Comprehensive Electronic Communication

Roy Blake

Niagara College of Applied Arts & Technology

West Publishing Company

Minneapolis / St. Paul   New York   Los Angeles   San Francisco

WEST'S COMMITMENT TO THE ENVIRONMENT

In 1906, West Publishing Company began recycling materials left over from the production of books. This began a tradition of efficient and responsible use of resources. Today, 100 percent of our legal bound volumes are printed on acid-free, recycled paper consisting of 50 percent new fibers. West recycles nearly 27,700,000 pounds of scrap paper annually—the equivalent of 229,300 trees. Since the 1960s, West has devised ways to capture and recycle waste inks, solvents, oils, and vapors created in the printing process. We also recycle plastics of all kinds, wood, glass, corrugated cardboard, and batteries, and have eliminated the use of polystyrene book packaging. We at West are proud of the longevity and the scope of our commitment to the environment.

West pocket parts and advance sheets are printed on recyclable paper and can be collected and recycled with newspapers. Staples do not have to be removed. Bound volumes can be recycled after removing the cover.

Production, Prepress, Printing and Binding by West Publishing Company.

PRINTED ON 10% POST CONSUMER RECYCLED PAPER

PRODUCTION CREDITS

Text Design: *John Edeen*
Copyeditor: *Pamela McMurray*
Artwork: *Illustrious, Inc.*
Composition: *G & S Typesetters*
Production Management: *Elm Street Publishing Services, Inc.*

British Library Cataloguing-in-Publication Data. A catalogue record for this book is available from the British Library.

Printed in the United States of America

04 03 02 01 00 99 8 7 6 5 4 3

Library of Congress Cataloging-in-Publication Data

Blake, Roy.
Comprehensive electronic communication / Roy Blake.
p. cm.
Includes index.
ISBN 0-314-20140-8
1. Telecommunication. I. Title.
TK5101.B556 1997
621.382—dc20 96-41373
CIP

For my wife Penny, son Adam,
and daughter Shira, who arrived
too late for mention
in my previous book.

Brief Contents

Contents

*Denotes optional sections

Preface

This text is intended for students in two-, three-, and four-year programs in electronics technology. It begins with traditional analog communications—amplitude and frequency modulation and their variations—and then discusses modern developments in digital and data communications and networks. The final section emphasizes radio and optical transmission and concludes with chapters on optical and personal communication systems.

There is enough material here for a three-course sequence. Schools with less time to spend on the subject can pick and choose the topics most important to their students. The material covered in earlier chapters is reviewed in later ones, so that in many cases the order in which chapters are covered can be easily changed, and some chapters can be omitted without harming the students' ability to understand the remainder of the book.

Most chapters have one or more optional sections, which are indicated with asterisks in the table of contents. As many of these sections may be included as time and the interests of the teacher and students permit. Omitting optional sections should not cause any difficulties in later chapters.

This book balances a "systems" orientation with enough detail to allow the student to do useful work. Essential theory is covered, using just enough mathematics to make the material clear, but the emphasis is on applying the theory to practical systems. Many representative systems are introduced, complete with photographs and manufacturers' specifications wherever possible. Complete circuit analyses are not attempted; rather, overviews are given, using a block diagram approach, and actual circuits are provided for representative sections of most interest.

Many chapters include a section on test equipment and measurement techniques. This material is intended to introduce the student to some real-world procedures and should both help to increase motivation and be useful in the laboratory.

This text assumes a basic knowledge of analog and digital electronics. It is expected that students may need to be reminded of the ways in which high-frequency circuits differ from those operating at audio frequencies, so these differences are reviewed. Class C and other tuned amplifiers are also explained because students may not have covered them in previous courses. The text does not assume that students are familiar with mixing and modulation or with techniques for making measurements at radio frequencies, so these topics are covered in the text. Phase-locked loops are explained at a basic level, and no prior knowledge of these devices is assumed.

Frequency-domain analysis is essential in any book of this type, and Fourier analysis is introduced in the text. No prior knowledge is assumed.

It is assumed that the student's mathematical background includes algebra and basic trigonometry. Calculus is not required, but knowledge of logarithms and decibels is assumed in the text; these topics are reviewed in an appendix.

Objectives are included at the beginning of each chapter, so that the student can easily refer to them while reading the material. Each chapter also contains a summary, a glossary, and a list of important equations, as well as numerous questions and problems. Answers to odd-numbered problems are provided at the back of the book, and an instructor's manual with complete solutions to all the problems is available. A set of transparency masters and a laboratory manual can also be obtained.

Students should find that the book is written in a clear and readable style. Frequent references to the work of communications pioneers give a sense of history, and these are also numerous glimpses into future possibilities. Readers should find that this text helps to make the field of electronic communications understandable and enjoyable.

Acknowledgments

Many peoples' efforts, in addition to mine, contributed to this book. Editor Chris Conty saw the project through from its beginning until it went into production three years later. Without his efforts there would have been no book. He made many suggestions that resulted in improvements to the book. Liz Riedel worked on the ancillaries: lab manual, solutions manual, and overhead transparency masters, as well as gathering reviews.

Thanks is also due to my colleagues at Niagara College, especially Joe Kopec and Greg Swick, both of whom made many excellent suggestions.

Once the book went into production, many others worked on it. Paul O'Neill at West Publishing coordinated the whole project. Michele Heinz of Elm Street Publishing Services made sure that text, artwork, and photographs came together in a smooth and error-free manner. She coordinated the efforts of the copy editor, typesetter, artist, and myself, and found and fixed errors that all of us had missed. The art and typesetting were very ably handled by Illustrious, Inc. and G & S Typesetters, respectively. Thanks also to copy editor Pam McMurray for straightening out some of my more convoluted sentences and making sure all the commas were in the right places.

Many reviewers contributed valuable suggestions to the book. A list of reviewers follows.

Richard Bain
DeVry Institute of Technology, Kansas City, Missouri

Ray Burns
Red River Community College, Manitoba

Bruce Koller
Diablo Valley College, California

Ronald Mackie
DeVry Institute of Technology, Weston, Ontario

Susan Meardon
Wake Technical Community College, North Carolina

Thomas G. Minnich
West Virginia Institute of Technology

Dennis K. Morland
Northern Alberta Institute of Technology

Gary J. Mullet
Springfield Technical Community College, Massachusetts

Vicki Price
DeVry Institute of Technology, Columbus, Ohio

James Stewart
DeVry Technical Institute, Woodbridge, New Jersey

Thanks also to the reviewers of the previous *Basic Electronic Communication* (1993), whose comments are still appreciated and often still included here.

PART

I

Basic Communication Systems

1 Introduction to Communication Systems

Objectives After studying this chapter, you should be able to:

1. Describe the essential elements of a communication system
2. Explain the need for modulation in communication systems
3. Distinguish between baseband, carrier, and modulated signals and give examples of each
4. Write the equation for a modulated signal and use it to list and explain the various types of continuous-wave modulation
5. Describe time-division and frequency-division multiplexing
6. Explain the relationship between channel bandwidth, baseband bandwidth, and transmission time
7. List the requirements for distortionless transmission and describe some of the possible deviations from this ideal
8. Use frequency-domain representations of signals and convert simple signals between time and frequency domains
9. Use a table of Fourier series to find the frequency-domain representations of common waveforms
10. Describe several types of noise and calculate the noise power and voltage for thermal noise
11. Calculate signal-to-noise ratio, noise figure, and noise temperature for single and cascaded stages
12. Use the spectrum analyzer for frequency, power, and signal-to-noise ratio measurements

1.1 Introduction

Communication was one of the first applications of electrical technology. Today, in the age of fiber optics and satellite television, facsimile machines and cellular telephones, communication systems remain at the leading edge of electronics. Probably no other branch of electronics has as profound an effect on people's everyday lives.

This book will introduce you to the study of electronic communication systems. After a brief survey of the history of communications, this chapter will consider the basic elements that are common to any such system: a **transmitter**, a **receiver**, and a communication **channel**. We will also begin discussions of signals and noise that will continue throughout the book.

Chapter 2 will review some of the fundamentals that you will need for later chapters and introduce you to some of the circuits that are commonly found in radio-frequency systems and with which you may not be familiar. We will then proceed, in succeeding chapters, to investigate the various ways of implementing each of the three basic elements.

We begin with analog systems. The reader who is impatient to get on to digital systems should realize that many digital systems also require analog technology to function. The modern technologist needs to have at least a basic understanding of both analog and digital systems before specializing in one or the other.

It is often said that we are living in the information age. Communication technology is absolutely vital to the generation, storage, and **transmission** of this information.

1.1.1 The Past, Present, and Future of Communications

Practical electrical communication began in 1837 with Samuel Morse's telegraph system. It was not the first system to use electricity to send messages, but it was the first to be commercially successful. Though not electronic, it had all the essential elements of the communication systems we will study in this book. There was a transmitter, consisting of a telegraph key and a battery, to convert information into an electrical signal that could be sent along wires. There was a receiver, called a *sounder*, to convert the electrical signal into sound that could be perceived by the operator (a variant of this system used marks on a paper tape). There was also a transmission channel, which consisted of wire on poles. Later, beginning in 1866, telegraph cables were run under water. By 1898, there were twelve transatlantic cables in operation.

Voice communication by electrical means began with the invention of the telephone by Alexander Graham Bell in 1876. Since that time, telephone systems have steadily expanded and have been linked together in complex ways. It is now possible to talk to someone almost anywhere in the world from nearly anywhere else.

The electronic content of telephony has also increased. In its early days, the telephone system did not involve any electronics at all. Eventually the vacuum tube and later the transistor allowed the use of amplifiers to increase the distance over which signals could be sent. Radio links were installed to allow communication where cables were impractical. Changes to digital electronics in both signal transmission and switching systems are now under way.

Radio is an important medium of communication. The theoretical framework was constructed by James Clerk Maxwell in 1865 and was verified experimentally by Heinrich Rudolph Hertz in 1887. Practical radiotelegraph systems were in use by the turn of the century, mainly for ship-to-shore and ship-to-ship service. The first transatlantic communication by radio was accomplished by Guglielmo Marconi in 1901.

Early radio transmitters used spark gaps to generate radiation and were not well suited to voice transmission. By 1906, some transmitters were using specially designed high-frequency alternators, and one of these was used experimentally to transmit voice. The one-kilowatt transmitter was modulated by connecting a microphone in series with the antenna. The microphone was allegedly water-cooled. Regular radio broadcasting did not begin until 1920, by which time transmitters and receivers used vacuum-tube technology. The diode tube had been invented by Sir John Ambrose Fleming in 1904, and the triode, which could work as an amplifier, by Lee De Forest in 1906.

By the late 1920s, radio broadcasting was commonplace and experiments with television were under way. The United States and several European countries had experimental television services before World War II, and after the war it became a practical worldwide reality.

Satellite communication is a relatively new form of radio. The first artificial satellite, Sputnik I, was launched by the Soviet Union in 1957, but practical, reliable communication via satellite really began with the launch of Intelsat I in 1965. This satellite was still partly experimental; the first modern commercial geostationary satellite, Anik A-1, was launched by Canada in 1972.

Modern communication systems have benefited from the invention of the transistor in 1948 and the integrated circuit in 1958. Modern equipment makes extensive use of both analog and digital solid-state technology, often in the same piece of equipment. The current cellular telephone system, for instance, uses analog FM radio coupled with an elaborate digital control system that would be completely impractical without microprocessors in the phones themselves and in the fixed stations in the network.

The coming years are likely to bring more exciting developments. Optical fiber is already a very important communication medium, and its use is bound to increase. The trend toward digital communication, even for such "analog" sources as voice and video, will continue.

It is impossible to predict accurately what will be developed in the future. Both cellular telephone and facsimile machines are improvements on systems that had been in use for many years and were regarded as not particularly exciting, yet their use has grown very quickly in the last several years. About the only thing that is certain about the future of communication technology is that it will be interesting.

1.2 Elements of a Communication System

Any communication system moves information from a source to a destination through a channel. Figure 1.1 illustrates this very simple idea, which is at the heart of everything that will be discussed in this book. The information from the source will generally not be in a form that can travel through the channel, so a device called a *transmitter* will be employed at one end, and a *receiver* at the other.

1.2.1 The Source

The source or information signal can be analog or digital. Common examples are analog audio and video signals and digital data. Sources are often described in terms of the frequency range that they occupy. Telephone-quality analog voice signals, for instance, contain frequencies from about 300 Hz to 3 kHz, while analog high-fidelity music needs a frequency range of approximately 20 Hz to 20 kHz.

Video requires a much larger frequency range than audio. An analog video signal of television-broadcast quality needs a frequency range from dc to about 4.2 MHz.

Digital sources can be derived from audio or video signals or can consist of data (alphanumeric characters, for example). Digital signals can have almost any bandwidth depending on the number of bits transmitted per second, and the method used to convert binary ones and zeros into electrical signals.

1.2.2 The Channel

A communication channel can be almost anything: a pair of conductors or an optical fiber, for example. Much of this book deals with radio communication, in which free space serves as the channel.

Sometimes a channel can carry the information signal directly. For example, an audio signal can be carried directly by a twisted-pair telephone cable. On the other hand, a radio link through free space cannot be used directly for voice signals. An antenna of enormous length would be required, and it would not be possible to transmit more than one signal without interference. Such situations require the use of a **carrier** signal whose frequency is such that it will travel, or propagate, through the channel. This carrier wave will be altered, or **modulated**, by the information signal in such a way that the information can be recovered at the destination. When a carrier is used, the information signal is also known as the *modulating signal*. Since the carrier frequency is generally much higher than that of the information signal, the frequency spectrum of the information signal is often referred to as the **baseband**. Thus, the three terms *information signal*, *modulating signal*, and *baseband* are equivalent in communication schemes involving modulated carriers.

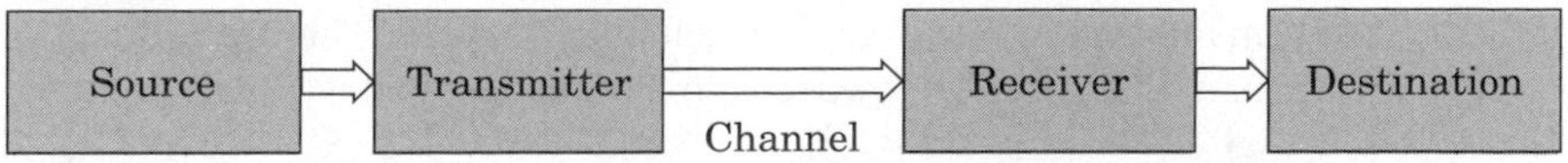

Figure 1.1 Elements of a communication system

1.2.3 Types of Modulation

All systems of modulation are variations on a small number of possibilities. A carrier is generated at a frequency much higher than the highest baseband frequency. Usually, the carrier is a sine wave. The instantaneous amplitude of the baseband signal is used to vary some parameter of the carrier.

A general equation for a sine-wave carrier is:

$$e(t) = E_c \sin (\omega_c t + \theta) \tag{1.1}$$

where

$e(t)$ = instantaneous voltage as a function of time
E_c = peak voltage
ω_c = frequency in radians per second
t = time in seconds
θ = phase angle in radians

Using radians per second in the mathematics dealing with modulation makes the equations simpler. Of course, frequency is usually given in hertz, rather than in radians per second, when practical devices are being discussed. It is easy to convert between the two systems by recalling from basic ac theory that $\omega = 2\pi f$.

In modulation, the parameters that can be changed are amplitude E_c, frequency ω_c, and phase θ. Combinations are also possible; for example, many schemes for transmitting digital information use both amplitude and phase modulation.

The modulation is done at the transmitter. An inverse process, called *demodulation* or *detection*, takes place at the receiver to restore the original baseband signal.

1.2.4 Signal Bandwidth

An unmodulated sine-wave carrier would exist at only one frequency and so would have zero bandwidth. However, a modulated signal is no longer a single sine wave, and it therefore occupies a greater bandwidth. Exactly how much bandwidth is needed depends on the baseband frequency range (or data rate, in the case of digital communication) and the modulation scheme in use. Hartley's Law is a general rule that relates bandwidth and information capacity. It states that the amount of information that can be transmitted in a given time is proportional to bandwidth for a given modulation scheme.

$$I = ktB \tag{1.2}$$

where

I = amount of information to be sent
k = a constant that depends on the type of modulation
t = time available
B = channel bandwidth

Some modulation schemes use bandwidth more efficiently than others. The bandwidth of each type of modulated signal will be examined in detail in later chapters.

1.2.5 Frequency-Division Multiplexing

One of the benefits of using modulated carriers, even with channels that are capable of carrying baseband signals, is that several carriers can be used at different frequencies. Each can be separately modulated with a different information signal, and filters at the receiver can separate the signals and demodulate whichever one is required.

Multiplexing is the term used in communications to refer to the combining of two or more information signals. When the available frequency range is divided among the signals, the process is known as *frequency-division multiplexing* (FDM).

Radio and television broadcasting, in which the available spectrum is divided among many signals, are everyday examples of FDM. There are limitations to the number of signals that can be crowded into a given frequency range because each requires a certain

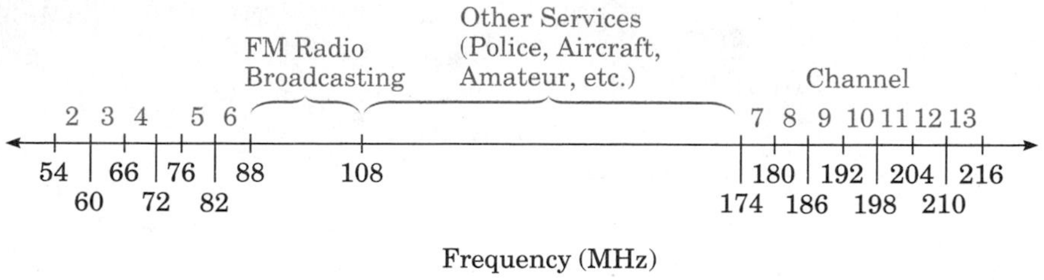

Figure 1.2 Frequency-division multiplexing in the VHF television band

bandwidth. For example, a television channel occupies a bandwidth of 6 MHz. Figure 1.2 shows how FDM applies to the VHF television band.

1.2.6 Time-Division Multiplexing

An alternative method for using a single communication channel to send many signals is to use *time-division multiplexing* (TDM). Instead of dividing the available bandwidth of the channel among many signals, the entire bandwidth is used for each signal, but only for a small part of the time. A glance at Equation (1.2) will confirm that time and bandwidth are equivalent in terms of information capacity. A nonelectronic example is the division of the total available time on a television channel among the various programs transmitted. Each program uses the whole bandwidth of the channel, but only for part of the time. Examples of TDM in electronic communication are not as common in everyday experience as FDM, but TDM is used extensively, especially with digital communication. The digital telephone system is a good example.

It is certainly possible to combine FDM and TDM. For example, the available bandwidth of a communication satellite is divided among a number of transmitter–receiver combinations called *transponders*. This is an example of FDM. A single transponder can be used to carry a large number of digital signals using TDM.

1.2.7 Frequency Bands

Hertz's early experiments in the laboratory used frequencies in the range of approximately 50 to 500 MHz. When others, such as Marconi, tried to apply his results to practical communication, they initially found that results were much better at lower frequencies. At that time, little was known about radio propagation (or about antenna design, for that matter). It is now known that frequencies ranging from a few kilohertz to many gigahertz all have their uses in radio communication systems.

In the meantime, a system of labeling frequencies came into use. The frequencies most commonly used in the early days, those from about 300 kHz to 3 MHz, were called *medium frequencies* (MF). Names were assigned to each order of magnitude of frequency, both upward and downward, and these names persist to this day. Thus we have *low frequencies* (LF) from 30 to 300 kHz and *very low frequencies* (VLF) from 3 to 30 kHz. Going the other way, we have *high frequencies* (HF) from 3 to 30 MHz, *very high frequencies* (VHF) from 30 to 300 MHz, and so on. Figure 1.3 shows the entire usable spectrum, with the appropriate labels.

Radio waves can also be described according to their wavelength, which is the distance a wave travels in one period. The general equation that relates frequency to wavelength for any wave is:

$$v = f\lambda \tag{1.3}$$

where

v = velocity of the wave in meters per second
f = frequency of the wave in hertz
λ = wavelength in meters

Figure 1.3 The radio-frequency spectrum

FREQUENCY		WAVELENGTH	SAMPLE USES
300 GHz		1 mm	
	Extremely High Frequencies (EHF)	Millimeter Waves	
30 GHz		1 cm	
	Super High Frequencies (SHF)		Radar
3 GHz		10 cm Microwaves	Communications Satellites
	Ultra High Frequencies (UHF)		Microwave Ovens Cellular Telephones
300 MHz		1 m	
	Very High Frequencies (VHF)		TV Broadcast FM Broadcast
30 MHz		10 m	
	High Frequencies (HF)	Short Waves	Shortwave Broadcast Commercial
3 MHz		100 m	
	Medium Frequencies (MF)	Medium Waves	AM Broadcast
300 kHz		1 km	
	Low Frequencies (LF)	Long Waves	Navigation
30 kHz		10 km	Submarine Communications
	Very Low Frequencies (VLF)		
3 kHz		100 km	
	Voice Frequencies (VF)		
300 Hz		1000 km	Audio
	Extremely Low Frequencies (ELF)		
30 Hz		10,000 km	Power Transmission

For a radio wave in free space, the velocity is the same as that of light, which is approximately 3×10^8 meters per second. The usual symbol for this quantity is c. Equation (1.3) then becomes

$$c = f\lambda \tag{1.4}$$

where

c = speed of light, 3×10^8 m/s
f = frequency in hertz
λ = wavelength in meters

It can be seen from Equation (1.4) that a frequency of 300 MHz corresponds to a wavelength of 1 m and that wavelength is inversely proportional to frequency. Low-frequency signals are sometimes called *long wave*, high frequencies correspond to *short wave*, and so on. The reader is no doubt familiar with the term *microwave* to describe signals in the gigahertz range.

Example 1.1

Calculate the wavelength in free space corresponding to a frequency of:

(a) 1 MHz (AM radio broadcast band)
(b) 27 MHz (CB radio band)
(c) 4 GHz (used for satellite television)

Solution

Rearranging Equation (1.4),

$$c = f\lambda$$
$$\lambda = \frac{c}{f}$$

(a)
$$\lambda = \frac{3 \times 10^8 \text{ m/s}}{1 \times 10^6 \text{ Hz}} = 300 \text{ m}$$

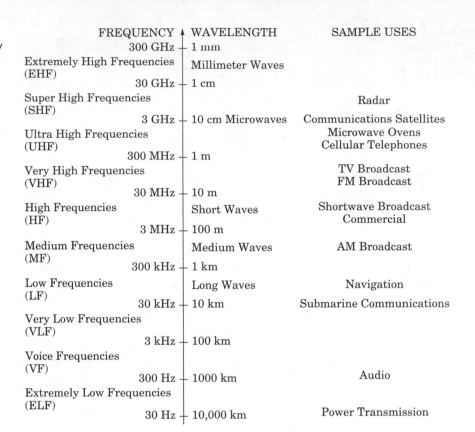

(b) $$\lambda = \frac{3 \times 10^8 \text{ m/s}}{27 \times 10^6 \text{ Hz}} = 11.1 \text{ m}$$

(c) $$\lambda = \frac{3 \times 10^8 \text{ m/s}}{4 \times 10^9 \text{ Hz}} = 0.075 \text{ m} = 7.5 \text{ cm}$$

1.2.8 Distortionless Transmission

The receiver should restore the baseband signal exactly. Of course, there will be a time delay due to the distance over which communication takes place, and there will likely be a change in amplitude as well. Neither effect is likely to cause problems, though there are exceptions. The time delay involved in communication via geostationary satellite can be a nuisance in telephone communication. Even though radio waves propagate at the speed of light, the great distance over which the signal must travel (about 70 thousand kilometers) causes a quarter-second delay.

Any other changes in the baseband signal reflect **distortion**, which has a corrupting effect on the signal. There are many possible types of distortion. Some types are listed below, but not all of them will be immediately clear. Much of the rest of this book will be devoted to explaining them in more detail. Some possible types of distortion are:

- Harmonic distortion: harmonics (multiples) of some of the baseband components are added to the original signal.
- Intermodulation distortion: additional frequency components generated by combining (mixing) the frequency components in the original signal. Mixing will be described in Chapter 2.
- Nonlinear frequency response: some baseband components are amplified more than others.
- Nonlinear phase response: phase shift between components of the signal.
- Noise: both the transmitter and the receiver add **noise**, and the channel is also noisy. This noise adds to the signal and masks it. Noise will be discussed later in this chapter.
- Interference: if more than one signal uses the same transmission medium, the signals may interact with each other.

One of the advantages of digital communication is the ability to regenerate a signal that has been corrupted by noise and distortion, provided that it is still identifiable as representing a one or a zero. In analog systems, however, noise and distortion tend to accumulate. In some cases distortion can be removed at a later point. If the frequency response of a channel is not flat but is known, for instance, equalization in the form of filters can be used to compensate. However, harmonic distortion, intermodulation, and noise, once present, are impossible to remove completely from an analog signal. A certain amount of immunity can be built into digital schemes, but excessive noise and distortion levels will be reflected in increased error rates.

1.3 Time and Frequency Domains

The reader is probably familiar with the **time-domain** representation of signals. An ordinary oscilloscope display, showing amplitude on one scale and time on the other, is a good example.

Signals can also be described in the **frequency domain**. In a frequency-domain representation, amplitude or power is shown on one axis and frequency is displayed on the other. A **spectrum analyzer** gives a frequency-domain representation of signals.

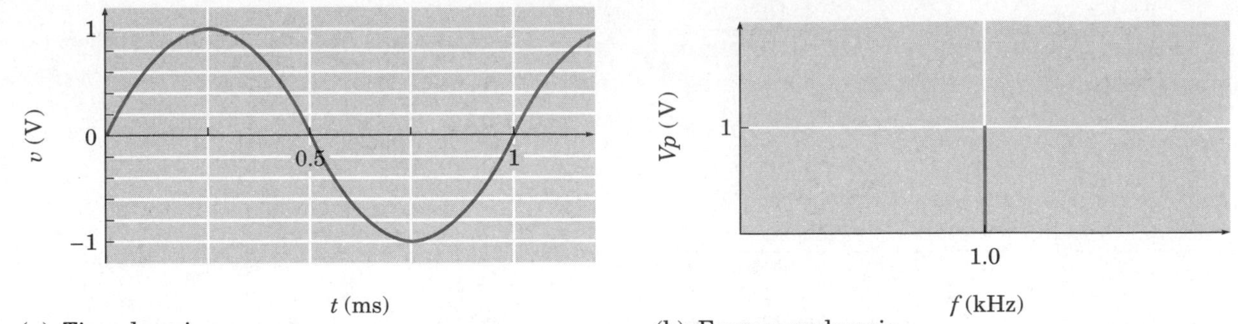

(a) Time domain (b) Frequency domain

Figure 1.4
Sine wave in time and frequency domains

Any signal can be represented either way. For example, a 1 kHz sine wave is shown in both ways in Figure 1.4. The time-domain representation should need no explanation. As for the frequency domain, a sine wave has energy only at its fundamental frequency, so it can be shown as a straight line at that frequency.

Notice that our way of representing the signal in the frequency domain fails to show one thing: the phase of the signal. The signal in Figure 1.4(b) could be a cosine wave just as easily as a sine wave.

Frequency-domain representations are very useful in the study of communication systems. For instance, the bandwidth of a modulated signal generally has some fairly simple relationship to that of the baseband signal. This bandwidth can easily be found if the baseband signal can be represented in the frequency domain. As we proceed, we will see many other examples in which the ability to work with signals in the frequency domain will be required.

1.3.1 Fourier Series

Any well-behaved periodic waveform can be represented as a series of sine and/or cosine waves at multiples of its fundamental frequency plus (sometimes) a dc offset. This is known as a **Fourier series**. This very useful (and perhaps rather surprising) fact was discovered in 1822 by Joseph Fourier, a French mathematician, in the course of research on heat conduction. Not all signals used in communications are strictly periodic, but they are often close enough for practical purposes.

Fourier's discovery, applied to a time-varying signal, can be expressed mathematically as follows:

$$f(t) = \frac{A_0}{2} + A_1 \cos \omega t + B_1 \sin \omega t + A_2 \cos 2\omega t + B_2 \sin 2\omega t + A_3 \cos 3\omega t + B_3 \sin 3\omega t + \cdots \tag{1.5}$$

where

$f(t)$ = any well-behaved function of time as described above. For our purposes, $f(t)$ will generally be either a voltage $v(t)$ or a current $i(t)$.

A_n and B_n = real-number coefficients; that is, they can be positive, negative, or zero.

ω = radian frequency of the fundamental.

The radian frequency can be found from the time-domain representation of the signal by finding the period (that is, the time T after which the whole signal repeats exactly) and using the equations:

$$f = \frac{1}{T} \tag{1.6}$$

and

$$\omega = 2\pi f \tag{1.7}$$

The simplest ac signal is a sinusoid. The frequency-domain representation of a sine wave has already been described and is shown in Figure 1.4(b) for a voltage sine wave with a period of 1 ms and a peak amplitude of 1 V. For this signal, all the coefficients of Equation (1.5) are zero except for B_1, which has a value of 1 V. The equation becomes

$$v(t) = \sin(2000\pi) \text{ V}$$

which is certainly no surprise.

The examples below use the equations for the various waveforms shown in Table 1.1.

Table 1.1 Fourier series for common repetitive waveforms

1. Half-wave rectified sine wave

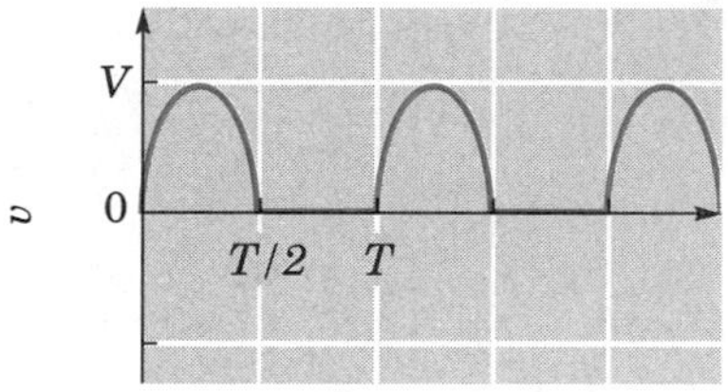

$$v(t) = \frac{V}{\pi} + \frac{V}{2}\sin\omega t - \frac{2V}{\pi}\left(\frac{\cos 2\omega t}{1 \times 3} + \frac{\cos 4\omega t}{3 \times 5} + \frac{\cos 6\omega t}{5 \times 7} + \cdots\right)$$

2. Full-wave rectified sine wave

(a) With time zero at voltage zero

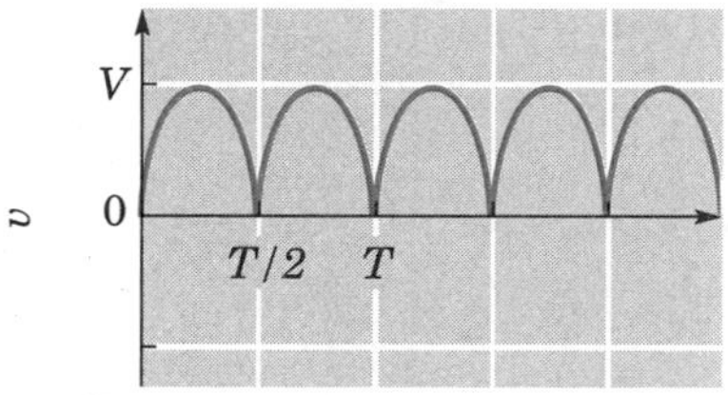

$$v(t) = \frac{2V}{\pi} - \frac{4V}{\pi}\left(\frac{\cos 2\omega t}{1 \times 3} + \frac{\cos 4\omega t}{3 \times 5} + \frac{\cos 6\omega t}{5 \times 7} + \cdots\right)$$

(b) With time zero at voltage peak

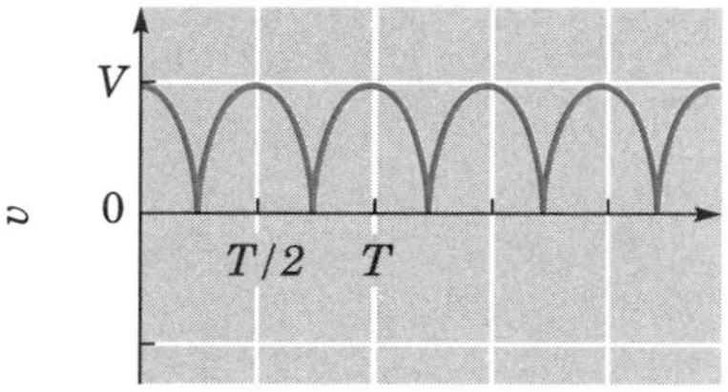

$$v(t) = \frac{2V}{\pi}\left(1 + \frac{2\cos 2\omega t}{1 \times 3} - \frac{2\cos 4\omega t}{3 \times 5} + \frac{2\cos 6\omega t}{5 \times 7} - \cdots\right)$$

3. Square wave

(a) Odd function

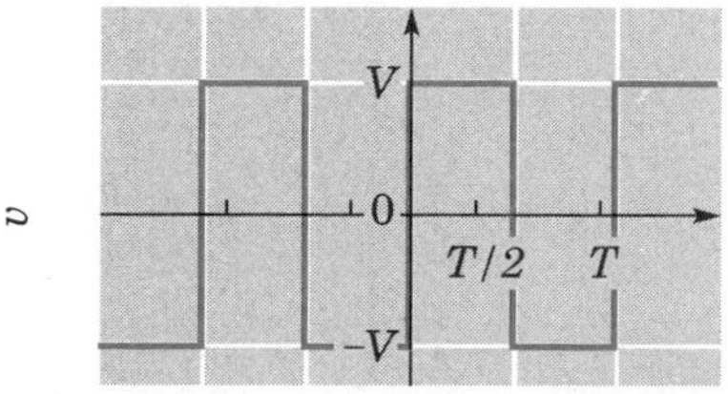

$$v(t) = \frac{4V}{\pi}\left(\sin\omega t + \frac{1}{3}\sin 3\omega t + \frac{1}{5}\sin 5\omega t + \cdots\right)$$

(b) Even function

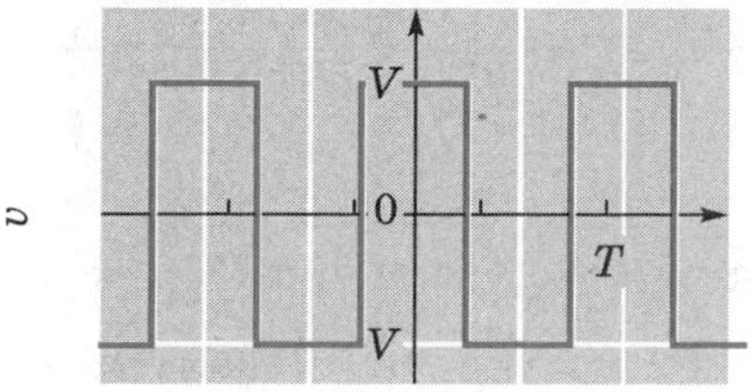

$$v(t) = \frac{4V}{\pi}\left(\cos\omega t - \frac{1}{3}\cos 3\omega t + \frac{1}{5}\cos 5\omega t - \cdots\right)$$

(continues)

Table 1.1 Continued

4. Pulse train

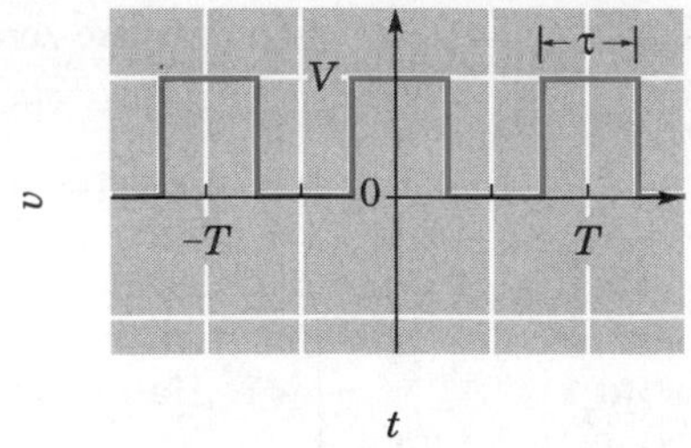

$$v(t) = \frac{V\tau}{T} + \frac{2V\tau}{T}\left(\frac{\sin \pi\tau/T}{\pi\tau/T}\cos \omega t + \frac{\sin 2\pi\tau/T}{2\pi\tau/T}\cos 2\omega t + \frac{\sin 3\pi\tau/T}{3\pi\tau/T}\cos 3\omega t + \cdots\right)$$

5. Triangle wave

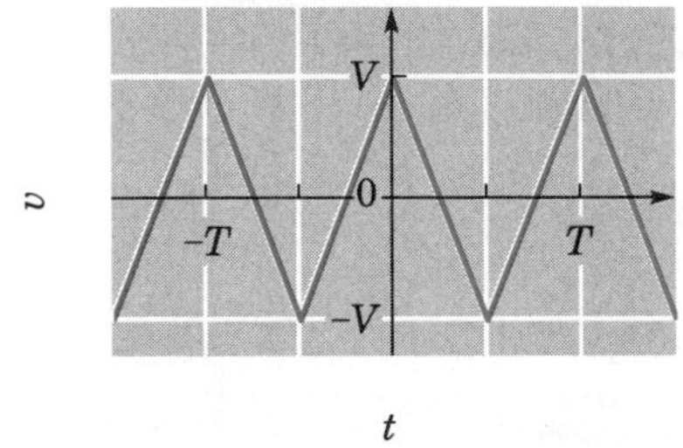

$$v(t) = \frac{8V}{\pi^2}\left(\cos \omega t + \frac{1}{3^2}\cos 3\omega t + \frac{1}{5^2}\cos 5\omega t + \cdots\right)$$

6. Sawtooth wave
(a) With no dc offset

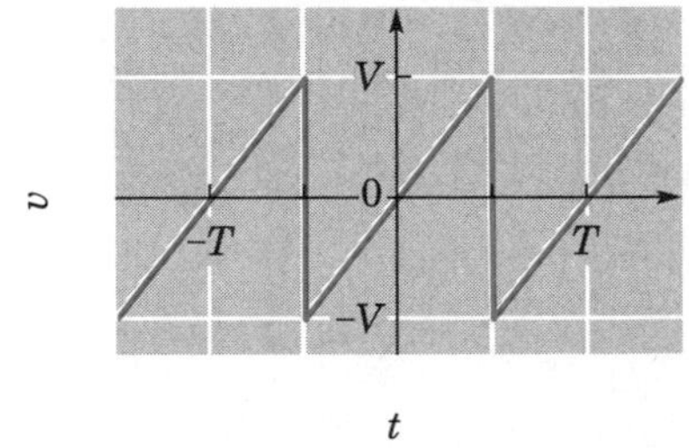

$$v(t) = \frac{2V}{\pi}\left(\sin \omega t - \frac{1}{2}\sin 2\omega t + \frac{1}{3}\sin 3\omega t - \cdots\right)$$

(b) Positive-going

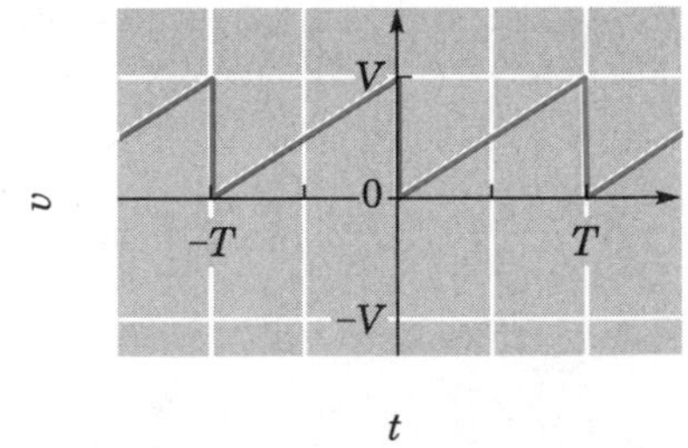

$$v(t) = \frac{V}{2} - \frac{V}{\pi}\left(\sin \omega t + \frac{1}{2}\sin 2\omega t + \frac{1}{3}\sin 3\omega t + \cdots\right)$$

Example 1.2 Find and sketch the Fourier series corresponding to the square wave in Figure 1.5(a).

Solution A square wave is another signal with a simple Fourier representation, although the representation is not quite as simple as that for a sine wave. For the signal shown in Figure 1.5(a), the frequency is 1 kHz, as before, and the peak voltage is 1 V.

According to Table 1.1, this signal has components at an infinite number of frequencies: all odd multiples of the fundamental frequency of 1 kHz. However, the amplitude decreases with frequency, so that the third harmonic has an amplitude one-third that of the fundamental, the fifth harmonic an amplitude of one-fifth the fundamental, and so on. Mathematically, a square wave of voltage with a rising edge at $t = 0$ and no dc offset can be expressed as follows:

$$v(t) = \frac{4V}{\pi}\left(\sin \omega t + \frac{1}{3}\sin 3\omega t + \frac{1}{5}\sin 5\omega t + \cdots\right) \tag{1.8}$$

where

V = peak amplitude of the square wave
ω = radian frequency of the square wave
t = time in seconds

For this example, Equation (1.8) becomes:

$$v(t) = \frac{4}{\pi}\left[\sin(2\pi \times 10^3 t) + \frac{1}{3}\sin(6\pi \times 10^3 t) + \frac{1}{5}\sin(10\pi \times 10^3 t) + \cdots\right] \text{V}$$

This equation shows that the signal has frequency components at odd multiples of 1 kHz, that is, at 1 kHz, 3 kHz, 5 kHz, and so on. The 1 kHz component has a peak amplitude of

$$V_1 = \frac{4}{\pi} = 1.27 \text{ V}$$

The component at three times the fundamental frequency (3 kHz) has an amplitude one-third that of the fundamental, that at 5 kHz has an amplitude one-fifth that of the fundamental, and so on.

$$V_3 = \frac{4}{3\pi} = 0.424 \text{ V}$$

$$V_5 = \frac{4}{5\pi} = 0.255 \text{ V}$$

$$V_7 = \frac{4}{7\pi} = 0.182 \text{ V}$$

Figure 1.5

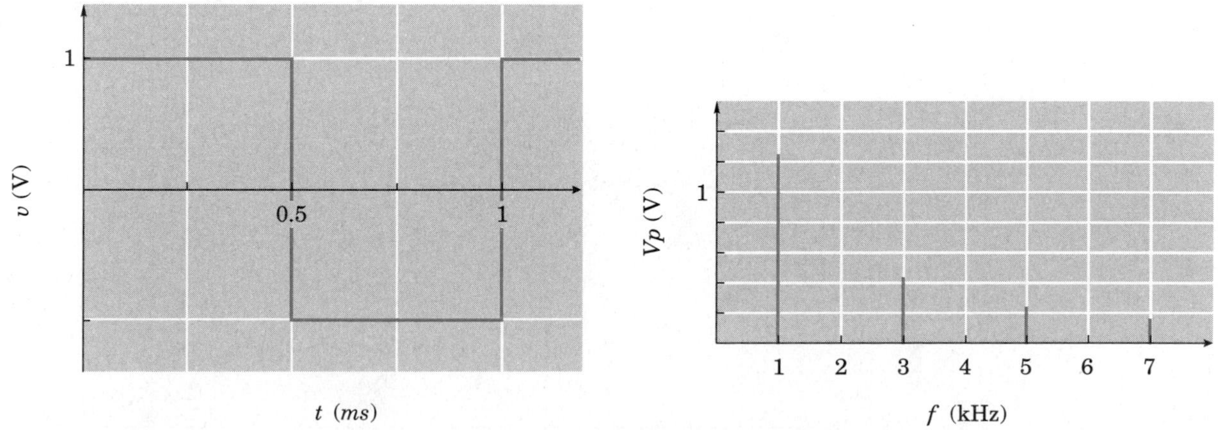

(a) Time domain (b) Frequency domain

The result for the first four components is sketched in Figure 1.5(b). Theoretically, an infinite number of components would be required to describe the square wave completely, but, as the frequency increases, the amplitude of the components decreases rapidly.

The representations in Figures 1.5(a) and 1.5(b) are not two different signals but merely two different ways of looking at the same signal. This can be shown graphically by adding the instantaneous values of several of the sine waves in the frequency-domain representation. If enough of these components are included, the result begins to look like the square wave in the time-domain representation. Figure 1.6 shows the results for two, four, and ten components. It was created by simply taking the instantaneous values of all the components at the same time and adding them algebraically. This was done for a number of time values, resulting in the graphs in Figure 1.6. Doing these calculations by hand would be simple but rather tedious, so a computer was used to perform the calculations and plot the graphs. A perfectly accurate representation of the square wave would require an infinite number of

Figure 1.6
Construction of a square wave from Fourier components

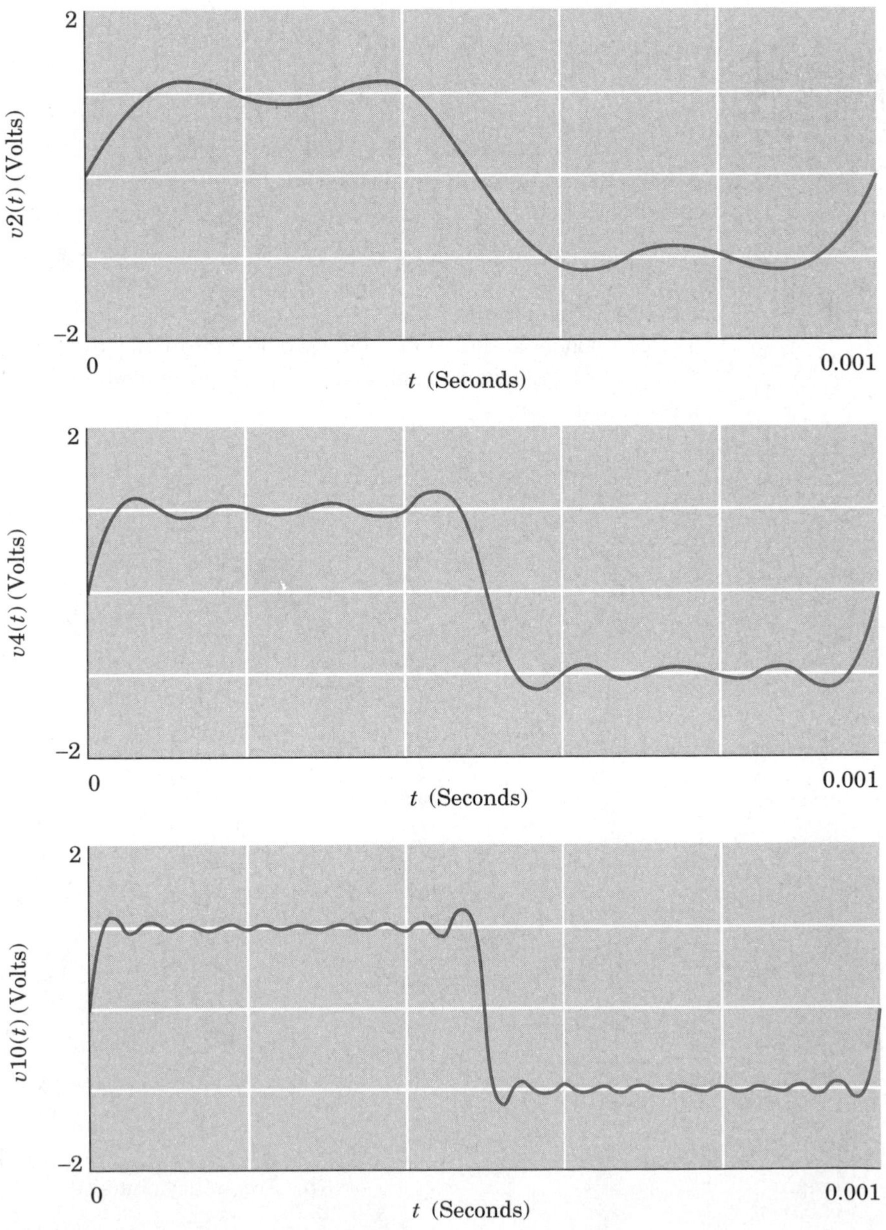

(a) Two components

(b) Four components

(c) Ten components

Figure 1.7
Addition of square-wave Fourier components with wrong phase angles

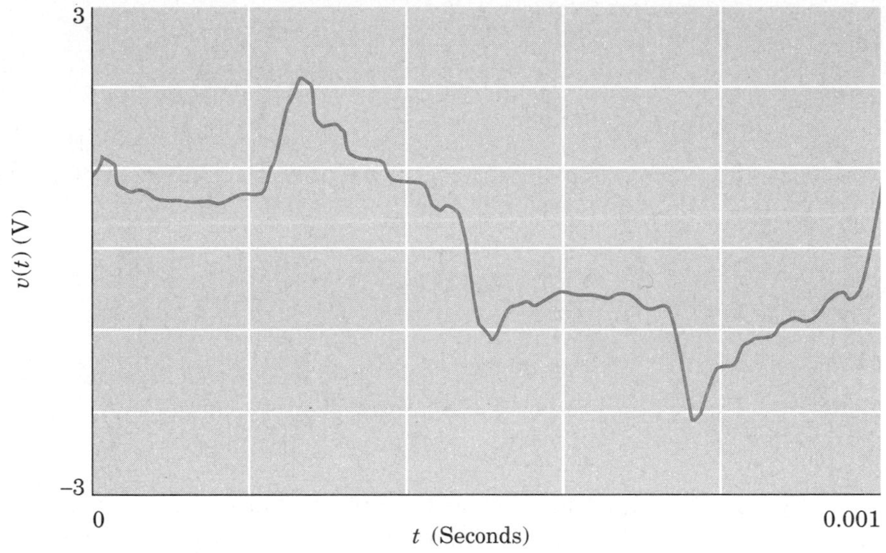

components, but we can see from the figure that using ten components gives a very good representation because the amplitudes of the higher-frequency components of the signal are very small.

It is possible to go back and forth at will between time and frequency domains, but it should be apparent that information about the relative phases of the frequency components in the Fourier representation of the signal is required to reconstruct the time-domain representation. The Fourier equations do have this information, but the sketch in Figure 1.5(b) does not. If the phase relationships between frequency components are changed in a communication system, the signal will be distorted in the time domain.

Figure 1.7 illustrates this point. The same ten coefficients were used as in Figure 1.6(c), but this time the waveforms alternated between sine and cosine: sine for the fundamental, cosine for the third harmonic, sine for the fifth, and so on. The result is a waveform that looks the same on the frequency-domain sketch of Figure 1.5(b) but very different in the time domain.

Example 1.3 Find the Fourier series for the signal in Figure 1.8(a).

Solution The positive-going sawtooth wave of Figure 1.8(a) has a Fourier series with a dc term and components at all multiples of the fundamental frequency. From Table 1.1, the general

Figure 1.8

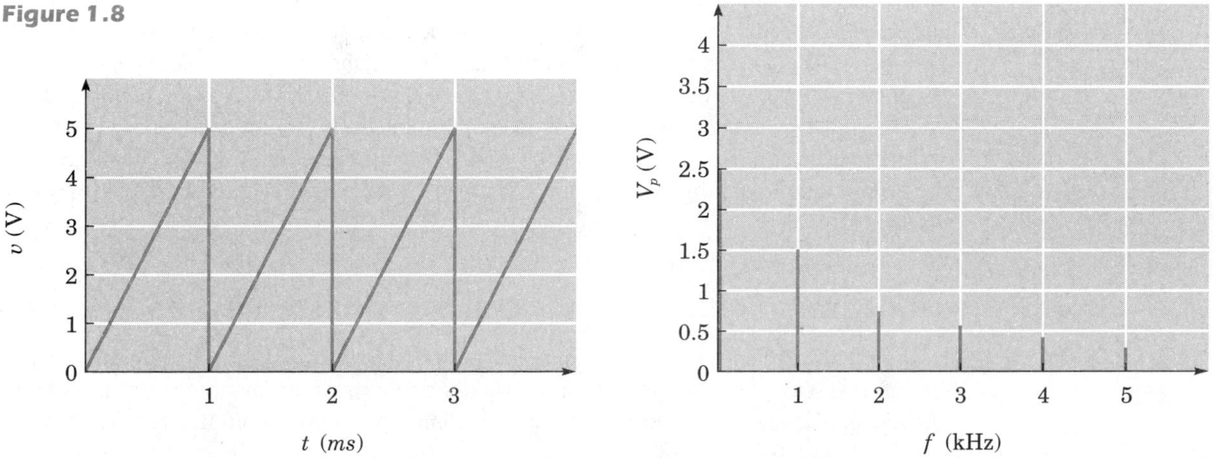

(a) Time domain (b) Frequency domain

equation for such a wave is

$$v(t) = \frac{A}{2} - \left(\frac{A}{\pi}\right)\left(\sin \omega t + \frac{1}{2} \sin 2\,\omega t + \frac{1}{3} \sin 3\,\omega t + \cdots\right) \tag{1.9}$$

The first (dc) term is simply the average value of the signal.

For the signal in Figure 1.8, which has a frequency of 1 kHz and a peak amplitude of 5 V, Equation (1.9) becomes:

$$v(t) = 2.5 - 1.59\,[\sin(2\pi \times 10^3 t) + 0.5 \sin(4\pi \times 10^3 t) + 0.33 \sin(6\pi \times 10^3 t) + \cdots]\text{ V}$$

1.3.2 Effect of Filtering on Signals

As we have seen, many signals have a bandwidth that is theoretically infinite. Limiting the frequency response of a channel removes some of the frequency components of these signals and causes the time-domain representation to be distorted. An uneven frequency response will emphasize some components at the expense of others, again causing distortion. Nonlinear phase shift will also affect the time-domain representation. For instance, shifting the phase angles of some of the frequency components in the square-wave representation of Figure 1.7 changed the signal to something other than a square wave.

However, Figure 1.6 shows that while an infinite bandwidth may theoretically be required, for practical purposes quite a good representation of a square wave can be obtained with a band-limited signal. In general, the wider the bandwidth, the better, but acceptable results can be obtained with a band-limited signal. This is welcome news, because practical systems always have finite bandwidth.

Although real communication systems have a solid theoretical base, they also involve many practical considerations. There is always a trade-off between fidelity to the original signal (that is, the absence of any distortion of its waveform) and such factors as bandwidth and cost. Increasing the bandwidth often increases the cost of a communication system. Not only is the hardware likely to be more expensive, but bandwidth itself may be in short supply. In radio communication, for instance, spectral allocations are assigned by national regulatory bodies in keeping with international agreements. This task is handled in the United States by the Federal Communications Commission (FCC) and in Canada by the communications branch of Industry and Science Canada. The allocations are constantly being reviewed, and there never seems to be enough room in the usable spectrum to accommodate all potential users.

In communication over cables, whether by electrical or optical means, the total bandwidth of a given cable is fixed by the technology employed. The more bandwidth used by each signal, the fewer signals can be carried by the cable.

Consequently, some services are constrained by cost and/or government regulation to use less than optimal bandwidth. For example, telephone systems use about 3 kHz of baseband bandwidth for voice. This is obviously not distortionless transmission: the difference between a voice heard "live," in the same room, and the same voice heard over the telephone is obvious. High-fidelity audio needs at least 15 kHz of baseband bandwidth. On the other hand, the main goal of telephone communication is the understanding of speech, and the bandwidth used is sufficient for this. Using a larger bandwidth would contribute little or nothing to intelligibility and would greatly increase the cost of nearly every part of the system.

1.4 Noise and Communications

Noise in communication systems originates both in the channel and in the communication equipment. Noise consists of undesired, usually random, variations that interfere with the desired signals and inhibit communication. It cannot be avoided completely, but its effects can be reduced by various means, such as reducing the signal bandwidth, increasing the transmitter power, and using low-noise amplifiers for weak signals.

It is helpful to divide noise into two types: *internal noise*, which originates within the communication equipment, and *external noise*, which is a property of the channel. (In this book, the channel will usually be a radio link.)

1.4.1 External Noise

If the channel is a radio link, there are many possible sources of noise. The reader is no doubt familiar with the "static" that afflicts AM radio broadcasting during thunderstorms. Interference from automobile ignition systems is another common problem. Domestic and industrial electrical equipment, from light dimmers to vacuum cleaners to computers, can also cause objectionable noise. Not so obvious, but of equal concern to the communications specialist, is noise produced by the sun and other stars.

Equipment Noise Noise is generated by equipment that produces sparks. Examples include automobile engines and electric motors with brushes. Any fast-risetime voltage or current can also generate interference, even without arcing. Light dimmers and computers fall into this category.

Noise of this type has a broad frequency spectrum, but its energy is not equally distributed over the frequency range. This type of interference is generally more severe at lower frequencies, but the exact frequency distribution depends upon the source itself and any conductors to which it is connected. Computers, for instance, may produce strong signals at multiples and submultiples of their clock frequency and little energy elsewhere.

Man-made noise can propagate through space or along power lines. It is usually easier to control it at the source than at the receiver. A typical solution for a computer, for instance, involves shielding and grounding the case and all connecting cables and installing a low-pass filter on the power line where it enters the enclosure.

Atmospheric Noise Atmospheric noise is often called *static* because lightning, which is a static-electricity discharge, is a principal source of atmospheric noise. This type of disturbance can propagate for long distances through space. Most of the energy of lightning is found at relatively low frequencies, up to several megahertz.

Obviously nothing can be done to reduce atmospheric noise at the source. However, circuits are available to reduce its effect by taking advantage of the fact that this noise has a very high peak-to-average power ratio; that is, the noise occurs in short intense bursts with relatively long periods of time between bursts. It is often possible to improve communication by simply disabling the receiver for the duration of the burst. This technique is called *noise blanking*.

Space Noise The sun is a powerful source of radiation over a wide range of frequencies, including the radio-frequency spectrum. Other stars also radiate noise called *cosmic*, *stellar*, or *sky noise*. Its intensity when received on the earth is naturally much less than for solar noise because of the greater distance.

Solar noise can be a serious problem for satellite reception, which becomes impossible when the satellite is in a line between the antenna and the sun. Space noise is more important at higher frequencies (VHF and above) because atmospheric noise dominates at lower frequencies and because most of the space noise at lower frequencies is absorbed by the upper atmosphere.

1.4.2 Internal Noise

Noise is generated in all electronic equipment. Both passive components (such as resistors and cables) and active devices (such as diodes, transistors, and tubes) can be noise sources. Several types of noise will be examined in this section, beginning with the most important, thermal noise.

Thermal Noise Thermal noise is produced by the random motion of electrons in a conductor due to heat. The term "noise" is often used alone to refer to this type of noise, which is found everywhere in electronic circuitry.

The power density of thermal noise is constant with frequency, from zero to frequencies well above those used in electronic circuits. That is, there is equal power in every hertz of bandwidth. Thermal noise is thus an equal mixture of noise of all frequencies. It is sometimes called *white noise*, in analogy with white light, which is an equal mixture of all colors.

The noise power available from a conductor is a function of its temperature, as shown by the equation:

$$P_N = kTB \tag{1.10}$$

where

P_N = noise power in watts
k = Boltzmann's constant, 1.38×10^{-23} joules/kelvin (J/K)
T = absolute temperature in kelvins (K); this can be found by adding 273 to the Celsius temperature
B = noise power bandwidth in hertz

This equation is based on the assumption that power transfer is maximum. That is, the source (the resistance generating the noise) and the load (the amplifier or other device that receives the noise) are assumed to be matched in impedance.

Equation (1.10) shows that noise power is directly related to bandwidth over which the noise is observed (the noise power bandwidth). Filters are often specified in terms of their bandwidth at half power, that is, at the points where the gain of the filter is 3 dB down from its gain in the center of the passband. An ideal bandpass filter would allow no noise to pass at frequencies outside the passband, so its noise power bandwidth would be equal to its half-power bandwidth. For a practical filter, the noise power bandwidth is generally greater than the half-power bandwidth, since some contribution is made to the total noise power by frequencies that are attenuated by more than 3 dB. An ordinary *LC* tuned circuit has a noise power bandwidth about 50% greater than its half-power bandwidth.

The power referred to above is *average* power. The instantaneous voltage and power are random, but over time the voltage will average out to zero, as it does for an ordinary ac signal, and the power will average to the value given by Equation (1.10). Figure 1.9 shows this effect.

Figure 1.9
Noise voltage and power

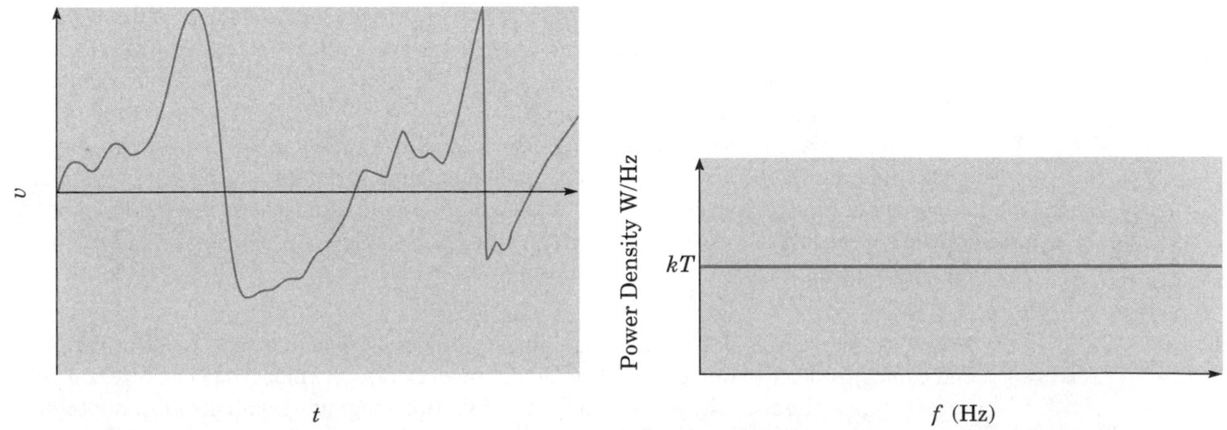

(a) Voltage as a function of time

(b) Power density as a function of frequency

Example 1.4 A receiver has a noise power bandwidth of 10 kHz. A resistor that matches the receiver input impedance is connected across its antenna terminals. What is the noise power contributed by that resistor in the receiver bandwidth, if the resistor has a temperature of 27°C?

Solution First, convert the temperature to kelvins:

$$\begin{aligned} T(\text{K}) &= T(°\text{C}) + 273 \\ &= 27 + 273 \\ &= 300\ \text{K} \end{aligned}$$

According to Equation (1.10), a conductor at a temperature of 300 K would contribute the following noise power in that 10 kHz bandwidth:

$$\begin{aligned} P_N &= kTB \\ &= (1.38 \times 10^{-23}\ \text{J/K})\ (300\ \text{K})\ (10 \times 10^3\ \text{Hz}) \\ &= 4.14 \times 10^{-17}\ \text{W} \end{aligned}$$

This is obviously not a great deal of power, but it can be significant at the signal levels dealt with in sensitive receiving equipment.

Thermal noise power exists in all conductors and resistors at any temperature above absolute zero. The only way to reduce it is to decrease the temperature or the bandwidth of a circuit (or both). Amplifiers used with very low level signals are often cooled artificially to reduce noise. This technique is called *cryogenics* and may involve, for example, cooling the first stage of a receiver for radio astronomy by immersing it in liquid nitrogen. The other method, *bandwidth reduction*, will be referred to many times throughout this book. Using a bandwidth greater than required for a given application is simply an invitation to noise problems.

Thermal noise, as such, does not depend on the type of material involved or the amount of current passing through it. However, some materials and devices also produce other types of noise that do depend on current. Carbon-composition resistors are in this category, for example, and so are semiconductor junctions.

Noise Voltage Often we are more interested in the *noise voltage* than in the power involved. The noise power depends only on bandwidth and temperature, as stated above. Power in a resistive circuit is given by the equation:

$$P = \frac{V^2}{R} \tag{1.11}$$

From this, we see that:

$$\begin{aligned} V^2 &= PR \\ V &= \sqrt{PR} \end{aligned} \tag{1.12}$$

Equation (1.12) shows that the noise voltage in a circuit must depend on the resistances involved, as well as on the temperature and bandwidth.

Figure 1.10 shows a resistor that serves as a noise source connected across another resistor, R_L, that will be considered as a load. The noise voltage is represented as a voltage source V_N, in series with a noiseless resistance R_N. We know from Equation (1.10) that the noise power supplied to the load resistor, assuming a matched load, is:

$$P_N = kTB$$

In this situation, one-half the noise voltage appears across the load, and the rest is across the resistor that generates the noise.

Figure 1.10 Thermal noise voltage

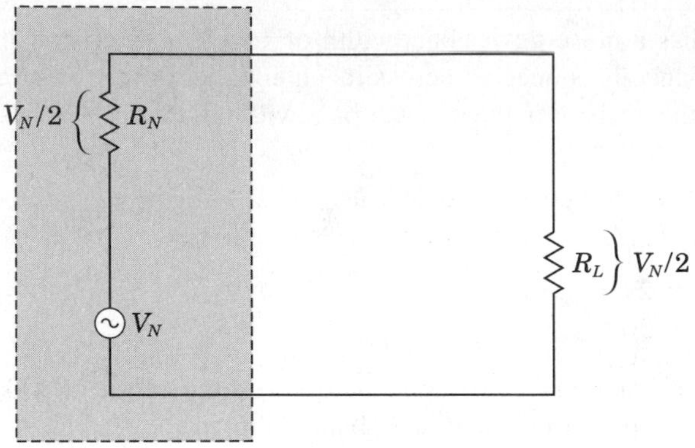

The root-mean-square (RMS) noise voltage across the load is given by:

$$\begin{aligned} V_L &= \sqrt{PR} \\ &= \sqrt{kTBR} \end{aligned}$$

An equal noise voltage will appear across the source resistance, so the noise source shown in Figure 1.10 must have double this voltage or

$$\begin{aligned} V_N &= 2\sqrt{kTBR} \\ &= \sqrt{4kTBR} \end{aligned} \tag{1.13}$$

Example 1.5

A 300 Ω resistor is connected across the 300 Ω antenna input of a television receiver. The bandwidth of the receiver is 6 MHz, and the resistor is at room temperature (293 K or 20°C or 68°F). Find the noise power and noise voltage applied to the receiver input.

Solution

The noise power is given by Equation (1.10):

$$\begin{aligned} P_N &= kTB \\ &= (1.38 \times 10^{-23}\ \text{J/K})\ (293\ \text{K})\ (6 \times 10^{6}\ \text{Hz}) \\ &= 24.2 \times 10^{-15}\ \text{W} \\ &= 24.2\ \text{fW} \end{aligned}$$

The noise voltage is given by Equation (1.13):

$$\begin{aligned} V_N &= \sqrt{4kTBR} \\ &= \sqrt{4\ (1.38 \times 10^{-23}\ \text{J/K})\ (293\ \text{K})\ (6 \times 10^{6}\ \text{Hz})\ (300\ \Omega)} \\ &= 5.4 \times 10^{-6}\ \text{V} \\ &= 5.4\ \mu\text{V} \end{aligned}$$

Of course, only one-half this voltage appears across the antenna terminals; the other half appears across the source resistance. Therefore the actual noise voltage at the input is 2.7 μV.

Shot Noise This type of noise has a power spectrum that resembles that of thermal noise, having equal energy in every hertz of bandwidth, at frequencies from dc into the gigahertz region. However, the mechanism that creates shot noise is different. Shot noise is due to random variations in current flow in active devices such as tubes, transistors, and semiconductor diodes. These variations are caused by the fact that current is a flow of carriers (electrons or holes), each of which carries a finite amount of charge. Current can thus be considered as a series of pulses, each consisting of the charge carried by one electron.

The name "shot noise" describes the random arrival of electrons arriving at the anode of a vacuum tube, like individual pellets of shot from a shotgun.

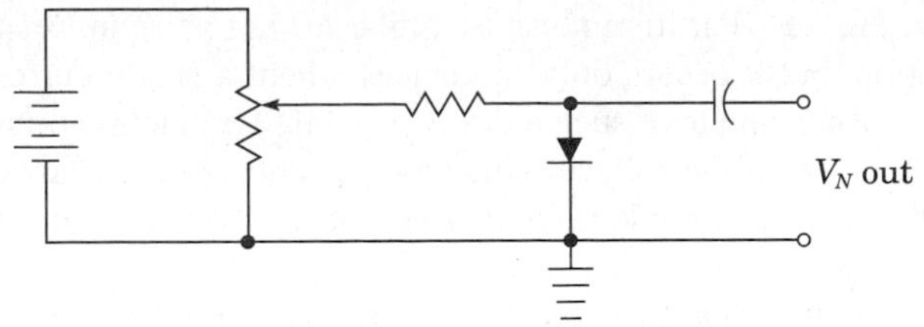

Figure 1.11 Diode noise generator

One would expect that the resulting noise power would be proportional to the device current, and this is true both for vacuum tubes and for semiconductor junction devices.

Shot noise is usually represented by a current source. The noise current for either a vacuum or a junction diode is given by the equation:

$$I_N = \sqrt{2qI_0B} \tag{1.14}$$

where

I_N = *RMS noise current, in amperes*
q = magnitude of the charge on an electron, equal to 1.6×10^{-19} coulomb
I_0 = dc bias current in the device, in amperes
B = bandwidth over which the noise is observed, in hertz

This equation is valid for frequencies much less than the reciprocal of the carrier *transit time* for the device. The transit time is the time a charge carrier spends in the device. Depending on the device, Equation (1.14) may be valid for frequencies up to a few megahertz or several gigahertz.

Since shot noise closely resembles thermal noise in its random amplitude and flat spectrum, it can be used as a substitute for thermal noise whenever a known level of noise is required. Such a noise source can be very useful, particularly when conducting measurements on receivers. A calibrated, variable noise generator can easily be constructed using a diode with adjustable bias current. Figure 1.11 shows a simplified circuit for such a generator.

Example 1.6 A diode noise generator is required to produce 10 μV of noise in a receiver with an input impedance of 75 Ω, resistive, and a noise power bandwidth of 200 kHz. (These values are typical of FM broadcast receivers.) What must the current through the diode be?

Solution First, convert the noise voltage to current, using Ohm's Law:

$$I_N = \frac{V_N}{R} = \frac{10\ \mu\text{V}}{75\ \Omega} = 0.133\ \mu\text{A}$$

Next, rearrange Equation (1.14) and solve for I_0:

$$I_N = \sqrt{2qI_0B}$$

$$I_N^2 = 2qI_0B$$

$$I_0 = \frac{I_N^2}{2qB} = \frac{(0.133 \times 10^{-6}\ \text{A})^2}{2\,(1.6 \times 10^{-19}\ \text{C})\,(200 \times 10^3\ \text{Hz})} = 0.276\ \text{A} \quad \text{or} \quad 276\ \text{mA}$$

Partition Noise Partition noise is similar to shot noise in its spectrum and mechanism of generation, but it occurs only in devices where a single current separates into two or more paths. An example of such a device is a bipolar junction transistor, where the emitter current is the sum of the collector and base currents. As the charge carriers divide into one stream or the other, a random element in the currents is produced. A similar effect can occur in vacuum tubes.

The amount of partition noise depends greatly on the characteristics of the particular device, so no equation for calculating it will be given here. The same is true of shot noise in devices with three or more terminals. Device manufacturers provide noise figure information on their data sheets when a device is intended for use in circuits where signal levels are low and noise is important.

Partition noise is not a problem in field-effect transistors, where the gate current is negligible.

Excess Noise Excess noise is also called *flicker noise* or $1/f$ noise (because the noise power varies inversely with frequency). Sometimes it is called *pink noise*, because there is proportionately more energy at the low-frequency end of the spectrum than with white noise, just as pink light has a higher proportion of red (the low-frequency end of the visible spectrum) than does white light. Excess noise is found in tubes but is a more serious problem in semiconductors and in carbon resistors. It is not fully understood, but it is believed to be caused by variations in carrier density. Excess noise is rarely a problem in communication circuits, because it declines with increasing frequency and is usually insignificant above approximately 1 kHz.

You may meet the term "pink noise" again. It refers to any noise that has equal power per octave rather than per hertz. Pink noise is often used for testing and setting up audio systems.

Transit-Time Noise Many junction devices produce more noise at frequencies approaching their cutoff frequencies. This high-frequency noise occurs when the time taken by charge carriers to cross a junction is comparable to the period of the signal. Some of the carriers may diffuse back across the junction, causing a fluctuating current that constitutes noise. Since most devices are used well below the frequency at which transit-time effects are significant, this type of noise is not usually important (except in some microwave devices).

1.4.3 Addition of Noise from Different Sources

All of the noise sources we have looked at have random waveforms. The amplitude at a particular time cannot be predicted, though the average voltage is zero and the RMS voltage or current and average power are predictable (and *not* zero). Signals like this are said to be uncorrelated; that is, they have no definable phase relationship. In such a case, voltages and currents do not add directly, but the total voltage (of series circuits) or current (of parallel circuits) can be found by taking the square root of the sum of the squares of the individual voltages or currents.

Mathematically, we can say, for voltage sources in series:

$$V_{Nt} = \sqrt{V_{N1}^2 + V_{N2}^2 + V_{N3}^2 + \cdots} \tag{1.15}$$

and similarly, for current sources in parallel:

$$I_{Nt} = \sqrt{I_{N1}^2 + I_{N2}^2 + I_{N3}^2 + \cdots} \tag{1.16}$$

Example 1.7

The circuit in Figure 1.12 shows two resistors in series at two different temperatures. Find the total noise voltage and noise power produced at the load, over a bandwidth of 100 kHz.

Figure 1.12

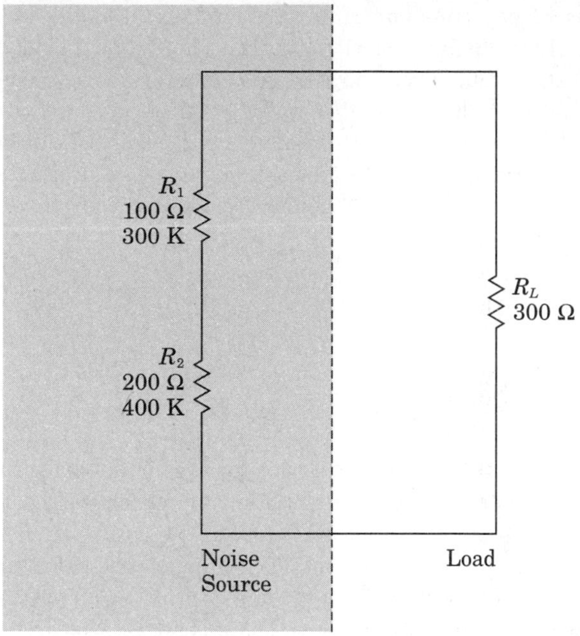

Solution The open-circuit noise voltage can be found from Equation (1.15).

$$\begin{aligned}
V_{Nt} &= \sqrt{V_{R_1}^2 + V_{R_2}^2} \\
&= \sqrt{(\sqrt{4kTBR_1})^2 + (\sqrt{4kTBR_2})^2} \\
&= \sqrt{4kT_1BR_1 + 4kT_2BR_2} \\
&= \sqrt{4kB(T_1R_1 + T_2R_2)} \\
&= \sqrt{4\ (1.38 \times 10^{-23}\ \text{J/K})\ (100 \times 10^3\ \text{Hz})\ [(300\ \text{K})\ (100\ \Omega) + (400\ \text{K})\ (200\ \Omega)]} \\
&= 779\ \text{nV}
\end{aligned}$$

This is, of course, the open-circuit noise voltage for the resistor combination. Since in this case the load is equal in value to the sum of the resistors, one-half of this voltage, or 390 nV, will appear across the load.

The easiest way to find the load power is from Equation (1.11):

$$\begin{aligned}
P &= \frac{V^2}{R} \\
&= \frac{(390\ \text{nV})^2}{300\ \Omega} \\
&= 0.506 \times 10^{-15}\ \text{W} \\
&= 0.506\ \text{fW}
\end{aligned}$$

1.4.4 Signal-to-Noise Ratio

The main reason for studying and calculating noise power or voltage is the effect that noise has on the desired signal. In an analog system, noise makes the signal unpleasant to watch or listen to and, in extreme cases, difficult to understand. In digital systems, noise increases the error rate. In either case, it is not really the amount of noise that concerns us but rather the amount of noise compared to the level of the desired signal. That is, it is the *ratio* of signal to noise power that is important, rather than the noise power alone. This **signal-to-noise** ratio, abbreviated *S/N* and usually expressed in decibels, is one of the most important specifications of any communication system. Typical values of S/N range from about 10 dB for barely intelligible speech to 90 dB or more for compact-disc audio systems.

Note that the ratio given above is a power ratio, not a voltage ratio. Sometimes calculations result in a voltage ratio that must be converted to a power ratio in order to make meaningful comparisons. Of course, the value in decibels will be the same either way, provided that both signal and noise are measured in the same circuit, so that the impedances are the same. The two equations below are used for power ratios and voltage ratios, respectively:

$$S/N \text{ (dB)} = 10 \log \frac{P_S}{P_N} \tag{1.17}$$

$$S/N \text{ (dB)} = 20 \log \frac{V_S}{V_N} \tag{1.18}$$

At this point, the reader who is not familiar with decibel notation should refer to Appendix A. Decibels are used constantly in communication theory and practice, and it is important to understand their use.

Although the signal-to-noise ratio is a fundamental characteristic of any communication system, it is often difficult to measure. For instance, it may be possible to measure the noise power by turning off the signal, but it is not possible to turn off the noise in order to measure the signal power alone. Consequently, a variant of S/N, called $(S+N)/N$, is often found in receiver specifications. It stands for the ratio of signal-plus-noise power to noise power alone.

There are other variations of the signal-to-noise ratio. For instance, there is the question of what to do about distortion. Though it is not a random effect like thermal noise, the effect on the intelligibility of a signal is rather similar. Consequently, it might be a good idea to include it with the noise when measuring. This leads to the expression SINAD which stands for $(S+N+D)/(N+D)$, or signal-plus-noise and distortion, divided by noise and distortion. SINAD is usually used instead of S/N in specifications for FM receivers.

Both $(S+N)/N$ and SINAD are power ratios, and they are almost always expressed in decibels. Details on the measurement of $(S+N)/N$ will be provided in Chapter 5, and SINAD will be described in Chapter 8.

Example 1.8

A receiver produces a noise power of 200 mW with no signal. The output level increases to 5 W when a signal is applied. Calculate $(S+N)/N$ as a power ratio and in decibels.

Solution

The power ratio is

$$\begin{aligned}(S+N)/N &= \frac{5 \text{ W}}{0.2 \text{ W}} \\ &= 25\end{aligned}$$

In decibels, this is

$$\begin{aligned}(S+N)/N \text{ (dB)} &= 10 \log 25 \\ &= 14 \text{ dB}\end{aligned}$$

1.4.5 Noise Figure

Since thermal noise is produced by all conductors and since active devices add their own noise as described above, it follows that any stage in a communication system will add noise. An amplifier, for instance, will amplify equally both the signal and the noise at its input, but it will also add some noise. The signal-to-noise ratio at the output will therefore be lower than at the input.

Noise figure (abbreviated *NF* or just *F*) is a figure of merit, indicating how much a component, stage, or series of stages degrades the signal-to-noise ratio of a system. The noise figure is, by definition:

$$NF = \frac{(S/N)_i}{(S/N)_o} \tag{1.19}$$

where

$(S/N)_i$ = input signal-to-noise power ratio (not in dB)
$(S/N)_o$ = output signal-to-noise power ratio (not in dB)

Very often both *S/N* and *NF* are expressed in decibels, in which case we have:

$$NF\ (\text{dB}) = (S/N)_i\ (\text{dB}) - (S/N)_o\ (\text{dB}) \tag{1.20}$$

Since we are dealing with power ratios, it should be obvious that:

$$NF\ (\text{dB}) = 10 \log NF\ (\text{ratio}) \tag{1.21}$$

Noise figure is occasionally called *noise factor*. Sometimes the term *noise ratio* (*NR*) is used for the simple ratio of Equation (1.19), reserving the term "noise figure" for the decibel form given by Equation (1.20).

Example 1.9

The signal power at the input to an amplifier is 100 μW and the noise power is 1 μW. At the output, the signal power is 1 W and the noise power is 30 mW. What is the amplifier noise figure, as a ratio?

Solution

$$(S/N)_i = \frac{100\ \mu\text{W}}{1\ \mu\text{W}} = 100$$

$$(S/N)_o = \frac{1\ \text{W}}{0.03\ \text{W}} = 33.3$$

$$NF\ (\text{ratio}) = \frac{100}{33.5} = 3$$

Example 1.10

The signal at the input of an amplifier has an *S/N* of 42 dB. If the amplifier has a noise figure of 6 dB, what is the *S/N* at the output (in decibels)?

Solution

The *S/N* at the output can be found by rearranging Equation (1.20):

$$\begin{aligned} NF\ (\text{dB}) &= (S/N)_i\ (\text{dB}) - (S/N)_o\ (\text{dB}) \\ (S/N)_o\ (\text{dB}) &= (S/N)_i\ (\text{dB}) - NF\ (\text{dB}) \\ &= 42\ \text{dB} - 6\ \text{dB} \\ &= 36\ \text{dB} \end{aligned}$$

1.4.6 Equivalent Noise Temperature

Equivalent **noise temperature** is another way of specifying the noise performance of a device. Noise temperature has nothing to do with the actual operating temperature of the circuit. Rather, as shown in Figure 1.13, it is the absolute temperature of a resistor that, connected to the input of a noiseless amplifier of the same gain, would produce the same noise at the output as the device under discussion. Figure 1.13(a) shows a real amplifier which, of course, produces noise at its output, even with no input signal. Figure 1.13(b) shows a hypothetical noiseless amplifier with a noisy resistor at its input. The combination produces the same amount of a noise at its output as does the real amplifier of Figure 1.13(a).

Figure 1.13 Equivalent noise resistance

(a) Real (noisy) amplifier

(b) Equivalent with noise resistance and ideal (noiseless) amplifier

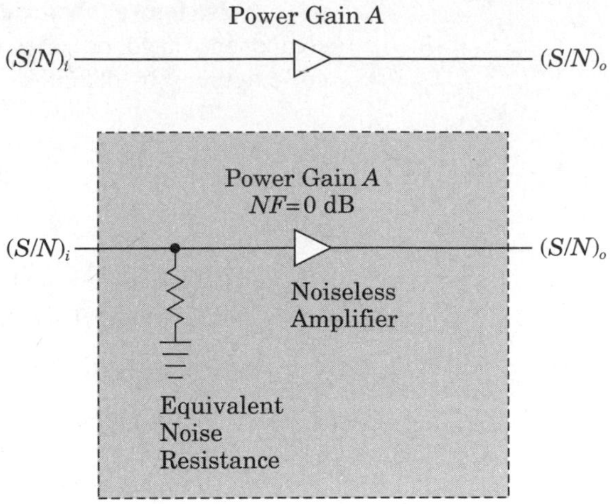

From Equation (1.19), the noise figure is given by:

$$NF = \frac{(S/N)_i}{(S/N)_o} = \frac{(S_i/N_i)}{(S_o/N_o)} = \frac{S_iN_o}{N_iS_o} = \frac{S_i}{S_o}\frac{N_o}{N_i} \quad (1.22)$$

where

S_i = signal power at input
N_i = noise power at input
S_o = signal power at output
N_o = noise power at output

If we are going to refer all the amplifier noise to the input, then this noise, as well as the noise from the signal source, will be amplified by the gain of the amplifier. Let the amplifier power gain be A. Then:

$$A = \frac{S_o}{S_i} \quad (1.23)$$

Substituting this into Equation (1.22), we get

$$NF = \frac{N_o}{N_iA} \quad (1.24)$$

or

$$N_o = (NF)N_iA \quad (1.25)$$

Thus the total noise at the input is $(NF)N_i$.

Assuming that the input source noise N_i is thermal noise, it has the expression, first shown in Equation (1.10):

$$N_i = kTB$$

The equivalent input noise generated by the amplifier must be

$$\begin{aligned} N_{eq} &= (NF)N_i - N_i \\ &= (NF)kTB - kTB \\ &= (NF - 1)kTB \end{aligned} \tag{1.26}$$

If we suppose that this noise is generated in a (fictitious) resistor at temperature T_{eq}, and if we further suppose that the actual source is at a reference temperature T_0 of 290 K, then we get:

$$\begin{aligned} kT_{eq}B &= (NF - 1)kT_0B \\ T_{eq} &= (NF - 1)T_0 \\ &= 290(NF - 1) \end{aligned} \tag{1.27}$$

This may seem a trivial result, because it turns out that T_{eq} can be found directly from the noise figure and thus contains no new information about the amplifier. However, T_{eq} is very useful when we look at microwave receivers connected to antennas by transmission lines. The antenna has a noise temperature due largely to space noise intercepted by the antenna. The transmission line and the receiver will also have noise temperatures. It is possible to get an equivalent noise temperature for the whole system by simply adding the noise temperatures of the antenna, transmission line, and receiver. From this, we can calculate the system noise figure merely by rearranging Equation (1.27):

$$\begin{aligned} T_{eq} &= 290(NF - 1) \\ NF - 1 &= \frac{T_{eq}}{290} \\ NF &= \frac{T_{eq}}{290} + 1 \end{aligned} \tag{1.28}$$

Equivalent noise temperatures of low-noise amplifiers are quite low, often less than 100 K. Note that this does not mean the amplifier operates at this temperature. It is quite common for an amplifier operating at an actual temperature of 300 K to have an equivalent noise temperature of 100 K.

Example 1.11 An amplifier has a noise figure of 2 dB. What is its equivalent noise temperature?

Solution First, we need the noise figure expressed as a ratio. This can be found from Equation (1.21):

$$NF\text{ (dB)} = 10 \log NF\text{ (ratio)}$$

From which we get:

$$\begin{aligned} NF\text{ (ratio)} &= \text{antilog} \frac{NF\text{ (dB)}}{10} \\ &= \text{antilog } 0.2 \\ &= 1.585 \end{aligned}$$

Now we can use Equation (1.27) to find the noise temperature:

$$\begin{aligned} T_{eq} &= 290(NF - 1) \\ &= 290(1.585 - 1) \\ &= 169.6\text{ K} \end{aligned}$$

1.4.7 Cascaded Amplifiers

When two or more stages are connected in *cascade*, as in a receiver, the noise figure of the first stage is the most important in determining the noise performance of the entire system because noise generated in the first stage is amplified in all succeeding stages. Noise pro-

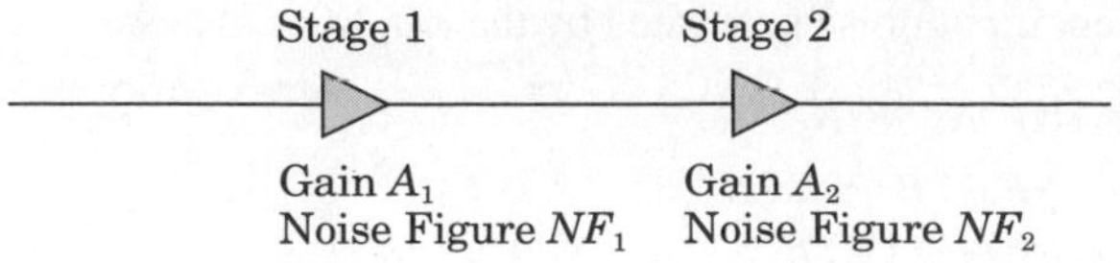

Figure 1.14
Noise figure for a two-stage amplifier

duced in later stages is amplified less, and noise generated in the last stage is amplified least of all.

While the first stage is the most important noise contributor, the other stages cannot be neglected. It is possible to derive an equation that relates the total noise figure to the gain and noise figure of each stage. In the following discussion, the power gains and noise figures are expressed as ratios, not in decibels.

We will start with a two-stage amplifier like the one shown in Figure 1.14. The first stage has a gain A_1 and a noise figure NF_1, and the second-stage gain and noise figure are A_2 and NF_2, respectively.

Let the noise power input to the first stage be

$$N_{i1} = kTB$$

Then, according to Equation (1.25), the noise at the output of the first stage is:

$$\begin{aligned} N_{o1} &= NF_1 N_{i1} A_1 \\ &= NF_1 kTBA_1 \end{aligned} \tag{1.29}$$

This noise appears at the input of the second stage. It is amplified by the second stage and appears at the output multiplied by A_2. However, the second stage also contributes noise, which can be expressed, according to Equation (1.26), as:

$$N_{eq2} = (NF_2 - 1)(kTB) \tag{1.30}$$

This noise also appears at the output, amplified by A_2. Therefore the total noise power at the output of the two-stage system is:

$$\begin{aligned} N_{o2} &= N_{o1}A_2 + N_{eq2}A_2 \\ &= NF_1 A_1 kTBA_2 + (NF_2 - 1)kTBA_2 \\ &= kTBA_2(NF_1 A_1 + NF_1 - 1) \end{aligned} \tag{1.31}$$

The noise figure for the whole system (NF_T) is, according to Equation (1.24):

$$NF_T = \frac{N_{o2}}{N_{i1}A_T}$$

Substituting the value for N_{o2} found in Equation (1.31) and making use of the fact that the total gain A_T is the product of the individual stage gains, we get

$$\begin{aligned} NF_T &= \frac{kTBA_2(NF_1A_1 + NF_2 - 1)}{kTBA_1A_2} \\ &= \frac{NF_1A_1 + NF_2 - 1}{A_1} \\ &= NF_1 + \frac{NF_2 - 1}{A_1} \end{aligned} \tag{1.32}$$

This equation, known as *Friis' formula*, can be generalized to any number of stages:

$$NF_T = NF_1 + \frac{NF_2 - 1}{A_1} + \frac{NF_3 - 1}{A_1A_2} + \frac{NF_4 - 1}{A_1A_2A_3} + \cdots \tag{1.33}$$

This expression confirms our intuitive feeling by showing that the contribution of each stage is divided by the product of the gains of all the preceding stages. Thus the effect of the first stage is usually dominant. Note once again that the noise figures given in this section are ratios, not decibel values, and that the gains are power gains.

Example 1.12 A three-stage amplifier has stages with the following specifications:

Stage	Power gain	Noise figure
1	10	2
2	25	4
3	30	5

Calculate the power gain, noise figure, and noise temperature for the entire amplifier, assuming matched conditions.

Solution The power gain is simply the product of the individual gains.

$$A_T = 10 \times 25 \times 30 = 7500$$

The noise figure can be found from Equation (1.33):

$$\begin{aligned} NF_T &= NF_1 + \frac{NF_2 - 1}{A_1} + \frac{NF_3 - 1}{A_1 A_2} \\ &= 2 + \frac{4 - 1}{10} + \frac{5 - 1}{10 \times 25} \\ &= 2.316 \end{aligned}$$

If answers are required in decibels, they can easily be found:

$$\begin{aligned} \text{gain (dB)} &= 10 \log 7500 = 38.8 \text{ dB} \\ NF \text{ (dB)} &= 10 \log 2.316 = 3.65 \text{ dB} \end{aligned}$$

The noise temperature can be found from Equation (1.27):

$$\begin{aligned} T_{eq} &= 290(NF - 1) \\ &= 290(2.316 - 1) \\ &= 382 \text{ K} \end{aligned}$$

1.5 Test Equipment: The Spectrum Analyzer

In this chapter, we have discovered that there are two general ways of looking at signals: the time domain and the frequency domain. The ordinary oscilloscope is a convenient and versatile way of observing signals in the time domain, providing as it does a graph of voltage with respect to time. It would be logical to ask whether there is an equivalent instrument for the frequency domain: that is, something which could display a graph of voltage (or power) with respect to frequency.

The answer to the question is yes: observation of signals in the frequency domain is the function of **spectrum analyzers**, which use one of two basic techniques. The first technique uses a number of fixed-tuned bandpass filters, spaced throughout the frequency range of interest. The display indicates the output level from each filter, often with a bar graph consisting of light-emitting diodes (LEDs). The more filters and the narrower the range, the better the resolution that can be obtained. This method is quite practical for audio frequencies, and in fact a crude version, with five to twenty filters, is often found as part of a stereo system. More elaborate systems are often used in professional acoustics and sound-reinforcement work. Devices of this type are called *real-time spectrum analyzers*, as all of

Figure 1.15
Real-time spectrum analyzer

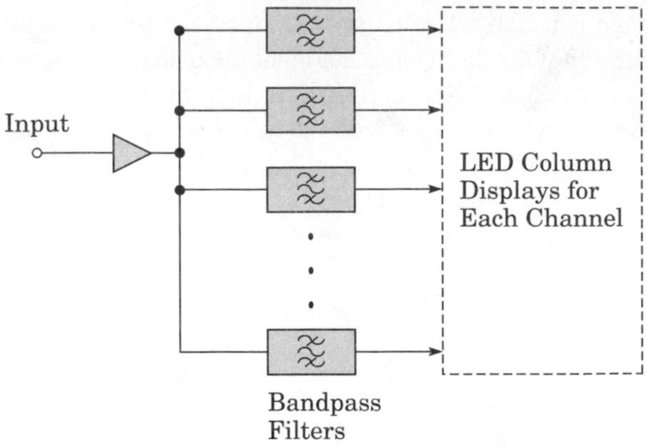

the filters are active all the time. Figure 1.15 shows a block diagram of a real-time spectrum analyzer.

When radio-frequency (RF) measurements are required, the limitations of real-time spectrum analyzers become very apparent. It is often necessary to observe signals over a frequency range with a width of megahertz or even gigahertz. Obviously, the number of filters required would be too large to be practical. For RF applications, it would be more logical to use only one, a tunable filter, and sweep it gradually across the frequency range of interest. This sweeping of the filter could coincide with the sweep of an electron beam across the face of a cathode-ray tube (CRT), and the vertical position of the beam could be made a function of the amplitude of the filter output. This is the essense of a *swept-frequency spectrum analyzer* like the one shown in Figure 1.16.

In practice, narrow-band filters that are rapidly tunable over a wide range with constant bandwidth are difficult to build. It is easier to build a single, fixed-frequency filter with adjustable bandwidth and sweep the signal through the filter. Figure 1.17 shows how this can be done. The incoming signal is applied to a mixer, along with a swept-frequency signal generated by a local oscillator in the analyzer. Mixers will be described in the next chapter; for now, it is sufficient to know that a mixer will produce outputs at the sum and the difference of the two frequencies that are applied to its input.

The filter can be arranged to be at either the sum or the difference frequency (usually the difference). As the oscillator frequency varies, the part of the spectrum that passes

Figure 1.16
Swept-frequency spectrum analyzer
Photograph courtesy of Hewlett-Packard Company.

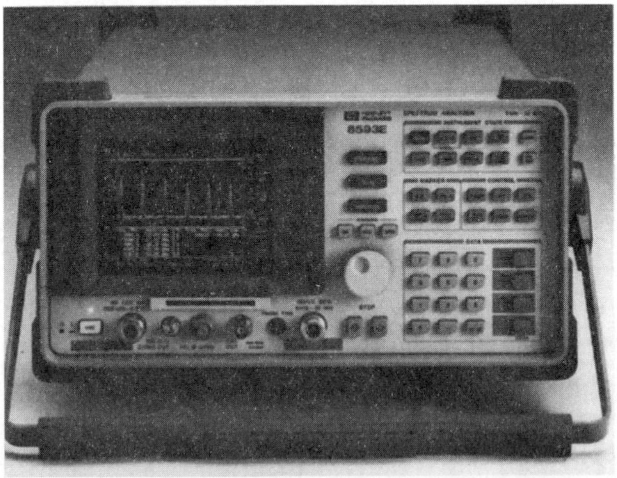

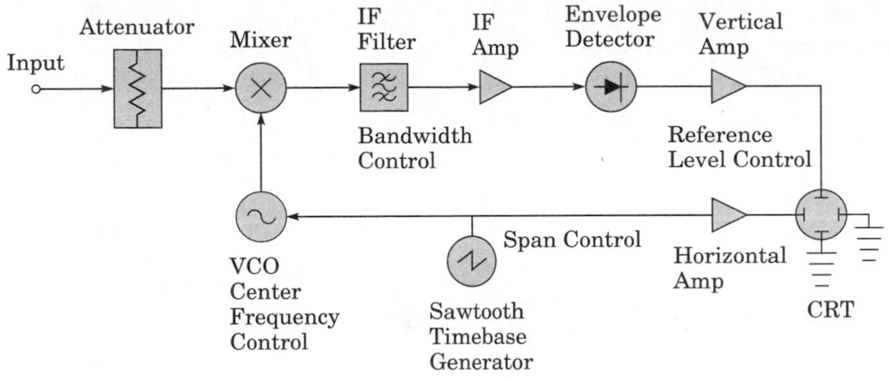

Figure 1.17 Simplified block diagram for a swept-frequency spectrum analyzer

through the filter changes. The oscillator is a voltage-controlled oscillator (VCO) whose frequency is controlled by a sawtooth wave generator that also provides the horizontal sweep signal for the CRT.

The filter output is an ac signal that must be rectified and amplified before it is applied to the vertical deflection plates of the CRT. If the amplification is linear, the vertical position of the trace will be proportional to the signal voltage amplitude at a given frequency. It is more common, however, to use logarithmic amplification, so that the display can be calibrated in decibels, with a reference level in decibels referenced to one milliwatt (dBm).

1.5.1 Using a Spectrum Analyzer

A typical swept-frequency spectrum analyzer has four main controls, which are shown on the block diagram of Figure 1.17. Their functions are described in the following paragraphs.

The *frequency control* sets either the center frequency of the sweep or the frequency at which it begins, called the start frequency (this is the frequency corresponding to the left edge of the display). Some analyzers provide a switch so that the display can show either the start or the center frequency.

The *span control* adjusts the range of frequencies on the display. There are two methods of specifying the span. Sometimes it is calibrated in kilohertz or megahertz per division. Other models give the span as the width of the whole display, which is normally ten divisions. Dividing this number by ten will give the span per division.

The frequency of a displayed signal can be found by adjusting the frequency control (also called *tuning control*) until the signal is in the center of the display (assuming the start/center control, if present, is at center). The frequency of a signal not at the center can be found by counting divisions and multiplying by the span per division.

The *bandwidth control* adjusts the filter bandwidth at the 3 dB points. Since the smaller the bandwidth, the better the potential resolution of the analyzer, it might seem logical to build the filter with as narrow a bandwidth as possible and use it that way at all times. However, sweeping a signal rapidly through a high-Q circuit (such as a narrow bandpass filter) gives rise to transients, which will reduce the accuracy of amplitude readings. If the frequency span being swept is small, a narrow bandwidth is necessary so that good resolution is obtained, but when a wide span is in use, narrow bandwidth is unnecessary. This is fortunate, for the sweep will naturally be faster, in terms of hertz per second, for the wider span. For wide spans, then, a wide bandwidth is appropriate, with narrow bandwidth being reserved for narrow span. Very often, the bandwidth and span controls are locked together, so that the correct filter bandwidth for a given span will be selected automatically. Provision is made for unlocking the span and bandwidth controls on the rare occasions when this is desirable.

It is important to realize that as viewed on a spectrum analyzer, a sine wave or a component of a more complex signal does not always appear as a single vertical line. If the

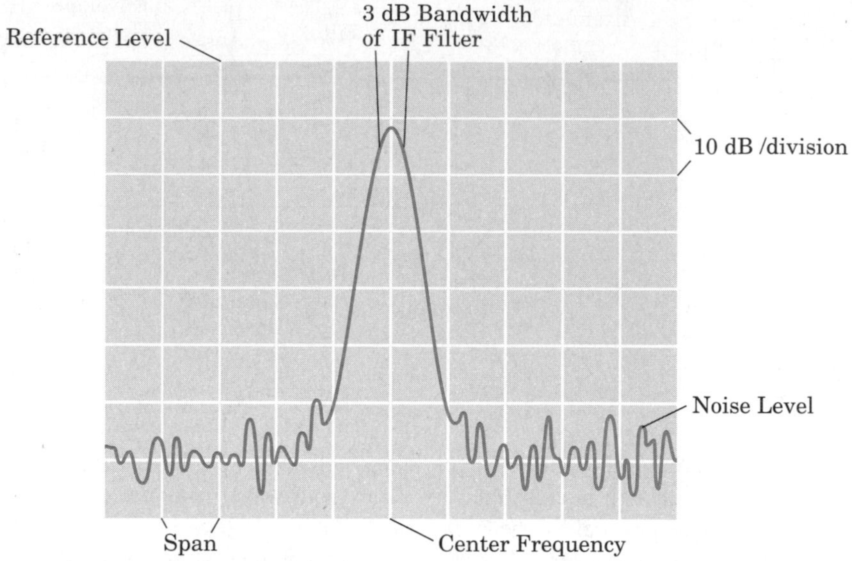

Figure 1.18 Spectrum analyzer display

span is sufficiently narrow that the filter bandwidth is a significant portion of a division, the signal will appear to have a finite width. Such a display is shown in Figure 1.18. What is seen in the display is the width and shape of the filter passband, not that of the signal.

The fourth major adjustment is the *reference level control*. Spectrum analyzers are set up so that amplitudes are measured with respect to a reference level at the top of the display (not the bottom, as one might assume). This level is set by the reference level control, which is normally calibrated in dBm.

There will also be an adjustable *attenuator* at the input to reduce strong signals to a level that can be handled by the input mixer. The attenuator control is normally coupled with the reference level setting, so that changing the attenuation automatically alters the displayed reference level.

A switch that sets the *vertical scale factor* is often closely associated with the reference level control. This switch typically has three settings: 10 dB/division, 1 dB/division, and linear. In the first position, one division vertically represents a 10 dB power difference at the input. The second position gives an expanded scale, with 1 dB per division. With either of these settings, the power of a signal can be measured in either of two ways: by setting the signal to the top of the display with the reference level controls and then reading the reference level from the controls, or by counting divisions down from the top of the screen to the signal, multiplying by 1 or 10 dB according to the switch setting, and subtracting the result from the reference level setting.

The linear position is a little more complicated. The reference level is at the top of the screen, as before. However, this time the screen is a linear voltage scale. Zero voltage is at the bottom, and the voltage corresponding to the reference level is at the top. This can easily be calculated from the power in dBm and the spectrum analyzer impedance, or you can use the graph in Figure 1.19. Please note, however, that this graph is valid only if the analyzer has an input impedance of 50 Ω, the most common value. Some analyzers allow the reference level to be set directly in volts.

Once the voltage that corresponds to the reference level is known, the signal voltage can be found. If the reference level has been adjusted so that the signal is just at the top of the screen, then the signal voltage is that corresponding to the reference level. Otherwise the signal voltage can be found by using proportion. The signal voltage will be:

$$V_{sig} = \frac{V_{ref} U_{sig}}{N_{div}} \tag{1.34}$$

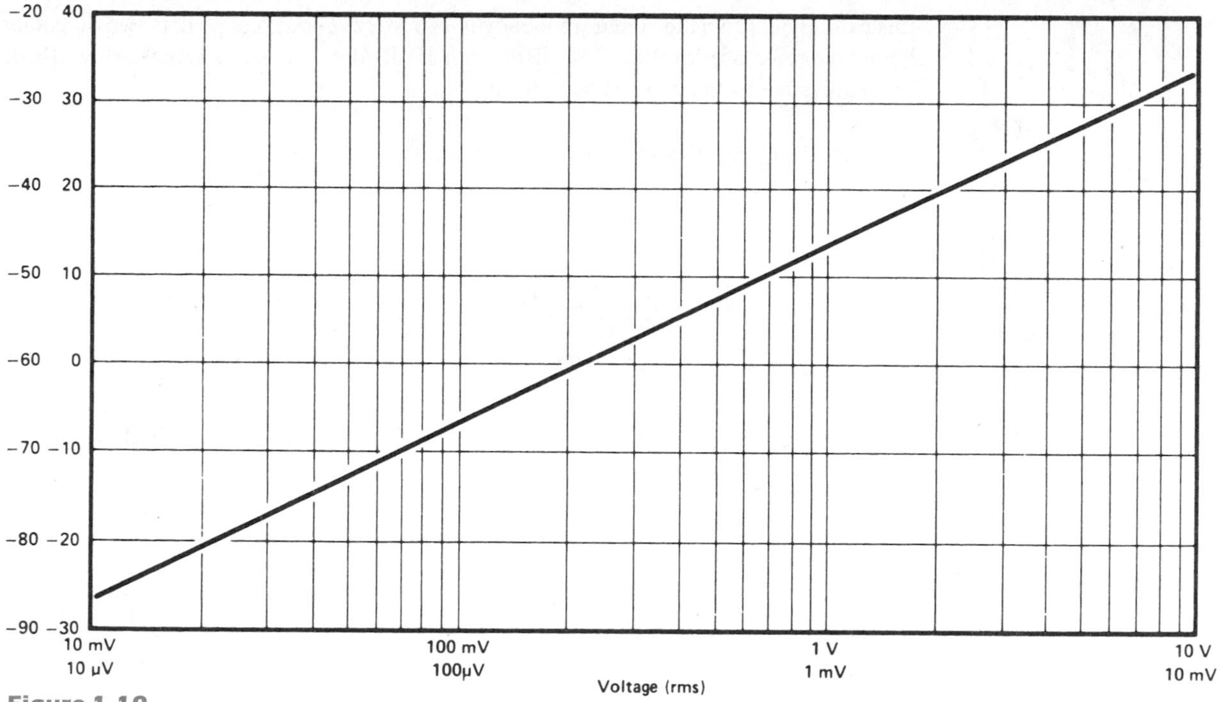

Figure 1.19
Conversion chart, dBm to voltage (for 50 Ω)
Courtesy Hewlett-Packard Company.

where

V_{sig} = signal voltage in RMS volts
V_{ref} = RMS voltage corresponding to the reference level
U_{sig} = position of the signal in divisions up from the bottom of the screen
N_{div} = number of divisions vertically on the display (usually 8)

The linear scale is not used very often with spectrum analyzers, but it can be handy on occasion.

Example 1.13

For each of the spectrum-analyzer displays in Figure 1.20, find the frequency, power level (in dBm and watts), and voltage level of the signals. The input impedance of the analyzer is 50 Ω.

Solution

(a) The signal is in the center of the display horizontally, so its frequency is displayed on the analyzer's frequency readout, assuming that it is set to read the center frequency. In

Figure 1.20

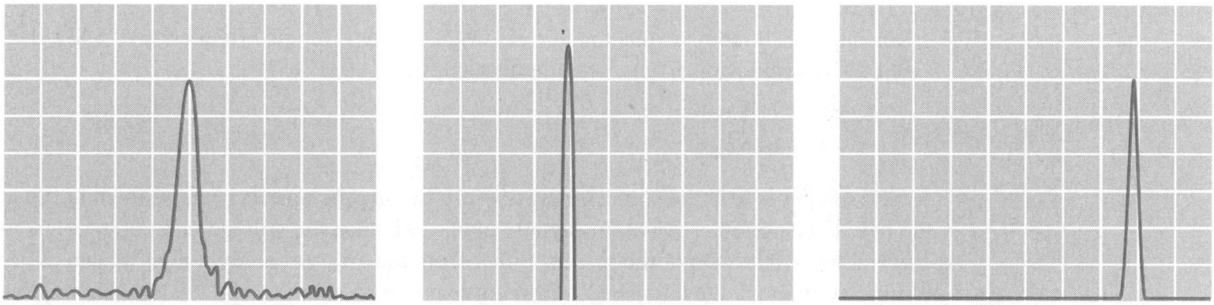

(a) Reference level −10 dBm
Vertical: 10 dB/division

Center frequency 110 MHz
Span: 10 kHz/division

(b) Reference level 10 dBm
Vertical: 1 dB/division

Center frequency 7.5 MHz
Span: 100 kHz/division

(c) Reference level −20 dBm
Vertical: linear

Center frequency 543 MHz
Span: 1 MHz/division

this example it is, so the signal frequency is 110 MHz. The signal peak is two divisions below the reference level of −10 dBm, with 10 dB/division, so its level is −30 dBm. The equivalent power can be found from:

$$P\ (\text{dBm}) = 10 \log \frac{P}{1\ \text{mW}}$$

$$\begin{aligned} P\ (\text{mW}) &= \text{antilog} \frac{\text{dBm}}{10} \\ &= \text{antilog} \frac{-30}{10} \\ &= 1 \times 10^{-3}\ \text{mW} \\ &= 1\ \mu\text{W} \end{aligned}$$

The voltage can be found from the graph in Figure 1.19, or it can be found more accurately from:

$$P = \frac{V^2}{R}$$

$$\begin{aligned} V &= \sqrt{PR} \\ &= \sqrt{1\ \mu\text{V} \times 50\ \Omega} \\ &= 7.07\ \text{mV} \end{aligned}$$

(b) The signal is one division to the left of center, with 100 kHz/division. The frequency is 100 kHz less than the reference frequency of 7.5 MHz, so

$$f = 7.5\ \text{MHz} - 0.1\ \text{MHz} = 7.4\ \text{MHz}$$

With regard to the amplitude, note that the scale is 1 dB/division and the signal is one division below the reference level, so the signal has a power level of

$$P\ (\text{dBm}) = 10\ \text{dBm} - 1\ \text{dB} = 9\ \text{dBm}$$

This can be converted to watts and volts in the same way as in part (a):

$$\begin{aligned} P\ (\text{mW}) &= \text{antilog} \frac{\text{dBm}}{10} \\ &= \text{antilog} \frac{9}{10} \\ &= 7.94\ \text{mW} \end{aligned}$$

$$\begin{aligned} V &= \sqrt{PR} \\ &= \sqrt{7.94\ \text{mW} \times 50\ \Omega} \\ &= 630\ \text{mV} \end{aligned}$$

(c) The signal is three divisions to the right of the center reference frequency of 543 MHz, with 1 MHz/division. Therefore, the frequency is:

$$\begin{aligned} f &= 543\ \text{MHz} + 3 \times 1\ \text{MHz} \\ &= 546\ \text{MHz} \end{aligned}$$

It is easier to find the signal voltage before finding the power, because the amplitude scale here is linear. To arrive at the signal voltage, we need the voltage that corresponds to the power level at the top of the screen. Either from the graph or analytically, we can easily find that the reference level of −20 dBm corresponds to a power of 10 μW, which represents 22.4 mV. The level on the display is $\frac{6}{8}$ of this, because the scale is linear with 8 divisions representing the reference level voltage. Therefore the signal level is:

$$V = 22.4\ \text{mV} \times \frac{6}{8} = 16.8\ \text{mV}$$

Now it is easy to find the power. In watts, it is:

$$P = \frac{V^2}{R} = \frac{(16.8\text{ mV})^2}{50\ \Omega} = 5.64\ \mu\text{W}$$

Now we find the power in dBm:

$$P\ (\text{dBm}) = 10 \log \frac{P}{1\text{ mW}} = 10 \log \frac{5.64\ \mu\text{W}}{1\text{ mW}} = -22.5\text{ dBm}$$

Summary

Here are the main points to remember from this chapter:

1. Any communication system has three essential elements: the transmitter, the receiver, and the communication channel.
2. The design of the transmitter and receiver must take into account the characteristics of the channel, particularly noise, distortion, and limited bandwidth.
3. Modulation is necessary with many types of communication channels. In modulation some characteristic of a carrier waveform is changed in accordance with the amplitude of a lower-frequency signal known as the baseband, information signal, or modulating signal.
4. The carrier characteristics that can be modulated are amplitude, frequency, and phase.
5. Time-division and frequency-division multiplexing are two systems for sharing a channel among several information signals.
6. For a given modulation scheme, the amount of information that can be transmitted is proportional to the time taken and the channel bandwidth employed.
7. Signal transmission is said to be distortionless if the information signal at the receiver output is identical to that at the transmitter input, except for time delay and change in amplitude. Any other change is known as distortion.
8. Signals can be represented in either the time or frequency domain. Fourier series can be used to find the frequency-domain representation of a periodic signal. Signals can be observed in the frequency domain using a spectrum analyzer.
9. Noise is present in all communication systems and has a degrading effect on the signal. There are many types of noise, but the most common is thermal noise, which is a characteristic of all electronic equipment.

Important Equations

Note: For handy reference, each chapter will contain a brief list of the most commonly used equations. These equations must not be used blindly: you must understand how and where to use them. Refer to the text when necessary. These are not the only equations you will need to solve the problems in each chapter.

$$e(t) = E_c \sin(\omega_c t + \theta) \tag{1.1}$$

$$v = f\lambda \tag{1.3}$$

$$f(t) = A_0/2 + A_1 \cos \omega t + B_1 \sin \omega t + A_2 \cos 2\omega t + B_2 \sin 2\omega t + A_3 \cos 3\omega t + B_3 \sin 3\omega t + \cdots \tag{1.5}$$

$$f = 1/T \tag{1.6}$$

$$\omega = 2\pi f \tag{1.7}$$

$$P_N = kTB \tag{1.10}$$

$$V_N = \sqrt{4kTBR} \tag{1.13}$$

$$I_N = \sqrt{2qI_0B} \tag{1.14}$$

$$V_{Nt} = \sqrt{V_{N1}^2 + V_{N2}^2 + V_{N3}^2 + \cdots} \tag{1.15}$$

$$S/N\,(\text{dB}) = 10 \log (P_S/P_N) \tag{1.17}$$

$$S/N\,(\text{dB}) = 20 \log (V_S/V_N) \tag{1.18}$$

$$NF = \frac{(S/N)_i}{(S/N)_o} \tag{1.19}$$

$$NF\,(\text{dB}) = (S/N)_i\,(\text{dB}) - (S/N)_o\,(\text{dB}) \tag{1.20}$$

$$T_{eq} = 290(NF - 1) \tag{1.27}$$

$$NF_T = NF_1 + \frac{NF_2 - 1}{A_1} + \frac{NF_3 - 1}{A_1A_2} + \frac{NF_4 - 1}{A_1A_2A_3} + \cdots \tag{1.33}$$

Glossary

baseband the band of frequencies occupied by an information signal before it modulates the carrier

carrier a signal that can be modulated by an information signal

channel a path for the transmission of signals

distortion any undesirable change in an information signal

Fourier series a way of representing periodic functions as a series of sinusoids

frequency domain a representation of a signal's power or amplitude as a function of frequency

modulation the process by which some characteristic of a carrier is varied by an information signal

multiplexing the transmission of more than one information signal over a single channel

noise any undesired disturbance that is superimposed on a signal and obscures its information content

noise figure ratio of the input and output signal-to-noise ratios for a device

noise temperature equivalent temperature of a passive system having the same noise-power output as a given system

receiver device to extract the information signal from the signal propagating along a channel

signal-to-noise ratio (S/N) ratio of signal to noise power at a given point in a system

spectrum analyzer device for displaying signals in the frequency domain

time domain representation of a signal's amplitude as a function of time

transmission transfer of an information signal from one location to another

transmitter device that converts an information signal into a form suitable for propagation along a channel

Questions

1. Identify the band of each of the following frequencies.
   (a) 10 MHz (used for standard time and frequency broadcasts)
   (b) 2.45 GHz (used for microwave ovens)
   (c) 100 kHz (used for the LORAN navigation system for ships and aircraft)
   (d) 4 GHz (used for satellite television)
   (e) 880 MHz (used for cellular telephones)
2. Suppose that a voice frequency of 400 Hz is transmitted on an AM radio station operating at 1020 kHz. Which of these frequencies is
   (a) the information frequency?
   (b) carrier frequency?
   (c) the baseband frequency?
   (d) the modulating frequency?
3. Which parameters of a sine-wave carrier can be modulated?

4. Explain briefly the concept of frequency-division multiplexing and give one practical example of its use.
5. Explain briefly the concept of time-division multiplexing and give one practical example of its use.
6. Describe what is meant by each of the following types of distortion:
   (a) harmonic distortion
   (b) nonlinear frequency response
   (c) nonlinear phase response
7. (a) What is the theoretical bandwidth of a 1 kHz square wave?
   (b) Suggest a bandwidth that would give reasonably good results in a practical transmission system.
   (c) Suppose that a 1 kHz square wave is transmitted through a channel that cannot pass frequencies above 2 kHz. What would the shape of the wave be after passing through the channel?
8. Describe the way in which each of the following types of noise is generated:
   (a) equipment
   (b) atmospheric
   (c) space
   (d) thermal
   (e) shot
   (f) partition
   (g) excess
   (h) transit-time
9. What is the difference between white and pink noise?
10. What is meant by the noise power bandwidth of a system?
11. Why is the noise power bandwidth greater than the half-power bandwidth for a typical bandpass filter?
12. Why are amplifiers that must work with very weak signals sometimes cooled to extremely low temperatures? What is this process called?
13. What is meant by the signal-to-noise ratio? Why is it important in communication systems?
14. Give two ratios that are similar to the signal-to-noise ratio and are easier to measure.
15. Explain how noise figure is related to signal-to-noise ratio.
16. What is the relationship between noise figure and noise temperature?
17. Explain why the first stage in an amplifier is the most important in determining the noise figure for the entire amplifier.
18. Name the two basic types of spectrum analyzer, and briefly describe how each works.
19. Name the four most important controls on a swept-frequency spectrum analyzer, and describe the function of each.
20. What happens to the noise level displayed on a spectrum analyzer as its bandwidth is increased? Why?

Problems

SECTION 1.2

21. Visible light consists of electromagnetic radiation with wavelengths between 400 and 700 nanometers (nm). Express this range in terms of frequency.
22. Equation (1.3) applies to any kind of wave. The velocity of sound waves in air is about 344 m/s. Calculate the wavelength of a sound wave with a frequency of 1 kHz.

SECTION 1.3

23. Sketch the spectrum for the half-wave rectified signal in Figure 1.21, showing harmonics up to the fifth. Show the voltage and frequency scales, and indicate whether your voltage scale shows peak or RMS voltage.

Figure 1.21

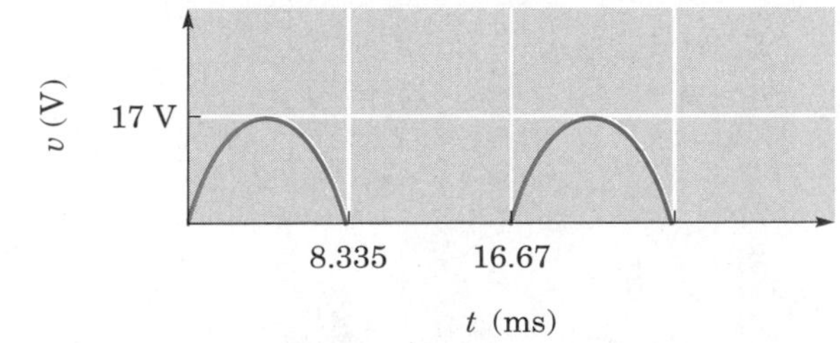

24. Sketch the frequency spectrum for the triangle wave shown in Figure 1.22, for harmonics up to the fifth. Show the voltage and frequency scales.

Figure 1.22

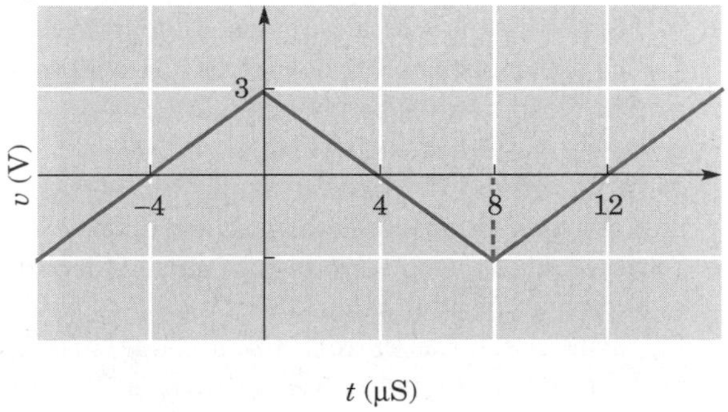

25. A 1 kHz square wave passes through each of three communication channels whose bandwidths are given below. Sketch the output in the time domain for each case.

(a) 0 to 10 kHz
(b) 2 kHz to 4 kHz
(c) 0 to 4 kHz

26. Sketch the spectrum for the pulse train shown in Figure 1.23.

Figure 1.23

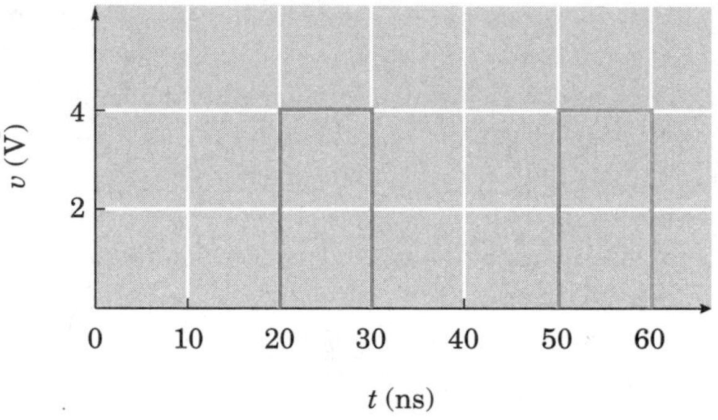

27. Sketch the spectrum for the sawtooth waveform in Figure 1.24. Explain why this waveform has no dc component, unlike the sawtooth waveform in Example 1.3.

Figure 1.24

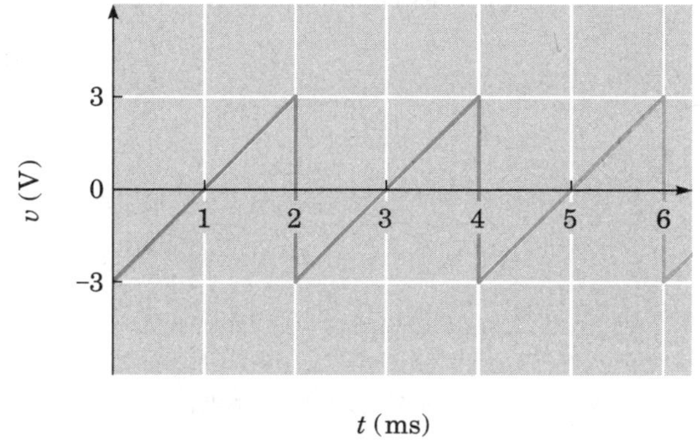

SECTION 1.4

28. A 50 Ω resistor operates at room temperature (21°C). How much noise power does it provide to a matched load over the bandwidth of
(a) a CB radio channel (10 kHz)?
(b) a TV channel (6 MHz)?
Express your answers in both watts and dBm.

29. What would be the noise voltage generated for each of the conditions in Problem 28?

30. Calculate the noise current for a diode with a bias current of 15 mA, observed over a 25 kHz bandwidth.

31. The signal-to-noise ratio is 30 dB at the input to an amplifier and 27.3 dB at the output.
(a) What is the noise figure of the amplifier?
(b) What is its noise temperature?

32. (a) The signal voltage at the input of an amplifier is 100 μV, and the noise voltage is 2 μV. What is the signal-to-noise ratio in decibels?
(b) The signal-to-noise ratio at the output of the same amplifier is 30 dB. What is the noise figure of the amplifier?

33. (a) A receiver has a noise temperature of 100 K. What is its noise figure in decibels?
(b) A competing company has a receiver with a noise temperature of 90 K. Assuming its other specifications are equal, is this receiver better or worse than the one in part (a)? Explain your answer.

34. Suppose the noise power at the input to a receiver is 1 nW in the bandwidth of interest. What would be the required signal power for a signal-to-noise ratio of 25 dB?

35. A three-stage amplifier has the following power gains and noise figures (as ratios, not in decibels) for each stage:

Stage	Power gain	Noise figure
1	10	3
2	20	4
3	30	5

Calculate the total gain and noise figure, and convert both to decibels.

SECTION 1.5

36. The display in Figure 1.25 shows the fundamental of a sine-wave oscillator and its second harmonic, as seen on a spectrum analyzer. The analyzer has an input impedance of 50 Ω. Find:
(a) the oscillator frequency
(b) the amplitude of the signal, in dBm and in volts
(c) the amount in decibels by which the second harmonic is less than the fundamental (usually expressed as "x dB down")

Figure 1.25
Reference level 10 dBm
Vertical: 10 dB/division
Center frequency 100 MHz
Span: 20 MHz/division

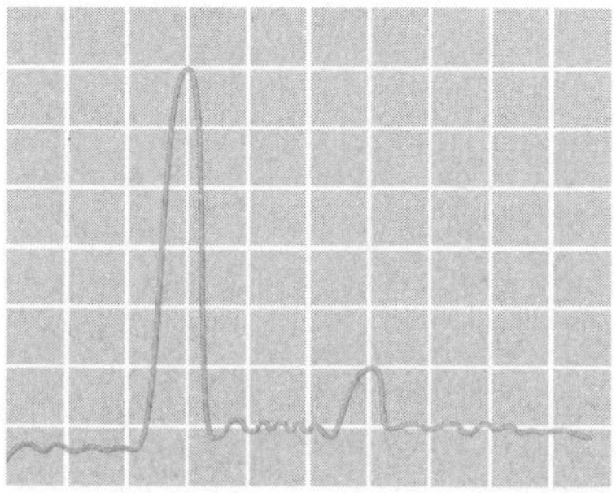

Figure 1.26

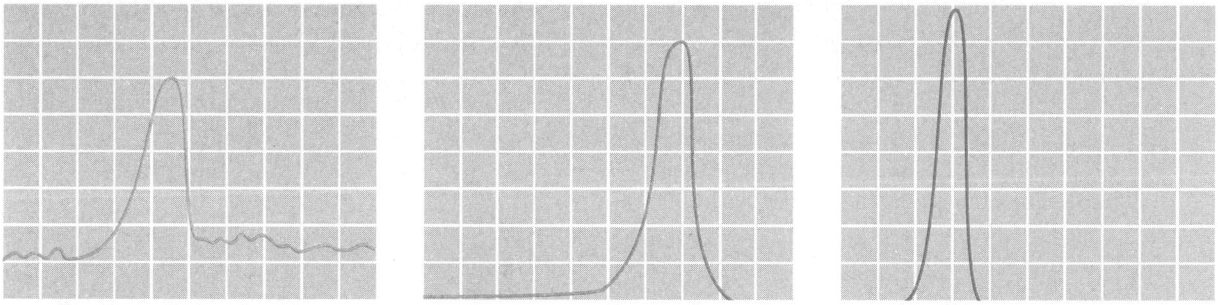

(a) Reference level −30 dBm
Vertical: 10 dB/division

Center frequency 872 MHz
Span: 10 MHz/division

(b) Reference level −18 dBm
Vertical: 1 dB/division

Center frequency 79 MHz
Span: 500 kHz/division

(c) Reference level 12 dBm
Vertical: linear

Center frequency 270 kHz
Span: 5 kHz/division

37. Find the frequency, power (both in dBm and in watts), and voltage for each of the signals shown in Figure 1.26.

COMPREHENSIVE

38. The four stages of an amplifier have gains and noise figures (in decibels) as follows:

Stage	Power gain	Noise figure
1	12 dB	2 dB
2	15 dB	4 dB
3	20 dB	6 dB
4	17 dB	7 dB

Calculate the overall noise figure in decibels.

39. A three-stage amplifier is to have an overall noise temperature no greater than 70 K. The overall gain of the amplifier is to be at least 45 dB. The amplifier is to be built by adding a low-noise first stage to an existing two-stage amplifier that has the gains and noise figures shown below. (The stage numbers refer to locations in the new amplifier.)

Stage	Power gain	Noise figure
2	20 dB	3 dB
3	15 dB	6 dB

(a) What is the minimum gain (in decibels) that the first stage can have?
(b) Using the gain you calculated in part (a), calculate the maximum noise figure (in decibels) that the first stage can have.
(c) Suppose the gain of the first stage could be increased by 3 dB without affecting its noise figure. What would be the effect on the noise temperature of the complete amplifier?

40. The $(S+N)/N$ ratio at the output of an amplifier can be measured by measuring the output voltage with and without an input signal. When this is done for a certain amplifier, it is found that the output is 2 V with the input signal switched on and 15 mV with it switched off. Calculate the $(S+N)/N$ ratio in decibels.

41. The frequency response of an amplifier can be measured by applying white noise to the input and observing the output on a spectrum analyzer. When this was done for a certain amplifier, the display shown in Figure 1.27 was obtained. Find the upper and lower cutoff frequencies (at 3 dB down) and the bandwidth of the amplifier.

Figure 1.27
Reference level −4 dBm
Vertical: 1 dB/division
Center frequency 210 MHz
Span: 10 MHz/division

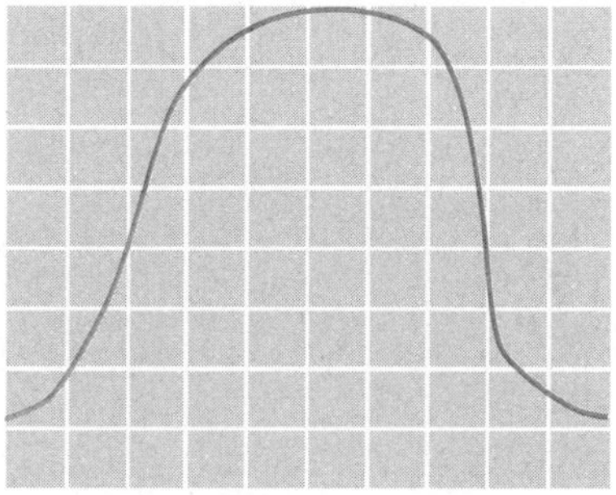

42. A receiver has a noise bandwidth of 200 kHz. A resistor is connected across its input. What is the noise power at the input if the resistor temperature is 20 degrees Celsius? Give your answer in dBf.

43. An amplifier has a noise figure of 3 dB. If the signal to noise ratio at its input is 30 dB, what is the signal to noise ratio at its output, in dB?
44. An amplifier consists of two stages. Stage one has a gain of 12 dB and a noise figure of 2 dB. Stage two has a gain of 20 dB and a noise figure of 5 dB. Calculate the noise figure, in decibels, and the equivalent noise temperature, in kelvins, for the amplifier.
45. What is the effect on the signal to noise ratio of a system, in dB, of doubling its bandwidth, other things being equal? Consider thermal noise only.
46. What is the effect on the signal to noise ratio of a system, in dB, of doubling the transmitter power?

2 Radio-Frequency Circuits

Objectives After studying this chapter, you should be able to:

1. Explain the differences in operation and construction between circuits that operate at low frequencies and those that operate at radio frequencies
2. Describe the characteristics of Class A, B, and C amplifiers and decide which type is the most suitable for a given application
3. Analyze radio-frequency amplifier circuits, both narrowband and broadband, and choose the correct configuration for a given application
4. Describe, draw circuits for, and analyze the most common types of radio-frequency oscillator circuits and discuss their relative stability
5. Explain the operation of varactor-tuned voltage-controlled oscillators and calculate the variation of frequency with tuning voltage
6. Describe the operation of crystal-controlled oscillators and explain their advantages and disadvantages compared with *LC* oscillators; perform frequency-stability calculations for crystal oscillators
7. Describe the function of a mixer and analyze several types of circuits, explaining how and where they are used and calculating output frequencies
8. Analyze frequency synthesizers employing phase-locked loops

2.1 Introduction

In the previous chapter, we looked at some of the signals found in a communication system. In particular, we saw the need to modulate a carrier signal using an information (baseband) signal. The carrier must be much higher in frequency than the baseband signal. Carrier frequencies can be as low as a few kilohertz but are typically much higher: megahertz or hundreds of megahertz. Microwave communications use carrier frequencies in the gigahertz range.

You are probably already familiar with amplifier and oscillator circuits that operate at audio frequencies. In this chapter, we will explore some of the differences in design and construction that permit these circuits to work at radio frequencies. We will also look at some techniques that are used in radio-frequency (RF) circuits but impossible or impractical to implement at lower frequencies. In addition, we will discuss devices such as frequency multipliers and mixers, which allow the frequency of a signal to be changed.

The purpose of this chapter is to give some insight into RF circuit design and to provide the reader with some electronic "building blocks" that can be used in later chapters as we look into the design and construction of practical transmitters and receivers.

2.2 High-Frequency Effects

When you first began to study electronics, you probably divided the frequency spectrum into two parts: ac and dc. A capacitor, for instance, would be considered an open circuit for dc and a short circuit for ac. This simplifying assumption works well in arriving at a general understanding of an audio amplifier circuit and in calculating its gain at midband frequencies.

Later on, you needed to be more careful in considering the effects of frequency. The bypass capacitor that looked like a short circuit at 1 kHz was no longer so simple at 20 MHz. In order to find the low-frequency response of a simple audio amplifier, you found it necessary to consider capacitive reactance. Similarly, other capacitances had to be taken into account to calculate the high-frequency response. Some of these capacitances involved actual components in the circuit, while others were incidental parts of components, for example, junction capacitances in transistors.

Perhaps an amplifier contains a transformer or other inductive element. Again, a first approximation might simplify an inductor as an open circuit for ac and a short circuit for dc. A transformer would be represented as ideal. Further investigation, however, especially at the extremes of the amplifier's frequency range, might reveal the need to consider not only inductive reactance but also losses in the iron core of the inductor or transformer.

As we extend our study of electronic circuits to higher frequencies, we have to be more careful to include reactive effects, not only those that are included deliberately as circuit elements but also the "stray" reactances in components and even within and between wires and circuit board traces. As we get still higher in frequency, into the UHF range, we find that conventional devices and construction methods become inefficient and innovative approaches to circuit design become important. At microwave frequencies, many circuits seem to bear very little physical resemblance to those used at lower frequencies.

Microwave-circuit design will be discussed in a later chapter. For the present, we shall look at more conventional circuitry operating in the range between approximately 300 kHz and 300 MHz, that is, in the MF, HF, and VHF frequency bands.

2.2.1 The Effect of Frequency on Device Characteristics

All electronic devices, whether active or passive, have capacitances and inductances that are not included intentionally but are an inevitable result of the design and construction of the component. A capacitor, for instance, will exhibit inductance and resistance as well as capacitance. It can be represented by the equivalent circuit in Figure 2.1(a). The series inductive component L_S is mainly due to the leads. The resistive component can be divided

Figure 2.1 High-frequency effects

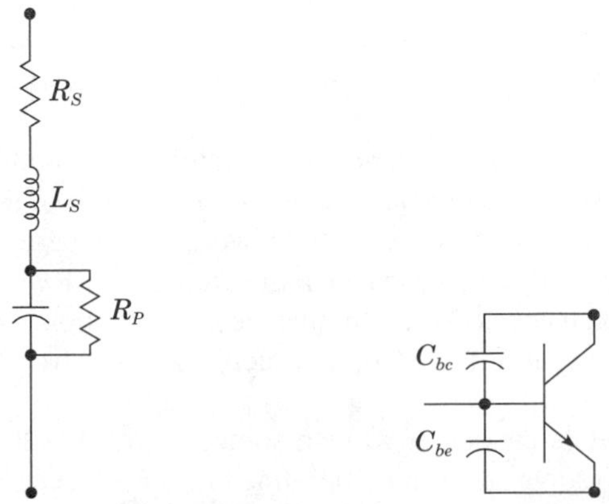

(a) Equivalent circuit for a capacitor (b) Transistor junction capacitance into two parts, a small series component R_S due to lead resistance and a large parallel resistance R_P representing dielectric losses.

As the frequency increases, so does the inductive reactance. Meanwhile, the capacitive reactance of the component decreases with increasing frequency. Eventually, a point will be reached where the two reactances are equal and the capacitor becomes a series-resonant circuit. This point is called the **self-resonant frequency.** Above this point, the magnitude of the inductive reactance becomes greater than that of the capacitive reactance, and our so-called capacitor behaves like an inductor.

Similar effects occur in transistors. Consider an ordinary bipolar transistor such as the 2N3904, for instance. Each of the two junctions has capacitance that can be represented by capacitors drawn in between base and emitter and between collector and base, as shown in Figure 2.1(b). The size of these capacitors depends partly on the physical structure of the transistor and partly on its operating point. As frequency increases, the capacitive reactances will decrease until the performance of the transistor is degraded. The base-to-collector capacitance, for instance, will cause feedback from output to input in an ordinary common-emitter amplifier circuit. The feedback can lower the gain of the amplifier or cause it to become unstable.

As the frequency increases into the gigahertz range, transit-time effects also become important. The transit time is the time it takes a charge carrier to cross a device. In an NPN transistor, it is the time taken for electrons to cross the base; in a PNP transistor, the holes exhibit transit time. In general, free electrons move more quickly than holes, so NPN transistors are preferred to PNP for high-frequency operation.

Transit time can be reduced by making devices physically small, but this causes problems with heat dissipation and breakdown voltage. It is difficult to remove large quantities of heat from a small area, and of course the dielectric strength of insulators is proportional to their thickness. There are limits to how small devices can be made, particularly if considerable power must be dissipated. For this reason, some devices include transit time as part of their design. Rather than simply making transit time as small as possible, for instance, it might be designed to be one or a number of periods at the operating frequency. Techniques such as these are used mainly in the microwave region of the spectrum and are discussed in Chapter 19.

2.2.2 Lumped and Distributed Constants

At low frequencies, we generally assume that capacitors have capacitance, resistors have resistance, and short sections of good conductors (for example, the traces on a circuit board) have neither. We saw in the previous section that this assumption is really a simplification that becomes less accurate as the frequency increases. For instance, a circuit board trace has a small amount of inductance in addition to resistance. There will also be capacitance between this trace and every other trace on the board.

A little thought will show that the inductance and capacitance related to this trace cannot be shown exactly by a single capacitor and a single inductor. In fact, no finite number of "lumps" of capacitance and inductance will do, since they are both distributed along the entire length of the trace. We can often approximate these distributed constants by a few lumped constants with reasonable accuracy, but you need to remember what is really happening, because some problems can be solved only using distributed constants. Figure 2.2 gives you an idea of how an ordinary circuit board trace might look when ana-

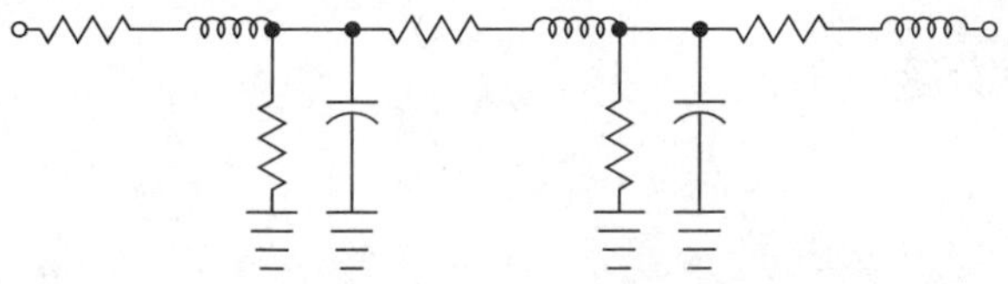

Figure 2.2
A circuit-board trace using distributed constants

lyzed this way. Of course, there is really no accurate way to sketch this trace, since the number of capacitors, resistors, and inductors should be infinite.

At frequencies in the UHF range and up, conductors even a few centimeters in length can no longer be ignored or have only their lumped capacitance and inductance taken into account. They must be analyzed as transmission lines. This process includes the distributed constants that are really there all the time but can usually be ignored at lower frequencies. At microwave frequencies, almost all conductors must be analyzed in this way. Transmission lines will be covered in detail in Chapter 16.

2.2.3 High-Frequency Construction Techniques

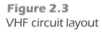

It is possible to design circuitry to reduce the effect of "stray" capacitance and inductance resulting from the wiring and circuit board traces themselves. In general, keeping wires and traces short reduces inductance, and keeping them well separated reduces capacitance between them. Inductive coupling can be reduced by keeping conductors and inductors that are in close proximity at right angles to each other. The use of toroidal cores for inductors and transformers also helps to reduce stray magnetic fields.

There may seem to be contradictions here. For instance, how do we keep components far away from each other while simultaneously keeping connections short? Obviously, compromise is necessary.

Another way to reduce interactions between components is to use *shielding*. Coupling by way of electric fields can be reduced by shielding sensitive circuits with any good conductor, such as copper or aluminum. Ideally, the shielding should form a complete enclosure and be connected to an earth ground, but even a piece of aluminum foil glued to the inside of a plastic cabinet and connected to the circuit common point will sometimes provide adequate shielding. This "cheap and dirty" technique is often found in consumer electronics. When double-sided circuit boards are used, most of the copper is often left on one side and connected to ground. This **ground plane** can provide useful shielding. Figure 2.3 shows a typical VHF circuit, in this case a cable-television converter, that uses the techniques outlined here.

Generally, conductors in RF circuits are kept as short and as far away from others as possible. The neat right-angle circuit-board layouts common at low frequencies are often avoided in favor of more direct routing. Such layouts may not be as pretty, but they can avoid problems of inductive or capacitive coupling.

Anyone attempting to service RF equipment should remember that in high-frequency circuits the placement of components and wiring can be critical. Moving one component lead may require a complete realignment. In fact, sometimes a length of solid insulated

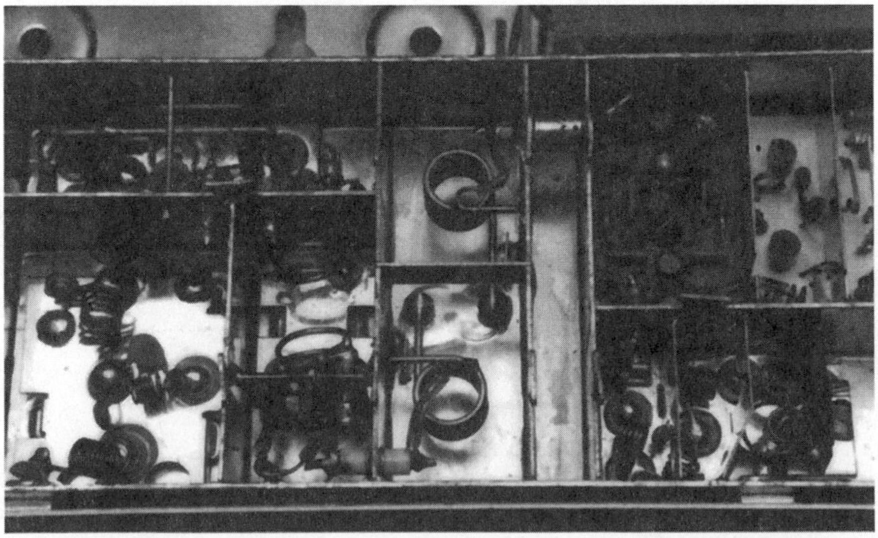

Figure 2.3
VHF circuit layout

hookup wire (a few centimeters long) will be seen, connected at only one end. This wire is called a **gimmick**; it provides a small capacitance to ground that is adjusted during circuit alignment by bending the wire slightly in one direction or another.

In order to prevent RF currents from traveling from one part of the circuit to another, careful **bypassing** is necessary. For instance, RF energy may travel from one stage to another stage that shares the same power supply. At first glance, this may seem unlikely, since power supplies typically contain large electrolytic capacitors that should have very low reactance at radio frequencies. Thus it would be expected that the power supply would look like a short circuit at high frequencies. Unfortunately, this assumption ignores the inductance in the leads from the circuit to the power supply and in the electrolytic capacitors themselves. To prevent energy from traveling from one circuit to another via this route, it is necessary to provide small capacitors to ground right at the power connection to each stage. Small capacitors are actually better than large electrolytics for this application, because they have less inductance. In difficult cases, either an inductance or a resistance can be added in the lead from the power supply to further discourage the transfer of RF energy. Using an inductor reduces the dc voltage drop, of course, and the inductor must be chosen with care because inductors, like capacitors, can exhibit self-resonance. Figure 2.4(a) shows a typical decoupling circuit between an amplifier stage and a power supply. If the power leads pass through a shielded enclosure, a feed-through capacitor like the one pictured in Figure 2.4(b) can be used to maintain the shielding.

Figure 2.4
Decoupling

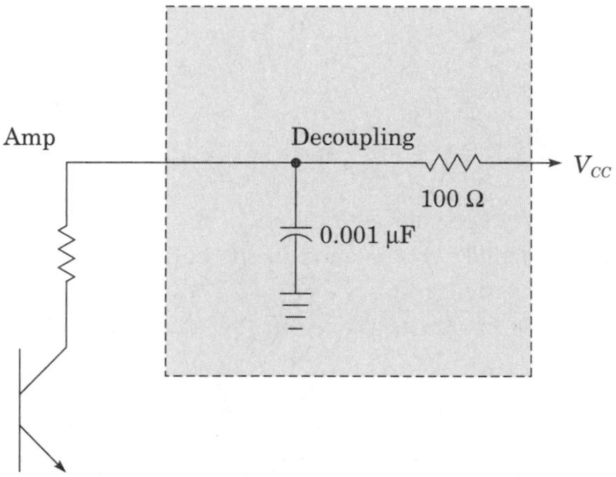

(a) Decoupling circuit

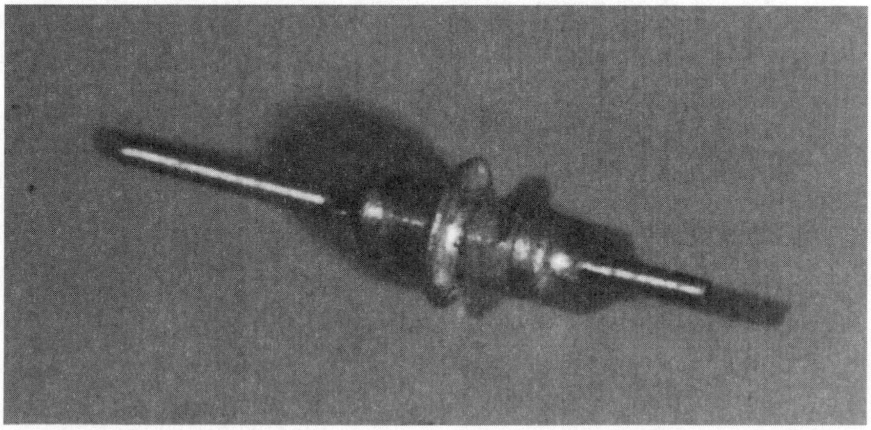

(b) Feed-through capacitor

2.3 Radio-Frequency Amplifiers

Amplifiers for RF signals can be distinguished from their audio counterparts in several important ways. Wide bandwidth may or may not be required. If it is not, gain can be increased and distortion reduced with the use of tuned circuits. Depending on the type of signal to be amplified, linearity of output with respect to input amplitude may or may not be required. If linearity is not necessary, efficiency can be improved by operating amplifiers in Class C, which will be described shortly. Impedance matching is likely to be more important than at lower frequencies because of the possibility of trouble caused by signal reflections.

2.3.1 Narrowband Amplifiers

Often the signals in an RF communication system are restricted to a relatively narrow range of frequencies. In such circumstances it is unnecessary and, in fact, undesirable to use an amplifier with a wide bandwidth. Doing so invites problems with noise and interference. Consequently, many of the amplifiers found in both receivers and transmitters incorporate filters to restrict their bandwidth. In many cases these filters also increase the gain of the amplifier.

The simplest form of bandpass filter is, of course, a resonant circuit, and these are very common in RF amplifiers. Consider the bipolar common-emitter amplifiers shown in Figure 2.5, for instance. Figure 2.5(a) shows a conventional amplifier using *RC* coupling. As you will recall, this amplifier is generally biased so that the emitter voltage is about 10% of V_{CC}. The collector resistor will drop another 40% or so, leaving the voltage between collector and emitter, V_{CE}, at about one-half the supply voltage. The voltage gain is given approximately by

$$A_V = \frac{-(R_C \parallel R_L)}{r_e'} \tag{2.1}$$

where

A_V = voltage gain as a ratio: $A_V = v_o/v_i$
$R_C \parallel R_L$ = parallel combination of the collector resistance and the load resistance
r_e' = ac emitter resistance of the transistor

Figure 2.5 Common-emitter amplifiers

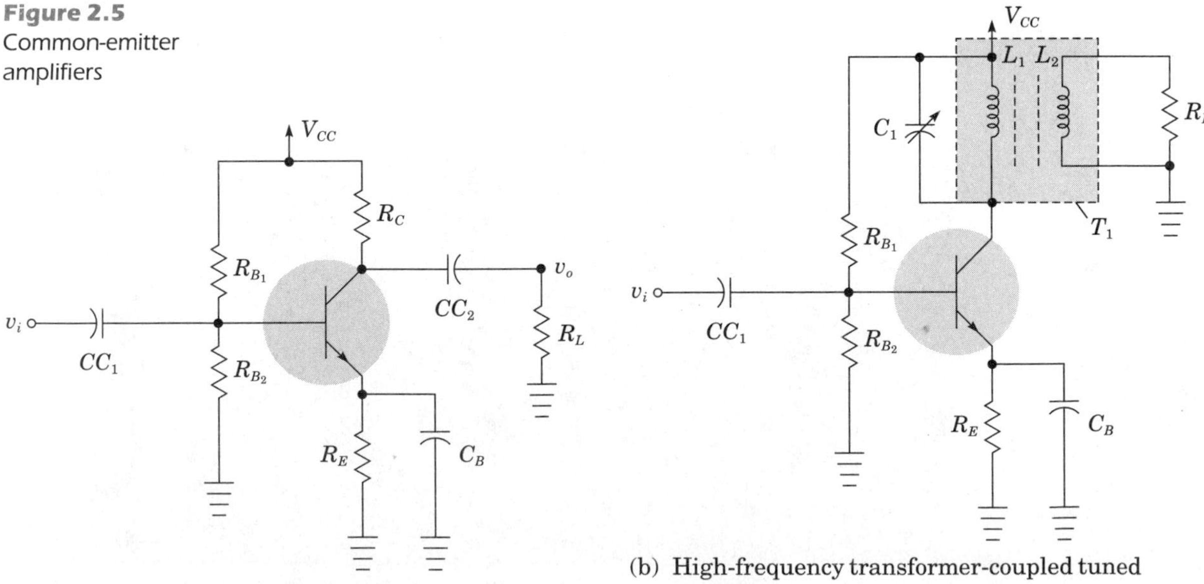

(a) Low-frequency *RC*-coupled amplifier

(b) High-frequency transformer-coupled tuned amplifier

Do not confuse r'_e, a transistor parameter, with R_E, which is part of the external circuit. The value of r'_e depends on the bias current. It is given, very approximately, by

$$r'_e = \frac{26\text{ mV}}{I_E}$$

where

r'_e = emitter resistance in ohms
I_E = dc emitter current in amperes

The tuned amplifier in Figure 2.5(b) is similar but not identical. For instance, the bias point is different. In the absence of a collector resistor, the dc collector voltage will obviously be equal to the supply voltage, less the very small drop due to the dc resistance of the transformer primary. The gain is still equal to the ratio of collector-circuit impedance to emitter-circuit impedance, but the collector-circuit impedance is now very much a function of frequency.

Once again, it is necessary to recall some basic electrical theory. The collector tuned circuit will be parallel resonant at a frequency that is given approximately by

$$f_o = \frac{1}{2\pi\sqrt{L_1 C_1}} \qquad (2.2)$$

where

f_o = resonant frequency in hertz
L_1 = primary inductance in henrys
C_1 = primary capacitance in farads

Equation (2.2) is reasonably accurate as long as the loaded quality factor (Q) of the tuned circuit is greater than about ten. At resonance, the impedance of the tuned circuit is resistive and its magnitude is at a maximum. On either side of resonance, the impedance is lower in magnitude and is reactive: inductive below resonance and capacitive above. The higher the impedance, the higher the gain. Therefore, the gain is greatest at resonance.

Once the Q has been found, the bandwidth at 3 dB down from the resonant-frequency gain is given very simply by

$$B = \frac{f_o}{Q} \qquad (2.3)$$

where

B = bandwidth
f_o = resonant frequency
Q = loaded Q at resonance

The gain goes down away from resonance because the impedance of the resonant circuit is greatest at resonance and drops rapidly at higher and lower frequencies. The loaded Q depends on the load resistance, the transformer turns ratio, and the coupling between the transformer windings. The more heavily loaded the amplifier, the lower its Q and the wider its bandwidth.

An example may clarify this analysis.

Example 2.1

An RF amplifier has the circuit shown in Figure 2.6. Find

(a) the operating frequency
(b) the bandwidth

Assume the loaded Q of the transformer primary is 15.

Figure 2.6

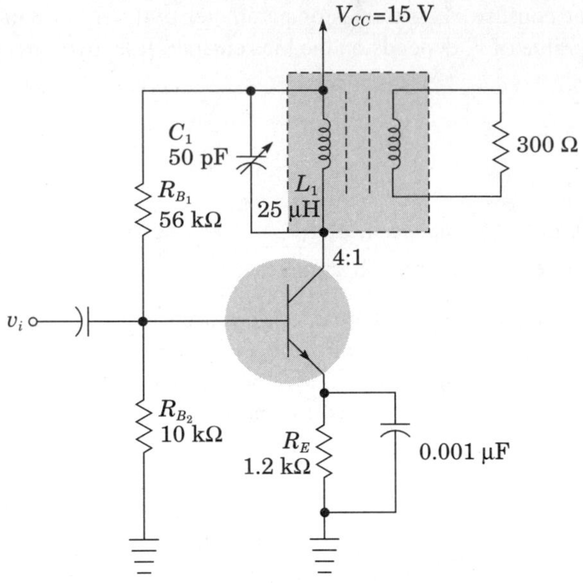

Solution (a) We can find the resonant frequency from Equation (2.2):

$$\begin{aligned} f_o &= \frac{1}{2\pi\sqrt{L_1 C_1}} \\ &= \frac{1}{2\pi\sqrt{(25 \times 10^{-6})(50 \times 10^{-12})}} \\ &= 4.50 \text{ MHz} \end{aligned}$$

(b) The bandwidth can be found from Equation (2.3).

$$\begin{aligned} B &= \frac{f_o}{Q} \\ &= \frac{4.5 \text{ MHz}}{15} \\ &= 300 \text{ kHz} \end{aligned}$$

There are some problems with the simple circuit in Figure 2.6. The transistor output impedance is connected across the tuned circuit. For a bipolar transistor operating at radio frequencies, this impedance may be only a few thousand ohms. Consequently, the transistor is usually connected across only part of the coil, as shown in Figure 2.7. The transformer primary acts as an autotransformer, increasing the effective impedance of the transistor and increasing the Q of the circuit. Note that while the transistor is connected across only part of L_1, the capacitor C_1 tunes the whole inductor. Remember that the top end of the coil is effectively connected to ground for ac through the filter capacity of the power supply. A decoupling network consisting of R_D and C_D has also been added.

The capacitance between the collector and base of the transistor is also likely to cause trouble. By feeding back some of the output signal to the input, it reduces the gain of the circuit at high frequencies. In fact, for the common-emitter circuit shown in Figure 2.7, the effect of this capacitance, often called the **Miller effect**, is the same as if a much larger capacitance had been connected across the input.

The influence of the Miller effect can be reduced somewhat by transformer-coupling the input as well as the output and tuning the secondary of the input transformer. The Miller capacitance then becomes part of the capacitance necessary to tune the input circuit to

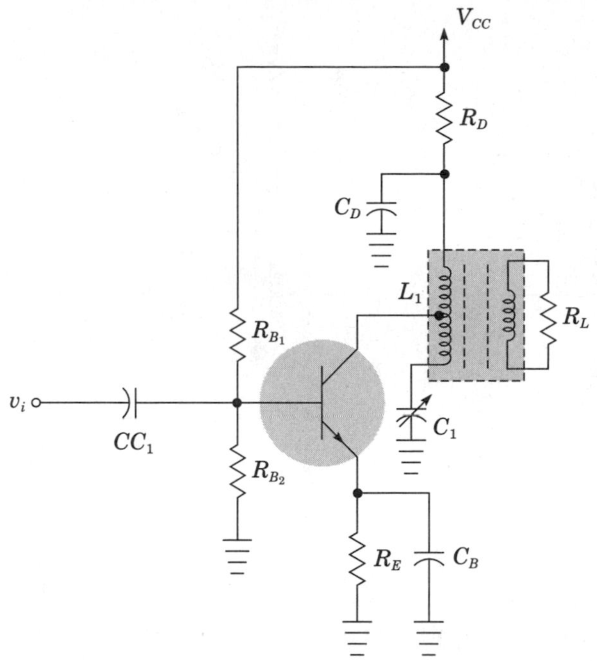

Figure 2.7 Practical common-emitter amplifier with tapped primary

resonance. Such a circuit is shown in Figure 2.8. The input is tapped down on the transformer secondary to reduce the loading effect of the transistor and its biasing circuit on the Q of the tuned circuit. C_1 tunes the input tuned circuit, and in this circuit C_2 is the output tuning capacitor.

The practice of tuning both the input and output of an amplifier can lead to instability. This instability can be corrected by neutralization, which will be discussed later in this chapter.

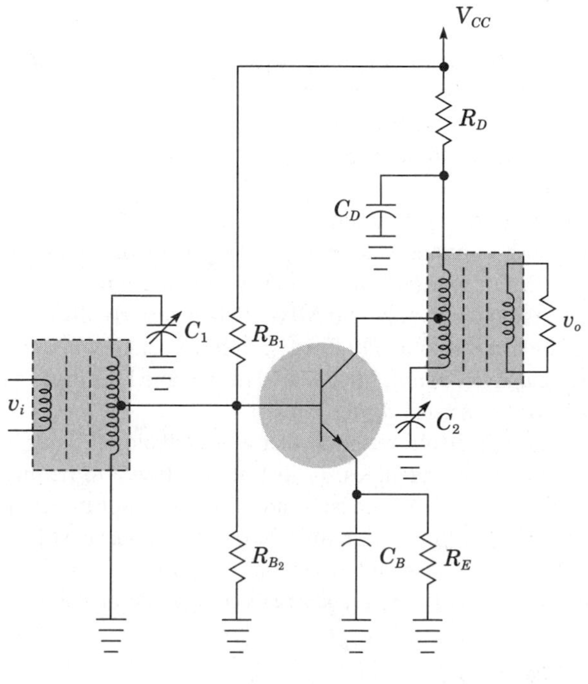

Figure 2.8 Narrowband RF amplifier with tuned input and output

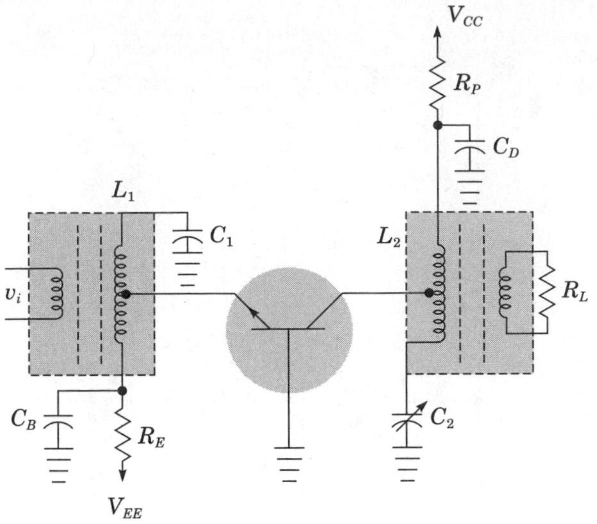

Figure 2.9
Common-base RF amplifier

Another way to avoid the problem of the Miller effect is to use a common-base amplifier, as shown in Figure 2.9. This circuit shows transformer coupling at both input and output. Common-base amplifiers are rare in low-frequency applications but quite common at radio frequencies. The capacitance between collector and base appears across the output tuned circuit, where it simply reduces the external capacitance needed to achieve resonance. This will greatly extend the useful frequency range of the transistor.

In Figure 2.9, the input tuned circuit consists of L_1 and C_1, and the output tuned circuit is L_2 and C_2. Emitter bias is shown, but voltage-divider bias is also suitable and does not require a negative supply voltage.

The loading effect of the transistor impedances on tuned circuits can be greatly reduced by using field-effect, rather than bipolar, transistors. Field-effect transistors (FETs) are famous for their very high input impedance, of course, but their output impedance is also much higher than for bipolar transistors.

2.3.2 Wideband Amplifiers

Not all amplifiers used in communications have tightly restricted bandwidth. For instance, the amplifiers used for the baseband part of the system are usually *wideband* (also called *broadband*). You are no doubt already familiar with audio amplifiers. Those used for baseband video are similar, though their bandwidth is larger (about 4.2 MHz for broadcast television signals).

In this section, we will consider amplifiers designed for higher-frequency operation, where the response is required to extend over a relatively wide range of frequencies. For instance, an amplifier for a cable television system might be required to amplify frequencies from 50 MHz to approximately 400 MHz. This is a fairly difficult requirement, especially if the system is expected to be *flat*, that is, to have equal gain across the entire bandwidth. (An actual cable television amplifier would be designed for more gain at higher frequencies to compensate for greater loss in the cable.)

In most cases, it is not required that a broadband RF amplifier have response down to dc or even to audio frequencies. In fact, this is likely to be undesirable. We would not want our hypothetical cable television amplifier to amplify 60 Hz hum, for example. In fact, we would prefer it to ignore both AM radio broadcasts at about 1 MHz and CB radio transmitters at 27 MHz as well. Broadband amplifiers, then, generally incorporate some form of filtering so that the frequency response, while broad, is restricted to the range of interest.

Wideband RF amplifiers, like their narrowband counterparts, typically use transformer coupling. This technique was once popular for audio amplifiers as well, but the size, the weight, and especially the high cost of audio transformers has led to their virtual elimination from audio circuitry. For RF amplifiers they retain some advantages, however. Transformers are very convenient for impedance matching, which is likely to be more important in RF than in audio designs. The isolation between input and output is useful in helping to keep spurious signals at frequencies greatly different from the desired signal frequency from propagating through the system. Transformer coupling also makes it easy to couple balanced inputs or loads to the amplifier. Balanced lines have equal impedance from each conductor to ground. They are often used with antennas; ordinary television twin-lead is an example of a balanced line that should be transformer-coupled to the first stage in a television receiver.

The transformers used in wideband amplifiers need careful design to avoid self-resonance and to maintain relatively constant gain across the frequency range of interest. In particular, the reactance of the windings must be large compared to the impedances that are connected to them. Toroidal transformers with ferrite cores are common. In the case of an amplifier that must have a very wide bandwidth, the ferrite, having much higher permeability than air, allows for sufficient reactance at the low end of the frequency range. At higher frequencies, the ferrite is likely to be less effective, but the reactance of the winding will be sufficient, since reactance increases with frequency.

Figure 2.10 is the circuit of a broadband RF amplifier using a bipolar transistor. Notice that it is the same as Figure 2.5(b), except that C_1, the tuning capacitor, is gone and the input as well as the output is transformer-coupled.

It is also possible to use conventional resistance-capacitance coupling between broadband amplifier stages. Such circuits can be analyzed in the same way as their low-frequency counterparts.

Whatever design is used, if an amplifier operates over a sufficiently wide frequency range, it will tend to have higher gain at the low end of its range than at the high end because the gain of the transistor falls off with increasing frequency. Negative feedback is often used to overcome this, with the amount of feedback decreasing as the frequency increases. Another method uses a high-pass filter at the input, so that as the frequency goes down, the amount of input signal reaching the amplifier is reduced.

Figure 2.10
Broadband RF amplifier

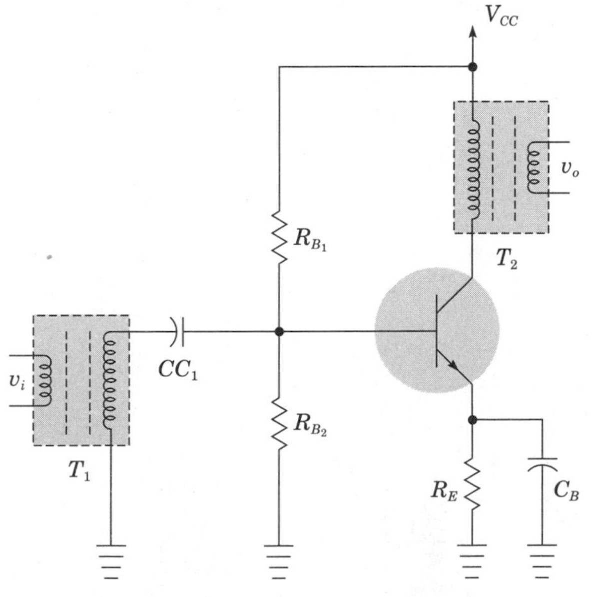

2.3.3 Amplifier Classes

Amplifiers are classified according to the portion of the input cycle during which the active device conducts current. This is called the *conduction angle* and is expressed in degrees. The active device may be a bipolar transistor, FET, or tube; for now we will assume it is a bipolar transistor. For a push-pull amplifier, the conduction angle for one of the two amplifying devices is used.

Single-ended audio amplifiers are generally operated in Class A, where the transistor conducts current at all times for a conduction angle of 360°. This is the only way to achieve linear operation, in which the output is a reasonably faithful copy of the input, except for amplitude. The small-signal RF amplifiers described earlier in this chapter were also biased for Class A operation.

Push-pull amplifiers can be linear if at least one of the two transistors is conducting at all times. In a Class B amplifier, each transistor is biased at cutoff, so that each conducts for 180° of the input cycle. This results in greater efficiency than for Class A amplifiers, but the distortion is larger due to nonlinearity of the transistors near cutoff.

Most audio power amplifiers use Class AB, which is a compromise between Class A and Class B, in terms of both distortion and efficiency. Each transistor conducts for slightly more than 180° of the input cycle, thus reducing the "crossover" distortion that occurs as the input signal passes through zero, where both transistors are near cutoff.

Figure 2.11 shows a simple Class B amplifier for RF operation. Like the Class A amplifiers shown earlier, this amplifier uses transformer coupling, which, while rare in modern audio amplifiers, is still the most common way to design amplifiers for high-frequency operation. The example shown is broadband, but the circuit could just as easily be tuned.

Both transistors are biased near cutoff by the voltage drop across D_1. When the top end of the input transformer secondary goes positive, Q_1 turns on and Q_2 is cut off. The reverse happens in the other half of the input cycle; thus each transistor conducts for half the cycle.

Audio signals, as well as some radio-frequency signals, have complex waveforms and require linear amplification to avoid distortion. Suppose, however, that we have to amplify an RF signal that we know is a sine wave. It might be possible to do this with an amplifier that created a great deal of distortion, provided that we had some means to remove that distortion after amplification and restore the signal to its original sinusoidal shape. This extra trouble would be worthwhile if the resulting amplifier were more efficient than either Class A or Class B amplification.

Such an amplifier class exists: it is called Class C. In it the active device conducts for less than 180° of the input cycle. The amplifier can be single-ended or push-pull; either way, it is apparent that the output for a sinusoidal input will resemble a series of pulses more than it does the original signal. These current pulses can be converted back into sine

Figure 2.11
Class B RF amplifier

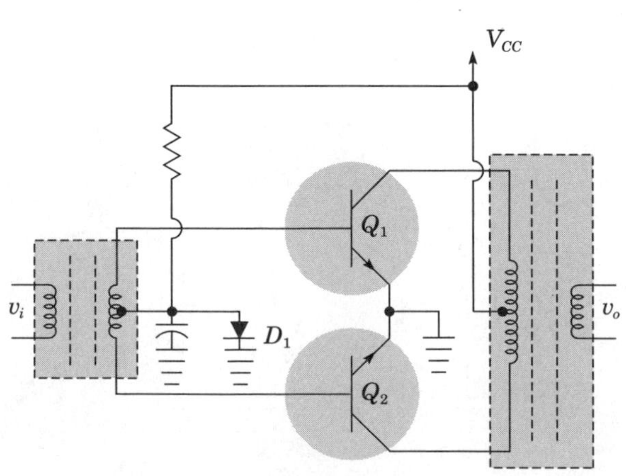

Table 2.1 Review of Amplifier Classes

Class	A	B	C
Conduction angle	360°	180°	<180°
Maximum efficiency	50%	78.5%	100%
Likely practical efficiency	25%	60%	75%

waves by an output tuned circuit. The resulting gain in efficiency can be quite worthwhile, as can be seen in Table 2.1.

Class C amplifiers achieve their maximum efficiency when the amplifying device almost saturates at peaks of the input cycle. It is, of course, cut off for most of the cycle. This type of operation minimizes power dissipation in the transistor, which is zero when the transistor is cut off, low when it is saturated, and much higher when it is in the normal, linear operating range. Since Class C amplifiers are biased beyond cutoff, they would be expected to have zero power dissipation with no input. This is indeed the case, provided the bias is independent of the input signal.

There remains the problem of distortion. Figure 2.12(a), a simplified circuit for a Class C amplifier, shows how the distortion is kept to a reasonable level. The process can be explained using either the time or the frequency domain. In the time domain, the output tuned circuit is excited once per cycle by a pulse of collector current. This keeps oscillations going in the resonant circuit. They are damped oscillations, of course, but the Q of the circuit is high enough to ensure that the amount of damping that takes place in one cycle is negligible, and the output is a reasonably accurate sine wave. The frequency-domain explanation is even simpler: the resonant circuit constitutes a bandpass filter that passes the fundamental frequency and attenuates harmonics and other spurious signals. In many cases, especially when the amplifier is the final stage of a transmitter, additional filtering after the amplifier further reduces the output of harmonics.

From the foregoing description, it might seem that Class C amplifiers would have to be narrowband. Many of them are, but this is not a requirement as long as the output is connected to a low-pass filter that will attenuate all the harmonics that are generated.

Let us look at the circuit in Figure 2.12(a) a little more closely. We begin with the bias circuit, which is unconventional. Far from being biased in the middle of the linear operating range, as for Class A, a Class C amplifier must be biased beyond cutoff. For a bipolar transistor, that can mean no base bias at all, since a base-to-emitter voltage of about 0.7 V is needed for conduction. With no signal at its input, the transistor will have no base current and no collector current and will dissipate no power.

If a signal with a peak voltage of at least 0.7 V is applied to the input of the amplifier, the transistor will turn on during positive peaks. It might seem at first glance that an input signal larger than this would turn on the transistor for more and more of the cycle, so that with large signals the amplifier would approach Class B operation, but this is not the case. The base-emitter junction rectifies the input signal, charging C_B to a negative dc level that increases with the amplitude of the input signal. This self-biasing circuit will allow the amplifier to operate in Class C with a fairly large range of input signals.

The collector circuit can be considered next. At the peak of each input cycle, the transistor turns on almost completely. This effectively connects the bottom end of the tuned circuit to ground. Current flows through the coil L_1.

When the input voltage decreases a little, the transistor turns off. For the rest of the cycle, the transistor represents an open circuit between the lower end of the tuned circuit and ground. The current will continue to flow in the coil, decreasing gradually until the stored energy in the inductor has been transferred to capacitor C_1, which becomes charged. The process then reverses, and oscillation takes place. Energy is lost in the resistance of the inductor and capacitor, and of course energy is transferred to the load, so the amplitude of the oscillations would gradually be reduced to zero, except for one thing: once each cycle,

Figure 2.12
Class C amplifier

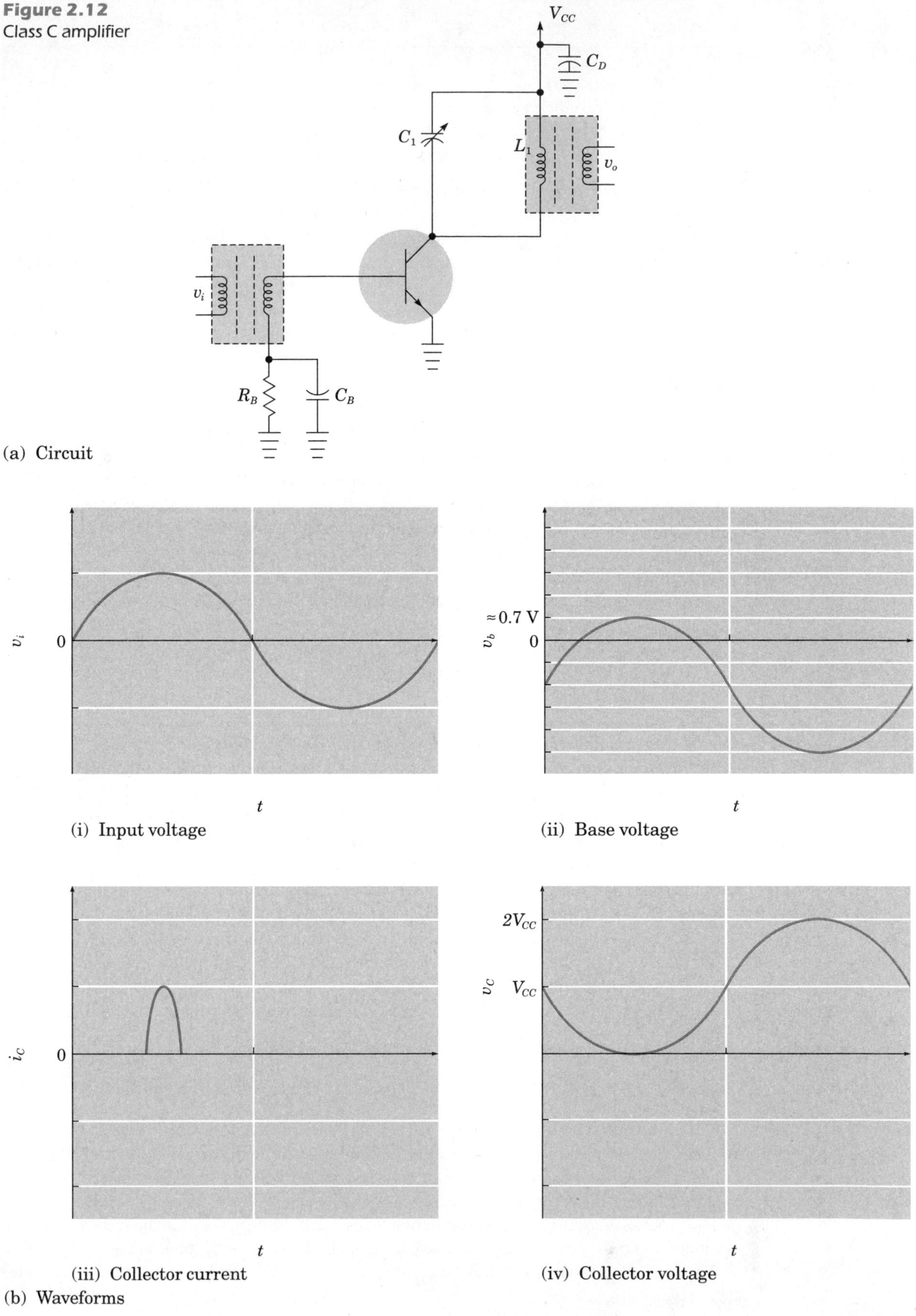

the transistor turns on, another current pulse is injected; because of this, oscillations continue indefinitely at the same level.

Figure 2.12(b) shows some of the waveforms associated with the Class C amplifier. Part (i) shows the input signal, and part (ii) shows the actual signal applied to the base. Note the bias level that is generated by the signal itself. Part (iii) shows the pulses of collector current, and part (iv), the collector voltage. It should not be surprising that this reaches a peak of almost $2V_{CC}$. The peak voltage across the inductor must be nearly V_{CC}, since, when the transistor is conducting, the top end of the coil is connected to V_{CC} and the bottom end is almost at ground potential ("almost" because, even when saturated, there will be a small voltage across the transistor). At the other peak of the cycle, the inductor voltage will have the same magnitude but opposite polarity and will add to V_{CC} to make the peak collector voltage nearly equal to $2V_{CC}$.

This description implies that the output tuned circuit must be tuned fairly closely to the operating frequency of the amplifier, and that is indeed the case. Since the transistor must swing between cutoff and something close to saturation for Class C operation, it is also implicit in the design that this amplifier will be nonlinear; that is, doubling the amplitude of the input signal will not double the output. These two limitations restrict the use of Class C amplifiers to RF signals, and in fact, only some RF applications can make use of this circuit.

Class C amplification can be used with either field-effect or bipolar transistors. It is also quite commonly used with vacuum tubes in large transmitters.

2.3.4 Neutralization

Device and stray capacitance tend to reduce gain and cause instability as frequency increases. Care must be taken to separate inputs and outputs to avoid feedback, but often a transistor or tube will itself introduce sufficient feedback to cause oscillations to take place. Sometimes this type of feedback can be cancelled by a process called **neutralization**.

Neutralization is accomplished by deliberately feeding back a portion of the output signal to the input in such a way that it has the same amplitude as the unwanted feedback but the opposite phase. Since the device capacitances vary from component to component, careful adjustment is necessary.

Figure 2.13 shows one type of neutralization. The basic circuit is that of Figure 2.8, a narrowband RF amplifier with tuned input and output. As mentioned earlier, such an

Figure 2.13 Neutralized RF amplifier

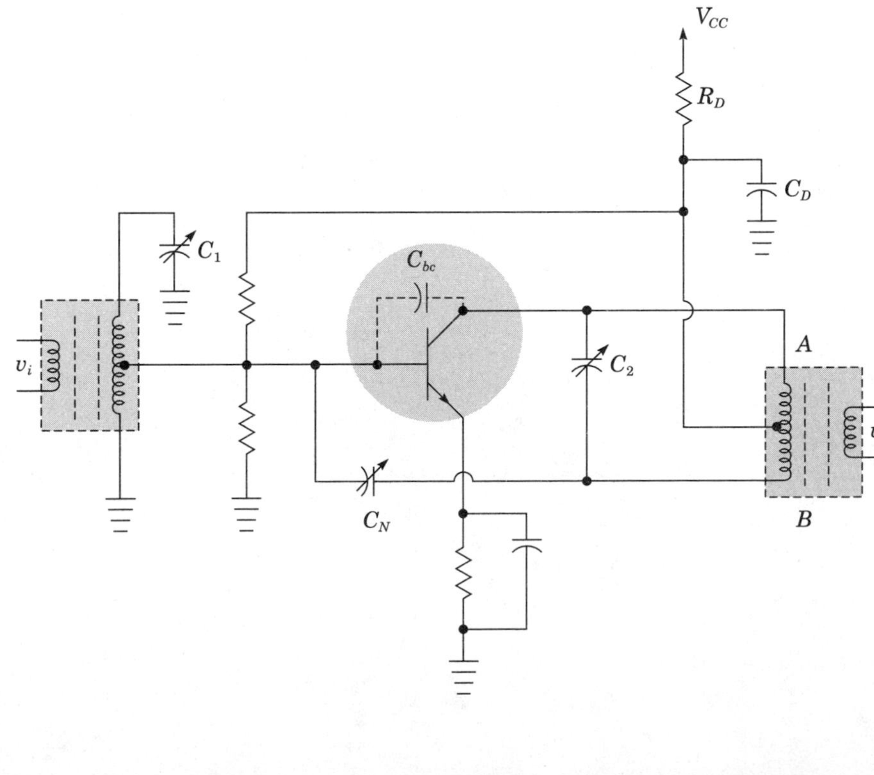

amplifier has a tendency to oscillate because of the feedback through the base-collector capacitance of the transistor. The method used in Figure 2.13 to provide feedback for neutralization is to rearrange the primary of the output transformer so that the center is connected to V_{CC}. The power supply is, of course, at ground potential for ac (especially when the decoupling network consisting of R_D and C_D is used). The transistor is still connected across only part of the output transformer, to reduce its loading effect as before, and there is no difference as far as the operation of the circuit is concerned.

The reason for rearranging the transformer connections is that now the top and bottom ends of the winding have opposite polarities with respect to ground for the ac signal. The voltages may or may not be equal in magnitude: there is no requirement that the transformer be tapped in the center as shown. The internal capacitance C_{bc}, shown dotted, feeds a signal from the top end of the coil, point A, to the transistor base. A small variable capacitor, C_N, is adjusted to feed a signal of equal magnitude but opposite polarity from the bottom of the coil, point B, to the base. When C_N is adjusted correctly, the signals will cancel and the amplifier will be stable.

2.4 Frequency Multipliers

The output circuit of a Class C amplifier can be tuned to a harmonic of the input signal. The amplifier will still operate, though with reduced efficiency. For instance, by providing energy to the circuit during alternate cycles the pulses of collector current will sustain oscillations in an output tuned circuit that is tuned to twice the input frequency. Another way to look at the process is to recognize that the collector-current pulse in a Class C amplifier is rich in harmonics. Any one of these harmonics can be chosen as the output by tuning the output bandpass filter to the appropriate frequency. This harmonic operation can be very useful when a signal is required at a higher frequency than can be conveniently generated. Figure 2.14 shows a circuit for a *frequency doubler* and its input, output, and collector-current waveforms.

Because **frequency multipliers** operate at lower efficiencies than straight-through amplifiers, they are used at low power levels. Most multipliers operate at the second or third harmonic of the input frequency and are known as **doublers** or **triplers**, respectively. They are more efficient than multipliers operating at higher-order harmonics.

Multipliers can be used in cascade if greater multiplication is required. Figure 2.15 shows one way to get a multiplication of 18 times. This is a common arrangement in VHF transmitters, though it is not used in new designs as much as it once was.

Figure 2.14
Frequency multiplier

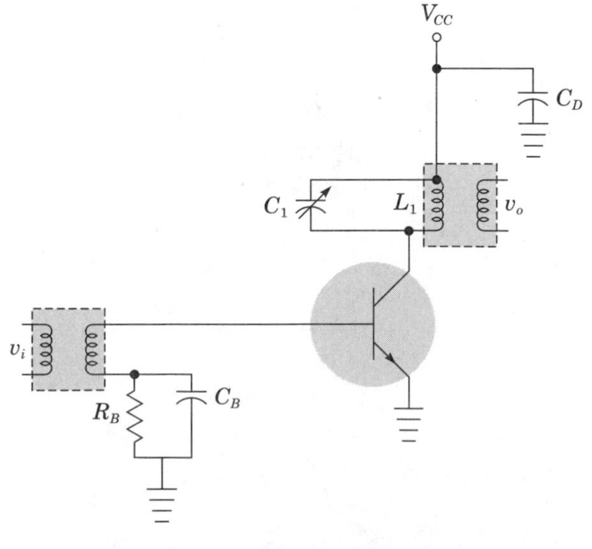

(a) Circuit

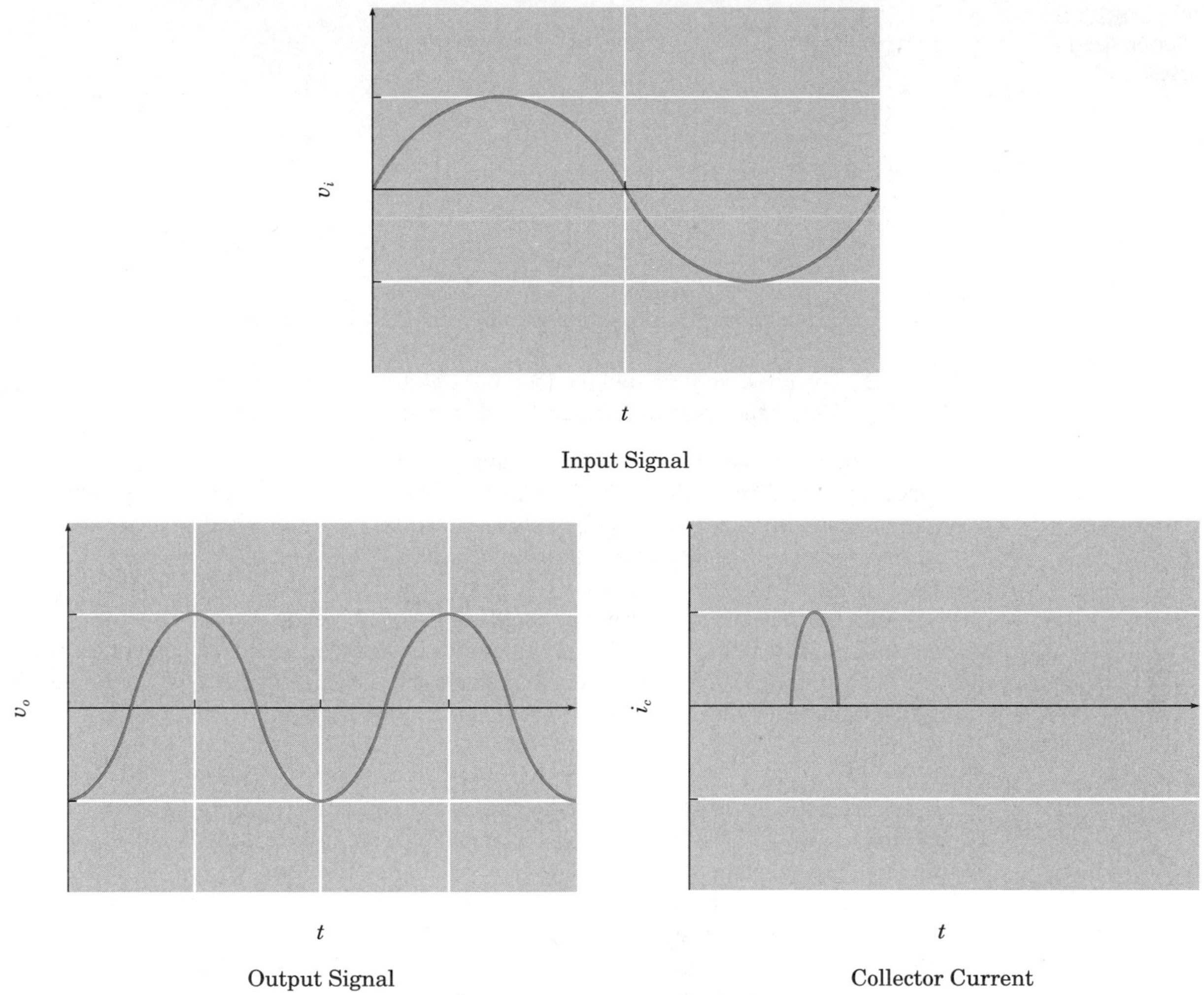

(b) Doubler waveforms

Figure 2.14
Continued

Figure 2.15
Frequency multiplication

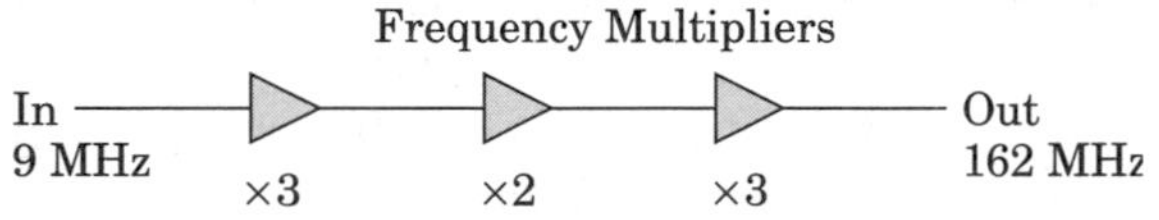

2.5 Radio-Frequency Oscillators

RF oscillators do not differ in principle from those used at lower frequencies, but the practical circuits are quite different. While low-frequency oscillators usually use *RC* circuits in the frequency-determining section, *LC* circuits are more common at radio frequencies. In addition, many RF oscillators are crystal controlled.

Any amplifier can be made to oscillate if a portion of the output is fed back to the input in such a way that the following criteria, known as the *Barkhausen criteria*, are satisfied:

1. The gain around the loop must be equal to one. (If it is initially greater than one, it will become equal to one when oscillations start, due to some process such as

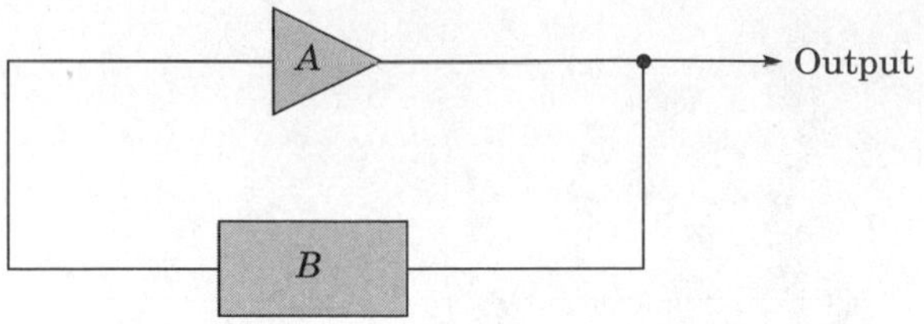

Figure 2.16 Generalized oscillator

transistor saturation; otherwise the output voltage would continue to increase without limit.)

2. The phase shift around the loop must total either 0° or some integer multiple of 360° at the operating frequency (and not at other frequencies).

Taken together, these statements simply mean that at the operating frequency, an input signal will be amplified then fed back in phase and with sufficient amplitude that it will maintain its value at the output without any further input. The initial signal needed to start the process can be noise or a transient caused by switching on the power to the oscillator circuit.

Figure 2.16 shows a "generic" oscillator consisting of an amplifier with gain A and a feedback network with "gain" B. B will usually be less than 1, and only part of the output will be fed back to the input. B is often called the *feedback fraction*. According to the Barkhausen criteria, we will have:

$$AB = 1 \tag{2.4}$$

when oscillations are in progress. The phase shift around the loop will be zero or some multiple of 360°. This means that if the amplifier is an inverting amplifier, B must also invert the signal phase.

The foregoing must be true at only one frequency, so B has to have some form of frequency dependence. Either its phase shift or its amplitude response or both must vary with frequency. There are many types of networks that meet this requirement, but one of the simplest (and the one most commonly used in radio-frequency oscillators) is the LC resonant circuit. Both series- and parallel-resonant circuits have amplitude and phase responses that are functions of frequency, and both have applications in oscillators.

2.5.1 *LC* Oscillators

It is now time to look at some practical oscillator configurations. We will begin with oscillators whose frequency is controlled by a resonant circuit using inductance and capacitance. Crystal-controlled oscillators will be described in Section 2.5.3.

All the circuits shown in this section have been in use for many years and have been implemented with vacuum tubes, bipolar and field-effect transistors, and integrated circuits. In each case, we will begin by showing the amplifier as a simple gain block. All of these oscillator types can be implemented with either inverting or noninverting amplifiers and will be shown both ways.

Hartley Oscillator This oscillator type can be recognized by its use of a tapped inductor, part of a resonant circuit, to provide feedback.

Figure 2.17(a) shows a Hartley oscillator using a noninverting amplifier and Figure 2.17(b) shows the same oscillator using an inverting amplifier. In either case, the resonant frequency is that of the tuned circuit, including the whole inductor,

that is

$$f_o = \frac{1}{2\pi\sqrt{LC}} \tag{2.5}$$

Figure 2.17
Hartley oscillators

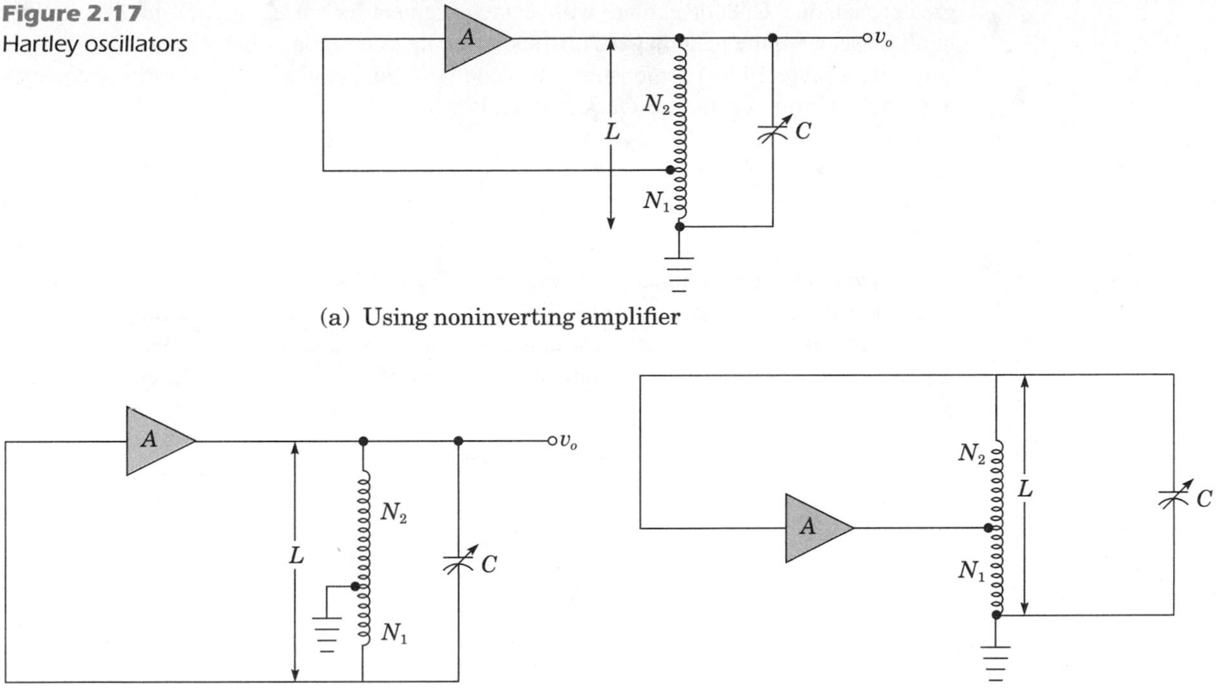

(a) Using noninverting amplifier

(b) Using inverting amplifier

(c) Using an amplifier with A_v<1 (but A_i >1)

where

f_o = frequency of oscillation in hertz
L = inductance of the whole coil in henrys
C = capacitance across the coil in farads

The feedback fraction is easy to find if the assumption is made that the inductor is a single coil with unity coupling. It is then given simply by the turns ratio (input turns to output turns). Thus there are two different equations, depending on whether the circuit uses an inverting or noninverting amplifier.

For the noninverting circuit of Figure 2.17(a),

$$B = \frac{N_1}{N_1 + N_2} \tag{2.6}$$

This circuit can be implemented with transistors connected in the common-base or common-gate configuration.

In the oscillator using the inverting amplifier, as in Figure 2.17(b),

$$B = \frac{-N_1}{N_2} \tag{2.7}$$

The negative sign indicates that the feedback has the opposite polarity to that of the output signal. This is required for an oscillator using an inverting amplifier, such as would be provided by a transistor connected for common-emitter or common-source operation.

Both of these circuits assume that the amplifier has a voltage gain greater than one. It might seem that this is a necessary condition for oscillation, but this is not the case. *AB* can be equal to one if *A* is less than one, provided that *B* is greater than one. The real amplifier requirement for oscillation is power gain, not voltage gain. The tapped coil of the Hartley oscillator can be used as an autotransformer to give a voltage feedback fraction greater than one. Of course, there will be a requirement for current gain from the amplifier in this case, a requirement that can be filled by using a common-collector or common-drain amplifier. Figure 2.17(c) demonstrates this configuration, in which the operating frequency remains as before, but the feedback fraction is now

$$B = \frac{N_1 + N_2}{N_1} \tag{2.8}$$

Example 2.2

Two practical Hartley oscillators are shown in Figure 2.18. The one in Figure 2.18(a) uses a junction field-effect transistor (JFET) in the common-source configuration, and in Figure 2.18(b) uses the same transistor connected as a common-drain amplifier (source follower). Find the operating frequency and the required gain for the amplifier part of each oscillator.

Figure 2.18

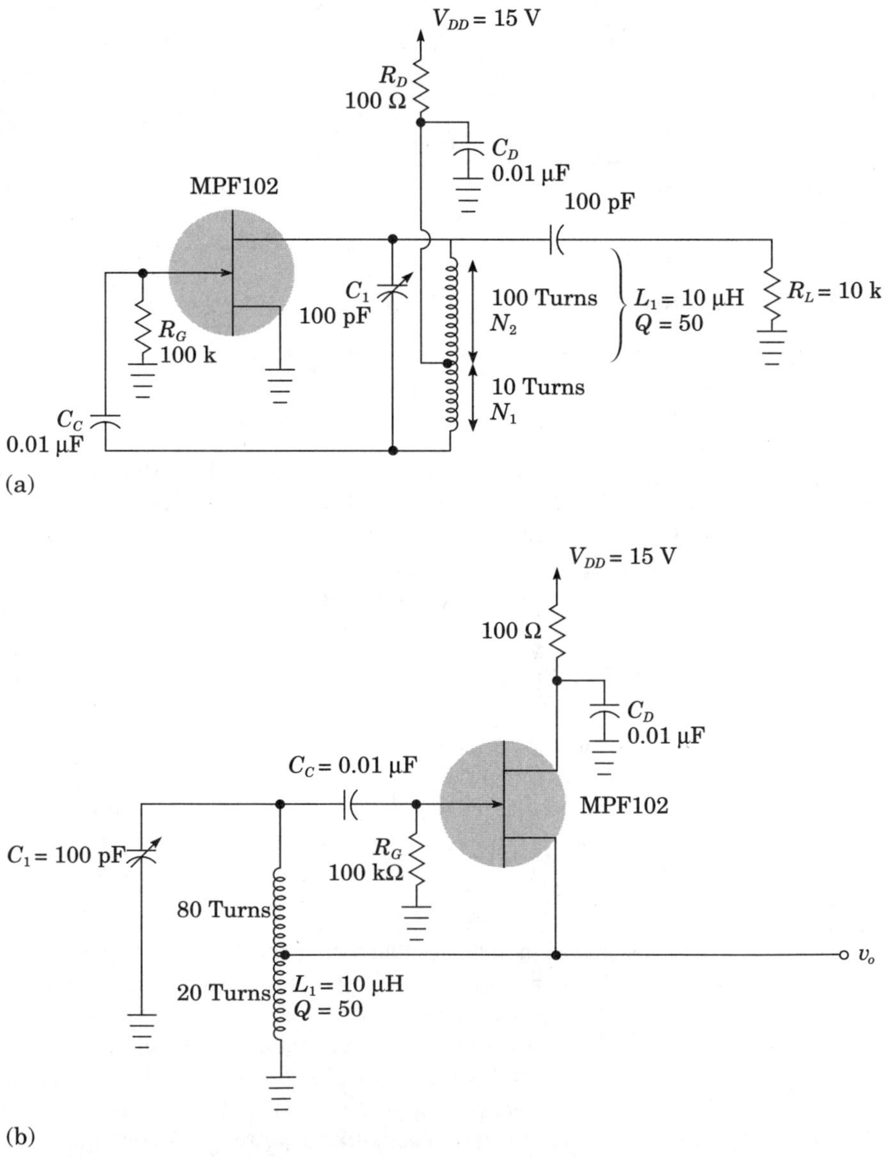

Solution (a) The operating frequency is given by Equation (2.5):

$$\begin{aligned} f_o &= \frac{1}{2\pi\sqrt{LC}} \\ &= \frac{1}{2\pi\sqrt{10 \times 10^{-6})(100 \times 10^{-12})} \\ &= 5.03 \text{ MHz} \end{aligned}$$

The feedback fraction is given by Equation (2.7):

$$\begin{aligned} B &= \frac{-N_1}{N_2} \\ &= \frac{-10}{100} \\ &= -0.1 \end{aligned}$$

To determine the minimum amplifier gain for oscillation, begin with Equation (2.4):

$$\begin{aligned} AB &= 1 \\ A &= \frac{1}{B} \\ &= \frac{1}{-0.1} \\ &= -10 \end{aligned}$$

As usual, the negative sign denotes a phase inversion.

(b) The operating frequency is the same as in part (a), but now the feedback fraction is given by Equation (2.8):

$$\begin{aligned} \mathrm{B} &= \frac{N_1 + N_2}{N_1} \\ &= \frac{80 + 20}{20} \\ &= 5 \end{aligned}$$

This requires an amplifier voltage gain of at least 0.2. Since the actual gain will be only slightly less than one, the circuit will work.

Colpitts Oscillator The Colpitts oscillator uses a capacitive voltage divider instead of a tapped inductor to provide feedback. Once again, the configuration of the feedback network depends on whether the amplifier is noninverting, as in Figure 2.19(a), or inverting, as in Figure 2.19(b).

The operating frequency is determined by the inductor and the series combination of C_1 and C_2.

$$f_o = \frac{1}{2\pi\sqrt{LC_T}} \tag{2.9}$$

where

$$C_T = \frac{C_1 C_2}{C_1 + C_2} \tag{2.10}$$

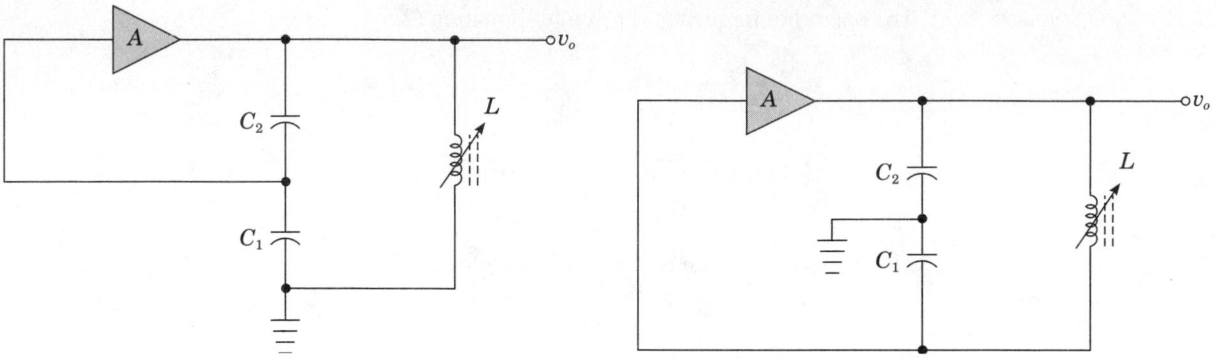

(a) Using noninverting amplifier
(b) Using inverting amplifier

Figure 2.19
Colpitts oscillators

The feedback fraction is given by the ratio of reactances between output and input circuits. This is, of course, the reciprocal of the ratio of capacitances, because

$$X_C = \frac{1}{2\pi f C} \tag{2.11}$$

From this it follows that the ratio of the reactances of two capacitors C_1 and C_2 is

$$\begin{aligned} \frac{X_{C_1}}{X_{C_2}} &= \frac{\frac{1}{2\pi f C_1}}{\frac{1}{2\pi f C_2}} \\ &= \frac{2\pi f C_2}{2\pi f C_1} \\ &= \frac{C_2}{C_1} \end{aligned} \tag{2.12}$$

For the noninverting version of the oscillator, shown in Figure 2.19(a), the output is across the series combination of C_1 and C_2, which corresponds to C_T in Equation (2.9), and the input is the voltage across C_1. It is easy to see that the feedback fraction is

$$\begin{aligned} B &= \frac{X_{C_1}}{X_{C_T}} \\ &= \frac{C_T}{C_1} \\ &= \frac{\frac{C_1 C_2}{C_1 + C_2}}{C_1} \\ &= \frac{C_2}{C_1 + C_2} \end{aligned} \tag{2.13}$$

The feedback fraction is even easier to determine for the inverting circuit of Figure 2.19(b). Since the output is applied across C_2 and the input is taken across C_1, the feedback fraction is

$$\begin{aligned} B &= -\frac{X_{C_1}}{X_{C_2}} \\ &= -\frac{C_2}{C_1} \end{aligned} \tag{2.14}$$

As before, the negative sign indicates that the feedback signal is 180° out of phase with the output.

Since changing either C_1 or C_2 to tune the oscillator will change the feedback fraction, it is quite common to use a variable inductor for tuning instead.

Like the Hartley, the Colpitts oscillator can be configured for an amplifier with power gain but no voltage gain. The derivation of the circuit and the feedback fraction is left as an exercise for the reader.

Example 2.3 Determine the feedback fractions and operating frequencies for the oscillators whose circuits are shown in Figure 2.20.

Solution Our Colpitts oscillator examples use bipolar transistors. Figure 2.20(a) uses a common-emitter circuit, while the transistor is connected common-base in Figure 2.20(b). These two examples use inverting and noninverting amplifiers, respectively, both with voltage gain greater than one. The use of a common-collector circuit is also possible.

In Figure 2.21(a), C_C is a coupling capacitor that prevents a dc short circuit from occurring between collector and base through the coil. The radio-frequency choke RFC takes the place of a collector resistor and keeps the ac at the collector from being short-circuited by the power supply. A collector resistor can be used, but the choke, because of its lower dc resistance, increases the output voltage and improves the efficiency of the circuit. R_{B_1} and R_{B_2} are bias resistors, of course. This leaves the frequency of the oscillator

Figure 2.20

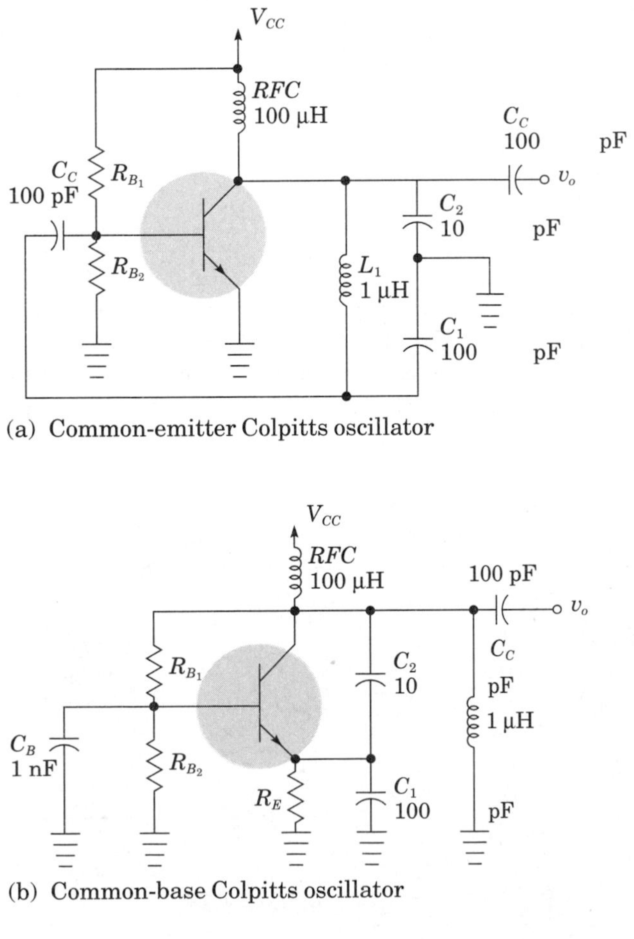

(a) Common-emitter Colpitts oscillator

(b) Common-base Colpitts oscillator

to be determined by L_1, C_1, and C_2. The effective capacitance for determining the frequency of operation is given by Equation (2.10):

$$C_T = \frac{C_1 C_2}{C_1 + C_2} = \frac{10 \times 100}{10 + 100}\ \text{pF} = 9.09\ \text{pF}$$

The operating frequency can be found from Equation (2.9):

$$f_o = \frac{1}{2\pi\sqrt{LC_T}} = \frac{1}{2\pi\sqrt{(1 \times 10^{-6})(9.09 \times 10^{-12})}} = 52.8\ \text{MHz}$$

The feedback fraction is given approximately by Equation (2.14).

$$B = -\frac{C_2}{C_1} = -\frac{10}{100} = -0.1$$

For the common-base circuit, the operating frequency will be the same but the feedback fraction will be different. Of course, the sign will be positive because the amplifier is noninverting, but the magnitude is also slightly different. From Equation (2.13),

$$B = \frac{C_2}{C_1 + C_2} = \frac{10}{100 + 10} = 0.0909$$

Clapp Oscillator The Clapp oscillator is a variation of the Colpitts circuit, designed to swamp device capacitances for greater stability. In the oscillators of Figure 2.21, the frequency of oscillation is determined by the inductor and the series combination of C_1, C_2, and C_3. In practice, the total capacitance is determined almost entirely by C_3, which is

Figure 2.21
Clapp oscillators

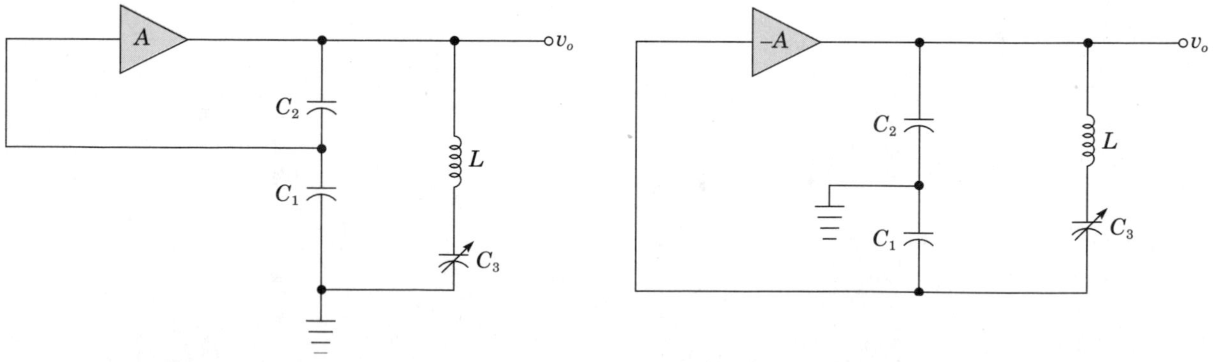

(a) Noninverting (b) Inverting

chosen to be much smaller than either C_1 or C_2. The total effective capacitance of the three capacitors in series is given by

$$C_T = \frac{1}{\frac{1}{C_1} + \frac{1}{C_2} + \frac{1}{C_3}} \tag{2.15}$$

After finding C_T from Equation (2.15), the operating frequency can easily be found from Equation (2.9), as for the Colpitts oscillator.

The feedback fraction is found in the same way as for the Colpitts oscillator.

One circuit will serve to illustrate the Clapp oscillator, since it is so similar to the Colpitts.

Example 2.4 Calculate the feedback fraction and oscillating frequency of the circuit in Figure 2.22.

Solution Figure 2.22 is the same as Figure 2.20(a) except for the addition of C_3 in series with the coil and the increase in the values of C_1 and C_2. Any transistor capacitances will appear across C_1 and C_2, where they will have little effect on the frequency.

Before calculating the oscillating frequency, it is necessary to calculate the effective total capacitance from Equation (2.15):

$$\begin{aligned} C_T &= \frac{1}{\frac{1}{C_1} + \frac{1}{C_2} + \frac{1}{C_3}} \\ &= \frac{1}{\frac{1}{100} + \frac{1}{1000} + \frac{1}{10}} \text{ pF} \\ &= 9.01 \text{ pF} \end{aligned}$$

Note the relatively small effect of C_1 and C_2.

Now the operating frequency can be found in the usual way from Equation (2.9).

$$\begin{aligned} f_o &= \frac{1}{2\pi\sqrt{LC_T}} \\ &= \frac{1}{2\pi\sqrt{(1 \times 10^{-6})(9.01 \times 10^{-12})}} \\ &= 53.02 \text{ MHz} \end{aligned}$$

Figure 2.22

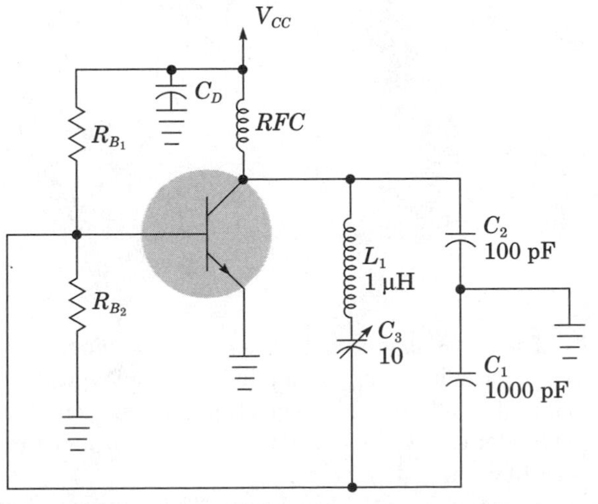

Figure 2.23
Variable capacitors and inductors

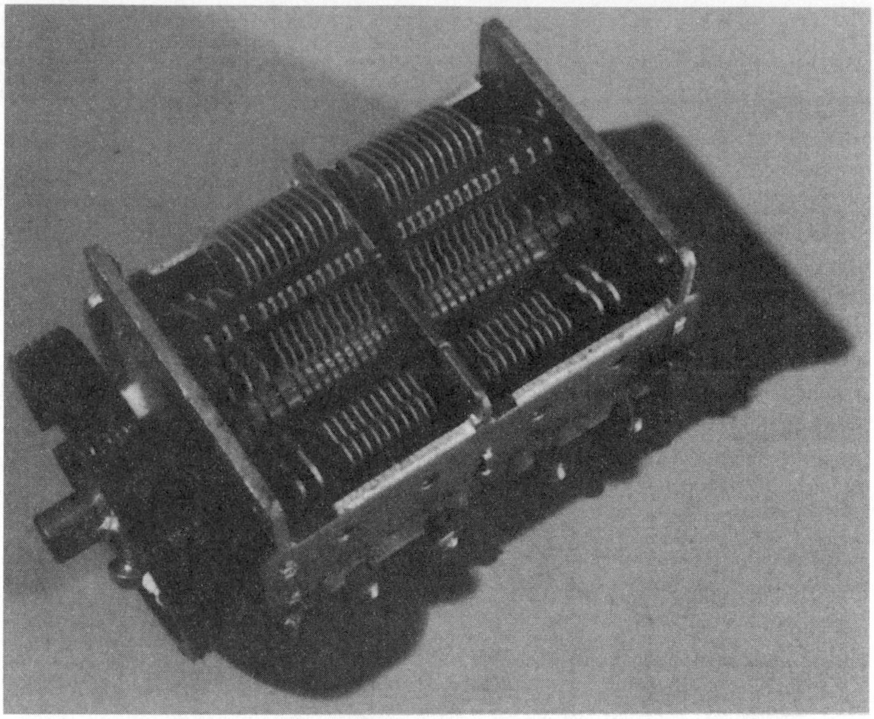

(a) Variable capacitor

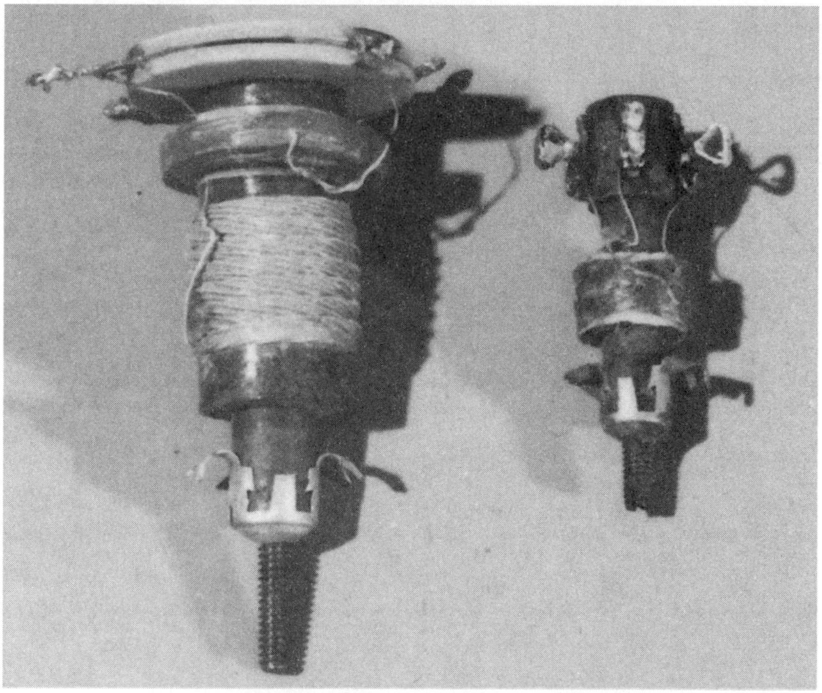

(b) Variable inductors

2.5.2 Varactor-Tuned Oscillators

The frequency of an *LC* oscillator can be changed by varying, or tuning, either the inductive or the capacitive element in a tuned circuit. Inductors are typically tuned by moving a ferrite core into or out of the coil; this is known as *slug tuning*. Variable capacitors usually have two sets of plates that can be interleaved to a greater or lesser extent. Figure 2.23 shows typical examples of each.

Mechanical tuning tends to be awkward. The components are bulky, expensive, and subject to accidental detuning, for instance, in the presence of vibration. Variable capacitors and inductors are mechanical devices that have to be moved physically. This makes remote or automatic frequency control rather cumbersome.

Varactors are a more convenient substitute for variable capacitors in many applications. Essentially, a varactor is a reverse-biased silicon diode. As the reverse voltage increases, so does the width of the diode's depletion layer. As a result, the junction capacitance decreases. If this junction capacitance is made part of a resonant circuit, that circuit can be tuned simply by varying the dc voltage on the varactor. This can be done in many ways and is well adapted to remote or automatic control. The resulting circuit is often called a voltage-controlled oscillator (VCO).

It is, of course, necessary to separate the dc control voltage from the ac signal voltages. This is quite straightforward: Figure 2.24 shows one way to do it. The noninverting Clapp oscillator of Figure 2.21(a) has been adapted for use as a VCO by using a varactor for C_3. Resistor R prevents the RF in the circuit from being short-circuited by the circuit that provides the tuning voltage, and the extra capacitor, C_4, keeps the dc control voltage out of the rest of the circuit. C_4 is made much larger than C_3 so that its reactance will be negligible and C_3 will still control the operating frequency.

The variation of capacitance with voltage is not linear for a varactor. It is given approximately by

$$C = \frac{C_o}{\sqrt{1 + 2V}} \tag{2.16}$$

where

C = capacitance at reverse voltage V
C_o = capacitance with no reverse voltage

From this equation it can be seen that for relatively large reverse-bias voltages, the capacitance is approximately inversely proportional to the square root of the applied voltage. A number of varactors, with maximum capacitances varying from a few picofarads to more than 100 pF are available. From Equation (2.23) it can be seen that the minimum capacitance will be limited by the breakdown voltage of the diode and, of course, by the tuning voltage available. In practice, a variation of about 5 : 1 in capacitance is quite practical. The magnitude of control voltage required for this can be found by letting $C_o/C = 5$ in Equation (2.16):

$$C = \frac{C_o}{\sqrt{1 + 2V}}$$

$$\sqrt{1 + 2V} = \frac{C_o}{C}$$

$$\sqrt{1 + 2V} = 5$$

$$1 + 2V = 25$$

$$2V = 24$$

$$V = 12 \text{ V}$$

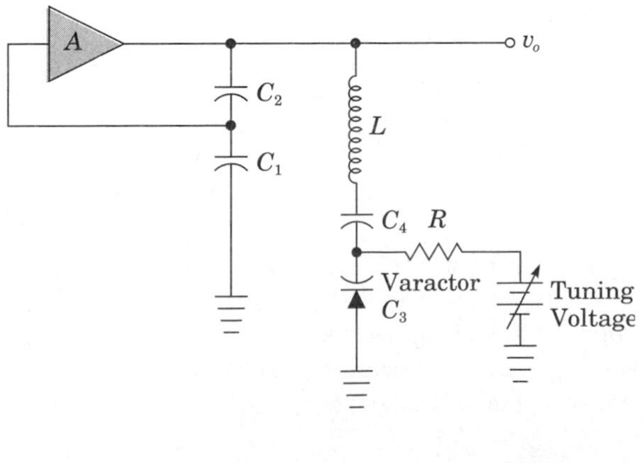

Figure 2.24 Varactor-tuned oscillator

Example 2.5

A varactor has a maximum capacitance of 80 pF and is used in a tuned circuit with a 100 μH inductor.

(a) Find the resonant frequency with no tuning voltage applied.
(b) Find the tuning voltage necessary for the circuit to resonate at double the frequency found in part (a).

Solution

(a) The maximum capacitance of the varactor occurs for zero bias voltage, so the capacitance in this case will be 80 pF. The resonant frequency can be calculated from Equation (2.2).

$$\begin{aligned} f_o &= \frac{1}{2\pi\sqrt{LC}} \\ &= \frac{1}{2\pi\sqrt{(100 \times 10^{-6})(80 \times 10^{-12})}} \\ &= 1.78 \text{ MHz} \end{aligned}$$

(b) From Equation (2.2) it is apparent that the resonant frequency is inversely proportional to the square root of capacitance, so that doubling the frequency will require reducing the capacitance by a factor of four. Alternatively we can use Equation (2.2) directly.

$$\begin{aligned} f_o &= \frac{1}{2\pi\sqrt{LC}} \\ f_o^2 &= \frac{1}{4\pi^2 LC} \\ C &= \frac{1}{4\pi^2 f_o^2 L} \\ &= \frac{1}{4\pi^2(2 \times 1.78 \times 10^6)^2(100 \times 10^{-6})} \\ &= 20 \times 10^{-12} \text{ F} \\ &= 20 \text{ pF} \end{aligned}$$

Now we can find the required tuning voltage from Equation (2.16).

$$\begin{aligned} C &= \frac{C_o}{\sqrt{1 + 2V}} \\ \sqrt{1 + 2V} &= \frac{C_o}{C} \\ 1 + 2\text{V} &= \left(\frac{C_o}{C}\right)^2 \\ V &= \frac{\left(\frac{C_o}{C}\right)^2 - 1}{2} \\ &= \frac{\left(\frac{80}{20}\right)^2 - 1}{2} \\ &= 7.5 \text{ V} \end{aligned}$$

2.5.3 Crystal-Controlled Oscillators

The frequency stability of any *LC* oscillator depends on that of its resonant circuit, including any stray or device reactances that may be present. Even with careful design, *LC* oscillators are subject to frequency change from such diverse sources as voltage variations, changes in load impedance, temperature changes, and mechanical vibration. All these problems can be reduced, but only at great expense.

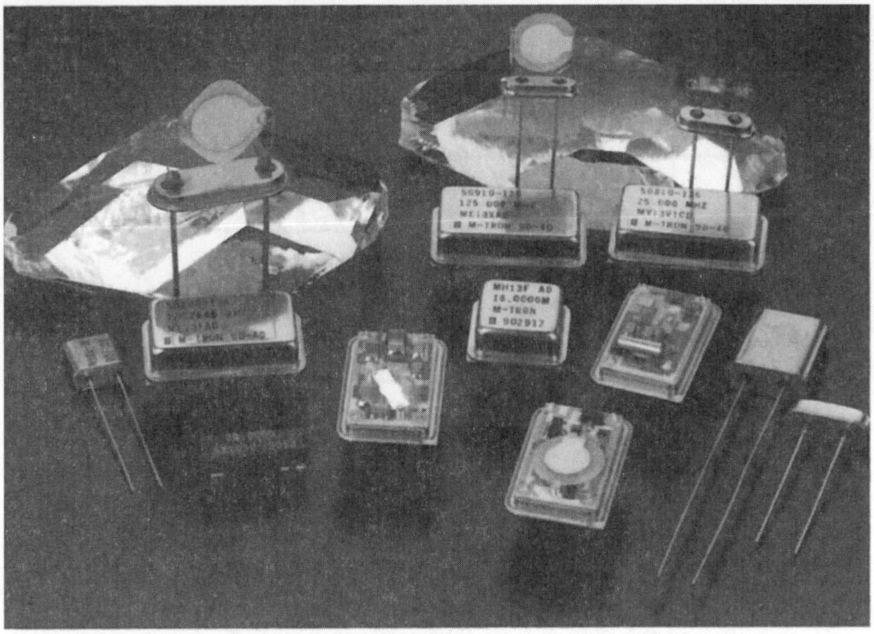

Figure 2.25
Quartz crystals
Photo courtesy of Yorkton, SD, USA.

Crystal oscillators achieve greater stability by using a small slab of quartz as a mechanical resonator, in place of an *LC* tuned circuit. Quartz is a **piezoelectric** material: deforming it mechanically causes the crystal to generate a voltage, and applying a voltage to the crystal causes it to deform. Like any rigid body, the crystal slab has a mechanical resonant frequency. If it is pulsed with voltage at that frequency, it will vibrate. From the outside, the crystal will have the appearance of an electrical resonant circuit.

The photograph in Figure 2.25 shows several typical crystals, some with the packaging removed to show the crystal slab. The actual quartz crystals from which RF crystals are made are shown in the background. Figure 2.26(a) is the schematic symbol for a crystal, and Figure 2.26(b) shows the equivalent circuit. Note that a crystal does not actually have

Figure 2.26

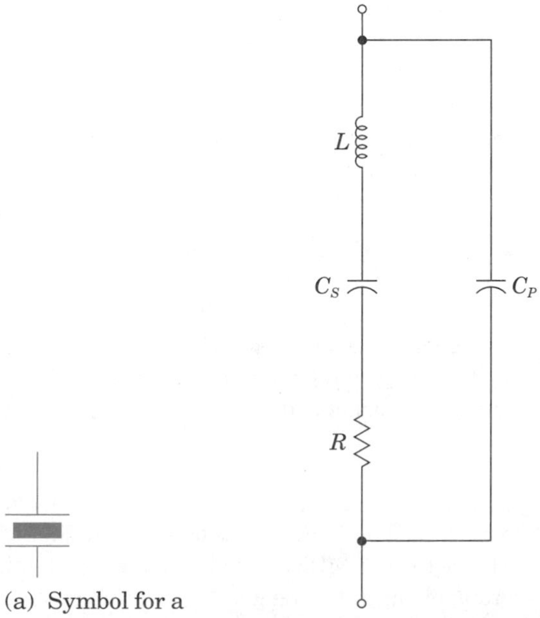

(a) Symbol for a quartz crystal

(b) Equivalent circuit for a crystal

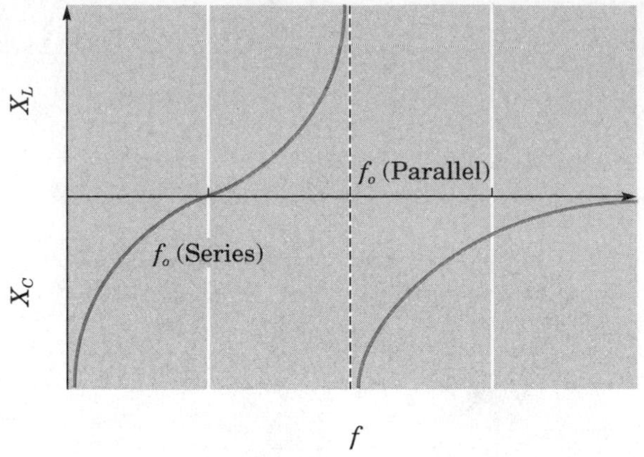

Figure 2.27 Variation of crystal reactance with frequency

this circuit. A crystal is just a slab of quartz with electrodes. This circuit simply models its behavior, as seen from outside. The only part of the equivalent circuit that resembles the actual device is C_P, the parallel capacitance, which actually is the capacitance of the holder and the crystal itself. The series inductor and capacitor are electrical analogs of the mechanical properties of the crystal.

In electrical terms, the inductor in the equivalent circuit is very large (on the order of henrys), while both capacitors are very small (picofarads or less), with the parallel capacitance much larger than the series. The resistance is small, and the Q of crystals is very high (in the range from 10^4 to 10^7). From the equivalent circuit, it is apparent that the crystal will actually have not one but two resonant frequencies: one series and one parallel. Figure 2.27 shows graphically how the reactance varies with frequency. The two resonances will vary by an amount that is on the order of 1% of the operating frequency, and when ordering a crystal, it is necessary to know at which of these frequencies the circuit operates. For parallel resonance, it is also necessary to have a constant, known circuit capacitance across the crystal. It is also possible to operate the crystal at a *harmonic*, or *overtone*, of the fundamental frequency. This is necessary when frequencies above approximately 20 MHz are required.

Crystal oscillators offer great stability at the price of fixed-frequency operation. The operating frequency is typically accurate to $\pm 0.005\%$, and the stability can be much better if a temperature-controlled "oven" is used to keep the crystal at a constant temperature a little higher than the ambient temperature. Crystals are not immune to temperature changes, since their temperature coefficients can be either positive or negative depending on the way the crystal is cut and can be as high as approximately 100 parts per million (ppm) per degree Celsius. The frequency is given by

$$f_T = f_o + kf_o(T - T_o) \tag{2.17}$$

where

f_T = operating frequency at temperature T
f_o = operating frequency at reference temperature T_o
k = temperature coefficient per degree

Example 2.6

A portable radio transmitter has to operate at temperatures from -5°C to $+35$°C. If its signal is derived from a crystal oscillator with a temperature coefficient of $+10$ ppm/degree C and it transmits at exactly 146 MHz at 20°C, find the transmitting frequency at the two extremes of the operating temperature range.

Solution From Equation (2.17), the frequency at a temperature of 35°C will be

$$f_T = f_o + kf_o(T - T_o)$$
$$f_{max} = 146 \text{ MHz} + (146 \text{ MHz})(10 \times 10^{-6})(35 - 20)$$
$$= 146.0219 \text{ MHz}$$

By similar reasoning, the operating frequency at the lower temperature limit will be

$$f_{min} = 146 \text{ MHz} + (146 \text{ MHz})(10 \times 10^{-6})(-5 - 20)$$
$$= 145.9635 \text{ MHz}$$

In other words, varying the temperature over a total range of 40°C caused the transmitter frequency to change by (21.9 + 36.5) = 58.4 kHz. This would not usually be acceptable in a practical transmitter, so some form of frequency compensation would be needed.

Figure 2.28 shows two typical circuits for crystal oscillators. The Colpitts circuit in Figure 2.28(a) uses the parallel-resonance mode of the crystal, with capacitors C_1 and C_2 providing feedback. R_D and C_D **decouple** the drain from the power supply. The bias is generated from the signal itself using the gate-leak technique previously described for FET oscillators. The circuit in Figure 2.28(b), known as the *Pierce circuit*, uses the crystal in place of the inductor in a series-resonant circuit consisting of C_1, C_2, and the crystal. It operates in the inductive region of the reactance curve of Figure 2.27, close to but slightly above the series resonant frequency.

Crystal oscillators are often supplied in convenient modules that include temperature compensation circuitry. Typical modules are shown in Figure 2.25.

The frequency of a crystal oscillator can be adjusted slightly by placing a variable capacitance in series or in parallel with the crystal, depending on the type of circuit. An oscillator of this sort is known as a *variable crystal oscillator* (VXO). If the capacitor is actually a varactor, the frequency can be adjusted by a thermistor for temperature compensation.

Figure 2.28 Crystal oscillator circuits

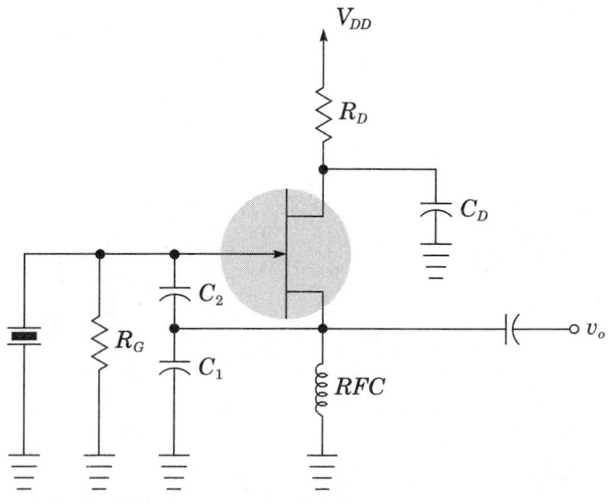

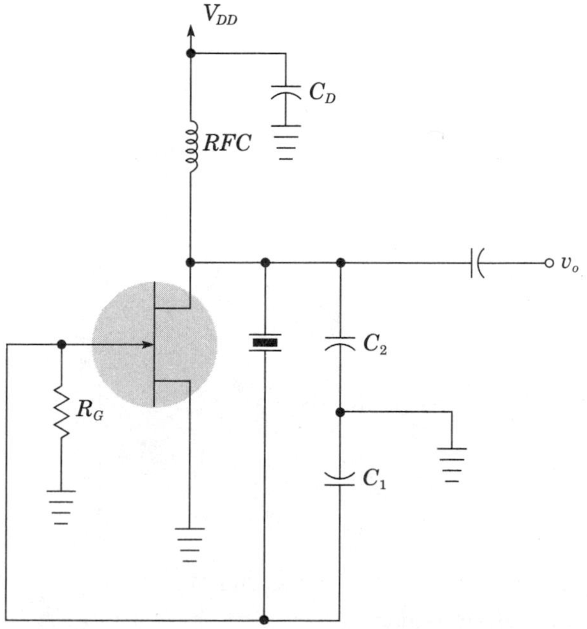

2.6 Mixers

A **mixer**, in communications, is a nonlinear circuit that combines two signals in such a way as to produce the sum and difference of the two input frequencies at the output. Sometimes the original input frequencies, and perhaps other frequencies as well, are also present. Mixers provide a way of moving a signal, complete with any modulation that may be present, from one frequency to another.

It is important to distinguish mixing from *linear summing*, which produces only the two input signal frequencies. Unfortunately the term *mixer* is often used in audio electronics to designate a linear summer. The "mixer" that regulates the sound in a recording studio is really a good example of a summer. Figure 2.29 illustrates the difference between mixing and summing.

Any nonlinear device can operate as a mixer. Familiar examples include a semiconductor diode and a Class C amplifier. In general, a nonlinear device produces a signal at its output that can be represented by a power series, for example

$$v_o = Av_i + Bv_i^2 + Cv_i^3 + \cdots \tag{2.18}$$

where

v_o = instantaneous output voltage
v_i = instantaneous input voltage
$A, B, C, \cdots$ = constants

With a single input frequency, the output will contain all the harmonics as well as the fundamental. As the order of the harmonics increases, the magnitude decreases. This is the principle of the frequency multiplier and also the cause of harmonic distortion.

Figure 2.29
Output spectra for mixer and summer

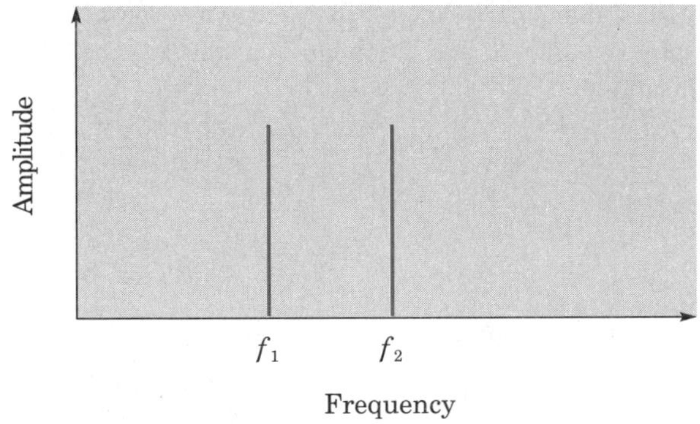

(a) Input frequencies

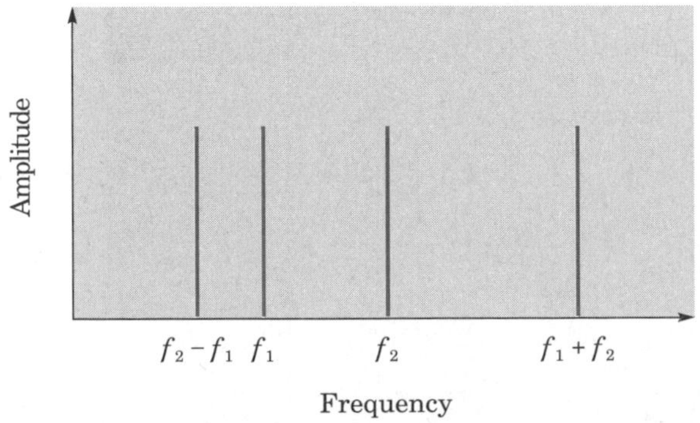

(b) Output frequencies: mixer

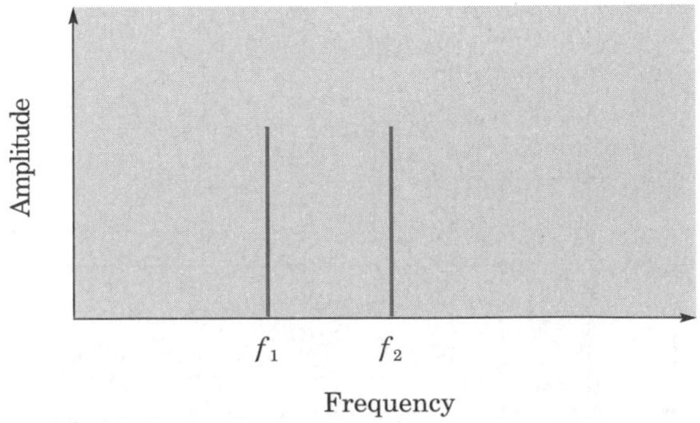

(c) Output frequencies: summer

When the input contains two different frequencies, we get cross products that represent sum and difference frequencies. If the two frequencies are f_1 and f_2, we get $mf_1 \pm nf_2$, where m and n are integers. Usually the most important are $f_1 + f_2$ and $f_1 - f_2$, where f_1 is assumed to be higher than f_2. When the cross products are not wanted, they are called *intermodulation distortion.*

2.6.1 Square-Law Mixers

The *square-law mixer* is the simplest to understand mathematically, and it closely models the actual performance of mixers using FETs. Its output is given by a simplified form of Equation (2.25):

$$v_o = Av_i + Bv_i^2 \tag{2.19}$$

Let us apply two signals to this circuit, summing them at the input. The signals will be sine waves of different frequencies. For convenience, we can let each signal have an amplitude of 1 V peak. Then

$$v_i = \sin \omega_1 t + \sin \omega_2 t \tag{2.20}$$

and

$$\begin{aligned} v_o &= Av_i + Bv_i^2 \\ &= A(\sin \omega_1 t + \sin \omega_2 t) + B(\sin \omega_1 t + \sin \omega_2 t)^2 \\ &= A \sin \omega_1 t + A \sin \omega_2 t + B \sin^2 \omega_1 t + B \sin^2 \omega_2 t \\ &\quad + 2B \sin \omega_1 t \sin \omega_2 t \end{aligned} \tag{2.21}$$

The first two terms in Equation (2.21) are simply the input signals multiplied by A, which is a gain factor. The next two terms involve the squares of the input signals, which are signals at twice the input frequency (plus a dc component). This can be seen from the trigonometric identity

$$\sin^2 A = \frac{1}{2} - \frac{1}{2} \cos 2A$$

This means that the third term in Equation (2.21) becomes

$$\frac{B}{2} - \frac{B}{2} \cos 2\omega_1 t$$

and the fourth term is

$$\frac{B}{2} - \frac{B}{2} \cos 2\omega_2 t$$

The final term is the interesting one. We can use the trigonometric identity

$$\sin A \sin B = \frac{1}{2} [\cos (A - B) - \cos (A + B)] \tag{2.22}$$

to expand this term.

$$\begin{aligned} 2B \sin \omega_1 t \sin \omega_2 t &= \frac{2B}{2} [\cos (\omega_1 - \omega_2)t - \cos (\omega_1 + \omega_2)t] \\ &= B [\cos (\omega_1 - \omega_2)t - \cos (\omega_1 + \omega_2)t] \end{aligned}$$

As predicted, there are output signals at the sum and difference of the two input frequencies, in addition to the input frequencies themselves and their second harmonics. In a practical application, either the sum or the difference frequency would be used, and the others would be removed by filtering.

Example 2.7 Sine-wave signals with frequencies of 10 MHz and 11 MHz are applied to a square-law mixer. What frequencies appear at the output?

Solution Let $f_1 = 11$ MHz and $f_2 = 10$ MHz. Then the output frequencies are as follows:

$$f_1 = 11 \text{ MHz} \quad 2f_1 = 22 \text{ MHz} \quad f_1 + f_2 = 21 \text{ MHz}$$
$$f_2 = 10 \text{ MHz} \quad 2f_2 = 20 \text{ MHz} \quad f_1 - f_2 = 1 \text{ MHz}$$

2.6.2 Diode Mixers

There are two senses in which we can say that a diode is a nonlinear device. First, consider an ideal diode, which has zero reverse current for any reverse voltage and zero forward voltage for any forward current. This voltage-current relationship is shown in Figure 2.30(a). There is a very obvious nonlinearity at zero voltage.

Now consider a real diode with its forward voltage-current characteristics. A *V-I* curve for a typical silicon signal diode is shown in Figure 2.30(b). It is obviously not a straight line; that is, a real diode is nonlinear even when operated in the forward-bias region. Of course, the larger the forward current, the more the diode approaches linearity.

Diode mixers can use either type of nonlinearity. In the first type, one of the signals is strong enough to switch the diode between the reverse-biased and forward-biased states. The diode ring balanced mixer, to be described later, operates in this way. There is also another possibility, where the diode operates with a small forward bias, in the "knee" of the curve of Figure 2.30(b). The input signal is small, so that the diode stays in this highly nonlinear region of its characteristic curve. Figure 2.31 shows a circuit for a mixer of this type. Such mixers are not used very often because of their poor noise figure.

2.6.3 Transistor Mixers

All transistors, both bipolar and field-effect types, have nonlinearities. In fact, much of the work in an introductory electronics course usually involves biasing these devices in such a way as to reduce the nonlinearity to a minimum. The success of these efforts is then measured in terms of percent distortion.

On the other hand, nonlinearities are desirable if a mixer is required. Either a bipolar transistor or an FET can be operated in such a way that the input causes the device to enter nonlinear regions. The FET is especially convenient for use as a mixer because the parabolic shape of its transconductance curve gives it approximately a square-law response. The bipolar transistor will produce more spurious frequencies at its output, frequencies that will probably have to be removed by filtering. However, both types of transistor are in common use in mixer circuits.

Figure 2.30
Diode curves

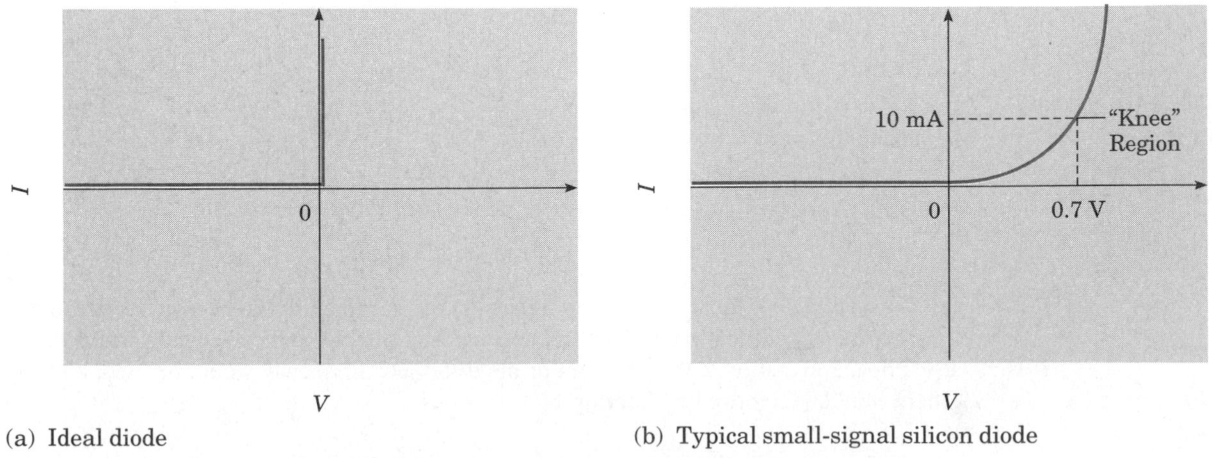

Figure 2.31
Diode mixer

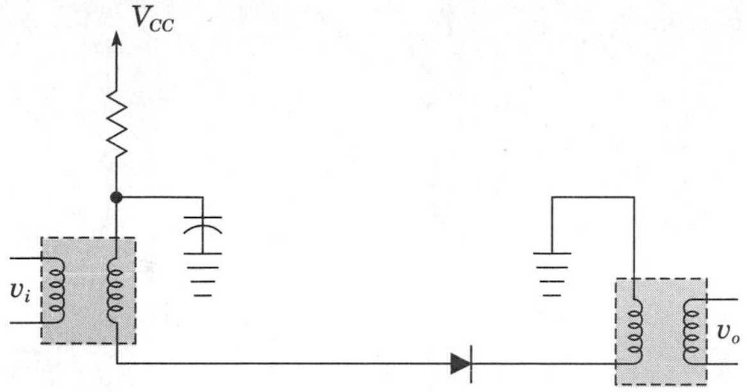

Figure 2.32 includes circuits for three typical transistor mixers. Figure 2.32(a) uses a bipolar transistor, and Figure 2.32(b) uses a JFET. The circuit in Figure 2.32(c) uses a dual-gate metal oxide semiconductor FET (MOSFET) with one of the two signals to be mixed applied to each gate.

Figure 2.32
Transistor mixers

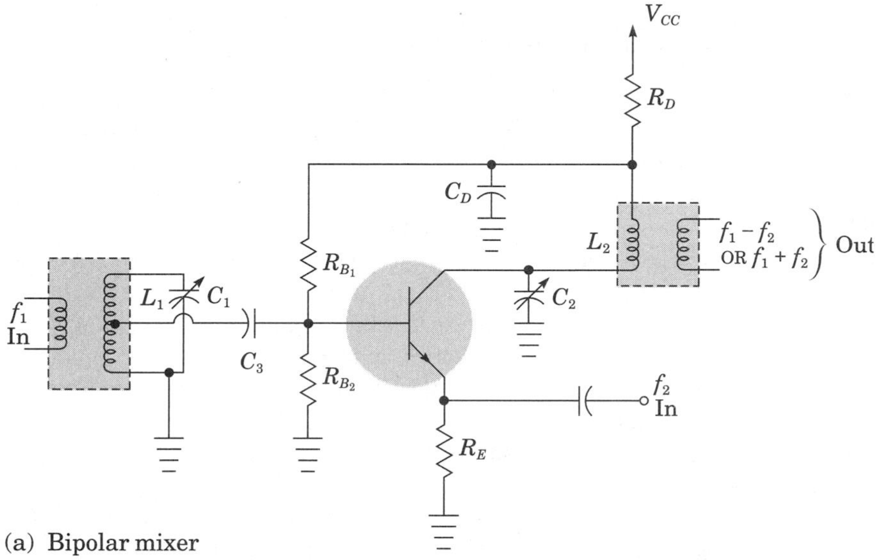

(a) Bipolar mixer

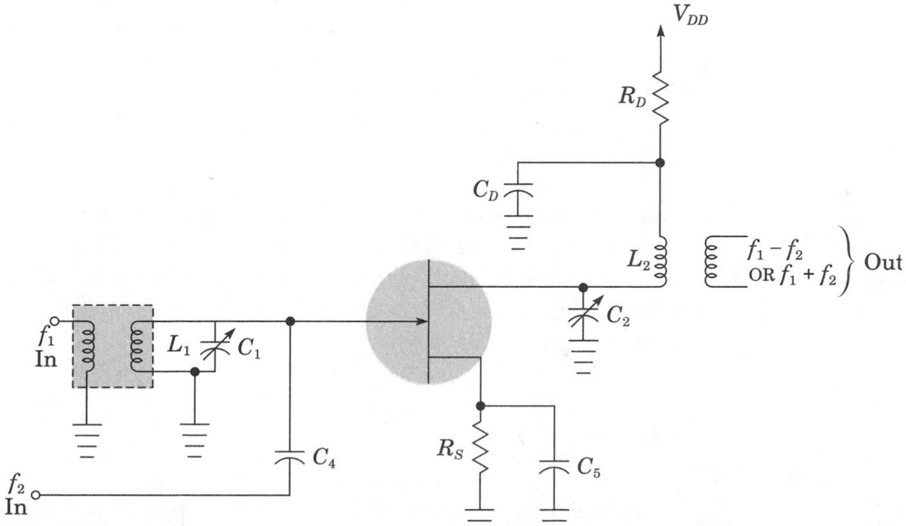

(b) JFET mixer

continues

Figure 2.32 Continued

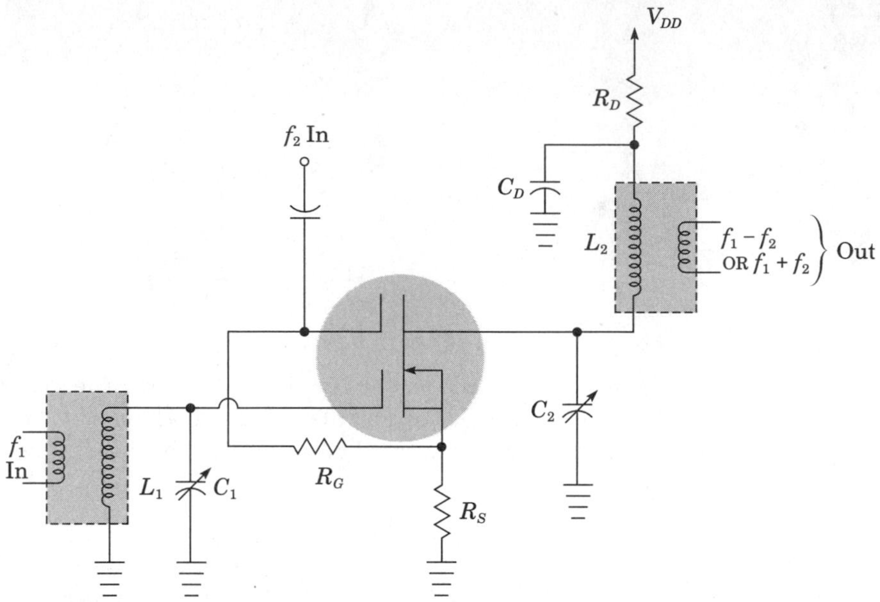

(c) Dual-gate MOSFET mixer

The bipolar-transistor mixer of Figure 2.32(a) resembles a conventional tuned RF amplifier, except that it has two inputs, one at the base and one at the emitter. One of the two signals to be mixed will be applied to each input, with the stronger of the two going to the emitter. The output can be tuned either to the sum or the difference of the two input signal frequencies, whichever is required. In this circuit, the resonant circuits L_1–C_1 and L_2–C_2 are tuned to f_1 and the required output frequency, respectively. R_{B_1}, R_{B_2}, and R_E form the usual voltage-divider bias circuit, C_3 is a coupling capacitor that keeps L_1 from short-circuiting the base bias, and C_D and R_D form a decoupling network that keeps RF out of the power supply line. It is also possible to apply f_2, as well as f_1, to the base, summing them at this point.

The JFET circuit of Figure 2.32(b) illustrates the technique of summing the two input signals at a single input terminal. Both f_1 and f_2 are applied to the gate. The tuned circuits L_1–C_1 and L_2–C_2 have the same functions as before, and so do C_D and R_D. C_4 couples the f_2 input into the circuit, C_5 bypasses the source resistor, and R_S provides self-bias. In some JFET mixers, one of the input signals is applied to the source.

Dual-gate MOSFETs are very easy to use as mixers, as one input signal can be applied to each gate. This is demonstrated in Figure 2.32(c). Dual-gate MOSFETs make excellent mixers. They have better dynamic range than mixers using bipolar transistors; that is, they can operate satisfactorily over a wider range of signal amplitudes. They also produce fewer unwanted output frequencies.

2.6.4 Balanced Mixers

A **balanced mixer** is one in which the input frequencies do not appear at the output. Ideally, the only frequencies that are produced are the sum and difference of the input frequencies.

A **multiplier** circuit, where the output amplitude is proportional to the product of two input signals, can be used as a balanced mixer. This is easy to show mathematically. Let the multiplier have the equation

$$v_o = A v_{i_1} v_{i_2} \tag{2.23}$$

where

v_o = the instantaneous output voltage
v_{i_1} and v_{i_2} = the instantaneous voltages applied at two input terminals
A = a constant

Let the input consist of two sine waves of different frequencies, as before. For simplicity, assume each signal has a peak amplitude of 1 V.

$$v_{i_1} = \sin \omega_1 t$$
$$v_{i_2} = \sin \omega_2 t$$

Then the output will be

$$\begin{aligned} v_o &= A v_{i_1} v_{i_2} \\ &= A \sin \omega_1 t \sin \omega_2 t \end{aligned}$$

We can use the same trigonometric identity, Equation (2.22), that we used earlier with the square-law mixer

$$\sin A \sin B = \frac{1}{2} [\cos (A - B) - \cos (A + B)]$$

to show that the output is

$$v_o = \frac{A}{2} [\cos (\omega_1 - \omega_2)t - \cos (\omega_1 + \omega_2)t]$$

As predicted, this mixer produces only the sum and difference of the input frequencies. Thus it is an excellent choice for many applications. In fact, the usual block-diagram representation of a mixer is the same as for a multiplier (see Figure 2.33).

Though general-purpose integrated-circuit multipliers exist, there are a number of balanced mixers designed especially for communications use. One of these is the MC1496, whose abridged data sheet is shown in Figure 2.34. These devices are capable of operating over a very wide frequency range and are extremely well balanced; that is, the input signal frequencies are reduced to very low levels.

It is also possible to build a balanced mixer using discrete components. Figure 2.35 shows a traditional circuit using four diodes in a ring configuration. For best results, the diodes should be carefully matched. The input signals should not have the same amplitude: the input at f_2 should have an amplitude large enough to turn the diodes on completely, and the other input signal should be of much lower voltage.

In order to understand the operation of this mixer, first consider an instant when the f_2 signal causes the secondary of transformer T_3 to have the polarity shown in Figure 2.35, that is, the left side is positive. This will cause diodes D_1 and D_2 to turn on and D_3 and D_4 to turn off. D_1 and D_2 will then connect the much weaker signal at the secondary of T_1 directly to the primary of T_2 without any polarity change.

Now consider what happens when the polarity of the secondary of T_3 reverses. D_1 and D_2 will now be off, and D_3 and D_4 will be on. The signal from T_1 will still be connected to T_2, but this time the polarity will be reversed.

From this description, it can be seen that the circuit multiplies the signal at f_1 by a factor of +1 or −1, alternating at the f_2 rate. This is equivalent to multiplying the signal at

Figure 2.33
Mixer symbol

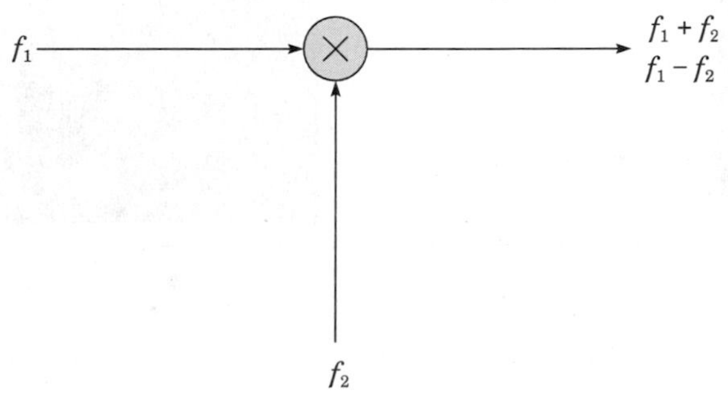

MOTOROLA

SEMICONDUCTORS

P.O. BOX 20912 • PHOENIX, ARIZONA 85036

MC1496
MC1596

Specifications and Applications Information

BALANCED MODULATOR/DEMODULATOR

SILICON MONOLITHIC INTEGRATED CIRCUIT

BALANCED MODULATOR/ DEMODULATOR

. . . designed for use where the output voltage is a product of an input voltage (signal) and a switching function (carrier). Typical applications include suppressed carrier and amplitude modulation, synchronous detection, FM detection, phase detection, and chopper applications. See Motorola Application Note AN-531 for additional design information.

- Excellent Carrier Suppression – 65 dB typ @ 0.5 MHz
  – 50 dB typ @ 10 MHz
- Adjustable Gain and Signal Handling
- Balanced Inputs and Outputs
- High Common Mode Rejection – 85 dB typ

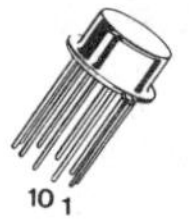

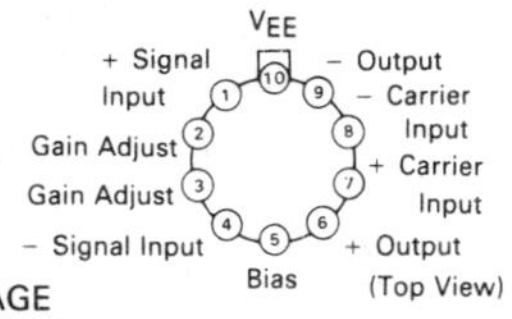

G SUFFIX
METAL PACKAGE
CASE 603-04

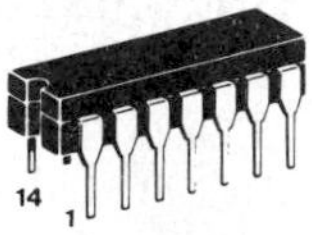

L SUFFIX
CERAMIC PACKAGE
CASE 632-08

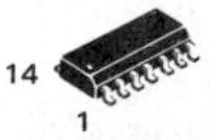

D SUFFIX
PLASTIC PACKAGE
CASE 751A-02
SO-14

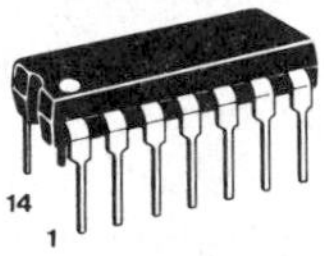

P SUFFIX
PLASTIC PACKAGE
CASE 646-06

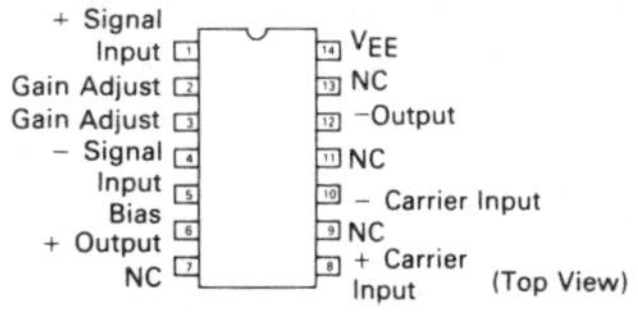

ORDERING INFORMATION

Device	Temperature Range	Package
MC1496D	0°C to +70°C	SO-14
MC1496G		Metal Can
MC1496L		Ceramic DIP
MC1496P		Plastic DIP
MC1596G	–55°C to +125°C	Metal Can
MC1596L		Ceramic DIP

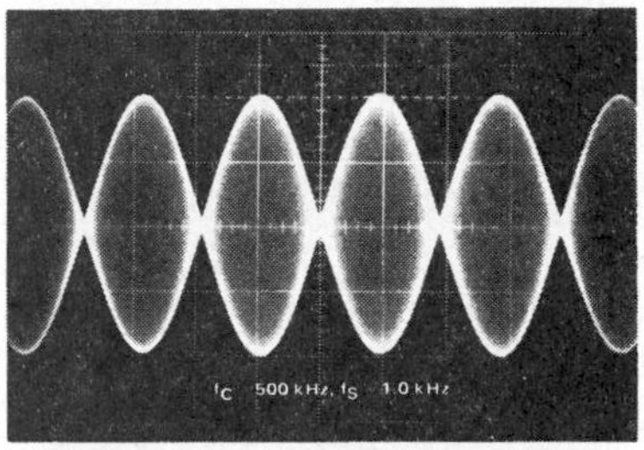

FIGURE 1 – SUPPRESSED-CARRIER OUTPUT WAVEFORM

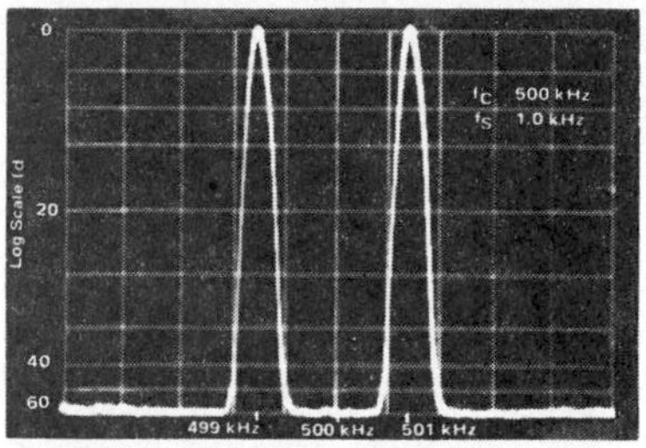

FIGURE 2 – SUPPRESSED-CARRIER SPECTRUM

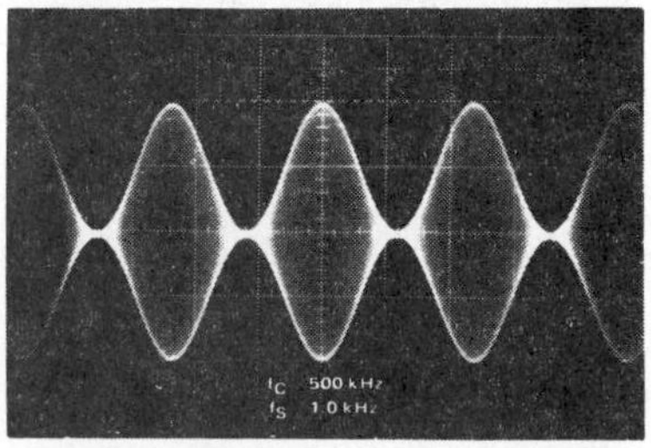

FIGURE 3 – AMPLITUDE-MODULATION OUTPUT WAVEFORM

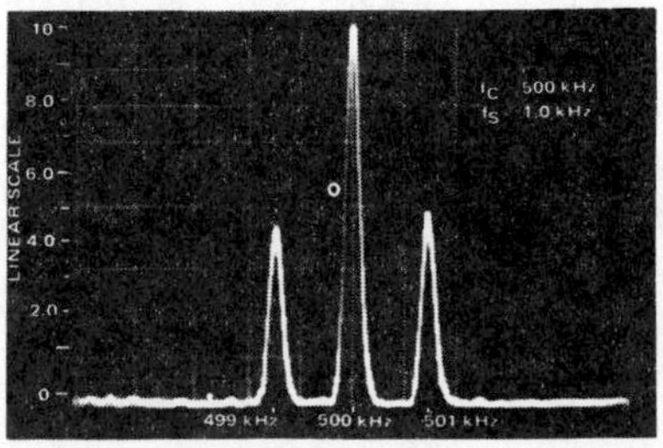

FIGURE 4 – AMPLITUDE-MODULATION SPECTRUM

Figure 2.34
Integrated-Circuit Balanced Mixer
Copyright of Motorola. Used by permission.

MAXIMUM RATINGS* (T_A = +25°C unless otherwise noted)

Rating	Symbol	Value	Unit
Applied Voltage ($V_6 - V_7, V_8 - V_1, V_9 - V_7, V_9 - V_8, V_7 - V_4, V_7 - V_1,$ $V_8 - V_4, V_6 - V_8, V_2 - V_5, V_3 - V_5$)	ΔV	30	Vdc
Differential Input Signal	$V_7 - V_8$ $V_4 - V_1$	+5.0 $\pm(5 + I_5R_e)$	Vdc
Maximum Bias Current	I_5	10	mA
Thermal Resistance, Junction to Air Ceramic Dual In-Line Package Plastic Dual In-Line Package Metal Package	$R_{\theta JA}$	 100 100 160	°C/W
Operating Temperature Range MC1496 MC1596	T_A	 0 to +70 −55 to +125	°C
Storage Temperature Range	T_{stg}	−65 to +150	°C

ELECTRICAL CHARACTERISTICS* (V_{CC} = +12 Vdc, V_{EE} = −8.0 Vdc, I_5 = 1.0 mAdc, R_L = 3.9 kΩ, R_e = 1.0 kΩ, T_A = +25°C unless otherwise noted) (All input and output characteristics are single-ended unless otherwise noted.)

Characteristic	Fig.	Note	Symbol	MC1596 Min	MC1596 Typ	MC1596 Max	MC1496 Min	MC1496 Typ	MC1496 Max	Unit
Carrier Feedthrough	5	1	V_{CFT}							μV(rms)
V_C = 60 mV(rms) sine wave and f_C = 1.0 kHz				—	40	—	—	40	—	
offset adjusted to zero f_C = 10 MHz				—	140	—	—	140	—	
V_C = 300 mVp-p square wave:										mV(rms)
offset adjusted to zero f_C = 1.0 kHz				—	0.04	0.2	—	0.04	0.4	
offset not adjusted f_C = 1.0 kHz				—	20	100	—	20	200	
Carrier Suppression	5	2	V_{CS}							dB
f_S = 10 kHz, 300 mV(rms)										
f_C = 500 kHz, 60 mV(rms) sine wave				50	65	—	40	65	—	
f_C = 10 MHz, 60 mV(rms) sine wave				—	50	—	—	50	—	k
Transadmittance Bandwidth (Magnitude) (R_L = 50 ohms)	8	8	BW_{3dB}							MHz
Carrier Input Port, V_C = 60 mV(rms) sine wave				—	300	—	—	300	—	
f_S = 1.0 kHz, 300 mV(rms) sine wave										
Signal Input Port, V_S = 300 mV(rms) sine wave				—	80	—	—	80	—	
$\lvert V_C \rvert$ = 0.5 Vdc										
Signal Gain V_S = 100 mV(rms), f = 1.0 kHz; $\lvert V_C \rvert$ = 0.5 Vdc	10	3	A_{VS}	2.5	3.5	—	2.5	3.5	—	V/V
Single-Ended Input Impedance, Signal Port, f = 5.0 MHz	6	—								
Parallel Input Resistance			r_{ip}	—	200	—	—	200	—	kΩ
Parallel Input Capacitance			c_{ip}	—	2.0	—	—	2.0	—	pF
Single-Ended Output Impedance, f = 10 MHz	6	—								
Parallel Output Resistance			r_{op}	—	40	—	—	40	—	kΩ
Parallel Output Capacitance			c_{oo}	—	5.0	—	—	5.0	—	pF
Input Bias Current	7	—								μA
$I_{bS} = \frac{I_1 + I_4}{2}$; $I_{bC} = \frac{I_7 + I_8}{2}$			I_{bS}	—	12	25	—	12	30	
			I_{bC}	—	12	25	—	12	30	
Input Offset Current	7	—								μA
$I_{ioS} = I_1 - I_4$; $I_{ioC} = I_7 - I_8$			$\lvert I_{ioS} \rvert$	—	0.7	5.0	—	0.7	7.0	
			$\lvert I_{ioC} \rvert$	—	0.7	5.0	—	0.7	7.0	
Average Temperature Coefficient of Input Offset Current (T_A = −55°C to +125°C)	7	—	$\lvert TC_{Iio} \rvert$	—	2.0	—	—	2.0	—	nA/°C
Output Offset Current ($I_6 - I_9$)	7	—	$\lvert I_{oo} \rvert$	—	14	50	—	14	80	μA
Average Temperature Coefficient of Output Offset Current (T_A = −55°C to +125°C)	7	—	$\lvert TC_{Ioo} \rvert$	—	90	—	—	90	—	nA/°C
Common-Mode Input Swing, Signal Port, f_S = 1.0 kHz	9	4	CMV	—	5.0	—	—	5.0	—	Vp-p
Common-Mode Gain, Signal Port, f_S = 1.0 kHz, $\lvert V_C \rvert$ = 0.5 Vdc	9	—	ACM	—	−85	—	—	−85	—	dB
Common-Mode Quiescent Output Voltage (Pin 6 or Pin 9)	10	—	V_{out}	—	8.0	—	—	8.0	—	Vp-p
Differential Output Voltage Swing Capability	10	—	V_{out}	—	8.0	—	—	8.0	—	Vp-p
Power Supply Current	7	6								mAdc
$I_6 + I_9$			I_{CC}	—	2.0	3.0	—	2.0	4.0	
I_{10}			I_{EE}	—	3.0	4.0	—	3.0	5.0	
DC Power Dissipation	7	5	P_D	—	33	—	—	33	—	mW

* Pin number references pertain to this device when packaged in a metal can. To ascertain the corresponding pin numbers for plastic or ceramic packaged devices refer to the first page of this specification sheet.

Figure 2.34
(Continued)

f_1 by a square wave with frequency f_2 and amplitude 1 V peak. The result will be the sum and difference frequencies described earlier plus higher-frequency components at $3f_2 \pm f_1$, $5f_2 \pm f_1$, and so on. Generally, f_2 is the higher frequency, and these spurious signals will be at such high frequencies compared to the desired frequency that they will be very easy to filter out.

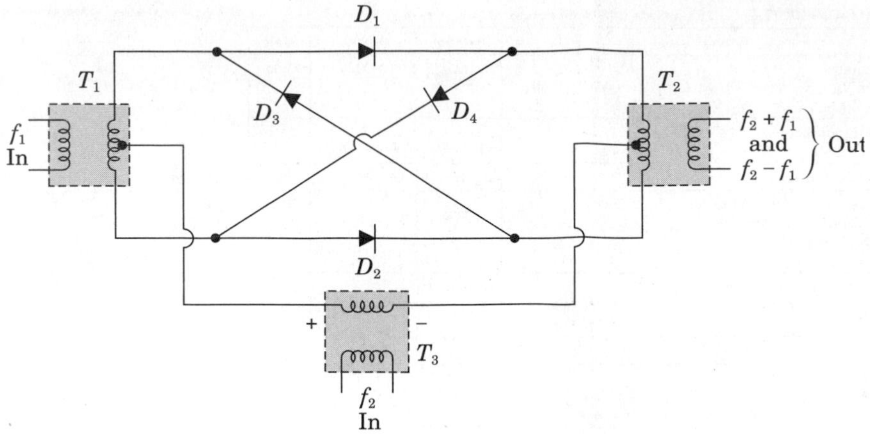

Figure 2.35 Double-balanced diode mixer

Balanced mixers are also called *balanced modulators* and will be seen again under that name when we discuss the generation of suppressed-carrier signals in Chapter 6.

2.7 Frequency Synthesizers

In section 2.5, we saw that a free-running *LC* oscillator can be easily tuned to different frequencies because its operating frequency is usually determined by tuned circuits. For that reason, they are often referred to as variable-frequency oscillators (VFOs). Unfortunately, they also exhibit undesired frequency changes as a result of vibration, voltage or

Early Frequency Synthesizers

Before phase-locked-loop frequency synthesizers became practical, attempts were made to generate multiple frequencies from a smaller number of crystals by using mixers and frequency multiplication. A simplified example of a synthesizer that uses banks of switched crystals is shown in Figure 2.36. There are two banks with four crystals in each. (In practice, there would likely be more.) There are two oscillators, each of which can work with any crystal in its associated bank. The mixer combines the two oscillator output frequencies, and the switchable bandpass filter can select either of the two crystal frequencies, their sum, or their difference. The range could be enhanced by using more mixers.

Figure 2.36

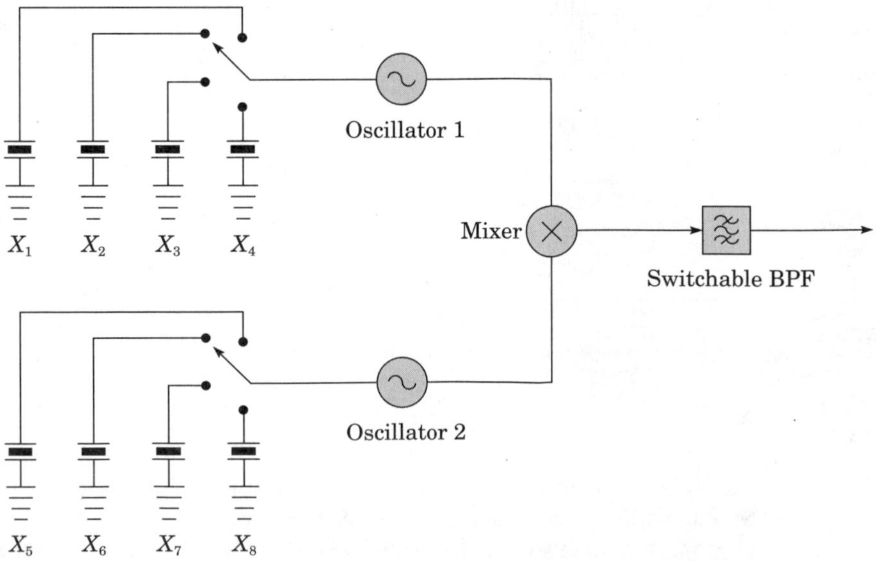

temperature changes, component aging, and so on. In addition, accurately setting these oscillators to a particular frequency is tricky, requiring precision-built variable capacitors or inductors and expensive dials using complicated arrangements of gears and pulleys.

Crystal oscillators, on the other hand, have very good stability. With voltage regulation and temperature control or compensation, frequency drift can be reduced to a few parts per million over long time periods. The disadvantage, for many applications, is that the frequency of a crystal oscillator can be changed only a very small amount by adjusting series or parallel capacitors. This makes its use awkward or impossible for any application that requires continuous frequency variation or even operation at more than a few discrete frequencies.

Most receivers and many transmitters require the frequency agility of the VFO coupled with the stability and accuracy of the crystal oscillator. For many years, it was customary to use VFOs for applications where tuning had to be continuous or where there were large numbers of frequencies in use and to use crystal oscillators with switchable crystals when operation was required on only a relatively small number of different frequencies. In recent years, however, the phase-locked frequency synthesizer has become very popular. It is now the preferred method of frequency generation in most modern receivers and transmitters. In fact, it is often possible to save money as well as improve performance when a synthesizer is used instead of a VFO because of reduced requirements for mechanical precision.

2.7.1 Phase-Locked Loops

Before looking more closely at synthesizers, it is necessary to know something about the **phase-locked loop** (PLL), because it is the basis of practically all modern synthesizer design. Though the PLL was actually invented in 1932, it is only since 1970—when it was first produced on an IC—that it has been much more than a laboratory curiosity. A PLL synthesizer can certainly be constructed using tubes or discrete transistors, but so many devices would be needed that the technique is not practical.

Figure 2.37 shows the essentials of a simple PLL. The loop consists of a phase detector, a **voltage-controlled oscillator (VCO)**, and a low-pass filter (LPF) called the "loop filter." An external reference signal is compared with the VCO signal in the **phase detector**. An error voltage is produced whose amplitude varies with the phase difference between the two signals. After filtering, this error signal is applied as a control voltage to the VCO.

The purpose of the PLL is to lock the VCO to the reference signal. That is, the two signals will have the same frequency. Since the frequency of the two signals is exactly the same, the phase angle between them will remain constant, hence the term *phase-locked*.

When the loop starts operating, the VCO will operate at its **free-running frequency** (that is, the frequency at which it operates when the control voltage is zero). This will probably not be the same as the reference frequency. The loop is said to be *unlocked*. The phase detector will generate a control voltage, which will cause the VCO frequency to change until it is exactly that of the external input signal. This is called the *acquisition of phase lock*. Once phase lock has been acquired, the loop will remain locked indefinitely. Any tendency of the VCO to drift in frequency will result in a change in the control voltage in the direction required to bring the loop back into a locked condition.

A practical loop will likely have some gain for the control signal. This is the purpose of the loop amplifier shown in Figure 2.37. There will also be some limits on how far apart

Figure 2.37
Phase-locked loop

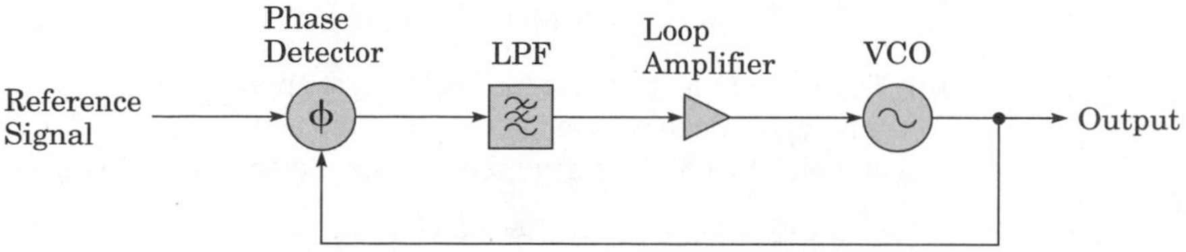

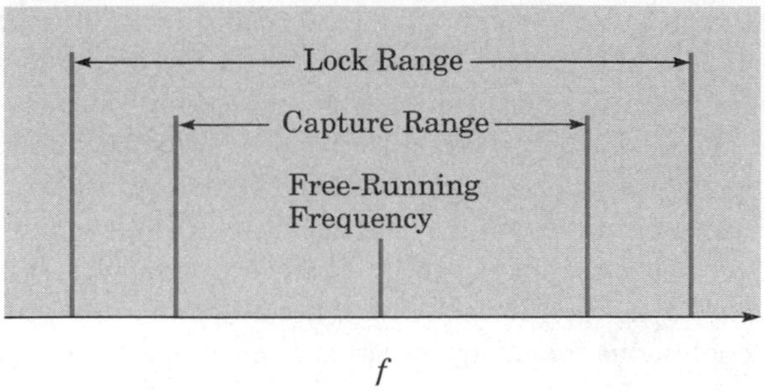

Figure 2.38
PLL frequency specifications

the free-running VCO frequency and the reference frequency can be for lock to be acquired or maintained. Of course, it will also take a small but finite amount of time to achieve lock.

Phase-Locked Loop Operation Suppose that the external reference frequency and the VCO frequency are initially very far apart. If the reference is gradually brought closer to that of the VCO, there will come a point where the VCO frequency will suddenly change to that of the external signal and the loop will lock. The range over which the reference frequency can be varied and still achieve phase lock is called the **capture range**. This is an important PLL specification, as it determines how far apart the external and internal frequencies can initially be for the loop to achieve lock.

Suppose that lock has been achieved, so that the VCO is synchronized to the reference frequency. Now we change the reference frequency, gradually moving it further from the free-running frequency of the VCO. For a while, the VCO will *track*; that is, its frequency will change to follow the external signal. Eventually, however, phase lock will be lost, and the VCO will return to its free-running frequency. The total frequency range within which lock, once achieved, can be maintained, is called the **lock range**.

Figure 2.38 should help to make all this clear. Note the difference between capture range and lock range. The lock range is almost always larger than the capture range.

Example 2.8

A phase-locked loop has a VCO with a free-running frequency of 12 MHz. As the frequency of the reference input is gradually raised from zero, the loop locks at 10 MHz and comes out of lock again at 16 MHz.

(a) Find the capture range and lock range.
(b) Suppose that the experiment is repeated, but this time the reference input begins with a very high frequency and steadily moves downward. Predict the frequencies at which lock would be achieved and lost.

Solution

(a) The capture range is approximately twice the difference between the free-running frequency and the frequency at which lock is first achieved. For this example,

$$\begin{aligned}\text{capture range} &= 2(12\text{ MHz} - 10\text{ MHz})\\ &= 4\text{ MHz}\end{aligned}$$

The lock range is approximately twice the difference between the frequency where lock is lost and the free-running frequency. Here,

$$\begin{aligned}\text{lock range} &= 2(16\text{ MHz} - 12\text{ MHz})\\ &= 8\text{ MHz}\end{aligned}$$

(b) The PLL frequency response is (at least approximately) symmetrical; that is, the free-running frequency is in the center of the lock range and capture range. The frequency at which lock will be acquired, moving downward in frequency, is

$$12\text{ MHz} + 2\text{ MHz} = 14\text{ MHz}$$

Figure 2.39

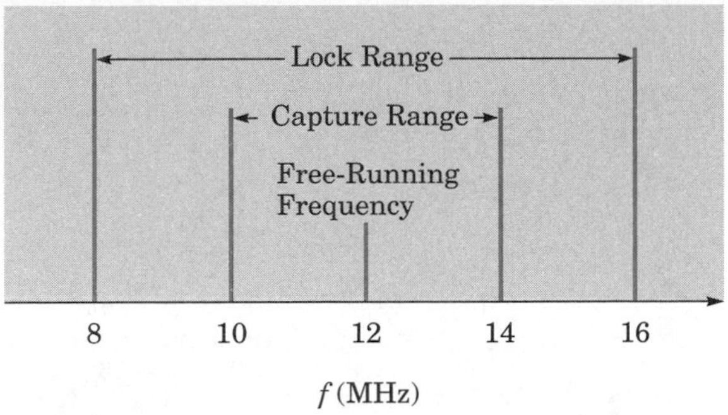

Lock will be lost on the way down at

$$12\text{ MHz} - 4\text{ MHz} = 8\text{ MHz}$$

Figure 2.39 shows these relationships for this example.

2.7.2 Simple Frequency Synthesizers

It may not be obvious how a PLL can fulfill our original goal of creating an oscillator with crystal-controlled stability and VFO agility without using a great number of crystals. To achieve this goal, we need to add one more component to the loop: a programmable divider. See Figure 2.40 for a sketch of a **frequency synthesizer** reduced to its simplest terms.

In this circuit, the phase detector still compares two frequencies and produces a control voltage that results in the two becoming locked together. However, while one of these is still an external reference frequency that could be generated by a crystal oscillator, the other is no longer the VCO frequency itself. That frequency is divided by some integer number N and then compared with the reference frequency. Using a programmable divider allows N to be varied. It is easy to see that, assuming the PLL is locked,

$$f_{ref} = \frac{f_{VCO}}{N}$$

so

$$f_{VCO} = Nf_{ref}$$

Generally, the VCO generates the output frequency f_o. Then

$$f_o = Nf_{ref} \tag{2.24}$$

This means that a large number of different output frequencies, all locked to a single crystal-controlled reference frequency, can be generated simply by changing the **modulus** (the value of N). The modulus can be changed by altering the voltages on some of the pins of the divider chip. Therefore the technique lends itself very easily to computer control

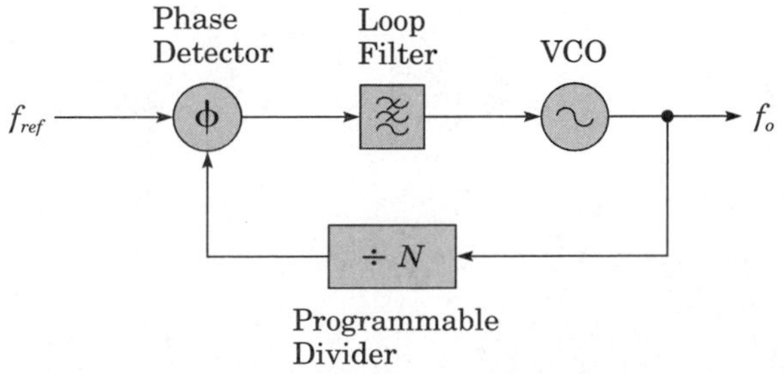

Figure 2.40
A simple frequency synthesizer

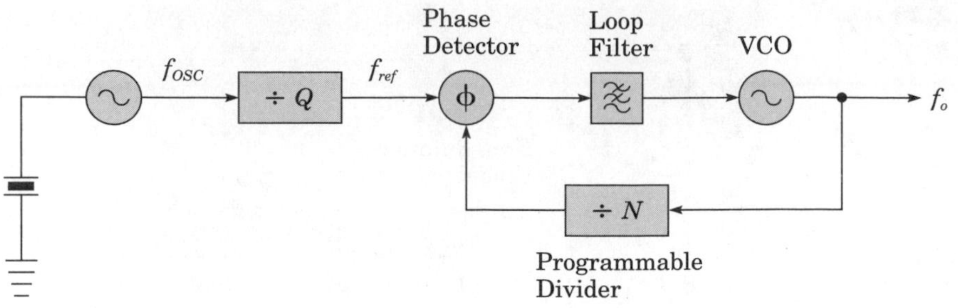

Figure 2.41 Frequency synthesizer with divider for the reference frequency

and/or remote control. Expensive variable capacitors and inductors are also eliminated. The resulting cost savings can actually make a frequency synthesizer cheaper than a conventional VFO, in spite of its greater complexity.

There is one problem, however. This circuit cannot generate just any frequency but only those that are multiples of f_{ref}. For instance, if f_{ref} is 100 kHz, the circuit can, at least in theory, generate any multiple of 100 kHz. This would be satisfactory in an FM broadcast receiver, since FM broadcast channels are 200 kHz apart, but it would have its limitations in AM broadcasting, where the channel spacing is only 10 kHz.The minimum frequency step is called the **resolution** of the synthesizer.

The obvious solution is to reduce f_{ref}. Crystals with frequencies much below 100 kHz are impractical, but a fixed-modulus divider can be used to divide down the reference frequency, as shown in Figure 2.41.

Example 2.9

Configure a simple PLL synthesizer using a 10 MHz crystal so that it will generate the AM broadcast frequencies from 540 to 1700 kHz.

Solution

Our synthesizer will have the block diagram shown in Figure 2.41. Since the channel spacing in AM broadcasting is 10 kHz and all channels are at integer multiples of 10 kHz, it would be logical to use this value for f_{ref}; in that case, each time N is incremented or decremented by 1, the output frequency will move to the adjacent channel.

Since we are using a 10 MHz crystal, it will be necessary to divide it by a factor (shown as Q in Figure 2.41) to get 10 kHz. This factor can easily be found as follows:

$$\begin{aligned} Q &= \frac{f_{OSC}}{f_{ref}} \\ &= \frac{10 \text{ MHz}}{10 \text{ kHz}} \\ &= 1000 \end{aligned}$$

Next, we should specify the range of values of N that will be required. We have already seen that changing N by 1 changes channels. All we need to do then is find N at each end of the tuning range. We can rearrange Equation (2.24)

$$f_o = N f_{ref}$$

to get

$$N = \frac{f_o}{f_{ref}}$$

At the low end of the band, we have

$$\begin{aligned} N &= \frac{540 \text{ kHz}}{10 \text{ kHz}} \\ &= 54 \end{aligned}$$

At the high end,

$$N = \frac{1700 \text{ kHz}}{10 \text{ kHz}} = 170$$

2.7.3 Prescaling

There is a problem with the basic synthesizer when output frequencies in the VHF range and higher are required. Programmable dividers are simply not available at frequencies much above 100 MHz. With current technology, it is not possible to build a UHF synthesizer with the simple topology of Figure 2.41. The simplest way to get a synthesizer to work at frequencies beyond those at which programmable dividers operate is to add a fixed-modulus divider in front of the programmable one, as illustrated by Figure 2.42. This divider could employ *emitter-coupled logic* (ECL), a digital technology that can be used at frequencies above 1 GHz. For even higher frequencies, discrete transistors using gallium arsenide could be used. This will work well into the microwave region.

There is one drawback to this idea. The VCO frequency is now divided by the fixed modulus M, then by the programmable modulus N, before it is compared with the reference frequency. Thus

$$f_{ref} = \frac{f_o}{MN}$$

or

$$f_o = MNf_{ref} \tag{2.25}$$

Since only N can be changed, the minimum amount by which the frequency can be changed is now Mf_{ref}. For example, if f_{ref} is 10 kHz, and a 10:1 **prescaler** is used, the minimum step by which the frequency can be changed is 100 kHz. We seem to have taken one step forward and one back.

A truly elegant solution to the problem, one that is often used in synthesizers for VHF and UHF, is to use a *two-modulus prescaler*. This is a divider that can be programmed to divide by either of two consecutive integers (for instance, 10 and 11 or 15 and 16). We can let these integers be P and $(P + 1)$. The frequency limitations of such dividers are not nearly as severe as for fully programmable dividers: using ECL, they can work up to at least 1.2 GHz. In addition, they are much more flexible than single-modulus counters.

Figure 2.43 shows a synthesizer with a two-modulus prescaler. The main counter divides by N, as before, but the prescaler can divide by either P or $(P + 1)$. In addition, one more programmable counter is needed. Let it divide by M. The output of this counter switches the modulus of the two-modulus counter between P and $P + 1$. Remember that M, N, and P are all integers. M and N can be changed by writing to registers on the counters, but P cannot be changed. In addition, N must be greater than M, as will shortly become obvious.

Figure 2.42 Frequency synthesizer with fixed prescaler

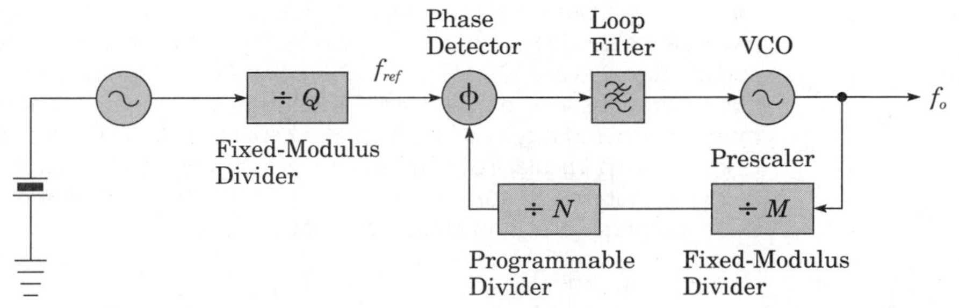

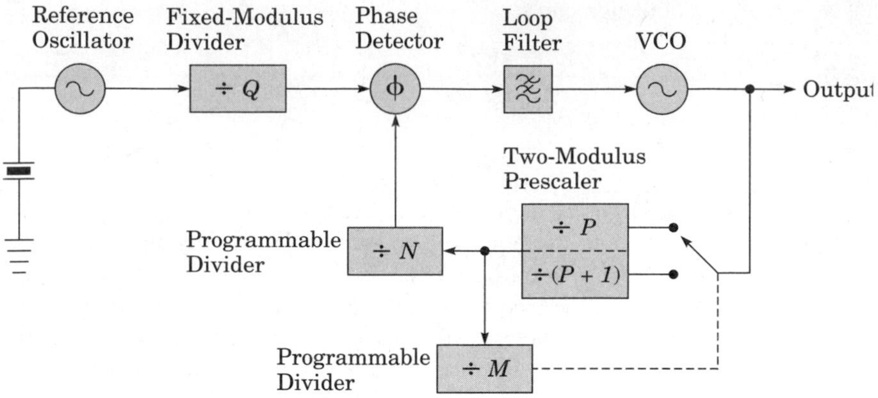

Figure 2.43 Frequency synthesizer with two-modulus prescaler

In order to understand the operation of this synthesizer, suppose that initially the prescaler is set to divide by $(P + 1)$ and the divide-by-M counter has just been reset. The two-modulus counter will switch to the P mode when the divide-by-M counter changes state (after it has received M transitions from the VCO). Then the two-modulus counter will divide by P until the divide-by-M counter changes state again. The trick here is that the M counter remains inactive until reset; that happens when the divide-by-N programmable counter reaches zero. That is, on reaching zero, the divide-by-N counter resets both itself and the divide-by-M counter, setting each to the value stored in its respective programming register.

Since initially the two-modulus counter divides by $(P + 1)$, it will take $(P + 1)$ transitions from the VCO to apply one count to the divide-by-M counter. Therefore, this counter will reach zero after $M(P + 1)$ transitions on the VCO output line. The same pulses that go to the divide-by-M counter are input to the divide-by-N counter. By the time the divide-by-M counter has reached zero, the divide-by-N counter has reached $(N - M)$. This value must be greater than zero, which accounts for the condition imposed above that N must be greater than M.

At this point, the prescaler switches to its other mode. The next P transitions from the VCO produce one transition at the prescaler output. This continues until the divide-by-N counter reaches zero. Since this counter begins this section of the process with a count of $(N - M)$, it must receive $(N - M)$ output transitions from the prescaler—this accounts for $P(N - M)$ state changes of the VCO output. At this point, both programmable counters reset to their programmed values, and the two-modulus counter switches back to its $(P + 1)$ mode.

Crystal Calibrators

One indication of the great improvement in accuracy achieved by synthesizers is the virtual disappearance of crystal calibrators from communications receivers. Crystal calibrators are (or were) crystal oscillators that operated at a low frequency (usually 100 kHz) but produced a waveform rich in harmonics. Coupled into the antenna circuit, the calibrator provided an audible signal every 100 kHz throughout the receiver tuning range. The idea was that the dial calibrations would be accurate enough so that the operator would know which harmonic was being received; the calibrator could then be used to calibrate the dial more precisely for the particular 100 kHz segment being used. Modern synthesized receivers have no need for crystal calibrators, as the crystal oscillator that controls the synthesizer is as accurate as the calibrator and is, of course, operative at all frequencies.

If we look at the whole sequence, we see that the total number of pulses emanating from the VCO for one output pulse to the phase detector is

$$\begin{aligned} M(P + 1) + P(N - M) &= MP + M + NP - MP \\ &= M + NP \end{aligned}$$

That is, the whole system including both the prescaler and the main counter has divided the VCO output by a factor of $(M + NP)$. Therefore

$$f_o = (M + NP)f_{ref} \tag{2.26}$$

The reader who has followed this rather convoluted argument may feel that this is not a very spectacular result from so many counters and so much head-scratching. It is actually worth all the trouble, however, as an example will show.

Example 2.10 The synthesizer in Figure 2.43 has $P = 10$ and $f_{ref} = 10$ kHz. Find the minimum frequency step size and compare it with that obtained using a fixed divide-by-10 prescaler.

Solution With a fixed-modulus prescaler, the minimum frequency step would be

$$\begin{aligned} \text{step size} &= Mf_{ref} \\ &= 10 \times 10 \text{ kHz} \\ &= 100 \text{ kHz} \end{aligned}$$

To find the step size with the two-modulus system, let the main divider modulus N remain constant and increase the modulus M to $(M + 1)$ to find how much the frequency changes. For the first case, the output frequency would be

$$\begin{aligned} f_o &= (M + NP)f_{ref} \\ &= (M + NP)10 \text{ kHz} \end{aligned}$$

If we now leave N alone but change M to $(M + 1)$, the new frequency is

$$\begin{aligned} f_o' &= (M + 1 + NP)f_{ref} \\ &= (M + 1 + NP)10 \text{ kHz} \end{aligned}$$

The difference is

$$\begin{aligned} f_o' - f_o &= (M + 1 + NP)10 \text{ kHz} - (M + NP)10 \text{ kHz} \\ &= (M + 1 + NP - M - NP)10 \text{ kHz} \\ &= 10 \text{ kHz} \end{aligned}$$

This is the same step size as would have been obtained without prescaling.

Thus, the two-modulus prescaler achieves the benefit of prescaling (increased high-frequency capability) without the disadvantage of poorer resolution.

2.7.4 Frequency Translation

All of the synthesizers shown so far are capable of generating very low frequencies, right down to f_{ref}. When such a synthesizer is used to generate a high frequency, a very large value of N is required. Dividing the VCO output by N divides the loop gain by the same amount, and very large values of N can cause instability.

In many practical applications, it is completely unnecessary for the synthesizer to operate at very low frequencies. For instance, an FM broadcast receiver must tune at 200 kHz intervals, but only between 88.1 and 107.9 MHz. One way to produce the local oscillator signal for such a receiver would be to generate it at a relatively low frequency and then raise it by mixing. Such a system is shown in Figure 2.44. The final output does not cover the FM band but rather a range of frequencies 10.7 MHz higher, since it is a local

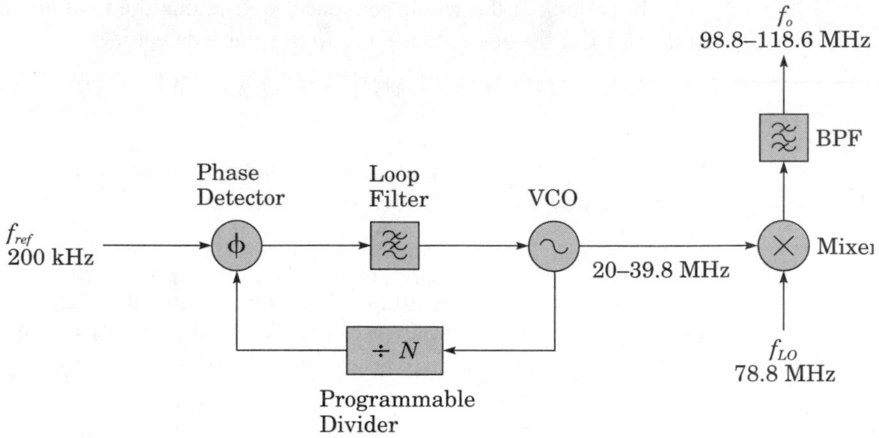

Figure 2.44 Synthesizer with frequency shifting

oscillator signal that is being generated and 10.7 MHz is the usual IF in FM broadcast receivers. Local oscillators are discussed in detail in Chapter 5. The VCO output varies from 20 MHz to 39.8 MHz in 200 kHz steps. It is mixed with a fixed-frequency oscillator at 78.8 MHz, and the sum of the two frequencies is used for the output, which varies from 98.8 to 118.6 MHz, as required. A bandpass filter is used to remove the difference component as well as the VCO and crystal oscillator frequencies from the mixer output. This movement of a block of frequencies is called a **frequency translation**.

Another possibility is to include the mixer and local oscillator within the loop, as shown in Figure 2.45. In this case, the VCO provides the output frequency directly, but its output is mixed down before being applied to the programmable divider.

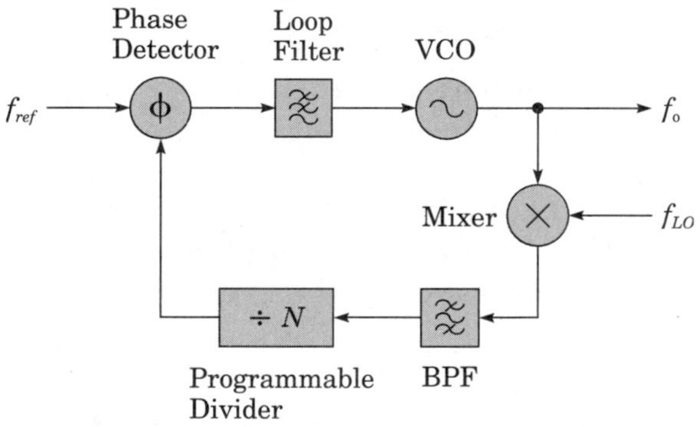

Figure 2.45 Synthesizer with mixer in the loop

Example 2.11 A synthesizer of the type shown in Figure 2.45 has $f_{ref} = 20$ kHz and a local oscillator operating at 10 MHz. Find the frequency range of the output as the value of N ranges from 10 to 100. Also find the minimum amount by which the frequency can be varied.

Solution The frequency at the input to the divider must be Nf_{ref}, by the same logic that was used with the simple synthesizer. This is the output of a mixer that subtracts the fixed 10 MHz frequency from the VCO frequency. Other mixer outputs are removed by the bandpass filter.

Therefore the VCO frequency, which is also the output frequency, is

$$f_o = Nf_{ref} + f_{LO}$$

For $f_{LO} = 10$ MHz, $f_{ref} = 20$ kHz, and $N = 10$, we have

$$f_o = 10 \times 20 \text{ kHz} + 10 \text{ MHz}$$
$$= 10.2 \text{ MHz}$$

If N changes to 100, the output frequency will be

$$f_o = 100 \times 20 \text{ kHz} + 10 \text{ MHz}$$
$$= 12 \text{ MHz}$$

By now we know that the step size will be constant, so we can simply divide the total frequency range by the number of steps to find it.

$$\text{step size} = \frac{12 \text{ MHz} - 10.2 \text{ MHz}}{100 - 10}$$
$$= \frac{1.8 \text{ MHz}}{90}$$
$$= 20 \text{ kHz}$$

As you would expect, the step size is simply equal to f_{ref}.

It is not always necessary to use a second crystal oscillator for f_{LO}; with careful choice of frequencies it may be possible to use the local-oscillator frequency, suitably divided down, as the reference frequency. This technique is shown in Figure 2.46. In this case, the only additional stage required is the mixer itself.

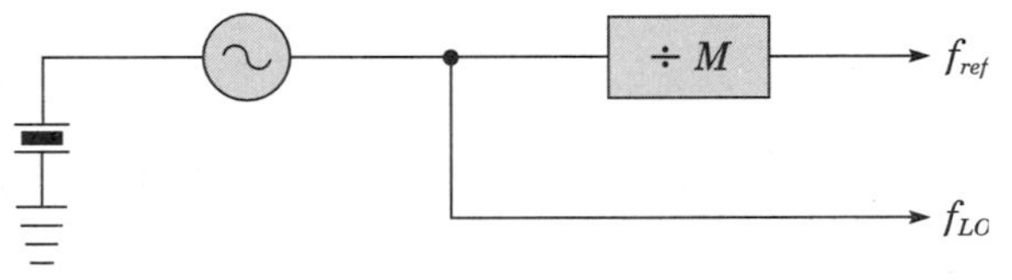

Figure 2.46 Derivation of f_{LO} and f_{ref} from the same source

Summary

Here are the main points to remember from this chapter.

1. As the frequency increases, the effects of distributed capacitance and inductance become more significant.
2. Interactions between components in high-frequency circuits can often be avoided by the use of shielding and bypassing.
3. Many RF amplifiers are designed for a relatively narrow range of frequencies to avoid problems with noise and spurious signals. Such amplifiers are called *narrowband amplifiers*. *Wideband amplifiers* are also used.
4. Depending on the application, an RF amplifier may operate in Class A, B, or C. Class A has the least distortion but is very inefficient. Class C is the most efficient, but its extreme nonlinearity makes it unsuitable for many signals.
5. A frequency multiplier can be created by tuning the output circuit of a Class C amplifier to a multiple of the input frequency.
6. Any oscillator requires a feedback loop with unity gain and zero phase shift around the loop, at the operating frequency.
7. Most variable-frequency RF oscillators use a resonant circuit for the feedback network. Crystal-controlled oscillators, in which a quartz crystal is used as the feedback element, are also common.
8. When two signals at different frequencies are applied to a nonlinear circuit, signals at the sum and difference of the frequencies are produced. A circuit designed to accomplish this is called a *mixer*.

9. PLL frequency synthesizers can provide the stability of a crystal oscillator with the flexibility of a VFO.
10. Simple single-loop frequency synthesizers are limited as to frequency resolution and maximum frequency. These limitations can be overcome by using mixing, prescalers, and/or multiple loops.

Important Equations

For handy reference, each chapter has a brief list of the most commonly used equations. These equations must not be used blindly: you must understand how and where to use them. Refer to the text as necessary. These are not the only equations you will need to solve the problems in this chapter.

$$A_V = \frac{-(R_C \parallel R_L)}{r'_e} \tag{2.1}$$

$$f_o = \frac{1}{2\pi\sqrt{L_1 C_1}} \tag{2.2}$$

$$B = \frac{f_o}{Q} \tag{2.3}$$

$$AB = 1 \tag{2.4}$$

$$f_o = N f_{\text{ref}} \tag{2.24}$$

$$f_o = M N f_{\text{ref}} \tag{2.25}$$

$$f_o = (M + NP) f_{\text{ref}} \tag{2.26}$$

Glossary

balanced mixer a mixer in which the input frequencies are cancelled and are therefore not present at the output

bypassing removal of an unwanted signal by providing a low-impedance path to ground

capture range the total frequency range over which a PLL can become locked to a signal

crystal a small slab of quartz with attached electrodes; used as a resonant circuit

decouple to prevent the undesired passage of signals between circuits

doubler a frequency multiplier whose output frequency is twice that of the input signal

free-running frequency the frequency at which a VCO operates when its control voltage is zero

frequency multiplier a circuit whose output frequency is a small integer multiple of the input signal frequency

frequency synthesizer a device that can produce a large number of output frequencies from a smaller number of fixed-frequency oscillators

frequency translation movement of a signal from one frequency to another using a mixer-oscillator combination

gimmick a small length of wire, connected at only one end and used as a capacitance to ground

ground plane an artificial ground, often consisting of an area of foil left on one side of a circuit board

lock range total range of frequencies over which a PLL, once locked, can remain locked

Miller effect In some amplifiers, the internal capacitance of the active device can cause feedback that produces the same effect on the circuit as a much larger capacitance across the amplifier input. This is called the Miller effect.

mixer a nonlinear circuit designed to generate sum and difference frequencies when two or more frequencies are present at its input(s)

modulus the number by which a digital divider chain divides

multiplier a circuit whose output is proportional to the product of the instantaneous amplitudes of two input signals

neutralization a means of avoiding instability in amplifiers by using negative feedback

phase detector a device whose output voltage is a function of the phase difference between two input signals

phase-locked loop (PLL) a device that locks the frequency of a VCO exactly to that of an input signal

piezoelectric effect an effect that occurs with some materials, such as quartz and some ceramics, whereby a voltage is produced across the material when it is deformed; the converse is also true: applying a voltage to the material causes mechanical deformation.

prescaler a divider that precedes the main programmable divider in a frequency synthesizer

resolution in a frequency synthesizer, the smallest amount by which the output frequency can be changed

self-resonant frequency the frequency at which a single component becomes a resonant circuit, because of the presence of stray capacitance or inductance, or both

tripler a frequency multiplier whose output frequency is three times that of the input signal

varactor a reverse-biased diode used as a voltage-variable capacitor

voltage-controlled oscillator (VCO) an oscillator whose frequency can be controlled by changing an external control voltage

Questions

1. What is the difference in the behavior of a resistor at low and high frequencies?
2. Explain what is meant by *self-resonance*.
3. Distinguish between lumped and distributed constants, and explain why both concepts are useful.
4. Explain how a circuit board can be made to incorporate shielding.
5. Why are bypass capacitors necessary at power-supply connections?
6. Why are narrowband amplifiers often preferred for RF applications?
7. Give one example of an RF application in which a wideband amplifier would be preferred.
8. State the characteristics of Class A, AB, B, and C amplifiers in terms of conduction angle, efficiency, and distortion level.
9. Why are Class C amplifiers unsuitable for use as audio amplifiers?
10. What is neutralization? Why is it sometimes needed in RF amplifiers?
11. How does a frequency multiplier differ from an ordinary RF amplifier?
12. State the Barkhausen criteria and explain them in your own words.
13. Draw the circuit for each *LC* oscillator type listed in this chapter, and show the feedback path.
14. What are the advantages of varactor tuning compared with variable capacitors or inductors?
15. Explain why a quartz crystal has two resonant frequencies. Name the two types of resonance. Which one occurs at the lower frequency?
16. Give two advantages of crystal-controlled over *LC* oscillators. Do crystal-controlled oscillators have any disadvantages?
17. What is the function of a mixer? Distinguish a mixer from a summer.
18. List all the frequency components produced when two sine waves of different frequencies are applied to a square-law mixer.
19. List all the frequency components produced when two sine waves of different frequencies are applied to a balanced mixer.
20. Compared with a bipolar transistor, what are the advantages of a dual-gate MOSFET when used as a mixer?
21. Sketch the block diagram of a PLL. Label all the components.
22. Sketch the block diagram of a basic PLL frequency synthesizer.
23. What is the smallest amount by which the frequency of a basic synthesizer can be changed?
24. Why are prescalers necessary when synthesizers are used at UHF frequencies?
25. Why is a mixer often incorporated into a frequency synthesizer?
26. How are low-frequency reference signals generated in a synthesizer?

Problems

SECTION 2.3

27. The narrowband RF amplifier in Figure 2.47 has an input signal with a frequency of 12 MHz. Calculate:
    (a) the value to which C_1 should be adjusted for best performance at the signal frequency,
    (b) the bandwidth of the circuit, assuming the tuned circuit has a loaded Q of 20.
28. Redraw the circuit of Figure 2.47 so that the transistor loads the circuit to a lesser extent. Suppose that this raises the loaded Q of the tuned circuit to 30.

Figure 2.47

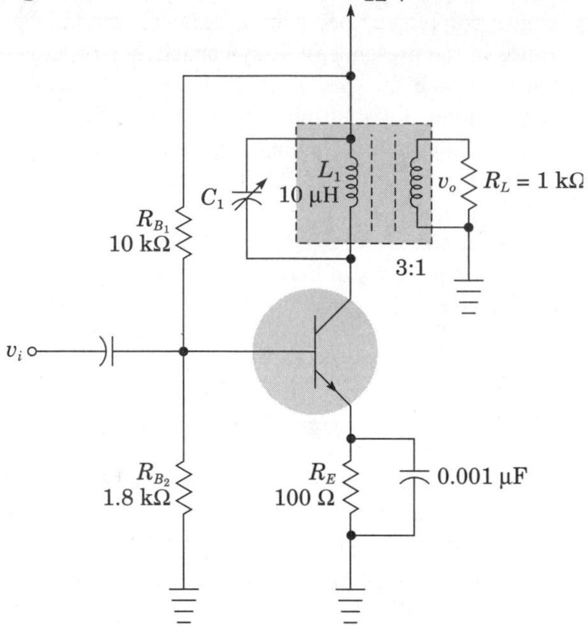

By what percentage does this reduce the bandwidth of the amplifier?

29. The circuit in Figure 2.48 represents an RF amplifier.
    (a) What class of operation is it designed for? How can you tell?
    (b) What is the collector current with no input signal?
    (c) Sketch the input voltage and collector-current waveforms with an input signal at 15 MHz. Sketch them one above the other using the same time scale. Current and voltage scales are not required.
    (d) Assuming an ideal transformer, what would be the maximum load voltage?
30. An amplifier is required to deliver 100 W to a load. Calculate the power it would draw from the power supply if it were to operate:
    (a) Class A with an efficiency of 30%
    (b) Class B with an efficiency of 55%
    (c) Class C with an efficiency of 80%

Figure 2.48

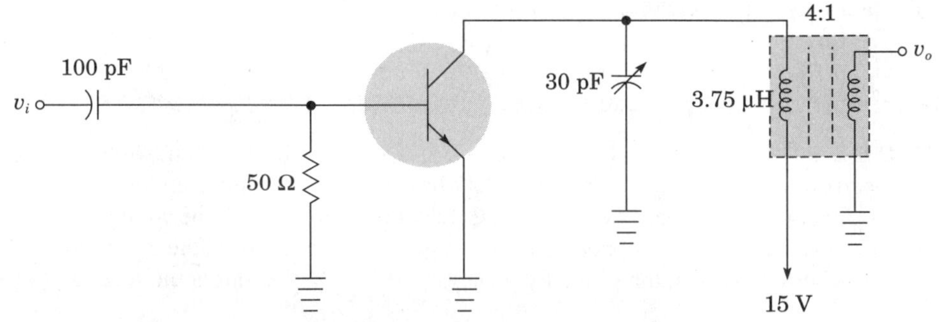

31. A transistor has a power dissipation rating of 25 W. Assume that the transistor is the only element that dissipates power in the circuit. Calculate the power an amplifier using this transistor could deliver to the load if it operates:
    (a) Class A with an efficiency of 30%
    (b) Class C with an efficiency of 80%

SECTION 2.4

32. A transmitter uses a frequency multiplication circuit consisting of two triplers and three doublers to get a frequency of 540 MHz from a crystal oscillator. What frequency should the crystal have?
33. The circuit in Figure 2.49 represents a frequency multiplier. If the input frequency is 10 MHz and the circuit operates as a tripler, calculate the value of L_1.

Figure 2.49

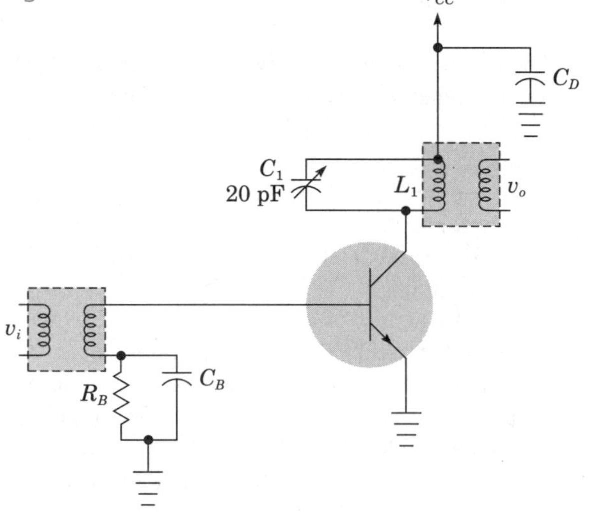

SECTION 2.5

34. The circuit in Figure 2.50 shows an oscillator.
    (a) Name the type of oscillator.
    (b) Should the amplifier be inverting or noninverting? Explain why.
    (c) For the circuit to oscillate, what is the minimum gain the amplifier could have?
    (d) What value should L have if the operating frequency is to be 5 MHz?

Figure 2.50

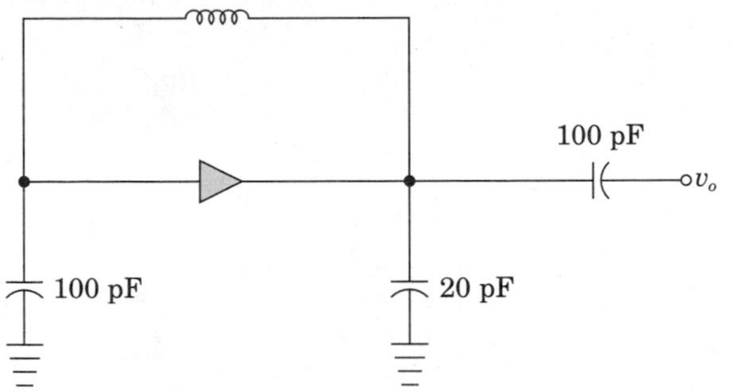

35. Suppose L is replaced by a quartz crystal in the circuit of Figure 2.50.
    (a) What is the name of this type of oscillator?
    (b) Sketch the curve of reactance as a function of frequency for a quartz crystal, and indicate where on the curve this oscillator would work.
36. (a) Calculate the operating frequency for the Clapp oscillator shown in Figure 2.51.

Figure 2.51

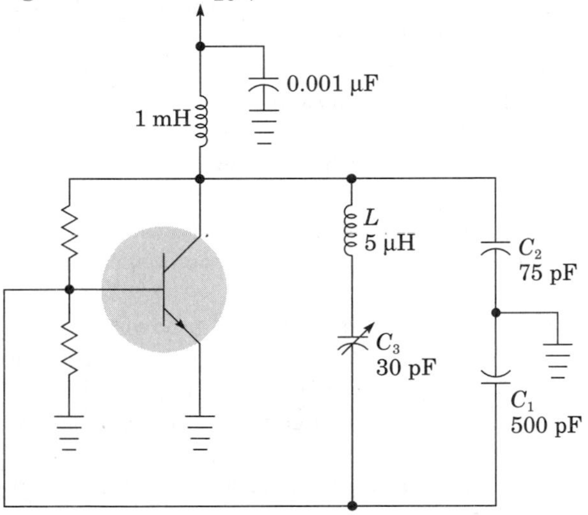

    (b) Suppose that transistor loading doubles the value of C_1. By what percentage does this change the operating frequency?
37. A varactor-tuned oscillator uses an inductance of 25 μH with a varactor having a maximum capacitance of 45 pF in the frequency-determining circuit. Calculate the range of frequencies for this oscillator, with a tuning voltage that varies from 1 V to 10 V.
38. A 10 MHz crystal oscillator has a temperature coefficient of −5 ppm/degree C. It operates over a frequency range from +10°C to +30°C and has been set to exactly 10 MHz at 20°C.
    (a) What are its output frequencies at 10° and 30°?
    (b) What is its percent accuracy over the operating temperature range?
    (c) Suppose the output of this oscillator is applied to a doubler. Recalculate the range of output frequencies and percent accuracy.
39. A quartz watch is guaranteed accurate to 15 seconds per month. Assuming a month has 30 days, calculate the accuracy of the crystal oscillator in the watch in parts per million.

SECTION 2.6

40. Signals at 12 MHz and 9 MHz are applied to the input of a mixer. What frequencies will be found at the output if the mixer is:
    (a) a square-law type?
    (b) a multiplier?

SECTION 2.7

41. A PLL has a free-running frequency of 10 MHz, with a capture range of 1 MHz and a lock range of 2 MHz. Graph the VCO frequency as a function of the reference input frequency as the latter varies from 5 to 15 MHz.
42. A PLL has the following characteristics:
    free-running frequency: 3 MHz
    lock range: 200 kHz
    capture range: 50 kHz

    What is the highest input frequency that
    (a) will cause an unlocked loop to lock?
    (b) can be tracked by a locked loop?
43. Figure 2.52 shows a simple frequency synthesizer.
    (a) What frequency range does it generate as N varies from 100 to 200?
    (b) What is the smallest possible frequency step with this synthesizer?

Figure 2.52

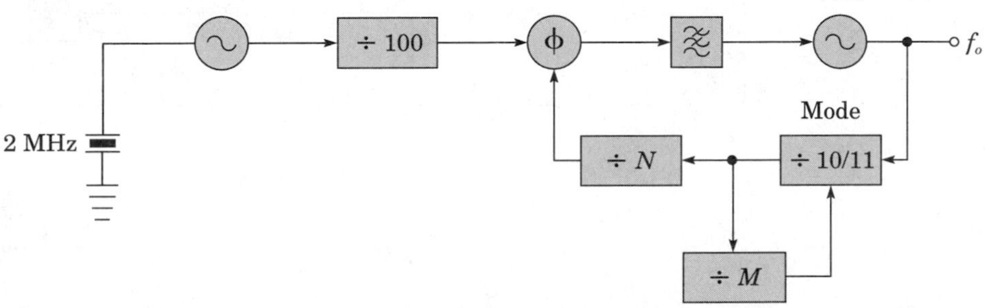

Figure 2.53

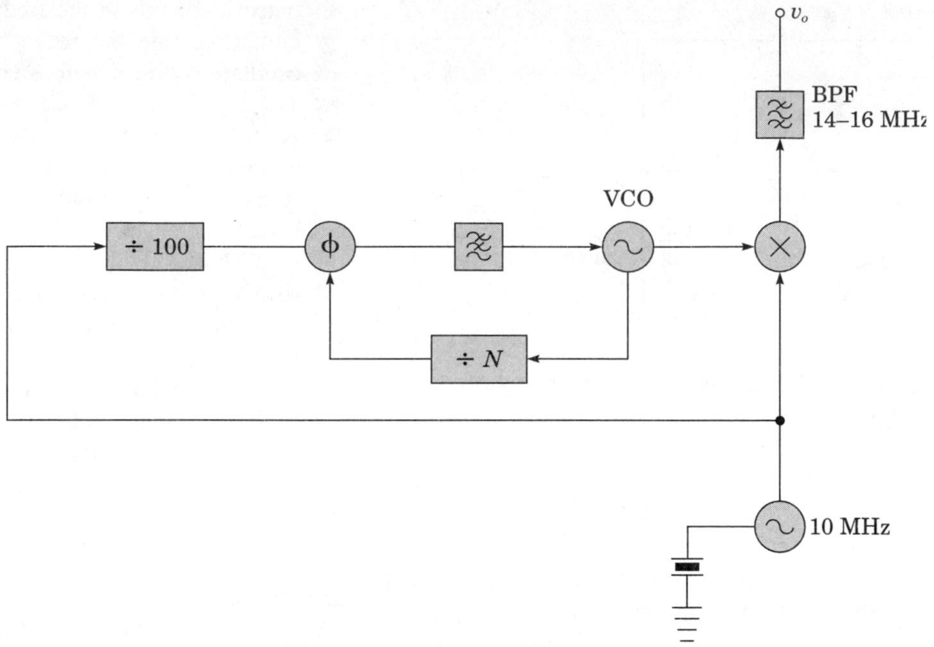

44. Draw a block diagram of a simple frequency synthesizer that will generate frequencies from 1 to 10 MHz in 500 kHz steps. Find the range of values of N that will be needed for the programmable divider.
45. A simple frequency synthesizer can function up to 50 MHz with a frequency step size of 10 kHz. Calculate new values for these specifications if a fixed 10:1 prescaler is used. The prescaler itself is capable of responding to frequencies up to 1 GHz.
46. Calculate the output frequency for the synthesizer shown in Figure 2.52, for $N = 50$ and $M = 10$.
47. Figure 2.53 shows a synthesizer with external frequency translation. Calculate the output frequency for $N = 50$.
48. The synthesizer in Figure 2.54 has frequency translation within the loop. Calculate its output frequency for $N = 30$.
49. The 10 MHz crystal oscillator in a PLL frequency synthesizer has been calibrated to a reference standard known to be accurate to 5 ppm. The oscillator is guaranteed to drift no more than 10 ppm per month.
    (a) A year later, what are the maximum and minimum frequencies that the oscillator could be producing?
    (b) Suppose this oscillator is used in a simple single-loop PLL to generate a frequency of 45 MHz. What are the maximum and minimum values that the output frequency could have after one year?
50. Draw a circuit for a Colpitts oscillator that will op-

Figure 2.54

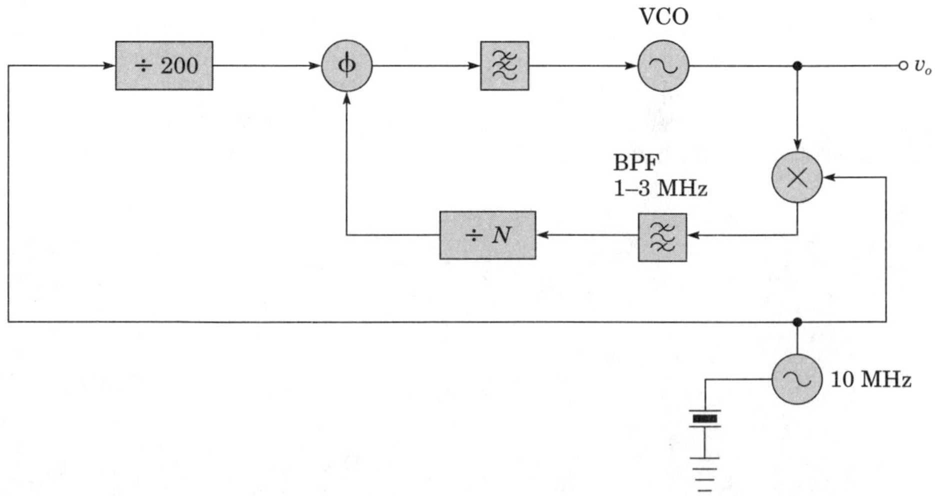

erate with a noninverting amplifier with a voltage gain of less than 1. Design it with a feedback fraction of 2, and an operating frequency of 16 MHz. The inductor should have a value of 5 μH.

51. An AM broadcast radio receiver uses high-side injection and an IF of 455 kHz. Design a frequency synthesizer to serve as the local oscillator for this receiver.

52. Draw a block diagram for a synthesizer with a two-modulus divide-by-10/divide-by-11 prescaler, and choose suitable values so that it will generate frequencies in the range from 100 to 200 MHz, at 1 MHz intervals.

53. Design a synthesizer with frequency translation within the loop, to generate the frequencies between 144 and 148 MHz, at 10 kHz intervals.

3 Amplitude Modulation

Objectives After reading this chapter, studying the examples, and solving the problems, you should be able to:

1. Write the time-domain equation for an AM signal and describe how the equation relates to the signal itself
2. Sketch an AM signal in both the time and frequency domains
3. Define the modulation index, calculate it, and measure it using either an oscilloscope or a spectrum analyzer
4. Describe the effects of overmodulation and explain why it must be avoided
5. Sketch and explain the operation of simple AM modulators and demodulators
6. Calculate the bandwidth of an AM signal and explain why bandwidth is an important factor in a communications system
7. Calculate power and voltage for an AM signal and for each of its components

3.1 Introduction

In Chapter 1, we mentioned the several ways in which an information signal can modulate a carrier wave to produce a higher-frequency signal that carries the information. The most straightforward of these, and historically the first to be used, is **amplitude modulation** (AM). AM has the advantage of being usable with very simple modulators and demodulators. It does have some disadvantages, including poor performance in the presence of noise and inefficient use of transmitter power. However, its simplicity and the fact that it was the first system to become established have ensured its continued popularity. Applications include broadcasting in the medium- and high-frequency bands, aircraft communications in the VHF frequency range, and CB radio, among others.

The basic technique of amplitude modulation can also be modified to serve as the basis for a variety of more sophisticated schemes that are found in applications as diverse as television broadcasting and long-distance telephony. Thus, it is essential to understand the process of amplitude modulation in some detail, both for its own sake and as a foundation for further study.

An AM signal can be produced by using the instantaneous amplitude of the information signal (the baseband or modulating signal) to vary the peak amplitude of a higher-frequency signal. Figure 3.1(a) shows a 1 kHz sine wave, which can be combined with the 10 kHz signal shown in Figure 3.1(b) to produce the AM signal in Figure 3.1(c). If the peaks of the individual waveforms of the modulated signal are joined, the resulting **envelope** resembles the original modulating signal. It repeats at the modulating frequency, and the shape of each "half" (positive or negative) is the same as that of the modulating signal.

The higher-frequency signal that is combined with an information signal to produce the modulated waveform is called the *carrier*. Figure 3.1(c) shows a case in which there

Figure 3.1
Amplitude modulation

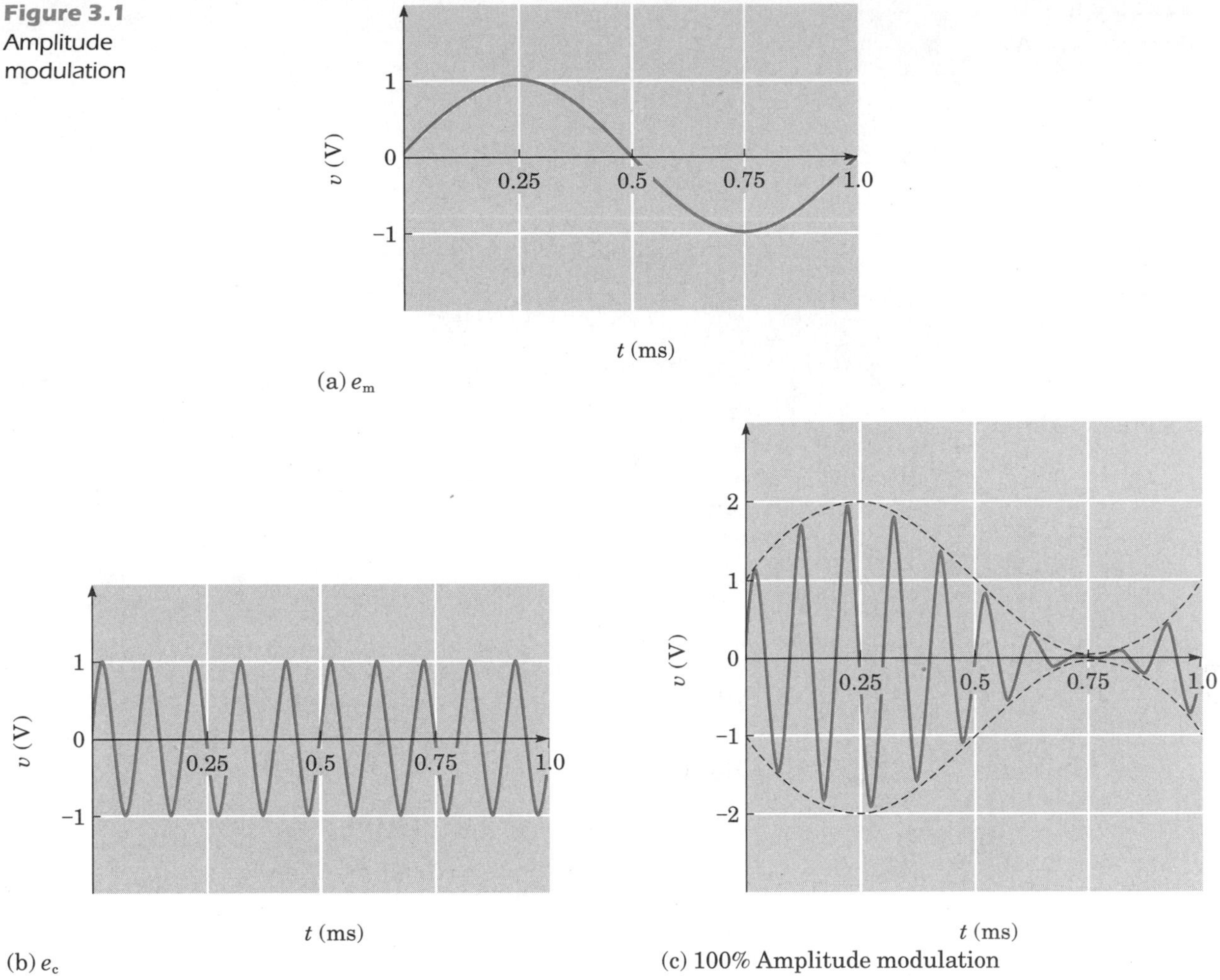

(c) 100% Amplitude modulation

are only 10 cycles of the carrier for each cycle of the modulating signal. In practice, the ratio between carrier frequency and modulating frequency is usually much greater. For instance, an AM broadcasting station might have a carrier frequency of 1 MHz and a modulating frequency on the order of 1 kHz. A waveform like this is shown in Figure 3.2. Since there are 1000 cycles of the carrier for each cycle of the envelope, the individual RF cycles are not visible, and only the envelope can be seen.

Note that amplitude modulation is not the simple linear addition of the two signals. Linear addition would produce the waveforms shown in Figure 3.3. Figure 3.3(a) shows a low-frequency signal, Figure 3.3(b) shows a higher-frequency signal, and Figure 3.3(c) shows the result of adding the two signals.

Amplitude modulation is essentially a nonlinear process. As in any nonlinear interaction between signals, sum and difference frequencies are produced that, in the case of amplitude modulation, contain the information to be transmitted. Another interesting thing about AM is that even though we seem to be varying the amplitude of the carrier (in fact, this is what is implied by the term *amplitude modulation*), a look at the frequency domain shows that the signal component at the carrier frequency survives intact, with the same amplitude and frequency as before! This mystery is easily unravelled with the aid of a little mathematics, as we shall see shortly; for the moment, just remember that AM is a bit of a misnomer, since the amplitude of the carrier remains constant. The amplitude of the entire signal does change with modulation, however, as is very clearly shown in Figure 3.1.

Figure 3.2
Envelope of an AM signal

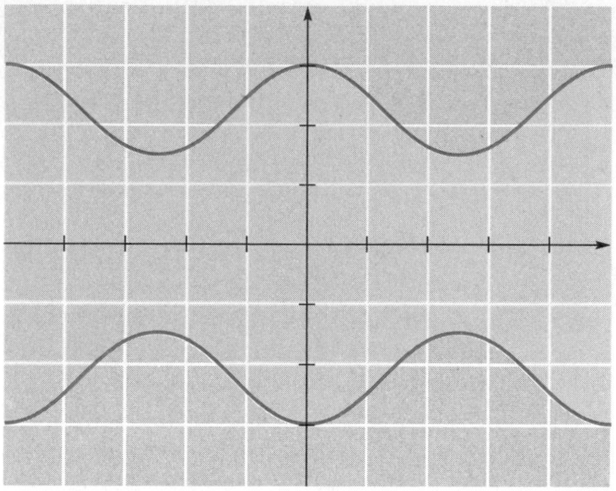

Vertical: 25 mV/division
Horizontal: 200 μs/division

Figure 3.3
Linear addition of two signals

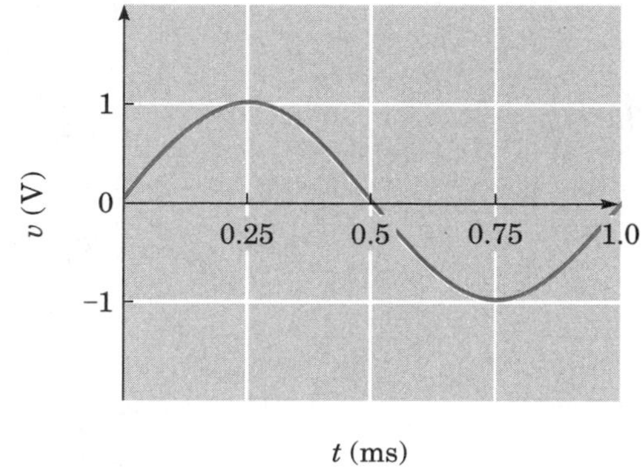

(a) e_m

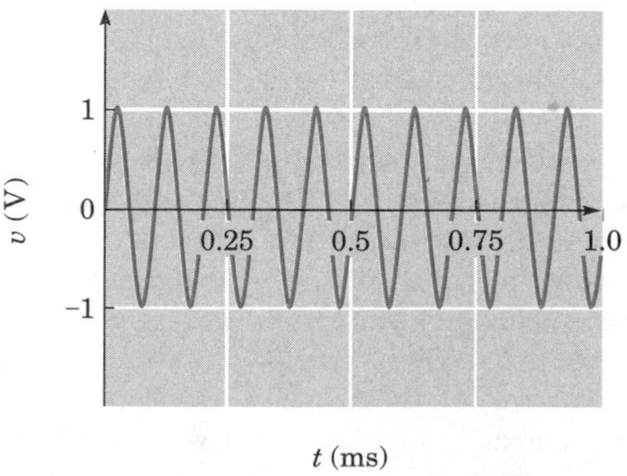

(b) e_c

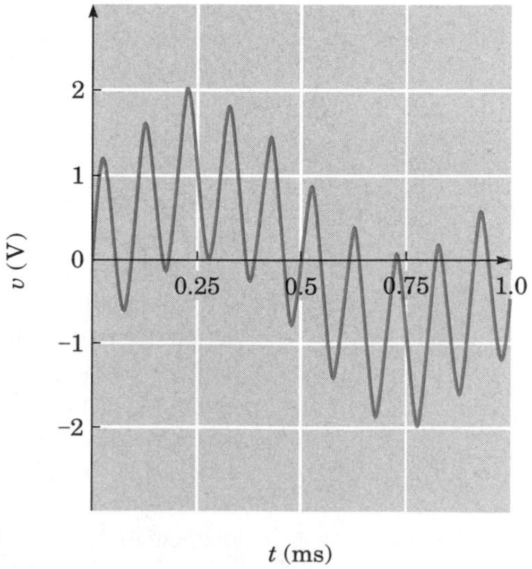

(c) $e_c + e_m$

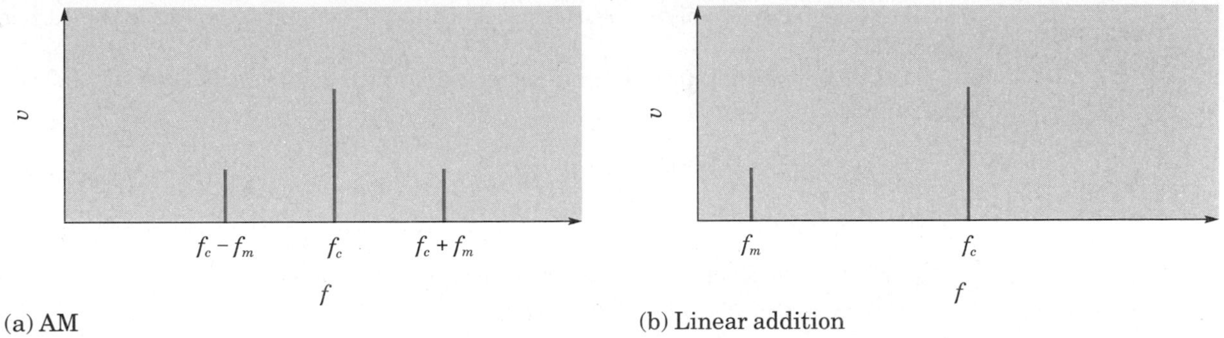

Figure 3.4
AM and linear addition in the frequency domain

Figure 3.4(a) shows frequency-domain representations for amplitude modulation, and the linear addition of two signals is shown in Figure 3.4(b). The AM signal has no component at the modulating frequency: all the information is transmitted at frequencies near that of the carrier. In contrast, linear addition has accomplished nothing: the frequency-domain sketch shows that the information and carrier signals remain separate, each at its original frequency.

3.2 Time-Domain Analysis

Now that the general idea of AM has been described, it is time to examine the signal in greater detail. We will look at the modulated signal in both the time and the frequency domains, since each approach emphasizes some of the important characteristics of AM. The time domain is probably more familiar, so let us begin here.

Amplitude modulation is created by using the instantaneous modulating signal voltage to vary the amplitude of the modulated signal. The carrier is almost always a sine wave. The modulating signal can be a sine wave but is more often an arbitrary waveform, such as an audio signal. However, an analysis of sine-wave modulation is very useful, since Fourier analysis often allows complex signals to be expressed as a series of sinusoids.

We can express the above relationship in the equation

$$v(t) = (E_c + e_m) \sin \omega_c t \tag{3.1}$$

where

$v(t)$ = instantaneous amplitude of the modulated signal in volts

E_c = peak amplitude of the carrier in volts

e_m = instantaneous amplitude of the modulating signal in volts

ω_c = radian frequency of the carrier

t = time in seconds

The addition of E_c and e_m is algebraic. That is, the peak amplitude of the modulated signal is both increased and decreased by the modulation.

Once again, let us note the difference between amplitude modulation and simple addition. Just adding the carrier and modulating signals would give the equation

$$v'(t) = E_c \sin \omega_c t + e_m \tag{3.2}$$

which is definitely not the same as Equation (3.1). The summing of two waveforms simply adds their instantaneous values at all times. Amplitude modulation, on the other hand, involves the addition of the instantaneous baseband amplitude to the peak carrier amplitude.

If the modulating (baseband) signal is a sine wave, Equation (3.1) has the following form:

$$v(t) = (E_c + E_m \sin \omega_m t) \sin \omega_c t \tag{3.3}$$

where

E_m = peak amplitude of the modulating signal in volts
ω_m = radian frequency of the modulating signal

and the other variables are as defined in Equation (3.1).

Example 3.1

A carrier wave with an RMS voltage of 2 V and a frequency of 1.5 MHz is modulated by a sine wave with a frequency of 500 Hz and amplitude of 1 V RMS. Write the equation for the resulting signal.

Solution

First, note that Equation (3.3) requires peak voltages and radian frequencies. We can easily get these as follows:

$$E_c = \sqrt{2} \times 2 \text{ V} = 2.83 \text{ V}$$

$$E_m = \sqrt{2} \times 1 \text{ V} = 1.41 \text{ V}$$

The equation also requires radian frequencies:

$$\omega_c = 2\pi \times 1.5 \times 10^6 = 9.42 \times 10^6 \text{ rad/s}$$

$$\omega_m = 2\pi \times 500 = 3.14 \times 10^3 \text{ rad/s}$$

So the equation is

$$v(t) = (E_c + E_m \sin \omega_m t) \sin \omega_c t = [2.83 + 1.41 \sin (3.14 \times 10^3 t)] \sin (9.42 \times 10^6 t) \text{ V}$$

3.2.1 The Modulation Index

The amount by which the signal amplitude is changed in modulation depends on the ratio between the amplitudes of the modulating signal and the carrier. For convenience, this ratio is defined as the **modulation index** m. It is expressed mathematically as

$$m = \frac{E_m}{E_c} \tag{3.4}$$

Modulation can also be expressed as a percentage, with percent modulation found by multiplying m by 100. For example, $m = 0.5$ corresponds to 50% modulation.

Substituting m into Equation (3.3) gives:

$$v(t) = E_c(1 + m \sin \omega_m t) \sin \omega_c t \tag{3.5}$$

Example 3.2

Calculate m for the signal in Example 3.1, and write the equation for this signal in the form of Equation (3.5).

Solution To avoid an accumulation of round-off errors, we should go back to the original voltage values to find m.

$$m = \frac{E_m}{E_c}$$

$$= \frac{1}{2},$$

$$= 0.5$$

It is all right to use the RMS values for calculating this ratio, as the $\sqrt{2}$ factors, if used to find the peak voltages, will cancel.

Now we can rewrite the equation:

$$v(t) = E_c(1 + m \sin \omega_m t) \sin \omega_c t$$
$$= 2.83[1 + 0.5 \sin (3.14 \times 10^3 t)] \sin (9.42 \times 10^6 t)$$

It is worthwhile examining what happens to Equation (3.5) and to the modulated waveform as m varies. To start with, when $m = 0$, $E_m = 0$ and we have the original, unmodulated carrier. As m varies between 0 and 1, the changes due to modulation become more pronounced. Resultant waveforms for several values of m are shown in Figure 3.5. Note es-

Figure 3.5
AM envelopes for various values of m

(a) $m = 0$

(b) $m = 0.3$

(c) $m = 0.8$

(d) $m = 1$

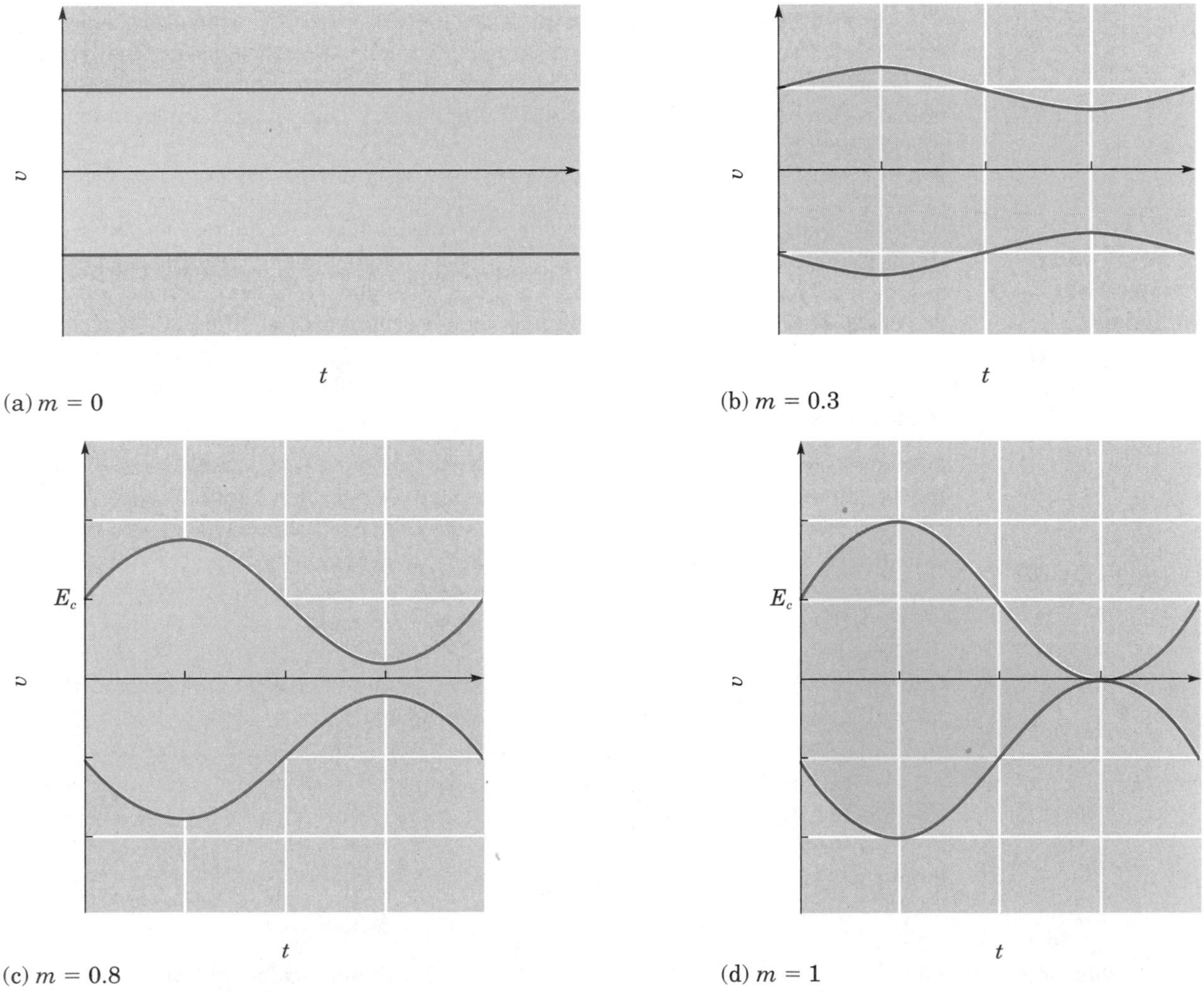

pecially the result for $m = 1$ ($m = 100\%$). Under these conditions, the peak signal voltage will vary between zero and twice the unmodulated carrier amplitude.

3.2.2 Overmodulation

When the modulation index is greater than 1, **overmodulation** is said to be present. There is nothing in Equation (3.3) that would seem to prevent E_m from being greater than E_c and m from being greater than 1. There are practical difficulties, however. Figure 3.6(a) shows the result of simply substituting $m = 2$ into Equation (3.5). As you can see, the envelope no longer resembles the modulating signal. Therefore, m must be kept less than or equal to 1.

Whenever we work with mathematical models, we must remember to keep checking against physical reality. This situation is a good example. It is possible to build a circuit that does produce an output that agrees with Equation (3.5) for m greater than 1. However, most practical AM modulators produce the signal shown in Figure 3.6(b) under these conditions. This is not the waveform predicted by Equation (3.5), but it does have the characteristic that the modulation envelope is no longer an accurate representation of the modulating signal. In fact, if subjected to Fourier analysis, the sharp "corners" on the waveform as the output goes to zero on negative modulation peaks would be found to represent high-frequency components added to the original baseband signal. This type of overmodulation creates side frequencies further from the carrier than would otherwise be the case. These spurious frequencies are known as **splatter**, and they cause the modulated signal to have increased bandwidth.

From the foregoing, we can conclude that for full-carrier AM, m must be in the range from 0 to 1. Overmodulation creates distortion in the demodulated signal and may result in the signal occupying a larger bandwidth than normal. Since spectrum space is tightly controlled by law, overmodulation of an AM transmitter is actually illegal, and means must be provided to prevent it.

3.2.3 Modulation Index for Multiple Modulating Frequencies

Practical AM systems are seldom used to transmit sine waves, of course. The information signal is more likely to be a voice signal, which contains many frequencies. Though a typical audio signal is not strictly periodic, it is close enough to periodic that we can use the idea of Fourier series to consider it as a series of sine waves of different frequencies.

When there are two or more sine waves of different, uncorrelated frequencies (that is, frequencies that are not multiples of each other) modulating a single carrier, m is calculated by using the equation

$$m_T = \sqrt{m_1^2 + m_2^2 + \cdots} \tag{3.6}$$

Figure 3.6
Overmodulation

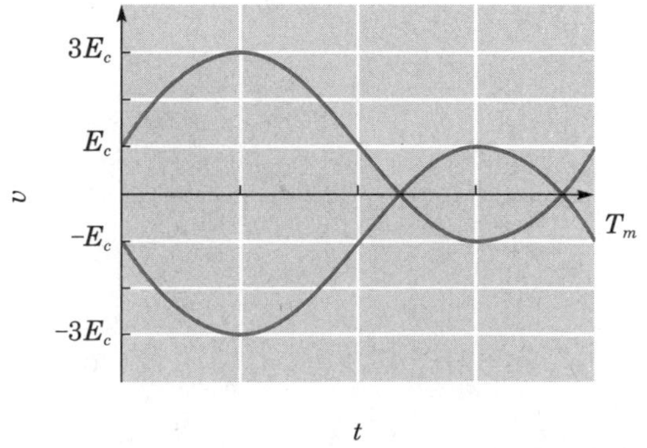

(a) $m = 2$ in Equation (3.5)

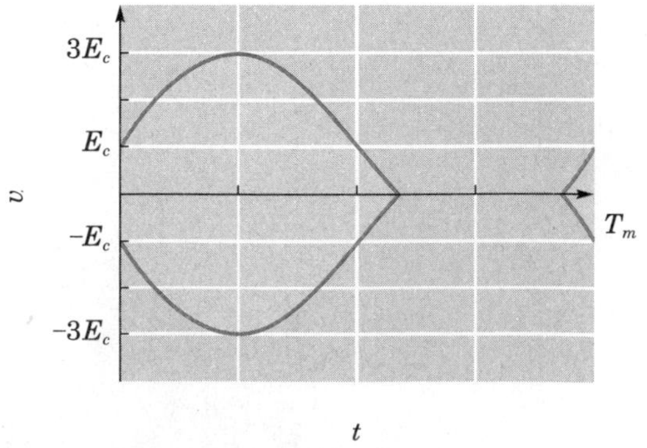

(b) $m = 2$ with practical modulation

where

$$m_T = \text{total resultant modulation index}$$
$$m_1, m_2, \ldots = \text{modulation indices due to the individual modulating components}$$

Example 3.3

Find the modulation index if a 10 V carrier is amplitude-modulated by three different frequencies with amplitudes of 1 V, 2 V, and 3 V, respectively.

Solution

The three separate modulation indices are:

$$m_1 = \frac{1}{10} = 0.1 \qquad m_2 = \frac{2}{10} = 0.2 \qquad m_3 = \frac{3}{10} = 0.3$$

$$\begin{aligned} m_T &= \sqrt{m_1^2 + m_2^2 + m_3^2} \\ &= \sqrt{0.1^2 + 0.2^2 + 0.3^2} \\ &= 0.374 \end{aligned}$$

3.2.4 Measurement of Modulation Index

If we let E_m and E_c be the peak modulation and carrier voltages, respectively, then we can see, either by using Equations (3.3) and (3.5) or by inspecting Figure 3.7, that the maximum envelope voltage is simply

$$E_{max} = E_c + E_m$$

or

$$E_{max} = E_c(1 + m) \tag{3.7}$$

and the minimum envelope voltage is

$$E_{min} = E_c - E_m$$

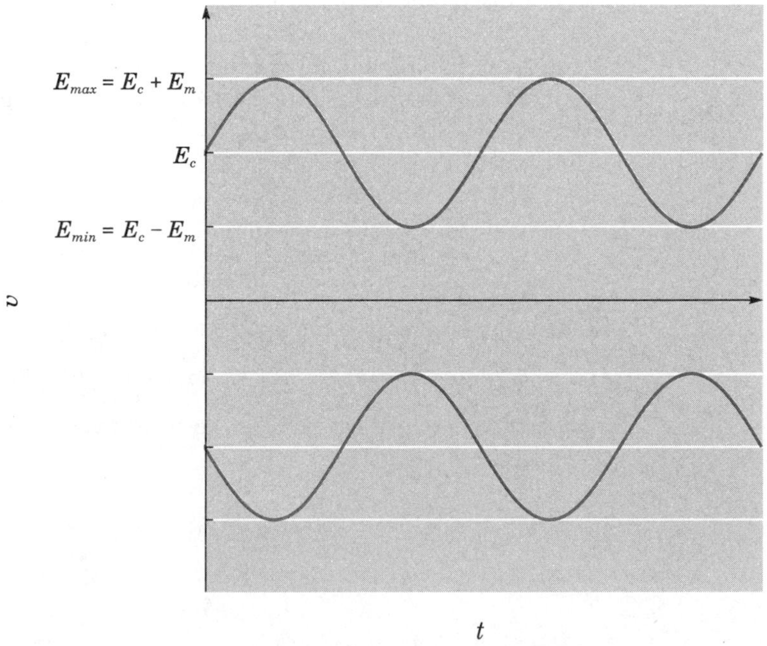

Figure 3.7 Voltage relationships in an AM signal

or

$$E_{min} = E_c(1 - m) \tag{3.8}$$

Note, by the way, that these results agree with the conclusions expressed earlier: for $m = 0$, the peak voltage is E_c, and for $m = 1$, the envelope voltage ranges from $2E_c$ to zero.

Applying a little algebra to the above expressions, it is easy to show that

$$m = \frac{E_{max} - E_{min}}{E_{max} + E_{min}} \tag{3.9}$$

Of course, doubling both E_{max} and E_{min} will have no effect on this equation, so it is quite easy to find m by displaying the envelope on an oscilloscope and measuring the maximum and minimum peak-to-peak values for the envelope voltage. Other time-domain methods for measuring the amplitude modulation index will be described along with AM transmitters, in the next chapter.

Example 3.4 Calculate the modulation index for the waveform shown in Figure 3.2.

Solution It is easiest to use peak-to-peak values with an oscilloscope. From the figure we see that:

$$E_{max} = 150 \text{ mV} \qquad E_{min} = 70 \text{ mV} \qquad m = \frac{E_{max} - E_{min}}{E_{max} + E_{min}}$$

$$= \frac{150 - 70}{150 + 70}$$

$$= 0.364 \quad \text{or} \quad 36.4\%$$

3.3 AM in the Frequency Domain

So far, we have looked at the AM signal exclusively in the time domain, that is, as it can be seen on an ordinary oscilloscope. In order to find out more about this signal, however, it is necessary to consider its spectral makeup. We could use Fourier methods to do this, but for a simple AM waveform it is easier and just as valid to use trigonometry.

To start, we should note that although both the carrier and the modulating signal may be sine waves, the modulated AM waveform is *not* a sine wave. This can be seen from a simple examination of the waveform of Figure 3.1(c). It is important to remember that the modulated waveform is not a sine wave when, for instance, trying to find RMS from peak voltages. The usual formulas, so laboriously learned in fundamentals courses, do not apply here!

If an AM signal is not a sine wave, then what is it? We already have a mathematical expression, given by Equation (3.5):

$$v(t) = E_c(1 + m \sin \omega_m t) \sin \omega_c t$$

Expanding it and using a trigonometric identity will prove useful. Expanding gives

$$v(t) = E_c \sin \omega_c t + mE_c \sin \omega_m t \sin \omega_c t$$

The first term is just the carrier. The second can be expanded using two trigonometric identities:

$$\sin A \sin B = \frac{1}{2}[\cos (A - B) - \cos (A + B)]$$

and

$$\cos A = \cos (-A)$$

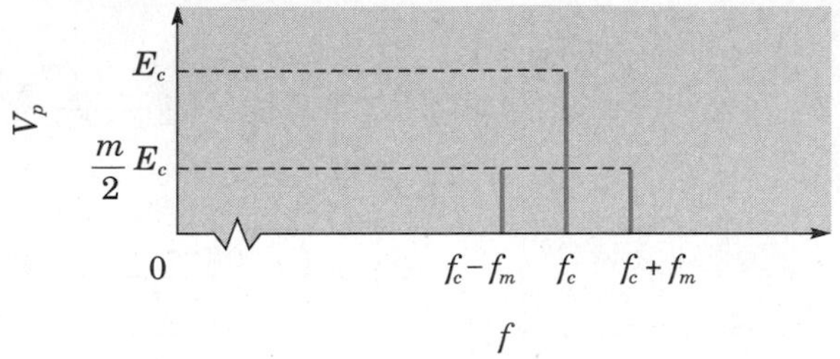

Figure 3.8 AM in the frequency domain

to give

$$v(t) = E_c \sin \omega_c t + \frac{mE_c}{2}[\cos(\omega_c - \omega_m)t - \cos(\omega_c + \omega_m)t]$$

which can be separated into three distinct terms:

$$v(t) = E_c \sin \omega_c t + \frac{mE_c}{2}\cos(\omega_c - \omega_m)t - \frac{mE_c}{2}\cos(\omega_c + \omega_m)t \tag{3.10}$$

We now have, besides the original carrier, two additional sinusoidal waves, one above the carrier frequency and one below. When the complete signal is sketched in the frequency domain, as in Figure 3.8, we see the carrier and two additional frequencies (one to each side), which are called, logically enough, **side frequencies**. The separation of each side frequency from the carrier is equal to the modulating frequency, and the relative amplitude of the side frequency, compared with that of the carrier, is proportional to m, becoming half the carrier voltage for $m = 1$.

In a realistic situation there is generally more than one set of side frequencies, because there is more than one modulating frequency. Each modulating frequency produces two side frequencies. Those above the carrier can be grouped into a band of frequencies called the upper **sideband**, and the lower sideband looks like a mirror image of the upper, reflected in the carrier.

From now on, we will generally use the term *sideband*, rather than *side frequency*, even for the case of single-tone modulation, because it is more general and more commonly used in practice.

Mathematically, we have:

$$f_{usb} = f_c + f_m \tag{3.11}$$

$$f_{lsb} = f_c - f_m \tag{3.12}$$

$$E_{lsb} = E_{usb} = \frac{mE_c}{2} \tag{3.13}$$

where

f_{usb} = frequency of the upper sideband
f_{lsb} = frequency of the lower sideband
E_{usb} = peak voltage of the upper-sideband component
E_{lsb} = peak voltage of the lower-sideband component

Example 3.5

(a) A 1 MHz carrier with an amplitude of 1 V peak is modulated by a 1 kHz signal with $m = 0.5$. Sketch the voltage spectrum.
(b) An additional 2 kHz signal modulates the carrier with $m = 0.2$. Sketch the voltage spectrum.

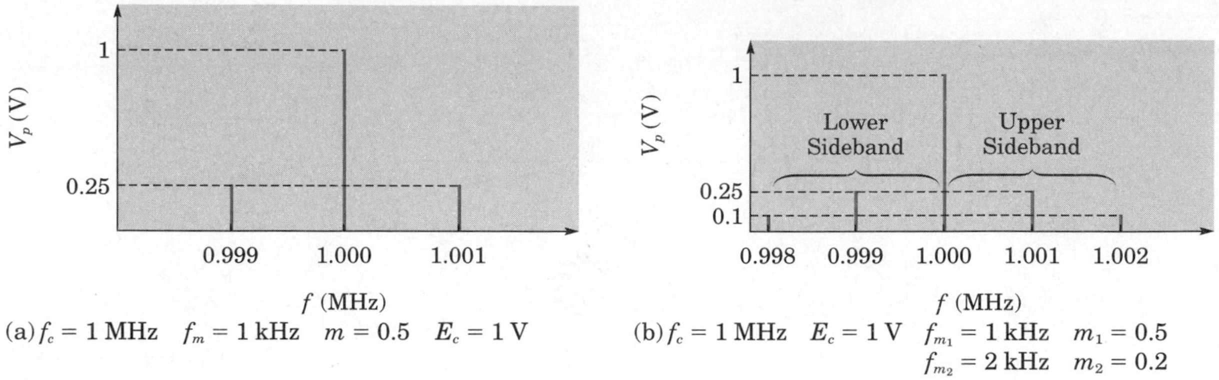

Figure 3.9

Solution (a) The frequency scale is easy. There will be three frequency components. The carrier will be at:

$$f_c = 1 \text{ MHz}$$

The upper sideband will be at:

$$\begin{aligned} f_{usb} &= f_c + f_m \\ &= 1 \text{ MHz} + 1 \text{ kHz} \\ &= 1.001 \text{ MHz} \end{aligned}$$

The lower sideband will be at:

$$\begin{aligned} f_{lsb} &= f_c - f_m \\ &= 1 \text{ MHz} - 1 \text{ kHz} \\ &= 0.999 \text{ MHz} \end{aligned}$$

Next, we have to determine the amplitudes of the three components. The carrier is unchanged with modulation, so it remains at 1 V peak. The two sidebands will have the same peak voltage:

$$\begin{aligned} E_{lsb} = E_{usb} &= \frac{mE_c}{2} \\ &= \frac{0.5 \times 1}{2} \\ &= 0.25 \text{ V} \end{aligned}$$

Figure 3.9(a) shows the solution.

(b) The addition of another modulating signal at a different frequency simply adds another set of side frequencies. It does not change anything that was done in part (a). The new frequency components will be at 1.002 and 0.998 MHz, and their amplitude will be 0.1 V. The result is shown in Figure 3.9(b).

3.3.1 Bandwidth

Signal bandwidth is one of the most important characteristics of any modulation scheme. In general, a narrow bandwidth is desirable. In any situation where spectrum space is limited, a narrow bandwidth allows more signals to be transmitted simultaneously than does a wider bandwidth. It also allows a narrower bandwidth to be used in the receiver. Since ordinary thermal noise is evenly distributed over the frequency domain, using narrower bandwidth in receivers will include less noise, thereby increasing the signal-to-noise ratio. (There is one major exception to the general rule that reducing the bandwidth improves the signal-to-noise ratio. The exception is for wideband frequency modulation, which will be described in Chapter 7.) However, the receiver must have a wide enough bandwidth to pass the complete signal including all the sidebands, or distortion will result. Consequently, we will have to calculate the signal bandwidth for each of the modulation schemes we consider.

A glance at Equation (3.10) and Figure 3.8 will show that this calculation is very easy for AM. The signal extends from the lower side frequency, which is at the carrier frequency less the modulation frequency, to the upper side frequency, at the carrier frequency plus the modulation frequency. The difference between these is simply twice the modulation frequency.

If we have a complex modulating signal with more than one modulating frequency, as in Figure 3.9(b), the bandwidth will be twice the *highest* modulating frequency. For telephone-quality voice, for instance, a bandwidth of about 6 kHz would suffice, while a video signal with a 4 MHz maximum baseband frequency would need 8 MHz of bandwidth if transmitted in this way. (Since a television channel is only 6 MHz wide, we can surmise, correctly, that television must actually be transmitted by a more complex modulation scheme that uses less bandwidth.)

Mathematically, the relationship is:

$$B = 2F_m \tag{3.14}$$

where

B = bandwidth in hertz
F_m = highest modulating frequency in hertz

Example 3.6

CB radio channels are 10 kHz apart. What is the maximum modulation frequency that can be used if a signal is to remain entirely within its assigned channel?

Solution

From Equation (3.14) we have

$$B = 2F_m$$

or

$$F_m = \frac{B}{2} = \frac{10 \text{ kHz}}{2} = 5 \text{ kHz}$$

By the way, many people assume that because AM radio broadcast channels are also assigned at 10 kHz intervals, there is a similar limitation on the audio frequency response for broadcasting. This is not the case: AM broadcast transmitters typically have an audio frequency response extending to about 10 kHz, giving a theoretical signal bandwidth of 20 kHz. This works because adjacent channels are not assigned in the same locality, so some overlap is possible. For instance, if your city has a station at 1000 kHz, there will not be one at 990 or 1010 kHz. In order to reduce interference from distant stations, many AM receivers do have narrow bandwidth and consequently limited audio frequency response.

3.3.2 Power Relationships

Power is important in any communications scheme, because the crucial signal-to-noise ratio at the receiver depends as much on the signal power being large as on the noise power being small. The power that is most important, however, is not the total signal power but only that portion that is used to transmit information. Since the carrier in an AM signal remains unchanged with modulation, it contains no information. Its only function is to aid in demodulating the signal at the receiver. This makes AM inherently wasteful of power, compared with some other modulation schemes we will describe later.

The easiest way to look at the power in an AM signal is to use the frequency domain. We can find the power in each frequency component and then add them to get total power. We will assume that the signal appears across a resistance R, so that reactive volt-amperes can be ignored. We will also assume that the power required is average power.

Suppose that the modulating signal is a sine wave. Then the AM signal consists of three sinusoids, the carrier and two side frequencies (sidebands), as shown in Figure 3.8.

The power in the carrier is easy to calculate, since the carrier by itself is a sine wave. The carrier is given by the equation

$$e_c = E_c \sin \omega_c t$$

where

e_c = instantaneous carrier voltage
E_c = peak carrier voltage

Since E_c is the peak carrier voltage, the power P_c developed when this signal appears across a resistance R is simply

$$\begin{aligned} P_c &= \frac{(E_c/\sqrt{2})^2}{R} \\ &= \frac{E_c^2}{2R} \text{ W} \end{aligned}$$

The next step is to find the power in each sideband. The two frequency components have the same amplitude, so they will have equal power. Assuming sine-wave modulation, each sideband is a cosine wave whose peak voltage is given by Equation (3.13):

$$E_{lsb} = E_{usb} = \frac{mE_c}{2}$$

Since the carrier and both sidebands are part of the same signal, the sidebands will appear across the same resistance R as the carrier. The two sidebands will have equal power. Looking at the lower sideband,

$$\begin{aligned} P_{lsb} &= \frac{E_{lsb}^2}{2R} \\ &= \frac{(mE_c/2)^2}{2R} \\ &= \frac{m^2 E_c^2}{4 \times 2R} \\ &= \frac{m^2}{4} \times \frac{E_c^2}{2R} \\ &= \frac{m^2}{4} P_c \end{aligned} \tag{3.15}$$

The same reasoning will show that

$$P_{usb} = \frac{m^2}{4} P_c$$

Since the two sidebands have equal power, the total sideband power is given by

$$P_{sb} = \frac{m^2}{2} P_c \tag{3.16}$$

The total power P_t in the whole signal is just the sum of the power in the carrier and the sidebands, so it is

$$P_t = P_c + \left(\frac{m^2}{2}\right) P_c$$

or

$$P_t = P_c\left(1 + \frac{m^2}{2}\right) \tag{3.17}$$

These latest equations tell us several useful things:

1. The total power in an AM signal increases with modulation, reaching a value 50% greater than that of the unmodulated carrier for 100% modulation. As we shall see, this result has implications for transmitter design.
2. The extra power with modulation goes into the sidebands: the carrier power does not change with modulation.
3. The useful power, that is, the power that carries information, is rather small, reaching a maximum of one third of the total signal power for 100% modulation and much less at lower modulation indices. For this reason, AM transmission is more efficient when the modulation index is as close to 1 as practicable.

Example 3.7

An AM broadcast transmitter has a carrier power output of 50 kW. What total power would be produced with 80% modulation?

Solution

$$\begin{aligned} P_t &= P_c\left(1 + \frac{m^2}{2}\right) \\ &= (50\text{ kW})\left(1 + \frac{0.8^2}{2}\right) \\ &= 66\text{ kW} \end{aligned}$$

3.3.3 Measuring the Modulation Index in the Frequency Domain

Since the ratio between sideband and carrier power is a simple function of m, it is quite possible to measure the modulation index by observing the spectrum of an AM signal. The only slight complication is that spectrum analyzers generally display power ratios in decibels. The power ratio between sideband and carrier power can easily be found from the relation:

$$\frac{P_{lsb}}{P_c} = \text{antilog } (dB/10) \tag{3.18}$$

where

P_c = carrier power
P_{lsb} = power in one sideband (the lower has been chosen for this example, but of course the upper sideband has the same power)
dB = difference between sideband and carrier signals, measured in decibels (this number will be negative)

Once the ratio between carrier and sideband power has been found, it is easy to find the modulation index from Equation (3.15):

$$P_{lsb} = \frac{m^2}{4} P_c \tag{3.19}$$

$$m^2 = \frac{4P_{lsb}}{P_c}$$

$$m = 2\sqrt{\frac{P_{lsb}}{P_c}}$$

Figure 3.10 is a handy graph for finding m directly from the difference in decibels between sideband and carrier powers.

Although the time-domain measurement described earlier is simpler and uses less-expensive equipment, frequency-domain measurement enables much smaller values of m to be found. A modulation level of 5%, for instance, would be almost invisible on an oscilloscope, but it is quite obvious and easy to measure on a spectrum analyzer. The spectrum

Figure 3.10
Courtesy Hewlett-Packard.

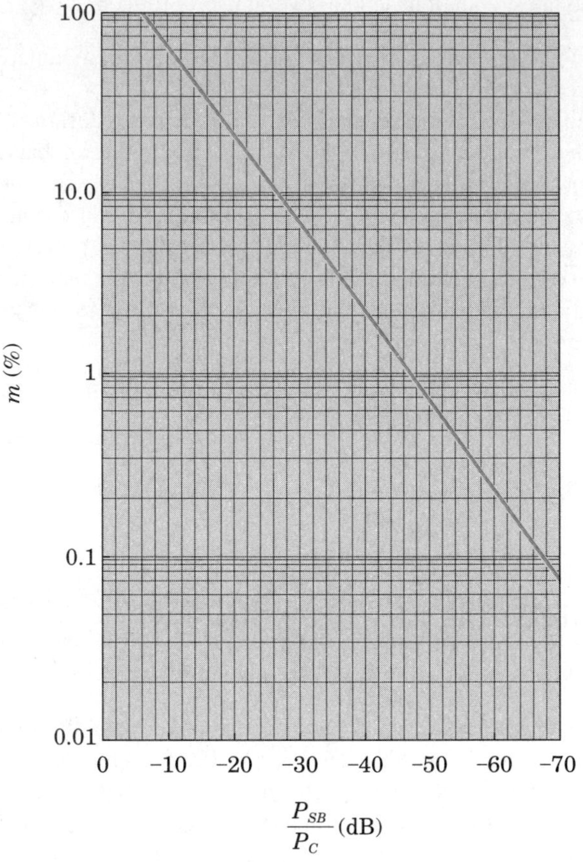

analyzer also allows the contribution from different modulating frequencies to be observed and calculated separately.

Example 3.8 Calculate the modulation frequency f_m and modulation index m for the spectrum analyzer display shown in Figure 3.11.

Figure 3.11

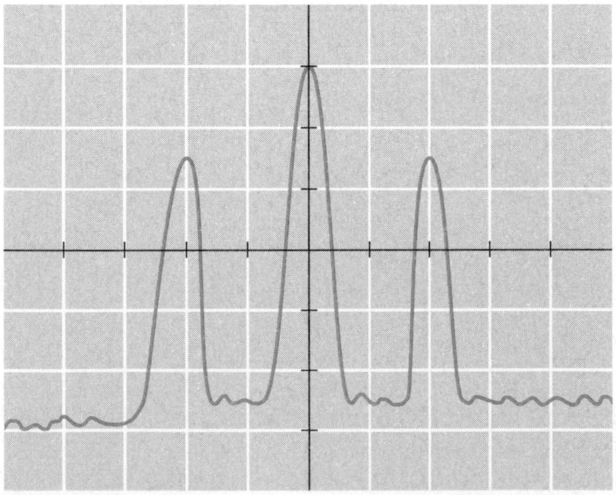

Ref. Level: 10 dBm Center Frequency: 10 MHz
Vert: 10 dB/division Span: 5 kHz/division

Solution First let us find f_m. The difference between the carrier and either sideband is 2 divisions at 5 kHz/division, or 10 kHz. So f_m is 10 kHz.

Next, we need to find the modulation index. The two sidebands have the same power, so we can use either. Let us use the lower one. The spectrum analyzer is set for 10 dB/division, and the sideband is 1.5 divisions or 15 dB below the carrier. This corresponds to a power ratio, as given in Equation (3.17), of

$$\frac{P_{lsb}}{P_c} = \text{antilog}\left(\frac{-15}{10}\right) = 0.0316$$

From Equation (3.19),

$$m = 2\sqrt{\frac{P_{lsb}}{P_c}} = 2\sqrt{0.0316} = 0.356$$

We could have used the nomograph instead, with the same result. Try it.

3.4 RMS and Peak Voltages and Currents

The well-known power equation

$$P = \frac{V^2}{R}$$

is valid for any kind of waveform, provided that P represents average power and V represents RMS voltage. To find the RMS voltage when the power is known, rearrange this equation to get

$$V = \sqrt{PR} \tag{3.20}$$

The total power in an AM waveform was found earlier. It is given by Equation (3.17) as

$$P_t = P_c\left(1 + \frac{m^2}{2}\right)$$

Substituting this into Equation (3.20) gives

$$V = \sqrt{P_c\left(1 + \frac{m^2}{2}\right)R} \tag{3.21}$$

Similarly, from the fundamental equation

$$P = I^2R$$

we can easily find the RMS current:

$$I = \sqrt{\frac{P}{R}} \tag{3.22}$$

For a modulated signal, the power required is the total power, so the equation becomes

$$I = \sqrt{\frac{P_c(1 + m^2/2)}{R}} \tag{3.23}$$

Example 3.9 Find the RMS voltage and current at the output of a transmitter that has a carrier power of 500 W into a load impedance of 50 Ω:

(a) without modulation
(b) with 70% modulation

Solution (a) Without modulation, the signal is a sine wave and the RMS voltage is given by Equation (3.20):

$$\begin{aligned} V &= \sqrt{PR} \\ &= \sqrt{(500\ \text{W})(50\ \Omega)} \\ &= 158\ \text{V} \end{aligned}$$

The current is found from Equation (3.22):

$$\begin{aligned} I &= \sqrt{\frac{P}{R}} \\ &= \sqrt{\frac{500\ \text{W}}{50\ \Omega}} \\ &= 3.16\ \text{A} \end{aligned}$$

(b) With modulation, the RMS voltage and current are found in the same way, except that the total power now includes the power in the sidebands. For voltage, we can use Equation (3.21):

$$\begin{aligned} V &= \sqrt{P_c\left(1 + \frac{m^2}{2}\right)R} \\ &= \sqrt{(500\ \text{W})\left(1 + \frac{0.7^2}{2}\right)(50\ \Omega)} \\ &= 176\ \text{V} \end{aligned}$$

Similarly the RMS current is given by Equation (3.23):

$$\begin{aligned} I &= \sqrt{\frac{P_c(1 + m^2/2)}{R}} \\ &= \sqrt{\frac{(500\ \text{W})(1 + 0.7^2/2)}{50\ \Omega}} \\ &= 3.53\ \text{A} \end{aligned}$$

It is not necessary to remember Equations (3.21) and (3.23). The only thing to remember is that to find RMS voltage or current, you use total average power. This is true for an AM signal just as for any other signal.

Sometimes an RF ammeter in the transmission line from transmitter to antenna is used to measure the modulation index for an AM station. It is easy to derive an expression to show how this can be done. Once again, it is not necessary to memorize this equation; the derivation is presented as one more example of the many things that can be learned about an AM signal once its basic structure is understood. From Equation (3.23) we know that

$$I = \sqrt{\frac{P_c(1 + m^2/2)}{R}}$$

We can rearrange this equation to express m in terms of the current:

$$1 + \frac{m^2}{2} = \frac{I^2 R}{P_c}$$

$$\frac{m^2}{2} = \frac{I^2 R}{P_c} - 1$$

$$m = \sqrt{2\left(\frac{I^2 R}{P_c} - 1\right)}$$

In this equation, of course, I is the current with modulation, as measured by the RF ammeter. The ammeter will not measure P_c directly, but we do know that

$$P_c = I_0^2 R$$

where I_0 = current without modulation

If we substitute this into our equation for m, we get

$$m = \sqrt{2\left(\frac{I^2 R}{I_0^2 R} - 1\right)}$$

R cancels out, so we find

$$m = \sqrt{2\left(\frac{I^2}{I_0^2} - 1\right)} \qquad (3.24)$$

Example 3.10 An RF ammeter in the transmission line from transmitter to antenna measures 5 A without modulation and 5.5 A with modulation. What is the modulation index?

Solution From Equation (3.24), we know that:

$$m = \sqrt{2\left(\frac{I^2}{I_0^2} - 1\right)}$$

$$= \sqrt{2\left(\frac{5.5^2}{5^2} - 1\right)}$$

$$= 0.648$$

Now, how about peak values for voltage and current? We may be tempted to use the formula

$$V_p = \sqrt{2}V_{RMS}$$

but we had better resist the temptation, because that equation is valid only for sine waves. Instead, let us go back to Equation (3.7), which shows us that the peak voltage is

$$E_{max} = E_c(1 + m)$$

Thus, we can find the peak voltage of an AM signal by first finding the peak carrier voltage and the modulation index. We are more likely to know the carrier power than the carrier voltage, but since the carrier is a sine wave we should have no difficulty.

Example 3.11 Find the peak voltage at the output of a CB transmitter with 4 W carrier power output, operating into a 50 Ω resistive load, with $m = 0.8$.

Solution From the carrier power, we can get the carrier voltage. The RMS voltage will be given by Equation (3.20):

$$V_c = \sqrt{P_c R}$$

but we want the peak carrier voltage, which, since the carrier is a sine wave, will be the RMS voltage multiplied by $\sqrt{2}$ or

$$\begin{aligned} E_c &= \sqrt{2P_c R} \\ &= \sqrt{2(4\text{ W})(50\ \Omega)} \\ &= 20\text{ V} \end{aligned}$$

Now, the peak voltage for the signal is, from Equation (3.7),

$$\begin{aligned} E_{max} &= E_c(1 + m) \\ &= 20(1 + 0.8) \\ &= 36\text{ V} \end{aligned}$$

3.5 Quadrature AM and AM Stereo

It is possible to send two separate information signals using amplitude modulation at one carrier frequency. This can be accomplished by generating two carriers at the same frequency but separated in phase by 90°. Then, each is modulated by a separate information signal and the two resulting signals are summed. Because of the 90° phase shift involved, the scheme is called **quadrature AM** (QUAM or QAM). Figure 3.12(a) shows how it can be implemented.

Recovery of the two information signals requires *synchronous detection* using two balanced demodulators. The demodulators must be supplied with reference carriers having exactly the same frequency and phase as the original carriers. Otherwise the output from each detector will be some combination of the two baseband signals. Figure 3.12(b) gives the general idea. Balanced demodulators are essentially the same as the balanced mixers described in Chapter 2. Their operation as detectors will be covered in detail in Chapter 6, in connection with single-sideband systems.

Quadrature AM is a much more complex scheme than ordinary AM. Until recently, its main applications have been in data transmission and in color television. The main

Improved Quality for AM

There is no inherent reason why AM radio has to have poor audio quality. Much of the problem rests with receivers whose bandwidth is too narrow to include all the sidebands of the broadcast signal. Many AM receivers, including those included with high-fidelity audio systems, have an audio response that extends only to about 4 kHz.

In an attempt to compensate for this, most AM broadcasters boost the higher audio frequencies before modulation. This is called pre-emphasis. The problem is that until now there has been no standard for the amount of boost used.

Recent decisions by the FCC in the United States have standardized the amount of pre-emphasis and have also determined that stations may transmit an audio signal up to 10 kHz in frequency (20 kHz total bandwidth). The AM band has now been extended to 1700 kHz. At the same time, the Electronics Industry Association in the United States has developed a voluntary standard for receivers requiring reasonably flat audio response to at least 7.5 kHz, with an option for selectable reduced bandwidth to avoid interference.

It will be interesting to see whether these improvements, plus the standardizing of AM stereo to Motorola's C-Quam™ system (discussed in Section 3.5), will reverse the decline in popularity of AM compared to FM broadcasting.

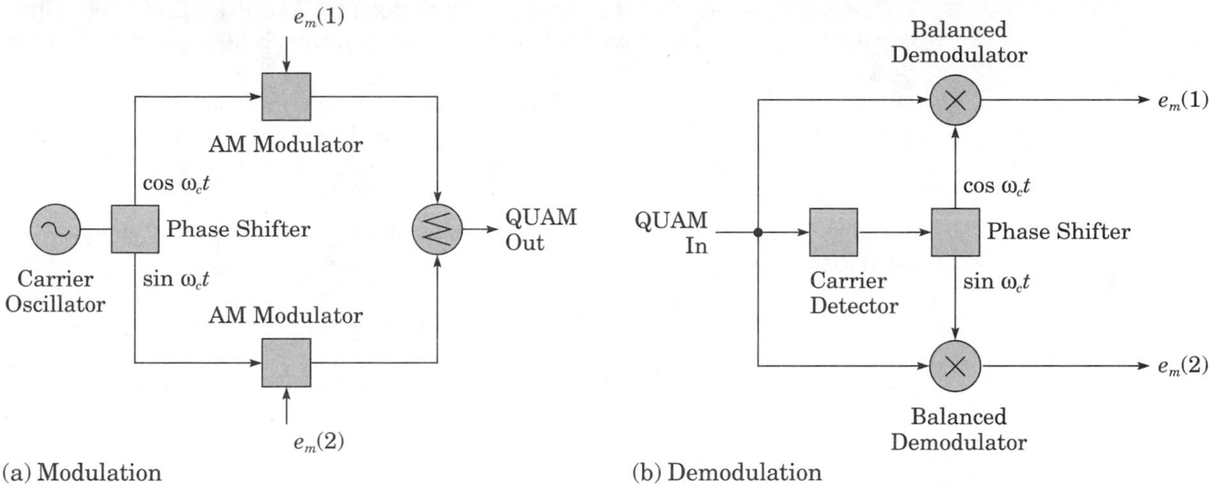

Figure 3.12
Quadrature AM

reason for including a brief introduction to it here is that a variation of QUAM is also used for AM stereo broadcasting. The scheme used is called C-Quam™ and was developed by Motorola, Inc.

Quadrature AM could be used to transmit stereo signals without any increase in bandwidth by modulating each carrier with one channel (left or right). Unfortunately, ordinary quadrature AM would not be very suitable for stereo broadcasts because it would not be compatible with the envelope detectors used in ordinary AM radios. For compatibility, such a detector should produce a monaural (mono) signal consisting of the sum of the two channels. Using the system just described would result in distorted signals in an ordinary (mono) radio. Mathematically, the signal produced by an ideal envelope detector would be

$$v(t) = \sqrt{[1 + v_L(t) + v_R(t)]^2 + [v_L(t) - v_R(t)]^2}$$

where $v_L(t)$ and $v_R(t)$ represent the left and right signal voltages, respectively. It is easy to see that this represents a sum $(v_L + v_R)$ signal (plus a dc component) only when $v_L - v_R = 0$, that is, when the original signal is monaural. Otherwise, there will terms of the form $v_L v_R$, which represent intermodulation distortion.

The C-Quam™ system solves this problem by predistorting the QUAM signal so that the envelope of the resulting signal represents the sum of the two stereo channels (referred to as the L+R signal). Thus a mono receiver, with a conventional envelope detector, will receive an L+R signal. A stereo receiver will have the necessary circuitry to convert the C-Quam™ signal to an ordinary QUAM signal and extract the L-R information.

In addition to the predistorted QUAM signal, a stereo AM signal has a 25 Hz pilot carrier that indicates the presence of stereo to the receiver. Single-chip C-Quam™ decoders that sense the 25 Hz carrier and automatically switch to stereo operation in the presence of a stereo signal are available.

Summary

Here are the main points to remember from this chapter.

1. Amplitude modulation is the simplest form of modulation.
2. In the time domain, the process of amplitude modulation creates a signal with an envelope that closely resembles the original information signal.
3. In the frequency domain, an AM signal consists of the carrier, which is unchanged from its unmodulated state, and two sidebands. The total bandwidth of the signal is twice the maximum modulating frequency.
4. The peak voltage of an AM signal varies with the modulation index, becoming twice that of the unmodulated carrier for the maximum modulation index of 1.

5. The power in an AM signal increases with modulation. The extra power goes into the sidebands. At maximum modulation, the total power is 50% greater than the power in the unmodulated carrier.
6. The modulation index can easily be measured using either an oscilloscope or a spectrum analyzer. The oscilloscope method requires less expensive equipment but the spectrum analyzer can measure much lower modulation indices.
7. When calculating power and voltage values for an AM signal, it is important to take into account the actual makeup of the signal and avoid reliance on formulas derived for sine waves.

Important Equations

$$v(t) = (E_c + e_m) \sin \omega_c t \quad (3.1)$$

$$v(t) = (E_c + E_m \sin \omega_m t) \sin \omega_c t \quad (3.3)$$

$$m = \frac{E_m}{E_c} \quad (3.4)$$

$$v(t) = E_c(1 + m \sin \omega_m t) \sin \omega_c t \quad (3.5)$$

$$m_T = \sqrt{m_1^2 + m_2^2 + \cdots} \quad (3.6)$$

$$E_{max} = E_c(1 + m) \quad (3.7)$$

$$E_{min} = E_c(1 - m) \quad (3.8)$$

$$m = \frac{E_{max} - E_{min}}{E_{max} + E_{min}} \quad (3.9)$$

$$v(t) = E_c \sin \omega_c t + \frac{mE_c}{2} \cos (\omega_c - \omega_m)t - \frac{mE_c}{2} \cos (\omega_c + \omega_m)t \quad (3.10)$$

$$f_{usb} = f_c + f_m \quad (3.11)$$

$$f_{lsb} = f_c - f_m \quad (3.12)$$

$$E_{lsb} = E_{usb} = \frac{mE_c}{2} \quad (3.13)$$

$$B = 2F_m \quad (3.14)$$

$$P_{lsb} = \frac{m^2}{4} P_c \quad (3.15)$$

$$P_{sb} = \frac{m^2}{2} P_c \quad (3.16)$$

$$P_t = P_c\left(1 + \frac{m^2}{2}\right) \quad (3.17)$$

$$m = 2\sqrt{\frac{P_{lsb}}{P_c}} \quad (3.19)$$

$$V = \sqrt{P_c\left(1 + \frac{m^2}{2}\right)R} \quad (3.21)$$

$$I = \sqrt{\frac{P_c(1 + m^2/2)}{R}} \tag{3.23}$$

$$m = \sqrt{2\left(\frac{I^2}{I_0^2} - 1\right)} \tag{3.24}$$

Glossary

amplitude modulation (AM) a modulation scheme in which the amplitude of a high-frequency signal is varied in accordance with the instantaneous amplitude of an information signal

envelope the curve produced by joining the tips of the individual RF cycles of a modulated wave

modulation index measure of the extent of modulation of a signal

overmodulation modulation to a greater depth than allowed (for AM, this means a modulation index greater than one)

quadrature AM transmission of two separate information signals using two amplitude-modulated carriers at the same frequency but differing in phase by 90°

side frequency a signal component in a modulated signal, at a frequency different from that of the carrier

sideband all of the side frequencies to one side of the carrier frequency

splatter colloquial term used to describe additional side frequencies produced by overmodulation or distortion in an AM system

Questions

1. What is meant by the *envelope* of an AM waveform? What is its significance?
2. Although amplitude modulation certainly involves changing the amplitude of the signal, it is not correct to say that the amplitude of the carrier is modulated. Explain this statement.
3. Why is it desirable to have the modulation index of an AM signal as large as possible without overmodulating?
4. Describe what happens when a typical AM modulator is overmodulated, and explain why overmodulation is undesirable.
5. Explain the difference between amplitude modulation and linear addition of the carrier and information signals.
6. How does the bandwidth of an AM signal relate to the information signal?
7. Describe three different ways in which the modulation index of an AM signal can be measured.
8. By how much does the power in an AM signal increase with modulation, compared to the power of the unmodulated carrier?
9. Suppose a modulation system similar to AM but with half the bandwidth could be devised. Would this be an improvement? Explain your answer.
10. How can the peak and RMS voltages of an AM signal wave be calculated?
11. Explain why the peak voltage of an AM waveform does not have the same relation to its RMS voltage as a sine wave does.
12. How is it possible for AM broadcast stations to transmit audio frequencies greater than 5 kHZ?
13. What is the advantage of keeping the modulation index of an AM signal as high as possible without exceeding one?
14. Briefly describe the C-Quam system for generating AM stereo.
15. How is the AM stereo signal made compatible with monaural receivers?
16. How is the modulation index determined when there are several modulating signals of different frequencies?
17. How is the bandwidth of an AM signal determined when there are several modulating signals of different frequencies?

Problems

SECTION 3.2

18. An AM signal has the equation:

$$v(t) = [15 + 4 \sin(44 \times 10^3 t)] \sin(46.5 \times 10^6 t) \text{ V}$$

(a) Find the carrier frequency.
(b) Find the frequency of the modulating signal.
(c) Find the value of m.
(d) Find the peak voltage of the unmodulated carrier.
(e) Sketch the signal in the time domain, showing voltage and time scales.

19. An AM signal has a carrier frequency of 3 MHz and an amplitude of 5 V peak. It is modulated by a sine wave with a frequency of 500 Hz and a peak voltage of 2 V. Write the equation for this signal and calculate the modulation index.
20. An AM signal consists of a 10 MHz carrier modulated by a 5 kHz sine wave. It has a maximum positive envelope voltage of 12 V and a minimum of 4 V.
    (a) Find the peak voltage of the unmodulated carrier.
    (b) Find the modulation index and percent.
    (c) Sketch the envelope.
    (d) Write the equation for the signal voltage as a function of time.
21. An AM transmitter is modulated by two sine waves at 1 kHz and 2.5 kHz, with modulations of 25% and 50%, respectively. What is the effective modulation index?
22. For the AM signal sketched in Figure 3.13, calculate:
    (a) the modulation index
    (b) the peak carrier voltage
    (c) the peak modulating-signal voltage

Figure 3.13

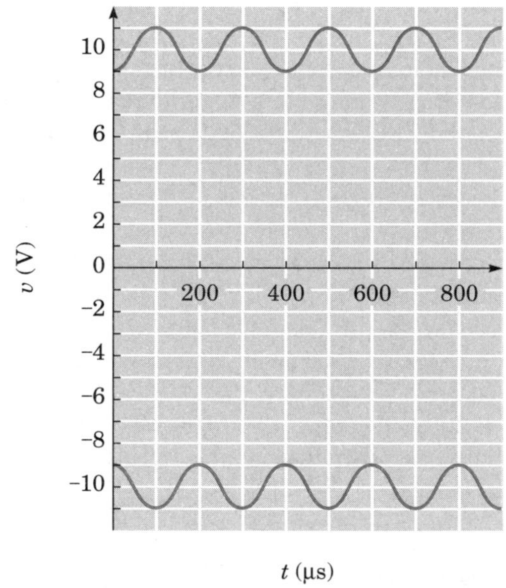

23. For the signal of Figure 3.14, calculate:
    (a) the index of modulation
    (b) the RMS voltage of the carrier without modulation

Figure 3.14

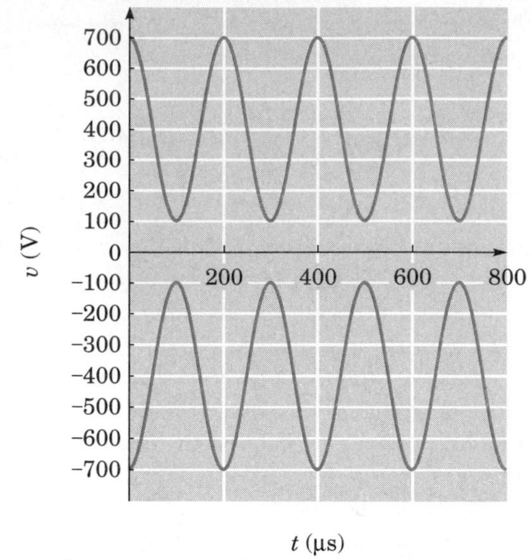

SECTION 3.3

24. An audio system requires a frequency response from 50 Hz to 15 kHz for high fidelity. If this signal were transmitted using AM, what bandwidth would it require?
25. A transmitter operates with a carrier frequency of 7.2 MHz. It is amplitude-modulated by two tones with frequencies of 1500 and 3000 Hz. What frequencies are produced at its output?
26. Sketch the signal whose equation is given in problem 18, in the frequency domain, showing frequency and voltage scales.
27. An AM signal has the following characteristics:

$$f_c = 150 \text{ MHz} \qquad f_m = 3 \text{ kHz}$$
$$E_c = 50 \text{ V} \qquad E_m = 40 \text{ V}$$

For this signal, find:
    (a) the modulation index
    (b) the bandwidth
    (c) the peak voltage of the upper side frequency

SECTION 3.4

28. An AM signal observed on a spectrum analyzer shows a carrier at +12 dbm, with each of the sidebands 8 dB below the carrier. Calculate:
    (a) the carrier power in milliwatts
    (b) the modulation index
29. An AM transmitter supplies 10 kW of carrier power to a 50 Ω load. It operates at a carrier frequency of 1.2 MHz and is 80% modulated by a 3 kHz sine wave.
    (a) Sketch the signal in the frequency domain, with frequency and power scales. Show the power in

dBW. (If necessary, refer to the discussion of dBW in Appendix A.)

(b) Calculate the total average power in the signal, in watts and dBW.

(c) Calculate the RMS voltage of the signal.

(d) Calculate the peak voltage of the signal.

30. A transmitter with a carrier power of 10 W at a frequency of 25 MHz operates into a 50 Ω load. It is modulated at 60% by a 2 kHz sine wave.

(a) Sketch the signal in the frequency domain. Show power and frequency scales. The power scale should be in dBm.

(b) What is the total signal power?

(c) What is the RMS voltage of the signal?

COMPREHENSIVE

31. The signal in Figure 3.15 is the output of an AM transmitter with a carrier frequency of 12 MHz and a carrier power of 150 W.

(a) Calculate the percent modulation.

(b) Calculate the RMS voltage of the carrier and use it to find the transmitter's load resistance.

(c) Sketch the signal in the frequency domain, with power in dBm on the vertical axis.

Figure 3.15

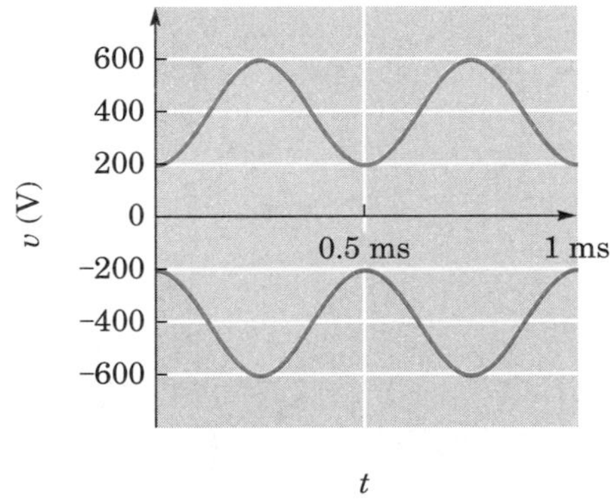

32. An AM transmitter with a carrier power of 20 kW is connected to a 50 Ω antenna. In order to design a lightning arrestor for the antenna, it is necessary to know the maximum instantaneous voltage that will appear at the antenna terminals. Calculate this voltage, assuming 100% modulation for the transmitter.

33. The spectrum analyzer display in Figure 3.16 represents the output of an AM transmitter. The analyzer has an input impedance of 50 Ω and is connected to the transmitter output through a 60 dB attenuator. Sketch the envelope of the signal in the time domain, as it would appear at the transmitter output terminals. Be sure to show both time and voltage scales.

Figure 3.16

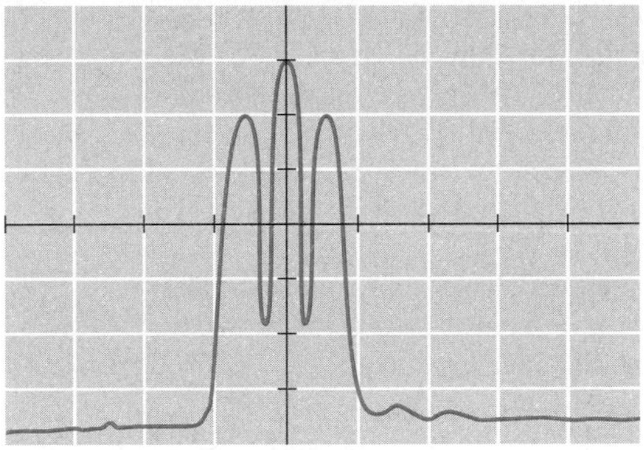

Reference level: − 10 dBm
Vertical: 10 dB/division

Center frequency: 21.200 MHz
Span: 5 kHz/division

34. An AM CB transmitter with a carrier frequency of 27.005 MHz is connected as shown in Figure 3.17(a). The oscilloscope display is shown in Figure 3.17(b). Sketch the same signal as it would appear on the spectrum analyzer.

Figure 3.17

Transmitter

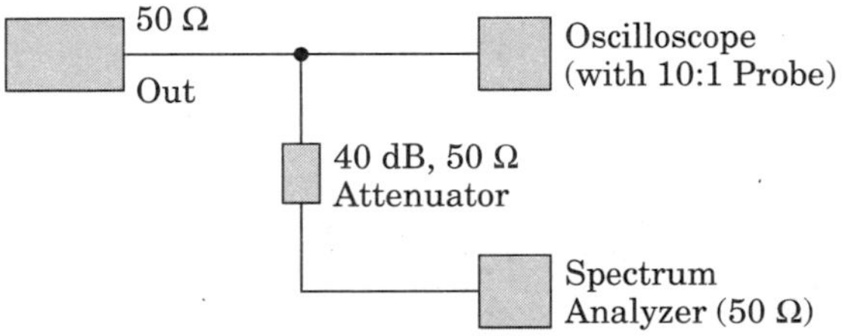

(a) Test Setup

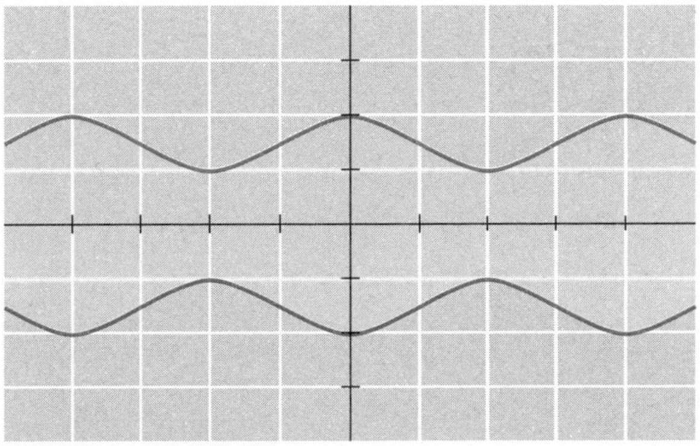

Vertical: 1 V/division
Horizontal: 100 μs/division

(b) Oscilloscope display (The probe has *not* been taken into account.)

35. Write the equation for an AM waveform with a carrier frequency of 5 MHz, a modulation frequency

of 3 kHz, and a modulation index of 40%. The peak-to-peak voltage of this signal is 10 volts.

36. A sine wave with a frequency of 1 MHz and an amplitude of 20 V peak is amplitude modulated at 80% by another sine wave with a frequency of 1 kHz.
   (a) Sketch the envelope of this signal in the time domain, with voltage and time scales.
   (b) Sketch this signal in the frequency domain, with voltage and frequency scales.

37. An AM signal appears across a 50 ohm load and has the following equation:

$$v(t) = 12\ (1 + \sin 12.566 \times 10^3 t) \sin 18.85 \times 10^6 t \text{ V}$$

   (a) Sketch the envelope of this signal in the time domain.
   (b) Calculate the following quantities:
      (i) carrier frequency
      (ii) modulating frequency
      (iii) total power
      (iv) bandwidth

38. An AM waveform can be described by the following equation. It appears across a 50 ohm resistor.

$$v(t) = [15 + 6 \sin (6 \times 10^3 t)] \sin (200 \times 10^6 t) \text{ V}$$

   (a) What is the modulation index for this signal?
   (b) Sketch this signal in the frequency domain showing frequency and voltage scales.
   (c) Find the total average power in this signal if it appears across a 50 ohm resistor.

4 AM Transmitters

Objectives After studying this chapter, you should be able to:

1. Discuss the requirements and specifications of AM transmitters and determine whether a given transmitter is suitable for a particular application
2. Draw block diagrams for several types of transmitters and explain their operation
3. Analyze the operation of transmitter circuits
4. Perform measurements on AM transmitters

4.1 Introduction

Chapter 3 dealt with the theory behind amplitude modulation and showed some simple modulator and demodulator circuits. This chapter covers the design of AM transmitters, using the theory from Chapter 3 and many of the circuit elements, such as oscillators and amplifiers, of Chapter 2. AM transmitter designs range from unlicensed toys developing 100 mW or less to transmitters used in international shortwave radio broadcasting, with powers in the megawatt range. Representative AM transmitters are shown in Figure 4.1. Though not a particularly efficient mode of radio communication, AM has remained popular, largely because of the simplicity of receiver design associated with it.

Later in this book, transmitters for other communication systems, such as single-sideband and FM, will be described. Many of the principles and a good number of the circuit features in these transmitters resemble those in AM transmitters, so much of the material in this chapter will be useful beyond its application to AM.

4.2 Transmitter Requirements

Before looking at actual circuits, we need to consider what a transmitter has to do. We already know the characteristics of the AM signal that it must generate. The signal must be generated with sufficient power, at the right frequency, with reasonable efficiency, and be coupled into an antenna. The demodulated signal must be a reasonably faithful copy of the original modulating signal.

4.2.1 Frequency Accuracy and Stability

The accuracy and stability of the transmitter frequency are essentially fixed by the oscillator. The exact requirements vary with the use to which the transmitter is put and are set by government regulatory bodies: the Federal Communications Commission in the United States and Industry and Science Canada in Canada, for example. Depending on the application, frequency accuracy and stability are specified in hertz or as a percentage of the operating frequency. It is easy to convert between the two methods, as the following example shows.

Figure 4.1
AM transmitters
(a) Courtesy of Radio Shack, a division of Tandy Corporation. (b) Continental Electronics Corporation, Dallas, Texas, USA.

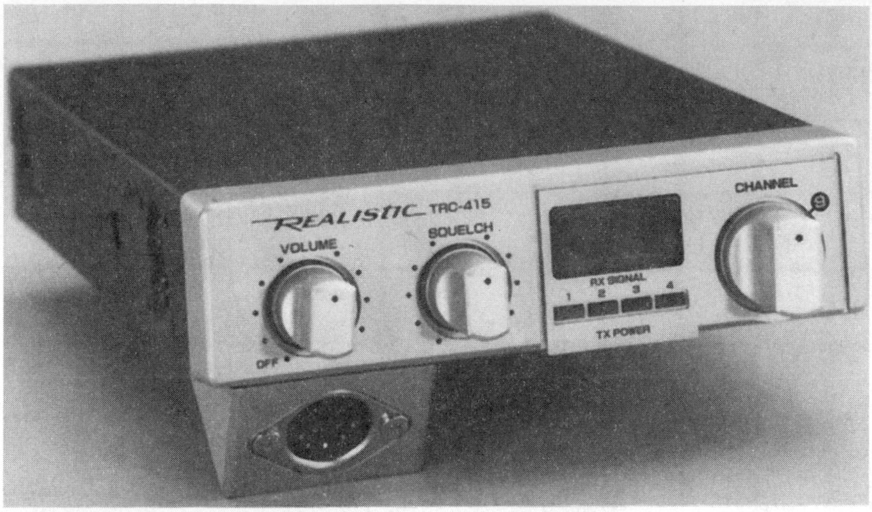

(a) CB transceiver

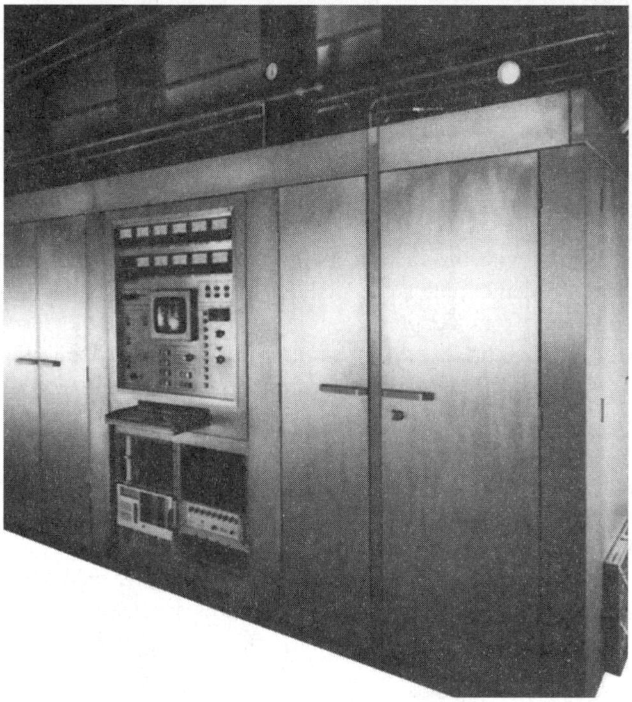

(b) Broadcast transmitter

Early Transmitters

The very first transmitters used a spark gap connected directly to an antenna. The only "tuned circuit" was the antenna itself. Needless to say, the bandwidth of such a transmitter was rather wide. It resembled more the interference from automobile ignition than any radio signal transmitted today. In fact, the consensus in the earliest days of radio, up to about the turn of the century, was that only one signal could be received in a given area. In other words, frequency-division multiplexing was unknown.

The next step was to connect one or more tuned circuits between the spark gap and the antenna. The original name for this idea was "syntony." This greatly reduced the bandwidth and made it possible to transmit more than one signal in the same area, at different frequencies.

Spark-gap transmitters became more and more complex, involving multiple gaps, rotating gaps, ac supplies, and elaborate tuning systems, until they were eventually superseded by the development of vacuum tubes.

Example 4.1 A crystal oscillator is accurate within 0.005%. How far off frequency could its output be at 27 MHz?

Solution The frequency could be out by 0.005% of 27 MHz, which is

$$27 \times 10^6 \text{ Hz} \times \frac{0.005}{100} = 1350 \text{ Hz}$$
$$= 1.35 \text{ kHz}$$

4.2.2 Frequency Agility

Frequency agility refers to the ability to change operating frequency rapidly, without extensive retuning. In a broadcast transmitter, this is not a requirement, since such stations rarely change frequency. When they do, it is a major, time-consuming operation involving extensive changes to the antenna system as well as the retuning of the transmitter.

With other services, for example, CB radio, the situation is different. Rapid retuning to any of the 40 available channels is essential to any modern CB transceiver. In addition to a frequency synthesizer for setting the actual transmitting frequency, such transmitters are required to use broadband techniques throughout so that frequency changes can be made instantly without retuning.

4.2.3 Spectral Purity

All transmitters produce spurious signals. That is, they emit signals at frequencies other than those of the carrier and the sidebands required for the modulation scheme in use. Spurious signals are often harmonics of the operating frequency or of the carrier oscillator if it operates at a different frequency. Any amplifier will produce harmonic distortion. Class C amplifiers, which are very common in AM transmitters, produce a large amount of harmonic energy. All frequencies except the assigned transmitting frequency must be filtered out to avoid interference with other transmissions.

The filtering of harmonics can never be perfect, of course, but in modern, well-designed transmitters it is very effective. For example, Figure 4.2 shows the spectrum produced by a typical CB transmitter. Note the fundamental at about 27 MHz and the har-

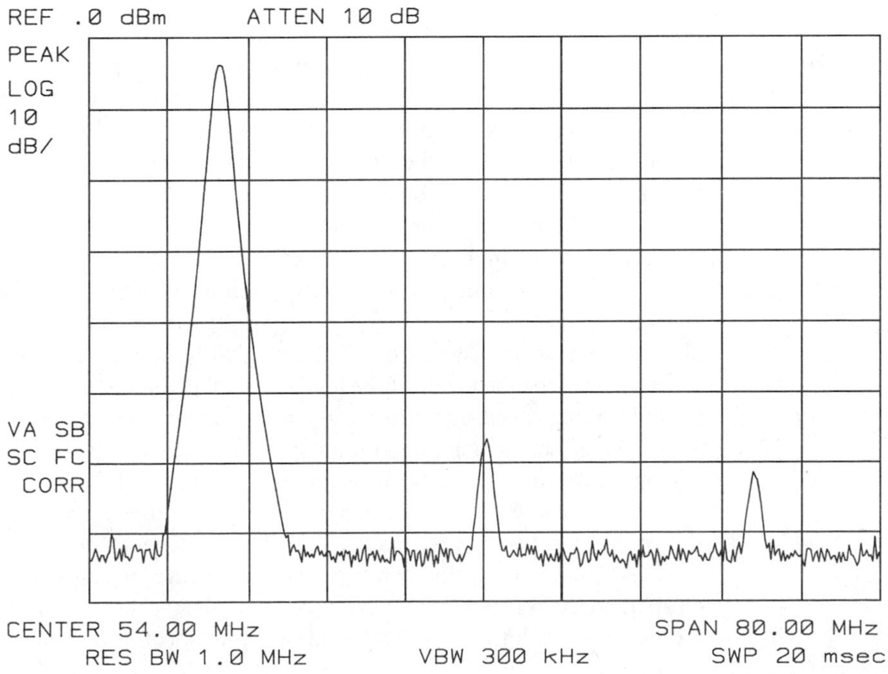

Figure 4.2 Output of a CB transmitter in the frequency domain

monics at twice and three times the fundamental frequency. How far down from the fundamental are these harmonic emissions?

4.2.4 Power Output

As pointed out in Chapter 3, there are a number of ways to measure the power of an AM signal. AM transmitters are generally rated in terms of carrier power output. Sometimes dc power input to the RF power amplifier stage is used because it is easier to measure. However, different amplifier efficiencies result in a considerable difference in output power for the same power input, so output power is a more useful rating.

When testing communications transmitters, the technologist should be aware of the rated **duty cycle** of the transmitter. Many transmitters designed for two-way voice communication are not rated for continuous operation at full power, since it is assumed that the operator will normally be talking for less than half the time and probably only for a few seconds to a minute at a time. This is easy to forget when testing, and the result can be overheated transistors or other components. Broadcast transmitters, of course, are rated for continuous duty (24 hours a day at full power).

4.2.5 Efficiency

Transmitter efficiency is important for two reasons. The most obvious one is *energy conservation*. This is especially important when very large power levels are involved, as in broadcasting, or, at the other extreme of the power-level range, when hand-held operation using batteries is required. Another reason for achieving high efficiency becomes apparent when we consider what happens to the power that enters the transmitter from the power supply but does not exit via the antenna: it is converted into heat in the transmitter, and this heat must be dissipated. Large amounts of heat require large components—heat sinks, fans, and in the case of some high-powered transmitters, even water cooling. All of these add to the cost of the equipment.

When discussing efficiency, it is important to distinguish between the efficiency of an individual stage and that of the entire transmitter. Knowing the efficiency of an amplifier stage is useful in designing cooling systems and sizing power supplies. In calculating energy costs, on the other hand, what is important is the **overall efficiency** of the system. The overall efficiency is the ratio of output power to power input from the primary power source, whether it be the ac power line or a battery. Overall efficiency is reduced by factors such as tube-heater power and losses in the power supply.

4.2.6 Modulation Fidelity

As mentioned in Chapter 1, an ideal communication system allows the original information signal to be recovered exactly, except for a time delay. Any distortion introduced at the transmitter is likely to remain—in most cases, it is not possible to remove it at the receiver. It might be expected, then, that an AM transmitter would be capable of modulating any baseband frequency onto the carrier, at any modulation level from 0% to 100%, to preserve the information signal as much as possible. In practice, however, the baseband spectrum often must be restricted in order to keep the transmitted bandwidth (which is twice the highest frequency in the baseband) within legal limits.

In addition, some form of **compression**, in which low-level baseband signals are amplified more than high-level signals, is often used to keep the modulation percentage high. Compression distorts the original signal by reducing the **dynamic range**, which is the ratio between the levels of the loudest and the quietest passages in the audio signal. The result is an improved signal-to-noise ratio at the receiver, at the expense of some distortion.

The effects of compression on dynamic range can be removed by applying an equal and opposite expansion at the receiver. Such expansion involves giving more gain to signals at higher levels. Compression-expansion combinations are quite common in communication systems but are not normally used with full-carrier AM.

AM communication transmitters tend to use a relatively simple **automatic-level-control (ALC) circuit** that (as far as possible) keeps the modulation at a level approaching but never exceeding 100%. The result is a great deal of dynamic range reduction, but this

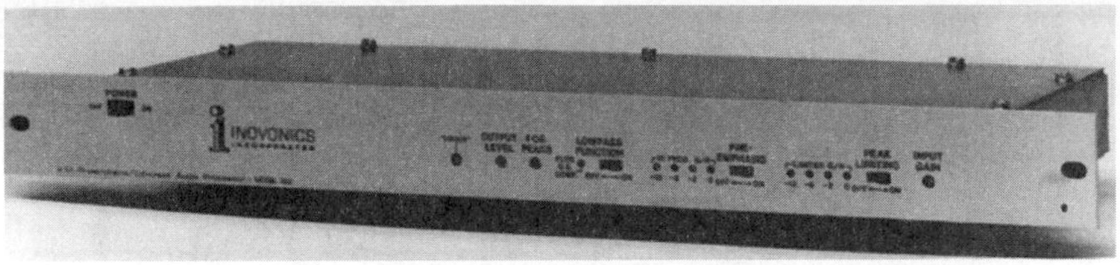

Figure 4.3
Audio processor for AM broadcasting
Courtesy Inovonics, Inc.

does not affect the intelligibility of speech. Broadcast transmitters, on the other hand, are often used with quite elaborate processors that apply varying amounts of compression to different frequency ranges. See Figure 4.3 for a photograph of one of these devices. They are set up very carefully to achieve the best compromise between natural-sounding music and good signal-to-noise ratio (which also means increased transmitter range).

Other kinds of distortion, such as harmonic and intermodulation distortion, also have to be kept within reasonable limits. As would be expected, low distortion levels are more important in the broadcast service than in the mobile-radio service.

4.3 Typical Transmitter Topologies

AM transmitters can take several different forms, as shown in the block diagrams in Figure 4.4. In Figure 4.4(a), an oscillator is modulated and connected directly to an antenna. This is not a practical circuit for several reasons. Oscillators are susceptible to frequency changes due to variation in any of the operating conditions of the active device used. Amplitude modulation represents such a change and, when applied to an oscillator, is likely to result in frequency modulation as well. In addition, changes in load impedance (such as would be created by ice on the antenna, for example) will also result in frequency changes. Finally, oscillators should run at very low power levels to reduce heating effects that can also cause frequency changes. Even with crystal oscillators, low power levels are necessary to avoid damaging the crystal. In spite of all these disadvantages, "transmitters" with this topology are built, but only as toys. The "wireless microphones" that allow children to talk over an AM radio are sometimes this simple.

A more practical transmitter is shown in Figure 4.4(b). A Class C amplifier stage has been added to provide isolation and power gain. The amplifier, rather than the oscillator, is modulated.

Most practical transmitters use more than one stage between oscillator and antenna, as in Figure 4.4(c), where two extra stages are shown. The buffer isolates the oscillator from any load changes caused by modulation of the power amplifier, and the **driver** supplies the power needed at the input to the power amplifier.

If an AM transmitter is to operate in the HF or VHF range or higher, one or more frequency multipliers will probably be used between the oscillator and the driver, as shown in Figure 4.4(d). Operation at a relatively low frequency allows for a more stable oscillator design.

All of the designs shown so far have the RF power stage modulated. This allows all of the RF amplifying stages to operate Class C for maximum efficiency. However, it is possible to apply the modulation to an earlier stage, as shown in Figure 4.4(e). The drawback is that all amplifier stages following the modulator must be linear (Class A or AB). This is so wasteful of power that this topology is seen only in very low power transmitters or in those designed mainly for another modulation scheme. Many transmitters that are designed for single-sideband suppressed-carrier operation fit this description. Single-sideband transmitters will be studied in Chapter 6.

In the balance of this chapter, it will be assumed that the modulation is applied to the RF power amplifier stage. Sometimes the driver is modulated as well, to reduce distortion.

Figure 4.4
AM transmitter topologies

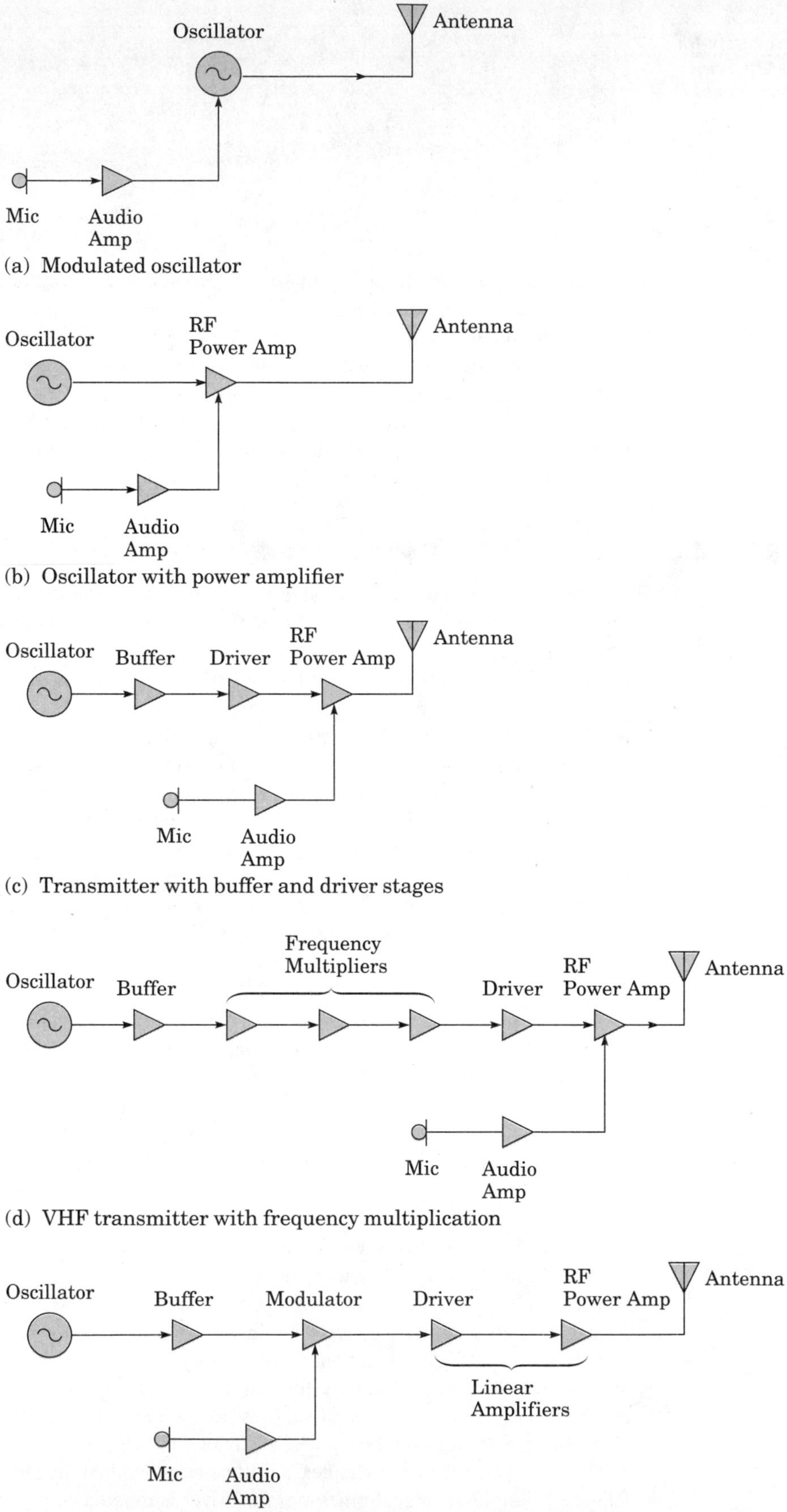

(a) Modulated oscillator

(b) Oscillator with power amplifier

(c) Transmitter with buffer and driver stages

(d) VHF transmitter with frequency multiplication

(e) Transmitter with low-level modulation and linear amplifier

4.3.1 Low-Level and High-Level Modulation

By definition, **high-level modulation** is accomplished in the output circuit of the RF power amplifier stage. Usually this output circuit is the collector of a transistor or the anode of a tube. Anything else is **low-level modulation** and is much less efficient. On the other hand, low-level modulation requires a great deal less power from the modulating amplifier. The configuration in Figure 4.4(e), in which the modulated stage is followed by a linear amplifier, is an example of low-level modulation, but so is a transmitter in which the modulation is accomplished in the base or grid circuit of the power amplifier.

4.4 Transmitter Stages

All of the stages in a transmitter (except the power amplifier and possibly the driver) operate at low power levels and are quite similar to the circuits examined in Chapter 2. Often this part of the transmitter, exclusive of the power-handling stages, is called the **exciter**.

4.4.1 The Oscillator Stage

Good stability in modern transmitter design generally requires a crystal-controlled oscillator. Where variable-frequency operation is required the usual practice is to use a frequency synthesizer locked to a crystal-controlled master oscillator.

4.4.2 The Buffer and Multiplier Stages

Whatever type of oscillator or synthesizer is used to generate the operating frequency, it must be isolated from any changes in load impedance in order to maintain good stability. The **buffer** stage accomplishes this. In modern designs it is likely to be a wideband amplifier to minimize adjustments when changing frequency. The buffer stage operates at low power, where efficiency is less important than spectral purity, so it is likely to operate Class A.

Multiplier stages, on the other hand, are required to generate harmonics as part of their design. They will operate Class C to ensure this, with tuned output circuits designed to pass the desired frequency and attenuate **spurious signals**.

4.4.3 The Driver Stage

Depending on the output power of the transmitter, its power-amplifier stage may require considerable power at its input. For instance, if the power amplifier of a 10 kW transmitter has a power gain of 20 dB, it will need an input power of 100 W. This calls for a power amplifier, probably operating Class C, to drive the final stage. This stage may be referred to as the **intermediate power amplifier** (IPA).

4.4.4 The Power Amplifier/Modulator

In small transmitters, a single transistor operating in Class C is likely to be used in the final amplifier, with collector modulation accomplished by means of a push-pull Class AB audio amplifier. For powers much above 100 W, it is necessary to use more than one output transistor. Simply putting transistors in parallel will not do, since the internal capacitances will add to unacceptable levels. Accordingly, it is necessary to use several complete amplifiers whose output power is then combined before being delivered to the antenna. An alternative that is still much used for high-power transmitters is to use vacuum tubes in the power amplifier, modulating amplifier, and sometimes the driver stage as well.

Figure 4.5 shows a simplified circuit for the modulated power amplifier of a low-powered AM transmitter. It is based on the Class C amplifier of Chapter 2. The essential difference is that the modulating signal is used to vary the collector supply voltage. The most straightforward—though certainly not the only—way of doing this is to use a transformer. Using a transformer also provides a means of transforming the audio amplifier's optimum load impedance to the impedance of the modulated atage. Ideally, for 100% modulation, the voltage at point A, the transistor side of the modulation transformer secondary, will vary between zero and twice the supply voltage, V_{CC}.

The power supply voltage does not change with modulation (assuming a properly designed supply), and neither does the average collector current. In fact, a milliammeter in the collector supply lead should not show any change in reading, since the variations with modulation are too rapid for the meter to follow. From this it appears that the power provided directly by the power supply to the final amplifier does not change with modulation.

Figure 4.5
Modulated Class C amplifier

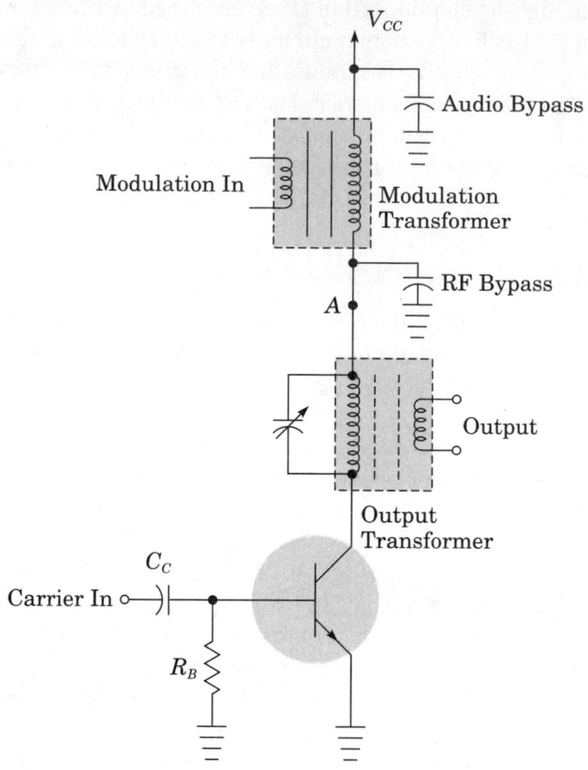

Basic AM theory tells us that modulation will increase the power output of the amplifier by 50% for 100% modulation. Therefore, the power input will have to increase by the same percentage. Where does the extra power come from, if not directly from the supply? There is only one other possible source, and that is the modulating amplifier, in this case assumed to be an audio amplifier. It is apparent that this amplifier will have to provide 50% of the dc carrier power input to the final amplifier. That is,

$$P_a = 0.5P_i \tag{4.1}$$

where

P_a = power required from the audio amplifier
P_i = dc power input to the final amplifier

Example 4.2

A transmitter has a carrier power output of 10 W at an efficiency of 70%. How much power must be supplied by the modulating amplifier for 100% modulation?

Solution The dc power input to the amplifier can be found from

$$\eta = \frac{P_o}{P_i}$$

$$P_i = \frac{P_o}{\eta} = \frac{10}{0.7} = 14.3 \text{ W}$$

The audio power required is

$$P_a = 0.5P_i$$
$$= 7.14 \text{ W}$$

The impedance looking from the transformer secondary into the modulated stage is easy to calculate. For 100% modulation, we know from Equation (4.1) that

$$P_a = 0.5P_i$$

P_i is just the dc power from the supply, so

$$P_i = V_{CC}I_c \tag{4.2}$$

where

V_{CC} = power supply voltage
I_c = average collector current

Therefore,

$$P_a = 0.5V_{CC}I_c \tag{4.3}$$

For 100% modulation, we also know that the voltage from the modulating amplifier, when added to the supply voltage, must be capable of reducing it to zero on one peak and doubling it on the other. In other words, at the transformer secondary

$$V_{a(pk)} = V_{CC} \tag{4.4}$$

where

V_a = audio voltage measured across the modulation transformer secondary

Assuming the modulating signal is a sine wave,

$$V_{a(RMS)} = \frac{V_{a(pk)}}{\sqrt{2}} = \frac{V_{CC}}{\sqrt{2}} \tag{4.5}$$

Now, letting Z_a be the impedance seen looking into the power amplifier from the modulation transformer secondary and using the power equation for RMS voltage and impedance:

$$P = \frac{V^2}{Z}$$

which can be rearranged to get

$$Z = \frac{V^2}{P} \tag{4.6}$$

$$Z_a = \frac{V^2_{a(RMS)}}{P_a} = \frac{\left(\frac{V_{CC}}{\sqrt{2}}\right)^2}{0.5V_{CC}I_c} = \frac{0.5V^2_{CC}}{0.5V_{CC}I_c} = \frac{V_{CC}}{I_c}$$

Example 4.3 A transmitter operates from a 12 V supply, with a collector current of 2 A. The modulation transformer has a turns ratio of 4:1. What is the load impedance seen by the audio amplifier?

Solution From Equation (4.6), the impedance at the transformer secondary is

$$\begin{aligned} Z_a &= \frac{V_{CC}}{I_c} \\ &= \frac{12\text{ V}}{2\text{ A}} \\ &= 6\ \Omega \end{aligned}$$

This impedance is multiplied by the square of the transformer turns ratio. Letting the impedance looking into the primary be Z_p, we have:

$$\begin{aligned} Z_p &= Z_a\left(\frac{N_1}{N_2}\right)^2 \\ &= (6\ \Omega)(4^2) \\ &= 96\ \Omega \end{aligned}$$

As noted previously, the collector voltage in an unmodulated Class C amplifier can vary between almost zero (the "almost" is due to the fact that there is a small voltage, typically less than 1 V, across the transistor when it is saturated) to twice the supply voltage. With 100% modulation, the supply voltage can double, so the collector voltage now reaches a maximum of

$$V_{c(max)} = 4V_{CC} \tag{4.7}$$

This fact must be borne in mind when choosing transistors and other components, such as capacitors, for the output circuit. For instance, for the amplifier in Example 4.2, the collector voltage can reach 48 V, even though the circuit operates from a 12 V supply.

Another factor to consider is power dissipation. Though the theoretical maximum efficiency of a Class C amplifier is 100%, the actual efficiency is unlikely to be much greater than 75%. Most of the remaining power is dissipated in the transistor, though some will be converted to heat in the output circuit and other components. Though the efficiency is likely to remain about the same with modulation, the power dissipation goes up.

Example 4.4 A collector-modulated Class C amplifier has a carrier output power P_c of 100 W and an efficiency of 70%. Calculate the input power and the transistor power dissipation with 100% modulation.

Solution Assume all the power dissipation occurs in the transistor. This will give a conservative answer that provides some safety margin. The output power with 100% modulation is

$$\begin{aligned} P_o &= 1.5P_c \\ &= 1.5 \times 100\text{ W} \\ &= 150\text{ W} \end{aligned}$$

The input power is

$$\begin{aligned} P_i &= \frac{P_o}{\eta} \\ &= \frac{150\text{ W}}{0.7} \\ &= 214\text{ W} \end{aligned}$$

The power dissipated P_D is the difference between the input and the output power.

$$\begin{aligned} P_D &= P_i - P_o \\ &= 214 \text{ W} - 150 \text{ W} \\ &= 64 \text{ W} \end{aligned}$$

By the way, it was stated above that the supply voltage should not change with modulation. Sometimes it does: since the supply is required to supply up to 50% more power with modulation than without, a poorly regulated supply may provide a lower voltage when the transmitter is fully modulated. If this is the case, the carrier will have a lower amplitude with modulation than without. The phenomenon is called **carrier shift**, and it is undesirable.

4.4.5 Audio Circuitry

Audio circuitry is required to amplify the very small signal (on the order of 1 mV) from a microphone to a sufficient level to modulate the transmitter. For a large transmitter of 50 kW or more, that can be a considerable amount of power.

In a communication transmitter, the audio circuitry will be straightforward. There will probably be some form of automatic level control to prevent overmodulation while keeping the modulation as close to 100% as possible. There will likely be some frequency-response shaping as well, to emphasize those frequencies in the voice range, from about 300 Hz to 3 kHz, that are most important for intelligibility. Lower frequencies contribute little to intelligibility and waste energy if transmitted, while higher frequencies waste bandwidth.

Most of the audio circuitry involved in broadcast radio is actually not in the transmitter. Studio equipment amplifies the tiny signals from microphones, phonograph cartridges, and tape heads, *mixes* them (electrically, the process is really *summation*), and applies any required equalization. The transmitter must be located in an area where there is a considerable amount of land for the antenna system; the studio may be located in the same building, but it is often at a more central location. In that case, the audio signal is sent to the transmitter over a leased telephone line or a microwave radio link. The transmitter itself receives a *line level* signal, often either +4 or +8 dBm into 600 Ω, and amplifies that to the very considerable power levels needed by the modulator. The final audio stage is undoubtedly push-pull, operating Class B, and very often uses vacuum tubes.

4.5 Output Impedance Matching

Most practical transmitters are designed to operate into a 50 Ω resistive load to match the characteristic impedance of the coaxial cable that is generally used to carry the transmitter power to the antenna. The antenna itself may not have this impedance, but in that case there will be matching circuitry at the antenna (or as close to it as practical). Often the match will not be exact, especially if the system must operate over a range of frequencies, and many transmitters have some latitude for adjustment.

The transmitter output circuitry must be designed to transform the standard load resistance at the output terminal to whatever is required by the active device or devices. Here, the requirements vary dramatically, depending on the type of amplifier. Transistors usually require a load impedance less than 50 Ω, while tubes are suited for much higher impedances.

The simplest way to find the appropriate required load impedance for a Class C amplifier is to note that for an unmodulated carrier, the collector voltage varies between approximately zero and $2V_{CC}$. The peak value of the output voltage is then

$$V_{o(pk)} = V_{CC}$$

$$V_{o(RMS)} = \frac{V_{CC}}{\sqrt{2}}$$

Assuming that the amplifier is being designed for a specific power output at a given supply voltage, we can use the usual power equation for RMS voltage and a resistive load:

$$P = \frac{V^2}{R}$$

Since we know P and V and require R, we rearrange this to

$$R = \frac{V^2}{P}$$

Now, substitute the output voltage and power into this equation to get

$$R_L = \frac{\left(\frac{V_{CC}}{\sqrt{2}}\right)^2}{P_c} = \frac{V_{CC}^2}{2P_c} \tag{4.8}$$

where

R_L = load resistance seen at the collector (or plate, for a tube)
P_c = carrier power output, without modulation
V_{CC} = collector (or plate) supply voltage

Example 4.5 An AM transmitter is required to produce 10 W of carrier power when operating from a 15 V supply. What is the required load impedance as seen from the collector?

Solution From Equation (4.8),

$$R_L = \frac{V_{CC}^2}{2P_c} = \frac{15^2}{2 \times 10} = 11.25\ \Omega$$

For many years, the *pi network*, which is shown in Figure 4.6(a), has been a popular output circuit. Its name stems from the resemblance of its schematic diagram to the shape of the Greek letter π. The pi network can be used to transform impedances either up or down, but it is best suited to active devices that require a fairly high load impedance, such as tubes. With devices that require a very low load impedance, such as bipolar transistors with more than a few watts of output, the design can be accomplished in theory, but some of the required component values are likely to be too small to be practicable.

High-power transistor amplifiers, with their requirement for small values of load impedance, can use a variety of narrowband coupling circuits. One common network is shown in Figure 4.6(b). For obvious reasons, it is called a *T network*. There are equations and charts available to aid in the design of these circuits, but it is easier to use one of the many computer programs available. One handy one is called APPCAD™; it is available free from Hewlett-Packard.

To be practical, both of these circuits need an extra capacitor (labeled C_c in the figures) to keep the dc voltage on the collector or plate away from the antenna. Some antenna systems represent a short circuit for dc. An RF choke is used to allow power to reach the output stage.

Besides providing impedance transformation, the pi and T networks also act as low-pass filters, aiding in the reduction of harmonic levels. They are very often followed by additional low-pass filtering to further attenuate harmonics and other spurious signals.

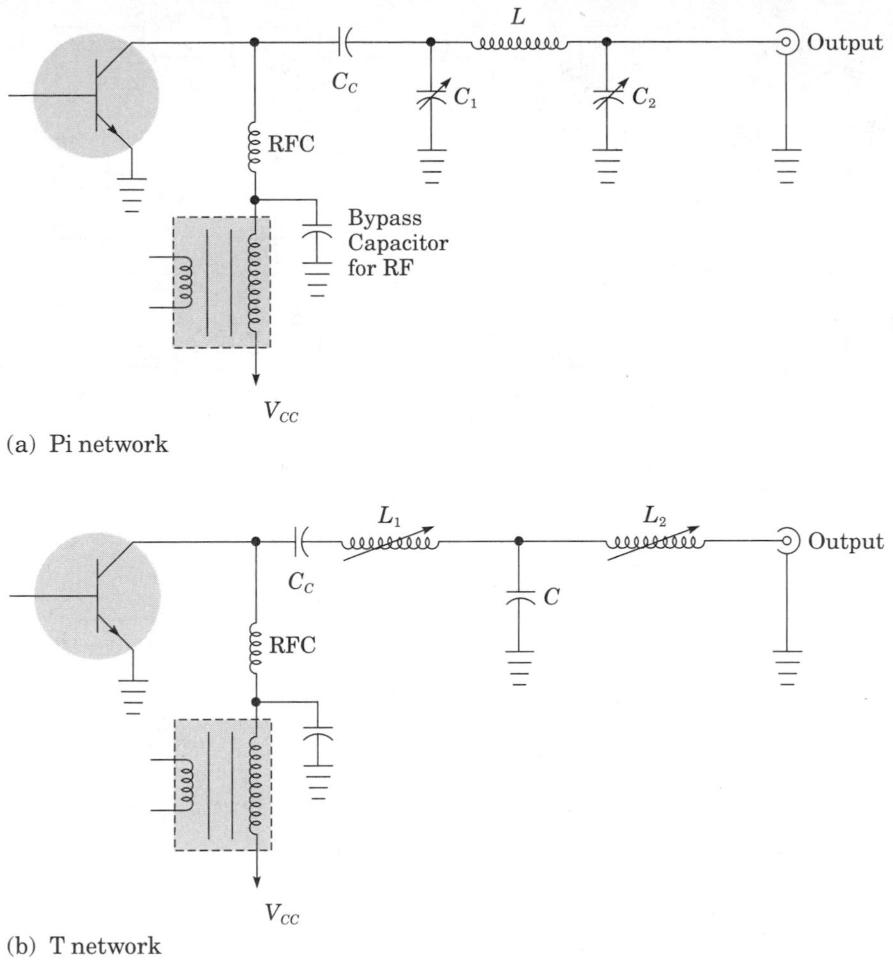

Figure 4.6 Narrowband output circuits

Any transmitter should operate into an impedance close to its rated load at all times. Severe mismatches, such as an open or short circuit at the load, can cause power-amplifier voltages or currents to be much higher than normal. Destruction of output transistors or other components is quite likely under these circumstances. Many solid-state transmitters have circuits to sense these conditions and automatically protect the power amplifier, either by reducing power or by shutting it off completely. However, not all transmitters are so equipped, and in any case it is poor practice to tempt fate. Whether on the bench or in the field, a transmitter should always operate into a matched load. When conducting tests, one should use a noninductive resistor of the correct value (usually 50 Ω) that is capable of dissipating the transmitter's rated power. Such a resistor is called a **dummy load**.

4.6 An AM Citizens' Band Transmitter

A CB radio transmitter is always found as part of a transceiver, that is, a combination receiver and transmitter. Compared with a separate transmitter and receiver, transceivers are compact, convenient to install and operate, and economical due to the use of some of the same components for both the transmit and receive functions.

Figure 4.7 is the complete block diagram of a typical CB transceiver using amplitude modulation. This transceiver is also shown in the photograph in Figure 4.1. Figure 4.8 shows only those stages that are used in transmitting. Some of these, such as the oscillator and the audio amplifier, are also used for receiving. (The operation of the receiver section will be described in Chapter 5.)

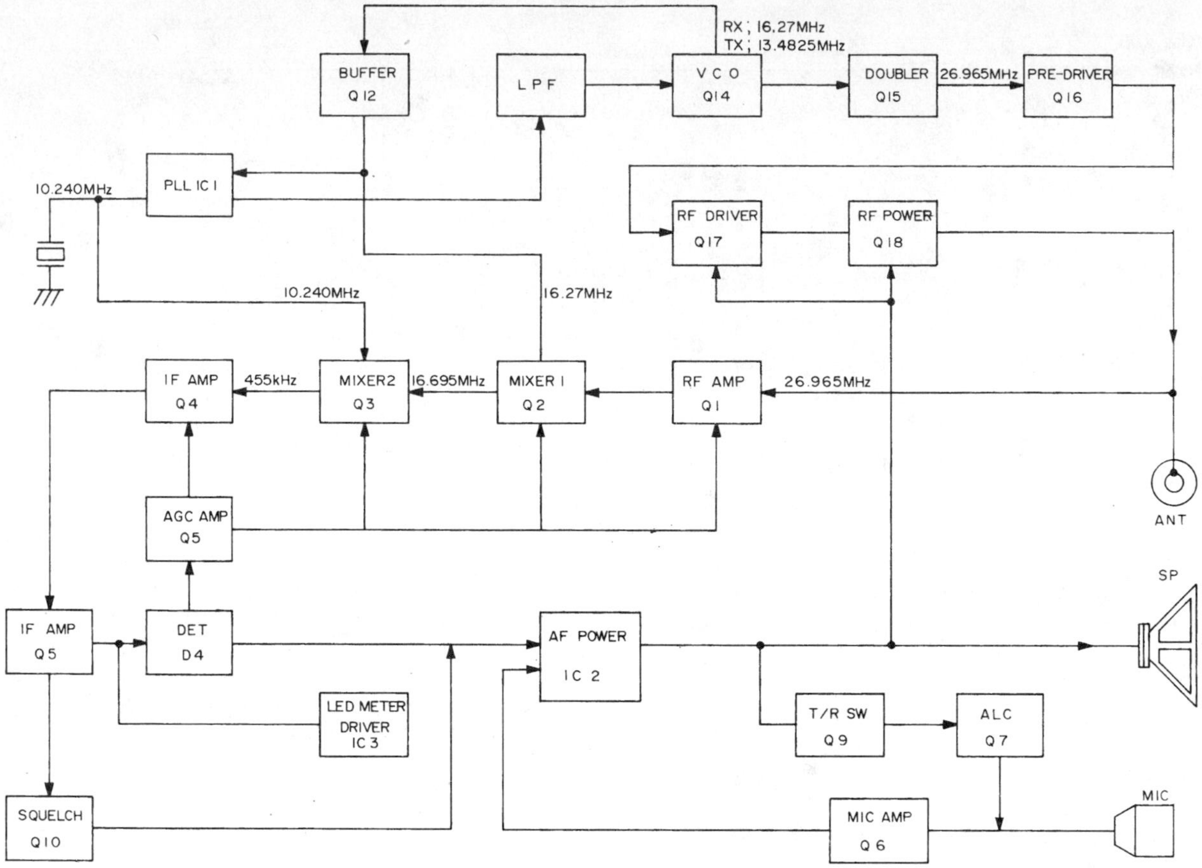

Figure 4.7
Block diagram of CB transceiver
Courtesy of Radio Shack, a division of Tandy Corporation.

The oscillator is actually a frequency synthesizer, to maintain crystal-controlled frequency accuracy and stability. It operates at half the output frequency, so the next stage is a doubler. The doubler is followed by two more gain stages before the RF power amplifier.

The audio circuitry consists of a microphone preamplifier followed by an integrated-circuit amplifier, which also provides the power to the loudspeaker when the transceiver is in the receive mode. The automatic level control is designed to keep the modulation index as high as possible without permitting overmodulation. Thus there is no need for the operator to use an audio gain control while transmitting or to monitor the modulation percentage.

The complete circuit of the transceiver is shown in Figure 4.9. If it seems a bit intimidating at first, here are a few hints to put some order into it. First, locate the antenna

Figure 4.8
Transmitter section of CB transceiver

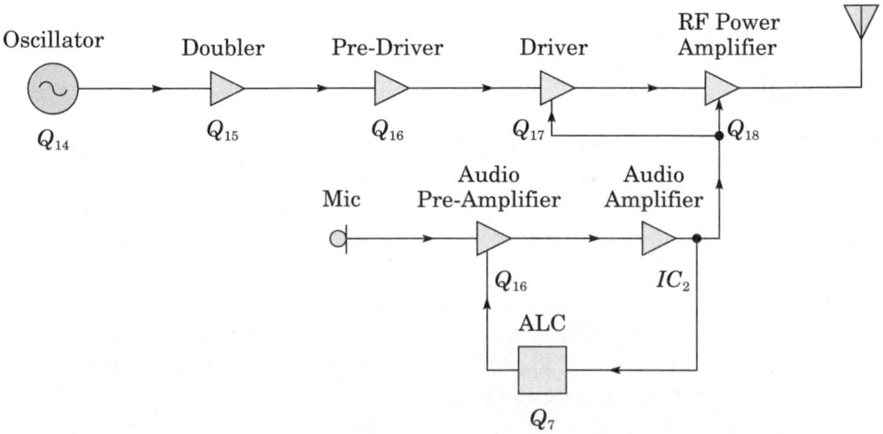

Figure 4.9
Schematic diagram of CB transceiver
Courtesy of Radio Shack, a division of Tandy Corporation.

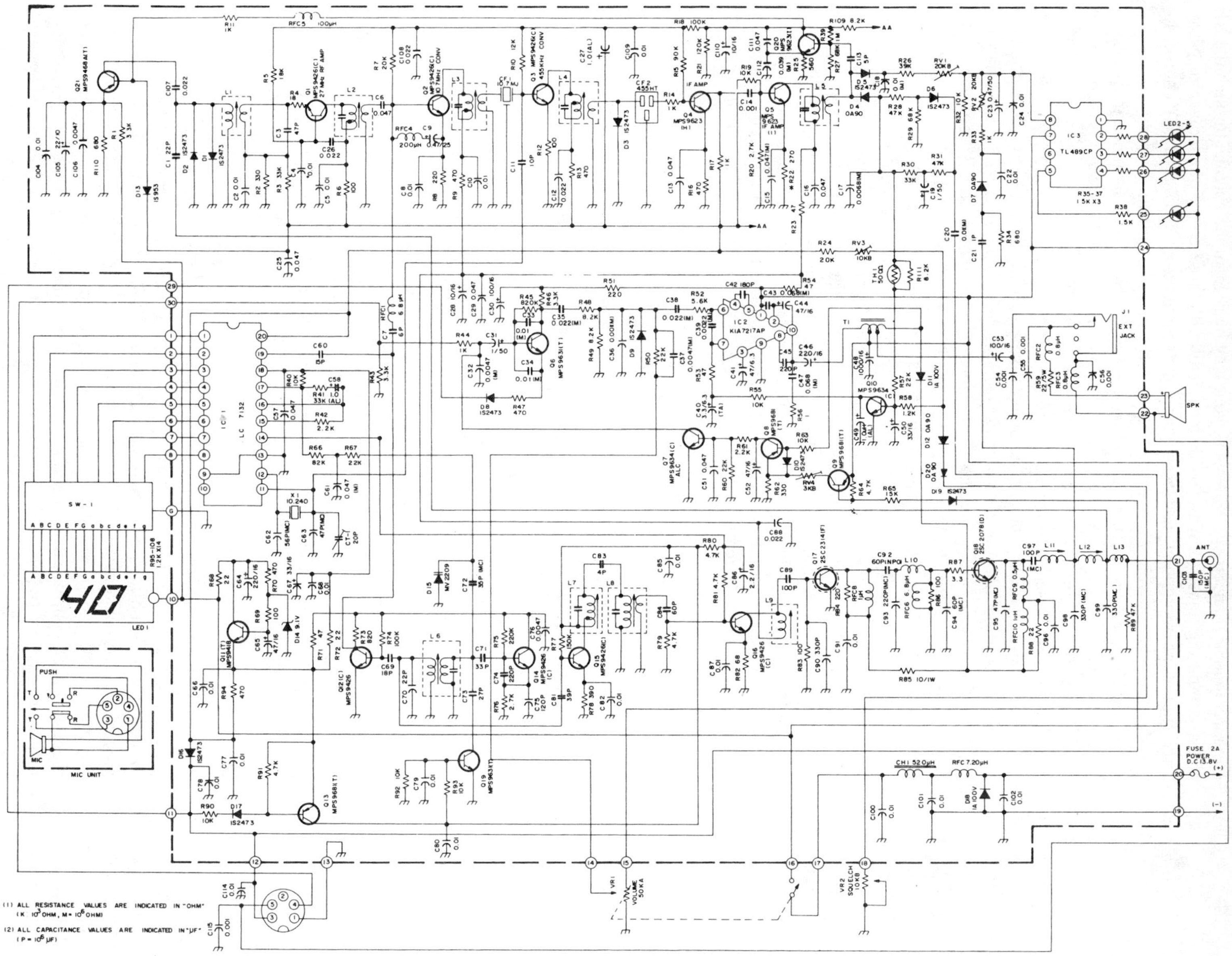

NOTES (1) ALL RESISTANCE VALUES ARE INDICATED IN "OHM" (K = 10^3 OHM, M = 10^6 OHM)
(2) ALL CAPACITANCE VALUES ARE INDICATED IN "μF" (P = 10^{-6} μF)

and microphone connections. Now work your way back from the antenna through the RF chain, comparing transistor numbers on the circuit with those on the block diagram. Similarly, start at the microphone and work your way through the audio circuitry. Of course, the fact that this is a transceiver makes the circuit a little more complicated.

Some of the circuit details should be familiar by now. Note the tuned transformer-coupled output on Q_{15}, the frequency doubler. Also note that the collector is tapped down on the primary winding so as not to reduce the Q of the tuned circuit too much. (This technique was discussed in Chapter 2.)

The output circuit for the final amplifier is very similar to the T network described in the previous section, but it has an extra LC section. Although this is essentially a narrow-band amplifier chain, the bandwidth is sufficient to cover the entire CB frequency range from 26.965 to 27.405 MHz when the transmitter is tuned for maximum output on Channel 18 (27.175 MHz), near the center of the band.

The audio circuitry is also interesting. The output of IC_2 modulates both the driver and the power amplifier via T_1, which is an autotransformer rather than the conventional transformer previously discussed. Modulation of the driver as well as the power amplifier is very common with transistor modulators, because otherwise the saturation voltage of the power-amplifier transistor prevents 100% modulation from being achieved. A sample of the output signal from the modulation transformer goes to the ALC circuit, which adjusts the gain of the microphone preamplifier Q_6 to keep the audio output relatively constant.

4.7 Modern AM Transmitter Design

AM transmitters have been built since the invention of the vacuum tube, and in some ways, their design has changed little since then. Recently, however, some fresh approaches have been tried. Of course, the best approach technically might be to abandon full-carrier AM altogether in favor of some of the more efficient systems to be described shortly. For historical and economic reasons, however, that is not likely to happen in the near future.

High-power AM transmitters are necessarily large and expensive because of the amount of power to be handled by the power amplifier and modulating amplifier, the high-voltage power supplies required for the tubes, and the need for a modulation transformer that is capable of handling very large amounts of audio-frequency power. Recent efforts at improving AM transmitters have involved the development of high-power solid-state power amplifiers and the use of pulse-duration modulation and switching amplifiers in the modulation process. A representative example of each technique will be presented here.

4.7.1 Solid-State Radio-Frequency Power Amplifiers

Transistors have been slow to take over in transmitter power amplifiers because of the greater simplicity of using one or two large tubes instead of many separate transistor amplifiers and an elaborate power combiner. The use of solid-state components can provide greater reliability and higher efficiency, however, as well as reducing the physical size of the transmitter by about 50%. Using many output modules allows operation at only slightly reduced power in the event of the failure of an output transistor and also allows instantaneous and efficient switching of power levels. This is often required of broadcast transmitters, since many station licenses require reduced power at night to avoid interference. Signals in the standard AM broadcast band tend to travel farther at night than in the daytime (see Chapter 17 for an explanation of this phenomenon).

Figure 4.10 shows the block diagram of a typical solid-state transmitter. Both the RF and audio power amplifiers are made up of a number of solid-state modules, typically six to twelve. The outputs of the RF modules must be combined before being applied to the antenna.

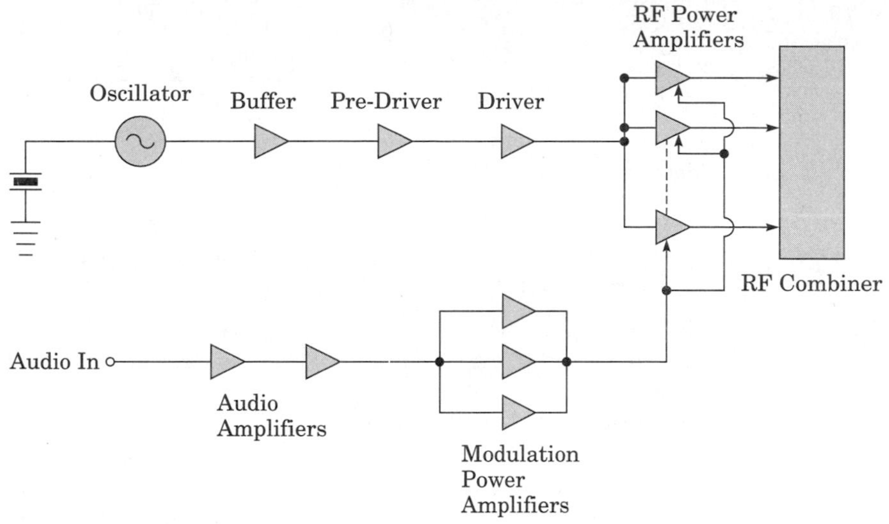

Figure 4.10
High-power solid-state AM transmitter

4.7.2 Pulse Duration-Modulators

The traditional Class B (or AB) modulator described earlier becomes very expensive when large power levels are involved. Class C, while more efficient, cannot be used for audio because of the extreme amount of distortion that would be generated. However, there is another way to amplify audio linearly and yet efficiently. Sometimes it is called Class D or *switching* amplification, but it is essentially pulse-duration modulation.

Suppose that the audio signal in Figure 4.11(a) is sampled at frequent intervals. (In theory, it must be sampled at least twice per cycle, but in practice the rate must be somewhat higher.) The result will be the discrete analog samples shown in Figure 4.11(b). Then suppose that an electronic switch is turned on for a length of time that is proportional to the amplitude of the sample. The result is *pulse-duration modulation* (PDM), or *pulse-width modulation*, which is shown in Figure 4.11(c). The pulses are produced by switching, so the theoretical efficiency so far can be 100%. (Since an ideal switch has zero current

Figure 4.11
Pulse-duration modulation

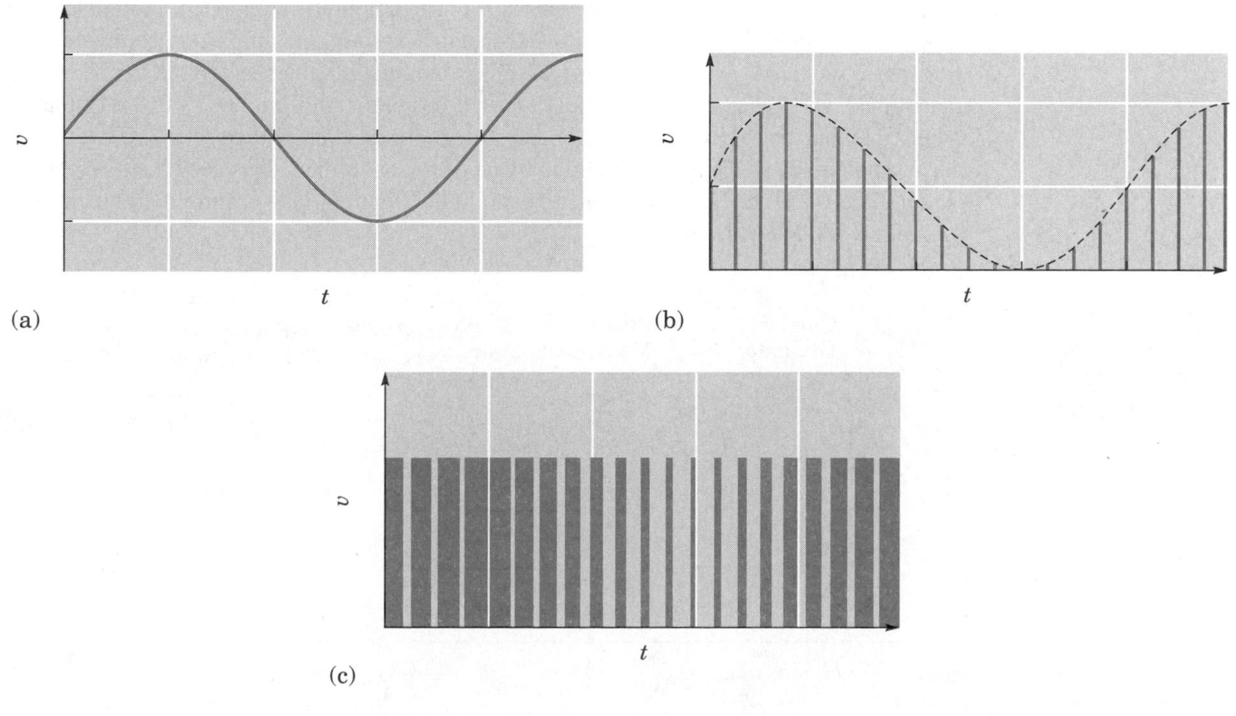

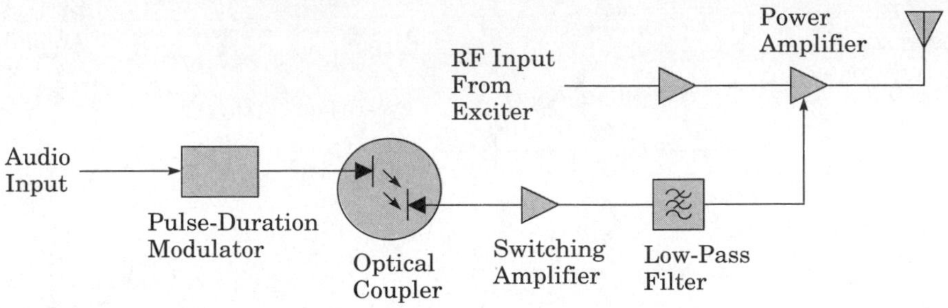

Figure 4.12 AM transmitter using pulse-duration modulation

through it when it is open and zero voltage across it when it is closed, it never dissipates any power.)

This PDM signal can be demodulated by simply low-pass filtering it. That process could again be 100% efficient in theory, though not, of course, in practice. Still, the efficiency could be considerably greater than that of a Class B amplifier.

This configuration has been used occasionally in the relatively low-powered audio amplifiers found in home and commercial sound systems, but it is relatively complicated and has not become very popular. In large transmitters, however, the gain in efficiency is worth almost any amount of complication. Figure 4.12 shows one possible implementation, in which a solid-state PDM modulator is followed by a tube-type switching amplifier that is used to modulate a vacuum-tube RF power amplifier. Note the optical coupler, which isolates the solid-state circuitry from the high voltages in the vacuum-tube part of the transmitter.

4.7.3 Transmitters for AM Stereo

When AM stereo was introduced, there were several incompatible systems, and adoption was slow. Now that the Motorola C-Quam™ system has become the standard, there is likely to be more interest in the mode.

A major concern of broadcasters who are considering improvements must always be cost. In particular, replacing a transmitter, especially the high-power portion, represents a major capital investment. Luckily, AM stereo can be implemented without replacing the transmitter power amplifier and modulator. This is done by separating the quadrature AM signal into a phase and an amplitude component. The amplitude component modulates the signal in the usual way; in fact, with the C-Quam™ system, the amplitude component is simply the monaural left-plus-right signal. Phase modulation must be applied to one of the transmitter's early stages. Phase modulation will be studied in Chapter 7, but we can note at this time that since phase modulation has no effect on the amplitude of the RF signal, a phase-modulated signal can be amplified by a Class C amplifier. Thus only the low-power exciter circuitry of the transmitter must be modified. Figure 4.13 is a simplified block diagram of an AM stereo transmitter.

Figure 4.13 Simplified block diagram of AM stereo transmitter

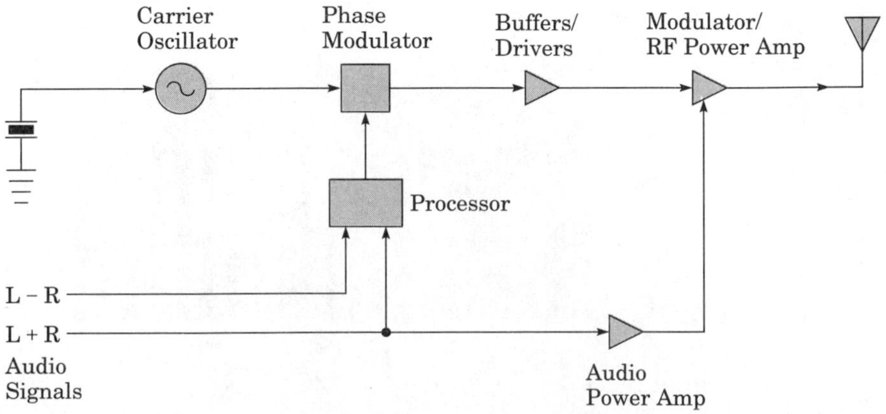

4.8 AM Transmitter Measurements and Troubleshooting

Many of the measurements that must be made in transmitter circuitry are quite ordinary, involving the measurement of dc or low-frequency ac quantities. No special techniques are required, but a few precautions are in order.

First, there are safety requirements. Technologists who are used to solid-state circuits may be intimidated by the voltage levels (in the kilovolts) found in tube-type transmitters. They should be. These voltages often come from supplies capable of providing several amperes of current, and mishaps are very likely to be fatal. High-voltage compartments are interlocked, so that the power is removed when they are opened, and the temptation to defeat the interlocks must be resisted. The beginner should follow the manual exactly and should work with a more experienced person.

There is also the possibility of RF burns. The output from a multikilowatt transmitter is very obviously dangerous, but even a small transmitter is capable of inflicting painful burns if the antenna leads are touched during operation. The author found this out the hard way many years ago with a 30 W transmitter! Any operating transmitter must be treated with respect and care.

Speaking of RF, strong radio-frequency fields can cause problems when seemingly unrelated measurements are made. All electronic test equipment must be properly grounded and shielded to avoid interference.

Transmitters should be tested off-the-air whenever possible to avoid creating interference. Suitable loads for such testing are described in the next section.

4.8.1 Measurement of Radio-Frequency Power

There are many situations in which the power in an RF circuit must be measured. Conventional wattmeters, such as those used at power-line frequencies, are useless because of the reactance of their coils. In this section, we will look at a few of the many ways in which power can be measured at high frequencies.

It is possible to measure power by measuring the voltage across a known resistance, and this can certainly be done using an oscilloscope or a specially designed high-frequency ac rectifier-type voltmeter. The problem here is to be sure that the resistance is just that: resistive and free from inductive or capacitive reactance. Ordinary carbon resistors can be used for low power at frequencies into the VHF range, but wire-wound resistors have far too much inductance. Noninductive resistors can be made for RF use; an example is shown in the photograph of Figure 4.14. Resistors of this sort are often called *dummy loads*, since they can be connected to a transmitter instead of an antenna for testing.

One way of measuring true power is to measure its heating effect. This has the advantages of ignoring any reactive power and not requiring any connections of test equipment to high-frequency or high-voltage points. It is a common technique to measure the power of large transmitters operating into dummy loads. Perhaps surprisingly, a variation

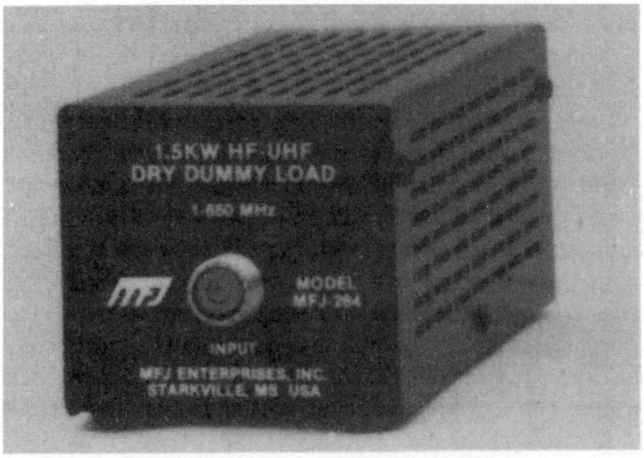

Figure 4.14 Transmitter dummy load

MFJ Enterprises, Inc., P. O. Box 494, Mississippi State, MS 39762.

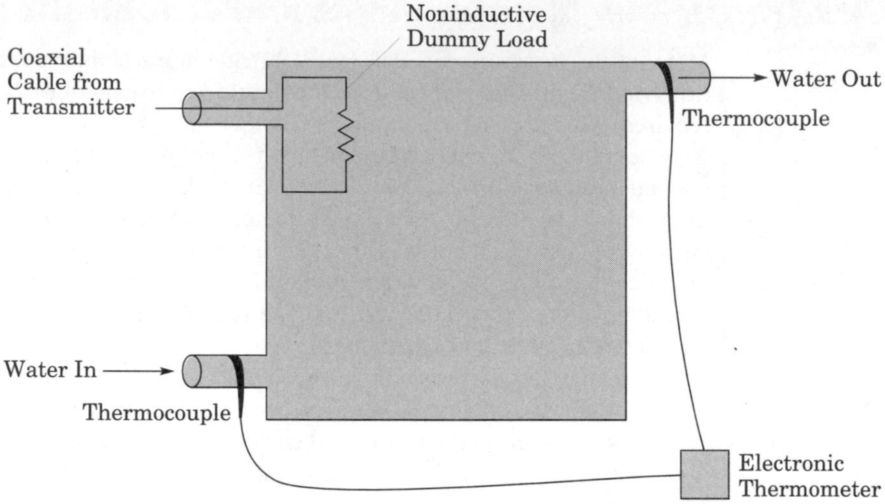

Figure 4.15 Calorimeter technique for RF power measurement

of the same idea can be used to measure quite small amounts (milliwatts) of power at microwave frequencies.

Figure 4.15 illustrates one way to measure power using a **calorimeter wattmeter**. The dummy load heats the circulating water. If the rate of flow is known and controlled, the amount of heat given to the water can be calculated from the rise in temperature between input and output. Notice that the only actual measurements involved are temperature measurements.

When the transmitter load is a real antenna, power measurement becomes more complicated. The antenna may look like something approaching a pure resistance, but this is certainly not always the case. Antenna impedance will be discussed in detail later in the book; for now, let us just remark that it varies with frequency and may very well have a reactive component that can be inductive at one frequency and capacitive at another. Furthermore, if the load is not exactly matched to the transmission line that feeds it, some power will be reflected from the load and measurements of voltage or current at the transmitter end of the line can be very deceptive. This effect is due to *standing waves*, which will be described in the discussion of transmission lines in Chapter 16. Broadcast stations use an RF ammeter at the antenna; it gives an accurate indication of carrier power provided that the resistive component of the antenna impedance is known.

Some meters have **directional couplers** that allow them to distinguish between power flowing from the transmitter toward the antenna and reflected power from the antenna. Subtracting the two readings gives the actual power delivered to the antenna. Figure 4.16 is a photograph of one such meter. The plug-in elements allow for the measurement of a wide variety of power levels at different frequencies. The direction of the power flow to be measured is given by the arrow on the plug-in element and can be changed by simply rotating the element. The meter movement is a dc microammeter and can easily be removed from the box for remote mounting.

It should be noted that these meters measure carrier power and do not show the increase in power output with amplitude modulation; their internal circuits sense rectified average voltage, which does not change with modulation even though average power does.

4.8.2 Measurement of Modulation Percentage

The use of an envelope display on an oscilloscope to measure AM modulation depth was discussed in Chapter 3. That technique works quite well in a laboratory situation where the modulating signal is a sine wave with constant amplitude and frequency, but it is inconvenient in actual operation where the baseband signal is more likely to be voice or music. With a random signal, the oscilloscope's horizontal sweep will not be triggered consistently. Actual overmodulation can still be seen if the display is observed carefully, most obviously

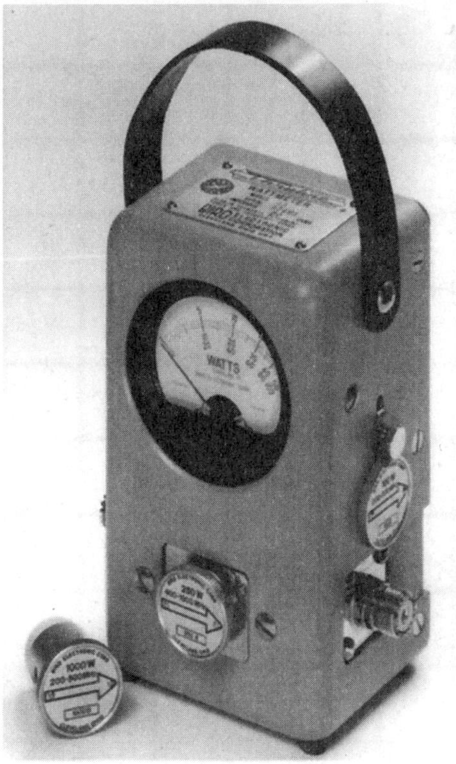

Figure 4.16
Bird Thruline™ wattmeter
Courtesy Bird Electronic Corporation.

as brief flashes of light at the x axis when overmodulation occurs in the negative direction. Another drawback to this system is that it is difficult to tell from an envelope pattern whether the modulator is distorting the signal. With a constant sine wave, deviation from the sinusoidal shape may be noticeable if it is severe, but there is little hope of seeing distortion with normal, rapidly varying program material.

The setup shown in Figure 4.17(a) solves some of these problems. This time the oscilloscope is set to the x–y mode. The vertical input samples the modulated waveform as before. The horizontal input, however, is connected to the output of the modulating amplifier; that is, it receives the baseband signal just before modulation. This creates a **trapezoidal pattern** on the screen, as shown in Figure 4.17(b). This is easier to observe than the envelope pattern, because the top and bottom edges of the trapezoid are (or should be) straight lines, regardless of the shape of the modulating waveform—both the vertical and the horizontal amplitudes change in proportion to the amplitude of the modulating signal.

Figure 4.17
Trapezoidal method of AM modulation monitoring

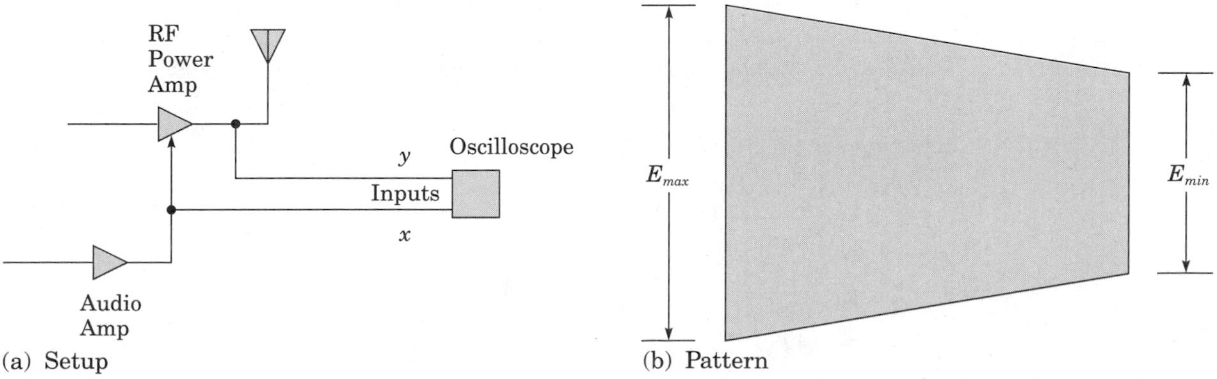

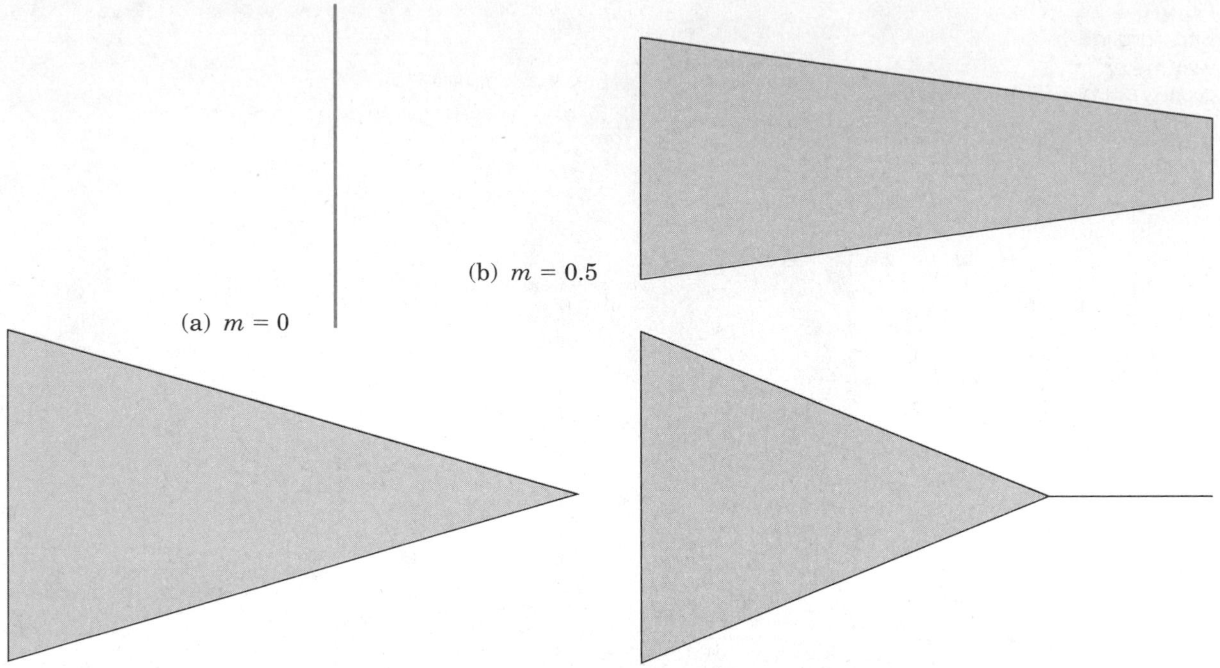

Figure 4.18
Trapezoidal patterns

Calculating the modulation index from this pattern is the same as calculating it for the envelope pattern. That is, as shown in Chapter 3,

$$m = \frac{E_{max} - E_{min}}{E_{max} + E_{min}} \tag{4.9}$$

Figure 4.18 shows the appearance of some representative values of m on the trapezoidal display. Note that for $m = 0$ the trapezoid becomes a vertical line, while for $m = 1$ it is actually a triangle.

Detecting overmodulation is easy with this setup. In the positive direction, it will simply result in the long side of the trapezoid being too long, which may not be obvious unless the oscilloscope has been properly calibrated in advance with an unmodulated carrier. In the negative direction, however, the results will be very obvious, as in Figure 4.18(d). The horizontal line extending to one side of the trapezoid (actually a triangle) indicates that there is no RF output for part of the baseband cycle. By the way, it is *negative* overmodulation that is most important, because it produces spurious signals and severe distortion of the envelope. Slight overmodulation in the positive direction does not create these problems and is, in fact, sometimes employed deliberately.

Example 4.6 Calculate the modulation index for the trapezoidal pattern shown in Figure 4.19.

Solution From Equation (4.9),

$$\begin{aligned} m &= \frac{E_{max} - E_{min}}{E_{max} + E_{min}} \\ &= \frac{10 - 4}{10 + 4} \\ &= 0.428 \end{aligned}$$

Figure 4.19

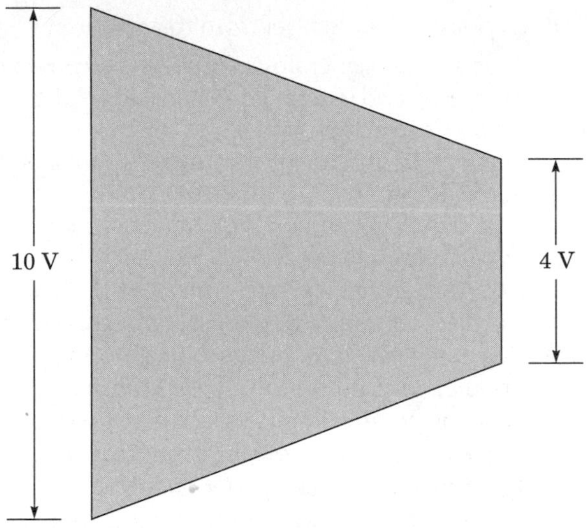

The trapezoidal pattern can indicate nonlinearity in the modulation stage. If the peak output signal level varies exactly in accordance with the instantaneous amplitude of the modulating signal, the top and bottom of the trapezoid will be straight lines, regardless of the waveshape of the modulating signal. Conversely, any deviation from straightness represents distortion in the modulated stage. Obviously it is easier to see deviations from a straight line than from a sinusoidal shape, and it is much easier than trying to notice deviations from the unknown waveform of program material. If the audio output as heard on a receiver is distorted but the trapezoidal pattern is normal, then the fault must be in the audio portion of the transmitter and not in the modulator.

While very useful, oscilloscope displays require interpretation. For continuous use, it is more practical to have a device that reads percent modulation on a meter and indicates overmodulation with a visible or audible alarm. One way to accomplish this is to demodulate the signal and measure the peak amplitude of the output. The detector will have to be calibrated, of course, and this can easily be done with an oscilloscope. After that, the meter will measure modulation directly. A comparator can be used with a calibrated reference voltage to operate an alarm whenever the modulation reaches a preset level. A highly simplified circuit for this type of modulation meter is shown in Figure 4.20.

If the calibrated detector is incorporated into a receiver, the modulation check can easily be done remotely and the audio output from the receiver can be used to get a qualitative indication of signal quality. Commercial units that accomplish this are available.

Figure 4.20
Modulation monitor (simplified)

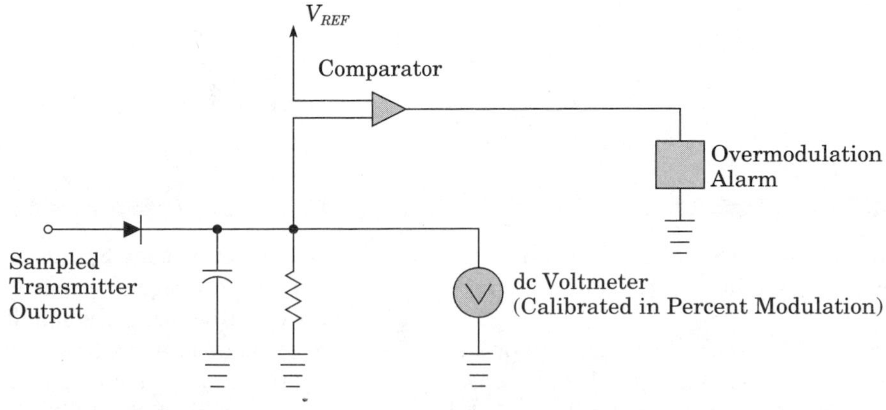

Summary

Here are the main points to remember from this chapter.

1. An AM transmitter must generate a signal at the correct carrier frequency, with the right power level, and with a modulation envelope that is an accurate reflection of the original information signal.
2. It is important in the design of any transmitter to minimize the presence of spurious signals at the output.
3. The efficiency of a transmitter is an important specification, particularly in view of the large power levels often used.
4. A typical AM transmitter has a crystal-controlled or frequency-synthesized oscillator followed by several stages of amplification. Usually the final amplifier stage, called the power amplifier, is modulated. This allows all stages to operate Class C.
5. High-level modulation is used in most AM transmitters for greater efficiency. It involves modulation at the output element of the output stage of the transmitter. Any other type of modulation is low-level modulation.
6. To achieve 100% modulation for conventional high-level modulation, the modulating amplifier must have a power output equal to one-half the dc power input to the transmitter's output stage.
7. Transmitters require an impedance-matching circuit between the output of the power amplifier and the antenna. A properly designed matching circuit can also act as a filter to prevent spurious frequencies from being radiated.
8. In an effort to improve the overall efficiency of AM transmitters, designers have used innovative techniques such as pulse-width modulation in modulator circuits.
9. Techniques for the measurement of transmitter output power include calorimeter wattmeters, directional couplers, and RF ammeters, among others.
10. Oscilloscopes can be used for modulation monitoring, but in a practical situation a calibrated detector is easier to use.
11. Though solid-state devices are used in most modern transmitter circuits, high-powered amplifier stages still make use of vacuum tubes.

Important Equations

$$P_a = 0.5P_i \tag{4.1}$$

$$P_a = 0.5V_{CC}I_c \tag{4.3}$$

$$V_{a(pk)} = V_{CC} \tag{4.4}$$

$$Z_a = \frac{V_{CC}}{I_c} \tag{4.6}$$

$$V_{c(max)} = 4V_{CC} \tag{4.7}$$

$$R_L = \frac{V_{CC}^2}{2P_c} \tag{4.8}$$

$$m = \frac{E_{max} - E_{min}}{E_{max} + E_{min}} \tag{4.9}$$

Glossary

automatic-level-control (ALC) circuit a circuit for keeping the amplitude of a signal within prescribed limits

buffer an amplifier stage used to isolate two stages. Often used to prevent changes in load impedance from affecting the frequency of an oscillator

calorimeter wattmeter device for measuring power by sensing a change in temperature

carrier shift change in amplitude of the carrier component of an AM signal with modulation

compression system that provides more gain for low-level signals than for higher-level signals

directional coupler device which allows a signal moving along a transmission line in one direction to be measured

driver amplifier which supplies the required input power for a power amplifier

dummy load a noninductive power resistor used to simulate an antenna or loudspeaker when testing a transmitter or audio power amplifier respectively

duty cycle ratio of time in use to total time for an electronic system

dynamic range ratio between largest and smallest signal present at a point in a system; usually expressed in decibels

exciter the stages of a transmitter that operate at low power levels

frequency agility ability of a transmitter to tune rapidly from one operating frequency to another

high-level modulation amplitude modulation of the output element of the output stage of a transmitter

intermediate power amplifier (IPA) the driver stage of a large transmitter

low-level modulation modulation of a transmitter at any point before the output element of the output stage

overall efficiency ratio of the power output of a device such as a transmitter to the total power required from the primary power source

spectral purity absence of spurious signals in the output of a transmitter

spurious signal any emission from a transmitter other than the carrier and the sidebands required by the modulation scheme in use (in suppressed-carrier systems the carrier is also a spurious signal)

trapezoidal pattern a means of measuring amplitude modulation using an oscilloscope

Questions

1. What is frequency agility? Under what circumstances is it desirable?
2. What spurious signals are commonly generated by transmitters?
3. Why is it necessary to suppress the emission of harmonics and other spurious signals by a transmitter?
4. What is meant by the overall efficiency of a transmitter?
5. Why is compression used with many AM transmitters?
6. Why is it poor practice to apply amplitude modulation directly to an oscillator?
7. What is the most common way of rating the power output of an AM transmitter?
8. What is meant by the duty cycle of a transmitter?
9. Is it possible to use Class C amplifiers to amplify an AM signal? Explain your answer.
10. What advantages does the use of a frequency synthesizer for the oscillator stage of a transmitter have over:
    (a) a crystal-controlled oscillator?
    (b) a VFO?
11. What advantages does high-level modulation have over low-level modulation? Does low-level modulation have any advantages?
12. It is possible to apply a modulating signal to the control grid of a tube or the base of a transistor used as the power amplifier (output) stage of a transmitter. Is this procedure considered to be high- or low-level modulation? Explain your answer.
13. Explain the function of each of the following stages in a transmitter: buffer, multiplier, and driver.
14. What is carrier shift? How is it caused?
15. Why is it undesirable to connect a large number of transistors in parallel in an RF power amplifier to achieve a greater power-dissipation capability?
16. What is meant by a "switching" audio amplifier?
17. Explain how pulse-duration modulation can be used to improve the efficiency of an AM transmitter.
18. Explain the operation of a calorimeter-type power meter.
19. What is a dummy load? How is it useful?
20. What advantages does the trapezoidal method of modulation measurement have over the envelope method?

Problems

SECTION 4.2

21. The oscillator of a CB transmitter is guaranteed accurate to $\pm 0.005\%$. What are the maximum and minimum frequencies at which it could actually transmit if it is set to transmit on channel 20, with a nominal carrier frequency of 27.205 MHz?
22. A CB transmitter is rated to supply 4 W of carrier power to a 50 Ω load while operating from a power supply that provides 13.8 V. The nominal supply current is 1 A without modulation and 1.8 A with 80% modulation. Calculate the overall efficiency of this transmitter, with and without modulation.

23. The audio frequency response of an AM transmitter, measured from microphone input to the secondary of the modulation transformer, is 3 dB down at 3 kHz from its level at 1 kHz. If a 1 kHz signal modulates the transmitter to 90%, what will be the modulation percentage due to a 3 kHz signal with the same level at the input?

SECTION 4.3

24. Draw a block diagram for an AM transmitter using high-level modulation, with an oscillator, buffer, driver, and power amplifier. Indicate the probable class of operation of each amplifier stage.
25. Draw a block diagram for an AM transmitter with an oscillator, buffer, pre-driver, driver, and power amplifier. Modulation is applied to the pre-driver stage. Indicate the probable class of operation of each amplifier stage.

SECTION 4.4

26. The RF power amplifier stage of a transmitter has an output of 50 kW and a gain of 15 dB. How much power must be supplied to this stage by the previous stage?
27. The power amplifier of an AM transmitter has an output carrier power of 25 W and an efficiency of 70% and is collector-modulated. How much audio power will have to be supplied to this stage for 100% modulation?
28. If the transmitter in the previous question operates from a 24 V supply, what will be the impedance seen looking into the power amplifier from the modulation transformer secondary?
29. A transistor RF power amplifier operating Class C is designed to produce 30 W output with a supply voltage of 50 V.
    (a) If the efficiency of the stage is 70%, what is the average collector current?
    (b) Assuming high-level modulation, what is the impedance seen by the modulation transformer secondary?
    (c) What power output would be required from the audio stages for 100% modulation of the amplifier?
    (d) What is the maximum voltage that appears between the collector and emitter of the transistor?

SECTION 4.5

30. Calculate a suitable load impedance for the amplifier in Problem 29. Sketch a suitable matching network to allow this amplifier to drive a 50 Ω load (component values not required).

SECTION 4.7

31. A transmitter uses twelve modules in its solid-state output stage. Calculate the power reduction (in decibels) that would occur if one module fails.
32. What would be the minimum allowable sampling rate for a pulse-duration modulator if it is required to handle a baseband frequency range from 50 Hz to 10 kHz?

SECTION 4.8

33. An in-line wattmeter connected in the transmission line between a transmitter and its antenna reads 50 W in the forward direction and 30 W in the reverse direction. How much output power is actually being produced?
34. Calculate the percent modulation for each of the trapezoidal patterns in Figure 4.21.

Figure 4.21

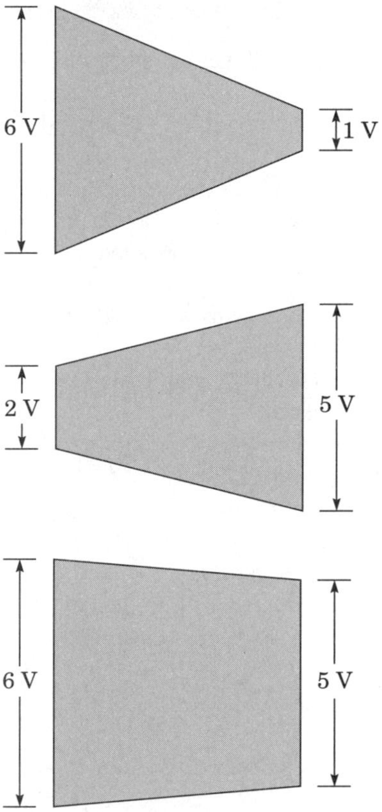

35. The circuit in Figure 4.22 can be used to indicate overmodulation.
    (a) Explain how it works.
    (b) Why will it not show very small amounts of overmodulation?

Figure 4.22

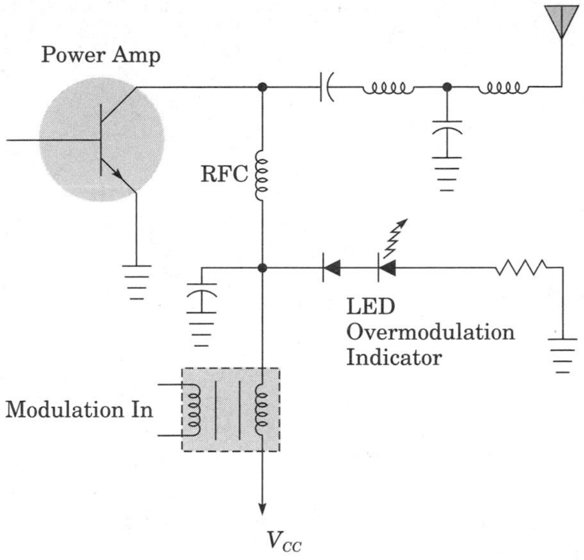

COMPREHENSIVE

36. A transmitter operates into a 50 Ω resistive load. The RMS voltage measured at the output is 250 V without modulation and 300 V with modulation, using a true-RMS reading meter. Find:
    (a) the power with modulation
    (b) the power without modulation
    (c) the modulation index
    (d) the peak voltage with modulation
    (e) the overall efficiency with modulation if the transmitter draws 3 kW from the ac line when modulated
    (f) the amount of power that will be drawn from the ac line when the transmitter is unmodulated, if the efficiency is the same as in part (e)
37. Use a computer program such as APPCAD™ to solve the following problems. In each case, try several configurations, and choose one with reasonable component values.
    (a) Match a 10 ohm source to a 50 ohm load at 1 MHz.
    (b) Match a 1000 ohm source to a 50 ohm load at 10 MHz.
38. An AM transmitter uses high-level modulation. The RF power amplifier runs from a 12 volt source, putting out a carrier power of 75 watts, at an efficiency of 75%.
    (a) What is the dc current drawn by this stage from the power supply?
    (b) What load impedance is required by this amplifier in order for it to deliver rated power?
    (c) How much power must be provided by the modulating amplifier for 100% modulation?
    (d) What impedance would be seen at the modulation transformer secondary?
39. Draw a block diagram for a moderate-power (100W) AM transmitter using high-level modulation. Show the probable class of operation of each stage.
40. An AM transmitter uses high-level modulation. The RF power amplifier draws 10 amperes from a 20 volt supply, putting out a carrier power of 125 watts.
    (a) What is the efficiency of this stage?
    (b) What load impedance is required by this amplifier in order for it to deliver rated power?
    (c) How much power must be provided by the modulating amplifier for 90% modulation?
    (d) What impedance would be seen at the modulation transformer secondary?
41. An AM radio station operates with a power of 10 kW at a carrier frequency of 1 MHz. It is modulated at 75% by a 1 kHz sine wave. The output is connected to a 50 ohm load.
    (a) Sketch the transmitter output in the frequency domain, showing a frequency scale, and a power scale in watts.
    (b) Sketch the transmitter output in the time domain (envelope display). Include time and voltage scales.
42. A transmitter power amplifier has a gain of 30 dB and an efficiency of 75%. Its output power is 100 watts.
    (a) What signal power must be provided at the input?
    (b) How much current does the amplifier draw from a 100 volt power supply?
    (c) Suppose this amplifier is to be amplitude modulated using high level modulation. How much power must be supplied by the audio amplifier, for 100% modulation?
    (d) What will be the output power with 100% modulation, assuming the efficiency is the same with and without modulation?

5 AM Receivers

Objectives After studying this chapter, you should be able to:

1. Describe the basic superheterodyne system and explain why it is the preferred design for most receivers
2. Distinguish between single- and multiple-conversion receivers and decide which type would be more suitable for a given application
3. Choose suitable intermediate frequencies and calculate image rejection for a receiver
4. Explain the requirements for each stage in a receiver and suggest suitable types of circuits to fulfill the requirements
5. Test and troubleshoot receivers
6. Analyze specifications for receivers and use them to determine suitability for a given application
7. Analyze a circuit diagram for a receiver or transceiver to determine the function of each stage

5.1 Introduction

Most of the emphasis in this chapter is on receivers for AM signals because that is the only modulation scheme that has been described so far. However, the principles introduced here will be used again and again, with all types of modulation. In the chapters that follow, we will only have to look at the particular modifications that are necessary for different applications.

Two important specifications are fundamental to all receivers. **Sensitivity** is a measure of the signal strength required to achieve a given signal-to-noise ratio, and **selectivity** is the ability to reject unwanted signals at frequencies different from that of the desired signal. These concepts will help you understand the following discussion of receiver types. The exact mathematical definitions and the techniques for measuring these parameters differ depending on the application and will be described later in this chapter, along with several other important specifications.

5.2 AM Demodulation

Every receiver requres a demodulator (also known as a detector). This circuit restores the baseband signal as it was before it was combined with the carrier wave to create the transmitted signal. The fact that the envelope of an AM wave reproduces the modulating signal means that an AM demodulator can be very simple.

An envelope detector (see Figure 5-1(a)) consists of a diode that rectifies the AM signal, discarding one-half of the envelope, followed by a simple low-pass filter to remove

Early Receivers

The very first receiver was used by Heinrich R. Hertz for his experiments in 1887. It was a loop of wire with a small spark gap in the center and was obviously not sensitive enough to be useful outside the laboratory. The first practical receiving device was the *coherer,* which was first used by Edouard Branly (1844–1940) in 1890. It consisted of a tube of metal filings through which the RF signal and a dc current were passed. The signal caused the filings to adhere together, or "cohere," reducing the resistance. This change in resistance increased the dc current, which activated a telegraph sounder. Unfortunately, the particles remained stuck together after the RF signal was removed, and the tube had to be tapped periodically to ascertain whether the signal was still present. Nonetheless, the coherer could be used to detect radiotelegraph signals. An improved version of the coherer was used for Guglielmo Marconi's first marine radio installations around the turn of the century. It was connected across a tuned circuit and had an automatic "tapper" to separate or "decohere" the metal filings between the dots and dashes of the Morse code.

The coherer could not demodulate AM signals, and it was quickly replaced by solid-state detectors made of galena, a semiconductor, and a thin wire "cat's whisker" that was carefully adjusted to touch a sensitive spot on the crystal, forming, in effect, a point-contact diode. Vacuum tubes soon replaced these early "crystal" detectors, since they could provide gain. This may have been the only time in history when a vacuum tube replaced a solid-state device. One can only wonder what would have happened if, instead of devoting all its resources to the new vacuum tube, the electronics industry had developed the potential inherent in the early and unreliable "cat's whisker" detector. Perhaps the transistor would have been invented thirty years sooner.

the radio-frequency component while preserving the baseband modulating signal. The effects of this circuit are shown in Figure 5-1(b). The modulated signal is applied at point A. Sketch (i) shows the envelope for a 1 kHz sinusoidal modulating signal. The diode removes half of the envelope. The result is shown in sketch (ii). Note, however, that this sketch ignores the low-pass filter connected to point B.

The capacitor C_1 charges to the peak value of the RF waveform and is slowly discharged by the resistor. If the RC_1 time constant is much longer than the period of the RF waveform and also much shorter than that of the modulating signal, the output will be a reasonably faithful reproduction of the modulating signal. There will be an additional dc component, which can easily be removed by blocking capacitor C_2. Since the carrier and modulating signal frequencies usually differ by several orders of magnitude, the choice of time constant is not difficult.

For best results, the signal voltage should be sufficient to fully turn on the diode, even when the envelope has its lowest amplitude. Germanium diodes are often used for detectors because of their low turn-on voltage of about 300 mV. Even so, it can be seen that as modulation approaches 100%, there will inevitably be distortion due to the nonlinearity of the diode.

5.3 Receiver Types

Almost all modern receiver designs use the **superheterodyne** principle, which will be described shortly. However, in order to recognize its advantages, we should first look at some more straightforward methods.

The simplest conceivable receiver would be an envelope detector connected directly to an antenna, as in Figure 5.2(a). Any AM signal arriving at the antenna would be demodulated, and the detector output would be connected to sensitive headphones. Since this is a passive circuit, only strong signals received by a good antenna could be heard at all. In addition, this receiver would have no ability to discriminate against unwanted signals and noise and so would receive all local AM stations at once. Obviously the results would not be at all satisfactory.

Figure 5.1
Envelope detector

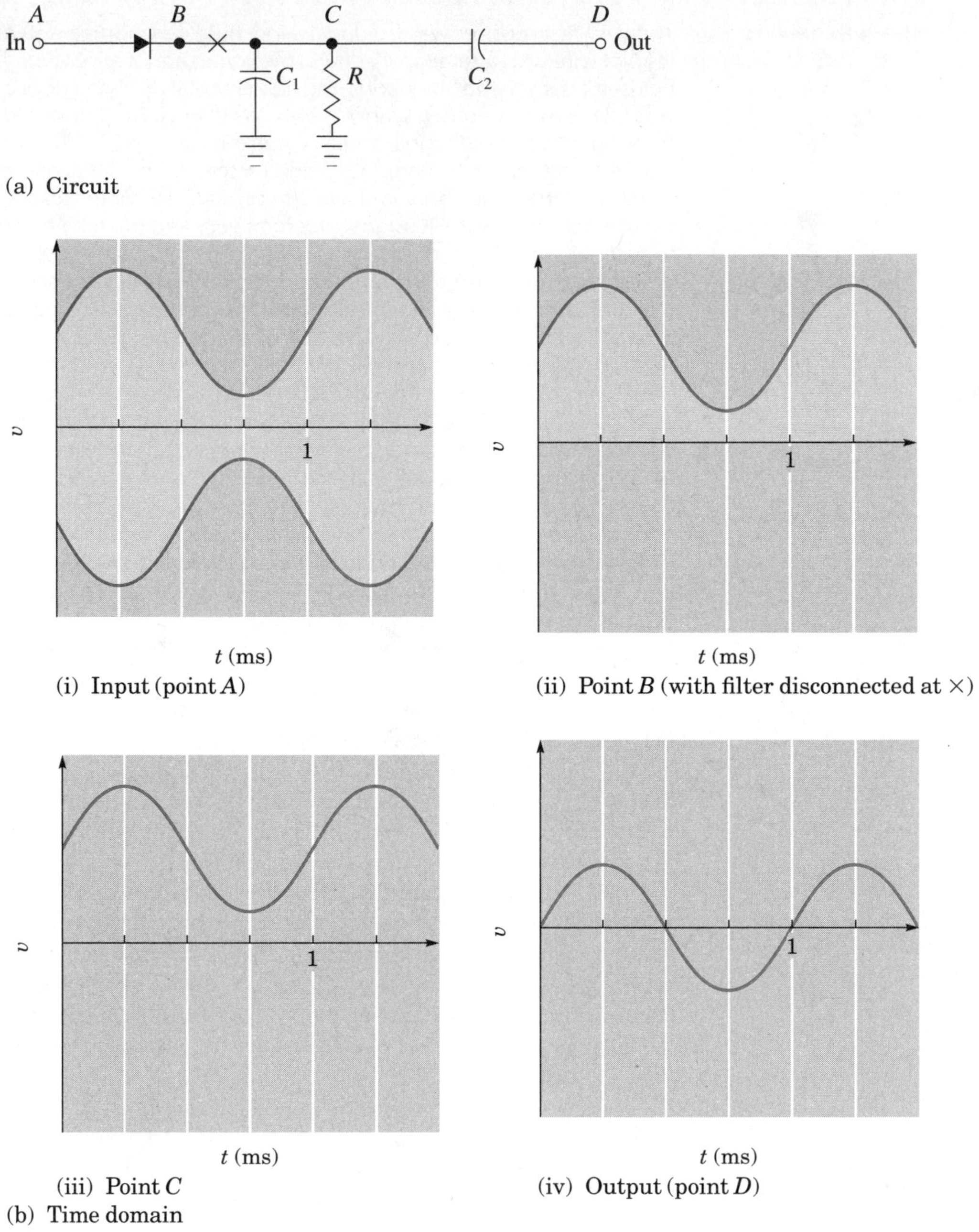

This receiver could be improved by adding a tuned circuit at the input, as shown in Figure 5.2(b). This would provide some selectivity, that is, the receiver could be tuned to a particular station. Signals at the resonant frequency of the tuned circuit would be passed to the detector, and those at other frequencies would be attenuated. However, there is still no gain. Receivers of this type are often called *crystal radios* ("crystal" being an early term for a semiconductor diode). They can still be found as toys.

The addition of an audio amplifier, as shown in Figure 5.2(c), could provide enough output power to operate a speaker. However, the selectivity would remain poor because of the single tuned circuit, and the receiver would not be sensitive enough to receive weak signals because a diode detector needs a relatively large input voltage to operate efficiently,

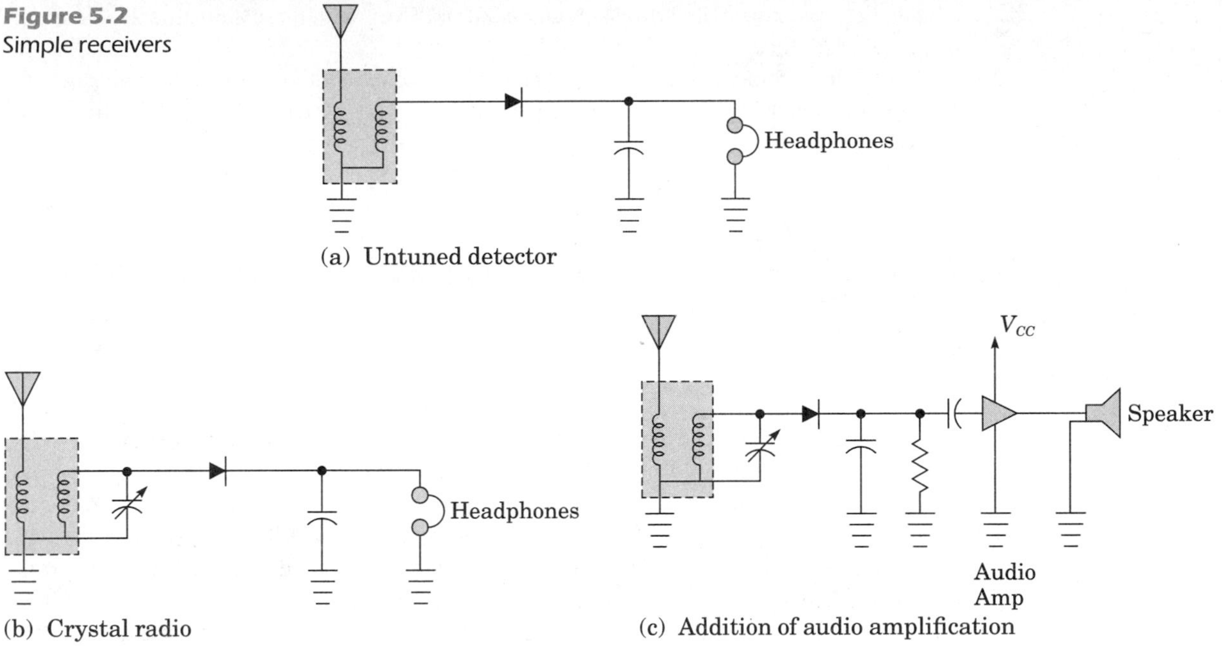

Figure 5.2
Simple receivers

with low noise and distortion. More sensitive detectors can be devised, but a better solution is to provide gain before the detector.

5.3.1 The Tuned-Radio-Frequency Receiver

Figure 5.3 shows a block diagram for a **tuned-radio-frequency (TRF) receiver**. Several RF amplifiers, each tuned to the signal frequency, provide gain and selectivity before the detector. An audio amplifier after the detector supplies the necessary power amplification to drive the speaker.

The problems with this receiver are in the RF stages. To achieve satisfactory gain and selectivity, several stages will probably be needed. All of their tuned circuits must tune together to the same frequency (or **track** very closely); this tends to cause both electrical and mechanical problems. Having several high-gain tuned amplifier stages in close physical proximity is likely to lead to oscillation, since it is difficult to prevent feedback. In addition, some means must be provided to tune all the circuits simultaneously. The usual way, when TRF receivers were popular, was to use several variable capacitors connected together mechanically—either *ganged* (mounted on the same shaft) or linked by belts or gears. However, component tolerances cause the circuits to tune to slightly different frequencies when their capacitors are at the same position, unless the frequencies are corrected with small

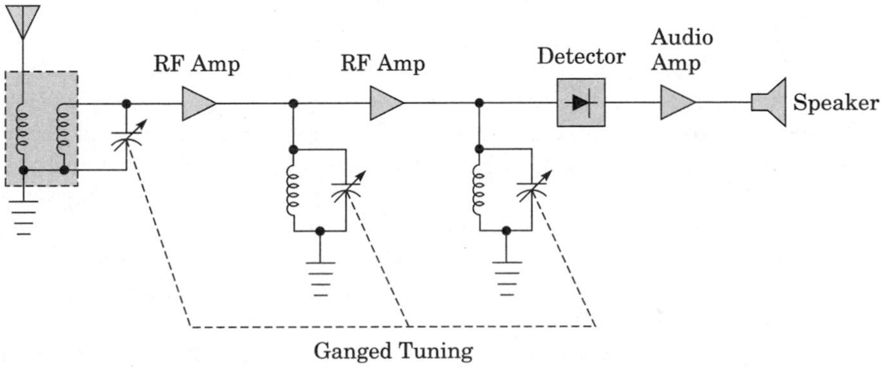

Figure 5.3
TRF receiver

adjustable capacitors called *trimmers* and *padders*. Even with these adjustments, the tracking is never perfect.

Another problem arises from the fact that the bandwidth of a tuned circuit does not remain constant when its resonant frequency is changed. At first glance this may seem surprising since for an inductor,

$$Q = \frac{X_L}{R} \tag{5.1}$$

and X_L increases directly with frequency. The required Q to achieve a given bandwidth also varies directly with frequency, as can be seen from the equation

$$B = \frac{f_o}{Q} \tag{5.2}$$

However, in practice the resistance of a coil also increases with frequency due to the **skin effect**. At higher frequencies, internal magnetic fields in the wire cause the current to flow mainly in the region near the surface of the conductor. This decreases the effective cross-sectional area of the conductor and increases its resistance. The resistance varies with the square root of frequency. Therefore, the bandwidth of a tuned circuit increases approximately with the square root of frequency.

Thus a receiver with the correct selectivity at the low end of its tuning range will have too wide a bandwidth at the high end. An example will make this clear.

Example 5.1

A tuned circuit tunes the AM radio broadcast band (from 540 to 1700 kHz). If its bandwidth is 10 kHz at 540 kHz, what is it at 1700 kHz?

Solution

The bandwidth varies with the square root of frequency. Therefore, at the high end the bandwidth will increase to:

$$\begin{aligned} B &= 10\text{ kHz} \times \sqrt{\frac{1700}{540}} \\ &= 17.7\text{ kHz} \end{aligned}$$

Assuming that the bandwidth is correct at the lower end, there may well be interference from adjacent stations at the top end.

For high signal frequencies and narrow bandwidths, it may not be possible to obtain satisfactory results with a reasonable number of RF stages, because of limitations on the Q of conventional tuned circuits. Circuits with higher Q, such as **crystal filters**, are not practical because of the difficulty in operating them over a wide range of frequencies. In addition, most active devices show a reduction of gain with increasing frequency.

The TRF system is now used only for simple, fixed-frequency receivers, where most of its disadvantages do not apply.

5.3.2 The Superheterodyne Receiver

The superheterodyne receiver or *superhet* was invented by Edwin H. Armstrong (1890–1954) in 1918 and is still almost universally used, in many variations. Figure 5.4 shows its basic layout. It may contain one or more stages of RF amplification, and the RF stage can be either tuned, as in the TRF receiver, or broadbanded. This stage should have a good noise figure as being the first stage in the receiver, it is largely responsible for the noise performance of the entire system. Low-cost receivers sometimes omit the RF amplifier, but they still include some sort of input filter, such as a tuned circuit. The input filter and RF stage (or mixer, if there is no RF stage) are sometimes referred to as the **front end** of a receiver.

The next stage is a mixer. The signal frequency is mixed with a sine-wave signal generated by an associated stage called the **local oscillator**, creating a difference frequency

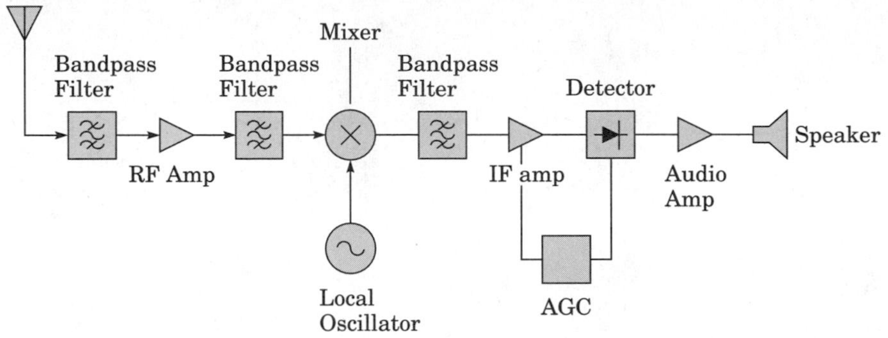

Figure 5.4 Basic superheterodyne receiver

called the **intermediate frequency (IF)**. The local oscillator is tunable, so the IF is fixed regardless of the signal frequency. The combination of mixer and local oscillator is known as a **converter**.

Figure 5.5 shows the signal and local oscillator frequencies for a typical AM broadcast receiver with an IF of 455 kHz. In Figure 5.5(a), the frequency of the desired signal is 740 kHz and the local oscillator is set to 1195 kHz. The difference frequency is 455 kHz, and this is passed to the next stage. The mixer will also produce a sum frequency of 1935 kHz, but this is easily removed by filtering. In Figure 5.5(b), the receiver is tuned to 1020 kHz. The local oscillator is returned to 1475 kHz so that the difference frequency is still 455 kHz.

Receivers with conventional variable-capacitor tuning usually use a two- or three-gang tuning capacitor. One section tunes the local oscillator, and the other section or sections tune the mixer input circuit and the input circuit for the RF amplifier (if present).

The mixer is followed by the IF amplifier, which provides most of the receiver's gain and selectivity. Generally there are two or more IF stages, with selectivity provided either by resonant circuits or, in more advanced designs, by a crystal filter or a **ceramic filter**. The use of a fixed IF greatly simplifies the problem of achieving adequate gain and selectivity.

The remainder of the receiver is straightforward and resembles the TRF design. There is a detector to demodulate the signal and an audio amplifier to increase the signal power to the level required to operate a loudspeaker.

The **automatic gain control (AGC)** adjusts the gain of the IF—and sometimes the RF—stages in response to the strength of the received signal, providing more gain for weak signals. It thus allows the receiver to cope with the very large variation in signal levels found in practice. This range may be more than 100 dB for a communications receiver.

The receiver in Figure 5.4 is for AM audio use, but the superheterodyne system is also used with other modulation schemes (for instance, FM) and other modulating signals

Figure 5.5 Signal and local-oscillator frequencies

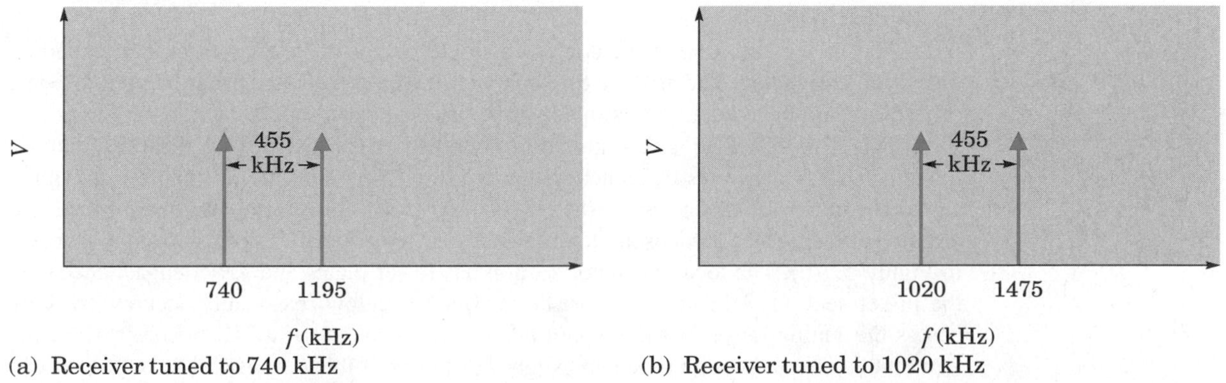

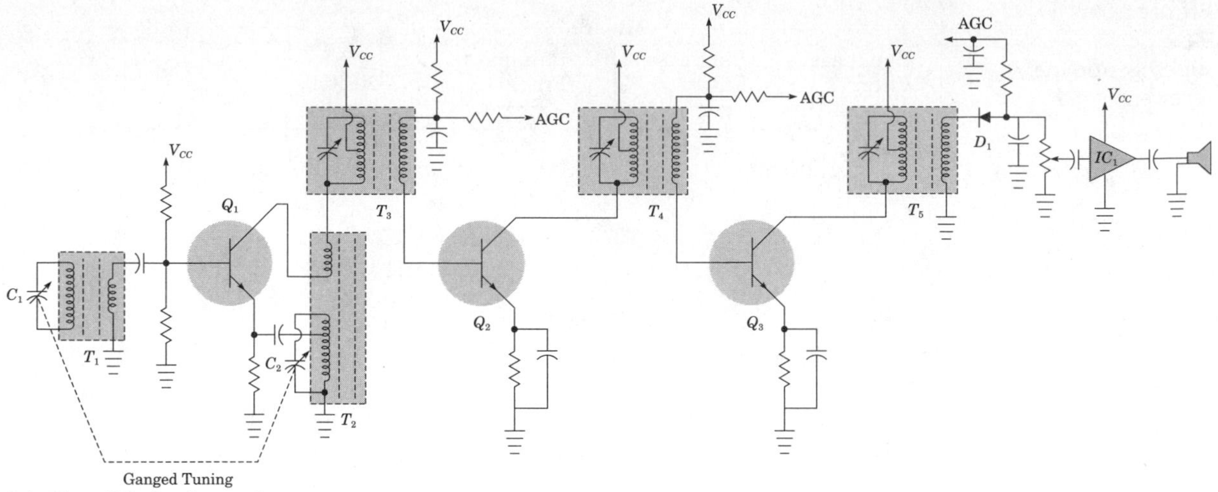

(a) Simplified schematic

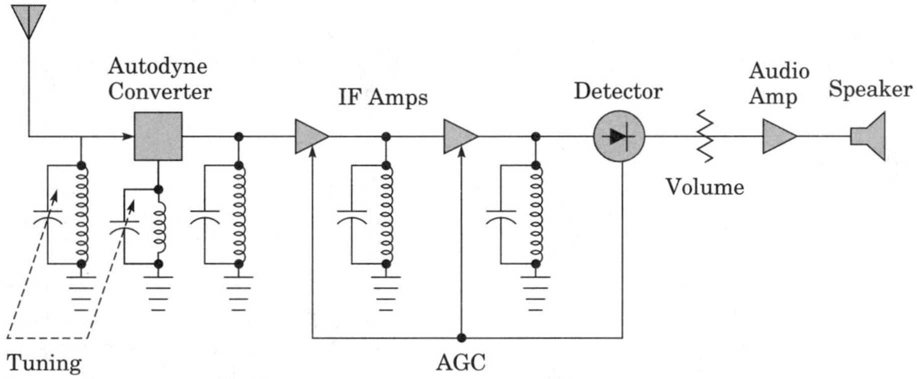

(b) Block diagram

Figure 5.6
AM broadcast receiver

(such as video). Once you are thoroughly familiar with the basic system, it will be easy to understand the variations.

5.3.3 Example: An AM Broadcast-Band Receiver

Figure 5.6(a) shows the schematic of a typical AM broadcast-band receiver, and Figure 5.6(b) shows its block diagram. Note the similarity to the generic block diagram in Figure 5.4.

For reasons of economy, the RF stage has been omitted from the receiver in Figure 5.6. The received signal is coupled to the mixer through a single tuned circuit. In fact, the inductive element of this tuned circuit is actually the receiving antenna, which is a ferrite *loopstick* (that is, a coil wound around a ferrite rod). An example is shown in the photograph in Figure 5.7(a).

The mixer and local oscillator use a single transistor in a configuration called an **autodyne converter**. This is done for economy: it would be better from the point of view of stability and the reduction of **spurious responses** to use a separate local oscillator.

The IF is 455 kHz, and **high-side injection** is used, that is, the local oscillator is tuned so that it generates a frequency that is always 455 kHz higher than the incoming signal frequency. This is accomplished by using two variable capacitors, one for the input circuit and one for the local oscillator, mounted on a single shaft. Such a capacitor is shown in Figure 5.7(b). The local oscillator section has fewer plates and lower capacitance than the mixer section. Adjustments must be provided to allow reasonably accurate tracking across the tuning range, but this is not nearly as critical as for a TRF receiver, since the input tuned circuit is quite broad and slight mistuning of that circuit will have little effect. The actual selection of the desired station is done by the local oscillator tuning.

Figure 5.7
Receiver tuning components

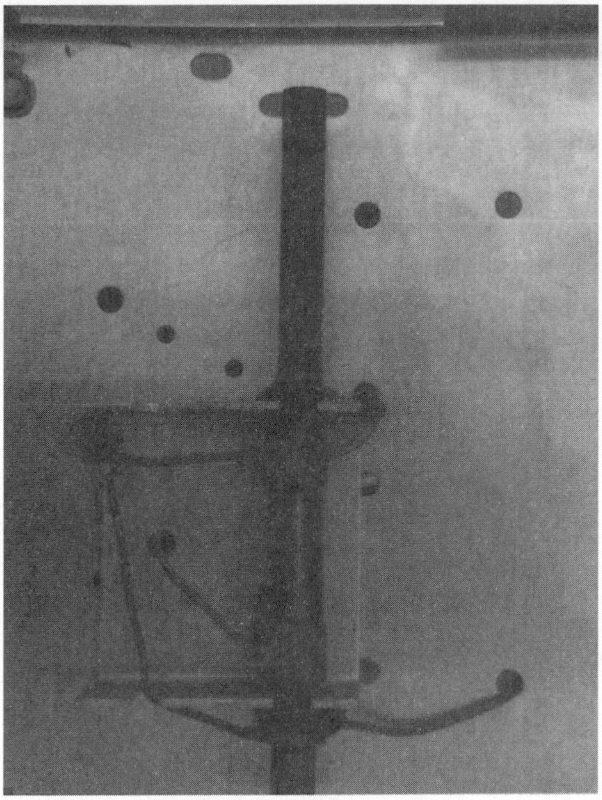

(a) Ferrite "loopstick" antenna

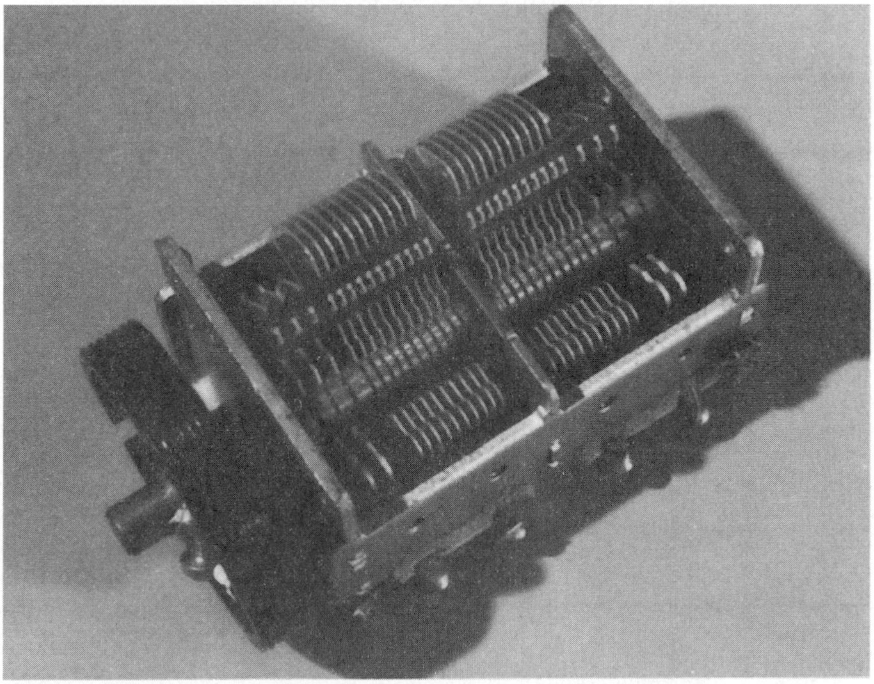

(b) Dual-gang variable capacitor

The mixer will, of course, produce the sum and difference of the local oscillator and signal frequencies. The sum is not used in this application; the difference of 455 kHz is amplified by the IF stages. Conventional tuned circuits are used here to give the desired bandwidth, which, for a small AM radio, is about 7 to 10 kHz.

Low-side injection—in which the local oscillator frequency is lower than that of the received signal—could have been used, but it would require the oscillator to have a wider tuning range, in percentage terms, which would make the mechanical design of the tuning system more difficult. An example will make this clear. For high-side injection, the oscillator must tune from:

$$f_l = 540\text{ kHz} + 455\text{ kHz} = 995\text{ kHz}$$

to:

$$f_h = 1600\text{ kHz} + 455\text{ kHz} = 2055\text{ kHz}$$

The ratio of highest to lowest frequency is:

$$\frac{f_h}{f_l} = \frac{2055}{995} = 2.065$$

With low-side injection, on the other hand, the oscillator would have to tune from 85 to 1145 kHz, a ratio of 13.47. Larger capacitance and inductance values would also require the components to be physically larger. Nonetheless, low-side injection is sometimes used for receivers operating at higher frequencies, as will be seen later in this chapter.

The AGC in this receiver is quite simple. Low-pass filtering the detector output (with a time constant of about a second) swamps the variations due to modulation and leaves a dc voltage proportional to the carrier level. This voltage is used to alter the bias on the IF amplifier transistors, which in turn changes the gain of the IF stages.

In addition to AGC, there is a manual volume control that varies the level of signal provided from the detector to the audio amplifier.

Practically all of the active circuitry for a radio such as this can be incorporated into a custom integrated circuit (IC). Figure 5.8 shows a typical example, the National Semiconductor LM1863. The circuit shown needs only an audio amplifier, which could be another IC, to complete it.

5.4 Receiver Characteristics

At the beginning of our discussion of receivers, we introduced two important receiver parameters, *sensitivity* and *selectivity*. We will now look more closely at these parameters and introduce some other receiver specifications.

5.4.1 Sensitivity

The transmitted signal may have a power level ranging from milliwatts to hundreds of kilowatts at the transmitting antenna. However, the losses in the path from transmitter to receiver are so great that the power of the received signal is often measured in dBf (that is, decibels relative to one femtowatt: 1 fW = 1×10^{-15} W). A great deal of amplification is needed to achieve a useful power output. To operate an ordinary loudspeaker, for instance, requires a power on the order of 1 W, or 150 dB greater than 1 fW.

Because received signals are often quite weak, noise added by the receiver itself can be a problem. Diode detectors are inherently noisy and also operate best with fairly large signals (hundreds of millivolts), so some of the amplification must take place before demodulation.

The ability to receive weak signals with an acceptable signal-to-noise ratio (S/N) is called *sensitivity*. It is expressed in terms of the voltage or power at the antenna terminals necessary to achieve a specified signal-to-noise ratio or some more easily measured equivalent. One common specification for AM receivers is the signal strength required for a 10 dB signal-plus-noise to noise [$(S+N)/N$] ratio at a specified output power level. Procedures for measuring sensitivity will be discussed later in this chapter.

Figure 5.8
Data sheet for the LM1863
Courtesy National Semiconductor.

National Semiconductor

LM1863 AM Radio System for Electronically Tuned Radios

General Description

The LM1863 is a high performance AM radio system intended primarily for electronically tuned radios. Important to this application is an on-chip stop detector circuit which allows for a user adjustable signal level threshold and center frequency stop window. The IC uses a low phase noise, level-controlled local oscillator.

Low phase noise is important for AM stereo which detects phase noise as noise in the L-R channel. A buffered output for the local oscillator allows the IC to directly drive a phase locked loop synthesizer. The IC uses a RF AGC detector to gain reduce an external RF stage thereby preventing overload by strong signals. An improved noise floor and lower THD are achieved through gain reduction of the IF stage. Fast AGC settling time, which is important for accurate stop detection, and excellent THD performance are achieved with the use of a two pole AGC system. Low tweet radiation and sufficient gain are provided to allow the IC to also be used in conjunction with a loopstick antenna.

Features

- Low supply current
- Level-controlled, low phase noise local oscillator
- Buffered local oscillator output
- Stop circuitry with adjustable stop threshold and adjustable stop window
- Open collector stop output
- Excellent THD and stop time performance
- Large amount of recovered audio
- RF AGC with open collector output
- Meter output
- Compatible with AM stereo

Block Diagram

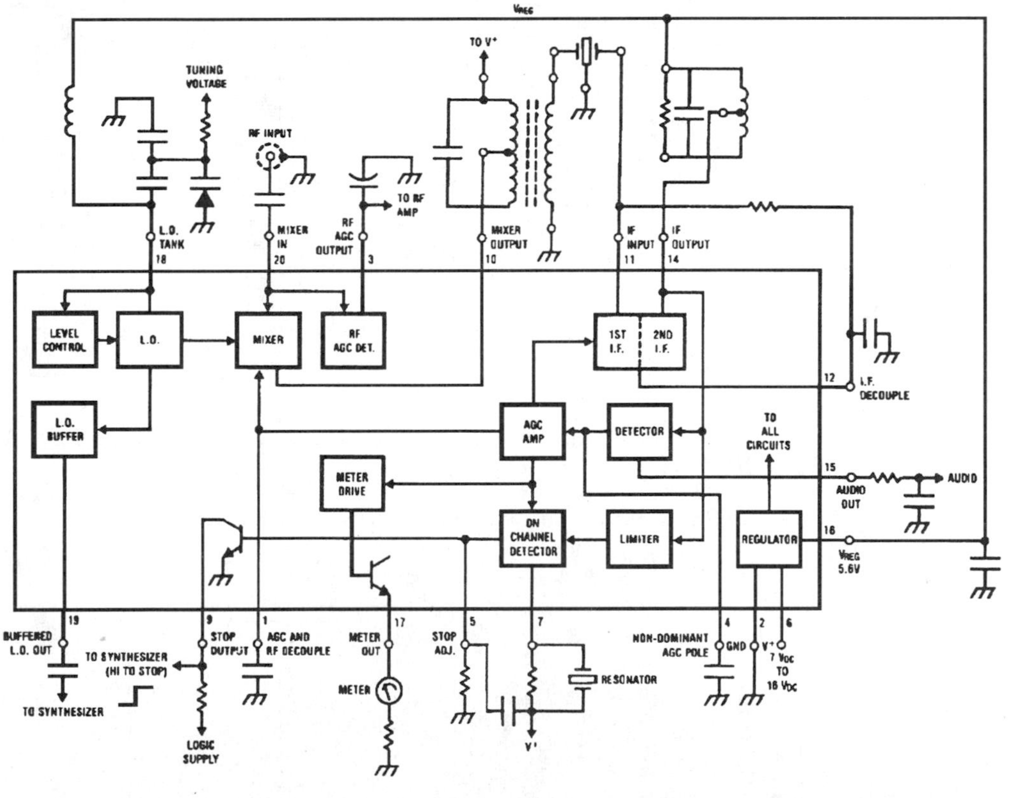

5.4.2 Selectivity

In addition to the noise generated within the receiver, there will be noise coming in with the signal, along with interfering signals with frequencies different from that of the desired signal. All of these problems can be reduced by limiting the receiver bandwidth to that of the signal, including all its sidebands. When interference is severe, an even smaller bandwidth can be used in an AM receiver, at the expense of reducing the response to high frequencies in the original modulating signal. The ability to discriminate against interfering signals is known as *selectivity*.

Selectivity can be expressed in various ways. The bandwidth of the receiver at two different levels of attenuation can be specified. The bandwidth at the points where the signal is 3 or 6 dB down is helpful in determining whether all the sidebands of the desired signal will be passed without attenuation. To indicate the receiver's effectiveness in rejecting interference, a bandwidth for much greater attenuation, for example 60 dB, should also be given.

The frequency-response curve for an ideal IF filter would have a square shape, with no difference between its bandwidths at 6 dB and 60 dB down. The closer the two bandwidths are, the better the design. The ratio between these bandwidths is called the **shape factor**. That is,

$$SF = \frac{B_{-60\text{ dB}}}{B_{-6\text{ dB}}} \tag{5.3}$$

where

SF = shape factor
$B_{-60\text{ dB}}$ = bandwidth at 60 dB down from maximum
$B_{-6\text{ dB}}$ = bandwidth at 6 dB down from maximum

The shape factor should be as close to one as possible. The following example shows the effect of changing the shape factor on the rejection of interfering signals.

Example 5.2 Calculate the shape factors for the two IF response curves shown in Figure 5.9, and calculate the amount by which the interfering signal shown would be attenuated in each case.

Figure 5.9

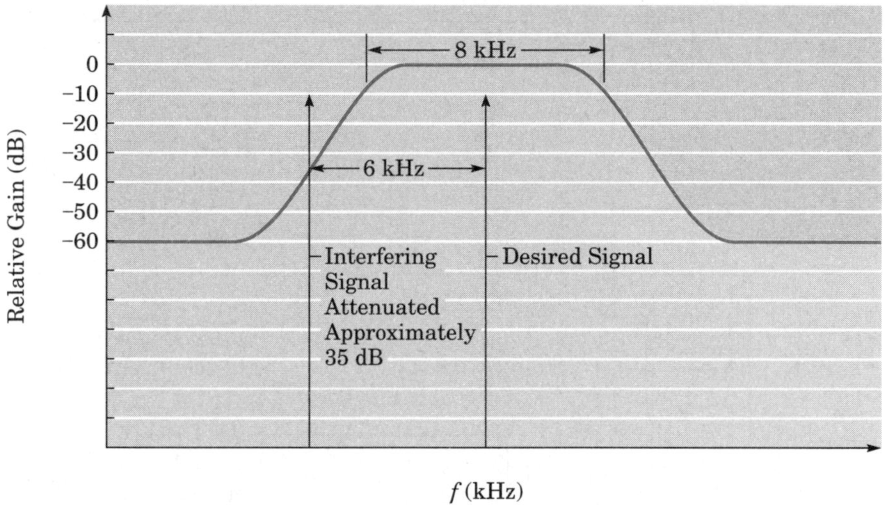

(a) Shape factor = 1 Bandwidth (6 dB) = 8 kHz

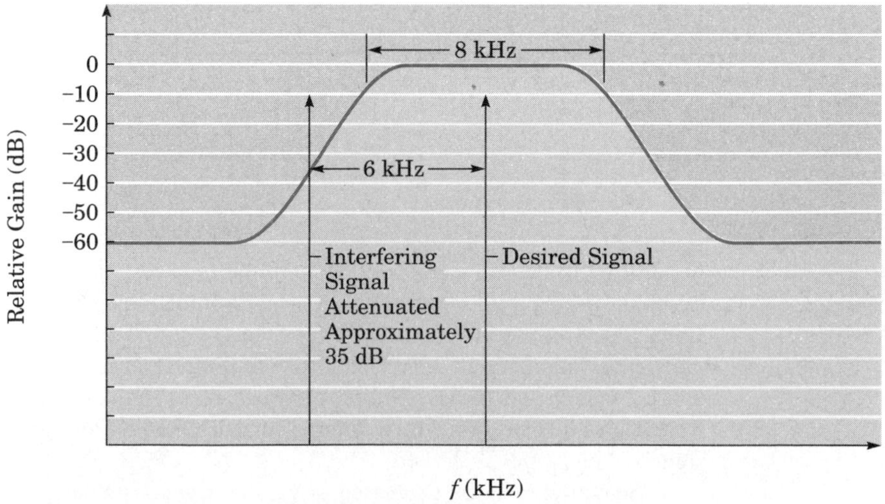

(b) Shape factor = 2 Bandwidth (6 dB) = 8 kHz

Solution Figure 5.9(a) shows an ideal filter. Since the −6 dB and −60 dB bandwidths are equal, $SF = 1$. The interfering signal is attenuated by 60 dB.

In Figure 5.9(b), the −6 dB bandwidth is 8 kHz and the −60 dB bandwidth is 16 kHz, so $SF = 2$. The interfering signal is attenuated approximately 35 dB compared to the desired signal.

Adjacent channel rejection is another way of specifying selectivity; it is commonly used with channelized systems, such as CB radio. It is defined as the number of decibels by which an **adjacent channel** signal must be stronger than the desired signal for the same receiver output.

Alternate channel rejection is also used in systems, such as FM broadcasting, where stations in the same locality are not assigned to adjacent channels. The **alternate channel** is two channels removed from the desired one. It is also known as the *second adjacent channel*. Figure 5.10 shows the relation between adjacent and alternate channels. For example, if 94.1 MHz were assigned to a station, the adjacent channel at 94.3 MHz would not be assigned in the same community, but the alternate channel, at 94.5 MHz, might be. Any signal at 94.3 MHz, then, will be relatively weak, and a receiver's ability to reject strong adjacent channel signals will be less important than its ability to reject the alternate channel (in this case, the one at 94.5 MHz).

Figure 5.10
Adjacent and alternate channels

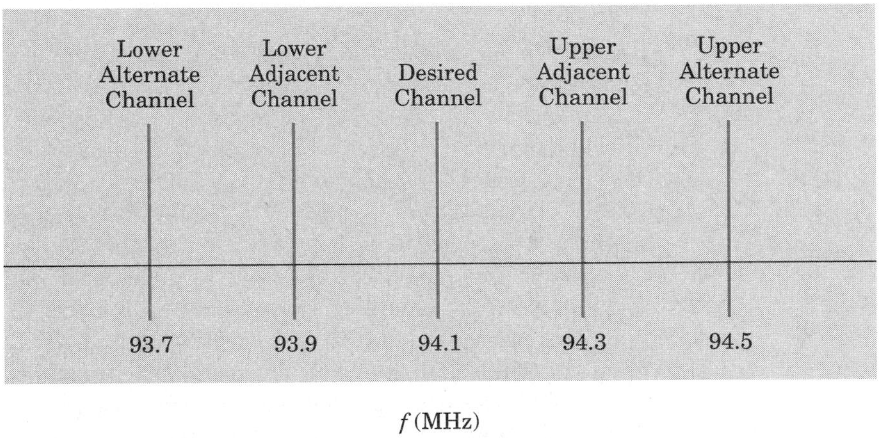

5.4.3 Distortion

In addition to having good sensitivity and selectivity, an ideal receiver would reproduce the original modulation exactly. A real receiver, however, subjects the signal to several types of *distortion*. They are the same types of distortion encountered in other analog systems: harmonic and intermodulation distortion, uneven frequency response, and phase distortion.

Harmonic distortion occurs when the frequencies generated are multiples of those in the original modulating signal. Envelope detectors produce significant harmonic distortion because they operate part of the time in the square-law portion of the diode curve.

Intermodulation takes place when frequency components in the original signal mix in a nonlinear device, creating sum and difference frequencies. A type of intermodulation peculiar to receivers consists of mixing between the desired signal and an interfering one that is outside the IF passband of the receiver but within the passband of the RF stage and is therefore present in the mixer. This can result in interference from a strong local station that is not at all close in frequency to the desired signal.

Example 5.3 A receiver is tuned to a station with a frequency of 1 MHz. A strong signal with a frequency of 2 MHz is also present at the amplifier. Explain how intermodulation between these two signals could cause interference.

Solution See Figure 5.11. Assume that the tuned circuit at the mixer input has insufficient attenuation at 2 MHz to block the interfering signal completely. (Remember that the interfering signal

Figure 5.11

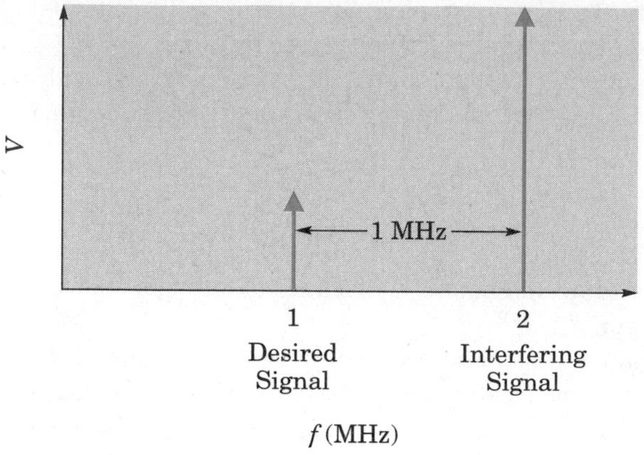

may be much stronger than the desired signal and therefore may still have strength similar to that of the desired signal, even after being attenuated by the input tuned circuit.) The two signals could mix in the mixer stage to give a difference frequency of 1 MHz, at the same frequency as the original signal. This interfering signal will then pass through the rest of the receiver in the same way as the desired signal.

In addition, any nonlinearities in the detector or any of the amplification stages can cause intermodulation between components of the original baseband signal.

The frequency response of an AM receiver is related to its IF bandwidth. Restricting the bandwidth to less than the full width of the signal reduces the upper limit of response, since the sidebands corresponding to higher modulating frequencies are further away from the carrier. For example, a receiver bandwidth of 10 kHz will allow a modulation frequency of 5 kHz to be reproduced. Reducing the bandwidth to 6 kHz reduces the maximum audio frequency that can be reproduced to 3 kHz. In addition, any unevenness within the IF passband will affect the audio frequency response. Of course, as would be expected, the audio amplifier also has an effect on frequency response.

Phase distortion is slightly more difficult to understand than frequency response. Of course, the signal at the receiver output will not be in phase with the input to the transmitter: there will be some time delay, which can be translated into a phase shift that increases linearly with frequency. Phase distortion consists of irregular shifts in phase and is quite a common occurrence when signals pass through filters.

The impact of distortion depends to a great extent on the application. For instance, phase distortion is unimportant for voice communications, of questionable importance for music (some audiophiles claim to be able to hear it), and fatal to some other applications, such as color television (where it results in incorrect colors) and certain types of data transmission. For voice communications, an audio bandwidth of more than about 3 kHz is simply a waste of the radio spectrum, while for high-fidelity music, an audio bandwidth of at least 15 kHz is essential. In a CB radio at maximum output power, 10% harmonic distortion may be acceptable, while 1% might be considered excessive in a quality FM broadcast tuner.

5.4.4 Dynamic Range

As mentioned above, a receiver must operate over a considerable range of signal strengths. The response to weak signals is usually limited by noise generated within the receiver. On the other hand, signals that are too strong will overload one or more stages, causing unacceptable levels of distortion. The ratio between these two signal levels, expressed in decibels, is the *dynamic range* of the receiver.

The above description is actually a bit simplistic. A well-designed AGC system can easily vary the gain of a receiver by 100 dB or so. Thus, almost any receiver can cope with very wide variations of signal strength, provided only one signal is present at a time. This range of signal strengths is sometimes referred to as *dynamic range* but should properly be called *AGC range*.

The hardest test of receiver dynamic range occurs when two signals with slightly different frequencies and very different power levels are applied to the antenna input simultaneously. This can cause overloading of the receiver input stage by the stronger signal, even though the receiver is not tuned to its frequency. The result can be **blocking** (also called *desensitization* or *desense*), which is a reduction in sensitivity to the desired signal. Intermodulation between the two signals is also possible.

Example 5.4

A receiver has a sensitivity of 0.5 μV and a blocking dynamic range of 70 dB. What is the strongest signal that can be present along with a 0.5 μV signal without blocking taking place?

Solution

Since both signal voltages are across the same impedance, the input impedance of the receiver, the general equation

$$\frac{P_1}{P_2}\text{ (dB)} = 20 \log \frac{V_1}{V_2} \tag{5.4}$$

can be used.

Here, we can let $V_2 = 0.5\ \mu\text{V}$ and

$$\frac{P_1}{P_2}\text{ (dB)} = 70$$

and solve for V_1 as follows. From Equation (5.4),

$$\frac{V_1}{V_2} = \text{antilog}\ \frac{P_1/P_2\text{ (dB)}}{20}$$

$$V_1 = V_2\ \text{antilog}\ \frac{P_1/P_2\text{ (dB)}}{20}$$

$$= (0.5\ \mu\text{V})\ \text{antilog}\ \frac{70}{20}$$

$$= 1581\ \mu\text{V}$$

$$= 1.58\ \text{mV}$$

There is yet another type of dynamic range, *audio dynamic range*, which is essentially the usable range of modulation depth with a given carrier level. Generally, a strong signal is specified for this measurement, which relates to the sound quality that can be expected with a good signal. This type of dynamic range specification is more common with FM than with AM receivers.

5.4.5 Spurious Responses

The superheterodyne receiver has many important advantages over simpler receivers, but it is not without its problems. In particular, it has a tendency to receive signals at frequencies to which it is not tuned and sometimes to generate signals internally, interfering with reception. Careful design can reduce these *spurious responses* almost to insignificance, but they will still be present.

Image Frequencies The intermediate-frequency signal in a superheterodyne receiver is the difference between the received signal and local oscillator frequencies. It does not matter whether the local oscillator is higher in frequency than the received signal (high-side injection) or lower (low-side injection).

It follows that in any receiver there will be two frequencies, one below and one above the local oscillator frequency, that will mix with it to produce a signal at the intermediate frequency f_{IF}. One of these will be the frequency f_{sig} to which the radio is tuned; the other is called the **image frequency** f_{image}. The image is an equal distance from the local oscillator frequency f_{LO} on the other side of it from the signal. This can be shown mathematically as follows. Assuming high-side injection,

$$f_{IF} = f_{LO} - f_{sig}$$

so

$$f_{LO} = f_{IF} + f_{sig} \tag{5.5}$$

For the image,

$$f_{IF} = f_{image} - f_{LO}$$

therefore

$$f_{LO} = f_{image} - f_{IF} \tag{5.6}$$

Combining Equations (5.5) and (5.6),

$$\begin{aligned} f_{IF} + f_{sig} &= f_{image} - f_{IF} \\ f_{image} &= f_{sig} + 2f_{IF} \end{aligned} \tag{5.7}$$

In a similar way, it can be shown that for low-side injection

$$f_{image} = f_{sig} - 2f_{IF} \tag{5.8}$$

Figure 5.12 shows the similarity to reflection in a mirror. Think of the local oscillator signal as the mirror, with the desired signal and the image equidistant from it, on opposite sides.

An image must be *rejected* prior to mixing: once it has entered the IF chain, the image will be indistinguishable from the desired signal and impossible to filter out. Image rejection is accomplished by tuned circuits or other filters before the mixer. Using a higher IF will improve image rejection by placing the image further away in frequency from the desired signal, where it can more easily be removed by tuned circuits before the mixer.

Image rejection is defined as the ratio of voltage gain at the input frequency to which the receiver is tuned to gain at the image frequency. Image rejection *IR* is usually expressed in decibels. For a single tuned circuit:

$$IR = \frac{A_{sig}}{A_{image}} = \sqrt{1 + Q^2x^2} \tag{5.9}$$

where

$$\begin{aligned} Q &= Q \text{ of the tuned circuit} \\ A_{image} &= \text{voltage gain at image frequency} \\ A_{sig} &= \text{voltage gain at signal frequency} \\ x &= \frac{f_{image}}{f_{sig}} - \frac{f_{sig}}{f_{image}} \end{aligned}$$

Figure 5.12
Image response

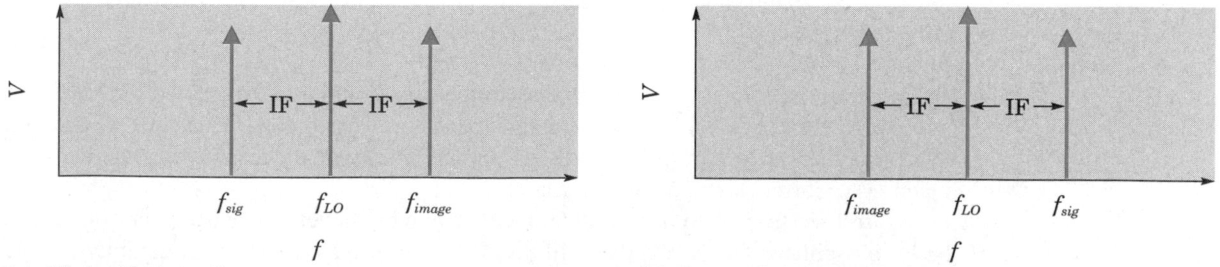

(a) High-side injection

(b) Low-side injection

Since we are dealing with voltage ratios across the same impedance, it is easy to convert to decibels:

$$IR\ (\text{dB}) = 20 \log \frac{A_{sig}}{A_{image}} \tag{5.10}$$

For multiple uncoupled tuned circuits, multiply the image rejection ratios (or add, if they are given in decibels). Many receivers have a single RF amplifier stage with a tuned circuit at its input, followed by a mixer, also with a tuned input. These two resonant circuits are uncoupled, and the above equation applies.

Example 5.5 The receiver in Figure 5.6 is tuned to a station at 590 kHz.

(a) Find the image frequency.
(b) Calculate the image rejection in decibels, assuming that the input filter consists of one tuned circuit with a Q of 40.

Solution

(a) This receiver uses high-side injection of the local oscillator. From Equation (5.7),

$$\begin{aligned} f_{image} &= f_{sig} + 2f_{IF} \\ &= 590\text{ kHz} + 2(455\text{ kHz}) \\ &= 1500\text{ kHz} \end{aligned}$$

Since 1500 kHz is also within the AM broadcast band, a strong local signal at that frequency is a real possibility.

(b) The image rejection can be found using Equation (5.9). First, find

$$\begin{aligned} x &= \frac{f_{image}}{f_{sig}} - \frac{f_{sig}}{f_{image}} \\ &= \frac{1500\text{ kHz}}{590\text{ kHz}} - \frac{590\text{ kHz}}{1500\text{ kHz}} \\ &= 2.149 \end{aligned}$$

Now,

$$\begin{aligned} \frac{A_{sig}}{A_{image}} &= \sqrt{1 + Q^2x^2} \\ &= \sqrt{1 + 40^2 \times 2.149^2} \\ &= 85.97 \end{aligned}$$

In decibels, the image rejection is

$$\begin{aligned} IR\ (\text{dB}) &= 20 \log 85.97 \\ &= 38.7\text{ dB} \end{aligned}$$

The use of a higher IF in Example 5.5 would have improved image rejection but at the expense of requiring more elaborate filtering in the IF amplifier to achieve the desired bandwidth. Choice of IF is always a compromise. Over the years, 455 kHz has come to be a very common frequency for the IF in receivers designed for the standard AM broadcast band.

Image response is usually undesirable, but in certain cases (for example, some of the "scanners" used to receive police and fire calls) image reception is used deliberately to increase the frequency coverage of a receiver.

The term *double-spotting* is sometimes used to describe the same phenomenon as image response, but seen from a different point of view. For example, suppose there is a strong AM station at 1500 kHz. It may be possible to tune a receiver to a frequency lower than this by twice the IF and receive the signal. For a typical AM radio with an IF of 455 kHz, this would require tuning it to 1500 kHz − 455 kHz × 2 = 590 kHz. Obviously, what has been done is to tune the receiver to a frequency such that this station is an image response. However, we could look at it the other way around and say that the station comes in at *two spots* on the dial, hence the term *double-spotting*.

Other Spurious Responses Besides images, superheterodyne receivers are subject to other problems. The local oscillator will have harmonics, for instance, and these can mix with incoming signals to produce spurious responses. The incoming signal may also have harmonics created by distortion in the RF stage. In fact, it is possible for a receiver to respond to any frequency given by the equation

$$f_s = \left(\frac{m}{n}\right) f_{LO} \pm \frac{f_{IF}}{n} \tag{5.11}$$

where

f_s = frequency of the spurious response
f_{LO} = local oscillator frequency
f_{IF} = intermediate frequency
m, n = any integers

Example 5.6 An AM high-frequency receiver has an IF of 1.8 MHz using high-side injection. If it is tuned to a frequency of 10 MHz, calculate the frequencies that can cause an IF response, for values of *m* and *n* ranging up to 2.

Solution First, we find f_{LO}:

$$\begin{aligned} f_{LO} &= f_{sig} + f_{IF} \\ &= 10 \text{ MHz} + 1.8 \text{ MHz} \\ &= 11.8 \text{ MHz} \end{aligned}$$

Now the problem can be easily solved using Equation (5.11) and a table of values. All frequencies in the table are in MHz.

m	n	$\left(\frac{m}{n}\right) f_{LO}$	$\frac{f_{IF}}{n}$	$\left(\frac{m}{n}\right) f_{LO} + \frac{f_{IF}}{n}$	$\left(\frac{m}{n}\right) f_{LO} - \frac{f_{IF}}{n}$
1	1	11.8	1.8	13.6	10.0
1	2	5.9	0.9	6.8	5.0
2	1	23.6	1.8	25.4	21.8
2	2	11.8	0.9	12.7	10.9

Rearranging the results in the last two columns in order of ascending frequency gives us the frequencies to which the receiver can respond. In MHz, they are:

5.0	12.7
6.8	13.6
10.0	21.8
10.9	25.4

In spite of the problems listed above, however, the superheterodyne remains the preferred arrangement for almost all receiving applications. It proves to be easier to improve the design to reduce its problems than to go to a different system.

5.5 Receiver Variations

Now that we have looked at the basic superheterodyne receiver topology, we will consider some of the variations that appear in receivers for different applications. This discussion will not be exhaustive, since practically every receiver, from the simplest AM radio to the most complex radio telescope, uses the superheterodyne technique. We will discuss some of the more common variations, progressing from the antenna towards the loudspeaker. For our purposes we will assume the baseband information signal is audio, though superheterodyne receivers are also used for video and data signals.

5.5.1 The Radio-Frequency Amplifier

Inexpensive receivers often omit the RF amplifier stage, especially if they are designed for operation at low to medium frequencies, where atmospheric noise entering the receiver with the signal is likely to be more significant than noise generated within the receiver itself. That is why our example AM broadcast-band receiver in Figure 5.6 did not have an RF amplifier. On the other hand, weak-signal reception at microwave frequencies requires very careful design of the first RF amplifier for the best possible noise figure. The low-noise amplifier (LNA) used with television satellite receivers is just an RF amplifier using a gallium arsenide field-effect transistor (GaAsFET) located right at the antenna in order to provide gain before the attenuation of the antenna cable can degrade the signal-to-noise ratio.

Between these extremes, we find high-frequency communications receivers and FM broadcast receivers, for instance. These receivers generally have one RF amplifier stage, located in the receiver, not at the antenna. The stage may be tuned to the signal frequency, in which case it has to track the local oscillator tuning. This can be accomplished by adding a third section to the dual-gang tuning capacitor shown earlier in Figure 5.7(b).

When a varactor-tuned or frequency-synthesized local oscillator is used, it is often inconvenient to tune the RF amplifier. A fixed or switchable bandpass filter is often used in such cases. The bandwidth of the filter must be wide enough to pass the range of frequencies to be received but small enough to exclude image frequencies.

There are two main reasons why FETs are more common than bipolar transistors for RF amplifiers in modern receiver designs. Their very high input impedance provides less loading to the input tuned circuit, improving its Q and allowing better image rejection. Perhaps more important, FETs exhibit lower levels of third-order intermodulation distortion than bipolar transistors. This reduces the possibility of interference between signals that are close enough in frequency to pass through the input filter together and improves the receiver's dynamic range. In the event that AGC is required, using a dual-gate MOSFET provides a handy extra element to which the AGC voltage can be applied.

5.5.2 The Mixer/Converter

Any of the mixer circuits described in Chapter 2 will work. Diode mixers are generally rejected as too noisy and too lossy, except in the simplest receivers (for example, some simple devices used to detect police radar signals have used a diode mixer connected directly to the antenna). Either bipolar transistors or FETs can be used as mixers, but the latter are preferred because they create fewer intermodulation distortion components. The problem here, as in the RF amplifier, is not so much distortion of the modulating signal as the creation of spurious responses due to interactions between desired and interfering signals.

The mixer and local oscillator can be combined for economy. The combination is called an autodyne converter or a *self-excited mixer*, and it is very common in simple AM broadcast receivers like the one discussed above. However, better designs use a separate local oscillator.

It is extremely important that the local oscillator be stable, as any frequency change will result in the receiver drifting away from the station to which it is tuned. *LC* oscillators must be carefully designed for stability, following the methods described in Chapter 2. This is expensive, of course, and some designs have used various types of *automatic frequency control* (AFC), which reduces drift by feeding an error signal back to the local oscillator. FM broadcast receivers and television receivers have made extensive use of AFC (often called AFT, for *automatic fine tuning*, when used in television receivers), and these methods will be discussed later when we cover those types of receivers. The higher the local oscillator frequency, the worse the drift (in kilohertz) for a given percentage frequency change; thus AFC is more commonly seen at VHF and up than at lower frequencies.

Another approach to local oscillator stability is to use crystal control. The local oscillator can be a simple crystal oscillator with a switch to change crystals for different channels. This becomes unwieldy with more than a few channels, so crystal-controlled frequency synthesizers have become very popular in new designs. The use of a synthesizer also allows remote control, direct frequency or channel number entry, and other similar conveniences. AFC is not necessary with a properly designed synthesizer. (See Chapter 2 for a discussion of synthesizers.)

Oscillator *spectral purity* is also important. Any frequency components at the oscillator output have the possibility of mixing with incoming signals and creating undesirable spurious responses. Noise generated by the local oscillator will degrade the noise performance of the receiver and should be minimized. Some designs use a bandpass filter to remove spurious signal components and noise from the local oscillator signal before it is applied to the mixer.

5.5.3 The Intermediate-Frequency Amplifier

The intermediate-frequency amplifier is basically a linear, fixed-tuned amplifier. The IF amplifier of an AM receiver must be Class A to avoid distorting the signal envelope.

The IF amplifier accounts for most of the receiver's gain and selectivity. The classical way to provide this is to use several stages, to give sufficient gain, coupled by tuned transformers that provide the selectivity. These transformers can be single-tuned. In that case, either the primary or secondary winding will be part of a *resonant circuit*. Alternatively, both windings can be included in resonant circuits, creating a double-tuned transformer. In either case, the capacitors that are needed for resonance are generally combined with the inductors in a shielded enclosure. Tuning is done with a nonmetallic screwdriver or "alignment tool." Figure 5.13(a) is a photograph of two of these IF transformers.

The choice of transformer depends on the frequency, bandwidth, and shape factor required. When narrow bandwidth is required, single- or double-tuned transformers with loose coupling may be used. For single-tuned transformers, tighter coupling means more gain but broader bandwidth, as shown in Figure 5.13(b).

The ideal IF passband has a flat top and steep sides. This is often achieved by using overcoupled double-tuned transformers. *Overcoupling* means that k is greater than the critical coupling factor k_c given by

$$k_c = \frac{1}{\sqrt{Q_p Q_s}} \tag{5.12}$$

where

Q_p, Q_s = primary and secondary Q, respectively

As a rule of thumb, a good compromise between steep skirts and flat passband is given by using the optimum coupling factor

$$k_{opt} = 1.5k_c \tag{5.13}$$

The effect of varying k for an overcoupled double-tuned transformer is illustrated in Figure 5.13(c). The bandwidth for a double-tuned amplifier with $k = k_{opt} = 1.5k_c$ is given approximately by:

$$B = kf_o \tag{5.14}$$

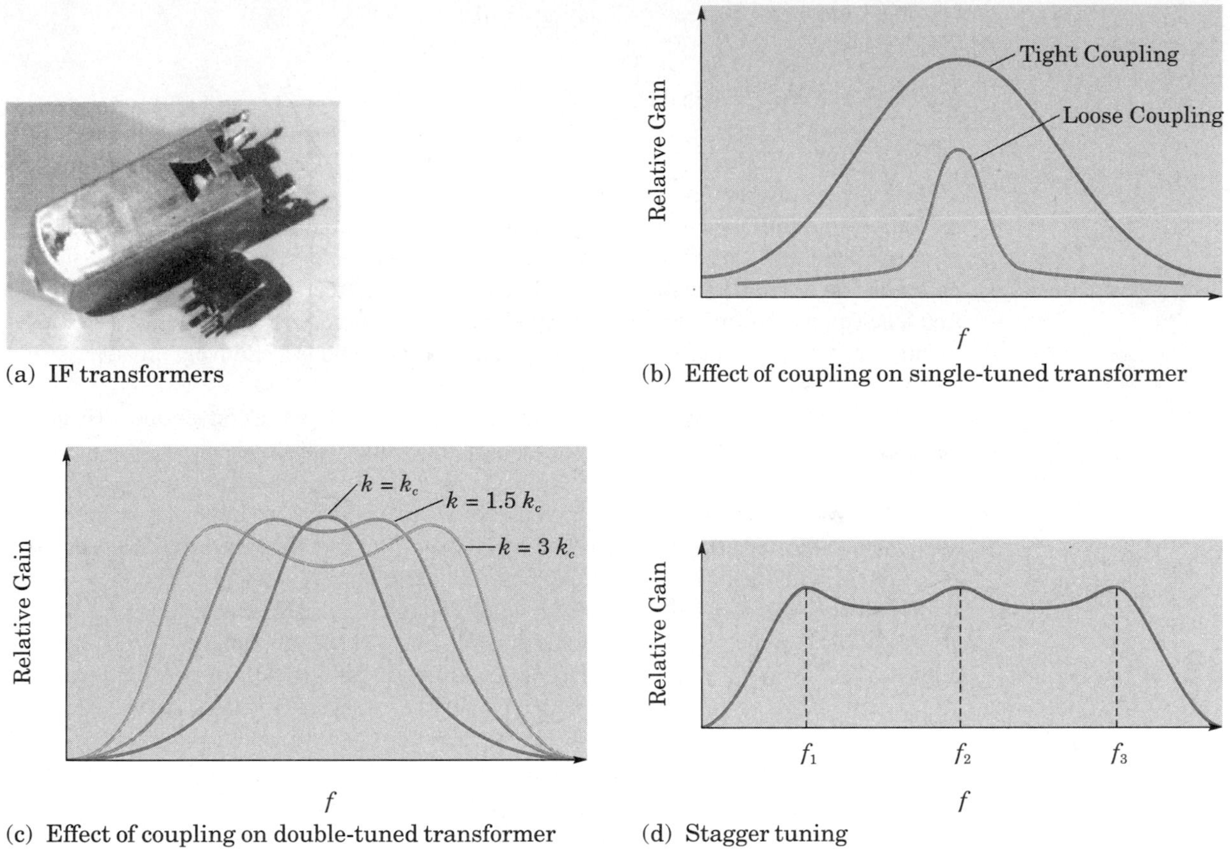

Figure 5.13
IF transformers

Example 5.7 An IF transformer operates at 455 kHz. The primary circuit has a Q of 40 and the secondary has a Q of 30. Find

(a) the critical coupling factor
(b) the optimum coupling factor
(c) the bandwidth using the optimum coupling factor

Solution (a) The critical coupling factor is given by Equation (5.12):

$$\begin{aligned} k_c &= \frac{1}{\sqrt{Q_p Q_s}} \\ &= \frac{1}{\sqrt{40 \times 30}} \\ &= 0.0289 \end{aligned}$$

(b) From Equation (5.13),

$$\begin{aligned} k_{opt} &= 1.5k_c \\ &= 1.5 \times 0.0289 \\ &= 0.0433 \end{aligned}$$

(c) The bandwidth is given by Equation (5.14):

$$B = kf_o$$
$$= 0.0433 \times 455 \text{ kHz}$$
$$= 19.7 \text{ kHz}$$

When more than one IF transformer is used, as is practically always the case, their responses are multiplied (or added, if expressed in decibels). If all are tuned to the same frequency, this will cause a reduction in bandwidth. If a wide passband with steep sides is required, two or more transformers may be *stagger tuned*, (that is, tuned to slightly different frequencies), as shown in Figure 5.13(d), where three single-tuned transformers are tuned to three slightly different frequencies, f_1, f_2, and f_3. The result is a relatively broad passband with three peaks.

The broadcast receiver we looked at earlier (Figure 5.6) has a conventional IF amplifier with two stages of amplification and three single-tuned transformers.

Crystal and Ceramic Filters The use of quartz crystals as frequency-determining components in oscillators was described in Chapter 2. It should be apparent that the properties of a crystal that make it useful in an oscillator, namely very high Q and excellent stability, could also be used to advantage in the design of bandpass filters. Using only one crystal (instead of a tuned circuit) would actually make the bandwidth too narrow for most purposes, but several crystals, tuned to slightly different frequencies, can be used in a *crystal lattice* to make a bandpass filter with a bandwidth of several kilohertz and an excellent shape factor. Figure 5.14 shows the schematic diagram for a typical crystal filter.

Some ceramics exhibit a piezoelectric effect similar to that of quartz and can also be used as resonators and filters. The Q of ceramic filters is lower than for quartz, and the bandwidth is wider. They are quite useful for wideband signals, such as broadcast FM.

Crystal and ceramic filters have the additional advantage, compared with conventional IF amplifier circuitry, of requiring no adjustment. IF amplifiers with multiple tuned transformers require a fairly elaborate process of alignment to set all the tuned circuits to the correct frequencies. Due to slight differences between different components, this must be done by hand *after* the receiver is constructed. This labor-intensive process is not required with crystal and ceramic filters, nor with mechanical and surface-acoustic-wave filters (to be described next).

It is quite possible to include more than one crystal filter in a receiver, for use with signals having different bandwidths. The correct filter can be selected with an ordinary switch, or (more elegantly) diode switching can be used.

Mechanical and Surface-Acoustic-Wave Filters An older technique, no longer much used, is to use mechanical resonators (consisting of discs and rods) as elements in a bandpass filter. Transducers at the input and output convert electrical energy to mechanical, kinetic energy and back again. These **mechanical filters** work very well, but only at frequencies up to about 500 kHz.

The **surface-acoustic-wave** (SAW) filter is a modern filter type based on similar principles. Again, transducers are used, and electrical energy is converted into mechanical

Figure 5.14
Crystal filter

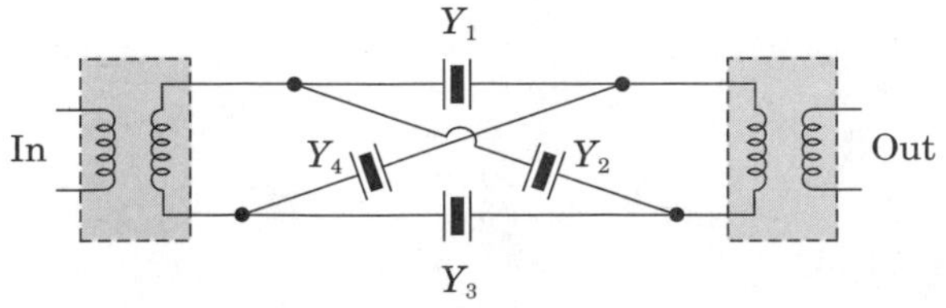

vibrations. This time, however, the vibrations are transverse waves on the surface of a piezoelectric substrate. The transducers are simply electrodes on the surface. By careful shaping of the substrate and the transducers, a wide variety of responses can be recreated at frequencies well into the megahertz range. For instance, television receivers, with an IF of 45.75 MHz for the picture signal, often use SAW filters. The problem in a television receiver is not to achieve a narrow bandwidth, since the signal is almost 6 MHz wide, but to produce a carefully shaped (not flat) response. The SAW filter is very well suited to this application.

Double Conversion The choice of IF usually involves a compromise. It should be high for image rejection, but satisfactory gain and selectivity are more easily obtained at lower frequencies. High and low are relative, of course, and over the years standard designs that achieve reasonable compromises have evolved. For instance, 455 kHz gives fair image rejection at AM broadcast frequencies (on the order of 1 MHz), and the relatively narrow bandwidth of most AM broadcast receivers (about 10 kHz) can easily be obtained at that frequency. FM broadcasting, which has a signal frequency 100 times as high, requires a higher IF for reasonable image rejection, but luckily the receiver bandwidth need not be as narrow; in fact, about 150 to 200 kHz is needed. A frequency of 10.7 MHz has become very common for the IF in such receivers.

Suppose, however, that the signal frequency is high and the required bandwidth is narrow. Such a condition can be found in high-frequency and VHF communications receivers. There may be no satisfactory compromise for an IF that provides both excellent image rejection and satisfactory selectivity. There are two solutions to this problem. One is to use a high IF for image rejection and to achieve narrow bandwidth by using more exotic filters than the simple tuned circuits referred to above. Crystal filters can be used well into the megahertz range, and so they allow the construction of excellent single-conversion receivers for narrow-bandwidth signals into the VHF range.

Another approach, which actually is often combined with the first, is to use double conversion. (See Figure 5.15 for a block diagram.) The first mixer is conventional; it uses a tunable local oscillator and converts all incoming signals to a first IF, which will be at a relatively high frequency for excellent image rejection. In fact, in some designs, the first IF is actually higher than the incoming signal frequency, and the sum, rather than the difference, of the signal and local oscillator frequencies is used. This type of mixing is called *up-conversion*.

The first mixer is followed by a relatively wideband filter and perhaps some gain. Then the signal enters another mixer whose local oscillator operates at a fixed, crystal-controlled frequency. The signal is converted to a lower second IF, where most of the gain and selectivity are provided. The filter in the first IF stage is there just to avoid image responses in the second mixer; it is the second IF filter that sets the bandwidth of the

Figure 5.15 Double-conversion receiver

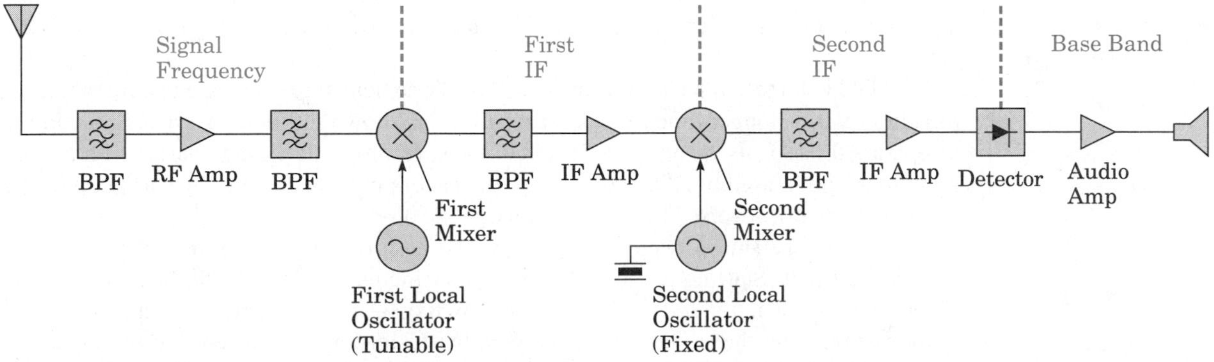

receiver. Double-conversion receivers require careful design because the extra mixer and local oscillator can give rise to additional spurious signals.

Of course, there is no reason that the number of conversions has to be limited to two. Some communications receivers use triple and even quadruple conversion.

It is also possible to make the first local oscillator fixed and the second variable in frequency. This is often done in systems where the first conversion takes place in a separate unit (often called, logically, a *converter*). The *block converters* used in some cable and satellite television systems (to be described in Chapters 9 and 21) are examples of this method, which has the advantage of requiring no adjustments on the converter unit.

Example 5.8 Find the intermediate frequencies for the receiver whose block diagram is shown in Figure 5.16, and state whether each local oscillator uses high-side or low-side injection.

Figure 5.16

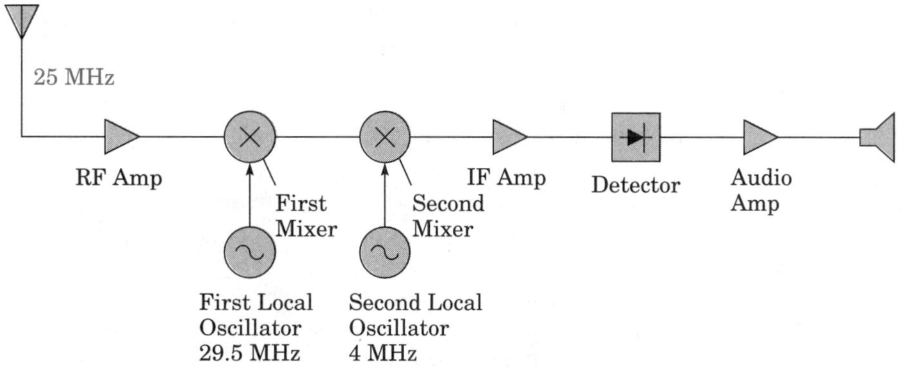

Solution The signal input frequency is 25 MHz, and the first local oscillator frequency is 29.5 MHz. Therefore the first local oscillator uses high-side injection, and the first IF is

$$\begin{aligned} f_{IF_1} &= f_{LO_1} - f_{sig} \\ &= 29.5\text{ MHz} - 25\text{ MHz} \\ &= 4.5\text{ MHz} \end{aligned}$$

The second local oscillator operates at 4 MHz. Therefore it uses low-side injection, and the second IF is

$$\begin{aligned} f_{IF_2} &= f_{IF_1} - f_{LO_2} \\ &= 4.5\text{ MHz} - 4\text{ MHz} \\ &= 0.5\text{ MHz} \\ &= 500\text{ kHz} \end{aligned}$$

5.5.4 Automatic Gain Control

Some form of gain control is necessary before the detector to reduce gain with strong signals and prevent overloading. This is usually done automatically using a feedback circuit.

AGC voltage can be derived from an AM diode detector by using an additional low-pass filter with a longer time constant (about 1 s), as shown in Figure 5.17(a). Sometimes, a separate diode is used for the AGC circuit. The resulting dc voltage varies with carrier amplitude and is used to adjust the bias on transistors in IF and sometimes RF amplifiers. Its polarity can be reversed by simply reversing the diode.

Figure 5.18 shows approximately how the gain of a bipolar transistor varies with collector current. Starting at the current I_c, either increasing or reducing the current will reduce the gain. Circuits that reduce the gain by increasing the current are called *forward AGC systems*, and those that reduce the gain by decreasing the current are *reverse AGC systems*. The circuit in Figure 5.17 uses reverse AGC: an increase in signal strength causes

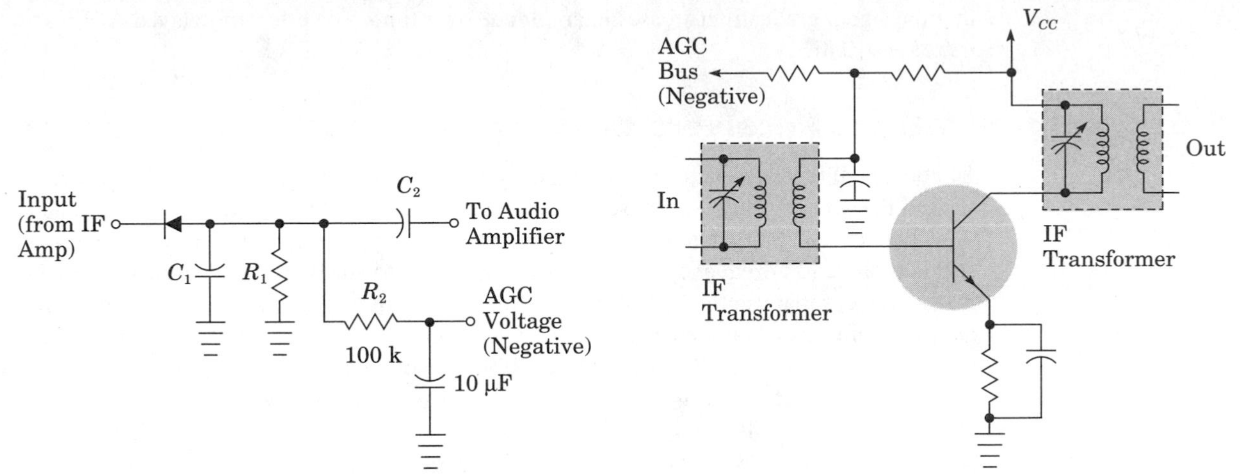

(a) Derivation of AGC voltage

(b) Application of AGC to NPN bipolar IF amplifier

Figure 5.17
Automatic gain control

the cathode of the detector diode to become more negative, resulting in the bases of the controlled transistors becoming less positive. Since the transistors are NPN, the bias is reduced with increasing signal strength.

Application of AGC to a representative IF amplifier stage is shown in Figure 5.17(b). The amplifier circuit uses conventional voltage-divider bias, except that the negative end of the voltage divider is connected to the AGC bus instead of to ground. As the signal becomes stronger at the detector, the AGC bus becomes more negative, and the voltage at the transistor base becomes lower. This reduces the transistor emitter current, lowering the gain of the stage.

Many communications receivers allow for a manual override of the AGC using a control misleadingly labeled "RF gain," although it is usually the gain of the IF stages that is being adjusted.

Many modern IF amplifiers use integrated circuits, in which case there will be a gain-control terminal to which an AGC voltage can be applied.

Delayed AGC Ordinary AGC reduces gain even for quite weak signals. More advanced AGC systems operate only when the signal reaches a threshold value set sufficiently low to avoid amplifier overload. Thus the "delay" in the name is really based on the assumption

Figure 5.18
Variation of transistor gain with collector current

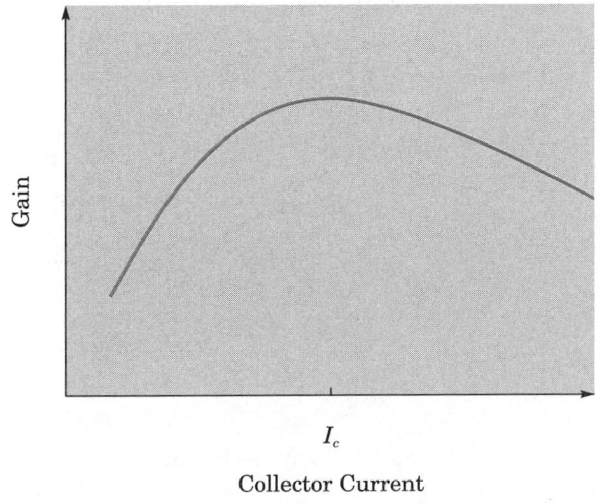

that the signal gradually increases in amplitude over time. The effect of delayed AGC is to increase sensitivity.

5.6 Communications Receivers

Of course, all receivers are used for communication, but the term *communications receiver* is used mainly for general-purpose receivers that cover a relatively wide range of frequencies. For instance, a typical receiver might cover the range from 100 kHz to 30 MHz. It would include low-frequency navigation signals, the AM standard broadcast band, and the many services that use the high-frequency range, including shortwave broadcasting, military and commercial radioteletype, amateur radio, citizens' band, and others. Other similar receivers are available for the VHF and UHF frequency ranges.

Communications receivers generally divide their frequency coverage into several bands. They are usually equipped to receive several types of modulation, including AM. In addition, they will have some of the special features listed below, which are included to make reception easier under difficult conditions.

5.6.1 Squelch

A **squelch** circuit disables the receiver audio in the absence of a signal. It can be implemented using the AGC voltage. When the voltage is very low, the audio amplifier is biased off. Squelch is very convenient for two-way mobile radio, as it eliminates channel noise in the absence of a carrier.

More elaborate squelch systems that shut off the audio unless a particular station is being received are also possible. These systems respond to a transmitted signal, which can vary from a simple low-frequency (sub-audible) tone to a complex digital code sent at the beginning of each transmission. This type of squelch is useful in mobile radio systems, as it enables a central dispatcher to address the mobile units individually or in any desired combination.

5.6.2 Noise Limiters and Blankers

Impulse noise, unlike thermal noise, can be removed from a signal to some extent in the receiver. This type of noise derives from sources such as lightning and automobile ignition systems and is characterized by short pulses with relatively large amplitudes and fast rise times.

The simplest way to deal with these noise pulses is simply to use a diode *limiter* or *clipper* in the audio section of the receiver. The limiting threshold should be set just above the peak level of the audio signal. Noise pulses, then, will be limited to about the same peak amplitude as the signal itself. Since the noise pulses are generally of large amplitude and short duration, clipping them to the same peak level as the rest of the signal removes much of their energy, rendering them less disturbing. Figure 5.19(a) shows the circuit of a simple clipper that will remove peaks greater than the diode threshold voltage.

Figure 5.19
Noise limiters for AM receivers

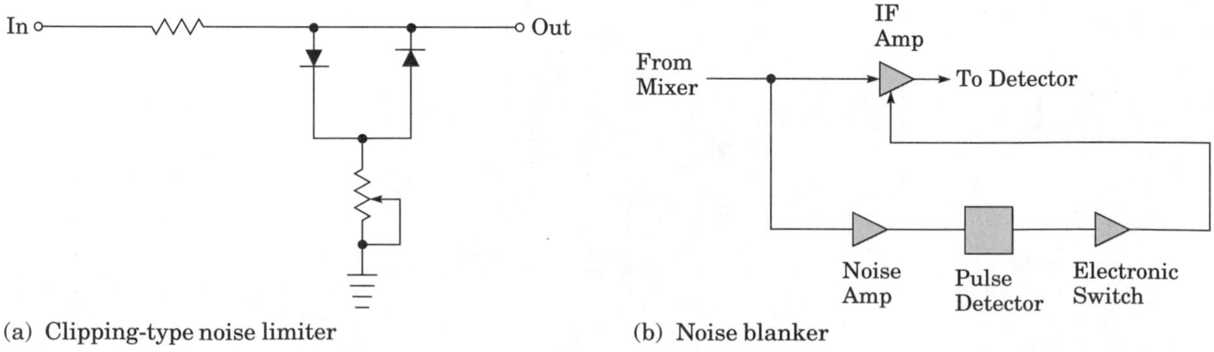

A better system is to *blank* noise pulses by muting the receiver output for the duration of the pulse. A block diagram of such a system is shown in Figure 5.19(b). It works by switching off the IF stage rather than the audio stages. This is more effective because it avoids any broadening of the noise pulse by the IF bandpass filter.

Both of these systems sense noise pulses by their large amplitude and fast rise time. Neither is effective in removing thermal noise, which is generally at a level below that of the signal.

5.6.3 Notch Filters

One of the most annoying types of interference occurs when two stations have almost the same carrier frequency. If an interfering signal has a carrier frequency 1 kHz above that of the desired signal, for example, the two carriers will mix in the detector to produce a difference frequency signal at 1 kHz, which will cause a very audible whistle.

Some communications receivers incorporate a tunable band-reject (*notch*) filter in the IF stage. This filter can be tuned to the frequency of the interfering carrier, greatly attenuating it and avoiding the audible whistle. Of course, this solution is not perfect: the notch filter will not attenuate the sidebands of the interfering signal. However, as shown in Chapter 3, these sidebands have an amplitude that is much smaller than that of the interfering carrier, so their potential for interference is less. The notch filter will also attenuate some of the sidebands of the desired signal, causing distortion, but if the filter is narrow, the distortion created will be less objectionable than the interference eliminated.

5.6.4 The S-Meter

The **S-meter** is designed to indicate signal strength for comparison and to aid in tuning the receiver and sometimes in adjusting the antenna. It is usually calibrated in units from 1 to 9 and then in "decibels above S-9." According to one standard, S-9 represents 50 μV at the antenna terminals, and each S-number represents 6 dB. However, there is really very little standardization, and S-meter readings are almost useless, except as an aid in tuning.

The AGC line provides a handy source for a voltage that varies with signal strength, and it is often used as a source for the S-meter circuit.

Example 5.9

An S-meter of the type described above reads S-6. Calculate the signal strength at the receiver input.

Solution

Since each S-number represents 6 dB, the signal will be 18 dB less than 50 μV. Therefore, using the equation

$$dB = 20 \log \frac{V_1}{V_2}$$

we can let $dB = 18$. We know that

$$V_1 = 50\ \mu\text{V}$$

so we can solve for V_2:

$$\frac{V_1}{V_2} = \text{antilog} \frac{dB}{20}$$

$$\begin{aligned} V_2 &= \frac{V_1}{\text{antilog}\ (dB/20)} \\ &= \frac{50\ \mu\text{V}}{\text{antilog}\ (18/20)} \\ &= 6.29\ \mu\text{V} \end{aligned}$$

5.7 Transceivers

A transceiver is essentially just a transmitter and a receiver in the same box. Transceivers are convenient and also allow certain economies to be made.

Most two-way radio schemes do not require *full-duplex operation*, which allows the operator to talk and listen simultaneously, as on an ordinary telephone. *Half-duplex* communication, in which the station transmits and receives alternately, is more common. Since the transmitter and receiver are never used at the same time, it is possible to use some of the same circuitry for both transmitting and receiving.

Audio circuitry is one example. The transmitter needs an audio amplifier to boost the microphone output to sufficient power to modulate the transmitter. The receiver also needs an audio amplifier to amplify the detector output to an adequate level to drive a loudspeaker. With suitable switching, the same amplifier can be used for both purposes.

5.7.1 Example: A Citizens' Band Transceiver

Figure 5.20 is a block diagram of a typical CB transceiver. The frequencies shown are present for an operating frequency of 26.965 MHz (channel 1). We looked at the transmitter section of this unit in Chapter 4. Now we will look at the receiver section and see how certain economies have been made by using some stages for both transmitting and receiving.

A look at the schematic in Figure 5.21 shows that the actual circuitry is quite conventional. In particular, the receiver and transmitter both use discrete, bipolar technology, except for the frequency synthesizer and the audio power amplifier, which employ integrated circuits.

In Chapter 3, we found that the transmitter section has five stages: oscillator (which is a frequency synthesizer), frequency doubler, and three amplifier stages. The driver and

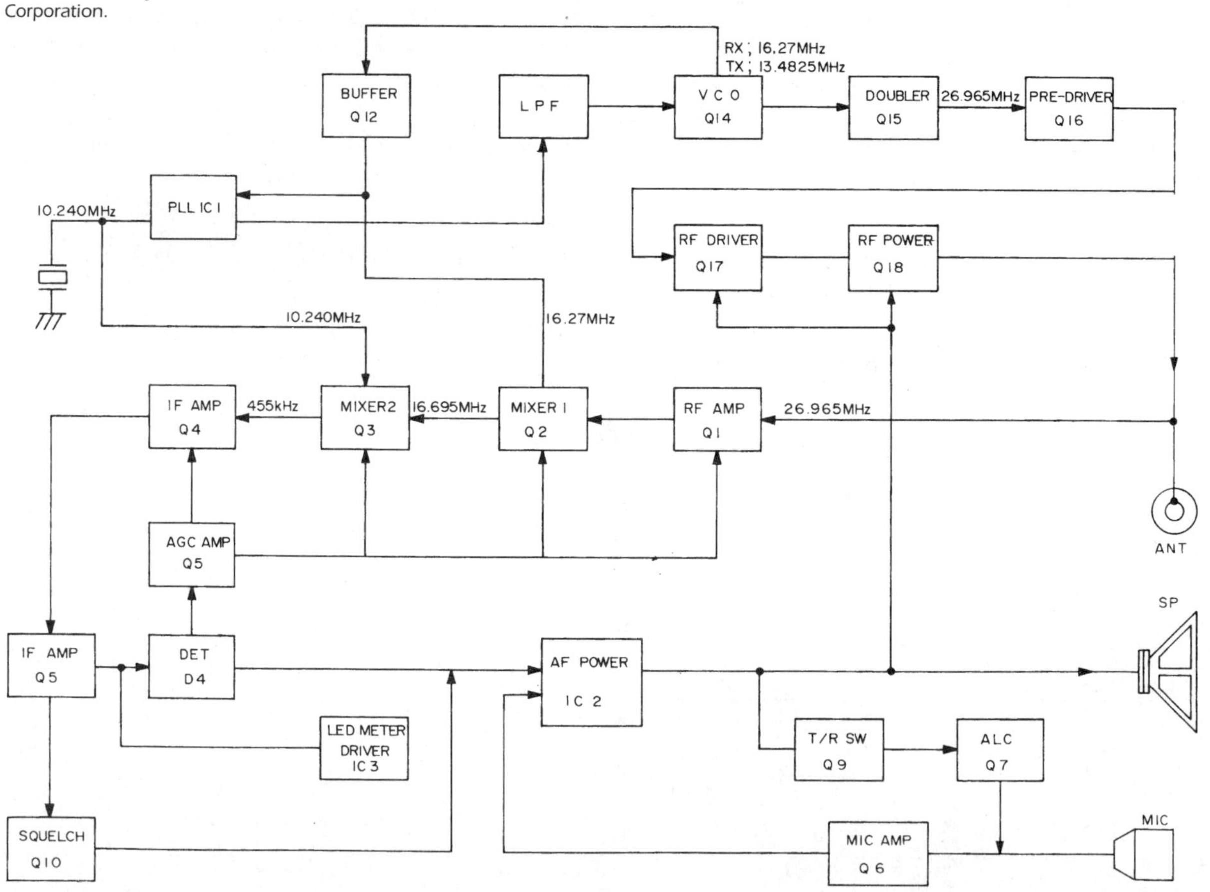

Figure 5.20
Block diagram of CB transceiver
Courtesy of Radio Shack, a division of Tandy Corporation.

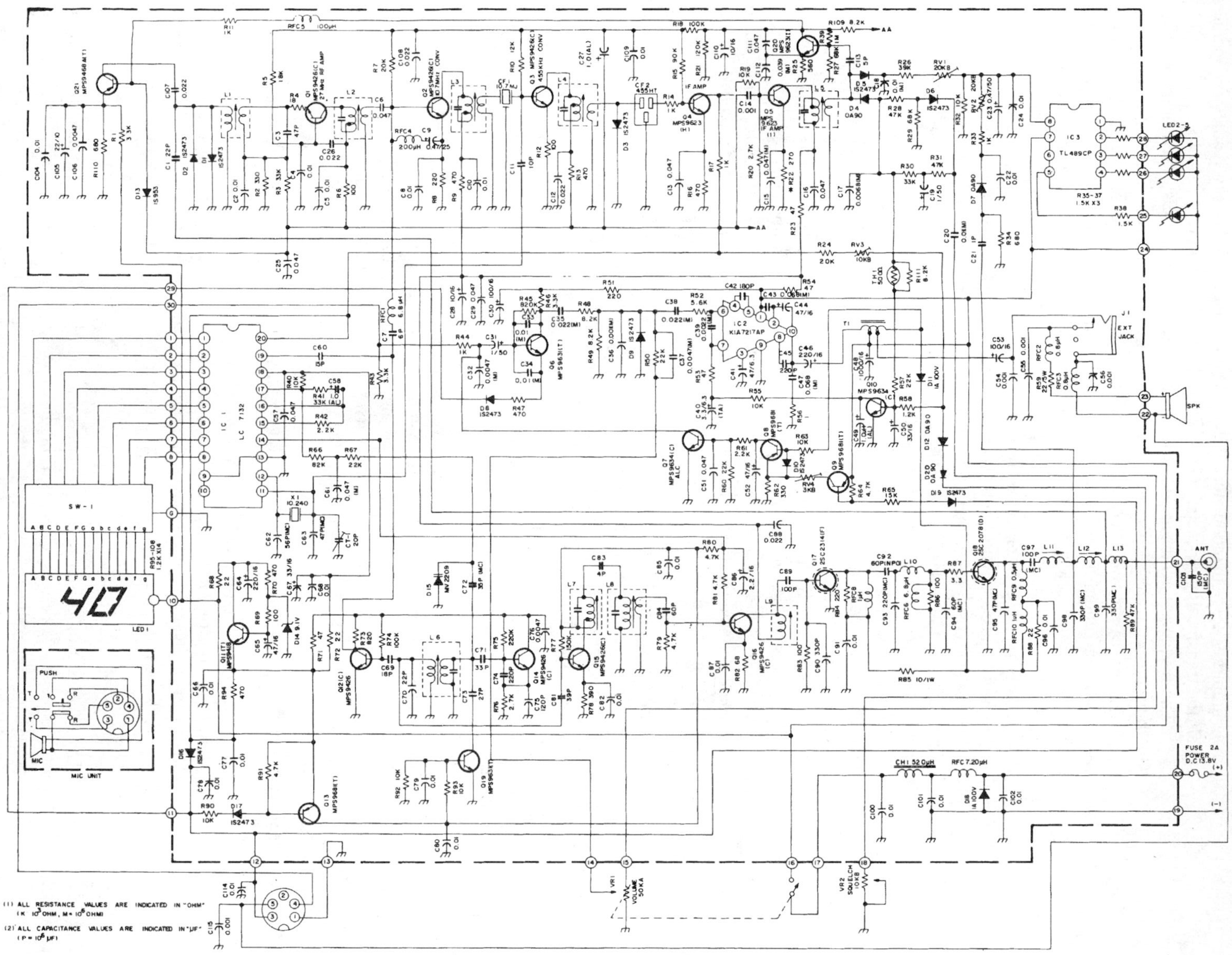

Figure 5.21
Schematic diagram of CB transceiver
Courtesy of Radio Shack, a division of Tandy Corporation.

NOTES. (1) ALL RESISTANCE VALUES ARE INDICATED IN "OHM" (K 10^3 OHM, M = 10^6 OHM)
(2) ALL CAPACITANCE VALUES ARE INDICATED IN "μF" (P = 10^{-6} μF)

power amplifier are both modulated. When transmitting, the frequency synthesizer operates at half the transmitting frequency (13.4825 MHz in this example).

The frequency synthesizer and audio power amplifier are also used in the receiver. The transceiver thus has economies compared with a separate receiver and transmitter.

The receiver is double conversion. It has a single RF stage, followed by a mixer that converts the incoming frequency to a 10.695 MHz first IF. This is high enough to achieve the specified image rejection of at least 60 dB. This particular choice of frequency also allows the use of components such as IF transformers and a ceramic filter designed for the 10.7 MHz IF used in FM broadcast receivers.

The synthesizer that sets the transmitting frequency also serves as the first local oscillator for the receiver: it switches frequencies between transmit and receive. Note that the first mixer uses low-side injection. This requires it to operate at the receive frequency less 10.695 MHz (16.270 MHz in this example).

The synthesizer is based on a 10.240 MHz crystal oscillator, which also provides a fixed local oscillator signal for the receiver's second mixer. That mixer also uses low-side injection. The difference between the first IF of 10.695 MHz and 10.240 MHz is 455 kHz, which is the second IF. You will recall that this is the usual IF for AM broadcast receivers. Again, it offers the designer a choice of available, low-cost components.

The rest of the receiver is conventional, with two stages of IF amplification, an envelope detector, and an audio amplifier. Squelch and S-meter circuits are provided.

Looking more closely at the schematic diagram, we see that Q_1, the RF amplifier, is in the common-base configuration. The pair of diodes at its input (D_1 and D_2) are to protect Q_1 from damage while the transceiver is transmitting. D_1 and D_2 limit the voltage at the receiver input to approximately 700 mV peak.

The receiver RF stage is fixed-tuned, so no tracking with the local oscillator is necessary. The output of the first mixer Q_2 is coupled by a single-tuned transformer L_3 to a 10.7 MHz ceramic filter CF_1 and then directly to the second mixer. There is no amplifier stage at the first IF. The second mixer Q_3 sends its output through another single-tuned transformer L_4 to another ceramic filter CF_2, which operates at 455 kHz. Two stages of IF amplification follow.

The detector is a conventional envelope type, using diode D_4. D_5 provides a voltage proportional to carrier strength for the S-meter, which is a row of LEDs driven by IC_3. The detected signal is low-pass filtered and amplified by Q_{20} before being used as an AGC voltage to control the bias on the RF, mixer, and IF stages. The audio is amplified by an integrated amplifier IC_2 and routed to the speaker.

5.8 AM Receiver Test Procedures

At the beginning of this chapter, we looked at some of the major criteria for receiver quality, such as sensitivity and selectivity. In the meantime, we have found that there are other important receiver characteristics, some of which (like image rejection) are unique to the superheterodyne system. We will now consider how some of these characteristics can be measured.

5.8.1 Sensitivity

Any specification of *sensitivity* requires some mention of noise level to have much meaning. Signal-to-noise ratio (*S/N*) is difficult to measure directly because of the difficulty of completely separating the two quantities. Usually, signal-plus-noise to noise [$(S+N)/N$] or SINAD (signal-plus-noise-and-distortion to noise-and-distortion) is measured instead. Manufacturers usually use the $(S+N)/N$ method, to be described here, to specify the sensitivity of AM receivers. (The SINAD method is more commonly used with FM receivers, so it will be studied in Chapter 8.)

AM receiver sensitivity is usually specified as the minimum signal level, with 30% modulation, that will give a 10 dB $(S+N)/N$ ratio with at least 500 mW of audio output. Measuring it requires a calibrated RF signal generator and an audio voltmeter. Figure 5.22

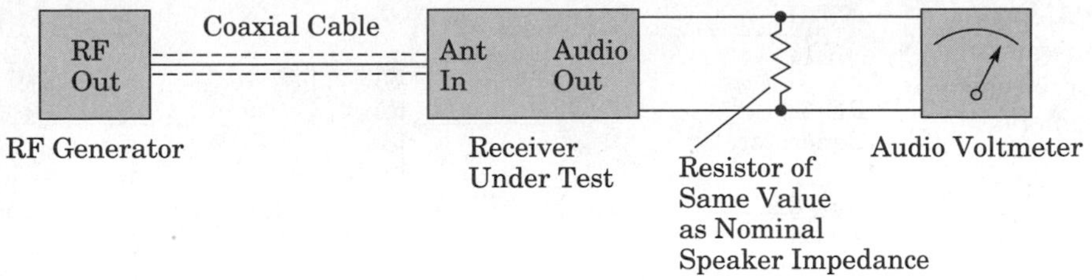

Figure 5.22 Test setup for measurement of receiver sensitivity

shows a typical test setup. The receiver audio gain (and RF gain, if present) should be at maximum, and any squelch circuit must be off. The RF generator output impedance should be the same as the receiver input impedance. If it is not, an impedance-matching network must be used, and any losses in the network must be taken into account. All covers should be on the receiver, and all cables should be shielded to prevent the entry of noise other than that generated in the receiver.

To measure $(S+N)/N$, first measure the receiver audio output with the signal modulated to a depth of 30% by a 1 kHz tone. This measurement will include the signal plus the noise. Then switch the modulation off at the generator, leaving the carrier on. The remaining output from the receiver is the noise. The carrier is left on so that the receiver AGC level will not change. Otherwise, the AGC is likely to increase the receiver gain and hence its noise output when the carrier is turned off.

If both of the above measurements are taken in decibels with respect to the same reference (in dBV, for example), then simple subtraction will yield $(S+N)/N$ in decibels. If the result is greater than 10 dB, the RF generator carrier level should be reduced and the measurements repeated. If the ratio is less than 10 dB, the generator level must be increased.

When the 10 dB ratio is reached, the audio output level should be checked to make sure it is at least 500 mW or whatever is specified by the manufacturer. This is rarely a problem. The power can, of course, easily be calculated from the voltage across a known load resistance. A loudspeaker does not make a very satisfactory load, as its impedance may vary widely with frequency and may not be at all close to its nominal value. Besides, even 500 mW of a 1 kHz tone is annoyingly loud! Using a noninductive resistor of the same value as the nominal speaker impedance (in place of the speaker) is better.

Assuming that at least the minimum audio output power is achieved, the carrier level necessary to achieve 10 dB $(S+N)/N$ is the sensitivity; otherwise, the carrier level is increased until the specified audio output is obtained, and that carrier level is the sensitivity.

5.8.2 Selectivity

There are as many ways to measure selectivity as there are to specify it. Two representative methods will be described; others can be found in manufacturers' service manuals.

There is a straightforward method using only one RF generator. The test setup is the same as for sensitivity measurements (see Figure 5.22). The generator is tuned to the same frequency as the receiver and set for the level that provides a 10 dB $(S+N)/N$ reading, just as for sensitivity measurements. Then either the receiver or the generator is detuned, and the generator level is increased until the 10 dB $(S+N)/N$ ratio is restored. The ratio in decibels between the two generator levels is the amount by which the off-frequency signal is attenuated. With this method, it helps to switch off or otherwise disable the AGC, so that the receiver gain will remain constant.

This method is easy to understand and carry out, but its results may not predict the receiver's performance under conditions of actual interference very accurately, because a strong interfering signal may overload the receiver's front end, causing blocking. Figure 5.23 shows a test setup for selectivity measurement using two RF generators that approximates real-world conditions more closely. The attenuation of the combiner network must be taken into account in the measurements. Generator A is set to the receiver's frequency and is modulated to a depth of 30% by a 1 kHz tone. Generator B is set to the

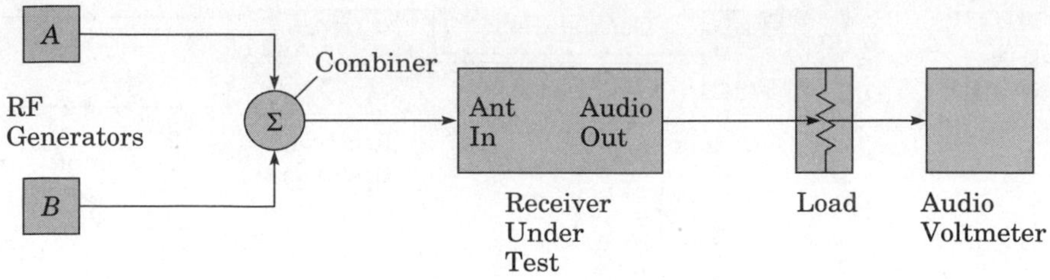

Figure 5.23 Two-generator test setup for selectivity measurements

adjacent channel and is modulated 30% by a 400 Hz tone. With generator *B*'s output off, generator *A* is adjusted to give an $(S+N)/N$ reading of 10 dB, as for a sensitivity measurement. Then generator *B* is switched on, and its carrier level is increased until the $(S+N)/N$ ratio drops to 6 dB. The ratio between the output power levels of the two generators, measured in decibels, is the adjacent channel selectivity.

This second method gives reasonably good results, as it approximates actual operating conditions in which the adjacent channel signal interferes with the reception of a signal on the channel to which the receiver is tuned.

5.8.3 Intermediate Frequency and Image Rejection

Intermediate frequency and image rejection can be measured in the same way as selectivity: instead of being on the adjacent channel, the interfering signal is inserted at the intermediate frequency or at an image frequency. Generally the first method described above, using one generator, is used.

5.8.4 Trouble-shooting

The tests above will identify some receiver problems such as low sensitivity and poor selectivity and will allow the comparison of receivers. Assuming that a receiver met its specifications when new, any deviation from specified performance represents a fault that can be found and corrected. A reasonably detailed knowledge of the particular receiver is required to correct subtle faults like poor sensitivity, which often indicates a need for realignment of the tuned circuits. Information on alignment can be found in the receiver service manual.

Much of the time, receiver faults are not subtle enough to require detailed testing. For instance, the unit may not work at all, or its output may be so distorted as to be unintelligible. Or the receiver may work but may be so insensitive that it requires an input signal measured in millivolts rather than microvolts.

The power supply should always be suspected when a radio is completely dead. Assuming all is in order there, cases like this can often be solved by using either of two time-honored and rather straightforward techniques known as *signal injection* and *signal tracing*.

In signal tracing, the object is to follow the signal through the receiver and see where it disappears or becomes severely distorted. Beginning at the antenna input, a sensitive detector can be moved through the receiver, checking the signal at the input and output of each stage. The detector can be an oscilloscope or a sensitive amplifier with a speaker. For RF and IF signals, a *demodulator probe* can be used. A demodulator probe is simply a diode detector that provides a demodulated audio signal to the amplifier or oscilloscope.

Signal injection is similar, except that a signal is applied to the device under test, starting at a point close to the output and working back towards the antenna. The receiver's loudspeaker can be used as a monitoring device. Once again, the point where the signal disappears indicates the defective stage. Of course, the signal must be appropriate, in both amplitude and frequency, to the point in the circuit to which it is applied.

Once the problem has been localized to a single stage, other techniques must be used. In-circuit voltage measurements can be used to form an opinion as to the defective component. Then individual components can be tested. Bear in mind that power-supply problems can sometimes cause the failure of only one stage (where a regulated supply is used for a critical circuit, for example).

Signal tracing and signal injection each have their difficulties. There may be a problem finding a device sensitive enough to detect a signal in the RF stage of a receiver, for instance, while injecting such a signal is no problem: an ordinary RF generator will do. It should be coupled through a capacitor to the circuit to avoid any problems with dc voltages that may be present.

Signal injection requires a variety of signals. For an ordinary AM receiver, AM signals must be available at the carrier and the intermediate frequencies. A baseband audio signal must also be available, and the correct signal must be applied at each test point. Care must also be taken not to overload sensitive circuits too greatly (to the point where damage could occur).

An objection might be made that in modern receivers much of the circuitry is likely to be on one integrated circuit, and it makes very little difference which part of the chip is defective. That is certainly true; once a defect is localized to an integrated circuit, the whole chip must be replaced. Before doing so, however, especially if the IC is soldered in place, it is a good idea to make sure that the chip is receiving all the correct supply voltages and signals.

Most AM broadcast receivers, except those incorporated in high-fidelity tuners, are so inexpensive that it is not cost-effective to repair them. A radio that sells new for twenty dollars is not worth very much of a technician's time. Nevertheless, the basic techniques of signal tracing and signal injection are applicable to many more complex and expensive systems, and a simple AM radio is a good place to learn these techniques.

Example 5.10 Find suitable test points for signal injection or tracing in the circuit shown in Figure 5.6.

Solution This is a typical AM broadcast-band receiver. Suitable test points are at the input and output of each stage. Some easily identifiable points in the mixer and IF stages are the transistor bases and collectors. The tap of the volume control is easy to find and will provide access to the audio-amplifier input.

Summary

Here are the main points to remember from this chapter.

1. A receiver must separate the desired signal from other signals and noise and then demodulate the signal. A considerable amount of gain is also necessary in a practical receiver.
2. By far the most common receiver type is the superheterodyne, which uses a mixer/local oscillator combination to transfer all incoming signal frequencies to a common IF.
3. Better-quality receivers, particularly at the higher frequencies, use at least one stage of RF amplification before the mixer. The RF amplifier is principally responsible for setting the noise figure for the receiver.
4. Superheterodyne receivers can receive signals at frequencies other than that to which the receiver is tuned. The most important such signal is called the image frequency.
5. Images and other spurious responses must be rejected before the mixer. This requires bandpass filtering in the receiver before the mixer.
6. The IF must be high enough to provide good image rejection but low enough to allow the required selectivity to be obtained with the type of filter in use.
7. The stability of a superheterodyne receiver depends directly on that of the local oscillator. Changing the local oscillator frequency tunes the receiver.

8. The IF amplifier is principally responsible for the selectivity of the receiver and it also provides most of the predetection gain.
9. Most AM receivers use envelope detectors, which have the advantage of great simplicity and the disadvantage of relatively high levels of distortion.
10. Receivers require some form of AGC to compensate for the very great range in signal strength at the antenna.
11. The most important specifications for a receiver are sensitivity and selectivity. *Sensitivity* refers to the signal strength required for a satisfactory signal-to-noise ratio, and *selectivity* refers to the ability of the receiver to reject interference and out-of-channel noise.
12. Signal injection and signal tracing are useful troubleshooting techniques for receivers and for many other types of electronic systems.

Important Equations

$$Q = \frac{X_L}{R} \tag{5.1}$$

$$B = \frac{f_o}{Q} \tag{5.2}$$

$$SF = \frac{B_{-60\text{ dB}}}{B_{-6\text{ dB}}} \tag{5.3}$$

$$f_{image}\text{ (high-side injection)} = f_{sig} + 2f_{IF} \tag{5.7}$$

$$f_{image}\text{ (low-side injection)} = f_{sig} - 2f_{IF} \tag{5.8}$$

$$IR = \frac{A_{sig}}{A_{image}} = \sqrt{1 + Q^2x^2} \tag{5.9}$$

$$\text{where } x = \frac{f_{image}}{f_{sig}} - \frac{f_{sig}}{f_{image}}$$

$$IR\text{ (dB)} = 20 \log \frac{A_{sig}}{A_{image}} \tag{5.10}$$

$$f_s = \left(\frac{m}{n}\right) f_{LO} \pm \left(\frac{f_{IF}}{n}\right) \tag{5.11}$$

$$k_c = \frac{1}{\sqrt{Q_p Q_s}} \tag{5.12}$$

$$k_{opt} = 1.5k_c \tag{5.13}$$

$$B = kf_o \tag{5.14}$$

Glossary

adjacent channel the communications channel immediately above or below the desired channel in frequency

alternate channel the next communications channel beyond the adjacent channel

autodyne converter a combined mixer and local oscillator that uses one transistor or tube for both

automatic gain control (AGC) a circuit to adjust the gain of a system in accordance with the input signal strength

blocking reduction of gain for a weak signal due to a strong signal close to it in frequency

ceramic filter a bandpass filter using piezoelectric ceramic elements

converter a combination of a mixer and a local oscillator that is used to move a signal from one frequency to another

crystal filter a bandpass filter that uses piezoelectric quartz elements

front end the first stage of a receiver

high-side injection application to a mixer of a signal from a local oscillator that operates at a frequency above that of the incoming signal

image frequency in a frequency converter, a second input frequency that produces the same output frequency

intermediate frequency (IF) a frequency to which a signal is shifted as an intermediate step in reception or transmission

local oscillator an oscillator used in conjunction with a mixer to shift a signal to a different frequency

low-side injection application to a mixer of a signal from a local oscillator that operates at a frequency below that of the incoming signal

mechanical filter a bandpass filter that uses mechanical resonators

S-meter a meter on a receiver that indicates the strength of the received signal

selectivity the ability of a receiver to reject signals of frequencies other than the frequency to which the receiver is tuned

sensitivity the ability of a receiver to receive weak signals with a satisfactory signal-to-noise ratio

shape factor for a bandpass filter, the ratio between the bandwidths for two specified amounts of attenuation

skin effect reduction in effective cross-sectional area of a conductor with increasing frequency

spurious response reception of signals at frequencies other than that to which a receiver is tuned

squelch a system that disables the output of a receiver in the absence of a suitable signal

superheterodyne receiver a receiver in which the signal is moved, using a mixer, to an intermediate frequency before demodulation

surface-acoustic-wave (SAW) filter a filter that uses acoustic waves on the surface of a substrate to achieve the desired response

tracking adjustment of two or more tuned circuits so that they can be tuned simultaneously with one adjustment

tuned-radio-frequency (TRF) receiver a receiver in which the signal is amplified at its original frequency before demodulation

Questions

1. Explain the operation of an envelope detector.
2. Why do all envelope detectors add distortion?
3. Compare the TRF and superheterodyne receiver types. Which is better? Why?
4. Why does the resistance of an inductor increase with frequency?
5. Explain the purpose and operation of trimmer and padder capacitors.
6. Distinguish between low- and high-side injection of the local oscillator signal.
7. Explain how image-frequency signals are received in a superheterodyne receiver. How can these signals be rejected?
8. Why do some superheterodyne receivers use an RF stage while others do not?
9. What are the main characteristics of a well-designed RF stage?
10. What is an autodyne converter?
11. Why is the stability of the local oscillator important?
12. In addition to images, what spurious responses are possible with superheterodyne receivers? How are they caused?
13. What is AGC? Why is it required in a practical receiver?
14. Why do designers of communications receivers often use a switchable RF stage and a switchable input attenuator instead of applying AGC to the RF stage?
15. Describe four types of filters that can be used in the IF stages of a receiver. Give an application for each.
16. What is the shape factor of a filter? Explain why a small value for the shape factor is better for the IF filter of a receiver.
17. What causes the relatively high distortion that is a characteristic of envelope detectors?
18. What is the advantage of delayed AGC in a communications receiver?
19. What advantage is gained by using double conversion in a receiver? Are there any disadvantages?
20. Describe a system for reducing the effect of lightning-caused noise in a receiver. Will the same system work for thermal noise? Explain your answer.
21. Describe one technique for measuring the sensitivity of a receiver.

22. Describe two techniques for measuring the selectivity of a receiver.

23. List and describe two troubleshooting techniques that are useful with receivers.

Problems

SECTION 5.3

24. A tuned circuit has a Q of 60 at 5 MHz. Find its bandwidth at 5 MHz and 20 MHz.
25. A superheterodyne receiver is tuned to a frequency of 5 MHz when the local oscillator frequency is 6.65 MHz.
    (a) What is the IF?
    (b) Which type of injection is in use?
26. A superheterodyne receiver has an IF of 9 MHz and tunes the frequency range from 50 to 60 MHz. The mixer uses low-side injection of the local oscillator signal. Calculate the range of local oscillator frequencies.
27. An FM broadcast-band receiver tunes from 88 to 108 MHz. The IF is 10.7 MHz, and the receiver uses high-side injection. Calculate the range of local oscillator frequencies.

SECTION 5.4

28. One receiver has a sensitivity of 1 μV and another has a sensitivity of 10 dBf under the same measurement conditions. Both receivers have an input impedance of 50 Ω. Which receiver is more sensitive?
29. A receiver has a sensitivity of 0.3 μV. The same receiver can handle a signal level of 75 mV without overloading. What is its AGC range in decibels?
30. The receiver in problem 29 has a blocking dynamic range of 80 dB. If the desired signal has a level of 10 μV, what is the maximum signal level that can be tolerated within the receiver passband?
31. A receiver uses low-side injection for the local oscillator, with an IF of 1750 kHz. The local oscillator is operating at 15.750 MHz.
    (a) To what frequency is the receiver tuned?
    (b) What is the image frequency?
32. An AM broadcast receiver with high-side injection and an IF of 455 kHz is tuned to a station at 910 kHz.
    (a) What is the local oscillator frequency?
    (b) What is the image frequency?
    (c) Find four other spurious frequencies that could be picked up by this receiver.
33. A receiver's IF filter has a shape factor of 2.5 and a bandwidth of 6 kHz at the 6 dB down point. What is its bandwidth at 60 dB down?
34. What is the shape factor of the filter sketched in Figure 5.24?

Figure 5.24

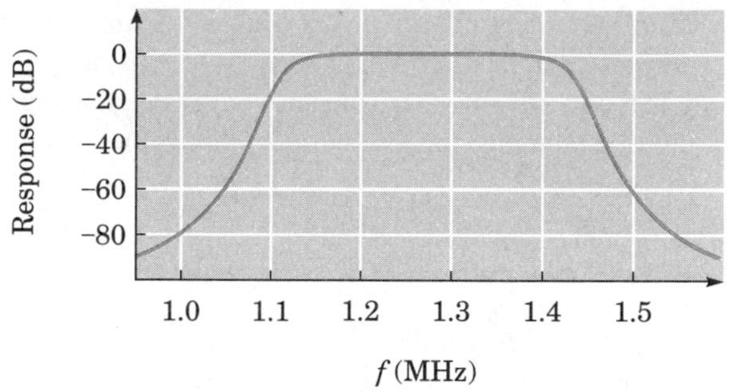

SECTION 5.5

35. A receiver has two uncoupled tuned circuits before the mixer, each with a Q of 75. The signal frequency is 100.1 MHz, and the IF is 10.7 MHz. The local oscillator uses high-side injection.
    (a) Calculate the image rejection ratio in decibels.
    (b) To show the advantage of a high IF, recalculate the image rejection, assuming an IF of 455 kHz.
36. An AM broadcast receiver tunes from 530 to 1700 kHz. The IF is 455 kHz and the local oscillator uses high-side injection. The local oscillator uses a variable capacitor with a maximum value of 365 pF. Calculate:
    (a) the range of frequencies that must be generated by the local oscillator
    (b) the value of inductor needed so that the local oscillator will tune the receiver to the lowest frequency on the band when the capacitor is at maximum
    (c) the required minimum value for the variable capacitor so that the receiver will be tuned to the highest frequency on the band when the capacitor is set at minimum
37. A double-tuned IF transformer has $k = 1.5k_c$. The primary Q is 50 and the secondary Q is 40. Calculate the bandwidth of the transformer at a frequency of 10.7 MHz.
38. The block diagram of Figure 5.25 shows a double-conversion receiver.
    (a) Does the first mixer use low- or high-side injection?
    (b) What is the second IF?
    (c) Suppose that the input signal frequency is changed to 17.000 MHz. What would the frequencies of the two local oscillators be now?

Figure 5.25

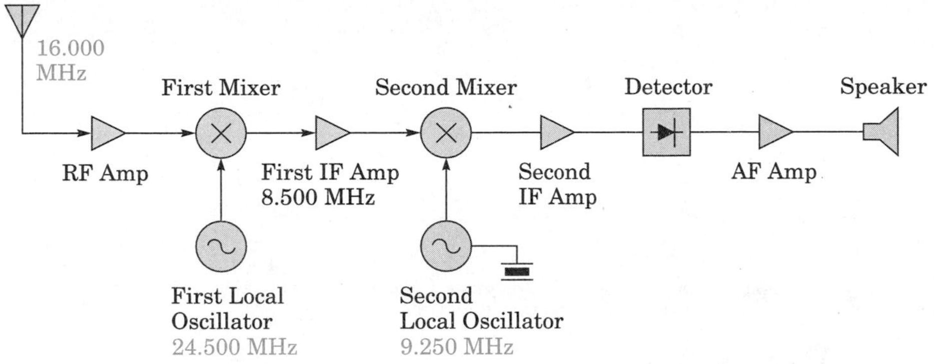

39. Suppose the receiver of Problem 27 is tuned by a variable capacitor with a minimum capacitance of 10 pF.
    (a) What is the required inductance of the coil which resonates with this capacitor?
    (b) What is the required maximum capacitance for the variable capacitor?
40. A tuned transformer operating at 5 MHz has a primary Q of 60 and a secondary Q of 40.
    (a) Calculate the value of the coupling factor for critical coupling.
    (b) Calculate the value of the coupling factor for optimum coupling.
    (c) Calculate the bandwidth of the circuit for optimum coupling.
41. A double-conversion receiver covers the frequency range from 100 to 200 MHz. The first IF is 30 MHz, and the first mixer uses high-side injection of the local-oscillator signal. The second IF is 1 MHz, and the second mixer uses low-side injection.
    (a) Calculate the operating frequency or frequency range for both local oscillators.
    (b) Draw a block diagram for this receiver.

SECTION 5.6

42. How much stronger, in decibels, is an S-9 signal compared with an S-1 signal?
43. A receiver is equipped with an S-meter as described in the text. What would it read for the following signal levels?
    (a) 12.5 μV
    (b) 100 μV
    (c) 2 mV

SECTION 5.7

44. Refer to the schematic diagram for the transceiver described in Section 5.7, and determine how the unit is switched from receive to transmit; that is, electrically, what happens when the push-to-talk button is pressed?

SECTION 5.8

45. What frequency and type of signal would be applied to each of the following points during signal injection with a typical AM broadcast-band receiver?
    (a) audio-amplifier input
    (b) detector input
    (c) mixer output
    (d) mixer input

COMPREHENSIVE

46. Suppose a receiver has the following gain structure (and that all stages are operating at maximum gain):

    RF amp: 12 dB gain
    mixer: 1 dB loss
    IF amp: 76 dB gain
    detector: 3 dB loss
    audio amp: 35 dB gain

    (a) What is the total power gain of the receiver from antenna to speaker?
    (b) What would be the minimum signal required at the antenna in order to get a power of 0.5 W into the speaker? Express your answer both in watts and in microvolts, assuming a 50 Ω input impedance.
    (c) Now suppose a 100 mV signal is applied to the antenna. Calculate the output power. Is this a reasonable answer? Explain why or why not. What would actually happen in a real receiver?

6 Suppressed-Carrier AM Systems

Objectives After studying this chapter, you should be able to:

1. Explain the advantages of suppressed-carrier and single-sideband transmission over standard AM
2. Calculate the improvements in signal-to-noise ratio that result from the use of suppressed-carrier and single-sideband techniques
3. Describe the operation of a balanced modulator and sketch the output in both time and frequency domains for typical input signals
4. Describe the filter method of single-sideband generation and calculate suitable filter and oscillator parameters to produce a given sideband at a given carrier frequency
5. Analyze the operation of a single-sideband transmitter, calculating the output frequency and sideband
6. Describe the operation of communications receivers that are capable of single-sideband reception and show how they differ from receivers designed only for full-carrier AM
7. Trace the signal path for transmit and receive operations in a block diagram of a single-sideband transceiver and determine which components are used for both operations
8. Perform tests on single-sideband transmitters to measure peak envelope power and the suppression of harmonics and other spurious signals
9. Describe several practical communication systems that use suppressed-carrier and/ or single-sideband techniques

6.1 Introduction

Although full-carrier AM is simple, it is not a particularly efficient form of modulation in terms of bandwidth or signal-to-noise ratio. We have seen that the transmission bandwidth is twice the highest modulating frequency because there are two sidebands containing the same information. We have also noticed that two-thirds or more of the transmitted power is found in the carrier, which contains no information and merely serves as an aid to demodulation.

Over the years, many variations on the basic AM system have been developed, and some have become quite popular for specific applications. This chapter looks at a number of these schemes from both the theoretical and practical points of view.

6.1.1 Efficiency Improvement by Carrier Suppression

As we have just noted, at least two-thirds of the power in an ordinary double-sideband (DSB) AM signal is in the carrier.

Removing the carrier before power amplification takes place would allow all of the transmitter power to be devoted to the sidebands, resulting in a substantial increase in sideband power. Removing the carrier from a fully modulated AM signal would change the power available for the sidebands from one-third of the total to all of it. The power increase in the sidebands would be the total available power divided by the power in the sidebands with full carrier:

$$A_P = \frac{P_t}{(1/3)P_t} = 3$$

where

A_P = power advantage gained by suppressing the carrier

This can easily be expressed in decibels:

$$A_P\ (\text{dB}) = 10 \log A_P = 4.77\ \text{dB} \tag{6.1}$$

The power advantage of almost 5 dB given by Equation (6.1) is a minimum value, since a practical AM system generally operates at less than 100% modulation.

Figure 6.1 shows the effect of removing the carrier from a fully modulated AM signal, in both the frequency and time domains. Figure 6.1(a) shows a 1 MHz carrier 100% modulated by a 1 kHz sine wave and applied to a 50 Ω load. The carrier power is 1 W (30 dBm). Each sideband has one-quarter the carrier power, or 24 dBm. In Figure 6.1(b), we see the result of using the same total signal power (1.5 W) to produce a *double-sideband suppressed-carrier* (DSBSC) signal. Since there is no carrier, each sideband has half the total power: 0.75 W (28.8 dBm).

Obviously, the envelope of the signal is no longer a faithful representation of the modulating signal. In fact, it is merely the sum of the upper and lower sideband signals. When these two sine waves (one at 0.999 MHz and the other at 1.001 MHz) are added, there is reinforcement when the two signals are in phase and cancellation when they are out of phase. The result is an envelope with a frequency equal to the difference between the frequencies of the two sidebands; that is, the envelope frequency is twice the modulating frequency.

Demodulation cannot be accomplished by the method used for full-carrier AM. An envelope detector would produce a distorted signal at twice the baseband frequency. To demodulate this signal, it would be necessary to reinsert the missing carrier at the receiver.

The peak amplitude of this signal can be found as follows. Each sideband has a power of 0.75 W. In a 50 Ω load, the RMS voltage corresponding to one sideband can be found from

$$P = \frac{V_{RMS}^2}{R}$$

$$V_{RMS} = \sqrt{PR} = \sqrt{0.75\ \text{W} \times 50\ \Omega} = 6.12\ \text{V}$$

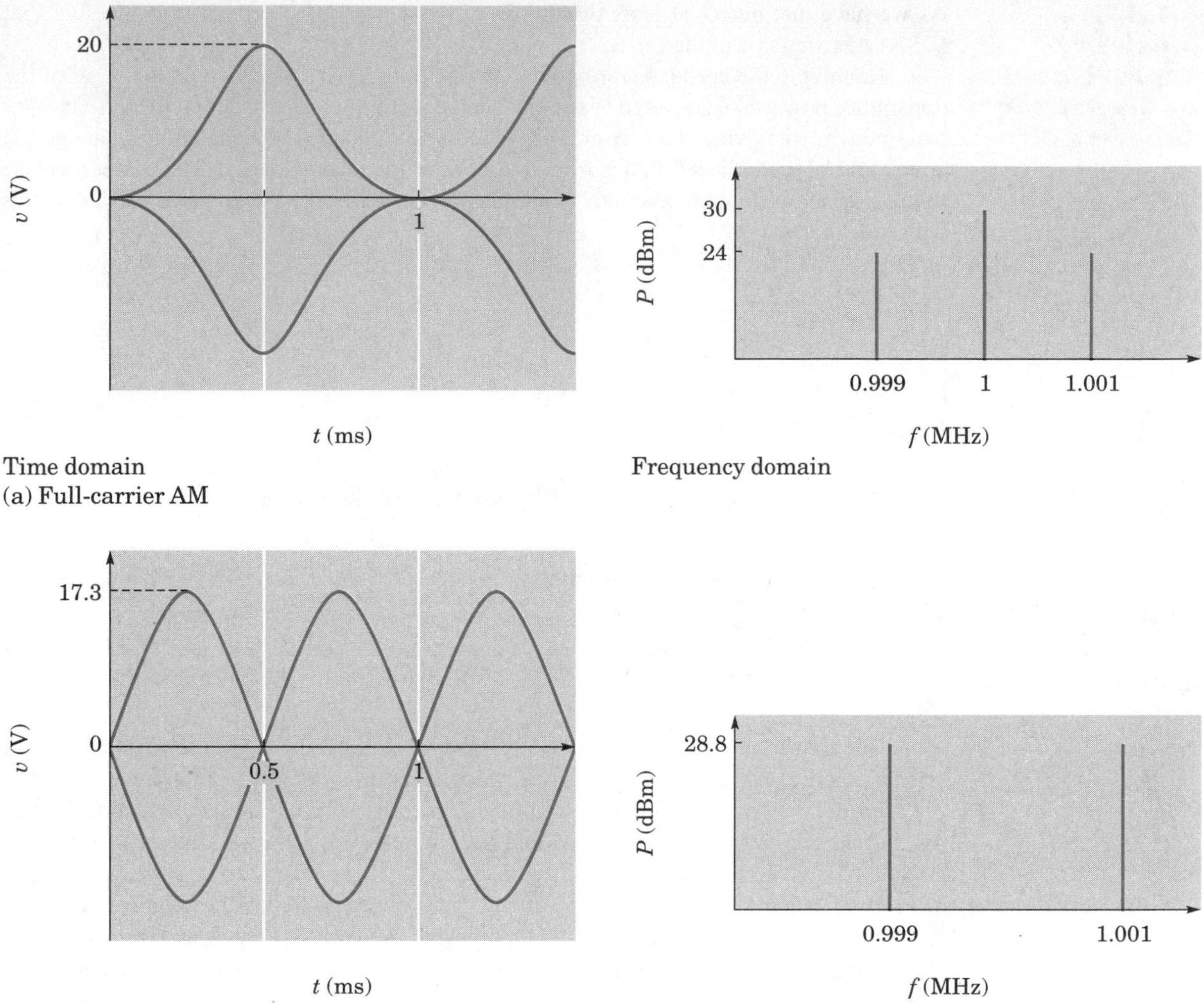

Time domain
(a) Full-carrier AM

Frequency domain

Time domain
(b) DSBSC AM

Frequency domain

Figure 6.1
DSBSC in time and frequency domains

We need the peak voltage. Since one individual sideband is a sine wave, this is given by

$$
\begin{aligned}
V_p \text{ (for one sideband)} &= \sqrt{2}V_{RMS} \\
&= \sqrt{2} \times 6.12 \text{ V} \\
&= 8.66 \text{ V}
\end{aligned}
$$

When the two signals are in phase, the peak envelope voltage will be the sum of the individual peak voltages, or

$$
\begin{aligned}
V_p \text{ (for the whole signal)} &= 8.66 \text{ V} + 8.66 \text{ V} \\
&= 17.3 \text{ V}
\end{aligned}
$$

Example 6.1

A spectrum analyzer connected to a DSBSC signal has the display shown in Figure 6.2. Calculate:

(a) the total signal power
(b) the modulating frequency
(c) the carrier frequency

Figure 6.2

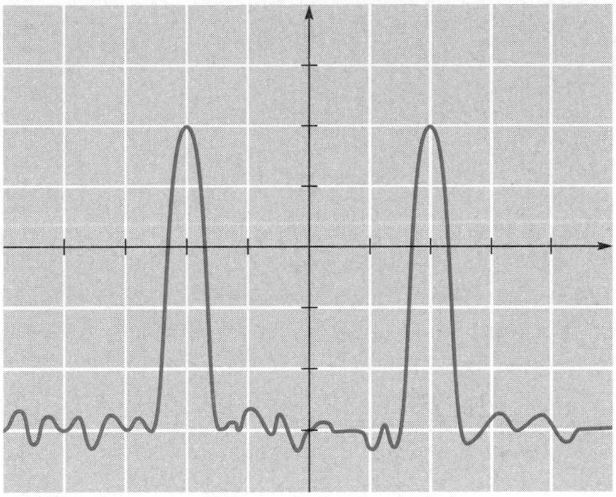

Reference level 10 dBm
Vertical: 10 dB/division
Center frequency 10 MHz
Span: 1 kHz/division

Solution

(a) Each of the two sidebands is two divisions (20 dB) below the reference level of 10 dBm. Therefore each sideband has a power of −10 dBm (100 μW). The total power is twice this, or 200 μW.
(b) The two sidebands are separated by four divisions at 1 kHz per division, or 4 kHz total. The separation between the sidebands is twice the modulating frequency, which must therefore be 2 kHz.
(c) Even though this signal has no carrier, it still has a carrier frequency. This is the frequency the carrier had before it was suppressed. The carrier frequency is midway between the two sidebands, so it is 10 MHz.

DSBSC AM is not often found on its own as a modulation scheme. It is used as the basis for generating **single-sideband suppressed-carrier** (SSBSC, or just SSB) signals, which will be discussed in the next section, and it is also found as a component in some rather complex multiplexed signals, such as color television and stereo FM signals, which will be described later in this book.

6.1.2 Bandwidth Reduction Using Single-Sideband Transmission

In Chapter 3, it was discovered that the two sidebands of an AM signal are mirror images of each other, since one consists of the sum of the carrier and modulation frequencies and the other is the difference. Thus one sideband is redundant, assuming the carrier frequency is known, and it should not be necessary to transmit both in order to communicate.

Removing one sideband obviously reduces the bandwidth by at least a factor of two. Since the modulating signal rarely extends right down to dc, the bandwidth improvement will usually be greater than two.

Figure 6.3 illustrates this effect. The baseband, shown in Figure 6.3(a), is a voice signal extending over a frequency range from 300 Hz to 3 kHz. Figure 6.3(b) shows this signal transmitted by DSBSC AM with a carrier frequency of 1 MHz. The bandwidth will be

$$\begin{aligned} B &= 2\, f_{m(max)} \\ &= 2 \times 3 \text{ kHz} \\ &= 6 \text{ kHz} \end{aligned}$$

With SSB transmission, as shown in Figure 6.3(c), the bandwidth of one sideband is

$$\begin{aligned} B &= f_{m(max)} - f_{m(min)} \\ &= 3 \text{ kHz} - 0.3 \text{ kHz} \\ &= 2.7 \text{ kHz} \end{aligned}$$

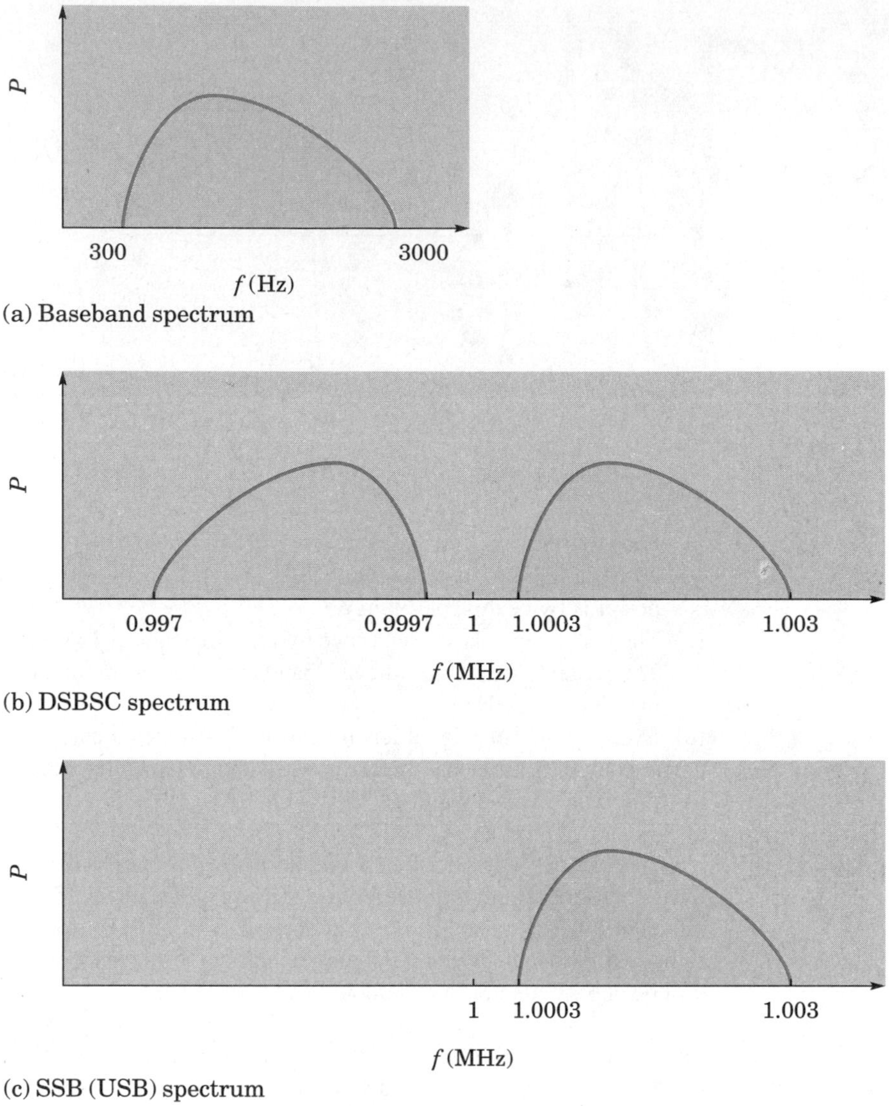

Figure 6.3 DSBSC and SSB transmission of a voice signal

This bandwidth reduction has two benefits. Perhaps the more obvious one is that the signal takes up less spectrum. This allows twice as many signals to be transmitted in a given spectrum allotment. At least as important, however, is the improvement in the signal-to-noise ratio that can be achieved by reducing bandwidth. If the bandwidth of the transmitted signal is reduced by 50%, the receiver bandwidth can be reduced by an equivalent amount. Since noise power is proportional to bandwidth, reducing the receiver bandwidth by one-half eliminates one-half the noise. Assuming the signal power remains constant, this represents a 3 dB improvement in the signal-to-noise ratio.

The signal-to-noise improvement resulting from bandwidth reduction is in addition to that achieved by increasing the transmitted power in the sidebands. Combining the 3 dB improvement from bandwidth reduction with the 4.77 dB improvement calculated in Equation (6.2) gives a total improvement in signal-to-noise ratio of

$$\begin{aligned} S/N \text{ improvement} &= 4.77 \text{ dB} + 3 \text{ dB} \\ &= 7.77 \text{ dB} \end{aligned} \quad (6.2)$$

compared with full-carrier AM with 100% modulation.

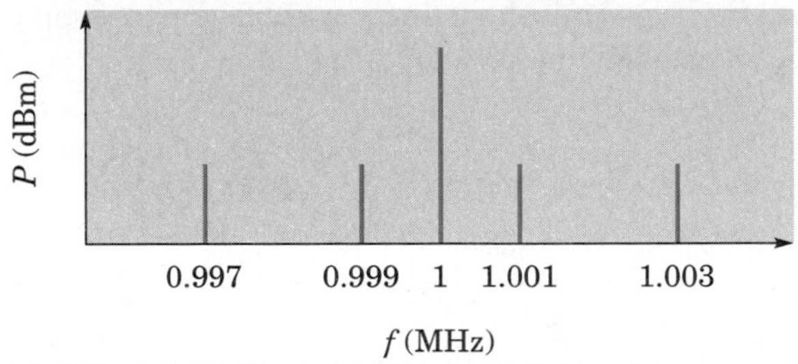

(a) AM with 1 kHz and 3 kHz modulation (frequency domain)

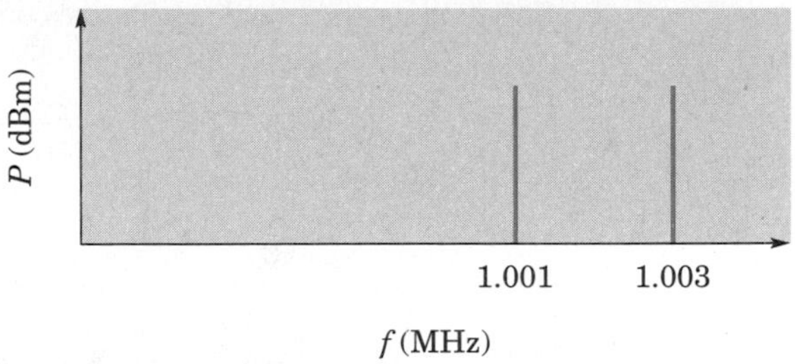

(b) USB with two-tone modulation (frequency domain)

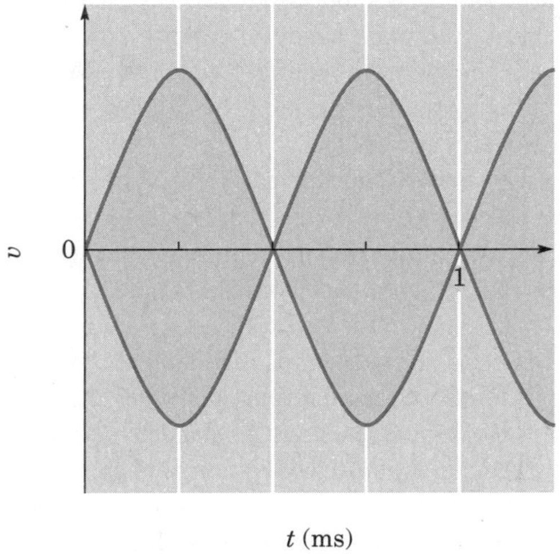

(c) USB with two-tone modulation (time domain)

Figure 6.4 Two-tone modulation

Since the two sidebands of an AM signal contain identical information, either upper-sideband (USB) or lower-sideband (LSB) transmission can be used, and both are found in practice.

SSB transmissions are even more unlike full-carrier AM than are DSBSC signals. In fact, if we look at a single-tone SSB signal, we find that there is no envelope at all. For instance, consider a USB signal in which a 1 MHz carrier is modulated by a 1 kHz baseband signal. The USB will be simply a sinusoid at a frequency given by

$$\begin{aligned} f_{usb} &= f_c + f_m \\ &= 1.001 \text{ MHz} \end{aligned}$$

Another form of SSB modulation that is often used for transmitter measurements is the **two-tone test signal.** For an example, see Figure 6.4. A 1 MHz carrier is modulated by two baseband frequencies, 1 kHz and 3 kHz, producing the AM signal shown in Figure 6.4(a). The carrier and the LSB are suppressed, leaving the two USB components at 1.001 and 1.003 MHz. This is quite obvious in the frequency domain, as shown in Figure 6.4(b). In the time domain, illustrated in Figure 6.4(c), the summation of the two frequency components creates an envelope that is identical to that of the DSBSC signal with single-tone modulation, as shown in Figure 6.1. As before, an envelope detector would give a distorted 2 kHz signal having no relation to the original baseband signal.

Example 6.2

A transmitter generates an LSB signal with a carrier frequency of 8 MHz. What frequencies will appear at the output with a two-tone modulating signal with frequencies of 2 kHz and 3.5 kHz?

Solution Since the signal is LSB, the components are found by subtracting the modulating frequency from the carrier frequency. Therefore the output frequencies are

$$8 \text{ MHz} - 2 \text{ kHz} = 7.998 \text{ MHz}$$

and

$$8 \text{ MHz} - 3.5 \text{ kHz} = 7.9965 \text{ MHz}$$

6.2 Single-Sideband Transmitters

A typical single-sideband transmitter has the simplified block diagram shown in Figure 6.5. First a double-sideband suppressed-carrier signal is generated. Next, one sideband is removed, usually by using a bandpass filter to pass one sideband and reject the other (it is also possible to do this without filters using complex phase-shift networks). Then the signal is moved to the operating frequency by mixing. Finally the SSBSC signal is amplified to the required output power level.

Note the similarity to the receivers studied in Chapter 5. An SSB transmitter looks almost like a receiver in reverse because it is easiest to generate an SSB signal at one frequency and then move it to the required frequency.

Let us look in turn at each of these steps.

6.2.1 Balanced Modulators for Double-Sideband Suppressed-Carrier Generation

Balanced mixers were introduced in Chapter 2. A balanced modulator is the same thing used for a different purpose. In fact, the type MC1496 IC shown in Chapter 2 is widely used as both a balanced mixer and a balanced modulator.

Mixing and amplitude modulation are essentially the same process. Both amplitude modulators and ordinary (unbalanced) mixers are provided with two input frequencies. They produce sum and difference signals at their output, as well as the two original signals. In a transmitter modulator, the two inputs are the baseband and the carrier signals, and the sum and difference signals are, of course, the USB and LSB. The carrier signal feeds through the modulator to the output. The baseband input does not feed through in a practical circuit because of the choice of output components. In the usual case, the baseband is audio and is completely rejected by the output tuned circuits, which are tuned to the carrier frequency several orders of magnitude higher.

In a receiver, on the other hand, the two inputs to the mixer are the incoming signal and the local oscillator signal. Usually the difference is used by the IF amplifier, and the other outputs are rejected by filters. The only differences between a mixer and a modulator, then, are practical considerations: modulators generally operate at higher power levels and use one RF and one audio-frequency signal, while receiver mixers operate at low power and use RF signals for all inputs and outputs.

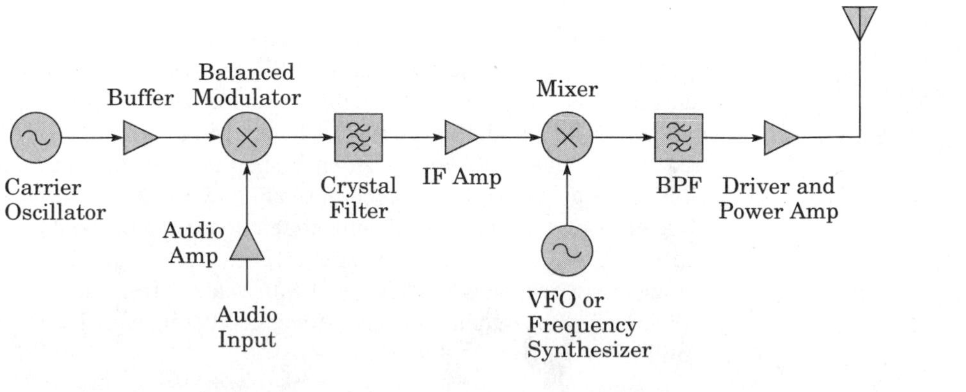

Figure 6.5
Basic SSB transmitter

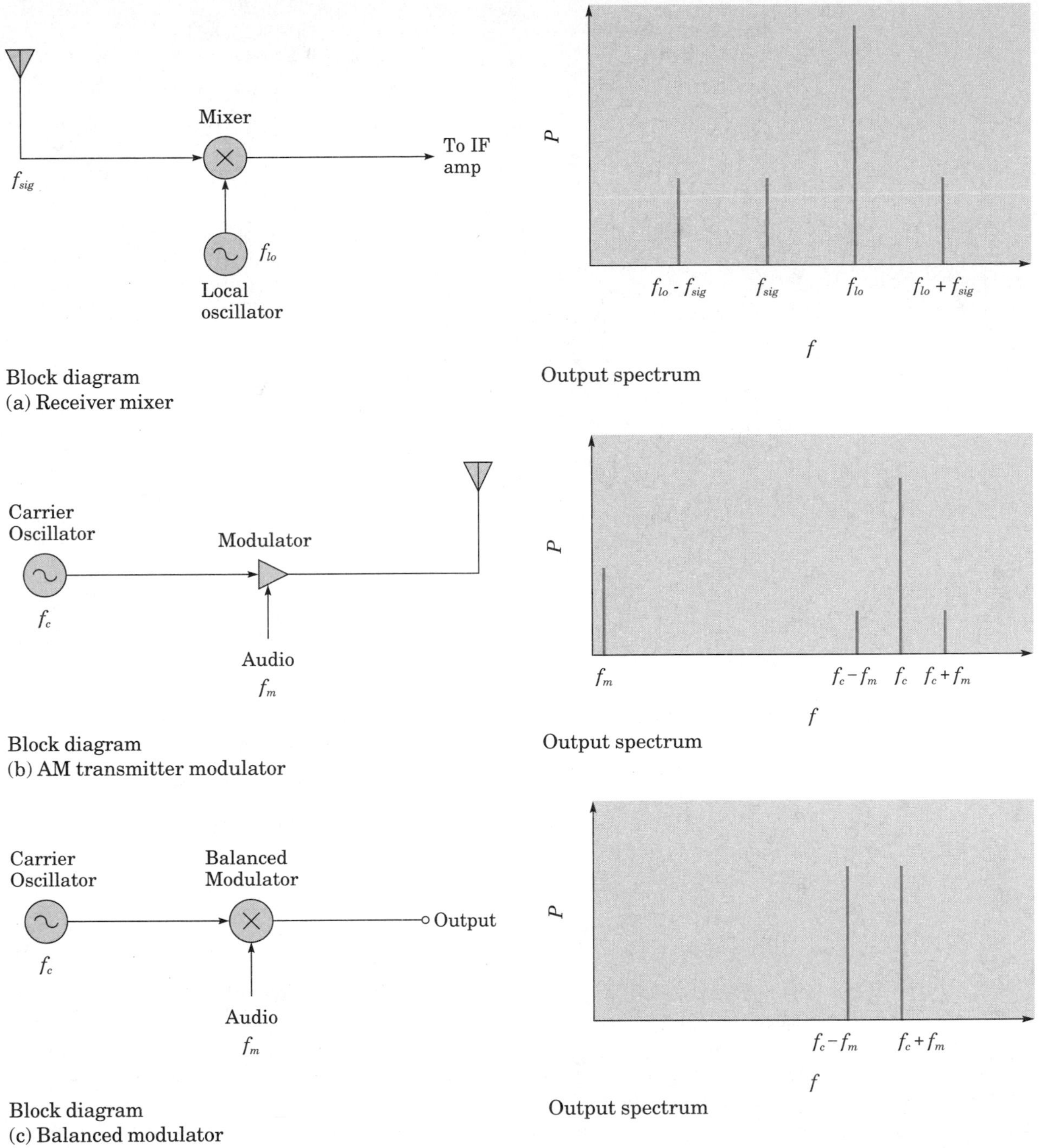

Figure 6.6
Mixing and modulation

Figure 6.6 should help to clarify the similarities between mixers and modulators. Figure 6.6(a) shows a receiver mixer and its output, and Figure 6.6(b) illustrates a transmitter modulator.

When discussing balanced mixers, we noted that they produce only the sum and difference of the two input signal frequencies that are applied. The input frequencies themselves are cancelled, or *balanced*, in the circuitry. Applying the same idea to an amplitude modulator, we would end up with a signal that had upper and lower sidebands, but no carrier; in other words, a DSBSC signal. In fact, that is how it is done. The baseband and

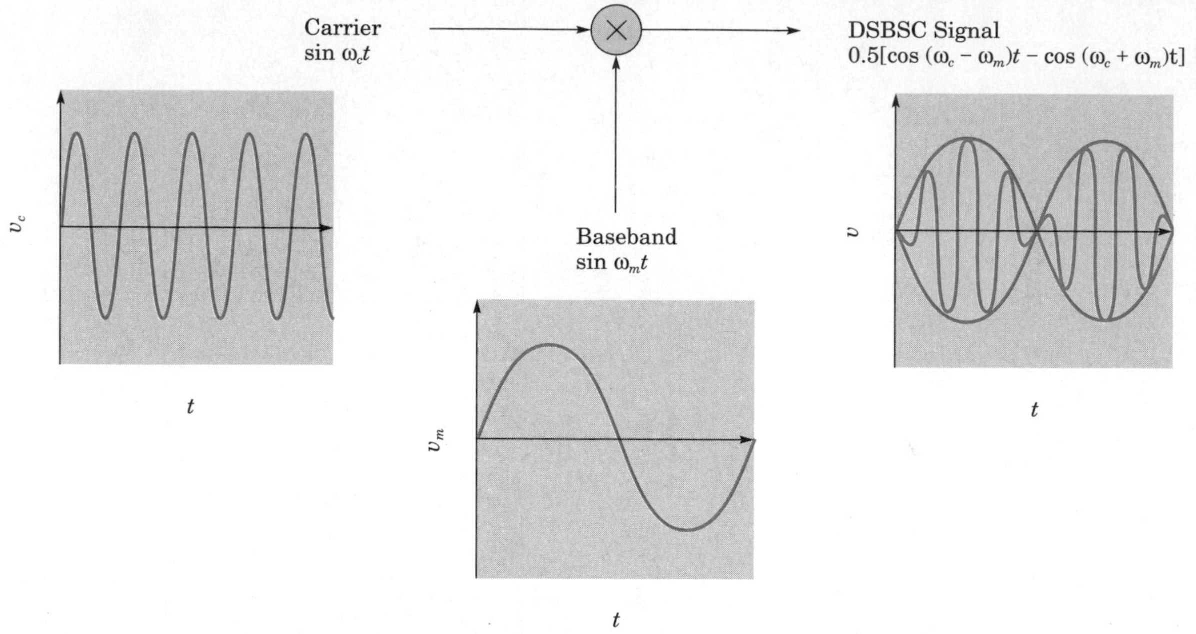

Figure 6.7
DSBSC generation

carrier signals are applied to the inputs of a balanced modulator, and the output is DSBSC, as shown in Figure 6.6(c).

Any of the balanced mixer configurations described in Chapter 2 can be employed, but, in modern designs, specialized ICs (such as the MC1496) are generally used. In theory, there is no reason why balanced modulation could not be accomplished at high power, but for practical reasons—which will very shortly become apparent—the DSBSC signal is always generated at low level, at an early stage in the transmitter.

We showed previously that an ideal balanced modulator is mathematically equivalent to a multiplier. Figure 6.7 shows such a modulator fed with sine-wave carrier and baseband signals. The result (ignoring amplitude factors) is simply

$$\begin{aligned} v &= \sin \omega_m t \sin \omega_c t \\ &= 0.5[\cos (\omega_c - \omega_m)t - \cos (\omega_c + \omega_m)t] \end{aligned} \tag{6.3}$$

This is indeed a DSBSC signal, with two components representing the upper and lower sidebands.

6.2.2 Generating Single-Sideband Signals

The obvious way to eliminate the unwanted sideband, once the carrier has been suppressed, is to use a bandpass filter. You might wonder why we do not simply filter out the carrier as well. The problem is that in an AM signal, the carrier is both very large in amplitude and very near the wanted sideband. There would be practical difficulties in removing the carrier while leaving the sideband untouched. It is more convenient to null the carrier with a balanced modulator and then remove one of the sidebands by filtering.

Figure 6.8 shows the filter method of SSB generation. The block diagram in Figure 6.8(a) is the same for either LSB or USB: which sideband is passed depends on the position of the filter passband relative to the signal. If the carrier frequency is at or near the low end of the filter passband, the USB passes through the filter and the LSB is suppressed. On the other hand, if the carrier frequency is at or near the high end of the passband, the filter passes the LSB and suppresses the USB.

The sidebands are shown as triangles, not because that is the shape of the spectrum but in order to emphasize that they are *mirror images* of each other and to make clear which sideband is present when a single-sideband signal is shown.

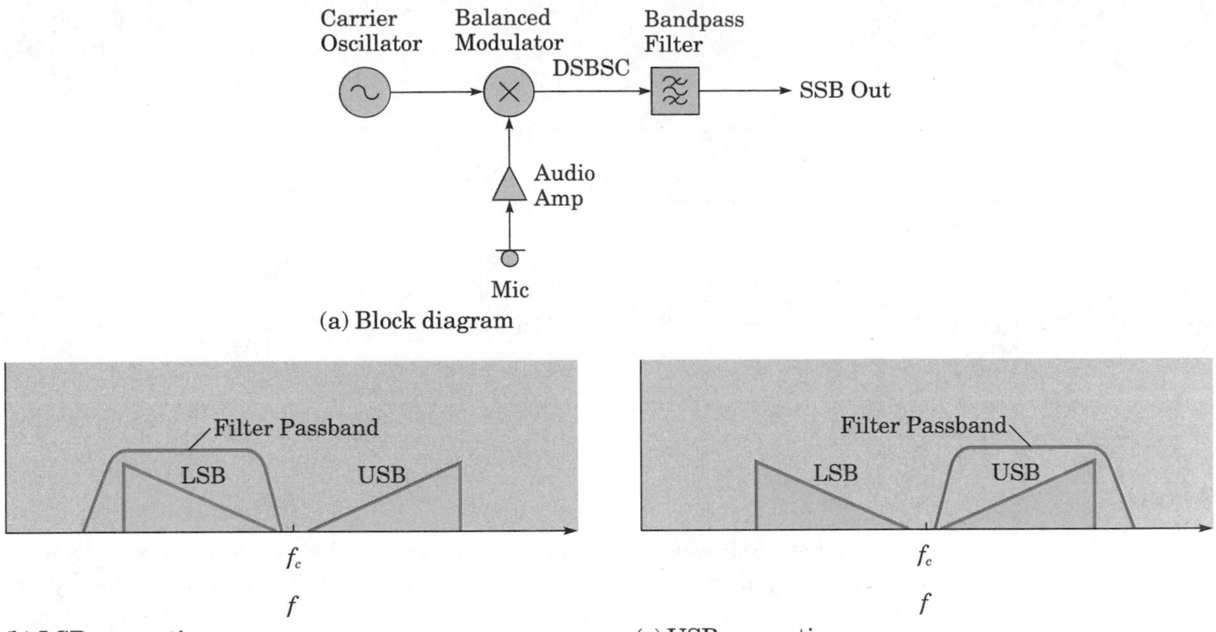

(b) LSB generation

(c) USB generation

Figure 6.8
SSB generation by filter method

Even with the carrier suppressed, the cutoff of the filter must be quite sharp. The lowest baseband frequency transmitted is likely to be about 300 Hz in a communications-quality voice signal. Thus there is a gap of 600 Hz between the two sidebands. The attenuation of the filter must change from zero to at least 60 dB within this span.

The generation of SSB requires a high-order filter with high Q. It can be done with *LC* filters, but only at low frequencies, and it is not an economical proposition. *Mechanical filters* have also been used.

By far the most common device for suppressing the unwanted sideband is the *crystal filter*. These circuits can provide a Q of about 100,000. Crystal filters were introduced in Chapter 5, where they were used in receivers. Later we will see that it is feasible to use one crystal filter in an SSB transceiver, switching it between receiver and transmitter as required.

Example 6.3

A filter-method SSB generator of the type shown in Figure 6.8 has the following specifications:

Filter center frequency: 5.000 MHz
Filter bandwidth: 3 kHz
Carrier-oscillator frequency: 4.9985 MHz

(a) Which sideband will be passed by the filter?
(b) What frequency should the carrier oscillator have if it is required to generate the other sideband?

Solution

(a) Since the carrier frequency is at the low end of the filter passband, the USB will be passed.
(b) To generate the LSB, the carrier frequency should be moved to the high end of the filter passband, at 5.0015 MHz.

6.2.3 Mixing

When the filter method is used, the SSB signal must be generated at a single carrier frequency, since a crystal filter cannot be retuned. If the transmitter is to operate at more than one frequency, the output frequency can be changed by using one or more mixers.

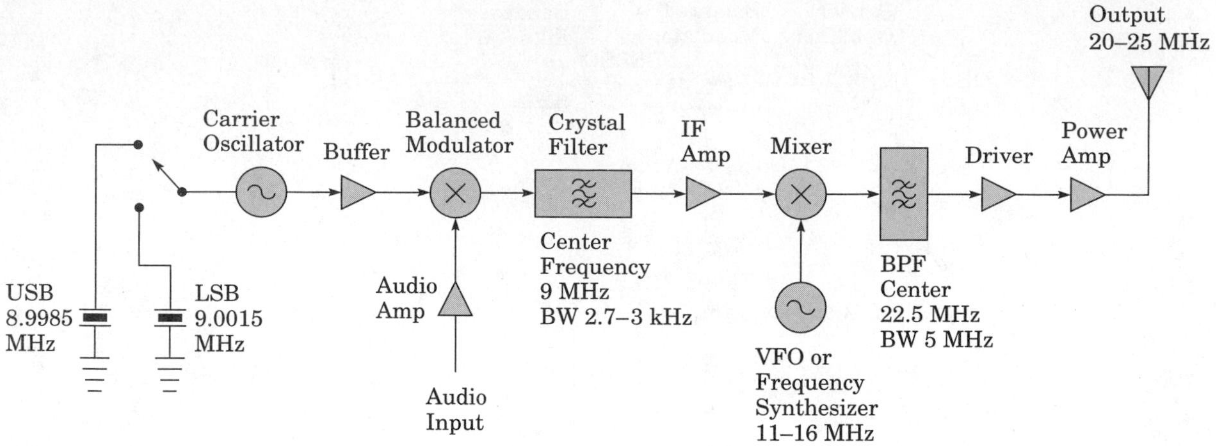

Figure 6.9
An SSB transmitter using the filter method

In the past, SSB was generated at low frequencies (100 to 500 kHz) because of the problem of designing a high-frequency filter with sufficient *Q*. The signal frequency was then raised by two or more up-conversions, using mixers and local oscillators. Modern practice is to use crystal filters and generate SSB at higher frequencies (9 to 11 MHz is common). One stage of mixing is generally all that is required for a high-frequency transmitter.

The block diagram of a typical SSB transmitter is shown in Figure 6.9. This transmitter uses a carrier oscillator and a balanced modulator to generate DSBSC, followed by a crystal filter to remove the unwanted sideband. By changing the frequency of the carrier oscillator, either sideband can be generated using the same crystal filter. This method is less expensive than using two crystal filters. The transmitter in the figure is designed to transmit either sideband over a frequency range of approximately 20 to 25 MHz, using a single crystal filter with a center frequency of 9 MHz. Once the SSB signal has been generated, it is mixed with a local oscillator signal. The sum of the two input frequencies is chosen at the output of the mixer and amplified by the driver and power amplifier.

Filter-method SSB transmitters require that the carrier frequency be positioned near one edge or the other of the filter passband. Figure 6.10 shows one possibility. Here, the center frequency of the filter is 9.0000 MHz, and it has a bandwidth of 3 kHz. For sim-

Figure 6.10
Use of a single crystal filter for both USB and LSB signals

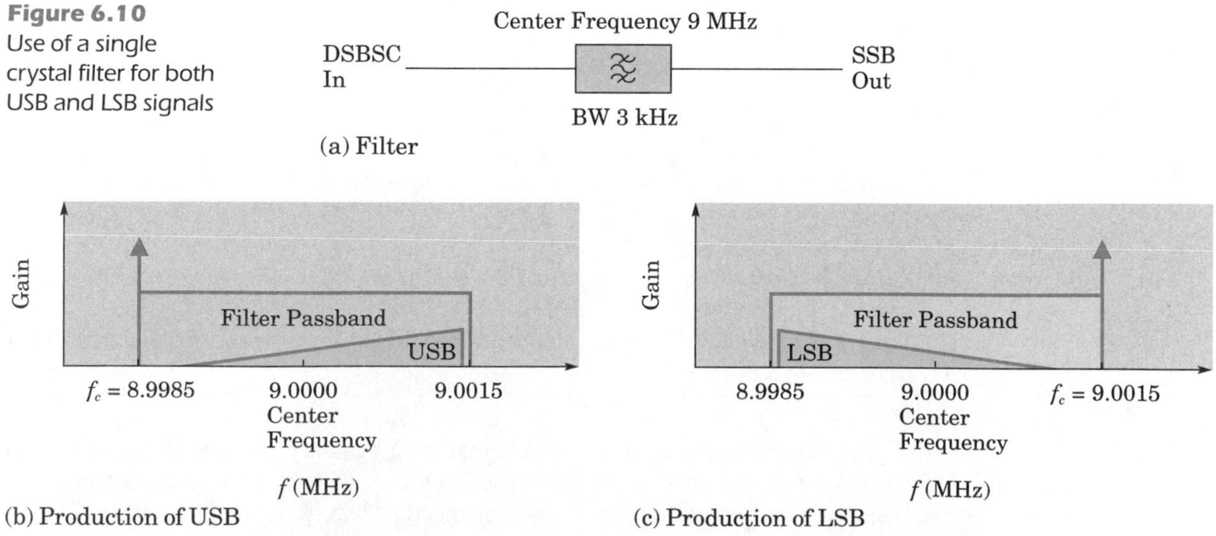

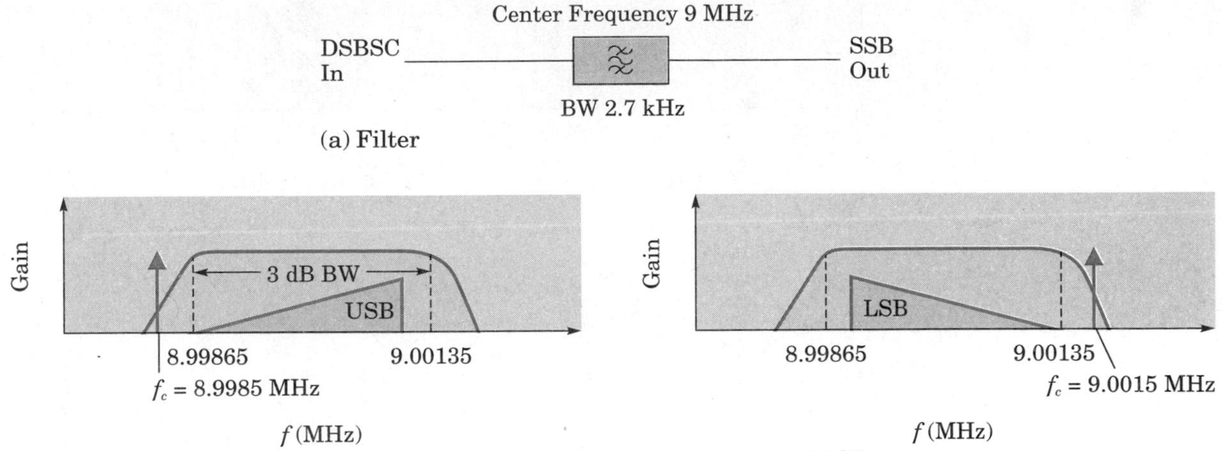

(b) Production of USB

(c) Production of LSB

Figure 6.11
SSB generation using a practical crystal filter

plicity, let us assume that the filter is ideal, so that it passes everything in its passband with no attenuation and blocks all other frequencies completely. In that case, when the carrier frequency is at the low end of the passband (8.9985 MHz), all of the USB and none of the LSB will be passed. The same filter can be used to generate an LSB signal by simply placing the carrier frequency at the top end of the passband (9.0015 MHz).

Note that the term *carrier frequency* does not imply that the carrier signal actually exists. By this point in the transmitter, the carrier has already been suppressed.

A real crystal filter does not have an infinite slope between passband and stopband. Consequently, placing the carrier frequency exactly at the filter's cutoff frequency allows some of the unwanted sideband to pass through the filter. Therefore, it is more usual to place the carrier frequency very slightly outside the passband. This provides more attenuation of the unwanted sideband and also attenuates that part of the desired sideband that is very close to the carrier in frequency. However, these sideband frequencies represent low-frequency components of the baseband signal. As long as baseband frequencies above 300 Hz are passed, there is no problem for voice communications. In our example, this modification could be made by leaving the carrier frequencies as they are and using a filter with a nominal bandwidth of, for example, 2.7 kHz. See Figure 6.11 for the results.

These filter considerations make it difficult to build a filter-method SSB transmitter that will handle very low frequency baseband components. This is one reason why SSB is not used for broadcasting, which must accommodate music signals with frequencies as low as 50 Hz or so.

After filtering, the SSB signal is moved to the required operating frequency by means of a mixer and local oscillator combination. In this case, the transmitted frequency is the sum of the carrier oscillator and local oscillator frequencies. That is,

$$f_o = f_{CO} + f_{LO} \tag{6.4}$$

where

f_o = carrier frequency of the transmitted signal
f_{CO} = carrier oscillator frequency
f_{LO} = local oscillator frequency

After mixing, the signal is amplified to the required power level. The power amplifier must be linear in order to preserve the amplitude variations in the signal.

Example 6.4

The SSB transmitter shown in Figure 6.12 is required to transmit a USB signal at a carrier frequency of 21.5 MHz. What must be the frequency of the local oscillator?

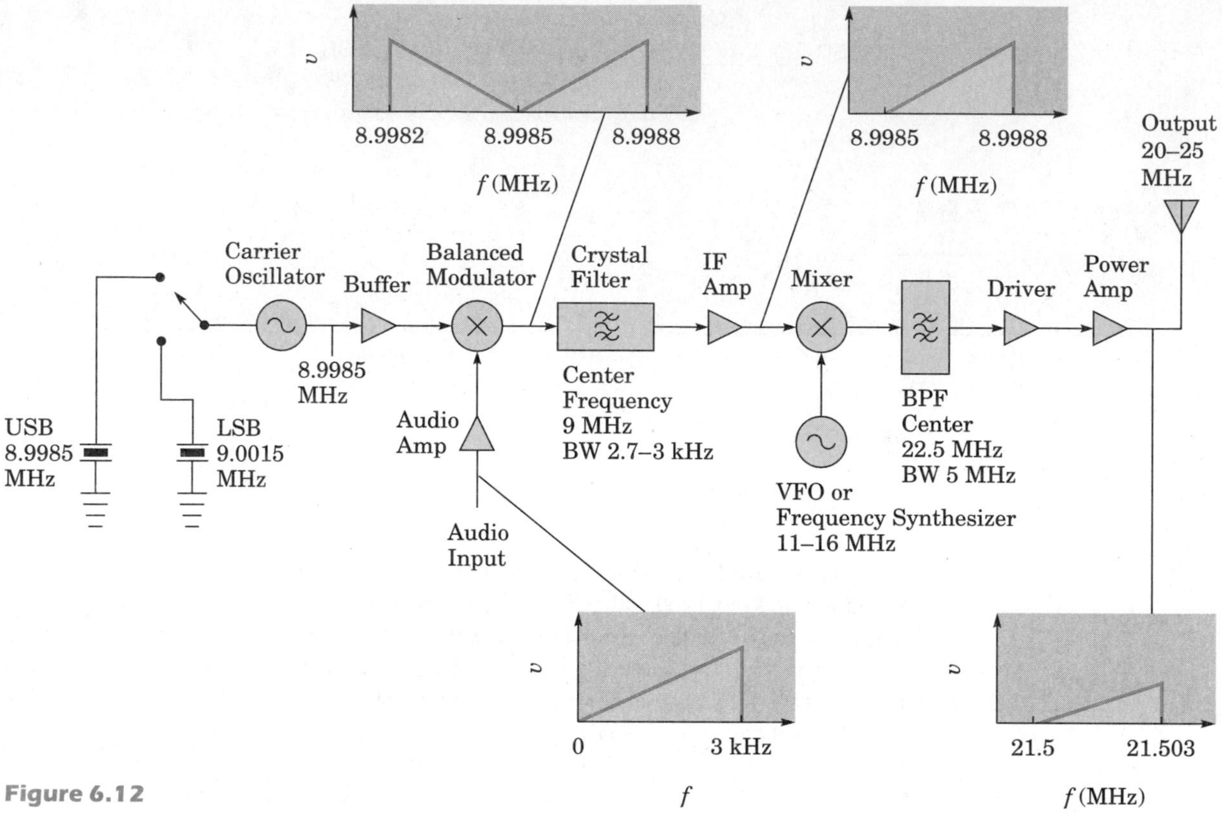

Figure 6.12

Solution In this transmitter, the carrier oscillator and local oscillator frequencies are summed to get the transmitting frequency. For USB operation with this transmitter, the carrier oscillator frequency is 8.9985 MHz. We can easily find the local oscillator frequency from Equation (6.4).

$$\begin{aligned} f_o &= f_{CO} + f_{LO} \\ f_{LO} &= f_o - f_{CO} \\ &= 21.5 \text{ MHz} - 8.9985 \text{ MHz} \\ &= 12.5015 \text{ MHz} \end{aligned}$$

Assuming that the baseband spectrum extends from 0 to 3 kHz, the frequency spectra at various points in the transmitter will be as shown in Figure 6.12.

Older SSB transmitter designs practically always require the SSB signal emerging from the filter to be raised in frequency and often need two or more mixer stages to do so. In modern designs, it may be necessary to move the signal either up or down, or perhaps both, depending on the frequency of operation. For instance, an amateur radio transmitter that generates an SSB signal at 9 MHz will have to use up-conversion for the 14 MHz band and down-conversion for the 7 MHz band.

Mixers for SSB transmitters are straightforward and resemble their receiver counterparts, except that the power levels involved are somewhat higher. Obviously, for down-conversion, the difference between the signal from the filter and the local oscillator signal must be used. The local oscillator frequency may be either higher or lower than that of the carrier oscillator. For up-conversion, either the sum or the difference can be used, but if the difference is used, the local oscillator must use high-side injection; that is, it must be higher in frequency than the carrier oscillator. Figure 6.13 shows a few possibilities. In each case, a USB signal with a carrier frequency of 9 MHz is shown at the input to the mixer.

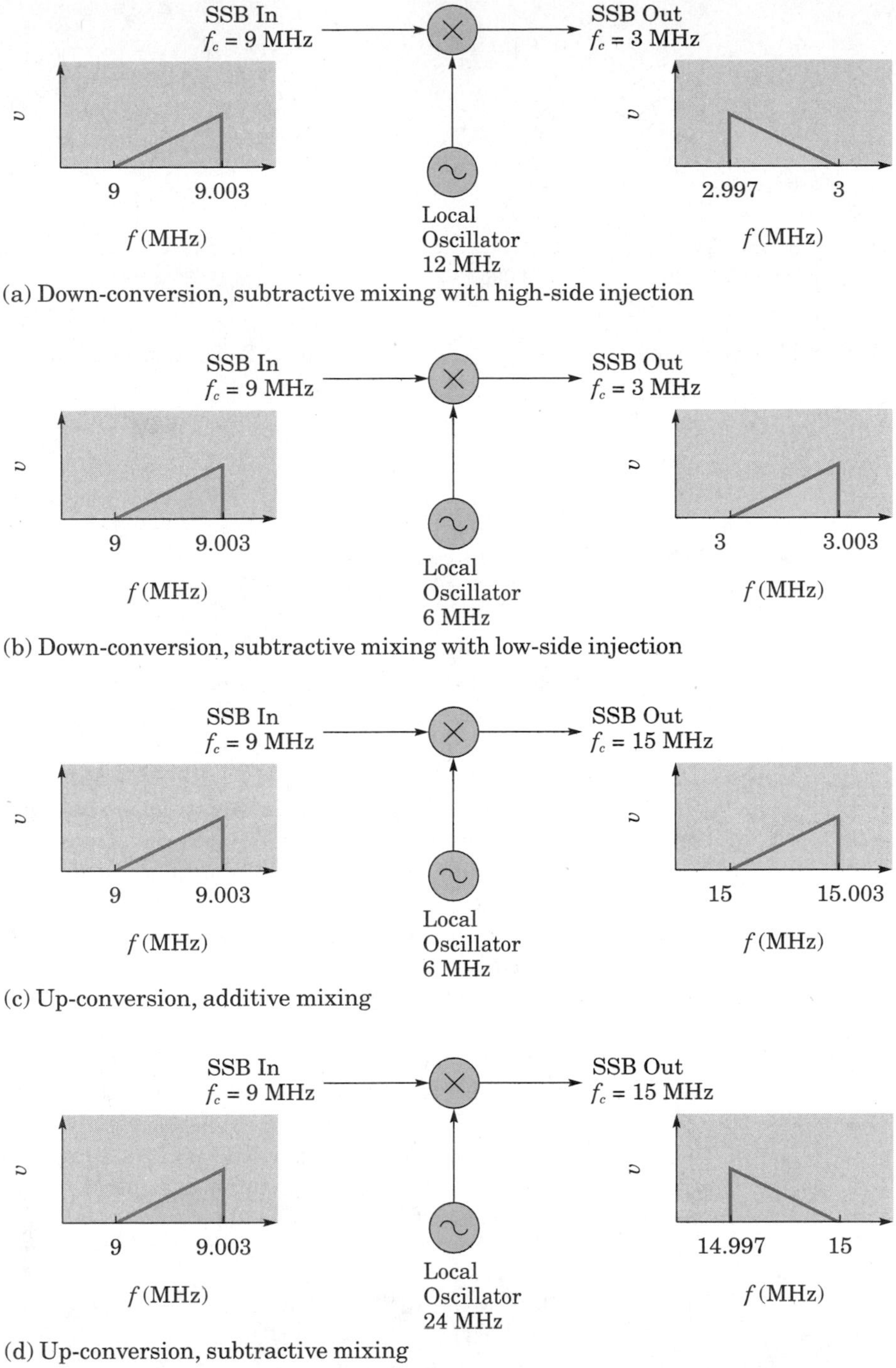

Figure 6.13 Examples of frequency conversion in SSB transmitters

It may not be immediately obvious that some types of mixing interchange sidebands, that is, a USB signal becomes LSB, and vice versa. This can be seen by referring to Figure 6.13(a). Suppose the incoming signal is a USB signal with a (suppressed) carrier frequency of 9 MHz, modulated with 1 kHz and 3 kHz tones. Then it has the spectrum shown in Figure 6.14(a), which has components at 9.001 and 9.003 MHz. The mixing process produces the difference between the local oscillator frequency and each of these two components. These difference-frequency signals will be found at

$$12 \text{ MHz} - 9.001 \text{ MHz} = 2.999 \text{ MHz}$$

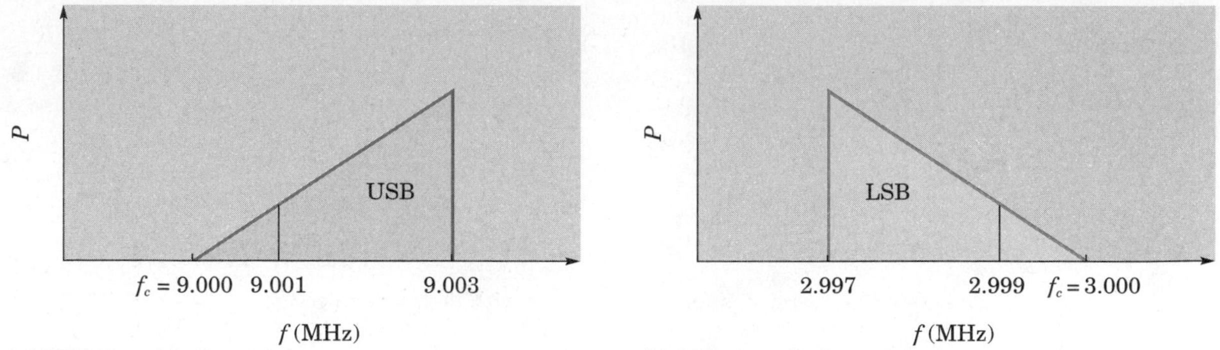

(a) USB signal before mixing (b) LSB signal after mixing

Figure 6.14 Effect of subtractive mixing with high-side injection on SSB signals

and

$$12\text{ MHz} - 9.003\text{ MHz} = 2.997\text{ MHz}$$

while the new carrier frequency is

$$12\text{ MHz} - 9\text{ MHz} = 3\text{ MHz}$$

It is clear from Figure 6.14(b) that what has happened is that the USB signal has been changed into an LSB signal by the mixing process. This is based on simple arithmetic: subtracting a larger number gives a smaller result. The effect occurs only for subtractive mixing with high-side injection of the local oscillator: verify it with the other examples in Figure 6.13.

Exactly the same thing happens in a receiver mixer. It goes unnoticed in an AM receiver because the two sidebands of a DSB AM signal are mirror images, so interchanging them makes no difference. With SSB transmitters and receivers, this effect must be taken into account, however.

Note that changing from one sideband to the other by switching the frequency of the carrier oscillator also changes the carrier frequency at the output. This will require a change in the frequency readout for the transmitter.

Example 6.5 Suppose that the transmitter in Example 6.4 is switched to LSB by simply switching carrier oscillator crystals. What is the new carrier frequency of the transmitted signal?

Solution The new carrier oscillator frequency is 9.0015 MHz (see Figure 6.9). The local oscillator frequency remains 12.5015 MHz, as found in Example 6.4. The new output carrier frequency can easily be found from Equation (6.4):

$$\begin{aligned} f_o &= f_{CO} + f_{LO} \\ &= 9.0015\text{ MHz} + 12.5015\text{ MHz} \\ &= 21.503\text{ MHz} \end{aligned}$$

This is 3 kHz higher than before. In order to maintain an accurate frequency readout, the dial display would have to change when sidebands are changed or the local oscillator frequency would have to change automatically when sidebands are switched. Either method is quite easy when frequency synthesis is used for the local oscillator.

6.2.4 Power Amplification

The power-amplification stages of an SSB transmitter must be linear. SSB is a variation of AM, and any AM signal must be amplified in such a way that the amplitude variations are preserved; if not, the demodulated signal will be distorted. As we saw in Chapter 4, the need for linear amplification can be avoided in a full-carrier AM transmitter by modulating the RF power amplifier. This is not possible in a filter-type SSB transmitter, because the crystal filter must operate at low power.

The enormous power levels that are often found in AM broadcast transmitters are rare with SSB. Since SSB is usually used for point-to-point communication with reasonably effective receivers and receiving antennas, transmitter power levels can be much lower than for broadcasting. Typical high-frequency SSB transmitters operate at power levels from 100 W to a few kilowatts.

6.2.5 Single-Sideband Transmitter Specifications

Speaking of power, you might wonder how an SSB transmitter would be rated for power output. Since there is no carrier, carrier power is obviously useless as a specification. The actual transmitter output power varies greatly with modulation, from zero with no modulation to some maximum level with full modulation. That maximum is referred to as the **peak envelope power** (PEP). It is simply the power at modulation peaks, calculated using the RMS formula:

$$PEP = \frac{(V_p/\sqrt{2})^2}{R_L} = \frac{V_p^2}{2R_L} \tag{6.5}$$

where

PEP = peak envelope power in watts
V_p = peak signal voltage in volts
R_L = load resistance in ohms

Figure 6.15 illustrates this calculation. The signal shown represents an SSB signal with two-tone modulation, as previously discussed in Section 6.1.2. The peak voltage is 25 V, and let us assume that the load is 50 Ω, resistive.

Then the peak envelope power is

$$\begin{aligned} PEP &= \frac{V_p^2}{2R_L} \\ &= \frac{(25\text{ V})^2}{2 \times 50\ \Omega} \\ &= 6.25\text{ W} \end{aligned}$$

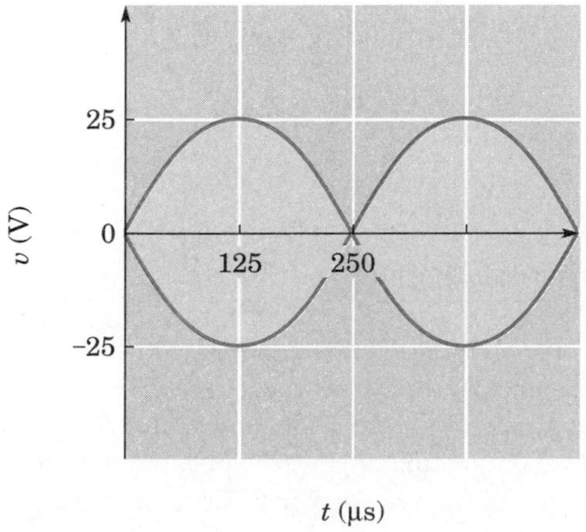

Figure 6.15 PEP calculation

PEP is not the same as instantaneous peak power. In fact, it is one-half the maximum instantaneous power. This can easily be seen by recalling that instantaneous power is simply

$$P = \frac{v^2}{R}$$

where

v = instantaneous voltage across resistance R

The maximum instantaneous voltage is simply V_p, so the maximum instantaneous power is

$$P_{max} = \frac{V_p^2}{R_L} \tag{6.6}$$

which is just twice the PEP given by Equation (6.5).

It is also important to distinguish PEP from average power. For a well-defined test signal, a relation can be found between them. With single-tone modulation, the PEP and the average power are equal, since the signal is a sine wave with the same peak voltage for every cycle. With the two-tone signal of Figure 6.15, the relationship is a little more complicated but still easily found. There are a number of ways to do this, of which one is illustrated in the following example.

Example 6.6 Find the average power of the signal illustrated in Figure 6.15.

Solution It helps to remember that this waveform is created from the algebraic addition of two sine waves of equal amplitude, separated in frequency by a small amount (2 kHz, in this example). When the signals are exactly 180° out of phase, they will cancel completely, as shown at $t = 0$ and $t = 250$ μs. On the other hand, when they are exactly in phase, the peak amplitudes will add, as occurs in the figure at $t = 125$ μs. Since the two sine waves have equal amplitude, the peak amplitude of each must be one-half the peak voltage of the combined signal, or 12.5 V each.

The total average power in the signal will be the sum of the average powers of the two components. Each component, being a sine wave, will have a power equal to

$$\begin{aligned} P &= \frac{V_{RMS}^2}{R_L} \\ &= \frac{(V_p/\sqrt{2})^2}{R_L} \\ &= \frac{V_p^2}{2R_L} \\ &= \frac{(12.5\text{ V})^2}{2 \times 50\ \Omega} \\ &= 1.56\text{ W} \end{aligned}$$

The total average signal power will be twice this, or 3.125 W. This is one-half the PEP of 6.25 W, which we already found.

There is no simple relationship between PEP and average power for a random voice signal. In general the average power with voice modulation varies from PEP/4 to PEP/3.

The PEP rating for a transmitter is generally limited by clipping, while the average power is limited by heat dissipation or power-supply capacity. The duty cycle of many SSB transmitters is less than 100% even for voice, since they are designed for two-way communication, where short transmissions alternate with longer idle periods. Consequently,

TABLE 6.1
SSB transmitter specifications

AM carrier power	3.8 W
PEP, two-tone SSB	12 W
Harmonic suppression	65 dB
Carrier suppression, SSB	55 dB
Unwanted sideband suppression at 2500 Hz	55 dB

care must be taken to avoid overheating when using or testing such transmitters with single-tone or two-tone signals.

Other important specifications for SSB transmitters relate to output-frequency stability and accuracy and to spectral purity (that is, the absence of harmonics and other unwanted signals). The former are the same as for full-carrier AM transmitters. Spectral purity figures are a little more complex for SSB transmitters because the carrier and unwanted sideband are among the spurious signals that must be eliminated.

Table 6.1 shows some nominal specifications for a typical small SSB transmitter, in this case the transmitter part of a CB transceiver. This particular transmitter is also capable of operating in the full-carrier AM mode.

Note that the PEP rating is considerably greater than the AM carrier power. Also note that the carrier and unwanted-sideband suppression are not quite as good as the harmonic suppression. The unwanted-sideband suppression is given at a point 2500 Hz from the carrier; it would undoubtedly be worse for lower modulating frequencies.

Measurement techniques for SSB transmitters will be discussed later in this chapter.

6.3 Single-Sideband Receivers

The block diagram of a basic SSB receiver (see Figure 6.16) is similar to that for an AM receiver. There are some differences, however. We have already seen that SSB signals require a carrier to be reinserted at the receiver. This necessitates a different sort of detector from that of a receiver designed for full-carrier AM. Since SSB signals have much smaller bandwidth than full-carrier AM transmissions, we might also expect a narrower IF filter response. Finally, since tuning is very critical with SSB, an SSB receiver must have a very stable local oscillator. This is desirable in any case, but it is more critical with SSB than with full-carrier AM.

The receiver in Figure 6.16 is designed to cover the same frequency range as the transmitter in Figure 6.9. In fact, the receiver is almost a mirror image of the transmitter.

Figure 6.16
Block diagram of SSB receiver

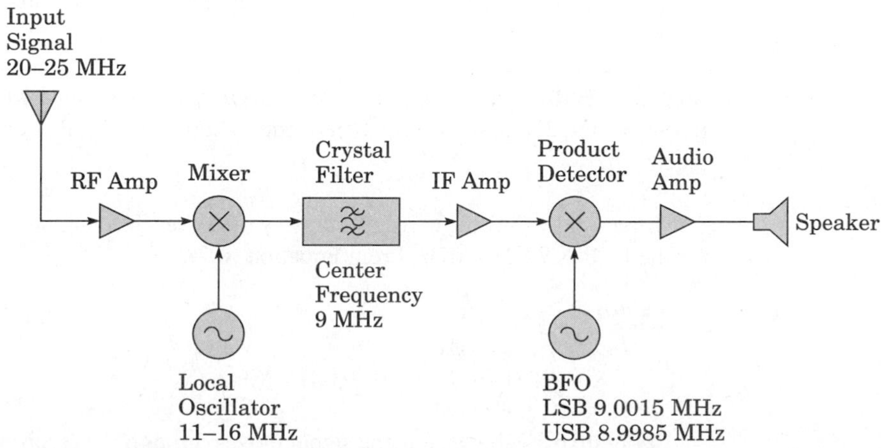

The BFO and CW Reception

Early communications receivers were usually designed for two types of signals: full-carrier AM and CW, which stands for *continuous waves*. The CW carrier was not actually transmitted continuously—it was switched on and off in the dot and dash patterns of the Morse code.

The BFO was used for CW reception. In that application, the BFO signal mixed with the transmitted carrier in the detector to produce an audible difference (beat) frequency. CW reception remains quite possible with any receiver designed for SSB, though it can be optimized by using a narrower IF filter and a different BFO frequency. Though gradually falling into disuse, CW is still quite popular with radio amateurs and sees limited use in marine and military communications. It is especially useful for emergency communication—in the presence of severe noise and interference, slow-speed CW is much more intelligible than voice.

After RF amplification and filtering to reject image frequencies, the signal from the antenna is applied to a mixer. The one in the illustration uses low-side injection, but high-side injection works equally well. A crystal filter provides IF selectivity; other than that, the IF amplifier is conventional. At the **product detector**, a signal is injected at the carrier frequency of the incoming signal. The oscillator that produces this signal is usually called the **beat-frequency oscillator** (BFO), largely for historical reasons.

Of course, at this point in the receiver, the carrier frequency has been moved by the mixer from its original value to an IF of approximately 9 MHz. Mathematically, for any receiver using low-side injection, we have

$$f_{IF} = f_{sig} - f_{LO}$$

where

f_{IF} = carrier frequency in the IF amplifier
f_{sig} = carrier frequency of the input signal
f_{LO} = local oscillator frequency

For this example, f_{IF} is equal to the BFO frequency f_{BFO}, so

$$f_{BFO} = f_{sig} - f_{LO} \tag{6.7}$$

Just as in the SSB transmitter studied earlier, the carrier is injected at one of two frequencies near either edge of the IF passband, depending on whether the received signal is USB or LSB. For the USB, the BFO frequency is near the lower edge of the passband, so that the USB signal passes through the filter. In the receiver, the purpose of the filter is not to reject the LSB of the received signal (that has already been done at the transmitter)—rather, the crystal filter rejects noise and interfering signals outside the frequency range occupied by the desired signal.

Example 6.7

The receiver illustrated in Figure 6.16 is tuned to an LSB signal with a (suppressed) carrier frequency of 23 MHz. To what frequency should the local oscillator be tuned for best reception?

Solution

The mixer must shift the carrier frequency of the received signal to that of the BFO, which for the LSB is 9.0015 MHz. From Equation (6.7),

$$\begin{aligned} f_{BFO} &= f_{sig} - f_{LO} \\ f_{LO} &= f_{sig} - f_{BFO} \\ &= 23.0000 \text{ MHz} - 9.0015 \text{ MHz} \\ &= 13.9985 \text{ MHz} \end{aligned}$$

Figure 6.17 shows how the signal moves through the receiver.

Figure 6.17

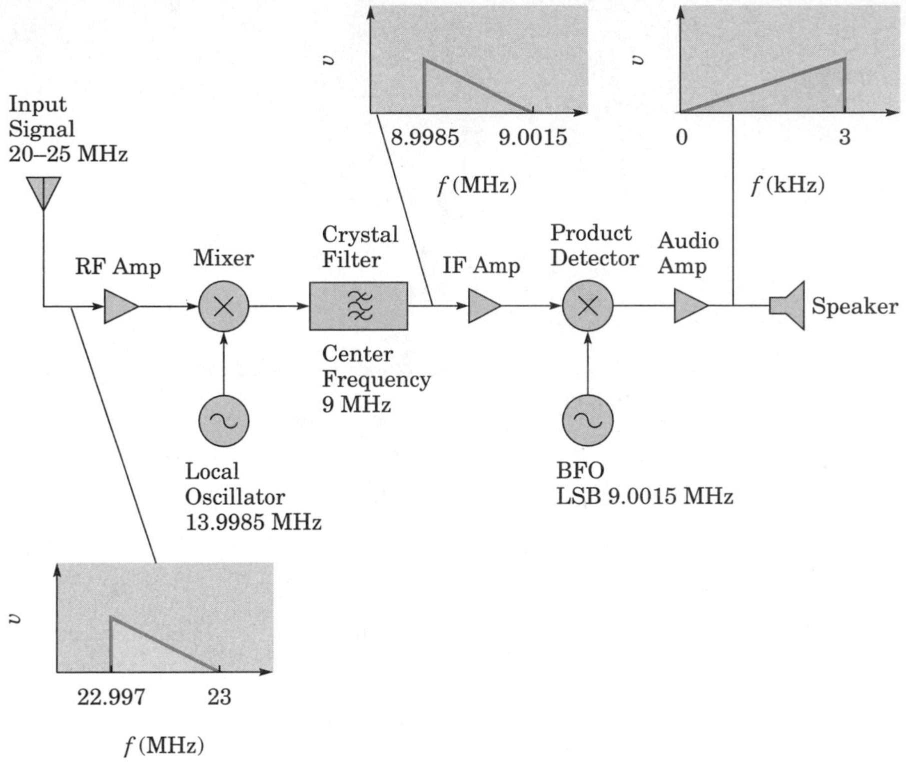

Just as with transmitters, receivers that use subtractive mixing and high-side injection will switch sidebands in the mixing process; that is, a USB signal at the antenna will become LSB in the IF amplifier, and an LSB signal will be changed to USB.

6.3.1 Detectors for Suppressed-Carrier Signals

A diode detector used alone will not work for SSB or DSBSC because the envelope is different from that of AM. A locally generated carrier can be injected along with the SSB signal into a diode detector, but it is better to use an arrangement similar to the one used to generate suppressed-carrier signals (that is, a balanced modulator). When used in receivers, such a device is called a *product detector*. Any of the balanced modulator circuits previously studied will work, but most contemporary designs use ICs such as the MC1496.

It may not be obvious that SSB and DSBSC can be demodulated by essentially the same process that created them, but it is easy to show mathematically that this is the case. Consider the detector in Figure 6.18, and remember that both balanced modulators and product detectors are actually multiplier circuits, that is, they have a single output whose instantaneous amplitude is proportional to the product of the amplitudes of two input signals.

The two inputs to the detector consist of the signal to be demodulated and a locally generated carrier produced by the BFO. In modern receivers, the BFO is crystal-controlled and switchable so that the carrier can be injected at the exact frequency of the (suppressed) signal carrier in the IF. This differs depending on whether the signal is USB or LSB. Since DSBSC is seldom used by itself, communications receivers do not make special provision for it; a DSBSC signal would be treated as SSB, with one of its sidebands removed in the IF before detection.

Consider an SSB signal with single-tone modulation. Let us choose USB, leaving the LSB case as an exercise for the reader. Assuming that this signal was originally produced by modulating a sine-wave carrier with radian frequency ω_c with another sine wave at ω_m,

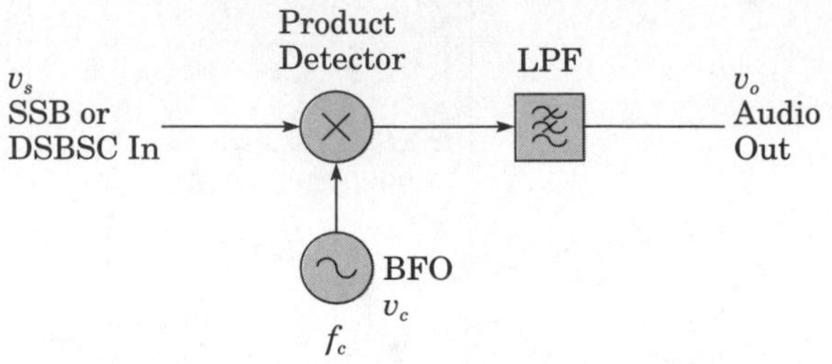

Figure 6.18 Product detector for DSBSC and SSB

it will have the following equation (neglecting amplitude factors that are not relevant to this discussion):

$$v_{sig} = -\cos(\omega_c + \omega_m)t \tag{6.8}$$

This is simply the USB component of the DSBSC signal shown in Equation (6.3).

Let the locally generated carrier be

$$e_c = \sin \omega_c t \tag{6.9}$$

The product detector will multiply these two signals to produce the output

$$\begin{aligned} v_o &= \sin \omega_c t[-\cos(\omega_c + \omega_m)t] \\ &= -\sin \omega_c t \cos(\omega_c + \omega_m)t \end{aligned}$$

Using the trigonometric identity

$$\sin a \cos b = 0.5[\sin(a - b) + \sin(a + b)]$$

we get

$$\begin{aligned} v_o &= -0.5[\sin(-\omega_m t) + \sin(2\omega_c + \omega_m)t] \\ &= 0.5 \sin \omega_m t - 0.5 \sin(2\omega_c + \omega_m)t \end{aligned} \tag{6.10}$$

This signal has two components. The first is simply the demodulated baseband signal. The second component can be removed by passing the output signal through a low-pass filter. This is quite easy because the unwanted component has a frequency more than twice the carrier frequency and therefore is several orders of magnitude higher than the baseband frequency.

The same method can be used to show that the product detector will work for multiple modulation frequencies, whether the signal is USB, LSB, or DSBSC.

In the foregoing, it was assumed that the carrier inserted at the receiver was exactly the same as the original carrier in both frequency and phase. This is *not* a trivial matter. Because there is no transmitted carrier to act as a reference and because the receiver is generally tuned by ear for the most natural-sounding voice, the phase relationship between the original and locally generated carriers in most communications receivers will be random, and the two frequencies will not be exactly equal. The effect can be determined by repeating the mathematical analysis performed earlier for a BFO frequency not quite equal to ω_c. However, the effect on an SSB signal can be seen more easily by looking at Figure 6.19. Figure 6.19(a) shows an example of a USB signal with a (suppressed) carrier frequency of 1 MHz and two modulating signals of 1 kHz and 3 kHz, resulting in an RF spectrum with components at 1001 and 1003 kHz.

Figure 6.19(b) shows this signal being detected correctly, with the BFO at exactly the carrier frequency of 1 MHz. The output, after low-pass filtering, consists of the difference signals that result from subtracting the BFO frequency from each of the transmitted components. This gives the original 1 kHz and 3 kHz baseband signals at the output.

Figure 6.19
Effect of receiver mistuning on SSB reception

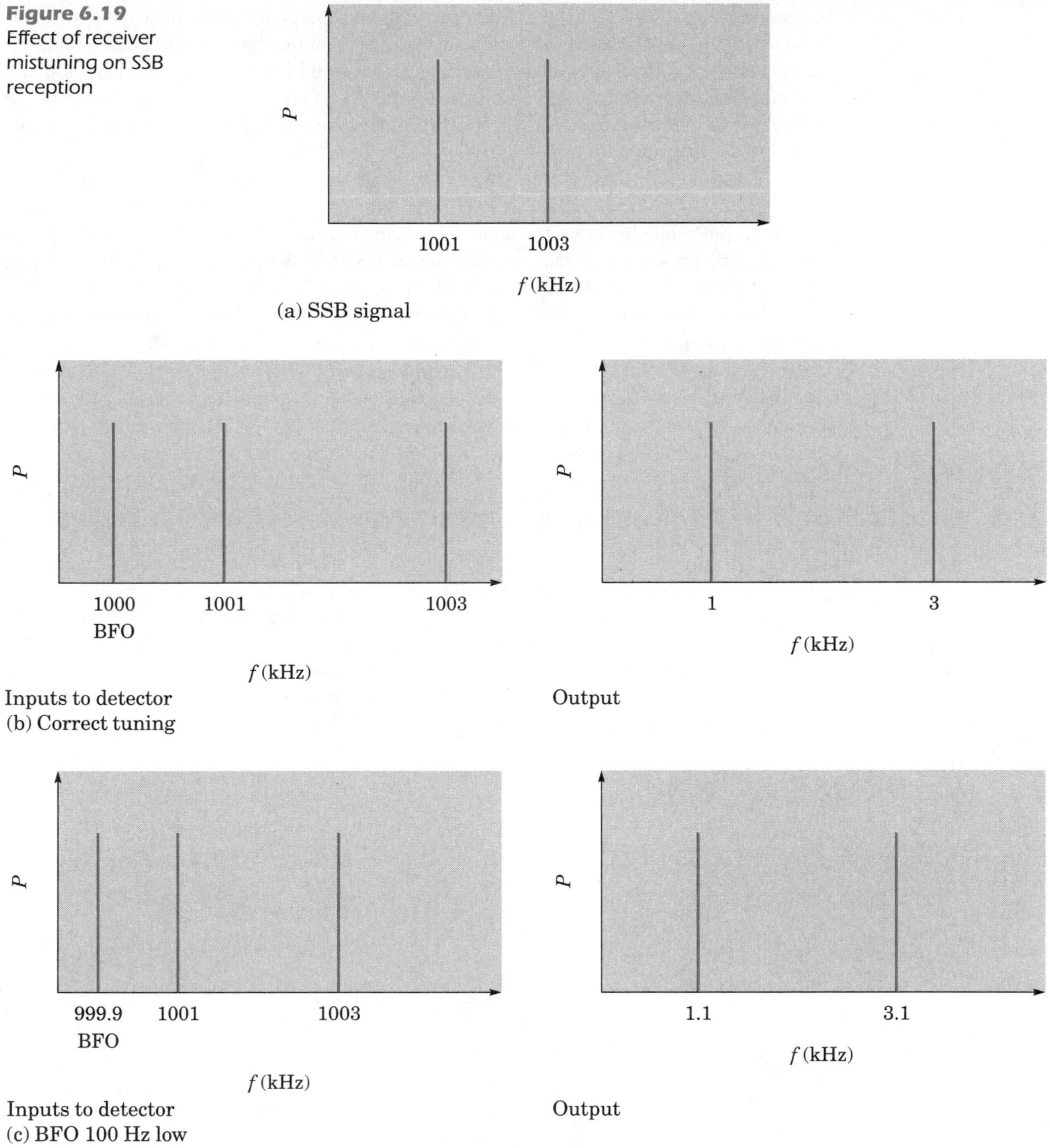

Figure 6.19(c) shows what happens when the BFO signal is 100 Hz too low, at 999.9 kHz. The difference signals are now

$$1001 \text{ kHz} - 999.9 \text{ kHz} = 1.1 \text{ kHz}$$

and

$$1003 \text{ kHz} - 999.9 \text{ kHz} = 3.1 \text{ kHz}$$

That is, a BFO that is set 100 Hz low causes each component of the demodulated baseband signal to be 100 Hz higher than it should be. It is obvious that this will have a distorting effect on voice signals, but if the difference is small (less than 100 Hz or so), the information signal will still be intelligible. It is not obvious from the figure, but the phase relationships between components of the baseband signal will also be disrupted. This is of no consequence for voice-grade communications. In fact, experts disagree as to whether phase coherence is important even for high-fidelity audio systems. In any case, as explained earlier, SSB is never used for high fidelity.

The situation is more critical with a DSBSC signal, since a BFO frequency offset will simultaneously raise one of the two difference frequencies between sideband and inserted carrier while lowering the other. This causes severe distortion. Phase relationships must also be preserved in many applications of DSBSC. In color television, for example, the hue of the color is determined by the phase angle of the color signal.

We can thus conclude that a product detector with a BFO that is at almost the original carrier frequency is acceptable for SSB voice but that DSBSC or phase-critical signals require a locally generated carrier that is phase-locked to the original carrier. This process is called *coherent detection*. There are several ways of accomplishing this, the simplest of which is to transmit a **pilot carrier** with reduced amplitude to which the receiver BFO can be locked. Other methods will be described later in the appropriate sections.

6.4 Single-Sideband Transceivers

SSB is not currently used for broadcasting, though there has been some talk of using it in shortwave radio broadcasting. Most of the applications for SSB are in two-way half-duplex communication, in which each station transmits and receives alternately. Combining the transmitter and receiver into a single piece of equipment (called a *transceiver*) provides greater convenience and a considerable reduction in cost.

Figure 6.20 is a simplified block diagram of a typical SSB transceiver. The dotted lines show the transmit signal path, and the dashed lines show the received signal. The

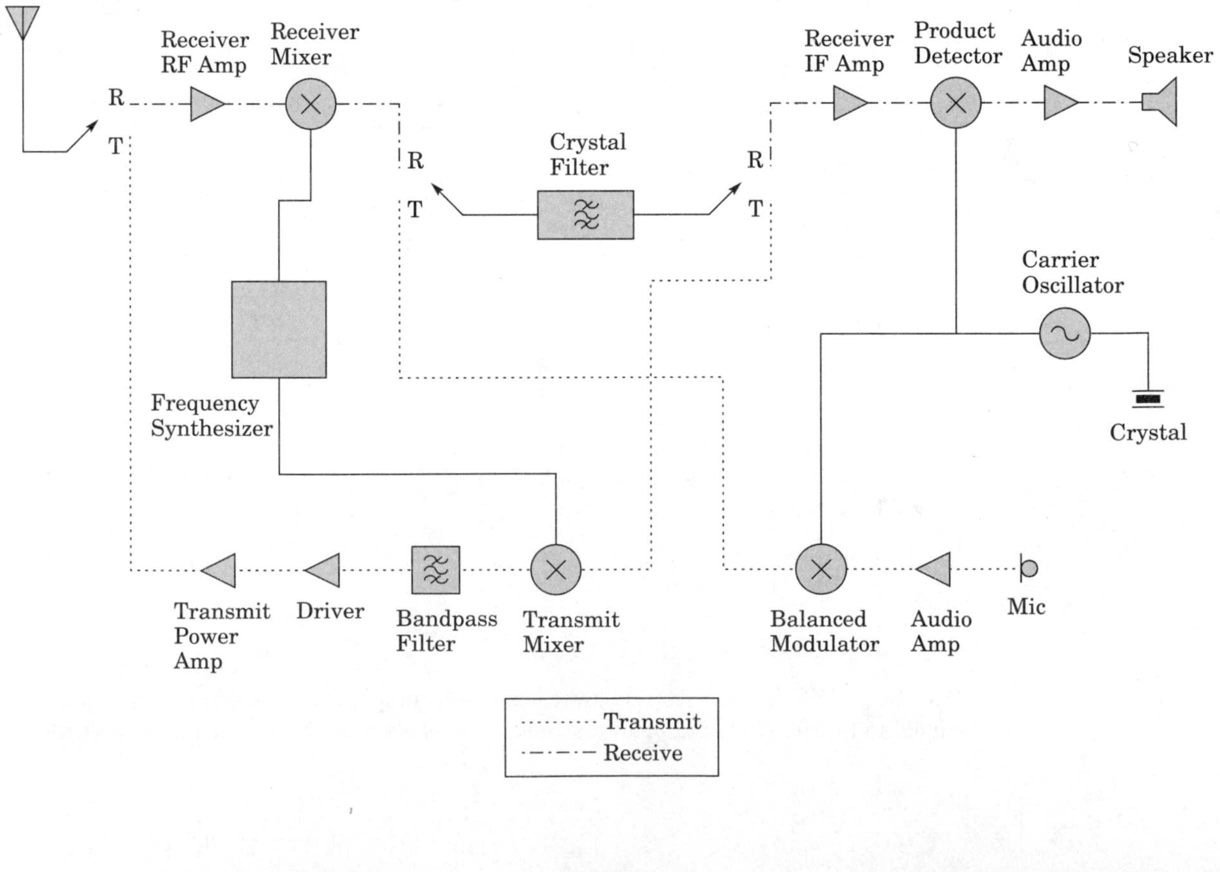

Figure 6.20
Block diagram of SSB transceiver

expensive crystal filter and frequency synthesizer are used for both transmitting and receiving. There is also a single crystal oscillator to provide the carrier signal for the balanced modulator when transmitting and to act as a BFO while receiving.

The transmitting process begins at the microphone. The audio signal is amplified and applied to a balanced modulator, along with the carrier oscillator signal. The output of the balanced modulator is a DSBSC signal at the IF frequency. It is sent through a crystal filter to remove the unwanted sideband and then to a mixer. The transmit mixer combines the SSB signal with a signal from the frequency synthesizer to produce an output signal at the operating frequency. This signal is then amplified to produce the required output power.

The received signal goes first to a stage of RF amplification, then to a mixer, where it is converted to the same IF as is used for transmitting. The crystal filter provides the receiver selectivity and is followed by IF amplification. A product detector, with the carrier oscillator serving as BFO, recovers the modulation. The baseband signal is then amplified and sent to a loudspeaker.

Example 6.8

An SSB transceiver has the block diagram shown in Figure 6.20. It is designed to operate at a frequency of approximately 30 MHz. When the unit is switched to receive, the synthesizer produces a frequency of 21.000 MHz. The crystal filter has a center frequency of 9.000 MHz and a bandwidth of 2.8 kHz.

(a) Determine which sideband will be received.
(b) What should the carrier frequency of the received signal be for proper reception?
(c) What frequency should the synthesizer produce when transmitting, in order to transmit the same sideband with the same carrier frequency?

Solution

(a) First, notice that the receive local oscillator operates at 21 MHz, which is approximately 9 MHz lower than the received signal frequency. This means that the mixer is using subtractive mixing with low-side injection. Therefore there will be no sideband inversion, that is, the sideband that is present in the IF amplifier will be the same sideband that is received. Since the BFO is near the high end of the IF filter, the lower sideband will be received.

(b) The carrier frequency f_i at the antenna must satisfy the equation

$$f_i - f_{LO} = f_{IF}$$

because low-side injection and subtractive mixing are used. Therefore

$$\begin{aligned} f_i &= f_{IF} + f_{LO} \\ &= 9.0015 \text{ MHz} + 21.000 \text{ MHz} \\ &= 30.0015 \text{ MHz} \end{aligned}$$

The frequency of the carrier oscillator, rather than the filter center frequency, is used for f_{IF}, so that the carrier frequency is moved to the same frequency as the BFO. This is a requirement for proper demodulation.

(c) The transmitted SSB signal is generated at the same IF as is used in the receiver and must be raised to the transmitting frequency. Therefore the transmit mixer must be additive, so that

$$f_o = f_{IF} + f_{LO}$$

The same carrier oscillator can be used as before. This will generate an LSB signal with a carrier frequency of 9.0015 MHz. The output frequency must be 30.0015 MHz, as calculated above. Rearranging the above equation gives

$$\begin{aligned} f_{LO} &= f_o - f_{IF} \\ &= 30.0015 \text{ MHz} - 9.0015 \text{ MHz} \\ &= 21.000 \text{ MHz} \end{aligned}$$

This is the same frequency as used for transmitting. If the transmit and receive frequencies are the same, it is not even necessary to retune the synthesizer when switching between transmit and receive operations.

6.5 Two-Tone Tests of Single-Sideband Transmitters

Measurement of output power is somewhat more complicated with SSB than with AM transmitters. Obviously, carrier-power measurements are useless, since SSB transmitters are not intended to produce a carrier. We have already noted that PEP (peak envelope power) is the usual way to measure the power output of SSB transmissions. This power is often limited by clipping in the transmitter output stage. On the other hand, such limitations as power-supply capacity and heat dissipation tend to put limits on the average power output. The ratio between PEP and average power depends on the waveform of the modulating signal and the duty cycle of the transmitter.

Figure 6.21 shows the envelope of an SSB signal for several situations. Continuous modulation with a single tone (a sine wave), as shown in Figure 6.21(a), is the hardest test, since PEP and average power output are equal. As shown in Figure 6.21(b), voice signals have much higher peak-to-average power ratios, about 3 : 1 or 4 : 1, but of course, this varies with the voice. It was shown previously that a transmitter modulated by a *two-tone test signal* consisting of two sine waves of equal amplitude but different frequencies, as in Figure 6.21(c), has a PEP-to-average-power ratio of 2 : 1. This is a little more severe than voice operation.

Figure 6.21
SSB signal envelopes

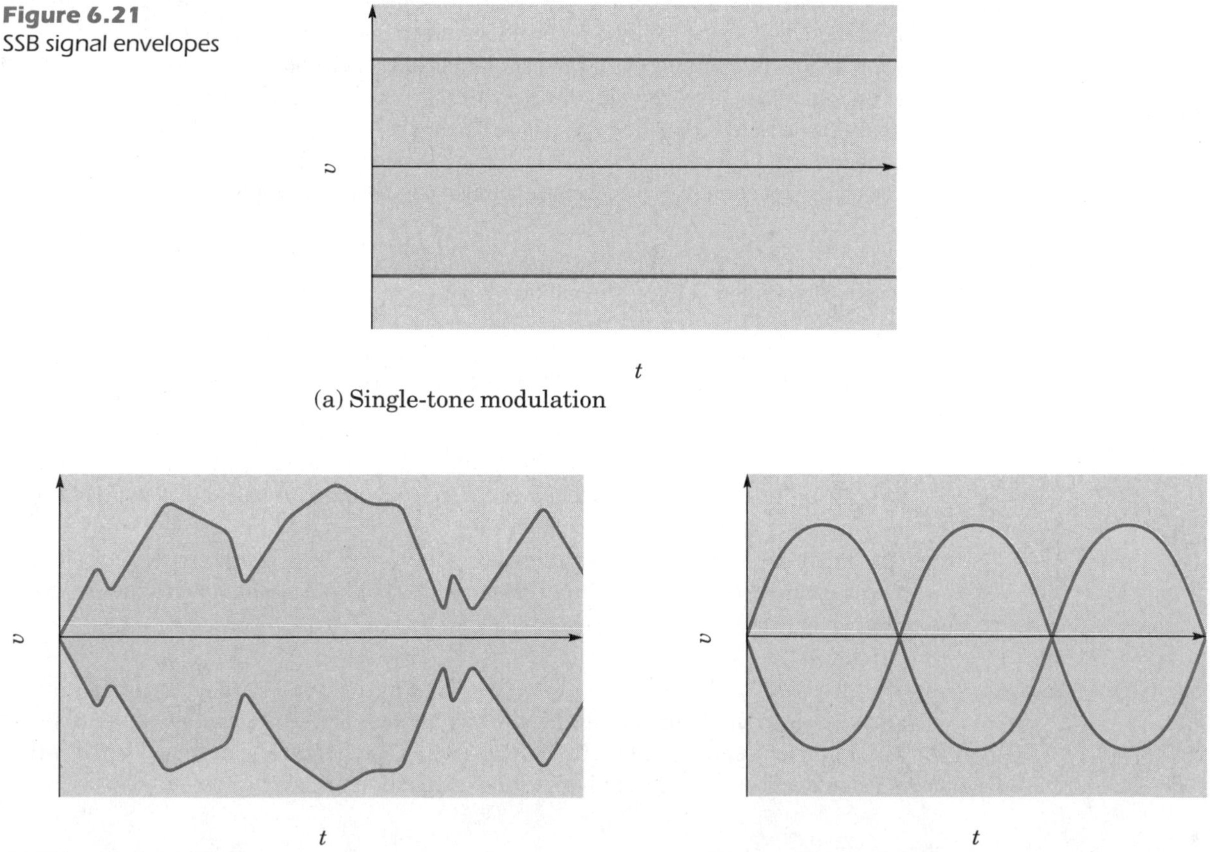

(a) Single-tone modulation

(b) Voice modulation

(c) Two-tone modulation

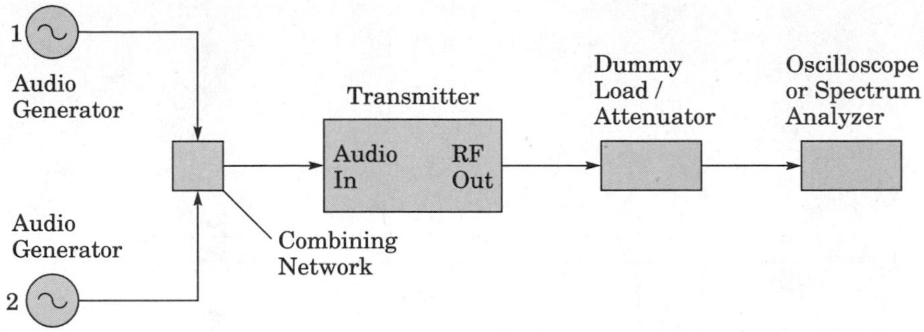

(a) Equipment setup

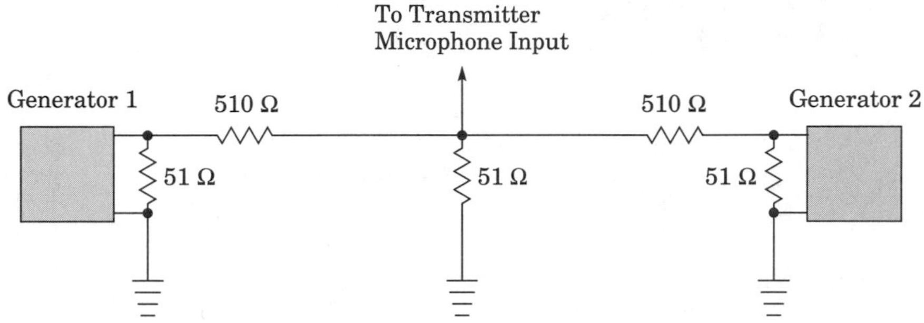

(b) Combining network

Figure 6.22 Two-tone transmitter tests

Any of the modulating signals shown in Figure 6.21 can be continuous or intermittent. Two-way half-duplex communication, the commonest form of SSB service, usually represents a duty cycle of 50% or less, that is, the operator spends at least as much time listening as talking. A 50% duty cycle reduces transmitter cooling requirements for a given PEP rating. In fact, many transmitters have PEP ratings that vary according to the use to which the transmitter will be put.

PEP can easily be measured by using a two-tone test signal and the test setup sketched in Figure 6.22(a). Two audio generators are needed, and they should be well isolated from each other by a combining network. A resistive network like the one shown in Figure 6.22(b) can be used with generators having a 50 Ω output impedance. The frequencies of the audio generators should not be harmonically related.

The transmitter power output can be measured using an oscilloscope or a spectrum analyzer. Figure 6.23(a) shows normal displays for each type of instrument. The scope allows severe distortion, such as clipping, to be seen as a change in the shape of the waveform, while the spectrum analyzer displays spurious signals if distortion is present. See Figure 6.23(b) for examples of distorted signals.

Besides power output, operators of SSB transmitters are also very concerned with carrier and unwanted-sideband suppression and intermodulation products. All of these are easily measured with a two-tone test signal and a spectrum analyzer. Figure 6.24 shows a typical spectrum analyzer display. The carrier and the unwanted sideband can be clearly seen, even though they are very much weaker than the desired sideband. There will always be some carrier output, because no balanced modulator is ever perfectly balanced, and there will always be some unwanted-sideband energy, because no bandpass filter has infinite rejection.

There are also spurious signals that result from the addition and subtraction of the desired frequency components and their harmonics. These are due to undesired nonlinear-

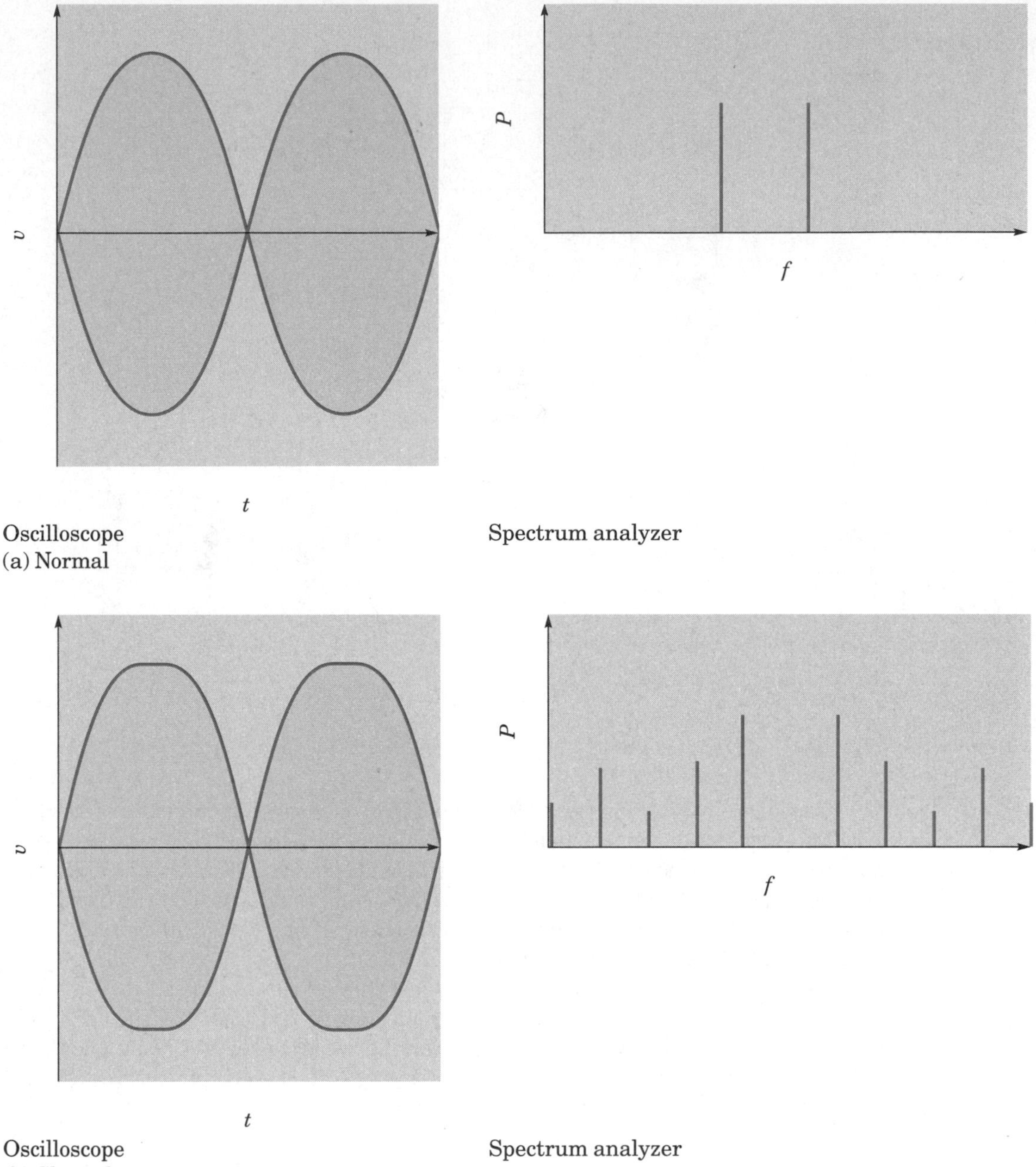

Figure 6.23
Two-tone SSB test results

ities in the amplifier stages in the transmitter and can never be removed entirely. Most, but unfortunately not all, of the possible combinations will be so far removed in frequency from the main signal that they will not be transmitted. Two of these spurious signals are shown in the display in Figure 6.24. In order to measure the effectiveness of the suppression of unwanted frequency components, it is only necessary to note the number of decibels by which the level of each component is less than that of each of the desired signal components (which should be at the same level).

The most bothersome of these intermodulation distortion (IMD) products are the third-order products. If the two transmitted frequencies with a two-tone test signal are f_A

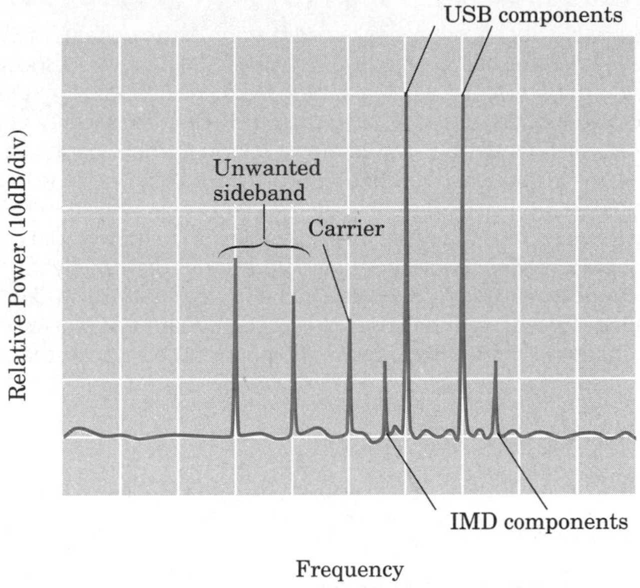

Figure 6.24
Two-tone test of SSB transmitter

and f_B, the third-order intermodulation products have frequencies of $(2f_A - f_B)$ and $(2f_B - f_A)$. Since f_A and f_B are very close together, these difference frequencies will be close in frequency to f_A and f_B and will be in the passband of any filter that is applied to the transmitter output.

Example 6.9 An SSB transmitter operating on the USB with a carrier frequency of 26.9 MHz is modulated with two tones at 1 kHz and 1.6 kHz, respectively. Find the output frequencies and the frequencies of the third-order intermodulation distortion products.

Solution The two desired output frequencies are:

$$\begin{aligned} f_A &= 26.9 \text{ MHz} + 1 \text{ kHz} \\ &= 26.901 \text{ MHz} \\ f_B &= 26.9 \text{ MHz} + 1.6 \text{ kHz} \\ &= 26.9016 \text{ MHz} \end{aligned}$$

The third-order IMD products are at:

$$\begin{aligned} 2f_A - f_B &= 2 \times 26.901 \text{ MHz} - 26.9016 \text{ MHz} \\ &= 26.9004 \text{ MHz} \\ 2f_B - f_A &= 2 \times 26.9016 \text{ MHz} - 26.901 \text{ MHz} \\ &= 26.9022 \text{ MHz} \end{aligned}$$

Figure 6.24 shows how these distortion products relate to the transmitted signal. Both of them are within the range of voice frequencies that would be transmitted by this transmitter. Once generated, these distortion products would be impossible to remove by filtering.

Single-Sideband in Amateur Radio

Amateur radio operators have taken advantage of the benefits of SSB for many years. In fact, very few amateurs use full-carrier AM at present. Figure 6.25 shows a typical modern amateur radio transceiver for use in the high-frequency bands. Anyone with a communications receiver can listen to amateur radio signals. Tune in the vicinity of 3.8, 7.2, or 14.2 MHz to start. You will find that LSB is used on the lowest of these bands and USB on the others.

Many portable radios with shortwave bands receive these frequencies but do not give intelligible results with SSB signals. By now you know why: they lack the essential BFO.

After listening for a while, many people decide they would like to transmit as well. This requires a license, of course, but getting one is not difficult. By the time you finish this book, you will know the required theory. It is also necessary to study the relevant government regulations and, for some classes of license, to learn the Morse code. More details can be obtained from a local amateur radio club (many colleges have them) or, in the United States, from

American Radio Relay League
Newington, CT 06111

In Canada, contact

Radio Amateurs of Canada, Inc.
614 Norris Court, Unit 6
Kingston, ON
K7P 2R9

Amateur radio is an enjoyable hobby and a good way to study practical communication systems.

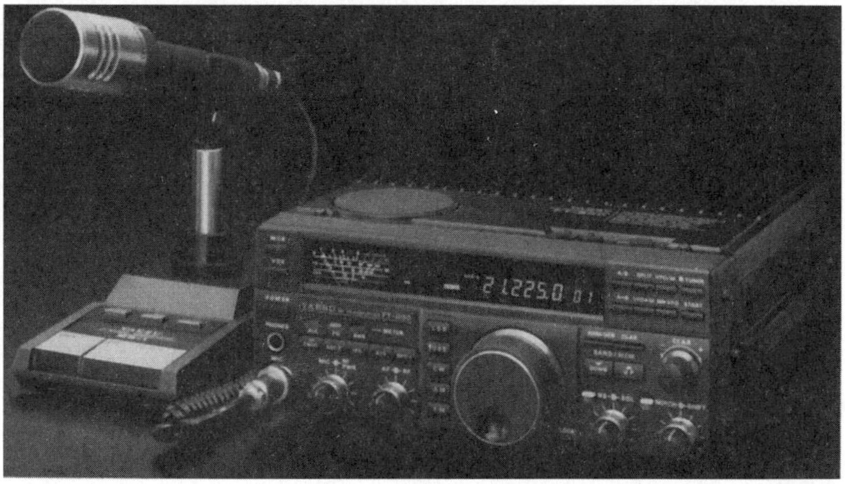

Figure 6.25
Amateur radio SSB transceiver
Courtesy Yaesu U.S.A.

Summary

Here are the main points to remember from this chapter.

1. Suppressing the carrier of an AM signal leads to greater efficiency with no loss of information content.
2. Removing one of the two sidebands of an AM signal reduces its bandwidth by approximately 50%, leading to an improved signal-to-noise ratio and allowing more signals to occupy a given frequency range.
3. A DSBSC AM signal can be generated by means of a balanced modulator, to which baseband and carrier-frequency signals are applied.

4. The most popular method for generating an SSB signal is to use a crystal filter to remove the unwanted sideband from a DSBSC signal. This is called the *filter method* of SSB generation.
5. Once an SSB signal has been generated, it can be moved to the transmitting frequency by means of a mixer–local oscillator combination.
6. SSB and DSBSC signals, like all AM signals, must be amplified using linear amplifiers in both transmitters and receivers.
7. In order to demodulate an SSB or DSBSC signal, it is necessary to reinsert the carrier at the receiver. The locally generated carrier frequency must be accurate to within approximately 100 Hz for intelligible voice reception.
8. When half-duplex communication is required, considerable reductions in cost and complexity are possible by combining a transmitter and receiver into a transceiver. Generally, a single crystal filter, local oscillator, and frequency converter can be used for both transmission and reception.

Important Equations

$$\sin \omega_m t \sin \omega_c t = 0.5[\cos (\omega_c - \omega_m)t - \cos (\omega_c + \omega_m)t] \quad (6.3)$$

$$PEP = \frac{V_p^2}{2R_L} \quad (6.5)$$

$$v_o \text{ (product detector)} = 0.5 \sin \omega_m t - 0.5 \sin (2\omega_c + \omega_m)t \quad (6.10)$$

Glossary

beat-frequency oscillator (BFO) an oscillator that reinserts the carrier signal in a single-sideband or continuous-wave receiver

peak envelope power (PEP) the power measured at modulation peaks in an AM or single-sideband signal

pilot carrier a signal transmitted in a suppressed-carrier communication system in order to supply carrier frequency and phase information to the receiver

product detector a balanced modulator used as a detector for suppressed-carrier signals

single-sideband (SSB) any AM scheme in which only one of the two sidebands is transmitted

suppressed-carrier signal an AM signal in which the carrier-frequency component is eliminated and only one or both sidebands are transmitted

two-tone test signal a signal consisting of two audio frequencies, not harmonically related, used to test single-sideband transmitters

Questions

1. Give two advantages of SSB operation compared with full-carrier AM.
2. Does full-carrier AM have any advantages over SSB? Explain your answer.
3. Why and by how much does suppressing the carrier improve the efficiency of an AM signal?
4. Why and by how much does eliminating one sideband improve the efficiency of an AM signal?
5. Why is the transmitted bandwidth of a typical SSB signal actually less than one-half that of a full-carrier AM signal transmitting the same information signal?
6. What is the audible effect of a slight mistuning of an SSB receiver?
7. What is the audible effect of setting an SSB receiver to the wrong sideband?
8. Describe one method of generating DSBSC signals.
9. Describe one method of generating SSBSC signals.
10. Why are Class C amplifiers unsuitable for use in SSB transmitter power amplifiers?
11. How is the power output of an SSB transmitter specified?
12. What is a BFO? Why is it necessary in an SSB receiver?
13. Why is local-oscillator stability more important in SSB receivers than in those for full-carrier AM?
14. Suggest suitable bandwidths for the AM, SSB, and CW filters in a communications receiver.

15. What is the most common means of achieving the required receiver IF bandwidth?
16. Which stages are typically shared between the receiver and transmitter sections of an SSB transceiver?
17. How does peak envelope power compare with average power for a typical voice signal?
18. By how much are the carrier and unwanted sideband suppressed in a typical SSB transmitter?
19. What problems might arise if SSBSC AM were to be used in shortwave broadcasting?
20. Selective fading, a problem in which the received signal strength varies rapidly with the frequency of the received signal, plagues high-frequency radio. Explain how SSB can at least partly overcome the effects of selective fading.
21. What do the balanced mixer, balanced modulator, and product detector circuits have in common?
22. Explain, using a block diagram, how a two-tone test of an SSB transmitter can be performed.

Problems

SECTION 6.1

23. A 5 MHz carrier is modulated by a 5 kHz sine wave. Sketch the result in both the time and frequency domains for each of the following modulation types. Time and frequency scales are required but amplitude scales are not.
    (a) DSB full-carrier AM
    (b) DSBSC AM
    (c) SSBSC AM (USB)
24. If a transmitter power of 100 W is sufficient for reliable communication over a certain path using SSB, approximately what power level would be required using each of the following?
    (a) DSBSC
    (b) full-carrier AM
25. (a) An AM transmitter has a carrier power of 50 W at a carrier frequency of 12 MHz. It is modulated at 80% by a 1 kHz sine wave. How much power is contained in the sidebands?
    (b) Suppose the transmitter in part (a) can also be used to transmit a USB signal with an average power level of 50 W. By how much (in decibels) will the signal-to-noise ratio be improved when the transmitter is used in this way, compared with the situation in part (a)?
26. Sketch the output signal from each of the transmitters described in the previous problem, in both the time and frequency domains. Assume the load impedance is 50 Ω.
27. CB radio was developed using full-carrier DSB AM. It operates using 40 channels, each 10 kHz wide. When SSBSC is used, the same carrier frequencies are used as with conventional AM, but the transceivers are switchable to either lower or upper sideband. How many channels are available when SSB is used?
28. A baseband video signal has a spectrum that extends from dc to 4.2 MHz.
    (a) What bandwidth would be required to transmit this signal using full-carrier DSB AM?
    (b) What bandwidth would be required to transmit this signal using SSBSC AM?
    (c) Can you suggest any difficulties that could arise when attempting to transmit this signal using SSBSC AM?
29. Sketch the following SSBSC signal in the frequency domain. Show a frequency scale, and a power scale in dBm. Assume perfect suppression of unwanted frequency components.
    Carrier frequency: 20 MHz
    Modulating frequencies: 2 kHz and 5 kHz with equal amplitude for each
    Peak envelope power: 10 W
    Modulation: upper sideband
30. (a) Sketch each of the following signals in the frequency domain:
        (i) A DSBSC signal with single-tone modulation. The carrier frequency is 10 MHz and the modulating frequency is 1 kHz. The average power is 5 W.
        (ii) An SSBSC USB signal with two-tone modulation. The carrier frequency is 9.998 MHz and the modulating frequencies are 1 kHz and 3 kHz with equal amplitude for each. The average power is 5 W.
    (b) Compare the two sketches. What do you notice about them?
    (c) How would a receiver distinguish between these signals?

Figure 6.26

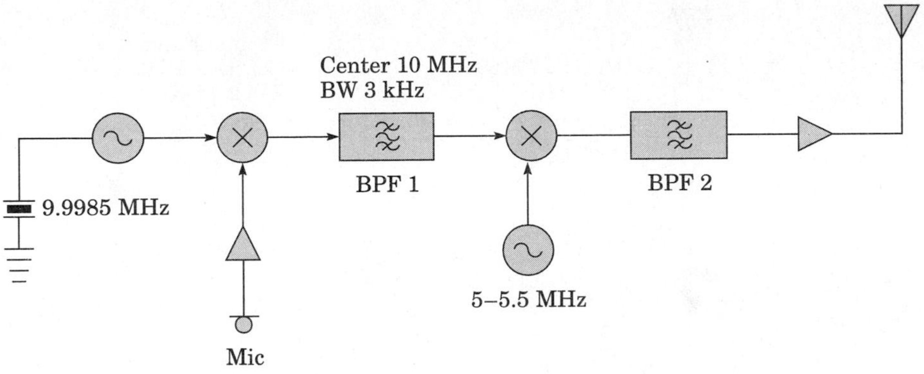

SECTION 6.2

31. An SSB transmitter has a PEP of 250 W. It is modulated equally by two audio frequencies at 400 Hz and 900 Hz. The carrier frequency is 14.205 MHz and the transmitter generates the LSB. Sketch the output in the time and frequency domains for a load impedance of 50 Ω.
32. A carrier at 10 MHz and a sine-wave modulating signal at 2 kHz are applied to a balanced modulator. Sketch the output in the time and frequency domains showing time and frequency scales, respectively.
33. A filter-type SSB generator uses an ideal bandpass filter with a center frequency of 5.000 MHz and a bandwidth of 2.7 kHz. What frequency should be used for the carrier oscillator if the generator is to produce a USB signal with a baseband frequency response having a lower limit of 280 Hz?
34. An SSB transmitter has the block diagram shown in Figure 6.26.
    (a) Using a variable-frequency oscillator (VFO) that tunes from 5.0 to 5.5 MHz, the transmitter will operate over two frequency ranges that can be selected by a suitable choice of bandpass filter BPF2. What ranges are they?
    (b) Which sideband would be produced at the output for each of the frequency ranges specified in part (a)?
    (c) The other sideband could be generated by changing the carrier oscillator frequency. Choose a suitable value for this frequency.
35. The block diagram of an SSB transmitter is shown in Figure 6.27. The local oscillator frequency is higher than the frequency at which the SSB signal is generated, and the difference between the two frequencies is used at the output.
    (a) Choose a suitable frequency for the carrier oscillator if the transmitter is to produce a USB signal.
    (b) What should be the frequency of the local oscillator if the (suppressed) carrier frequency at the antenna is to be exactly 30 MHz?
    (c) Suppose that the transmitter is modulated by a single sine-wave tone at 1 kHz. It is operating with a PEP of 100 W into a 50 Ω load. Sketch the output in the time and frequency domains, showing all appropriate scales.
36. Consider a full-carrier AM signal with $m = 1$. Show mathematically that the PEP of this signal is equal to four times the carrier power.

Figure 6.27

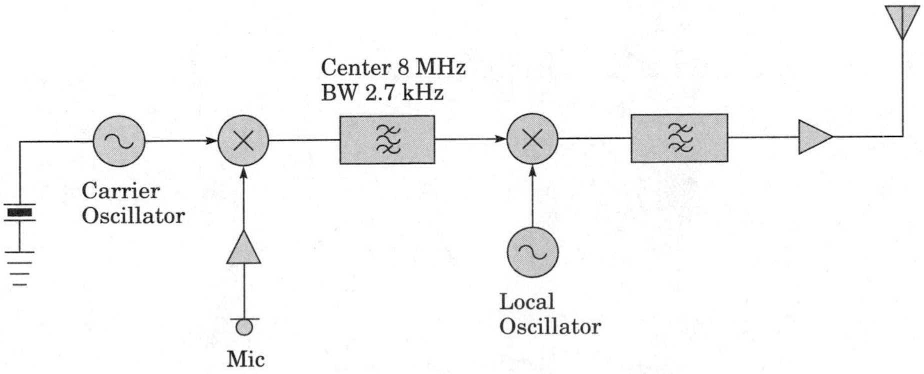

Figure 6.28

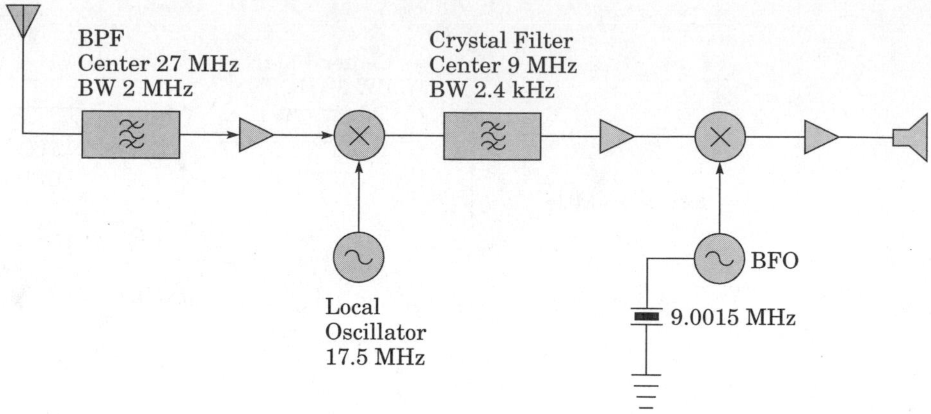

SECTION 6.3

37. The block diagram of an SSB receiver is shown in Figure 6.28.
    (a) Which sideband will be received with the values shown?
    (b) What will be the carrier frequency of the received signal?
    (c) What changes would have to be made to set up this receiver to receive the other sideband at the same carrier frequency?
38. Sometimes an AM receiver is made capable of SSB reception by simply adding a BFO, with the IF bandwidth left as for DSB AM. What would be the effect of this on the signal-to-noise ratio, compared with a receiver with the proper IF filter?
39. A CW signal is being transmitted at 7.100 MHz. Explain how it can be received by an SSB receiver, and suggest suitable frequencies to which the receiver could be tuned. Does it matter whether the receiver is set to USB or LSB?
40. Draw a block diagram for an SSB receiver with the following characteristics:
    (a) The receiver is single conversion with one RF stage.
    (b) The IF uses a single filter with a center frequency of 10 MHz and a bandwidth of 2.8 kHz.
    (c) The receiver can be used to receive either sideband.
    (d) The mixer uses high-side injection of the local oscillator signal.
    (e) The receiver tunes the range from 20 to 30 MHz.
41. Suppose the receiver described in the previous problem were used to receive a full-carrier double-sideband AM signal. Could this be made to work? Explain your answer. Can you think of any difficulties that could arise?

SECTION 6.4

42. A block diagram of an SSB transceiver is given in Figure 6.29. Redraw it to show the transmitter and receiver functions in two separate diagrams. Note that some blocks are used for both transmitting and receiving.

Figure 6.29

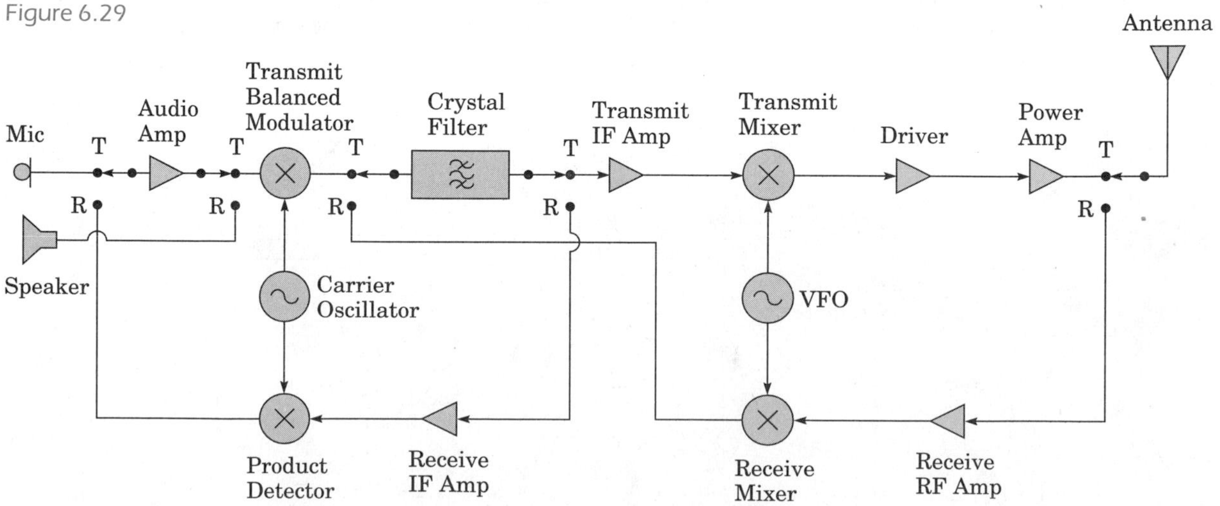

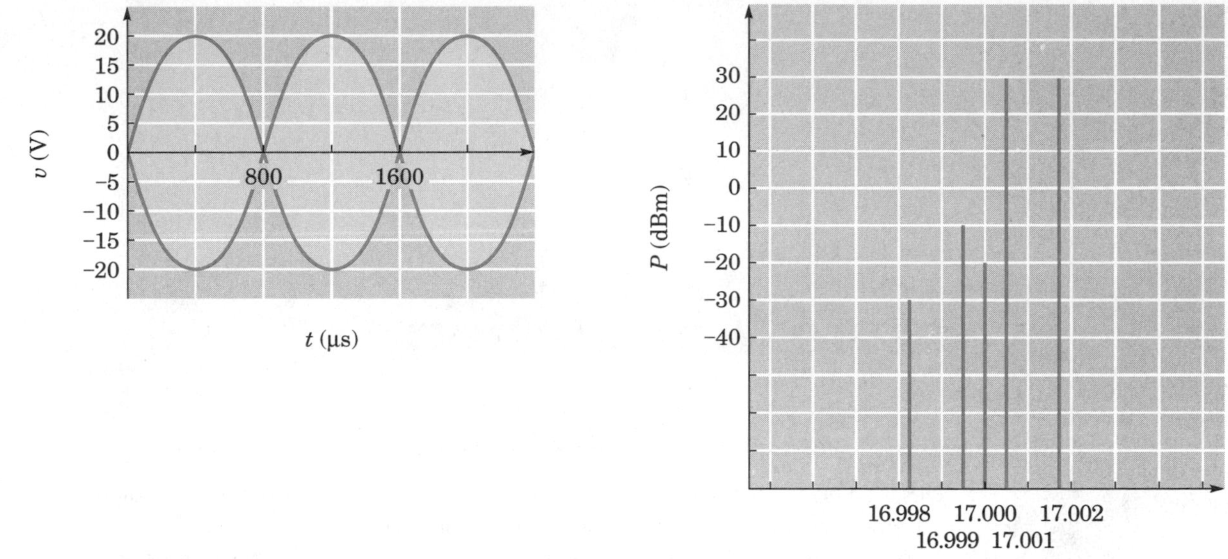

Figure 6.30

SECTION 6.5

43. The results of a two-tone test on an SSB transmitter with a carrier frequency of 17 MHz are shown in both the time and frequency domains in Figure 6.30. Use the information provided to find:
    (a) the two tones
    (b) whether the transmitter is USB or LSB
    (c) the PEP (assume the transmitter has a 50 Ω load)
    (d) the carrier suppression
    (e) the unwanted-sideband suppression at each of the test frequencies

COMPREHENSIVE

44. SSB transmitters use Class AB power amplifiers, while AM transmitters generally use more efficient Class C circuits. Suppose that 100 W total dc power is used to power the final stages of each of two transmitters as follows (for the AM transmitter, the 100 W includes the power supplied to the modulating amplifier):

    Transmitter 1: Full-carrier AM. The modulating amplifier is Class AB with 70% efficiency, the final amplifier is Class C with 85% efficiency, and the modulation index is 75%.

    Transmitter 2: SSB. The RF power amplifier is Class AB with 65% efficiency.

    Which transmitter will produce a better signal-to-noise ratio at the receiver? By how much?
45. An SSB transmitter transmits on the USB with a (suppressed) carrier frequency of 7.2 MHz. It is modulated by two tones, with frequencies of 1 kHz and 2.5 kHz and equal amplitude. The transmitter PEP is 75 W into a 50 Ω load. The carrier and unwanted sideband are each suppressed by 60 dB.
    (a) Calculate the peak voltage across the load.
    (b) Calculate the average power into the load.
    (c) Sketch the signal envelope in the time domain, showing voltage and time scales.
    (d) Sketch the signal in the frequency domain, showing frequency and power scales.
46. An SSB transmitter is modulated by tones of 1, 2, and 3 kHz. A receiver is tuned to this signal, but it is slightly misadjusted: it is tuned 100 Hz too high. What frequencies will be present at the output of this receiver if the transmitter is:
    (a) transmitting on the USB?
    (b) transmitting on the LSB?

7 Angle Modulation

Objectives After studying the material in this chapter, you should be able to:

1. Describe and explain the differences between amplitude and angle modulation schemes and the advantages and disadvantages of each
2. Describe and explain the differences between frequency and phase modulation and show the relationship between the two
3. Calculate bandwidth, sideband frequencies, carrier and sideband voltage and power levels, and modulation index for frequency- and phase-modulated signals
4. Explain the capture effect and noise threshold level for FM signals and calculate the signal-to-noise ratio for simple situations
5. Relate deviation, bandwidth, and signal-to-noise improvement for FM systems
6. Explain the use of pre-emphasis and de-emphasis in FM systems and calculate component values for pre-emphasis and de-emphasis circuits
7. Describe the system used for FM stereo broadcasting and draw a diagram showing the spectrum of an FM stereo signal
8. Perform measurements on FM signals using a spectrum analyzer

7.1 Introduction

At the beginning of this book, we observed that only three parameters of a carrier wave can be changed or modulated in order for it to carry information: amplitude, frequency, and phase. The last two are closely related, since frequency (expressed in radians per second) is the rate of change of phase angle (in radians). If either frequency or phase is changed in a modulation system, the other will change as well. Consequently, it is useful to group frequency and phase modulation together in the term **angle modulation**.

Both *frequency modulation* (FM) and *phase modulation* (PM) are widely used in communications systems. FM is more familiar in daily life, since it is used extensively for radio broadcasting. FM is also used for the sound signal in television, for two-way fixed and mobile radio systems, for satellite communications, and for cellular telephone systems, to name only a few of its more common applications.

While PM may be less familiar, it is used extensively in data communications. It is also used in some FM transmitters as an intermediate step in the generation of FM. FM and PM are closely related mathematically, and it is quite easy to change one to the other.

The most important advantage of FM or PM over AM is the possibility of a greatly improved signal-to-noise ratio. A penalty is paid for this in increased bandwidth: an FM signal may occupy several times as much bandwidth as that required for an AM signal. There may seem to be a contradiction here, as we found that for AM, decreasing the bandwidth improved the signal-to-noise ratio. This seeming contradiction will be resolved shortly.

Historical Development of FM

FM was considered very early in the development of radio communications. At first, it was thought that FM might permit a reduced transmission bandwidth compared with AM. This was refuted by experimental tests and also mathematically by John Renshaw Carson (1887–1940) in 1922.

Carson failed to notice that FM does have an advantage over AM in terms of signal-to-noise ratio. Edwin Armstrong (yes, the same Armstrong who invented the superheterodyne receiver) did notice this, and in 1936 he proposed a practical FM system. FM broadcasting began in the United States in 1939 but suffered a setback in 1944 when its frequency allocation was abruptly shifted from 42–50 MHz to its present range of 88–108 MHz. FM broadcasting gradually became popular because of its noise and fidelity advantages over AM. There are now actually more FM than AM listeners.

Unfortunately, Armstrong did not benefit from the success of FM broadcasting. He spent the remainder of his life involved in lawsuits in an attempt to receive royalties from his inventions, and finally, a broken man, he committed suicide in 1954.

In our discussion of amplitude modulation, we found that the amplitude of the modulated signal varied in accordance with the instantaneous amplitude of the modulating signal. In FM, the *frequency* of the modulated signal varies with the amplitude of the modulating signal. In PM, the *phase* varies directly with the modulating-signal amplitude. It is important to remember that in all types of modulation, it is the amplitude of the modulating signal that varies the carrier wave.

In contrast to AM, the amplitude and the power of an FM or PM signal do not change with modulation. Thus, an FM signal does not have an envelope that reproduces the modulation. This is actually an advantage: an FM receiver does not have to respond to amplitude variations, and thus it can ignore noise to some extent. Similarly, FM transmitters can use Class C amplifiers throughout, since amplitude linearity is not important. Modulation can be accomplished at low power levels.

7.2 Frequency Modulation

Figure 7.1 shows FM with a square wave modulating a sine-wave carrier. Figure 7.1(a) shows the unmodulated carrier and the modulating signal. Figure 7.1(b) shows the modulated signal in the time domain, as it would appear on an oscilloscope. The amount of frequency change has been exaggerated for clarity. The amplitude remains as before, and the frequency changes can be seen in the changing times between zero crossings for the waveforms.

The next two sections are interesting. Figure 7.1(c) shows how the signal frequency varies with time in accordance with the amplitude of the modulating signal. Figure 7.1(d) shows how the phase changes with time. For this figure, the phase angle of the unmodulated carrier is used as a reference. When the frequency is greater than the **carrier frequency**, the phase angle gradually moves ahead, and when the frequency is lower than the carrier frequency, the phase begins to lag.

One further inference can be drawn from Figure 7.1. While the unmodulated carrier is a sine wave, the modulated signal is not—a "sine" wave that changes frequency is not really a sine wave at all. In our study of AM, we found that changing the amplitude of a sine wave generated extra frequencies called *side frequencies* or *sidebands*. This happens in FM, too. In fact, for an FM signal, the number of sets of sidebands is theoretically infinite.

There are many ways to generate FM, some of which will be described in the next chapter. The simplest method is to use a voltage-controlled oscillator (VCO) to generate the carrier frequency and to apply the modulating signal to the oscillator's control signal input, as indicated in Figure 7.2. As might be expected, this is a little too simple for most

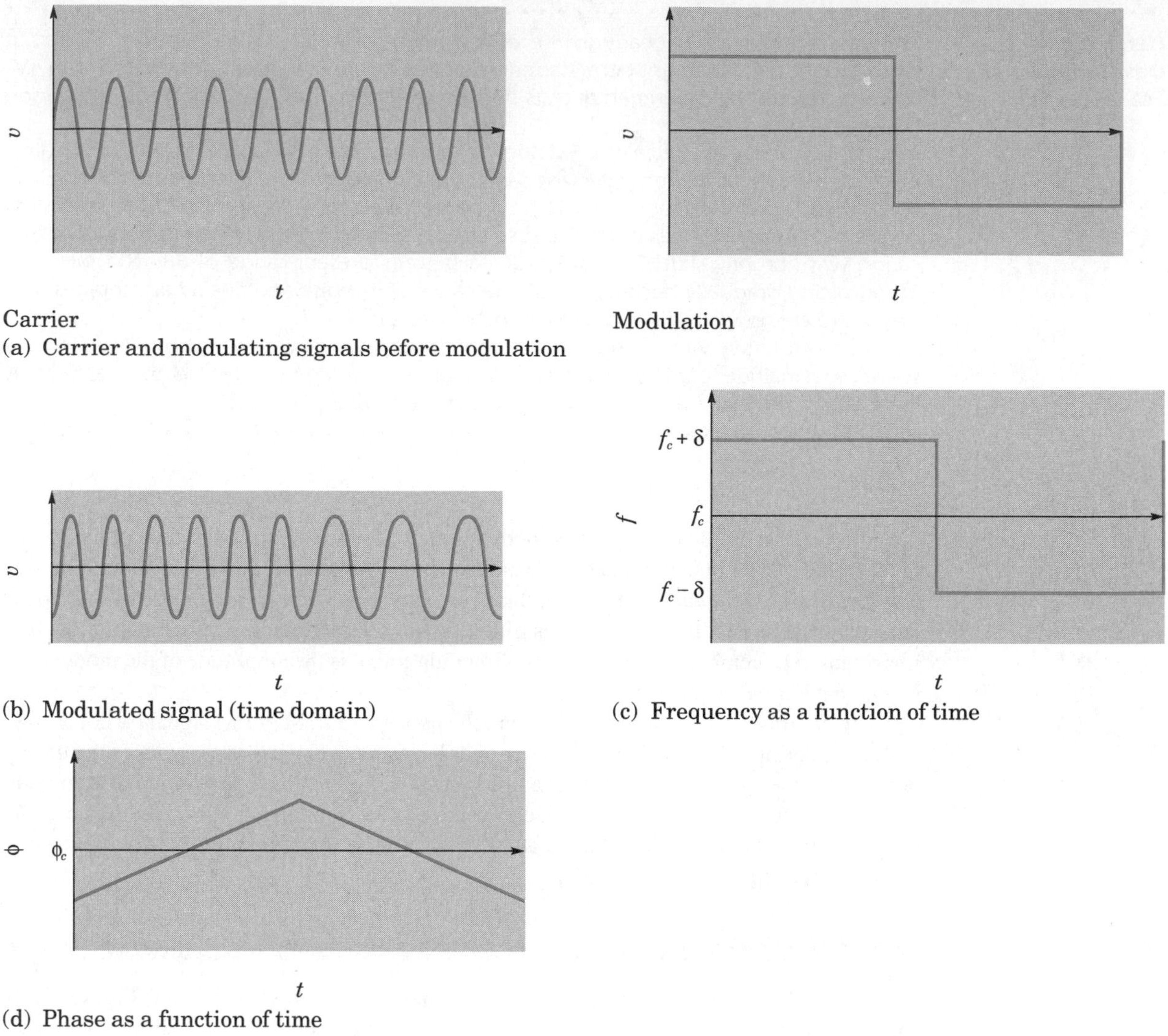

Figure 7.1 Frequency modulation of a sine-wave carrier by a square wave

practical transmitters; in particular, the stability of a free-running VCO is not likely to be good enough. It does, however, give us a conceptual model to use for the time being.

7.2.1 Frequency Deviation

Assume that the carrier frequency is f_c. Modulation will cause the signal frequency to vary (deviate) from its resting value. If the modulation system is properly designed, this deviation will be proportional to the amplitude of the modulating signal. Sometimes this is referred to as *linear modulation*, though no modulation process is really linear. FM can be called linear only in the sense that the graph relating instantaneous modulating-signal amplitude e_m to instantaneous frequency deviation Δf is a straight line. The slope of this line

Figure 7.2 Simplified FM generator

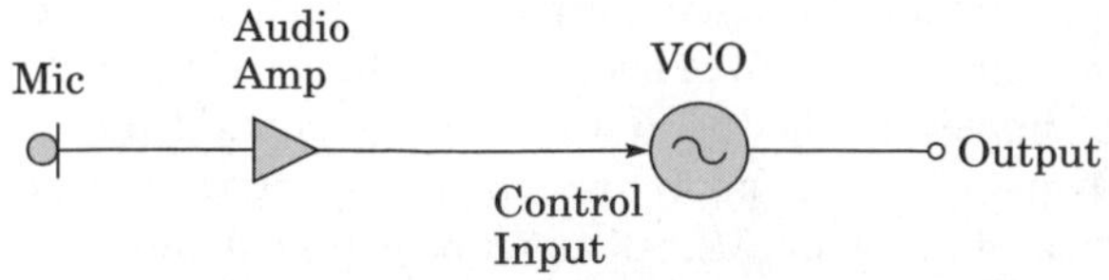

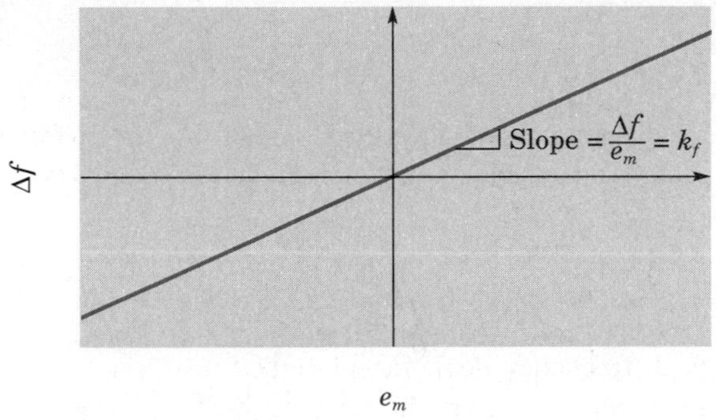

Figure 7.3 Deviation sensitivity of FM modulator

is the ratio $\Delta f/e_m$, and it represents the deviation sensitivity of the modulator, in units of hertz per volt. Let us call this constant k_f. Then

$$k_f = \frac{\Delta f}{e_m} \tag{7.1}$$

Figure 7.3 demonstrates this.

Once again, remember that the **frequency deviation** is proportional to the amplitude, not the frequency, of the modulating signal. The number of times per second that the frequency varies from its lowest to its highest value is, of course, equal to the modulating-signal frequency.

It is possible to write an equation for the signal frequency as a function of time:

$$f_{sig}(t) = f_c + k_f e_m(t) \tag{7.2}$$

where

$f_{sig}(t)$ = signal frequency as a function of time
f_c = unmodulated carrier frequency
k_f = modulator deviation constant
$e_m(t)$ = modulating voltage as a function of time

Example 7.1 An FM modulator has $k_f = 30$ kHz/V and operates at a carrier frequency of 175 MHz. Find the output frequency for an instantaneous value of the modulating signal equal to:

(a) 150 mV (b) −2 V

Solution Equation (7.2) can be used for both parts of the question.

(a)
$$\begin{aligned} f_{sig} &= (175 \times 10^6\ \text{Hz}) + (30 \times 10^3\ \text{Hz/V})(150 \times 10^{-3}\ \text{V}) \\ &= 175.0045 \times 10^6\ \text{Hz} \\ &= 175.0045\ \text{MHz} \end{aligned}$$

(b)
$$\begin{aligned} f_{sig} &= (175 \times 10^6\ \text{Hz}) + (30 \times 10^3\ \text{Hz/V})(-2\ \text{V}) \\ &= (175 \times 10^6\ \text{Hz}) - (30 \times 10^3\ \text{Hz/V})(2\ \text{V}) \\ &= 174.94 \times 10^6\ \text{Hz} \\ &= 174.94\ \text{MHz} \end{aligned}$$

In our study of AM, we found it convenient to assume a sine wave for the modulating signal. This treatment can then be generalized to cover any periodic signal using Fourier techniques. If the modulating signal is a sine wave with the equation

$$e_m(t) = E_m \sin \omega_m t \tag{7.3}$$

then Equation (7.2) becomes

$$f_{sig}(t) = f_c + k_f E_m \sin \omega_m t \tag{7.4}$$

and the peak frequency deviation (on each side of the carrier frequency) will be $k_f E_m$ Hz. Usually, the peak frequency deviation is given the symbol δ. Then

$$\delta = k_f E_m \tag{7.5}$$

where

δ = peak frequency deviation in hertz
k_f = modulator sensitivity in hertz per volt
E_m = peak value of the modulating signal in volts

Example 7.2

The same FM modulator as in the previous example is modulated by a 3 V sine wave. Calculate the deviation.

Solution

Unless otherwise stated, ac voltages are assumed to be RMS. On the other hand, δ is a peak value. Therefore, the modulating voltage must be converted to a peak value before Equation (7.5) can be used.

$$\begin{aligned} E_m &= 3\sqrt{2}\ \text{V} \\ &= 4.24\ \text{V} \\ \delta &= k_f E_m \\ &= 30\ \text{kHz/V} \times 4.24\ \text{V} \\ &= 127.2\ \text{kHz} \end{aligned}$$

Using frequency deviation, Equation (7.4) becomes

$$f_{sig}(t) = f_c + \delta \sin \omega_m t \tag{7.6}$$

7.2.2 Frequency Modulation Index

Another basic term related to FM is the frequency **modulation index** m_f (not to be confused with f_m, which is the modulating frequency). By definition, for sine-wave modulation,

$$m_f = \frac{\delta}{f_m} \tag{7.7}$$

The reason for this rather peculiar definition, which includes not only the frequency deviation but also the modulating frequency, will become apparent very soon. Meanwhile, note one other peculiarity of m_f: unlike the amplitude modulation index, which cannot exceed one, there are no theoretical limits on m_f. It can exceed one and often does.

Example 7.3

An FM broadcast transmitter operates at its maximum deviation of 75 kHz. Find the modulation index for a sinusoidal modulating signal with a frequency of:

(a) 15 kHz (b) 50 Hz

Solution

(a)

$$\begin{aligned} m_f &= \frac{\delta}{f_m} \\ &= \frac{75\ \text{kHz}}{15\ \text{kHz}} \\ &= 5.00 \end{aligned}$$

(b) $$m_f = \frac{\delta}{f_m} = \frac{75 \times 10^3 \text{ Hz}}{50 \text{ Hz}} = 1500$$

Substituting Equation (7.7) into Equation (7.6) gives

$$f_{sig}(t) = f_c + m_f f_m \sin \omega_m t \tag{7.8}$$

as an equation for the frequency of an FM signal with sine-wave modulation.

7.3 Phase Modulation

In phase modulation, it is the phase shift, rather than the frequency deviation, that is proportional to the instantaneous amplitude of the modulating signal. As we did for FM, we can define a constant for a phase modulator that relates the change in phase angle to the amplitude of the modulating signal:

$$k_p = \frac{\phi}{e_m} \tag{7.9}$$

where

k_p = phase modulator sensitivity in radians per volt
ϕ = phase deviation in radians
e_m = modulating-signal amplitude in volts

An equation can be written for the phase of a PM signal as a function of time (similar to Equation (7.2) for the frequency of an FM signal):

$$\theta(t) = \theta_c + k_p e_m(t) \tag{7.10}$$

Once again it would be useful to express this in terms of sine-wave modulation. If

$$e_m(t) = E_m \sin \omega_m t$$

then

$$\phi = k_p E_m \sin \omega_m t \tag{7.11}$$

and

$$\theta(t) = \theta_c + k_p E_m \sin \omega_m t$$

The peak phase deviation, in radians, is defined as m_p, the *phase modulation index*. Then

$$\theta(t) = \theta_c + m_p \sin \omega_m t \tag{7.12}$$

Example 7.4 A phase modulator has $k_p = 2$ rad/V. What RMS voltage of a sine wave would cause a peak phase deviation of 60°?

Solution Remembering that a circle has 360° or 2π rad, we see that

$$360° = 2\pi \text{ rad}$$

$$60° = \frac{2\pi \text{ rad} \times 60}{360} = \frac{\pi}{3} \text{ rad}$$

The voltage to cause this deviation can be found from Equation (7.9):

$$k_p = \frac{\phi}{e_m}$$

$$e_m = \frac{\phi}{k_p}$$

$$= \frac{(\pi/3)\text{ rad}}{2\text{ rad/V}}$$

$$= \frac{\pi}{6}\text{ V}$$

$$= 0.524\text{ V}$$

This is peak voltage. We can find its RMS value in the usual way:

$$V_{RMS} = \frac{V_{\text{peak}}}{\sqrt{2}}$$

$$= \frac{0.524}{\sqrt{2}}$$

$$= 0.37\text{ V}$$

7.3.1 The Relationship between Frequency Modulation and Phase Modulation

As mentioned earlier, either FM or PM results in changes in both the frequency and phase of the modulated waveform. It has also been pointed out that frequency (in radians per second) is the rate of change of phase (in radians). That is, frequency is the derivative of phase. This leads to a relatively simple relationship between FM and PM that can make it easier to understand both and to perform calculations with either.

For any angle-modulated signal with sine-wave modulation, the modulation index m_p or m_f represents the peak phase deviation from the phase of the unmodulated carrier, in radians. This is obvious for PM but not quite so obvious for FM. We know that

$$m_f = \frac{\delta}{f_m}$$

That is, the modulation index (which corresponds to peak phase deviation) is proportional to frequency deviation and inversely proportional to modulating frequency. The first part of the statement sounds very reasonable. Suppose that the frequency increases. Then the higher frequency represents a phase angle that changes more quickly than before. The greater the frequency change, the greater the increase in phase angle.

The second part of the statement needs more explanation. Why should an increased modulating frequency result in a reduced change in phase angle? Consider a low modulating frequency, say 1 Hz, that causes a frequency deviation of 1 kHz. For simplicity, suppose that the modulating signal is a square wave. Then the signal frequency will increase to a point 1 kHz above the carrier frequency, stay there for one-half second, then decrease by 2 kHz to a point 1 kHz below the carrier frequency for the next one-half second. Figure 7.4(a) shows the way the instantaneous signal frequency varies.

When the signal frequency is higher than normal, the phase angle with respect to that of the unmodulated carrier steadily increases as the modulated wave gets further and further ahead of the unmodulated signal. This continues until the frequency decreases. At that point, the signal's phase angle starts to lag, letting the carrier phase angle catch up to and then overtake the angle of the modulated signal.

The amount of phase change is proportional to the length of time the instantaneous frequency stays above the carrier frequency, that is, the phase deviation is proportional to the period of the modulating signal. Another way of saying this is that phase deviation is inversely proportional to the modulating-signal frequency. Figures 7.4(c) and (d) show the

Figure 7.4
Relation between modulating frequency and modulation index for FM

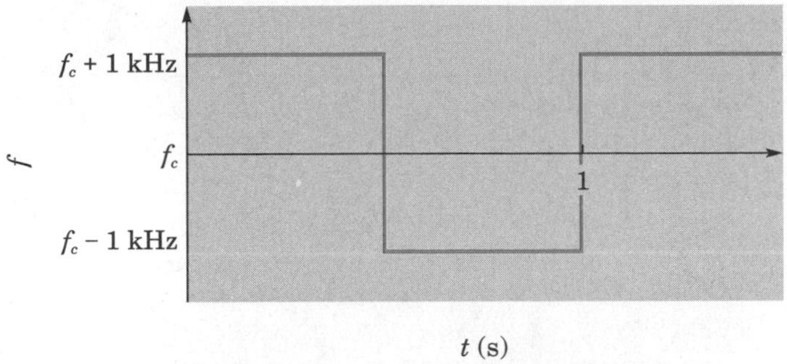

(a) Frequency shift for 1 Hz modulating signal

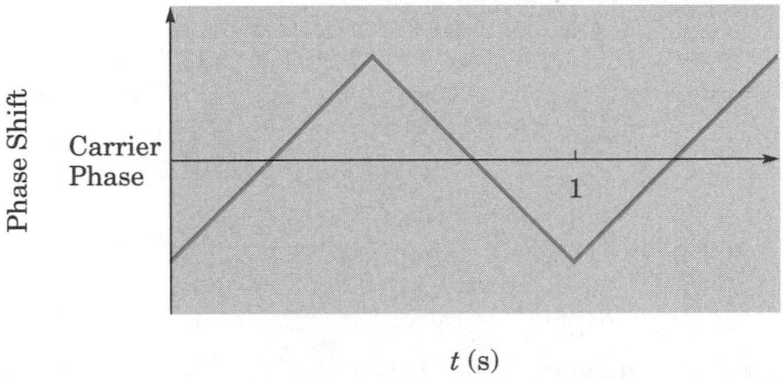

(b) Phase shift for 1 Hz modulating signal

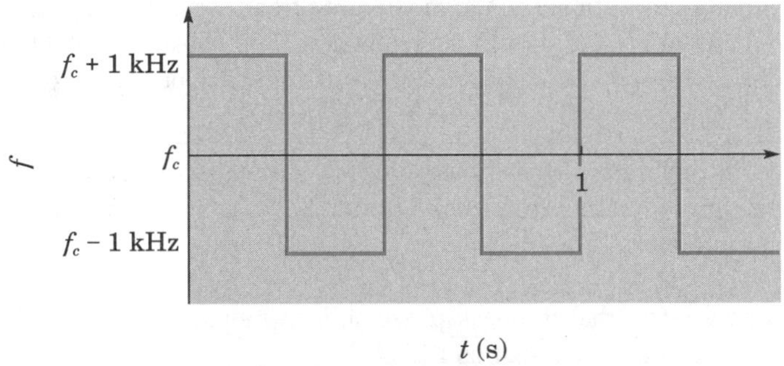

(c) Frequency shift for 2 Hz modulating signal

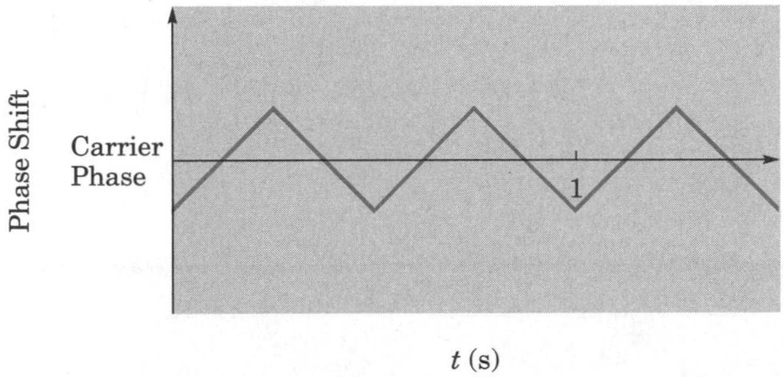

(d) Phase shift for 2 Hz modulating signal

frequency and phase change, respectively, for a modulating signal with the same amplitude as before but twice the frequency. The phase deviation is only one-half as much as before.

This simple relationship between FM and PM suggests that it would be easy to convert one to the other, and this is true. For instance, a phase modulator can be used to generate FM. The baseband signal is passed through a low-pass filter with the frequency re-

Figure 7.5
Use of an integrator to convert PM to FM

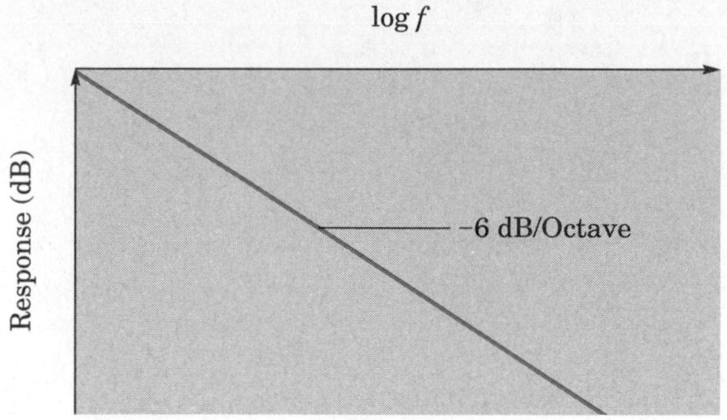

(a) Filter response

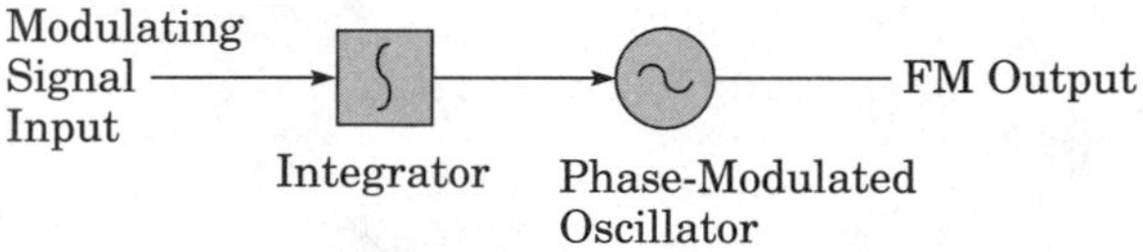

(b) Use of filter

sponse shown in Figure 7.5(a). This type of filter is often referred to as an *integrator*. For equal amplitudes at the modulator input, the amplitude of the output from the filter will be inversely proportional to the modulating frequency. Figure 7.5(b) shows the low-pass filtered signal applied to a phase modulator. The phase-modulated output has a modulation index inversely proportional to the modulating frequency, that is, it is identical to FM!

Example 7.5

An FM communications transmitter has a maximum frequency deviation of 5 kHz and a range of modulating frequencies from 300 Hz to 3 kHz. What is the maximum phase shift that it produces?

Solution

We now know that the peak phase shift in radians is equal to the frequency modulation index m_f. Since, by Equation (7.7),

$$m_f = \frac{\delta}{f_m}$$

m_f will be largest for the lowest possible value of f_m, in this case 300 Hz. For this frequency, the phase shift is

$$\begin{aligned}\phi_{max} &= m_f \\ &= \frac{\delta}{f_m} \\ &= \frac{5000}{300} \\ &= 16.7 \text{ rad}\end{aligned}$$

Example 7.6

A phase modulator has a sensitivity of $k_p = 3$ rad/V. How much frequency deviation does it produce with a sine-wave input of 2 V peak at a frequency of 1 kHz?

Solution

The maximum phase shift is easily found from Equation (7.11):

$$\phi = k_p E_m \sin \omega_m t$$

The maximum value of ϕ is m_p, and it occurs for the peak modulating voltage.

$$\begin{aligned} m_p &= \phi_{max} \\ &= k_p E_m \\ &= 3 \text{ rad/V} \times 2 \text{ V} \\ &= 6 \text{ rad} \end{aligned}$$

This has the same value as m_f, if the signal is considered as frequency modulation. From Equation (7.7),

$$m_f = \frac{\delta}{f_m}$$

$$\begin{aligned} \delta &= m_f f_m \\ &= 6 \times 1 \text{ kHz} \\ &= 6 \text{ kHz} \end{aligned}$$

7.4 The Angle Modulation Spectrum

Angle modulation produces an infinite number of sidebands, even for single-tone modulation. These sidebands are separated from the carrier by multiples of f_m, but their amplitude tends to decrease as their distance from the carrier frequency increases. Sidebands with amplitudes less than about 1% of the total signal voltage can usually be ignored, so for practical purposes an angle-modulated signal can be considered to be band-limited. In most cases, though, its bandwidth is much larger than that of an AM signal.

7.4.1 Bessel Functions

FM and PM signals have similar equations. In fact, without a knowledge of the baseband signal, it would be impossible to tell them apart. For the modulation of a carrier with amplitude A and radian frequency ω_c by a single-frequency sinusoid, the equation is of the form

$$v(t) = A \sin(\omega_c t + m \sin \omega_m t) \tag{7.13}$$

The factor m can represent m_f for FM or m_p for PM.

This equation cannot be simplified by ordinary trigonometry, as is the case for amplitude modulation. About the only useful information that can be gained by inspection is the fact that the signal amplitude remains constant regardless of the modulation index. This observation is significant, since it demonstrates one of the major differences between AM and FM or PM, but it provides no information about the sidebands.

This signal can be expressed as a series of sinusoids by using Bessel functions of the first kind. Proving this is beyond the scope of this text, but it can be done. The Bessel functions themselves are rather tedious to evaluate numerically, but that, too, has been done. Some results are presented in Table 7.1 and Figure 7.6.

The table and graph of Bessel functions represent normalized voltages for the various frequency components of an FM or PM signal, that is, the numbers in the tables represent actual voltages if the unmodulated carrier has an amplitude of 1 V. Here J_0 represents the amplitude of the component at the carrier frequency (sometimes called the **rest frequency**). The variable J_1 represents the amplitude of each of the first set of sidebands, which have frequencies of $(f_c + f_m)$ and $f_c - f_m)$. J_2 represents the amplitude of each of the second pair of sidebands, at $(f_c + 2f_m)$ and $(f_c - 2f_m)$, and so on. Figure 7.7 shows this on a frequency-domain plot. All of the Bessel terms should be multiplied by the voltage of the unmodulated carrier to find the actual sideband amplitudes. Of course, the Bessel coefficients are equally valid for peak or RMS voltages, but the user should be careful to keep track of which type of measurement is being used. There certainly will be a difference when sideband power calculations are involved.

Table 7.1

m	J_0	J_1	J_2	J_3	J_4	J_5	J_6	J_7	J_8	J_9	J_{10}	J_{11}	J_{12}	J_{13}	J_{14}	J_{15}	J_{16}	J_{17}	J_{18}	J_{19}	J_{20}
0	1.00																				
0.25	0.98	0.12																			
0.5	0.94	0.24	0.03																		
0.75	0.86	0.35	0.07	0.01																	
1	0.77	0.44	0.11	0.02																	
1.25	0.65	0.51	0.17	0.04	0.01																
1.5	0.51	0.56	0.23	0.06	0.01																
1.75	0.37	0.58	0.29	0.09	0.02																
2	0.22	0.58	0.35	0.13	0.03	0.01															
2.25	0.08	0.55	0.40	0.17	0.05	0.01															
2.4	0.00	0.52	0.43	0.20	0.06	0.02															
2.5	−0.05	0.50	0.45	0.22	0.07	0.02															
2.75	−0.16	0.43	0.47	0.26	0.10	0.03	0.01														
3	−0.26	0.34	0.49	0.31	0.13	0.04	0.01														
3.5	−0.38	0.14	0.46	0.39	0.20	0.08	0.03	0.01													
4	−0.40	−0.07	0.36	0.43	0.28	0.13	0.05	0.01													
4.5	−0.32	−0.23	0.22	0.42	0.35	0.20	0.08	0.03	0.01												
5	−0.18	−0.33	0.05	0.36	0.39	0.26	0.13	0.05	0.02	0.01											
5.5	0.00	−0.34	−0.12	0.26	0.40	0.32	0.19	0.09	0.03	0.01											
6	0.15	−0.28	−0.24	0.11	0.36	0.36	0.25	0.13	0.06	0.02	0.01										
6.5	0.26	−0.15	−0.31	−0.03	0.28	0.37	0.30	0.18	0.09	0.04	0.01										
7	0.30	−0.01	−0.30	−0.17	0.16	0.35	0.34	0.23	0.13	0.06	0.02	0.01									
7.5	0.27	0.14	−0.23	−0.26	0.02	0.28	0.35	0.28	0.17	0.09	0.04	0.01	0.01								
8	0.17	0.24	−0.11	−0.29	−0.11	0.19	0.34	0.32	0.22	0.13	0.06	0.03	0.01								
8.5	0.04	0.27	0.02	−0.26	−0.21	0.07	0.29	0.34	0.27	0.17	0.09	0.04	0.02	0.01							
8.65	0.00	0.27	0.06	−0.24	−0.23	0.03	0.27	0.34	0.28	0.18	0.10	0.05	0.02	0.01							
9	−0.09	0.25	0.14	−0.18	−0.27	−0.06	0.20	0.33	0.30	0.21	0.13	0.06	0.03	0.01							
10	−0.25	0.04	0.26	0.06	−0.22	−0.23	−0.01	0.22	0.32	0.29	0.21	0.12	0.06	0.03	0.01						
11	−0.17	−0.18	0.14	0.23	−0.01	−0.24	−0.20	0.02	0.23	0.31	0.28	0.20	0.12	0.06	0.03	0.01	0.01				
12	0.05	−0.22	−0.08	0.20	0.18	−0.07	−0.24	−0.17	0.04	0.23	0.30	0.27	0.20	0.12	0.07	0.03	0.01	0.10			
13	0.21	−0.07	−0.22	0.00	0.22	0.13	−0.12	−0.24	−0.14	0.07	0.23	0.29	0.26	0.19	0.12	0.07	0.03	0.01	0.01		
14	0.17	0.13	−0.15	−0.18	0.08	−0.15	−0.23	−0.11	0.08	0.24	0.29	0.25	0.19	0.12	0.07	0.03	0.02	0.01			
15	−0.01	0.20	0.04	−0.19	−0.12	0.13	0.21	0.03	−0.17	−0.22	−0.09	0.10	0.24	0.28	0.25	0.18	0.12	0.07	0.03	0.02	0.01
16	−0.17	0.09	0.19	−0.04	−0.20	−0.06	0.17	0.18	−0.01	−0.19	−0.21	−0.07	0.11	0.24	0.27	0.24	0.18	0.11	0.07	0.03	0.02
17	−0.17	−0.10	0.16	0.14	−0.11	−0.19	0.00	0.19	0.15	−0.04	−0.20	−0.19	−0.05	0.12	0.24	0.27	0.23	0.17	0.11	0.07	0.04

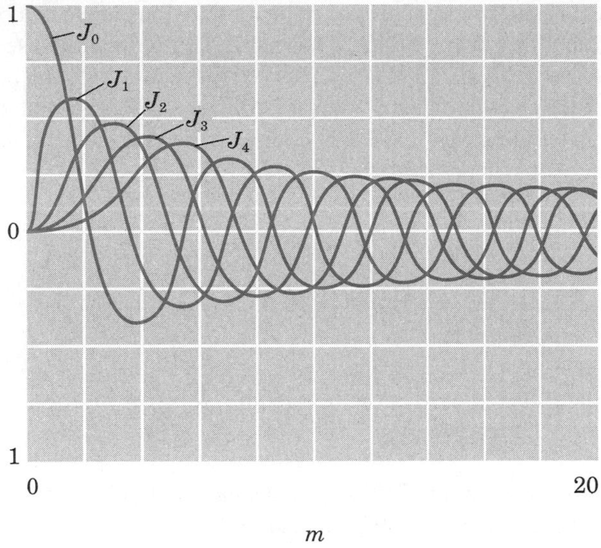

Figure 7.6
Bessel functions

When Bessel functions are used, the signal of Equation (7.13) becomes

$$
\begin{aligned}
v(t) &= A \sin(\omega_c t + m \sin \omega_m t) \\
&= A\{J_0(m) \sin \omega_c t \\
&\quad - J_1(m)[\sin(\omega_c - \omega_m)t - \sin(\omega_c + \omega_m)t] \\
&\quad + J_2(m)[\sin(\omega_c - 2\omega_m)t + \sin(\omega_c + 2\omega_m)t] \\
&\quad - J_3(m)[\sin(\omega_c - 3\omega_m)t + \sin(\omega_c + 3\omega_m)t] \\
&\quad + \cdots\}
\end{aligned}
\tag{7.14}
$$

With angle modulation, the total signal voltage and power do not change with modulation. Therefore, the appearance of power in the sidebands indicates that the power at the carrier frequency must be reduced below its unmodulated value in the presence of modulation. In fact, the carrier-frequency component disappears for certain values of m (for example, 2.4 and 5.5).

Figure 7.7
FM in the frequency domain, for an unmodulated carrier amplitude of 1 V

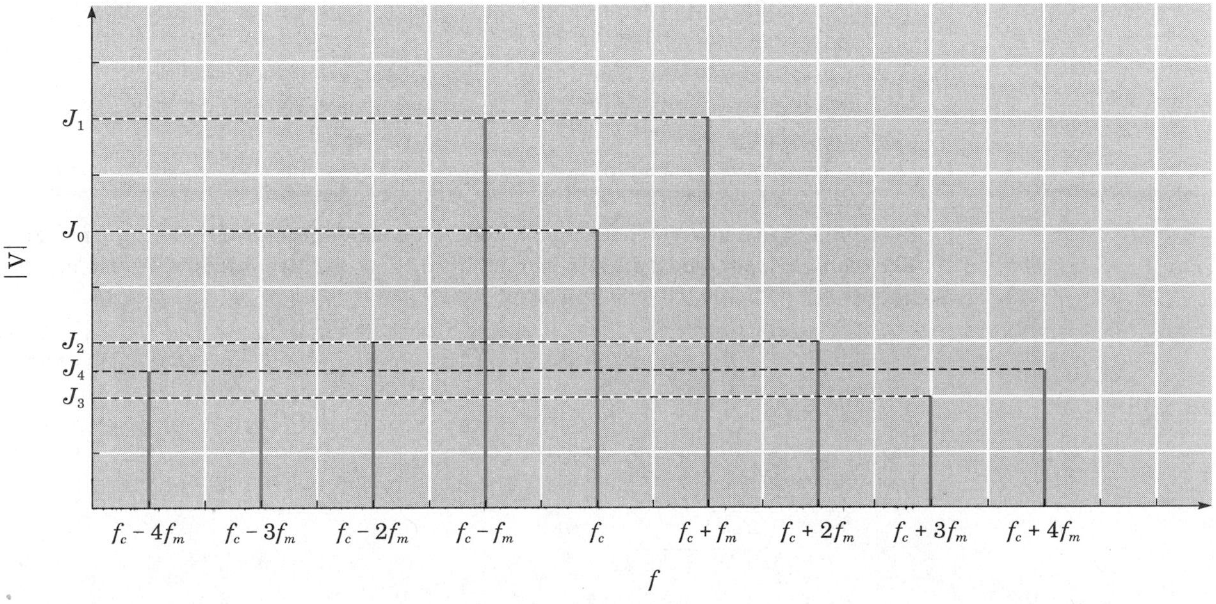

This constant-power aspect of angle modulation can be demonstrated using the table of Bessel functions. For simplicity, normalized values can be used. Let the unmodulated signal have a voltage of 1 V RMS across a resistance of 1 Ω. Its power is, of course, 1 W. When modulation is applied, the carrier voltage is reduced and sidebands appear. From Table 7.1, J_0 will represent the RMS voltage at the carrier frequency, and the power at the carrier frequency will be

$$\begin{aligned} P_c &= \frac{V_c^2}{R} \\ &= \frac{J_0^2}{1} \\ &= J_0^2 \end{aligned}$$

Similarly, the power in each of the first set of sidebands will be

$$P_{SB_1} = J_1^2$$

The combined power in the first set of sidebands will be twice as much as this, of course. The power in the whole signal will then be

$$P_T = J_0^2 + 2J_1^2 + 2J_2^2 + \cdots \tag{7.15}$$

If the series is carried on far enough, the result will be equal to 1 W, regardless of the modulation index.

For a signal with a power different from 1 W, it is only necessary to multiply the terms in Equation (7.15) by the total signal power.

The bandwidth of an FM or PM signal is to some extent a matter of definition. The Bessel series is infinite, but as we can see from the table or the graph, the amplitude of the components gradually diminishes until at some point they can be ignored. The process is slower for large values of m, so the number of sets of sidebands that has to be considered is greater for larger modulation indices. A practical rule of thumb is to ignore sidebands with a Bessel coefficient whose absolute value is less than 0.01. The bandwidth, for practical purposes, is equal to twice the number of the highest significant Bessel coefficient, multiplied by the modulating frequency.

Example 7.7

An FM signal has a deviation of 3 kHz and a modulating frequency of 1 kHz. Its total power P_T is 5 W, developed across a 50 Ω resistive load. The carrier frequency is 160 MHz.

(a) Calculate the RMS signal voltage V_T.
(b) Calculate the RMS voltage at the carrier frequency and each of the first three sets of sidebands.
(c) For the first three sideband pairs, calculate the frequency of each sideband.
(d) Calculate the power at the carrier frequency and at each of the sideband frequencies found in part (c).
(e) Determine what percentage of the total signal power is unaccounted for by the components described above.
(f) Sketch the signal in the frequency domain, as it would appear on a spectrum analyzer. The vertical scale should be power in dBm, and the horizontal scale should be frequency.

Solution

(a) The signal power does not change with modulation, and neither does the voltage, which can easily be found from the power equation.

$$P_T = \frac{V_T^2}{R_L}$$

$$\begin{aligned} V_T &= \sqrt{P_T R_L} \\ &= \sqrt{5 \text{ W} \times 50 \ \Omega} \\ &= 15.8 \text{ V (RMS)} \end{aligned}$$

(b) The modulation index must be found in order to use Bessel functions to find the carrier and sideband voltages.

$$\begin{aligned} m_f &= \frac{\delta}{f_m} \\ &= \frac{3\text{ kHz}}{1\text{ kHz}} \\ &= 3 \end{aligned}$$

From the Bessel function table, the coefficients for the carrier and the first three sideband pairs are:

$$J_0 = -0.26 \qquad J_1 = 0.34 \qquad J_2 = 0.49 \qquad J_3 = 0.31$$

These are normalized voltages, so they will have to be multiplied by the total RMS signal voltage to get the RMS sideband and carrier-frequency voltages.
For the carrier,

$$V_c = J_0 V_T$$

J_0 has a negative sign. This simply indicates a phase relationship between the components of the signal. It would be required if we wanted to add together all the components to get the resultant signal. For our present purpose, however, it can simply be ignored, and we can use

$$\begin{aligned} V_c &= |J_0| V_T \\ &= 0.26 \times 15.8\text{ V} \\ &= 4.11\text{ V} \end{aligned}$$

Similarly, we can find the voltage for each of the three sideband pairs. Note that these are voltages for individual components. There will be a lower and an upper sideband with each of these calculated voltages.

$$\begin{aligned} V_1 &= J_1 V_T \\ &= 0.34 \times 15.8\text{ V} \\ &= 5.37\text{ V} \end{aligned}$$

$$\begin{aligned} V_2 &= J_2 V_T \\ &= 0.49 \times 15.8\text{ V} \\ &= 7.74\text{ V} \end{aligned}$$

$$\begin{aligned} V_3 &= J_3 V_T \\ &= 0.31 \times 15.8\text{ V} \\ &= 4.9\text{ V} \end{aligned}$$

(c) The sidebands are separated from the carrier frequency by multiples of the modulating frequency. Here, $f_c = 160$ MHz and $f_m = 1$ kHz, so there are sidebands at each of the following frequencies:

$$\begin{aligned} f_{USB_1} &= 160\text{ MHz} + 1\text{ kHz} \\ &= 160.001\text{ MHz} \end{aligned}$$

$$\begin{aligned} f_{USB_2} &= 160\text{ MHz} + 2\text{ kHz} \\ &= 160.002\text{ MHz} \end{aligned}$$

$$\begin{aligned} f_{USB_3} &= 160\text{ MHz} + 3\text{ kHz} \\ &= 160.003\text{ MHz} \end{aligned}$$

$$\begin{aligned} f_{LSB_1} &= 160\text{ MHz} - 1\text{ kHz} \\ &= 159.999\text{ MHz} \end{aligned}$$

$$f_{LSB_2} = 160 \text{ MHz} - 2 \text{ kHz}$$
$$= 159.998 \text{ MHz}$$
$$f_{LSB_3} = 160 \text{ MHz} - 3 \text{ kHz}$$
$$= 159.997 \text{ MHz}$$

(d) Since each of the components of the signal is a sinusoid, the usual equation can be used to calculate power. All the components appear across the same 50 Ω load.

$$P_c = \frac{V_c^2}{R_L}$$
$$= \frac{4.11^2}{50}$$
$$= 0.338 \text{ W}$$

$$P_1 = \frac{V_1^2}{R_L} = \frac{5.37^2}{50} = 0.576 \text{ W} \qquad P_2 = \frac{V_2^2}{R_L} = \frac{7.74^2}{50} = 1.2 \text{ W} \qquad P_3 = \frac{V_3^2}{R_L} = \frac{4.9^2}{50} = 0.48 \text{ W}$$

(e) To find the total power P_T in the carrier and the first three sets of sidebands, it is only necessary to add the powers calculated above, counting each of the sideband powers twice, because each of the calculated powers represents one of a pair of sidebands. We only count the carrier once, of course.

$$P_T = P_c + 2(P_1 + P_2 + P_3)$$
$$= 0.338 + 2(0.576 + 1.2 + 0.48) \text{ W}$$
$$= 4.85 \text{ W}$$

This is not quite the total signal power, which was given as 5 W. The remainder is in the additional sidebands. To find how much is unaccounted for by the carrier and the first three sets of sidebands, we can subtract. Call the difference P_x:

$$P_x = 5 - 4.85$$
$$= 0.15 \text{ W}$$

As a percentage of the total power this is

$$P_x\ (\%) = \left(\frac{0.15}{5}\right)100$$
$$= 3\%$$

(f) All the information we need for the sketch is on hand, except that the power values have to be converted to dBm using the equation

$$P \text{ (dBm)} = 10 \log \frac{P}{1 \text{ mW}}$$

This gives

$$P_c \text{ (dBm)} = 10 \log 338$$
$$= 25.3 \text{ dBm}$$
$$P_1 \text{ (dBm)} = 10 \log 576$$
$$= 27.6 \text{ dBm}$$
$$P_2 \text{ (dBm)} = 10 \log 1200$$
$$= 30.8 \text{ dBm}$$
$$P_3 \text{ (dBm)} = 10 \log 480$$
$$= 26.8 \text{ dBm}$$

The sketch is shown in Figure 7.8.

Figure 7.8

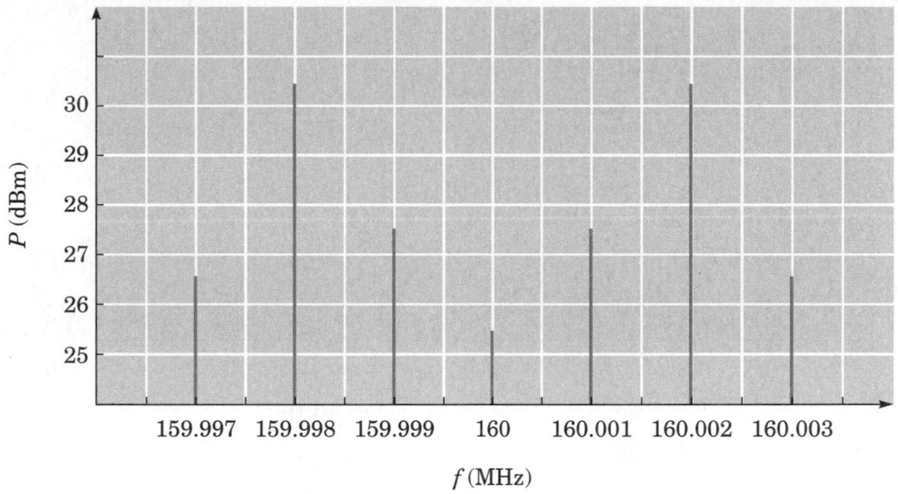

7.4.2 Bandwidth

For PM, the bandwidth varies directly with the modulating frequency, since doubling the frequency doubles the distance between sidebands. It is also roughly proportional to the maximum phase deviation, since increasing m_p increases the number of sidebands. For FM, however, the situation is complicated by the fact that

$$m_f = \frac{\delta}{f_m}$$

For a given amount of frequency deviation, the modulation index is inversely proportional to the modulating frequency. Recall that the frequency deviation is proportional to the amplitude of the modulating signal. Then, if the amplitude of the modulating signal remains constant, increasing the frequency reduces the modulating index. Reducing m_f reduces the number of sidebands with significant amplitude. On the other hand, the increase in f_m means that the sidebands will be further apart in frequency, since they are separated from each other by f_m. These two effects work in opposite directions. The result is that the bandwidth does increase somewhat with increasing modulating-signal frequency, but the bandwidth is not directly proportional to the frequency. Sometimes FM is called a *constant-bandwidth* communications mode for this reason, though the bandwidth is not really constant. Figure 7.9 provides a few examples that show the relationship between modulating frequency and bandwidth. For this example, the deviation remains constant at 10 kHz as the modulating frequency varies from 2 kHz to 10 kHz.

One other point must be made about sidebands: with AM, restricting the bandwidth of the receiver has a very simple effect on the signal. Since the side frequencies furthest from the carrier contain the high-frequency baseband information, restricting the receiver bandwidth reduces its response to high-frequency baseband signals, leaving all else unchanged. When reception conditions are poor, bandwidth can be restricted to the minimum necessary for intelligibility. For FM, the situation is more complicated, since even low-frequency modulating signals can generate sidebands that are far removed from the carrier frequency. FM receivers must be designed to include all the significant sidebands that are transmitted; otherwise, severe distortion, not just limited frequency response, will result.

7.4.3 Carson's Rule

Calculating the bandwidth of an FM signal from Bessel functions is easy enough, since the functions are available in a table, but it can be a bit tedious. There is an approximation, known as *Carson's rule*, that can be used to find the bandwidth of an FM signal. It is not as accurate as using Bessel functions, but it can be applied almost instantly, without using tables or even a calculator.

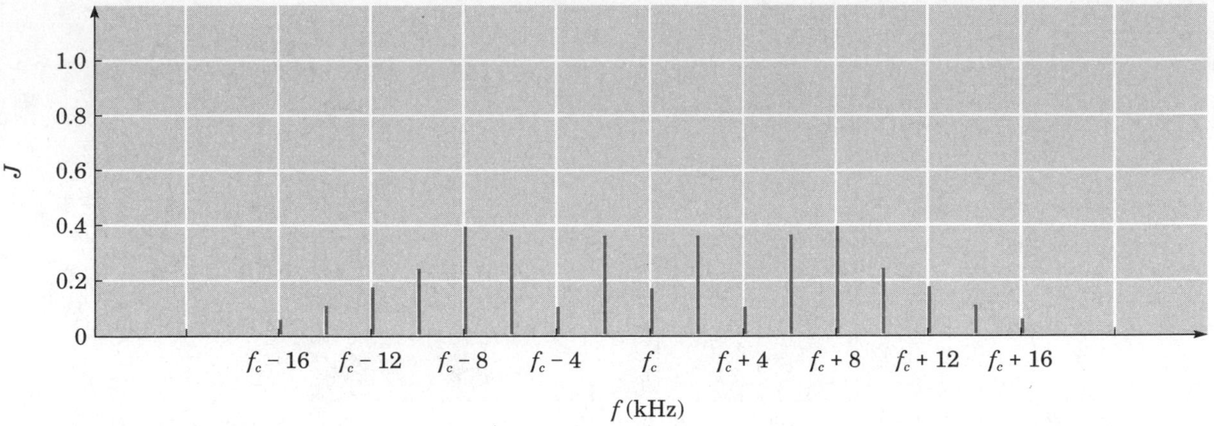

(a) $f_m = 2$ kHz $m_f = 5.0$

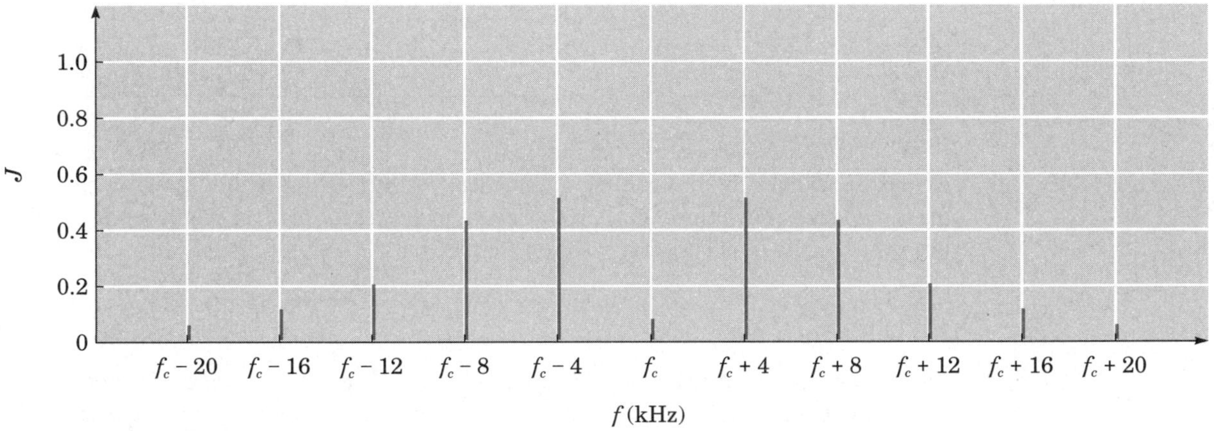

(b) $f_m = 4$ kHz $m_f = 2.5$

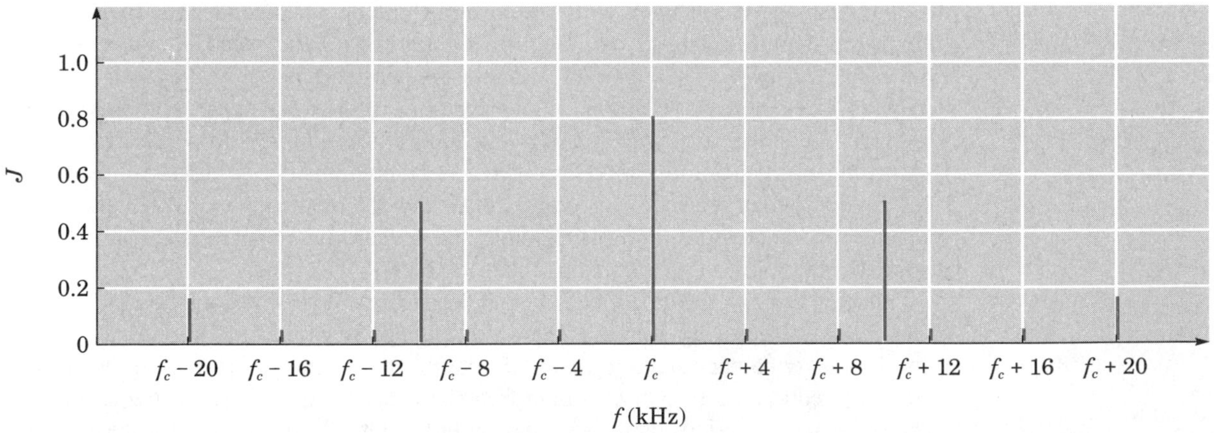

(c) $f_m = 10$ kHz $m_f = 1.0$

Figure 7.9
Variation of effective FM signal bandwidth with modulating frequency (deviation is 10 kHz in all cases)

Here is Carson's rule:

$$B = 2(\delta_{max} + f_{m\ (\max)}) \tag{7.16}$$

Equation (7.16) assumes that the bandwidth is proportional to the sum of the deviation and the modulating frequency. This is not strictly true. Carson's rule also makes the assumption that maximum deviation occurs with the maximum modulating frequency. Sometimes this leads to errors in practical situations, where often the highest baseband frequencies have much less amplitude than lower frequencies and therefore do not produce as much deviation.

Example 7.8 Use Carson's rule to calculate the bandwidth of the signal used in Example 7.7.

Solution Here there is only one modulating frequency, so

$$\begin{aligned} B &= 2(\delta + f_m) \\ &= 2(3 \text{ kHz} + 1 \text{ kHz}) \\ &= 8 \text{ kHz} \end{aligned}$$

In the previous example, we found that 97% of the power was contained in a bandwidth of 6 kHz. An 8 kHz bandwidth would contain more of the signal power. Carson's rule gives quite reasonable results in this case, with very little work.

7.4.4 Narrowband and Wideband FM

We mentioned earlier that there are no theoretical limits to the modulation index or the frequency deviation of an FM signal. The limits are practical and result from a compromise between signal-to-noise ratio and bandwidth. In general, larger values for the deviation result in an increased signal-to-noise ratio, while also resulting in greater bandwidth. The former is desirable, but the latter is not, especially in regions of the spectrum where frequency space is in short supply. It is also necessary to have some agreement about deviation, since receivers must be designed for a particular signal bandwidth.

For these reasons, the bandwidth of FM transmissions is generally limited by government regulations that specify the maximum frequency deviation and the maximum modulating frequency, since both of these affect bandwidth. In general, relatively narrow bandwidths (on the order of 15 kHz) are used for voice communication, with wider bandwidths for such services as FM broadcasting (about 200 kHz) and satellite television (36 MHz for one system).

It seems logical to distinguish between **narrowband FM** (NBFM) signals, used for voice transmission, and **wideband FM** (WBFM), used for most other transmissions. This is, in fact, the terminology that is in daily use. The reader should be aware that there is a more restrictive use of the term *narrowband FM*, found mostly in textbooks, to refer to a signal with m_f less than about 0.5. Such a signal has a bandwidth about the same as that of an AM signal and only one pair of sidebands with significant power. Even communications transmitters that are normally called NBFM do not qualify under this definition. For example, a typical communications transmitter has a maximum deviation of 5 kHz, with a voice-frequency response from about 300 Hz to 3 kHz. The maximum value of m_f occurs for maximum deviation with minimum modulating frequency, in this case,

$$\begin{aligned} \text{Maximum } m_f &= \frac{5000}{300} \\ &= 16.7 \end{aligned}$$

This hardly qualifies as NBFM in the strict, textbook sense, though it is universally called NBFM in practice.

7.5 FM and Noise

The original reason for developing FM was to give improved performance in the presence of noise, and that is still one of its main advantages over AM. This improved noise performance can actually result in a better signal-to-noise ratio at the output of a receiver than is found at its input.

One way to approach the problem of FM and noise is to think of the noise voltage as a phasor having random amplitude and phase angle. The noise will add to the signal, causing random variations in both the amplitude and phase angle of the signal as seen by the receiver. Figure 7.10 shows this vector addition.

Figure 7.10 Effect of noise on an FM signal

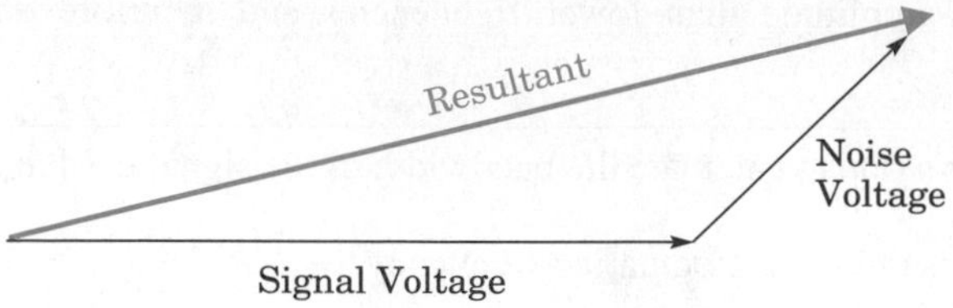

The amplitude component of noise is easily dealt with in a well-designed FM system. Since FM signals do not depend on an envelope for detection, the receiver can employ limiting to remove any amplitude variations from the signal. That is, it can use amplifiers whose output amplitude is the same for a wide variety of input signal levels. The effect of this on the amplitude of a noisy signal is shown in Figure 7.11. As long as the signal amplitude is considerably larger than the noise to begin with, the amplitude component of the noise will not be a problem.

It is not possible for the receiver to ignore phase shifts, however. A PM receiver obviously must respond to phase changes, but so will an FM receiver because, as we have seen, phase shifts and frequency shifts always occur together. Therefore, phase shifts due to noise are associated with frequency shifts that will be interpreted by the receiver as part of the modulation.

Figure 7.12 shows the situation at the input to the receiver. The circle represents the fact that the noise phasor has a constantly changing angle with respect to the signal. Its greatest effect, and thus the peak phase shift to the signal, occurs when the noise phasor is perpendicular to the resultant, E_R. At that time, the phase shift due to noise is

$$\phi_N = \sin^{-1} (E_N/E_S) \tag{7.17}$$

where E_N/E_S is the reciprocal of the voltage signal-to-noise ratio at the input. A little care is needed here, as *S/N* is usually given as a power ratio, in decibels, and has to be converted to a voltage ratio before being used in Equation (7.17).

Equation (7.17) can be simplified as long as the signal is much larger than the noise. In this case the phase deviation is small, and for small angles the sine of the angle is approximately equal to the angle itself, in radians. Thus in a practical situation we can use

$$\phi_N \approx E_N/E_S \tag{7.18}$$

The effect of noise can be reduced by making the signal voltage as large as possible relative to the noise voltage. Figure 7.12 demonstrates this. This, of course, requires increased transmission power, a better receiver noise figure, or both. Perhaps less obvious is the fact that the relative importance of phase shifts due to noise can be reduced by having the phase shifts in the signal as large as possible. This is accomplished by keeping the value of m_f high, since m_f represents the peak phase shift in radians. It would seem that the ratio

Figure 7.11 Removal of amplitude noise component by limiting

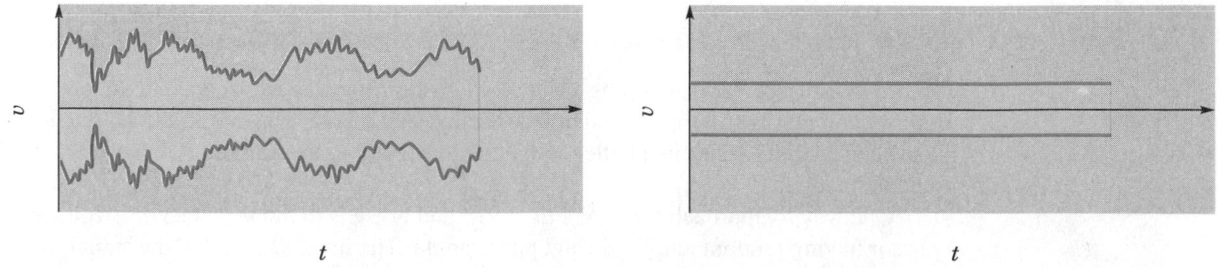

(a) Signal before limiting (b) Signal after limiting

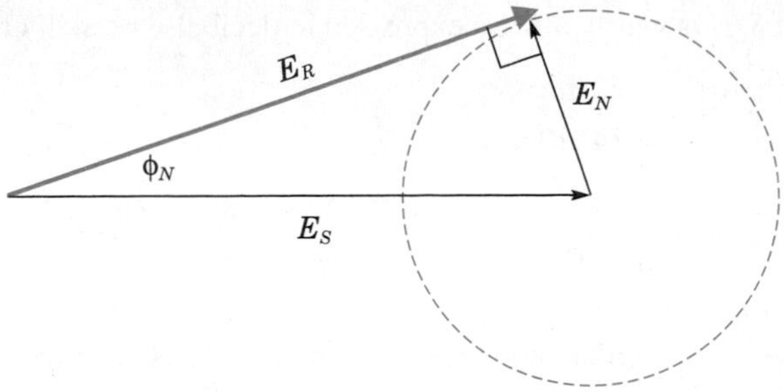

Figure 7.12 Phase shift due to noise

of signal voltage to noise voltage at the output would be proportional to m_f, and this is approximately true for wideband FM under strong-signal conditions.

Example 7.9 An FM signal has a frequency deviation of 5 kHz and a modulating frequency of 1 kHz. The signal-to-noise ratio at the input to the receiver detector is 20 dB. Calculate the approximate signal-to-noise ratio at the detector output.

Solution First, notice the word "approximate." Our analysis is obviously a little simplistic, since noise exists at more than one frequency. We are also going to assume that the detector is completely unresponsive to amplitude variations and that it adds no noise of its own. Our results will not be precise, but they will show the process that is involved.

Let us first convert 20 dB to a voltage ratio:

$$\begin{aligned} E_S/E_N &= \log^{-1} \frac{(S/N)\ (\text{dB})}{20} \\ &= \log^{-1} \frac{20}{20} \\ &= 10 \end{aligned}$$

$$\begin{aligned} E_N/E_S &= 1/10 \\ &= 0.1 \end{aligned}$$

Since $E_S >> E_N$, we can use Equation (7.18).

$$\begin{aligned} \phi &\approx E_N/E_S \\ &= 0.1 \text{ rad} \end{aligned}$$

Remembering that the receiver will interpret the noise as an FM signal with a modulation index equal to ϕ_N, we find

$$m_{fN} = 0.1$$

This can be converted into an equivalent frequency deviation δ_N due to the noise.

$$\begin{aligned} \delta_N &= m_f f_m \\ &= 0.1 \times 1 \text{ kHz} \\ &= 100 \text{ Hz} \end{aligned}$$

The frequency deviation due to the signal is given as 5 kHz, and the receiver output voltage is proportional to the deviation. Therefore the output S/N as a voltage ratio is equal to the ratio between the deviation due to the signal and that due to the noise.

$$\begin{aligned} (E_S/E_N)_o &= \delta_S/\delta_N \\ &= 5 \text{ kHz}/100 \text{ Hz} \\ &= 50 \end{aligned}$$

Since S/N is nearly always expressed in decibels, we will change this to decibels.

$$(S/N)_o \text{ (dB)} = 20 \log 50$$
$$= 34 \text{ dB}$$

This is an improvement of 14 dB over the S/N at the input.

7.5.1 The Threshold Effect and the Capture Effect

An FM signal can produce a better signal-to-noise ratio at the output of a receiver than an AM signal with a similar input S/N, but this is not always the case. The superior noise performance of FM depends on there being a sufficient input S/N ratio. There exists a threshold S/N below which the performance is no better than AM. In fact, it is actually worse, because the greater bandwidth of the FM signal requires a wider receiver noise bandwidth.

The FM **threshold effect** can easily be observed by driving away from an FM broadcast transmitter while listening to it on a car radio. The reception remains essentially noise-free for a considerable distance, then degrades very quickly as the limit of the station's coverage is reached. Shortly before reception of the station is lost completely, there will be some very loud clicks in the audio. This effect is caused by noise that causes a rapid 360° phase shift in the signal. AM reception, in contrast, degrades gradually and continuously as the signal strength is reduced.

When the signal strength is above the threshold, the noise performance of FM can be more than 20 dB better than for AM, as shown in Figure 7.13. The numbers in this figure are approximate: the exact values of the threshold and S/N improvement depend on the modulation index. Because wideband FM allows more noise to enter the receiver than narrowband FM, the threshold is higher for WBFM. For signal strengths above the threshold, the performance of WBFM is superior to that of NBFM; in fact, the S/N improves with the square of the modulation index.

The analysis of noise that was conducted in the previous section would apply equally well to an interfering signal. As long as the desired signal is considerably stronger than the interference, the ratio of desired to interfering signal strength will be greater at the output of the detector than at the input. We could say that the stronger signal *captures* the receiver, and in fact this property of FM is usually called the **capture effect**. It is very easy to demonstrate with any FM system. For example, it is the reason why there is less interfer-

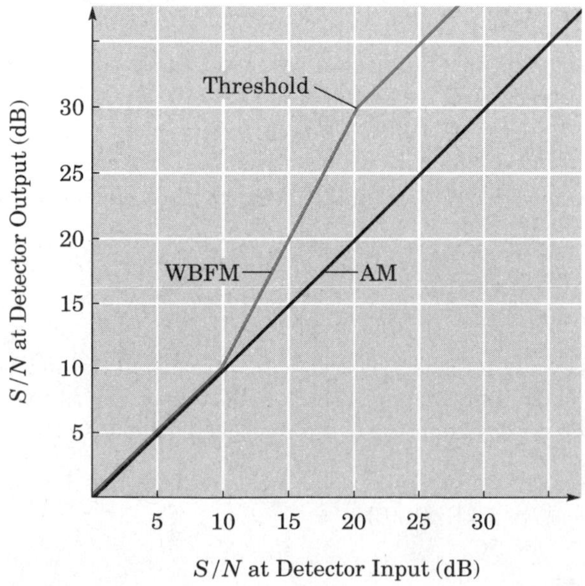

Figure 7.13 S/N improvement with wideband FM

ence between cordless telephones, which all share a few channels in the 46–49 MHz bands, than one might expect.

7.5.2 Pre-emphasis and De-emphasis

The phase component of noise creates phase modulation of the signal. If the noise power is evenly distributed across the channel at the receiver input (*white noise*), the phase modulation will be evenly distributed over the receiver bandwidth. With a phase modulation system, the resulting noise after demodulation would also be evenly distributed over the baseband spectrum.

With an FM receiver, the situation is different. The phase modulation due to noise is interpreted as frequency modulation by the receiver. As we have already discovered, phase and frequency deviation are related by Equation (7.7):

$$m_f = \frac{\delta}{f_m}$$

which can be restated as

$$\delta = m_f f_m$$

Remember that the modulation index m_f is simply the peak phase deviation in radians. The frequency deviation is proportional to the modulating frequency. This tells us that if the phase deviation due to thermal noise is randomly distributed over the baseband spectrum, then the amplitude of the demodulated noise will be proportional to frequency. This relationship between noise voltage and frequency is shown in Figure 7.14. Since power is proportional to the square of voltage, the noise power has the parabolic spectrum shown in Figure 7.14.

To improve the noise performance of an FM system, the deviation can be increased. There are limits to how much this can be done, however, because of considerations of channel width and threshold carrier-to-noise ratio. However, the foregoing shows that a large deviation is more important for high modulating frequencies. Unfortunately, for most common analog signals (including voice, music, and video signals), the high-frequency components generally have lower amplitudes than low and medium frequencies. An improvement in S/N can be made by boosting (*pre-emphasizing*) these high frequencies before modulation, with a corresponding cut in the receiver after demodulation.

Obviously, it is necessary to use similar filter characteristics for **pre-emphasis** and **de-emphasis**. In FM broadcasting, for instance, a 75 μs standard is used. The number refers to the time constants of the high-pass filter at the transmitter and the low-pass filter at the receiver. This gives a turnover frequency of 2.12 kHz. Depending on the program

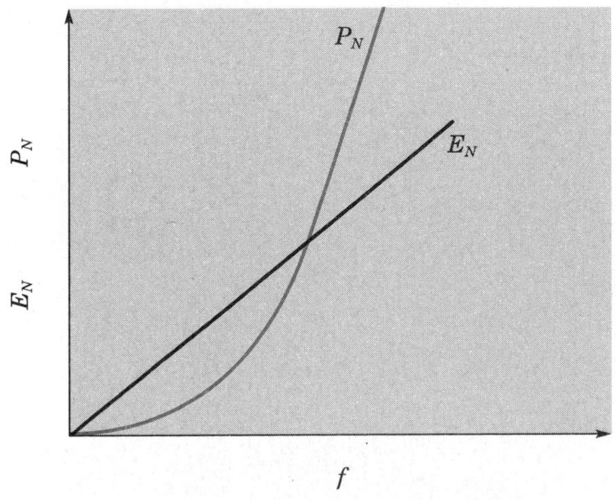

Figure 7.14 Spectrum of demodulated noise in an FM system (without de-emphasis)

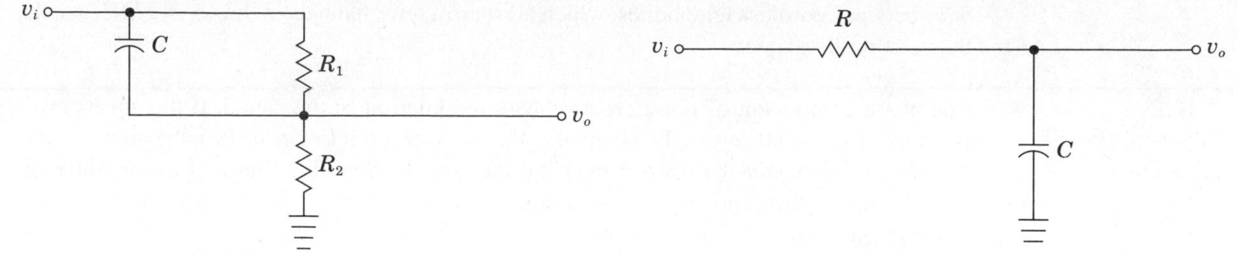

$$\tau = R_1C = 75\ \mu\text{s}$$

$$f_1 = \frac{1}{2\pi R_1 C} = 2120\ \text{Hz}$$

$$f_2 = \frac{1}{2\pi}\left(\frac{R_1 + R_2}{R_1 R_2 C}\right) > 15\ \text{kHz}$$

(a) Pre-emphasis (in transmitter)

$$\tau = RC = 75\ \mu\text{s}$$

$$f_1 = \frac{1}{2\pi RC} = 2120\ \text{Hz}$$

(b) De-emphasis (in receiver)

Figure 7.15 Typical pre-emphasis and de-emphasis circuits

material, pre-emphasis can make a considerable difference. The improvement in *S/N* is about 12 dB for FM broadcasting.

Figure 7.15 gives examples of simple circuits that provide pre-emphasis and de-emphasis, and Figure 7.16 shows the frequency response of these circuits.

7.6 FM Stereo

The introduction of FM stereo in 1961 was accomplished in such a way as to maintain compatibility with the monaural system already in use. It was essential that the mono receiver should reproduce not the left channel or the right channel, but a combination of the two, formed by adding the two signals to get a left-plus-right (L+R) signal. It was also required that the existing 200 kHz channel spacing be preserved. This was accomplished by using a form of frequency-division multiplexing that was modified to maintain compatibility.

Figure 7.17(a) shows the baseband spectrum of an FM stereo signal. This is the signal that is applied to the transmitter. It is also the signal that appears at the output of the receiver

Figure 7.16 Frequency response of pre-emphasis and de-emphasis circuits

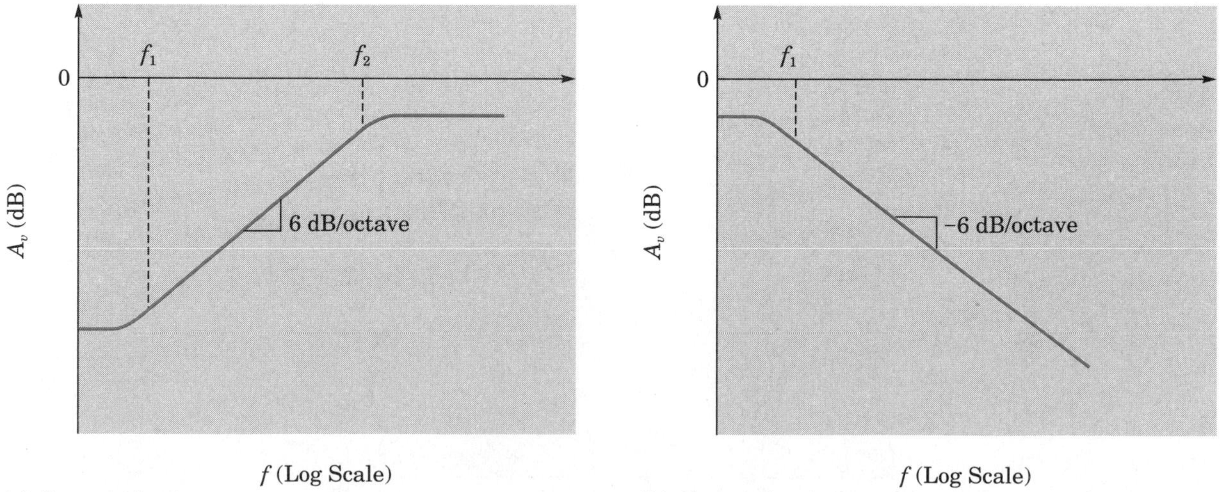

(a) Pre-emphasis

(b) De-emphasis

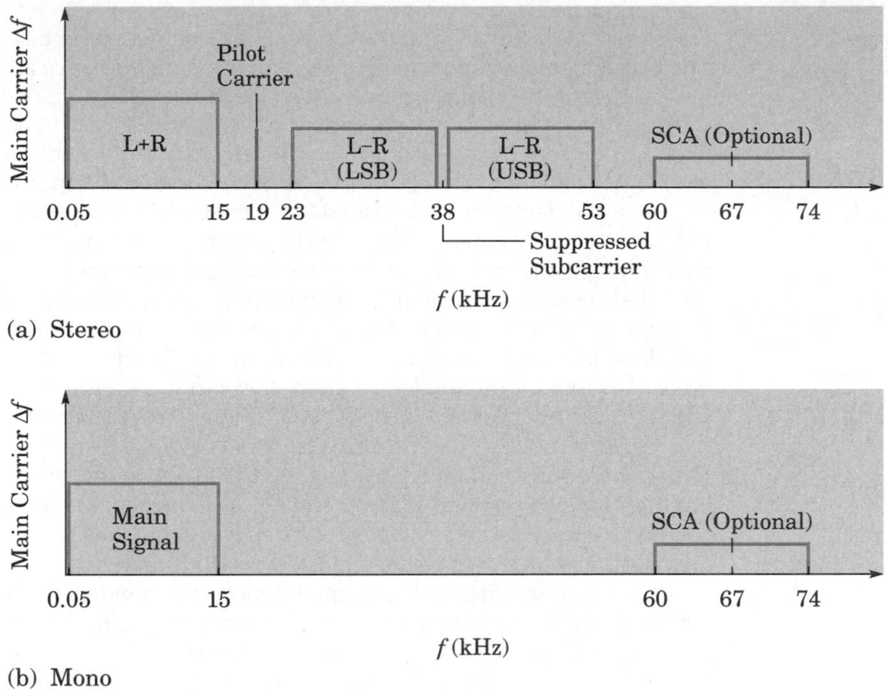

Figure 7.17 FM broadcasting: baseband spectra

detector. For comparison, Figure 7.17(b) shows the baseband of a monaural FM broadcast transmission.

The mono L+R signal occupies the frequencies from about 50 Hz to 15 kHz in the stereo signal. This maintains the compatibility spoken of above, since the monaural receiver will ignore the rest of the stereo signal. Even if frequencies above 15 kHz make their way through the audio amplifier, they are not likely to be reproduced by the loudspeaker, and if they are, 19 kHz is at or outside the limits of hearing of most people.

The two stereo channels can be reproduced at the receiver if, in addition to the sum (L+R) signal, a difference (L − R) signal is transmitted. This signal is sent using DSBSC AM at a (suppressed) **subcarrier** frequency of 38 kHz. The DSBSC AM requires a pilot carrier for proper detection; it is sent at half the subcarrier frequency, at 19 kHz. The pilot carrier causes a 10% (7.5 kHz) deviation of the main carrier frequency.

It might be expected that a stereo FM signal would have a much wider bandwidth than would the monaural version. This is what would be predicted by Carson's rule. The maximum deviation in both cases is 75 kHz, but the maximum baseband frequency is 15 kHz for monaural and 53 kHz for stereo FM. Carson's rule would predict, for monaural FM, a bandwidth of

$$\begin{aligned} B &= 2(\delta_{max} + f_{m\ (max)}) \\ &= 2(75\text{ kHz} + 15\text{ kHz}) \\ &= 180\text{ kHz} \end{aligned}$$

For stereo, the prediction from Carson's rule would be

$$\begin{aligned} B &= 2(\delta_{max} + f_{m\ (max)}) \\ &= 2(75\text{ kHz} + 53\text{ kHz}) \\ &= 256\text{ kHz} \end{aligned}$$

In practice, the increase is not as drastic. It is unlikely that maximum frequency deviation will result from L − R signals with maximum frequency. The L − R signal usually

FM Stereo Broadcasting

FM is now the preeminent mode of local radio broadcasting because of its better performance compared with AM. In part, this enhanced performance is due to the *S/N* improvement inherent in FM, but there are other advantages that are unrelated to the modulation scheme. The higher carrier frequencies used in FM broadcasting allow it to avoid nighttime interference from distant stations, which often ruins AM reception. The FM broadcast system, as implemented, also allows a wider audio frequency response.

Another factor that established FM as the preferred broadcast medium for music was the availability of FM stereo. AM stereo is a relative newcomer that so far shows little sign of challenging the predominance of FM for music programming.

FM stereo broadcasting, using the present multiplexing system, began in 1961. Before that time, and even after, there were a number of experimental stereo broadcasts using two stations, one for each channel. Most used one FM and one AM station, since few households had more than one FM receiver. This was an interesting novelty but, beyond the inherent waste of spectrum space, the differences in the two transmission systems and the two receivers usually resulted in stereo sound that would be rather disconcerting to listeners today. There are still some 1960s vintage AM/FM tuners that attempt to make the best of a bad idea by incorporating separate tuning dials and separate outputs so that AM and FM can be heard on separate channels of a stereo system. At least in this case, the audio amplifiers and speakers would be the same for the two channels.

The current system allows high fidelity by the standards of the early 1960s. Present-day audiophiles would prefer a wider frequency range and a better *S/N*. The only way to get a wider frequency range now would be to move the pilot carrier and the L-R signal to a higher frequency. Since that would make all current FM receivers obsolete, such a move is unlikely. An attempt to improve the *S/N* by using DOLBY™ noise reduction was made in the 1970s, but since the result was only partially compatible with ordinary receivers, the system never became popular.[1] Interestingly, the new television stereo sound system does incorporate noise reduction, using the dBx™ system for the L-R signal only.[2] You'll learn more about that in Chapter 9.

[1] DOLBY is a trademark of Dolby Laboratories Licensing Corporation.
[2] dBx is a trademark of dBx Technology Licensing.

has a lower amplitude than the L+R signal, and in any case, frequencies at the upper end of the audio range are likely to have low amplitude, even after pre-emphasis has been applied.

There is a noise penalty with stereo FM. Pre-emphasis is applied to the left and right signals. This is effective with the L+R signal, but the L−R signal does not get the full advantage of the pre-emphasis, as is made clear in Figure 7.18. The upper sideband receives some benefit from pre-emphasis, but the pre-emphasis actually works in reverse on the lower sideband. Matters are made even worse by the fact that the L−R signal has been

Figure 7.18
Noise performance of FM stereo

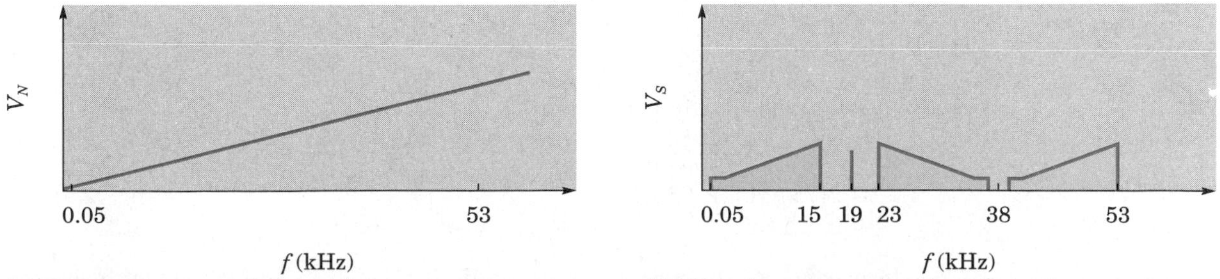

(a) Noise spectrum at baseband

(b) Signal spectrum showing effect of pre-emphasis

shifted upward in frequency before modulation onto the main FM carrier. The noise voltage in the demodulated signal is proportional to frequency, as we have seen. This means that the L−R signal will be noisier than the L+R signal. When the DSBSC L−R signal is demodulated, this noise is shifted down in frequency into the audible range. The result of all this is an *S/N* degradation of about 22 dB for stereo FM compared to the monaural signal. The difference can easily be heard by tuning in a relatively weak station and then switching a stereo receiver between stereo and monaural reception. In fact, many receivers switch automatically to monaural mode in the presence of weak signals.

There is one part of the baseband signal in Figure 7.17(a) that has not yet been explained. The Subsidiary Carrier Authorization (SCA) signal, centered on a subcarrier at 67 kHz, is used for services such as background music for stores and offices. The SCA signal is monaural, with a maximum audio frequency of 7 kHz. The subcarrier can be modulated using either AM or (usually) NBFM, but the total deviation of the main carrier due to the SCA subcarrier is limited to 10% (7.5 kHz). Not all stations use SCA.

7.7 Measurement of FM Index and Deviation

The maximum frequency deviation of FM transmitters is restricted by law, not by any physical constraint (as is the AM index). Setting a particular value of deviation is not quite as simple as determining 100% modulation from an oscilloscope display of an AM signal. In fact, oscilloscope displays of FM signals are not very enlightening at all—there is no envelope variation, and the frequency variations are usually such a small percentage of the carrier frequency that they are difficult to see (let alone measure) with an oscilloscope.

The spectrum analyzer is a more useful instrument for the observation of FM signals. It allows the power in the carrier and each sideband to be measured, and it allows the modulating frequency to be found by measuring the separation between sidebands. At least this is true for single-tone modulation; with several modulating frequencies, the pattern of sidebands becomes very confusing, as each tone produces multiple sets of sidebands.

It would, in theory, be possible to measure the value of m_f for single-tone modulation by measuring the amplitude of each sideband pair and comparing it with a table of Bessel functions. This process is tedious and seldom attempted. There are several values of m_f, however, that are very easily recognized—the values where the term J_0 is equal to zero. As m_f increases from zero, this happens for m_f equal to 2.4, 5.5, 8.65, 11.8, and 14.9, to list the first few points. At each of these values of m_f, the carrier-frequency component of the spectrum disappears, an occurrence that is easy to observe on a spectrum analyzer. This provides several reference points for measuring m_f.

In adjusting a transmitter, for instance, it is usually deviation rather than modulation index that must be measured. This can be done using the relation between deviation, modulation index, and modulating frequency given in Equation (7.7):

$$m_f = \frac{\delta}{f_m}$$

An example will show how this is done. The method is often called the *Bessel-zero method*, for obvious reasons.

Example 7.10 Show how the deviation of a voice-communications FM transmitter can be set to 5 kHz using the Bessel-zero method with a spectrum analyzer.

Solution Set up the equipment as in Figure 7.19. The frequency counter is not an absolute requirement, but it is recommended because the accuracy of the deviation measurement depends on the accuracy with which the modulating frequency can be set. The dial calibrations on many function generators are not very accurate.

Figure 7.19

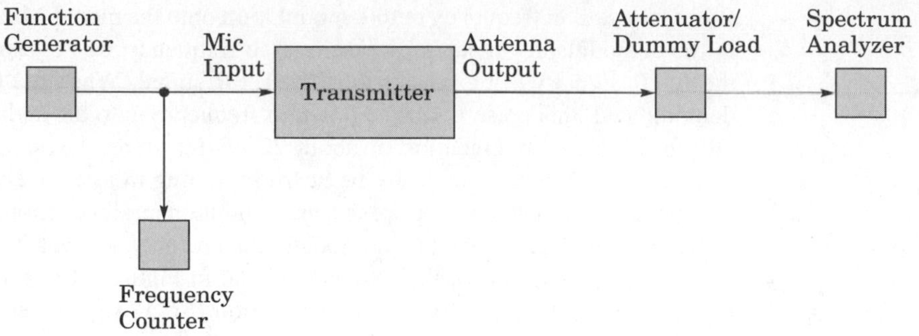

Since the transmitter is designed for voice frequencies, it is necessary to find a frequency within the range of 300 Hz to 3 kHz, approximately, that will produce a carrier null for a deviation of 5 kHz. This can be done by trial and error, starting with the first null for $m_f = 2.4$. It is easy to rearrange Equation (7.7) to find f_m.

$$
\begin{aligned}
m_f &= \frac{\delta}{f_m} \\
f_m &= \frac{\delta}{m_f} \\
&= \frac{5\text{ kHz}}{2.4} \\
&= 2.08\text{ kHz}
\end{aligned}
$$

This frequency is within the acceptable range. If it had not been, the next carrier null, for $m_f = 5.5$, could have been tried.

The function generator is set to 2.08 kHz, and the deviation is increased gradually from zero until the carrier disappears for the first time. The deviation is now 5 kHz.

This method of measuring deviation is elegant and simple, but it does require an expensive spectrum analyzer and complete control over the modulating frequency. Thus, it is unsuitable for continuous monitoring of transmitter operation. For that purpose, the fact that the output of an FM detector is proportional to deviation can be used. A detector can have a meter attached and then be calibrated (using a spectrum analyzer, for instance). The result is an instrument that can provide instant readings, with reasonable accuracy, for any type of modulating signal. Figure 7.20 shows an example of this type of instrument, which is used for continuous monitoring of FM broadcast transmitters.

Figure 7.20
FM deviation meter
Courtesy Belar Electronics Laboratory, Inc., Devon, Pennsylvania

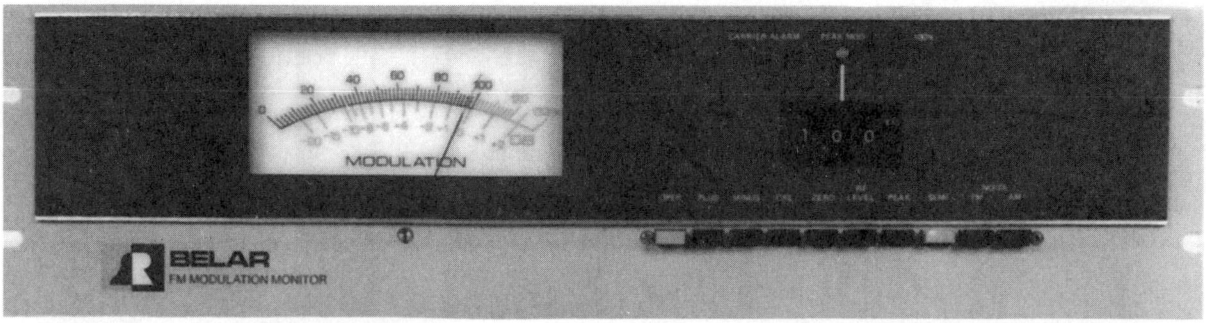

Summary

Here are the main points to remember from this chapter.

1. Angle modulation includes frequency and phase modulation, which are closely related.
2. FM is widely used for analog communications, while PM sees greatest application in data communications.
3. The power of an angle modulation signal does not change with modulation, but the bandwidth increases due to the generation of multiple sets of sidebands.
4. The voltage and power of each sideband can be calculated using Bessel functions. An approximate bandwidth is given by Carson's rule.
5. FM has a significant advantage compared with AM in the presence of noise or interference, provided the deviation is relatively large and the signal is reasonably strong.
6. The *S/N* for FM can be improved considerably by using pre-emphasis and deemphasis, which result in greater gain for the higher baseband frequencies before modulation, with a corresponding reduction after demodulation.
7. Stereo FM transmission is accomplished using a multiplexing scheme that preserves compatibility with monaural receivers and requires only a slight increase in bandwidth. A noise penalty is paid with stereo FM, however.
8. A spectrum analyzer is much more useful than an oscilloscope for the analysis of FM signals. It can also be used to calibrate other FM test equipment, such as deviation meters.

Important Equations

$$k_f = \frac{\Delta f}{e_m} \tag{7.1}$$

$$f_{sig}(t) = f_c + k_f e_m(t) \tag{7.2}$$

$$f_{sig}(t) = f_c + k_f E_m \sin \omega_m t \tag{7.4}$$

$$\delta = k_f E_m \tag{7.5}$$

$$f_{sig}(t) = f_c + \delta \sin \omega_m t \tag{7.6}$$

$$m_f = \frac{\delta}{f_m} \tag{7.7}$$

$$f_{sig}(t) = f_c + m_f f_m \sin \omega_m t \tag{7.8}$$

$$k_p = \frac{\phi}{e_m} \tag{7.9}$$

$$\theta(t) = \theta_c + k_p e_m(t) \tag{7.10}$$

$$\theta(t) = \theta_c + m_p \sin \omega_m t \tag{7.12}$$

$$\begin{aligned} v(t) &= A \sin (\omega_c t + m \sin \omega_m t) \\ &= A\{J_0(m) \sin \omega_c t \\ &\quad - J_1(m)[\sin (\omega_c - \omega_m)t - \sin (\omega_c + \omega_m)t] \\ &\quad + J_2(m)[\sin (\omega_c - 2\omega_m)t + \sin (\omega_c + 2\omega_m)t] \\ &\quad - J_3(m)[\sin (\omega_c - 3\omega_m)t + \sin (\omega_c + 3\omega_m)t] \\ &\quad + \cdots\} \end{aligned} \tag{7.14}$$

$$P_T = J_0^2 + 2J_1^2 + 2J_2^2 + \cdots \tag{7.15}$$

$$B = 2(\delta_{max} + f_{m\ (max)}) \tag{7.16}$$

$$\phi = \sin^{-1} (E_N/E_S) \tag{7.17}$$

$$\phi_n \approx E_N/E_S \tag{8.18}$$

Glossary

angle modulation a general term that includes frequency and phase modulation

capture effect the ability of an FM receiver to receive the stronger of two signals, ignoring the weaker

carrier frequency the frequency of a signal before modulation is applied; in contrast to AM signals, the power transmitted at the carrier frequency varies with modulation for an FM signal

de-emphasis use of low-pass filter in a receiver to remove the effect of pre-emphasis on the frequency response

frequency deviation the amount by which the frequency of an FM signal shifts to each side of the carrier frequency

modulation index in FM and PM, the peak amount in radians by which the phase of a signal deviates from its resting value

narrowband FM (NBFM) FM with a relatively low modulation index

pre-emphasis use of a high-pass filter in an FM transmitter to improve the signal-to-noise ratio; always used with de-emphasis at the receiver

rest frequency the frequency of the unmodulated carrier of an FM signal; a synonym for *carrier frequency*

subcarrier a secondary carrier that can carry an additional modulating signal and is itself modulated onto the main carrier

threshold effect the noise-reduction effect that occurs with strong FM signals

wideband FM (WBFM) FM with a relatively large modulation index

Questions

1. What two types of modulation are included in the term *angle modulation*?
2. What is the relationship between phase and frequency for a sine wave?
3. Define and compare the modulation index for FM and PM.
4. Compare, in general terms, the bandwidth and signal-to-noise ratio of FM and AM.
5. Describe and compare two ways to determine the practical bandwidth of an FM signal.
6. What is pre-emphasis? How is it used to improve the *S/N* of FM transmissions?
7. For FM, what characteristic of the modulating signal determines the instantaneous frequency deviation?
8. What is the capture effect?
9. Draw a diagram showing the baseband spectrum of an FM stereo signal, and explain why this system is used.
10. What is the purpose of the pilot carrier in an FM stereo signal?
11. Where is PM used?
12. Explain why the signal-to-noise ratio of FM can increase with the bandwidth. Is this always true for FM? Compare with the situation for AM.
13. Compare the effects of modulation on the carrier power and the total signal power in FM and AM.
14. What is the threshold effect?
15. Why is the *S/N* for FM stereo transmissions worse than for mono?
16. Explain how limiting reduces the effect of noise on FM signals.
17. Explain how noise affects FM signals even after limiting.
18. Explain the fact that there is no simple relationship between modulating frequency and bandwidth for an FM signal.
19. Define NBFM in two different ways, and explain the use of each.
20. Why does limiting the receiver bandwidth to less than the signal bandwidth cause more problems with FM than with AM?

Problems

SECTION 7.2

21. An FM signal has a deviation of 10 kHz and a modulating frequency of 2 kHz. Calculate the modulation index.

22. Calculate the frequency deviation for an FM signal with a modulating frequency at 5 kHz and a modulation index of 2.
23. A sine-wave carrier at 100 MHz is modulated by a 1 kHz sine wave. The deviation is 100 kHz. Draw a graph showing the variation of signal frequency with time.
24. An FM modulator has $k_f = 50$ kHz/V. Calculate the deviation and modulation index for a 3 kHz modulating signal of 2 V (RMS).
25. When a positive dc voltage of 5 V is applied to an FM modulator, its output frequency drops by 100 kHz. Calculate the deviation sensitivity of the modulator.
26. A 10 kHz signal modulates a 500 MHz carrier, with a modulation index of 2. What are the maximum and minimum values for the instantaneous frequency of the modulated signal?

SECTION 7.3

27. A phase modulator with $k_p = 3$ rad/V is modulated by a sine wave with an RMS voltage of 4 V at a frequency of 5 kHz. Calculate the phase modulation index.
28. A PM signal has a modulation index of 2, with a modulating signal that has an amplitude of 100 mV and a frequency of 4 kHz. What would be the effect on the modulation index of:
    (a) changing the frequency to 5 kHz?
    (b) changing the voltage to 200 mV?
29. A sine wave of frequency 1 kHz phase-modulates a carrier at 123 MHz. The peak phase deviation is 0.5 rad. Calculate the maximum frequency deviation.
30. What is the maximum phase deviation that can be present in an FM radio broadcast signal, assuming it transmits a baseband frequency range of 50 Hz to 15 kHz, with a maximum deviation of 75 kHz?
31. A phase modulation system operates with a modulation index of 1.5. What is the maximum phase shift in degrees?
32. An FM signal uses a deviation of 100 kHz and a modulating frequency of 25 kHz.
    (a) Calculate the modulation index.
    (b) What is the bandwidth predicted by Carson's rule?
    (c) Use Bessel functions to find the bandwidth of the signal, ignoring all sidebands with an amplitude more than 40 dB below the amplitude of the unmodulated carrier.
    (d) If the total power of this signal is 10 W, how much power is transmitted at the carrier frequency for the conditions described above?

SECTION 7.4

33. An FM signal has a deviation of 10 kHz and is modulated by a sine wave with a frequency of 5 kHz. The carrier frequency is 150 MHz, and the signal has a total power of 12.5 W, operating into an impedance of 50 Ω.
    (a) What is the modulation index?
    (b) How much power is present at the carrier frequency?
    (c) What is the voltage level of the second sideband below the carrier frequency?
    (d) What is the bandwidth of the signal, ignoring all components that have less than 1% of the total signal voltage?
34. An FM transmitter operates with a total power of 10 W, a deviation of 5 kHz, and a modulation index of 2.
    (a) What is the modulating frequency?
    (b) How much power is transmitted at the carrier frequency?
    (c) If a receiver has a bandwidth sufficient to include the carrier and the first two sets of sidebands, what percentage of the total signal power will it receive?
35. An FM transmitter has a carrier frequency of 220 MHz. Its modulation index is 3 with a modulating frequency of 5 kHz. The total power output is 100 W into a 50 Ω load.
    (a) What is the deviation?
    (b) Sketch the spectrum of this signal, including all sidebands with more than 1% of the signal voltage.
    (c) What is the bandwidth of this signal, according to the criterion used in part (b)?
    (d) Use Carson's rule to calculate the bandwidth of this signal, and compare it with the result found in part (c).
36. An FM transmitter has a carrier frequency of 160 MHz. The deviation is 10 kHz and the modulation frequency is 2 kHz. A spectrum analyzer shows that the carrier-frequency component of the signal has a power of 5 W. What is the total signal power?
37. Use Carson's rule to compare the bandwidth that would be required to transmit a baseband signal with a frequency range from 300 Hz to 3 kHz using:
    (a) NBFM with maximum deviation of 5 kHz
    (b) WBFM with maximum deviation of 75 kHz

SECTION 7.5

38. An FM receiver operates with an S/N of 30 dB at its detector input and with $m_f = 10$.
    (a) If the received signal has a voltage of 10 mV, what is the amplitude of the noise voltage?

(b) Find the maximum phase shift that could be given to the signal by the noise voltage.
(c) Calculate the S/N at the detector output, assuming the detector is completely insensitive to amplitude variations.

39. The circuits shown in Figure 7.21 are 75 μs pre-emphasis and de-emphasis networks. Identify which is which, and calculate values for the missing components.

Figure 7.21

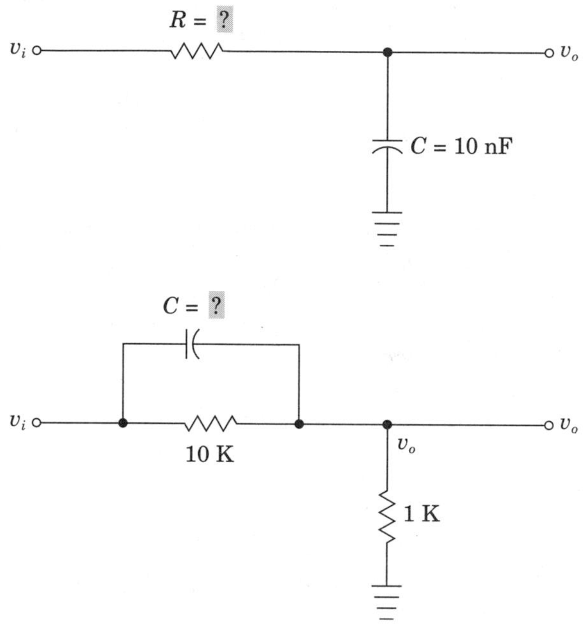

40. Using a pre-emphasis circuit with a 75 μs time constant, by approximately how much, in decibels, will a 10 kHz signal be boosted compared with a 5 kHz signal?

41. A frequency-modulation system requires a signal-to-noise ratio of 40 dB at the detector output. The modulating frequency is 2 kHz, and the deviation is 50 kHz. Calculate the required minimum S/N at the detector input.

SECTION 7.6

42. Sketch the baseband spectrum for each of the following stereo FM signals. A frequency scale is required, but an amplitude scale is not.
(a) a 1 kHz sine wave, of equal amplitude and phase on the left and right channels
(b) as in part (a), except that the left and right channels are 180° out of phase
(c) a 1 kHz sine wave on the left channel, no signal on the right channel

43. Sketch the baseband spectrum for a stereo FM signal modulated by a 1 kHz sine wave on the left channel only.

44. (a) Sketch the baseband spectrum of an FM stereo multiplex signal with both left-plus-right and left-minus-right signals containing frequencies ranging from 50 Hz to 15 kHz.
(b) Use Carson's rule to calculate the bandwidth of the signal described in part (a), assuming a maximum deviation of 75 kHz.
(c) The signal-to-noise ratio for the left-minus-right signal will not be as good as for the left-plus-right signal in a stereo multiplex signal. Explain why.

SECTION 7.7

45. Using a function generator and spectrum analyzer, how could the deviation of an FM generator be set to 75 kHz to simulate a broadcast FM signal?

COMPREHENSIVE

46. A certain full-carrier DSB AM signal has a bandwidth of 20 kHz. What would be the approximate bandwidth required if the same information signal were to be transmitted using:
(a) DSBSC AM
(b) SSB
(c) FM with 10 kHz deviation
(d) FM with 50 kHz deviation

47. Using the table of Bessel functions, demonstrate that the total power in an FM signal is equal to the power in the unmodulated carrier for $m = 2$. Compare with the situation for full-carrier AM and for SSB AM.

48. Suppose you were called upon to recommend a modulation technique for a new communication system for voice frequencies. State which of the techniques studied so far you would recommend and why, in each of the following situations:
(a) simple, cheap receiver design is of greatest importance
(b) narrow signal bandwidth is of greatest importance
(c) immunity to noise and interference is of greatest importance

8 FM Equipment

Objectives After studying the material in this chapter, you should be able to:

1. Describe the ways in which FM transmitters and receivers differ from those for AM and explain why they differ
2. Explain the difference between direct- and indirect-FM generation and draw block diagrams for both systems
3. Analyze the operation of both direct and indirect frequency modulators
4. Explain the use of frequency multipliers and mixers in FM transmitters
5. Calculate the carrier frequency and deviation for FM transmitters
6. Analyze and compare the operation of each of the following types of FM detector: discriminator, ratio detector, phase-locked loop, and quadrature
7. Explain why limiters are used in FM receivers and describe the operation of a limiter
8. Explain the operation of FM transceivers, including those with split-frequency operation
9. Perform sensitivity and channel-separation measurements on FM receivers
10. Describe the operation of FM stereo transmitters and receivers

8.1 Introduction

Now that you are familiar with the basics of FM signals, it is time to consider how such signals are created and employed. The fact that both transmitters and receivers are covered in a single chapter is certainly not intended to imply that these circuits are of less importance than those for AM. If anything, the contrary is true. However, there are many similarities between transmitters for different modulation types, and the same is true for receivers. In this chapter, it will be necessary to describe only the differences between FM and AM equipment.

The processes of modulation and demodulation are very different for FM than for AM. In addition, there is more variety in the circuits used for the modulation and demodulation of FM than there is for AM. Consequently, most of this chapter will actually be devoted to modulators and demodulators, and their associated circuitry.

8.2 FM Transmitters

The differences between FM and AM transmitters are easier to understand if we start by considering the differences between FM and AM signals. To begin, FM requires the transmitter frequency to be varied. That in turn implies that the modulation must be applied early, probably at the carrier-oscillator stage (the "probably" is intended to leave room for *indirect* FM, which will be described shortly).

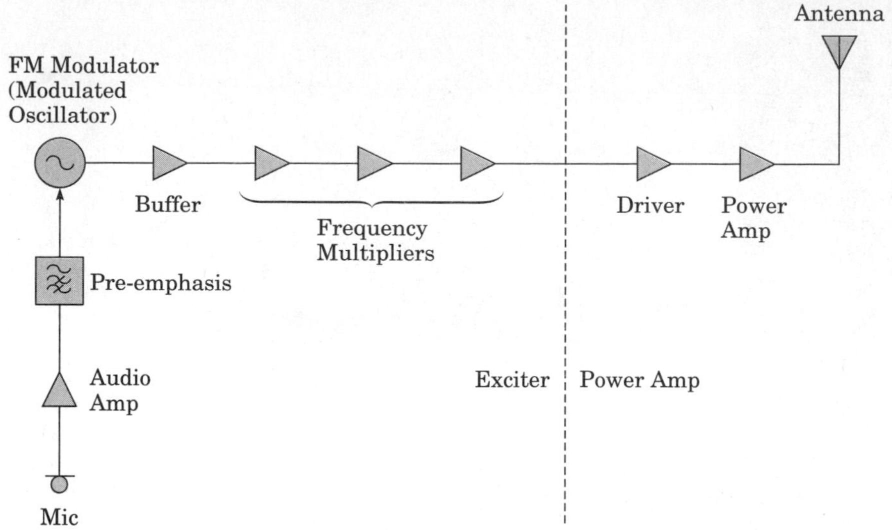

Figure 8.1 Typical FM transmitter

Another difference is that FM signals have no amplitude variations. This means that Class C amplifiers can be used with FM signals. Since Class C cannot be used to amplify AM signals, most AM transmitters use high-level modulation, which takes place in the final stage of power amplification. High-level modulation allows Class C to be used for all RF amplifier stages. FM transmitters can employ Class C throughout, even after modulation.

Frequency multipliers are used before the modulator stage in AM transmitters when very high carrier frequencies are required. Multiplier stages can be used for the same reason in FM transmitters, but they have another function as well: frequency multiplication of an FM signal also multiplies the deviation. Since the amount of frequency deviation that can be achieved with some types of FM modulators is quite small, frequency multiplication is often used as a way of increasing deviation.

Figure 8.1 is a simplified block diagram of a typical FM transmitter. There are many variations on this, of course, and some of them will be described in the following sections. For instance, frequency multiplication is not always necessary, depending on the modulator design and the amount of deviation required.

You will notice that the transmitter in Figure 8.1 is divided into two sections, the *exciter* and the *power amplifier*. This is typical of large transmitters such as those used in broadcasting, where the exciter is a separate unit containing the audio and low-power RF stages. The power amplifier stages—involving high voltages, large amounts of power, forced-air or water cooling, and sometimes vacuum-tube technology—are in a separate cabinet. Small transmitters like those used in mobile communications are, of course, self-contained. Those in current production are completely solid-state.

8.2.1 Direct-FM Modulators

FM signals can be generated either *directly*, by varying the frequency of the carrier oscillator, or *indirectly*, by converting phase modulation to frequency modulation. In this section, we will discuss techniques for **direct FM**; **indirect FM** will be covered later in the chapter.

Direct FM requires that the frequency of the carrier oscillator be varied in accordance with the instantaneous amplitude of the modulating signal (after any required pre-emphasis has been added). Using a *reactance modulator* is the simplest way to do this. It works by using the modulating signal to vary a reactance in the frequency-determining circuit. One common way to build a reactance modulator is to put a varactor into the frequency-determining circuit of the carrier oscillator. Figure 8.2 shows two simple circuits: in Figure 8.2(a), an *LC* oscillator is used, while the circuit in Figure 8.2(b) is crystal-controlled.

Figure 8.2
Direct FM varactor modulators

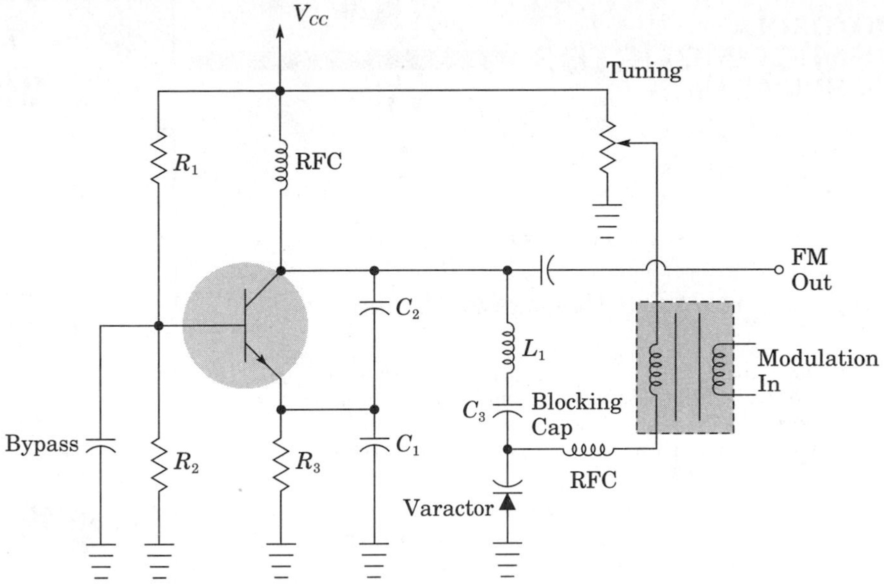

(a) Using an *LC* oscillator

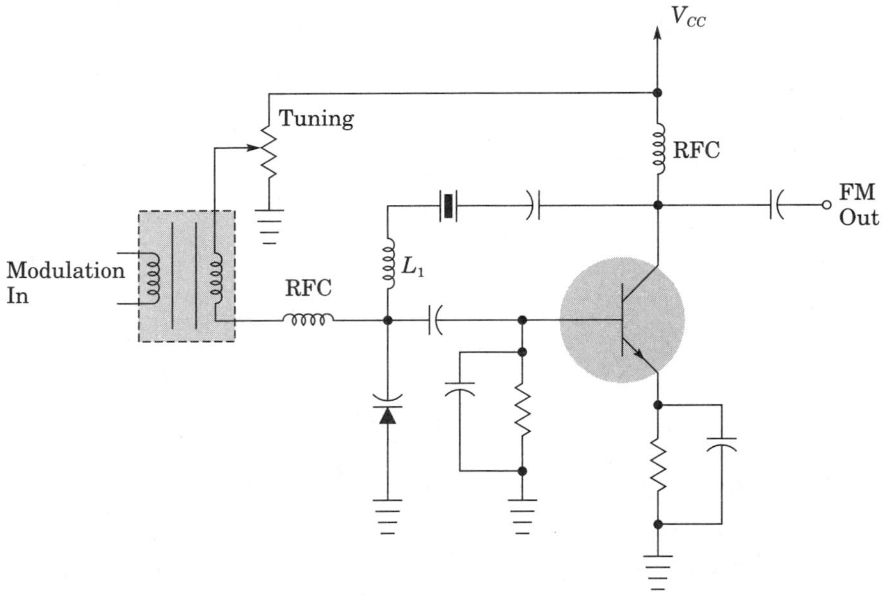

(b) Using a crystal-controlled oscillator

In both circuits, the varactor is reverse-biased and the bias is varied by the modulating signal.

The circuit in Figure 8.2(a) should be familiar from Chapter 2 as a Clapp oscillator. The frequency is determined mainly by L_1 and the varactor. Capacitor C_3 has a large value and exists to isolate the varactor from V_{CC}; C_1 and C_2 determine the feedback fraction and have some small effect on the frequency. The tuning potentiometer adjusts the dc bias on the varactor. To this is added the ac modulating signal, causing the bias voltage, and therefore the frequency, to vary around its nominal value.

Quite large values of frequency deviation are possible with varactor modulation of an *LC* oscillator. In fact, there are broadcast FM transmitter designs that generate the FM signal at the carrier frequency, achieving the full 75 kHz of deviation without any frequency

MOTOROLA
SEMICONDUCTOR
TECHNICAL DATA

MC2833

Advance Information

LOW POWER FM TRANSMITTER SYSTEM

MC2833 is a one-chip FM transmitter subsystem designed for cordless telephone and FM communication equipment. It includes a microphone amplifier, voltage controlled oscillator and two auxiliary transistors.

- Wide Range of Operating Supply Voltage (2.8–9.0 V)
- Low Drain Current (I_{CC} = 2.9 mA Typ)
- Low Number of External Parts Required
- −30 dBm Power Output to 60 MHz Using Direct RF Output
- +10 dBm Power Output Attainable Using On-Chip Transistor Amplifiers

LOW POWER FM TRANSMITTER SYSTEM

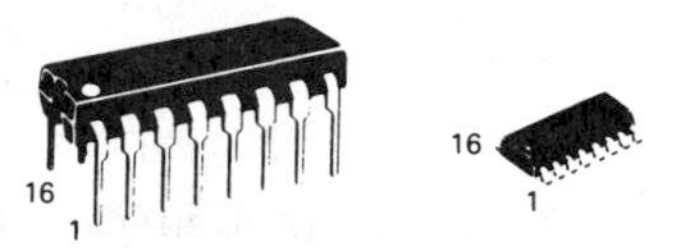

P SUFFIX
PLASTIC PACKAGE
CASE 648

D SUFFIX
PLASTIC PACKAGE
CASE 751B
(SO-16)

FUNCTIONAL BLOCK DIAGRAM

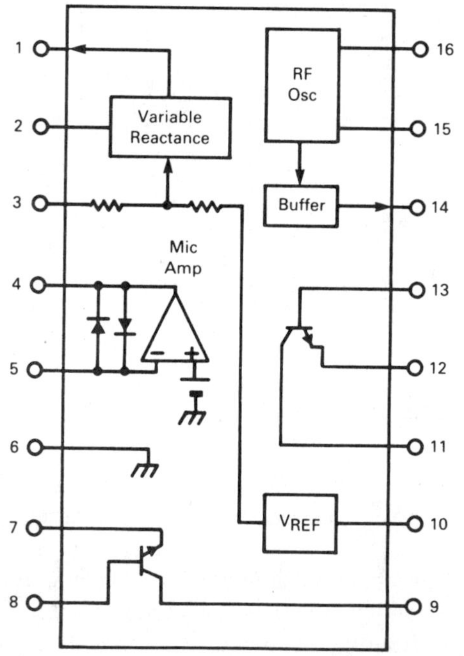

PIN ASSIGNMENTS

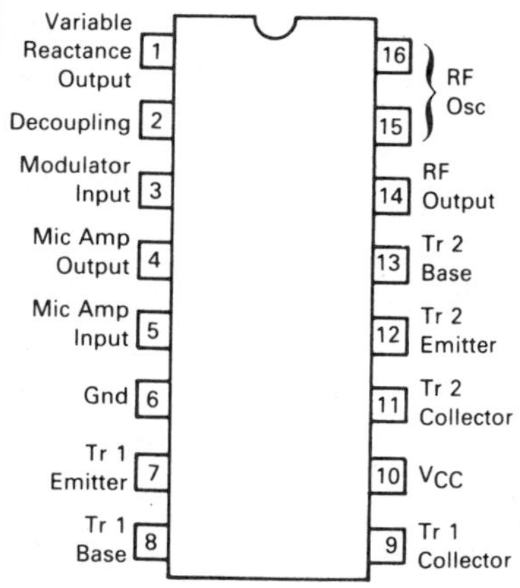

Figure 8.3
An IC transmitter: the MC2833

multiplication at all. On the other hand, an *LC* oscillator at VHF frequencies will be quite unstable and will require some form of **automatic frequency control (AFC)**.

As for the crystal-controlled version in Figure 8.2(b), its frequency can be changed only very slightly (usually only a few tens of hertz) by modulation. A great deal of frequency multiplication is always needed with crystal-controlled direct-FM modulators.

Another, more modern way to generate direct FM is to use either an integrated-circuit VCO or an oscillator chip with an external varactor. For low-power applications such as cordless phones, it is even possible to get an entire transmitter on a chip. The MC2833, shown in Figure 8.3, is an example.

8.2.2 Frequency Multipliers

Frequency multipliers were discussed in Chapter 2. You should recall that they are essentially Class C amplifiers with the output circuit tuned to a harmonic of the input frequency. Doublers and triplers are the most common types of frequency multiplier, though greater multiplication is possible at the expense of reduced efficiency.

Frequency multipliers are particularly useful in FM transmitters because they multiply the deviation of an FM signal by the same factor as the carrier frequency. It is easy to show how this is done. First, suppose that an unmodulated carrier with frequency f_c is applied to the input of a circuit that multiplies the frequency by N. The frequency of the output signal will of course be Nf_c. Now, suppose that due to modulation the frequency of the input changes by an amount δ to $(f_c + \delta)$. The output signal will have frequency

$$\begin{aligned} f_o &= N(f_c + \delta) \\ &= Nf_c + N\delta \end{aligned} \tag{8.1}$$

which corresponds to a carrier at N times the original carrier frequency, with a frequency deviation N times the original deviation. Thus a frequency multiplier can increase the deviation obtained at the modulator by any required amount.

Example 8.1

A direct-FM transmitter has a varactor modulator with $k_f = 2$ kHz/V and a maximum deviation of 300 Hz. This modulator is followed by a buffer and three stages of frequency multiplication: a tripler, a doubler, and another tripler, followed by a driver and power amplifier.

(a) Draw a block diagram of this transmitter.
(b) Will this transmitter be capable of 5 kHz deviation at the output?
(c) What should the oscillator frequency be if the transmitter is to operate at a carrier frequency of 150 MHz?
(d) What audio voltage will be required at the modulator input to obtain full deviation?

Solution

(a) See Figure 8.4 for a block diagram of this transmitter.
(b) The frequency multipliers multiply the deviation by a factor of

$$3 \times 2 \times 3 = 18$$

The maximum deviation at the oscillator must then be

$$\begin{aligned} \delta_{OSC} &= \frac{\delta_o}{18} \\ &= \frac{5 \text{ kHz}}{18} \\ &= 278 \text{ Hz} \end{aligned}$$

Since the modulator is capable of 300 Hz deviation, it will enable the output deviation to reach 5 kHz.

Figure 8.4

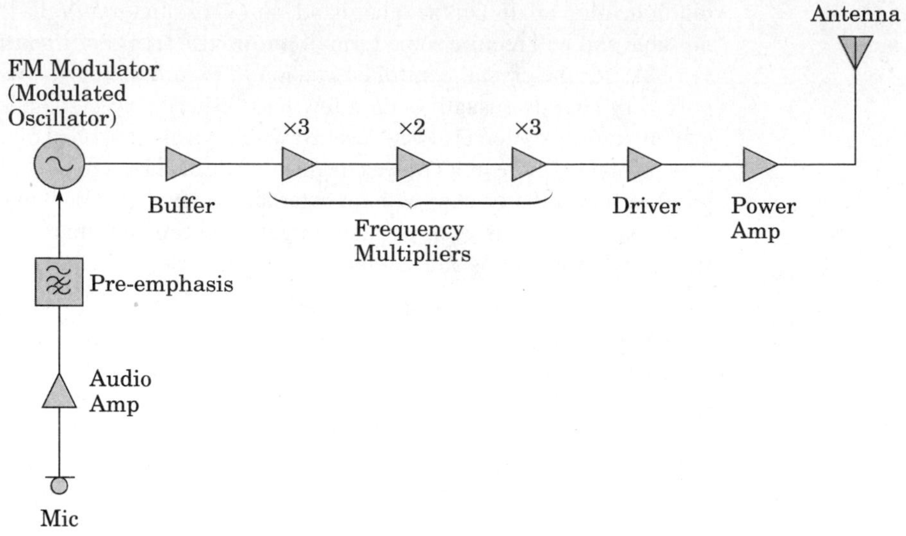

(c) The oscillator frequency will be multiplied by 18, so

$$
\begin{aligned}
f_{OSC} &= \frac{f_o}{18} \\
&= \frac{150 \text{ MHz}}{18} \\
&= 8.33 \text{ MHz}
\end{aligned}
$$

(d) In Part (b) we found that the full 5 kHz deviation at the antenna requires a deviation at the oscillator of 278 Hz. Therefore the required input modulating-signal level can be found from

$$
\begin{aligned}
k_f &= \frac{\delta}{V_i} \\
V_i &= \frac{\delta}{k_f} \\
&= \frac{278 \text{ Hz}}{2000 \text{ Hz/V}} \\
&= 0.139 \text{ V}
\end{aligned}
$$

Since the deviation is a peak value, so is the input voltage. If an RMS voltage is required, it will be

$$
\begin{aligned}
V_i \text{ RMS} &= \frac{V_i \text{ peak}}{\sqrt{2}} \\
&= \frac{0.139 \text{ V}}{\sqrt{2}} \\
&= 0.098 \text{ V} \\
&= 98 \text{ mV}
\end{aligned}
$$

Of course, increasing the deviation by frequency multiplication also increases the carrier frequency. What about the case where sufficient multiplication to achieve the required amount of deviation results in a carrier frequency that is too high? The solution to that problem is actually quite simple: the carrier frequency can be lowered by mixing, as shown in Figure 8.5.

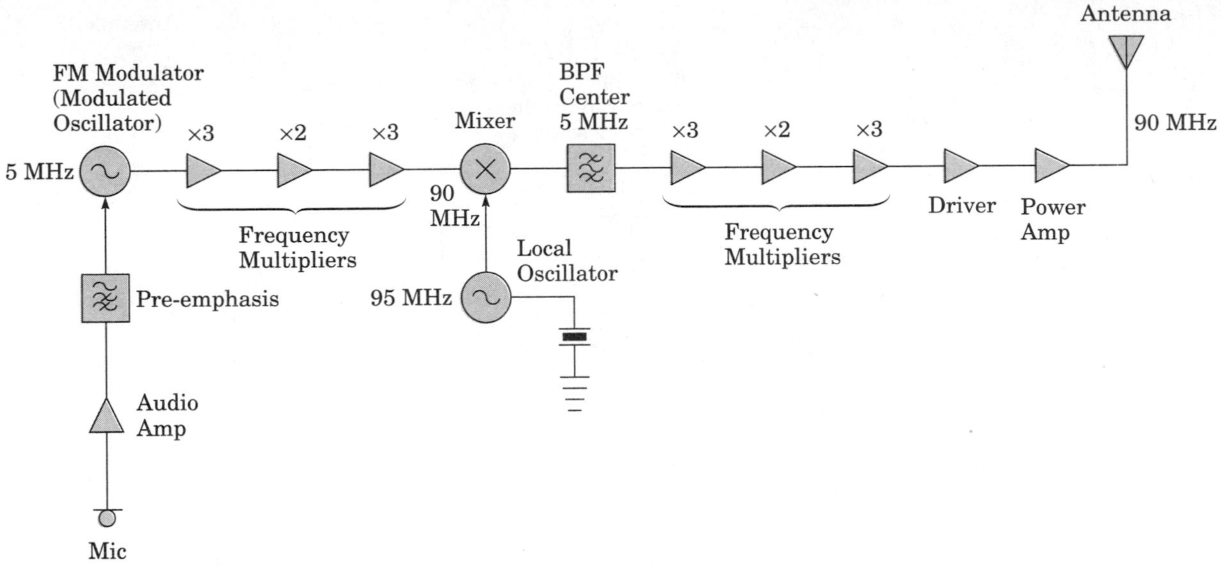

Figure 8.5
FM transmitter with multiplication and mixing

The mixing process can change the carrier frequency to any required value, but it has no effect on the deviation. Again, this is easy to show. Let one input to the mixer have frequency ($f_c + \delta$), and let the other input frequency be f_{LO}, provided by a local oscillator. The sum and difference frequencies will be produced by the mixer, but in this case, it is the difference that is used, since the object is to lower the carrier frequency. Here there are two possibilities, depending on which frequency is higher. If the carrier frequency is higher than the local oscillator frequency, the output will be

$$\begin{aligned} f_o &= (f_c + \delta) - f_{LO} \\ &= (f_c - f_{LO}) + \delta \end{aligned}$$

The carrier frequency has been reduced, but the deviation has not changed. Alternatively, if $f_{LO} > f_c$,

$$\begin{aligned} f_o &= f_{LO} - (f_c + \delta) \\ &= (f_{LO} - f_c) - \delta \end{aligned}$$

This time, the carrier frequency has been decreased, and the deviation has changed in sign but not in magnitude. The change in sign is equivalent to a 180° phase shift in the modulating signal and makes no practical difference.

Example 8.2 For the transmitter in Figure 8.5, calculate the required frequency deviation of the oscillator if the output frequency is to deviate by 75 kHz.

Solution In this transmitter, a modulated oscillator operating at 5 MHz has its frequency multiplied by 18, just as in the previous example. Then the signal is mixed back down to 5 MHz and multiplied another 18 times. The net multiplication is 18 for the carrier frequency, but the deviation has been multiplied by $18 \times 18 = 324$. A deviation of 75 kHz at the antenna would require the oscillator frequency to deviate by only (75 kHz)/324 = 232 Hz.

8.2.3 Automatic Frequency Control Systems

As an alternative to the use of crystal-controlled oscillators and a great deal of frequency multiplication, it is possible to achieve a reasonable amount of deviation at the oscillator frequency by using an *LC* oscillator operated as a VCO. As mentioned before, however, stability is a serious problem with such oscillators, and some form of AFC is always required.

Figure 8.6
The Crosby AFC system

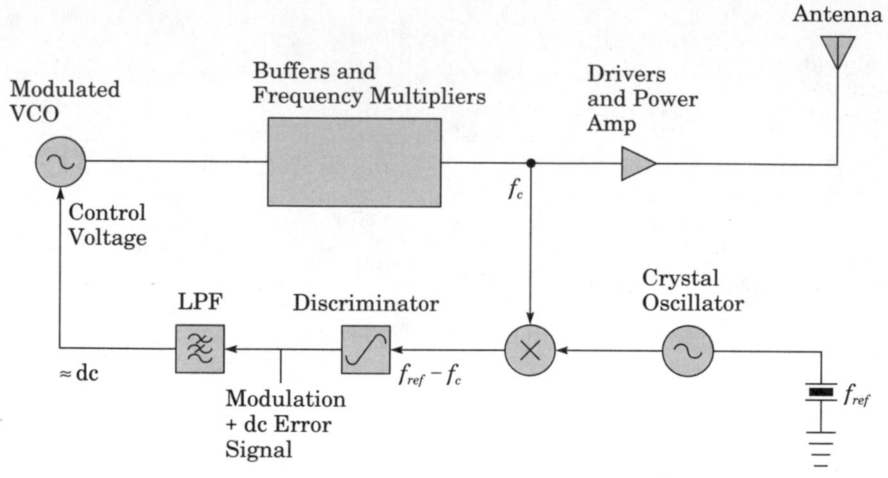

The classical way to do this is known as the **Crosby system**, and it is illustrated in Figure 8.6. Essentially, the transmitter carrier frequency (or some submultiple of it) is mixed with a crystal-controlled reference signal. The difference frequency is sent to a frequency discriminator. The operation of a discriminator circuit will be discussed in connection with receivers; for now, all we need to know is that it produces an output voltage proportional to the difference between the frequency at its input and the center frequency to which the circuit has been tuned. The center frequency of the discriminator is set so that the transmitter will have the correct carrier frequency when the mixer output is at that frequency.

The output of the discriminator is sent to a low-pass filter so that rapid changes of voltage due to modulation will be removed, leaving only dc levels due to slow drift of the carrier oscillator away from its correct frequency. When the carrier is at exactly the correct frequency, the output voltage from the discriminator is zero. The dc level is applied to the carrier oscillator, where it causes the oscillator frequency to move toward the correct value. The whole network thus exhibits negative feedback. The AFC circuit does not give perfect stability, because the control voltage is present only when there is an error between the oscillator frequency and the correct value. In addition, the discriminator itself contains an *LC* tuned circuit whose components are subject to some variation.

Example 8.3

A Crosby-type FM transmitter has the block diagram shown in Figure 8.6. The transmitter is to have a carrier frequency of 99.9 MHz. The crystal oscillator has a frequency of 105 MHz. What should the center frequency of the discriminator be?

Solution

The difference frequency output from the mixer will be

$$\begin{aligned} f_{ref} - f_c &= 105.0 \text{ MHz} - 99.9 \text{ MHz} \\ &= 5.1 \text{ MHz} \end{aligned}$$

This is the frequency to which the discriminator should be tuned. By the way, 105 MHz is not a very practical frequency for a crystal oscillator. Frequency multipliers would actually have to be used between the oscillator and the mixer. Alternatively, the transmitter signal could be brought out to the mixer at a lower frequency, partway along the chain. (See Problem 26 at the end of this chapter for a situation like this.)

8.2.4 Phase-Locked Loop FM Generators

Another, more modern way to solve the stability problem is to make the modulator VCO part of a phase-locked loop (PLL). See Figure 8.7 for an example of this method of stabilization. As in a frequency synthesizer, the VCO is locked to some multiple *N* of a crystal-controlled reference frequency. In fact, the VCO can be part of a frequency synthesizer for

Figure 8.7
PLL FM transmitter

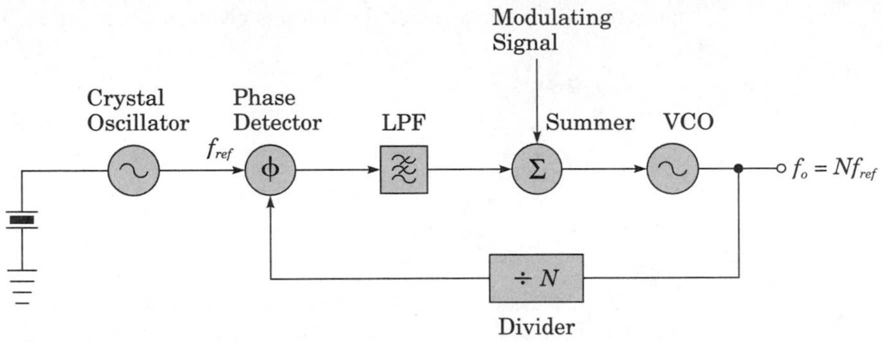

the carrier frequency. The loop filter allows the system to ignore the rapid variations of frequency associated with modulation, while preventing drift of the center frequency of the VCO away from its nominal value. Unlike the Crosby AFC system, the PLL FM transmitter is capable of locking the transmitting frequency exactly to the crystal-controlled reference frequency. With a well-designed VCO, the PLL system is also capable of producing wideband FM without frequency multiplication. These advantages make the use of PLLs very popular in new designs for both communications and broadcast transmitters.

The modulator sensitivity with this system is determined by the VCO. The VCO in a PLL has a constant k_f that is defined as

$$k_f = \frac{\Delta f}{\Delta v} \tag{8.2}$$

where

k_f = VCO sensitivity in hertz per volt
Δf = change in frequency
Δv = change in control voltage required for the change in frequency

It is no coincidence at all that k_f was also used in Chapter 7 for the deviation sensitivity of a frequency modulator and was defined in a similar way. In fact, k_f for the VCO becomes k_f for the modulator when a VCO is used as a frequency modulator.

Example 8.4

A PLL FM generator has the block diagram shown in Figure 8.7, with f_{ref} = 100 kHz, N = 200, and k_f = 50 kHz/V.

(a) Calculate the carrier frequency of the output signal.
(b) What RMS modulating voltage will be required for a deviation of 10 kHz at the carrier frequency?

Solution (a)

$$\begin{aligned} f_c &= Nf_{ref} \\ &= 200 \times 100 \text{ kHz} \\ &= 20 \text{ MHz} \end{aligned}$$

(b)

$$\begin{aligned} k_f &= \frac{\delta}{V_p} \\ V_p &= \frac{\delta}{k_f} \\ &= \frac{10 \text{ kHz}}{50 \text{ kHz/V}} \\ &= 0.2 \text{ V} \end{aligned}$$

This is the peak voltage, since deviation is given as a peak value. The RMS voltage will be

$$V_{RMS} = \frac{V_p}{\sqrt{2}} = \frac{0.2 \text{ V}}{\sqrt{2}} = 0.141 \text{ V} = 141 \text{ mV}$$

Many, perhaps most, FM transmitters now use this technique, which makes frequency multiplication unnecessary. The modulator is simply followed by enough stages of amplification to achieve the desired power output.

8.2.5 Indirect-FM Modulators

We devoted considerable time in Chapter 7 to showing that FM and PM are closely related. To review, the modulation index in frequency modulation with a sinusoidal modulating signal is

$$m_f = \frac{\delta}{f_m}$$

where

δ = maximum frequency deviation
f_m = frequency of the modulating signal
m_f = modulation index, and also the maximum phase deviation in radians

Therefore, there is more phase deviation in an FM signal for lower modulating frequencies. Frequency modulation can be produced using a phase modulator if the modulating signal is passed through a suitable low-pass filter before it reaches the modulator, so that lower modulating frequencies will produce greater phase deviation.

For readers familiar with the terminology of calculus, another way to look at the situation is to note that since phase is essentially the integral of frequency, it is possible to integrate the baseband signal and then apply it to a phase modulator. The resulting signal will be exactly the same as if the original baseband signal had been applied to a frequency modulator. An integrator circuit is in fact a low-pass filter, so the two explanations lead to the same result. Figure 8.8 shows the general idea of the *indirect FM* transmitter.

Figure 8.8 Indirect FM transmitter

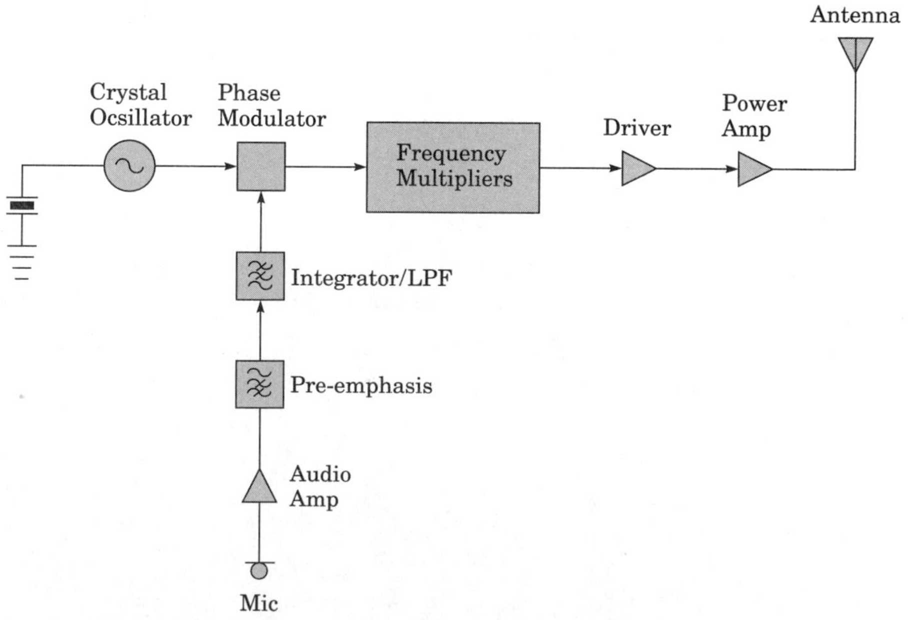

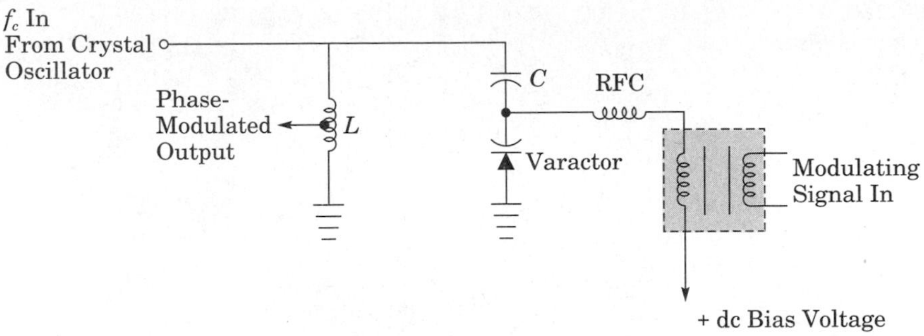

Figure 8.9
Phase modulator

One reason for using indirect FM is that it is easier to change the phase than the frequency of a crystal oscillator. Figure 8.9 shows one circuit that can do this. The changing voltage across the varactor (due to the modulating signal) changes the resonant frequency of the tuned circuit. This causes the constant-frequency output from the crystal oscillator to be alternately below and above the changing resonant frequency of the tuned circuit. The phase shift due to the tuned circuit will be zero at resonance and will vary from that value for a nonresonant condition. Thus, phase modulation is achieved without any interference with the stability of the crystal oscillator.

Unfortunately, the amount of phase shift that can be achieved in this way is relatively small. With more than a few degrees of phase shift, the variation of phase with frequency becomes nonlinear, and the variation of varactor capacitance with voltage is not linear either. To avoid serious distortion of the modulating signal, the phase shift must be small. This means that only a small modulation index is possible, and frequency multipliers will be needed with this system.

8.3 FM Receivers

Although FM receivers are similar to AM receivers in basic design, this discussion will emphasize the differences, which are found mainly in the IF and detector stages. Of course, the demodulator must be based on different principles from that in an AM receiver, since it must respond to changes in frequency rather than amplitude. In addition, the IF amplifiers need not respond linearly to amplitude changes; in fact, it would be better if they did not, for the purpose of reducing the effect of noise. Those in an AM receiver, on the other hand, must be as linear as possible, since any amplitude distortion in the RF or IF stages appears directly as distortion of the demodulated baseband signal.

8.3.1 Demodulators

The FM demodulators must convert frequency variations of the input signal into amplitude variations at the output. To demodulate FM properly, the amplitude of the output must be proportional to the frequency deviation of the input. This results in a characteristic *S-curve* for many FM detectors (see Figure 8.10). The output voltage is proportional to frequency deviation over a range at least equal to 2δ, since the deviation is the distance the signal frequency moves above and below the carrier frequency. Beyond that, the shape of the curve is not important. This type of curve explains why the tuning is relatively critical with FM receivers: once the detector begins to operate in the nonlinear portion of the S-curve, severe distortion results.

The sensitivity of an FM detector can be given as

$$k_d = \frac{V_o}{\delta} \tag{8.3}$$

where

k_d = detector sensitivity in volts per hertz
V_o = output voltage
δ = frequency deviation required for the output voltage

Figure 8.10
S-curve characteristic of FM detectors

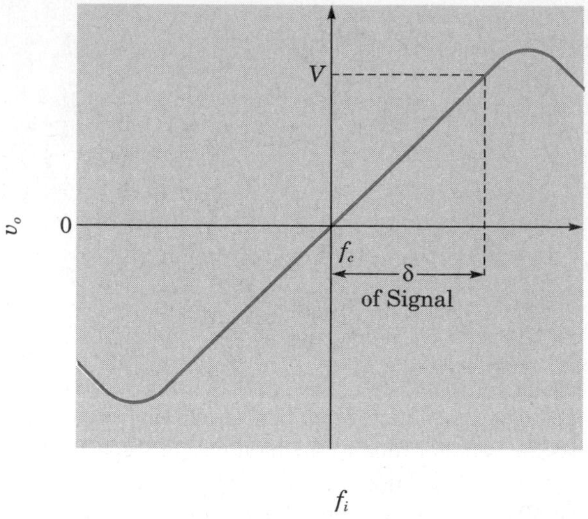

Note the similarity to the sensitivity of a frequency modulator, which is given in hertz per volt. The sensitivity of a detector is the slope of the straight-line portion of the S-curve of Figure 8.10.

Example 8.5 An FM detector produces a peak-to-peak output voltage of 1.2 V from an FM signal that is modulated to 10 kHz deviation by a sine wave. What is the detector sensitivity?

Solution First, the peak-to-peak voltage must be changed to peak, since that is the way deviation is specified.

$$V_o \text{ peak} = \frac{V_o \text{ peak-to-peak}}{2} = \frac{1.2 \text{ V}}{2} = 0.6 \text{ V}$$

Now the detector sensitivity is

$$k_d = \frac{0.6 \text{ V}}{10 \text{ kHz}} = 60\ \mu\text{V/Hz}$$

There are three major types of FM detectors. The Foster-Seeley **discriminator**, along with its variation, the **ratio detector**, are obsolescent but still very commonly found in older receivers and new receivers built to older designs. These detectors are effective but require a good number of discrete components, including a specially designed transformer, and must be adjusted by hand. More modern types of demodulators include the **quadrature detector** and the PLL, both of which are well adapted to IC construction.

Foster-Seeley Discriminator and Ratio Detector Both of these circuits convert frequency changes first to phase shifts and then to amplitude variations. The resulting AM signal is then demodulated by a combination of two diode detectors. The conversion from frequency to phase modulation is achieved by using the fact that the phase angle between voltage and current in a tuned circuit changes as the applied frequency goes through resonance. Recall that a series-tuned circuit is capacitive for frequencies below resonance, resistive at resonance, and inductive above resonance. At resonance, the current through the

Using an AM Receiver for FM: Slope Detection

Although hardly a practical system, the slope detector is perhaps the easiest type of FM detector to understand. Its operation can be demonstrated by tuning an AM receiver to an FM signal, or rather, mistuning it slightly. As shown in Figure 8.11, this mistuning can put the FM carrier frequency about halfway down the skirt of the IF amplifier frequency-response curve. In that case, as the frequency deviates in one direction, the amplitude of the output signal from the IF amplifier increases; in the other direction, it decreases. Frequency variations have thus produced amplitude variations, which can be demodulated by an envelope detector. Of course, the output is likely to be distorted, since the side of the IF amplifier frequency-response curve is not likely to be a straight line. This type of reception can be demonstrated by using a VHF receiver (such as a scanner) set to AM to receive FM signals (mobile-radio signals, for example).

Figure 8.11

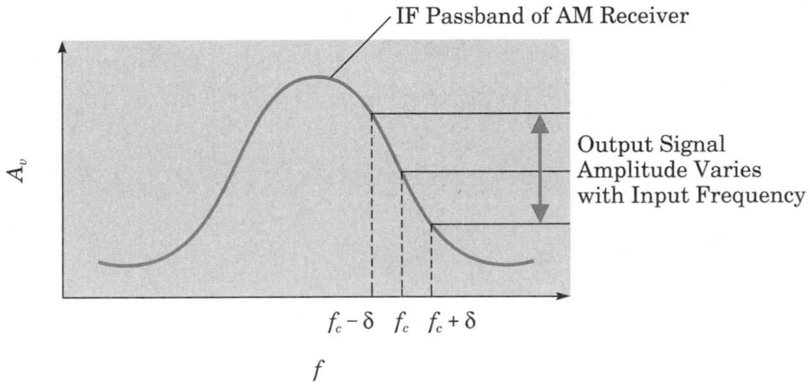

circuit is in phase with the voltage across it. Below resonance, the current leads the voltage, and above resonance, the current lags the voltage.

In the Foster-Seeley discriminator, shown in Figure 8.12(a), the transformer is double-tuned, with both the primary (L_1–C_1) and secondary (L_2–C_2) circuits resonant at the carrier frequency. The secondary voltage at resonance is 90° out of phase with the primary voltage. The primary voltage is also applied, in phase, to the center of the secondary winding through capacitor C_3, which is chosen to have low reactance at the carrier frequency. Capacitors C_4 and C_5 are chosen in the same way as the filters for AM detectors: they should have low reactance at the carrier frequency but high reactance at the modulating frequency, so that D_1–C_4 and D_2–C_5 act as peak detectors. Inductor L_3 is an RF choke with high reactance at the carrier frequency and low reactance at the modulating frequency. It can be seen that the low-reactance connection made by C_3 causes virtually the entire primary voltage to appear across L_3.

Figure 8.12(b) shows how the primary and secondary voltages add. The vector sum of the primary voltage and one-half the secondary voltage is applied to one diode detector (consisting of D_1 and C_4), and the vector difference between the primary voltage and one-half the secondary voltage goes to the other detector (D_2 and C_5). The output will be the difference between the rectified and filtered voltages at the output of the two detectors, so it will be the difference between the amplitudes of the two phasors representing the voltages applied to the two detectors.

When the incoming signal frequency is equal to the resonant frequency of the tuned circuits, the voltages applied to the two detectors are equal in magnitude. This can be seen by observing that the vectors $\mathbf{V}_4$ and $\mathbf{V}_5$ in Figure 8.12(b)(i) are the same length. Therefore, the net output voltage is zero. If the frequency increases, the secondary voltage has a leading phase angle, and the relative lengths of these vectors change. Vector $\mathbf{V}_4$ is now greater than $\mathbf{V}_5$, as shown in Figure 8.12(b)(ii). The output voltage becomes positive. Reducing the

Figure 8.12
Foster-Seeley discriminator

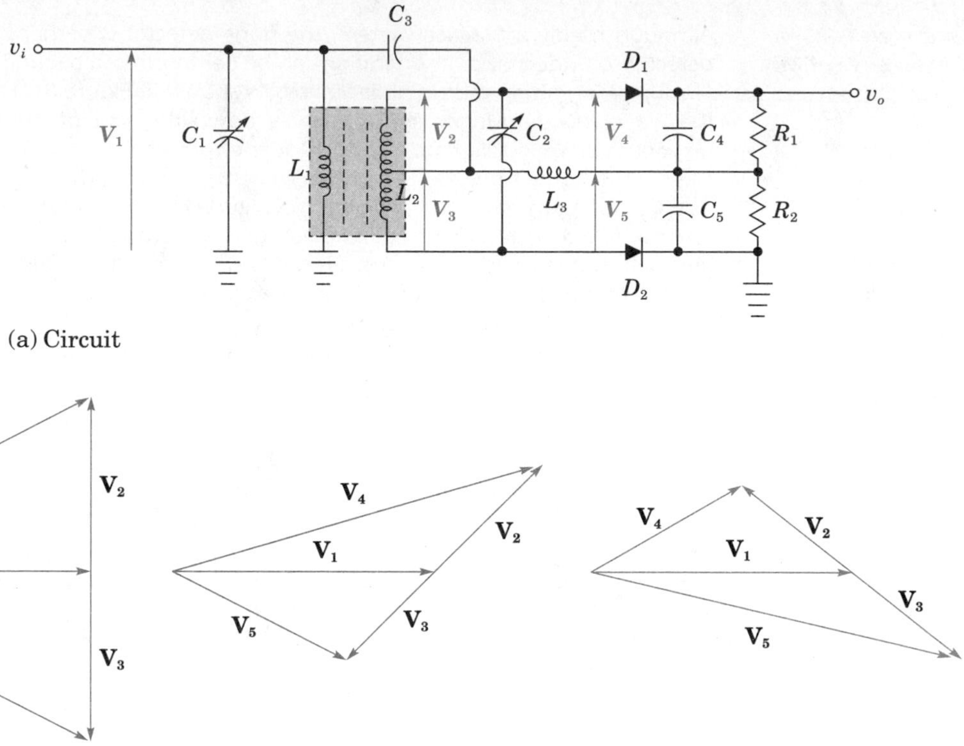

(i) Input frequency = f_c
(ii) Input frequency > f_c
(iii) Input frequency < f_c
(b) Phasor diagrams

frequency below resonance causes a similar but opposite phase change, causing $\mathbf{V_5}$ to be less than $\mathbf{V_4}$ and producing a negative output voltage. The result is an output voltage that follows the modulation.

A change in primary voltage amplitude (due to noise, for example) will also cause a change in output voltage. This undesirable effect can be reduced by using limiters before the detector. However, a detector that is less sensitive to amplitude variations would increase the effectiveness of the limiters.

The *ratio detector* is a variation on the discriminator. It greatly reduces sensitivity to amplitude variations, at the cost of a 50% reduction in output voltage. This 6 dB loss can easily be made up elsewhere in the receiver. Its circuit (see Figure 8.13) is similar to that of the discriminator but can be recognized at a glance by the fact that one of the diodes has been reversed, so that the two outputs across C_4 and C_5 add rather than subtract. Rather than going to the output, the sum of the two voltages is applied to an additional capacitor

Figure 8.13
Ratio detector

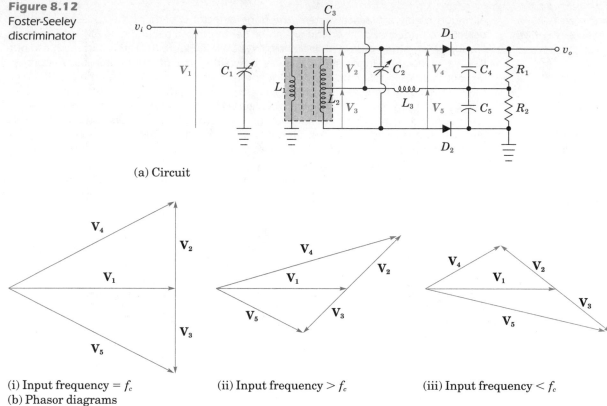

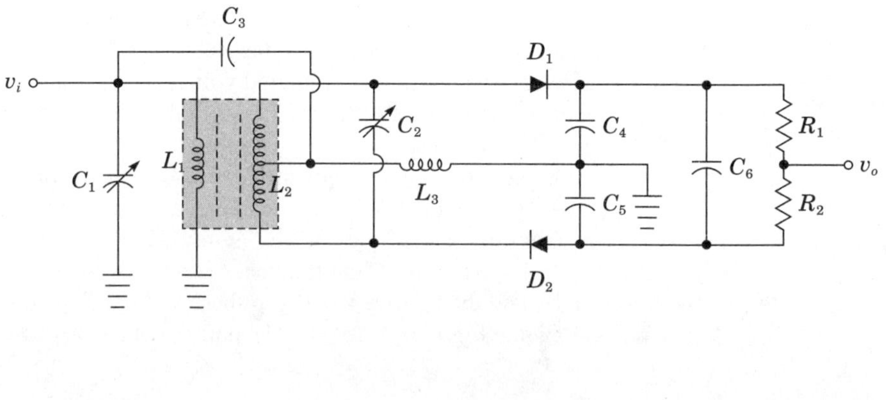

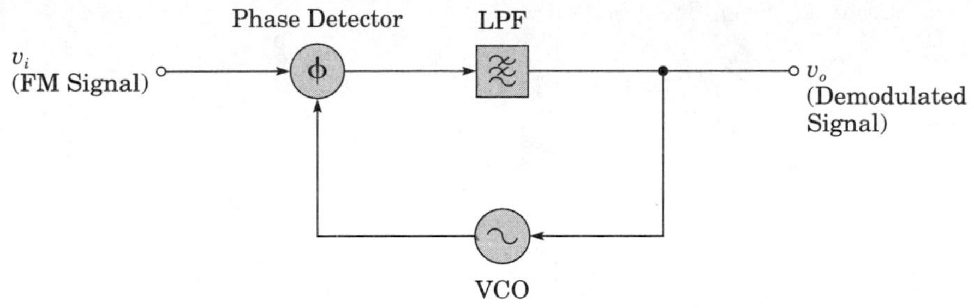

Figure 8.14
PLL FM detector

shown as C_6 in the diagram. This capacitor has a long discharge time constant in combination with R_1 and R_2 and provides a reference level. Its voltage changes only with long-term variations in signal strength and not with brief amplitude variations, such as those due to noise. As the vectors change in the way described above, the ratio between the two voltages produced by the diode detectors across C_4 and C_5 changes, but their sum is fixed by the capacitor voltage.

The output of this circuit cannot be taken between the two diodes as it is for the discriminator, since this voltage does not vary with modulation. Instead, it is taken between the center point of the load resistor and the junction of C_4 and C_5. The output voltage will be one-half of the difference between the voltages across C_4 and C_5.

PLL Method The use of a PLL to demodulate FM signals is very straightforward. See Figure 8.14 for a typical PLL detector. The incoming FM signal is used to control the frequency of the VCO. As the incoming frequency varies, the PLL generates a control voltage to change the VCO frequency, which follows that of the incoming signal. This control voltage varies at the same rate as the frequency of the incoming signal, and so it can be used directly as the output of the circuit. Unlike the PLLs used in transmitter modulator circuits, this PLL must have a short time constant so that it can follow the modulation. The capture range of the PLL is not important, since the free-running frequency of the VCO is set equal to the signal's carrier frequency at the detector (that is, to the center of the IF passband). The lock range must be at least twice the maximum deviation of the signal. If it is deliberately made wider, the detector will be able to function in spite of a small amount of receiver mistuning or local oscillator drift. Amplitude variations in the input signal do not affect the operation of this detector, unless they are so great that it stops working altogether.

It is easy to calculate the output voltage for a PLL detector, provided k_f for the VCO is known. Since the loop stays locked as it follows the modulation, the VCO frequency follows the signal frequency, varying over the range from $(f_c - \delta)$ to $(f_c + \delta)$. Therefore, the peak output voltage is the voltage necessary to move the local oscillator by the amount of the deviation δ. That is,

$$V_o \text{ peak} = \frac{\delta}{k_f} \tag{8.4}$$

where

V_o = output voltage from the detector
δ = deviation of the signal in hertz
k_f = VCO proportionality constant in hertz per volt

Example 8.6

A PLL FM detector uses a VCO with k_f = 100 kHz/V. If it receives an FM signal with a deviation of 75 kHz and sine-wave modulation, what is the RMS output voltage from the detector?

Solution From Equation (8.4),

$$V_o \text{ peak} = \frac{\delta}{k_f} = \frac{75 \text{ kHz}}{100 \text{ kHz/V}} = 0.75 \text{ V}$$

For sine-wave modulation,

$$V_o \text{ RMS} = \frac{V_o \text{ peak}}{\sqrt{2}} = \frac{0.75 \text{ V}}{\sqrt{2}} = 0.53 \text{ V}$$

Quadrature Detector Like the PLL detector, the *quadrature detector* is adapted to integrated circuitry. Like the discriminator, it uses the phase shift in a resonant circuit. In a quadrature detector (see Figure 8.15), the incoming signal is applied to one input of a phase detector. The signal is also applied to a phase-shift network that consists of a capacitor (C_1 in the figure) with high reactance at the carrier frequency and causes a 90° phase shift (this is the origin of the term *quadrature*). The tuned circuit consisting of L_1 and C_2 is resonant at the carrier frequency. Therefore, it causes no phase shift at the carrier frequency but does cause a phase shift at other frequencies that add to or subtract from the basic 90° shift caused by C_1.

The output of the phase-shift network is applied to the second input of the phase detector. When the input frequency changes, the angle of phase shift in the quadrature circuit also varies, as the resonant circuit becomes inductive or capacitive. The output from the phase detector varies at the signal frequency but has an average value proportional to the amount by which the phase angle differs from 90°. Low-pass filtering the output recovers the modulation. In Figure 8.15 this function is accomplished by a simple first-order filter consisting of R_2 and C_3. As is usual with detector low-pass filters, the cutoff frequency should be well above the highest modulating frequency and well below the receiver intermediate frequency.

The phase detector can be an analog multiplier (product detector) or a digital gate (either an AND or an exclusive-OR gate).

8.3.2 Limiters

FM obtains some immunity to noise because amplitude variations of the signal do not carry information. This advantage is lost if the receiver responds to variations in signal amplitude. Some detectors, like the ratio detector and PLL, are inherently insensitive to ampli-

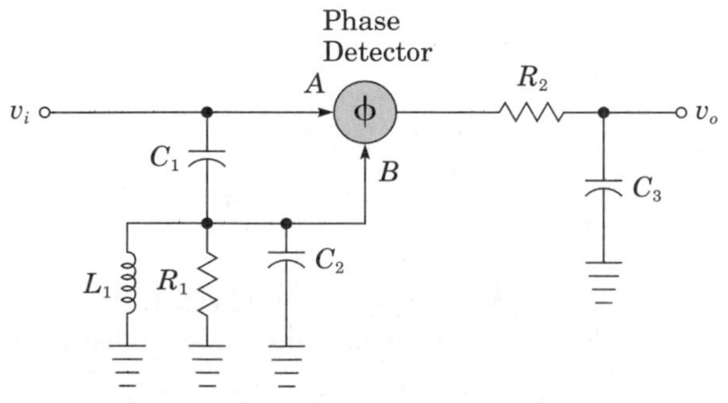

Figure 8.15 Quadrature FM detector

tude variations. The Foster-Seeley discriminator and quadrature detector, on the other hand, respond to changes in amplitude as well as in frequency.

Since only the frequency of the signal carries information, distortion of the waveform is of no consequence. In our study of Class C amplifiers, we noticed that the output signal level from such amplifiers is almost independent of the input signal amplitude, provided it is above the level needed to drive the transistor into saturation on peaks. It thus might seem that just as Class C amplifiers are satisfactory in FM transmitters, they might be useful as limiters in the receiver IF section.

In fact, a receiver limiter is usually designed to operate Class A with small signals and switch to Class C as the signal becomes larger. That is, it is biased in the active region, unlike a Class C amplifier, which is biased beyond cutoff.

The operating range is deliberately made narrow, however, so that large signals drive the transistor into saturation on one signal peak and cutoff on the other. This allows the receiver amplifier to amplify weak signals, whereas the straight Class C amplifier would not respond at all to these signals and would reduce the sensitivity of the receiver.

Figure 8.16 shows the difference between a straightforward Class C amplifier and a typical limiter stage. A Class C amplifier, as described in Chapter 2, is shown in Figure 8.16(a). With no signal, there is no transistor bias. The presence of a signal with sufficient peak voltage to cause base current to flow causes C_B to charge and put a negative bias on the base, which ensures that the transistor continues to conduct only on peaks. With a weak signal, the transistor is never driven into conduction, and there is no output.

The limiter in Figure 8.16(b) is very similar. The designation of R_B has been changed to R_{B_2}, and resistors R_{B_1} and R_E (bypassed by C_E) have been added to form a more conventional bias circuit that causes the transistor to conduct with no signal. Capacitor C_B is still present, so that large signals cause a negative bias as before, causing the transistor to be cut off except on signal peaks. In addition, resistor R_c in the collector circuit limits the current flow in the circuit, making it easier for the signal to saturate the stage. Of course, this resistor also reduces the efficiency of the amplifier, but that is not important in a receiver.

Figure 8.17(a) shows the transfer curve for a limiter. For an input voltage less than the threshold value, the output is proportional to input. Above the threshold, the output rises hardly at all. Another way of saying this is that the stage gain reduces sharply above the threshold. Since the output remains constant for input levels above the threshold, the gain, defined as the ratio of output to input voltage, decreases as the input increases. Fig-

Figure 8.16
Class C amplifier and limiter

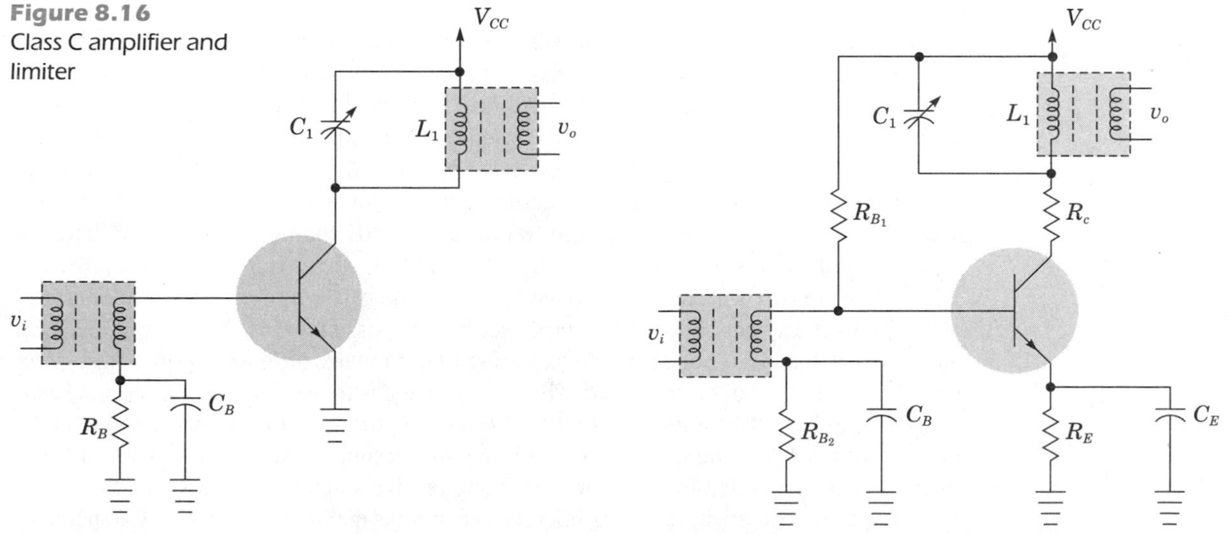

(a) Class C amplifier

(b) Typical limiter stage

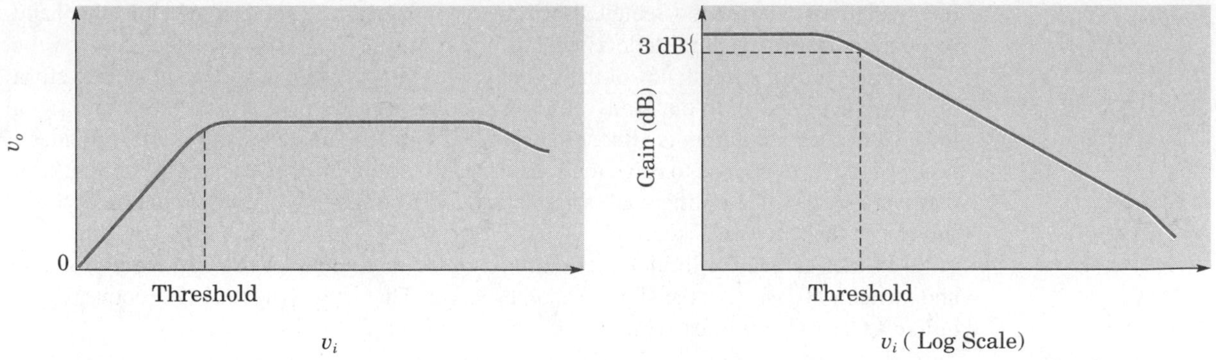

(a) Transfer curve

(b) Variation of gain with input-signal level

Figure 8.17
Limiters

ure 8.17(b) illustrates this idea. The threshold point is often defined as the input signal level where the gain is 3 dB down from its small-signal value.

For very large input signals, the output is actually reduced, because the transistor conducts for only a very short time during each cycle. This can be prevented by applying automatic gain control (AGC) to an earlier stage in the receiver. If a sharper transition is required, two or more limiting amplifiers can be used in cascade.

In many modern receiver designs, the limiting function is built into a specially designed FM IF integrated circuit. In fact, commonly available ICs include both the IF amplifier stages and a quadrature FM detector. The CA3089 is one such circuit. A block diagram of its internal structure is shown in Figure 8.18, along with some of its specifications.

It is even possible to obtain an entire FM receiver, except for a few passive components, in a single IC. The MC3362 (see Figure 8.19), for instance, is a complete double-conversion receiver on a chip.

One characteristic of limiting amplifiers is that they have large gain for small signals. This can be quite annoying, and for that reason *squelch* (called *muting* in broadcast receivers) is almost universal in FM receivers. Turning off the muting on an FM broadcast receiver (if this is possible) will demonstrate the effect of limiting. The loud hiss that is heard between stations is amplified noise. The hiss disappears when a strong station is received and the gain of the limiters is reduced. Squelch systems were described in connection with AM receivers. The same techniques are used with FM.

8.3.3 Automatic Frequency Control Systems

As in any receiver, it is important that the local oscillator frequency be stable so that the receiver does not drift away from the station to which it is tuned. A small amount of drift can cause the detector to operate in the nonlinear portion of its S-curve, resulting in distortion; a greater amount of mistuning will cause reception to be lost altogether. The need for accurate tuning with FM is generally not quite as critical as it is for SSB, but it is more critical than for AM. The problem of stability is made more difficult by the fact that FM is generally transmitted at relatively high frequencies (VHF and up). The *LC* oscillators at VHF and UHF are not likely to have sufficient stability.

There are two basic solutions to the problem. The first is to use some form of crystal control. Direct crystal control of the local oscillator is possible when the receiver has from one to a few channels. For receivers that must tune to many different frequencies, a frequency synthesizer can be employed. The other method is to use automatic frequency control (AFC) to keep the frequency of an *LC* local oscillator close to the correct value in the presence of a received signal. Most modern receiver designs use frequency synthesis, rendering AFC unnecessary, but there are still many receivers that incorporate AFC.

The AFC circuits in receivers operate in a similar manner to the Crosby transmitter system already discussed. One major difference is that there is no need to add a discriminator, since the receiver's FM detector circuit serves that purpose.

Figure 8.18
The CA3089
Reprinted with permission of Harris Semiconductor.

Features

- **For FM IF Amplifier Applications in High-Fidelity, Automotive, and Communications Receivers**
- **Includes: IF Amplifier, Quadrature Detector, AF Preamplifier, and Specific Circuits for AGC, AFC, Muting (Squelch), and Tuning Meter**
- **Exceptional Limiting Sensitivity .... 12μV (Typ.) @ -3dB Point**
- **Low Distortion: (with Double-Tuned Coil) .... 0.1% (Typ.)**
- **Single-Coil Tuning Capability**
- **High Recovered Audio ....400mV (Typ.)**
- **Provides Specific Signal for Control of Interchannel Muting (Squelch)**
- **Provides Specific Signal for Direct Drive of a Tuning Meter**
- **Provides Delayed AGC Voltage for RF Amplifier**
- **Provides a Specific Circuit for Flexible AFC**
- **Internal Supply-Voltage Regulators**

Description

Harris CA3089E is a monolithic integrated circuit that provides all the functions of a comprehensive FM-IF system. Figure 1 is a block diagram showing the CA3089E features, which include a three-stage FM-IF amplifier/limiter configuration with level detectors for each stage, a doubly-balanced quadrature FM detector and an audio amplifier that features the optional use of a muting (squelch) circuit.

The advanced circuit design of the IF system includes desirable deluxe features such as delayed AGC for the RF tuner, and AFC drive circuit, and an output signal to drive a tuning meter and/or provide stereo switching logic. In addition, internal power supply regulators maintain a nearly constant current drain over the voltage supply range of +8.5 to +16 volts.

The CA3089E is ideal for high-fidelity operation. Distortion in a CA3089E FM-IF System is primarily a function of the phase linearity characteristic of the outboard detector coil.

The CA3089E utilizes the 16-lead dual-in-line plastic package and can operate over the ambient temperature range of -40°C to +85°C.

Pinout

CA3089E (PLASTIC DIP)
TOP VIEW

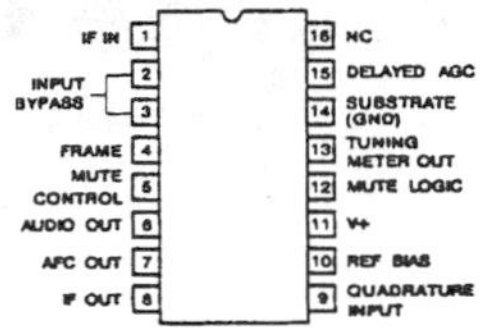

Block Diagram

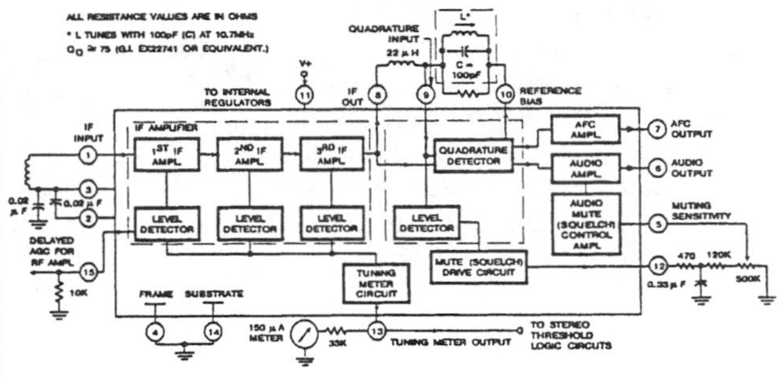

FIGURE 1. BLOCK DIAGRAM OF THE CA3089E

CAUTION: These devices are sensitive to electrostatic discharge. Proper I.C. handling procedures should be followed.

A block diagram for a receiver AFC is shown in Figure 8.20(a). The S-curve representing the discriminator output is shown in Figure 8.20(b). When the tuning is correct, the discriminator output varies symmetrically about the zero point, and there is no dc output. When the receiver is tuned too high, most of the signal-frequency variation is in the lower-frequency portion of the curve and there is a negative dc component to the output voltage. The opposite effect occurs for a receiver that is tuned to too low a frequency. This dc offset can be applied to a varactor diode in the frequency-determining circuit of the local oscillator in such a way that a negative voltage causes the frequency to which the receiver is tuned to decrease, and vice versa. This negative feedback results in the local oscillator moving to a frequency that is almost correct (though not exactly correct, since the needed control voltage for the oscillator can be generated only when there is a frequency error).

AFC is not needed with synthesized receivers, because their local oscillators are stable enough without it. It is found in virtually all receivers of reasonable quality that use VFO (rather than synthesizer) tuning. Since the effect of AFC is to *pull* the receiver to almost the correct frequency even when it is tuned a relatively long way from the station's frequency, AFC can make a receiver difficult to tune accurately. Therefore, AFC is sometimes made switchable, so that it can be disabled to allow precise tuning then switched on to counteract drift in the local oscillator circuit. Another reason for making the AFC switch-

MOTOROLA
■ SEMICONDUCTOR ■
TECHNICAL DATA

MC3362

Advance Information

**LOW-POWER
DUAL CONVERSION
FM RECEIVER**

**SILICON MONOLITHIC
INTEGRATED CIRCUIT**

LOW-POWER NARROWBAND FM RECEIVER

. . . includes dual FM conversion with oscillators, mixers, quadrature discriminator, and meter drive/carrier detect circuitry. The MC3362 also has buffered first and second local oscillator outputs and a comparator circuit for FSK detection.

- Complete Dual Conversion Circuitry
- Low Voltage: V_{CC} = 2.0 to 6.0 Vdc
- Low Drain Current (3.6 mA (Typ) @ V_{CC} = 3.0 Vdc)
- Excellent Sensitivity: Input Limiting Voltage —
 (− 3.0 dB) = 0.7 μV (Typ)
- Externally Adjustable Carrier Detect Function
- Low Number of External Parts Required
- Manufactured in Motorola's MOSAIC Process Technology
- See AN980 for Additional Design Information

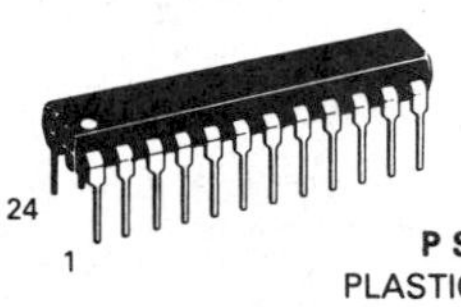

P SUFFIX
PLASTIC PACKAGE
CASE 724

DW SUFFIX
PLASTIC PACKAGE
CASE 751E
(SO-24L)

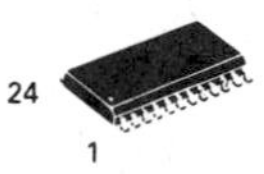

FIGURE 1 — TYPICAL APPLICATION IN A PLL FREQUENCY SYNTHESIZED RECEIVER

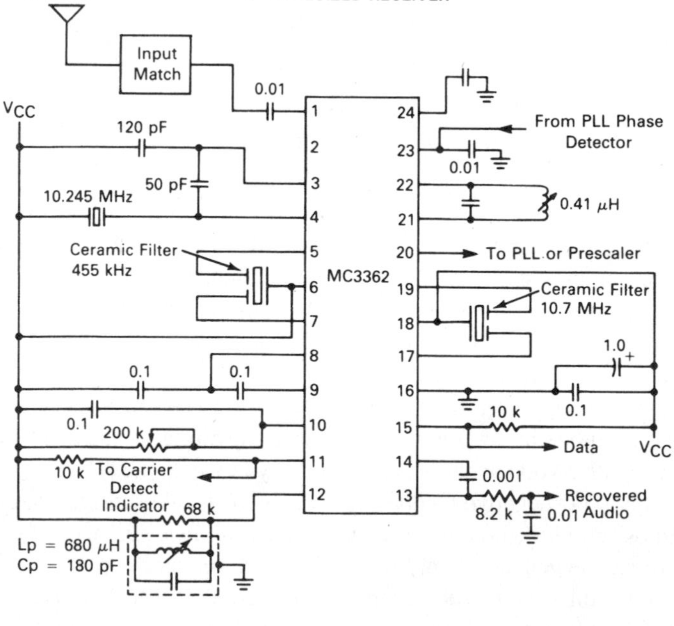

FIGURE 2 — PIN CONNECTIONS AND FUNCTIONAL BLOCK DIAGRAM

Figure 8.19
The MC3362
Copyright of Motorola.
Used by permission.

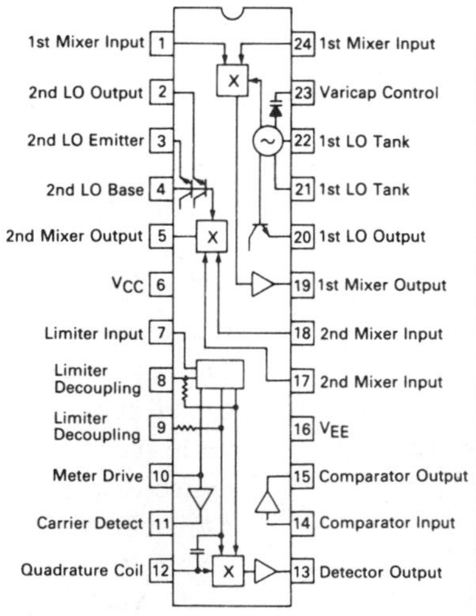

Figure 8.20
Receiver AFC System

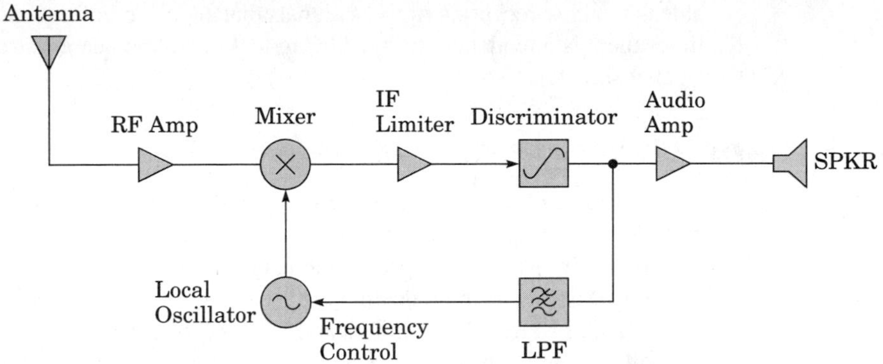

(a) Block diagram

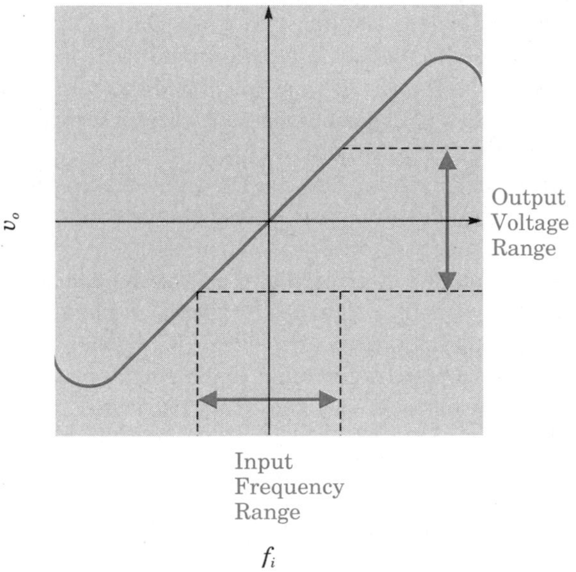

(i) Receiver tuned correctly

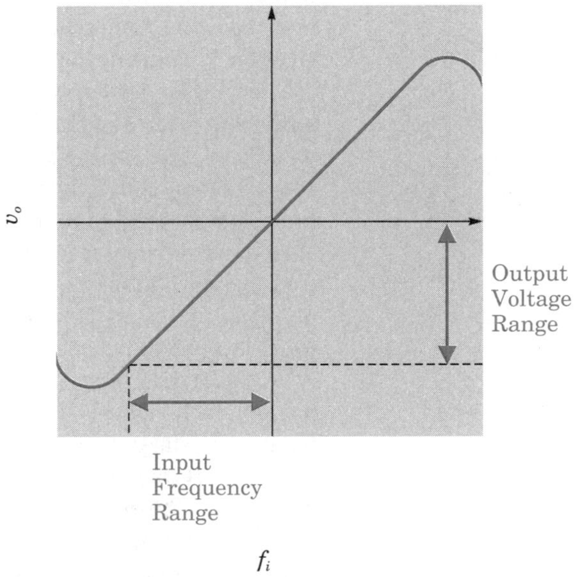

(ii) Receiver tuned too high

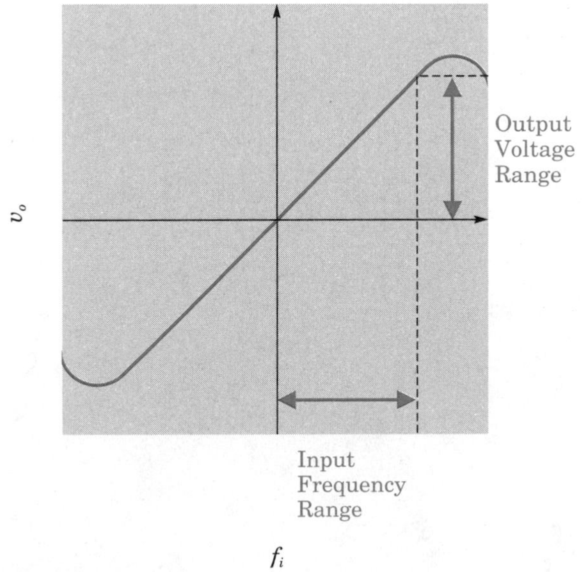

(iii) Receiver tuned too low

(b) Effect of mistuning on discriminator output

able is to allow reception of weak signals that are close in frequency to strong ones. Sometimes there is a tendency for the AFC to lock on to the nearest strong station, ignoring the weaker signals.

8.4 FM Transceivers

A major use for FM systems, like SSB systems, is in two-way communications, particularly mobile radio. In such applications, compactness and ease of installation and operation are important. Transmitter-receiver combinations called *transceivers* are therefore very common for FM voice communications.

Much FM mobile communication is done by means of *repeaters*, which are receiver-transmitter combinations connected to an antenna or set of antennas located well above local terrain, as illustrated in Figure 8.21. Communication between mobile units is by way of the repeater, which receives transmissions on one frequency and simultaneously retransmits them on another. For instance, the mobile units can transmit on frequency f_1 and receive on f_2, while the repeater receives on f_1 and simultaneously retransmits on f_2. The two frequencies are generally a few hundred kilohertz apart. It would be difficult to design the repeater to receive and transmit simultaneously on frequencies that were very close together without the receiver experiencing interference from the transmitter.

Consequently, while there is no need for the mobile transceivers (unlike the repeaters) to be able to both transmit and receive at the same time, it is often essential that they be able to transmit and receive on different, accurately determined frequencies. Often a single frequency synthesizer is used for the transmitter carrier oscillator and the receiver local oscillator; in that case, it must be able to move quickly from one frequency to another and to settle rapidly at the new frequency without instability.

Other than the main frequency synthesizer and some of the audio circuitry, there is not too much that can be shared between transmitter and receiver. The receiver may incorporate a crystal or ceramic filter, but there is no need to share it with the transmitter, as is done in SSB transceivers. The combination transmitter and receiver does offer convenience of operation, however.

Figure 8.22 shows a photograph and a block diagram of a typical FM communications transceiver. This one is used for VHF marine radio.

As in most modern designs, the central feature of the transceiver is a microprocessor-controlled frequency synthesizer. The same PLL controls two VCOs, one each for trans-

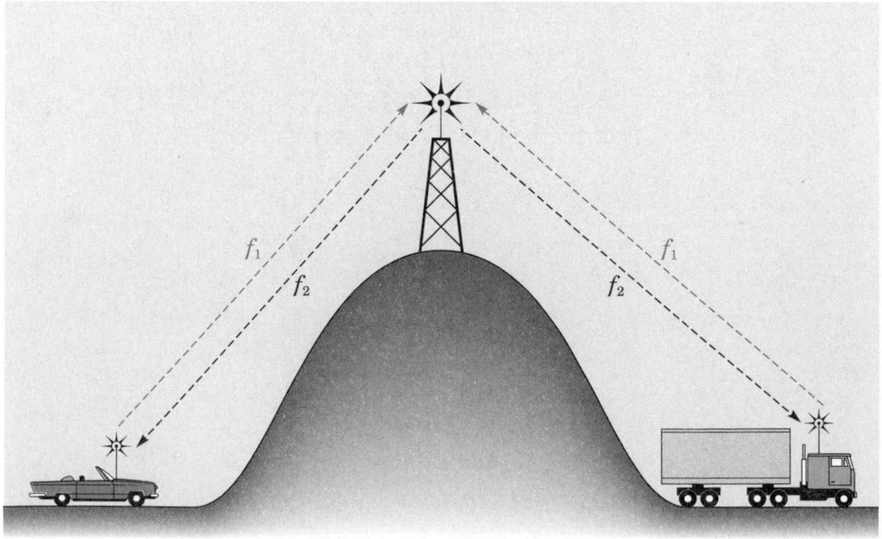

Figure 8.21
FM mobile radio system with repeater

mitting and receiving. The transmitting VCO is modulated directly at the carrier frequency; there are a total of four stages of transmitter amplification but no frequency multipliers. The transmitter output power is switchable between 1 W and 25 W, with the low-power setting being used for short-range communication, such as between boats in harbor.

The receiver is double conversion. There is one stage of RF amplification before the first mixer, which works with the receive VCO as its local oscillator. There is a crystal filter at the first IF of 16.9 MHz. A crystal-controlled local oscillator and second mixer lower the signal frequency to that of the second IF, which is the familiar 455 kHz. The second IF amplifier chain and the detector are contained in an IC. The receiver is also equipped with an adjustable squelch circuit.

8.5 FM Stereo Transmitters and Receivers

In the previous chapter, we looked at the signals associated with FM stereo but made no attempt to discuss hardware for producing or decoding these signals. For review, the baseband spectrum of an FM stereo signal is shown in Figure 8.23. The optional SCA signal has been removed for simplicity. Recall that this entire signal, spanning the frequency range from about 30 Hz to 53 kHz, is used to modulate the FM transmitter. No modifications to the transmitter are required, assuming that it is capable of handling this frequency range, as all FM broadcast transmitters now are. All the circuitry for generating the FM stereo signal is contained in the baseband (audio) portion of the system.

Figure 8.24 is a block diagram of a system that can be used to encode an FM stereo signal. Both channels are given pre-emphasis in the usual way. Then a simple matrix arrangement adds and subtracts the two signals. This process can easily be accomplished by a couple of operational amplifiers (op-amps). The L − R signal is applied to a balanced modulator, along with the 38 kHz subcarrier, to create a DSBSC signal. The L + R signal is passed through a delay network to compensate for the time taken for the L − R signal to pass through the balanced modulator, so that they arrive at the summing amplifier (summer) at the same time. In the summer, the L + R signal is added to the DSBSC L − R signal, and a 19 kHz pilot carrier is added as well. This last signal must be phase-coherent with the suppressed 38 kHz subcarrier, a requirement that is easily satisfied by deriving them from

Figure 8.22
VHF FM transceiver
Diagram and photo courtesy of Radio Shack, a division of Tandy Corporation.

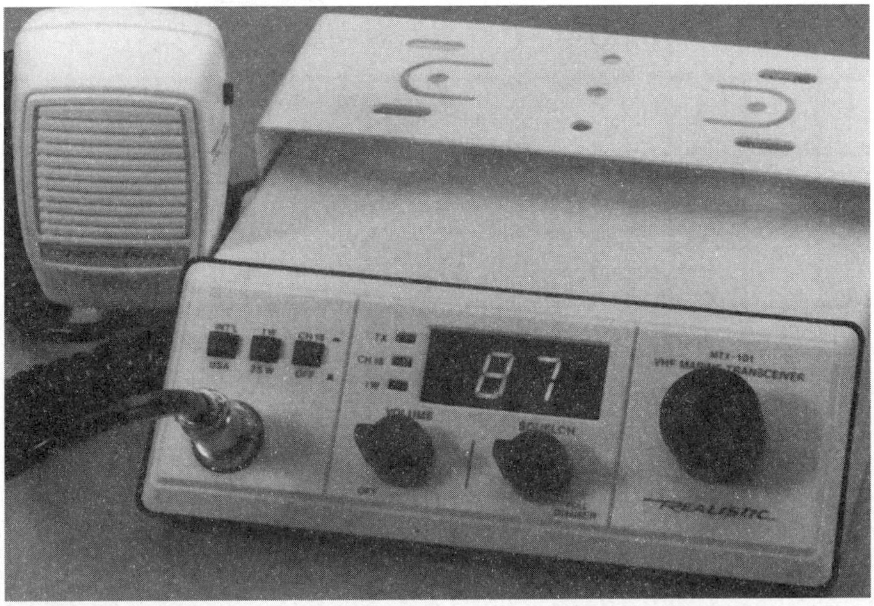

(a) Photograph

continues

Figure 8.22
Continued

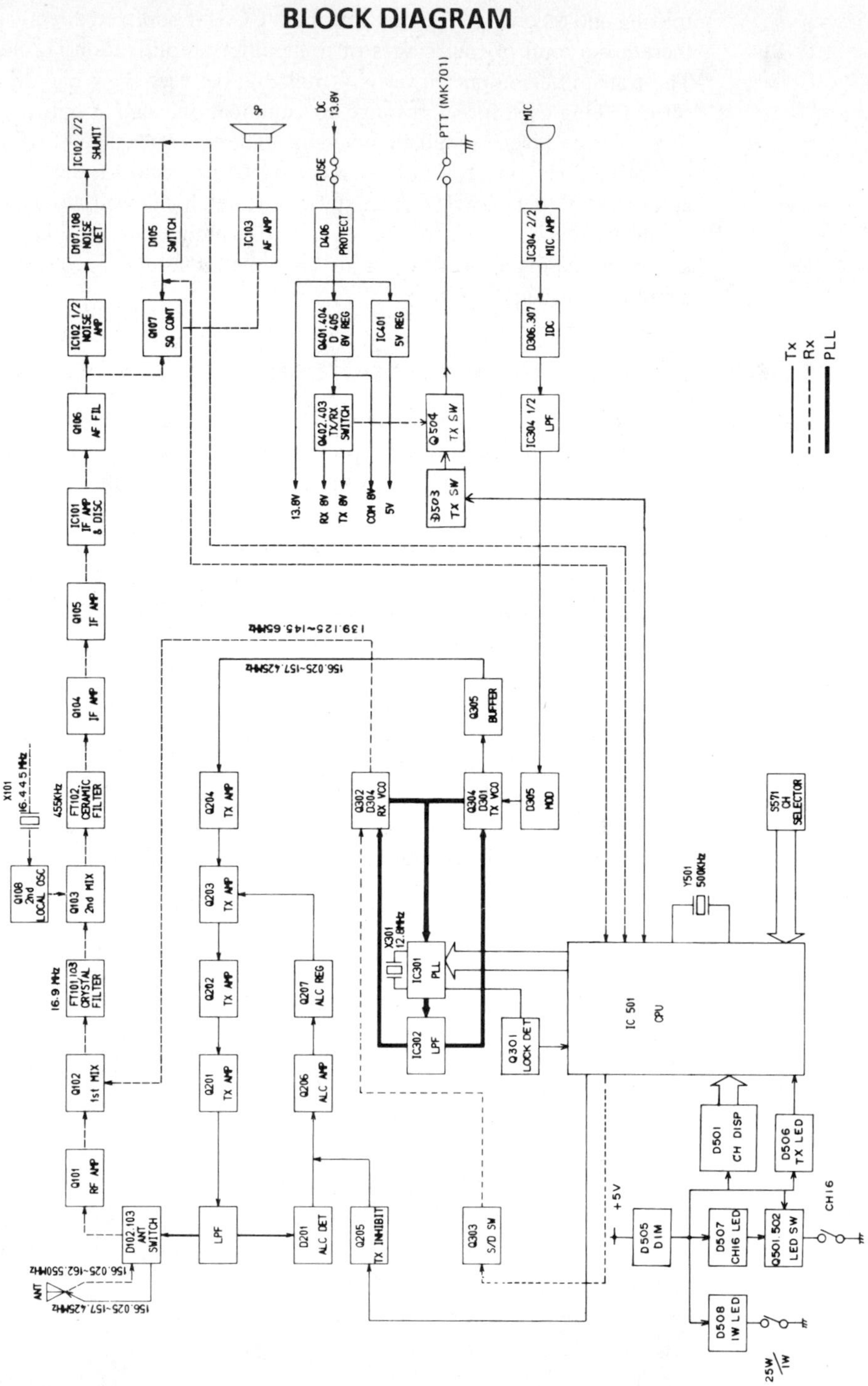

(b) Block diagram

the same source. Either the 38 kHz signal can be divided down to produce the 19 kHz pilot, as shown, or the 19 kHz signal can be sent through a frequency doubler to produce the 38 kHz subcarrier.

The entire baseband stereo signal is applied to the input of an FM broadcast transmitter. Figure 8.25 is a partial data sheet for a typical FM broadcast transmitter. Notice that the frequency response for the modulating signal is specified as flat to 53 kHz. Note also

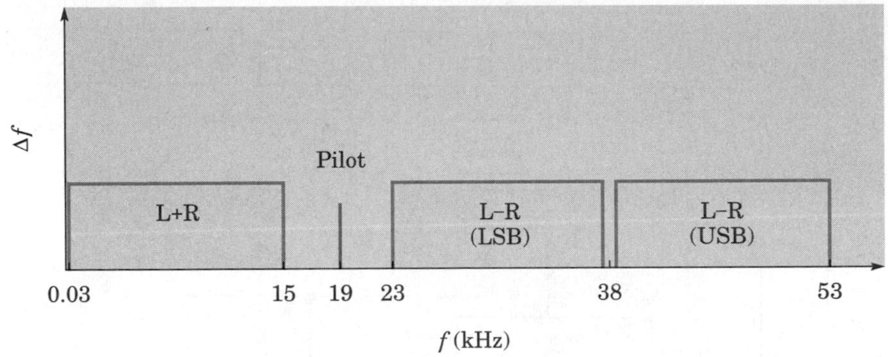

Figure 8.23 Baseband spectrum for FM stereo (without SCA)

that this transmitter uses direct FM at the operating frequency, followed by enough amplification to achieve its specified power output of 10 kW. All the amplification stages are solid-state, except for the power amplifier, which uses a tube. All-solid-state FM broadcast transmitters are also available, just as they are for AM.

Figure 8.26 shows one popular way to decode the stereo signal at the receiver. The stereo decoder receives a signal at a point after the FM detector in the receiver but before de-emphasis. The L+R and modulated L − R signals are easily separated by filtering, using a low-pass filter with a cutoff frequency of about 15 kHz for L+R and a bandpass filter with a passband extending from 23 to 53 kHz for the L − R signal. A small time delay is again added to the L+R signal to compensate for the time taken for the L − R signal to go through a demodulator.

The DSBSC L − R signal must be demodulated in a product detector before it can be used. To do this, the original 38 kHz subcarrier must be recovered. The 19 kHz pilot carrier can be recovered using a bandpass filter and then either sent through a frequency doubler or used to stabilize a PLL that provides the required subcarrier. The latter method is shown in Figure 8.26.

Once the L+R and demodulated L − R signals are available, they can simply be added and subtracted to produce left and right signals, which must then go through individual de-emphasis networks.

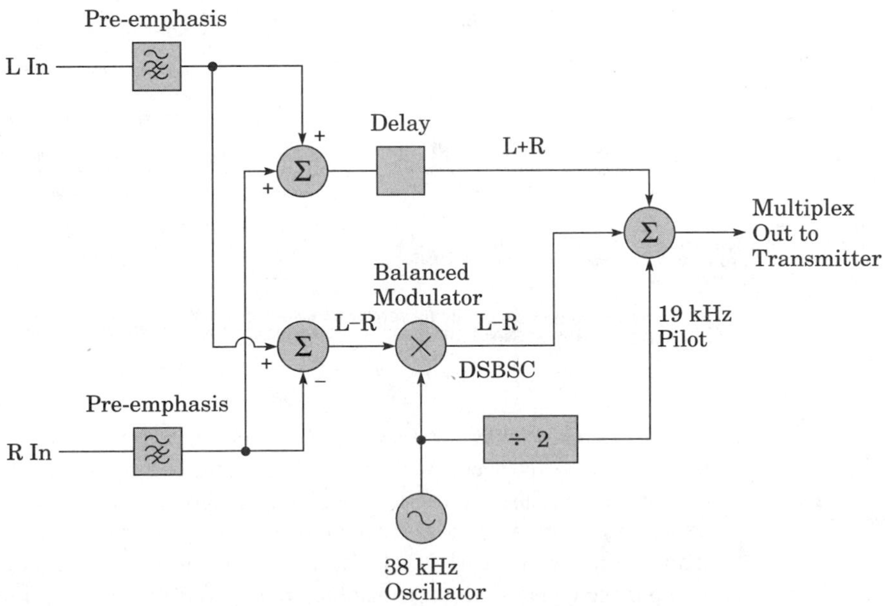

Figure 8.24 Generation of FM stereo signal

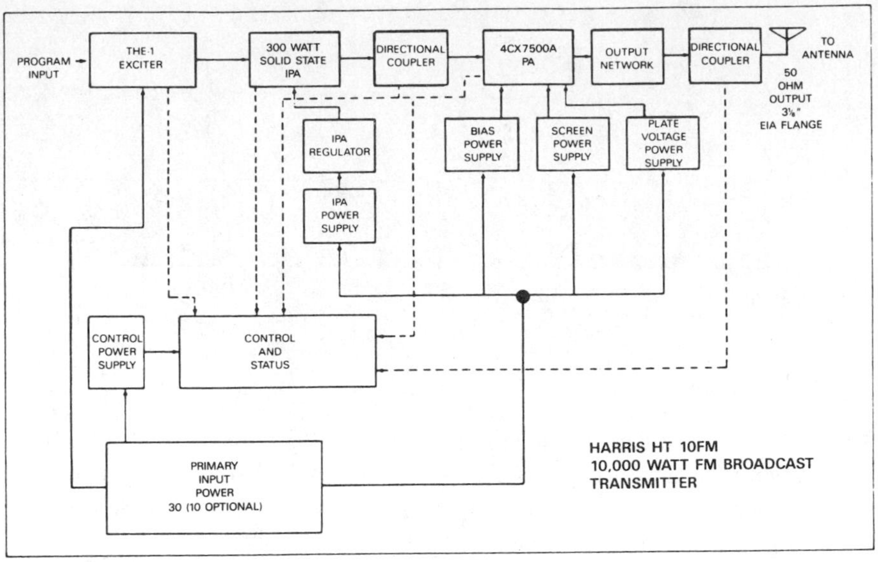

HARRIS HT 10FM SPECIFICATIONS

GENERAL

POWER OUTPUT: 5,000 to 10,000 watts (FCC type-notified range)
FREQUENCY RANGE: 87.5 to 108 MHz in 50 kHz steps. Tuned to single frequency.
EXCITATION: Harris THE-1 High Power FM Exciter
TYPE OF MODULATION: Direct carrier frequency modulation (DCFM)
MODULATION CAPABILITY: ±200 kHz
RF LOAD IMPEDANCE: 50 ohms
RF OUTPUT TERMINATION: 3⅛" EIA flange, female
PA MATCHING RANGE: 1.7:1 VSWR, maximum for full output power; automatic power reduction into high VSWR's
RF HARMONIC/SPURIOUS OUTPUT: Suppression meets or exceeds FCC/DOC/CCIR specifications
AC INPUT POWER: 197-250 VAC, 50/60 Hz, three phase, 3-wire closed delta or 360-415 VAC 4-wire WYE. OPTIONAL: 197-250 VAC, 50/60 Hz, single phase, 2-wire.
POWER CONSUMPTION: 15.7 kW typical at 10 kW RF output
POWER FACTOR: 0.95 (3-phase), 0.85 (1-phase)
AMBIENT TEMPERATURE RANGE: −20 to +50°C at sea level; derated 2°C/1000 ft altitude
MAXIMUM ALTITUDE: 10,000 ft (60 Hz), 7,500 ft (50 Hz)
MAXIMUM HUMIDITY: To 95% non-condensing
AIR REQUIREMENTS: 400 CFM (60 Hz), 350 CFM (50 Hz); no back pressure
CABINET SIZE: 33"W (84 cm) × 34"D (99 cm) × 72"H (183 cm)
WEIGHT/VOLUME: 1,125 lbs/46.7 cu. ft., domestic packed

WIDEBAND COMPOSITE OPERATION (STANDARD)

INPUTS: Two; one balanced, floating and one unbalanced.
INPUT IMPEDANCE: 2000 ohms resistive.
INPUT CONNECTORS: Female BNC (rear panel).
INPUT LEVEL: 1.0 volt RMS nominal for ±75 kHz deviation.
AMPLITUDE RESPONSE: ±0.1 dB, 30 Hz to 53 kHz; −0.2 dB at 100 kHz.
FM SIGNAL TO NOISE: 80 dB below 100% modulation (reference 400 Hz at ±75 kHz deviation with 75 microsecond de-emphasis, 20 Hz to 200 kHz bandwidth).
HARMONIC DISTORTION: 0.08%
INTERMODULATION DISTORTION: 0.02% (60 Hz/7 kHz 1:1 tone pair).
CCIF INTERMODULATION DISTORTION: All distortion products below 80 dB (reference 14 kHz/15 kHz test tone pair).

ASYNCHRONOUS AM SIGNAL TO NOISE: 55 dB below equivalent 100% amplitude modulation
SYNCHRONOUS AM SIGNAL TO NOISE: 50 dB below equivalent 100% amplitude modulation of output carrier with 75 microsecond de-emphasis (FM modulation ±75 kHz @ 400 Hz).
PHASE RESPONSE: +0.5/−1.0 degrees from linear phase, 20 Hz to 53 kHz.
TRANSIENT INTERMODULATION DISTORTION: 0.05%, 2.96 kHz square wave/14 kHz sine wave modulation.
TEST INPUT (FRONT PANEL): Nominal 1.0 volt for ±75 kHz deviation at 400 Hz (10,000 ohm input impedance, BNC female connector)
TEST OUTPUT (FRONT PANEL): Nominal 1.0 volt for ±75 kHz deviation at 400 Hz (200 ohm source impedance, BNC female connector)

MONAURAL OPERATION (STANDARD)

AUDIO INPUT IMPEDANCE: 600 ohms, balanced, resistive, transformerless.
AUDIO INPUT LEVEL: +10 dBm, ±1 dB for ±75 kHz deviation at 400 Hz.
AUDIO FREQUENCY RESPONSE: Standard 75 microsecond FCC pre-emphasis curve ±0.5 dB, 30 Hz-15 kHz. Selectable: flat, 25, 50 or 75 microsecond pre-emphasis.
HARMONIC DISTORTION: 0.08%, 30 Hz to 15 kHz, de-emphasized.
INTERMODULATION DISTORTION: 0.04%, 60 Hz/7 kHz test tone pair, 4:1 ratio.
CCIF INTERMODULATION DISTORTION: All distortion products down 70 dB (reference 14 kHz/15 kHz test tone pair).
TRANSIENT INTERMODULATION DISTORTION: 0.05%, 2.96 kHz square wave/14 kHz sine wave modulation
FM SIGNAL TO NOISE RATIO: At least 80 dB below 100% modulation (reference 400 Hz @ ±75 kHz deviation, measured 20 Hz to 200 kHz bandwidth, 75 microsecond de-emphasis).

SCA INPUTS (STANDARD)

EXTERNAL SCA GENERATOR INPUTS: Two
INPUT CONNECTORS: BNC female (rear panel).
INPUT IMPEDANCE: 10,000 ohms, unbalanced.
INPUT LEVEL: 0.1V (nominal) for 10% injection.
RANGE OF SUBCARRIER FREQUENCIES: 57 kHz to 92 kHz (25 kHz to 92 kHz in monaural operation).
AMPLITUDE RESPONSE: +0.1 dB, −0.2 dB; 20 kHz to 100 kHz.

SPECIFICATIONS ARE REFERENCED TO 10 kW OPERATION, EXCEPT WHEN NOTED.

HARRIS MAINTAINS A POLICY OF CONTINUOUS IMPROVEMENTS ON ITS EQUIPMENT AND THEREFORE RESERVES THE RIGHT TO CHANGE SPECIFICATIONS WITHOUT NOTICE.

Figure 8.25
Data sheet for the Harris FM broadcast transmitter

A stereo decoder generally has an output for a light or LED to indicate when a stereo signal is being received. A stereo signal is identified by the presence of the 19 kHz pilot carrier. There is also a provision for manual and/or automatic switching between mono and stereo reception. As previously mentioned, there is a noise penalty for stereo, so its use is inappropriate with weak signals. Recall that many stereo receivers automatically switch to mono if the signal strength is insufficient for good stereo reception. In addition, when the

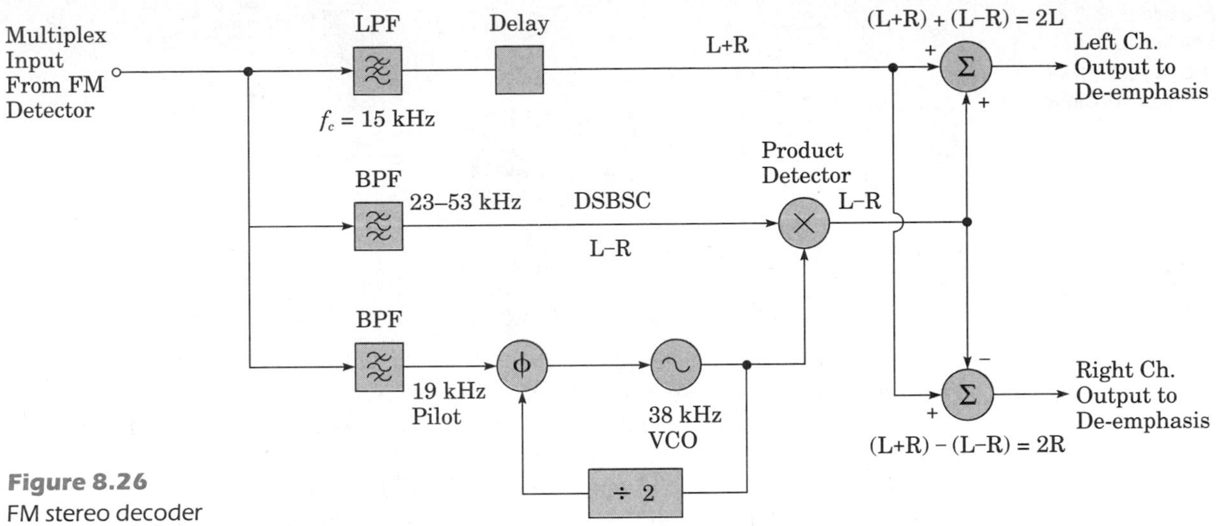

Figure 8.26
FM stereo decoder

S/N is marginal for stereo reception, some receivers are capable of operating in stereo at low and midrange audio frequencies, while blending the high-frequency audio components of the two channels. The theory here is that noise is more troublesome at high frequencies and that blending the two channels can reduce this noise, since noise signals that are different on the two channels tend to cancel each other. At the same time, the listener can experience some stereo effect from the lower frequencies.

Modern stereo decoders are generally contained, except for some filter and frequency-determining components, on a single IC. One example is the CA3090, which works in the way just described.

8.6 FM Receiver Measurements

Selectivity can be measured for FM receivers in a manner similar to that for AM. Measurement techniques for sensitivity are different, however. There are two popular methods for measuring the sensitivity of FM receivers. One of them, known as **usable sensitivity** or **SINAD** sensitivity, can also be used with AM receivers and occasionally is. The other technique, called **quieting sensitivity**, is suitable only for FM receivers.

8.6.1 Usable Sensitivity

Usable sensitivity is the signal level required for a given SINAD. SINAD, you probably remember, stands for the ratio of signal-plus-noise-and-distortion to noise-and-distortion. A value of 12 dB is common for communications systems, while FM broadcast receivers are rated for a SINAD of 30 dB. The reason for the difference is that while a 12 dB SINAD is sufficient for the understanding of speech, it is not enough for enjoyable music listening. The test signal is modulated at 60% of full deviation. The sensitivity may be expressed in terms of voltage (in microvolts) or power (in dBm or dBf, where the term dBf means decibels referenced to 1 femtowatt, which is 1×10^{-15} W). The receiver must be capable of producing at least one-half its rated audio output power with this signal; if not, the (higher) signal level required to produce one-half its rated power is the usable sensitivity.

Figure 8.27 shows a setup for measuring usable sensitivity. It is important that all cables be shielded and that all covers be on the receiver. Remember that any noise that gets into the receiver through faulty connections or unshielded leads detracts from the sensitivity reading. If the generator output impedance does not match the receiver input impedance, a suitable matching network must be used, and its loss must be taken into account. It will not do just to connect, for instance, a 50 Ω unbalanced generator output to the 300 Ω balanced input of an FM broadcast receiver. If this is done, any results obtained will be meaningless.

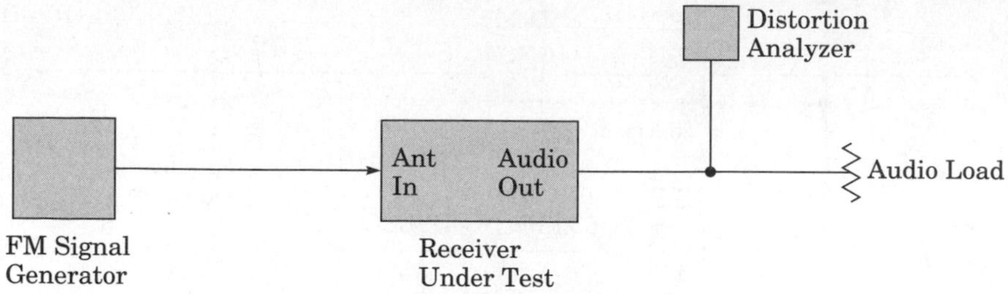

Figure 8.27 Test setup for measurement of receiver sensitivity

Similarly, the receiver output must be connected to a suitable load, otherwise the output power and distortion readings may be incorrect. For example, a receiver designed for use with an 8 Ω loudspeaker can be connected to an 8 Ω resistor with a power rating at least equal to the receiver's rated output power.

The generator and receiver must be tuned to the same frequency. Use a 1 kHz modulating signal that produces 60% of the rated deviation for the system in use. For example, with a communications system that uses 5 kHz maximum deviation, the deviation at the generator would be set to 3 kHz.

The distortion analyzer connected to the receiver audio output is used to measure SINAD. The procedure is identical to that used to measure harmonic distortion. The meter is first adjusted for a full-scale reading, then its notch filter is used to tune out as much of the fundamental 1 kHz signal as possible. What remains consists of noise and distortion, and the ratio between the two readings, expressed in decibels, is the SINAD. In most cases, the analyzer has a decibel scale that allows SINAD to be measured directly in decibels.

Normally, the nominal sensitivity of the receiver is known, so that is a logical RF signal level at which to begin. Measure the SINAD with this signal. If it is greater than 12 dB (for a communications receiver), reduce the RF level; if the SINAD is less than 12 dB, increase the level. Continue until a SINAD of 12 dB is obtained. Then measure the audio power output. This is easily done by using the voltmeter function of the distortion analyzer to measure the voltage across the load resistor, then calculating the power from the equation

$$P = \frac{V^2}{R} \tag{8.5}$$

Assuming that this power is at least one-half the rated power output specification, the generator RF output (after any matching networks are taken into account) is the usable sensitivity of the receiver. If not, simply increase the generator output until the audio output power increases to one-half the specified value, and use the generator output level required to achieve this as the usable sensitivity.

The procedure outlined above uses general-purpose test equipment. In applications in which a great deal of receiver testing is done, a dedicated SINAD meter is used. It is similar to a distortion analyzer, except that it operates only at 1 kHz and is self-nulling. It can read out the SINAD in decibels without any adjustment and therefore speeds up the process, though it does not improve the accuracy of the measurement.

8.6.2 Quieting Sensitivity

Quieting sensitivity is a measure of the effectiveness of FM limiting. The receiver is adjusted to produce noise power of 25% of rated output with no input signal (squelch or muting off). An unmodulated carrier is applied to the receiver, and its strength is increased until the noise output decreases by a specified amount (20 dB for communications receivers, 50 dB for broadcast receivers). The level of carrier required for this is the quieting sensitivity. Like usable sensitivity, quieting sensitivity can be specified in microvolts, dBm, or dBf.

The test setup for measuring quieting sensitivity can be the same as that described above for usable sensitivity. A distortion analyzer is not required, however—an audio voltmeter works just as well. For the best accuracy, a meter that reads true-RMS voltage for any waveshape is preferred, since the noise waveform will not be a sine wave. Since only a carrier is required, it is not necessary that the generator be capable of FM. Some older generators can produce only AM and CW signals; they are quite satisfactory as long as the output level can be set accurately and the amount of signal leakage through the cabinet of the generator is low. Most low-cost generators do not meet these requirements and cannot be used for sensitivity measurements, though they can be of some use in troubleshooting.

Summary

Here are the main points to remember from this chapter.

1. FM transmitters and receivers have many similarities with their AM counterparts, but there are some significant differences. The differences are mainly due to the fact that FM signals have constant amplitude.
2. FM can be generated in two basic ways. Direct FM requires that the carrier oscillator be frequency modulated. In indirect FM, the modulating signal is integrated and then applied to a phase modulator.
3. Frequency multipliers are used in some FM transmitters to increase the frequency deviation. Mixers can be used to move the signal to a different frequency without affecting the modulation.
4. PLLs can be used to achieve wideband FM directly at the operating frequency. When incorporated as part of a frequency synthesizer, they can have excellent stability.
5. Class C amplifiers can be used to amplify FM signals, as amplitude distortion is not important with FM.
6. FM demodulators typically convert frequency variations to phase shifts and then to amplitude variations. The exception is the PLL detector, in which the control voltage of a loop locked to the incoming signal becomes the output signal.
7. Limiters are used in FM receivers to remove amplitude variations due to noise, improving the output signal-to-noise ratio.
8. Both transmitters and receivers commonly use AFC circuits to achieve stability approximating that of crystal control with *LC* oscillator circuits. This is not necessary when PLL techniques are used in transmitter carrier oscillators and receiver local oscillators.
9. Transceivers are common for two-way mobile operation. Transceivers can use a common frequency synthesizer and can share audio components.
10. Stereo FM broadcasting is accomplished by synthesizing the composite baseband signal before modulation onto the main carrier. Similarly, stereo receivers first demodulate the FM signal, then demodulate the subcarrier, and finally combine the main and subcarrier signals.

Important Equations

$$k_f = \frac{\Delta f}{\Delta v} \tag{8.2}$$

$$k_d = \frac{V_o}{\delta} \tag{8.3}$$

$$V_o \text{ peak} = \frac{\delta}{k_f} \tag{8.4}$$

Glossary

automatic frequency control (AFC) a scheme for keeping a transmitter or a receiver tuned to the correct frequency by applying a correction voltage to a VCO when the operating frequency is incorrect

Crosby system a method of generating FM signals directly, with a discriminator for automatic frequency control

direct FM any system that generates FM without going through a preliminary stage of phase modulation

discriminator a circuit whose output is proportional to the difference between its input frequency and a predetermined center frequency

indirect FM any method that generates FM using a phase modulator and an integrator applied to the modulating signal

quadrature detector an FM detector circuit that makes use of the phase shift in a resonant circuit as a signal moves through resonance

quieting sensitivity the strength of an unmodulated carrier that reduces the noise output of an FM receiver by a specified amount

ratio detector a type of discriminator that has the advantage of being insensitive to amplitude variations of the signal

SINAD ratio of signal-plus-noise-and-distortion to noise-plus-distortion (closely related to signal-to-noise ratio)

usable sensitivity the FM signal strength with defined deviation, required to produce a specified SINAD in a receiver

Questions

1. Why can an FM transmitter use low-level modulation followed by Class C amplification, while this is impossible with AM?
2. What limits the amount of frequency deviation that can be obtained with a varactor modulator for direct FM?
3. How can the deviation of an FM signal be increased?
4. Draw a block diagram of an FM transmitter using the Crosby system, and explain how this system stabilizes the carrier frequency.
5. What is meant by indirect FM?
6. What method of direct FM allows wideband FM to be generated directly at the carrier frequency? Draw a block diagram for this type of modulator.
7. What is a limiter? Why and where are limiters used in FM receivers?
8. Why is AGC a requirement in an AM receiver but not always used with FM?
9. What is the function of an AFC circuit?
10. Why is AFC more common in FM than in AM broadcast receivers?
11. Why is AFC unnecessary with receivers using a frequency-synthesized local oscillator?
12. Compare the Foster-Seeley discriminator and the ratio detector, giving one advantage of each.
13. Draw and explain the S-curve for a typical FM detector.
14. Describe two modern types of FM detectors, and explain how each works. Use diagrams to help with the explanations.
15. How does an FM stereo decoder detect that a stereo signal is present?
16. What is the reason for high-frequency blending in FM stereo receivers?
17. Why are FM transceivers often designed to transmit and receive on different frequencies?
18. Define two types of sensitivity measurement for FM receivers, and describe the procedure for each.
19. Why are the sensitivity specifications defined in different ways for broadcast and for communications receivers?

Problems

SECTION 8.2

20. A direct-FM transmitter has a block diagram as shown in Figure 8.28.
    (a) What is the carrier frequency of the output signal?
    (b) If the modulator has a sensitivity of 5 kHz/V, what modulating voltage would be required for a deviation of 50 kHz at the output?
21. Calculate the carrier frequency and deviation at the output of the direct-FM transmitter shown in Figure 8.29.
22. Suppose it is required to transmit at a carrier frequency of 600 MHz using the same baseband signal, modulator, and deviation as in Problem 21. Draw a block diagram showing how the original transmitter configuration could be modified.

Figure 8.28

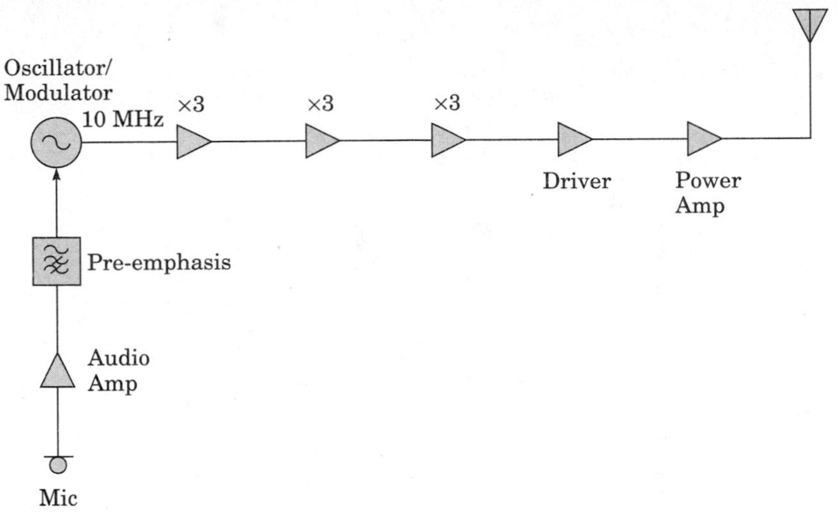

23. (a) In an indirect-FM transmitter, what phase shift, in degrees, corresponds to a frequency-modulation index of 4.5?
    (b) Suppose a phase modulator is available with a maximum phase shift of 45 degrees. How can this phase shift be increased to the amount calculated in part (a)? Draw a block diagram to illustrate.
24. An indirect-FM transmitter has the block diagram shown in Figure 8.30. The total output power is 1 W into 50 Ω.
    (a) Label the block that has been left blank.
    (b) What phase deviation will be necessary at the modulator in order for the output frequency to deviate by 5 kHz with a 1 kHz modulating signal?
    (c) Use Bessel functions to sketch the spectrum of the output signal. Be sure to include suitable scales. It is necessary to include only the carrier and the first three sidebands on each side.
25. A block diagram for an FM transmitter using indirect FM is shown in Figure 8.31.
    (a) What is the frequency of the carrier oscillator?
    (b) What maximum phase deviation must the phase modulator be capable of supplying if the transmitter is to produce FM with a maximum deviation of 25 kHz with a modulating-signal frequency of 10 kHz?
    (c) Why is the integrator necessary?
26. The block diagram of a Crosby-type transmitter is shown in Figure 8.32. Calculate the output frequency and deviation.

Figure 8.29

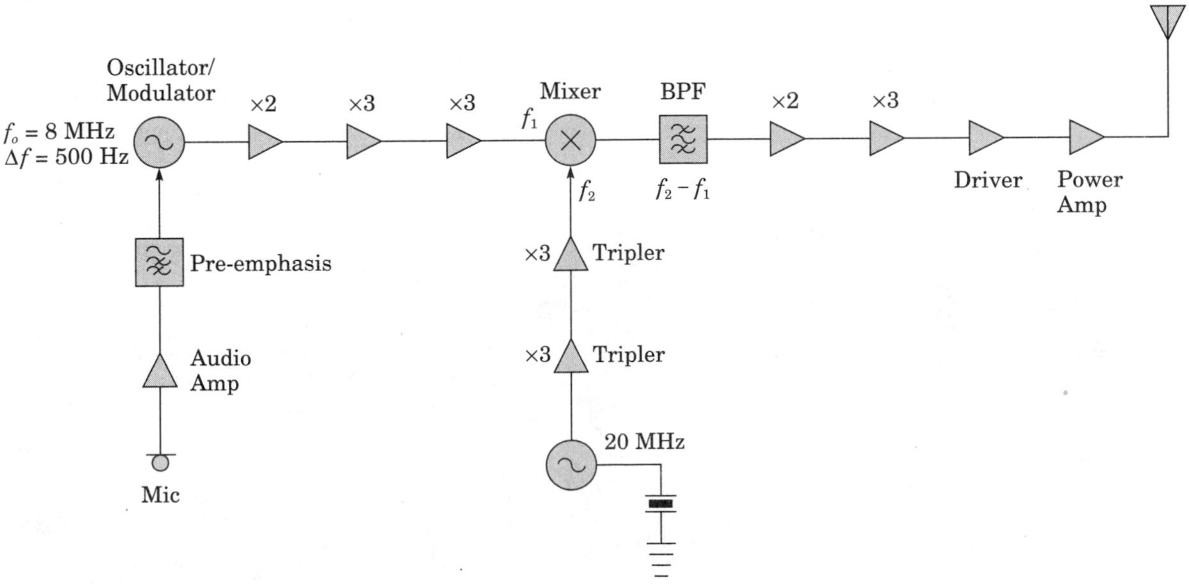

Figure 8.30

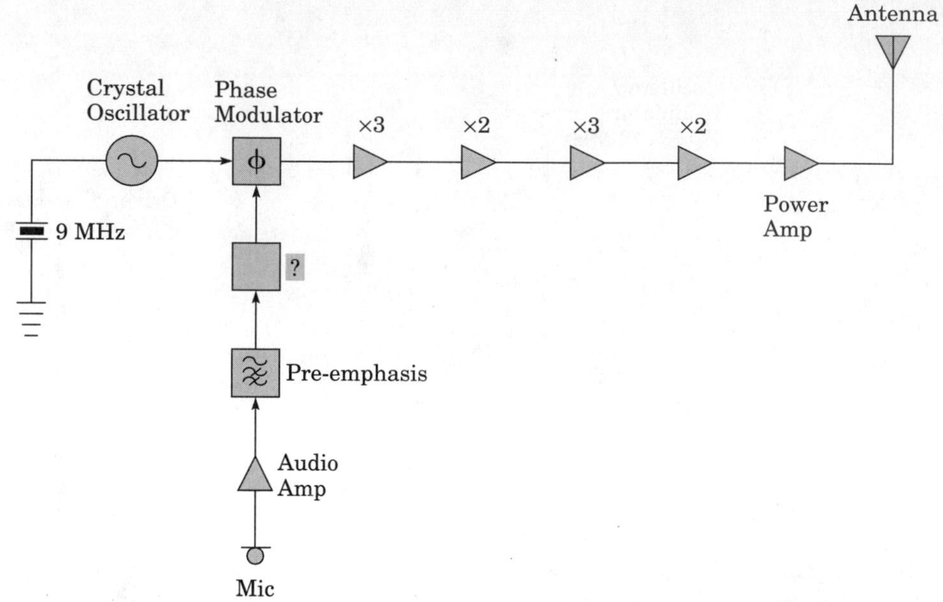

Figure 8.31

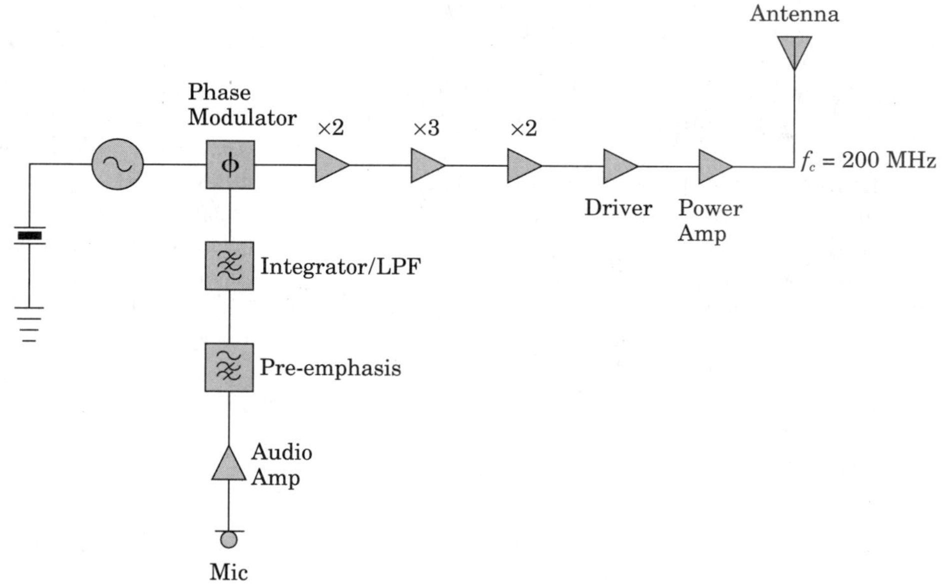

Figure 8.32

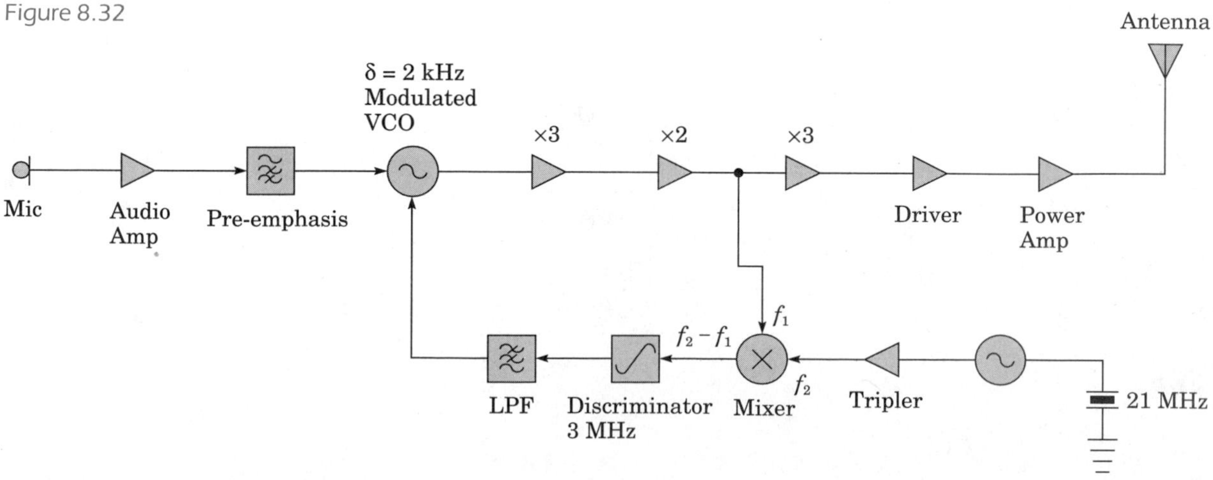

Figure 8.33

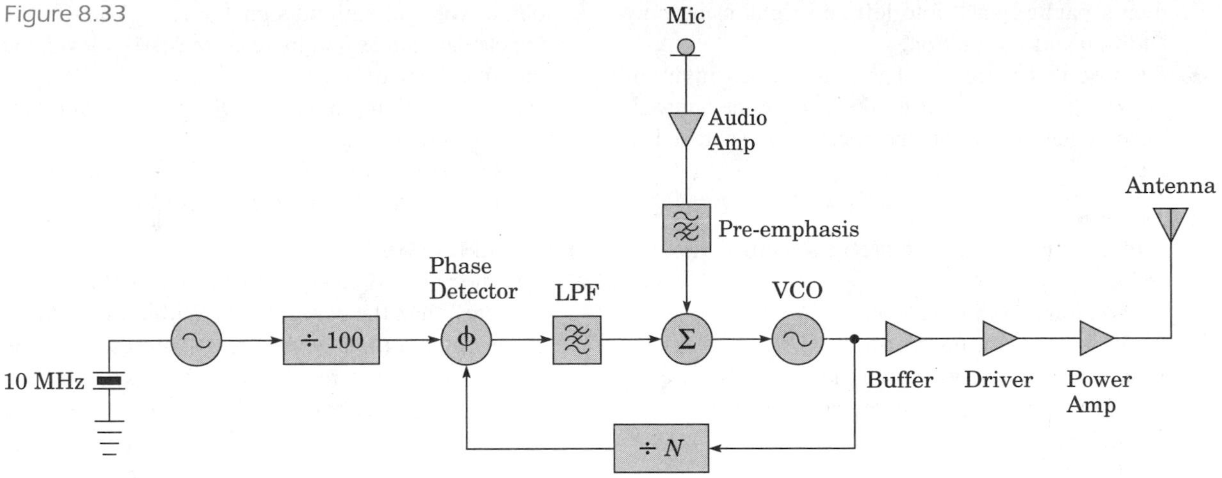

27. Figure 8.33 is the block diagram of an FM transmitter using a PLL.
    (a) Calculate the required value of N for a carrier frequency of 91.1 MHz.
    (b) Calculate the minimum acceptable value of k_f for the VCO if the transmitter is to produce a 75 kHz deviation with an input signal of 2 V RMS.

SECTION 8.3

28. (a) Draw the block diagram for an FM broadcast receiver. It is to have one RF stage and an IF of 10.7 MHz. The local oscillator will operate above the signal frequency. Indicate on the diagram the frequency or frequencies at which each stage operates when the receiver is receiving a station at 94.5 MHz.
    (b) What is the image frequency of the receiver described above?
    (c) Give two ways in which the image rejection of the receiver could be improved.
29. A PLL FM detector uses a VCO with $k_f = 20$ kHz/V. If it receives an FM signal with a deviation of 30 kHz and sine-wave modulation, what is
    (a) the output voltage from the detector?
    (b) the minimum value for the lock range of the loop, for satisfactory operation of the detector?
30. An FM detector has a sensitivity of 100 mV/kHz. How much deviation, assuming a sinusoidal modulating signal, is required for an output of 2 V RMS?
31. A limiter has a limiting threshold of 225 mV at the input.
    (a) How much voltage gain is needed before the limiter if the receiver is to start limiting with a signal of 0.4 μV at the antenna?
    (b) Would it be possible to receive weaker signals with this receiver? Explain why or why not.
32. An FM receiver is double conversion with a first IF of 10.7 MHz and a second IF of 455 kHz. High-side injection is used in both mixers. The first local oscillator is a VFO, while the second is crystal-controlled. There is one RF stage, one stage of IF amplification at the first IF, and three stages of combined IF amplification and limiting at the second IF. A Foster-Seeley discriminator is used as the detector. The receiver is tuned to a signal with a carrier frequency of 160 MHz.
    (a) Draw a block diagram for this receiver.
    (b) Calculate the Q that will be necessary in the input circuits to achieve a rejection of 50 dB at the first image frequency, assuming that there are two tuned circuits with no coupling between them before the first mixer. [*Hint*: you will find the necessary equations in Chapter 5.]
    (c) If AGC were used with this receiver, to which stages could it be applied?
    (d) If AFC were necessary, from which stage would it be derived and to which stage would it be applied?

SECTION 8.4

33. Refer to the block diagram for an FM transceiver shown in Figure 8.22.
    (a) What frequencies would the synthesizer produce while transmitting and receiving, respectively, on marine channel 14 (156.7 MHz)?
    (b) Explain the absence of frequency multipliers in the transmitter section.
    (c) The transmitter section uses a maximum deviation of 5 kHz and operates with modulating frequencies in the range of approximately 300 Hz to 3 kHz. Suggest an appropriate receiver bandwidth.

SECTION 8.5

34. Show mathematically how the left and right signals can be combined into L+R and L−R signals and

then separated back into left and right, using only addition and subtraction.

35. Suppose that a receiver has a usable sensitivity of 50 μV in stereo. What is likely to be the approximate sensitivity when the receiver is switched to mono?

SECTION 8.6

36. A receiver has a specified usable sensitivity listed as follows:

    Nominal: 0.5 microvolt
    Limit: 1.0 microvolt

    What can you conclude about the receiver for each of the following sensitivity measurements?
    (a) 0.4 microvolt
    (b) 0.7 microvolt
    (c) 1.1 microvolt
37. You are making a usable sensitivity test of an FM communications receiver and have just measured a SINAD of 10 dB with an input signal level of 0.5 μV. What do you do next?
38. In making a quieting sensitivity test on an FM broadcast receiver, you have measured the noise output voltage, with no signal, as 2 V. You turn on the carrier, and as you increase the carrier level, the output voltage decreases.
    (a) You continue to increase the carrier level until the output voltage declines to what value?
    (b) Once the output voltage level found in part (a) is reached, what do you do next?

COMPREHENSIVE

39. Commercial FM radio stations often try to sound "louder" than the competition. Within the primary coverage area (that is, the area where the signal is quite strong at the receiver antenna), which of the following approaches would be more effective in increasing loudness?
    (a) increasing the transmitter carrier power
    (b) using audio compression to increase the average deviation

    Explain your answer.
40. Draw a block diagram of a communications receiver that could be used to receive AM, SSB, and FM transmissions.

9 Television

Objectives After studying this chapter, you should be able to:

1. Describe the characteristics of an NTSC television signal (monochrome and color)
2. Calculate horizontal and vertical resolution for video signals
3. Explain the composite color video system and compare it with RGB and component systems
4. Describe the system used for terrestrial television broadcasting
5. Explain the system used for stereo sound in television and compare it with the system used in FM broadcasting
6. Draw block diagrams for monochrome and color television receivers and explain their operation
7. Diagnose faults in monochrome and color television receivers
8. Calculate signal levels at various points in a cable-television system
9. Explain the need for and operation of cable-television components, such as amplifiers, directional couplers, and converters
10. Describe some of the problems with the current television standard and suggest ways in which it could be improved

9.1 Introduction

Television and video systems form a very important part of the communications environment. This chapter will introduce the video system used in conventional television broadcasting and will also provide insights into the kinds of changes that are made for such applications as high-definition video and computer video displays. In addition, the hardware involved in television receivers and video monitors will be introduced.

Video systems form pictures by a scanning process. The image is divided into a number of horizontal lines, which are traced out in synchrony at the camera and receiver. The number of lines is arbitrary, but increasing it gives better resolution in the vertical direction. The North American standard uses 525 lines, compared to 625 in Europe. Not all the lines are visible on the television screen: in North America, the number of scan lines actually used to form the image is about 483. The remaining lines are transmitted during the time interval between images.

In order to make the image visible all at once, rather than as a series of consecutively drawn lines, the process must be completed quickly. Also, in order to provide the illusion of motion, many images must be drawn in quick succession. The more quickly the images follow one another, the less flicker is present. However, this requires more information to be transmitted, increasing the required bandwidth. In North America, the images, or **frames**, are sent at the rate of approximately 30 per second; the equivalent number for European systems is 25.

Historical Development of Television

It is perhaps surprising that the development of television followed closely after that of radio. Regular radio broadcasting began in 1920; by 1928 experimental television systems existed. Regular broadcasting began in the 1930s in both North America and Europe. It was suspended in most countries during World War II but became popular immediately after the war. Color television was introduced in North America in 1953.

Early television systems were about equally divided between electromechanical and all-electronic systems. The former used a single photodetector at the camera, which scanned the image to be transmitted using a complex mechanical system of disks containing holes and slots. At the receiver, another mechanical system, synchronized with the first, directed light from a single source to reproduce an image on a screen.

Electronic television, which with refinements became the system in use today, worked by scanning electron beams by means of magnetic fields. At the camera, the beam scanned a photosensitive surface, creating a current proportional to the intensity of light at a given time at a particular point on the image. At the receiver, the electron beam scanned a phosphorescent material, producing a moving beam of light that drew the image on the screen of a **cathode-ray tube** (CRT). The **luminance** (brightness) of the original image was reproduced by intensity-modulating the electron beam in the receiver CRT. Once again, the transmitter and receiver had to be synchronized.

Figure 9.1 shows an early electronic television system. Though they worked, the mechanical systems showed less detail, and they were also subject to mechanical failures. The electronic systems soon prevailed.

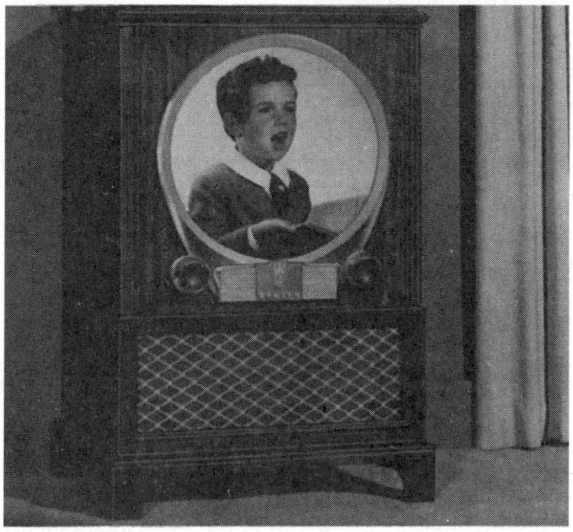

Figure 9.1
Photo courtesy Zenith Electronics Corporation.

Frame rates of 25 or 30 Hz cause noticeable flicker. To reduce this, television systems use a technique called *interlaced scan*, which involves transmitting alternate lines of the picture, then returning and filling in the missing lines. Figure 9.2 illustrates the idea. Each half of the picture thus sent is called a **field**, and of course the field rate is twice the frame rate, or 60 Hz in the North American system. The result is reduced flicker without increased bandwidth.

As Figure 9.2 shows, the picture is scanned from left to right and from top to bottom. The electron beam that traces the picture is blanked during the time intervals in which the beam **retraces** its path, from right to left and from bottom to top. These time intervals are called the *horizontal* and *vertical blanking intervals*, respectively.

The ratio of width to height, or **aspect ratio**, is 4 : 3. If the frame rate is 30 Hz, then the rate at which lines are sent is 525 multiplied by 30 or 15,750 Hz. These frequency

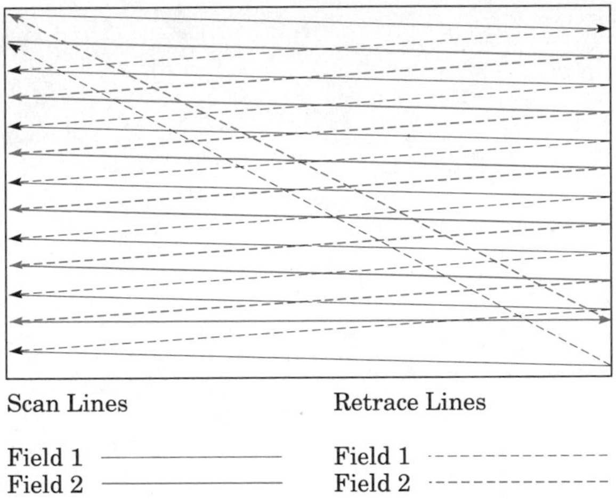

Figure 9.2
Interlaced video

specifications are somewhat arbitrary, but obviously the same standards must be in use at both ends of the communication path.

Good color reproduction can be achieved by mixing three primary colors: red, green, and blue. Thus considerably more information must be transmitted for color television than for monochrome. Color television also requires a rather elaborate picture tube containing three electron guns.

9.2 Baseband Video Signal

In this section, we shall look at the video signal as it might emerge from a camera and enter a video monitor. Later, we shall see how this signal can be modulated on a carrier and transmitted along with a sound signal by a television broadcasting station.

The signal we will examine most closely is called a **composite video signal**, because it combines all the picture information, along with synchronizing pulses. With a composite signal, the whole signal can be transmitted on the same cable or the same radio channel. It is also possible to separate the various components of a video signal, and this is often done when the transmission distance is not great—with computer monitors, for instance.

9.2.1 Luminance Signal

Figure 9.3 shows one line of a monochrome video signal that conforms to the North American standard. Note that the duration of the line is 63.5 μs. This is simply the period corresponding to the horizontal line frequency of 15.75 kHz. About 10 μs of this is used by the horizontal synchronizing (sync) pulse, which will be discussed later. The rest of the time is occupied by an analog signal that represents the variation of the *luminance* (brightness) level along the line.

The video signal shown in Figure 9.3 has negative sync, that is, the sync pulses are in the negative direction. An inverted version of this signal, called *positive sync*, is also possible. The polarity of the luminance portion depends on that of the sync pulses, which are always in the direction of black. Both polarities and a variety of amplitudes are used in video equipment, but for the interconnection of equipment, the standard is negative sync with an amplitude of 1 V peak-to-peak, into a 75 Ω terminating resistance.

Because of the variety of possible levels and polarities for the video signal, a scale of relative amplitudes is used for setting the correct proportions between the level of the synchronizing pulses and the luminance signal. It is called the *IRE scale*, for the Institute of Radio Engineers, a precursor to the Institute of Electrical and Electronic Engineers (IEEE), with which you may be familiar. On this scale, which is also shown in Figure 9.3, zero represents **blanking** level, and −40 represents the level of the sync pulses. The maximum

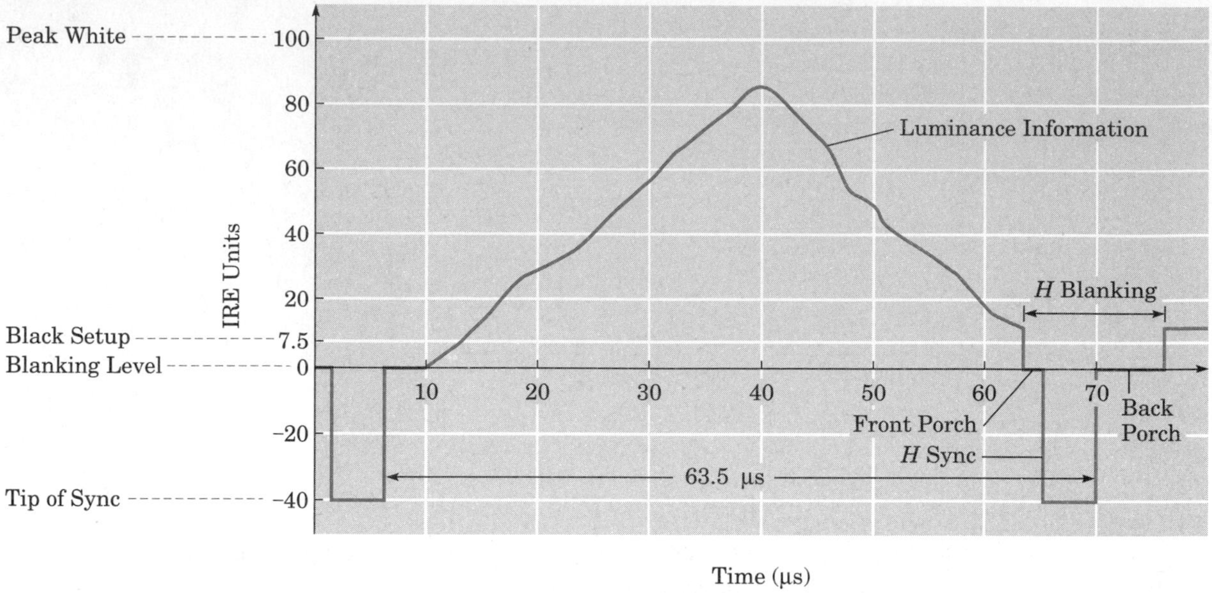

Figure 9.3 Monochrome video signal

luminance level, called **peak white**, is 100, and 7.5 is a level that should represent black at the receiver. From this, it can be seen that the blanking and synchronizing pulses are "blacker than black" and ensure that the CRT electron beam is completely turned off while it retraces from right to left.

Example 9.1 A video signal has 50% of the maximum luminance level. Find its level in IRE units.

Solution The **black setup** level of 7.5 IRE represents zero luminance, and 100 IRE is maximum brightness. Therefore, the range from minimum to maximum luminance has 100 − 7.5 = 92.5 units. We must add 50% of this to the setup level of 7.5. Therefore, the level is

$$\begin{aligned} \text{IRE} &= 7.5 + 0.5 \times 92.5 \\ &= 53.75 \text{ IRE units} \end{aligned}$$

9.2.2 Synchronizing Pulses

In order to synchronize the scanning of a scene at the camera and the receiver, two types of synchronizing pulses are used: horizontal sync at the end of each line and vertical sync at the end of each field. Interlaced scan is accomplished by using two slightly different types of vertical sync, one for odd-numbered and one for even-numbered fields.

The electron beam in the receiver CRT must be turned off, or *blanked*, as it returns from right to left and from bottom to top. To do this, the sync pulses extend into the blacker-than-black region of the video signal and are part of the blanking interval, which is also in the blacker-than-black region. Modern receivers also blank the electron beam automatically during retrace.

Figure 9.3 shows the horizontal blanking interval at the end of each line for a monochrome signal. The duration of the blanking pulse is approximately 10 μs, of which about one-half is used for the sync pulse itself. The period of blanking before the sync pulse is called the **front porch**, and naturally, the **back porch** follows the sync pulse. The back porch is longer to allow the electron beam time to move from the right to the left side of the screen.

The vertical blanking interval is shown in Figure 9.4. The two diagrams in Figure 9.4 correspond to the two video fields in each frame. The vertical blanking interval occupies the time required for approximately 21 horizontal lines: about 1.3 ms, compared with 10 μs for horizontal blanking. The time difference allows the receiver to distinguish between

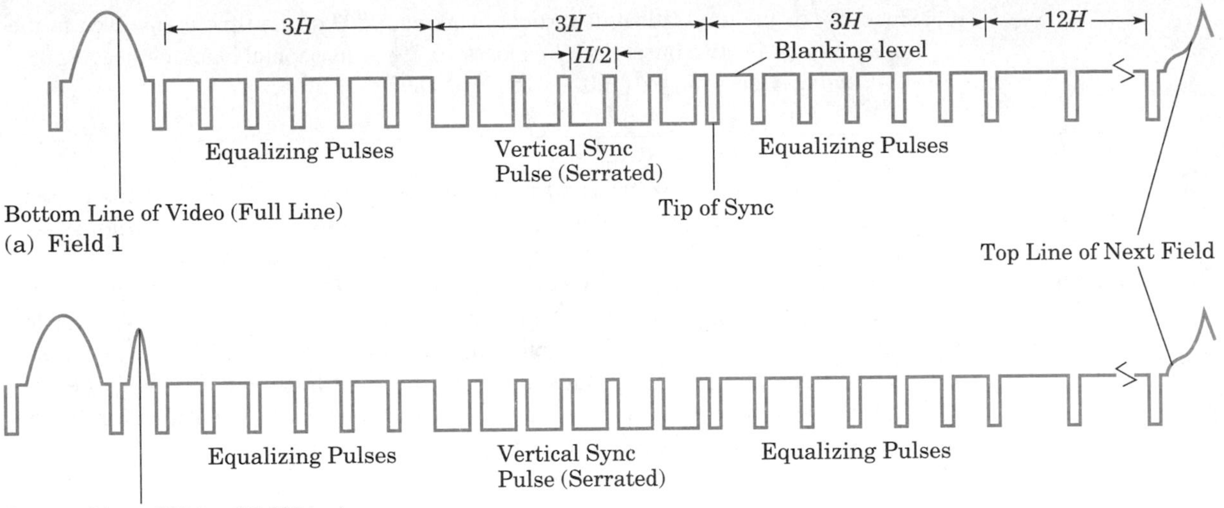

Figure 9.4
Vertical blanking interval

horizontal and vertical sync on the basis of duration. It also allows more time for the electron beam to move from the bottom to the top of the CRT, which simplifies the receiver deflection circuitry.

The structure of the vertical blanking interval is relatively complex. It begins with six **equalizing pulses** spaced $H/2$ apart, where H is the length of one horizontal line, or 63.5 μs. This takes a total time of $3H$, or 190.5 μs. **Interlace** is accomplished by putting the first equalizing pulse in the middle of a line for one field and at the end of a line for the next.

Since the vertical blanking interval is several lines long, it is necessary to continue to send horizontal sync information during the vertical blanking interval. Otherwise, the horizontal scanning in the receiver might lose synchronization during vertical retrace. For this reason, the vertical sync pulse is *serrated*, that is, the pulse is interrupted at intervals of $H/2$ throughout its duration of $3H$. Following the vertical sync pulse, there are six more equalizing pulses spaced $H/2$ apart, for a total duration so far of nine horizontal lines. This time may not be sufficient for the electron beam in the receiver CRT to return to the top of the screen, so a further period of about $12H$ consists of blanking level with normal horizontal sync pulses. Often, some of this extra blanking time is used for such purposes as the sending of test signals and closed-caption information.

Example 9.2

Calculate the total percentage of the signal time that is occupied by:

(a) horizontal blanking
(b) vertical blanking
(c) active video

Solution

(a) Horizontal blanking occupies approximately 10 μs of the 63.5 μs duration of each line. Therefore,

$$\text{horizontal blanking (\%)} = \frac{10}{63.5} \times 100 = 15.7\%$$

(b) Vertical blanking normally occupies 21 lines per field or 42 lines per frame. A frame has 525 lines altogether, so

$$\text{vertical blanking (\%)} = \frac{42}{525} \times 100 = 8.0\%$$

(c) Since 8% of the time is lost in vertical blanking, 92% of the time is involved in the transmission of active lines. Each line loses 15.7% to horizontal blanking, leaving

$$\text{active video (\%)} = (100 - 15.7) \times 0.92 = 77.6\%$$

9.2.3 Resolution and Bandwidth

The resolution of a video system is a function of the number of details that can be seen both horizontally and vertically. Horizontal and vertical resolution in a television system are determined quite differently. The vertical resolution is perhaps more straightforward, since it depends directly on the number of scanning lines. The maximum possible number of details in the vertical direction is the number of visible scan lines, about 483 for the North American system. However, in an ordinary picture, only about 70% of the vertical resolution can be used, because details in the scene do not line up exactly with the scan lines in the camera and receiver. This factor of about 0.7 is called the **utilization factor**. The number of details that can be seen vertically is thus about

$$N_V = 483 \times 0.7 = 338 \tag{9.1}$$

where

N_V = number of details in the vertical direction

Horizontal resolution in an analog video system depends on the bandwidth that is allowed, as can be seen by referring to Figure 9.5. Each cycle of the highest-frequency component of the video signal represents two picture details, assuming that a detail is any change in the luminance level.

For a baseband signal, there is no theoretical limit on the bandwidth. A limit is imposed on signals to be broadcast, however, because of the finite width of the broadcast channel. The upper limit for a broadcast video signal is 4.2 MHz. It is easy to find how many details this allows horizontally. First, multiply by the scan time for the visible portion of one line to find the number of cycles per line, then multiply by two. This gives the following result:

$$N_H = (4.2 \times 10^6)(53.5 \times 10^{-6}) \times 2 = 449 \text{ details} \tag{9.2}$$

where

N_H = number of details in the horizontal direction

Figure 9.5
Horizontal resolution

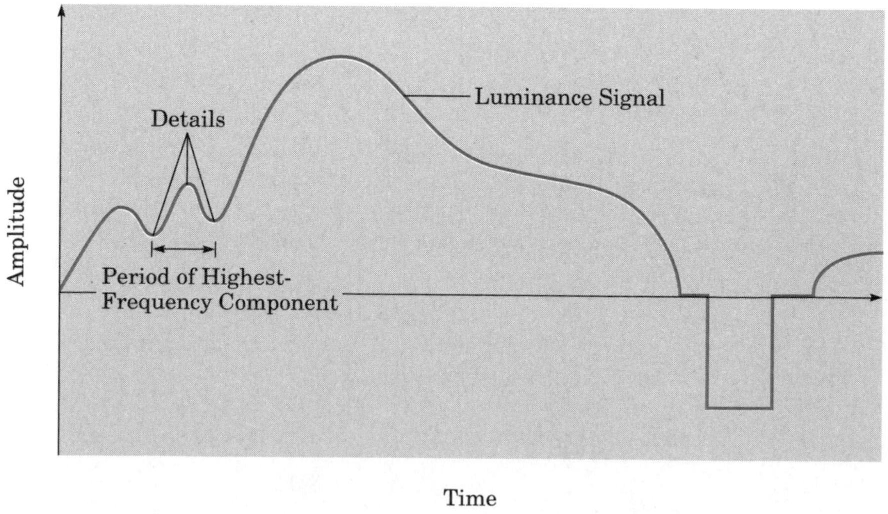

In practice, horizontal resolution for a television system is usually specified in lines rather than details. The aspect ratio of the picture is 4:3; that is, the picture is one-third wider than it is high. Picture elements will be the same size horizontally as they are vertically if the number of details horizontally is four-thirds of the vertical number. The horizontal resolution in lines is defined as the number of details across three-quarters of the screen. This gives numbers for horizontal and vertical resolution that can easily be compared. For a 4.2 MHz signal, the horizontal resolution in lines is

$$\begin{aligned} L_H &= 449 \times 0.75 \\ &= 337 \text{ lines} \end{aligned} \tag{9.3}$$

where

L_H = horizontal resolution in lines

This is almost exactly the same as the vertical resolution found above.

Horizontal resolution depends only on bandwidth once the other parameters of a video system, such as line rate and blanking time, are fixed. Therefore it is possible to derive a simple relation between **resolution** and bandwidth. Resolution is proportional to bandwidth; therefore, if 4.2 MHz of bandwidth gives 337 lines of resolution,

$$\begin{aligned} \frac{L_H}{B} &= \frac{337 \text{ lines}}{4.2 \text{ MHz}} \\ &= 80 \text{ lines/MHz} \end{aligned}$$

where

B = bandwidth in megahertz

This relation is usually expressed:

$$L_H = B\ (\text{MHz}) \times 80 \tag{9.4}$$

There is no low-frequency limit for a video signal. In fact, there will be a dc or near-dc component that represents the average brightness of the scene.

The maximum number of picture elements (**pixels**) in the broadcast television system can be found by multiplying the numbers of details horizontally and vertically. This gives a total number of details of

$$\begin{aligned} N_P &= N_H N_V \\ &= 449 \times 338 \\ &= 151{,}761 \end{aligned} \tag{9.5}$$

where

N_P = total number of pixels
N_H = number of details horizontally
N_V = number of details vertically

Example 9.3 A typical low-cost monochrome receiver has a video bandwidth of 3 MHz. What is its horizontal resolution in lines?

Solution From Equation (9.4),

$$\begin{aligned} L_H &= B\ (\text{MHz}) \times 80 \\ &= 3 \times 80 \\ &= 240 \text{ lines} \end{aligned}$$

The frequency spectrum of a monochrome video signal is not uniform. It should be obvious that there will be strong components at the horizontal and vertical scanning rates

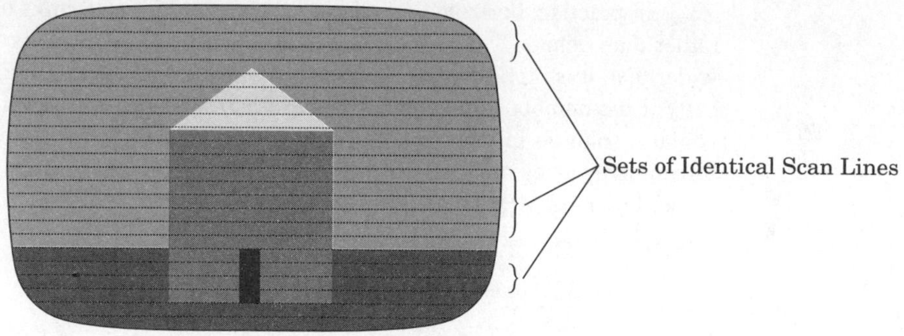

Figure 9.6
Components at multiples of horizontal frequency

and at harmonics of these frequencies, due to the synchronizing pulses. What is not quite so obvious is that most of the energy due to the picture information is also at multiples of these frequencies. This can be seen by referring to Figure 9.6, which shows a section of a picture and several horizontal scanning lines. For a typical image, one scan line is much like the next. Therefore, the whole signal repeats at the horizontal rate, and the smaller details along the lines cause frequency components at multiples of that rate. Diagonal lines and areas with motion do create components at one-half the horizontal line rate, as well as at odd multiples of that frequency.

A similar argument can be made for repetition at the vertical scanning rate. Usually, the two fields that make up one frame are very similar since they are two views of the same image shifted by one horizontal line. Therefore, we expect a spectrum like that shown in Figure 9.7, with peaks at multiples of the horizontal scan rate and with a finer line structure showing peaks at 60 Hz intervals. The pattern is not perfect, of course: horizontal lines in the picture, abrupt changes of scene, and motion in the picture all disturb it. Nonetheless, the frequency distribution of the signal follows the suggested pattern quite closely. This fact can be used to advantage in the transmission of color, which we will consider in the next section.

9.2.4 The NTSC Color System

In 1953, The National Television Systems Committee (NTSC) of the Electronics Industries Association (EIA), an American industry group, established the first color television standard. **NTSC video** is still used today in North and Central America, most of South America, and Japan. Two other standards, known as *PAL* (phase alternation by line) and *SECAM* (sequential color and memory) were developed in Germany and France, respectively. SECAM is used in France and eastern Europe, and PAL is employed in most of the rest of the world.

As was mentioned in the introduction to this chapter, it is not necessary for a color video system to actually reproduce all the colors found in nature. In fact, because of the way color is perceived by the human eye-brain combination, it is only necessary to transmit information for three colors: red, green, and blue. Even this is a rather daunting requirement, since at first glance it would appear to triple the bandwidth required for a color (as compared to a monochrome) video signal. Indeed, some video systems transmit red, green, and blue signals separately on three separate conductors—these systems are called **RGB video** systems. Most color computer monitors use three separate lines included in one cable for the three colors, with additional lines for synchronizing information.

RGB video is not suitable for conventional television broadcasting because of bandwidth limitations. Furthermore, when the NTSC color system was developed, it was required to maintain compatibility with monochrome, that is, an existing monochrome receiver had to be able to receive a color television signal.

In order to maintain compatibility, the composite color system includes a luminance signal. Two other signals are multiplexed onto it to provide color. Only two are needed because the original luminance signal is a combination of red, green, and blue. Two other

Figure 9.7
Baseband spectrum of a video signal

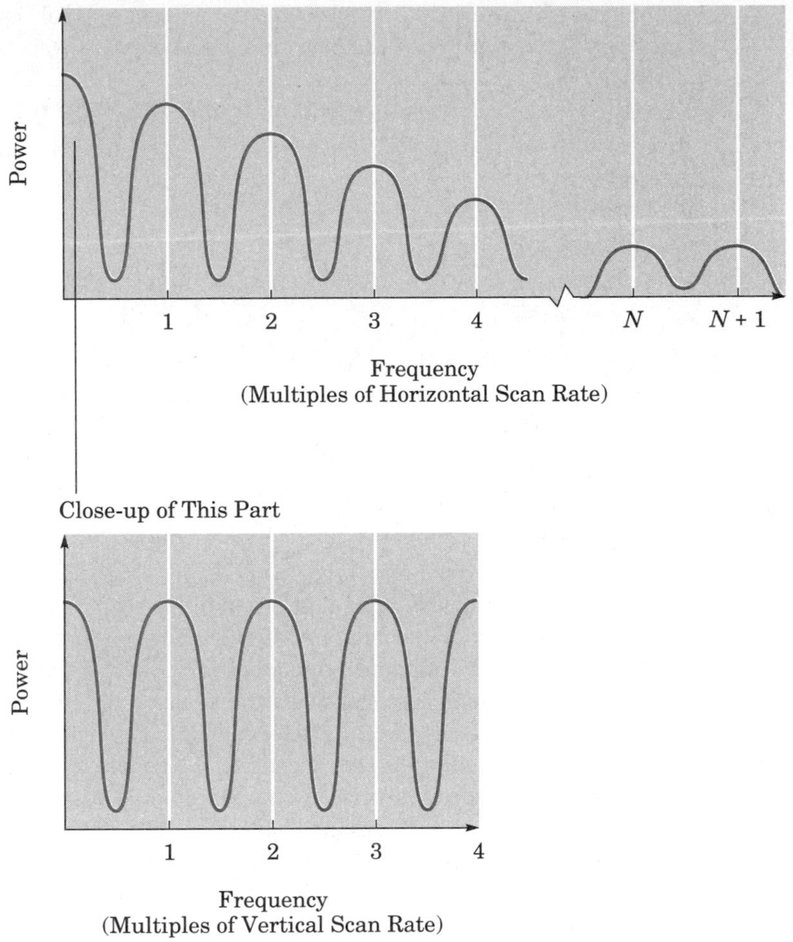

signals, each a linear, independent combination of the three colors, provide enough information to reconstruct the original red, green, and blue signals. If the red, green, and blue signals are normalized (that is, they have amplitude values between zero and one), then the equations for the three signals that make up the composite color signal are as follows.

$$\begin{aligned} Y &= 0.30R + 0.59G + 0.11B \\ I &= 0.60R - 0.28G - 0.32B \\ Q &= 0.21R - 0.52G + 0.31B \end{aligned} \tag{9.6}$$

where

Y = luminance signal
I = in-phase component of the color signal
Q = quadrature component of the color signal

The terms *in-phase* and *quadrature* refer to the process by which the **chrominance** (color) signal is combined with the luminance signal and will be made clear very shortly.

There are several noteworthy points about these equations. First, the luminance signal is not composed of equal amounts of red, green, and blue, as you might suppose. The frequency response of the human eye is not flat—we are much more sensitive to green light than to red or blue. By assigning the three primary colors the weights given above, the system designers ensured that monochrome pictures would replace the various colors with shades of grey whose luminance on the television screen corresponds to the brightness they would have when viewed "live." The two chrominance signals have zero amplitude when a grey or white object is viewed, that is, when the values of R, G, and B are all equal.

Example 9.4 An RGB video signal has normalized values of $R = 0.2$, $G = 0.4$, $B = 0.8$. Find the values of Y, I, and Q.

Solution From Equation (9.6),

$$\begin{aligned} Y &= 0.30R + 0.59G + 0.11B \\ &= (0.30 \times 0.2) + (0.59 \times 0.4) + (0.11 \times 0.8) \\ &= 0.384 \end{aligned}$$

$$\begin{aligned} I &= 0.60R - 0.28G - 0.32B \\ &= (0.60 \times 0.2) - (0.28 \times 0.4) - (0.32 \times 0.8) \\ &= -0.248 \end{aligned}$$

$$\begin{aligned} Q &= 0.21R - 0.52G + 0.31B \\ &= (0.21 \times 0.2) - (0.52 \times 0.4) + (0.31 \times 0.8) \\ &= 0.082 \end{aligned}$$

The I and Q signals are modulated onto a subcarrier at approximately 3.58 MHz, using suppressed-carrier quadrature AM. The subcarrier frequency is chosen to be an odd multiple of one-half the horizontal frequency ($227.5f_H$), so that the chrominance sidebands tend to fit between luminance sidebands and so that interference between color and luminance is less visible on the screen. The diagram in Figure 9.8 shows how this works. Recall from the previous section that the spectrum of a video signal has most of its energy at multiples of the horizontal line rate. Setting the color subcarrier at an odd multiple of one-half the horizontal frequency ensures that the maxima of the luminance and chrominance signals (often called **luma** and **chroma**) do not occur at the same frequencies. This allows the chroma and luma frequency ranges to overlap and reduces the total bandwidth required for the signal.

If the chroma signal were allowed to extend right down to zero frequency, there would be excessive interference between luma and chroma. On the other hand, if the signal is to be broadcast, its upper frequency limit must be limited to 4.2 MHz. Therefore, the chroma signal must be band-limited. The Q signal is cut off at 0.5 MHz and transmitted using both sidebands, while the I signal has a cutoff at about 1.3 MHz for the lower sideband and 0.5 MHz for the upper, that is, the I signal uses vestigial sideband (VSB) modulation. This allows the receiver to take advantage of the extra I bandwidth to improve the color resolution. The combination of colors used for the I signal places it in a region where the acuity of the eye for color, which is never as good as its acuity for monochrome, is maximum.

Even if the receiver takes advantage of the entire I bandwidth (and most do not), the

Figure 9.8
Spectrum of a color video signal

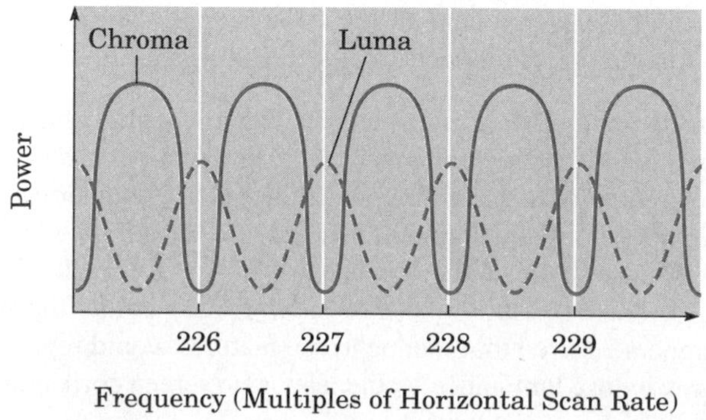

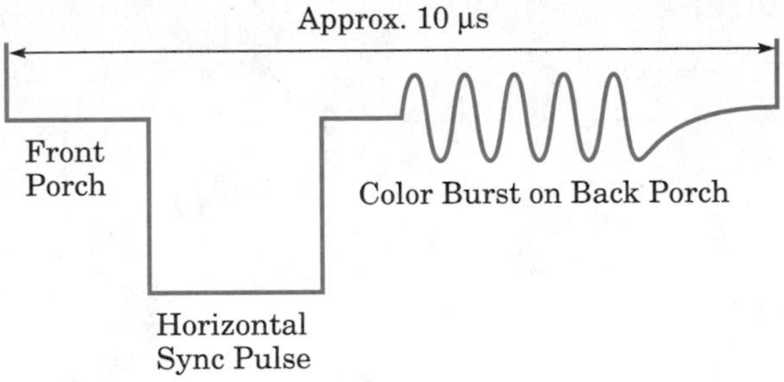

Figure 9.9
Color burst

horizontal resolution for color is much less than for the luminance signal. With a bandwidth of 1.3 MHz, the color resolution is

$$L_H \text{ (color)} = 80 \times 1.3 = 104 \text{ lines} \tag{9.7}$$

and, for the more common situation where the color bandwidth is restricted to 0.5 MHz, the resolution is

$$L_H \text{ (color)} = 80 \times 0.5 = 40 \text{ lines} \tag{9.8}$$

This is much lower than the luminance resolution; the difference can readily be seen when observing stationary alphanumeric characters on a television screen. The colors will appear to "bleed" between the characters and the background, assuming that the background color is different from that of the lettering. The lack of color resolution is usually less apparent with ordinary program material. The vertical resolution depends on the number of scanning lines, so of course it is the same for color as for luminance.

As in any suppressed-carrier system, the receiver is required to regenerate the color subcarrier. Thus information about the subcarrier frequency and phase must be transmitted. In color television, this is accomplished by adding a **color burst** consisting of eight to eleven cycles of the subcarrier to the back porch of each horizontal sync pulse, as shown in Figure 9.9. The receiver also uses the color burst to sense the presence of a color signal.

9.3 Television Broadcasting

In ordinary VHF and UHF television broadcasting, the picture and sound are transmitted using separate modulation schemes on separate carriers (sometimes with separate transmitters). The two carriers are 4.5 MHz apart, and the whole channel has a width of 6 MHz. See Figure 9.10 for the spectrum of channel 2, which has the lowest frequency. (See Appendix C for a complete list of television channel frequencies.)

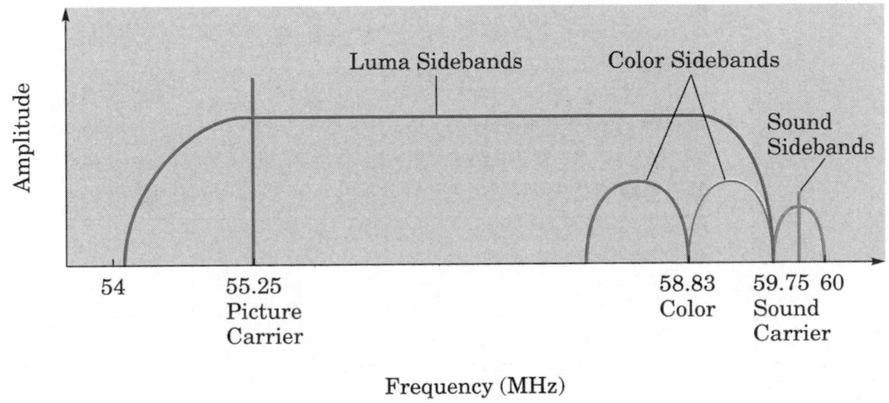

Figure 9.10
Spectrum of
channel 2

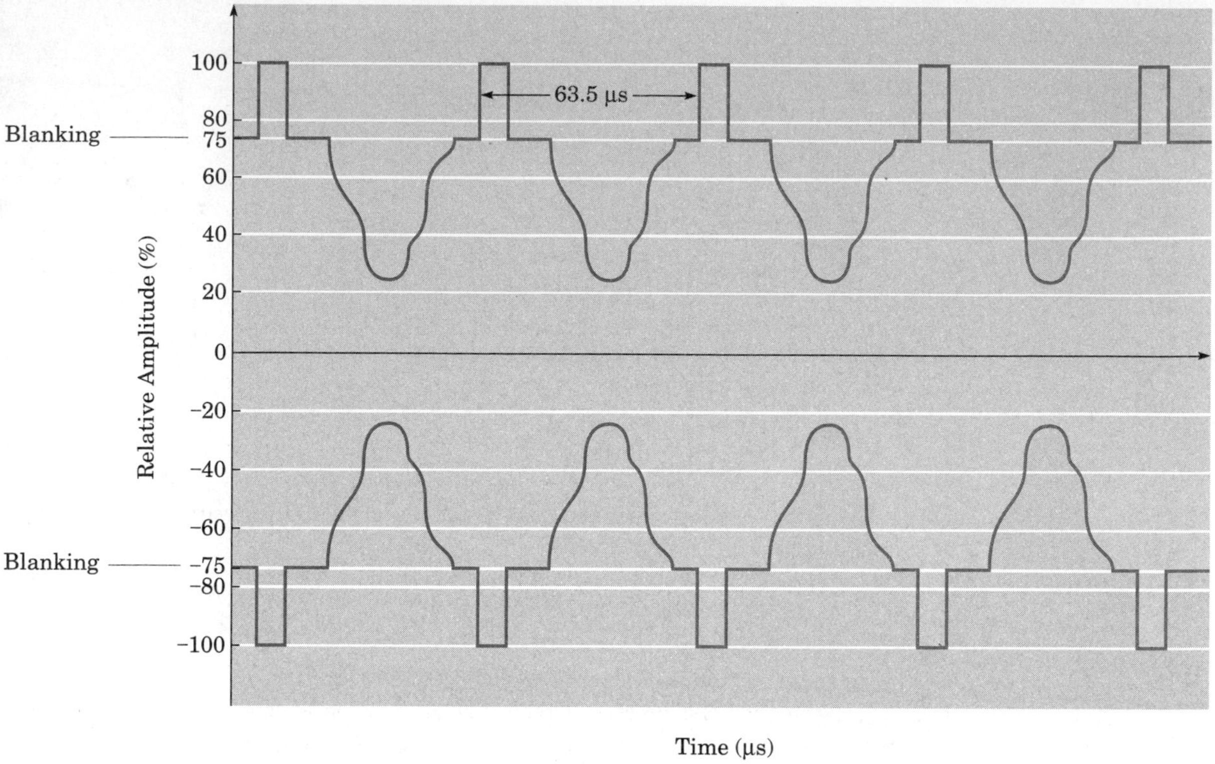

Figure 9.11 Television picture signal in the time domain

The picture signal uses vestigial-sideband AM, with a full upper sideband and a band-limited lower sideband. This reduces the bandwidth compared with double-sideband AM, while avoiding the problems that single-sideband systems have in transmitting low-frequency information. The picture carrier is not suppressed so that critical tuning is avoided and a simple envelope detector can be used for the video signal. The color modulation appears only on the upper sideband of the main picture carrier.

The sound signal uses wideband FM and employs a stereo system similar, but not identical, to that used for stereo FM radio broadcasting.

9.3.1 Picture Signal

Figure 9.11 shows the AM picture signal in the time domain. As can be seen in the figure, sync pulses are transmitted with the maximum signal amplitude and hence the maximum transmitter power, and black represents a higher amplitude than white. This form of transmission is called *negative picture transmission*. Table 9.1 shows the relative levels of the various parts of the signal, in terms of percentage of the maximum amplitude.

Negative picture transmission uses less transmitter power than would the reverse polarity, since the duration of the sync pulses is relatively short. It also ensures that the sync

Table 9.1 Broadcast video levels

Signal	IRE level	Carrier level (percent)
Tip of sync	−40	100
Blanking	0	75
Black setup	7.5	67.5
Peak white	100	12.5

pulses will be received in the presence of noise. This is important since, without sync, the picture is completely unwatchable. Impulse noise tends to cause black spots in the picture, which are thought to be less disturbing than white spots. Note that the signal level never reaches zero—this would disturb the operation of the intercarrier sound system in modern receivers, which will be described shortly.

Example 9.5 What proportion of the maximum transmitter power is used to transmit a black setup level?

Solution The proportions in Table 9.1 are voltage levels and have to be squared to get power. Therefore, the power level, as a fraction of the maximum, is

$$\begin{aligned} P \text{ (black setup)} &= 0.675^2 \\ &= 0.456 \quad \text{or} \quad 45.6\% \end{aligned}$$

9.3.2 Sound Signal

Television sound is wideband FM, similar to FM radio broadcasting. While radio broadcasting uses a total maximum deviation of 75 kHz for both mono and stereo, the television specification is 25 kHz for mono. For multichannel television sound (MTS), the television stereo sound system, the maximum total deviation is 75 kHz, with only 25 kHz in the mono (left-plus-right) signal. The same 75 μs pre-emphasis is used in both radio and television broadcasting.

The sound carrier must be exactly 4.5 MHz above the picture carrier: the difference frequency is used in television receivers, as will be explained shortly. There is a possibility of interference in receivers due to mixing of the sound carrier with the color signal on the same channel. The difference frequency is approximately

$$\begin{aligned} f_d &= 4.5 \text{ MHz} - 3.58 \text{ MHz} \\ &= 0.92 \text{ MHz} \end{aligned} \tag{9.9}$$

where

f_d = difference frequency between the sound and color signals

This 920 kHz signal will cause visible interference if it gets into the video signal path in the receiver. To reduce the visibility of this interference, the designers of the NTSC color system changed the horizontal scanning rate slightly so that the difference between the color subcarrier frequency and the sound carrier frequency would be an odd multiple of one-half the horizontal scanning frequency. The color subcarrier is set at 3.579545 MHz. Now the difference frequency is

$$\begin{aligned} f_d &= 4.5 \text{ MHz} - 3.579545 \text{ MHz} \\ &= 0.920455 \text{ MHz} \end{aligned}$$

The horizontal scanning frequency f_H is changed so that the color subcarrier is still at $227.5f_H$:

$$\begin{aligned} f_H &= \frac{3.579545 \times 10^6}{227.5} \\ &= 15.734 \times 10^3 \text{ Hz} \\ &= 15.734 \text{ kHz} \end{aligned}$$

The ratio of f_d to the horizontal scanning frequency is

$$\begin{aligned} \frac{f_d}{f_H} &= \frac{920.455 \times 10^3}{15.734 \times 10^3} \\ &= 58.5 \end{aligned}$$

That is, the interference frequency is an odd multiple of $f_H/2$.

Figure 9.12
Stereo FM radio and television sound

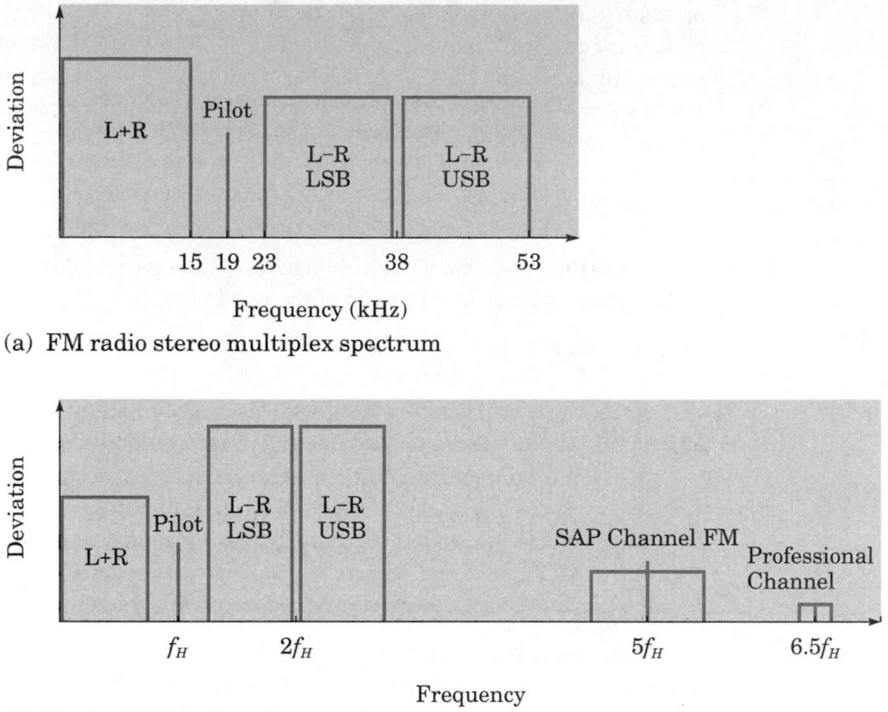

(a) FM radio stereo multiplex spectrum

(b) Typical MTS television sound spectrum

9.3.3 Multichannel Television Sound

The television stereo sound system is called multichannel television sound. It is similar to the system used for FM radio broadcasting with the following differences. The spectra in Figure 9.12 illustrates the differences. Figure 9.12(a) shows the spectrum of a stereo FM radio signal, before modulation, and Figure 9.12(b) shows a typical multichannel television sound signal.

Perhaps the most obvious difference between radio and television stereo sound is that the pilot frequency for television is equal to the horizontal scanning frequency of 15.734 kHz, instead of the 19 kHz frequency used for FM radio. As before, the left-minus-right (L − R) signal is modulated on a subcarrier at twice the pilot-carrier frequency, using double-sideband suppressed-carrier (DSBSC) modulation.

Another difference in television sound is that noise reduction is used for the L − R channel. Recall from Chapter 7 that stereo FM radio broadcasting has a severe noise penalty compared with mono as a result of shifting the L − R signal upward in frequency during the modulation process. With any FM system, the demodulated noise voltage is directly proportional to frequency. The additional noise is moved downward in frequency, into the audible range, during demodulation. A noise reduction scheme is used to reduce this problem in the MTS system.

The noise reduction scheme involves gain compression at the transmitter, followed by expansion at the receiver. A *compressor* is an amplifier that has more gain for low-level signals than for those with greater amplitude. The result of applying compression is to produce a greater modulation index for the low-level signals and therefore produce a greater signal-to-noise ratio for these signals. In order to restore the original relationship between loud and quiet sounds, the receiver must reverse the compression process. The amplifier that does this is called an *expander*. It has more gain for high-level than for low-level audio signals. The combination of compressor and expander should result in the same gain at all frequencies in the audio range.

Noise reduction is not used for the mono L+R channel in television sound, in order to maintain compatibility with receivers that are not equipped for MTS.

The multichannel television sound system also provides for a separate audio program (SAP) and one or more professional channels. The SAP can be received by the public and is intended to be used for services such as second-language translation. It can also be used for a completely separate audio program such as background music or weather information. The professional channels are for use by the television station for its own purposes, such as sending messages to employees in the field.

9.4 Television Receivers

As we have seen, the color television signal evolved from the monochrome signal. Similarly, color television receivers evolved from their monochrome predecessors. The easiest way to understand television receivers is therefore to begin with monochrome, then add the color circuitry once the monochrome receiver is understood. Our procedure with the monochrome receiver will be first to examine the stages that process the incoming signal, then to look at the circuits that scan and synchronize the CRT electron beam and that provide power to the receiver.

9.4.1 Signal Processing in a Monochrome Receiver

Figure 9.13 shows the block diagram of a monochrome receiver with the signal-processing circuitry highlighted. Note the similarity to the superheterodyne radio receivers studied earlier.

The incoming signal is applied to the *tuner*, which is a separate box that includes the RF amplifier, mixer, and local oscillator circuitry. In fact, the traditional method is to use two boxes, one each for VHF and UHF. In this type of system, which can be recognized by the set of two channel-switching knobs on the front panel, the UHF tuner has no RF stage,

Figure 9.13 Monochrome television receiver

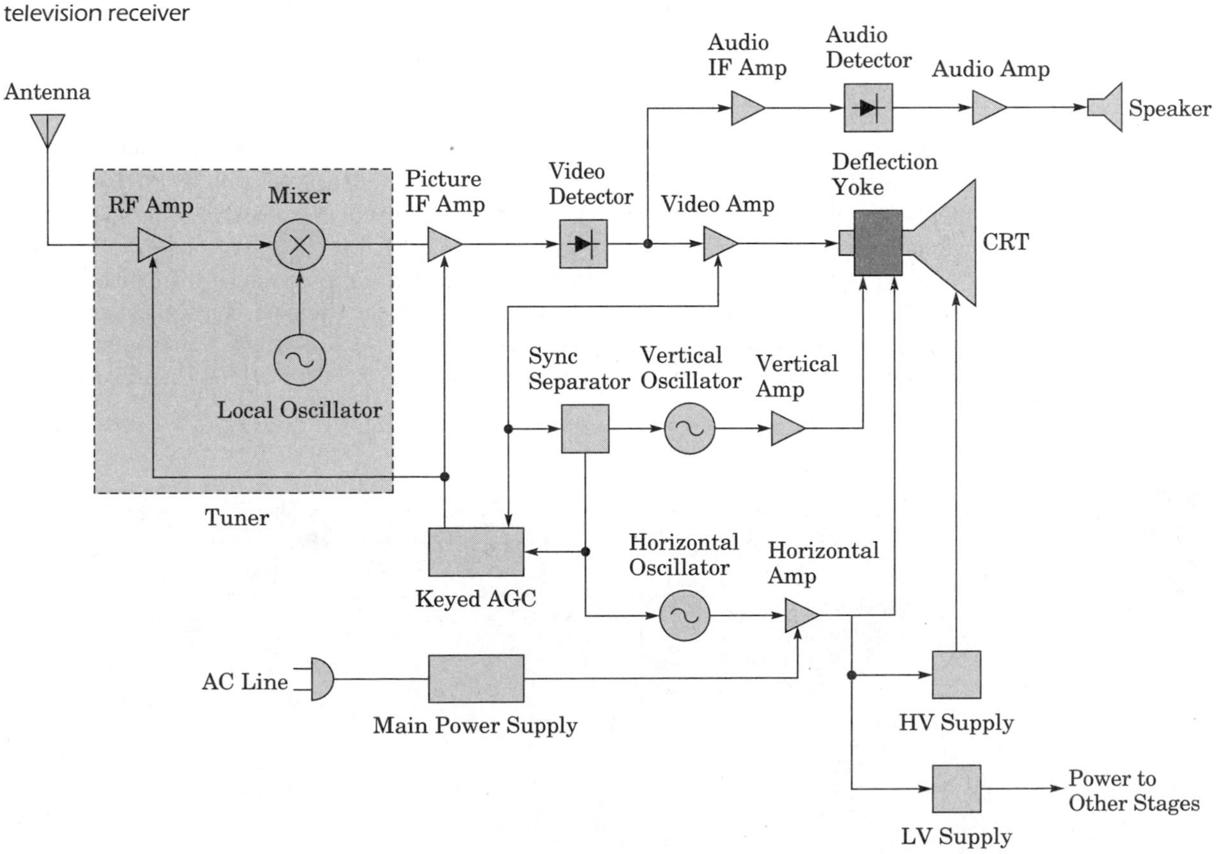

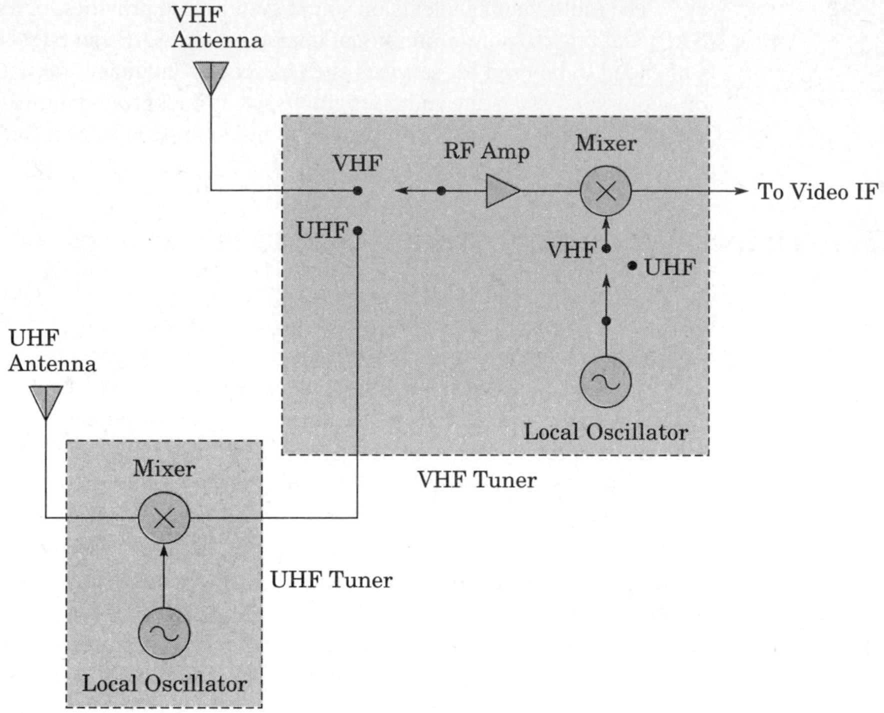

Figure 9.14
Television tuner

and the VHF tuner is configured as an additional stage of intermediate frequency (IF) amplification when the receiver is receiving a UHF channel. Figure 9.14 shows how this works. The IF is the same for VHF and UHF. The signal from the UHF tuner is fed to the VHF tuner for additional IF amplification but not for frequency conversion. The local oscillator in the VHF tuner is off when UHF is selected.

The standard IF is 45.75 MHz for the picture carrier. Since high-side injection of the local oscillator is used, the sound carrier will be converted to an IF of 41.25 MHz, 4.5 MHz below the picture carrier. Figure 9.15 shows the spectrum of a television signal in the IF circuitry.

Most modern receivers use some form of varactor tuning, rather than a mechanical switch, to change channels. Varactor tuning allows easy remote control and programmability, as all tuning can be accomplished using variable dc voltages. Varactor tuners also use diode switching to change among three frequency bands: VHF-low (channels 2–6),

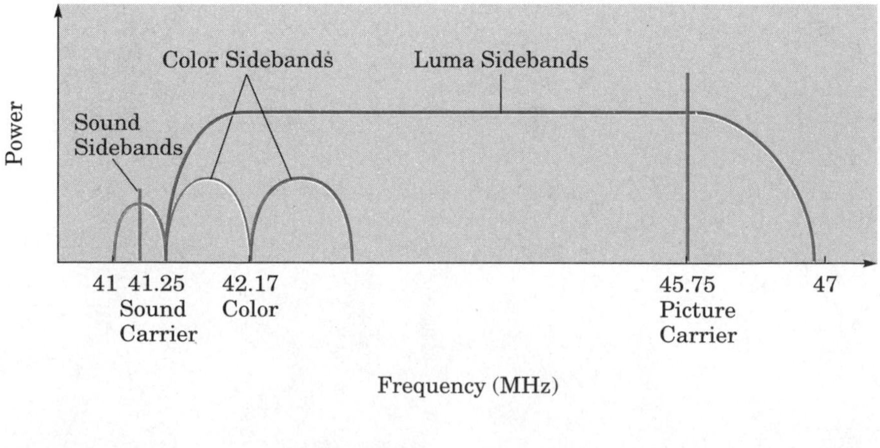

Figure 9.15
Television signal spectrum at tuner output

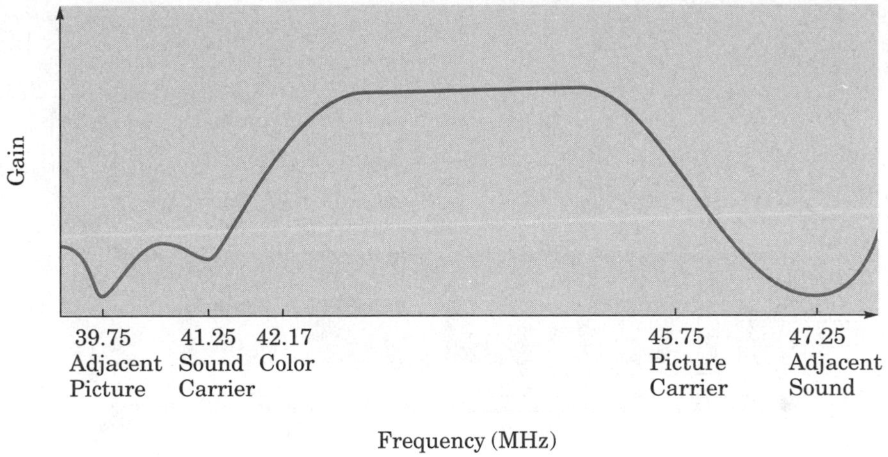

Figure 9.16 Television receiver picture IF amplifier frequency response

VHF-high (channels 7–13), and UHF. The most advanced tuners use frequency synthesis to generate the local oscillator signal.

Following the tuner, the signal passes to the picture IF amplifier. The term is actually a misnomer, for this section of the receiver also handles the sound signal. The IF amplifier looks much like those studied earlier, except that its bandwidth of about 6 MHz is much wider. The shape of the IF passband is not flat: the spectral response has the form of a "haystack curve" as shown in Figure 9.16. It has more gain for those parts of the video signal that are transmitted on only one sideband and less gain for the sound signal, to reduce interference between the sound and picture signals. Notch filters, or *traps*, are often present, tuned to the adjacent-channel picture and sound carrier frequencies. These filters reduce adjacent-channel interference.

Older designs have several stages using discrete transistors and employ several tuned circuits to shape the IF amplifier response, but newer versions often use a single integrated circuit with a surface-acoustic-wave (SAW) filter. A SAW filter consists of two sets of electrodes mounted on a piezoelectric substrate. Voltage applied to the input causes vibrations in the substrate that are transmitted acoustically through the crystal and converted back to electrical energy at the output. The frequency response of the filter depends both on the substrate and on the electrode configurations. Unlike multiple tuned circuits, SAW filters do not need alignment.

As in the receivers studied earlier, an automatic gain control (AGC) circuit allows the receiver to cope with the dynamic range of available signals. In television receivers, however, the AGC is described as *gated* or *keyed*, which simply means that the AGC looks at the signal level during the blanking interval, monitoring either the blanking or sync-pulse level and adjusting the IF gain accordingly. The advantage of this method is that the relative levels of blanking and sync are fixed and vary only with the strength of the signal. Other video levels are affected by picture content.

The stage following the video IF amplifier is the *video detector*. This is usually a diode envelope detector followed by a low-pass filter. As well as demodulating the video signal, the diode provides a 4.5 MHz sound IF signal because of the mixing between sound and picture carriers that occurs in the diode. This technique is called **intercarrier sound**. Since no separate local oscillator is required for the sound signal, the sound will automatically be correctly tuned whenever the receiver is tuned for the best picture.

Following the video detector, the sound and picture signals go their separate ways. The sound is still modulated and must go through a sound IF amplifier operating at 4.5 MHz, followed by an FM detector. The video, on the other hand, is now a baseband signal and can proceed to the video amplifier and then to the CRT.

The video amplifier provides a positive-sync video signal at the CRT cathode. It has

Figure 9.17
Monochrome CRT

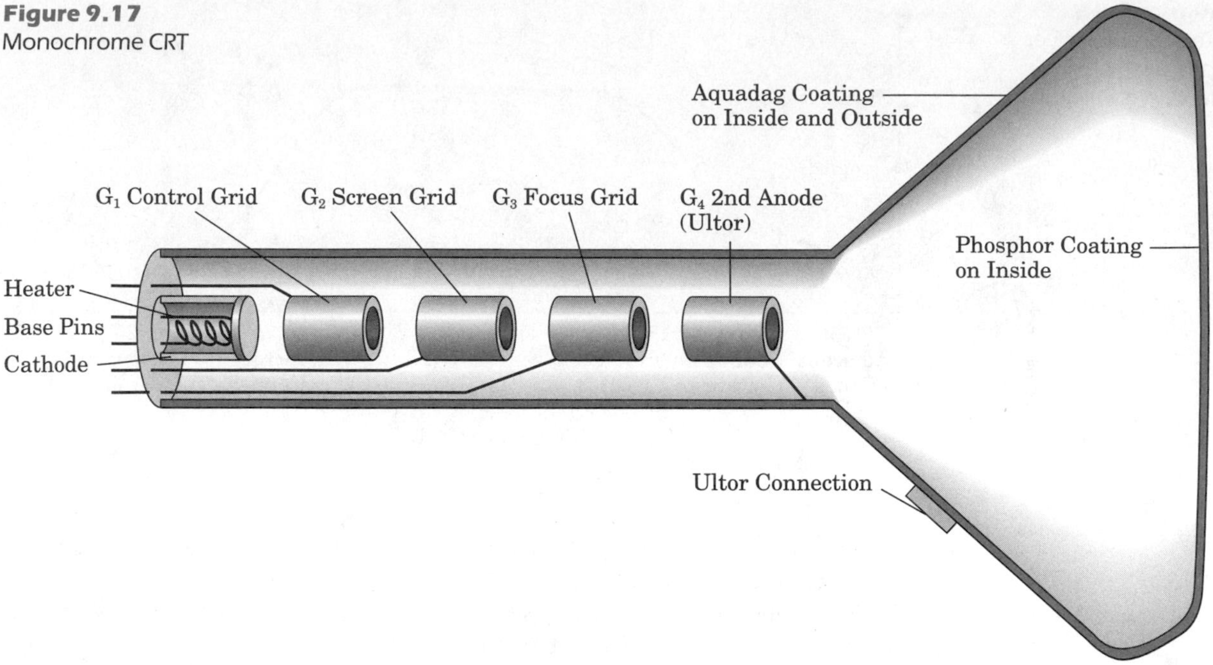

a 4.5 MHz trap at its input to reduce interference from the sound signal. The contrast control is a video gain control.

9.4.2 The Monochrome Cathode-Ray Tube

The basic structure of a monochrome CRT is shown in Figure 9.17. The electron gun emits a beam of electrons, the intensity of which is controlled by the video signal. Inside the electron gun, a hot cathode emits electrons, by thermionic emission. The control grid (G_1) is actually a cylinder maintained at a negative potential with respect to the cathode. In modern circuits, the grid is kept at ground potential, and the cathode is about 50 V positive. The brightness control adjusts the bias, and the positive-sync video signal is also applied to the cathode. As the video signal becomes more positive, the intensity of the electron beam is reduced.

The screen grid (G_2), sometimes called the *first anode*, is maintained at a positive voltage of about 400 V and serves to accelerate the electron beam. The screen grid and the focus grid (G_3) form an electrostatic lens to focus the electron beam on the screen. The second anode (G_4) is connected to the high-voltage supply, which in a monochrome receiver is about 10 to 12 kV, and accelerates the beam to its final velocity. The beam current is low, generally less than 1 mA.

The second anode is often called the **ultor**. Its connection is on the flared part of the tube, called the *bell*, to avoid arcing, which could occur if the high voltage were applied to the pins at the end of the tube. Both the inside and outside of the tube have a conductive coating called the *aquadag*. The inside coating is at the second anode potential, and the outside is grounded. The coatings provide shielding, and the capacitance between the two coatings is used as a filter in the high-voltage power supply.

The electron beam strikes a phosphor coating on the inside of the CRT faceplate. For a monochrome tube, the phosphor is white with a persistence of about 5 ms. There is a thin aluminum coating on the inside of the faceplate to provide a return path for electrons.

In order to trace out the lines that form an image on the screen of the tube, the beam is deflected by electromagnets that surround the tube and are collectively called the **yoke**. They are provided with sawtooth current waveforms by special circuits that will be described shortly. Permanent magnets are also used for beam centering and to correct distortion around the edges of the screen.

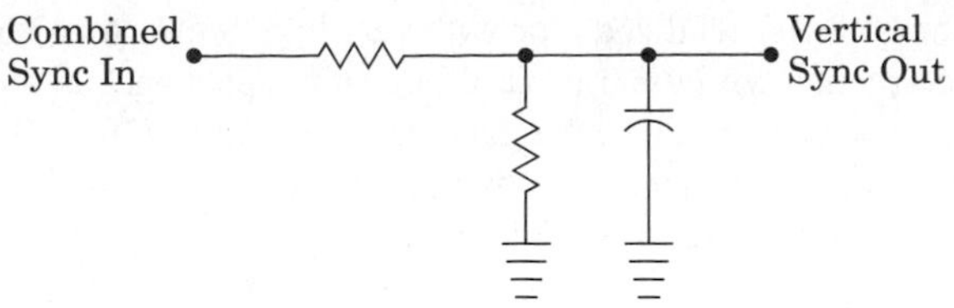

Figure 9.18 Integrator to separate vertical sync

9.4.3 Scanning and Synchronization

A television receiver needs some means of producing the required sawtooth current waveforms in the vertical and horizontal deflection coils. In addition, these waveforms must be synchronized with the incoming signal. The means of doing this differ from one receiver to another, with modern designs using one integrated circuit that separates the sync pulses from the rest of the signal, further separates horizontal from vertical sync, and generates horizontal and vertical scanning signals that are properly synchronized. All that remains are the horizontal and vertical output stages, which use too much power to be included in the integrated circuit. Older designs use a number of discrete transistors (or tubes, in still older circuits) to perform the same functions.

Whether part of an integrated circuit or made up of discrete components, the sync separator uses the difference in amplitude between the sync pulses and the rest of the video signal to detect sync pulses. Then it uses the difference in duration to separate horizontal from vertical sync. That can be accomplished by a simple integrator like the one shown in Figure 9.18. The capacitor does not have time to charge very far during the brief horizontal sync pulses, and it discharges completely between pulses. The vertical sync pulses are much longer, however, and allow the capacitor time to charge completely.

The separated horizontal and vertical sync pulses are used to control the frequency of the horizontal and vertical oscillators, which must also be free-running in the absence of sync pulses so that there will be a **raster** on the screen in the absence of a signal. (*Raster* is the term used to describe the pattern of scanning lines on the screen.)

The output from each oscillator must be amplified and shaped to produce the required sawtooth current waveform in the deflection coils. In the vertical coils, the current and voltage waveforms have roughly the same shape, and the amplifier circuitry resembles an ordinary push-pull Class B amplifier.

The horizontal output circuitry is less conventional. The deflection coils have considerable inductive reactance, causing the voltage waveform across them to resemble a train of pulses more than it does a sawtooth. There is also a good deal of power required, since the coils store energy in their magnetic fields. The circuitry used for horizontal deflection circuits is designed to increase efficiency by recycling some of the stored energy, using a resonant circuit to do so.

Figure 9.19 is a simplified schematic diagram of a typical horizontal output circuit.

Figure 9.19 Simplified horizontal output circuit

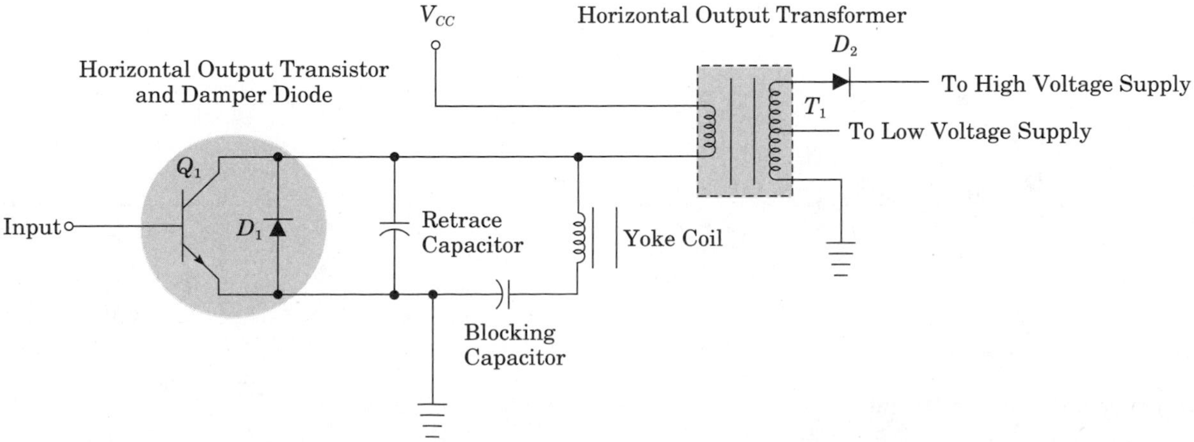

The transistor Q_1 is a specialized type with very high breakdown voltage. The damper diode D_1 is often combined with the transistor in a single package, as shown. T_1 is the horizontal output transformer, often called the *flyback transformer*. Often D_2, the high-voltage rectifier, is really a voltage-tripler circuit that is enclosed in the package containing T_1.

The input to this circuit is in the form of pulses that occur once per horizontal line. The output consists of a sawtooth current waveform in the horizontal deflection coils. The voltage pulses necessary to provide this are also used, suitably changed in amplitude, to provide both the high voltage for the CRT and low voltages for many of the other circuits in the receiver.

Figure 9.20 illustrates some of the important waveforms that occur in this circuit. The simplest way to analyze the circuit is to begin in the center of the trace. When the beam is in the center of the screen, there is no current flowing in the deflection coils. A current pulse flows into the base of Q_1 (Figure 9.19) at this time, saturating it. The current through the transistor and the deflection coils increases gradually because of the inductance of the coils. The graph of current with respect to time is exponential, of course, but it appears to be almost a straight line because we are looking at only the beginning of the curve.

As the deflection-coil current increases, the electron beam moves at a constant rate toward the right of the screen. When it gets there, the base current into Q_1 is reduced to zero, turning off the transistor. In fact, the current goes negative briefly to remove any stored charge in the base and turn Q_1 off as quickly as possible. This is necessary because the transistor must dissipate a good deal of heat when it is in the active region. When it is saturated, the power dissipation is small, and when cut off, it dissipates no power at all.

Of course, the current through the deflection coils cannot be reduced to zero instantaneously because of the energy stored in their magnetic fields. Current continues to flow, charging the retrace capacitor C_1. After several microseconds, the current has been reduced

Figure 9.20
Horizontal output circuit waveforms

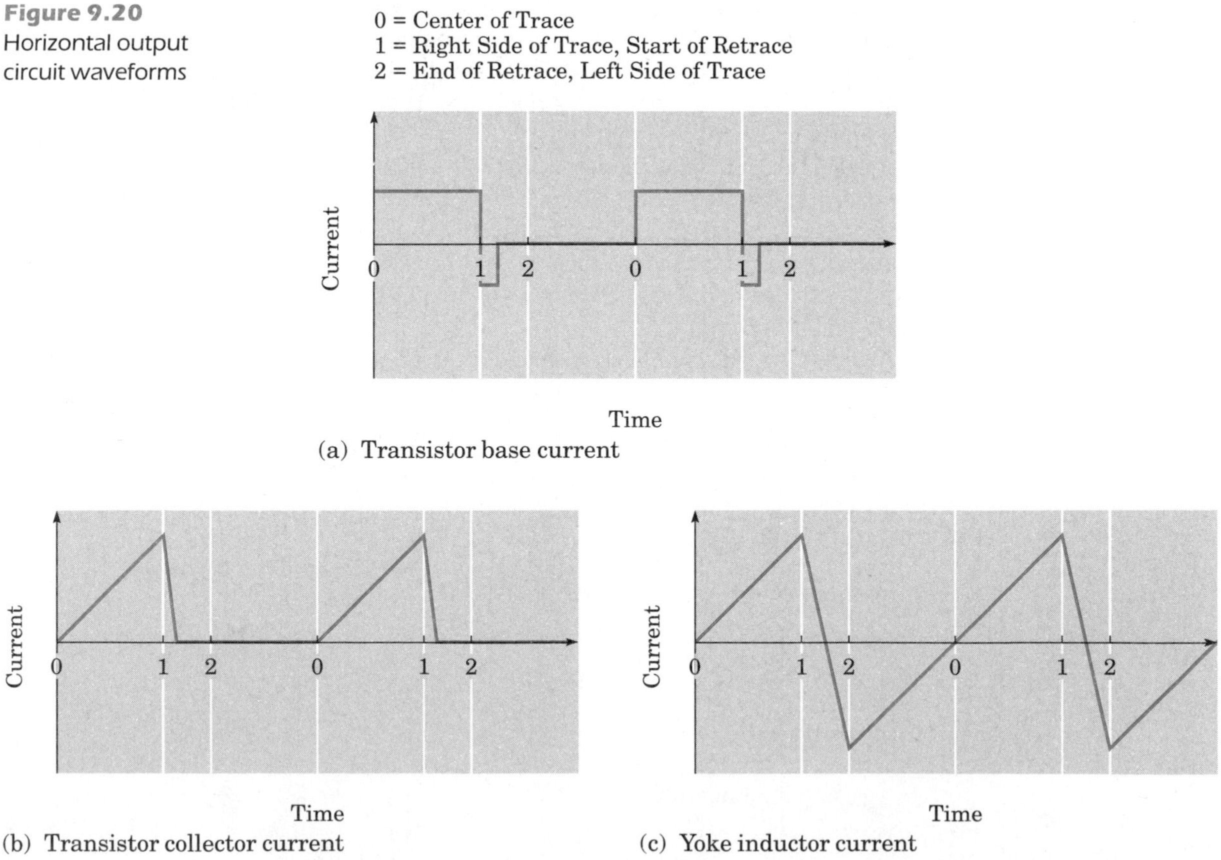

(a) Transistor base current

(b) Transistor collector current

(c) Yoke inductor current

Figure 9.21
Television receiver power supply

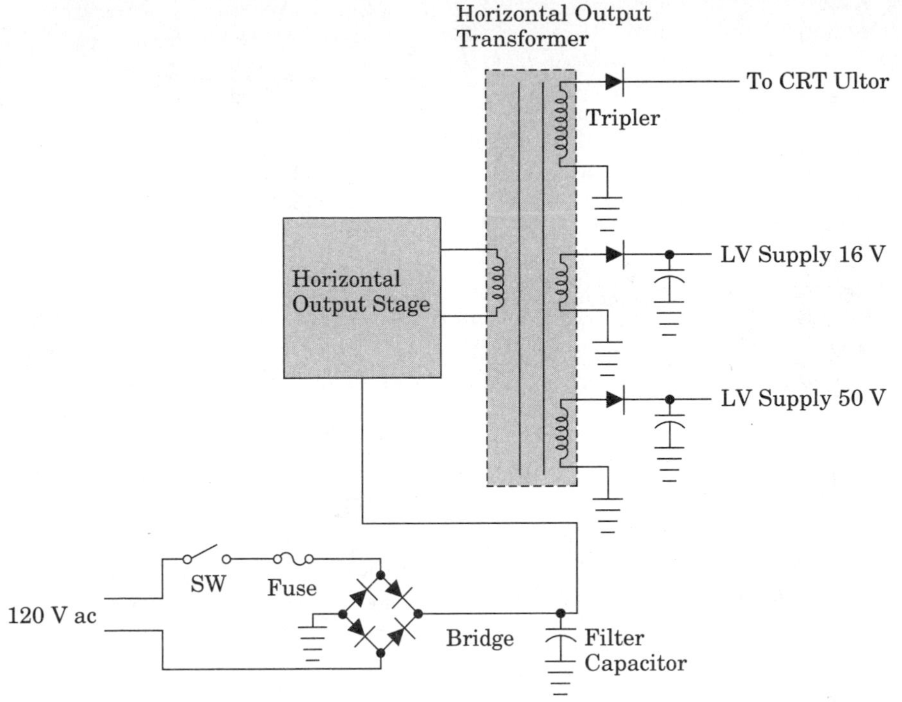

to zero, at which point the capacitor begins to discharge into the coil. In other words, there is a resonant circuit. Damped oscillations would continue for some time except for the fact that just as the capacitor finishes discharging and the electron beam reaches the left of the screen, the damper diode begins to conduct and the coils discharge the remainder of their energy through the diode. By the time this has occurred, the electron beam is once more in the center of the screen and the process is ready to repeat.

The presence of a large amount of power at a frequency of about 16 kHz is used by receiver manufacturers to simplify power-supply design. Figure 9.21 is a representative example. Typically, there is either a full-wave bridge or a half-wave rectifier operating directly from the ac line voltage, without a power transformer. After filtering, this produces a voltage on the order of 160 V, which can be used to power some stages, notably the horizontal output. The rest of the voltages are derived from the horizontal output transformer. The whole operation somewhat resembles a switching power supply, in that advantage is taken of the reduced requirements for transformer iron and filter capacitance at the higher frequency. The use of a transformerless supply does represent a safety hazard when working on television receivers, since one side of the power line is often connected directly to the chassis. An isolation transformer, which is simply a transformer with a 1 : 1 turns ratio and sufficient power capacity to supply the power required by the receiver, should always be used when working on television receivers.

Since the horizontal output circuit supplies much of the power used in the receiver, a failure in the horizontal circuitry is likely to result in a television that shows little if any sign of life. This is a useful thing to remember when troubleshooting a television receiver.

9.4.4 Color-Signal Processing

Thus far, we have been looking at monochrome receivers. They are still made, of course, but mainly for use as second receivers, for kitchens, as portables, and so on. However, we will see that what we have learned in our study of monochrome receivers is directly applicable to color television.

Figure 9.22 is the block diagram of a typical color receiver. Some similarities should be immediately apparent. The tuner and video IF sections are unchanged from those of the

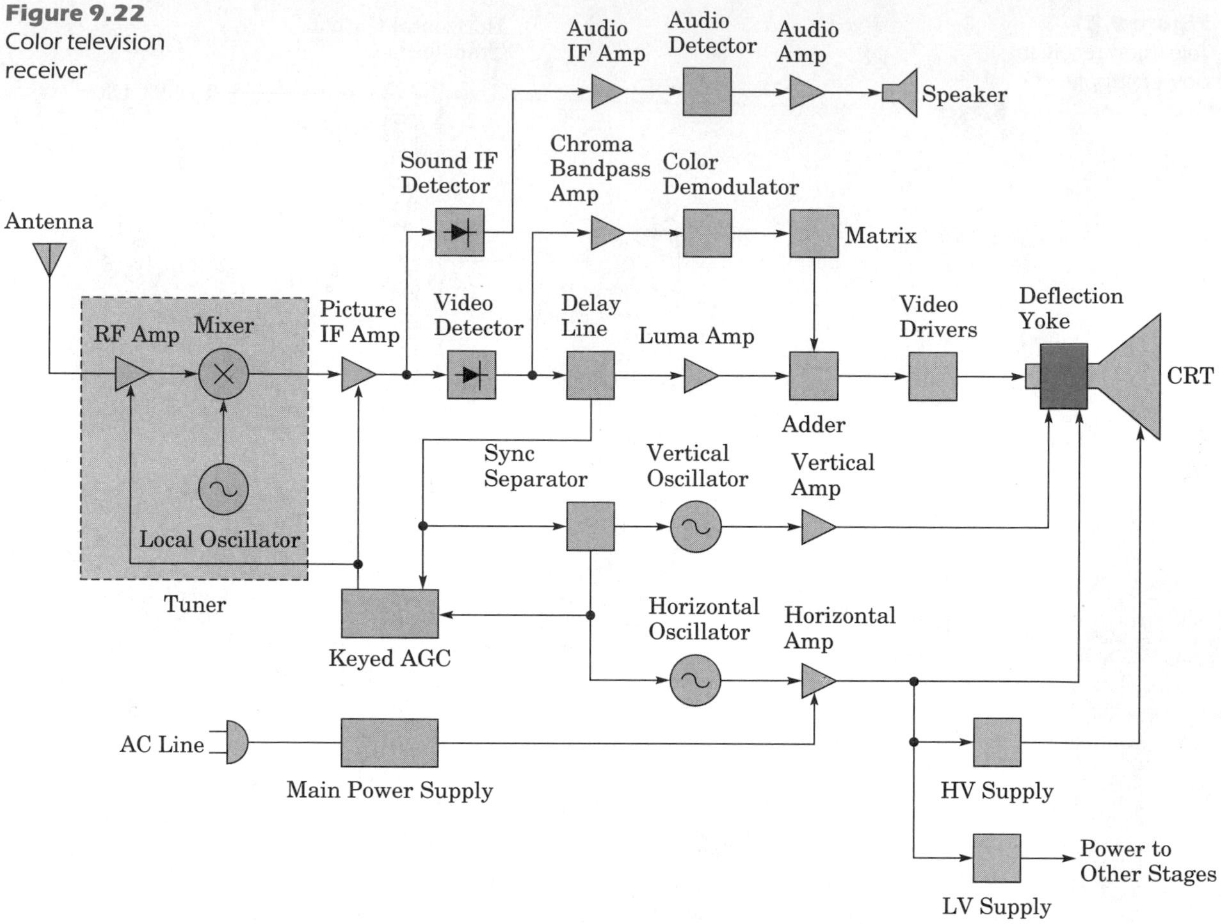

Figure 9.22
Color television receiver

monochrome receiver, on the block diagram. Actually there are likely to be a few differences in circuitry. Color receivers are far more likely to use more advanced technology such as varactor tuners, frequency synthesis, and remote control. Likewise, the audio section is similar, but color receivers are much more likely to incorporate stereo sound. Raster and synchronizing circuitry is the same for monochrome and color receivers, except that color receivers use a higher accelerating voltage for the CRT.

What, then, are the essential differences between monochrome and color receivers? The differences are mainly related to the way in which the chroma signal is demodulated and the result combined with the luma. The composite color signal must be converted back into an RGB signal before it is applied to the color CRT.

Proceeding along the signal chain from the antenna, the first notable difference is that there is a separate sound IF detector, which separates out the 4.5 MHz sound IF signal. In the monochrome receiver, this function was quite adequately served by the video detector, but this is inadvisable in a color receiver. If the sound signal were allowed to enter the video detector, undesirable interference could result from the mixing of the sound carrier and the color signal. This is avoided by taking off the sound at an earlier point and installing a 41.25 MHz trap to remove the sound signal before the video detector.

Following the video detector, the chroma signal must be separated from the luma and sent to the color demodulators. There are two common ways to do this. The simplest, shown in Figure 9.22, is to apply the signal to a bandpass amplifier with its center frequency at the color subcarrier frequency of approximately 3.58 MHz and a bandwidth of about 1 MHz. This will remove most of the luma, though of course the highest-frequency luma components will pass through the filter.

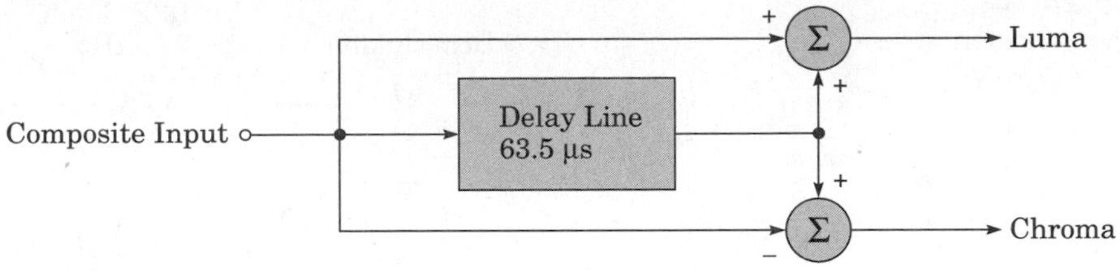

Figure 9.23
Comb filter

Similarly, there will be some chroma information in the luma signal. Because of the frequency interleaving described earlier, any interference will not be very serious. The delay line shown in the luma signal path is there to compensate for the delay that the chroma signal experiences in the bandpass filter, so that chroma and luma arrive at the CRT at the same time.

There is a better way to separate luma from chroma by using a **comb filter**. Figure 9.23 shows how this works. The composite signal is applied to the input of a $1H$ delay line, that is, one with a delay of exactly one horizontal line period, or 63.5 μs. The original and delayed signals are added to give the luminance output and subtracted to produce the chroma.

The operation of a comb filter can be understood by considering a signal with a period of $1H$ (63.5 μs). The delay results in a phase shift of 360°, and the delayed signal is indistinguishable from the original. Assuming that the amplitudes are made equal, the output of the summer is enhanced, while the subtractor has zero output. The same logic holds for any harmonic of the horizontal scanning frequency. If we recall that luminance signals have most of their power at or near the harmonics of the horizontal line rate, it is apparent that the luma signal emerges from the filter almost unchanged.

Now consider a signal with a frequency one-half that of the horizontal scanning frequency, that is, its period is $2H$ and the delay line shifts its phase by 180°. Summing the delayed and original signals results in cancellation, but subtracting results in reinforcement. It can easily be shown that the same is true for odd multiples of one-half the horizontal line rate. Recalling that the chroma signal is modulated on a subcarrier at $227.5f_H$, it is apparent that the subtractor output has most of the chroma and very little of the luma. Comb filters are used in most new designs.

Once the chroma signal has been separated from the luma, it must still be demodulated. The obvious solution would be to demodulate the QUAM color signal into the original I and Q signals using two balanced modulators, each supplied with a carrier of the correct phase. Of course, those carriers have to be synchronized with the original subcarrier, a relatively easy job since the color burst provides a sample once per line.

Unfortunately the reality of a demodulator using the I and Q axes is fairly complex. Figure 9.24 shows how it can be done. The extra filters take into account the different bandwidths of the I and Q signals, and the extra delay line compensates for the different delays in filters of different bandwidth. The initials AFPC, by the way, stand for *automatic frequency and phase control*, a system that phase-locks a crystal oscillator to the exact frequency and phase of the color burst. This in turn requires that the burst be separated from the rest of the signal. Its position right after horizontal sync makes this fairly easy. The block marked *color killer* does what it says: it turns off the color circuitry when there is no color burst.

It is possible to simplify the color circuitry somewhat by using two different axes to demodulate the color signal. This is possible as long as the axes chosen are at a phase angle of 90° to one another. Of course, the signals that result are combinations of I and Q, but by judicious choice of axes, it is possible to end up with the two signals $R-Y$ and $B-Y$.

Figure 9.25 shows a color demodulator using the $R-Y$ and $B-Y$ axes. The number of filters is reduced, and there is no need for a delay line—since both of the demodulated

Figure 9.24
Color demodulation using *I* and *Q* axes

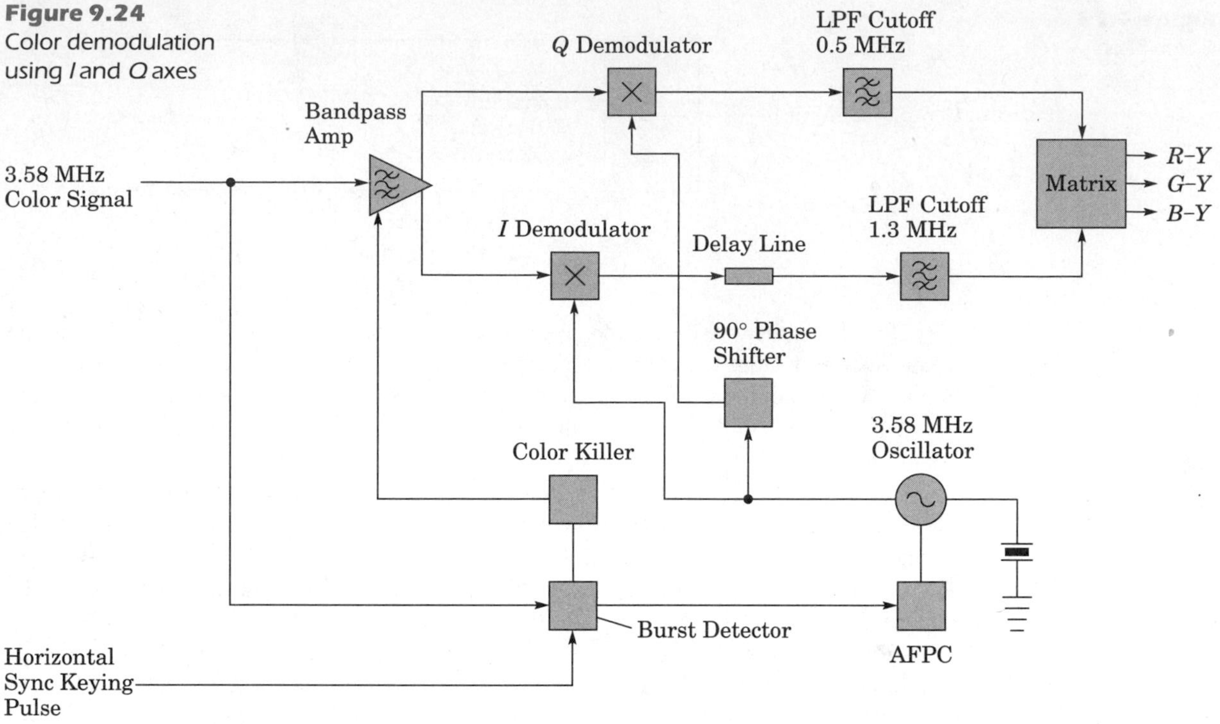

signals are combinations of *I* and *Q*, there is no point in using a bandwidth larger than that of the *Q* signal, which is 500 kHz. The relative simplicity of the $R-Y/B-Y$ demodulator is paid for in reduced horizontal resolution for color.

Whichever color demodulator system is chosen, the output will consist of three signals, $R-Y$, $B-Y$, and $G-Y$. These can simply be added to the luminance signal to produce the three primary signals (red, green, and blue) that are required by the CRT.

Figure 9.25
Color demodulation using R–Y and B–Y axes

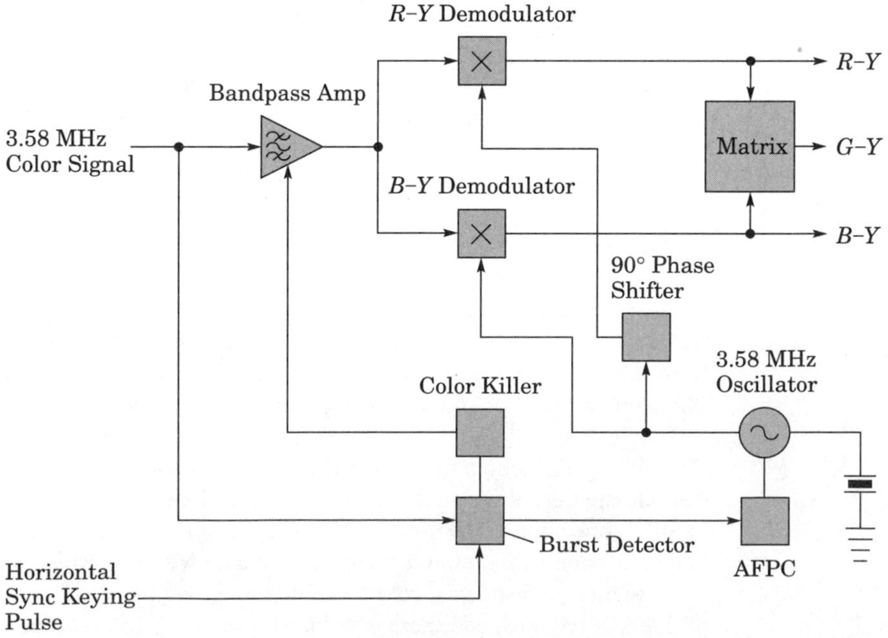

Figure 9.26
Color CRT

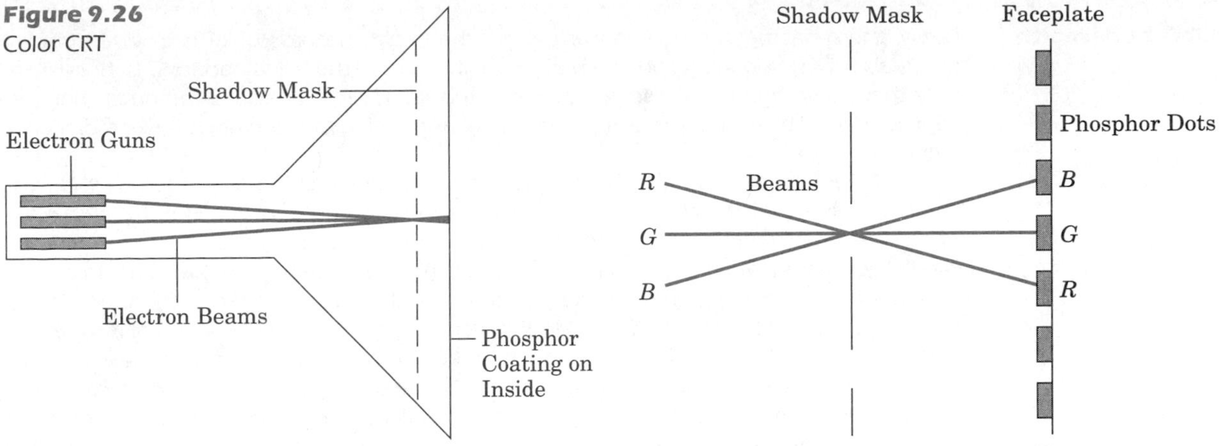

(a) Cross section

(b) Close-up showing convergence of beams

The color and tint controls adjust the color circuitry. The tint color alters the phase of the injected carriers to the demodulators, and the color control varies the gain of the color bandpass amplifier.

9.4.5 Color Cathode-Ray Tubes

In order to reproduce the three primary colors of the color television system, a color CRT must have three electron beams. Most tubes use three electron guns, arranged either in a triangle (delta guns) or in a straight line (in-line guns). Most modern tubes use the in-line system, which makes adjustment easier at the expense of requiring a large neck diameter. An exception is the Sony Trinitron™, which has a single large gun with three cathodes. Figure 9.26 shows a color CRT using in-line guns.

The three video signals are applied at the cathodes, one to each electron gun. The faceplate is covered with a dot pattern using three types of phosphor that glow red, green, and blue when bombarded with electrons. Each gun illuminates a different type of phosphor. Because the electron beams are large enough to illuminate more than one phosphor dot, a *shadow mask* consisting of a sheet of steel containing thousands of tiny holes is placed near the faceplate. Excess electrons that would have hit the wrong phosphor dot strike the shadow mask instead. Of course, the efficiency of the tube is much reduced, since about 70% of the electrons strike the mask rather than the tube. Consequently the accelerating voltage must be higher than for a monochrome tube—on the order of 20 to 30 kV for a typical color tube.

Color CRTs require careful adjustment. First, the beam from each gun must land on the proper color dots; this adjustment is called **purity**. If the purity is not adjusted correctly, patches of incorrect color can be seen. Second, all three beams must land on the same triad, or group of three dots. This adjustment is called **convergence**, and misadjusted convergence results in color fringes. The means of adjustment of purity and convergence varies according to whether the tube employs delta or in-line guns. For in-line guns, purity is adjusted using permanent magnets mounted around the neck of the tube. Convergence near the center of the screen is adjusted similarly, using a different set of magnets. Convergence near the center is called *static convergence*. The term *dynamic convergence* refers to convergence around the edges and in the corners; this is adjusted by tilting the entire yoke containing the deflection coils.

Because of the steel shadow mask, color CRTs can be affected by external magnetic fields, which can magnetize the shadow mask so that it deflects the electrons a small amount, causing them to land on phosphor dots of the wrong color. The problem is avoided by installing a coil around the faceplate of the tube; the coil is energized by 60 Hz ac for a few seconds each time the power to the receiver is turned on. The function of this *degaussing coil* is to remove any magnetism that has built up on the shadow mask since the last

CRTs and Health

Lately, there has been much concern about the health hazards of CRTs in video display terminals (VDTs) and television receivers. Actually, most of the talk is about VDTs, but electrically there is very little difference between a television receiver and a computer monitor. People do tend to sit closer to computers, increasing their exposure to whatever radiation is emitted.

It is certainly true that CRTs emit radiation. The most obviously dangerous and least controversial form is X-ray emission. X-rays are known as ionizing radiation because each photon has sufficient energy to ionize molecules. X-rays are known to cause cancer. They can be generated when high-energy electrons strike the screen or shadow mask in a color CRT. Only color receivers and monitors emit X-rays; the voltages used with monochrome CRTs are too low to generate them. Modern color CRTs emit very low levels of X-rays, provided the high voltage level is correct, and all modern monitors and receivers have circuitry to shut them down if the voltage gets too high due to a malfunction. Older receivers using tubes can also emit X-rays from the high-voltage rectifier and damper diode tubes. The levels of X-rays from modern equipment are generally considered so low as to represent negligible danger, but with ionizing radiation there really is no "safe" limit. Background radiation due to cosmic rays and uranium deposits is already "dangerous" in the sense that it causes some cancers and mutations.

CRTs also produce a strong dc electric field due to the accelerating voltage and fairly strong electric and magnetic fields at the vertical and horizontal sweep frequencies. Whether there is any danger from these fields is highly controversial. Some people have claimed that they contribute to everything from headaches to miscarriages to cancer. Studies have been and are being done, but nothing has really been proved as of this writing. Still, reputable scientists, who used to scoff at claims of any danger from nonionizing radiation, have begun to take these fields seriously. It is certainly true that reducing exposure to them would do no harm. Some monitor designers have already begun to take steps to improve electric and magnetic shielding.

One obvious solution to these problems is to move to solid-state liquid-crystal displays. They emit no X-rays at all, and the electric fields are very weak due to the low voltages involved. Unfortunately, so far the cost is higher, and the resolution and contrast are not as good as for CRTs.

A more mundane problem with CRTs is eyestrain, which can result from too low a resolution (and hence is worse with LCD displays) or from glare from the screen. Regardless of whether electromagnetic fields cause headaches, eyestrain certainly does, and the VDT user is wise to eliminate sources of glare. Screens with antireflective coating are also helpful.

time the set was used. Even the magnetic field of the earth can cause problems if the receiver is moved.

9.5 Cable Television

Community-antenna television (CATV) systems, usually known as cable television, have several advantages over individual rooftop antennas. Besides removing what some people (though not usually communications specialists) think of as unsightly antennas from rooftops, the system often allows better reception, since cable companies can choose a good location and erect a relatively elaborate antenna system. CATV systems can also provide services unavailable on regular broadcast television, for example, locally generated channels and specialized services delivered by satellite, such as news, weather, sports, and movie channels.

9.5.1 Structure

Figure 9.27 shows the structure of a typical cable-television system. The origination point for the signals, called the *head-end*, is located at a suitable antenna site usually just out of town and preferably on a hill. The structure of a cable-television system is treelike, with

Figure 9.27 Cable-television system

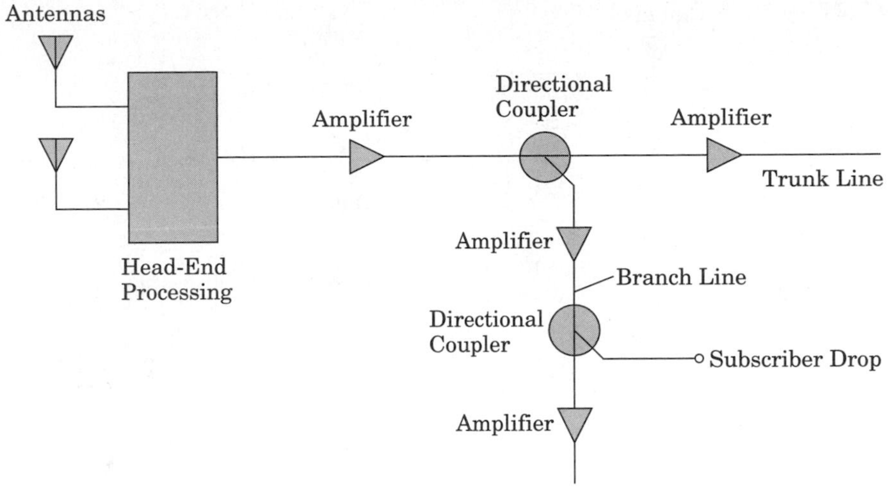

trunk lines using large-diameter, low-loss coaxial cable and smaller-diameter branches leading to *drop lines* that provide the signal to individual buildings.

Because of the losses in the cable, amplifiers are necessary at intervals of about 0.65 km. The actual distance is governed by the loss on the line, which varies with the maximum frequency used by the system and with the type of line. A typical cable system has amplifiers with approximately 20 dB gain, followed by a section of line with about 20 dB of loss. Since the loss on the cable increases with frequency, so does the gain of the amplifiers. Temperature also affects cable losses, so cable system amplifiers use a form of automatic gain control that monitors the levels of at least two pilot carriers sent down the cable from the head-end, keeping them at the appropriate value.

Signal levels on cable-television systems are usually given in dBmV (that is, decibels with respect to 1 mV). Mathematically, this is

$$\text{dBmV} = 20 \log \frac{V}{1 \text{ mV}} \tag{9.10}$$

where

V = voltage on the cable

Signal levels at the subscriber drop vary from 0 dBmV (1 mV) to about 10 dBmV (3 mV) or more; that is, the levels at the receiver are similar to what would be obtained from an antenna with a reasonably strong signal. On the trunk and branch cables there is somewhat more variation, in the range of about 10 to 40 dBmV. The standard impedance for cable used in CATV systems is 75 Ω.

Directional couplers are used where the cable branches, to avoid problems due to reflection. Mismatches that cause reflection of signals can be caused by defects in the cable or by faulty connections. The subscriber drops sometimes use directional couplers and sometimes simple resistive taps. In either case, there must be at least 18 dB of isolation between subscribers.

Example 9.6

Referring to the portion of a CATV system shown in Figure 9.28, calculate the signal level, in dBmV, at the input and output of each amplifier and at the subscriber drop. Also calculate the signal level at the subscriber drop in microvolts and in dBm.

Figure 9.28

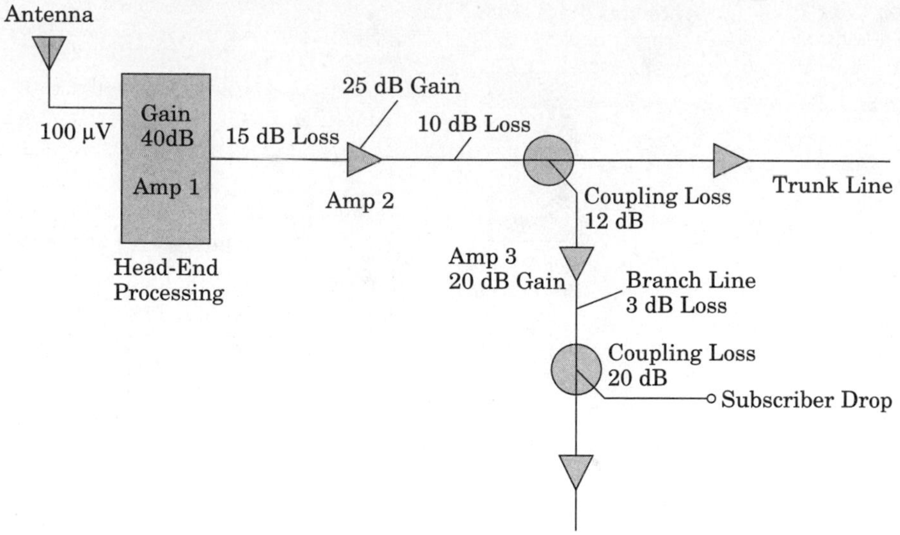

Solution The first step is to convert the 100 μV level from the antenna to dBmV. From Equation (9.10),

$$\begin{aligned} In_1 \text{ (dBmV)} &= 20 \log \frac{V}{1 \text{ mV}} \\ &= 20 \log 0.1 \\ &= -20 \text{ dBmV} \end{aligned}$$

This signal is applied to the input of the first amplifier, actually the head-end signal processing, which has a gain of 40 dB. One of the pleasant things about the decibel system is the fact that once inputs are converted to a decibel form, addition of gains and subtraction of losses are all that is required to find output levels. Thus, the output level of Amp 1 is

$$\begin{aligned} Out_1 \text{ (dBmV)} &= -20 + 40 \\ &= 20 \text{ dBmV} \end{aligned}$$

The input level of Amp 2 is this amount less the loss in the intervening cable:

$$\begin{aligned} In_2 \text{ (dBmV)} &= 20 - 15 \\ &= 5 \text{ dBmV} \end{aligned}$$

At the output of this amplifier, we have

$$\begin{aligned} Out_2 \text{ (dBmV)} &= 5 + 25 \\ &= 30 \text{ dBmV} \end{aligned}$$

This is followed by 10 dB of loss in the cable and another 12 in the directional coupler, for a total of 22 dB loss before the next amplifier.

$$\begin{aligned} In_3 \text{ (dBmV)} &= 30 - 22 \qquad & Out_3 \text{ (dBmV)} &= 8 + 20 \\ &= 8 \text{ dBmV} & &= 28 \text{ dBmV} \end{aligned}$$

There is a further 3 dB cable loss and 20 dB loss in the subscriber tap, leaving a signal strength at the subscriber drop of

$$\begin{aligned} V_{drop} \text{ (dBmV)} &= 28 - 23 \\ &= 5 \text{ dBmV} \end{aligned}$$

As a voltage, this is

$$V_{drop} = \text{antilog} \frac{5}{20} \text{ mV}$$
$$= 1.78 \text{ mV}$$
$$= 1780 \ \mu\text{V}$$

The power into 75 Ω is

$$P_{drop} = \frac{V_{drop}^2}{75}$$
$$= \frac{(1.78 \times 10^{-3})^2}{75}$$
$$= 42.16 \times 10^{-9} \text{ W}$$
$$= 42.16 \text{ nW}$$

In dBm, this is

$$P_{drop} \text{ (dBm)} = 10 \log \frac{P_{drop}}{1 \text{ mW}}$$
$$= 10 \log (42.16 \times 10^{-6})$$
$$= -43.8 \text{ dBm}$$

9.5.2 Midband and Superband Channels

The loss in any type of coaxial cable increases with frequency. It is certainly possible to transmit UHF television frequencies along a cable, but more amplifiers would be required. Since every amplifier adds noise, the signal-to-noise ratio would be impaired. A less expensive and more practical alternative is to take advantage of the gap in the spectrum between channels 6 and 7 and to use the frequencies above channel 13. Of course, these regions are occupied by other services, but there is no need to carry most of them on the cable. The one exception is the FM broadcast band, located immediately above channel 6, which most cable systems do carry. (Appendix D shows the most common way of numbering the cable-television channels.) The numbers 2 through 13 are used for the channels that coincide with the normal VHF broadcast frequencies—this allows an ordinary receiver to be used with the cable without modification. The midband channels 14 through 22 occupy the spectrum between the top of the FM broadcast band and the bottom of channel 13. The superband channels start with number 23 just above the normal broadcast channel 13 and continue as far as necessary to handle all the requirements of the system.

Obviously, UHF television stations must be converted to lower frequencies in order to be carried on cable. Often, VHF stations are moved from their broadcast channels as well. There are many reasons for this. Sometimes it is to avoid interference between the channel on cable and signals that may be picked up on the internal wiring of the television receiver. Some channels are also subject to local oscillator interference from other receivers connected to the cable. Another factor that influences channel allocation is the fact that cable loss increases with frequency. The loss is made up by amplifiers along the cable, but the signal-to-noise ratio becomes worse as the frequency gets higher.

For all these reasons and others, cable companies frequently move stations from one channel to another. This can be done either by mixing or by demodulating the signal and remodulating it on another carrier.

Example 9.7

Suppose a television receiver tuned to channel 6 radiates a local oscillator signal into the cable. What channel will be interfered with?

Solution All modern receivers use a picture IF of 45.75 MHz with high-side injection of the local oscillator signal. The picture carrier of channel 6 is at a frequency of 83.25 MHz, so

$$f_{LO} = 83.25 + 45.75$$
$$= 129 \text{ MHz}$$

This could interfere with midband cable channel 15 (126–132 MHz).

9.5.3 Converters and Cable-Ready Receivers

Now that we have seen how the midband and superband channels are configured, it is possible to make sense of the term *cable-ready* as it applies to receivers. Such a receiver simply has a tuner that is capable of tuning to the midband and at least some of the superband channels. Exactly how high the superband channels go varies both with the cable system and the receiver design.

When an ordinary, non-cable-ready receiver is used with a CATV system, it can tune in only channels 2 to 13. Cable converters allow the reception of midband and superband channels with such a receiver. Usually, these are tunable converters that can move any channel on the cable to a single VHF channel. Figure 9.29 shows a photograph of a typical converter, and Figure 9.30 is its block diagram. The signal is first converted up in frequency, then back down to the VHF output channel. This ensures that the first local oscillator frequency is higher than any frequency used by the cable. That way, any signal leakage into the cable from the local oscillator will cause no interference. Virtually all converters use varactor tuning, and some have quite elaborate remote control and programming features.

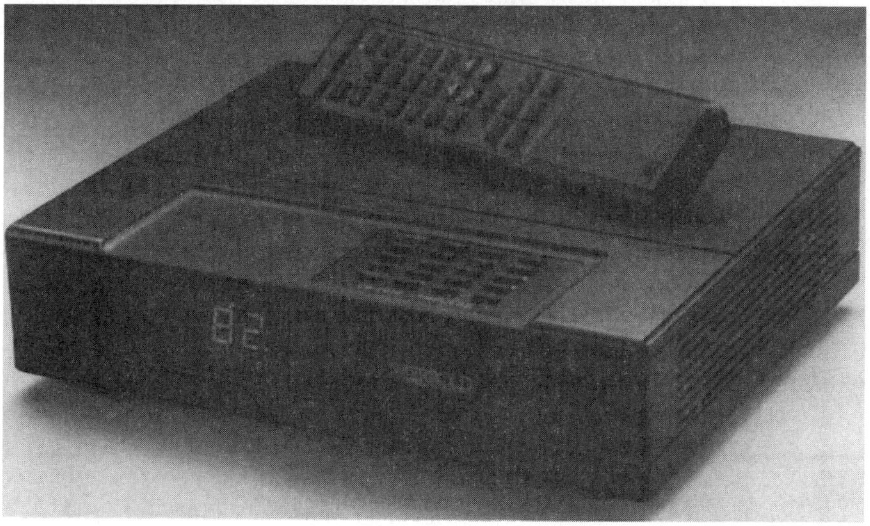

Figure 9.29 Cable-television converter
Photo courtesy of General Instrument Corp.

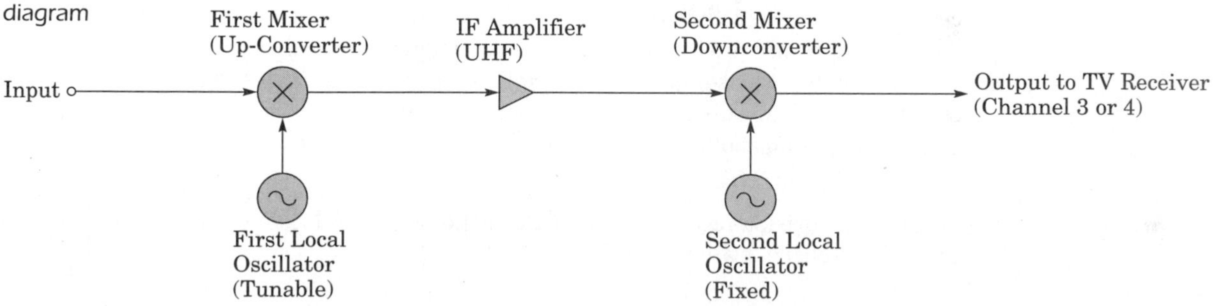

Figure 9.30 Cable-television converter block diagram

One feature of some cable converters is a volume control. It adds considerably to the complexity (and the price) because the only way to control the volume is to demodulate the sound, adjust the volume, and then remodulate it onto a new carrier.

9.6 Test Equipment and Signals

Television and video systems require a wide variety of specialized test equipment, only some of which will be described in this section. In the next section, we will briefly investigate troubleshooting techniques, and we will learn how some of these signals and generators are used.

9.6.1 Color Bar Pattern

Figure 9.31 shows a sketch of a color test signal. Figure 9.31(a) shows what the signal looks like on the television screen, and Figure 9.31(b) shows the appearance of the signal on an oscilloscope. There are seven color bars, from left to right: gray, yellow, cyan, green, magenta, red, and blue. All are modulated at 75% (that is, the color component is 75% of its maximum amplitude) except for gray, which is just 75% of maximum luminance. When sent in this order, the luminance levels form a signal whose amplitude decreases from left to right. Such a signal is called a *staircase signal*, because of its appearance when viewed on an oscilloscope at the horizontal rate.

When the color modulation is turned on, it appears on an oscilloscope as a high-frequency (approximately 3.58 MHz) ac signal superimposed on the luminance level. Because of the frequency interleaving used for color, the color modulation is not synchronized on the scope and appears as a blur, like the envelope of an AM signal.

Figure 9.32 shows a generator that generates this pattern. It also generates red, green, and blue rasters, which are useful for checking CRT purity, and a crosshatch pattern that can be used for convergence adjustment.

A more elaborate version of the color bar pattern is used in broadcasting studios. It has extra bars across the bottom of the picture, including 100% white, $-I$ and Q color signals, and black level.

Figure 9.31
Color bars
Courtesy Leader Instruments Corporation.

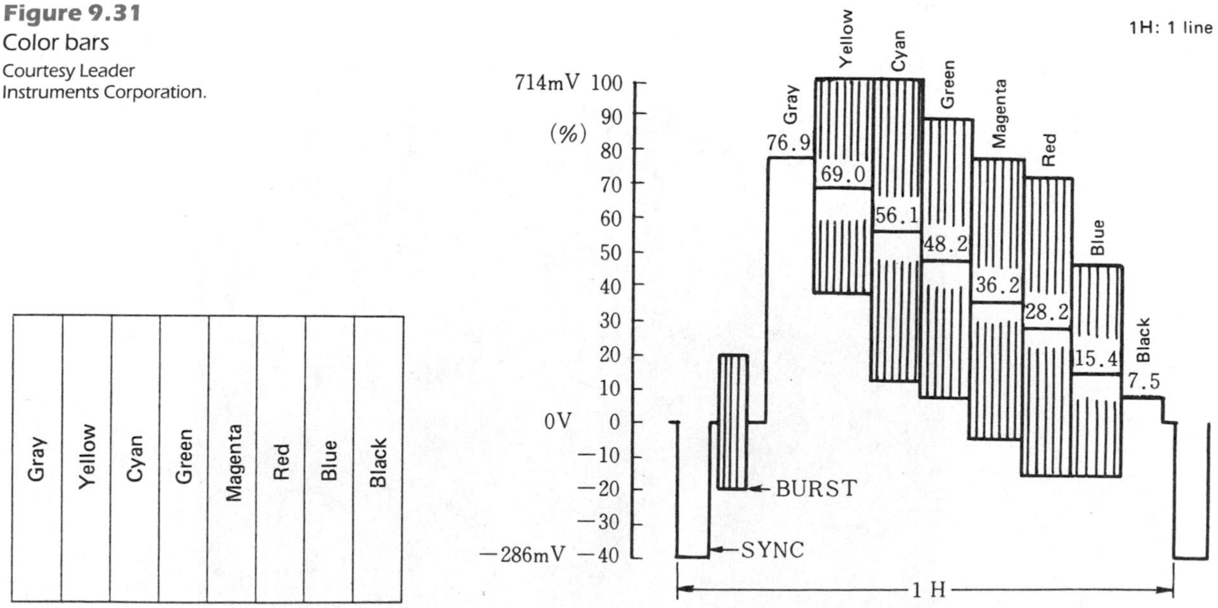

(a) Appearance on TV screen (b) Appearance on oscilloscope

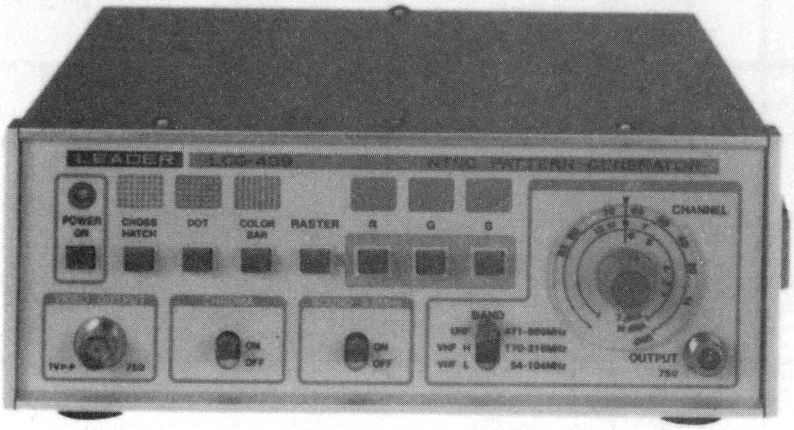

Figure 9.32
Color bar generator
Courtesy Leader Instruments Corporation.

Color bars can be used to set up the color response of video systems by eye, and, in fact, the eye is the most suitable instrument for overall evaluation of a video system. However, some quantitative results can be obtained by using a **vectorscope**, which will be described in the next section.

9.6.2 Vectorscope Display

The magnitude of the chroma signal represents the color intensity (saturation) and the phase represents the hue. If the color signal is displayed on an oscilloscope in such a way that two signals with a phase difference of 90° are applied, one to the vertical and one to the horizontal plates of the oscilloscope, then a color-bar signal with predetermined levels and phases appears and is the same each time. The screen can be marked with the correct values for use in adjustment. This is called a *vectorscope display*. Any axes 90° apart can be used, but $B - Y$ and $R - Y$ are common. Note that looking at the bars on an ordinary scope gives a check on amplitude but not phase. Figure 9.33 is a photograph of a combination vectorscope and waveform monitor. The latter is simply an oscilloscope calibrated with IRE levels and equipped with a triggering arrangement that can allow it to scan any line as required.

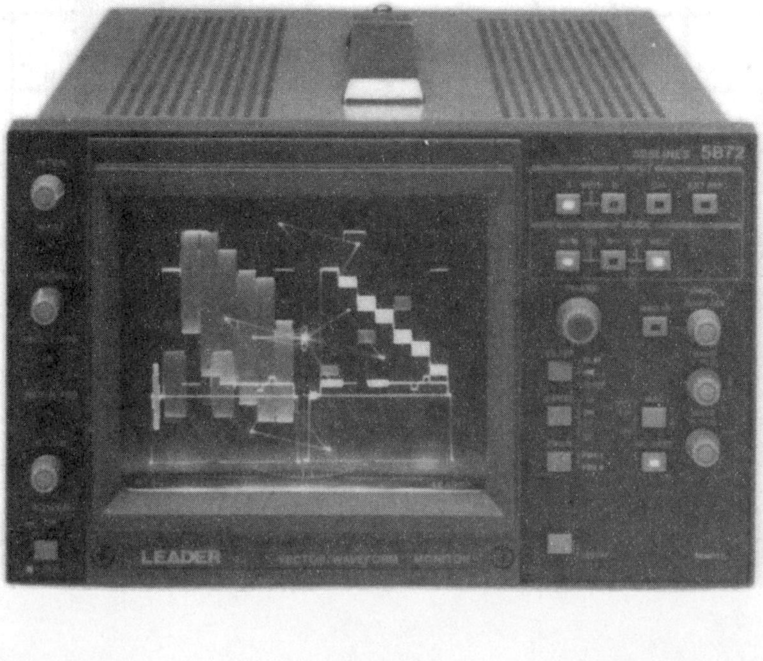

Figure 9.33
Vectorscope/ waveform monitor
Courtesy Leader Instruments Corporation.

9.7 Television Receiver Troubleshooting

A detailed discussion of television repair would merit a book in itself, and in fact, many books have been written on the subject. This brief section will point out how a knowledge of the basic operation of a color television receiver can be used to localize problems. Following that, signal injection or tracing can be used to narrow the search further. At this point, of course, reference to a complete circuit diagram, and preferably a service manual, is required.

The first step in the process should be to examine the symptoms of a fault carefully. A color television is a complex device, but it also has a complex display that can reveal a good deal of information about troubles. Attention should be paid to the presence or absence, normality or abnormality, of raster, picture, sound, and color. Next, referring to the block diagram of a typical color television receiver, an attempt can be made to localize the problem. The following sections will show how the symptoms can lead to at least an approximate diagnosis of the fault.

9.7.1 No Raster

Lack of any light on the screen points to a problem with either the main power supply or the horizontal deflection circuit. The former is obvious; the latter conclusion depends on a knowledge that the high-voltage supply for the CRT, as well as various low voltages for signal-processing stages, are derived from the horizontal output transformer.

If there is only one horizontal line on the screen, it is a safe guess that the problem is in the vertical oscillator or amplifier.

9.7.2 No Video and/or Audio

If there is light on the screen but no picture, a good deal can be learned from noticing whether there is audio. Recall that the video and audio are combined through the tuner and the picture IF amplifier. Therefore, if both video and audio disappear, the problem is probably in one of these circuits. On the other hand, if the audio is present but the video is not, the problem is after the point at which the sound is separated from the video. Check the video detector and the circuitry past that point. In a color receiver, the problem is likely to be between the video detector and the point at which the chroma and luma are separated. A failure in the luminance amplifier after that point will allow a picture to appear, but it will consist of only saturated colors, while a fault in the color circuitry may result in a monochrome picture or one with incorrect color. On the other hand, a fault in a monochrome receiver anywhere between the video detector and the CRT can cause the picture to be absent.

Another possibility is that the picture may be normal while the sound is absent. This points to the circuitry between the first sound detector (or the video detector in a monochrome receiver) and the loudspeaker.

9.7.3 Synchronization Problems

When the receiver displays a picture but synchronization is lacking, it is important to notice whether the problem is with vertical or horizontal synchronization or both. Lack of vertical synchronization is indicated by a picture that moves up or down. With no horizontal sync, the picture does not move sideways—rather, it breaks up into diagonal bars. Lack of both types of synchronization in the presence of an otherwise strong image points to the sync separator. On the other hand, the presence of one and absence of the other could result from a fault in either the sync separator or one of the oscillators (horizontal or vertical, depending on which is out of sync).

Color receivers have a third type of synchronization: color sync. This is what keeps all the colors stable, and its failure will probably result in a rainbow effect in the picture. The trouble could be in the color oscillator itself or in the AFPC circuit (in modern receivers, these are included in the same integrated circuit).

9.7.4 Signal Tracing and Injection

Once the trouble has been localized to a relatively small area of the receiver, and in the absence of any obvious faults (such as overheating components or loose wires), it is necessary to make further checks. The simplest test is to look for the proper signals and voltages at various test points, following manufacturers' instructions. However, even when the fault has been localized to some extent, there may be a great number of such points to check.

There are two basic methods for following a signal through a series of stages. One is signal tracing, in which a normal television signal is applied to the antenna terminals and followed through the receiver or, more likely, through that part of the receiver where the fault is suspected to lie. The instrument most commonly employed for signal tracing is the oscilloscope, though for audio circuits a sensitive amplifier with a small loudspeaker is sometimes used.

In television receivers, the oscilloscope can be used in all the baseband video circuitry, the audio IF and baseband circuitry, and the horizontal and vertical sync, oscillator, and amplifier circuits. Be careful in the horizontal output stage, however: because of the high voltages and brief pulses in this stage, there are often points on the circuit marked "do not measure." For your own safety and that of your equipment, obey that injunction.

Depending on the high-frequency response of the oscilloscope, it may detect the presence of a signal in the video IF circuits. As long as the signal level is high enough, a demodulator probe can eliminate the necessity for a high-frequency scope. In the tuner, however, the signals are likely to be too weak and at too high a frequency for measurement. For these areas, signal substitution is useful. For instance, applying a simulated video IF signal to the input of a video IF amplifier can determine whether a tuner is at fault. Tuners, by the way, are seldom repaired except by returning them to the manufacturer in exchange for a new or rebuilt tuner. The main exception is that conventional switch-type tuners sometimes become intermittent due to dirt and oxidation on the contacts. A liberal application of contact cleaner will often improve them considerably.

Other circuits where signal injection is useful include the horizontal and vertical sweep circuits. In fact, sometimes a horizontal drive signal is applied to the horizontal output transistor in order to get it working and allow the set to generate high and low power-supply voltages. Sometimes, this is the only way a problem that shuts down the horizontal output signal can be found, since without power it is very difficult to determine which stages are operable. Figure 9.34 shows a specialized test instrument that is designed to

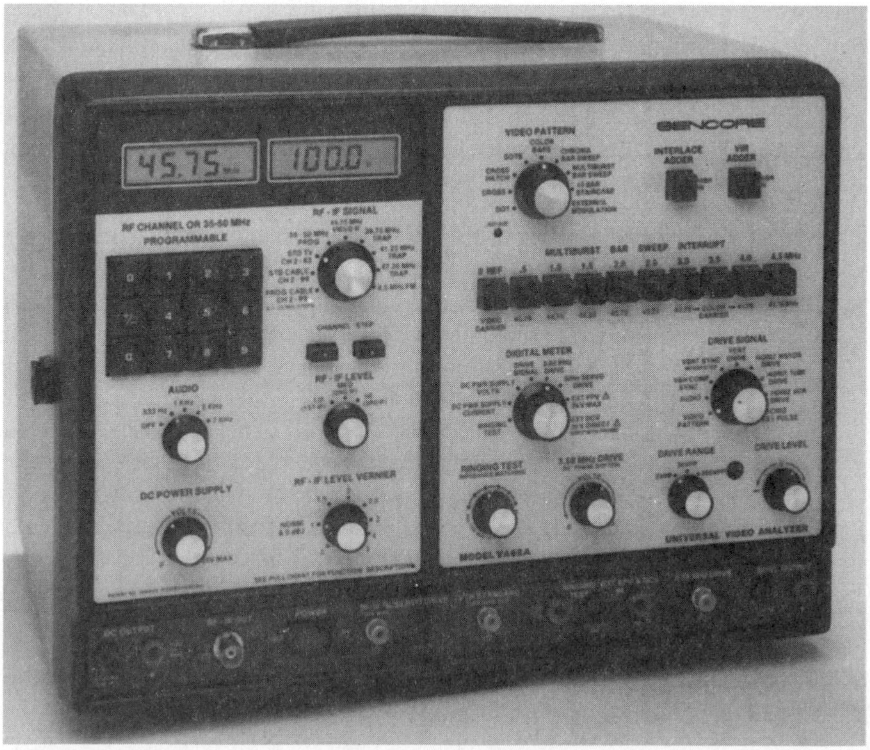

Figure 9.34
Video analyzer
Courtesy of Sencore Electronics Inc.

supply these types of signals, as well as RF, IF, and baseband video signals. It can also provide signals at specific frequencies for aligning the notch filters commonly found in television receivers.

9.8 High-Definition Television

The NTSC television system has served well for many years. However, it is far from ideal. The following are some of its more important deficiencies:

1. Luma resolution is inadequate for large-screen use.
2. Chroma resolution, while adequate in the vertical direction, is insufficient in the horizontal direction. This causes color smearing.
3. A wider aspect ratio, similar to that of commercial motion pictures, would be more suitable for many types of programming. The current 4:3 ratio was set in the early days of television and was compatible with the motion pictures of the day.
4. The interleaving of luma and chroma signals conserves bandwidth but causes interference between the two signals. Cross luminance results from the luminance component interfering with the chroma signal. It occurs whenever there is diagonal detail in the picture and gets worse with motion. It is visible as spurious color in the picture. Cross color is caused by color information being interpreted as luminance, causing a structure of fine dots.
5. Interlaced scanning conserves bandwidth for a given flicker rate with most pictures. However, certain types of images, alphanumerics for instance, still appear to flicker at a 30 Hz rate. Interlacing also appears to lower the resolution for motion and makes the scanning-line structure more obvious.

In recent years, there have been many attempts to solve these problems. These attempts can be divided roughly into three categories: systems designed to get the best performance possible from the existing NTSC system; systems that change the signal to achieve higher definition but change it in such a way as to maintain compatibility with existing receivers; and completely new schemes, requiring new equipment from beginning to end of the signal chain. Obviously, there are possible compromises; for instance, an incompatible system might be made compatible by adding a specially designed converter to an ordinary NTSC receiver.

Compatibility between any new system and the present one is very desirable for economic reasons, and many of the early proposals for high-definition television tried to maintain compatibility. Several systems did this by using two channels for the HDTV signal, transmitting an ordinary NTSC signal on one and the additional information required for HDTV on the other. For instance, the additional information required for the wider picture could be transmitted on the extra channel. However, such schemes, besides using twice the bandwidth of the NTSC system, have problems integrating the information from the two signals. They have fallen out of favor with the development of new ditigal methods for creating HDTV.

9.8.1 Improved Performance with the Current System

The cheapest and least disruptive approach to improving the television image is to work with the existing NTSC system to extract maximum performance. There are a few things that can be done at the transmitter without sacrificing compatibility and many more that can be done at the receiver. In general, techniques that work with the NTSC signal are known as *improved-definition television* (IDTV) rather than *high-definition television* (HDTV).

Comb filtering can be used to reduce interference between the luma and chroma signals. The use of a comb filter in the receiver has already been described. The signal can also be *pre-combed* by using a comb filter at the transmitter in the creation of the color signal.

The impression of greater vertical resolution can be achieved by deriving additional scan lines from those that are transmitted. For instance, with enough memory in the receiver to store two lines, it is possible to insert new lines that are the average of the two adjacent lines.

One way to create the illusion of greater horizontal resolution is to emphasize edges. Since vertical edges involve high-frequency components in the video signal, a simple peaking circuit that emphasizes the high frequencies in the luminance signal will have this effect. This has been done in most receivers for many years.

More elaborate signal processing can be accomplished if the video signal is first digitized in the receiver. Once the signal has been digitized, it is relatively easy to use digital signal processing to sharpen edges (giving the impression of greater resolution) and to reduce noise by correlating consecutive lines. Since adjacent lines tend to be quite similar, any sharp difference between one line and the next is likely to be caused by noise. Of course, it can also be caused by a horizontal edge in the picture, so some care is needed and it is better to compare more than two lines. Similarly, ghosts can be eliminated by looking for repeating patterns along a line.

Picture-in-picture is another possibility; for this, more than one source is digitized, and the sources can be combined on the screen.

Noninterlaced scanning, called **progressive scan**, can be implemented at the receiver. This can be accomplished by including a field store, that is, a memory that can store one complete video field, in the receiver. Recall that the complete television frame includes two fields, each sent in 1/60 s. Normally, the fields are shown as they are received, and no memory is required at the receiver. However, if one of the fields is stored, lines from that field could be interspersed with lines from the field currently being received, allowing the whole frame to be displayed at once. As lines are clocked out of memory and displayed, the vacated memory locations can be used to store the current field for use during the next field. This system requires that the receiver be able to convert the signal to digital form for storage and back to analog for display. The receiver also has to have at least one megabit of memory, assuming that data compression is used.

The only way to achieve a wider aspect ratio with the current system is to use fewer active scan lines. This process is called *letterboxing*. It is done occasionally by stations broadcasting widescreen movies and is also seen with some music videos. Of course, this leaves a blank band at the top and bottom of the screen with current receivers, and it also results in decreased vertical resolution, since some of the available scan lines are unused. It is therefore not a very satisfactory arrangement. See Figure 9.35 for an illustration of letterboxing.

9.8.2 Analog Schemes

All early methods for transmitting high-definition television were analog. Digital transmission requires greater bandwidth than analog methods, unless a great deal of signal processing is used. Since any practical video system has to do all the necessary processing in real time, that is, as the signal is transmitted, the required computations have to be done very quickly, and thus digital television was not practical until very recently. The next section will introduce digital television.

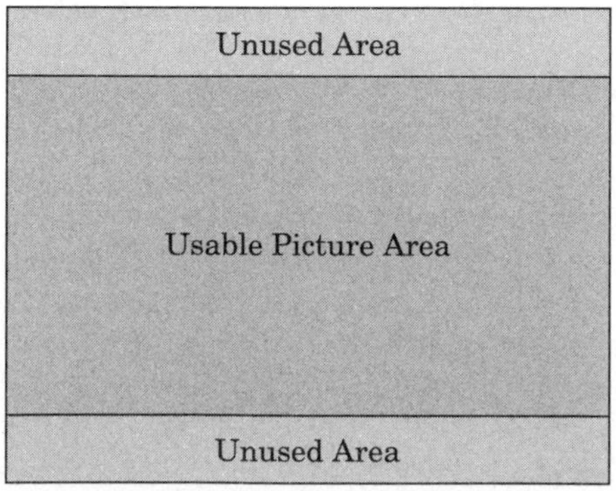

Figure 9.35 Letterboxed video

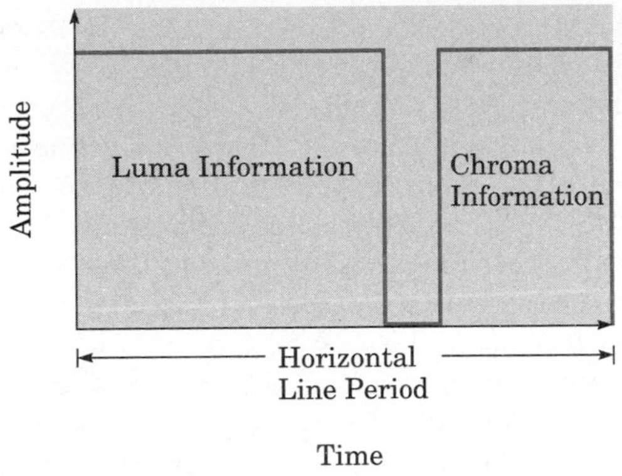

Figure 9.36 Multiplexed analog component video signal

In order to improve the color resolution and remove artifacts created by interference between the luma and chroma signals, practically all analog high-definition systems use **component color;** that is, they transmit the luma and chroma signals sequentially, one component after the other, rather than simultaneously using frequency interleaving, as is done by the NTSC system. Any such scheme is known as a MAC (multiplexed analog components) system—Figure 9.36 illustrates the basic idea. Note that the amount of time allocated to chroma is less than that for luma, so the chroma resolution is still not as good as for luma. Typically the chroma resolution is about half that of the luma signal, which is much better, in the horizontal direction, than NTSC. Each signal has to be time-compressed, so that one line of luma, followed by one line of chroma, fits into the time allocated for one horizontal line. This time compression necessitates an increase in bandwidth. Note also that the receiver needs to have enough memory to store at least one line of video so that it can combine the luma and chroma signals and display them together. MAC systems have been used for some time for satellite transmission of non-HDTV television, especially in Europe.

The MAC concept does not by itself imply a wider screen with more picture details. To implement that would generally require still more bandwidth. The Japanese have been broadcasting HDTV from satellites using a modified MAC scheme since 1990. The system is called MUSE (for multiple sub-Nyquist encoding). The picture has a 16:9 aspect ratio, with 1125 scan lines. The signal has a bandwidth of 8 MHz and does not fit in an ordinary television channel. Even so, this relatively narrow bandwidth is obtained only by applying compression to the signal.

Essentially, video compression works by transmitting only the changes from one frame to another, rather than transmitting the whole frame. Such a scheme obviously requires the receiver to have at least enough memory to store one complete frame of video. The Japanese HDTV system is quite expensive, and as of early 1995, only about 30,000 receivers, plus about 100,000 converters to convert MUSE into NTSC, had been sold. The Japanese are still working on improvements to MUSE, while simultaneously developing digital HDTV systems.

9.8.3 Digital Television and the Grand Alliance

North America was the first area to standardize its analog color television standard. It is arguable that it has suffered ever since, because the later PAL and SECAM standards have better performance. In HDTV, however, North America lags behind Japan and, to some extent, Europe, which has also experimented with analog MAC systems, though no high-definition system has been implemented commercially in Europe. Digital technology has improved so quickly that even while analog HDTV proposals were being evaluated, digital schemes became practical. Just in time, the analog proposals were withdrawn and a so-called "Grand Alliance" was set up to develop a digital HDTV scheme. The members of the alliance include AT&T, General Instrument Corporation, Massachusetts Institute of Technology (MIT), Philips Electronics North America Corporation, David Sarnoff Research Center, Thomson Consumer Electronics (known for its products under the RCA name), and Zenith Electronics Corporation.

The North American HDTV standard will be digital and will allow a complete HDTV signal to be transmitted in a single 6 MHz channel. The digital HDTV signal will not be compatible with NTSC receivers; however, it will be possible to transmit this signal in "taboo" channels that are presently unusable for broadcast because of cochannel or adjacent-channel interference. The plan is to introduce HDTV initially using these channels; eventually conventional NTSC broadcasting would be phased out, and all channels would be available for the new system.

The new HDTV standard will have an aspect ratio of 16:9, with a luminance resolution of 1920 by 1080 pixels at 30 frames per second, noninterlaced. Note that the vertical resolution is more than double the current 480 lines. The horizontal resolution for this digital system is the number of details across the whole screen, and it compares very favorably with about 449 for NTSC. The color resolution will be half that for luminance.

The broadcast version of HDTV requires a carrier-to-noise ratio of only 14.9 dB for satisfactory operation. When HDTV is transmitted by cable television systems, the carrier-to-noise ratio will be much better than this. In that case, it is possible to use less error correction and transmit a much higher data rate in the same bandwidth. For transmission over cable with a carrier-to-noise ratio of at least 28.3 dB, two HDTV programs can fit into one 6 MHz channel. That means that existing cable systems could actually carry twice as many channels using digital HDTV as with conventional analog television signals.

The new digital scheme will support other standards as well. For instance, much video material originates as motion-picture film, shot at 24 frames per second. Conventional NTSC television converts the frame rates by constructing three television fields, rather than two, from alternate film frames, but the new digital scheme will be able simply to reduce speed to 24 frames per second, allowing more bits per frame.

The digital HDTV system will use an elaborate data compression scheme to reduce its bit rate to 19.3 megabits per second, which can be transmitted in a conventional television channel with a bandwidth of 6 MHz using a vestigial sideband system. The audio will also be digital, compressed using the Dolby AC-3 system.

Summary

Here are the main points to remember from this chapter.

1. The North American NTSC television system transmits 30 frames per second, each consisting of 525 lines. A 2:1 interlace is used so that 60 fields are transmitted per second.
2. Vertical resolution depends on the number of scan lines, while horizontal resolution depends on the maximum frequency that can be transmitted.
3. Composite color video uses two color signals modulated on a subcarrier at 3.58 MHz using DSBSC QUAM. This results in no increase in bandwidth for color, but it also means that the horizontal resolution for color is much less than for monochrome.
4. Terrestrial television broadcasting uses a channel 6 MHz wide. The picture is transmitted using VSB AM with carrier, and the sound uses FM on a separate carrier.
5. Picture and sound travel together through the tuner and picture IF sections of a television receiver. The sound is then converted to a separate 4.5 MHz IF, in a system known as intercarrier sound.
6. Color receivers use synchronous demodulation to retrieve the color signal. The frequency and phase reference for the color demodulators are provided by a color burst on the back porch of the horizontal synchronizing signal.
7. The CRTs used in television receivers generate an electron beam that is intensity-modulated by the video signal and deflected horizontally and vertically by coils that surround the tube. The beam produces an image by striking a phosphor coating on the tube face. Color CRTs use three electron beams striking three different types of phosphors that glow red, green, and blue.

8. Cable-television systems reduce losses by moving UHF signals to the VHF range. They use frequencies between channels 6 and 7 and above channel 13. Amplifiers are used at frequent intervals along the cable to compensate for losses.
9. It is often possible to localize faults in television receivers by looking for the presence, absence, or abnormality of the raster, picture, color, and sound. Once the fault has been localized to one or more stages, there are many types of specialized test equipment that can be used.
10. Both analog and digital schemes have been proposed for high-definition television. A digital method is expected to be implemented in North America in the near future.

Important Equations

$$N_V = 338 \text{ for NTSC system} \tag{9.1}$$

$$N_H = 449 \text{ for NTSC system} \tag{9.2}$$

$$L_H = B\text{ (MHz)} \times 80 \tag{9.4}$$

$$\begin{aligned} Y &= 0.30R + 0.59G + 0.11B \\ I &= 0.60R - 0.28G - 0.32B \\ Q &= 0.21R - 0.52G + 0.31B \end{aligned} \tag{9.6}$$

$$\text{dBmV} = 20 \log \frac{V}{1 \text{ mV}} \tag{9.10}$$

Glossary

aspect ratio ratio of the width to the height of a television picture

back porch the portion of the horizontal blanking pulse after the sync pulse

black setup the video level corresponding to zero luminance

blanking the period of time when the electron beam in a CRT is cut off

cathode-ray tube (CRT) a vacuum tube that uses a moving electron beam to produce patterns or images on a phosphorescent screen

chrominance (chroma) the color signal

color burst several cycles of color subcarrier on the back porch of the horizontal sync, for color synchronization

comb filter a filter that can pass (or reject) a fundamental frequency and its harmonics

component color a video system in which color and luminance are sent separately, without frequency interleaving

composite video a video system in which luma, sync, and chroma signals (if present) are combined

convergence alignment of the three electron beams in a color CRT so that they land on the same triad of color phosphor dots

equalizing pulses the pulses in the vertical blanking interval of a video signal that create interlaced scan

field in an interlaced video system, one-half of a frame, consisting of alternate lines

frame one complete image in a video system

front porch the portion of horizontal blanking pulse before the sync pulse

intercarrier sound a television receiver design that uses mixing between the picture and sound carriers to generate the sound intermediate frequency

interlace a video scanning system that divides a frame into two fields, to reduce flicker

luminance (luma) the signal that provides brightness information in a video system

NTSC video North American television standard

peak white the video signal level representing maximum luminance

pixel picture element

progressive scan a video system that does not use interlace

purity in a color CRT, the adjustment of the three electron beams so that each lands on phosphor dots of the appropriate color

raster the pattern of scanning lines in a video system

resolution the amount of detail produced by a video system

retrace the return of the electron beam in a CRT from right to left or from bottom to top

RGB color a color video system in which the three primary colors are transmitted separately

ultor the main accelerating element in a CRT

utilization factor the proportion of scanning lines in a video system that, on average, can be used in determining vertical resolution

vectorscope a specialized oscilloscope designed for the observation of composite color signals

yoke the assembly that contains the deflection coils and is mounted on the neck of a CRT

Questions

1. Explain the difference between a frame and a field.
2. Describe interlaced scan and explain its advantages and disadvantages.
3. State the standard for video signals at the inputs and outputs of equipment, and sketch one line of a color signal that complies with these standards. Show an IRE scale on your sketch.
4. How do the vertical blanking intervals differ for odd and even fields?
5. Explain how it is possible to transmit a color signal using the same bandwidth as for monochrome.
6. What is meant by negative picture transmission? Why is it used for television broadcasting?
7. Explain briefly how the method used for television stereo sound differs from that used for FM radio broadcasting.
8. Sketch the block diagram of a monochrome television receiver, including only those blocks that process the signal. Above each block, note the frequency or range of frequencies that it handles.
9. (a) Sketch the frequency response for a typical video IF amplifier, showing the frequencies for picture carrier, sound carrier, and color subcarrier.
   (b) Explain why the sound carrier is at a lower frequency than the picture carrier in the video IF.
   (c) What frequency is used for the sound IF? Why?
10. Give two reasons for the use of the intercarrier sound system in television receivers.
11. Sketch the electron gun in a typical CRT. Explain briefly how the electron beam is focused.
12. What is the aquadag? What does it do?
13. Describe and compare two methods of color-signal demodulation.
14. Explain briefly the meaning of each of the following:
    (a) color killer
    (b) AFPC
    (c) luminance delay line
    (d) gated AGC
15. Explain briefly what is meant by each of the following, as it refers to a color CRT:
    (a) purity
    (b) static convergence
    (c) dynamic convergence
16. Sometimes a color monitor can be set for "green only" when viewing a monochrome image, as on a computer. Often this gives a sharper image than when white is used. Why?
17. Why is the accelerating voltage much higher in a color CRT than in a monochrome CRT of the same size?
18. Why is it necessary for CATV systems to have isolation between subscribers?
19. Why do CATV amplifiers need automatic gain and slope (frequency response) control?
20. Describe the picture improvements needed to create a high-definition television system.
21. What is a "taboo" channel, and why is it useful for digital HDTV?
22. Why is it possible for a digital HDTV signal to occupy a smaller bandwidth on a cable television system than it needs when broadcast?

Problems

SECTION 9.2

23. A certain receiver has a vertical retrace time of $0.025V$, where V is the time for one field.
    (a) How much time in seconds is this?
    (b) How many horizontal lines does this represent?
24. For the following pictures, sketch one line of the composite video signal:
    (a) an all-white frame
    (b) two vertical white bars and two black bars, equally spaced
    (c) five pairs of equally spaced vertical black and white bars (Why does this signal have more high-frequency content than that in part (b)?)
25. Suppose the signal from a color camera has $R = 0.8$, $G = 0.4$, and $B = 0.2$, where 1 represents the maximum signal possible.
    (a) Calculate values for Y, I, and Q for the transmitted signal.
    (b) Express the luminance component in IRE units.
26. A certain TV receiver displays 475 scanning lines on the screen and has a frequency response of 3.2 MHz for luminance and 0.4 MHz for color. Cal-

culate its horizontal and vertical resolution in lines for both monochrome and color components.

SECTION 9.3

27. Find the frequency of each of the following, for a television signal on channel 47.
    (a) picture carrier
    (b) sound carrier
    (c) color subcarrier

SECTION 9.4

28. Draw a block diagram for a monochrome television receiver, indicating beside each block the frequency or frequency range at which that block will operate when the receiver is tuned to channel 9.
29. Draw a block diagram for a comb filter that will remove a 2 kHz signal and all its harmonics from a signal.

SECTION 9.5

30. A CATV system is sketched in Figure 9.37. Find the signal levels at each of the subscriber drops shown.
31. Convert 4.5 dBmV in 75 Ω to:
    (a) volts (b) watts
32. Suppose an ordinary TV receiver (without a converter) is connected to a cable system and that its local oscillator sends some of its signal back into the cable. Find three channels that could be interfered with, and state which channels the TV will be tuned to when it causes the interference.

SECTION 9.6

33. What would be the effect of each of the following on the performance of a monochrome television receiver?
    (a) mixer stage in VHF tuner not working (assume the set has conventional switch-type tuners)
    (b) gain in video IF amplifier is much too low
    (c) sync separator not working
    (d) horizontal oscillator not functioning
34. What area of a receiver would you suspect first for each of the following symptoms? Give a reason for your choice.
    (a) no sync (neither horizontal nor vertical)
    (b) no sound, picture OK
    (c) no picture, sound OK
35. What area(s) of a color receiver would you suspect for each of the following symptoms?
    (a) no color sync
    (b) no color at all, monochrome picture OK
    (c) no red in the picture

SECTION 9.8

36. Suppose that a high-resolution television signal used the NTSC standard, except that both vertical and horizontal resolution were doubled. What would be the required bandwidth for the baseband video signal?
37. Suppose that the aspect ratio of a television system were changed from 4:3 to 2:1, with the horizontal resolution keeping the same ratio to vertical resolution as at present. Assume the number of scan lines remains as at present.
    (a) How many details would have to be shown on a horizontal line?
    (b) What would be the bandwidth of the baseband signal?
38. Express the horizontal resolution of the Grand Alliance digital HDTV scheme in lines.
39. By what percentage is the transmitted bit rate reduced when a digital HDTV system transmits material derived from film, compared with the bit rate for material originating as video?

Figure 9.37

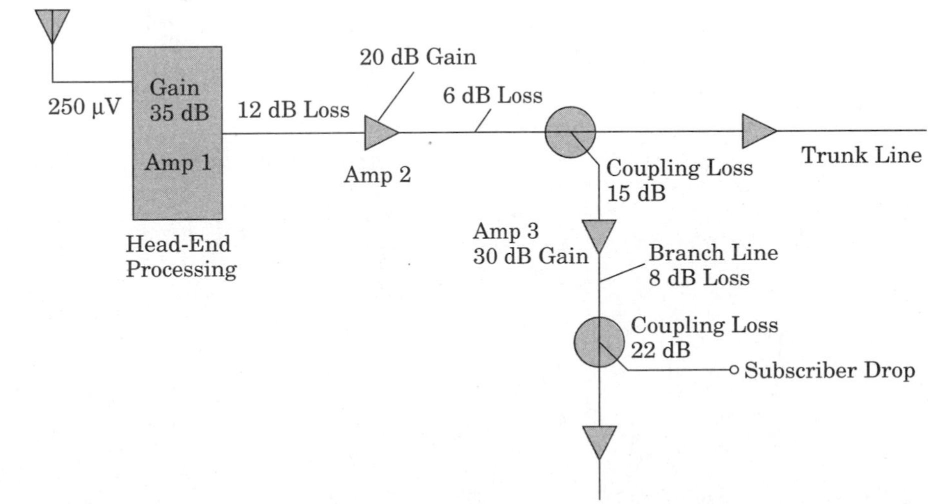

COMPREHENSIVE

40. Suppose a television signal has 10.3 μs of horizontal blanking per line and 21 lines of vertical blanking per field. Calculate the percentage of total time that is occupied by blanking.
41. Under ideal conditions, the eye can resolve two details subtending 0.5 minutes of arc. Suppose a monitor screen 30 cm wide is located 1.5 m from a viewer. What video bandwidth would be required in order to separate horizontal details in the center of the screen by 0.5 minutes of arc, assuming NTSC video?
42. Assume that the horizontal deflection coils in a receiver can be modeled as a pure inductance of 2 mH and that the current waveform is a sawtooth with a peak current of 2 A. Sketch the voltage waveform, including scales.
43. Express the R signal algebraically in terms of Y, I, and Q.

10 The Telephone System

Objectives After studying this chapter, you should be able to:

1. Describe the topology of the switched telephone network
2. Explain the operation of basic and electronic telephone equipment
3. Describe the various signals present on a local-loop telephone line and explain the function of each
4. Perform calculations related to signal and noise levels in a telephone system
5. Describe several methods for switching telephone signals
6. Describe and compare in-band and out-of-band signaling systems for telephony
7. Explain the advantages of common-control signaling
8. Explain the use of frequency-division multiplexing in telephony and perform bandwidth calculations with FDM signals

10.1 Introduction

The public switched telephone system is undoubtedly the largest and probably the most important communication system in the world. The reasons for this are contained in the two words *public* and *switched*. The telephone system is public in the sense that anyone can connect to it. The fact that it is switched means that, in theory at least, it is possible for anyone to communicate with anyone else. This makes the telephone system conceptually very different from broadcasting systems and from private communication networks.

The telephone network employs many of the most interesting developments in communications, such as fiber optics and digital signal transmission, but it remains in many ways consistent with its origins in the nineteenth century—this will become obvious when we look at telephone signaling systems and the voltage and current levels that are found on subscriber lines. Compatibility has been maintained in most areas of the system, so that simple dial-type telephones can coexist with modern data-communication equipment. Though originally intended only for voice communication, the switched telephone network has been adapted to serve many other needs, including data communication, facsimile, and even video.

The Inventors of the Telephone

The first practical telephone was invented by Alexander Graham Bell (1847–1922) with the assistance of Thomas A. Watson (1854–1934) in 1876. Born in Scotland, Bell emigrated to Canada in 1870. He did most of his work in the United States, becoming an American citizen in 1882, though he maintained a summer home in Canada. He became interested in the possibility of telephony while working as a teacher of speech and simultaneously trying to invent an improved telegraph. Bell made a great deal of money from his telephone patents, and he used it to finance a long career of experimentation in other fields, including aircraft and hydrofoil boats. Watson went on to found a shipyard and organize tours of Shakespearean plays, among other things.

This chapter introduces the telephone system and describes the way it works for ordinary voice telephony using analog technology. The telephone network is now in a period of transition from an all-analog to a mostly-digital system. Digital telephony and the use of the telephone network to transfer digital data will be described in succeeding chapters, as the underlying techniques are discussed.

10.2 The Structure of a Switched Network

Figure 10.1 shows the basic structure or **topology** of a local calling area (known as a **local access and transport area** or **LATA**) in a typical switched telephone system. Each subscriber is connected via a separate line, called a **local loop**, to a **central office**, where switching is done. The central office represents one exchange: that is, in a typical seven-digit telephone number, all the lines connected to a single central office begin with the same three digits. Thus there can be ten thousand telephones connected to a central office. Actually the term *office* can be deceiving: in urban areas, it is quite possible for there to be more than one central office in the same building.

Subscribers connected to the same central office can communicate with each other by means of the central office switch, which can connect any line to any other line.

The central offices themselves are connected together by **trunk lines**, so that any subscriber can contact any other subscriber within a local calling area. There are not enough trunks or enough switching facilities to allow every subscriber to use the system at once, so there is the possibility of overload. When overload occurs, it is impossible for a subscriber to place a call—this situation is known as **call blocking**. The likely number of simultaneous conversations is predicted by statistical methods, and it is usually exceeded only during emergencies.

When there are no available trunks between two central offices, a connection can sometimes be made through a **tandem office**, which connects central offices without having any direct connection to individual telephones.

Long-distance calls in North America used to be routed through a hierarchy of **toll stations** similar to the system sketched in Figure 10.2. The local central office, also called an **end office**, was at the bottom of the hierarchy. Calls could be completed at various layers of the hierarchy. Calls from one side of the country to the other would have to move as many as five layers up and down the hierarchy and go through many switches.

The hierarchical concept works well with a minimum of computation and thus was suitable to electromechanical switches. With modern computer-controlled switches, however, it is more efficient to have a mesh of long-distance switching centers, each one

Figure 10.1
Local access and transport area

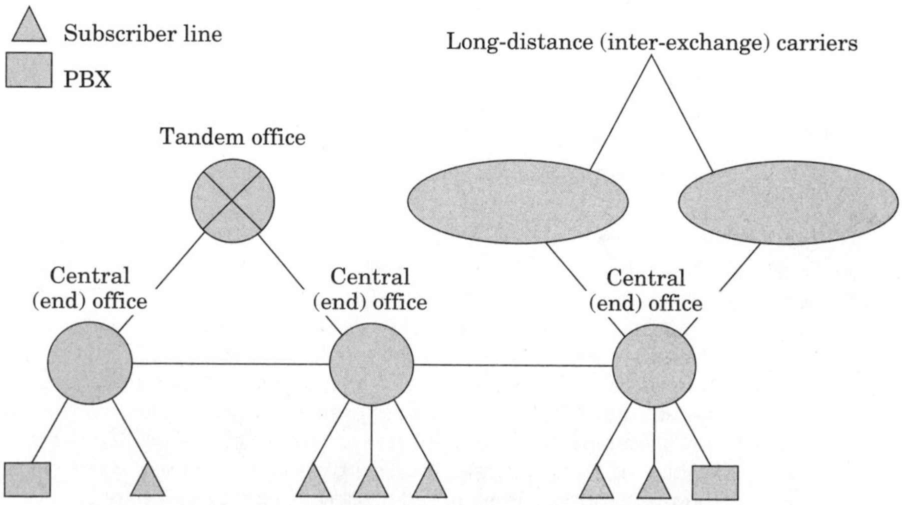

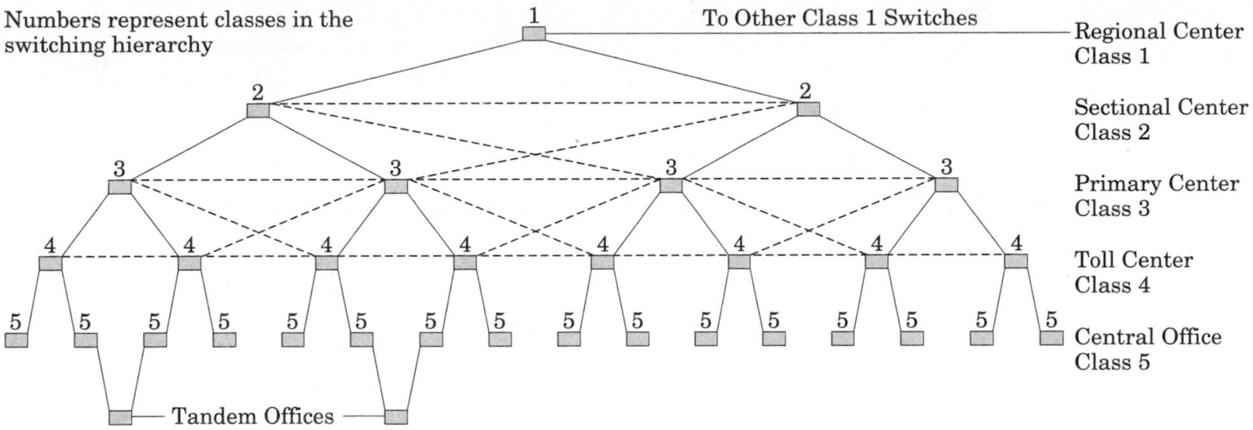

Figure 10.2
Hierarchical switched network

connected to every other one. This **flat network** usually lets the system find a direct route from one area of the country to the other, and there is never a need for more than one intermediate switch. Figure 10.3 illustrates this type of system, which is now used in North America, though the hierarchical system is still in use in many parts of the world.

The telephone system in the United States and Canada was formerly a monopoly, and it still is in much of the world. There is now competition for long-distance calls in North America, and it appears likely that there will be competition for the local subscriber loop in the near future. Each competing long-distance carrier (for example, AT&T, Sprint, and MCI) has its own connection to the local access and transport area. For simplicity, only two long-distance carriers are shown in Figure 10-3. Each carrier has its own connection, called a **point of presence** (**POP**), to the local telephone system.

New switching equipment and trunk lines are digital, with time-division multiplexing used to combine many signals on one line. This technology will be discussed in the next chapter. Fiber optics technology is increasingly used for trunk lines; other common media include terrestrial microwave links, geostationary satellites, and coaxial cable. Short distances between central offices are often covered using multipair cable, which has many twisted pairs in one protective sheath. So far, most local loops are still analog and use twisted-pair copper wire, but that seems likely to change in the future. Eventually

Figure 10.3
Nonhierarchical long-distance network with multiple providers

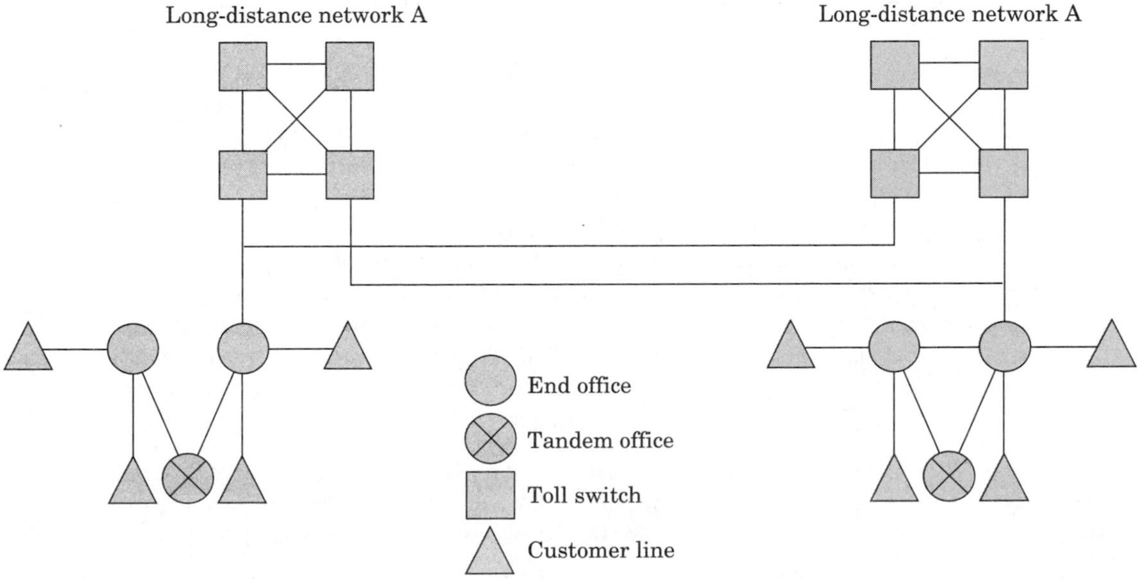

the system will be digital from one end to the other, and most of it will employ fiber-optic cable. However, telephone equipment is built for maximum reliability, and it lasts a long time. It is also expensive, so telephone companies quite naturally expect to use it until it wears out. It will be well into the next century before we have anything like a complete digital network.

10.3 Setting Up a Call

Let us begin our study of the operation of the telephone network by looking at an ordinary telephone call, which is sometimes called **POTS** (**plain old telephone service**) within the telephone industry. This requires us to look briefly at the central office switch, the subscriber telephone, and the connection between them, which is called the local loop. Since all the components of the system work together to establish communication, it is necessary to jump back and forth from one to the other to see what is happening. The local loop connects the telephone instrument to the rest of the network and carries the various control signals as well as the voice signal, so let us look at it first.

10.3.1 The Local Loop

Normally each individual subscriber telephone is connected to the central office by a single twisted pair of wires. The wires are twisted to help cancel their magnetic fields and reduce interference between circuits in the same cable. In telephone parlance, this interference is known as **crosstalk**. It is common practice to run a four-conductor cable to each residence, but only two of these wires (usually red and green) are used for a single line. The others (black and yellow) allow for the installation of a second line without running more cable.

Recently there has been a trend toward running multiplexed digital signals to junction boxes in neighborhoods and switching at that point to analog signals on individual pairs, in an effort to reduce the total amount of copper cable. Some optical fiber has been installed, though the last section of loop to the customer normally remains copper. In the future, **fiber-in-the-loop** (**FITL**) may be used. It will cost more because of the necessity of converting back and forth between electrical and optical signals at each subscriber location, but the bandwidth will be vastly increased and will thus allow a great number of additional services, such as cable television and data-based services like banking and shopping, to be carried on the same fiber.

The local loop performs several functions. It carries voice signals both ways. It must also carry signaling information both ways: dialing pulses or tones to the central office from the customer and dial tones, ringing, busy signals, and prerecorded messages from the network to the subscriber. In addition, the pair of wires in the local loop must transmit power from the central office to operate the telephone and ring the bell. Optical fiber cannot carry power, of course, so when it is used, the power to operate the telephone equipment at the subscriber location must be provided by the subscriber.

Twisted pairs attenuate signals because of their resistance and also because of reactive effects that cause the attenuation to increase with frequency. The dc resistance of the loop can be kept within allowable limits by using larger wire sizes for longer loops. The attenuation of voice frequencies can be greatly reduced by adding inductance in series with the line. This additional inductance takes the form of **loading coils**, which are small inductors. One common technique is to add 88 mH coils every 6000 ft (approximately 1.8 km). However, the loading coils greatly increase the attenuation of the line at high frequencies above the voice-frequency range, and when a wide frequency response is needed for high-speed data or high-fidelity audio, the loading coils must be removed. Figure 10.4 compares the frequency response of loaded and unloaded lines.

10.3.2 Signals on the Local Loop

Let us look at the process of making a call. Before we begin, assume the phone is not in use: it is **on hook**, in telephone parlance. The central office maintains a voltage of about 48 V dc across the line. Of the two wires in the twisted pair, one, normally the

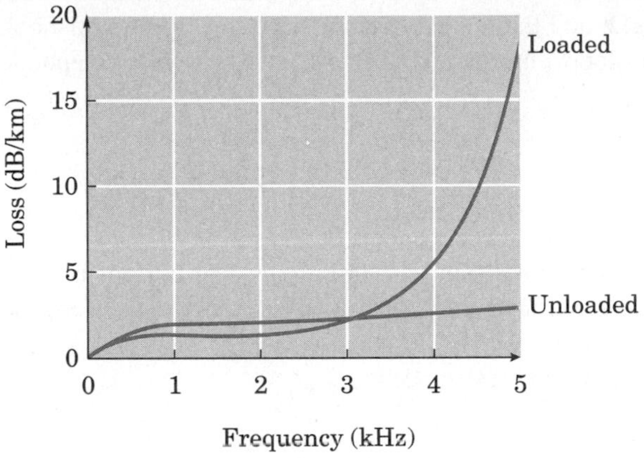

Figure 10.4
Effect of loading coils

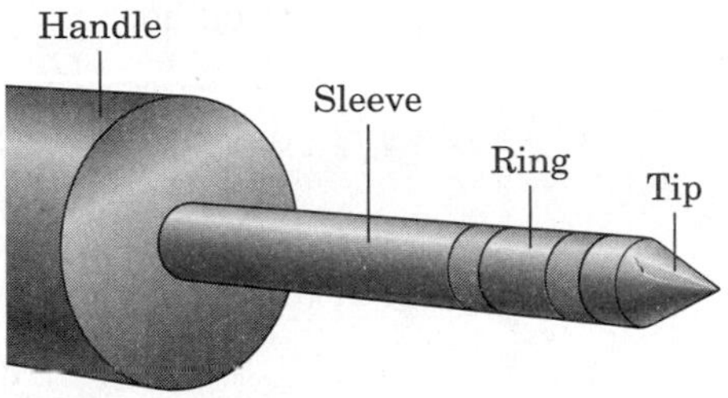

Figure 10.5
Telephone plug

green, is designated **tip** and the other (red), **ring**. The ring is connected to the negative side of the supply. A red wire in electronic equipment is usually positive, but not here! The central office supply is called the *battery*. In fact, the voltage does derive from a battery that is constantly under charge. Thus the telephone system can function during electrical power outages, whether they occur at the central office or at the customer's premises.

The "tip" and "ring" terminology dates from the days of manual switchboards, whose plugs looked like the one sketched in Figure 10.5. The same sort of plug is now used for stereo headphones. The telephone system is not stereo, of course, but the lines are balanced with respect to ground. The words *tip* and *ring* refer to the parts shown in the figure.

When the phone is on hook, it represents an open circuit to the dc battery voltage. The subscriber signals the central office that he or she wishes to make a call by lifting the receiver, placing the instrument **off hook**. The telephone has a relatively low resistance (about 200 ohms) when off hook, so it allows a dc current to flow in the loop. The presence of this current signals the central office to make a line available (in telephone parlance, the telephone is said to have *seized* the line). When the telephone is off hook, the voltage across it drops considerably, to about 5 to 10 volts, due to the resistance of the telephone line. Resistance can also be added at the central office, if necessary, to maintain the loop current in the desired range of approximately 20 to 80 mA. Figure 10.6 illustrates these ideas.

Figure 10.6
Local loop

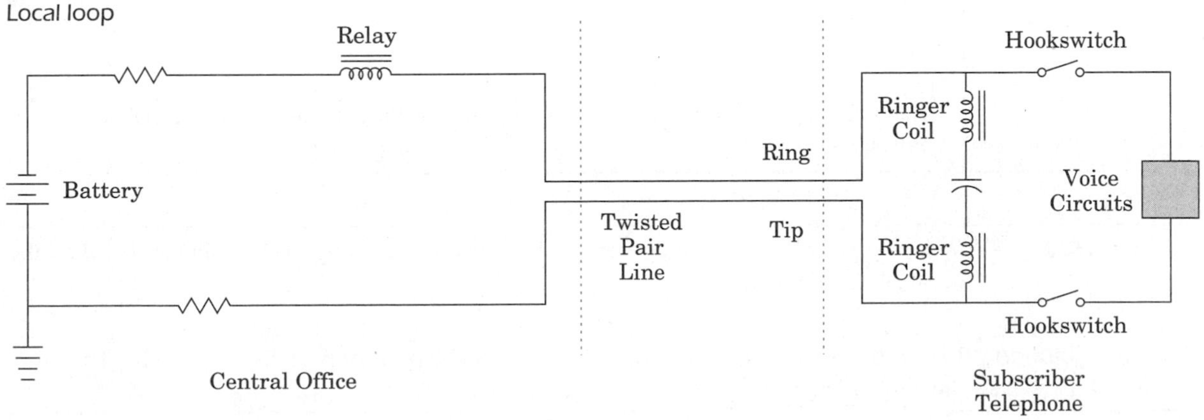

Example 10.1 A local loop has a resistance of 1 kΩ, and the telephone connected to it has an off-hook resistance of 200 Ω. Calculate the loop current and the voltage across the telephone when the phone is:

(a) on hook
(b) off hook

Solution (a) When the telephone is on hook, its dc resistance is infinite, so the current is zero. Since there will be no voltage drop around the loop, except at the phone itself, the full battery voltage will appear across the phone.

(b) When the phone is off hook, the total loop resistance is

$$R_T = 1000\ \Omega + 200\ \Omega = 1200\ \Omega$$

Then the loop current is

$$I = 48\ \text{V}/1200\ \Omega = 40\ \text{mA}$$

The voltage across the telephone is

$$V = IR = 40\ \text{mA} \times 200\ \Omega = 8\ \text{V}$$

Once a line has been assigned to a subscriber who lifts the receiver, the office signals the user to proceed by transmitting a dial tone, which consists of 350 Hz and 440 Hz signals added together.

Dialing can be accomplished in one of two ways. The old-fashioned rotary dial functions by breaking the loop circuit at a 10 Hz rate, with the number of interruptions equal to the number dialed. That is, dialing the number 5 causes five interruptions or pulses in the loop current. This technique is called **pulse dialing**, and it can be emulated by some electronic telephones. The second and much more efficient way is for the phone to transmit a combination of two tones for each number. This method is officially known as **dual-tone multifrequency (DTMF) dialing**, and it is commonly referred to as **Touch-Tone dialing** or just tone dialing (the term Touch-Tone is a registered trademark of AT&T). Table 10.1 shows the combinations of tones used for each digit. This system improves efficiency because digits can be transmitted in much less time than with pulse dialing.

TABLE 10.1 DTMF frequencies

Frequencies (Hz)	1209	1336	1477	1633
697	1	2	3	A
770	4	5	6	B
852	7	8	9	C
941	*	0	#	D

The letters A through D are included in the system specifications but are not present on ordinary telephones.

Example 10.2 What frequencies would be generated by a telephone using DTMF signaling, when the number 9 is pressed?

Solution Use Table 10.1. Go across from 9 to find 852 Hz; go up to find 1477 Hz. Therefore the output frequencies are 852 Hz and 1477 Hz.

10.3.3 The Central Office Switch

The earliest telephone switchboards were manually operated and used patch cords that were plugged into sockets. Any line could be connected to any other line in the same office or to a trunk line for long-distance calling. Automatic analog switches still work in much the same way.

The first automatic telephone switch, invented by Almon B. Strowger in 1893, worked in a step-by-step fashion, gradually finding a way from one subscriber line to another as the number was dialed. Modern electronic switches, however, wait until the entire number has been dialed before looking for a route to the destination line. This process requires memory and some computing power.

The Strowger step-by-step switch was superseded by the crossbar type of switch, which was invented in Sweden and first installed in the United States in 1938. It was capable of storing the whole telephone number before making the connection, thus reducing the amount of time the expensive switching equipment was tied up. The separation of control functions from actual signal switching, a process that began with the crossbar system, is known as **common control**.

The crossbar system used relays for "memory." The first large system to use electronic memory was the ESS-1 (1965) from Western Electric, though computerized small switches known as private automatic branch exchanges (PABXs) were available in 1963. The ESS-1 used electronic digital logic circuits for its control circuitry, but the actual switching of the telephone lines was done mechanically using reed relays. Some later analog switches used solid-state switching for the telephone lines.

Figure 10.7 shows the essence of a **crosspoint switch**. In typical systems, any incoming line can be connected to any outgoing line, but not all lines can be in use at the same time. For example, in Figure 10.7 there are ten incoming and ten outgoing lines but only three tie lines, so only three simultaneous connections are possible.

The most modern switches digitize the analog voice signal and manipulate the digital signals by assigning them to different time slots in a time-division multiplexed signal. Figure 10.8 shows a typical modern telephone switch. Some of these switches have very large capacities. The No. 4 ESS used by AT&T, for example, can handle 700,000 calls per hour. We shall have more to say about digital telephony later.

Assume for now that the called party is connected to the same central office as the calling party; that is, they have the same exchange and the first three numbers of their seven-digit telephone numbers are the same. When the switch connects to the called party, it must send an intermittent ringing signal to that telephone. The standard for the ringing voltage at the central office is 100 V ac at a frequency of 20 Hz, superimposed on the 48 V dc battery voltage. Of course, the voltage at the telephone will be less than this, due to the resistance of the wire in the local loop. In order to respond to the ac ringing signal when on hook, without allowing dc current to flow, the telephone ac-

Figure 10.7
Crosspoint switch

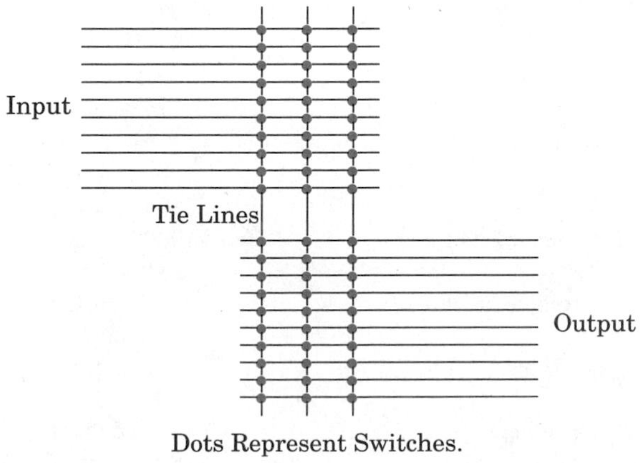

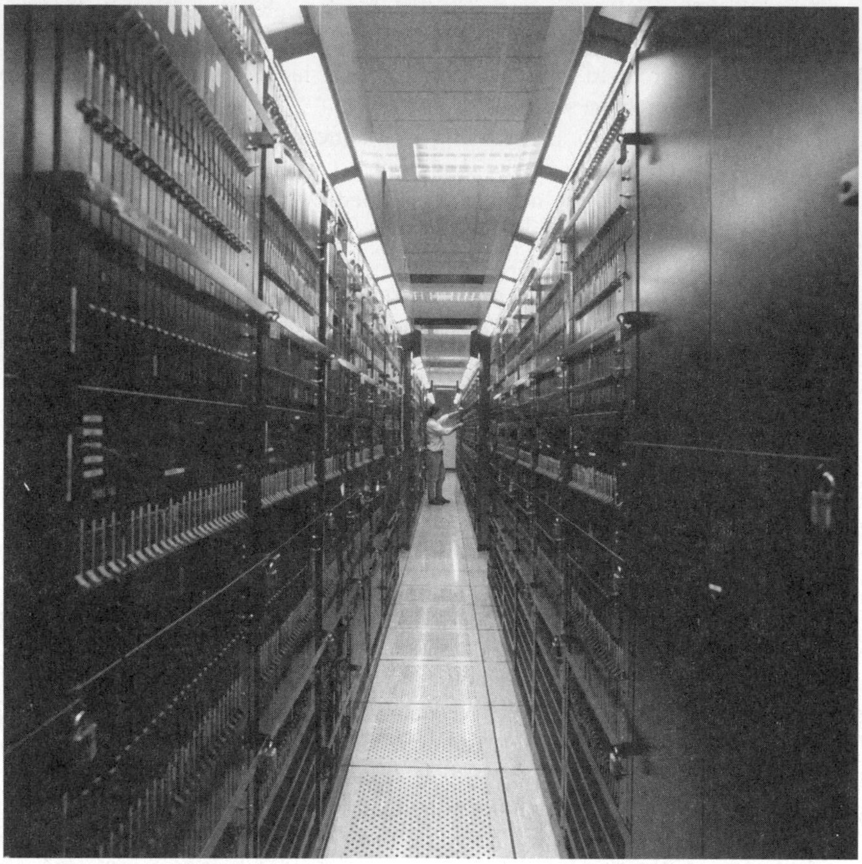

Figure 10.8
Photo of modern phone switch
Courtesy Northern Telecom. (Nortel).

couples the ringer to the line. In a conventional telephone with an electromechanical ringer, the ringer consists of two coils and a capacitor in series across the line.

While the called telephone is ringing, the central office switch sends a pulsed ac voltage called a **ringback signal** to the calling telephone. The ringback signal consists of 440 and 480 Hz signals added together. When the called phone goes off hook, the circuit is complete, the ringing voltages are switched off, and conversation can begin. If the circuit corresponding to the called telephone is in use, a busy signal is returned to the caller.

See Table 10.2 for a summary of the signals described so far.

TABLE 10.2
Local-loop voltages and currents

On-hook voltage	48 V dc
Off-hook voltage (at phone)	5–10 V dc, depending on loop resistance
Off-hook current	23–80 mA dc (approximately), depending on loop resistance
Dial tone	350 and 440 Hz
Ringing voltage (at office)	100 V ac, 20 Hz, superimposed on 48 V dc
Ringing voltage (at phone)	approximately 80 V ac, depending on loop resistance and number of ringers, superimposed on 48 V dc
Ringback voltage	440 and 480 Hz, pulsed
Busy signal	480 Hz and 620 Hz, pulsed

10.3.4 The Subscriber Line Interface Card

The local loop connects to the central office by means of a **subscriber line interface card** (abbreviated **SLIC** or just **line card**). The functions of this card can be remembered by using the mnemonic BORSCHT. The letters stand for:

Battery: the 48 V dc supply already described.

Overvoltage: protection against lightning and other high-voltage transients.

Ringing: the 100 V, 20 Hz ac ring voltage previously described is connected to the line by a relay on the line card, as required.

Supervision: monitoring the line for on- or off-hook conditions, and so forth.

Coding: for digital switches, analog-to-digital and digital-to-analog conversion take place here, at the interface between the analog loop and the digital switch. SLICs for analog switches do not have this function.

Hybrid: the local loop is a two-wire circuit with signals traveling in both directions on the same pair, and the rest of the network is usually four-wire (one pair for each direction). The conversion is done on the line card.

Testing: checking of the line for opens, shorts, and so forth.

10.3.5 The Telephone Instrument

Ordinary telephones use carbon microphones (traditionally called **transmitters**) and magnetic earphones (called **receivers**). The carbon microphone needs a dc bias current to operate. The sketch in Figure 10.9 illustrates its operating principle. As the diaphragm moves due to air pressure variations resulting from sound waves, the pressure on the carbon granules changes. When they are packed closer together, the resistance of the microphone decreases; when the pressure is less, the granules separate a little, and the resistance of the microphone increases. These resistance changes modulate the loop current.

Carbon microphones have the advantages of simplicity and the ability to generate a relatively large signal voltage without amplification. The audio quality is poor, however. In many modern telephones, carbon microphones have been replaced by electret condenser microphones. In these microphones, the vibrating diaphragm effectively changes the plate spacing in a permanently charged capacitor consisting of electrodes on both sides of a layer of plastic that has a permanently stored electrical charge. An electret condenser microphone produces a very small ac voltage (less than one millivolt) that requires amplification. In these telephones, the battery voltage powers an electronic amplifier instead of biasing the microphone directly.

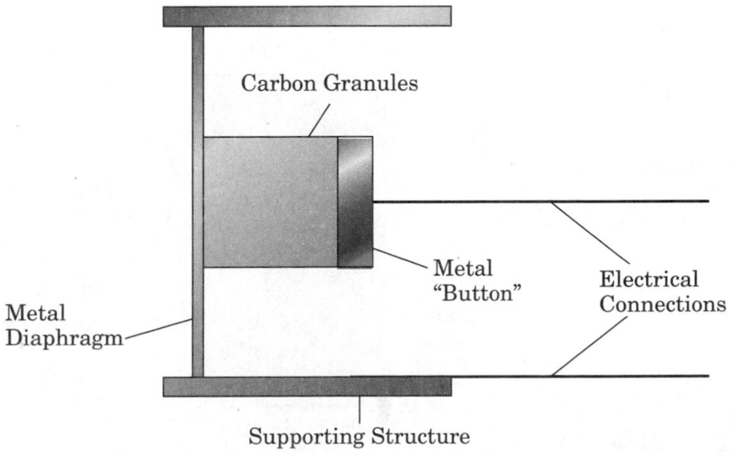

Figure 10.9 Carbon microphone (simplified cross-section)

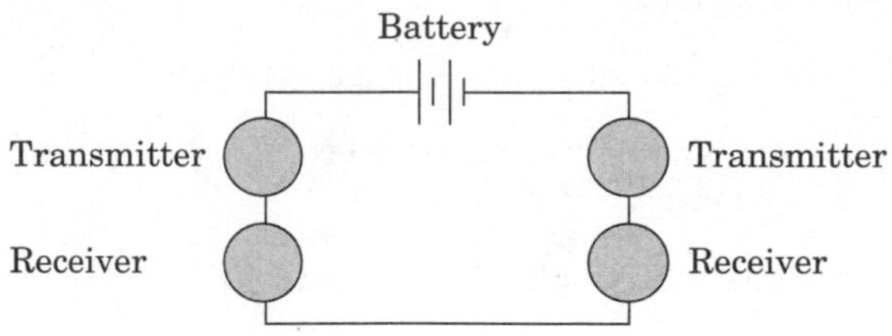

Figure 10.10
Simple telephone system

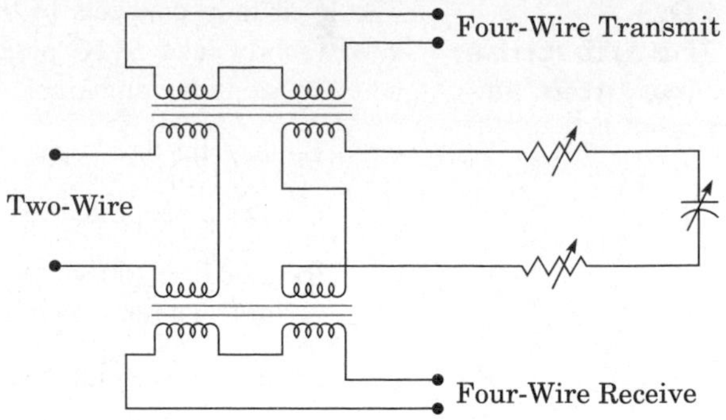

Figure 10.11
Hybrid coil

The single twisted-pair line is required to carry both sides of the conversation simultaneously, thus providing **full-duplex communication**. This could be accomplished by simply connecting both transmitters and both receivers in series, as shown in Figure 10.10. However, the two receivers would receive equal power, and some of the transmitter power would be wasted in the other transmitter. These problems can be solved using a circuit called a **hybrid coil**, which is shown in Figure 10.11. The same thing can be accomplished electronically. Signals from the transmitter will add at the line and cancel at the receiver. Similarly, signals coming in on the line will cancel at the transmitter and add at the receiver. Deliberately unbalancing the circuit allows a small portion of the transmitter signal to reach the receiver, creating a **sidetone** that lets the user know that the line is active and hear what is being transmitted. Hybrid coils are also used in the central office line cards, as already noted.

Figure 10.12 shows a simplified circuit for a standard dial telephone. The switch-hook, S_1, closes the line circuit when the receiver is lifted, alerting the central office to find an available line. S_2 is the dial; it interrupts the line at a 10 Hz rate during dialing. A portion of S_2 also short circuits the receiver when dialing to prevent the dial pulses from being heard. The balancing network works with the hybrid coil to provide a sidetone, as described in the preceding paragraph. The capacitor C_1 between the two ringer coils blocks dc so that there is no dc continuity when the phone is on hook.

Modern electronic telephones typically use one or two specialized integrated circuits. If two are used, one generates the DTMF signals and the other does pretty much everything else. The mechanical ringer is replaced with an electronic oscillator and either a piezoelectric transducer or an ordinary loudspeaker. Figure 10.13 shows a circuit for a typical electronic telephone.

Figure 10.12
Simplified circuit for dial telephone

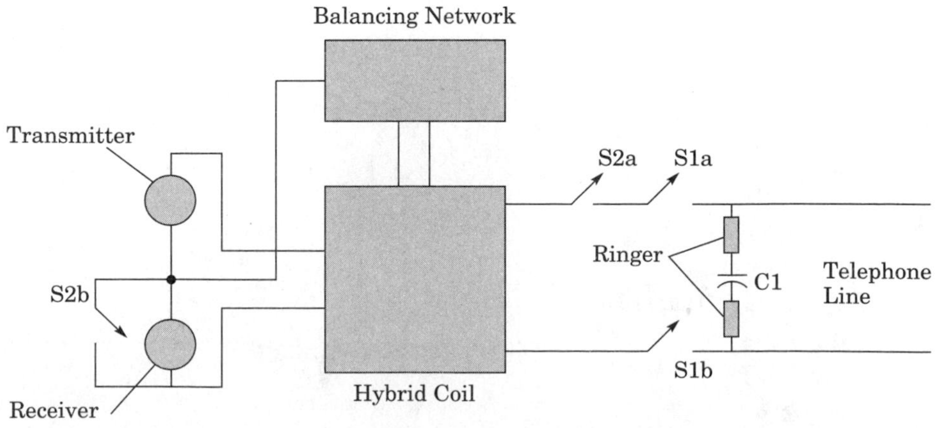

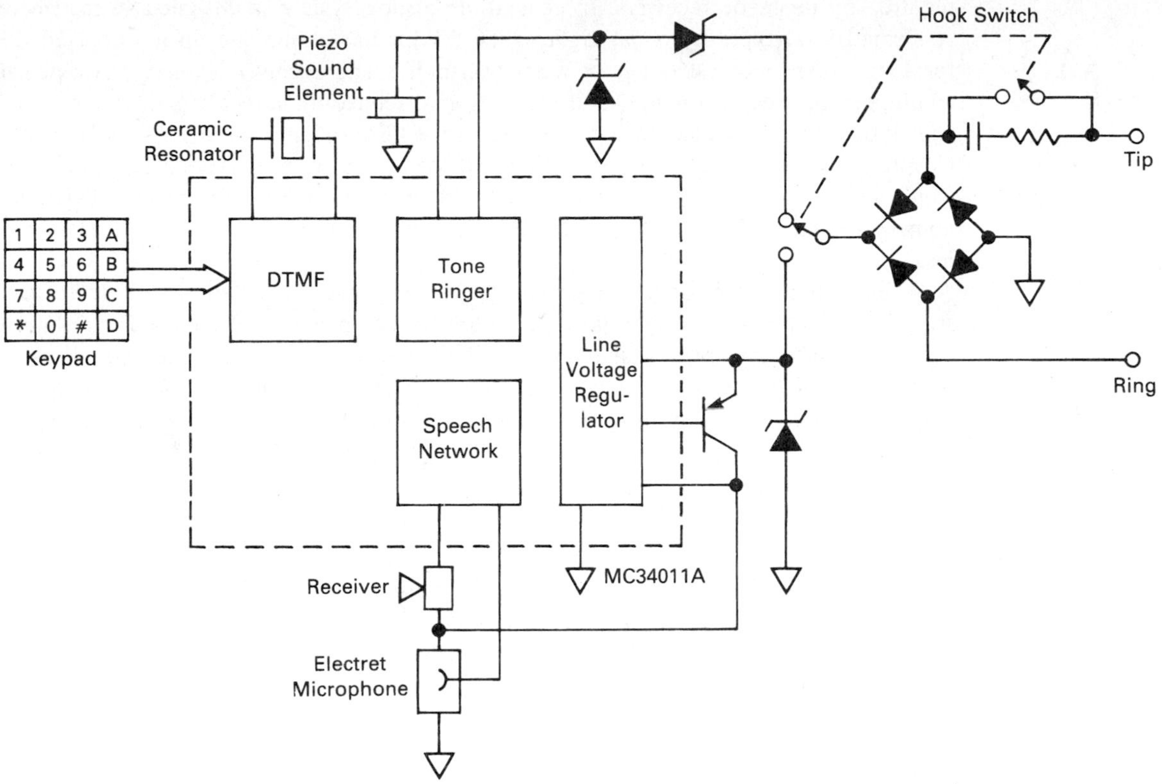

Figure 10.13
Circuit for 1-chip electronic telephone
Copyright of Motorola. Used by permission.

Just as the battery and ringing voltages on a local loop are compatible with the requirements of carbon microphones and electromechanical ringers, the audio signal levels on telephone lines were set to be those typically produced by carbon microphones. No amplification is required for communication over a distance of several kilometers between telephone sets.

10.4 Signals and Noise in the Telephone System

Two of the most important characteristics of any communication system are bandwidth and signal-to-noise ratio. The bandwidth of voice-grade telephone systems is deliberately restricted to about 3.2 kHz in order to allow the multiplexing of as many signals as possible in a channel of limited bandwidth. Restricting the bandwidth also reduces the amount of noise that gets into the voice channel. For special applications, such as studio-to-transmitter links for broadcast stations and high-speed data links for large businesses, wider-bandwidth "conditioned" lines are available.

The signal-to-noise ratio for an analog system is bound to degrade with increasing distance from the source. Losses in the system can be compensated for by using amplifiers called repeaters, but noise will continue to accumulate. In a properly designed digital system, on the other hand, repeaters can regenerate pulses with only a very slight increase in the bit error rate (the bit error rate is the digital equivalent of noise in terms of its degrading effect on the intelligibility of the signal)—this is one of the main advantages of digital communication, and it will be explored in more detail in the next chapter.

The signal levels in analog telephone systems have changed little in the past century, since they were determined by the levels produced by carbon microphones and required by magnetic receivers. In general, an analog system is designed so that there is a small loss (typically about eight to ten dB for local calls and up to about 20 dB for long-distance connections) between transmitter and receiver, even when repeater amplifiers are used. The purpose of this loss is to reduce the effect of reflections of the signal due to impedance mismatches. (See the discussion of transmission lines in Chapter 16 for more details on reflections.) If the system were to have a net gain, there would be a possibility of oscillation due to feedback from these reflections. This phenomenon is called "singing," and the loss introduced to avoid it is called the **via net loss (VNL)**.

Reducing the net gain to less than 0 dB will prevent oscillations, but there may still be an annoying echo as reflections return to the speaker and interfere with his or her speech. The amount of loss required to reduce echo to an acceptable level depends on the amount of the delay, since long-delayed echoes are more confusing. Delays on long-distance circuits can easily be several milliseconds (or much longer when satellites are used). The required loss is given by the equation

$$VNL = 0.2t + 0.4 \text{ dB} \tag{10.1}$$

where

VNL = minimum required via net loss in dB
t = time delay in ms for propagation one way along line

Example 10.3 A telephone signal takes 2 ms to reach its destination. Calculate the via net loss required for an acceptable amount of echo.

Solution

$$\begin{aligned} VNL &= 0.2t + 0.4 \text{ dB} \\ &= 0.2 \times 2 + 0.4 \text{ dB} \\ &= 0.8 \text{ dB} \end{aligned}$$

If reducing the level of reflections in this way requires an unacceptably large loss, as it would on very long circuits, echo suppressors can be used. Echo suppressors are electronic switches that sense the direction of travel in which the signal level is greater and switch off the other path, which is assumed to be the return (reflection) path. Essentially, they convert a full-duplex system to half-duplex with automatic switching. Echo suppressors can cause problems in data communication, so they must be switched off by tone signals sent by the modems. A tone with a frequency between 2010 and 2240 Hz and duration of 400 ms is used to disable the echo suppression.

Signal and noise levels are defined in a way that is unique to telephony: in decibels above reference noise when C-message weighted, abbreviated dBrnc. The reference noise level is one picowatt or -90 dBm when C-message weighted. This weighting requires a little explanation. While thermal noise has a flat frequency spectrum, the response of the telephone instrument does not. C-message weighting is an attempt to adjust the noise or signal level to the response of a typical telephone receiver. To do this, a correction factor is applied to frequencies at which the response is less than maximum. The resulting curve is shown in Figure 10.14.

The net loss described above is specified with respect to a reference point called the zero-transmission-loss point (0-TLP). This point is considered to be located at the output of a two-wire switch. If the switch is a four-wire type, its output is considered to be -2 dB TLP and the 0-TLP is not physically present. Any point in the network can be defined in terms of its gain (or more usually, loss) compared to the 0-TLP. For instance, at a telephone instrument, received signals might be at -8 dB TLP. If the signal were measured at this point, 8 dB would be added to get the equivalent level at the 0-TLP.

In telephony, signal and noise levels are often given separately with reference to the 0-TLP, rather than being expressed as a signal-to-noise ratio. For instance, a noise level for a long-distance circuit might be 34 dBrnc0, that is, 34 dB above one pico-

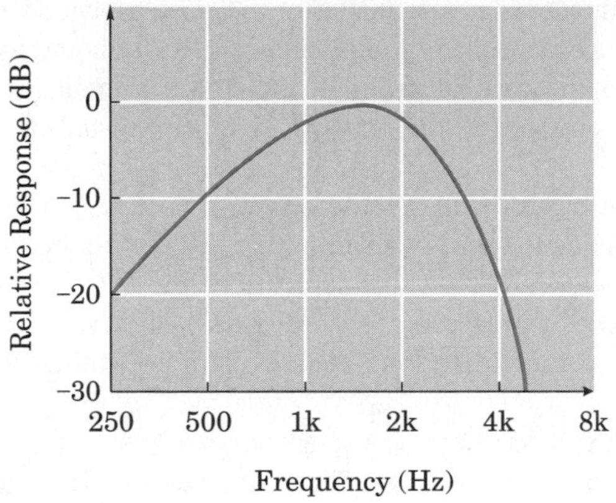

Figure 10.14 C-message weighting

watt, C-weighted, expressed in terms of the 0-TLP. The actual measured C-message weighted noise level at the receiver mentioned above would be 8 dB less than this, or 26 dBrnc. A typical signal level would be about 74 dBrnc0. The signal-to-noise ratio would be the difference:

$$S/N = 74 \text{ dBrnc0} - 34 \text{ dBrnc0} = 40 \text{ dB}$$

Example 10.4 A 1 kHz test tone is inserted at a local loop with an amplitude of +4 dBm and is transmitted toward the central office. In this direction the loop has a level of +10 dB TLP, because the signal will be attenuated as it moves toward the central office. Express the level of the tone in dBrnc0.

Solution First let us convert from dBm to dBrn. Since 0 dBrn = −90 dBm, a signal level of +4 dBm corresponds to 90 + 4 = 94 dBrn. Next, look at the C-weighting curve to find that at 1 kHz there is an attenuation of about 5 dB, which reduces our signal down to 94 − 5 = 89 dBrnc at the loop. At the 0-TLP point the signal will be 10 dB less than this, or 79 dBrnc TLP.

10.5 In-Channel and Common-Channel Signaling

We have already looked at some of the control and supervisory signals used with the telephone system. DTMF signals, the dial tone, the busy signal, and the ringback signal are examples. All these signals use the same channel as the voice, but not at the same time. Since they use the same channel, they are called **in-channel signals**. Their frequencies are also in the same range as voice frequencies so that they can be heard by the user, so they are also referred to as **in-band signals**.

We also noted that the telephone instrument communicates its off-hook or on-hook status to the central office by the presence or absence, respectively, of a dc current. This signal is in-channel because it uses the same pair of wires as the voice, but it is **out-of-band** because the dc current is not in the same frequency range as voice signals. Consequently, the central office can receive the off-hook signal continuously, even during the call.

Traditionally, similar methods have been used within the network for such purposes as communicating the number of the calling party to billing equipment and determining which trunk lines are idle (that is, ready for use). Earlier systems used dc currents and dial pulses, as in local loops. Later versions used either a switched single-frequency tone at 2600 Hz (called *SF signaling*) or, still later, combinations of tones similar to but not the same as the DTMF system, known as **MF (multifrequency) signaling**. For a time, that system was plagued by fraud as people ranging from amateur "hackers" to members of organized crime used so-called blue boxes to duplicate network signaling tones and make long-distance calls without paying for them and, even more importantly in the case of criminal activities, without there being any record of the calls. However, changes to the network soon eliminated or at least greatly reduced this problem.

In-channel but out-of-band signaling is also used. A tone at 3825 Hz can be sent along a long-distance network. It will pass through the allotted 4 kHz channel for each call but will be filtered out by voice-separation filters before it reaches the customer telephone. This type of signal can replace the dc loop current as a means of indicating whether the line is in use, since long-distance circuits, unlike local loops, do not have dc continuity.

Recently the trend has been to use a completely separate data channel to transmit control information between switches. This **common-channel signaling** completely eliminates fraud, since users have no access to the control channels, and it also allows a call to be set up completely before any voice channels are used. The state of the whole network can be known to the control equipment, and the most efficient routes for calls can be planned in advance. Common-channel signaling also makes services such as calling-number identification much more practical. The signaling channels can be either special data lines or conventional voice-grade lines with modems. They are arranged into a signaling network that is completely separate from the voice network.

10.6 Frequency-Division Multiplexing

A number of analog telephone signals can be combined (multiplexed) onto one channel using frequency-division multiplexing (FDM). The channel can be a twisted pair, a coaxial cable, or a microwave radio link, either terrestrial or satellite. (Optical fiber is used almost exclusively for digital signals employing time-division multiplexing (TDM) and will be discussed later in this book.)

The concept of FDM should be familiar from its use in ordinary radio broadcasting. The available spectrum is divided among a number of information signals. In ordinary AM broadcasting, each channel is 10 kHz wide and double-sideband full-carrier AM transmission is used. In FDM telephony, the modulation is usually single-sideband suppressed-carrier (SSB or SSBSC) and 4 kHz of spectrum is allocated to each conversation. A telephone conversation actually needs only about 3 kHz of bandwidth, so this scheme allows a small guard band between channels to make it easier to separate them. Either upper- or lower-sideband SSB can be used, but lower-sideband (LSB) modulation is standard in North America.

The number of simultaneous conversations that can be transmitted using FDM depends on the total bandwidth available, which in turn, varies with the medium. The channels are grouped according to a hierarchical structure, as shown in Figure 10.15. The lowest level of the hierarchy is the **group**, which consists of 12 LSB signals spanning a frequency range from 60 kHz to 108 kHz. A group is also known as an A (analog) channel bank. Five of these groups can be multiplexed together to form a **supergroup**, which occupies the spectrum from 312 kHz to 552 kHz and contains 60 voice channels or their equivalent.

If the bandwidth of the transmission medium permits, ten supergroups can be combined to form a **mastergroup** with a frequency range from 564 kHz to 3084 kHz and a capacity of 600 voiceband channels. The mastergroup incorporates guardbands between the supergroups to lessen the need for precision in frequency-shifting the signals. A typical terrestrial microwave link would carry three mastergroups. For wider-bandwidth applications such as satellite links, it is possible to form a **jumbogroup** (six mastergroups or 3600 voiceband channels) or a **superjumbogroup** (three jumbogroups or 10,800 voiceband channels.)

The process of generating an FDM telephone signal is similar to generating an ordinary SSBSC radio signal. It is illustrated by the block diagram in Figure 10.16.

Figure 10.15 FDM telephone system spectrum

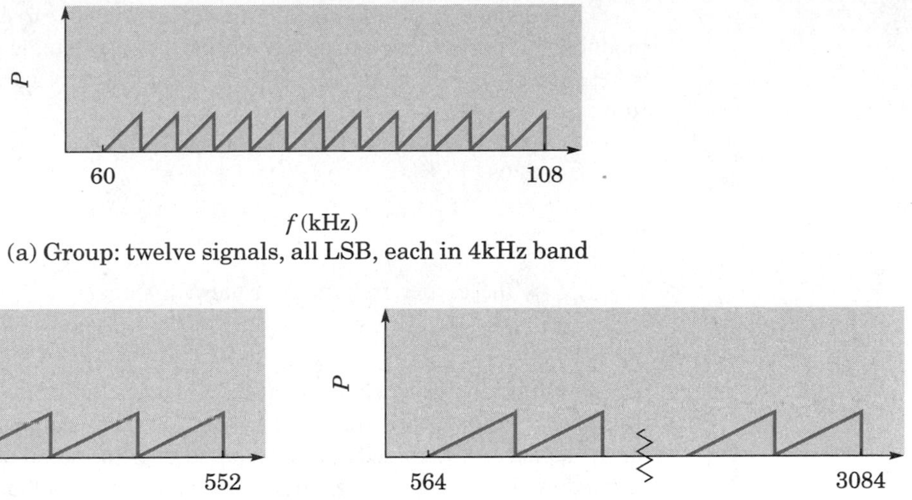

(a) Group: twelve signals, all LSB, each in 4kHz band

(b) Supergroup: five groups

(c) Mastergroup: ten supergroups, separated by guard bands

Figure 10.16 Generation of a group

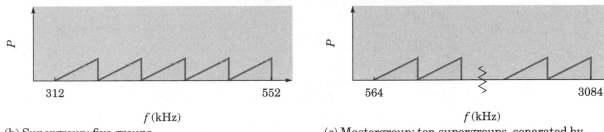

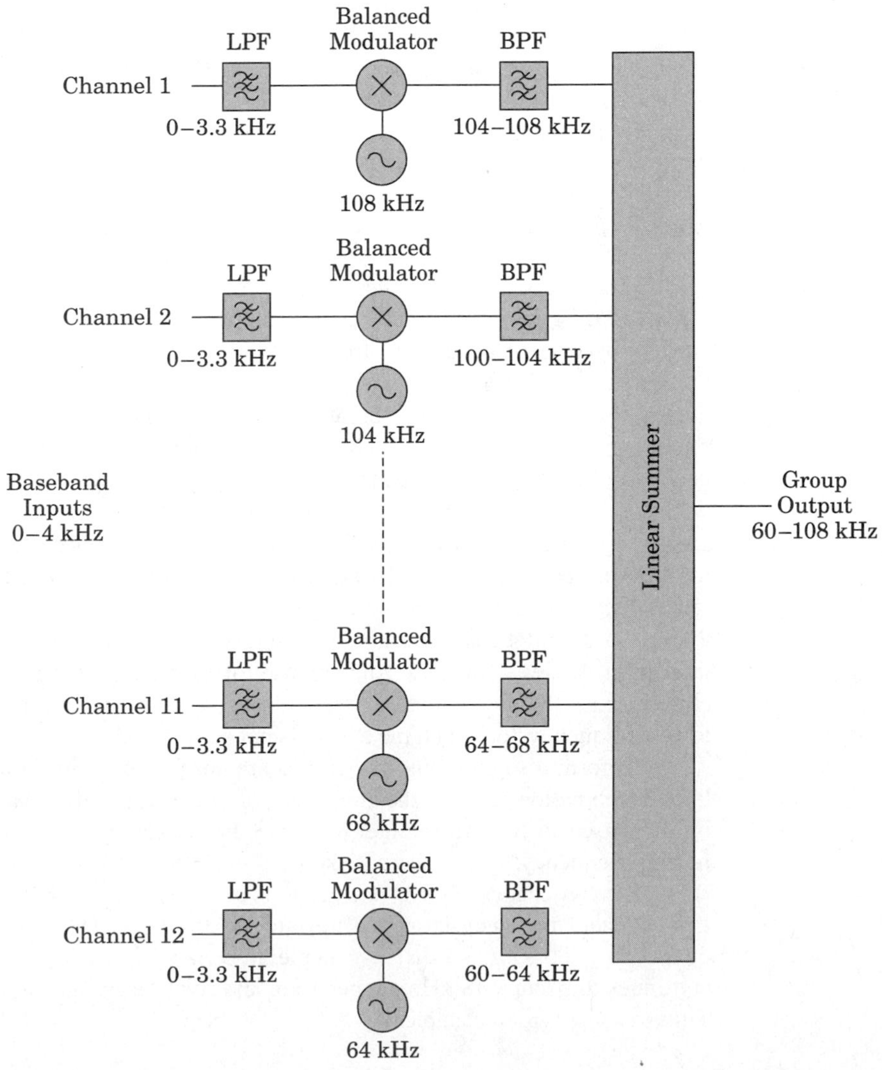

Each voice baseband signal and an associated carrier are applied to a balanced modulator. The carriers are all synthesized from a single oscillator and are spaced at 4 kHz intervals, beginning at 64 kHz. Therefore, any carrier frequency can be found from:

$$f_{cN} = 64 + 4\,(12 - N)\text{ kHz} \tag{10.2}$$

where

f_{cN} = the carrier frequency of the Nth signal
N = the number of the signal

The resulting double-sideband suppressed-carrier (DSBSC) signals are bandpass-filtered to pass only the desired sideband; in this case, the lower sideband is required. Finally, all the signals are summed and transmitted over the same channel. At the receiving end, the process is reversed. A bandpass filter first separates the signals. After the filter, each signal is applied to a balanced demodulator (also called a product detector) along with a locally generated carrier. The output of the product detector is the required baseband audio.

Example 10.5 Find the (suppressed) carrier frequency for channel 5 of a group.

Solution From Figure 10.15, note that channel 12 has the lowest carrier frequency, 64 kHz, and that the carrier frequency goes up 4 kHz per channel.

Therefore, using Equation (10.2) for channel 5:

$$f_c = 64\text{ kHz} + 4\,(12 - 5)\text{ kHz} = 92\text{ kHz}$$

There are some differences between SSB radio and FDM telephony. First, in order for there to be no interference between the signals that are multiplexed together, all the carriers used must be derived from the same master oscillator. That way, any drift in the carrier oscillator frequency will cause all the multiplexed signal frequencies to move together, by the same amount and in the same direction. Second, since the telephone system must operate automatically, without the skilled operator that is usually necessary for SSB voice radio, the carriers generated in the receiver must be synchronized with the transmitted carrier—otherwise there would be a frequency shift in the received signal. Synchronization can be accomplished by transmitting a pilot carrier. Since all the transmit carriers are synchronized, only one pilot carrier needs to be sent, and all the receiver carriers can be generated from it. All the carrier frequencies are multiples of 4 kHz, so the pilot carrier can be any multiple of 4 kHz. For a group, 64 kHz is a common pilot carrier frequency. Another way of looking at this is to say that all the voice channels use suppressed-carrier transmission except the one that is transmitted at the lowest frequency, for which the carrier is not suppressed.

To form a supergroup, five groups are combined by using a mixer and local oscillator combination to raise the frequency of each. You will recall that a similar process is often used in SSB transmitters to raise the frequency of an SSB signal to the transmitting frequency. This process is sketched in Figure 10.17.

The frequency range of one group is changed from 60–108 kHz to 312–360 kHz, that is, each frequency is moved upward by 252 kHz. The next group is raised in frequency by 300 kHz, so that it occupies the range from 360 to 408 kHz. Each group continues to occupy 48 kHz of spectrum, and the five groups are packed tightly together. Thus the supergroup occupies $48 \times 5 = 240$ kHz.

Figure 10.17 Generation of a supergroup

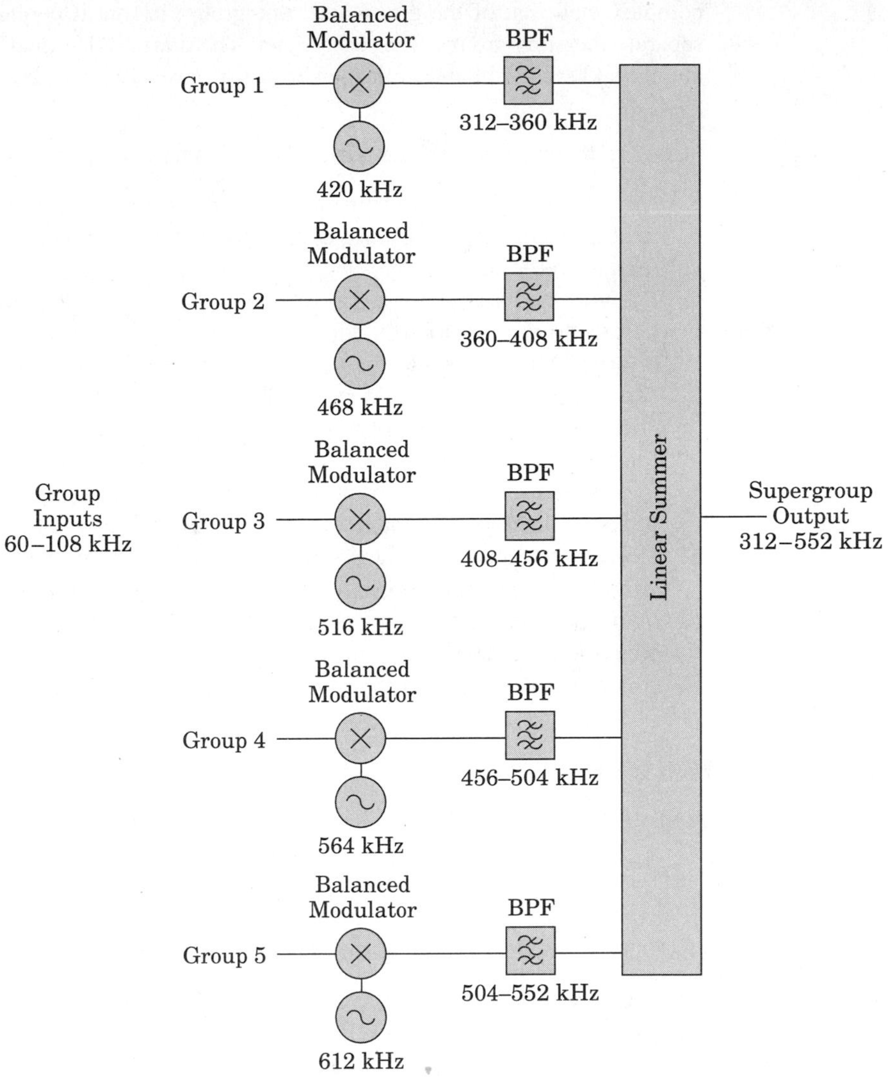

Example 10.6 A 2 kHz tone is present on channel 5 of group 3 of a supergroup. At what frequency does the tone appear in the supergroup output?

Solution In Example 10.5 we discovered that the carrier frequency for channel 5 of a group is 92 kHz. Since the signal is lower sideband, the 2 kHz baseband signal will be moved to

$$f_g = 92\text{ kHz} - 2\text{ kHz} = 90\text{ kHz}$$

It can be seen from Figure 10.15 that group 3 in the supergroup is moved to the range 408–456 kHz, with a suppressed carrier frequency of 516 kHz. Again the modulation is lower sideband, so the supergroup output frequency will be 90 kHz lower than the carrier frequency.

$$f_{sg} = 516\text{ kHz} - 90\text{ kHz} = 426\text{ kHz}$$

A similar process is used to generate the higher levels of FDM signals. The mastergroup shown in Figure 10.15 is known as a U600 mastergroup and can be further multiplexed into even larger conglomerations of signals. Its structure is somewhat more complex than that of the group and supergroup, in that it contains guard bands to help separate the supergroups from each other. There are 8 kHz guard bands between groups and a 56 kHz guard band in the center of the mastergroup, between 2044 and 2100 kHz.

Summary

Here are the main points to remember from this chapter.

1. The modern public switched telephone networks in North America have a flat structure with multiple long-distance carriers connected to a local network that remains a monopoly.
2. The voltages and currents on telephone lines, with the exception of tone-dialing signals, were set in the days of electromechanical ringers and carbon microphones. Modern equipment must remain compatible with these old standards.
3. The telephone network is gradually becoming digital, though most local loops and subscriber equipment are still analog.
4. Telephone connections generally have a net loss of a few decibels from one end to the other, even when amplification is used, in order to minimize the effects of line reflections, which otherwise could cause oscillations or annoying echoes.
5. Modern telephone networks use common-channel signaling, which uses data channels completely separate from the voice channels to set up calls.
6. Many analog voice channels can be combined using frequency-division multiplexing. Time-division multiplexing, which is used with digital telephony, will be covered in the next chapter.

Important Equations

$$\text{VNL} = 0.2t + 0.4 \text{ dB} \quad (10.1)$$

$$f_{cN} = 64 + 4\,(12 - N) \text{ kHz} \quad (10.2)$$

Glossary

call blocking inability of the network to complete a telephone call due to lack of free trunk lines or switching equipment

central office switch that handles one telephone exchange (up to 10,000 lines)

common-channel signaling a signaling system that sets up and monitors long-distance calls and is completely independent of the channels used for voice

common control a control system for telephone switching that sets up one call and then goes on to set up another call, without being tied up for the duration of the first call

crosspoint switch a switching system that uses a matrix arrangement of incoming and outgoing lines

crosstalk interference between two telephone conversations, usually caused by inductive and capacitive coupling between pairs in the same cable

dual-tone multifrequency (DTMF) dialing a dialing method that uses combinations of tones to represent each digit

end office see *central office*

fiber-in-the-loop (FITL) use of fiber-optic cable in the local loop to the individual subscriber

flat network a switching network that does not use a hierarchical structure but instead relies on a computer to find the most direct route between two points

full duplex a type of communication system that allows communication in both directions simultaneously

group a frequency-division multiplexing scheme that allows 12 voice signals to be transmitted on one channel

hybrid coil a transformer that allows two- and four-wire full-duplex systems to be connected together; also called a hybrid transformer

in-band signal a control or supervisory signal that is transmitted in the voice band on a voice channel, such as a dial tone

in-channel signal a control or supervisory signal that is transmitted in the voice channel; it may or may not be within the voice frequency range

jumbogroup a frequency-division multiplexed signal

consisting of six mastergroups or 3600 voiceband channels

line card see *subscriber line interface card*

loading coil inductance added to a twisted-pair telephone line to reduce its losses for voice frequencies

local access and transport area (LATA) an area consisting of several central offices and handled by a local carrier

local loop the link from the central office to an individual subscriber's premises

mastergroup a frequency-division multiplexed signal carrying 10 supergroups or 600 voice channels

multifrequency (MF) signaling a system for setting up long-distance calls using pairs of tones sent along voice channels; a form of in-band signaling

off hook a term used to describe a telephone instrument that is ready for use, with the handset removed from its cradle

on hook a term used to describe a telephone instrument whose handset is on its cradle, ready to receive a ring signal

out-of-band signal a control or supervisory signal that is transmitted on a voice channel, but at dc or at such a frequency that it will not be heard

point of presence (POP) connection of a long-distance carrier to the local telephone system

POTS (plain old telephone service) a term used to describe ordinary voice telephony

pulse dialing dialing that works by interrupting the dc loop current; used by dial-type telephones and some electronic phones

receiver the earpiece in a telephone

ring the red wire in a telephone circuit; it normally has negative polarity

ringback signal signal sent by the network to the calling telephone to indicate that the called telephone is ringing

sidetone a low-level voice signal sent to a telephone receiver from the transmitter in the same telephone

subscriber line interface card (SLIC) the circuit board that connects a local loop to the central office

supergroup a frequency-division multiplexed signal consisting of five groups or 60 voice channels

superjumbogroup a frequency-division multiplexed signal consisting of three jumbogroups or 10,800 voiceband channels

tandem office a switch that connects central offices together but is not directly connected to any subscriber lines

tip the green wire in a telephone loop; it normally has positive polarity

toll station a long-distance telephone switch

topology the layout of a system such as a telephone network

Touch-Tone dialing AT&T trademark for DTMF dialing

transmitter the microphone in a telephone

trunk line a connection between telephone offices

via net loss (VNL) the signal loss in decibels between the transmitting and receiving ends of a telephone connection

Questions

1. Explain briefly how the telephone network differs from a broadcasting network.
2. Describe the difference between flat and hierarchical switched networks. Explain why modern networks are flat.
3. What is a tandem office?
4. What is a LATA?
5. How has the breakup of the Bell monopoly changed the telephone network?
6. What is call blocking? How does it happen?
7. Why has fiber optics been much slower in penetrating the local loop part of the telephone system than in replacing copper in trunk lines?
8. How many wires are needed from the individual telephone set to the central office (for a single line)?
9. How many wires are normally contained in the cable from an individual residence subscriber to the network? Why is this number different from the answer to question 8?
10. What is the function of loading coils on a telephone line? Do they have any disadvantages?
11. Explain the meanings of the terms *tip* and *ring*. Which has negative polarity?
12. Explain how pulse dialing works.
13. What is meant by DTMF dialing? Why is it better than pulse dialing?
14. What is a crosspoint switch?
15. Give one advantage of common control over the step-by-step system in telephone switching.
16. Why is it not possible for all subscribers to use their telephones simultaneously with a typical crosspoint switch?
17. List and describe the functions commonly performed by the subscriber line interface card.

18. Explain the operation of a carbon microphone.
19. What is the function of the hybrid coil in a telephone instrument?
20. What is sidetone? Why is it used in a telephone instrument?
21. Approximately how much bandwidth, at baseband, is needed for one channel of telephone-quality audio?
22. Why is it usually necessary to provide some loss on a telephone circuit?
23. What is the purpose of echo suppressors?
24. Explain what is meant by the term dBrnc0.
25. Describe the difference between in-channel and common-channel signaling. Which is the more modern system?
26. Describe the difference between in-band and out-of-band signaling. Give an example of each.
27. Name two forms of multiplexing. Which one is more commonly used with analog signals?
28. Which of the FDM combinations described in the text carries the most voice-grade signals?
29. What type of modulation is used in FDM telephony?

Problems

SECTION 10.3

30. Suppose the voltage across a telephone line at the subscriber drops from 48 V to 10 V when the phone goes off hook. If the telephone instrument has a resistance of 200 ohms when off hook and represents an open circuit when on hook, calculate:
    (a) the current that flows when the phone is off hook
    (b) the combined resistance of the local loop and the power source at the central office
31. The local loop has a resistance of 650 ohms, and the telephone instrument has a ringer voltage of 80 volts when the voltage at the central office is 100 V. Calculate the impedance of the ringer in the telephone.
32. Find the DTMF frequencies for the number 8.
33. What number is represented by tones of 770 and 1209 Hz in the DTMF system?
34. Draw a schematic diagram for a crosspoint switch that will switch 20 inputs to 20 outputs, allowing 10 simultaneous conversations.
35. List the steps in making a local call and the local-loop voltages and frequencies associated with them.

SECTION 10.4

36. Calculate the required via net loss for a line 1000 km long on coaxial cable with a velocity factor of 0.66.
37. Calculate the required via net loss for a satellite link with a one-way delay of 250 ms. Is this loss a practical value? Explain your answer.
38. A 1 kHz tone has a level of 70 dBrnc at a point that is −9 dB TLP. What would be the maximum C-message weighted noise level at the 0 TLP for a signal-to-noise ratio of 30 dB?
39. A tone measures 80 dBrnc at the output of a four-wire central office. What will it be at a local loop that is at −8 dB TLP?

SECTION 10.6

40. A broadcast-quality video signal needs approximately 4.2 MHz of baseband bandwidth. If it is transmitted along a telephone-type microwave link, how many voice channels will it displace? How many mastergroups does this represent?
41. Find the frequency to which a 400 Hz baseband frequency will be translated, for channel 8 of a group.
42. The group in the previous problem is multiplexed into a supergroup as group 2. Find the frequency to which the tone is translated at the output of the supergroup.

PART

II

Digital and Data Communication

11 Digital Communications

Objectives On completing this chapter, you should be able to:

1. Compare analog and digital communication techniques and discuss the appropriate use of each
2. Calculate the information capacity of a channel
3. Calculate the minimum sampling rate for a signal and explain the necessity for sampling at that rate or above
4. Describe the common types of analog pulse modulation
5. Describe pulse-code modulation and calculate the number of quantizing levels, the bit rate, and the dynamic range for PCM systems
6. Perform calculations to show the effect of compression on a PCM signal
7. Describe the coding and decoding of a PCM signal
8. Describe delta modulation and explain the advantages of adaptive delta modulation
9. Describe and compare line codes in terms of frequency components and clock-information content
10. Show how time-division multiplexing can be used to send multiple digital signals over a single channel
11. Describe the use of PCM and TDM in the telephone system
12. Describe the Integrated Services Digital Network
13. Explain the use of compression in the coding of video signals

11.1 Introduction

Many of the signals used in modern communication are digital (for example, the codes for alphanumeric characters and the binary data used in computer programs). In addition, digital techniques are often used in the transmission of analog signals. Digitizing a signal often results in improved transmission quality, with a reduction in distortion and an improvement in signal-to-noise ratio. This chapter looks mainly at the digital transmission of analog signals such as voice and video signals. Data communication will be studied in the next chapter.

Figure 11.1 shows several possible types of signal transmission. In Figure 11.1(a) an analog signal is sent over a channel with no modulation. A typical example is an ordinary public-address system consisting of a microphone, an amplifier, and a speaker and using twisted-pair wire as a channel.

Figure 11.1(b) shows analog transmission using modulation and demodulation. Broadcast radio and television are good examples.

Figures 11.1(c) and 11.1(d) start with a digital signal (for example, a data file from a computer). In (c), the link can handle some kind of digital pulse signal directly. In (d), the channel cannot transmit pulses directly (a radio channel, for example, requires a

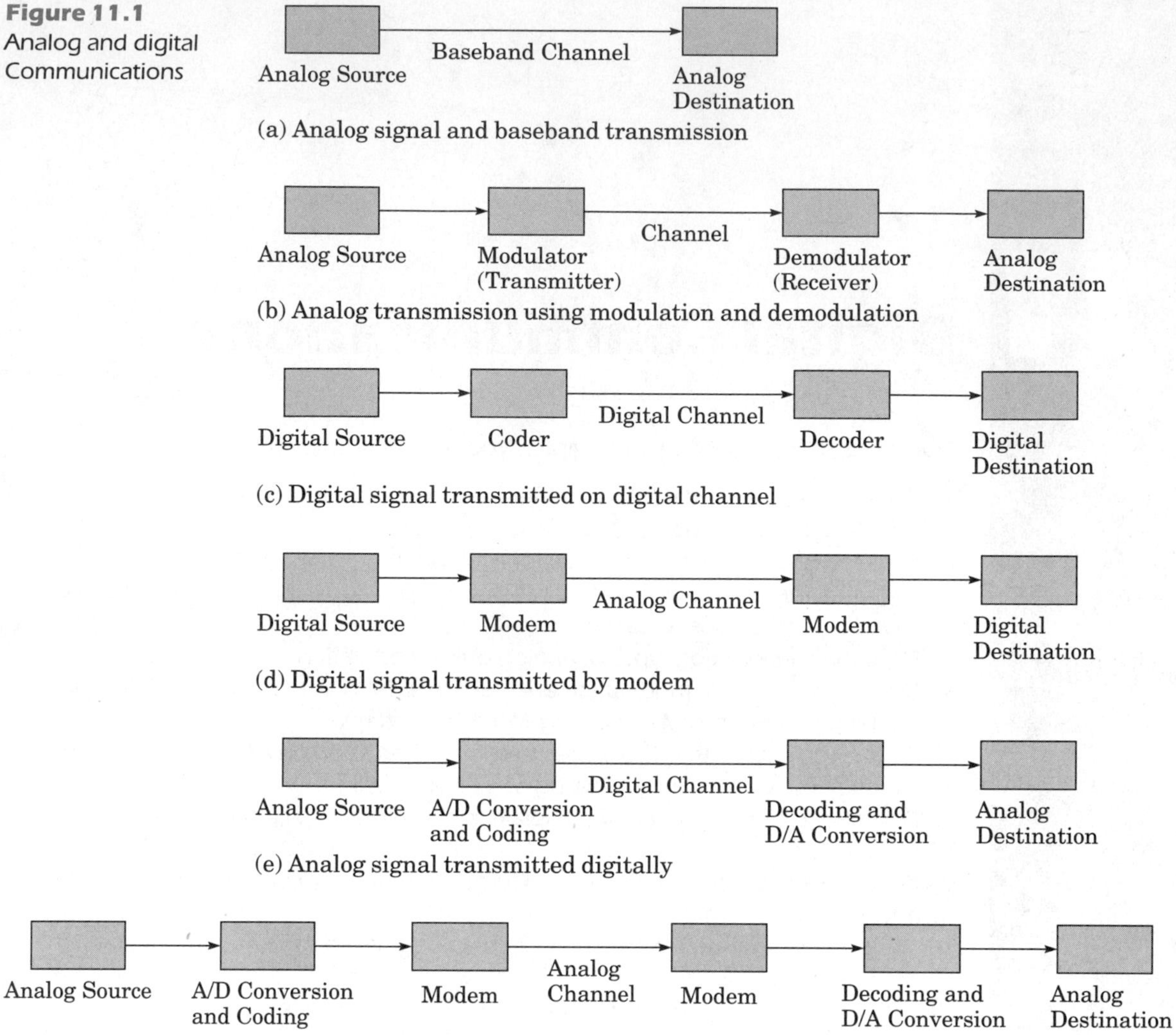

Figure 11.1 Analog and digital Communications

(a) Analog signal and baseband transmission

(b) Analog transmission using modulation and demodulation

(c) Digital signal transmitted on digital channel

(d) Digital signal transmitted by modem

(e) Analog signal transmitted digitally

(f) Analog signal digitized and transmitted by modem

modulation process, and an ordinary telephone connection cannot pass dc). In these cases the digital signal has to be modulated onto a carrier at one end and demodulated at the other. The *modem* shown is a combination modulator-demodulator. Transmission of digital data will be discussed in the next chapter.

Figures 11.1(e) and 11.1(f) show an analog signal that is digitized at the transmitter and converted back to analog form at the receiver. The difference between these two systems is that in (e) the transmission is digital, while in (f) the transmission channel cannot carry pulses, so modulation and demodulation are required.

Analog transmission of analog signals, such as voice, seems to make sense. It is certainly simpler than converting the signal to digital form and back again. Similarly, it seems obvious that signals that begin as digital, such as the contents of computer memories, should be kept in digital form as much as possible. What seems awkward at first glance is the idea of converting analog signals to digital form for transmission. Actually the use of digital techniques with analog signals is one of the fastest growing areas in communications, for several good reasons.

Throughout this book, beginning in Chapter 1, we have looked at the effects of noise and distortion on analog signals. Once noise and distortion are present, there is usually no way to remove them. Furthermore, the effects of these impairments are cumulative. Noise is added in the transmitter, the channel, and the receiver. If the communication system

Figure 11.2
Removal of noise and distortion from digital signal

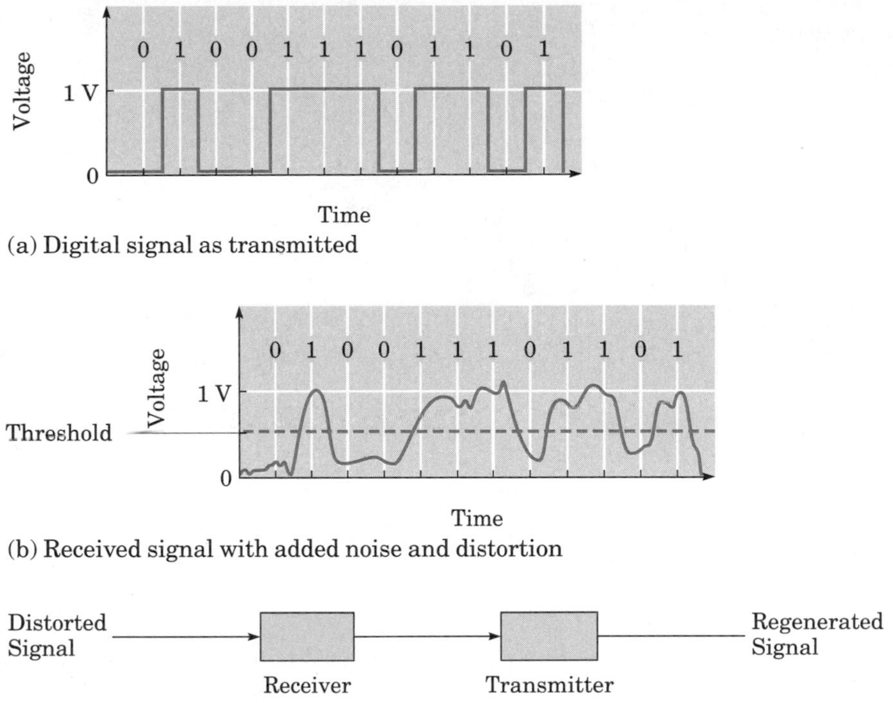

(a) Digital signal as transmitted

(b) Received signal with added noise and distortion

(c) Digital repeater

involves several trips through amplifiers and channels, as in a long-distance telephone system, the signal-to-noise ratio will gradually decrease with increasing distance from the source.

Digital systems are not immune from noise and distortion, but it is possible to reduce their effect. Consider the simple digital signal shown in Figure 11.2. Suppose that a transmitter generates 1 V for a binary one and 0 V for a binary zero. The receiver examines the signal in the middle of the pulse and has a decision threshold at 0.5 V; that is, it considers any signal with an amplitude greater than 0.5 V to be a one and any amplitude less than that to represent a zero. Figure 11.2(a) shows the signal as it emerges from the transmitter, and Figure 11.2(b) shows it after its passage through a channel that adds noise and distorts the pulse. In spite of the noise and distortion, the receiver has no difficulty deciding correctly whether the signal is a zero or a one. Since the binary value of the pulse is the only information in the signal, the distortion has had no effect on the transmission of information.

The perfectly received signal of Figure 11.2(b) could now be used to generate a new pulse train to send further down the channel. This receiver-transmitter combination, called a **regenerative repeater** and illustrated in Figure 11.2(c), has not only avoided adding any distortion of its own, but it has also removed the effects of noise and distortion that were added by the channel preceding the repeater. The elusive goal of distortionless transmission seems to have been achieved!

Unfortunately, as in most areas of life, absolute perfection is impossible. Since noise is random, a noise pulse can have any amplitude, including one that will cause a transition to the wrong level. Similarly, extreme distortion of pulses can cause errors. Figure 11.3 demonstrates these problems. Errors can never be eliminated completely, but by judicious choice of such parameters as signal levels and bit rates, it is possible to reduce the probability of error to a very small value. There are even techniques to detect and correct some of the errors.

The other major source of error in the digital transmission of analog signals appears in the conversion of the infinitely variable analog signal to digital form. This conversion

Figure 11.3 Excessive noise on a digital signal

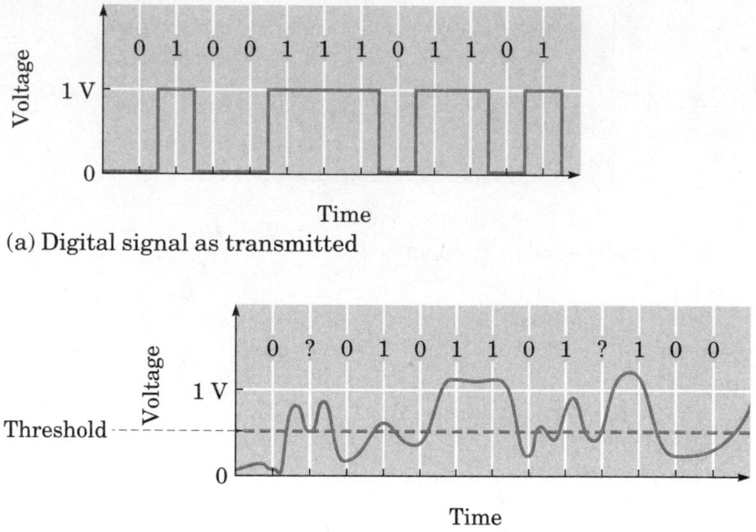

(a) Digital signal as transmitted

(b) Received signal with excessive noise and distortion

inevitably results in the loss of some information and the creation of a certain amount of noise and distortion. Once again, however, it is possible to predict quite accurately the amount of error that will be introduced and reduce it to any required value.

Other advantages of digital communication include convenience in multiplexing and switching. Time-division multiplexing (TDM) is quite easy with digital signals, and different types of signals (for example, voice and data) can be multiplexed together on the same channel.

The only real disadvantages of digital systems are their greater complexity and the larger transmission bandwidth they require. Large-scale, low-cost digital integrated circuits are reducing the difficulty and expense of constructing complex circuitry, and ingenious data-compression techniques, combined with wider-bandwidth media like fiber-optic cable, are beginning to decrease the bandwidth penalty. In general, the advantages outweigh the disadvantages.

11.1.1 Channels and Information Capacity

All practical communication channels are band-limited; either they pass frequencies from dc to some upper limit, or they have both lower and upper cutoff frequencies. In addition, all channels are noisy. There is always thermal noise, and there may be other kinds of noise as well (see Chapter 1 for a discussion of noise).

There are theoretical limits to the rate at which information can be sent along a channel with a given bandwidth and signal-to-noise ratio. Of course, there is no limit to the amount of information that can be sent if there is sufficient time to send it. The relationship between time, information capacity, and channel bandwidth is given by a simple equation called Hartley's Law:

$$I = ktB \tag{11.1}$$

where

I = amount of information to be sent
k = a constant
t = time available
B = channel bandwidth

Hartley's law tells us that the information rate (that is, the amount of information that can be sent in a given time) is proportional to the bandwidth of the channel. It is also

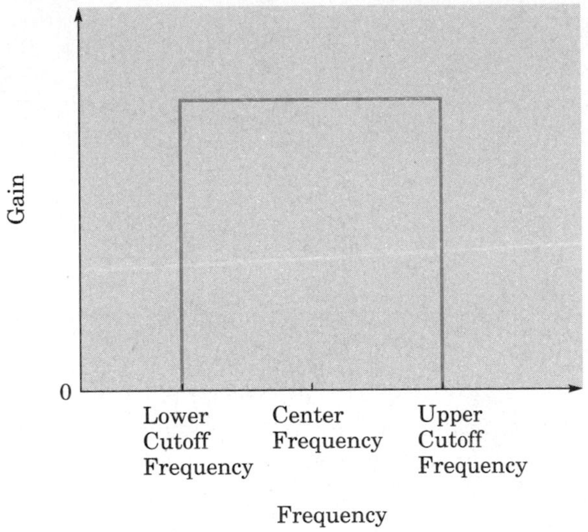

Figure 11.4
Ideal bandpass channel

proportional to the factor k, which depends on the type of data coding used and the signal-to-noise ratio of the channel.

If the noise is assumed to have the same flat spectrum and random amplitude as thermal noise (the amplitude probability function is of the type called *Gaussian*) and if the bandwidth of the channel has the square-sided shape shown in Figure 11.4, then the maximum theoretical data rate for a given channel can be found using two simple equations.

First, ignoring noise, there is a limit to the amount of data that can be sent in a given bandwidth. This limit is given by the Shannon-Hartley theorem:

$$C = 2B \log_2 M \tag{11.2}$$

where

C = information capacity in bits per second
B = the channel bandwidth in hertz
M = number of levels transmitted

In the above equation, C and B should be easy to understand. M is a little less obvious. The idea is perhaps best explained by looking at a low-pass (rather than a bandpass) channel. Suppose that the channel can pass all frequencies from zero to some maximum frequency B. Then, of course, the highest frequency that can be transmitted is B. Suppose that a simple binary signal consisting of alternate ones and zeros is transmitted through the channel. This time let a logic 1 be 1 V and a logic 0 be -1 V. The input signal will look like Figure 11.5(a): it will be a square wave with a frequency one-half the bit rate (since there are two bits, a one and a zero, for each cycle). Since it is a square wave, this signal has harmonics at all odd multiples of its fundamental frequency, with declining amplitude as the frequency increases. At very low bit rates, the output signal after passage through the channel will be similar to the input, but as the bit rate increases, the frequency of the square wave also increases and more of its harmonics are filtered out. Therefore, the output will become more and more distorted. Finally, for a bit rate of $2B$, the frequency of the input signal will be B, and only the fundamental of the square wave will pass through the channel, as shown in Figure 11.5(b). Nonetheless, the receiver will still be able to distinguish a one from a zero, and the information will be transmitted. Thus, with binary information, the channel capacity will be

$$C = 2B$$

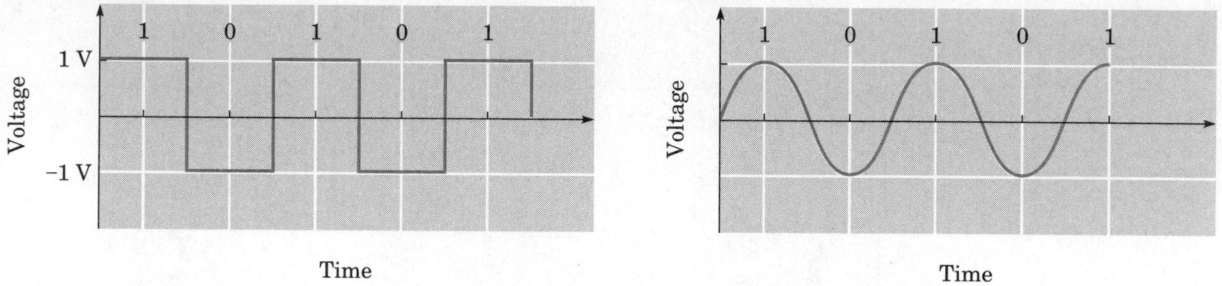

(a) Input square-wave signal
(b) Output signal at maximum rate

Figure 11.5
Digital transmission through a low-pass channel

Now suppose that instead of only two possible levels, several different levels, each corresponding to a different number, are possible. For instance, each level could represent one of four possibilities. The possible levels could be −1 V, −0.5 V, +0.5 V, and +1 V. Figure 11.6 shows this four-level code. With this code, measuring the voltage level once at the receiver would actually provide two bits of information, since it takes two bits to express four different possibilities. However, the maximum frequency of the signal would not change. We have, it seems, managed to transmit twice as much information in the same bandwidth. This idea can be expanded to any number of levels, in order to give Equation (11.2).

It might seem that any desired amount of information could be transmitted in a given channel by simply increasing the number of levels. This is not possible, however, because of noise. The more levels there are, the closer together they are, and the more likely it is that noise will cause the receiver to mistake one level for another (this can also be seen from Figure 11.6). Therefore, there will be, for a given noise level, a maximum data rate that cannot be exceeded without errors, no matter how elaborately the data is coded. This maximum rate is called the Shannon limit:

$$C = B \log_2(1 + S/N) \tag{11.3}$$

where

C = information capacity in bits per second
B = bandwidth in hertz
S/N = signal-to-noise ratio (as a power ratio, not in decibels)

There is no contradiction between Equations (11.2) and (11.3). Each represents a maximum rate, so the one that applies in a given situation is the one that gives the lower data rate. Remember also that these rates are theoretical maxima, and real, practical equipment is unlikely ever to reach these limits.

At this point, a small reminder may be in order. Not every scientific calculator is

Figure 11.6
Four-level code

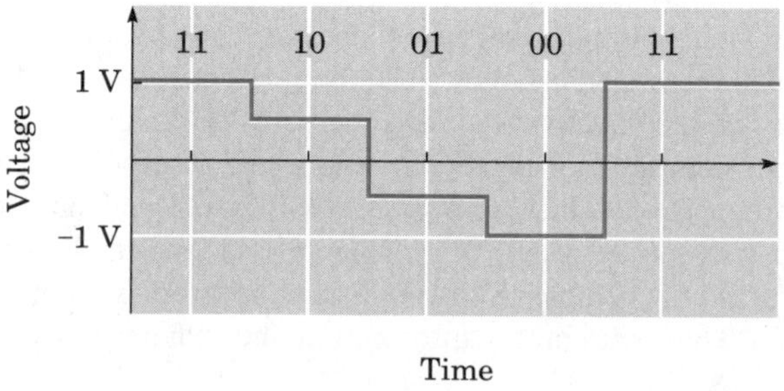

capable of finding logs to the base 2, but they can all find logs to the base 10. A handy way to find logs to the base 2 is to use the following equation:

$$\log_2 N = \frac{\log_{10} N}{\log_{10} 2} \tag{11.4}$$

Example 11.1 A telephone line has a bandwidth of 3.2 kHz and a signal-to-noise ratio of 35 dB. A signal is transmitted down this line using a four-level code. What is the maximum theoretical data rate?

Solution First, we use the Shannon-Hartley theorem to find the maximum data rate for a four-level code in the available bandwidth, ignoring noise. From Equation (11.2),

$$\begin{aligned} C &= 2B \log_2 M \\ &= 2\,(3.2 \times 10^3) \times \log_2 4 \\ &= 12.8 \times 10^3 \text{ b/s} \\ &= 12.8 \text{ kb/s} \end{aligned}$$

Next, we use the Shannon limit to find the maximum data rate for any code, given the bandwidth and signal-to-noise ratio. Remember that S/N is required as a power ratio.

$$\begin{aligned} S/N &= \text{antilog}_{10}\,(35/10) \\ &= 3162 \end{aligned}$$

From Equation (11.3)

$$\begin{aligned} C &= B \log_2 (1 + S/N) \\ &= (3.2 \text{ X } 10^3) \log_2 (1 + 3162) \\ &= 37.2 \text{ kb/s} \end{aligned}$$

Since both results are maxima, we take the lesser of the two, 12.8 kb/s. This means that it would be possible to increase the data rate over this channel by using more levels.

11.2 Pulse Modulation

In order to transmit an analog signal by digital means, it is first necessary to sample the signal at intervals. The amplitude of each sample can then be expressed as a binary number for transmission. At the receiver, the samples can be reconstituted and used to form a replica of the original signal.

11.2.1 Sampling

In 1928, Harry Nyquist showed mathematically that it is possible to reconstruct a band-limited analog signal from periodic samples, as long as the sampling rate is at least twice the frequency of the highest frequency component of the signal. This assumes that an ideal low-pass filter prevents higher frequencies from entering the sampler. In practice, the sampling frequency should be considerably greater than twice the maximum frequency to be transmitted. For example, in telephony, a sample rate of 8 kHz is used for a maximum audio frequency of 3.4 kHz, and compact disc systems have a 44.1 kHz sampling rate and a maximum audio frequency of 20 kHz.

Figure 11.7(a) demonstrates the simplest type of sampling, called **natural sampling**. The incoming analog signal is low-pass filtered and then multiplied by a pulse train. If we assume that the pulses have an amplitude of 1 V at the input to the multiplier, then the output voltage will be equal to the amplitude of the analog signal at the time of the pulse. Since a real pulse has a finite duration, the amplitude of the pulse varies during

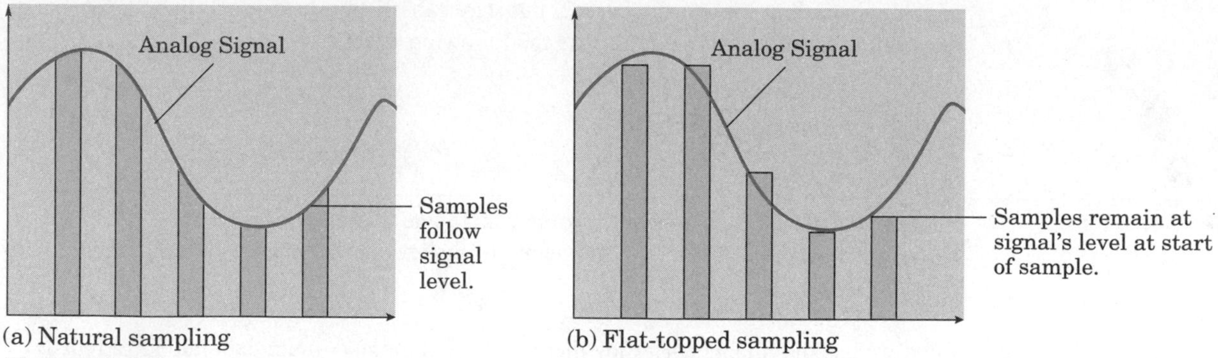

Figure 11.7
Natural and flat-topped sampling

the length of time it is present. In some situations where this is undesirable, a **sample-and-hold circuit** can be used to keep the pulse amplitude constant for the duration of the pulse. This variation, which is shown in Figure 11.7(b), is called **flat-topped sampling**.

The penalty for a sampling rate that is too low is called **aliasing** or **foldover distortion**. In this form of distortion, frequencies are translated downward. Figure 11.8 shows how aliasing develops. In Figure 11.8(a), the sampling rate is adequate and the signal can be reconstructed. In Figure 11.8(b), however, the rate is too low and the attempt to reconstruct the original signal results in a lower-frequency signal. Once aliasing is present, it cannot be removed.

The frequency of the interference generated by aliasing is easier to see by looking at the frequency domain. Recall from Chapter 6 that multiplying a modulating signal and a sine-wave carrier produces a double-sideband suppressed-carrier signal. We might

Figure 11.8
Aliasing

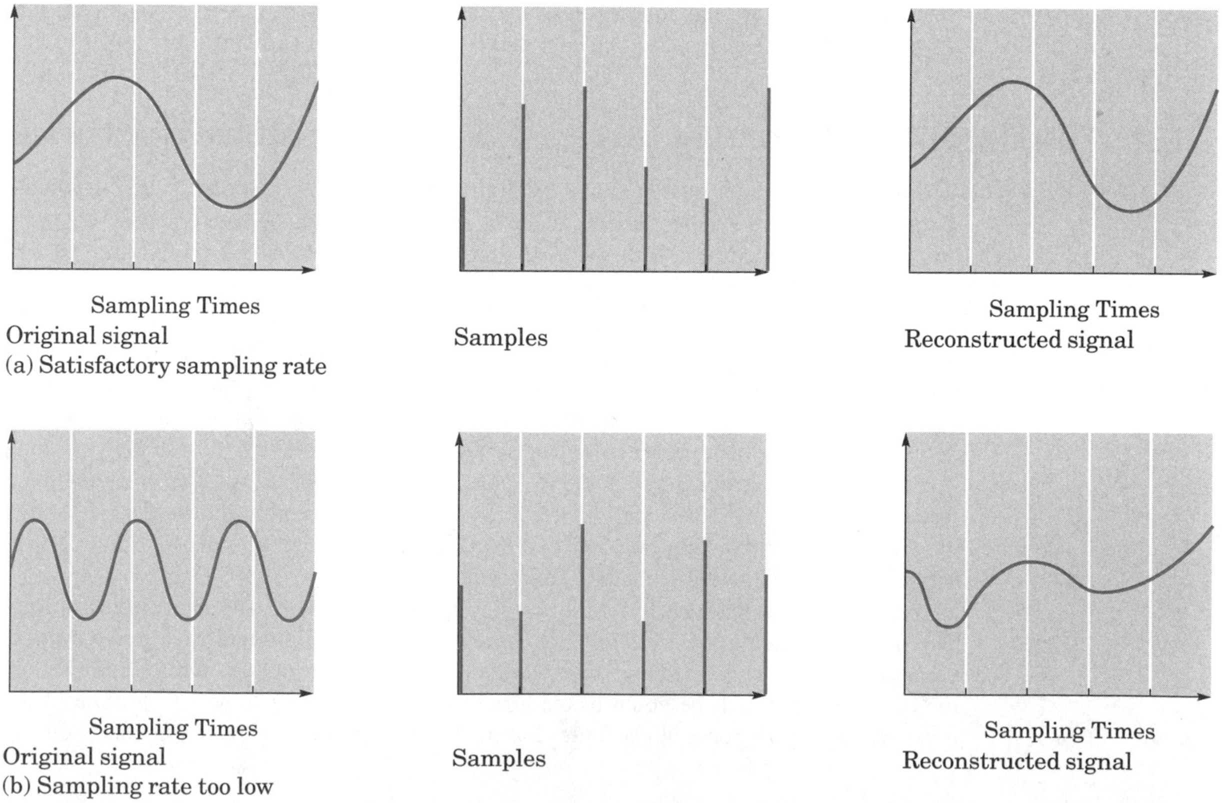

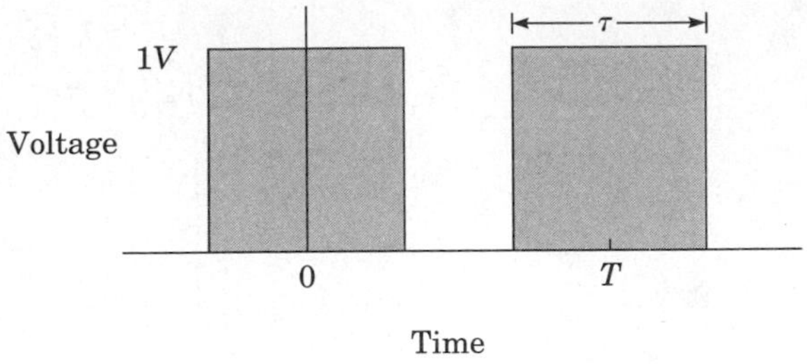

Figure 11.9
Pulse Train

expect similar results when the carrier is a pulse train. Let us find out. For simplicity, we shall assume that the modulating signal is a sine wave,

$$e_m = E_m \sin \omega_m t \tag{11.5}$$

where

e_m = the instantaneous modulating-signal voltage
E_m = the peak modulating-signal voltage
ω_m = the radian frequency of the modulating signal

The carrier signal is a pulse train like the one in Figure 11.9. Let us assume for simplicity that the pulses have an amplitude of 1 V. The spectrum for such a signal is given in Chapter 1 as:

$$e_s = \frac{\tau}{T} + 2\frac{\tau}{T}\left(\frac{\sin \pi\tau/T}{\pi\tau/T}\cos \omega_s t + \frac{\sin 2\pi\tau/T}{2\pi\tau/T}\cos 2\omega_s t + \frac{\sin 3\pi\tau/T}{3\pi\tau/T}\cos 3\omega_s t + \dots\right) \tag{11.6}$$

where

e_s = the instantaneous voltage of the sampling pulse
τ = the duration of a pulse
T = the period of the pulse train
ω_s = the radian frequency of the pulse train

Multiplying the two signals together gives the following output:

$$v(t) = E_m\frac{\tau}{T}\sin \omega_m t + 2E_m\frac{\tau}{T}\left(\frac{\sin \pi\tau/T}{\pi\tau/T}\sin \omega_m \cos \omega_s t + E_m\frac{\sin 2\pi\tau/T}{2\pi\tau/T}\sin \omega_m \cos 2\omega_s t + E_m\frac{\sin 3\pi\tau/T}{3\pi\tau/T}\sin \omega_m \cos 3\omega_s t + \dots\right) \tag{11.7}$$

As is often the case with equations of this type, we do not have to "solve" anything. We simply need to study the equation to see what is happening. In particular, look at the first term. It is just the modulating signal multiplied by a constant. This tells us that applying a low-pass filter to the pulse signal can recover the original modulation, provided that there are no other components of the signal occupying the baseband frequency range.

Next, look at the second term, which contains the product of sin ω_m and cos ω_s. We have seen many terms of this sort before, and we should remember that they represent sum and difference frequencies. In this case the frequencies produced will be $f_s - f_m$ and $f_s + f_m$. (We have changed from radian notation because while radian notation simplifies equations, most practical work is done with hertz, not radians per second.)

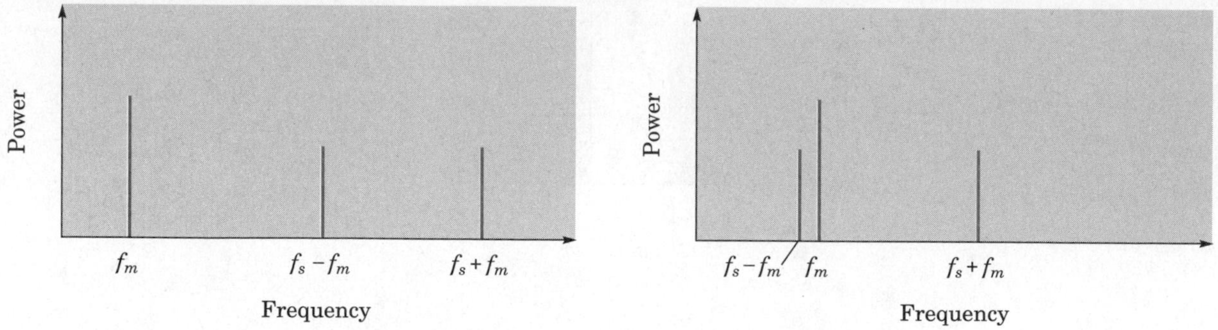

(a) Satisfactory sampling rate (b) Sampling rate too low

Figure 11.10 Aliasing in the frequency domain

Figure 11.10 shows sketches of the frequency spectrum for the first two terms in Equation (11.7). The subsequent terms produce still higher frequency components and are not of interest in this discussion. In Figure 11.10(a), the baseband frequency is lower than the component at $f_s - f_m$, so a low-pass filter can recover the baseband signal. In Figure 11.10(b), however, the difference term has a lower frequency than the original baseband signal, and the two cannot be separated by a low-pass filter. This is a frequency-domain representation of the aliasing that we saw earlier in the time domain.

It appears from Figure 11.10 that aliasing will take place if f_m is greater than $f_s - f_m$. If $f_s = 2f_m$, then $f_s - f_m = f_m$, and we are just on the edge of aliasing. In order to separate the signals with a real low-pass filter, the carrier frequency (which is also the sampling rate) must be greater than $2f_m$, as stated earlier. If aliasing does take place, the interfering component will be at a frequency

$$f_a = f_s - f_m \tag{11.8}$$

where

f_a = the frequency of the aliasing distortion
f_s = the sampling rate
f_m = the modulating (baseband) frequency

Example 11.2 An attempt is made to transmit a baseband frequency of 30 kHz using a digital audio system with a sampling rate of 44.1 kHz. What audible frequency would result?

Solution From Equation (11.8),

$$\begin{aligned} f_a &= f_s - f_m \\ &= 44.1 \text{ kHz} - 30 \text{ kHz} \\ &= 14.1 \text{ kHz} \end{aligned}$$

In addition, of course, the original baseband signal of 30 kHz would be present.

11.2.2 Analog Pulse-Modulation Techniques

Sampling alone is not a digital technique. The immediate result of sampling is a **pulse-amplitude modulation (PAM)** signal like the one in Figure 11.11(b). PAM is an analog scheme in which the amplitude of each pulse is proportional to the amplitude of the signal at the instant at which it is sampled. The example in Figure 11.11 uses flat-topped sampling (described earlier), in which a sample-and-hold circuit is used to keep the amplitude constant during each pulse.

For our purposes, the main use of PAM is as an intermediate step; before being

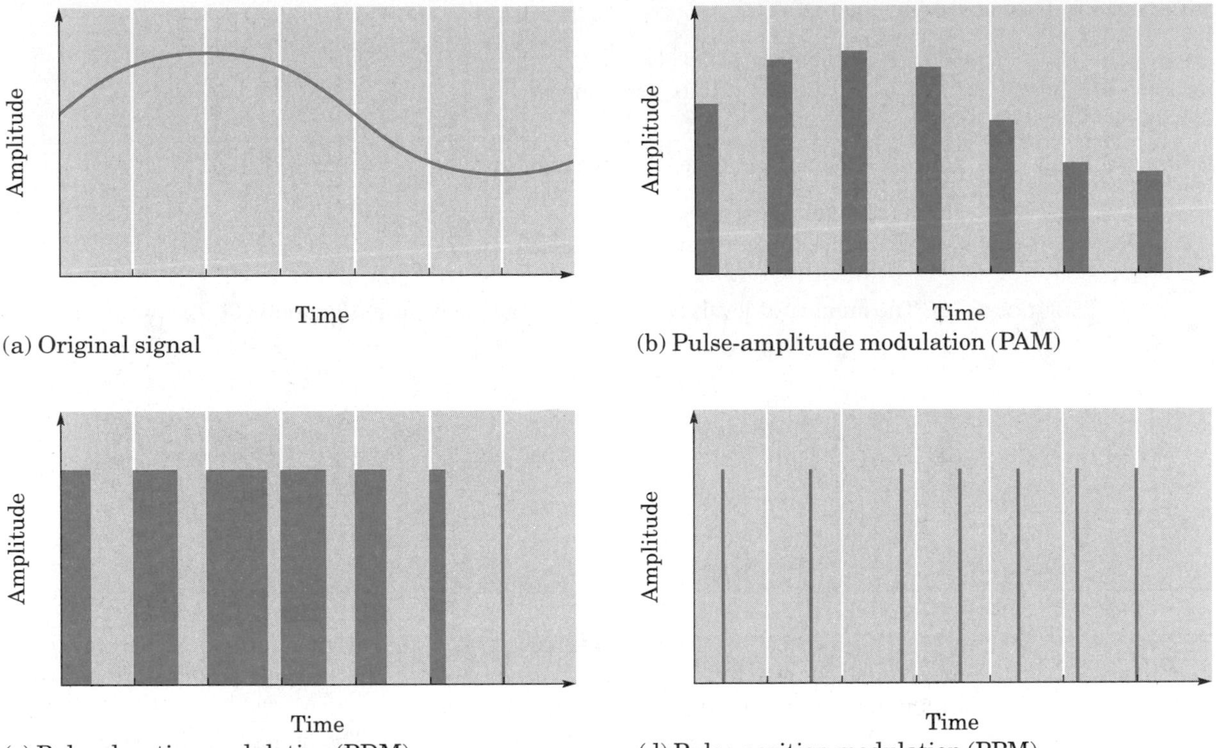

(a) Original signal (b) Pulse-amplitude modulation (PAM) (c) Pulse-duration modulation (PDM) (d) Pulse-position modulation (PPM)

Figure 11.11 Analog pulse modulation

transmitted, the PAM signal has to be digitized. Similarly, at the receiver, the digital signal must be converted back to PAM as part of the demodulation process. The original signal can then be recovered using a low-pass filter.

Before looking at digital schemes, a quick glance at two other analog pulse-modulation techniques is in order. **Pulse-duration modulation (PDM)**, which is shown in Figure 11.11(c), uses pulses that all have the same amplitude. The duration of each pulse depends on the amplitude of the signal at the time it is sampled. PDM has its communications uses; for instance, it is often used in the high-powered audio amplifiers used to modulate AM transmitters. It has also been used for telemetry systems. Though still an analog mode, it is more robust than PAM because it is insensitive to amplitude changes due to noise and distortion. Like PAM, PDM can be demodulated using a low-pass filter. Sometimes the term **pulse-width modulation (PWM)** is used for this method.

Pulse-position modulation (PPM), shown in Figure 11.11(d), is closely related to PDM. All pulses have the same amplitude and duration, but their timing varies with the amplitude of the original signal. PPM also sees some use in telemetry systems.

11.2.3 Pulse-Code Modulation

Pulse-code modulation (PCM) is the most commonly used digital modulation scheme. In PCM the available range of signal voltages is divided into levels, and each is assigned a binary number. Each sample is then represented by the binary number representing the level closest to its amplitude, and this number is transmitted in serial form. In *linear* PCM, levels are separated by equal voltage gradations.

The number of levels available depends on the number of bits used to express the sample value. The number of levels is given by

$$N = 2^m \qquad \textbf{(11.9)}$$

where

$$N = \text{number of levels}$$
$$m = \text{number of bits per sample}$$

Example 11.3 Calculate the number of levels if the number of bits per sample is:

(a) 8 (as in telephony)
(b) 16 (as in compact disc audio systems)

Solution (a) The number of levels with 8 bits per sample is, from Equation (11.9),

$$\begin{aligned} N &= 2^m \\ &= 2^8 \\ &= 256 \end{aligned}$$

(b) The number of levels with 16 bits per sample is, from the same equation,

$$\begin{aligned} N &= 2^m \\ &= 2^{16} \\ &= 65{,}536 \end{aligned}$$

This process is called **quantizing**. Since the original analog signal can have an infinite number of signal levels, the quantizing process will produce errors called **quantizing errors** or often **quantizing noise**.

Figure 11.12 shows how quantizing errors arise. The largest possible error is one-half the difference between levels. Thus the error is proportionately greater for small signals. This means that the signal-to-noise ratio varies with the signal level and is greatest for large signals. The level of quantizing noise can be decreased by increasing the number of levels, which also increases the number of bits that must be used per sample.

The dynamic range of a system is the ratio of the strongest possible signal that can be transmitted and the weakest discernible signal. For a linear PCM system, the maximum dynamic range in decibels is given approximately by

$$DR = (1.76 + 6.02m) \text{ dB} \tag{11.10}$$

where

$$DR = \text{dynamic range in decibels}$$
$$m = \text{number of bits per sample}$$

This equation ignores any noise contributed by the analog portion of the system.

Figure 11.12
Quantizing error

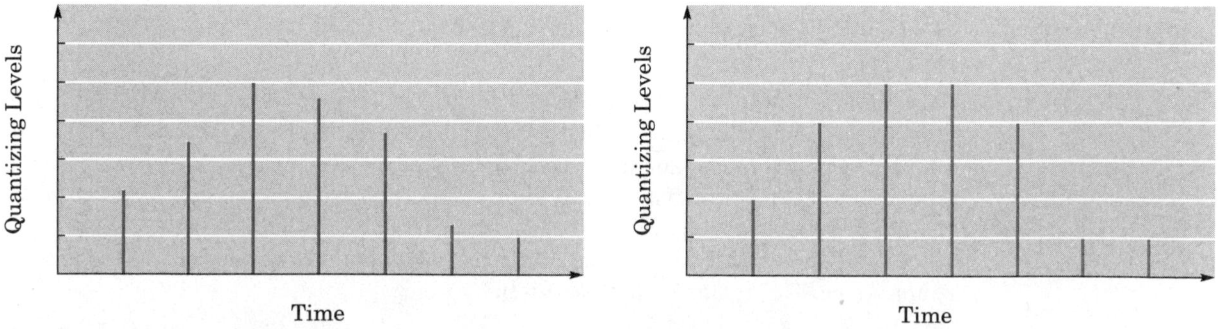

(a) Samples before quantizing
(b) Samples after quantizing

Example 11.4 Find the maximum dynamic range for a linear PCM system using 16-bit quantizing.

Solution From Equation (11.10)

$$\begin{aligned} DR &= 1.76 + 6.02m \text{ dB} \\ &= 1.76 + 6.02 \times 16 \\ &= 98.08 \text{ dB} \end{aligned}$$

Increasing the number of bits per sample increases the data rate, which is given very approximately by

$$D = f_s m \tag{11.11}$$

where

D = data rate in bits per second
f_s = sample rate in samples per second
m = number of bits per sample

Extra bits are needed to detect and correct errors. A few bits, called **framing bits**, are also needed to ensure that the transmitter and receiver agree on which bits constitute one sample. The actual bit rate will be somewhat higher than calculated above.

Example 11.5 Calculate the minimum data rate needed to transmit audio with a sampling rate of 40 kHz and 14 bits per sample.

Solution From Equation (11.11)

$$\begin{aligned} D &= f_s m \\ &= 40 \times 10^3 \times 14 \\ &= 560 \times 10^3 \text{ b/s} \\ &= 560 \text{ kb/s} \end{aligned}$$

11.2.4 Companding

The transmission bandwidth varies directly with the bit rate. In order to keep the bit rate and thus the required bandwidth low, **companding** is often used. Companding involves using a compressor amplifier at the input, with greater gain for low-level than for high-level signals. The compressor reduces the quantizing error for small signals. The effect of **compression** on the signal can be reversed by using expansion at the receiver, with a gain characteristic that is the inverse of that at the transmitter.

It is necessary to follow the same standards at both ends of the circuit so that the dynamics of the output signal are the same as at the input. The system used in the North American telephone system uses a characteristic known as the *mu* (μ) *law*, which has the following equation for the compressor:

$$v_o = \frac{V_o \ln (1 + \mu v_i / V_i)}{\ln (1 + \mu)} \tag{11.12}$$

where

v_o = output voltage from the compressor
V_o = maximum output voltage
v_i = actual input voltage
V_i = maximum input voltage
μ = a parameter that defines the amount of compression (contemporary systems use μ = 255)

Figure 11.13
Mu-law compression

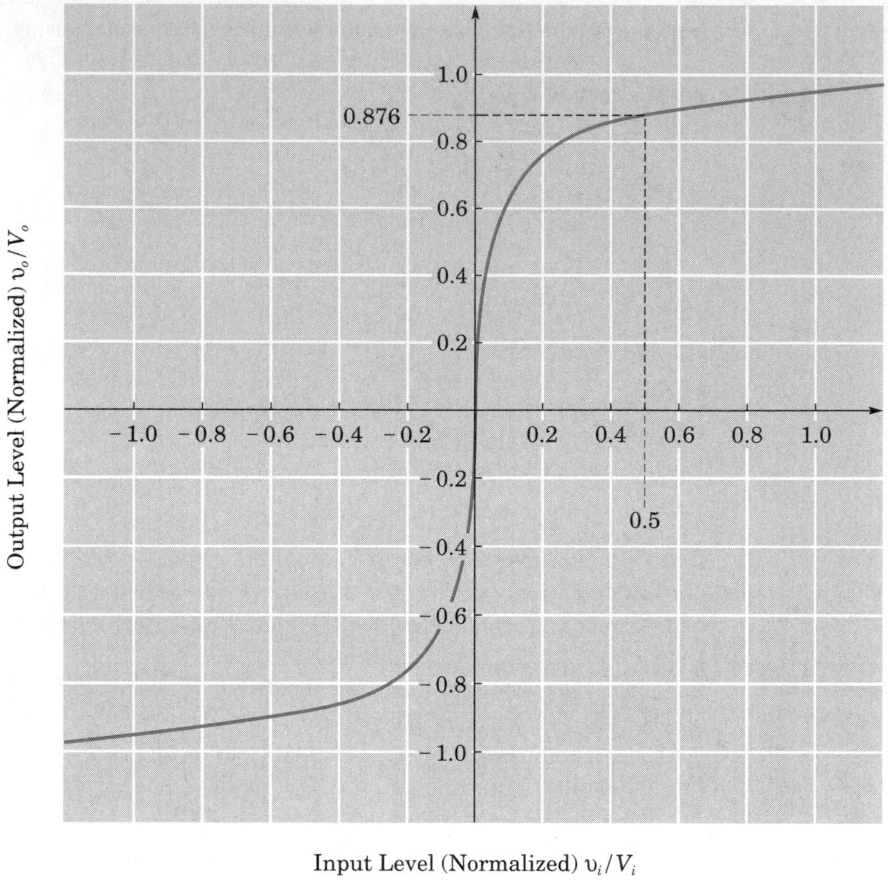

European telephone systems use a similar but not identical scheme called A-law compression.

Figure 11.13 shows the μ-255 curve. The curve is a transfer function for the compressor and relates the input and output levels. It has been normalized, that is, v_i/V_i and v_o/V_o are plotted, rather than v_i and v_o.

Example 11.6 A signal at the input to a mu-law compressor is positive, with its voltage one-half the maximum value. What proportion of the maximum output voltage is produced?

Solution From Equation (11.12)

$$v_o = \frac{V_o \ln (1 + \mu v_i/V_i}{\ln (1 + \mu)}$$

$$= \frac{V_o \ln (1 + 255 \times 0.5)}{\ln (1 + 255)}$$

$$= 0.876\ V_o$$

This problem can also be solved graphically, as shown in Figure 11.13.

11.2.5 Delta Modulation

Although PCM is the most commonly used system of digital transmission, there is one other important technique. **Delta modulation** uses an idea that is, at first glance, breathtakingly simple. Instead of transmitting complete information about the amplitude of

Figure 11.14
Delta modulation

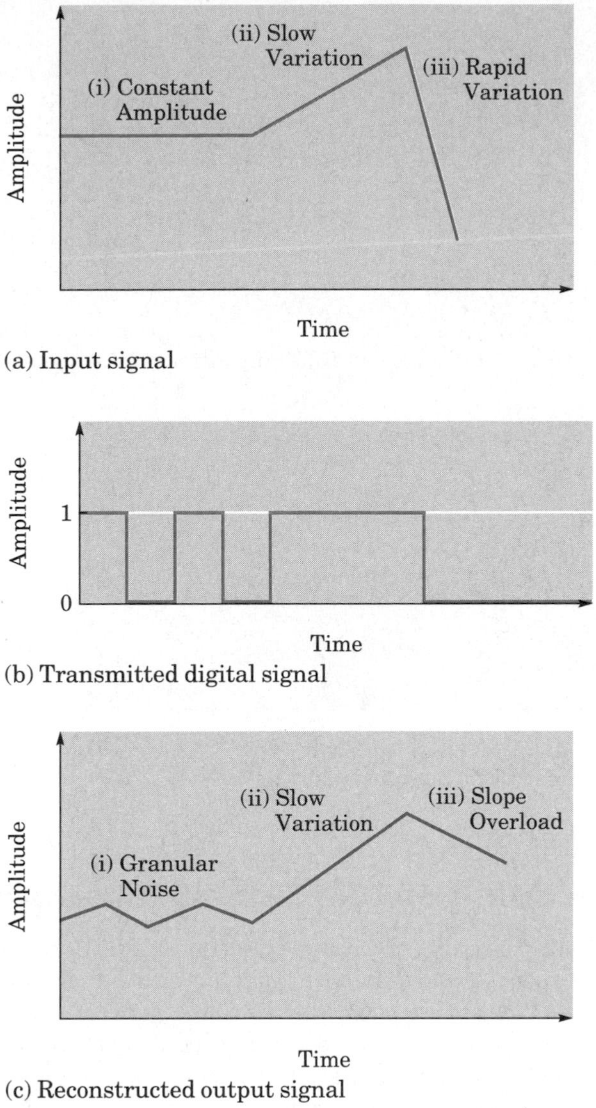

every sample, only one bit is transmitted. That bit will be a one if the current sample is more positive than the previous sample, zero if it is more negative.

If this seems too good to be true, it is. Since only a small amount of information about each sample is transmitted, delta modulation requires a much higher sampling rate than PCM for equal quality of reproduction. Nyquist did not say that transmitting samples at twice the maximum signal frequency would always give undistorted results, only that it could, provided the samples were transmitted accurately.

Figure 11.14 shows how delta modulation generates errors. In region (i), the signal does not vary at all; the transmitter can send only ones and zeros, however, so the output waveform has a triangular shape, producing a noise signal called granular noise. On the other hand, the signal in region (iii) changes more rapidly than the system can follow, creating an error in the output called **slope overload**.

Adaptive delta modulation, in which the step size varies according to previous values, is more efficient. Figure 11.15 shows how it works. After a number of steps in the same direction, the step size increases.

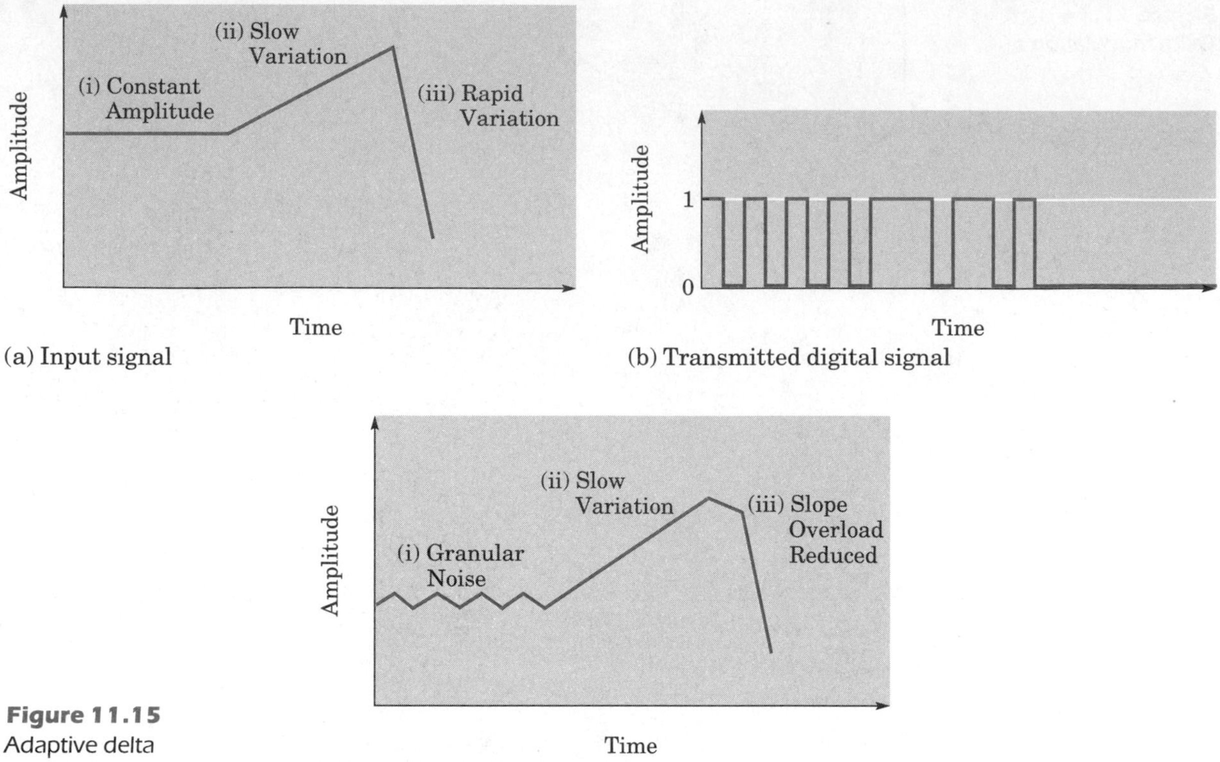

(c) Reconstructed output signal

Figure 11.15
Adaptive delta modulation

11.3 Coding and Decoding of PCM Signals

The process of converting an analog signal into a PCM or delta modulation signal is called **coding**, and the inverse operation, converting back from digital to analog, is known as **decoding**. In telephony, the two processes are usually accomplished in the line card, where the analog local loop meets the digital switch. There are a number of integrated circuits called **codecs** available; they combine the functions of coding and decoding. Before looking at one of these, however, let us examine the necessary steps in coding and decoding a PCM signal. We choose PCM because it is much more common than delta modulation, but the process is similar for delta modulation.

11.3.1 Coding and Decoding a PCM Signal

The block diagram in Figure 11.16 shows the steps in converting an analog signal into a compressed PCM code such as would be used in a digital telephone system. The first block is a low-pass filter, which is required to prevent aliasing. It must have a cutoff frequency below one-half the sampling rate, and it should be a high-order filter so that the signal is greatly attenuated above the cutoff frequency. Switched-capacitor filters are often used.

The next step is to sample the incoming waveform using a sample-and-hold circuit. There are many such circuits; a simple one is shown in Figure 11.17. The field-effect

Figure 11.16
PCM coding

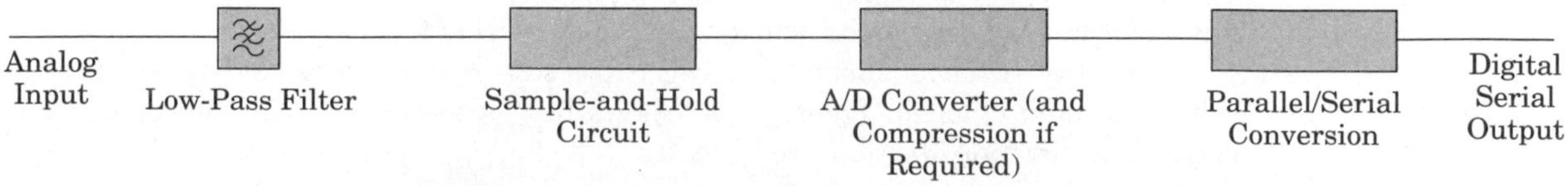

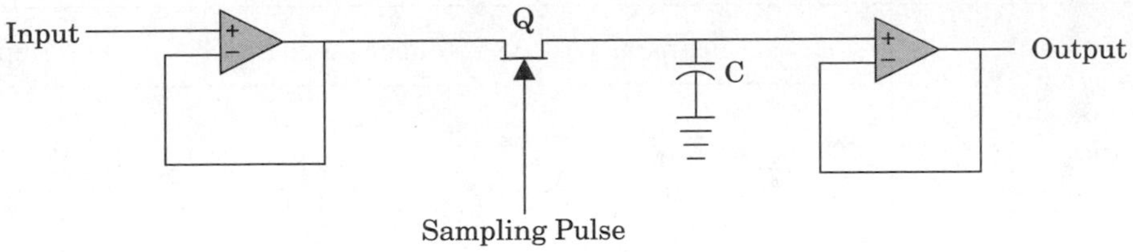

Figure 11.17 Sample-and-hold circuit

transistor Q turns on during the sample time, allowing the capacitor to charge to the amplitude of the incoming signal. The transistor then turns off, and the capacitor stores the signal value until the analog-to-digital converter has had time to convert the sample to digital form. The two operational amplifiers, connected as voltage followers, isolate the circuit from the other stages. The low output impedance of the first stage ensures that the capacitor quickly charges or discharges to the value of the incoming signal when the transistor conducts.

The samples must now be coded as binary numbers. If we are using linear PCM, all that is required is a standard analog-to-digital (A/D) converter. Compression, if required, can be applied to the analog signal, but it is more common to incorporate the compression into the coding process.

The codecs used in telephony generally accomplish compression by using a piecewise-linear approximation to the mu-law curve shown in Figure 11.13. The positive- and negative-going parts of the curve are each divided into seven segments, with an additional segment centered around zero, for a total of fifteen segments. Figure 11.18 shows the segmented curve. Segments 0 and 1 have the same slope and do not compress the signal. For each higher-numbered segment, the step size is double that of the previous segment. Each segment has sixteen steps. The result is a close approximation to the actual curve.

The binary number produced by the codec in a telephone system has eight bits. The first is a sign bit, one for a positive voltage and zero for negative. Bits 2, 3, and 4 represent the segment number, which ranges from zero to seven. The last four bits determine the step within the segment. If we normalize the signal, that is, set the maximum input level equal to one volt, the step sizes can easily be calculated as follows: let the step size for segments 0 and 1 be x mV. Then segment 2 has a step size of $2x$, segment 3 a step size of $3x$, and so on. Since each segment has 16 steps, the value of x can be found as follows.

$$16(x + x + 2x + 4x + 8x + 16x + 32x + 64x) = 1000 \text{ mV}$$
$$x = 0.488 \text{ mV}$$

The relationship between input voltage and segment is shown in Table 11.1.

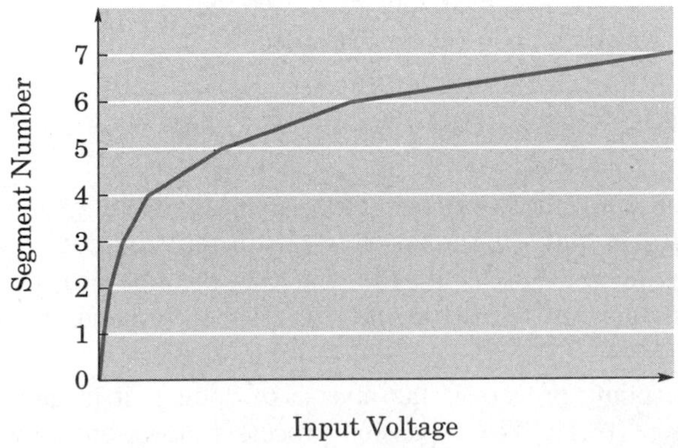

Figure 11.18 Segmented mu-law curve (positive half)

Table 11.1 Mu-law compressed PCM coding

Segment	Voltage range (mV)	Step size (mV)
0	0–7.8	0.488
1	7.8–15.6	0.9772
2	15.6–31.25	1.953
3	31.25–62.5	3.906
4	62.5–125	7.813
5	125–250	15.625
6	250–500	31.25
7	500–1000	62.5

Example 11.7 Code a signal whose amplitude is 30% of the maximum allowed as a PCM sample.

Solution The signal is positive, so the first bit is a one. On the normalized voltage scale, the amplitude is 300 mV. A glance at Table 11.1 shows that the signal is in segment 6, so the next three bits are 110 (6 in binary). This segment starts at 250 mV and increases 31.25 mV per step. The signal voltage is 50 mV above the lower limit, which translates into 50/31.25 = 1.6 steps. This is more than halfway from step 1 to step 2, so the signal will be quantized as step 2, making the last four bits 0010 (2 in binary). Therefore the code representing this sample is 11100010.

In operation, many modern codecs achieve compression by first encoding the signal using a 12-bit linear PCM code, then converting the 12-bit linear code into an 8-bit compressed code by discarding some of the bits. This is a simple example of **digital signal processing (DSP)**. Once an analog signal has been digitized, it can be manipulated in a great many ways simply by performing arithmetic with the bits that make up each sample. In the case of the 12-to-8 bit conversion described here, some precision will be lost for large-amplitude samples, but the data rate needed to transmit the information will be much less than for 12-bit PCM. Since most of the samples in an audio signal have amplitudes much less than the maximum, there is a gain in accuracy compared with 8-bit linear PCM.

Briefly, the conversion works as follows: The 12-bit PCM sample begins with a sign bit, which is retained. The other 11 bits describe the amplitude of the sample, with the most significant bit first. For low-level samples, the last few bits and the sign bit may be the only non-zero bits. The segment number for the 8-bit code can be determined by subtracting the number of leading zeros (not counting the sign bit) in the 12-bit code from 7. The next four bits after the first 1 give the level number within the segment. Any remaining bits are discarded.

Example 11.8 Convert the 12-bit sample 100110100100 into an 8-bit compressed code.

Solution Copy the sign bit to the 8-bit code. Next, count the leading zeros (2) and subtract from 7 to get 5 (101 in binary). The first four bits of the 8-bit code are thus 1101. Now copy the next four bits after the first 1 (not counting the sign bit) to the 8-bit code. Thus the next four bits are 1010. Discard the rest. The corresponding 8-bit code is 11011010.

The decoding process is the reverse of coding; it is illustrated in the block diagram in Figure 11.19. The expansion process follows an algorithm analogous to that

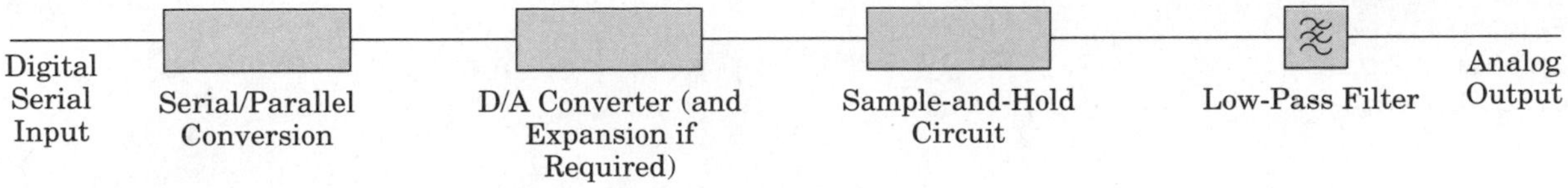

Figure 11.19
PCM decoding

Figure 11.20
Simplified block diagram of combo chip

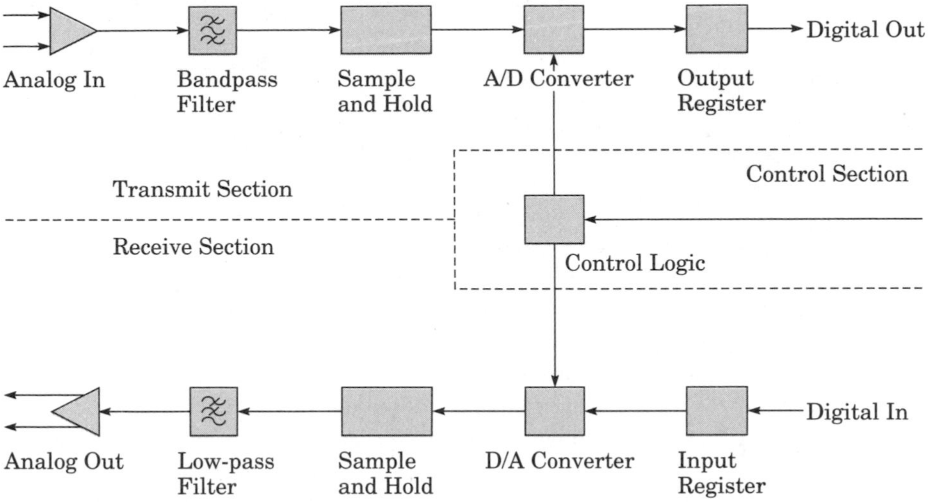

used in the compressor. The low-pass filter at the output removes the high-frequency components of the PAM signal that exits from the digital-to-analog converter.

11.3.2 Codecs

In telephony, the coding and decoding of PCM voice signals is currently done at the subscriber line interface card (SLIC) and is accomplished by a single integrated circuit called a *codec* (from *co*der-*dec*oder). Many of the integrated circuits used for this are known as **combo chips**, because they combine the coding and decoding with low-pass filtering and companding. Figure 11.20 shows a simplified block diagram for a typical combo chip. In the block diagram, "transmit" refers to the digital signal sent to the switch, and "receive" refers to the analog circuit sent to the local loop. A bandpass filter is used in the transmit (analog-to-digital) part of the chip, because it is unnecessary in telephony to transmit very low frequencies; of course, as pointed out earlier, it is also very important not to transmit frequencies high enough to cause aliasing. The compression and expansion are digital and are built into the A/D and D/A converters. Note also that the analog inputs and outputs are balanced for easy interface with the local loop.

11.4 Line Codes

So far, our discussion of digital signals has been in terms of zeros and ones. It is now time to convert these binary numbers into voltage or current levels on a line. This can be done in many ways. Probably the simplest **line code** is to use the presence of a current or voltage to represent one logic state, while its absence indicates the other logic state. For example, TTL voltage levels could be used. Figure 11.21 shows a binary sequence sent using TTL levels. **Positive logic**, in which a high level represents a logical one, was used for the example, but **negative logic** could be used just as well.

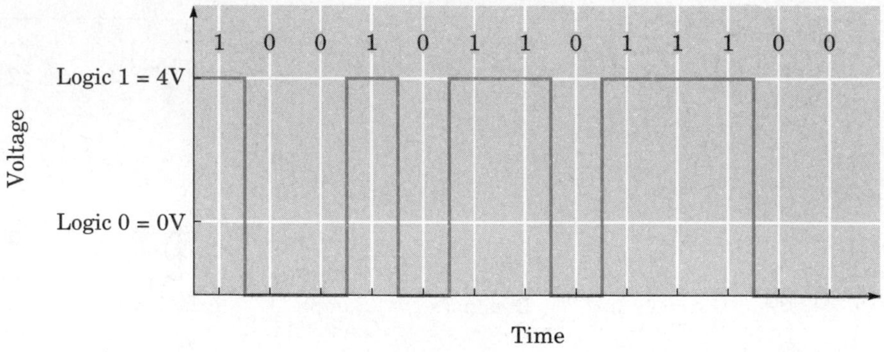

Figure 11.21 Unipolar NRZ code using TTL levels

This example requires a line with dc continuity, since the current, when present, always flows in the same direction. Codes of this sort are called **unipolar NRZ codes**. The term *unipolar* means that the voltage or current polarity is always the same, and **NRZ** (non-return-to-zero) means that there is no requirement for the signal level to return to zero at the end of each element. For example, the message 111 requires the voltage or current to go to the high state and stay there for three bit periods.

Sometimes, the condition of dc continuity cannot be met (for instance, when there are transformers or ac-coupled amplifiers on the line). Codes that have zero average dc content have been developed for ac-coupled lines. In the long run, the **bipolar NRZ code** shown in Figure 11.22 will accomplish this, provided the message contains equal numbers of ones and zeros. On the other hand, a long string of ones or zeros will result in a component at very low frequency, and for some systems there is no guarantee that there will be equal numbers of ones and zeros.

Low-frequency ac components and dc components can be eliminated completely with bipolar **RZ (return-to-zero) codes**. Figure 11.23 provides two examples. Figure 11.23(a) shows a system used in telephony; it is called *AMI*, for alternate mark inversion. A binary zero is coded as 0 V, and binary ones are recorded alternately by positive and negative voltages. This signal will have no dc or low-frequency ac content, even with long strings of ones or zeros. On the other hand, long strings of zeros must be avoided when synchronous communication takes place with this code. A string of zeros can cause the signal to disappear and the timing to be lost. There is some error detection built into this code: any time two consecutive pulses with the same polarity are received, an error must have occurred.

Figure 11.23(b) shows the *Manchester* code, which is a type of *biphase* code. Every bit has a level transition in the center of the bit period. For a one there is an upward transition; for a zero, a downward one. There is no dc or low-frequency energy regardless of the proportion of zeros and ones in the signal. The Manchester code also

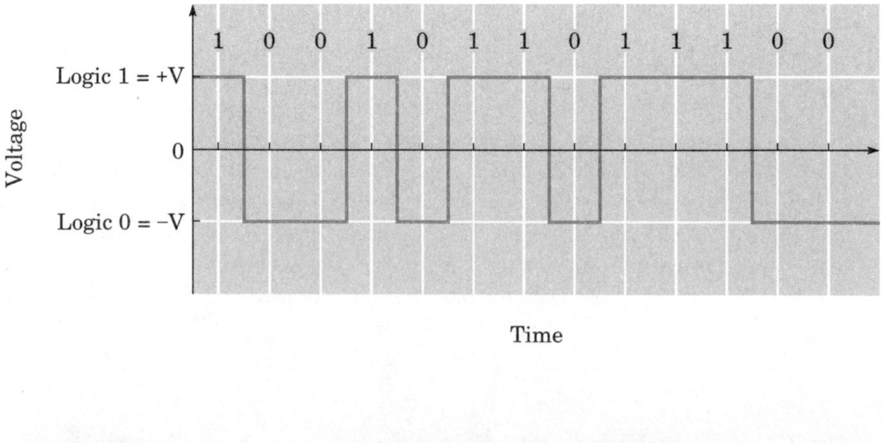

Figure 11.22 Bipolar NRZ code

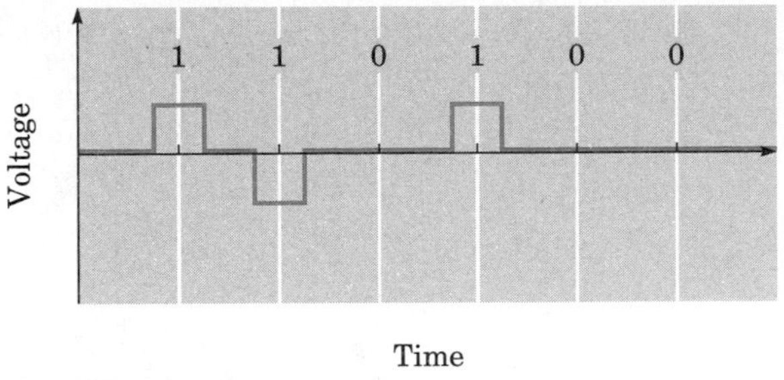

(a) AMI code

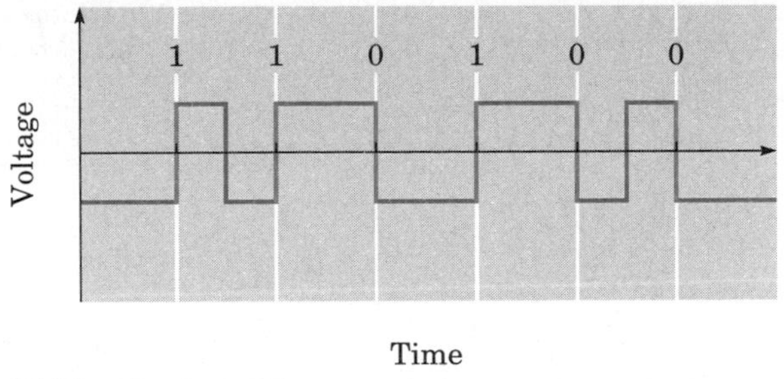

(b) Manchester code

Figure 11.23
RZ codes

provides strong timing information regardless of the pattern of ones and zeros. Its disadvantage is that it requires more bandwidth than the AMI code.

11.5 Time-Division Multiplexing

There are two basic types of multiplexing. We looked at frequency-division multiplexing (FDM) in the previous chapter. Time-division multiplexing (TDM) is used mainly for digital communication. In TDM, each information signal is allowed to use all the available bandwidth, but only for part of the time. From Hartley's Law (Equation 11.1), it can be seen that the amount of information transmitted is proportional to both bandwidth and time. Therefore, at least in theory, it is equally possible to divide the bandwidth or the time among the users of a channel. Continuously varying signals, such as analog audio, are not well adapted to TDM because the signal is present at all times. On the other hand, sampled audio is very suitable for TDM, as it is possible to transmit one sample from each of several sources sequentially, then send the next sample from each source, and so on. As already mentioned, sampling itself does not imply digital transmission, but in practice sampling and digitizing usually go together.

Many signals can be sent on one channel by sending a sample from each signal in rotation. Time-division multiplexing requires that the total bit rate be multiplied by the number of channels multiplexed. This means that the bandwidth requirement is also multiplied by the number of signals.

11.5.1 TDM in Telephony

TDM is used extensively in telephony. There are many different standards for TDM. One commonly used arrangement is the DS-1 signal, which consists of 24 PCM voice channels, multiplexed using TDM. Each channel is sampled at 8 kHz, with 8 bits per sample, as previously described. This gives a bit rate of $8 \text{ k} \times 8 = 64$ kb/s for each voice channel.

The DS-1 signal consists of frames, each of which contains the bits representing one sample from each of the 24 channels. One extra bit, called the *framing bit*, is added to each frame to help synchronize the transmitter and receiver. Each frame contains $24 \times 8 + 1 = 193$ bits.

The samples must be transmitted at the same rate as they were obtained in order for the signal to be reconstructed at the receiver without delay. This requires the multiplexed signal to be sent at a rate of 8000 frames per second. Thus the bit rate is 193×8000 b/s $= 1.544$ Mb/s. Figure 11.24 contains an illustration of a frame of a DS-1 signal. When this signal is transmitted over twisted-pair line using the AMI line code just discussed, the whole system is known as a T1 carrier. That is, the *signal* includes only the coding into ones and zeros, while the *carrier* also includes the voltage levels used.

The framing bits are used to enable the receiver to determine which sample and which bit in that sample are being received at a given time. In addition, the receiver must distinguish between frames in order to decode the signaling information that is

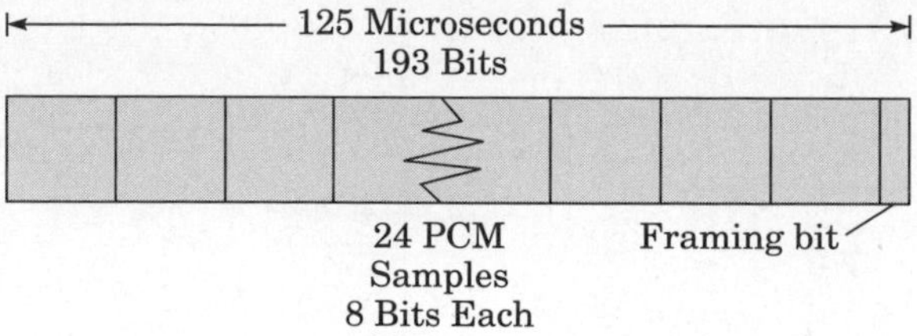

Figure 11.24
One frame of a DS-1 signal

Table 11.2
Digital signal hierarchy

Carrier	Signal	Voice channels	Bit rate (Mb/s)	Typical medium
T1	DS-1	24	1.544	twisted pair
T1C	DS-1C	48	3.152	twisted pair
T2	DS-2	96	6.312	low-capacitance twisted pair, microwave
T3	DS-3	672	44.736	coax, microwave
T4	DS-4	4032	274.176	coax, fiber optics
T5	DS-5	8064	560.16	fiber optics

sent along with the signal. In one frame out of every six, each of the least significant bits in the 24 samples is used for signaling information rather than as part of the PCM signal. This results in a very slight degradation of signal quality—for instance, the signal-to-noise ratio is degraded by about two decibels. The signaling information is different on the sixth and twelfth frames, known as the *A* and *B* frames, in a sequence. In effect, the receiver is required to count frames up to twelve. A group of twelve frames is called a *superframe*.

To allow the receiver to accomplish this, the framing bit alternates between two sequences, 100011 and 011100. The underlined bits indicate the *A* and *B* signaling frames, respectively.

The "stolen" signaling bits can be used to indicate such conditions as on-hook and off-hook, ringing, and busy signals.

11.5.2 Digital Signal Hierarchy

The DS-1 signal and T-1 carrier described above represent the lowest level in a hierarchy of TDM signals with higher bit rates. All of these signals contain PCM audio signals, each sampled 8,000 times per second. As the number of multiplexed voice signals increases, so does the bit rate. This requires the channel to have a wider frequency response and variations of time delay with frequency to be held to a low level. Twisted-pair lines can be specially conditioned for use as T1 and T2 carriers, but higher data rates require channels with greater bandwidth, such as coaxial cable, microwave radio, or optical fiber. See Table 11.2 for more details.

11.6 Digital Switching

One of the handiest things about digital communication is the ease with which switching can be accomplished and the variety of ways in which it can be done. The techniques described in the previous chapter can, of course, be used just as well with digital as with analog signals. Switching of signals from one line to another is known as **space switch-**

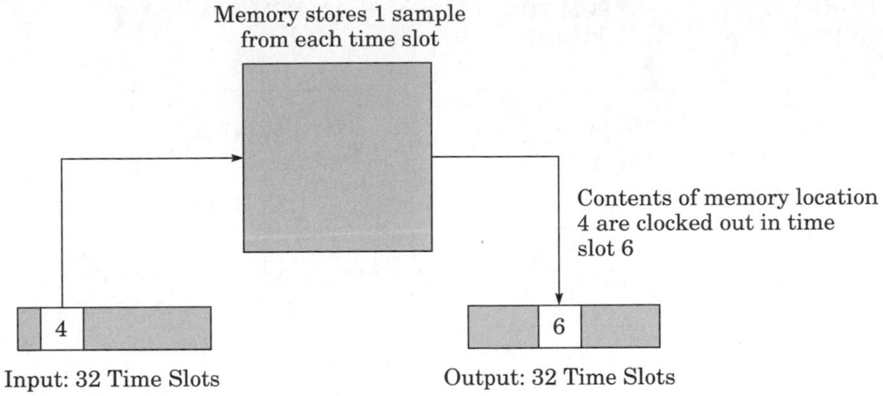
Memory stores 1 sample
from each time slot
Contents of memory location
4 are clocked out in time
slot 6
4
6
Input: 32 Time Slots
Output: 32 Time Slots

Figure 11.25
Time switching

ing, to distinguish it from **time switching**, which we shall look at next. Most digital switches use a combination of space and time switching.

11.6.1 Time Switching

A time switch moves PCM samples from one time slot to another in a TDM signal. Suppose a TDM signal has 32 time slots, as shown in Figure 11.25. Time switches normally use multiples of 32 time slots, rather than 24, as in the North American transmission hierarchy. A real switch might more typically have 128 time slots. Each time slot at the input to the switch could be associated with a particular subscriber line, and the output time slots could be similarly assigned. Thus, the switch can handle 32 subscribers for signals in one direction. Unlike an analog switch, digital switches are inherently "four-wire" devices and handle signals in one direction only. Another switch would be needed for the other direction.

Now let us suppose that subscriber 4 wants to talk to subscriber 6. Between the input and output of the switch is a memory capable of storing a sample from each input. Each input sample is clocked into this memory and remains there for the duration of one frame (125 microseconds). Then the contents of that memory location are replaced with the next sample from the same time slot. To switch a sample from slot 4 to slot 6, it is only necessary to read the contents of memory location 4 during the time period allotted to slot 6 and write the contents into the switch output. In such a manner any slot can be moved to any other slot, and any of the 32 subscribers can be connected to any other subscriber.

11.6.2 Space Switching

A real switch usually has to handle a much larger number of subscribers than the 32 shown in Figure 11.25. One way to arrange this is to use time and space switches together. A digital space switch is a crosspoint type of switch, but it differs from those described earlier in that it can switch much more quickly, connecting one line to another for the duration of one sample, rather than for an entire call. Of course, this means that electronic rather than mechanical switching must be used.

11.6.3 Time-Space-Time Switching

Figure 11.26 shows a time-space-time switch. Each of the time switches is just like the one in Figure 11.25, and each has a separate bus, often called a *highway*, at its output. The space switch connects two time switches at its input to two others at its output. A real switch would, of course, have many more than two inputs and outputs. The input switches are designated A and the output switches, B. Thus, for instance, time switch A1 can be connected to time switch B2 by the space switch. The space switching can be changed for each sample coming from the input time switches. The first sample from A1 can go to B2, the second to B1, and so on.

What the space switch cannot do is change the time slot of the samples on each bus. A second time switch at the output of the space switch does that. It is now pos-

Figure 11.26 Time-space-time switching

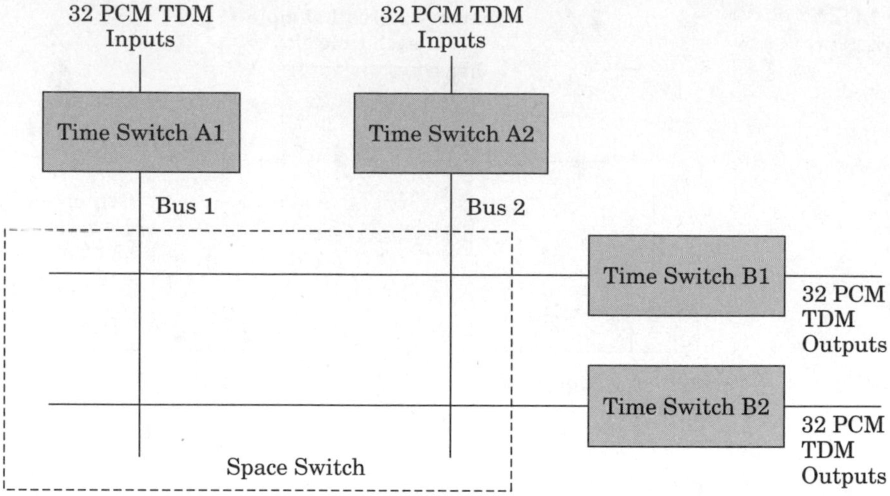

sible for any time slot of either input bus to be switched to any time slot of either output bus. Other combinations of time and space switching are also possible.

Example 11.9 Show how a signal on slot 4 of the input to time switch A1 can be moved to slot 1 of the output of time switch B2 in the switching setup shown in Figure 11.26.

Solution There are many ways to do this. The computer controlling the switch will choose a scheme that involves slots and connections that are not in use. One possibility is shown in Figure 11.27. Slot 4 is moved to slot 6 on time switch A1, as previously described. Note that any available slot would do. Then slot 6 on switch A1 is connected to slot 6 on switch B2 by the space switch. Had slot 6 on B2 not been free, it would have been necessary to switch slot 4 to some other free slot on switch A1. Finally, time switch B2 moves slot 6 to slot 1.

Figure 11.27

Input: Slot 4

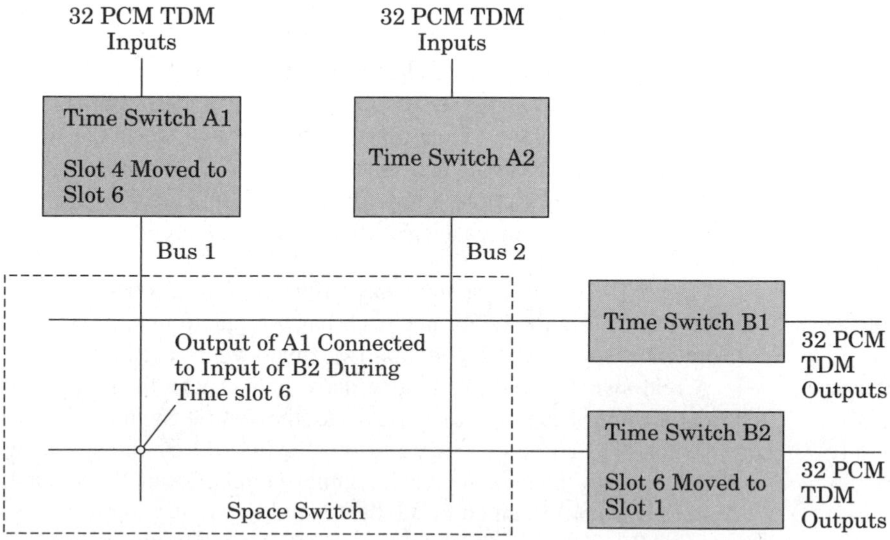

11.7 Integrated Services Digital Network

The concept behind the **integrated services digital network (ISDN)** is to allow voice and data to be sent in the same way along the same lines. Currently, most subscribers are connected to the switched telephone network by local loops and interface cards designed for analog signals. This system is reasonably well-suited to voice communications, but data can be accommodated only by using modems. We will consider the details of modems in Chapter 13, but for now, let us just note that they are restricted to relatively low data rates on ordinary dial-up lines, due both to the restricted bandwidth of the channel and to imperfections such as noise and phase shift. Currently the practical limit is about 28.8 kb/s. Large users of data communications can of course lease their own lines and create their own networks, but this is impractical for small- and medium-sized users.

As the telephone network gradually goes digital, it seems logical to send data directly over telephone lines, without using modems. If the local loop could be made digital, with the codec installed in the telephone instrument, there is no reason why the 64 kb/s data rate required for PCM voice could not be used for data instead, at the user's discretion. One voice channel could handle a much higher data rate than is practical using modems on dial-up lines.

The design of the integrated services digital network establishes a standard for this idea. The standard envisions two types of connections to the network. Large users connect at a primary-access point, with a data rate of 1.544 Mb/s. This, you will recall, is the same rate as for the DS-1 signal described earlier. It includes 24 channels with a data rate of 64 kb/s each. One of these channels is the **D (data) channel**, which is used for signaling, that is, for setting up and monitoring calls. The other 23 channels are called **B (bearer) channels** and can be used for voice or data or a combination of the two, for example, to handle high-speed data or digitized video signals.

Individual terminals connect to the network through a basic interface, at the basic access rate of 192 kb/s. Individual terminals in a large organization would use the basic access rate to communicate with a **private branch exchange (PBX)**, a small switch dedicated to that organization. Residences and small businesses would connect directly to the central office by way of a digital local loop. Two twisted pairs can be used for the local loop, but more use of fiber optics is expected in the future.

Basic-interface users have two 64 kb/s B channels for voice or data, one 16 kb/s D channel, and 48 kb/s for network overhead. The D channel is used to set up and monitor calls and can also be employed for low-data-rate applications such as remote meter-reading. It is important to note that all channels are carried on one physical line, using time-division multiplexing. Two pairs are used, one for signals in each direction.

Figure 11.28 shows typical connections to the ISDN. The primary interface is known as a T type interface, and the basic interface has the designation S. Terminal equipment, such as digital telephones and data terminals, designed especially for use with ISDN, is referred to as **TE1 (terminal equipment type 1)** and connects directly to the network at point S. The network termination equipment designated NT2 could

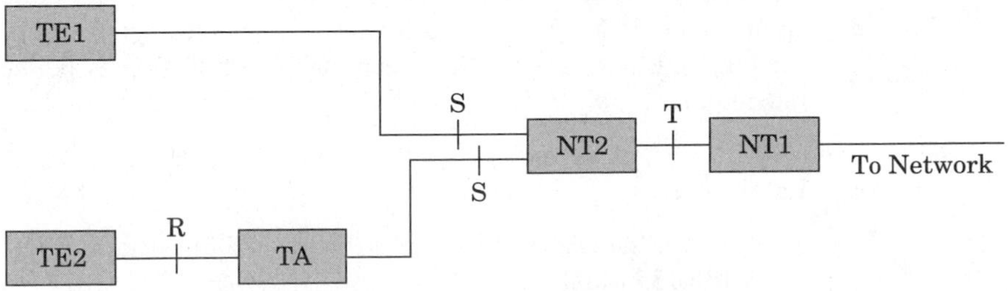

Figure 11.28
ISDN access

be a PBX, a small computer network called a **local area network**, or a central office. Terminal equipment not especially designed for ISDN is designated **TE2 (terminal equipment type 2)** and would need a **terminal adaptor (TA)** to allow it to work with the ISDN. Examples of type 2 equipment would be ordinary analog telephones, ordinary fax machines, and personal computers with serial ports. Each of these would need a different type of terminal adaptor.

Implementation of the ISDN has been slow, leading some telecommunications people to claim, tongue-in-cheek, that the abbreviation means "it still does nothing." Several reasons can be advanced for this. First, converting local loops to digital technology is expensive, and it is questionable whether the results justify the cost for most small users. Residential telephone users would not notice the difference (except that they would have to replace all their telephones or buy terminal adaptors), and analog lines with low-cost modems or fax machines are quite satisfactory for occasional data users. On the other hand, recent advances in the digital compression of video signals, which will be discussed in the next section, have led to a great deal of speculation about the practicality of running one digital line, probably using fiber optics (coaxial cable has also been proposed), to each residence. This single line could carry voice and data, as well as all the video signals now carried on cable television, with bandwidth to spare. Its data rate would have to be much higher than the basic ISDN data rate, however. Another possibility is to continue to use twisted-pair line to the residence. Using data compression, it is possible to transmit one video signal over a twisted pair, and the switched network could allow subscribers to choose which channel to watch.

Meanwhile, very large data users who often need a data rate well in excess of the primary interface rate for ISDN are already using other types of network. It appears possible that the ISDN standard is becoming obsolete before it can be implemented.

With this in mind, work has already begun on a revised and improved version of ISDN called **broadband ISDN (B-ISDN)**. The idea is to use much larger bandwidths and higher data rates, so that high-speed data and video can be transmitted. B-ISDN uses data rates between 100 and 600 Mb/s.

11.8 Digital Video Transmission

Digital video was mentioned briefly in Chapter 9, in connection with high-definition television. However, digital techniques are also used to transmit conventional video, as well as low-resolution, reduced-frame-rate video for videoconferencing. Digital video is also used in many multimedia computer applications. Though the subject of multimedia is beyond the scope of this book, it is worth noting that many of the techniques used in digitizing video (and audio) signals are equally useful whether the resulting signal is to be transmitted over a communications channel or stored on a hard disk or a CD-ROM.

Video signals can be encoded using PCM in the same way as audio signals, but the bit rate is usually much higher.

Example 11.10. The most popular digital studio standard for video is known as a D-1 or CCIR 601 signal. The luminance channel has 858 samples per line, and the chroma is coded as two components: R-Y and B-Y, each with 429 samples per line. Ten bits are used for each sample, and there are 525 lines per frame and 30 frames per second. Calculate the bit rate required.

Solution Luminance:

10 bits/sample × 858 samples/line × 525 lines/frame × 30 frames/s = 135.135 Mb/s

R-Y:

10 bits/sample × 429 samples/line × 525 lines/frame × 30 frames/s = 67.5675 Mb/s

B-Y (same as R-Y):

67.5675 Mb/s

Total:

135.135 Mb/s + 2 × 67.5675 Mb/s = 270.27 Mb/s

11.8.1 MPEG Video Standards

Many attempts have been made to get the bit rate for video signals down to more reasonable levels. All these schemes use one or both of two basic techniques. One technique is to reduce the data rate by reducing the amount of data, that is, either the number of picture elements (pixels) in each frame or the frame rate (number of frames per second) is reduced (or both). Either way, the result is a reduction in picture quality, which may be acceptable, depending on the application. Video conferencing, for instance, typically does not demand broadcast-quality video. The other technique takes advantage of the fact that in most cases, most of the pixels in one frame are the same as in the previous frame; thus, as far as possible, it transmits only the changes, rather than sending all the data for each frame. That is, sophisticated data-compression methods reduce the data rate without compromising picture quality.

The MPEG (Moving Picture Experts Group, a committee of the International Standards Organization (ISO)) set of video standards allows for data compression at a variety of frame rates and resolutions. Many people associate MPEG with the rather inferior video typically seen in multimedia computer applications, but the specification allows for broadcast-quality video as well and, in fact, can be adapted to high-definition television, as mentioned in Chapter 9.

There are two MPEG standards, or, more exactly, two groups of standards. The original version, MPEG-1, was completed in 1992 and is intended for progressive scan (noninterlaced) video. Its original intent was to allow a video signal and accompanying stereo audio (also digitally compressed) to be transferred at the data rate used by audio compact discs, which is about 1.5 Mb/s. Even with data compression, transmitting full-motion video at this data rate requires that the resolution be reduced to 352 by 240 pixels for luminance and half that in each direction for chroma. The resulting signal has approximately the quality of VHS videotape and is markedly below broadcast quality. With higher data rates, more detail can be included: the standard allows for resolutions as great as 4095 by 4095 pixels, at a bit rate of about 100 Mb/s.

A newer standard, MPEG-2, was completed in 1994 and allows for either progressive or interlaced scan and is therefore better adapted to the coding of broadcast-type television pictures. NTSC broadcast-quality video, MPEG-2 encoded, as employed in direct-broadcast satellite transmission, requires a data rate of about 3 Mb/s. A variation of MPEG-2 appears likely to be used for high-definition television broadcasting as well.

MPEG video is compressed in several ways. First it is necessary to transmit a complete single frame, then succeeding frames can to some extent be predicted from it. A complete frame has to be transmitted at relatively frequent intervals (about 0.4 seconds), so that the video signal can be acquired by a receiver. The completely-coded frames are called *I (intra)* frames. Even here, though, redundant data can be eliminated. For instance, if several successive pixels are the same, it is only necessary to tell the receiver this, rather than transmitting all the information for each pixel. This technique is called a *discrete cosine transform (DCT)*, and it works on an 8 × 8 block of pixels.

Next, motion is coded by comparing frames, using a 16 × 16 pixel block in the luminance channel. Rather than coding all the frames completely, those that lie between

I frames are coded as to the way they differ from the nearest frames. *P (predicted)* frames are derived from a previous *I* or *P* frame, and *B (bidirectional)* frames are derived from both previous and future frames. The result is a data compression of approximately 20:1 without very much loss in detail. The compression works better for video in which changes from scene to scene are slow. Some video (sports is a good example) has a lot of rapid change from frame to frame and cannot be compressed as much without losing some of the detail in fast-moving sequences. When several digital video signals are transmitted over a single wideband channel (for example, from a direct-broadcast satellite), it is possible to allocate greater bit rates to those signals that require them, on a moment-by-moment basis.

Summary

Here are the main points to remember from this chapter.

1. Modern communication systems are often a mixture of analog and digital sources and transmission techniques. The trend is to digital systems, which have better performance over long distances.
2. The information capacity of a channel is limited by its bandwidth and signal-to-noise ratio. Sometimes the theoretical capacity can be approached more closely by increasing the number of bits transmitted by each symbol.
3. An analog signal that is to be transmitted digitally must be sampled at least twice per cycle of its highest-frequency component.
4. PCM requires the amplitude of each sample of a signal to be converted to a binary number. The more bits used for the number, the greater the accuracy, but the greater the bit rate required.
5. Delta modulation transmits only one bit per sample, indicating whether the signal level is increasing or decreasing, but it needs a higher sampling rate than PCM for equivalent results.
6. The signal-to-noise ratio for either PCM or delta modulation signals can often be improved by using companding.
7. By appropriate choice of line codes, it is possible to transmit digital signals with no dc content and with sufficient clock information for synchronous communication.
8. Many digital signals can be transmitted on one channel using time-division multiplexing.
9. The integrated services digital network will allow combined voice and data on an all-digital telephone system.
10. The MPEG series of video-compression standards allows digital video to be transmitted using bandwidths even narrower than are required for analog video.

Important Equations

$$I = ktB \tag{11.1}$$

$$C = 2B \log_2 M \tag{11.2}$$

$$C = B \log_2 (1 + S/N) \tag{11.3}$$

$$\log_2 N = \frac{\log_{10} N}{\log_{10} 2} \tag{11.4}$$

$$f_a = f_s - f_b \tag{11.8}$$

$$N = 2^m \tag{11.9}$$

$$DR = (1.76 + 6.02m) \text{ dB} \tag{11.10}$$

$$D = f_s m \tag{11.11}$$

$$v_o = \frac{V_0 \ln (1 + \mu v_i/V_i)}{\ln (1 + \mu)} \tag{11.12}$$

Glossary

aliasing distortion caused by using too low a sampling rate

B (bearer) channel 64 kb/s ISDN channel used for voice or high-speed data

bipolar code a data code that uses both polarities of voltage or current

broadband ISDN (B-ISDN) variation of ISDN for use with higher data rates

codec data *co*der and *dec*oder combined into one integrated circuit

coding conversion of an analog signal into logic ones and zeros for digital transmission

combo chip an integrated circuit that combines coding and decoding with data compression and expansion

companding a combination of **compression** at the input to a system and **expansion** at its output

compression amplification of a signal in such a way that there is less gain for higher-level input signals than for lower-level input signals

D (data) channel a low data-rate ISDN channel used for low-speed data

decoding conversion of a digital data stream to an analog signal

delta modulation a digital pulse-modulation scheme that transmits changes in analog signal level, rather than the level itself

digital signal processing (DSP) manipulation of analog signals by first digitizing them and then performing computations with the digits

expansion amplification of a signal in such a way that there is more gain for higher-level input signals than for lower-level input signals

flat-topped sampling a sampling scheme that maintains the same amplitude for the duration of a given sample

foldover distortion see *aliasing*

framing bits extra bits used in a digital signal to ensure synchronization of transmitter and receiver

integrated services digital network (ISDN) a completely digital telephone system that can handle voice or data equally well

line code a system for translating logic ones and zeros into voltage or current levels for transmission

local area network (LAN) a small data network, usually confined to a company or campus

natural sampling a sampling scheme in which the amplitude of a sample varies during the sample time to follow the amplitude of the sampled signal

negative logic a logic system in which a low level represents logic one and a high level represents logic zero

NRZ (non-return-to-zero) code a data line code in which the voltage or current does not necessarily return to zero between bits

positive logic a logic system in which a high level represents logic one and a low level represents logic zero

private branch exchange (PBX) a small telephone switch operated independently of the public network, to which it appears as one or a number of subscriber lines

pulse-amplitude modulation (PAM) an analog pulse-modulation scheme in which one pulse is transmitted per sample of the original signal. The amplitude of each transmitted pulse is proportional to the amplitude of the original signal at the corresponding sampling time

pulse-code modulation (PCM) a digital pulse-modulation scheme in which sample amplitudes are expressed as binary numbers, which are transmitted as a series of pulses

pulse-duration modulation (PDM) an analog pulse-modulation scheme in which one pulse is transmitted per sample of the original signal; the duration of each transmitted pulse is proportional to the amplitude of the original signal at the corresponding sampling time

pulse-position modulation (PPM) an analog pulse-modulation scheme in which one pulse is transmitted per sample of the original signal; the time at which each pulse is transmitted is a function of the amplitude of the original signal at the corresponding sampling time

pulse-width modulation (PWM) see *pulse-duration modulation*

quantizing the division of a continuous range of values (for example, the possible voltages of an analog signal) into a number of discrete values

quantizing errors errors that occur when a continuously varying signal is quantized; these errors result from the fact that a range of values of the original signal must be represented by a single number

quantizing noise see *quantizing errors*

regenerative repeater a device that decodes and recodes a digital signal as well as amplifying it

RZ (return-to-zero) code a line code in which the voltage or current returns to zero at the end of each bit period

sample-and-hold circuit a device that detects the amplitude of an input signal at a particular time called the sampling time and maintains its output at or near that amplitude until the next sampling time

slope overload in delta modulation, the inability of the output to keep up with a rapidly changing input signal

space switching a method of telephone switching that provides a separate physical path for each call

terminal equipment type 1 (TE1) digital telephones and data terminals that are expressly designed for use with the ISDN system

terminal equipment type 2 (TE2) analog telephones and data modems that are designed to be used with a conventional analog telephone system

terminal adaptor (TA) a device that allows terminal equipment type 2 to be used with the ISDN system

time switching a method of telephone switching in which a call is moved from one time slot to another on a single time-division-multiplexed physical channel

unipolar code a line code in which the polarity of the voltage or the direction of the current remains the same at all times

Questions

1. Give four advantages and one disadvantage of using digital (rather than analog) techniques for the transmission of voice telephone signals.
2. What factors limit the theoretical maximum information rate on a channel?
3. What happens when a signal is sampled at less than the Nyquist rate?
4. (a) List three types of analog pulse modulation.
   (b) Which pulse modulation scheme is used as an intermediate step in the creation of PCM?
   (c) Which pulse modulation scheme also finds use in audio amplifiers and motor speed-control systems?
5. For a PCM signal, describe the effects of:
   (a) increasing the sampling rate
   (b) increasing the number of bits per sample
6. (a) Briefly explain what is meant by companding.
   (b) What advantage does companded PCM have over linear PCM for telephone communications?
7. Explain why the sample rate must be greater for delta modulation than for PCM.
8. What are the two functions of a codec? Where in a telephone system is it usually located?
9. Explain briefly how mu-law compression is implemented in a typical codec.
10. What advantages does the AMI bipolar code have over unipolar NRZ coding?
11. Compare the bipolar NRZ and Manchester codes:
    (a) Which of these codes is more suitable for synchronous communication? Why?
    (b) Which of these codes has more energy at low frequencies? Illustrate your answer with a bit pattern that contains a lot of low-frequency energy with this code.
    (c) Which of these codes requires the greater bandwidth? Illustrate your answer with a bit pattern that requires the greatest bandwidth with this code.
12. How many voice signals are combined into a DS-1 signal? How is this done?
13. Distinguish between a DS-1 signal and a T1 carrier.
14. How is signaling information incorporated into a DS-1 signal?
15. How is framing information incorporated into a DS-1 signal?
16. Explain the difference between space switching and time switching for a multiplexed signal.
17. Describe the integrated services digital network's basic and primary access points, including the number and types of channels and data rates.
18. Why is a terminal adapter needed when an ordinary telephone is used with the ISDN?
19. What is meant by the term *broadband ISDN*?
20. Describe how the MPEG video-compression schemes reduce the bandwidth required for digital video.

Problems

SECTION 11.1

21. A broadcast television channel has a bandwidth of 6 MHz.
    (a) Calculate the maximum data rate that could be carried in a TV channel using a 16-level code. Ignore noise.

(b) What would be the minimum permissible signal-to-noise ratio, in dB, for the data rate calculated above?

22. How many levels would be required to communicate at 9600 bits per second over a channel with a bandwidth of 4 kHz and a signal-to-noise ratio of 40 dB?

SECTION 11.2

23. A 1 kHz sine wave with a peak value of 1 volt and no dc offset is sampled every 250 microseconds. Assume the first sample is taken as the voltage crosses zero in the upward direction. Sketch the results over 1 ms using:
    (a) PAM with all pulses in the positive direction
    (b) PDM
    (c) PPM

24. The compact disc system of digital audio uses two channels with TDM. Each channel is sampled at 44.1 kHz and coded using linear PCM with sixteen bits per sample. Find:
    (a) the maximum audio frequency that can be recorded (assuming ideal filters)
    (b) the maximum dynamic range in decibels
    (c) the bit rate, ignoring error correction and framing bits
    (d) the number of quantizing levels

25. Suppose an input signal to a μ-law compressor has a positive voltage and an amplitude 25% of the maximum possible. Calculate the output voltage as a percentage of the maximum output.

26. A PCM/TDM system multiplexes 24 voice band channels. Each sample is encoded into seven bits, and a frame consists of one sample from each channel. A framing bit is added to each frame. The sampling rate is 9000 samples per second. What is the line speed in bits per second?

27. Suppose a composite video signal with a baseband frequency range from dc to 4 MHz is transmitted by linear PCM, using eight bits per sample and a sampling rate of 10 MHz.
    (a) How many quantization levels are there?
    (b) Calculate the bit rate, ignoring overhead.
    (c) What would be the maximum signal-to-noise ratio, in decibels?
    (d) What type of noise determines the answer to part (b)?

SECTION 11.3

28. How would a signal with 50% of the maximum input voltage be coded in eight-bit PCM, using digital compression?

29. Convert a sample coded as 11001100 using mu-law compression to a voltage, with the maximum sample voltage normalized as 1 V.

30. Convert the 12-bit PCM sample 110011001100 to an 8-bit compressed sample.

SECTION 11.4

31. Code the following data using each of the codes specified. The data is to be sent in time as it reads, from left to right.

    011000111

    (a) unipolar NRZ
    (b) AMI
    (c) Manchester code

32. Figure 11.29 shows a signal in AMI code. Decode it into ones and zeros.

Figure 11.29

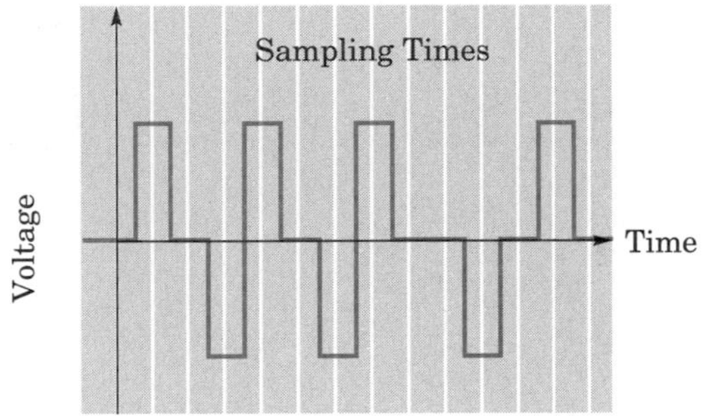

33. Figure 11.30 shows a signal in Manchester code. Decode it into ones and zeros.

Figure 11.30

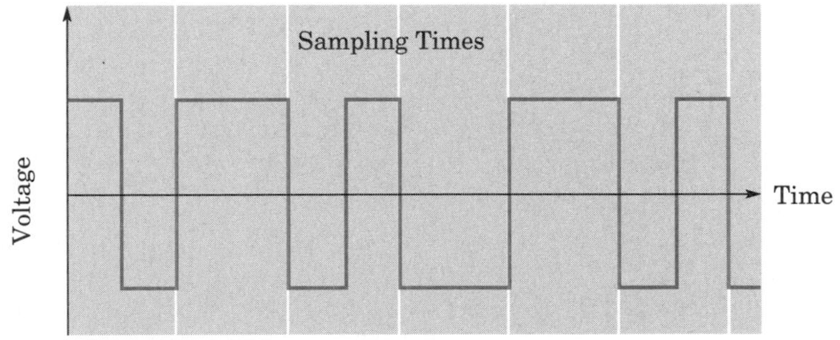

SECTION 11.5

34. (a) Calculate the bandwidth required to transmit a DS-1 signal using AMI code.
    (b) Suppose Manchester code were used instead. What would be the required bandwidth.

35. The basic European TDM telephone signal combines 30 voice channels with 2 signaling channels that have the same data rate as the voice channels. The sampling rate is 8 kHz and there are 8 bits per sample for each voice channel. Calculate the total bit rate for this signal.

SECTION 11.6

36. Draw a diagram showing how time switching could be used to move a voice signal from slot 4 to slot 22 of a DS-1 signal.
37. Draw diagrams showing how a voice signal could be moved from slot 6 to slot 12 using time-space-time switching.

SECTION 11.7

38. Draw a block diagram showing how an ordinary telephone could be connected to an ISDN primary interface using a PBX.

SECTION 11.8

39. Calculate the approximate compression ratio for MPEG by calculating the data rate required to encode the standard MPEG-1 signal (352 by 240 pixels) using eight-bit quantizing without compression. Don't forget to code the two color signals. Compare your result with the standard MPEG-1 rate (for CD video only) of 1.15 Mb/s.

COMPREHENSIVE

40. A certain cable television system has a usable bandwidth of 300 MHz. How many television signals could be transmitted in this bandwidth
    (a) using FDM and standard NTSC television channels, 6 MHz wide
    (b) using MPEG-2 with 3 Mb/s required per channel, time-division multiplexed and transmitted using an ordinary unipolar NRZ code

12 Data Transmission

Objectives After studying this chapter, you should be able to:

1. Explain the difference between text and binary data and the purpose of data codes
2. Describe the Baudot, ASCII, and ARQ codes and be able to encode and decode messages in these formats
3. Distinguish between synchronous and asynchronous communication and discuss the advantages and disadvantages of each
4. Explain the function of start and stop bits in an asynchronous system
5. Calculate the overhead and efficiency of a data communication system
6. Describe the conversion between parallel and serial data
7. Describe synchronous transmission and explain how clock and framing information is transmitted in synchronous communication systems
8. Describe several types of error control and perform sample calculations to show how errors can be detected and corrected
9. Explain the rationale behind data compression and describe several methods to accomplish it

12.1 Introduction

The previous chapter described how analog signals can be digitized and transmitted. In this chapter we will look at alphanumeric data and describe some of the ways it can be encoded. Then we will look more closely at data communication when the data may represent alphanumeric characters, binary information such as machine-language computer instructions, or digitized analog signals as described in Chapter 11.

In this chapter it will be assumed that the transmission channel is capable of carrying voltage pulses. Some transmission channels are not able to support a nonzero average level, that is, they are not dc-coupled. That problem can be dealt with by proper design of the line code, as discussed in the previous chapter. Channels such as radio links or **dial-up** telephone lines, which require modulation and demodulation, will be considered in the next chapter.

Either **serial** or **parallel transmission** can be used to transmit binary signals. In parallel transmission, multiple channels are used to transmit several bits simultaneously, while a single channel is used in serial transmission. Data communication over more than very short distances is generally serial to reduce the number of channels needed, so that is the only type of transmission we will study here. Keep the possibility of parallel communication in mind, however; later we will see that some local area networks employ a form of parallel communication using more than one pair of twisted-pair line to achieve higher data rates than would ordinarily be possible on such line.

12.2 Data Coding

Pulse-code modulation and delta modulation convert analog signals, such as voice and video, to binary numbers. The data coding techniques we will study in this section convert alphanumeric characters into binary numbers.

A **data code** is a standardized relationship between signaling elements and characters. Data codes are also called **character sets** or **character codes**.

Data communication uses three types of characters. Data link control characters facilitate the orderly flow of data; graphic control characters arrange the presentation of data at the receiving terminal; and alphanumeric characters represent the data itself.

12.2.1 Binary and Text Data

Alphanumeric characters are not the only source of data, of course. Digitized audio and video signals are examples of binary data that has nothing to do with the alphabet. Similarly, machine-language computer programs are not formed from alphanumeric characters. Such noncharacter data is usually called "binary" data.

Character codes are not needed for binary data, at least in principle. However, some communication systems designed primarily for the transmission of text insist on dividing all data into blocks that typically contain seven or eight bits; these small data blocks are then referred to as *characters*. This division would pose no problem, except that these systems define some characters as control characters; these control characters can change the way the system operates or even disconnect the communications link. In these systems, care must be taken that binary data does not accidentally form a control character. We will have more to say about this later.

12.2.2 Character Codes

Many codes have been used to convert alphanumeric and other characters into electrical signals. In this section we will look at two examples, the Baudot and ASCII codes.

Baudot Code The Baudot code dates from about 1870 and is still in use in telex and radioteletype services. It has been designated as International Telegraph Alphabet number two (ITA2) by the International Telegraph and Telephone Consultative Committee (CCITT), a European standards group, some of whose standards are also used in North America. Baudot was originally used with electromechanical teletype machines, but most modern systems use a computer for coding and decoding.

The Baudot code is listed in Table 12.1. Note that it is a five-bit code (sometimes called a five-level code). The number of possible combinations is therefore $2^5 = 32$. Additional combinations are created by using two characters called LTRS and FIGS, which act like the shift key on a typewriter to shift between letters and figures. The FIGS character set also includes control characters. The LTRS set includes only uppercase letters, which are satisfactory for telegraphic communication but have obvious drawbacks for other forms.

Another problem with Baudot is that an error that is interpreted as a FIGS character can cause all the following information to be interpreted as numerical until the next LTRS character occurs. To reduce this problem, most modern systems use a technique called *unshift-on-space* (USOS), in which the receiver switches back to LTRS whenever a space character is received, so that an error can cause at most one word to be garbled.

Table 12.1 lists the most significant bit (MSB) first, using standard binary notation. The code is actually transmitted with the least significant bit (LSB), bit 1, first. A few of the characters in the FIGS case have different interpretations in different variations of the code. The standard CCITT version is given first, followed by the variant(s).

The characters in parentheses are nonprinting control characters. Most of them are self-explanatory. WRU stands for "Who are you?" and is used to request the identity of the other station. BEL rings the bell on a teleprinter or beeps the speaker in a computer.

Table 12.1
Baudot code (CCITT ITA2)

Code 54321	LTRS case	FIGS case
11111	(LTRS)	
11011	(FIGS)	
00000	(BLANK)	
00100	(SPACE)	
01000	(LINE FEED)	
00010	(CARRIAGE RETURN)	
00011	A	-
11001	B	?
01110	C	:
01001	D	(WRU) or $
00001	E	3
01101	F	(unassigned) or ! or %
11010	G	(unassigned) or & or @
10100	H	(unassigned) or # or £
00110	I	8
01011	J	(BELL) or '
01111	K	(
10010	L	)
11100	M	.
01100	N	,
11000	O	9
10110	P	0
10111	Q	1
01010	R	4
00101	S	' or (BELL)
10000	T	5
00111	U	7
11110	V	= or ;
10011	W	2
11101	X	/
10101	Y	6
10001	Z	+ or "

A five-bit code has the advantage of faster data transfer for a given bit rate, compared to codes with more bits per character. Thus Baudot is useful for low-data-rate channels like HF radio.

ASCII Code The most common code for communication between microcomputers is known as ASCII, which stands for American Standard Code for Information Interchange. In its basic form, ASCII is a seven-bit code, so it allows for 128 possible combinations without shifting. As in Baudot, bit 1 (LSB) is sent first.

ASCII has certain regularities that make programming easier. For instance, switching from lower- to uppercase is accomplished by changing bit 6, and numbers are represented by their binary-coded decimal (BCD) values in bits 4 through 1. ASCII contains many control characters, which are normally generated on a keyboard by pressing the control key and another character simultaneously. Pressing the control key moves a character four columns to the left in Table 12.2 and is thus the same as subtracting 40 in hexadecimal notation from the character's value. For example, pressing CTRL-G gives the BEL character (the bell rings on a teletype machine or the speaker beeps on a computer).

Table 12.2
ASCII code
(binary notation)

Bit								
7	0	0	0	0	1	1	1	1
6	0	0	1	1	0	0	1	1
5	0	1	0	1	0	1	0	1
4321								
0000	NUL	DLE	SP	0	@	P	`	p
0001	SOH	DC1	!	1	A	Q	a	q
0010	STX	DC2	"	2	B	R	b	r
0011	ETX	DC3	#	3	C	S	c	s
0100	EOT	DC4	$	4	D	T	d	t
0101	ENQ	NAK	%	5	E	U	e	u
0110	ACK	SYN	&	6	F	V	f	v
0111	BEL	ETB	'	7	G	W	g	w
1000	BS	CAN	(	8	H	X	h	x
1001	HT	EM	)	9	I	Y	i	y
1010	LF	SUB	*	:	J	Z	j	z
1011	VT	ESC	+	;	K	[	k	{
1100	FF	FS	,	<	L	\	l	|
1101	CR	GS	–	=	M	]	m	}
1110	SO	RS	.	>	N	^	n	~
1111	SI	US	/	?	O	—	o	DEL

Many of the ASCII control characters are rather obscure; the ones we need to know about right now are:

BEL: ring the bell (or beep the speaker)

BS: backspace

LF: line feed (advance to the next line)

ESC: escape

FF: form feed (advance to the next page)

CR: carriage return (move to the beginning of the line)

DEL: delete

Example 12.1 Write the ASCII codes for the characters B and b.

Solution Find B in the table and proceed upward to find that bits 7,6,and 5 are 1,0, and 0, in that order. Then move from B to the left to find bits 4,3,2,1. They are 0010. Therefore the ASCII code for B is 1000010, with the most significant bit first. It would actually be transmitted with the least significant bit first.

The code for b can be found in the same way, as 1100010. Note that the only difference between the codes for B and b is the value of bit 6.

Sometimes ASCII codes are expressed in decimal or hexadecimal notation. The numbers are simply the decimal or hexadecimal translation of the binary numbers given in Table 12.2. Table 12.3 shows the decimal and hexadecimal numbers corresponding to the ASCII codes, along with the full names of all the control characters.

Table 12.3 ASCII Codes (decimal and hexadecimal notation)

Decimal	Hex	ASCII	Full name
0	00	NUL	null
1	01	SOH	start of heading
2	02	STX	start of text
3	03	ETX	end of text
4	04	EOT	end of transmission
5	05	ENQ	enquiry
6	06	ACK	acknowledge
7	07	BEL	bell
8	08	BS	backspace
9	09	HT	horizontal tab
10	0A	LF	line feed
11	0B	VT	vertical tab
12	0C	FF	form feed
13	0D	CR	carriage return
14	0E	SO	shift out
15	0F	SI	shift in
16	10	DLE	data link escape
17	11	DC1	device control 1
18	12	DC2	device control 2
19	13	DC3	device control 3
20	14	DC4	device control 4
21	15	NAK	negative acknowledge
22	16	SYN	synchronous idle
23	17	ETB	end of transmission block
24	18	CAN	cancel
25	19	EM	end of medium
26	1A	SUB	substitute
27	1B	ESC	escape
28	1C	FS	file separator
29	1D	GS	group separator
30	1E	RS	record separator
31	1F	US	unit separator
32	20	SP	space
33	21	!	
34	22	"	
35	23	#	
36	24	$	
37	25	%	
38	26	&	
39	27	'	
40	28	(	
41	29	)	
42	2A	*	
43	2B	+	
44	2C	,	
45	2D	-	
46	2E	.	
47	2F	/	
48	30	0	
49	31	1	
50	32	2	
51	33	3	
52	34	4	

continues

Table 12.3 Continued

Decimal	Hex	ASCII	Full name
53	35	5	
54	36	6	
55	37	7	
56	38	8	
57	39	9	
58	3A	:	
59	3B	;	
60	3C	<	
61	3D	=	
62	3E	>	
63	3F	?	
64	40	@	
65	41	A	
66	42	B	
67	43	C	
68	44	D	
69	45	E	
70	46	F	
71	47	G	
72	48	H	
73	49	I	
74	4A	J	
75	4B	K	
76	4C	L	
77	4D	M	
78	4E	N	
79	4F	O	
80	50	P	
81	51	Q	
82	52	R	
83	53	S	
84	54	T	
85	55	U	
86	56	V	
87	57	W	
88	58	X	
89	59	Y	
90	5A	Z	
91	5B	[	
92	5C	\	
93	5D	]	
94	5E	^	(circumflex)
95	5F	_	(underscore)
96	60	`	(grave accent)
97	61	a	
98	62	b	
99	63	c	
100	64	d	
101	65	e	
102	66	f	
103	67	g	
104	68	h	
105	69	i	
106	6A	j	

continues

Table 12.3
Continued

Decimal	Hex	ASCII	Full name
107	6B	k	
108	6C	l	
109	6D	m	
110	6E	n	
111	6F	o	
112	70	p	
113	71	q	
114	72	r	
115	73	s	
116	74	t	
117	75	u	
118	76	v	
119	77	w	
120	78	x	
121	79	y	
122	7A	z	
123	7B	{	
124	7C	|	
125	7D	}	
126	7E	~	(tilde)
127	7F	DEL	delete

Sometimes an eighth bit is added to the ASCII code to allow for the transmission of graphics characters, mathematical symbols, and foreign language characters. When the eighth bit is zero, the standard ASCII set described above is generated. Setting the eighth bit to one allows for 128 more characters. There is no single standard for these extra characters, but the set used by the IBM series of personal computers has become a de facto standard. These characters can be generated on the PC by holding the ALT key while typing the decimal number for the character, from 128 to 255, on the numeric keypad. These characters are listed in Table 12.4 below. Some software products (word processors, for example) can switch among several different character sets.

12.3 Asynchronous Transmission

A digital transmission format has to specify standard voltage ranges that will be interpreted as binary one and zero. There must also be a standard clock speed, and the transmitter and receiver clocks must be in phase with each other, so that the receiver checks the level of the line at the correct times. Framing is also needed. That is, the receiver has to be able to tell which bit represents the first bit of a character or a PCM sample. Of course, the same data code must be in use at both ends of the communications link.

Data communication schemes are designated as synchronous or asynchronous depending on how the timing and framing information is transmitted. The framing for asynchronous communication is based on a single character, while that for synchronous communication is based on a much longer block of data. This section will deal with asynchronous communication, which was originally intended for use with electromechanical teleprinters and is now the type of communication most commonly used with microcomputers. Both Baudot and ASCII codes are used, but ASCII is more popular and will be used in our examples.

In an asynchronous system, the transmit and receive clocks are free-running and are set to approximately the same speed. A **start bit** is transmitted at the beginning of each character, and at least one stop bit is sent at the end of the character. The **stop bit** leaves

Table 12.4
IBM PC ASCII Characters from 128 to 255

128	Ç	160	á	192	└	224	α
129	ü	161	í	193	┴	225	β
130	é	162	ó	194	┬	226	Γ
131	â	163	ú	195	├	227	π
132	ä	164	ñ	196	─	228	Σ
133	à	165	Ñ	197	┼	229	σ
134	å	166	ª	198	╞	230	μ
135	ç	167	º	199	╟	231	τ
136	ê	168	¿	200	╚	232	Φ
137	ë	169	⌐	201	╔	233	Θ
138	è	170	¬	202	╩	234	Ω
139	ï	171	½	203	╦	235	δ
140	î	172	¼	204	╠	236	∞
141	ì	173	¡	205	═	237	φ
142	Ä	174	«	206	╬	238	ε
143	Å	175	»	207	╧	239	∩
144	É	176	░	208	╨	240	≡
145	æ	177	▒	209	╤	241	±
146	Æ	178	▓	210	╥	242	≥
147	ô	179	│	211	╙	243	≤
148	ö	180	┤	212	╘	244	⌠
149	ò	181	╡	213	╒	245	⌡
150	û	182	╢	214	╓	246	÷
151	ù	183	╖	215	╫	247	≈
152	ij	184	╕	216	╪	248	°
153	Ö	185	╣	217	┘	249	∙
154	Ü	186	║	218	┌	250	·
155	¢	187	╗	219	█	251	√
156	£	188	╝	220	▄	252	^{n}
157	¥	189	╜	221	▌	253	^{2}
158	Pt	190	╛	222	▐	254	■
159	*f*	191	┐	223	▀	255	

the line or channel in the **mark** condition, which represents binary 1, and the start bit always switches the line to a **space** (binary 0). Figure 12.1 shows the arrangement of these bits.

In asynchronous communication, the framing is set by the start bit. The timing remains sufficiently accurate throughout the limited duration of the character as long as the clocks at the transmitter and receiver are reasonably close to the same speed.

Figure 12.1
Asynchronous communication

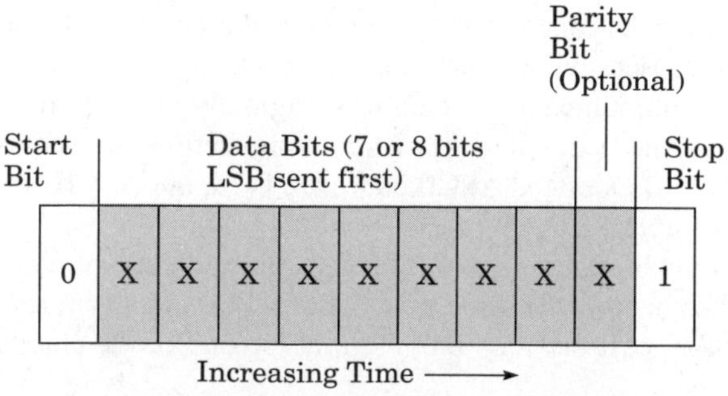

Historical Roots of Digital Communication

The terms *mark* and *space* date from the early days of telegraphy, when a pen linked to a solenoid marked a moving paper tape with the dots and dashes of Morse code. An energized line caused the pen to mark the tape, and a line with no current caused the pen to lift from the tape, creating a space.

Morse code itself represents an early form of digital communication. Morse code does not qualify as a binary system, however, since there are three meaningful symbols: dots, dashes and spaces. Some early telegraph systems that predated the Morse system used multiple wires in a form of parallel communication.

In addition to *mark* and *space*, many other terms and concepts used in telegraphy persist today. Telegraph operators sent their dots and dashes by means of a hand-operated momentary-contact switch called a *key* (see Figure 12.2(a)). To this day, digital modulation schemes are referred to as *keying*. In the next chapter, we will discuss frequency-shift keying, phase-shift keying, and so on.

Telegraph operators using hand keys were gradually replaced by teleprinters, electromechanical devices resembling electric typewriters (see Figure 12.2(b)). Teleprinters could code and decode characters, usually using Baudot code, and were also capable of transmitting previously coded messages from a punched paper tape. The crude speed control of their motors required asynchronous communication, which synchronizes transmitter and receiver at the start of each transmitted character.

Asynchronous communication remains popular today, even though most of the mechanical teleprinters have been replaced by computers with crystal-controlled clocks.

Figure 12.2

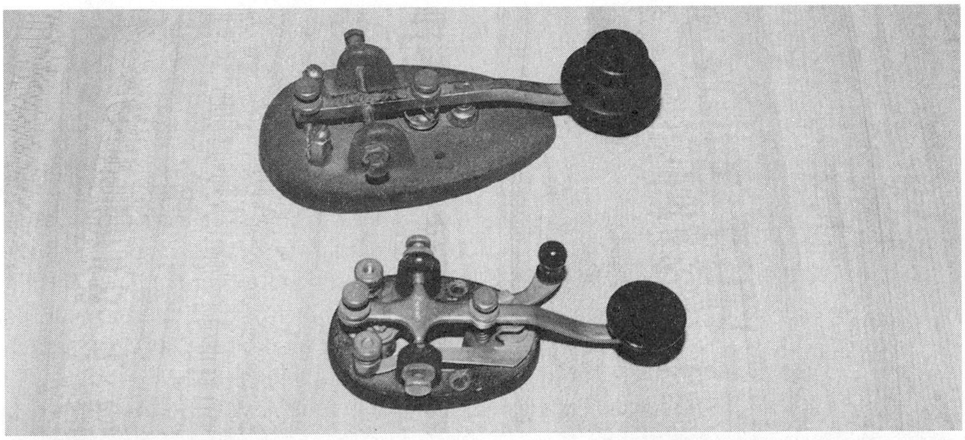

(a) Hand key

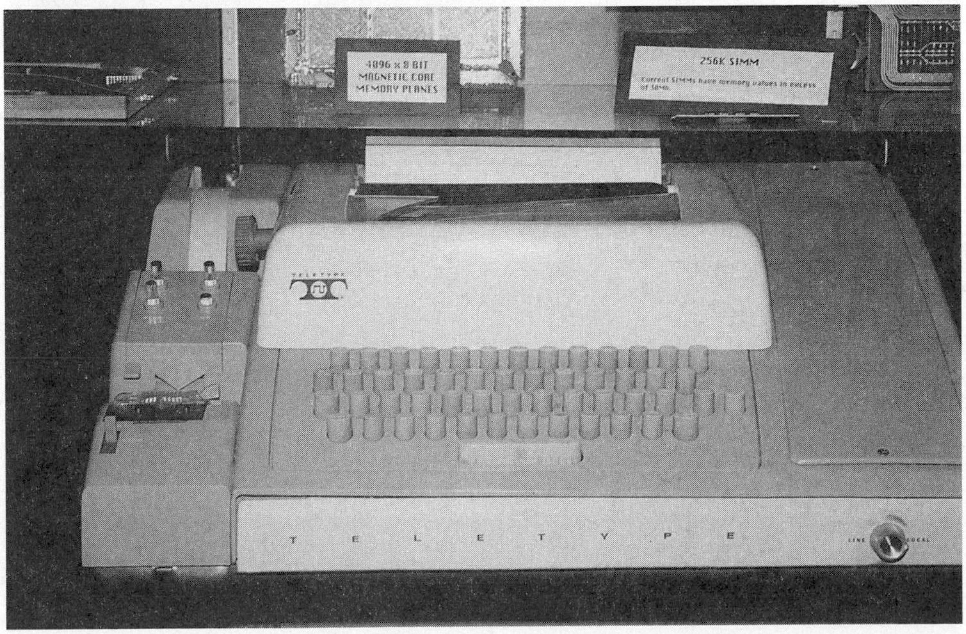

(b) Teletype machine

In asynchronous transmission, there is no set length of time between characters. The receiver monitors the line until it receives a start bit. It counts bits, knowing the character length being employed, and after the stop bit, it begins to monitor the line again, waiting for the next character. With ASCII, there may be an extra bit called the parity bit before the stop bit. Parity is a form of error control and will be described later in this chapter.

Not every bit that is transmitted contributes directly to the message. The start bit, stop bit(s), and parity bit, if present, are called **bit overhead**. The efficiency of the communication system can be defined as

$$\eta = N_D / N_T \tag{12.1}$$

where

η = efficiency
N_D = number of data bits
N_T = number of total bits

Example 12.2

Calculate the maximum efficiency of an asynchronous communication system using ASCII with seven data bits, one start bit, one stop bit, and one parity bit.

Solution

Assuming characters follow immediately one after another, there are ten bits per character, and seven of them are data bits. From Equation (12.1),

$$\begin{aligned} \eta &= N_D / N_T \\ &= 7/10 \\ &= 0.7 \text{ or } 70\% \end{aligned}$$

12.3.1 Parallel to Serial Conversion

Internally, computers use parallel communication, sending at least eight and often 16 or 32 bits along a bus. On the other hand, most data communication over longer distances is serial. Thus a device is needed to translate between parallel and serial data. (See Figure 12.3 for an illustration.) The line drivers and receivers shown in Figure 12.3 simply convert voltages from the levels used internally to those required on the communication link.

In an asynchronous communication system, this conversion between parallel and serial form is usually performed by an integrated circuit called a **Universal Asynchronous Receiver-Transmitter (UART)**. A typical UART is shown in block-diagram form in

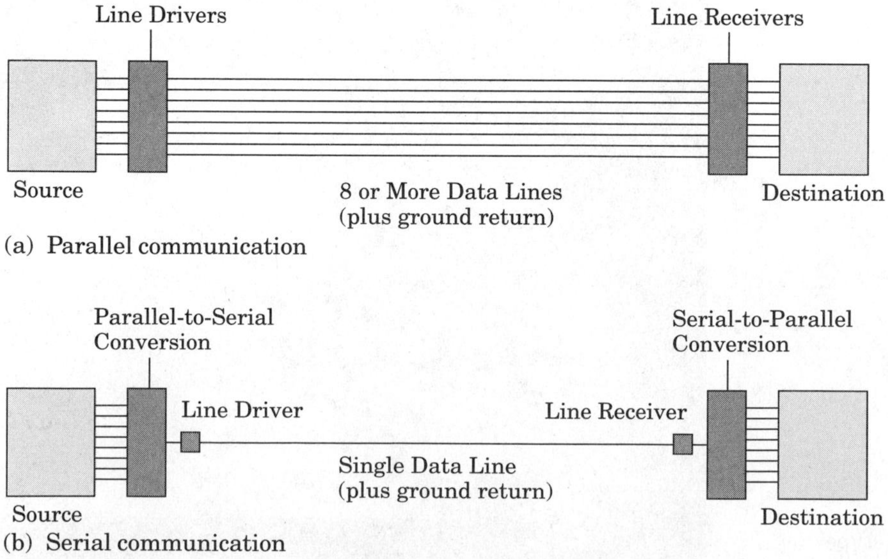

Figure 12.3
Parallel and serial communication

Figure 12.4
Simplified block diagram of a UART

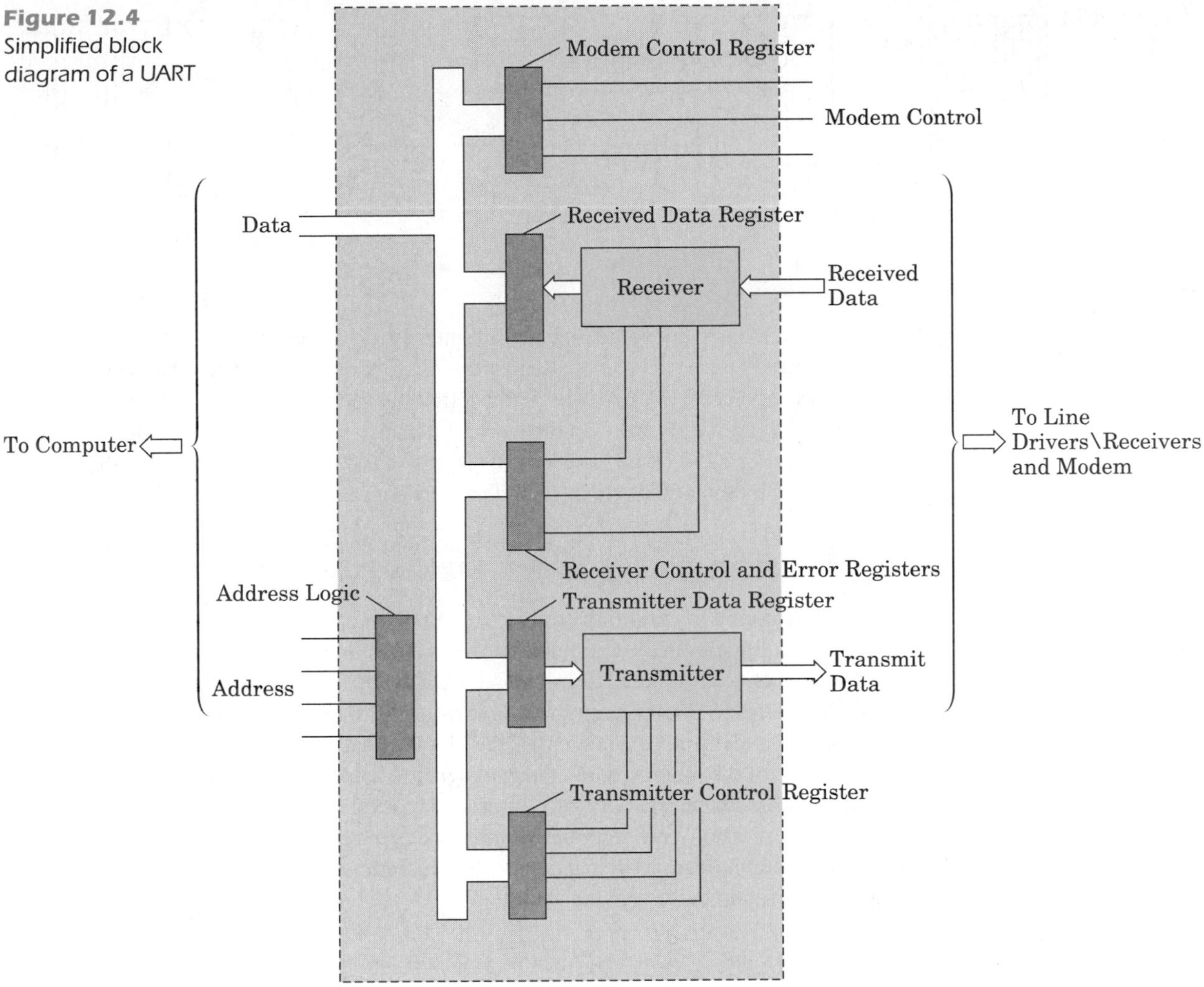

Figure 12.4. At the transmitting end of an asynchronous communication link, a UART converts parallel data from a computer or terminal into serial form, adds start and stop bits, and clocks it out at the correct rate. It may also add a parity bit for error control.

The receiver section of a UART must be able to detect the start-bit transition at the beginning of a character, before the receiver and transmitter clocks are in phase. The receiver must continuously monitor the line until it notes a one-to-zero transition, then it must look again at the signal a short time later to make sure it is still low. It then identifies this transition as a start bit rather than a noise pulse. It makes this identification by checking the line at a rate typically sixteen times the bit rate. When a transition is noted, the line is observed again eight clock cycles (one-half the time for one bit) later. If still low, the pulse is identified as a start bit. Figure 12.5 illustrates this process.

After the start bit has been identified, the signal is sampled once per bit time, approximately in the center of the bit, until the end of the character (the stop bit). If the stop bit is not a one, an error signal must be sent. An error signal is also sent if the parity bit (if present) is not correct. These error signals can be used by the computer to print an error message or request retransmission.

Generally, the UART will assemble the received character in a shift register, then transfer it to another register for access by the computer while it assembles the next character. A problem can arise if the computer is tardy in reading the incoming character: If the

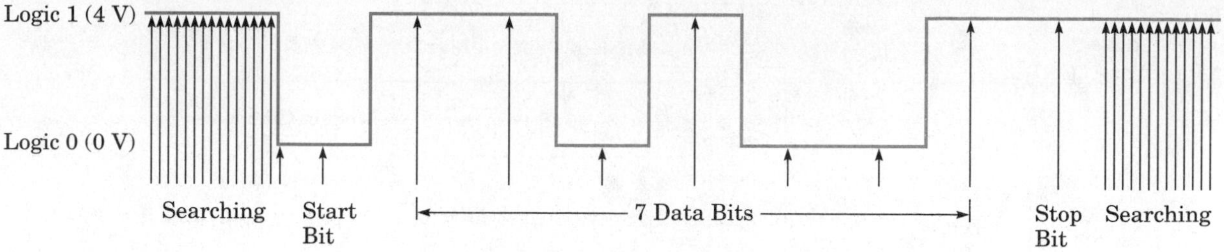

Figure 12.5 Asynchronous reception (arrows indicate sampling times)

processor is interrupted by something with higher priority than the UART, the character may remain unread until it is overwritten by the next incoming character, causing an error to occur. More advanced UARTs have a buffer that is capable of storing several characters to reduce the possibility of errors. In the world of IBM-compatible personal computers, for instance, the earlier 8250 (8-bit) and 16450 (16-bit) UARTs, which have no buffers, have been superseded by the 16550AFN, which has a 16-character buffer.

12.4 Synchronous Transmission

In synchronous communication, the transmitter and receiver are synchronized to the same clock frequency. Synchronous communication is more efficient than asynchronous because start and stop bits are not necessary. Instead, data is sent in blocks that are much longer than a single character. Blocks begin with an identifying sequence of bits that allows proper framing and often identifies the content of the block as well.

The price for the increased efficiency of synchronous communication is greater complexity in both hardware and software. In practice, asynchronous systems are commonly used for communication between microcomputers. Synchronous systems are used for the higher-speed communication typical of mainframe computers and for digitized analog signals such as are found in the telephone system.

12.4.1 Clock Recovery

Asynchronous digital systems can function with a receiver clock whose speed is only approximately the same as the transmitter clock, because the clock is reset at the start of each character. Simple crystal oscillators at each end, running at the same nominal frequency, provide more than sufficient accuracy. However, the longer message blocks used in synchronous communication make it necessary to lock the two clocks together exactly—otherwise the receiver clock would soon begin to sample the signal at the wrong times. This problem becomes worse as the data rate increases, because an increase in data rate reduces the duration of each transmitted symbol.

The most economical way for the receiver to get clock-synchronizing information is from the incoming data stream. Assuming the incoming data is a random collection of ones and zeros, the clock period can easily be found from practically any line code. For instance, consider the following data stream, and suppose it is sent using an ordinary NRZ code, such as is found at the output of an RS-232 serial port:

11101010011001010111

The first three ones tell us nothing about the clock. Following them, however, are some 1010 transitions that occur at the maximum possible rate. As long as there are not too many ones or zeros in a row, it is quite easy to derive synchronization information from the data stream, using a phase-locked loop.

Of course, long streams of zeros or ones are quite possible with many types of data. One way to avoid this problem is to use line codes that increase the number of level transitions, such as the Manchester code discussed in the previous chapter. You should recall that the Manchester code has a transition for every bit, so it creates no problems due to strings of identical bits. Manchester and similar codes are often used with local area networks; they are not useful when modems are required, however, because the increased number of level transmissions increases the bandwidth required for a given bit rate. Since the bandwidth available on a telephone line is very limited, the result would be a reduction in the maximum data rate.

Another way of ensuring that enough transitions take place is to "scramble" the bits using shift registers, which delay the data and combine it in such a way that sufficient one-zero and zero-one transitions occur. Scramblers are often used with synchronous modems.

Still another method, called eight-to-fourteen modulation, is used in the compact disc digital audio system. In this scheme, the data is first divided into 8-bit samples. For transmission, the 8 bits are converted to 14 bits. The number of possible 8-bit combinations is $2^8 = 256$, and the number of 14-bit sequences is $2^{14} = 16{,}384$. From this large number of possibilities, it is easy to find 256 combinations that each have the required number of transitions.

12.4.2 Framing

The use of framing bits in digital telephony was discussed in the previous chapter, and a very similar problem exists in synchronous data communication, where it is necessary to determine which bit represents the beginning of a character. In asynchronous communication, the problem is solved by using start and stop bits for each character, but they add a great deal of overhead. In synchronous communication, it is only necessary to locate the beginning of a block of data. As long as the transmitter and receiver clocks remain synchronized, the receiver can keep track of the remaining data without needing stop and start bits. Thus synchronous systems can be much more efficient than asynchronous systems.

12.4.3 Synchronous Data-Link Protocols

Synchronous communication protocols are either character- or bit-oriented. Character-oriented protocols predate bit-oriented ones and are becoming obsolete. Probably the most common example is BISYNC, an IBM product. These protocols begin a block with at least two synchronizing (SYN) characters, followed by control and data characters. The SYN characters are repeated at intervals during a long block of data to help the system maintain synchronization. See Figure 12.6 for an illustration.

Problems arise in character-oriented protocols when noncharacter data is transmitted. Eight consecutive bits of data could easily form a control character by accident. When character-oriented protocols are used with binary data, it is necessary to precede each real control character with a unique bit sequence, such as the DLE (data link escape) character. Any control character not preceded by DLE is actually part of the message and is ignored by the system. Of course, it is also possible that a sequence of data bits could form the DLE character. In that event, the transmitter inserts an extra DLE. When two consecutive DLE characters are received, the receiver ignores the first and interprets the second as binary data.

Bit-oriented protocols do not distinguish between characters and binary data. They divide the data into blocks that can have either a fixed or a variable length, depending on the protocol. Examples of bit-oriented protocols include SDLC (synchronous data link control), an IBM product, and HDLC (high-level data link control), a similar ISO standard.

Each block of data is preceded by an eight-bit pattern called a *flag*, which signals the start of a **frame**. In SDLC and HDLC the flag consists of the bit sequence 01111110.

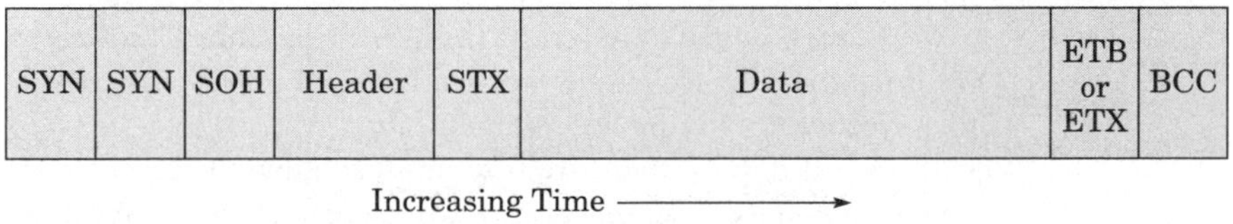

BCC = Block Check Character (used for error control)
For other control characters, see the ASCII table.

Figure 12.6 Character-oriented synchronous protocol (Bisync): one block with control and synchronizing characters

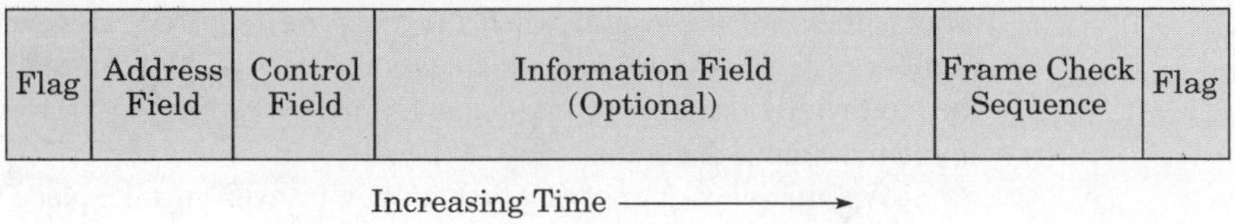

Figure 12.7 One frame of a bit-oriented synchronous protocol (HDLC)

Following the flag are control fields that perform the same functions of addressing and data link control as in character-oriented protocols. The data block follows the control fields, and the system examines it only for the presence of flags. The transmitter avoids accidental flags by inserting an extra zero after any series of five consecutive ones that occurs in the data. The extra zeros are removed by the receiver. This technique is called **bit stuffing**. After the data, there are frame check bits that can be used for error control and another flag. If another frame follows immediately, the flag that ends one frame starts the next. Figure 12.7 illustrates a data block in HDLC.

Bit-oriented protocols are more versatile than their character-oriented equivalents, because they are more easily adapted to different kinds of data. Modern communication systems carry a mixture of character-based information, computer programs, graphics, and video and audio information in various formats. New developments such as asynchronous transfer mode (ATM) and frame relay, which we will describe in our discussion of data networks, are bit-oriented.

12.5 Error Detection and Correction

Errors will occur whether data is sent by digital or analog means. Many of these errors are due to noise in the channel. The practical importance of errors varies with their number and with the type of information. For example, a bit error in a digital telephone signal will cause, at worst, a click in the receiver; an error in an exchange of financial data between banks could be much more serious. Usually, errors can be reduced by using a lower data rate, but it may be more efficient to accept a fairly high error rate in the channel and use error-correction techniques to remove the errors before they cause problems.

Thermal noise produces random errors, but other types of noise can cause burst errors, in which many bits in sequence are in error. Radio systems and tape and disk storage systems are examples of places where burst errors can occur. Burst errors are harder to correct than random errors, since the error can include bits added for error correction as well as data bits.

There are two stages in the control of errors. The first problem, of course, is to detect the fact that an error has occurred. Then the error must be corrected, if possible. If the error cannot be corrected, the receiver should at least be notified that the block of data it has just received contains errors. It might seem that with a binary system, detection and correction would be the same thing—if a bit received as a one is in error, then the bit must actually be a zero. However, in many cases an error-detection scheme only localizes an error to a character or a block of data. For instance, we may know that one of the seven bits in an ASCII character is in error, but we may not know which one.

There are many techniques for the detection and correction of errors. All of them involve adding some redundancy, that is, bits that do not actually contribute to the information sent. Error control therefore increases overhead.

Some systems rely on the receiver being able to correct errors from the redundant information. This technique is called **forward error correction (FEC)**. It is the only way to correct errors in simplex communication systems. Alternatively, if the communication is half- or full-duplex, there are schemes that allow the receiver to request the transmitter to repeat any data blocks that are known to contain errors. Systems of this type are known as **automatic request for retransmission (ARQ)** schemes. In general, ARQ methods do not require the transmission of as much redundant information as do FEC schemes. In this section we will look at a few representative systems for detecting and correcting errors in both of these situations.

12.5.1 Parity

Parity is a simple means of error detection. It involves the addition of one extra bit to the bits that encode a character. The parity bit is set to make the transmitted character (including the parity bit but not the start and stop bits) have either an even number of ones (even parity) or an odd number of ones (odd parity). The parity bit can easily be added by hardware using a simple logic circuit. The receiver checks the parity. The received parity will be incorrect if one error or an odd number of errors has occurred. It is not possible to use parity to determine which bit is in error; the whole character must be retransmitted to correct the error. Also, if two errors have occurred, the parity will be correct and the error will not be detected (the same is true for any even number of errors). Sometimes the parity bit is called a *vertical redundancy check* (*VRC*).

Example 12.3

Write the bit pattern for the letter b in ASCII with one start bit, odd parity, and one stop bit, in the order in which the bits would be transmitted.

Solution

From Table 12.2, the ASCII code for b is 1100010. However, this is given in the usual order in which binary numbers are written, with the most significant bit first. It will be transmitted in the reverse order: 0100011. Now we add a parity bit, which will be zero to give odd parity: 01000110. The start bit, which is sent first, is 0, and the stop bit, sent last, is 1. This gives the complete character, which will be sent as 0010001101.

12.5.2 Longitudinal Redundancy Check

The longitudinal redundancy check (LRC) is an extension of parity that can provide some error correction as well as detection. A block of seven-bit ASCII characters, starting with STX and ending with ETX or ETB, has a parity bit for each character. Another parity bit is made from the first bit of each character, another from the second bit, and so on, so that a *parity character* is made up from the block. The parity character is sent last. By checking the parity for each character, the fact that a single-bit error has occurred can be determined. Then the receiver can go to the LRC character to determine exactly which bit is in error. This allows the correction of single-bit errors.

The amount of redundancy and the power of the system depend on the length of the block. Longer blocks give less redundancy but poorer error correction, as only one bit per block can be corrected with complete reliability.

When there is more than one error per block, the errors may still be detected and an automatic request for retransmission (ARQ) can be sent. The optimum block length depends on the error rate of the channel. If there are few errors, a long block will reduce the amount of redundant information, but with a high channel error rate, a long block will result in many ARQ requests, greatly slowing the data rate as long blocks are sent over and over.

Example 12.4

Generate the vertical and longitudinal redundancy checks, using odd parity, for the block

STX L R C ETB

Solution

First, look up the code for each character and write it vertically:

Bit	STX	L	R	C	ETB
(MSB) 7	0	1	1	1	0
6	0	0	0	0	0
5	0	0	1	0	1
4	0	1	0	0	0
3	0	1	0	0	1
2	1	0	1	1	1
1	0	0	0	1	1

Next find the VRC (parity) bit for each character. (Now you can see why parity is called a *vertical* redundancy check.)

Bit	STX	L	R	C	ETB
7	0	1	1	1	0
6	0	0	0	0	0
5	0	0	1	0	1
4	0	1	0	0	0
3	0	1	0	0	1
2	1	0	1	1	1
1	0	0	0	1	1
VRC	0	0	0	0	1

Finally, find the LRC for each row.

Bit	STX	L	R	C	ETB	LRC
7	0	1	1	1	0	0
6	0	0	0	0	0	1
5	0	0	1	0	1	1
4	0	1	0	0	0	0
3	0	1	0	0	1	1
2	1	0	1	1	1	1
1	0	0	0	1	1	1
VRC	0	0	0	0	1	0

Example 12.5 The block below contains exactly one error. Find the error and decode the block. Odd parity is used.

0	1	1	1	1	0	1
0	0	1	1	1	0	0
0	0	0	0	0	1	1
0	0	0	0	0	0	1
0	0	0	1	1	1	0
1	1	0	0	0	1	0
0	0	1	1	0	1	0
0	1	0	0	0	1	1

Solution Looking at the VRC bits (the bottom row) reveals one column with incorrect parity. Similarly, looking at the LRC (the right-hand column) shows an error in one row.

0	1	1	1	1	0	1	
0	0	1	1	1	0	0	
0	0	0	0	0	1	1	—Error in this row
0	0	0	0	0	0	1	
0	0	0	1	1	1	0	
1	1	0	0	0	1	0	
0	0	1	1	0	1	0	
0	1	0	0	0	1	1	
			Error in this column				

Next, change the bit at the intersection of that row and column.

0	1	1	1	1	0	1
0	0	1	1	1	0	0
0	0	0	1	0	1	1
0	0	0	0	0	0	1
0	0	0	1	1	1	0
1	1	0	0	0	1	0
0	0	1	1	0	1	0
0	1	0	0	0	1	1

Now discard the check bits and decode the characters.

0	1	1	1	1	0
0	0	1	1	1	0
0	0	0	1	0	1
0	0	0	0	0	0
0	0	0	1	1	1
1	1	0	0	0	1
0	0	1	1	0	1
STX	B	a	u	d	ETB

12.5.3 Checksums

Another error-detection method consists of adding together all the data words in a block, for instance, by adding their ASCII values, then dividing by some fixed number. The remainder that results from this division is transmitted at the end of the block. The receiver performs the same division and should get the same remainder. If not, there is an error in the transmitted block.

XMODEM is a data-transfer protocol for microcomputers that, in its original and most basic form, uses this method. (There is also a version of XMODEN that uses cyclic redundancy checking, which will be described later in this section.) XMODEM uses a block of 128 bytes. The checksum is obtained by summing the ASCII values of the transmitted characters, dividing by 128, and using the remainder. The transmitter appends this remainder to the block, and the receiver compares it with its own calculated remainder. If the results agree, the receiver sends an ACK character back to the originating station, which then transmits the next block. If the two checksums are different, the receiver transmits a NAK character and the transmitter sends the entire block again.

12.5.4 ARQ and FEC Techniques in HF Radio

Traditionally, radioteletype (RTTY) has used the five-bit Baudot code, which does not have any provision for parity. More recently, a special seven-bit code has been adopted in which each valid character has four zeros and three ones. This allows 35 valid combinations out of 128 possibilities (only uppercase is supported). Any other received combination is rejected as an error. This system can detect two or more errors in a character, provided that they do not occur in complementary pairs (that is, if a one becomes a zero and a zero becomes a one in the same character, it will be interpreted as valid). Table 12.5 lists the valid combinations in the ARQ code. Note that the code is given in terms of B and Y rather than one and zero. B is the higher frequency and Y the lower. Also note that unlike the ASCII and Baudot tables, this code specifies the signals in the order they are transmitted. Like Baudot, the ARQ code has variations. Where two characters are listed for a given bit combination, the first is the CCITT standard and the second the U.S. convention.

Unlike Baudot, the ARQ code is used for synchronous communication, so no start and stop bits are needed. When this code is used in half-duplex communication between

Table 12.5 HF radio ARQ code

Code	LTRS case	FIGS case
BBBYYYB	A	-
YBYYBBB	B	?
BYBBBYY	C	:
BBYYBYB	D	WRU or $
YBBYBYB	E	3
BBYBBYY	F	undefined or !
BYBYBBY	G	undefined or &
BYYBYBB	H	undefined or #
BYBBYYB	I	8
BBBYBYY	J	bell or ′
YBBBBYY	K	(
BYBYYBB	L	)
BYYBBBY	M	.
BYYBBYB	N	,
BYYYBBB	O	9
BYBBYBY	P	0
YBBBYBY	Q	1
BYBYBYB	R	4
BBYBYYB	S	′ or BELL
YYBYBBB	T	5
YBBBYYB	U	7
YYBBBBY	V	= or ;
BBBYYBY	W	2
YBYBBBY	X	/
BBYBYBY	Y	6
BBYYYBB	Z	+ or "
YYYBBBB	carriage return	
YYBBYBB	line feed	
YBYBBYB	LTRS	
YBBYBBY	FIGS	
YYBBBYB	space	
YBYBYBB	unperforated tape	

Control and phasing signals

Code	ARQ mode	FEC mode
BYBYYBB	control signal 1	
YBYBYBB	control signal 2	
BYYBBYB	control signal 3	
BBYYBBY	idle signal beta	
BBBBYYY	idle signal alpha	phasing signal 1
YBBYYBB	signal repetition (RQ)	phasing signal 2

two stations, such as a ship at sea and a shore station, a sequence of three characters is transmitted, then the transmitting station waits for an acknowledgement from the other station. If the receiving station receives all three characters without detecting any errors, it sends a control character, alternating between control signals 1 and 2. If the receiving station detects an error, it sends the same control signal as last time. In the latter case the transmitter repeats all three characters and again pauses for an acknowledgement.

The same code can also be used in a forward error correction (FEC) system that is broadcasting a message to a number of stations. In that case, the transmitter sends each block of three characters twice, and the receiver checks each character for errors. If there are errors in one transmission of a character but not the other, the error-free one is used. If both versions of the character are in error, an error message is displayed. Of course, some errors get through this way, so the ARQ mode is better if it is available.

12.5.5 Hamming Codes

Hamming codes allow single errors in a block of data to be corrected without any need for retransmission. Double errors can be detected but not corrected. Extra check bits, called Hamming bits, are added to the block. The number required depends on the length of the block. The minimum number of Hamming bits is given by

$$2^n \geq m + n + 1 \tag{12.2}$$

where

n = number of Hamming bits
m = number of bits in the data block

Obviously, the longer the data block, the less the overhead due to the check bits. On the other hand, the longer the block, the more likely it is to contain more than one error. The Hamming bits can be placed anywhere within the block, provided the location is known to the receiver. In fact, it is better not to place them all together in order to reduce the chances of the Hamming bits themselves being in error. The check bits are generated by a fairly simple process involving a number of exclusive OR operations and are detected the same way.

Example 12.6 How many Hamming bits are required for a block length of 21 message bits?

Solution The correct number of check bits is the smallest number that satisfies Equation (12.2). This number is most easily found by trial and error. Try five check bits as a starting point. The equation becomes

$$2^n \geq m + n + 1$$
$$2^5 \geq 21 + 5 + 1$$
$$32 \geq 27$$

The equation is satisfied, but perhaps a smaller value of n would also satisfy it. Try $n = 4$.

$$2^n \geq m + n + 1$$
$$2^4 \geq 21 + 4 + 1$$
$$16 \geq 26$$

The equation is not satisfied, so five Hamming bits are required.

12.5.6 Cyclic Redundancy Checking

Cyclic redundancy checking (CRC) codes are used in many sophisticated digital systems, such as the compact disc system. This type of error-correcting code uses a form of feedback in which the state of each message bit depends on the state of the previous bits in the block. CRC codes are particularly good at detecting burst errors, in which several consecutive bits are missing or incorrect.

CRC codes are generated by computing a polynomial from the message bits. The whole message is considered as a polynomial and is divided by a fixed polynomial called the generator polynomial, yielding a quotient plus a remainder. The remainder is appended to the message and transmitted as a block check character (BCC). At the receiver, the calculation is repeated. If there are no errors, the remainder (the block check character) should be the same as transmitted. If it is different, an error has occurred.

CRC codes can be implemented either in hardware or software, but hardware is more common.

12.6 Data Compression

So far we have assumed that data is transmitted in the exact form in which it is presented to the transmitter, with the addition, where required, of extra bits for data link control, framing, and error control. However, it is often possible to reduce the number of bits transmitted without losing any of the information. This technique is called data compression.

Two concepts are fundamental to data compression. One is that not all bit patterns are equally probable. In character-based systems, for instance, the letter *e* appears much more often than the letter *q*. It is possible to use fewer bits for the most common patterns, and more for the less common ones. This technique is called **Huffman coding**. A Huffman-coded version of the alphabet uses strings of 4 to 11 bits to encode the characters, instead of using the same number of bits for every character, as in ASCII. Similar considerations can be applied to noncharacter bit patterns, such as the brightness of pixels in facsimile (fax) transmissions.

The second important data-compression technique is called **run-length encoding**. It takes advantage of the fact that information is often repeated. A typical spreadsheet, for instance, may have long strings of spaces and carriage returns. Similarly, an image may have a lot of white space represented by identical pixels. The transmission system can note this redundancy and transmit the character or pixel once, followed by the number of consecutive occurrences. Run-length encoding can achieve a great deal of compression when used with some types of files.

Summary

Here are the main points to remember from this chapter.

1. Data that consists of alphanumeric characters must first be encoded using a character code such as ASCII or Baudot.
2. Asynchronous communication actually involves synchronizing the transmitter and receiver clocks at the start of each character. It is simpler but less efficient than synchronous communication, in which the transmitter and receiver clocks are continuously locked together.
3. Computer data must be converted from parallel to serial form before being transmitted and back to parallel form at the receiver.
4. Since noise is present in all communication systems, errors will occur. Errors can be detected and corrected, within limits, by adding redundant information.
5. Data can be compressed by using shorter codes for patterns that appear more often and by coding repetitive data by indicating the number of times an element repeats.

Important Equations

$$\eta = N_D/N_T \tag{12.1}$$

$$2^n \geq m + n + 1 \tag{12.2}$$

Glossary

automatic request for retransmission (ARQ) an error-control system based on the repetition of data blocks that contain errors

bit overhead bits that do not carry the message, for example, those used for timing and error control

bit stuffing addition of extra bits to a data block to avoid the accidental generation of a flag pattern

character code a set of rules that translates alphanumeric characters into binary numbers

character set see *character code*

cyclic redundancy check (CRC) an error-detecting method in which the binary number corresponding to the group of bits to be checked is divided by a predetermined binary number, and the remainder is transmitted as a check

data code see *character code*

dial-up line a telephone connection via the public switched telephone network

forward error correction (FEC) an error-correcting system in which errors are corrected at the receiver using redundant transmitted data without using retransmission requests

frame a group of bits sent between framing signals in a bit-oriented synchronous communication system

Huffman coding a data-compression scheme that uses fewer bits to represent more frequently occurring characters or bit patterns and more bits to represent those that occur less frequently

mark a line condition corresponding to a binary one

parallel transmission simultaneous transmission of multiple data bits using several channels

run-length encoding a data-compression scheme that replaces repeated characters or bit patterns with a code indicating the character or pattern and the number of repetitions

serial transmission data transmission using only one channel

space a line condition corresponding to a binary zero

start bit in asynchronous communication, this bit alerts the receiver to the beginning of a transmitted character by changing the line from the mark to the space condition

stop bit in asynchronous communication, this bit marks the end of a transmitted character and returns the line to the mark condition

Universal Asynchronous Receiver-Transmitter (UART) a device, generally an integrated circuit, that converts from parallel to serial format when transmitting and from serial to parallel format when receiving; it also adds start, stop, and parity bits and checks for errors

Questions

1. What types of characters are used in data communication?
2. How can the transmission of binary data using a system primarily designed for characters cause confusion?
3. How are lowercase letters handled with the Baudot code?
4. Explain the difference between the seven-bit and eight-bit versions of the ASCII code.
5. Explain the use of start and stop bits in asynchronous transmission. Why are they unnecessary in synchronous systems?
6. Which bit is normally transmitted first using the Baudot and ASCII codes?
7. How is data converted from parallel to serial form and back in a microcomputer-based data communication system?
8. How does a UART determine the point at which a received character begins?
9. Explain the use of a receive buffer in a UART.
10. Why is the choice of line code important in synchronous systems?
11. How is the start of a block of data indicated in character-oriented protocols?
12. How do bit-oriented protocols indicate the start of a frame of data?
13. Compare the ways in which character-oriented and bit-oriented protocols deal with the problem of binary data patterns that resemble control characters.
14. What is the difference between burst and random errors? Which type of error is harder to correct?
15. Explain the difference between FEC and ARQ error-control systems. Why is ARQ unsuitable for simplex communications?
16. Can parity be used to correct errors? Explain your answer.
17. Explain why the combination of VRC and LRC is more effective than either alone.
18. How many bits in a block can be corrected using a Hamming code?
19. What is the difference between a checksum and a CRC code?
20. Which of the data-compression systems we have studied would be more useful for transmitting pages containing mostly blank space, with a few lines of text on each?
21. Which of the data-compression systems we have studied would be more useful for transmitting ordinary, closely spaced text?
22. Describe the use of scrambling and eight-to-fourteen modulation in ensuring reliable clock recovery in a synchronous system.

Problems

SECTION 12.2

23. Suppose that in the TELEX message (using CCITT Baudot code)

    BE HERE TONITE BY 8

    the space after HERE was changed by a noise pulse to a FIGS character. What would the received message read?
    (a) without USOS
    (b) with USOS
24. The message RYRYRYRYRYRYRYRYRY is often sent to test and set up radioteletype circuits. Look at the Baudot code table and suggest a reason why.
25. Encode the message HELLO THERE in Baudot.
26. Decode the message below, assuming it has been sent in ASCII with one start bit, seven data bits, one stop bit, and even parity. The bits are printed, from left to right, in the order in which they would be transmitted.

    1111001000010111111110100101
    1010001011101111111111111111

27. Encode the message *Hello There!* in ASCII, using one start bit, eight data bits, one stop bit, and no parity. Write the bits in the order in which they would be transmitted, from left to right.
28. Encode the message $R5 = 47\ \Omega$ in ASCII, using one start bit, eight data bits, one stop bit, and odd parity. Use the IBM extended ASCII set given in Table 12.4. Write the bits in the order in which they would be transmitted, from left to right.

SECTION 12.3

29. An asynchronous communication system uses ASCII at 9600 bits/sec., with eight data bits, one start bit, one stop bit, and no parity bits.
    (a) Explain what is meant by efficiency, and calculate the maximum percent efficiency in this system.
    (b) Express the data rate in characters per second and in words per minute (assume a word has five letters and one space).
    (c) Suppose a synchronous system with the same bit rate is used instead. Would you expect the number of characters per second to increase or decrease? Why?
30. A data-communication system operates at 28.8 kb/s (typical of current high-speed telephone modems). How long does the receiving UART have to transfer an incoming character to a buffer or to the main computer before the next character arrives?

SECTION 12.4

31. Suppose a data stream consists of a long sequence of ones. Sketch the corresponding signal using each of the following line codes, and state which ones would be suitable for synchronous transmission.
    (a) unipolar NRZ
    (b) bipolar NRZ
    (c) AMI
    (d) Manchester
32. Repeat problem 31 for a string of zeros.
33. Sketch the format of a block of data to be transmitted using BISYNC.
34. Sketch the format of a frame transmitted using SDLC.
35. Suppose part of a binary data block transmitted using ASCII consists of the bits 0000111 (shown MSB first).
    (a) What control character does this represent?
    (b) How would a character-oriented protocol tell the receiver that this is data and not a control character?
    (c) How would a bit-oriented protocol handle the situation?
36. Suppose part of a frame transmitted using a bit-oriented protocol happens to consist of the bits (shown from left to right in the order in which they are transmitted)

    1010110111111101101011
    00011010111010000111010

    (a) Identify the problem with this data.
    (b) How would the transmitter fix the problem? (Write down the sequence as it would be transmitted.)
    (c) What would the receiver do to the received bit sequence?

SECTION 12.5

37. (a) Calculate the VRC and LRC bits for the block of data below, assuming even parity for both. Indicate which are VRC and which are LRC bits.

0	1	1	1	0	1	0	1	1	0
1	0	1	0	1	0	1	0	1	0
0	1	0	1	0	1	0	1	0	1
1	1	1	1	0	0	0	0	1	0
1	1	0	0	1	1	0	0	1	1
0	1	0	1	0	1	0	1	0	1

38. Assume the LRC uses even parity and VRC uses odd parity in the data block below. There is a one-bit error in the block.

Characters				
1	2	3	4	
0	1	0	1	0
0	0	0	0	0
1	0	1	0	0
0	1	0	0	0
0	0	1	0	1
0	0	0	0	0
1	1	1	1	0
1	1	0	1	0

(a) Identify the LRC and VRC.
(b) Circle the erroneous bit.

39. A receiver receives the following partial transmission in ARQ code in FEC mode (from left to right). The transmission includes a group of three characters sent twice in succession. Identify the errors and decode the message.

YBBBBYYYBBYBYBBBYYYBY
YBBBYYYYBBYBYBBBYBYBY

40. How many Hamming bits would have to be added to a data block containing 128 bits?

13 Modems

Objectives After studying this chapter, you should be able to:

1. Explain the need for modems and give examples of situations where modems are required
2. Describe several modulation schemes for transmitting data on analog carriers and calculate the baud and bit rates for different schemes
3. Use RS-232 cables and connectors to connect modems with computers
4. Describe and compare several methods of error correction and data compression that are used with modems
5. Describe the operation of "intelligent" modems and configure them using the Hayes AT command language
6. Set up a connecton between two computers using modems
7. Compare several file-transmission protocols and choose the best one for a given situation
8. Explain the functions of communications software used with microcomputers
9. Configure and use fax modems to send and receive facsimile (fax) transmissions

13.1 Introduction

It is often necessary to transmit digital data over an analog channel. Some communication methods (transmission via radio, for example) require analog carriers, while other media, such as coaxial cable and optical fiber, can carry either analog FDM or digital TDM signals. Digital transmission is preferred when it is practical, but if a system has already been constructed using analog techniques, it is simpler and less expensive to convert data signals to analog form than to rebuild the whole system. This is especially true of the dial-up telephone system, which was set up for analog voice signals. Although many telephone central offices and long-distance trunk lines are now digital, nearly all the individual telephone lines, known as local loops, are still analog.

Analog channels have limited bandwidth, which prevents the transmission of short pulses. Often, as in the telephone system, neither dc nor high-frequency ac is allowed. Even over the limited frequency range of the channel, the amplitude response may vary with frequency. For example, see the frequency response of a typical voice-grade telephone channel, shown in Figure 13.1. The phase response may also be nonlinear.

All channels are subject to noise of various types. Thermal noise has already been discussed, and noise can also originate from switching, lightning, and so forth.

Channel characteristics may vary with time. For example, fading due to rain will reduce the signal-to-noise ratio of a microwave radio link. Dial-up telephone connections may take different routes each time a call is placed between the same two locations. High-frequency radio links are very susceptible to fading as well as to noise and interference.

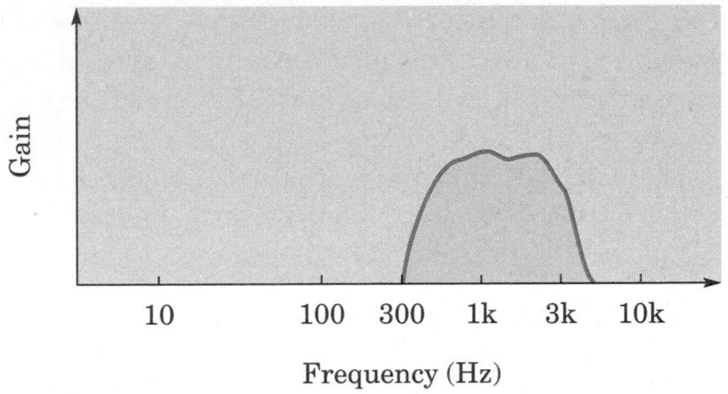

Figure 13.1
Voice-grade telephone channel

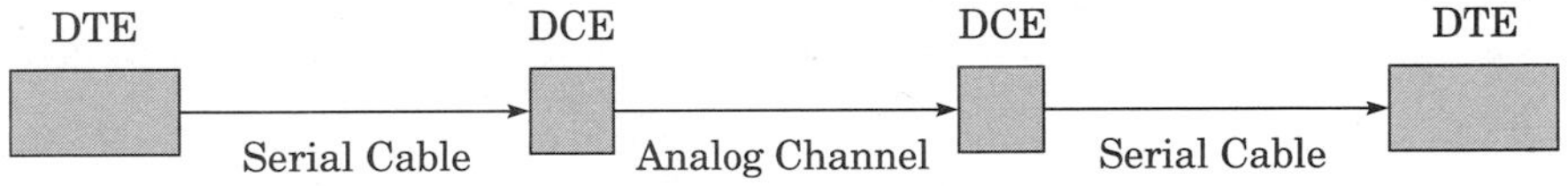

Figure 13.2
Data communication using modems

For transmission via analog channels, the digital information signal is used to modulate a sine-wave carrier. Demodulation at the other end restores the original baseband information. A **modem** is a single unit that contains both a *mo*dulator and a *dem*odulator.

Figure 13.2 shows a data communication setup using modems. The abbreviation **DTE** refers to **data terminal equipment** and **DCE** means **data communications equipment**. Normally a DTE is a terminal, a computer, or a teleprinter. DCE is another term for a modem.

13.2 Modulation Techniques

Whether the modulating signal is analog or digital, there are only three carrier parameters that can be varied: amplitude, frequency, and phase. In practice, frequency modulation, which is known as **frequency-shift keying (FSK)** when used with digital signals, is generally used at low data rates. Phase modulation (**phase-shift keying** or **PSK**) is used at moderate data rates, and **quadrature amplitude modulation (QAM)**, which is a combination of amplitude and phase modulation, is used in high-speed modems. The terms low, moderate, and high are relative, since the maximum possible data rate depends on the channel bandwidth and signal-to-noise ratio as well as on the modulation scheme.

Digital signals result in discrete, rather than continuous, changes in the modulated signal. The receiver examines the signal at specified times, and the state of the signal at each such time is called a **symbol**. The eventual output will be binary, but it is certainly possible to use more than two states for the transmitted symbol. Complex schemes using several levels can send more data in a given bandwidth. For example, four bits could be sent at once using sixteen combinations of amplitude and phase.

As we discussed in Chapter 11, there is a theoretical limit to the maximum data rate that can be transmitted in a channel with a given bandwidth and signal-to-noise ratio. These limits apply to communication via modem, just as they do to direct digital transmission. For convenience, we will repeat two important equations here. First, the Shannon-Hartley theorem:

$$C = 2B \log_2 M \qquad (13.1)$$

where

C = information capacity in bits per second
B = channel bandwidth in hertz
M = number of levels transmitted

The number of levels transmitted, in modem communications, corresponds to the number of different possible states that one signaling element (symbol) can have. For instance, if a modem can transmit either of two frequencies, $M = 2$. A more elaborate modem using four different phase angles, each of which can have two different amplitudes, has $M = 8$.

Unfortunately, the information capacity of a channel cannot be increased without limit by making the modem more complex, because noise makes it difficult to distinguish between signal states. The ultimate limit is called the Shannon limit:

$$C = B \log_2(1 + S/N) \tag{13.2}$$

where

C = information capacity in bits per second
B = bandwidth in hertz
S/N = signal-to-noise ratio (as a power ratio, not in decibels)

At this point, and before we begin any discussion of the speed at which information is transmitted by modems, we should distinguish between **bit rate** and **baud rate**. The bit rate is simply the number of bits transmitted per second, while the baud rate is the number of symbols. Therefore

$$\begin{aligned} \text{bit rate} &= \text{baud rate} \times \text{bits per symbol} \\ &= \text{baud rate} \times \log_2 M \end{aligned} \tag{13.3}$$

where

M = number of possible states per symbol

If each symbol has only two possibilities and therefore carries one bit of information, then the baud rate and the bit rate are equal. Confusion arises, however, when using more complex schemes that allow several bits to be transmitted at once. With such systems the bit rate is higher than the baud rate, but the two terms are often used interchangeably to represent the bit rate.

Example 13.1 A modem transmits symbols, each of which has 64 different possibilities, 10,000 times per second. Calculate the baud rate and bit rate.

Solution The baud rate is simply the symbol rate, or 10 kbaud. The bit rate is given by Equation (13.3):

$$\begin{aligned} \text{bit rate} &= \text{baud rate} \times \log_2 M \\ &= 10 \times 10^3 \times \log_2 64 \\ &= 10 \times 10^3 \times 6 \\ &= 60 \times 10^3 \text{ b/s} \\ &= 60 \text{ kb/s} \end{aligned}$$

13.2.1 Frequency-Shift Keying Modems

Frequency-shift keying (FSK) is a simple and reliable modulation scheme that is commonly used for low-data-rate applications. A carrier is switched between two frequencies, one defined as the mark frequency and the other as the space frequency. For full-duplex communication there are four frequencies, with two transmitted in each direction.

We will learn about FSK modems by studying some practical examples. First let us briefly consider the Bell 103 standard, intended for use with dial-up telephone lines. The frequencies used are shown in Figure 13.3. This standard allows full-duplex operation at 300 bits per second using frequency-division multiplexing. There are two pairs of mark and space frequencies. The modem that places the call is called the *originate modem*; it transmits with a mark frequency of 1270 Hz and a space frequency of 1070 Hz. The other modem is called the *answer modem*; it uses a mark frequency of 2225 Hz and a space

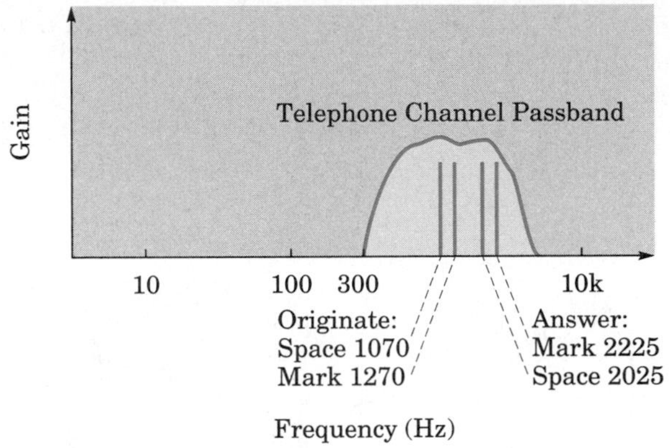

Figure 13.3
Bell 103 modem

frequency of 2025 Hz. All of these frequencies are well inside the telephone passband and sufficiently removed from each other to allow room for the sidebands that are generated by modulation.

The Bell 103 standard is reliable but slow and is found on current modems only as a **fallback** system for use with very noisy lines. Another FSK standard for use with the telephone system is known as the Bell 202 modem. This system, shown in Figure 13.4, allows a higher bit rate, but it is essentially a half-duplex standard. The mark frequency is 1200 Hz, and the space frequency is 2200 Hz. The use of only two more widely separated frequencies allows a data rate of 1200 bits per second. In addition there is provision for a slow-speed (5 bits per second) return channel using a carrier at 387 Hz that can be keyed on and off to acknowledge transmissions or request a retransmission. The Bell 202 standard is used for the transmission of call-display information to modern telephones.

FSK is also used for digital communication via both HF and VHF radio. Here the carrier frequency is much higher, of course, and depends on the frequency allotted to the station by government regulation. Digital VHF radio systems use two audio tones, one each for mark and space, to modulate a conventional FM transmitter. An FM receiver produces the two audio tones at its output, and these are sent to another modem for demodulation. The technique is called **audio frequency-shift keying** or **AFSK**. There is an extensive amateur digital packet radio network that communicates mainly with modems using the Bell 202 standard. See Figure 13.5 for a block diagram and Figure 13.6 for a typical VHF radio modem, known as a *terminal node controller (TNC)*.

High-frequency digital radio uses a different technique. Rather than defining the actual mark and space frequencies as is done in telephone modems, these systems specify the

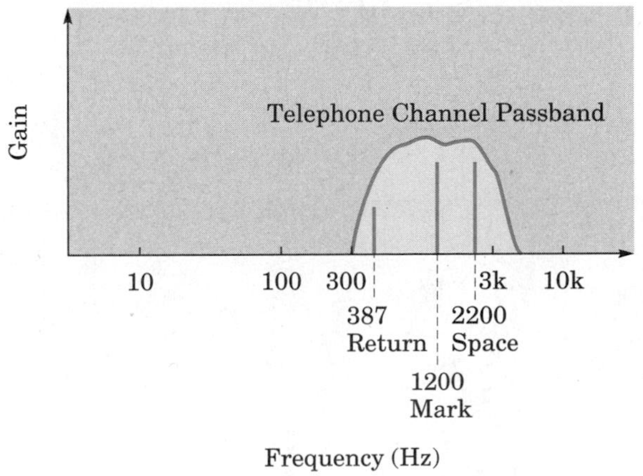

Figure 13.4
Bell 202 modem

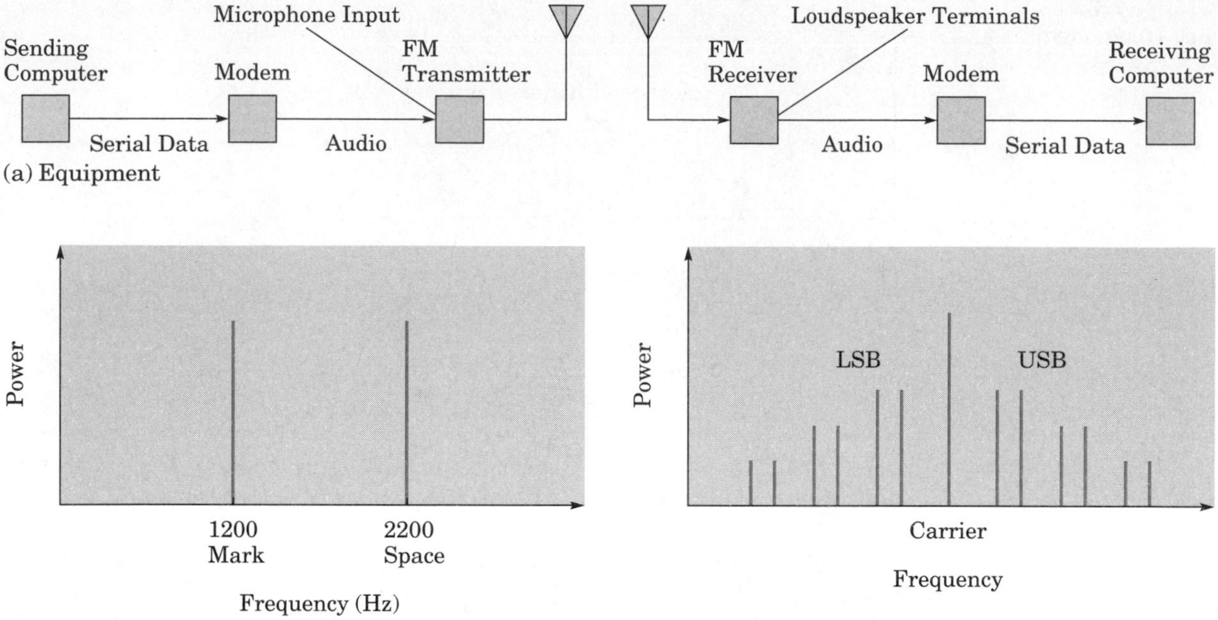

Audio spectrum

Radio spectrum

(b) Typical spectra (this example is for an amateur packet radio station)

Figure 13.5
Radio AFSK transmission using FM equipment

Figure 13.6
Terminal node controller

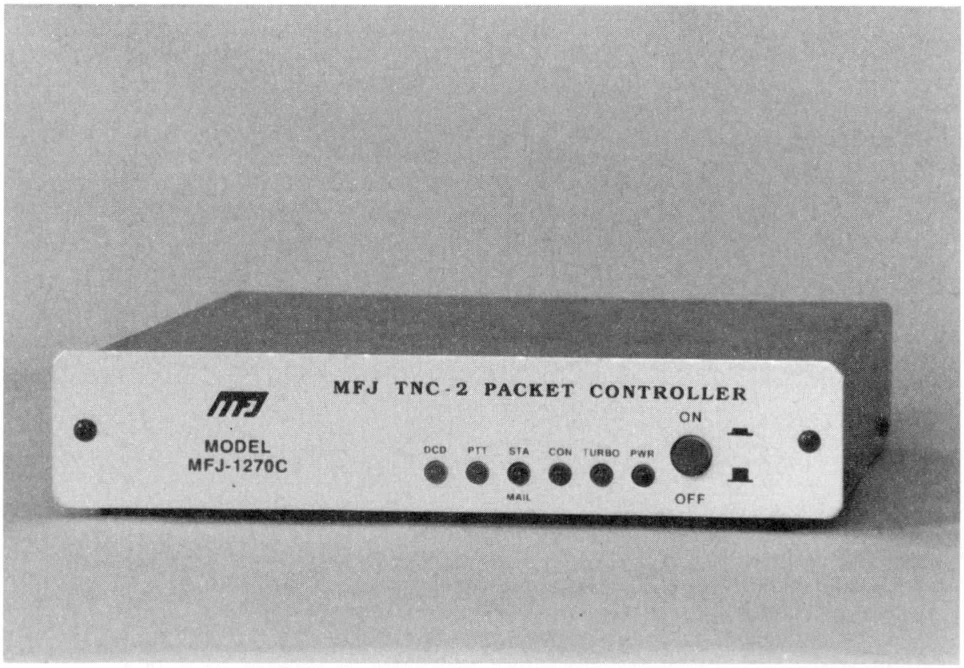

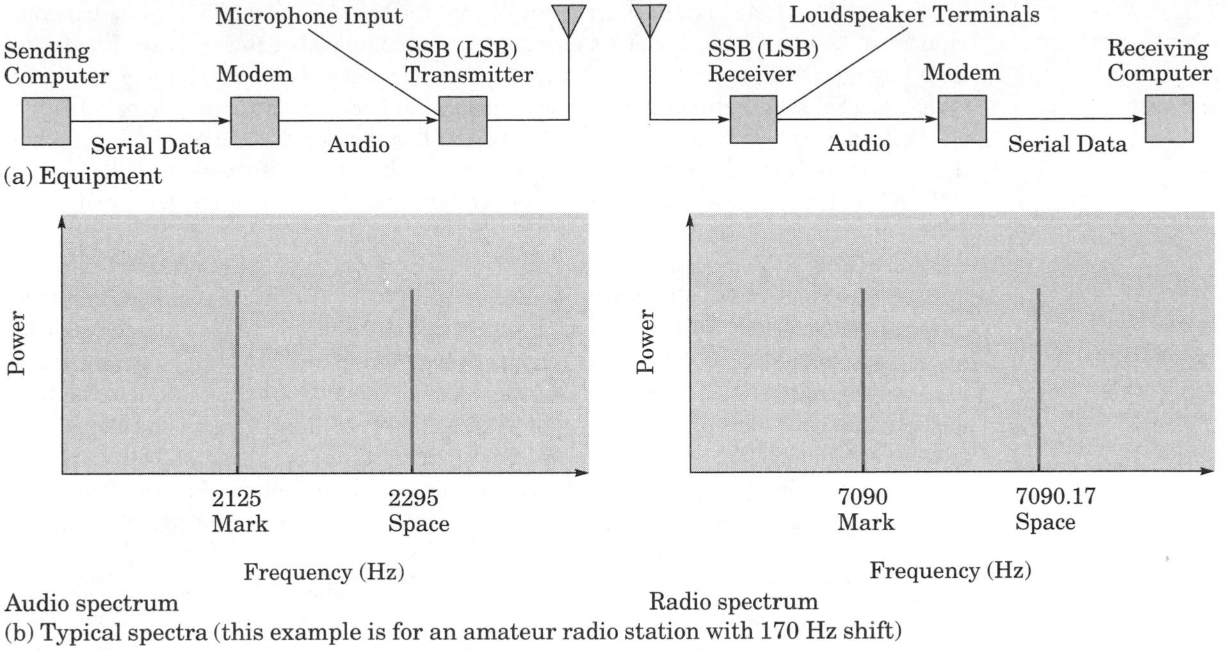

Figure 13.7
Radio FSK transmission using SSB equipment

frequency shift between mark and space. There are many possible shifts, with 170, 425, 800, and 850 Hz being the most common. A single-sideband (SSB) AM transmitter is modulated by an audio tone that switches between two frequencies differing by the required separation. Recall that when a single-sideband transmitter is modulated by a single frequency, its output is a sine wave at one frequency. For an upper-sideband signal, the output frequency is the sum of the (suppressed) carrier frequency and the modulating frequency, and for a lower-sideband signal, it is the difference. Thus, switching the modulation between the two tones switches the transmitter output between two RF frequencies that differ by the same amount.

Figure 13.7 shows how SSB equipment can be used to produce FSK. At the transmitting station, a modem produces two audio frequencies, one for mark and one for space, that differ by the required frequency shift. These tones modulate the SSB transmitter, whose output consists of a sine wave that shifts between two frequencies. For example, 2125 and 2295 Hz can be used to give a shift of 170 Hz. The receiver is also set up for SSB operation, and it produces two audio tones at its output. They may not be the same two tones as at the transmitter input, depending on the exact tuning of the receiver and its beat frequency oscillator (BFO), but they will differ in frequency by the same amount.

Example 13.2 A lower-sideband SSB transmitter has a suppressed carrier frequency of 14.08 MHz. It is modulated by frequencies of 1275 Hz (mark) and 1445 Hz (space). Calculate the frequency shift and the mark and space frequencies as transmitted.

Solution The frequency shift is

$$1445\text{ Hz} - 1275\text{ Hz} = 170\text{ Hz}$$

The transmitted mark frequency is

$$14.08\text{ MHz} - 1275\text{ Hz} = 14.078725\text{ MHz}$$

The transmitted space frequency is

$$14.08\text{ MHz} - 1445\text{ Hz} = 14.078555\text{ MHz}$$

13.2.2 Phase-Shift Keying Modems

When somewhat faster data rates are required than can be achieved with FSK, phase modulation is often used. Actually the system used is more completely described as **delta phase-shift keying (DPSK)** because what is measured at the receiver is not the phase of the signal with respect to some permanent reference (which would be awkward to maintain) but rather any shift in the phase of the signal with respect to that measured at the previous sampling time. Most DPSK modems use a four-phase system called **quadrature phase-shift keying (QPSK)**. In QPSK, each symbol represents two bits, and the bit rate is twice the baud rate. This is often called a **dibit system**.

The Bell 212A modem is an example of a system of this type. Its maximum data rate is 1200 bits per second, full duplex. Only a single carrier frequency is needed for each direction with a DPSK modem, and the Bell 212A standard uses 1200 Hz for the originating modem and 2400 Hz for the answering modem. Since four different phase shifts are used, this is a dibit system and 1200 bits per second is actually 600 baud. Modems built to this standard have fallback capability, that is, when line conditions are not good enough for reliable communication at 1200 bits per second, they can reduce speed or "fall back" to 300 bits per second using the Bell 103 standard. The vector diagram in Figure 13.8 shows how the four possible phase shifts from one symbol to the next are translated into four two-bit combinations.

Another DPSK system is known as the CCITT V.22 standard. The CCITT (International Telephone and Telegraph Consultative Committee) was an international standards group that is now officially the Telecommunications Standards Division of the International Telecommunications Union (ITU). It works under the auspices of the United Nations, but it is often referred to by its old name. V.22 modems also operate at 1200 bits per second, using a dibit system and the same carrier frequencies as for the Bell 212A. The difference is that V.22 modems fall back to 600 b/s using DPSK with only two phase angles.

13.2.3 Quadrature Amplitude Modulation Modems

The only way to achieve high data rates with a narrowband channel, such as a dial-up telephone line, is to increase the number of bits per symbol, and the most reliable way to do that is to use a combination of amplitude and phase modulation known as *quadrature amplitude modulation (QAM)*. For a given system, there is a finite number of allowable amplitude-phase combinations. Figure 13.9 is a **constellation diagram** that shows the pos-

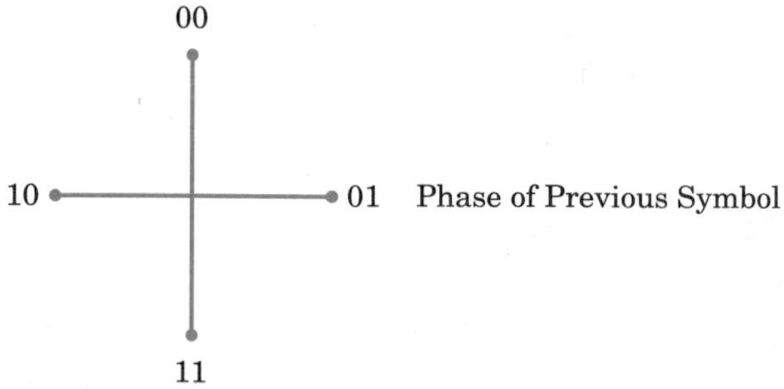

Figure 13.8
DPSK modem vector diagram

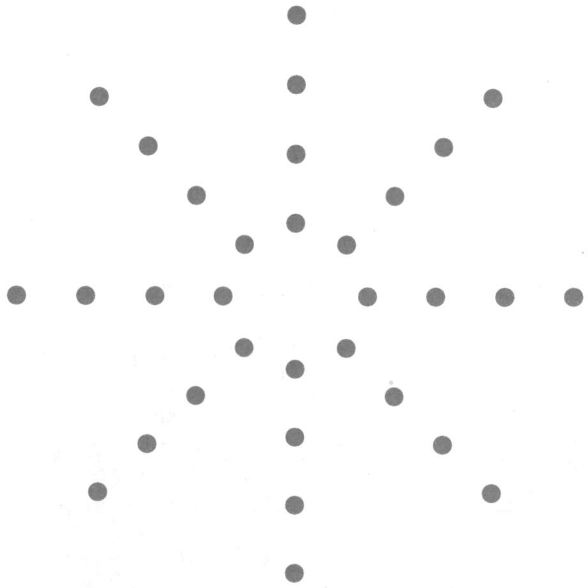

Figure 13.9
Constellation diagram

sibilities for a hypothetical system with eight phase angles and four amplitudes, for a total of 32 combinations. Thus each transmitted symbol represents five bits. This diagram is similar to the previous figure (the vectors have not been drawn here). Each circle represents a possible amplitude-phase combination or state. With a noiseless channel, the number of combinations could be increased indefinitely, but a practical limit is reached when the difference between adjacent states becomes too small to be detected reliably in the presence of noise and distortion.

Example 13.3 A modem uses 16 different phase angles and 4 different amplitudes. How many bits does it transmit for each symbol?

Solution The number of possible states per symbol is $16 \times 4 = 64$. The number of bits per symbol is $\log_2 64 = 6$.

Once modem speeds exceed 1200 b/s over dial-up lines, QAM is the modulation scheme of choice. The CCITT V.22 *bis* (*bis* is French for "again") operates at 2400 b/s using QAM and is still in common use. Another example of a QAM modem for telephone use is the CCITT V.32 standard. This system operates at 2400 baud and varies the number of bits per symbol according to the line conditions. There are three possibilities:

- 9600 b/s using 2 amplitudes and 8 angles, for 16 states or 4 bits per symbol
- 7200 b/s using 1 amplitude and 8 angles, for 8 states or 3 bits per symbol
- 4800 b/s using 1 amplitude and 4 angles, for 4 states or 2 bits per symbol

Both originate and answer modems occupy approximately the same bandwidth, and each modem incorporates circuitry to cancel its own transmitter signal at its receiver, in a manner similar to that of the hybrid transformer in an ordinary telephone. There is also a half-duplex, 9600 b/s standard known as V.29.

In 1991 the V.32 *bis* standard was promulgated by the CCITT. It allows for full-duplex data transfer at 14.4 kb/s, with fallback to the earlier standards. Finally (as of this writing), the V.34 standard (1993) allows 28.8 kb/s. This speed is getting quite close to the theoretical limit for a dial-up telephone line, which, as discussed earlier, depends on both the bandwidth and the signal-to-noise ratio of the channel.

Example 13.4 A typical dial-up telephone connection has a bandwidth of 3 kHz and a signal-to-noise ratio of 30 dB. Calculate the Shannon limit.

Solution The signal-to-noise ratio, given in decibels, must be converted to a power ratio:

$$\begin{aligned} S/N &= \text{antilog } (30/10) \\ &= 1000 \end{aligned}$$

From Equation (13.2)

$$\begin{aligned} C &= B \log_2 (1 + S/N) \\ &= 3 \times 10^3 \times \log_2 1001 \\ &= 29.9 \text{ kb/s} \end{aligned}$$

In order to communicate at rates near the Shannon limit, modems must apply equalization to the line to compensate for its uneven phase and frequency response. The transmitter applies an equalization designed for average lines, and the receiver uses variable equalization. The receiver equalization is adjusted automatically at the start of communication, by means of a series of bits called a **training sequence**.

13.3 Error Correction and Data Compression with Modems

High-speed modems often incorporate error correction and data compression. Error correction becomes more important as speeds increase, because states in the constellation pattern are closer together at higher speeds and thus the possibility of errors due to line noise increases. Data compression allows a modem to achieve data rates greater than would normally be possible over telephone lines by eliminating redundant data. Both error correction and data compression can be (and often are) implemented in software that is not part of the modem. There are, however, some types of error correction and data compression that are commonly built into modems, especially high-speed modems, and they will be considered here.

The error-correction protocols commonly used in modems include several developed by modem manufacturer Microcom, Inc., and known as MNP2, MNP3, MNP4, and MNP10. MNP10 is used mainly with modems that operate over very noisy channels, such as cellular telephone connections, which are subject to signal interruptions as a vehicle moves from one cell to another. In addition, there is now a CCITT standard known as V.42, which includes the MNP4 protocol as well as one called Link Access Procedure for Modems (LAPM). All of these protocols assemble the data into blocks called packets and apply a cyclic redundancy check (CRC) to each packet. For proper operation, the same error-control scheme must be in use at both ends of the link.

The data-compression schemes built into modems are limited by the fact that they must operate on data in real time, as it passes through the modem. They do not get the chance to examine a complete file to determine the best way to compress it, as do archiving programs such as the popular PKZIP program. On the other hand, archiving programs normally operate only on data files, while the built-in compression found in modems works with whatever is sent through the modem.

Currently the most popular data-compression schemes for modems are MNP5 from Microcom and CCITT standard V.42 *bis* (not to be confused with V.42, which, as mentioned above, is an error-correction protocol). Both data-compression schemes use a combination of Huffmann and run-length encoding. Of the two, V.42 *bis* is preferred. MNP5 tends to slow the transfer of any file that has already been compressed by other software, as it searches unsuccessfully for ways to compress it further.

Though most modems that are capable of error correction and data compression perform these operations internally, there are some, especially low-cost high-speed modems, that rely on software running on the associated computer to perform these operations. Such modems use a Rockwell chip set that supports the Rockwell Protocol Interface. They can be identified by sending the command ATI3 to the modem—it will then identify itself as an RPI or RPI+ (a later version) modem. Software to work with these modems is built into some (but not all) communications packages. A Microsoft Windows software driver for these modems, called WinRPI, can be obtained free of charge from Rockwell at their World Wide Web site, at this address:

http://www.nb.rockwell.com

13.4 The RS-232C Serial Communications Standard

The previous chapter described the role of a UART in interfacing between the parallel data of the computer and the serial data necessary for communication. Normally the UART is in the computer or terminal. The modem may also be inside the computer, but often it is a separate device located nearby. Figure 13.10 shows examples of internal and external modems. Serial communication between the computer and the modem involves data and also, in the case of the "smart" modems to be described shortly, commands and responses. There may be several other connections between computer and modem, to indicate to each the status of the other and to allow the computer to control the modem.

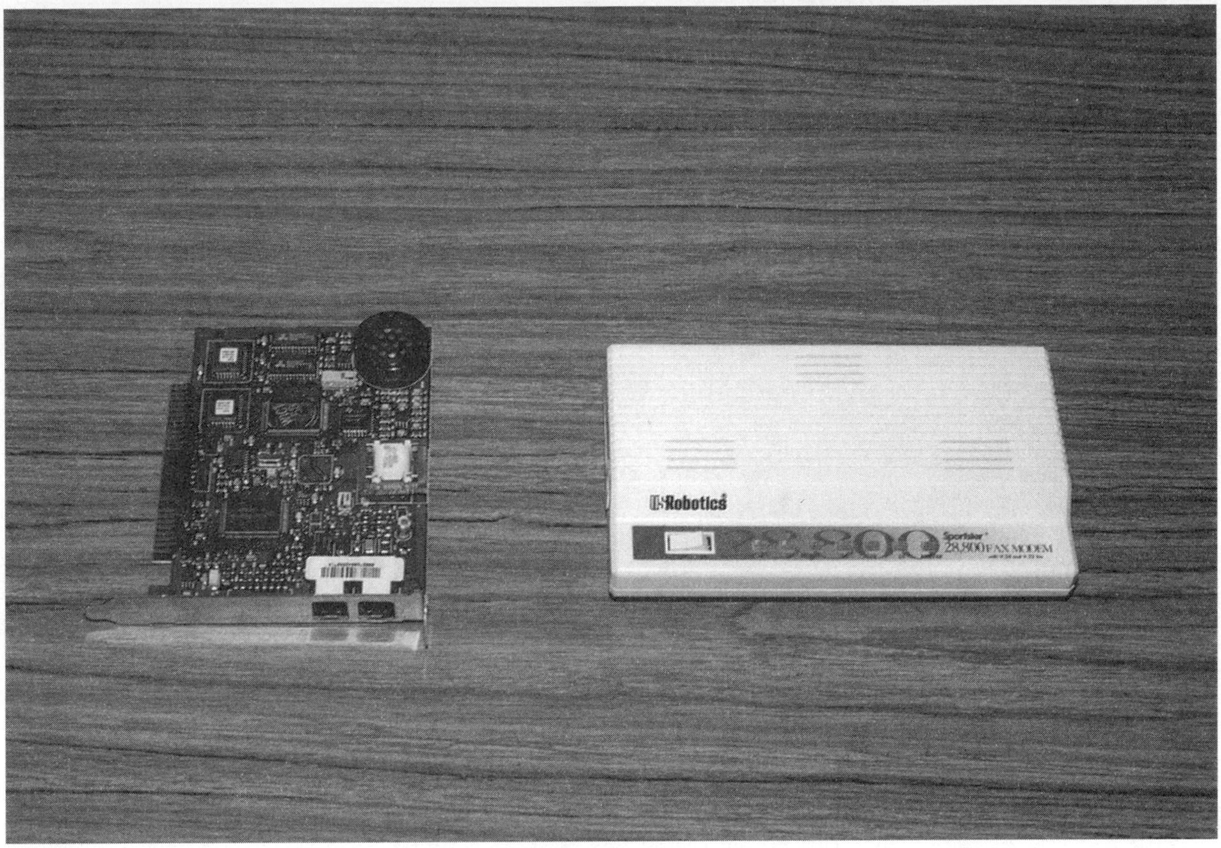

Figure 13.10
Internal and external modems

There are a number of standards for communication between the computer (DTE) and the modem (DCE), but the most popular is the Electronics Industries Association (EIA) standard popularly known as RS-232C, though its official name is now EIA 232D. The standard defines voltage levels and even connector pin numbers for both data and control lines. The *serial port* found on most personal computers follows this standard. Not all of the lines are used in all installations, and the port can also be used for nonmodem applications, for example, the connection of printers, plotters, and computer mice. The uses (and misuses) of the RS-232C standard could fill a book, so only the basics will be described here.

Table 13.1 shows the lines defined by RS-232C. Two different types of connectors are commonly used. Some computers use the 25-pin DB-25 connector, and some use a 9-pin DB-9 connector to save space and/or cost. (The author's no-name 80486 computer has one of each.) Obviously the 9-pin connector has room for only a subset of the RS-232C lines, but these are the only ones needed for asynchronous communication with microcomputers. See Figure 13.11 for the connectors.

The EIA has official letter designations for each line, but they are hard to remember and there are simple mnemonic designations for the most-often used lines. These are unofficial but reasonably standardized and are shown in the "common designation" column in Table 13.1.

The RS-232C interface has lines for data in each direction: transmit data (TD) and receive data (RD) describe the process as seen by the DTE. That is, the TD line carries data from the computer to the modem, and RD carries it from the modem to the computer. There are provisions for a second data channel using the secondary data lines, but these are seldom used.

Table 13.1
RS-232C connections

Connector pins		Designation		
9-Pin	25-Pin	EIA	Common	Function
	1	AA	GND	protective (chassis) ground
3	2	BA	TD	transmit data
2	3	BB	RD	receive data
7	4	CA	RTS	request to send
8	5	CB	CTS	clear to send
6	6	CC	DSR	data set ready
5	7	AB	SG	signal ground
1	8	CF	DCD	data carrier detect
	9			reserved
	10			reserved
	11			unassigned
	12	SCF		secondary received line signal detect
	13	SCB		secondary clear to send
	14	SBA		secondary transmit data
	15	DB		transmitter signal element timing (dce)
	16	SBB		secondary received data
	17	DD		receiver signal element timing
	18			unassigned
	19	SCA		secondary request to send
4	20	CD	DTR	data terminal ready
	21	CG	SQ	signal quality detector
9	22	CE	RI	ring indicator
	23	CH		data signal rate selector (DCE)
	24	DA		transmitter signal element timing (DTE)
	25			unassigned

Figure 13.11
DB-9 and DB-25 connectors

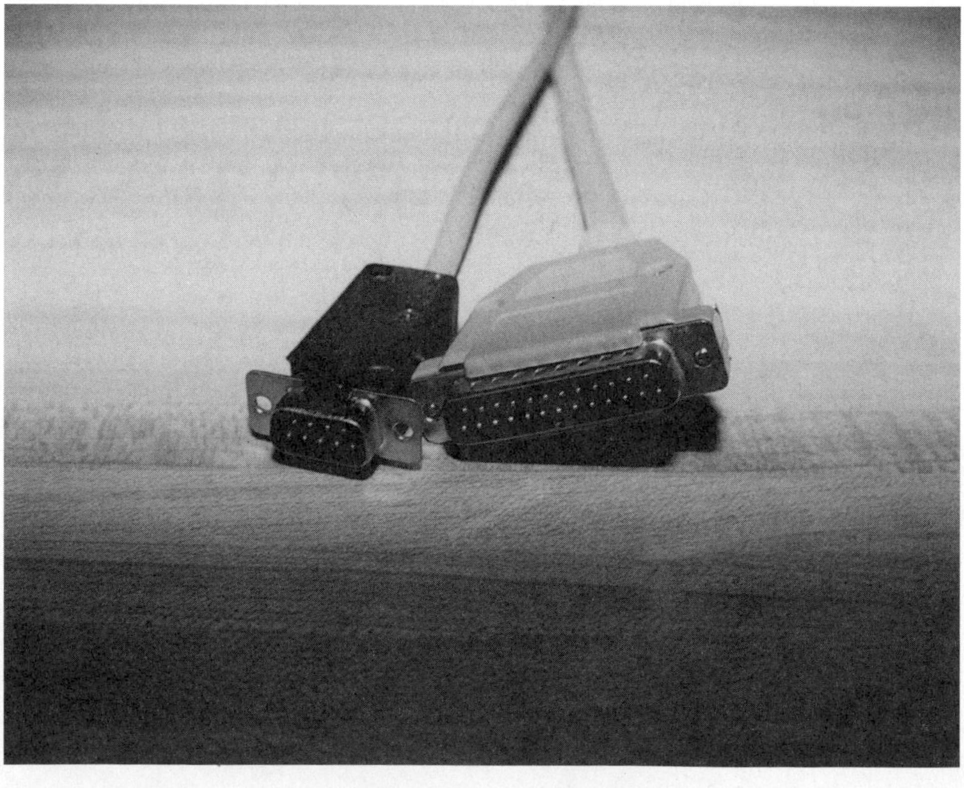

The RTS and CTS lines provide "handshaking" between the DTE and the DCE. That is, the modem uses CTS (clear to send) to signal the DTE that it is ready to transmit data. Normally this occurs after a connection has been made with a modem at the other end. RTS (ready to send) is a signal from the DTE that it has information to transmit. This type of signaling is called **flow control**. When flow control is accomplished by changing levels on the RS-232 lines, it is known as *hardware flow control*. With an intelligent modem, the CTS function can be accomplished by sending control signals (XON and XOFF in ASCII) along the data lines—this is known as *software flow control*. Of the two types, hardware flow control is preferred with high-speed modems, since it is faster. Of course, software flow control is required on the path between the two modems, because there are no extra lines between the modems to facilitate hardware flow control.

The DSR and DTR functions are related to RTS and CTS. DSR (data set ready) indicates the DCE (modem) is turned on, but it does not imply that a telephone connection has been made. Similarly, DTR (data terminal ready) indicates that the terminal is on and ready for communications.

The modem uses the DCD (data carrier detect) line to signal that the analog carrier from another modem is being received, and the RI (ring indicator) line is one way the modem can signal that the phone is ringing. Another way, used by intelligent modems, is to send the word "ring" along the data lines in command mode.

Thus it may seem that many of the features of RS-232C are redundant when an intelligent modem is used. This is true—it follows from the fact that the standard dates from 1966, long before modems with built-in microprocessors and memory were available. To say that confusion reigns in the world of serial interfacing with RS-232 is an understatement.

The RS-232C standard specifies voltage levels and logic type. For all lines, a *high* at the transmitter is in the range of $+5$ V to $+15$ V and a *low* is -5 V to -15 V; the receiver should count anything above $+3$ V as high and below -3 V as low. The voltage levels for data are in negative logic, that is, a positive voltage on a data signal line indicates a zero (space), while a negative voltage indicates a one (mark). Positive logic is used for the control lines, that is, a line is *asserted* (active) when it carries a positive voltage.

The official standard allows a maximum data rate of 20 kb per second over lines with not more than 2500 pF of capacitance. Thus its use is limited, with typical cable, to distances of approximately 15 m or less. The problem with line capacitance is that it increases the rise and fall time of pulses on the line, causing them to merge together. In practice, the RS-232C standard can be used over longer lines, especially if the data rate is reduced. It can also be used with higher data rates, provided the distance between DTE and DCE is short.

Since one or more RS-232C ports are provided on most personal computers, the interface is often used for purposes other than connection to a modem. For instance, data can be transferred between two computers by connecting their serial ports together directly. In that case, the TD and RD lines must be cross-connected in the connecting cable, as shown in Figure 13.12, so that information transmitted by one device will be received by

Figure 13.12
A typical null modem cable

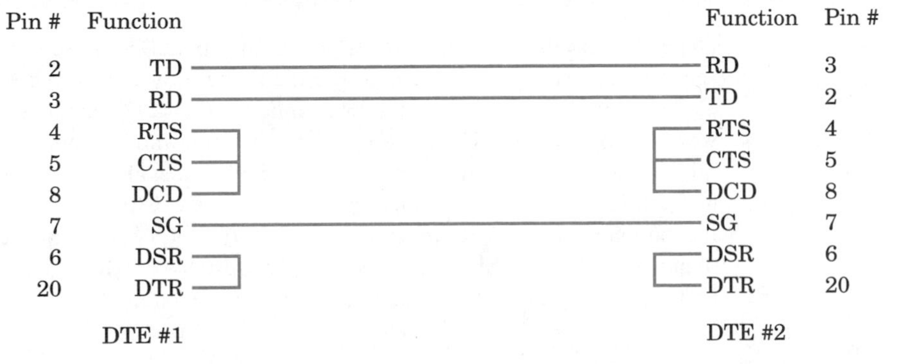

the other. Depending on the software in use, it may also be necessary to "fool" each computer into perceiving that a modem is connected and ready by connecting DTR to DSR and RTS to CTS and DCD. A cable designed for direct connection without a modem is called a **null modem** cable. In addition to these internal connections, a null modem cable must have a female connector on each end; a cable for use with a modem needs a female connector on the computer end and a male connector on the modem end.

RS-232C serial ports are frequently used to connect a serial printer or plotter or a mouse. The manufacturer generally provides specifications for the connecting cable, and a program called a *driver* is needed to ensure operation. Some of the control lines may be used for such functions as reporting whether the printer is ready or out of paper, whether its buffer is full, and so on. This type of application is really outside the scope of our discussion; the point to remember is that an RS-232C connection can be reconfigured by software to perform a variety of functions not envisioned by the original standard.

13.5 Establishing a Connection

The steps required to establish digital communication over an analog link such as a standard telephone line vary considerably depending on the type of modem and the type of connection. Some high-speed modems, for example, are permanently connected to leased lines. In this section we will look at the procedure for establishing digital communication between two computers over an ordinary dial-up telephone line (for example, as in connecting to a computer bulletin board). We will assume that asynchronous, full-duplex communication is being used, since that is the most common technique for microcomputer communications.

Full-duplex communication using a single line requires frequency-division multiplexing (FDM)—that is, the carrier frequency or frequencies vary depending on whether the modem is set to originate or answer the call. Formerly, the call would be dialed by hand and answered by a human, and then the modems would be connected, but current practice is to have a modem dial the phone. The modem that dials should be set to the originate mode. Almost all current modems are "smart," that is, they are programmable to a certain extent, using a language that will be described shortly.

The answer modem detects the ringing signal on the telephone line, and if it has been programmed to do so, it answers the phone. Alternatively, it can indicate to the computer that the phone is ringing and then await instructions as to whether to answer. Once it has taken the phone line off-hook, it sends a carrier down the line to indicate to the originating modem that the phone has been answered. The originating modem also sends a carrier, and both modems indicate to their associated terminals that a connection has been made. For high-speed modems and those with error correction and/or data compression, there must be more communication between the modems to establish common protocols and to send a training sequence that allows the receiver equalizers to adjust themselves for best results with the particular channel being used.

13.6 "Smart" Modems

Depending on the modem, various operations, including those discussed in the preceding section, are usually programmed into the modem by commands sent from the computer along the same serial lines used for communications. The language normally used for this was developed by the Hayes company for its Smartmodem but has been adopted by most other modem manufacturers. It is supported by almost all the software that has been written to facilitate modem communications. Sometimes it is called the "AT" language because its commands begin with those letters. A partial list of some of the more important commands is given in Table 13.2. Modems generally have a default setup that can be changed by using these commands, and sometimes they can store one or more user profiles in non-

Table 13.2. "AT" modem command language: sample commands

Command	Function
AT	attention: starts all command strings.
A	answer: modem answers call.
D	dial phone: followed by T (tone) or P (pulse) to indicate type of dialing, then the number to be dialed.
E	command echo: followed by 0 to inhibit, 1 to enable, echo of commands back to computer.
H	hang up phone, or go off hook if followed by 1.
L	set speaker volume: followed by 1 (low), 2 (medium), or 3 (high).
M	speaker control: followed by 0 (always off), 1 (on till connected, the usual situation), 2 (always on), or 3 (on during answer).
N	modulation handshake: followed by 0 for fixed speed as set in register S7 or 1 to allow the two modems to negotiate and work at the highest speed common to them (the usual setup).
&C	DCD options: followed by 0 to lock DCD on or 1 to allow DCD to follow the state of the carrier. The default setting is usually 0 and should be changed to 1 by the software initialization string.
&D	controls modem action after DTR is dropped: 0 to ignore, 2 to disconnect when DTR is dropped. The usual default setting of 0 should be changed to 2 by the modem initialization string.

volatile memory. However, the latter feature often remains unused because most communications software automatically sets up the modem by sending a series of commands called an **initialization string** whenever the software is loaded.

High-speed modems have commands to turn the internal data compression and error correction on and off. Unfortunately, this is an area that lacks standardization at present, and it is often necessary to consult the manuals for both the modem itself and the associated communications software in order to determine the proper procedure.

An example of a command string is

ATDT 1,855-1155

This command instructs the modem to dial the number given (the hyphen is ignored, but the comma is interpreted as a pause). Since commands should be interpreted by the modem rather than being sent down the telephone line, the modem must distinguish between commands and data to be transmitted. When the modem is turned on, it is in the local mode and is receptive to commands; once a connection is made, it goes on-line. To return the modem to local mode to receive more commands (a command to disconnect, for instance), the computer must send a sequence that is not likely to occur in normal communication. One commonly used sequence is a pause followed by three plus signs then another pause.

In addition to being controlled by AT commands, modem settings are affected by values stored in setting (S) registers in the nonvolatile memory of the modem. These registers control such characteristics as the delay in dropping a connection when the carrier disappears and the time the modem will wait for a carrier after dialing. It is seldom necessary to change the register settings, but when necessary it can be done using the command:

ATSn=x

where

n = the register number to be changed
x = the value to be stored in the register

The manufacturer's manual supplied with a modem generally lists the registers and their default values.

13.7 File Transfers via Modem

The preceding sections have outlined modem operations and the role of modems in making a connection between two computers. Once this connection has been made, communication can take place. In many cases, a live operator at one computer explores menus, reads messages, or runs programs stored on the other computer. For instance, a microcomputer user may access a mainframe or minicomputer by means of a modem, or two people at two different computers can carry on a conversation by typing at their respective terminals.

Modems are also used to transfer files between computers. For example, a microcomputer user at home can obtain a copy of a text file or a computer program that is stored on a computer at work or on a **computer bulletin board** run by another individual or a company. Transferring a file is not at all the same thing as running the program on the host computer. In fact, there is no need for the program to be one that would run on either computer. For example, an Apple Macintosh computer could download a Microsoft Windows program from a minicomputer running the Unix operating system, even though the program would not run on either of these computers.

As we mentioned in Chapter 12, files can be categorized in two main types. Text files consist of ASCII characters only, while binary files contain other material and include both computer programs and much of the data created by them. For instance, a data file created by a word-processing application probably contains normal ASCII characters as well as nonprinting characters that control such features as font size and line spacing. Often these nonprinting characters are eight-bit combinations that could be interpreted as the extended ASCII characters from decimal 128 to 255. The word processor, however, does not interpret them as graphics characters but uses them to set up the printer, format the page, and so on. Computer programs are even further removed from ASCII text. They contain binary information that would produce random ASCII characters if arbitrarily grouped in 7- or 8-bit blocks.

It is possible to transfer ASCII files from one computer to another by simply displaying them on the screen of the receiving computer while using a text-capture program to store the received information to disk. However, this is not a good procedure to use for binary files. For one thing, a single error that would pass almost unnoticed in a text file could prevent a computer program from running at all. Furthermore, non-ASCII files contain random "characters" that can often be interpreted as control characters (see Chapter 12).

Several **file transfer protocols** have been designed expressly for the transfer of binary files. Some of the more commonly used ones will be described here. Any of these protocols can also be used to transfer an ASCII file, but in most cases it is faster and simpler to capture the file from the screen as described above.

13.7.1 Xmodem

This is the oldest of the file-transfer protocols discussed here. The original version of Xmodem was developed by Ward Christensen in 1977. This version transmits data in packets that are only 128 bytes long, with a checksum for error control. The use of such short packets makes transmission relatively slow, since each packet must be acknowledged by the receiver before the next one is sent. Xmodem has been updated by the addition of cyclic redundancy checking, a more advanced form of error correction. This newer version is called Xmodem-CRC to distinguish it from the original. Another version known as Xmodem-1k uses both CRC error control and larger packets (1024 bytes). This version is much faster than the original, unless the line is very noisy. Larger packets are usually faster to transmit because there are fewer pauses for acknowledgement. If there are many errors, however, short packets are faster because there are fewer bits to repeat if an error is found in a packet. All three of these systems are built into most communications software packages.

13.7.2 Ymodem

Sometimes called Ymodem Batch, Ymodem CRC, or Ymodem 1k, the basic Ymodem protocol is similar to Xmodem-1k except that it allows the transfer of multiple files at one time and sends the file name with the file. This is an improvement over Xmodem, which requires files to be transferred one at a time with a separate command for each. A variation of Ymodem called Ymodem-G sends the entire file in one packet without error checking. This protocol is designed for modems in which error checking is built in.

13.7.3 Zmodem

This is the most advanced file-transfer protocol in common use. It can vary the size of data packets to suit the condition of the line and can send data continuously until interrupted by the receiver with a request for retransmission. Both 16-bit and 32-bit CRC error checking are supported.

13.7.4 Kermit

Named after the famous frog from Sesame Street, Kermit is especially designed to handle file transfers between microcomputers and minicomputers or mainframes. Packet size is variable, up to a maximum of 1024 bytes. Super Kermit, a more advanced version of Kermit, allows data to be transmitted continuously until the transmitter is interrupted by a receiver request for retransmission, as in Zmodem.

13.8 Communications Software

Modems are generally used with specialized communications software designed to automate many of the tasks associated with programming and using them. There are many such packages on the market. Several popular examples are Crosstalk, MicroPhone, Procomm Plus, QuickLink, and Smartcom. Communications programs also come bundled with some general-purpose "works" application packages. Microwsoft Windows 3.1 has a simple communications program called Terminal, and Windows 95 has a more elaborate communications application that is designed especially for the Microsoft Network but is also usable as a general-purpose communications program. There are also specialized programs supplied by and designed for accessing the major commercial information services like America Online and Compuserve.

All general-purpose communications packages have several things in common. When first started, they transmit an *initialization string* to the modem to set it up for the type of communication most commonly in use. Commands may be transmitted to set error control and data compression, the speed and type of dialing, the speaker volume (and whether the speaker is heard all the time, not at all, or only during the setup of calls), and so on. These communications packages also program the modem for each call to change the initial settings as required. In addition, the communications software programs the UART with the correct number of data and stop bits and parity (if applicable). The software also dials the phone and enables the modem to establish a connection. Most communication packages have provision for the user to store a list of phone numbers and the proper communication settings for each, so that connection can be made with one key press or mouse click.

Once a connection is made, the more elaborate packages allow such data as account numbers and passwords to be transmitted automatically, as requested by the computer at the other end of the connection. This is made possible by a short program, usually written in a proprietary "script language" specific to the software. Procomm's script language for example, is called *Aspect*. Such scripts make it possible to log on to an on-line service, download electronic mail messages or data files, and log off again, all without operator attention. Thus file transfers can be made late at night when telephone rates are lower.

Once a connection has been made and any necessary user names and passwords have been transmitted, it is often necessary for the personal computer to emulate a specific type of terminal for meaningful communication to take place. All modem packages can emulate the most popular **"dumb" terminals**; the two most common are ANSI-BBS, used on most bulletin boards, and DEC VT-100, which is required by many Internet sites. Virtually all communications programs emulate both types of terminals, but sometimes the VT-100 emulation, in particular, is confusing or incomplete. Real VT-100 terminals have special keys to control cursor movement, and not all packages get this part right.

Communications packages also have provisions for file upload to and download from the remote computer. Most support all the types of file transfer described earlier in this chapter.

Many communications packages have a remote-access capability that lets the user set up his or her microcomputer to answer the phone and connect to a remote computer that calls it. In this way, for instance, an office computer can be accessed from home.

13.9 Computer Bulletin Boards and Online Services

One of the easiest ways to experiment with a modem is to use it to connect to local computer bulletin boards. A bulletin board can be as simple as a personal computer running appropriate software, a modem, and a telephone line. Boards are run by individuals, by companies, and by associations of all sorts. Many public and college libraries have bulletin boards that can be used to access the library catalog. Some boards charge a fee, but many are free. Most towns and cities in North America have several bulletin boards within their local calling areas.

In addition to local bulletin boards, there are a number of **online services** that are, in effect, very large bulletin boards using a central computer that can be accessed by many people simultaneously. Well-known examples include Compuserve, America Online, Delphi, Prodigy, and the Microsoft Network. To reduce long-distance charges, these networks are set up to transfer digital data in blocks called packets; this method is less expensive than a continuous connection from the user to the service. Packet-switched networks will be discussed in Chapter 14.

Local bulletin boards can be accessed with any generalized communications software package. Most boards require the data to be formatted with no parity, eight data bits, and one stop bit. Maximum speed varies. Beginning with ANSI-BBS terminal emulation is a good idea; if something else is required, it will at least be possible to read enough of the information from the bulletin board to find out what is required. Most bulletin boards use ANSI; other services, such as some library systems, require DEC VT-100 terminal emulation. Both are available in all the major communications software packages.

13.10 Fax Modems

Whether a facsimile (fax) machine is a stand-alone device or part of a microcomputer, it contains a specialized modem at its heart. Stand-alone fax machines are more convenient for many applications because they contain an optical scanner that can convert a written or printed document into a digital bit stream and a printer that can reproduce a received fax on paper. **Fax modems** used with computers, on the other hand, can transmit only data that already resides in the computer—thus handwritten documents, for example, must first be scanned into the computer using a separate optical scanner. On the other hand, if the data is already contained in a file on the computer, the fax modem may actually be more convenient. Fax modem software packages allow the user to "print" the file to the fax rather than to an attached printer. Received files can be stored on the computer's hard drive or printed using its associated printer.

Not all modems are capable of fax operation, though most modern ones are. Fax software is available separately or as part of some general-purpose communications software packages.

13.10.1 Fax Groups and Classes

There are four CCITT standards, called groups, for fax transmissions. Groups one and two are obsolete analog standards. Group three, which is by far the most common, uses digital coding of the document, followed by analog transmission using QAM at up to 14.4 kb/s. Data compression is used, and it greatly reduces the time needed to transmit a document containing large amounts of "white space," since it does not transmit all the repetitive pixels. Group four is for digital lines, such as ISDN connections, and can transmit at up to 64 kb/s. These group specifications are common to fax modems and dedicated fax machines.

There are two fax modem classes, class 1 and class 2, as specified by the EIA. These classes define the communication between the software and the modem and are therefore not relevant to dedicated fax machines. With class 1 modems, the computer does most of the processing, including data compression and error correction. Class 2 modems can do this internally, freeing the associated computer to do other tasks while a fax is being transmitted or received.

Summary

Here are the main points to remember from this chapter.

1. Modems are required whenever digital signals must be transmitted over analog channels.
2. While many different modulation schemes can be used, most modern modems employ quadrature amplitude modulation, which combines phase and amplitude modulation.
3. Modems can be either internal to the computer or external. If external, they are generally connected to the computer's serial port with cables that follow the RS-232C standard.
4. Most current modems are programmable using variations of a language originally developed for the Hayes Smartmodem.
5. When binary files are transmitted by modem, they are divided into packets and sent using a protocol that checks for errors and asks for a retransmission of any flawed data block.
6. Some modems can perform some data compression and error correction internally. Others achieve this by working with software running on the computer.
7. Facsimile machines and fax modems are similar to data modems in their method of operation, but they always include data compression.

Important Equations

$$\begin{aligned} \text{bit rate} &= \text{baud rate} \times \text{bits per symbol} \\ &= \text{baud rate} \times \log_2 M \end{aligned} \tag{13.3}$$

Glossary

audio frequency-shift keying (AFSK) data transmission by varying the frequency of an audio tone used to modulate a higher-frequency carrier

baud rate number of symbols transmitted per second in a data transmission scheme

bit rate number of bits transmitted per second

computer bulletin board a computer set up for unattended operation, with one or more modems to enable remote users to read information and download and upload files

constellation diagram a pattern showing the possible states of a modem on a plane where amplitude is represented by distance from the center of the pattern and phase angle by the angle made by an imaginary line from the center to a dot representing a particular state (with respect to a reference angle)

data communications equipment (DCE) a modem

data terminal equipment (DTE) a terminal or computer that communicates via a modem

delta phase-shift keying (DPSK) a phase-shift keying system that uses the change in phase angle from one symbol to the next to avoid the need for a reference phase

dibit system a data communication system that transmits two bits per symbol

dumb terminal a terminal for a mainframe or minicomputer system that has a monitor, a keyboard, and the means to communicate keystrokes and sometimes cursor movements to the host computer while displaying data from the host; it does no computing itself

fallback a transmission speed that is less than the maximum of which a modem is capable and that the modem resorts to when line conditions do not permit transmission at its maximum speed

fax modem a device for the transmission and reception of facsimile documents; it may be used with a microcomputer or built into a stand-alone fax machine

file transfer protocol a system to allow the transmission of binary data files between computers, incorporating some form of flow control and error control

flow control a means of ensuring that a transmitter sends data only when the associated receiver is ready to receive it

frequency-shift keying (FSK) data transmission by shifting the transmitter frequency; two frequencies are generally used, one for mark (binary 1) and one for space (binary 0)

initialization string a series of commands transmitted to the modem whenever a communications software program is loaded

modem acronym for *mo*dulator-*dem*odulator, a device used to allow digital data to be transmitted over an analog channel

null modem a cable used to make a direct connection between two devices using their serial ports; it is wired to simulate the presence of a link via modem between the devices

online service a large computer that connects to many users simultaneously via modem and typically offers such services as electronic mail, databases, interest groups, and Internet access

phase-shift keying (PSK) a means of transmitting data by shifting the phase angle of the transmitted signal

quadrature phase-shift keying (QPSK) phase-shift keying that employs four different phases and allows two bits of information to be transmitted simultaneously

quadrature amplitude modulation (QAM) a means of transmitting data by shifting both the amplitude and the phase of the transmitted signal

symbol a transmitted signal that can have two or more possible states

training sequence a series of tones transmitted by a modem to allow the automatic adjustment of line equalization

Questions

1. What is the meaning of the term *modem*?
2. Describe the circumstances in which modems are required for data communications. Give examples of typical applications.
3. (a) What parameters of a sine-wave carrier can be varied in analog modulation?
   (b) In a data communications system, which parameter(s) are more likely to be varied at (i) low, (ii) moderate, and (iii) high data rates?
4. Explain the difference between *bit rate* and *baud rate*.
5. In the Bell 103 standard, explain the difference between the originate and answer modems in terms of function and frequency.
6. What modem scheme is used to transmit call-display data in the telephone system?
7. How does the digital transmission scheme usually used with VHF radio differ from that employed in HF radio communication?
8. What is DPSK? What advantage does it have over ordinary PSK?
9. The Bell 212A modem is often referred to as a 1200 baud modem. Is this correct? Explain your answer.
10. What is meant by *fallback*? How does it apply to a V.22 *bis* modem using North American standards?
11. Which modem standard is currently the fastest for dial-up telephone lines? How fast is it?
12. What is a training sequence? How is it used in modem communication?
13. Give one advantage and one disadvantage of modem data-compression schemes, compared with other file-compression methods.
14. Describe the function of each of the RS-232 lines that appear on a nine-pin connector.
15. Explain what is meant by *flow control*, and describe the differences between hardware and software flow control.
16. Even though the nominal maximum length of an RS-232 cable is 15 meters, much longer lengths can be used if the data rate is less than maximum. Explain why (that is, what limits the length, and why does a lower data rate make it less critical?).
17. Given that a terminal is a DTE and a modem is a DCE, what would you call a "serial printer" that plugs into an RS-232 port? Explain briefly why this could be confusing.
18. What is a null modem? Under what circumstances is a null modem required in an RS-232 link?
19. What steps are necessary to set up communications between modems?
20. This question refers to a Hayes-compatible modem.
    (a) What is local mode?
    (b) How can the modem be switched from on-line to local mode?
    (c) What command would be used to get the modem to dial 9 for an outside line, pause, then dial the number 555-1212, using pulse dialing?
21. Prepare a table comparing the file-transfer protocols

described in this chapter. Use the headings block length, error-correction method, speed, and suitability for multiple file transfers.

22. Explain the uses of a script language as found in communications software packages.
23. What is the difference between a class one and a class two fax modem?
24. Nearly all stand-alone fax machines and fax modems belong to the same one of the four groups defined by the CCITT. Which one? What is its maximum data transmission rate?

Problems

SECTION 2

25. A constellation diagram for a modem is shown in Figure 13.13.

Figure 13.13

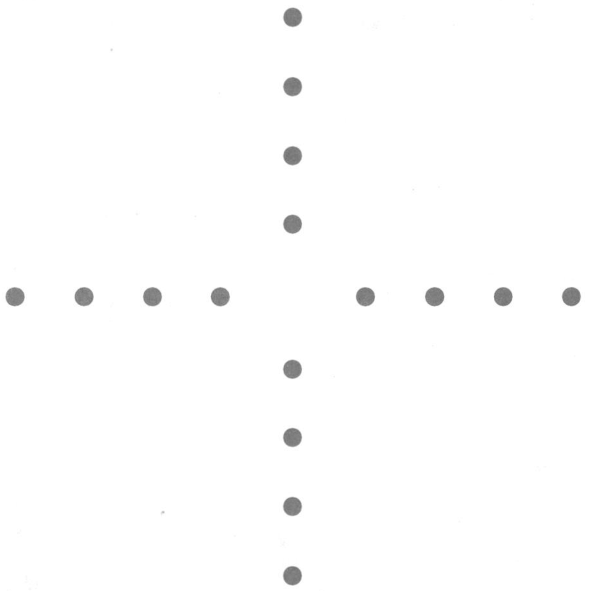

(a) What type of modulation is this?
(b) If the transmitted bit rate is 9600 b/s, what will be the baud rate using this modem?
(c) Suppose that a fallback mode at 4800 b/s, with the same baud rate, is required. Draw a constellation diagram showing how this could be accomplished. (There is more than one right answer.)
(d) When would a fallback mode of operation be useful?

26. Draw a constellation pattern for a modem that uses eight equally spaced phase angles and four equally spaced amplitude levels. If the modem operates at 4800 baud, what is its bit rate?

SECTION 4

27. Draw a circuit diagram for an adaptor that will enable a modem with a 25-pin connector to be used with a computer with a 9-pin connector.
28. Draw a connection diagram for a null-modem cable to connect two computers, with a 25-pin connector on one end and a 9-pin connector on the other. The cable should connect DTR on each computer to DSR on the other and connect RTS on each computer to CTS and DCD on the other.

SECTION 6

29. Write an initialization string to set up a modem as follows:
(a) Commands are to be echoed to the computer.
(b) The speaker is to be set to high volume and is to operate until a connection is established.
(c) The connection is to be at the highest speed common to the two communicating modems.
(d) The DCD line should follow the true state of the carrier.
(e) The modem should hang up the phone when the computer drops the DTR line.

COMPREHENSIVE

30. Calculate the maximum data rate that could be sent using simplex transmission in a voice-grade telephone channel having a bandwidth of 3 kHz and a modem that uses QAM modulation with four possible phase angles and two amplitude levels. Assume that the signal-to-noise ratio in the channel is 30 dB.
31. A microwave radio system uses 256-QAM, that is, there are 256 possible amplitude and phase combinations.
(a) How many bits per symbol does it use?
(b) If it has a channel with 40 MHz bandwidth, what would be its maximum data rate, ignoring noise?

14 Local-Area Networks

Objectives After studying this chapter, you should be able to:

1. Describe three basic topologies for local-area networks and discuss the advantages and disadvantages of each
2. Describe and compare CSMA-CD and token-passing protocols, as used with local-area networks
3. Describe and compare the Ethernet, token-ring, and Arcnet local-area network specifications
4. Explain the difference between baseband and broadband networks and discuss the application of each
5. Perform propagation-time calculations with LANs
6. Describe the operation of client-server networks
7. Explain the functions of the software that controls LANs

14.1 Introduction

We have already studied the switched telephone network and have noted that it can be used for data as well as voice. Sometimes a private branch exchange (PBX) is used to switch both data and voice lines within a building. This setup is quite practical if the PBX uses digital switching, though the data rate, typically 64 kb/s, is relatively low. In this and the next chapter, however, we will look at networks designed especially for data.

We will begin with **local-area networks (LANs)**, which, as the name implies, generally cover a small geographical area. In the next chapter, we will look at **wide-area networks (WANs)**, which can extend for much greater distances and can even be international. It is also possible to link data networks together to form a network of networks. Some of these are worldwide in scope. The largest and most famous of these is the *Internet*.

Local-area networks are intended for use within a building or group of buildings, such as an office complex or a college campus. Many colleges have one or more local-area networks. For instance, all of the individual workstations in a computer lab may be linked to a central computer called a **file server**, which contains the software to be used, and to a single printer. A well-constructed LAN can allow several people to use the same software at the same time and even to work on the same document (though the network must prevent one user from reading information while another is changing it, in order to avoid confusion). Information can also be moved from one kind of document to another. For instance, in a college, student grades could be compiled by professors in their offices using a spreadsheet program, submitted to the registrar over the network, and then used by the awards officer to determine who gets scholarships. The grades could also be incorporated into form letters prepared with a word processor to inform the lucky students. Meanwhile, statistics could be compiled by another person and then used by still another to determine how many students will be graduating at the end of the term. All this can happen without a flurry of paper memos, miscopied numbers, phone calls, and meetings.

14.2 Local-Area Network Topologies

There are many ways to build a network. We looked at one method when we studied the switched telephone network, which belongs to a type of network called a **circuit-switched network**. In such a system, switches connect two users in such a way that they have a dedicated channel, free of other traffic (messages), for the duration of the call. There is no reason why this cannot be done with data; in fact, as we have already discussed, the telephone network is often used for data transmission. However, the nature of digital data allows other possibilities. In this section we will look at some common data network architectures. Real networks, especially large ones, are often combinations of these topologies.

14.2.1 Star

In a **star** network, individual terminals, microcomputers, or workstations are connected directly to a central computer. The central computer is often a mainframe or minicomputer but could also be a dedicated microcomputer known as a file server. Each of the connections to a network, regardless of its type, can be referred to as a **node**.

A typical star network is shown in Figure 14.1. Note its similarity to the telephone network studied earlier. In fact, sometimes a private branch exchange is used to connect computers within a company. This method is quite practical when the PBX is digital, though the maximum data transmission rate is generally less than with a dedicated LAN.

The star is the oldest network topology. It is well suited to a centralized system, such as one in which many terminals or workstations communicate with and are largely controlled by a mainframe computer. A drawback of this system is that the failure of the central computer causes the whole network to "go down" (be disabled).

It is also possible for a star network to have an active or passive **hub**, rather than a computer, at the center of the star. The hub connects all of the nodes together but does not necessarily control the network. In such a case, we say that the network is a *physical star* but has some other *logical* topology.

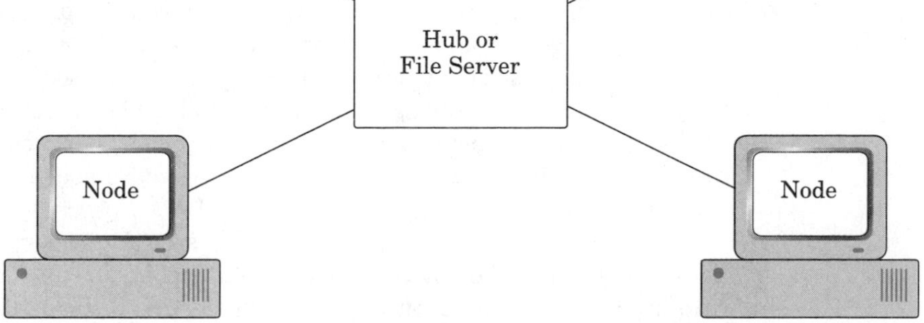

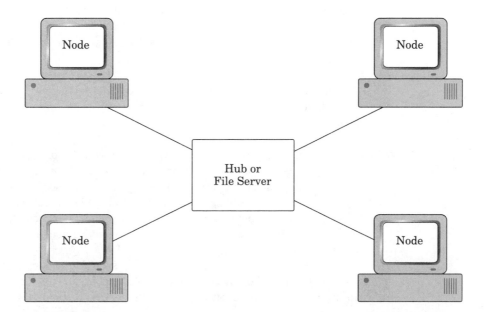

Figure 14.1
Star network

Figure 14.2
Ring network

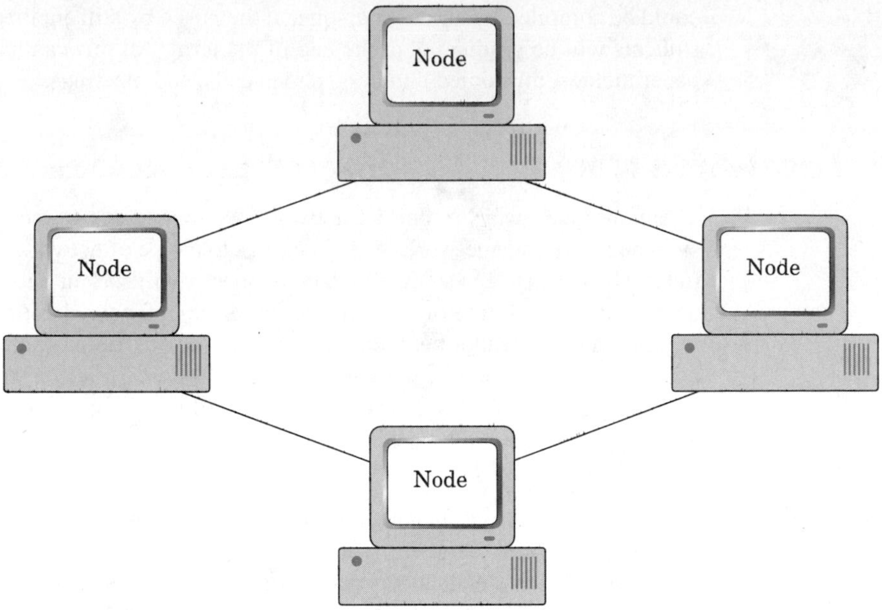

14.2.2 Ring

A typical **ring** network is shown in Figure 14.2. A cable runs from each node to the next. Each node receives every message and passes it on to the next. The signals move around the ring, usually in only one direction.

Ring networks require special provisions to keep the network operating if one of the terminals is not operating or is disconnected from the network.

Communication in a ring network is usually organized by means of a technique called **token passing**. A short message, called a free token, is sent around the ring to each node in turn. When the token reaches a node with data to send (often called **traffic**), the node changes the token to a busy token and sends it to the next station in order around the ring, followed by the message. Messages are divided into relatively short sections called **packets**. Each station examines the data packet in order to determine whether the message is intended for it. If so, it reads the packet. In any event, it sends the packet on to the next location in the ring. The process continues until the packet circulates all the way around the ring and reaches the originating node. At that time the originating station changes the token back to a free token and passes it to the next station in order. The process continues until the token reaches another node with traffic.

Token-passing is a very orderly way of making certain that there is only one message at a time on the network. However, it can result in inefficient use of the network when traffic is light, since a node has to wait until it receives the token before transmitting, even if no other node has traffic for the network.

14.2.3 Bus

Figure 14.3 shows a simpler network topology known as a bus. Each terminal connects to the same line, which can be twisted-pair wiring, coaxial cable, or even optical fiber. The wiring can be considerably simpler than with either the star or the ring topology. This system is much less centralized than the star, and unlike the ring, it needs no special provisions to allow it to continue to function when an individual node fails or is disconnected (the only exception is the situation in which a defective node sends a stream of incorrect data into the network).

One problem with the bus topology is that all traffic shares the same channel, so there must be some method of allocating time on the network to individual stations. One method is to use token-passing, as in ring networks. Instead of proceeding around the ring, the

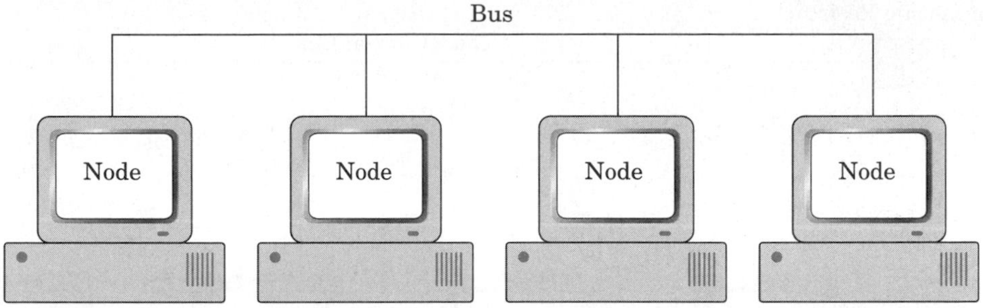

Figure 14.3
Bus network

token is sent to each node in turn, following a programmed table. It is even possible to send the token more often to some nodes than to others, if some nodes require faster service.

Another method that is more common with bus networks is called **CSMA/CD**, which is short for **carrier sense multiple access with collision detection**. With this system, network terminals monitor the line for data. When a terminal has a packet of data to transmit and the line is not busy, it simply transmits the data. Each station receives the data and determines from an address on the packet whether the data is intended for it. If it is not, the packet is ignored.

A problem arises when one node begins to transmit after another node has started its transmission but before the data from that transmission has reached it. The interval between the time that one node starts to transmit and the time that another node receives the transmission is equal to the propagation delay on the line between the two nodes. Since data moves along the network at an appreciable fraction of the speed of light, this time is short and depends on the distance between the nodes. If two or more nodes do transmit at once, a **collision** will occur, and both packets of data will be useless. Both transmitting stations will detect the resulting errors almost immediately, and each will stop sending, then wait a random amount of time before retransmitting the packet.

CSMA/CD works very well when the traffic is light, since stations with data can transmit at once, without waiting for a token, and collisions are rare. With heavy traffic, however, the effective data transmission rate is reduced because of the retransmissions made necessary by collisions. In such a situation, a token-passing scheme, which avoids collisions, may be more efficient.

14.3 Examples of Local-Area Networks

In order to understand how LANs work, it will be useful to look at some typical commercial systems. We will examine three popular systems: **Ethernet**, which is a bus system using CSMA/CD, the **IBM token-ring system**, and **Arcnet**, which uses a blend of bus and token-ring concepts.

14.3.1 Ethernet

Ethernet was originated by Xerox, with later participation by Intel and Digital Equipment Corporation. By 1980, when the IEEE began to draft standards for local-area networks, Ethernet was already a de facto standard. Thus while it is sometimes said that Ethernet follows IEEE standard 802.3, it is perhaps more accurate to say that the IEEE standard follows Ethernet! Currently the Ethernet standard is supported by many vendors.

A standard, "classical" Ethernet system uses a special "thick Ethernet" 50 Ω coaxial cable in a bus configuration. Up to 100 nodes are allowed over a bus length of 500 meters. Data is communicated along the bus using Manchester code at a rate of ten megabits per second. This system is commonly called 10Base5, meaning 10 Mb/s, baseband, 500 m length. Transceivers placed along the bus connect to the individual nodes with separate cables (see Figure 14.4).

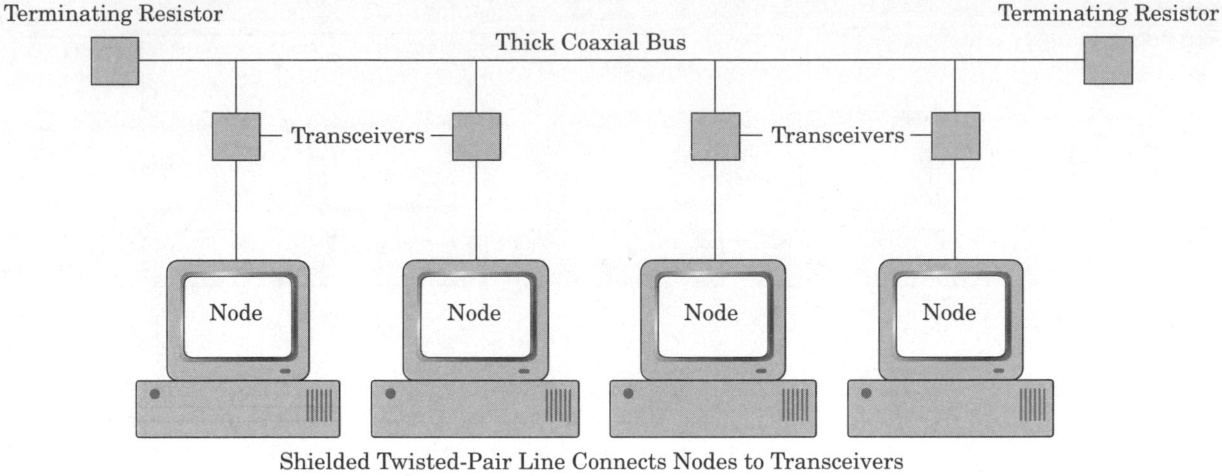

Figure 14.4
"Thick Ethernet" network

Thick Ethernet cable is now usually reserved for "backbone" cable runs that link groups (clusters) of computers. Small networks and clusters of nodes in larger networks can use "thin Ethernet" (colloquially known as "cheapernet"), which uses ordinary, inexpensive RG-58/U coaxial cable. In the IEEE standards, this setup is called 10Base2 (10 Mb/s, baseband, 200 meter length). Though the standard does specify a maximum 200 meter length, in practice the length can usually be extended to about 300 meters without giving rise to any problems. The cable can be connected to the computers using ordinary BNC tee connectors, as shown in Figure 14.5.

Ethernet can also be used with twisted-pair line, such as is often used for voice telephony. This 10BaseT (T for twisted) system is restricted to a length of 100 m, using two pairs of wires. Though logically it is a bus system, physically it is connected as a star with a central wiring hub, like the one shown in Figure 14.6. The hub can disconnect a defective node or one with shorted wiring.

Twisted-pair Ethernet can sometimes be implemented using spare telephone wiring already installed in an office building, so it can be an economical choice for a small LAN. The connectors used resemble standard modular telephone connectors, except that they can accept more wires.

There are also versions of the standard for higher data rates (100 Mb/s) on coaxial cable and for fiber-optic cable. Because of the difficulty of constructing fiber-optic tee connectors, fiber networks always use a physical star configuration. Ethernet can operate over greater distances with additional amplification (up to a total distance of 2.8 km under some conditions).

The reason for the limitation on the total length of an Ethernet LAN can be found in the CSMA/CD protocol used with Ethernet. Before transmitting, a node "listens" for a

Figure 14.5
"Thin Ethernet"

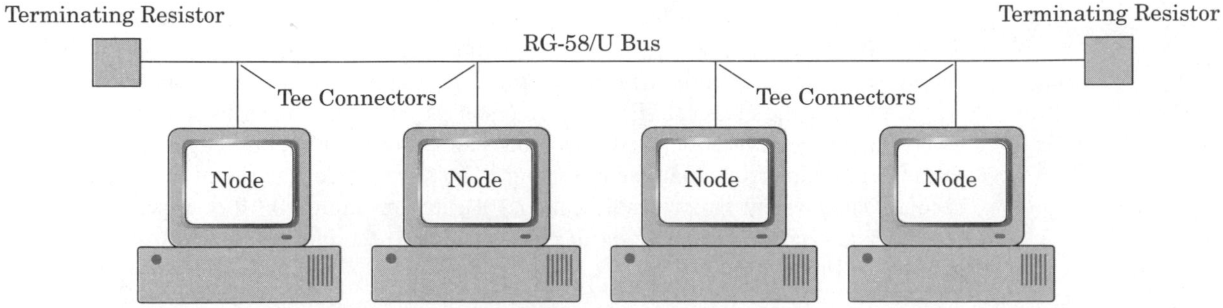

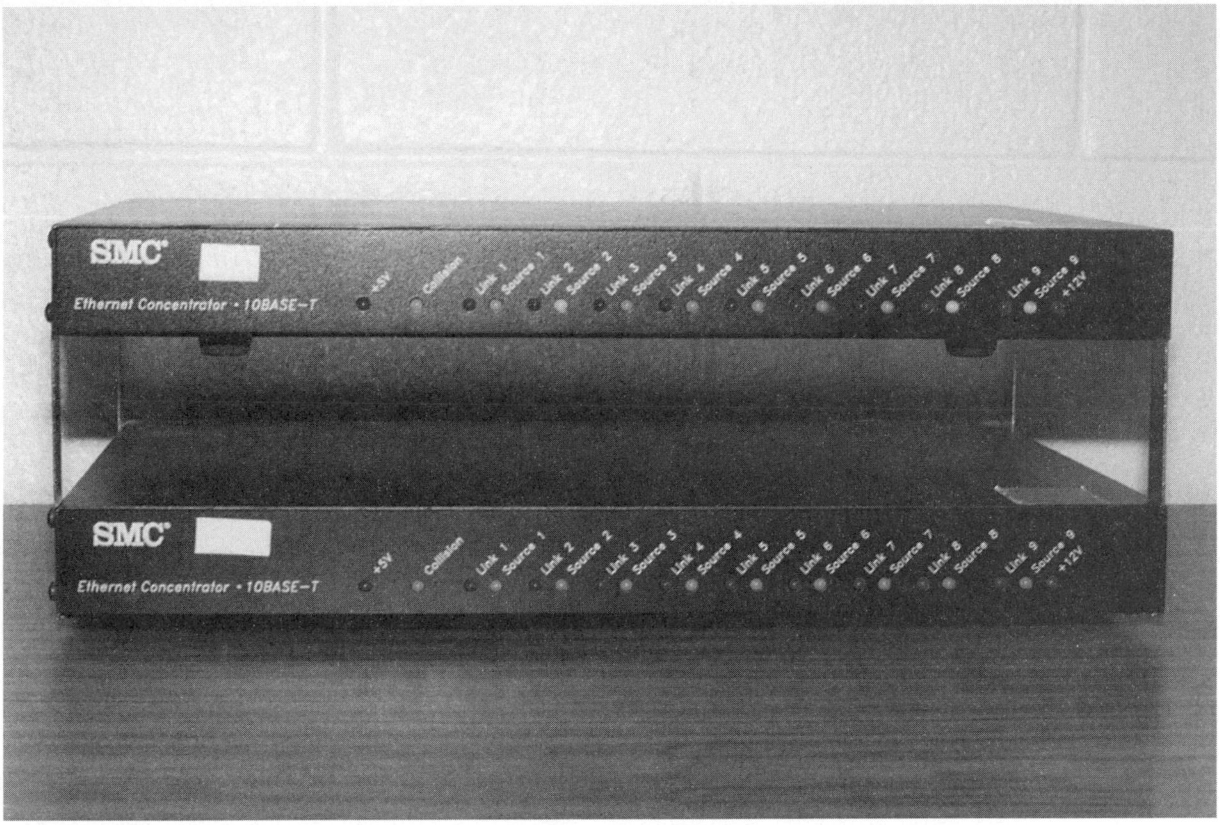

Figure 14.6
10BaseT hubs

period of 9.6 microseconds. If another station is transmitting at the beginning of this period, it will be detected by the end of the listening period, provided that its distance from the receiving station is less than the distance a signal can travel during the listening period. See Chapter 16 for a detailed discussion of transmission lines. For this example it is sufficient to realize that the speed of transmission along a line can be found by multiplying a number called the **velocity factor**, which is a characteristic of the cable, by the speed of light.

The velocity factor for typical coaxial cable is about 0.66, so the distance traveled is given by:

$$\begin{aligned} d &= vt \\ &= v_f ct \\ &= 0.66 \times 3 \times 10^8 \text{ m/s} \times 9.6 \ \mu\text{s} \\ &= 1900 \text{ m} \end{aligned}$$

where

d = distance traveled
v = velocity of propagation along the cable
t = propagation time
v_f = velocity factor
c = speed of light

Of course, it is possible that another station will begin transmitting while the first station is listening. If the station is too far away, the transmission will not be detected and a collision may result. In that case, the collision should be detected relatively early in the

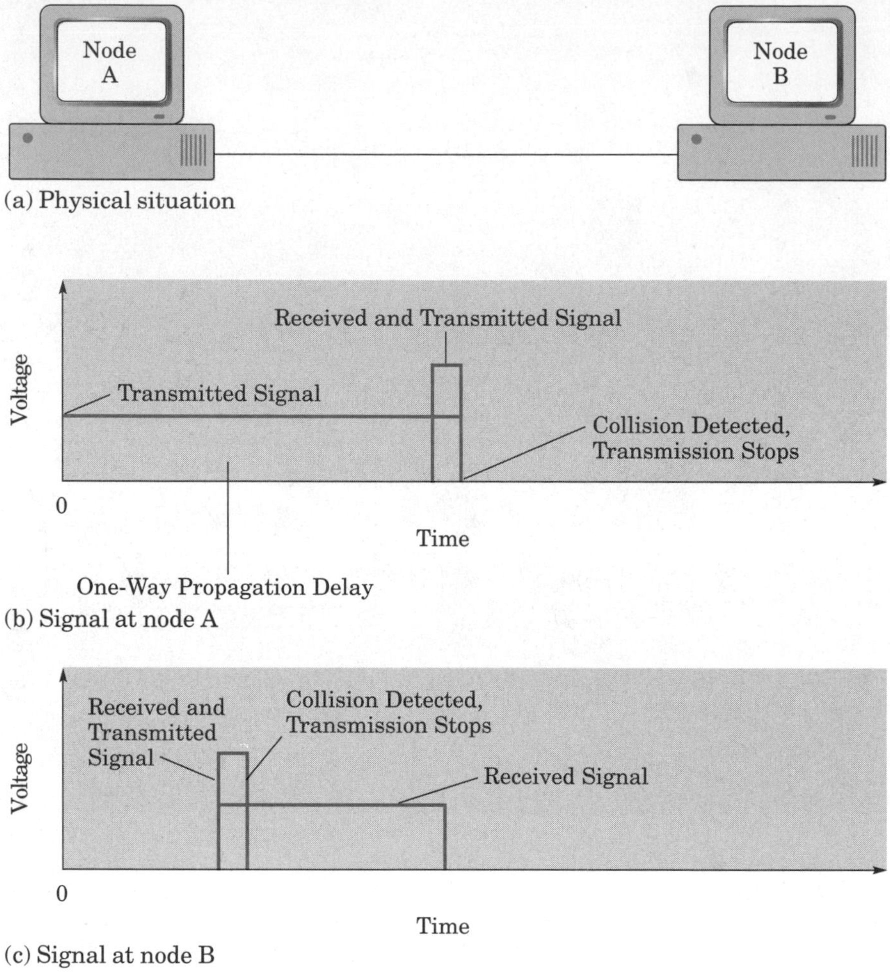

Figure 14.7
Collision in a CSMA/CD network

transmission, so that the process can be stopped and restarted with as little wasted time as possible. This means that the system is most efficient when the packets of data are relatively long.

To determine how long it can take to detect a collision, consider the worst-case scenario shown in Figure 14.7. Nodes A and B are at opposite ends of the bus. Suppose node A transmits. The signal from node A moves down the line toward node B, but it takes some time to get there. Meanwhile, node B listens and, hearing nothing, begins to transmit, just as the signal from A reaches it. Node B stops transmitting, but both transmissions now have errors and have to be discarded. Meanwhile, node A knows nothing about the collision until the signal from node B reaches it. Then it stops transmitting and the network is clear for one or the other to try again.

The total time used by this process is the time it takes for a signal to travel down the line and back. That is, it is two times the propagation delay along the line. Thus we have:

$$v = v_f c \tag{14.1}$$

The time for a signal to travel a given distance along the cable is given by

$$t = \frac{d}{v} \tag{14.2}$$

For an Ethernet system with a length of 2.5 km on coaxial cable with a velocity factor of 0.66, the propagation delay is:

$$\begin{aligned} t &= \frac{d}{v} \\ &= \frac{2.5\ \text{km}}{0.66 \times 3 \times 10^8\ \text{m/s}} \\ &= 12.6\ \mu\text{s} \end{aligned}$$

The maximum time it takes to detect a collision is twice this, or 25.3 μs. At 10 Mb/s, this corresponds to the time taken by 25.3 μs × 10 Mb/s = 253 bits. Ethernet packets are generally much longer than this, up to a maximum of about 12 kb.

14.3.2 IBM Token-Ring System

Logically, the IBM token-ring architecture is a ring, since each terminal communicates with the one next in order. Physically, however, the network looks like a star, with all the nodes connected to a central interface point called a **multistation access unit (MAU)**. Figure 14.8 illustrates the idea. Each MAU can support eight stations; for larger networks, a number of MAUs can be connected together in the form of a logical ring.

The IBM system uses an interesting method to prevent the network from being disrupted because a node is malfunctioning or simply switched off. The MAU contains a relay that bypasses the workstation unless it receives a 5V signal from that station to indicate that the power is on and the station is ready to connect to the network.

Figure 14.8 Token-ring network using multistation access unit

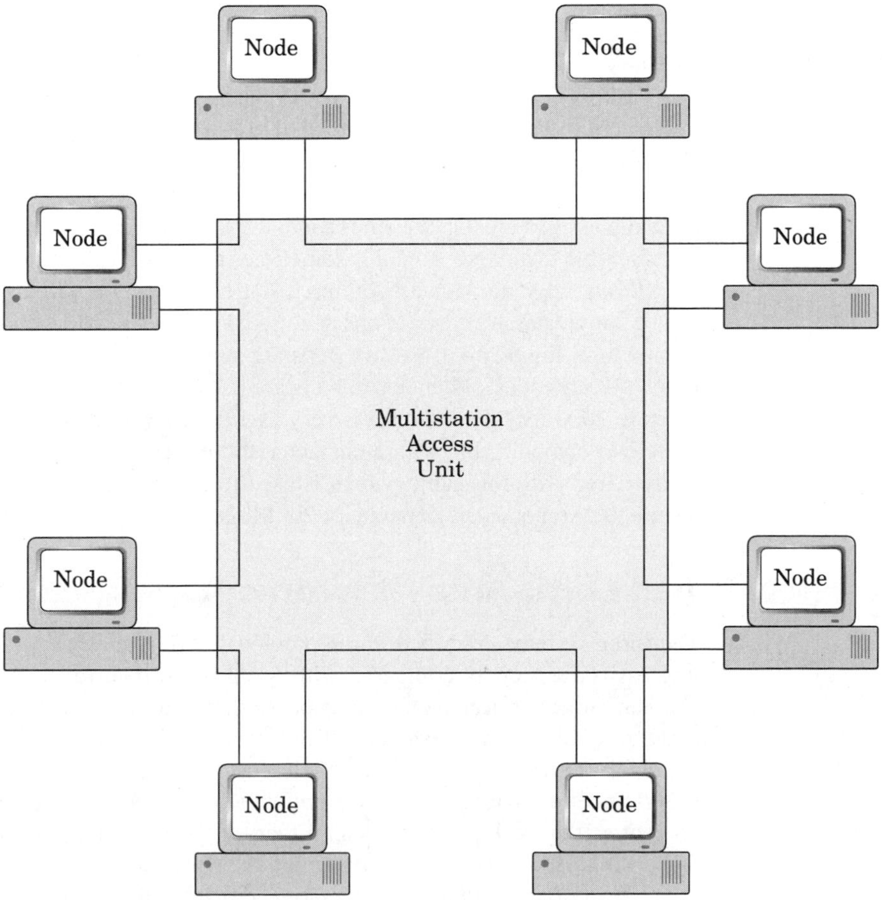

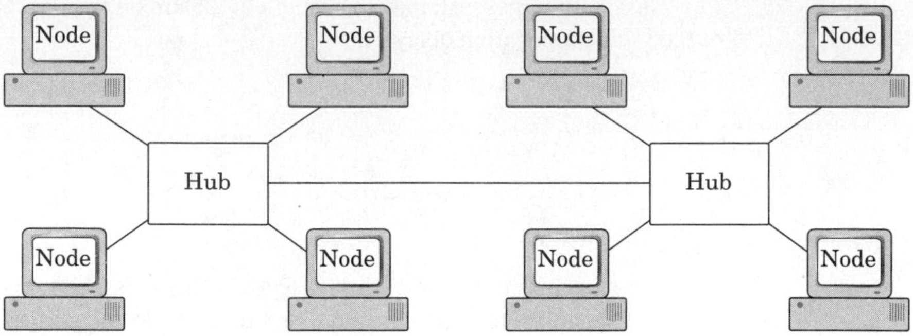

Figure 14.9 Arcnet

The IBM token-ring network uses twisted-pair wiring and can operate at either 4 Mb/s or 16 Mb/s. The network can support 72 nodes using ordinary telephone twisted-pair wiring or 260 nodes with special data-grade shielded twisted-pair wiring.

Token-ring architecture is important when microcomputers are required to communicate with an IBM mainframe as well as other microcomputers. Token-ring technology allows IBM mainframes to interact with microcomputers as equals, rather than as dumb terminals. Token-ring networks are considerably more expensive than Ethernet networks.

14.3.3 Arcnet

Arcnet was originated by Datapoint and popularized for microcomputers by Standard Microsystems. Arcnet networks use switches on the network adaptor card in each computer to assign a number from 0 to 255 to each station. Whenever a node connects to the network, the system reconfigures itself so that a token called an *invitation to transmit* always circulates around the network in numerical order from the lowest to the highest node number and then back to the lowest.

When a station with traffic to send receives the invitation to transmit, it first asks its intended destination node if it is ready to receive the packet of data. Then, on receiving the acknowledgement that the receiving node is ready, it transmits its data.

All traffic is transmitted over the whole network, as in Ethernet, so Arcnet has a logical bus structure. On the other hand, the transmission invitation messages used by Arcnet resemble tokens, so Arcnet is sometimes described as a token-ring system.

Physically, an Arcnet LAN has a star configuration with a central wiring hub. Hubs can be simple passive connections or active signal regenerators. Hubs can be interconnected as shown in Figure 14.9. Passive hubs are suitable only for short distances of 100 feet or less between nodes. When active hubs are used, however, the distance between nodes can be up to 2000 feet. Thus Arcnet is very useful for LANs that cover wide areas. Arcnet uses RG-62/U coaxial cable with a characteristic impedance of 93 Ω. Some buildings are already wired with this cable, which is used to connect IBM mainframe computers to their terminals. Arcnet uses a data rate of 2.5 Mb/s.

14.4 High-Speed Local-Area Networks

The three systems described in the previous section all operate at speeds that seem quite impressive, especially compared with the modem data rates described in the preceding chapter. However, it must be remembered that in all these systems, the specified data rate is the rate for the entire network. If only one node is transmitting data, that node can use almost the entire network bandwidth, less a small amount for overhead. On the other hand, if 100 nodes are trying to transmit simultaneously, each will get, on average, less than one percent of the total data rate. Thus, Ethernet's average data rate would be something less than 100 kb/s, not the 10 Mb/s that might be expected.

The efficiency of network topologies varies with the amount of network traffic. Ethernet, which allows stations to transmit without waiting for a token or a transmission permission message, lets a single transmitting station use the entire network bandwidth without additional overhead. On the other hand, when traffic is heavy, Ethernet experiences collisions and retransmissions, which reduce the data throughput very considerably. Token-ring and Arcnet systems are better performers under these conditions because of their tightly controlled, orderly way of assigning the right to transmit to each node in turn.

These traditional LAN systems are generally adequate provided the network is small and the amount of data transferred by each node is relatively small. If the number of users grows or the amount of data transferred by each becomes very large, noticeable delays will appear with any of these systems. Multimedia applications are especially problematic because they require very large amounts of data to be transferred quickly (for example, for live-action video).

There are two main ways to increase the speed of a local-area network. The first, and perhaps the most obvious, is to use a higher signaling speed. Ethernet, for instance, is available in a 100 Mb/s version. IBM's token-ring system comes in 4- and 16-Mb/s versions, and Arcnet has a 20 Mb/s version called Arcnet Plus.

Besides the additional cost of adaptors for the higher speeds, it is necessary to use more care in the wiring of high-speed LANs. This is especially important when twisted-pair wiring is used. Ethernet at 10 Mb/s sometimes works with existing telephone wiring, but for reliable performance at 100 Mb/s it is necessary to use higher-grade wire and special care in wiring it to connectors and junction blocks.

An alternative to upgrading premises cabling is to use more wires and send data in parallel. A system called 100BaseVG is designed to work with voice-grade twisted pair wiring (VG stands for voice grade). This system is sponsored by AT&T and Hewlett-Packard, among others, and uses four pairs of twisted-pair wires. The signaling system is also different from that used with conventional Ethernet. Nodes are assigned different priority levels, so that those with very time-sensitive traffic (real-time video, for example) receive a higher priority than those that can accept some delay. A **print server**, for instance, would likely be given a low priority.

Another way to improve communication speed on a network is to divide a single LAN into smaller segments, each with fewer nodes. The organization should follow a logical structure; for instance, a manufacturing company could have engineering in one LAN, accounting in another, sales in still another, and so on. The assumption is that most of the network traffic takes place among computers in the same area. When it is necessary for engineers to talk to sales or accounting people, this can be accomplished by providing internetwork connections that pass only those messages intended for members of the other network. These devices are called **routers**, and they will be discussed in the next chapter.

14.5 Broadband Networks

So far we have assumed that each cable in a LAN carries only one data signal at a given time. Such networks are called **baseband** systems, because no modulation is used. An ordinary public-address system is a good example of an analog baseband system.

In analog communications, it is very common to use frequency-division multiplexing to increase the capacity of a transmission medium. Cable television systems, for instance, can carry many television channels by assigning each a carrier frequency and using modulation. Exactly the same idea can be applied to data communications. In fact, a **broadband** LAN uses techniques very similar to those used in cable television. There are several channels, each with its own carrier frequency, and data signals are put onto these channels using special **radio-frequency (RF) modems**. A broadband network can carry much more traffic than a baseband network and can even be used simultaneously for data, video, and audio signals. On the other hand, broadband LANs are much more complex and difficult to install and adjust than baseband systems, so they are much less common.

RF modems have recently found a new application. Some cable-television companies have experimented with using RF modems to provide data services, especially Internet access, to customers. One television channel with a bandwidth of 6 MHz can provide a data rate of at least 10 Mb/s. Of course, this must be shared in some way among all the customers using the service at a given time. So far, the cable companies have used CSMA-CD systems. Cable television systems also work much better in one direction (from the central cable starting point or *head-end* toward the customer) than the other, so the data rate from subscriber to service provider is much lower than the other way around.

14.6 Software

A detailed discussion of LAN software merits a book in itself, and many such books have been written. In this section we will merely discuss some basic concepts common to this software, to give the reader an understanding of the ways in which networks can share resources.

Network software is generally independent of the network topology, that is, the same software can be used with Ethernet, token-ring, or Arcnet hardware. Packet assembly, token passing, and collision detection are accomplished in the network interface cards themselves. The interface cards also contain the address of the node. For most systems, the address is a unique serial number programmed into the card when it is built; for Arcnet cards, however, the address is set by the user when the network is constructed.

Once installed, the operation of most LANs is quite simple from the user's point of view. After logging on to the network using a procedure that usually involves entering a username and a password, the user finds that network resources appear as extensions to his or her computer's own capabilities. The hard drives and/or CD-ROM drives that are accessible through the network simply appear as extra disk drive letters on each station.

14.6.1 Client-Server and Peer-to-Peer Networking

There are two basic concepts in local-area network software. A **peer-to-peer network** treats all nodes as equal. Any node can share any of its resources, such as disk drives, printers, and modems, with the network. Which resources are shared is under the control of the individual node. Peer-to-peer networking is simple and inexpensive for small networks, as it does not require the purchase of an extra computer to be dedicated to running the network. All computers can be used as individual workstations while simultaneously using network resources and contributing resources to the network.

In networks using IBM microcomputers and compatibles, peer-to-peer networks tend to be DOS-based. Examples are Lantastic, Microsoft Windows for Workgroups and Windows 95, and PowerLAN.

The other major networking concept is the **client-server** model. One node of the network operates as a file server (occasionally there is more than one file server). Its disk drives contain all the data files and programs that will be used on the network. It may also connect to whatever printers, modems, and links to other networks are provided as network resources. Sometimes there are dedicated print servers and fax servers instead. Generally a server is dedicated to the network and is not used as a workstation.

The other nodes in a client-server network can be used as independent computers but can also share resources with the network by means of the file server. To avoid errors and confusion, LAN software makes sure that data on the server that is being changed by one station is not simultaneously read by another. This practice is called *record locking*. It is also possible to monitor software use so that, for instance, a program that is licensed for ten simultaneous users is not available to an eleventh user until one of the ten current users exits from the program.

The disadvantage of the client-server approach is that it ties up one computer as the network file server. Its advantage is that everything is more orderly and more amenable to central control. The client-server model becomes simpler to use once the size of the network exceeds a few nodes.

The software that runs client-server networks needs to have true **multitasking**. Much of this software is based on Unix or Unix-like systems: Banyan's Vines, Novell's NetWare, and IBM OS-2 fit into this category. Others, like Microsoft Windows NT, show the Unix heritage less clearly but are designed for multitasking from the beginning. That is, several programs can run at once without interfering with each other.

Summary

Here are the main points to remember from this chapter.

1. Local-area networks (LANs) are small data networks that can be arranged in star, ring, or bus configurations.
2. Many networks are arranged as a physical star for convenience in wiring and servicing, though their logical topology may be that of a ring or a bus.
3. Most ring networks use token-passing protocols, and most bus networks use carrier sense multiple-access with collision detection (CSMA/CD) protocols.
4. Ethernet is the most common example of a CSMA/CD network, while the IBM token-ring system is a popular token-passing network. Arcnet is an example of a token-bus configuration, though physically it is connected as a star.
5. LAN communication speeds have recently been increased by increasing the data rate possible with all three popular LAN configurations. Speed increases can also be achieved by dividing a network into segments.
6. Broadband networks can combine several local-area networks with voice and video on the same cable. They are more complex to configure than baseband LANs.
7. The simplest way to configure a small LAN is to use peer-to-peer networking, but larger networks use a dedicated server, with the other nodes operating as clients sharing files on the server.

Important Equations

$$v = v_f c \tag{14.1}$$

$$t = \frac{d}{v} \tag{14.2}$$

Glossary

Arcnet a type of LAN that has a physical star topology but uses token passing along a logical bus

baseband the information signal (data in the case of LANs)

broadband system a system in which the baseband signal is used to modulate a higher-frequency carrier signal

carrier sense multiple access with collision detection (CSMA/CD) a system for controlling network traffic; it allows any station to transmit without relying on central control but has provisions to recover from situations in which two or more stations transmit at once

circuit-switched network a network in which a physical connection from one end to the other of a data path is maintained for the duration of a period of communication; for example, the public switched telephone network

client-server network a network with one or more specialized nodes (servers) that contain files and operating software for the network; the other nodes, called clients, use the resources of the server

collision the loss of data that occurs when two stations transmit at the same time on a network

Ethernet a type of LAN that has a logical bus structure using CSMA/CD; the physical structure can be either a bus or a star

file server see *client-server network*

hub the central connecting point of a star network, to which all other nodes connect

IBM token-ring system a type of LAN that is a physical star and a logical token-ring

local-area network (LAN) a small data network that typically operates within one building or a localized group of buildings

multistation access unit (MAU) the hub of an IBM token-ring network

multitasking a term used to describe a computer operating system that allows multiple programs to run simultaneously without interfering with each other
node one station (terminal or computer) that is attached to a network
packet the smallest block of data transmitted over a network
peer-to-peer network a network in which all nodes can contribute network resources and also run local programs
print server a network node dedicated to interfacing between the network and one or more printers
radio-frequency (RF) modem a modem that modulates data onto a very high frequency carrier; used in broadband networks and recently in experimental systems using cable-television systems for data transmission
ring a network topology in which data circulates from one computer to the next in sequence
router a device that links two or more networks, passing only that information from one network that is intended for another
star a network topology in which all nodes are connected individually to a central point
token passing a method of network control that involves a short packet that circulates around the network, giving each node in turn permission to transmit
traffic messages to be transmitted over a network
velocity factor the ratio between the velocity of propagation of electromagnetic energy through a medium or along a transmission line, and the speed of light in a vacuum
wide-area network (WAN) a large network extending over an area greater than that of a city

Questions

1. Write out in words and explain the meanings of the terms LAN and WAN. How do they differ?
2. Explain how a telephone system can be used to switch data.
3. List three basic topologies for LAN use, and explain the differences among them.
4. How is it possible for a network to have different physical and logical topologies? When this is the case, what is the usual physical layout? Why?
5. Explain how a token-ring network determines which node can transmit.
6. How does a CSMA/CD protocol allow a node to decide whether to transmit?
7. What is a collision? With which topology are collisions likely? How are they dealt with?
8. What limits the total length of an Ethernet network?
9. What types of cable can be used with an Ethernet network? What differences does the choice of cable make in the capabilities of the network?
10. What is meant by a 10BaseT system?
11. Explain how the IBM token-ring system keeps operating in spite of the failure of one of its nodes.
12. What data rates are possible with an IBM token-ring system?
13. Sometimes Arcnet is said to have a logical ring topology and sometimes a logical bus. Clear up the confusion by explaining how Arcnet determines when each station signals and to whom.
14. What is the difference between an active and a passive Arcnet hub? What are the advantages and disadvantages of each type?
15. What is a broadband LAN? What advantages and disadvantages does it have compared with baseband LANs?
16. Explain the difference between peer-to-peer and client-server networks. Which is more economical for small networks? Which is preferred for larger networks?
17. What is record locking? When and why is it necessary?

Problems

18. Calculate the length of time it takes for a signal to pass down an Ethernet bus that is 300 m in length, assuming the velocity factor of the cable is 0.66.
19. For the Ethernet bus in problem 18 and a communication rate of 10 Mb/s, calculate the total number of bits that would be sent by each station before it detected a collision if:
    (a) both stations begin to transmit at the same time
    (b) one station begins to transmit at the first possible moment such that its transmission would not be detected by the other station before it began its transmission

15 Wide-Area Networks

Objectives On completing this chapter, you should be able to:

1. Describe and explain the differences between local-area, metropolitan-area, and wide-area networks
2. Describe and explain three ways of organizing a wide-area network and discuss the advantages and disadvantages of each
3. List the seven layers of the OSI protocol model and explain their meaning
4. Describe the X.25 packet-switching protocol
5. Explain the uses of frame relay and asynchronous transfer mode communications
6. Describe the Internet and explain how to use its major features

15.1 Introduction

The local-area networks we have studied so far differ in many details, but they have several things in common: limited distances, a relatively small number of nodes, and continuous, exclusive use of the communications medium, whether it is twisted pair line, coaxial cable, or optical fiber. We now turn our attention to larger networks: **metropolitan-area networks (MANs),** which can extend across a city, and **wide-area networks (WANs)**, which can cover a nation, a continent, or even the world. Many of these systems are actually networks of networks and thus link local-area networks at many locations.

15.2 Basic Network Structures

Unlike local-area networks, most wide-area networks use lines provided by telephone companies rather than by the network users. There are three basic ways of organizing the lines. Dedicated, leased lines can be used, or lines can be shared by using either circuit switching or packet switching.

15.2.1 Leased Lines

Perhaps the simplest system is for lines to be leased from telephone companies on a monthly basis. In most areas these lines are digital throughout, so it is not necessary to use modems. Leased lines are available with data rates from 56 kb/s up. The cost is considerable (often thousands of dollars per month) and depends on the data rate and the length of the line. Leased lines are most suitable when data must be transferred between two centers on a continuous basis.

15.2.2 Circuit Switching

The second way to organize a wide-area network is to use **circuit switching**. This method is actually very similar to placing a call on the ordinary switched telephone network. The connection is made only when a station has data to transmit. WAN circuit switching usually uses digital lines, making modems unnecessary, and can operate at much higher data rates than an ordinary dial-up line with a modem.

The ISDN system (see Chapter 10) is very well suited for use with circuit-switched data. The basic ISDN interface uses two 64 kb/s channels plus a 16 b/s channel for network signaling. The two 64 kb/s channels can be combined to give a data rate of 128 kb/s.

The implementation of ISDN is still in the early stages, and in the meantime there are several other ways to set up circuit-switched data connections. As on leased lines, the available data rates range from 56 kb/s up. Circuit-switched lines can be much less expensive than leased lines, as long as communication over a given path is not necessary for more than several hours a day. This is usually the case when a node has more than one other node with which communication must be maintained. The exact cost trade-off depends on the company, the data rate, the distance, and other factors.

15.2.3 Packet Switching

Packet switching is the third way to organize the lines of a wide-area network. Packets, as we saw in the preceding chapter, are short bursts of data. Each packet begins with a **header** that includes the address of the node to which the packet is being sent. Telephone companies and other service providers operate packet-switched networks. Each station of a WAN has an address, and packets of data can be addressed to it from any other station. The network automatically chooses the best route, which is not necessarily the shortest distance between the two nodes—if the shortest route is busy, a packet may be sent by a longer route. Each switching station in the network stores the packet briefly before sending it on to the next. For this reason, packet-switched networks are sometimes called **store-and-forward networks**. Packet-switched networks are often visualized as a "cloud" with various nodes connected to it. The network routes individual packets within the cloud in ways that are not apparent to the nodes (see Figure 15.1).

The nature of packet switching means that there is sometimes a short delay before packets are delivered. Usually this is much less than a second, but nonetheless it often makes real-time audio or video communication impractical. It is also quite possible that packets will be delivered out of order, but this should not cause a problem, since the packet headers contain information that lets the system rearrange the packets in the correct order.

One advantage of packet switching (compared with leased lines or even circuit switching) is redundancy. A failure in the network is not likely to prevent packets from being delivered, unless it occurs on the line between the originating or destination node and

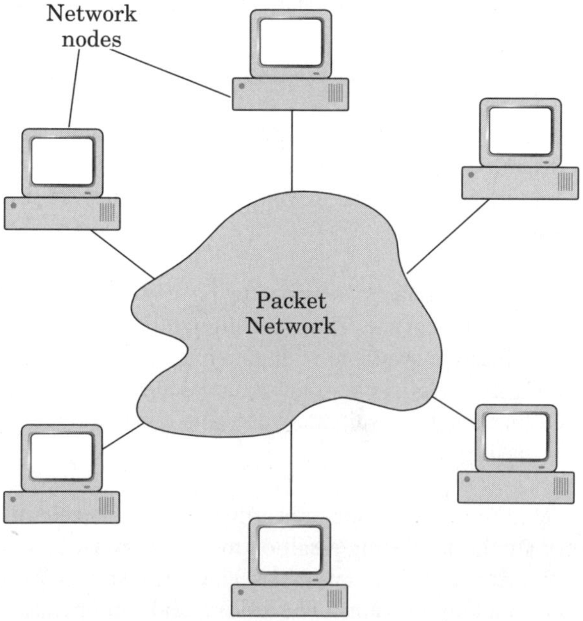

Figure 15.1 Packet-switching "cloud"

the nearest switching center. Anywhere else in the network, there should always be an alternate route in case of trouble. One of the first large packet-switched networks, ARPANET, was created by the U.S. military for just this reason. An organization can also use a public packet network to communicate with other institutions, given the proper permission (see the section on the Internet for an example).

Packet networks are very useful when many offices have to be connected. Running leased lines to several locations would be very expensive, and circuit switching has to be redone every time a message must be sent to a different destination. In contrast, a packet network can send an individual message to any station or combination of stations with a simple change in the address on the packet. Connections to packet services can be either full-period (like leased lines) or dial-in. Charges can be either flat rate, based on the capacity of the connection to the network, or usage-sensitive, based largely on the amount of data transferred.

15.3 Network Protocols

Any data communications network requires some system to organize the flow of data. The system is normally specified as a hierarchy of **protocols**. There are many ways, probably too many, of arranging these protocols, and there has been a historical tendency for each manufacturer to devise its own proprietary standard. As networks grow and link to other networks, however, it has become necessary to develop some standard protocols and to find ways to translate from one protocol to another.

15.3.1 The Open Systems Interconnection Model

The International Standards Organization (ISO) has proposed a standard way of arranging protocols known as the **Open Systems Interconnection (OSI) model**. OSI is not a protocol but a description of the way in which protocols should work. Adherence to this model would allow equipment from different vendors to work together. Unfortunately, such adherence is by no means universal, but many real-world products do follow the model at least in general, if not in every particular.

Two other protocol models that predate the OSI system are still in widespread use: IBM's **Systems Network Architecture (SNA)** and Digital Equipment's **DEC Network Architecture (DNA)**. Both of these systems are similar but not identical to the OSI model.

The OSI model has a total of seven layers. A graphic representation of the model is shown in Figure 15.2. Moving up the chart means moving to a more complex and abstract area. The three lowest levels are the ones that will interest us most.

The lowest level, the **physical layer**, concerns the way the hardware transmits data to the network. For example, it involves specifying the correct voltage levels and pulse timings to represent ones and zeros. The RS-232 standard described in Chapter 13 is an example of a physical layer protocol.

The **data link layer** detects and corrects errors within frames of data and also provides "flags" (the bit patterns that indicate the beginning and end of a frame).

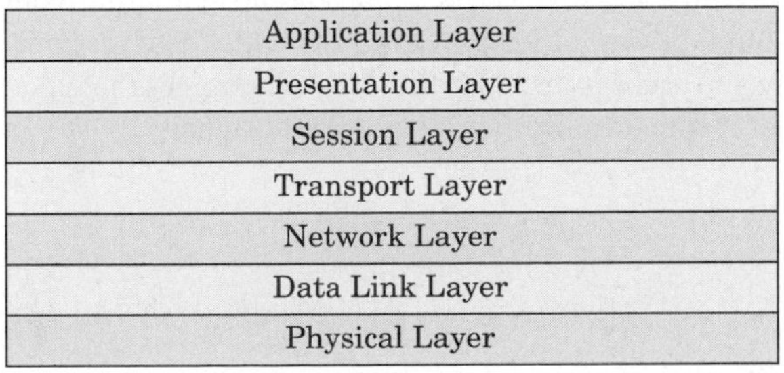

Figure 15.2
The OSI model

Figure 15.3
X.25 frame

Flag	Address	Control	Data	Frame Check Sequence	Flag

The **network layer** sets up the path to transmit data between terminals and arranges data into packets. Each packet needs to be supplied with a header containing the address used to route the data through the network.

At the higher levels, the network protocol may have to allow incompatible machines to talk to each other—IBM PCs to Apple Macintoshes or UNIX workstations to Digital Equipment VAX minicomputers, for example. The higher levels of the OSI model are concerned with such matters, as well as with network security (passwords and such) and billing details. These topics are beyond the scope of our discussion.

The Ethernet system discussed in Chapter 14 contains the bottom three layers of the OSI standard, as do the IBM token-ring system and the Arcnet protocol. In the next section, we will look at an implementation that is widely used in packet-switched wide-area networks.

15.3.2 The X.25 Packet-Network Protocol

The **X.25 protocol**, developed by the CCITT, is used in many packet networks to control the interface between an individual data terminal and the network. It corresponds to the bottom three layers of the OSI model. The lowest level of X.25 (the physical layer) can have several forms, one of which is identical to the RS-232 interface already discussed. The second layer of X.25, the frame layer, corresponds to the data link layer of the OSI model. The frame layer organizes data into frames, which are constructed as shown in Figure 15.3. Each frame is divided into six fields.

The first and last fields are identical bit patterns (01111110) that act as **flags** to mark the beginning and end of a frame. After the starting field comes the address, followed by a control field that contains information about the progress of the message, such as whether this is the starting or ending frame, whether errors have occurred, and whether they have been corrected. After the control field comes the actual information to be transmitted. The amount of information in one frame can vary between 16 and 1024 bytes. Finally, there is a frame check sequence for error control and the ending flag.

The third layer of the X.25 protocol is called the *packet layer*. It corresponds to the OSI network layer. This part of the protocol converts the frames described above into packets suitable for the network. Four fields are added between the control field and the data, as shown in Figure 15.4.

The general format identifier (GFI) includes, among other things, the number of the frame being sent. This number is important, because the routing in a packet network can change during the transmission of a message, and packets can get out of order. The logical channel group number (LCGN) and the logical channel number (LCN) give the number of a **virtual circuit** set up by the network. A virtual circuit is not a circuit in the sense that a circuit-switched network has circuits. The physical route of a connection is likely to be different every time it is made and can even change during communication. The logical channel is simply a way of keeping track of which two stations have messages for each other. Finally, the packet type identifier (PTI) determines the function of the packet, specifically, whether it contains information or control signals.

Figure 15.4
X.25 packet

Flag	Address	Control	GFI	LCGN	LCN	PTI	Data	Frame Check Sequence	Flag

15.3.3 Frame Relay and Asynchronous Transfer Mode

The X.25 packet-network protocol is well established in public data networks. It is famous for being robust, that is, its error-checking and error-correcting methods allow for very reliable communication, even over less-than-optimal channels. It is not the fastest possible protocol, however, and recently, two alternatives to X.25 have begun to gain popularity.

Frame relay is very similar to X.25, except that it has less built-in error correction. Removing some of the error correction reduces overhead and increases the net data-transmission rate, provided that the channel has a low bit-error rate. Modern communications methods, such as digital communication using fiber optics, tend to have very low error rates compared with, for example, a modem used with an ordinary dial-up telephone line. For digital channels, frame relay can provide faster transmission with acceptable error correction.

Both the X.25 and frame relay protocols allow for variable-length information packets. The maximum length is 4096 bytes, but the default is 128 bytes. The advantage of using a longer packet instead of several shorter packets is that it reduces overhead, since only one header is needed and fewer bits are required for error control. Long packets have the disadvantage that they require more memory at switching points, since the whole packet has to be stored before being forwarded to the next node in the network. Long packets can also cause synchronization problems when real-time transmission of audio or video signals is attempted, because other packets can be delayed for several milliseconds while a long packet is processed. When channels with high error rates are used, long packets have another disadvantage: if a packet must be repeated due to an uncorrectable error, a long packet takes longer to retransmit than a short packet.

While long and short packets each have advantages in certain situations, using a variable length packet also causes problems. Sorting, storing, and forwarding variable-length packets require more network intelligence than would be needed if all packets were the same length. Furthermore, transferring data between systems can require disassembling and reassembling packets. There would be definite advantages to a system in which all packets were required to be the same length.

There is such a system, called **asynchronous transfer mode (ATM)** (not to be confused with asynchronous communication, which we discussed earlier). With ATM, all packets have a relatively small size, just 53 bytes. This size is designed to allow an ATM system to handle real-time digital video and audio, and it represents a compromise for data communications.

Because the small packets result in considerable overhead, ATM is best suited to high-speed channels such as fiber optics. The main application for ATM is expected to be in connection with broadband ISDN (see Chapter 11). Eventually, individual homes and businesses could be connected with a single optical fiber that could deliver audio, video, and data, all using ATM. This is one way in which the much-discussed "electronic highway" could evolve.

15.4 Connecting LANs to WANs

A wide-area network can be configured in many different ways, depending on the needs of the organization that establishes it. A central mainframe computer can communicate with a large number of remote terminals, much as it would if the terminals were in the same building. Mainframes can also share data with each other over networks. Currently, however, one of the most popular ways to configure a WAN is as an extension of the local-area networks at each of the locations served by the WAN. When this is done, any microcomputer in any part of the organization can communicate with any other, just as if the two computers were part of the same LAN.

LANs can be connected to each other and to larger networks in a variety of ways. The connecting devices, in order of increasing complexity, are **repeaters, bridges, routers**, and **gateways**.

Repeaters are simple devices that regenerate and retransmit packets. They allow the

Figure 15.5
Use of bridges

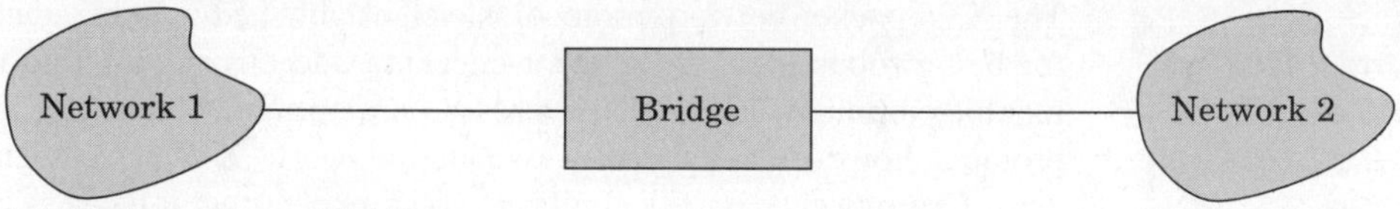

(a) Joining adjacent LANs

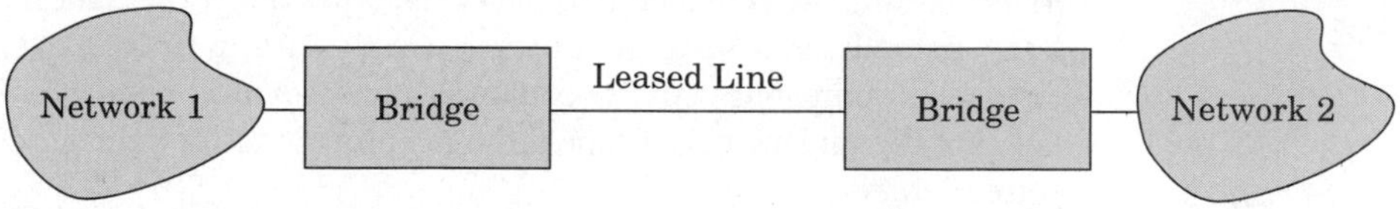

(b) Joining LANs linked by leased telephone line

distance spanned by a LAN to be increased substantially (from 500 m to 2500 m for Ethernet, for instance). Some repeaters can connect different media, such as coaxial cable, twisted pair, and optical fiber. Since repeaters do not do anything with the data but deal only with voltages and timing, they are considered to operate at the lowest level, the physical layer, of the OSI model.

Bridges are slightly more complex than repeaters. They examine the address field of each data packet and determine whether the packet should be restricted to the originating LAN or passed on to another section of the network. The decision can be based on a table programmed into the bridge by the network manager. More sophisticated bridges can develop the table themselves by broadcasting a message that generates a reply from all stations and then noting the location of each node from the replies.

Bridges operate at the second layer, the data link layer, of the OSI model. A single bridge can be used to connect two adjacent LANs, but if they are separated by long links that operate at a different (usually lower) data rate, two bridges are needed, one at each end of the long-distance link (see Figure 15.5.)

Routers, which are even more complex than bridges, can change packets from one protocol to another. For instance, an Ethernet packet can be converted to an X.25 packet or a TCP/IP (see section 15.5.1) datagram for transmission through a wide-area network and then converted back at the other end. Routers pass messages from one part of a network to another, guided by information in the address of each packet. Routers operate at the third level, the network layer, of the OSI protocol model.

Gateways are used to connect computers running otherwise incompatible systems, such as different types of mainframes or different electronic-mail systems. They operate at the higher levels of the OSI model and can disassemble packets of data and reassemble them according to the standards of different programs. See Figure 15.6 for an example of a network using routers and gateways.

Figure 15.6
Routers and gateways

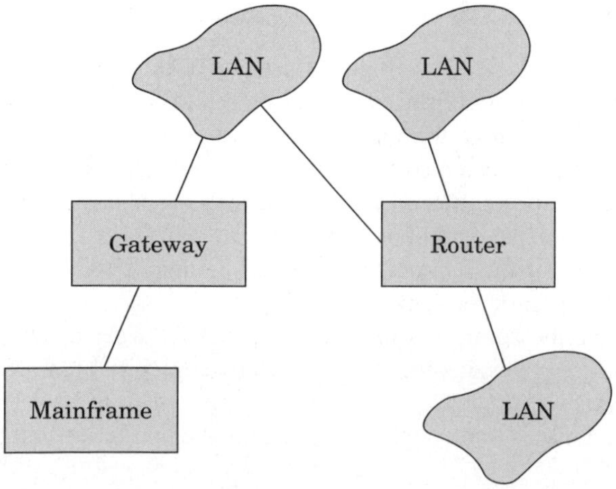

15.5 The Internet

The **Internet** is a good example of a network of networks. It is a packet-switched network that links many computers and computer networks. The Internet began as ARPANET, a network organized by the United States military with the intention of making it robust enough to survive nuclear war. To that end, the network was designed to be, and remains, as decentralized as possible. Unlike commercial information services like Compuserve or America Online, the Internet has no central host computer, since that would render the whole system inoperative if the central computer malfunctioned or were destroyed. Millions of computers are connected to the Internet, and there are many ways to get a message from one to another.

As time went on, the United States military found more secure ways to communicate, and the Internet became mainly an academic network linking universities, colleges, and government and other organizations dedicated to science and technology. Much of the funding came from the National Science Foundation. Recently, many commercial companies have also connected to the network, particularly through the World Wide Web, which will be described shortly.

The Internet is one of the great communications success stories. It is growing exponentially, at the rate of about 15% per month. At this writing, something like thirty million people have access to the Internet (no one knows for sure, since one Internet address often represents many potential users, only some of whom actually use the Net). No doubt that figure will be obsolete by the time this book is printed.

The great success of the Internet can actually cause problems. There is no single, central registry of available resources, for instance. Many books listing Internet addresses have been written, but they rapidly become obsolete. Various directories can be downloaded from the Internet itself, and while they tend to be more up-to-date than printed matter, they are never complete. Perhaps the best way to find information on the Net is to use the various search tools that will be described later in this section.

15.5.1 Connecting to the Internet

Individual computers and computer networks are connected to the Internet through routers, which convert the protocols used locally to a common network protocol called **Transmission Control Protocol/Internet Protocol (TCP/IP)**. The physical connections use leased telephone lines with data rates ranging from 56 kb/s to 1.544 Mb/s (T1), and even higher (T3 at 45 Mb/s) in the case of major "backbone" components.

Individual users with microcomputers can have different types of Internet access. The simplest is to have the user to operate as a dumb terminal on a computer with an Internet connection, which can be a mainframe or a minicomputer. The user can be local or can be connected to the host via a modem connection. The user logs on to the host in the usual way and then runs programs on the host to connect to the Internet.

A better way to access the Internet is as a node on a LAN whose server is connected to the Internet using a router. Then the network-access programs can be run on the individual microcomputer. Modem access to the server can use either of two protocols, SLIP (Serial Line Internet Protocol) or PPP (Point-to-Point Protocol), both of which allow graphics-oriented programs to operate.

Individuals not affiliated with an organization with a network connection can connect by way of commercial networks that are themselves linked to the Internet. This is done through gateways and routers, which are computers that are part of both networks. For example, subscribers to the Compuserve, America Online, GEnie and MCIMail networks, among others, can link to the Internet. There are also many local commercial access sites that specialize in Internet access via modem for individual users, and some noncommercial sites, known as **freenets**, also provide access.

Here is an example of a typical Internet connection. The author of this book is writing it on an IBM-compatible personal computer in his home. This computer can be connected by a modem and an ordinary dial-up telephone line to a Digital Equipment VAX computer at Niagara College in the small city of Welland, Ontario. The college computer is linked by a router and a 56 kb/s leased telephone line to Brock University in St. Catharines, a somewhat larger city about 20 km away. From the university there are two possible connections, to universities in either Toronto or London, Ontario. These links make the college part of ONET, an Ontario government network that is part of the Internet. As a result, the author can log on to computers and send mail to colleagues anywhere in the world.

15.5.2 Internet Structure and Protocols

The Internet is a very complex network capable of linking many normally incompatible types of computers, ranging from Apple and IBM-compatible microcomputers running MS-DOS, UNIX, and other various operating systems to DEC VAX minicomputers with their proprietary VMS operating system to IBM mainframes. Furthermore, there are many possible ways to get a message from one point to another. The network itself determines the most efficient route for each packet of data. Thus it can continue to operate even if one or more of its constituent computers or connecting lines are unavailable.

The Internet operates under a set of protocols known as **TCP/IP (Transmission Control Protocol/Internet Protocol)**. Originally created by the United States Defense Department in the 1970s, this protocol has proved to be a robust and adaptable means of allowing many different kinds of computers and computer networks to work together. Data is transmitted in the form of packets called *datagrams*, each of which contains a source address and a destination address, as well as data. Many other wide-area networks besides the Internet use the TCP/IP protocol set.

TCP/IP is currently being revised to increase its versatility still further; especially pressing, because of the rapid expansion of the Internet, is the need to increase the number of possible addresses. Currently each Internet address has 32 bits, but this will be quadrupled in the revised protocol.

Each Internet host has an address consisting of a network number, a subnetwork number, and a host number. Addresses are given as four groups of base 10 numbers, each group representing the value of eight of the address bits, for example, 123.45.67.89. Most addresses are translated into an easier-to-remember alphabetical form as well. Unlike Ethernet LANs in which all messages are transmitted to all nodes and each node determines whether it should read or ignore the message, Internet nodes have lookup tables that help them to route messages to their destinations. It would be impossible for each node to know the address of every other node, especially with the network doubling in size each year, but each node has enough information to route traffic in the right direction. For instance, the node at Niagara College sends all traffic to Brock University, so that is the only address it needs. As a message moves into larger network centers, it is routed by its domain name to any of various subnetworks, such as the ONET network. From the subnetwork, the message is routed to its final destination.

The TCP/IP protocol set supports many Internet applications. Some of the more common ones are described below. We will begin with relatively simple applications and conclude with the World Wide Web, which is the cause of much current excitement.

15.5.3 Electronic Mail

Electronic mail (e-mail) is the most basic way of communicating via the Internet. Many offices and schools already have in-house electronic mail based on a fairly simple system that allows messages to be left for other network users in files on a central computer or file server. The addressee is sent a message that mail exists and can read it at leisure.

Internet mail extends the range of electronic mail worldwide. Mail can be sent by addressing it to a specific person at a particular domain. The domain is the address of the Internet site. For example, e-mail can be sent to the author of this book at the address rblake@niagarac.on.ca. "rblake" refers to a specific person, "niagarac." refers to a computer at Niagara College, "on." indicates that the college is connected to ONET, a network that connects government and educational sites in the province of Ontario, and "ca" shows that the country is Canada. If the college were in the United States, the last part of the address would probably be "edu" (for education). Since the Internet started in the United States, American sites do not need a country address.

Internet mail uses an application called the **Simple Mail Transfer Protocol (SMTP)**, which is part of the TCP/IP suite and is available for many types of computers.

15.5.4 Mailing Lists

In addition to e-mail among individuals, there are many automated *listservers* that operate the electronic equivalent of mailing lists. They are located mostly on the *Bitnet* (*b*ecause *it*'s *t*ime) network, but since Bitnet is connected to the Internet, they can be accessed from Internet hosts. You can subscribe to one of these lists by sending an e-mail request to the appropriate address. In addition, it is often possible to "post" short articles to these lists for dissemination to all subscribers. Some mailing lists are open to contributions from anyone, while others are "moderated," that is, there is a person who reads postings and discards any that seem to be unsuitable.

In addition to the automated Bitnet listservers, there are many mailing lists that are operated by other organizations. Some are automated and some are not. The method of subscribing to these groups varies.

15.5.5 Usenet Newsgroups

Newsgroups are somewhat similar to mailing lists, but they operate on a computer-to-computer rather than a person-to-person basis. They form a system known as *Usenet.* Like Bitnet, Usenet is not actually part of the Internet, but the two networks are linked.

At this writing there are over 14,000 newsgroups covering practically every imaginable topic. A given site can subscribe to as many Usenet newsgroups as desired. Anyone wishing to read newsgroup messages simply accesses a computer that connects to Usenet, without having to be placed on a mailing list individually. This reduces the load on network resources, since all users of a site have access to all newsgroups that are maintained at that site, without having to individually obtain the material over the network. Users can read messages and can post their own messages on the newsgroup. They can also use e-mail to respond individually to a person who posted a message.

15.5.6 Telnet

Telnet allows a user to log on to a distant ("foreign" in Internet jargon) host computer just as if he or she were sitting at a terminal in the same building as the computer. Telnet is very useful for people with accounts on more than one computer. There are also many Internet sites that allow public access to some of their computing resources. For instance, many university computers allow public access to library catalogs and often to other information as well.

Probably the most famous Telnet sites are the Freenets. These computers are accessible to the public via modem for local use or via Internet Telnet for people out of the local telephone calling area. They use a relatively simple, menu-driven interface set up as an analog of a town. There may be a city hall menu for legal documents, a hospital for medical advice, a house of worship with material on religion, a school with bulletin boards for teachers and students, a meeting hall for discussion groups, a science and technology center, and so on. Often there is a "teleport" that allows access to other Internet sites.

15.5.7 File-Transfer Protocol

Another important use of the Internet is to transfer files, either text files or data (binary) files, from one computer to another. Transferring files can be done with the Internet **file-transfer protocol (FTP)**, which is part of the TCP/IP protocol suite. Many public-domain files can be obtained through *anonymous ftp*, in which the user connects with the user name *anonymous* and uses her or his e-mail address as a password. The user then moves through a file directory structure that is something like MS-DOS (and even more like UNIX). Upon locating a useful-looking file, the user can transfer it to his or her host system by simply issuing the command "get."

15.5.8 Internet Search Tools

Several programs have been developed to search the Internet for information, almost as if it were a very large database. Of course, the network is not a single, unified database but many different ones using different protocols, menu systems, directory trees, and so on.

This makes the search more complex and less thorough than it would be with a single well-organized database. The most useful search programs used are listed below.

- Archie: searches for files in FTP sites by file name or partial file name and gives the file name and the source where it may be found. Archie works very well for finding computer program files, provided the name is known.
- Gopher: organizes available information and presents it in a menu form, by topic. Information can be read from the screen after proceeding through a hierarchy of menus. Files can be transferred using FTP. Gophers can link with other gophers at other sites to provide worldwide gopher access from a single location. Among other things, gopher is useful for reading Usenet news from sites without direct access to Usenet.
- Veronica: similar to Archie, except that it uses a keyword to search the same resources gopher accesses.
- WAIS (Wide Area Information Server): allows a user-selected combination of available databases to be searched for keywords.

15.5.9 The World Wide Web

The **World Wide Web (WWW)** is a unique, linked database system begun at the European Particle Physics Laboratory (CERN) in Switzerland. The Web started as a method of finding scientific documents. Recently it has greatly expanded and is now used by many organizations of all types. The Web is especially popular with commercial companies, who have set up *home pages* on the Web, although some individuals have also done so.

Simplicity of use is the main reason for the sudden increase in the popularity of the World Wide Web. Equipped with the proper "Web browser" software, users can view text and graphics, hear sound, and even watch video clips. They can also download files as required, without needing any separate search tools or file-transfer protocols. The Web is much more user-friendly than gopher or telnet, since a standard interface is used for everything. Many documents on the Web are linked to other documents, which may be located on the same server or in another computer on the other side of the world. The user need not be concerned with and may never know the location of these files.

The user can begin an exploration of the Web by starting at an organization's *home page*, which is generally a hypertext document with links to other pages containing more detailed information. These pages are often lavishly illustrated and may even feature downloadable audio or video clips. From the home page, the user can move to other pages supplied by the same organization. Usually there are also convenient links to other organizations in the same or similar fields.

The Web uses a protocol known as **Hypertext Transport Protocol (HTTP)**, which allows quick transitions when the user simply clicks on a highlighted word or picture with a mouse or moves a cursor to the highlighted word and presses one key on the keyboard. The protocol determines the type of file and the way to display it, depending on the user's computer software. Some Web browsers, such as Netscape and Mosaic, can display text and graphics; others, notably Lynx, show text only and replace all graphics with the word "image." The latter gives faster results over slow Internet connections.

The Web has its own search tools. Several commercial companies allow free public use of search tools as an advertisement for their other products. Probably the best known such site is at the address: http://www.yahoo.com. Many other search engines can be reached by links from that site.

Summary

Here are the main points to remember from this chapter.

1. Wide-area networks (WANs) can extend over very large areas and can even be worldwide. The three major WAN topologies are leased lines, circuit switching, and packet switching.

2. The Open Systems International (OSI) model is a way of specifying communications protocols to make it easier to design them for compatibility among different systems. There are seven layers to the OSI model.
3. The most popular protocol for packet networks is X.25, but there are two competing systems, frame relay and asynchronous transfer mode (ATM).
4. Networks are increasingly being connected together to form very large communication systems. The largest of these, the Internet, is worldwide in scope.

Glossary

asynchronous transfer mode (ATM) a versatile data-transmisssion system using 53-byte packets and designed to enable various kinds of data, including live audio and video, to be multiplexed

bridge a network-to-network connection that passes only data addressed to a node in the other network

circuit switching a method of organizing a network in which a physical path is dedicated to communication between two nodes for the duration of the communication

data link layer the second layer of the OSI model, responsible for detecting and correcting errors within frames of data and providing the flags that indicate the beginning and end of frames

DEC Network Architecture (DNA) a data-communications protocol created by Digital Equipment Company for its minicomputer products

electronic mail (e-mail) a method whereby messages can be left for individual network users; messages are generally stored on the server in such a way that only the designated recipient can access them

file-transfer protocol (FTP) a program in the TCP/IP system that allows for the transfer of both binary and text files between computers with otherwise incompatible operating systems

flag a bit or sequence of bits that indicates the beginning or end of a packet

frame relay a packet-transport protocol, similar to X.25 but with less error correction, making it faster over low-error-rate channels

freenet a computer system, normally run by a nonprofit organization, that provides modem access to the public at no charge (though donations are encouraged); a typical freenet has bulletin-board areas for local messages, and many have Internet access

gateway a device used to connect computers or computer networks running incompatible operating systems

header a sequence of bits at the beginning of a packet, containing information about the type of packet and/or routing

Hypertext Transport Protocol (HTTP) a data-transfer method that allows quick transitions by simply clicking on a highlighted word or picture with a mouse or moving a cursor to the highlighted word and pressing one key on the keyboard; the protocol determines the type of file and the way to display it, depending on the user's computer software

Internet a worldwide public network of networks that connects a very wide variety of computers, applications, and users

metropolitan-area network (MAN) a network that encompasses a city and its environs

network layer the third layer of the OSI protocol model; the network layer sets up the path to transmit data between terminals and arranges data into packets

newsgroups electronic bulletin boards devoted to a wide variety of subjects, accessible via the Internet

Open Systems Interconnection (OSI) model a system for organizing data-transmission protocols developed by the International Standards Organization (ISO)

packet switching a way of organizing a network so that small blocks of data are routed individually from source to destination

physical layer the lowest level of the OSI protocol model, dealing with matters such as voltage and current levels

protocol a formal set of conventions governing the format and timing of messages

repeater a device used to extend a network or other digital communication system by regenerating bits and restoring voltage levels and timings to their original values

router a device used to interconnect networks; routers operate at level three, the network layer, of the OSI protocol model and can change packets from one protocol to another

Simple Mail Transfer Protocol (SMTP) the part of the TCP/IP suite that allows for e-mail over the Internet and similar networks by specifying the control messages used in mail transfer

store-and-forward network a network in which nodes receive a packet of data from the source or a node closer to the source and then transmit it to the destination or a node closer to the destination

Systems Network Architecture (SNA) an IBM system for transferring data between IBM mainframes and between IBM mainframes and other computers

Telnet a system to allow users to log on to a distant host by emulating a dumb terminal; part of the TCP/IP suite

Transmission Control Protocol/Internet Protocol (TCP/IP) a suite of protocols that allows a wide variety of computers to share the same network

virtual circuit a link between computers in which each recognizes a software connection to the other; the physical connection is not continuous but consists of packets routed as transmitted

wide-area network (WAN) any computer network that extends for more than a short distance such as a building or related group of buildings

World Wide Web (WWW) a system that allows users to access documents from widely separated sources on the Internet, using a common interface

X.25 protocol a very popular system for defining and switching data packets on computer networks

Questions

1. What are the advantages and disadvantages of using leased lines for a wide-area network (WAN)?
2. What type of data-communication requirement would lend itself particularly well to the use of leased lines for a WAN?
3. What is a circuit-switched network?
4. Describe ISDN and explain how it can be effectively used in the construction of a WAN.
5. What is the lowest data rate commonly available as a leased or circuit-switched data communications line?
6. What types of data communications particularly lend themselves to the use of circuit-switched lines?
7. What is a packet-switched network? What are the advantages and disadvantages of packet switching compared with dedicated lines and circuit switching?
8. What is a protocol? Why are protocols necessary in data communications?
9. List the seven levels of the OSI protocol model, and explain the meaning and function of the lowest three levels.
10. What level in the OSI system does the RS-232 interface represent?
11. What level(s) in the OSI system are represented by start bits, stop bits, and parity in an asynchronous communication system?
12. Explain how the X.25 protocol relates to the bottom three levels of the OSI model.
13. In X.25, what do the following terms mean: GFI, LCGN, and LCN.
14. What does the term *virtual circuit* mean?
15. Compare the frame relay and asynchronous transfer mode protocols with X.25, and suggest circumstances in which each would be preferred.
16. Explain the differences among the following: repeaters, bridges, routers, and gateways.
17. Briefly describe the origins and the present composition and extent of the Internet.
18. What is the name of the protocol used by the Internet?
19. Briefly describe how electronic mail is used on the Internet.
20. Describe two ways of disseminating news and opinion on the Internet. Which of these is generally more efficient?
21. Describe and explain the difference between Telnet and FTP on the Internet.
22. Describe and explain the differences between the Veronica and Archie programs for data searches on the Internet.
23. What is a gopher and what resources can it find on the Internet?
24. Explain briefly the meaning of the term *hypertext*.
25. How does the World Wide Web simplify the search for data on the Internet?
26. What is a *home page*?
27. Name at least one graphics-oriented Web browser and one text-only browser, and explain the difference. Are there any situations where the text-only browser would be preferred?
28. How are searches performed on the World Wide Web?

Problems

SECTION 3

29. Sketch one packet of an X.25 message. Note on your sketch the name and function of each of the various fields.
30. Discuss the advantages and disadvantages of shorter and longer packets for different link error rates and different types of data.

SECTION 4

31. Which of the types of hardware described in this section would you employ for each of the following applications? Explain your reasoning.
    (a) A LAN must be extended to reach an outlying campus building.
    (b) Two local-area networks in adjoining departments must be connected so that occasional messages can travel from one to the other.
    (c) Several local-area networks, all running the same operating system, must be connected to a wide-area network.
    (d) A wide-area network requires connections to several metropolitan-area networks and many local-area networks equipped with a variety of hardware and software configurations.

COMPREHENSIVE

32. What form of network topology would you recommend in each of the following situations? Refer to this chapter and the preceding one.
    (a) A central mainframe computer works with a number of *dumb terminals* (that is, machines consisting of a display and a keyboard, used only to input data into the mainframe and read results from the mainframe). The mainframe and the terminals are located in the same building.
    (b) A computer in a branch of a business needs to communicate continuously with head office.
    (c) A computer in a branch of a business needs to report once a day for about an hour.
    (d) Many offices of an international organization need to communicate with each other on a sporadic basis.
    (e) Ten computers on a college campus need to communicate with each other on an occasional basis to send large data files.

PART

III

Radio and Optical Transmission Systems

16 Transmission Lines

Objectives After studying this chapter, you should be able to:

1. Give several examples of transmission lines and explain what parameters of a transmission line must be considered as the frequency increases
2. Define characteristic impedance and calculate the impedance of a coaxial or open-wire transmission line
3. Describe the responses of a matched and a mismatched transmission line to a step or pulse input
4. Define reflection coefficient and standing-wave ratio and calculate them in practical situations
5. Explain the importance of impedance matching with respect to transmission lines and describe several methods of matching lines
6. Perform the necessary calculations to achieve an impedance match using a quarter-wave transformer and a single stub
7. Describe the function and use of several types of transmission-line test equipment

16.1 Introduction

So far in our study of communications systems, we have essentially looked at the two ends: the transmitter and the receiver. In the next few chapters, we will study the middle, which we referred to in Chapter 1 as the *channel*. A signal can proceed from transmitter to receiver by a variety of means, including metallic cable, optical fiber, and radio transmission. We begin with metallic cable, which is referred to as a **transmission line**.

Almost any configuration of two or more conductors can operate as a transmission line, but there are several types that have the virtues of being relatively easy to study as well as being very commonly used.

Figure 16.1 shows some **coaxial lines**, in which the two conductors are concentric and are separated by an insulating dielectric. Figure 16.1(a) shows a solid-dielectric cable; the cable in Figure 16.1(b) uses air (and occasional plastic spacers) for the dielectric. Sometimes the conductors are separated by a single helical (spiral) spacer, as shown in Figure 16.1(c). When high power is used, it is important to keep the interior of the line dry, so these lines are sometimes pressurized with nitrogen to keep out moisture. Coaxial cables are referred to as *unbalanced lines* because of their lack of symmetry with respect to ground (usually the outer conductor is grounded).

A number of parallel-line cables are sketched in Figure 16.2. Figure 16.2(a) shows television twin-lead. The two conductors are separated by a thin ribbon of plastic, but air actually forms a good part of the dielectric because the electric field fills the space around the wires, not just the region directly between them. Figure 16.2(b) illustrates an **open-wire line**, in which the amount of solid dielectric is greatly reduced, with only a few spacers being required to separate the conductors.

Figure 16.1
Coaxial cables
(a) Belden, Inc.
(b) and (c) Courtesy Andrew Corporation.

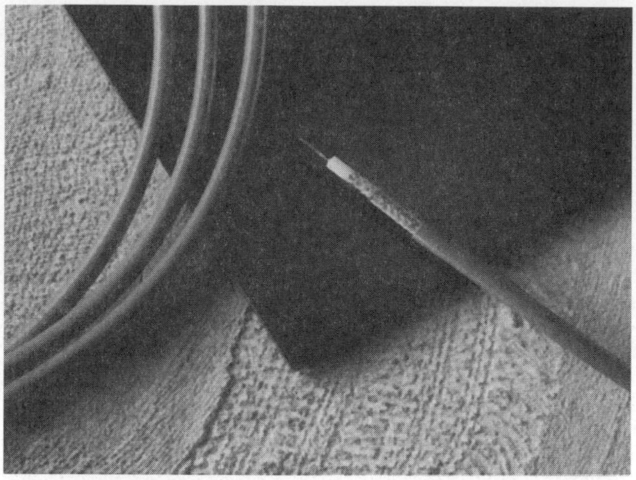

(a) Solid dielectric

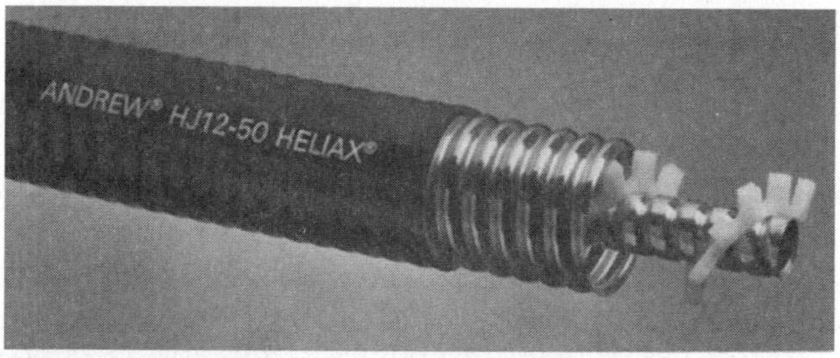

(b) Air dielectric

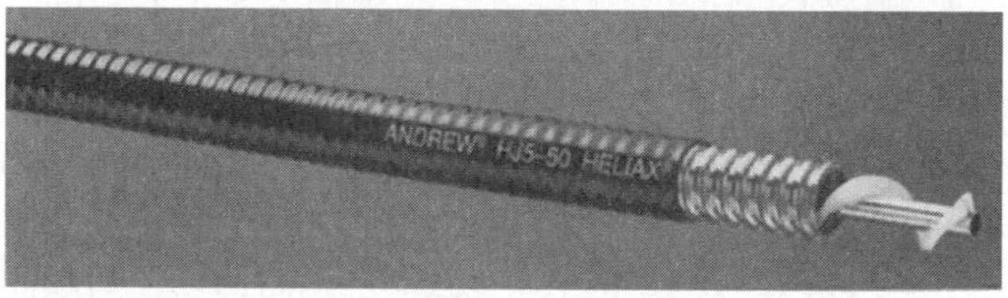

(c) Heliax®

Parallel lines are usually operated as *balanced lines*; that is, the impedance to ground from each of the two wires is equal. This ensures that the currents in the two wires will be equal in magnitude and opposite in sign, reducing both radiation from the cable and its susceptibility to outside interference. If required, a grounded shield can be placed around the cable, as shown in Figure 16.2(c). The shield plays no part in transferring the signal from source to destination; its only function is to reduce noise and interference.

At first glance, it may seem that there is very little to say about transmission lines. The wire connections that are familiar to us from everyday life (or from basic electricity and electronics) are often considered ideal. Like the lines representing wires on a schematic diagram, they are assumed to have no effect on the signals that pass through them.

When lines used at low frequencies are not considered ideal, they are usually represented as having only resistance. It is the resistance of a household extension cord that causes the voltage, under load, to be lower at the load than at the source. Resistance is easy to calculate and to measure, and its effects are very predictable.

Transmission lines become more complex in their behavior as frequency increases. Besides resistance in the conductors, inductive and capacitive effects become important. As

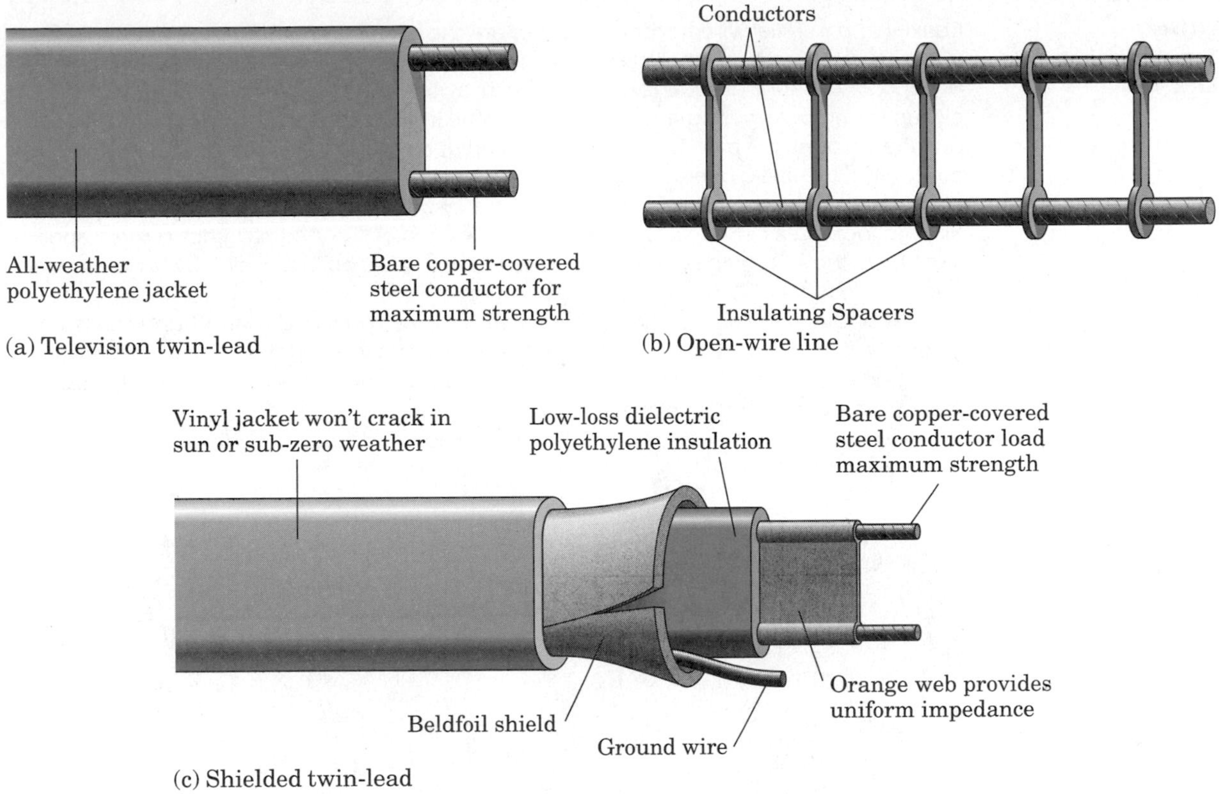

Figure 16.2
Parallel-line cables
(a) and (c) Belden, Inc.

we discussed in Chapter 2, these factors are more important with higher frequencies and longer lines.

Transmission lines are seldom the most interesting part of a communications system. In fact, the best transmission line is often the least interesting one: the one that approaches most closely the ideal of a line with no loss and no reactance. However, it is necessary to understand transmission lines in order to obtain this pleasant (if unexciting) result, which allows signals to move from source to load with minimal attenuation or distortion.

Transmission lines can have other uses as well. They can be used to provide phase shifts and time delays and to match impedances. A length of transmission line can even be used as a filter or as a tuned circuit in a receiver or transmitter stage. All of this makes the study of transmission lines important and necessary.

16.2 Electrical Model of a Transmission Line

In the analysis of transmission lines, it is necessary to use distributed rather than lumped constants. The factors that must be considered are as follows.

First, there is the resistance of the line. This idea is familiar from low-frequency applications, but there is one difference: resistance increases with frequency. Any current flow in a conductor is associated with a magnetic field, both within the conductor and in the space surrounding it. At high frequencies, the magnetic field within the conductor causes most of the current to flow near its surface. As the frequency increases, the region of high current density becomes thinner, reducing the effective cross-sectional area and increasing the resistance of the conductor. Because most of the current flows in a thin region resembling a "skin" near the surface of the wire, this phenomenon is called the *skin effect.* The skin effect explains why hollow tubing works just as well as a solid conductor for such applications as television antennas and the coils in radio transmitters for VHF and

Other Transmission Lines

Coaxial and parallel-wire transmission lines are the most common for high-frequency communications, but they are not the only possibilities. Twisted-pair line, shown in Figure 16.3(a), is popular for audio-frequency circuits, such as telephone and intercom cables. If hum pickup from power wiring and cross talk from nearby circuits are problems, a shield can be provided around the pair. The twisting results in quite good cancellation of interfering fields, especially if the line is operated in a balanced configuration with respect to ground. Twisted pairs are quite lossy at radio frequencies but are sometimes used at frequencies of several MHz (in local-area networks for computers, for example). Twisted-pair lines in these applications have the advantage of low cost; in fact, many office buildings already have extra telephone lines that can be used for this application.

A single wire can operate as a transmission line, provided that ground is used as a second, *return* connection. The early telegraph systems worked this way, with each circuit on a single wire. In fact, iron was used rather than copper, because the lower cost and greater strength of iron wire offset its lower conductivity.

Long-distance power cables, like the one shown in Figure 16.3(b), are also transmission lines.

Figure 16.3
(a) Belden, Inc.

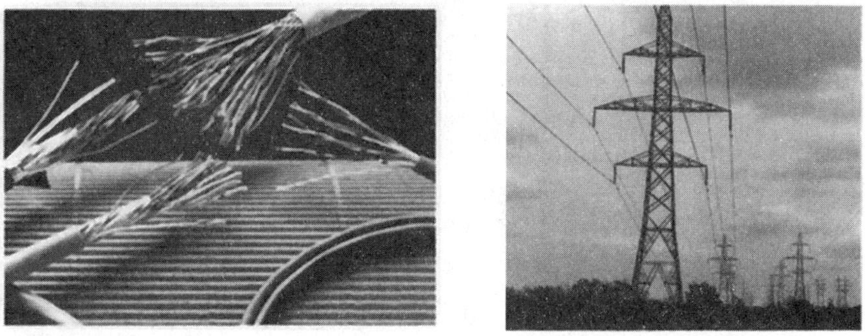

(a) Twisted-pair line (b) Electric power line

higher frequencies. We came across the skin effect once before, in Chapter 5, where we studied its effect on the variation of the Q-factor of inductors with frequency.

In addition to the resistance of the wires, we must consider the conductance of the dielectric. At low frequencies, this is very small and can be neglected. Dielectrics, however, tend to become more lossy as the frequency increases.

Any conductor or combination of conductors has inductance as well as resistance. There is also capacitance between any two conductors separated by a dielectric. Any transmission-line model must therefore include both inductance and capacitance.

It is difficult to visualize and to work with distributed constants. One way to approach the problem is to consider a short section of line and assign to it a number of lumped constants. Such a scheme is the basis for Figure 16.4, where models for both balanced (such as open-wire) and unbalanced (such as coaxial) line are illustrated. The figure allows for R, the resistance of the wire, G, the conductance of the dielectric, L, the series inductance, and

Figure 16.4
Model of a short transmission line section

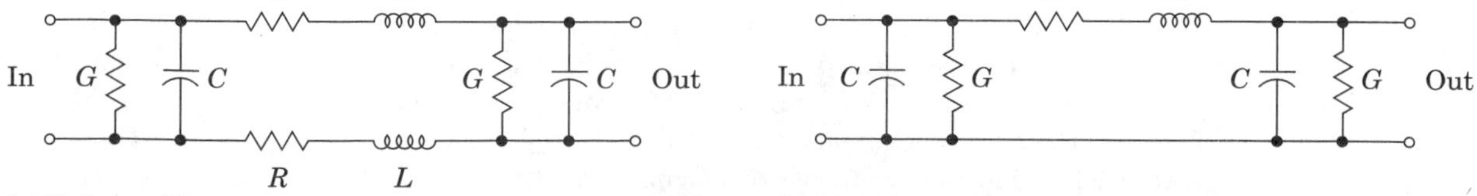

(a) Balanced line (b) Unbalanced line

C, the shunt capacitance. All of these constants will be given per unit length; for example, the resistance can be expressed in ohms per meter. However, the length of line we are considering is certainly much smaller than a meter. In fact, the idea is to allow the section to shrink until its length is infinitesimal.

At dc and low frequencies, the inductance has no effect because its reactance is very small compared with the resistance of the line. Similarly, the reactance of the shunt capacitance is very large, so the effect of the capacitance is negligible, as well. The line is characterized by its resistance and possibly by the conductance of the dielectric, though this can usually be neglected.

As the frequency increases, the inductance and capacitance begin to have an effect. The higher the frequency, the larger the series inductive reactance and the lower the parallel capacitive reactance. In fact, it is often possible at high frequencies to simplify calculations by neglecting the resistive elements and considering only the inductance and capacitance of the line. Such a line is called *lossless*, since the inductive and capacitive reactances store energy but do not dissipate it.

16.3 Step and Pulse Response of Lines

Although transmission lines are used for a wide variety of signals, it is useful to begin looking at the response of transmission lines by considering some very simple signals. It is also useful, as a starting assumption, to suppose that the line is lossless and infinite in length. That way, we will not have to worry about what happens to the signal when it reaches the end of the line. Let us consider such a line, shown schematically in Figure 16.5, and apply a step input to one end, that is, let the voltage applied to the line be zero until some time that we can arbitrarily call $t = 0$. At that time, the switch closes, connecting a source with voltage V to the line through a resistance R.

Current begins to flow into the line, charging the capacitances. A surge of energy moves along the line at some finite speed. It takes time for the current to build up in the inductances and for the capacitances to charge. If the line is infinitely long, the surge will continue along the line forever.

16.3.1 Characteristic Impedance

Returning to the source for a moment, we might ask what happens to the voltage and current at that point. Since the line has capacitance to be charged, the initial current will not be zero, as we might have expected. Even though the line is open-circuited, it does not look like an open circuit to the source. Nor does it look like a short circuit at any time, as a completely discharged capacitor would, because the inductance of the line limits the initial current. Rather, it appears that the current will have some definite, finite value that will not change as long as the surge continues to move down the line. Though all we have attempted here is to show that this is a plausible idea, it turns out to be true. There is a definite ratio between the voltage and the current for any transmission line under the conditions just described. Since the ratio of voltage to current is generally called impedance, and since the ratio we are talking about is a characteristic of the type of line used, we call it the **charac-**

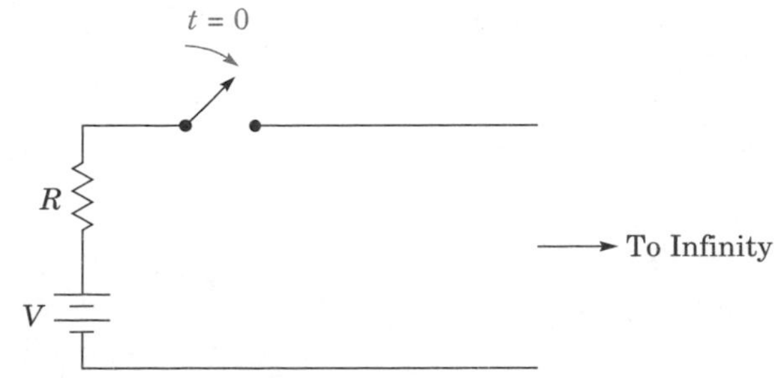

Figure 16.5
Step input applied to an infinitely long line

teristic impedance of the line. Sometimes it is also called the **surge impedance**. The impedance is a real number (that is, resistive) for a line with no losses. Note carefully that even though the characteristic impedance of a line has the units of ohms, it does not represent an actual resistance. That is, calling a transmission line a "50 Ω line" does not say anything about the resistance of the wire in the line. The hypothetical line that we are considering does not have any resistance, since it is lossless. However, since it has infinite length, energy put into the line continues to move along it forever, and looking into the line from the source, the line looks like a resistance. Instead of being dissipated as heat, the energy from the source continues to move along the line, away from the source, forever. The effect as seen from the source is the same: energy put into the line disappears.

If this is the case, then rather than continuing to deal with an infinite line that must remain a figment of the imagination, it should be possible to replace that line with one of finite length, terminated at the destination end with a resistance equal to the characteristic impedance of the line. Looking down the line from the source, it would be impossible to distinguish this finite, terminated line from the infinite line discussed earlier. Instead of moving down the line forever, the electrical energy would be converted into heat in the resistor, but no difference would be apparent from the source. A transmission line that is terminated in its characteristic impedance is called a *matched line*. Figure 16.6 shows a matched line of finite length.

The characteristic impedance depends on the electrical properties of the line. In particular, it can be shown that for any transmission line,

$$Z_0 = \sqrt{\frac{R + j\omega L}{G + j\omega C}} \tag{16.1}$$

where

Z_0 = characteristic impedance of the line in ohms
R = conductor resistance in ohms per unit length
$j = \sqrt{-1}$
L = inductance in henrys per unit length
G = dielectric conductance in siemens per unit length
C = capacitance in farads per unit length
ω = operating frequency in radians per second

This equation has several interesting mathematical properties. In general, the impedance is complex and is a function of frequency as well as the physical characteristics of the line. For a lossless line, however, R and G would be zero and Equation (16.1) would simplify to

$$Z_0 = \sqrt{\frac{j\omega L}{j\omega C}} \tag{16.2}$$

$$= \sqrt{\frac{L}{C}}$$

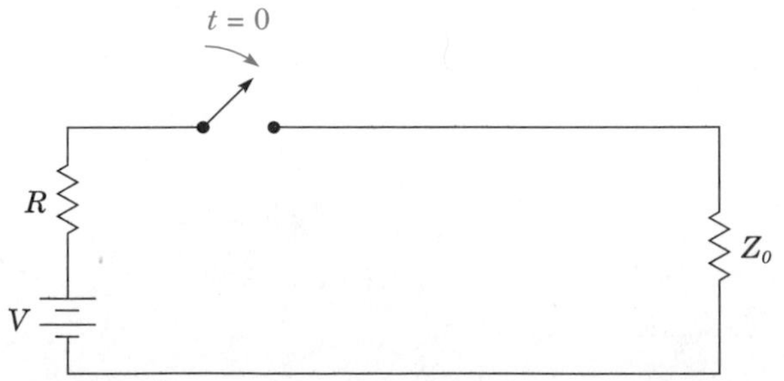

Figure 16.6
Step input applied to a matched line of finite length

Of course, there is no such thing as a completely lossless line, but many practical lines approach the ideal closely enough that the characteristic impedance can be approximated by Equation (16.2). This is especially true at high frequencies: as ω gets larger, the values of R and G become less significant in comparison with L and C. For this reason, Equation (16.2) is often referred to as the *high-frequency model of a transmission line*. Equation (16.2) gives a characteristic impedance that is a real number and does not depend on frequency but only on such characteristics as the geometry of the line and the permittivity of the dielectric. Remember that L and C are the inductance and capacitance per unit length. The unit of length does not matter, so long as it is the same for both L and C.

Example 16.1 A coaxial cable has a capacitance of 90 pF/m and a characteristic impedance of 50 Ω. Find the inductance of a 1 m length.

Solution We can do a little basic algebra with Equation (16.2):

$$Z_0 = \sqrt{\frac{L}{C}}$$

$$Z_0^2 = \frac{L}{C}$$

$$\begin{aligned} L &= Z_0^2 C \\ &= 50^2 \times 90 \times 10^{-12} \text{ H/m} \\ &= 225 \text{ nH/m} \end{aligned}$$

The characteristic impedance can be calculated for any type of transmission line by calculating the capacitance and inductance per unit length, but it can be a tedious business. Two equations presented here will allow the majority of practical cases to be dealt with. First, consider a parallel line with an air dielectric, as shown in cross section in Figure 16.7. The impedance of such a line can be found from

$$Z_0 \approx 276 \log \frac{D}{r} \tag{16.3}$$

where

Z_0 = characteristic impedance of the line
D = spacing between the centers of the conductors
r = conductor radius

Equation (16.3) is valid only for $D \gg r$. It also neglects the effect of spacers on the impedance. For open-wire ladder line, this is reasonable, but for open-wire line with a solid ribbon joining the conductors, such as TV twin-lead, the dielectric causes the characteristic impedance to be less than that given by Equation (16.3).

For a coaxial cable like the one shown in cross section in Figure 16.8, the impedance can be found from

$$Z_0 \approx \frac{138}{\sqrt{\epsilon_r}} \log \frac{D}{d} \tag{16.4}$$

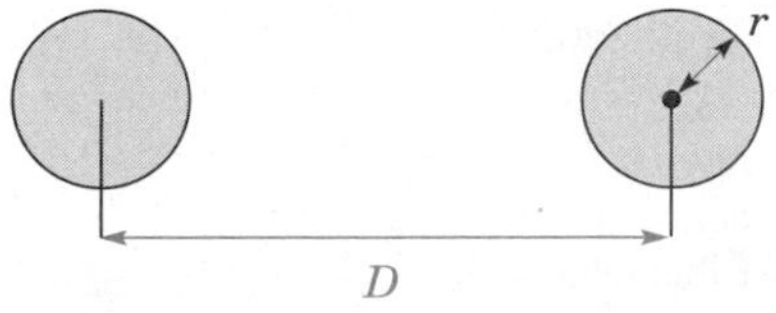

Figure 16.7
Air-dielectric parallel line: cross section

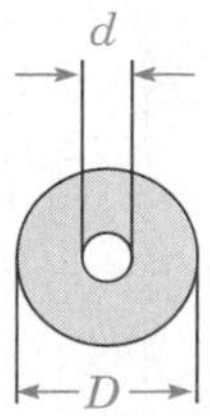

Figure 16.8
Coaxial cable: cross section

where

Z_0 = characteristic impedance of the line
D = inside diameter of the outer conductor
d = diameter of the inner conductor
ϵ_r = relative permittivity of the dielectric, compared with that of free space (ϵ_r is often called the *dielectric constant*.)

Here it is quite obvious that increasing the dielectric constant reduces the impedance of the cable.

Example 16.2 Find the characteristic impedance of each of the following lines:

(a) an open-wire line with conductors 3 mm in diameter separated by 10 mm
(b) a coaxial cable using a solid polyethylene dielectric having $\epsilon_r = 2.3$, with an inner conductor 2 mm in diameter and an outer conductor 8 mm in inside diameter

Solution (a) From Equation (16.3),

$$\begin{aligned} Z_0 &\approx 276 \log \frac{D}{r} \\ &= 276 \log \frac{10 \text{ mm}}{1.5 \text{ mm}} \\ &= 227\ \Omega \end{aligned}$$

(b) From Equation (16.4),

$$\begin{aligned} Z_0 &= \frac{138}{\sqrt{\epsilon_r}} \log \frac{D}{d} \\ &= \frac{138}{\sqrt{2.3}} \log \frac{8}{2} \\ &= 54.8\ \Omega \end{aligned}$$

In practice, it is usually not necessary to use Equation (16.4) to find the impedance of a coaxial cable, since the impedance is part of the cable specifications. In fact, the impedance is often marked right on the insulating jacket. For coaxial cable, there are a few standard impedances that fulfill most requirements. Some examples are shown in Table 16.1. The type numbers in Table 16.1 began as U.S. military specifications but have since come into general use and are employed by a number of manufacturers.

Practical open-wire lines usually have relatively high impedance. Television twin-lead is rated at 300 Ω, and open-wire lines are often 600 or 900 Ω.

16.3.2 Velocity Factor

We mentioned above that a signal moves down the line at some finite rate. This is obvious—even in free space, nothing can move at a velocity greater than the speed of light, which is given approximately by

$$c = 3 \times 10^8 \text{ m/s} \qquad (16.5)$$

Table 16.1 Coaxial cable applications

Impedance (ohms)	Application	Typical type numbers
50	radio transmitters	RG-8/U
	communications receivers	RG-58/U
75	cable television	RG-59/U
	TV antenna feedlines	
93	computer networks	RG-62/U

The speed at which energy is propagated along a transmission line is always less than the speed of light. The **propagation velocity** of a signal varies from about 66% of the speed of light for coaxial cable with solid polyethylene dielectric to 78% for polyethylene foam dielectric to about 95% for air-dielectric cable. Rather than specifying the actual velocity of propagation, it is normal for manufacturers to specify the **velocity factor**, which is given simply by

$$v_f = \frac{v_p}{c} \tag{16.6}$$

where

v_f = velocity factor, as a decimal fraction
v_p = propagation velocity on the line
c = speed of light in free space

The velocity factor is also commonly expressed as a percentage, which is found by simply multiplying the value found from Equation (16.6) by 100.

The velocity factor for a transmission line depends almost entirely on the dielectric used in the line. It is given by

$$v_f = \frac{1}{\sqrt{\epsilon_r}} \tag{16.7}$$

where

ϵ_r = line's dielectric constant

Example 16.3 Find the velocity factor and propagation velocity for a cable with a Teflon dielectric ($\epsilon_r = 2.1$).

Solution From Equation (16.7),

$$\begin{aligned} v_f &= \frac{1}{\sqrt{\epsilon_r}} \\ &= \frac{1}{\sqrt{2.1}} \\ &= 0.69 \end{aligned}$$

From Equation (16.6),

$$\begin{aligned} v_f &= \frac{v_p}{c} \\ v_p &= v_f c \\ &= 0.69 \times 3 \times 10^8 \text{ m/s} \\ &= 2.07 \times 10^8 \text{ m/s} \end{aligned}$$

16.3.3 Reflections

Suppose that we have a line of finite length but terminated in an open circuit (as shown in Figure 16.9) rather than in its characteristic impedance. This mismatched line will behave in exactly the same way as the matched line when the switch is first closed. A surge of energy will move down the line, and the ratio between voltage and current at the source will be equal to the characteristic impedance, Z_0. Since the source impedance is equal to the characteristic impedance of the line, one-half the source voltage will appear across the input end of the line. The other one-half appears across the source resistance.

Now let us follow the surge down the line. The voltage and current will continue to be related by Z_0 until the surge reaches the end of the line. At that point, the total current must be zero since there is an open circuit. On the other hand, the incoming surge of energy cannot simply disappear, because there is nothing capable of dissipating energy at this

Figure 16.9
Step input to an open-circuited line

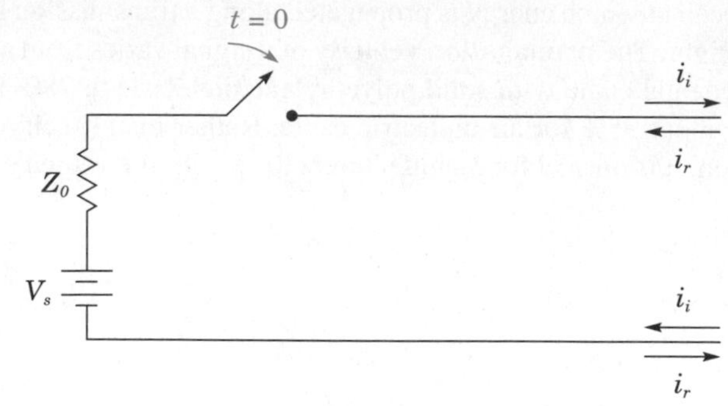

point. What happens is that the energy reflects from the open end of the line. The reflected voltage is the same as the incident voltage, but the current is equal in magnitude and opposite in direction to the incident current. Thus, the total voltage at the destination end of the line is twice the incident voltage, and the total current is zero. The incident voltage is one-half the source voltage, so the total voltage is simply equal to the source voltage.

As the reflected surge moves back toward the input, the current gradually becomes zero all along the line. Once the input is reached, the final conditions at the far end apply all along the line: the voltage is equal to the source voltage, and the current is zero. It is interesting that this final condition is exactly what anyone would predict: when a voltage is connected across an open-circuited line, the voltage across the line becomes equal to the source voltage, and the current becomes zero. The only difference is that with a long transmission line, the process takes time. Once the transient has died down, there are no surprises.

Figure 16.10 shows what happens. Figure 16.10(a) shows how the voltage at the source end of the line varies with time. The voltage rises from zero to one-half the supply voltage at $t = 0$, then rises again to the full supply voltage once the surge has had time to move down the line to the open end and back again. As shown in Figure 16.10(b), there is no voltage at the open end until the surge has reached it. At that instant, the voltage rises from zero to the full supply voltage. The time taken for the signal to move down the line is given simply by

$$T = \frac{L}{v_p} \tag{16.8}$$

where

T = time for a signal to travel the length of the line
L = length of the line
v_p = propagation velocity on the line

Figure 16.10
Voltage on an open-circuited line with step input

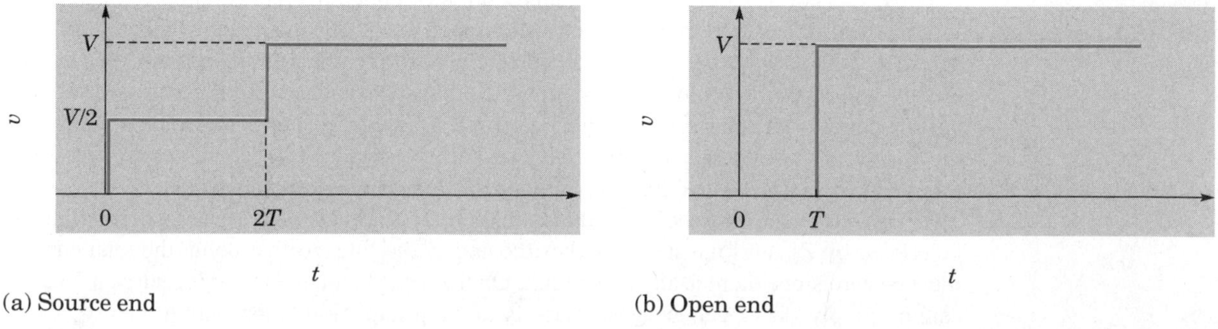

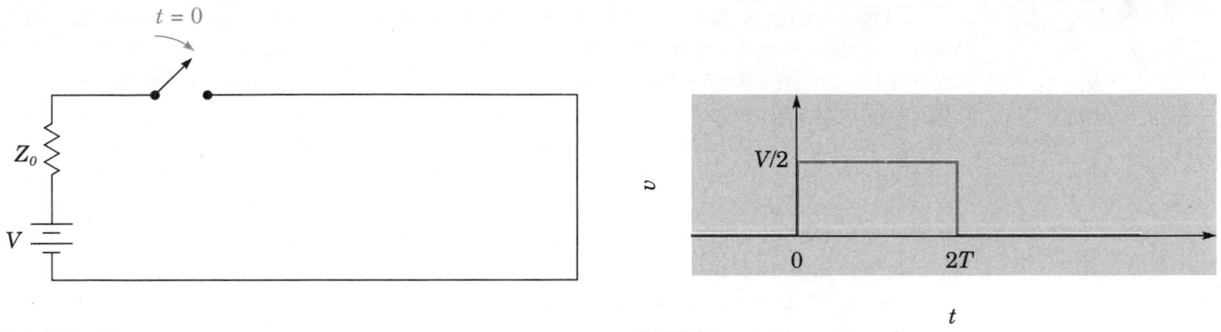

(a) Circuit

(b) Voltage at source end

Figure 16.11
Voltage step applied to a shorted line

Next, let us consider what happens if a step input is applied to a transmission line that is shorted at the far end, as shown in Figure 16.11(a). At first, there is no perceptible difference. While the incident surge is still moving along the line toward the shorted end, the voltage and current at the input end will be related by the characteristic impedance of the line, just as before. However, once the surge reaches the shorted end, a different boundary condition applies. With an open end, the total current was compelled to be zero, but there was no restriction on the voltage. A short circuit, on the other hand, must have a voltage of zero across it. The current can have any value, as determined by other constraints on the system. Once again, the signal will be reflected, but this time the current will be in the same direction as the incident current, and the voltage will have the opposite polarity. When the reflected surge reaches the input of the line, the voltage all along the line will be zero, and the current will be limited only by the source resistance. This is exactly what we would have expected previously, for a short-circuited line connected to a dc source. Figure 16.11(b) shows the sequence of events.

Instead of a step input, a pulse could be applied to the line, as in Figure 16.12. Once again, if the line were terminated in its characteristic impedance, the pulse would simply proceed down the line and be dissipated. With an open circuit at the far end, the pulse would be reflected with the same voltage polarity; with a shorted line, the reflected pulse would have opposite polarity.

Figure 16.12
Pulse input to transmission lines: voltages at the source end

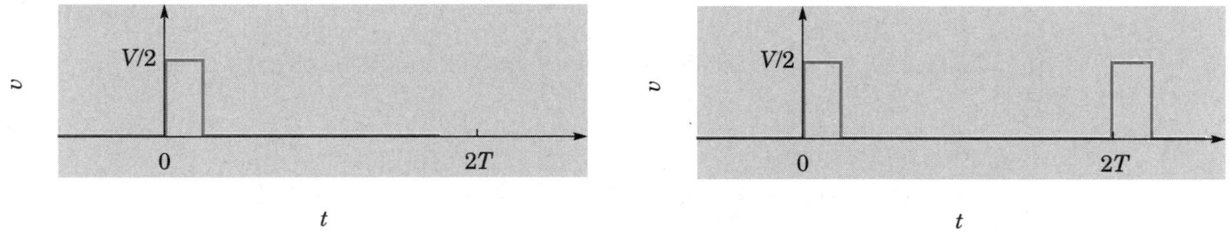

(a) Matched line

(b) Open-circuited line

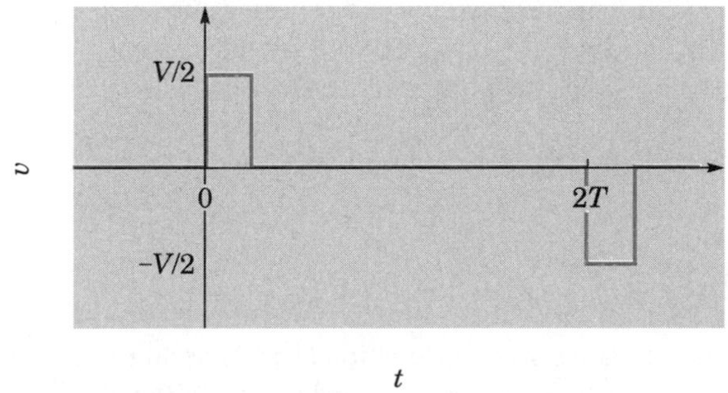

(c) Short-circuited line

The situations that have been described so far are simple, limiting cases in which there is either complete reflection or no reflection at all. Now suppose that the line is terminated in an impedance, but one that is not equal to the characteristic impedance of the line. For instance, consider a 50 Ω line terminated by a 25 Ω resistor. One might guess that there would be partial reflection of the incident surge or pulse, and that is in fact what happens. At the termination, the ratio of voltage to current is given by the load resistance Z_L, which in this case is 25 Ω. That is, Ohm's law must be satisfied at the termination. On the other hand, the voltage and current in the incident surge are related by the Z_0 of the line, in this case 50 Ω.

A line that is terminated by an impedance other than Z_0 is said to be *mismatched.* Mismatches, while usually undesirable, are very common with transmission lines. It is often difficult to design a load, such as an antenna, that presents precisely the right impedance to the line, especially when the frequency must vary. When performing calculations with mismatched lines, it is convenient to define a reflection coefficient

$$\Gamma = \frac{V_r}{V_i} \quad (16.9)$$

where

Γ = voltage reflection coefficient, a dimensionless number (in general, it is a complex number)
V_r = reflected voltage
V_i = incident voltage

Note that Γ is formally defined as the *voltage reflection coefficient.* This is to distinguish it from the reflection coefficient for power, which turns out to be equal to Γ^2. However, in common use, Γ is usually referred to simply as the **reflection coefficient**.

From the work done so far, it can be seen that a matched line has a reflection coefficient of zero, since there is no reflected voltage. For an open-circuited line, $\Gamma = 1$, since the reflected voltage has the same magnitude and sign as the incident voltage. On the other hand, a short-circuited line has $\Gamma = -1$ because the incident and reflected voltages are equal in magnitude but opposite in sign.

The reflection coefficient is determined by Z_L and Z_0. It is quite possible to derive a relation among these three parameters. First, note that at the load, the total voltage will be

$$V_T = V_r + V_i \quad (16.10)$$

and the total current, that is, the current into the load, will be given by Ohm's law:

$$I_T = \frac{V_T}{Z_L} \quad (16.11)$$

Also, the total current at the load will be

$$I_T = I_i + I_r \quad (16.12)$$

where I_i and I_r are the incident and reflected currents, respectively.

On the line, the incident voltage and current will be related by the characteristic impedance Z_0, and so will the reflected voltage and current:

$$I_i = \frac{V_i}{Z_0} \quad (16.13)$$

$$I_r = -\frac{V_r}{Z_0} \quad (16.14)$$

There is a negative sign in Equation (16.14) because the reference direction for the current is initially defined as toward the load on one of the conductors. A reflected current flowing away from the load will then be negative.

At the termination, combining Equations (16.11) and (16.12), we have

$$\frac{V_T}{Z_L} = I_i + I_r$$

Substituting the expression for V_T from Equation (16.10) gives

$$\frac{V_r + V_i}{Z_L} = I_i + I_r$$

Next, substitute the expressions for I_i and I_r found in Equations (16.13) and (16.14), respectively:

$$\frac{V_r + V_i}{Z_L} = \frac{V_i}{Z_0} - \frac{V_r}{Z_0} \tag{16.15}$$

$$\frac{V_r + V_i}{Z_L} = \frac{V_i - V_r}{Z_0}$$

Remembering that $\Gamma = V_r/V_i$, Equation (16.15) can be simplified by cross multiplication:

$$V_iZ_L - V_rZ_L = V_iZ_0 + V_rZ_0 \tag{16.16}$$

$$V_r(Z_0 + Z_L) = V_i(Z_L - Z_0)$$

$$\frac{V_r}{V_i} = \frac{Z_L - Z_0}{Z_L + Z_0}$$

$$\Gamma = \frac{Z_L - Z_0}{Z_L + Z_0}$$

Once again, it should be pointed out that any or all of the quantities in Equation (16.16) can be complex.

Example 16.4 The switch shown in Figure 16.13(a) closes at time $t = 0$, applying a 1 V source through a 50 Ω resistor to a 50 Ω line that is terminated by a 25 Ω resistor. The line is 10 m in length, with a velocity factor of 0.7. Draw graphs showing the variation of voltage with time at each end of the line.

Figure 16.13

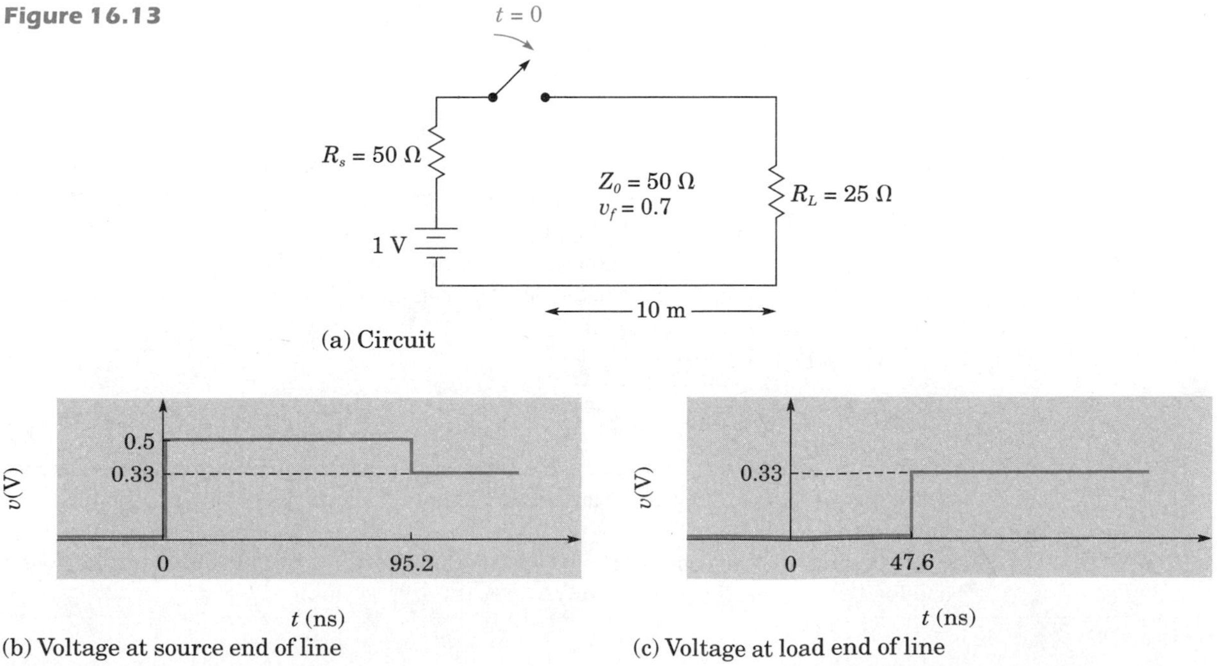

Solution At $t = 0$, the source voltage will divide between R_s and the Z_0 of the line. Since the two resistances are equal, the voltage will divide equally. Therefore, at $t = 0$, the voltage at the source end of the line will rise from zero to 0.5 V.

The voltage at the load will remain zero until the surge reaches it. The time for this is

$$\begin{aligned} T &= \frac{L}{v_p} \\ &= \frac{L}{v_f c} \\ &= \frac{10 \text{ m}}{0.7 \times 3 \times 10^8 \text{ m/s}} \\ &= 47.6 \text{ ns} \end{aligned}$$

After 47.6 ns, the voltage at the load will rise. The new value can be found by calculating the reflection coefficient. From Equation (16.16),

$$\begin{aligned} \Gamma &= \frac{Z_L - Z_0}{Z_L + Z_0} \\ &= \frac{25\ \Omega - 50\ \Omega}{25\ \Omega + 50\ \Omega} \\ &= \frac{-25\ \Omega}{75\ \Omega} \\ &= -0.333 \end{aligned}$$

The reflected voltage can be found from Equation (16.9):

$$\begin{aligned} \Gamma &= \frac{V_r}{V_i} \\ V_r &= \Gamma V_i \\ &= -0.333 \times 0.5 \text{ V} \\ &= -0.1665 \text{ V} \end{aligned}$$

The total voltage at the load is given by Equation (16.10):

$$\begin{aligned} V_T &= V_r + V_i \\ &= -0.1665 \text{ V} + 0.5 \text{ V} \\ &= 0.3335 \text{ V} \end{aligned}$$

The reflected voltage will propagate back along the line, reaching the source at time $2T =$ 95.2 ns. After this time, the voltage will be 0.3335 V all along the line.

The astute reader may have noticed that there is an easier way to calculate the final voltage. The final condition, after all transients have died down, must be the same as the dc situation, ignoring transmission-line characteristics. This situation is shown in Figure 16.13(a). It is simply a voltage divider. The voltage across the line, and the load, will be

$$\begin{aligned} V_L &= 1 \text{ V} \times \frac{25\ \Omega}{50\ \Omega + 25\ \Omega} \\ &= 0.3333 \text{ V} \end{aligned}$$

Allowing for round-off error, this is the same answer we obtained before. The reflection coefficient thus seems to provide an explanation of the mechanism by which the final condition is reached.

Figures 16.13(b) and (c) show the variation over time of the voltages at the source and load ends of the line, respectively.

16.4 Wave Propagation on Lines

We began our discussion of transmission-line characteristics with step and pulse inputs mainly to simplify matters, though both types of signals are in fact used with transmission lines. Pulses in particular are common in some digital applications, but we will now consider the effect of transmission lines on sine waves.

Let us begin, as before, with a matched line. A sinusoidal wave is applied to one end through a source resistance that is equal to Z_0. From our experience with step and pulse inputs, we would expect that the signal would simply move down the line and disappear into the load, and that is what does happen. Such a signal is called a *traveling wave*. We would also expect that this process would take time. Once the sine wave has been operating long enough for the first part of the signal to reach the far end, a steady-state situation will exist. The signal at any point along the line will then be the same as that at the source, except for a time delay.

With a sine wave, of course, a time delay is equivalent to a phase shift. A time delay of one period causes a phase shift of 360°, or one complete cycle; a wave that has been delayed that much is indistinguishable from one that has not been delayed at all. The length of line L that causes a delay of one period is known as a *wavelength*, for which the usual symbol is λ. If we could look at the voltage along the line at one instant of time, the resulting "snapshot" would look like the input sine wave, except that the horizontal axis would be distance rather than time, and one complete cycle of the wave would occupy one wavelength instead of one period. See Figure 16.14 for an example.

Figure 16.14
Traveling waves on a matched line

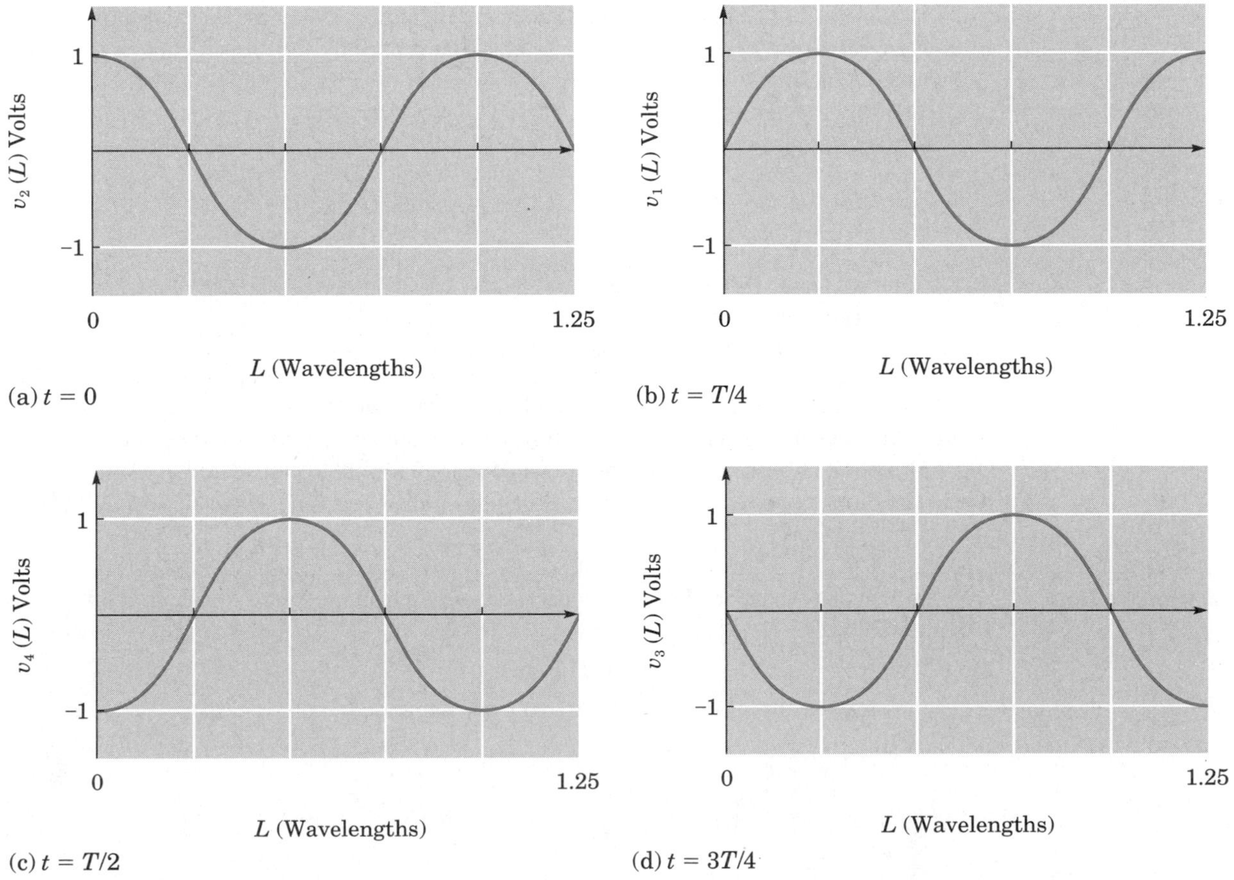

To arrive at an expression for wavelength on a line, we start with the definition of velocity

$$v = \frac{d}{t} \tag{16.17}$$

where

v = velocity in meters per second
d = distance in meters
t = time in seconds

Since we are trying to find wavelength, we substitute the period T of the signal for t, the wavelength λ for d, and the propagation velocity along the line v_p for v. This gives us

$$v_p = \frac{\lambda}{T} \tag{16.18}$$

More commonly, frequency is used rather than period. Since

$$f = \frac{1}{T}$$

Equation (16.18) becomes

$$v_p = f\lambda \tag{16.19}$$

By the way, Equation (16.19) is valid for all types of waves, in any medium.

Lengths of line that are not equal to a wavelength yield a phase delay proportional to their length. Since a length λ produces a phase shift of 360°, the phase delay produced by a given line is simply

$$\phi = (360)\,\frac{L}{\lambda} \tag{16.20}$$

where

ϕ = phase shift in degrees
L = length of the line
λ = wavelength on the line

In many applications, the time delay and phase shift due to a length of transmission line are of no concern. However, there are times, (for example, when two signals must arrive at a given point in phase with each other) when it is important to consider phase shift. Transmission lines can also be used to introduce phase shifts and time delays deliberately when they are required.

Example 16.5 What length of standard RG-8/U coaxial cable would be required to obtain a 45° phase shift at 200 MHz?

Solution The velocity factor for this line is 0.66, so from Equation (16.6),

$$\begin{aligned} v_p &= v_f c \\ &= 0.66 \times 3 \times 10^8 \text{ m/s} \\ &= 1.98 \times 10^8 \text{ m/s} \end{aligned}$$

The wavelength on the line of a 200 MHz signal is found from Equation (16.19):

$$v_p = f\lambda$$

$$\lambda = \frac{v_p}{f}$$

$$= \frac{1.98 \times 10^8 \text{ m/s}}{200 \times 10^6 \text{ Hz}}$$

$$= 0.99 \text{ m}$$

The required length for a phase shift of 45° would be

$$L = 0.99 \text{ m} \times \frac{45°}{360°}$$

$$= 0.124 \text{ m}$$

16.4.1 Standing Waves

It has been shown that a sine wave applied to a matched line results in an identical sine wave, except for phase, appearing at every point on the line as the incident wave travels down it. If the line is unmatched, we would probably expect a reflected wave from the load to add to the incident wave from the source. This is exactly what happens.

As we did for the step input, let us begin with a rather simple situation: a transmission line terminated in an open circuit. Assume that the line is reasonably long, say one wavelength. The situation is shown in Figure 16.15. The reflection of a sine wave is harder to visualize than that of a voltage pulse or step, but this figure shows the situation at several points in the cycle. As before, the reflected voltage has the same amplitude and polarity as the incident voltage at the load, since the open circuit cannot dissipate any power. The reflected voltage propagates down the line until it is dissipated in the source impedance, which has been chosen to match the line at the source. (If there is a mismatch at the two ends, the situation will be complicated by multiple reflections.) At every point on the line, the instantaneous values of incident and reflected voltage add algebraically to give the total voltage. From the figure, we can infer that the voltage at every point on the line varies sinusoidally but the amplitude of the voltage varies greatly due to constructive and destructive interference between the incident and reflected waves. At the open-circuited end of the line, the peak voltage is a maximum. One-quarter wavelength away from that end, the incident and reflected voltages exactly cancel because the two signals have equal amplitude and opposite phase. At a distance of one-half wavelength from the open-circuited end, there is another voltage maximum, and the situation repeats for every half-wavelength segment of line.

A sketch showing the variation of peak (or RMS) voltage along the line is shown in Figure 16.16. It is important to realize that this figure does not represent either instantaneous or dc voltages. There is no dc on this line at all. Rather, the figure shows how the amplitude of a sinusoidal voltage varies along the line.

The interaction between the incident and reflected waves, which are both traveling waves, causes what appears to be a stationary pattern of waves on the line. It is customary to call these *standing waves* because of this appearance.

For comparison, Figure 16.17 shows the standing waves of voltage on a line with a shorted end. Naturally, there is no voltage at the shorted end. A voltage maximum occurs one-quarter wavelength from the end, another null occurs at one-half wavelength, and so on.

The current responds in just the opposite way from the voltage. For the open line, the current must be zero at the open end. It will be a maximum one-quarter wavelength away, zero again at a distance of one-half wavelength, and so on. In other words, the graph of current as a function of position for an open-circuited line looks just like the graph of

Figure 16.15
Incident and reflected waves on an open-circuited line

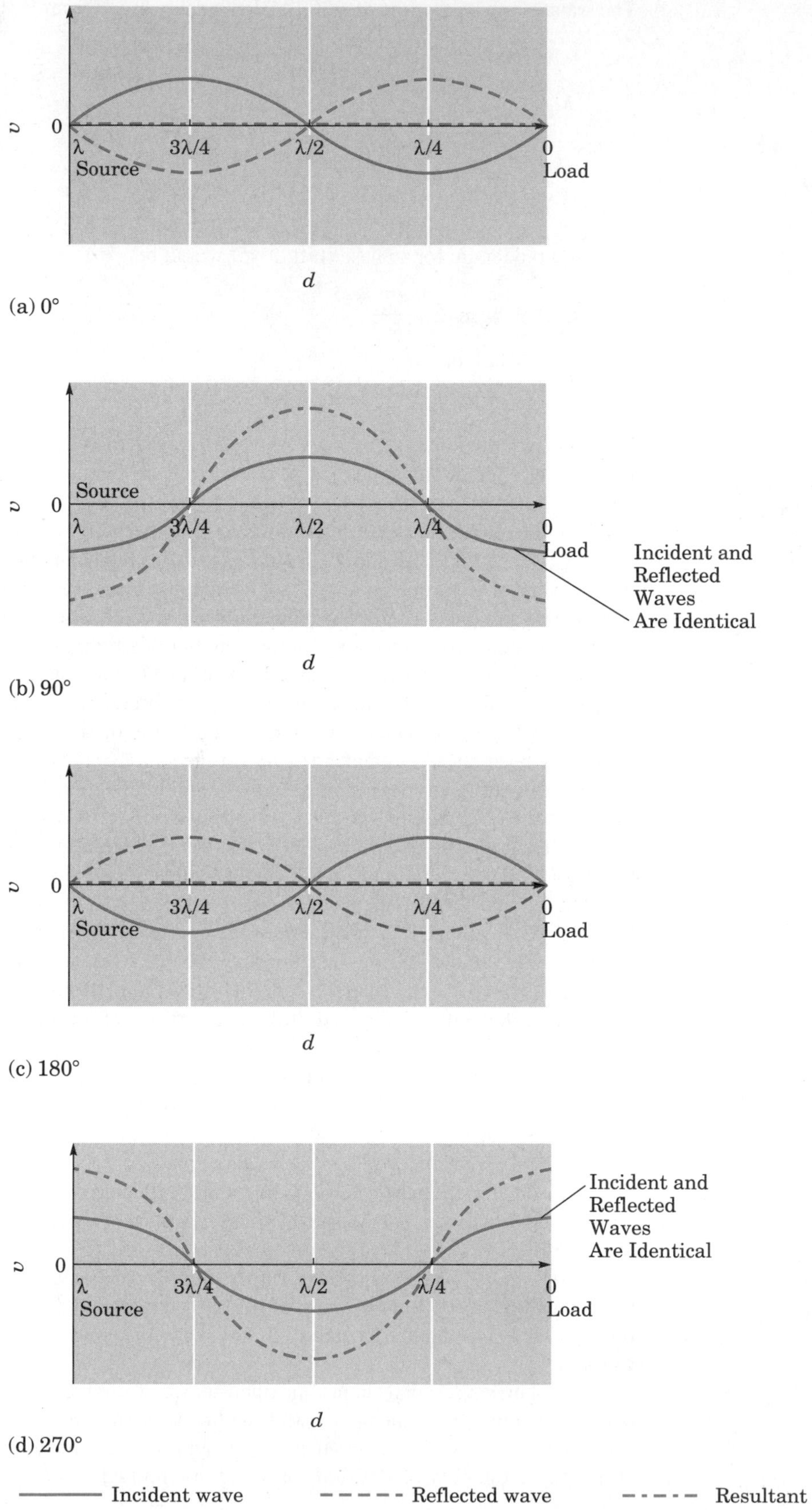

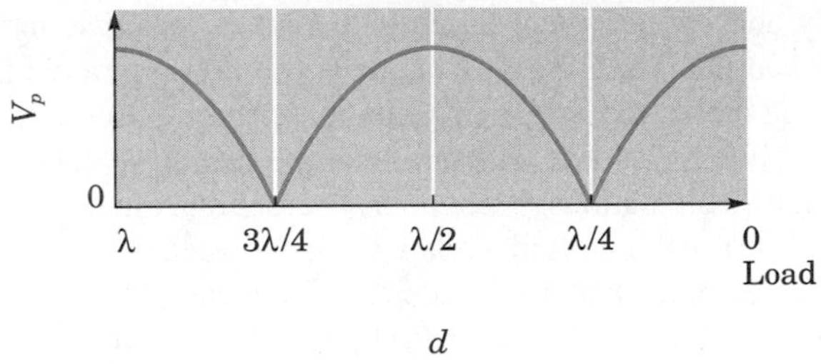

Figure 16.16
Standing waves on an open-circuited line

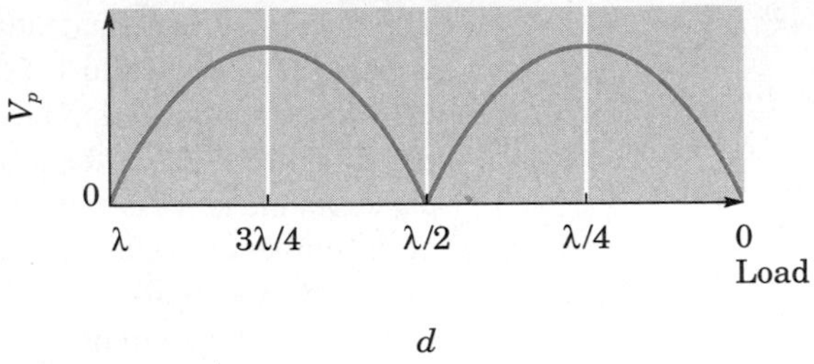

Figure 16.17
Standing waves on a short-circuited line

voltage versus position for a short-circuited line (see Figure 16.17). Similarly, the current on a short-circuited line has a maximum at the termination, a minimum one-quarter wavelength away, and another maximum at a distance of one-half wavelength, just like the voltage curve for the open-circuited line (Figure 16.16).

What happens when the line is mismatched, but not as drastically as discussed above? There will be a reflected wave, but it will not have as large an amplitude as the incident wave. Its phase angle will depend on the load impedance compared to that of the line. Recall from Equation (16.9) that the incident and reflected voltages are related by the coefficient of reflection:

$$\Gamma = \frac{V_r}{V_i}$$

and that, from Equation (16.16),

$$\Gamma = \frac{Z_L - Z_0}{Z_L + Z_0}$$

In general, Γ is complex, but for a lossless line it is a real number if the load is resistive. It is positive for $Z_L > Z_0$ and negative for $Z_L < Z_0$. For $Z_L = Z_0$, the reflection coefficient is of course zero. A positive coefficient means that the incident and reflected voltages are in phase at the load.

When a reflected signal is present but has a lower amplitude than the incident wave, there will be standing waves of voltage and current, but there will be no point on the line where the voltage or current remains zero over the whole cycle (see Figure 16.18 for an example). It is possible to define the voltage **standing-wave ratio** (VSWR, or just SWR) as follows:

$$\text{SWR} = \frac{V_{max}}{V_{min}} \tag{16.21}$$

Figure 16.18
Standing waves on a line with $R_L > Z_0$

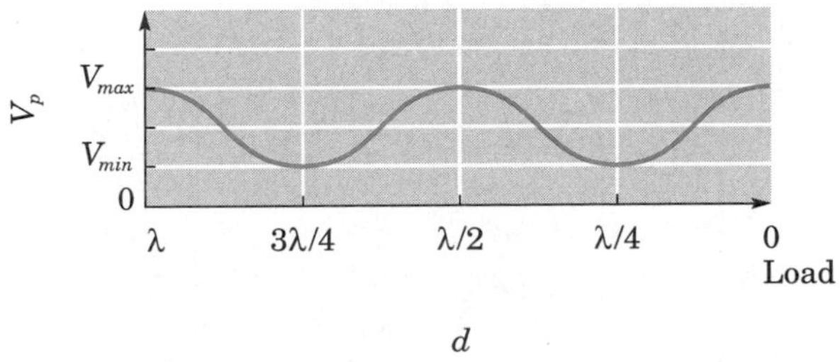

The SWR concerns magnitudes only and is thus a real number. It must be positive and greater than or equal to 1. For a matched line, the SWR is 1 (sometimes expressed as 1:1 to emphasize that it is a ratio), and the closer the line is to being matched, the lower its SWR. The SWR has the advantage of being easier to measure than the reflection coefficient, but the latter is more useful in many calculations. Since both are essentially measures of the amount of reflection on a line, it is possible to find a relationship between them.

The maximum voltage on the line occurs where the incident and reflected signals are in phase, and the minimum voltage is found where they are out of phase. Therefore, using absolute value signs to emphasize the lack of a need for phase information, we can write

$$V_{max} = |V_i| + |V_r| \tag{16.22}$$

and

$$V_{min} = |V_i| - |V_r| \tag{16.23}$$

Combining Equations (16.21), (16.22), and (16.23), we get

$$\begin{aligned} \text{SWR} &= \frac{V_{max}}{V_{min}} \\ &= \frac{|V_i| + |V_r|}{|V_i| - |V_r|} \\ &= \frac{1 + \dfrac{|V_r|}{|V_i|}}{1 - \dfrac{|V_r|}{|V_i|}} \\ &= \frac{1 + |\Gamma|}{1 - |\Gamma|} \end{aligned} \tag{16.24}$$

A little algebra will show that $|\Gamma|$ can also be expressed in terms of SWR:

$$|\Gamma| = \frac{\text{SWR} - 1}{\text{SWR} + 1} \tag{16.25}$$

For the special but important case of a lossless line terminated in a resistive impedance, it is possible to find a simple relationship between SWR and the load and line impedances. First, suppose that $Z_L > Z_0$. Then, from Equation (16.24),

$$\begin{aligned} \text{SWR} &= \frac{1 + |\Gamma|}{1 - |\Gamma|} \\ &= \frac{1 + \dfrac{Z_L - Z_0}{Z_L + Z_0}}{1 - \dfrac{Z_L - Z_0}{Z_L + Z_0}} \\ &= \frac{Z_L + Z_0 + Z_L - Z_0}{Z_L + Z_0 - Z_L + Z_0} \\ &= \frac{2Z_L}{2Z_0} \\ &= \frac{Z_L}{Z_0} \end{aligned} \tag{16.26}$$

It is easy to show, in a similar way, that if $Z_0 > Z_L$, then

$$\text{SWR} = \frac{Z_0}{Z_L} \tag{16.27}$$

Use of the appropriate equation will always give a positive SWR that is greater than or equal to one.

Example 16.6 A 50 Ω line is terminated in a 25 Ω resistance. Find the SWR.

Solution In this case, $Z_0 > Z_L$ so the solution is given by Equation (16.27):

$$\begin{aligned} \text{SWR} &= \frac{Z_0}{Z_L} \\ &= \frac{50}{25} \\ &= 2 \end{aligned}$$

The presence of standing waves causes the voltage at some points on the line to be higher than it would be with a matched line, while at other points the voltage is low but the current is higher than with a matched line. This situation results in increased losses. In a transmitting application, standing waves put additional stress on the line and can result in failure of the line or of equipment connected to it. For instance, if the transmitter happens to be connected at or near a voltage maximum, the output circuit of the transmitter may be subjected to a dangerous overvoltage condition. This is especially likely to damage solid-state transmitters, which for this reason are often equipped with circuits to reduce the output power in the presence of an SWR greater than about 2:1.

Reflections cause the power dissipated in the load to be less than it would be with a matched line, for the same source, because some of the power is reflected back to the source. Since power is proportional to the square of voltage, the fraction of the power that is reflected is Γ^2, that is,

$$P_r = \Gamma^2 P_i \tag{16.28}$$

where

P_r = power reflected from the load
P_i = incident power at the load
Γ = voltage reflection coefficient

Sometimes Γ^2 is referred to as the *power reflection coefficient.*

The amount of power absorbed by the load is the difference between the incident power and the reflected power, that is,

$$\begin{aligned} P_L &= P_i - \Gamma^2 P_i \\ &= P_i(1 - \Gamma^2) \end{aligned} \tag{16.29}$$

Example 16.7 A generator sends 50 mW down a 50 Ω line. The generator is matched to the line, but the load is not. If the coefficient of reflection is 0.5, how much power is reflected and how much is dissipated in the load?

Solution The amount of power that is reflected is, from Equation (16.28):

$$\begin{aligned} P_r &= \Gamma^2 P_i \\ &= 0.5^2 \times 50 \text{ mW} \\ &= 12.5 \text{ mW} \end{aligned}$$

The remainder of the power reaches the load. This amount is

$$\begin{aligned} P_L &= P_i - P_r \\ &= 50 \text{ mW} - 12.5 \text{ mW} \\ &= 37.5 \text{ mW} \end{aligned}$$

Alternatively, the load power can be calculated directly from Equation (16.29):

$$P_L = P_i(1 - \Gamma^2)$$
$$= 50 \text{ mW} \times (1 - 0.5^2)$$
$$= 37.5 \text{ mW}$$

Since SWR is easier to measure than the reflection coefficient, an expression for the power absorbed by the load in terms of the SWR would be useful. It is easy to derive such an expression by using the relationship between Γ and SWR given in Equation (16.25). The derivation is left as an exercise; the result is

$$P_L = \frac{4\text{SWR}}{(1 + \text{SWR})^2} P_i \tag{16.30}$$

Example 16.8 A transmitter supplies 50 W to a load through a line with an SWR of 2:1. Find the power absorbed by the load.

Solution From Equation (16.30),

$$P_L = \frac{4\text{SWR}}{(1 + \text{SWR})^2} P_i$$
$$= \frac{4 \times 2}{(1 + 2)^2} \times 50 \text{ W}$$
$$= 44.4 \text{ W}$$

Reflections on transmission lines can cause problems in receiving applications as well. For instance, reflections on a television antenna feedline can cause a double image or "ghost" to appear. In data transmission, reflections can distort pulses, causing errors.

16.4.2 Variation of Impedance Along a Line

A matched line presents its characteristic impedance to a source located any distance from the load. If the line is not matched, however, the impedance seen by the source can vary greatly with its distance from the load. At points where the voltage is high and the current low, the impedance is higher than at points with the opposite current and voltage characteristics. In addition, the phase angle of the impedance can vary. At some points, a mismatched line may look inductive, at others it may look capacitive, and at a few points, it may look resistive. Very near the load, the impedance looking into the line is very close to Z_L. This is one reason why transmission line techniques need to be used only with relatively high frequencies and/or long lines: a line shorter than about one-sixteenth of a wavelength can usually be ignored.

The impedance that a lossless transmission line presents to a source varies in a periodic way. We have already noticed that the standing-wave pattern repeats itself every one-half wavelength along the line; the impedance varies in the same fashion. At the load and at distances from the load that are multiples of one-half wavelength, the impedance looking into the line is that of the load.

The impedance at any point on a lossless transmission line is given by the equation

$$Z = Z_0 \frac{Z_L \cos\theta + jZ_0 \sin\theta}{Z_0 \cos\theta + jZ_L \sin\theta} \tag{16.31}$$

where

Z = impedance looking toward the load
Z_L = load impedance
Z_0 = characteristic impedance of the line
θ = distance to the load in degrees (for example, a quarter-wavelength would be 90°)

Provided that cos θ is not equal to zero, Equation (16.31) simplifies to

$$Z = Z_0 \frac{Z_L + jZ_0 \tan \theta}{Z_0 + jZ_L \tan \theta} \tag{16.32}$$

Example 16.9

Calculate the impedance looking into a 50 Ω line 1 m long, terminated in a load impedance of 100 Ω, if the line has a velocity factor of 0.8 and operates at a frequency of 30 MHz.

Solution

First, we need the length of the line in degrees. At 30 MHz with a velocity factor of 0.8, the wavelength is

$$\begin{aligned}\lambda &= \frac{v}{f}\\ &= \frac{v_f c}{f}\\ &= \frac{0.8 \times 3 \times 10^8 \text{ m/s}}{30 \times 10^6 \text{ Hz}}\\ &= 8 \text{ m}\end{aligned}$$

The length of the line in degrees is

$$\begin{aligned}\theta &= \frac{1 \text{ m}}{8 \text{ m}} \times 360^\circ\\ &= 45^\circ\end{aligned}$$

Substituting this value and the given values into Equation (16.32),

$$\begin{aligned}Z &= Z_0 \frac{Z_L + jZ_0 \tan \theta}{Z_0 + jZ_L \tan \theta}\\ &= 50\ \Omega \frac{100\ \Omega + j(50\ \Omega) \tan 45^\circ}{50\ \Omega + j(100\ \Omega) \tan 45^\circ}\\ &= 50\ \Omega \frac{100\ \Omega + j(50\ \Omega)}{50\ \Omega + j(100\ \Omega)}\\ &= \frac{100\ \Omega + j(50\ \Omega)}{1 + j2}\\ &= \frac{[100\ \Omega + j(50\ \Omega)](1 - j2)}{(1 + j2)(1 - j2)}\\ &= \frac{100\ \Omega + j(50\ \Omega) - j(200\ \Omega) + 100\ \Omega}{1 + 4}\\ &= \frac{200\ \Omega - j(150\ \Omega)}{5}\\ &= 40\ \Omega - j(30\ \Omega)\end{aligned}$$

The calculation of transmission-line impedances is usually done in one of two ways: graphically, with the aid of a **Smith chart** (see Section 16.6), or by computer.

16.4.3 Characteristics of Open and Shorted Lines

Though a section of transmission line that is terminated in an open or short circuit is useless for transmitting power, it can serve other purposes. Such a line can be used as an inductive or capacitive reactance or even as a resonant circuit. In practice, short-circuited sections are more common, because open-circuited lines tend to radiate energy from the open end.

The impedance of a short-circuited line can be found from Equation (16.32) by setting Z_L equal to zero. The impedance looking toward the short circuit is

$$Z = jZ_0 \tan \theta \tag{16.33}$$

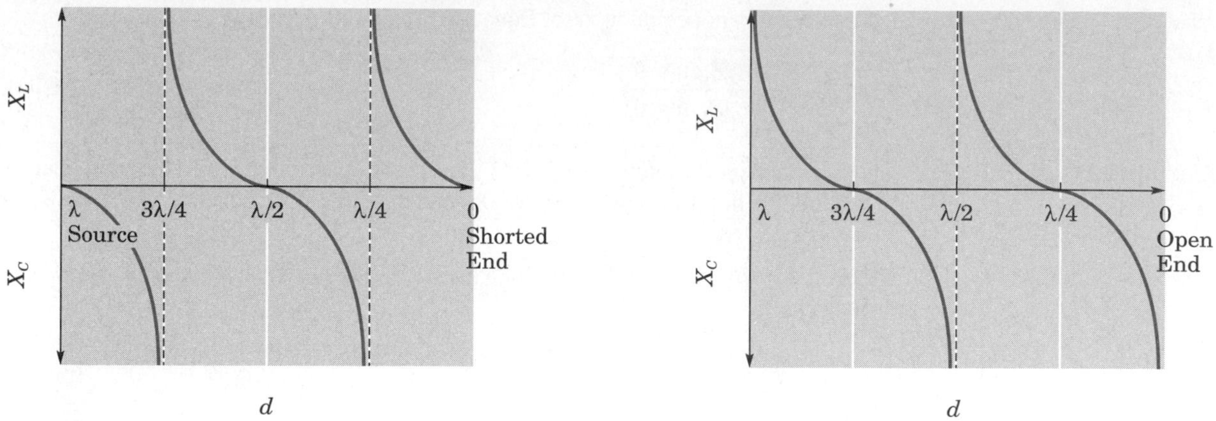

(a) Shorted line

(b) Open-circuited line

Figure 16.19 Variation of impedance with length on shorted and open lines

Note that the impedance has no resistive component. This is logical, since there is no resistive element that can dissipate any power. For short lengths (less than one-quarter wavelength or 90°), the impedance is inductive. At one-quarter wavelength, the line looks like an open circuit, and for lengths between one-quarter and one-half wavelength, the line is capacitive, since the tangent of the corresponding angle is negative. For a length of one-half wavelength or 180°, tan θ is zero, and the line once again behaves like a short circuit. For longer lines, the cycle repeats. Figure 16.19(a) shows graphically how the impedance varies with length. It is easy to remember that a short length of shorted line is inductive by visualizing the shorted end as a loop or coil and hence an inductance. At exactly one-quarter wavelength, the line behaves as a parallel-resonant circuit, that is, it has a very high impedance. As the line is made still longer, it becomes capacitive, finally becoming series-resonant at a length of one-half wavelength. For still longer lines, the process repeats.

The open-circuited line, on the other hand, is capacitive in short lengths. You can remember this by thinking of the two parallel lines at the end of the transmission line as a capacitor. It is series-resonant (that is, it behaves like a short circuit) at a length of one-quarter wavelength, inductive between one-quarter and one-half wavelength, and parallel-resonant at a length of one-half wavelength. For longer lines, the cycle repeats. Figure 16.19(b) shows graphically how the impedance varies with length for this line.

Short transmission line sections, called *stubs*, can be substituted for capacitors, inductors, or tuned circuits in applications where lumped-constant components (conventional inductors or capacitors) would be inconvenient or impractical. At VHF and UHF frequencies, the required values of inductance and capacitance are often very small. It would be difficult to build a 1 pF capacitor, for instance; the capacitance between its leads might well be more than that. In addition, physically small components are difficult to use where large amounts of power are involved, as in transmitters, because of the large voltages and currents that must be handled. A short section of transmission line can avoid all these problems. In addition, an air-dielectric transmission line can have much higher Q than a typical lumped-constant resonant circuit.

Though either short-circuited or open-circuited stubs could be used, in practice the short-circuited version is much more useful. Open-circuited lines tend to radiate energy from the end of the line, and furthermore, it is easier to adjust a short-circuited stub by means of a moving short circuit. Any extra line that extends past the short circuit will have no effect on the operation of the stub.

Example 16.10 A series tuned circuit operating at a frequency of 1 GHz is to be constructed from a shorted section of air-dielectric coaxial cable. What length should be used?

Solution The velocity factor of an air-dielectric line is about 0.95, so the propagation velocity is, from Equation (16.6),

$$\begin{aligned} v_p &= v_f c \\ &= 0.95 \times 3 \times 10^8 \text{ m/s} \\ &= 2.85 \times 10^8 \text{ m/s} \end{aligned}$$

The wavelength on the line is given by Equation (16.19).

$$\begin{aligned} v_p &= f\lambda \\ \lambda &= \frac{v_p}{f} \\ &= \frac{2.85 \times 10^8 \text{ m/s}}{1000 \times 10^6 \text{ Hz}} \\ &= 0.285 \text{ m} \end{aligned}$$

Since this is a shorted stub, a half-wavelength section will be series-resonant. Therefore the length will be

$$\begin{aligned} L &= \frac{\lambda}{2} \\ &= \frac{0.285 \text{ m}}{2} \\ &= 0.143 \text{ m} \end{aligned}$$

16.5 Transmission Line Losses

Up to this point, we have concentrated on lossless lines. No real transmission line is completely lossless, of course, but the approximation is often valid, particularly when the section of line is short. In that case, a matched line can be ignored, and a mismatched line is noteworthy mainly for the reactance that it introduces, the reflections (which can cause distortion), and the standing waves (which increase losses and can damage components).

In order to simplify our discussion of lossy lines, we will begin with matched lines. Then we will look briefly at the general case in which a line is both mismatched and lossy.

16.5.1 Loss Mechanisms

The most obvious form of loss in a transmission line is that due to the resistance of the conductors. Sometimes this is called *I^2R loss* because it is proportional to the square of the current or *copper loss* because copper is the most common material for the conductors. All other things being equal, this loss will be less with higher characteristic impedances, because the same power can be delivered with less current when the impedance is increased. I^2R loss increases with frequency due to the skin effect, which causes the resistance of the conductors to rise as the frequency increases.

The dielectric of a transmission line also has some loss. Usually, the conductance of the dielectric increases with frequency. Dry air has low loss, so air-dielectric lines, like open-wire line and air-dielectric coaxial cable, should have better performance than solid-dielectric coaxial cable in this regard. Because open-wire line is exposed to the weather, it does not always have the lower loss that might be expected. Foam-dielectric coaxial lines typically have lower loss than those that use solid polyethylene.

Open-wire lines can radiate energy. This type of loss becomes more significant as the frequency increases and is worse with greater spacing between the conductors. The tendency of these cables to radiate can be reduced by carefully balancing the line to ground.

Radiation is not a problem for coaxial cables that are properly terminated and grounded. The signal currents flow near the outside surface of the inner conductor and the inside surface of the outer conductor. The outside of the outer conductor acts as a shield.

This assumes, of course, that the shield completely covers the cable. Some low-cost cables use a twisted or braided shield with less than 100% coverage, with the result that there can be some radiation, especially at VHF and UHF frequencies.

Radiation of energy not only contributes to cable losses but can also be a source of interference and crosstalk. Any cable that radiates can also absorb radiation.

16.5.2 Loss in Decibels

Transmission line losses are usually given in decibels per 100 feet or per 100 meters. When choosing a transmission line for a given application, attention must be paid to losses. Remember, for instance, that a 3 dB loss in a feedline between a transmitter and its antenna means that only one-half the transmitter power actually reaches the antenna. The rest of the power goes to heat up the transmission line. Losses are just as important in receivers, where a low noise figure depends on minimizing the losses before the first stage of amplification. In some receiving applications, notably satellite receivers, it is common to install a low-noise amplifier right at the antenna to reduce the adverse effect of the transmission line on noise performance. In fact, with satellite receivers, it is very common to down-convert the signal to a lower frequency at the antenna in order to take advantage of the fact that cable losses are less at lower frequencies.

Example 16.11

A transmitter is required to deliver 100 W to an antenna through 45 m of coaxial cable with a loss of 4 dB/100 m. What must be the output power of the transmitter, assuming the line is matched?

Solution

The loss in decibels is

$$\text{loss (dB)} = 45\text{ m} \times \frac{4\text{ dB}}{100\text{ m}}$$
$$= 1.8\text{ dB}$$

The relation between input and output power is

$$\frac{P_{in}}{P_{out}} = \text{antilog}\,\frac{1.8}{10}$$
$$= 1.51$$

The transmitter power must be

$$P_{in} = 1.51 \times 100\text{ W}$$
$$= 151\text{ W}$$

16.5.3 Mismatched Lossy Lines

The technologist is often spared the problem of dealing with losses and mismatches simultaneously. Most transmission lines that are long enough to have significant losses are reasonably well matched. One thing that is worth noting, though, is the effect that losses have on SWR. When the line is lossy, the SWR at the source is lower than that at the load. Figure 16.20 shows why. Incident and reflected signals are subject to the same amount of loss in decibels, so the ratio between them is greater at the source than at the load. Therefore, the reflection coefficient and standing-wave ratio both have larger magnitudes at the load. Consequently, when using SWR measurements to aid in matching a load to a line, it is better to take the measurements at the load, rather than at the source, particularly if the line is long.

Any of the computer programs that are available for transmission-line analysis are capable of handling losses and mismatches simultaneously. Recourse to one of these programs is suggested for such calculations.

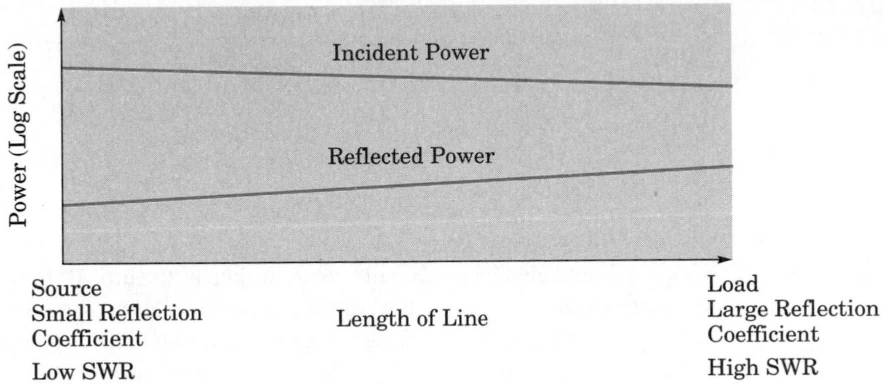

Figure 16.20
SWR on a lossy line

16.5.4 Power Ratings

The maximum power that can be applied to a transmission line is limited by one of two things: power dissipation in the line, which causes heating and can raise the temperature of the line to the point at which damage occurs, and voltage, which, when it exceeds a maximum value, can cause breakdown of the dielectric. Both power dissipation and maximum voltage increase with the standing-wave ratio, so a transmission line is capable of handling more power when it is properly matched than when it is mismatched. Published power ratings for commercial cables generally specify a maximum SWR at which they are valid. Since losses increase with frequency, a derating curve that reduces the maximum power-handling capability as the operating frequency increases may apply.

This is a good point at which to consider the effect of changing the characteristic impedance of a transmission line. Increasing the impedance increases the voltage required for a given power level and makes dielectric breakdown more likely. On the other hand, increasing the impedance reduces the current for a given power level, and consequently the I^2R losses are less with high-impedance lines. The two most common impedances for coaxial cables in RF circuits are 50 and 75 Ω. An impedance of 50 Ω is more common in transmitting applications, while an impedance of 75 Ω is used with television antennas and cable-TV systems, where power levels are low. This makes sense, since (other things being equal) the 75 Ω cable has lower loss, while the 50 Ω line has greater power-handling capability.

16.6 Impedance Matching

In most transmission-line applications, it is highly desirable that the load match the impedance of the line, in order to avoid the problems associated with reflections and a high SWR. Transmission lines are readily available with only a few standard impedance values, and most equipment is designed to work with only a narrow range of impedances. For example, a typical transmitter requires a 50 Ω load impedance with an SWR of 2:1 or less. If the antenna connected to such a transmitter has an impedance close to 50 Ω, resistive, there is no problem. Often, however, the impedance of the load differs, being neither resistive nor equal in magnitude to that of the line. In that case, some means must be provided for impedance matching.

Sometimes this is done by adding inductance or capacitance at the load as required to remove the reactance and using a conventional or toroidal transformer to change the magnitude of the impedance. Lumped-constant matching networks, such as the *T*, *L*, and pi networks, are also common at relatively low frequencies. It is also possible, however, to use transmission line sections as matching networks. The transmission line matching sections themselves are not usually matched and may have quite a high SWR, which must be taken into account whenever large power levels are present.

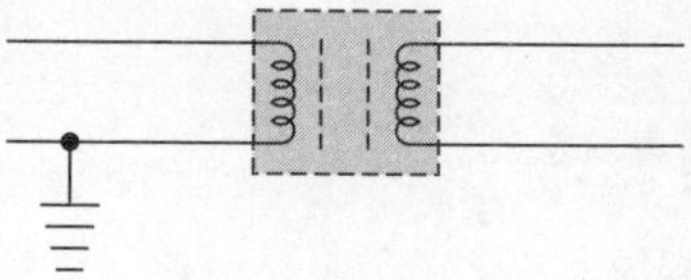

Figure 16.21
Balun transformer

Another requirement for proper transmission line operation is that balanced and unbalanced lines should be terminated by balanced and unbalanced loads, respectively. The sources should also have the same configuration as the line. If, for instance, a balanced load is connected to a coaxial cable, signal current is likely to flow on the outer surface of the shield, leading to radiation from the cable or the pickup of interference by the cable. On the other hand, connecting an unbalanced source or load to a balanced line will unbalance the entire line, causing similar problems. Balanced and unbalanced lines, loads, and sources can be connected together by using a transformer called a **balun** (see Figure 16.21). The balun can have a 1 : 1 turns ratio, or it can also be used to match impedances. An example of the latter is the common television-receiver balun. It is used to match a 75 Ω unbalanced coaxial cable to an antenna input designed for 300 Ω balanced twin-lead. Baluns can also be constructed from sections of transmission line, as described later in this chapter.

16.6.1 Quarter-Wave Transformers

A section of transmission line that is one-quarter wavelength long can act as an impedance transformer. This can be useful when the load impedance is resistive but has the wrong magnitude. This can be seen by looking at Equation (16.31):

$$Z = Z_0 \frac{Z_L \cos\theta + jZ_0 \sin\theta}{Z_0 \cos\theta + jZ_L \sin\theta}$$

For a quarter-wave line, $\theta = 90°$, so $\cos\theta = 0$ and $\sin\theta = 1$, so Equation (16.31) simplifies to

$$Z = Z_0 \frac{0 + jZ_0}{0 + jZ_L} \tag{16.34}$$

$$= Z_0 \frac{jZ_0}{jZ_L} \tag{16.35}$$

$$= \frac{Z_0^2}{Z_L}$$

The impedance looking into the quarter-wave section is Z, and Z_0 is the characteristic impedance of the **quarter-wave transformer**. To avoid confusion when the transformer is inserted into the main line, let us replace Z with the term Z_0, since the idea is to have the input impedance of the transformer equal to the characteristic impedance of the main line. The term Z_0 in the above equation can be replaced by Z_0' to emphasize that this is the impedance of the transformer section. Equation (16.35) then becomes

$$Z_0 = \frac{Z_0'^2}{Z_L}$$

This means that in order to match a load impedance Z_L to a line impedance Z_0, the quarter-wave section must have an impedance given by

$$Z_0' = \sqrt{Z_0 Z_L} \tag{16.36}$$

This technique usually requires a custom-made line section. It can be used with either open-wire or coaxial lines, though it is easier to construct with the former. A match can be obtained only over a small range of frequencies, since the length of the matching section will no longer be a quarter-wavelength if the frequency changes.

The quarter-wave transformer is only useful, by itself, in matching impedances that are resistive. If the load is reactive, however, it may be possible to add the quarter-wave section at a point along the line, at such a distance from the load that the impedance, looking down the line to the load, is real. Such a point can be found by using a Smith chart, a method to be shown later.

Example 16.12

Find the impedance and length of a quarter-wave transformer to match a 75 Ω load to a 50 Ω line at a frequency of 300 MHz.

Solution

The impedance of the matching section can be found from Equation (16.36):

$$\begin{aligned} Z_0' &= \sqrt{Z_0 Z_L} \\ &= \sqrt{50\ \Omega \times 75\ \Omega} \\ &= 61.2\ \Omega \end{aligned}$$

The length of the line section will depend on the dielectric. Assuming an air dielectric with a velocity factor of 0.95, the propagation velocity will be, from Equation (16.6),

$$\begin{aligned} v_p &= v_f c \\ &= 0.95 \times 3 \times 10^8 \text{ m/s} \\ &= 2.85 \times 10^8 \text{ m/s} \end{aligned}$$

The wavelength is given by Equation (16.19).

$$\begin{aligned} v_p &= f\lambda \\ \lambda &= \frac{v_p}{f} \\ &= \frac{2.85 \times 10^8 \text{ m/s}}{300 \times 10^6 \text{ Hz}} \\ &= 0.95 \text{ m} \end{aligned}$$

The line must be one-quarter wavelength long, so its length is given by

$$\begin{aligned} L &= 0.25 \times 0.95 \text{ m} \\ &= 0.238 \text{ m} \end{aligned}$$

16.6.2 Other Series Matching Sections

Though it is the simplest to analyze, the quarter-wave matching section is not the only possibility. It is possible to match complex impedances by using a section of a different length or two sections with different impedances. In order to avoid tedious calculations, it is logical to use a computer to design such sections. In fact, there is a free program called APPCAD, available from Hewlett-Packard, that handles these problems as well as matching using conventional *LC* networks. In addition, it performs calculations in several other areas as diverse as noise figure and thermal analysis of semiconductors, among others. The author recommends it highly, especially considering the price!

16.6.3 Transmission-Line Baluns

A transmission line balun, an alternative to a balun transformer, can be used within a relatively narrow frequency range. A diagram is given in Figure 16.22. The half-wavelength section has a phase shift of 180° along its length, causing the currents in the two wires of the balanced line to be equal in magnitude and opposite in phase, as required. By Kirchhoff's current law, the sum of the magnitudes of the two currents on the balanced line must be equal to the current on the center conductor of the unbalanced line at Node *A*. That is, the current on the balanced line is one-half that on the unbalanced line. Assuming a lossless line section is used to make the balun, the power on the two lines must be the same. Since

$$P = I^2 R$$

Figure 16.22
Transmission-line balun

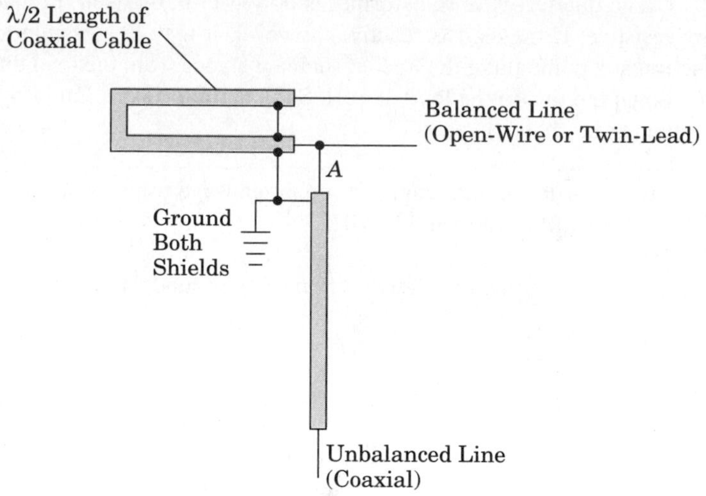

then

$$R = \frac{P}{I^2}$$

That is, for constant power, the impedance is inversely proportional to the square of current. Dividing the current by two will therefore multiply the impedance by four. This balun, then, effects a 4 : 1 impedance transformation as well as conversion from unbalanced to balanced lines. It would be quite suitable for matching 75 Ω coaxial cable to 300 Ω twin-lead, for example. It would not be a satisfactory replacement for the television-receiver matching transformer described earlier, because it would not operate over the wide frequency range required. It would, however, be a simple and economical balun for narrowband use at VHF and higher frequencies.

16.6.4 Stub Matching

Stub matching makes use of two properties of transmission lines that we have already noted. The impedance varies with position along a mismatched line, and a shorted section of line, called a **stub**, can simulate a capacitance or an inductance, depending on its electrical length. Figure 16.23(a) shows the idea. Theoretically, an open stub could also be used, but this is not practical because of radiation from the open end.

Single-stub matching consists of two steps. First, a point is found where the resistive component of the impedance looking into the line, toward the load, is equal to the characteristic impedance of the line. In general, the impedance at this point will be complex, of course, but the reactive component can be cancelled by adding an appropriate reactance to the line. Actually, admittance and susceptance are generally used in these calculations instead of impedance and reactance because it is easier to connect the stub in parallel with the line than to connect it in series. The second step, then, is to calculate the length of a stub with the correct susceptance. These calculations can be performed using a *Smith chart*, a calculator, or a computer. Smith chart calculations will be described in the next section.

Figure 16.23
Stub matching

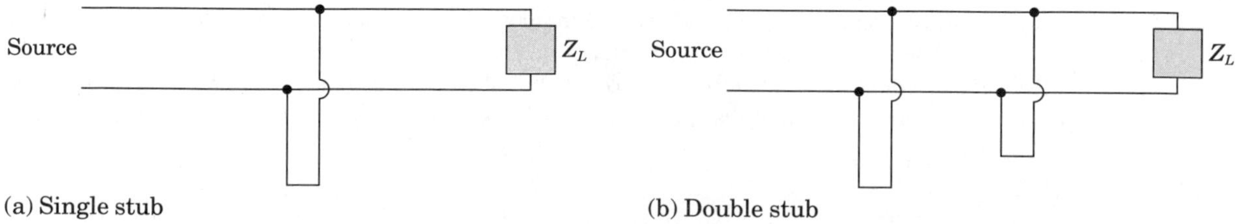

Single-stub matching requires that the position of the stub, as well as its length, be changed if any adjustments are required. This is relatively simple with open-wire line, but it is difficult for coaxial cable. Another technique uses two stubs in fixed positions and allows all adjustments to be made by changing the lengths of the stubs. This allows easier adjustment, but it is somewhat less flexible in the range of impedances that can be matched.

Figure 16.23(b) shows a double-stub tuner. The spacing of the stubs is usually an odd number of eighth-wavelengths. The stubs work together to achieve the impedance match.

16.6.5 The Smith Chart

All of the techniques for impedance matching described above require some way of calculating impedances and/or admittances at various points on a transmission line. This is not particularly difficult—it is only necessary to solve Equation (16.31) or its simplified form, Equation (16.32).

In general, of course, Z_L is complex, so the equation becomes quite messy. If, in addition, the line is very lossy, Equation (16.32) will no longer be sufficiently accurate, and it will be necessary to use a more complicated equation involving hyperbolic functions. However, the more complex equations are no problem for a computer or even a programmable calculator. These tools were unavailable in 1938, though, when P. H. Smith developed a graphical transmission-line calculator. The improved form shown in Figure 16.24 dates from 1944 and is still in use. It is not as accurate as a good computer program, but it does provide a great deal of insight into transmission-line operation. In fact, many computer programs, as well as test instruments for RF components and systems, are capable of providing output in Smith chart format, because many people find this graphical presentation easier to interpret than a list of numbers. Many technical papers also use Smith charts to illustrate the properties of transmission lines and other RF components. Therefore, while graphical computations are passing from general use, a brief study of the Smith chart will prove useful to anyone working with transmission lines.

The most obvious thing about the Smith chart is that it is circular. We noticed earlier that the impedance of a lossless transmission line repeats every one-half wavelength, that is, the impedance one-half wavelength from the load is the same as that of the load itself. On the Smith chart, a distance of one-half wavelength on the line corresponds to one revolution around the chart. Clockwise rotation represents movement toward the generator, and counterclockwise rotation represents progress toward the load, as shown on the chart itself. For convenience, there are two scales around the outside of the chart, one in each direction, in decimal fractions of a wavelength. Each scale runs from zero to 0.5 wavelength.

The body of the chart is made up of families of orthogonal circles; that is, they intersect at right angles. The impedance or admittance at any point on the line can be plotted by finding the intersection of the real component (resistance or conductance), which is indicated along the horizontal axis, with the imaginary component (reactance or susceptance), shown above the axis for positive values and below for negative. Because of the wide variation of transmission-line impedances, the Smith chart uses *normalized* impedance and admittance to reduce the range of values that must be shown.

To normalize an impedance, simply divide it by the characteristic impedance of the line:

$$z = \frac{Z}{Z_0} \tag{16.37}$$

where

z = normalized impedance at a point on the line
Z = actual impedance at the same point
Z_0 = characteristic impedance of the line

Since z is actually the ratio of two impedances, it is dimensionless.

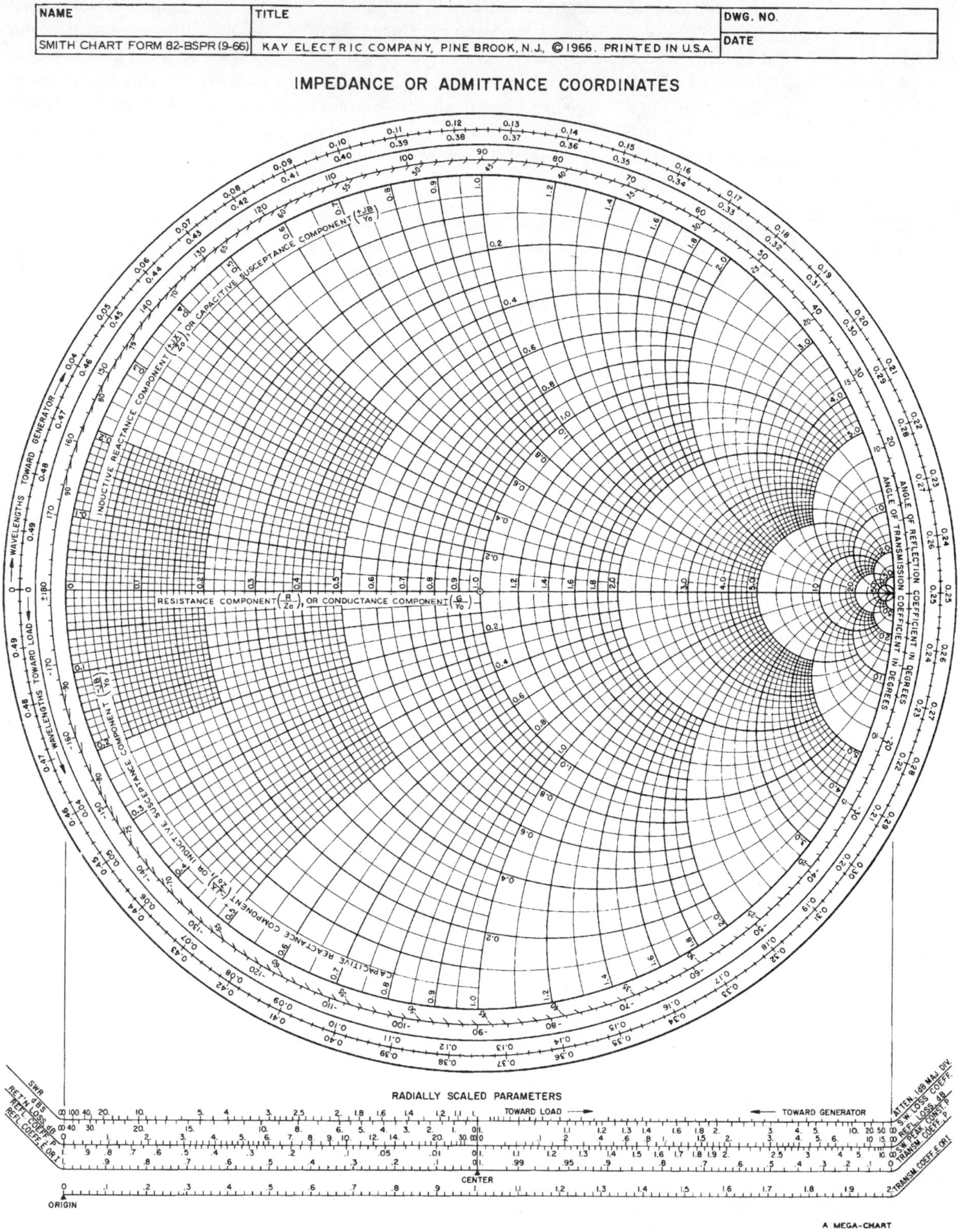

Figure 16.24

Smith chart

"Smith" is a trademark, and the Smith chart is copyright by Analog Instruments Co., Box 950, New Providence, NJ, 07974. Used by permission.

For example, a load impedance of $100 + j25\ \Omega$ on a $50\ \Omega$ line would be normalized to

$$z = \frac{Z_L}{Z_0} = \frac{100 + j25\ \Omega}{50\ \Omega} = 2 + j0.5$$

Figure 16.25 shows how this would be plotted on the chart.

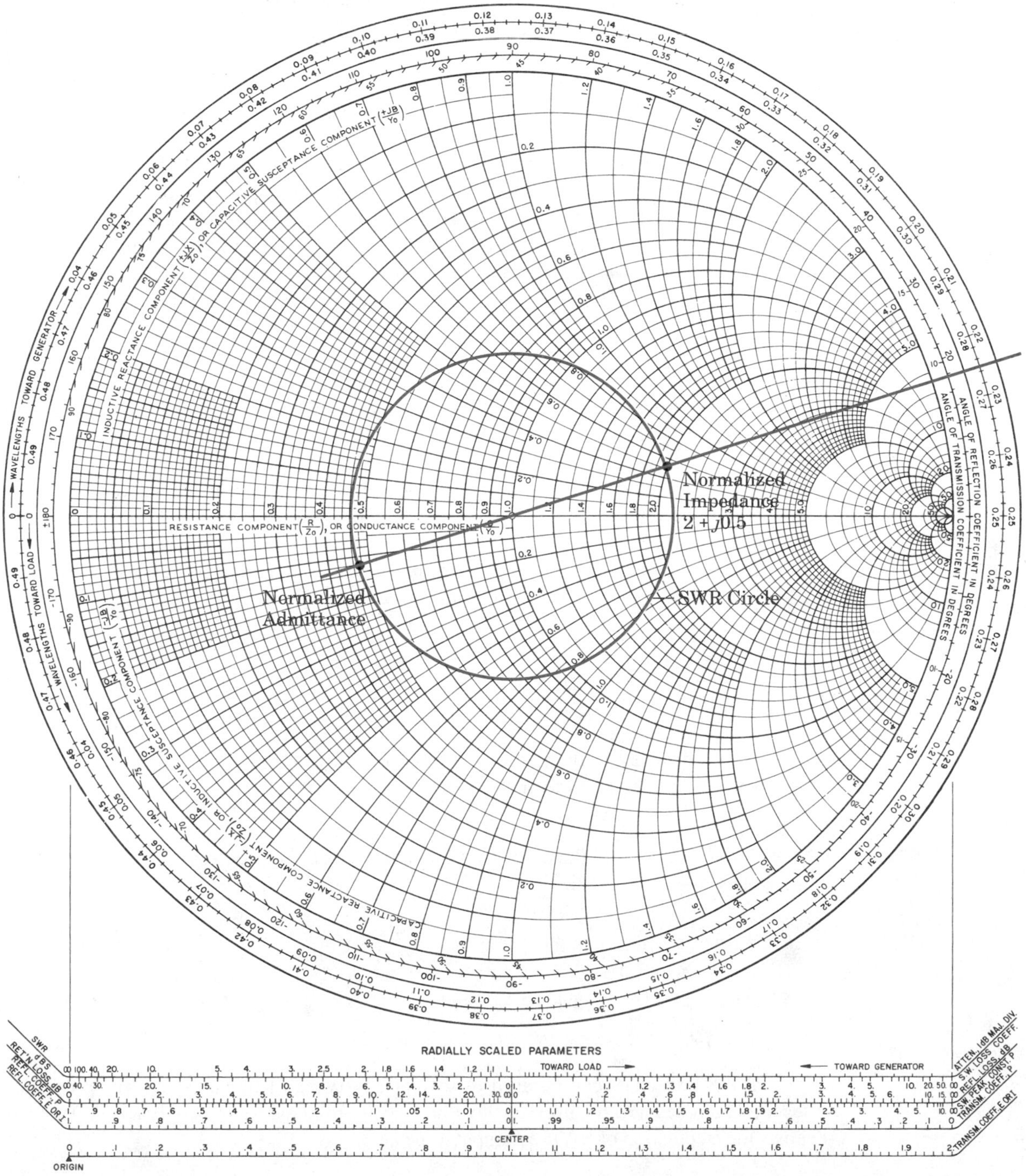

Figure 16.25
Smith chart operations
"Smith" is a trademark, and the Smith chart is copyright by Analog Instruments Co., Box 950, New Providence, NJ, 07974. Used by permission.

Once the normalized impedance at one point on the line has been plotted, that at any other point can be found very easily. Draw a circle with its center in the center of the chart, which is the point on the horizontal axis where the resistive component is equal to 1. Set the radius so that the circle passes through the point just plotted. Then draw a radius to that point and extend it to the outside of the chart, as shown in Figure 16.25. Move around the outside in the appropriate direction, using the wavelength scale as a guide. Just follow the arrows. If the first point plotted is the load impedance, then move in the direction of the generator. On the other hand, if the impedance looking into the line from the generator was plotted, move toward the load. Once the new location on the line has been found, draw another radius. The normalized impedance at the new position is the intersection of the radius with the circle. In other words, the circle is the locus of the impedance of the line. Every point on the circle represents the impedance at some point on the line.

The radius of the circle represents the SWR on the line. In fact, it is usually referred to as the *SWR circle*. The SWR can easily be found by reading the normalized resistance value where the circle crosses the horizontal axis to the right of the center of the chart. Another way is to mark the radius of the circle on the standing-wave voltage ratio scale at the bottom of the chart. If required, the SWR can also be found in decibels by using the adjoining scale. For this example, the SWR is read as approximately 2.16.

We noted before that the SWR depends on the magnitude of the reflection coefficient. Another scale on the chart allows this magnitude to be found directly. Find the reflection coefficient voltage scale at the lower left of the chart. Marking off the radius of the SWR circle on this scale gives the magnitude of Γ, which can be read as 0.37. The power reflection coefficient, which is simply Γ^2, is also provided, as are the reflection and transmission losses in decibels.

Many transmission line calculations are easier to perform using admittance. Converting from impedance to admittance is easy enough using a calculator, especially one that does complex arithmetic, but it is even easier on the Smith chart. All that is necessary is to draw a diameter that intersects the SWR circle at the impedance that must be converted to admittance. Follow the diameter across the circle to the other point at which it intersects the SWR circle—this point represents the normalized admittance at the same point on the transmission line. From Figure 16.25, the normalized admittance is found to be approximately $0.47 - j0.12$.

Normalized admittance, by the way, has the same relationship to the characteristic admittance of the line as normalized impedance has to characteristic impedance. That is,

$$y = \frac{Y}{Y_0} \tag{16.38}$$

where

y = normalized admittance at a point on the line
Y = admittance at the same point
Y_0 = characteristic admittance

Normalized admittance, like normalized impedance, is dimensionless. Since admittance is the reciprocal of impedance,

$$Y_0 = \frac{1}{Z_0}$$

and Equation (16.38) can also be written

$$y = YZ_0 \tag{16.39}$$

The Smith chart can be used to solve many types of transmission line problems. Several examples will be given to show how such problems can be approached. The first shows the calculation of SWR and input impedance.

Example 16.13 A 50 Ω line operating at 100 MHz has a velocity factor of 0.7. It is 6 m long, and is terminated with a load impedance of $50 + j50\ \Omega$. Find:

(a) the load admittance Y_L
(b) the SWR
(c) the input impedance of the line Z_i

Solution (a) Refer to Figure 16.26. First, normalize the load impedance, using Equation (16.37):

$$\begin{aligned} z_L &= \frac{Z_L}{Z_0} \\ &= \frac{50 + j50\ \Omega}{50\ \Omega} \\ &= 1 + j1 \end{aligned}$$

Enter this on the chart as shown, and draw a circle with this radius. This is the SWR circle.

To find the normalized admittance, draw a diameter at the z_L point. It intersects the circle, on the opposite side, at the point $0.5 - j0.5$. This is the normalized admittance. To find the actual admittance, use Equation (16.39):

$$\begin{aligned} y &= YZ_0 \\ Y &= \frac{y}{Z_0} \\ &= \frac{0.5 - j0.5}{50\ \Omega} \\ &= 0.01 - j0.01\ \text{S} \end{aligned}$$

(b) To find the SWR, simply note where the SWR circle crosses the horizontal axis to the right of the center of the chart, or scale the radius on the VSWR scale at the bottom of the chart. Read the SWR as 2.6.

(c) In order to find the input impedance, it is necessary to know the length of the line in wavelengths. To do that, first find the wavelength on the line.

$$\begin{aligned} \lambda &= \frac{v_p}{f} \\ &= \frac{v_f c}{f} \\ &= \frac{0.7 \times 3 \times 10^8\ \text{m/s}}{100 \times 10^6\ \text{Hz}} \\ &= 2.1\ \text{m} \end{aligned}$$

The length of the line in wavelengths is

$$\begin{aligned} L &= \frac{6\ \text{m}}{2.1\ \text{m}} \\ &= 2.857\ \lambda \end{aligned}$$

Mentally go five times around the Smith chart toward the generator, to account for 2.5 wavelengths. That is, any whole number of half-wavelengths can be subtracted from the length of the line before starting. Now go the remaining distance of 0.357 wavelength toward the generator. There are two scales around the outside of the chart. The one closer to the center increases as we move toward the load, and the other one, at the very outside of the chart, increases as we go toward the generator. Choose the outside scale, since we are going toward the generator. At the load, this scale reads 0.162. Add 0.357, the number of wavelengths remaining before we get to the generator, to get 0.519. The maximum value

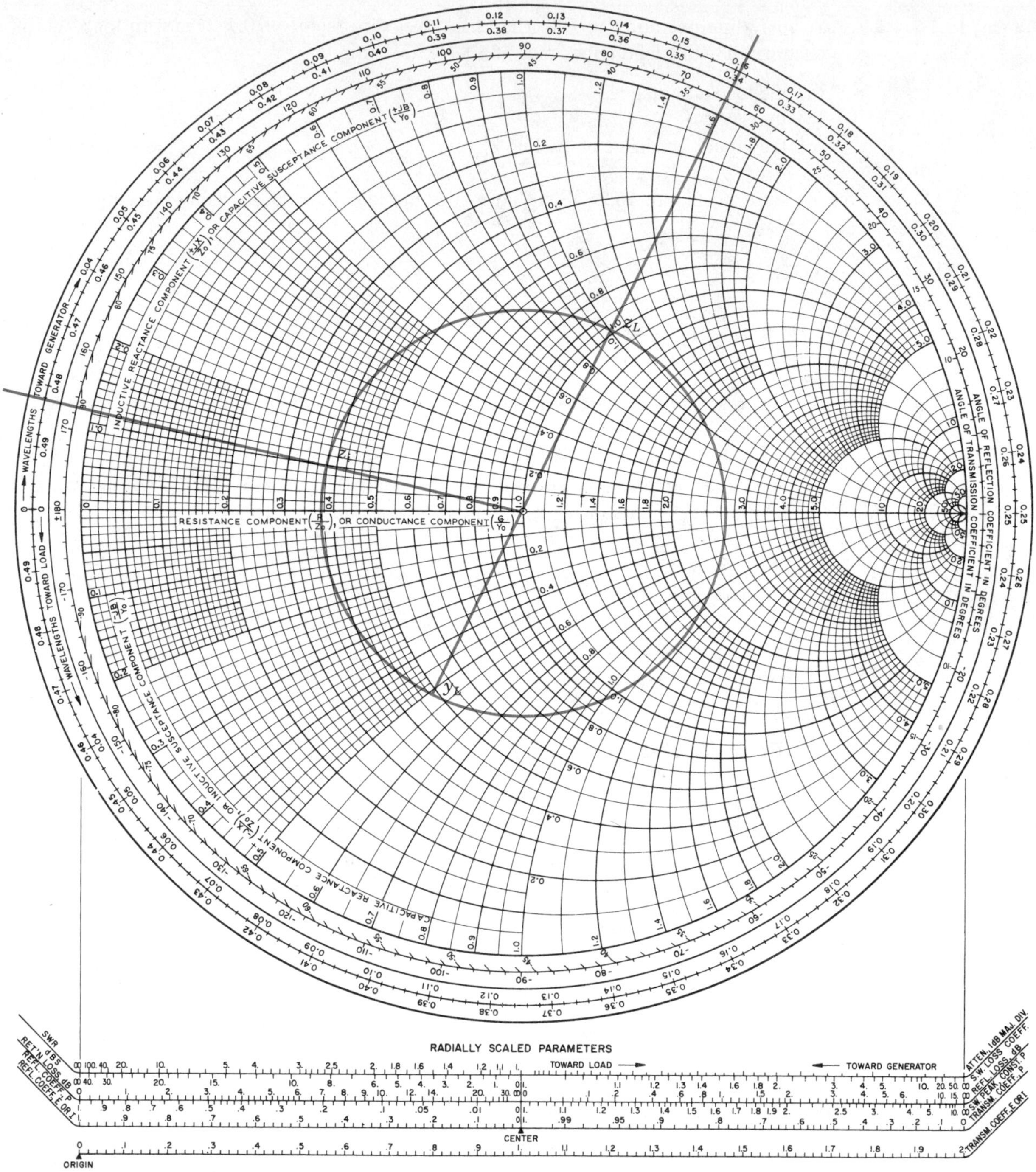

Figure 16.26
Smith chart calculations for Example 16.13

"Smith" is a trademark, and the Smith chart is copyright by Analog Instruments Co., Box 950, New Providence, NJ, 07974. Used by permission.

for the scale is 0.5, so go to that point and then go on a further distance equal to the distance

$$0.519 - 0.5 = 0.019$$

Draw a radius at this point. The place where it intersects the SWR circle represents the input impedance. Read the normalized impedance from the chart as

$$z_i = 0.39 + j0.1$$

To de-normalize this, multiply by Z_0:

$$
\begin{aligned}
Z_i &= z_i Z_0 \\
&= (0.39 + j0.1)50\ \Omega \\
&= 19.5 + j5\ \Omega
\end{aligned}
$$

The next example concerns the placement of a quarter-wave transformer. Recall that this must be added to the line at a point where the input impedance is resistive. It is easy to find such a point on a Smith chart by noting where the SWR circle crosses the horizontal axis. At such points, the reactive component of the impedance is zero, so the impedance is resistive. The value of the impedance can be calculated from the chart and used to calculate the impedance of the quarter-wave transformer section. This example also demonstrates that it is possible to move along the Smith chart in either direction. For example, the load impedance can be found when the input impedance of the line is known.

Example 16.14 The measured input impedance of a 10 m length of 300 Ω line is 120 − j650 Ω. The velocity factor is 0.9, and the frequency is 50 MHz. Find:

(a) the load impedance
(b) a suitable place to install a quarter-wave transformer
(c) the characteristic impedance that the transformer section should have

Solution (a) The first step is to normalize the input impedance.

$$
\begin{aligned}
z_i &= \frac{Z_i}{Z_0} \\
&= \frac{120 - j650\ \Omega}{300\ \Omega} \\
&= 0.4 - j2.17
\end{aligned}
$$

Plot this on the chart and draw the SWR circle, as shown in Figure 16.27.

Next, find the wavelength on the line. This can be found as before, or a slight shortcut can be taken:

$$
\begin{aligned}
\lambda &= \frac{v_p}{f} \\
&= \frac{v_f c}{f} \\
&= \frac{0.9 \times 3 \times 10^8\ \text{m/s}}{50 \times 10^6\ \text{Hz}} \\
&= 5.4\ \text{m}
\end{aligned}
$$

The length of the line in wavelengths is 10/5.4 = 1.85 wavelengths.

To find the load impedance, start at the point already plotted, which corresponds to the input, and go around the circle three times, then a further 0.35 wavelength. Be sure to go counterclockwise: follow the arrow marked "wavelengths toward load." Following the scale that increases toward the load, the source is at 0.182. Adding 0.35 to this gives 0.532, so go to 0.5 and then keep going in the same direction for the difference of 0.032.

Now the normalized impedance can be read from the point where a line drawn from the center to 0.032 on the wavelength scale intersects with the SWR circle. The

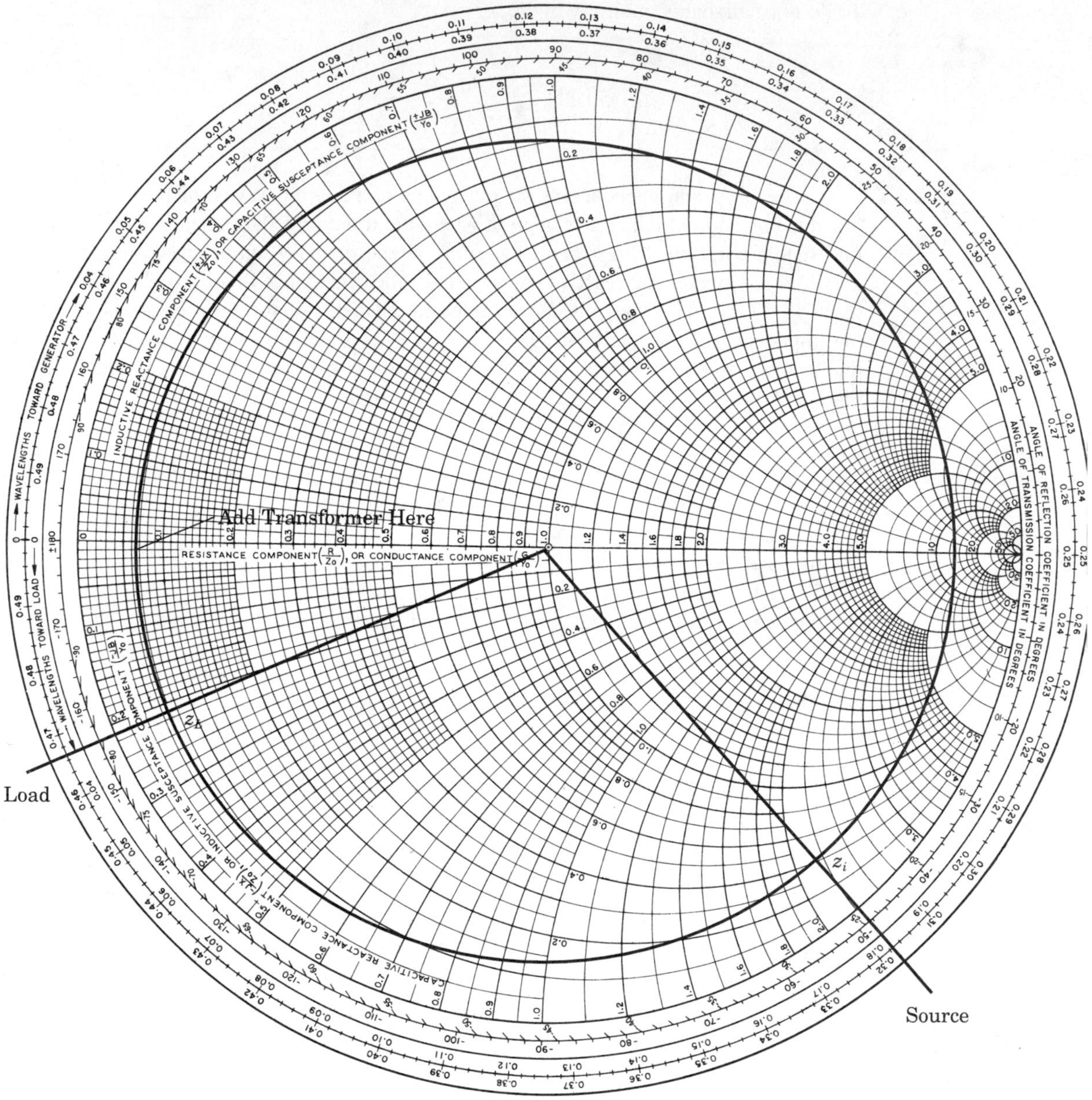

Figure 16.27
Smith chart calculations for Example 16.14

"Smith" is a trademark, and the Smith chart is copyright by Analog Instruments Co., Box 950, New Providence, NJ, 07974. Used by permission.

normalized impedance is $z_L = 0.075 - j0.21$. This can easily be converted to the actual load impedance.

$$\begin{aligned} Z_L &= z_L Z_0 \\ &= (0.075 - j0.21)300\ \Omega \\ &= 22.5 - j63\ \Omega \end{aligned}$$

(b) The quarter-wave transformer must be installed in a place where the impedance is real. It is easy to find such a place on a Smith chart: the impedance is real wherever the SWR circle crosses the horizontal axis, which it does once every quarter-wavelength. Normally, the closest such place to the load would be used, to minimize the length of mismatched line in the system. Remember that the transformer itself and the line between the transformer and the load will not be matched.

Proceeding from the load in the direction of the generator, it is necessary to go a distance of 0.032 wavelength before the circle crosses the x axis. In meters, the distance will be

$$0.032 \times 5.4 = 0.173 \text{ m}$$

(c) The impedance of the line at this point is real and equal to 0.07 in normalized form. The actual impedance is

$$\begin{aligned} Z_L &= z_L Z_0 \\ &= 0.07 \times 300\ \Omega \\ &= 21\ \Omega \end{aligned}$$

The required impedance of the quarter-wave transformer section can be found from Equation (16.36):

$$\begin{aligned} Z_0' &= \sqrt{Z_0 Z_L} \\ &= \sqrt{300\ \Omega \times 21\ \Omega} \\ &= 79.4\ \Omega \end{aligned}$$

For our final Smith chart example, let us look at single-stub matching. Since the stub cannot change the resistive part of the impedance, we look for a spot on the line where the resistive value is correct (that is, equal to Z_0), so that when normalized, it appears as a value of 1 on the chart. Then, we use the stub to cancel the reactive component. For practical reasons, we will use a shorted stub connected in parallel with the main line.

Example 16.15 Match a load with impedance $Z_L = 120 - j100\ \Omega$ to a line with $Z_0 = 72\ \Omega$, using a shorted stub.

Solution As usual, the first step is to normalize Z_L:

$$\begin{aligned} z_L &= \frac{Z_L}{Z_0} \\ &= \frac{120 - j100\ \Omega}{72\ \Omega} \\ &= 1.67 - j1.39 \end{aligned}$$

Draw the SWR circle (see Figure 16.28).

Since the stub will be in parallel with the line, it is easier to use admittances than impedances, since admittances in parallel add. Therefore, the next step is to change the load impedance to an admittance. Do this by drawing a diameter on the SWR circle, from the load impedance through the center. The point where this line intersects the circle on the opposite side represents the load admittance. Call it y_L. From the chart, $y_L = 0.35 + j0.3$.

The next step is to move toward the generator until the real part of the admittance has the correct normalized value of 1. Move from y_L toward the generator until the SWR circle intersects the $g = 1$ circle. At this point, the normalized admittance is $1 + j1.18$. We have moved a distance of 0.116 wavelength. The stub will have to have a normalized admittance of $-j1.18$ so that it will cancel the susceptance of the line.

The stub will be entirely reactive and will have infinite SWR. Its SWR circle, therefore, is the outside of the chart. To find the correct stub length, we have to start at the "load" end of the stub (the short-circuited end) and proceed around the outside of the chart in the direction of the generator until the susceptance has the required value. The shorted end will, of course, have infinite admittance, so we start at the right end of the chart. Moving

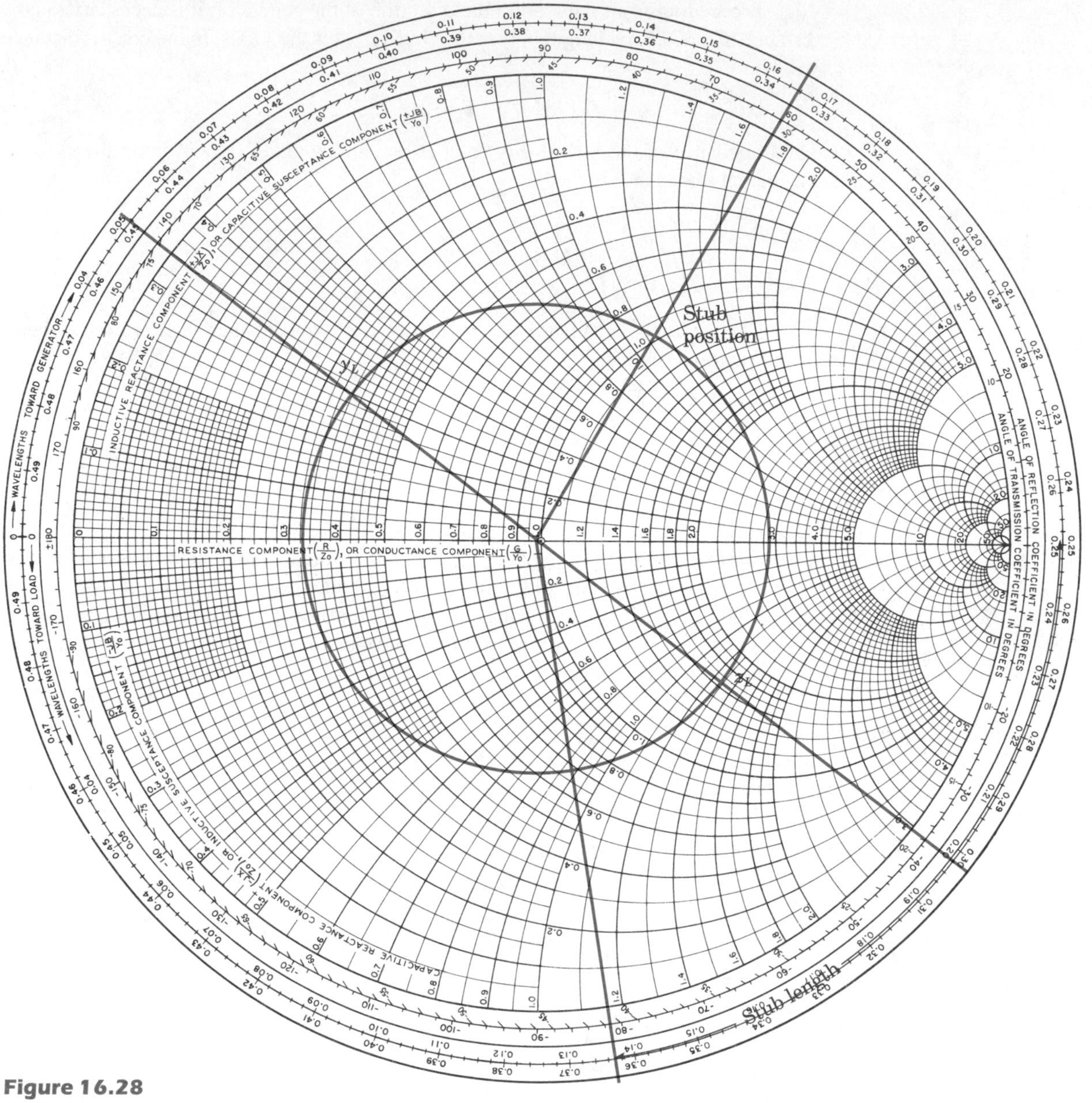

Figure 16.28
Smith chart calculations for Example 16.15
"Smith" is a trademark, and the Smith chart is copyright by Analog Instruments Co., Box 950, New Providence, NJ, 07974. Used by permission.

toward the generator a distance of 0.112 wavelength will bring us to a point with a normalized admittance of $-j1.18$.

Therefore the load can be matched by inserting a stub with a length of 0.112 wavelength at a point 0.116 wavelength from the load.

16.7 Transmission Line Measurements

There is a good deal of specialized test equipment for transmission line measurements in both the time and the frequency domains. The time domain is represented in this section by *time-domain reflectometry*. In the frequency domain, we will examine the *slotted line* and look briefly at two simpler methods for measuring the SWR on a transmission line.

16.7.1 Time-Domain Reflectometry

The techniques of time-domain reflectometry (TDR) are essentially practical applications of the methods used at the beginning of the chapter. A step input or a pulse is applied to the input of a transmission line. By examining the reflected signal, much information can be gained about such things as the length of the line, the way in which it is terminated, and the type and location of any impedance discontinuities on the line. These discontinuities can be caused by such things as faulty splices or connectors, water leaking into the cable, kinks, and so forth.

Figure 16.29 shows a typical TDR setup. Figure 16.29(a) is a photograph of the equipment, and Figure 16.29(b) is a block diagram showing its operation. In reality, the step input will be a low-frequency pulse, so that it can be repetitive to allow easier display. However, this will not affect measurements made near the leading edge of the step. If a dedicated TDR unit is not available, a pulse generator and a high-speed oscilloscope can be connected as shown in Figure 16.29(b).

The most important use of TDR is to determine the position and type of defects on a line. Often, these are not easy to locate by other means, especially in an underground or underwater cable. With TDR, the distance to the defect is easy to determine from the time taken for the reflection to return to the source. The type of defect can be gauged to some extent from the nature of the reflection.

Figure 16.29 Time-domain reflectometry
Photo courtesy Tektronix, Inc.

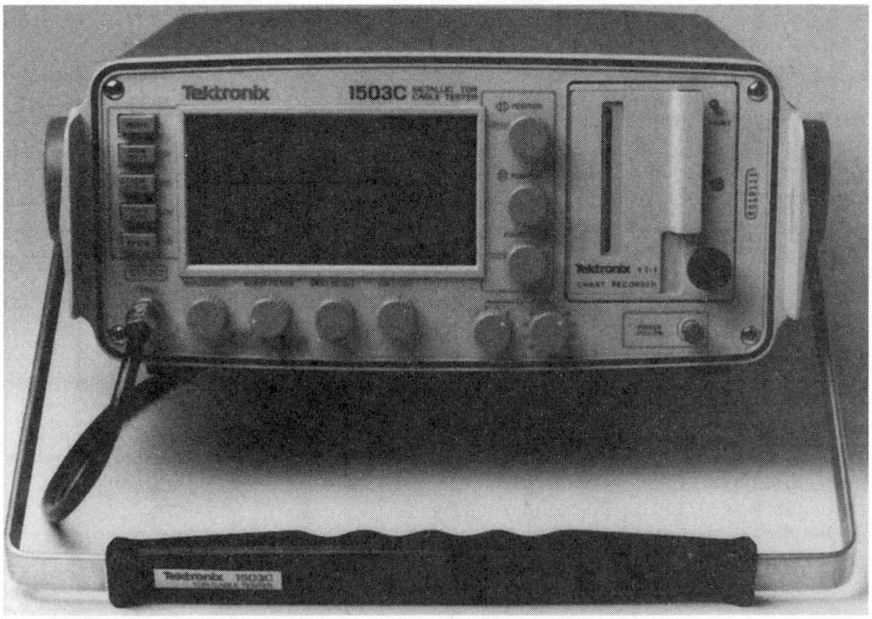

(a) TDR setup

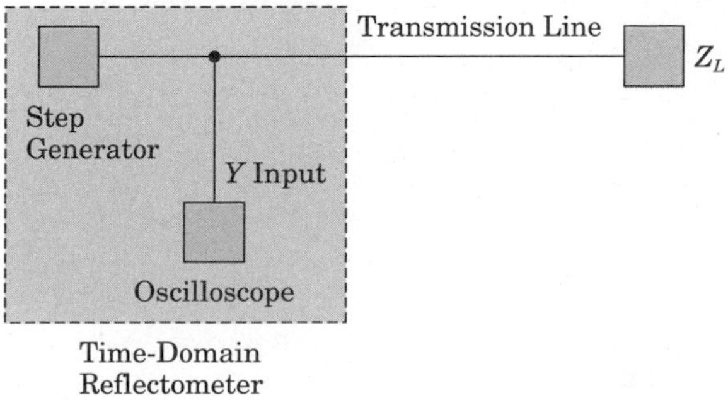

(b) TDR operation

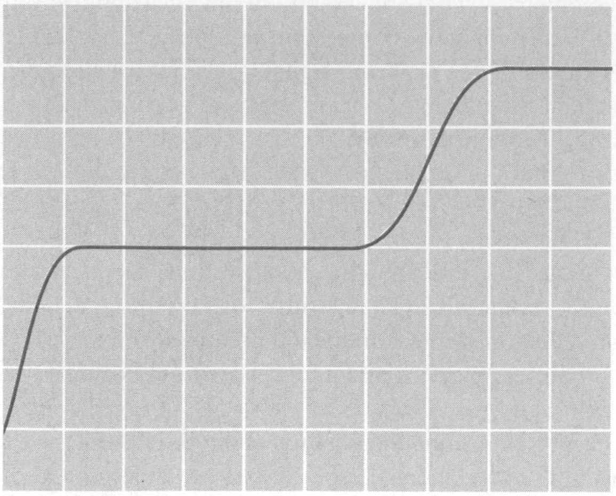

(a) Open-circuited line

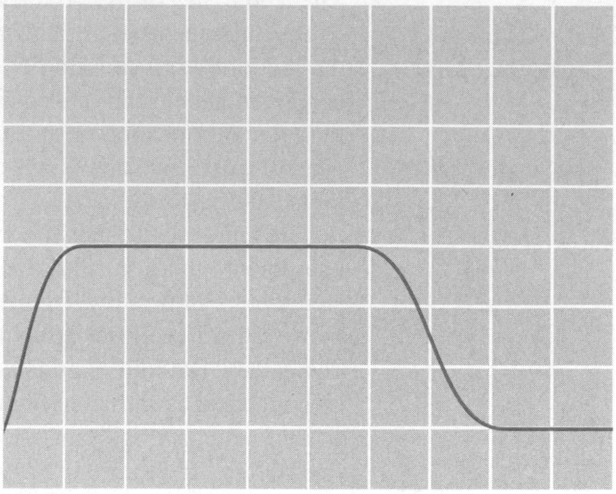

(b) Short-circuited line

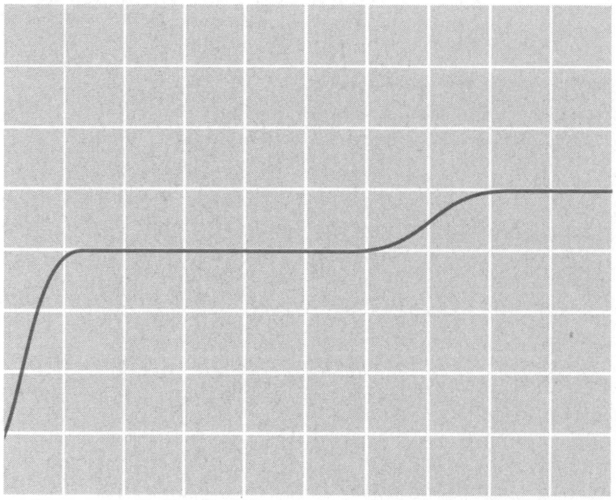

(c) Z_L resistive and $Z_L > Z_0$

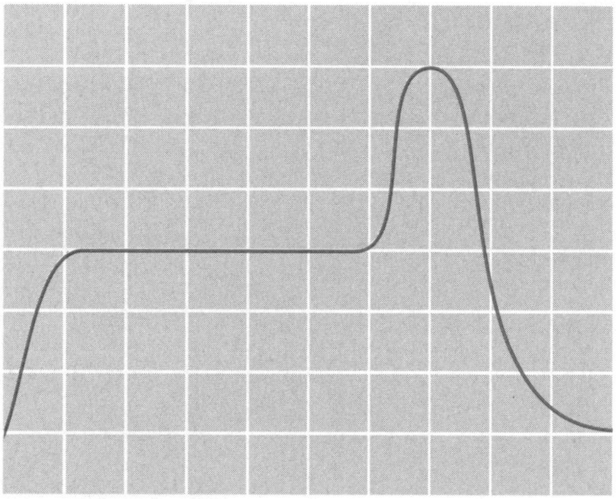

(d) Z_L inductive

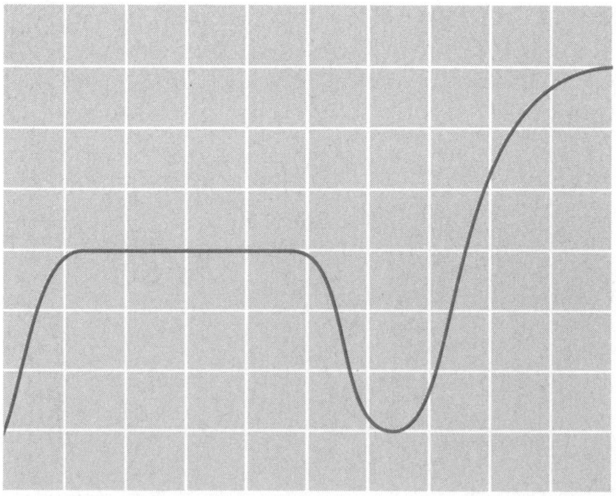

(e) Z_L capacitive

Figure 16.30
Typical waveform displays using TDR

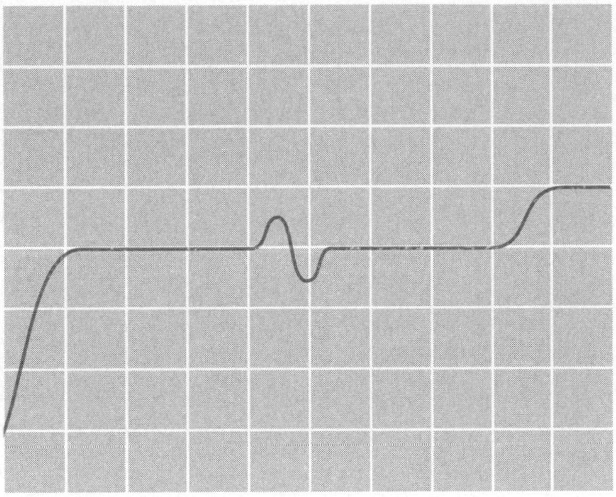

Figure 16.31
TDR display showing an impedance discontinuity

Figure 16.30 shows several types of reflections. The line in Figure 16.30(a) is open-circuited, and the voltage rises to double the initial value. In Figure 16.30(b), the termination is a short circuit, which causes the final voltage to fall to zero. Figure 16.30(c) shows $Z_L > Z_0$.

Reactive terminations can also be distinguished using TDR. An inductance will initially look like an open circuit, then it will gradually draw more current until after a time it appears as a short. Figure 16.30(d) shows how an inductive termination would appear using TDR. On the other hand, a capacitor would initially appear as a short circuit and become more like an open circuit as it charges, as can be seen in Figure 16.30(e).

Figure 16.31 shows two reflections. The first is from an impedance discontinuity in the line, the second from a mismatched load.

Example 16.16 A TDR display shows a discontinuity 1.4 μs from the start. If the line has a velocity factor of 0.8, how far is the fault from the reflectometer?

Solution The time shown on the screen is the time taken for the signal to travel to the defect and back. Therefore, one-half that time should be used to calculate the distance.

$$\begin{aligned} d &= \frac{v_p t}{2} \\ &= \frac{v_f c t}{2} \\ &= \frac{0.8(3 \times 10^8 \text{ m/s})(1.4 \times 10^{-6} \text{ s})}{2} \\ &= 168 \text{ m} \end{aligned}$$

16.7.2 The Slotted Line

The slotted line is a very straightforward way of conducting transmission line measurements (see Figure 16.32 for a photograph). Essentially, it is a short section of air-dielectric coaxial line, with a slot in the outer conductor through which a probe is inserted. The probe does not actually touch the inner conductor; it is capacitively coupled, so as to disturb the signal on the line as little as possible. The probe is connected to a diode detector that rectifies and filters the RF voltage. The resulting signal is sent to a sensitive voltmeter. Usually, an amplitude-modulated signal is used with the slotted line, so the output from the diode will be an audio-frequency signal that is easy to amplify and measure. A means is provided of accurately positioning the probe at a given point on the line.

Figure 16.32 A slotted line

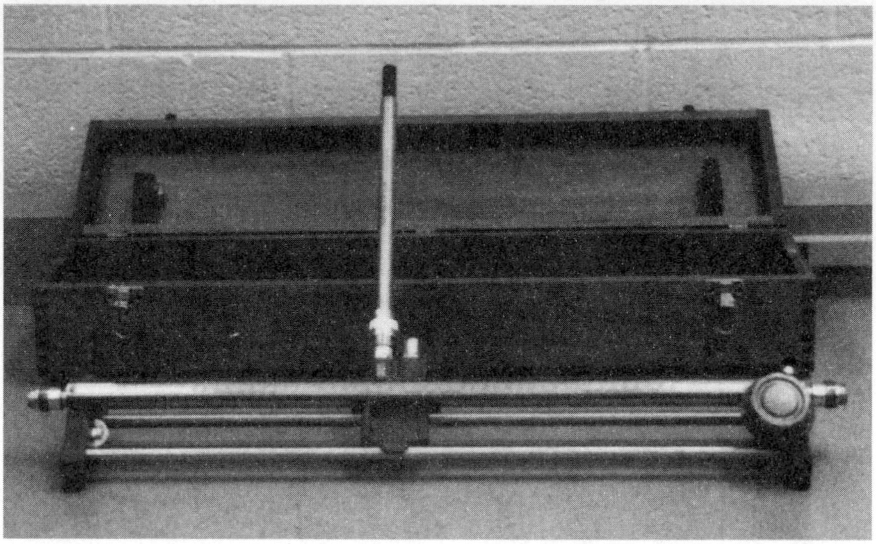

The length of a slotted line must be at least one-half wavelength. Due to the limited physical length of practical slotted lines, they are most commonly used at UHF and higher frequencies. The line must have the same impedance as the transmission line with which it is used; generally this will be 50 Ω.

In operation, the slotted line is inserted into the transmission line. It allows the wavelength on the line and the SWR to be measured very easily. Since the SWR is, by definition, V_{max}/V_{min}, it is easily measured by sliding the probe along the line and recording the maximum and minimum voltages found. The meters used with slotted lines allow the maximum to be adjusted to a set point. Then, the SWR can be read directly from the meter when the line is adjusted for a minimum reading. The measurement of wavelength is equally simple; since voltage minima occur at half-wavelength intervals on a line, all that is required is to note the position of two adjacent minimum voltage readings. Voltage maxima could be used instead, but in practice the minima can be located more precisely.

Example 16.17 Two adjacent minima on a slotted line are 23 cm apart. Find the wavelength and the frequency, assuming a velocity factor of 95%.

Solution The minima are separated by one-half wavelength so

$$\begin{aligned} \lambda &= 2 \times 23 \text{ cm} \\ &= 46 \text{ cm} \end{aligned}$$

The frequency is given by

$$\begin{aligned} v_p &= f\lambda \\ f &= \frac{v_p}{\lambda} \\ &= \frac{v_f c}{\lambda} \\ &= \frac{0.95 \times 3 \times 10^8 \text{ m/s}}{0.46 \text{ m}} \\ &= 620 \times 10^6 \text{ Hz} \\ &= 620 \text{ MHz} \end{aligned}$$

16.7.3 Standing-Wave Ratio Meters and Directional Wattmeters

Though the slotted line allows the SWR to be measured in a very straightforward way, it is a cumbersome piece of equipment for field use and is unsuitable for frequencies below UHF. There are other methods that are more convenient.

The direct measurement of SWR requires a *directional coupler*. This device allows the measurement of power moving along the line in each direction, that is, it is possible to measure incident and reflected power separately. Once this has been done, it is easy to calculate the reflection coefficient and from that, the SWR.

There are several ways to build a directional coupler. Figure 16.33 shows one method. The idea is to insert a short section of air-dielectric coaxial line with the same characteristic impedance into the transmission line on which measurements are to be made. An additional conductor is placed between the center conductor and the shield, as a sensing element. The current on the transmission line is sensed by using the mutual inductance between the center conductor of the line and the added conductor, and the voltage is sensed by capacitive coupling between the added conductor and the center conductor. The phase of the current relative to the voltage depends on the direction in which the wave is traveling along the line. It is possible to sum the current-derived and voltage-derived signals across a resistor in such a way that they cancel for a signal moving in one direction and add for a signal moving the other way. The resulting signal is rectified and filtered and can then be observed with an ordinary dc microammeter or milliammeter. With careful design and calibration, the power moving along the line can be measured fairly accurately.

Probably the best-known example of an in-line directional wattmeter is the Bird Thruline meter, which was pictured in Chapter 4. This device allows power to be measured in either direction by simply turning a removable element. The power traveling in the direction of the arrow is measured. A number of interchangeable elements calibrated for different power and frequency ranges can be used with this meter.

In order to measure SWR with a meter of this type, the power moving in each direction is measured, and the SWR is calculated from the power ratio. The SWR is given by Equation (16.24):

$$\text{SWR} = \frac{1 + |\Gamma|}{1 - |\Gamma|}$$

The ratio of reflected power to incident power is the square of the voltage ratio, or Γ^2. Therefore,

$$\text{SWR} = \frac{1 + \sqrt{P_r/P_i}}{1 - \sqrt{P_r/P_i}} \tag{16.40}$$

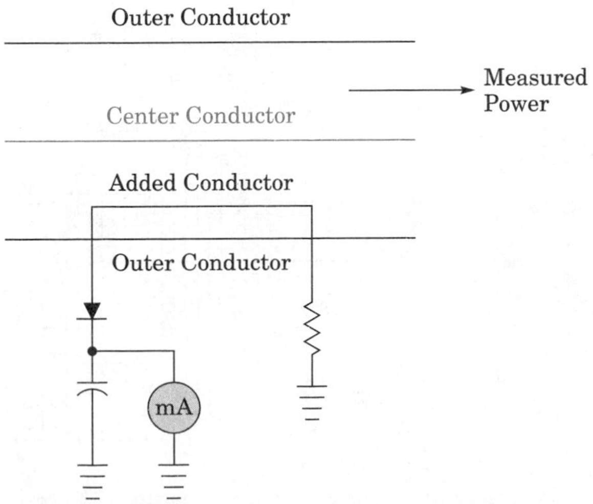

Figure 16.33
Directional coupler

Example 16.18 The forward power in a transmission line is 150 W, and the reverse power is 20 W. Calculate the SWR on the line.

Solution The easiest way is to first calculate the quantity $\sqrt{P_r/P_i}$.

$$\sqrt{\frac{P_r}{P_i}} = \sqrt{\frac{20\ \text{W}}{150\ \text{W}}}$$
$$= 0.365$$

Then, substituting into Equation (16.40),

$$\text{SWR} = \frac{1 + 0.365}{1 - 0.365}$$
$$= 2.15$$

Figure 16.34
SWR nomograph
Courtesy Bird Electronic Corp.

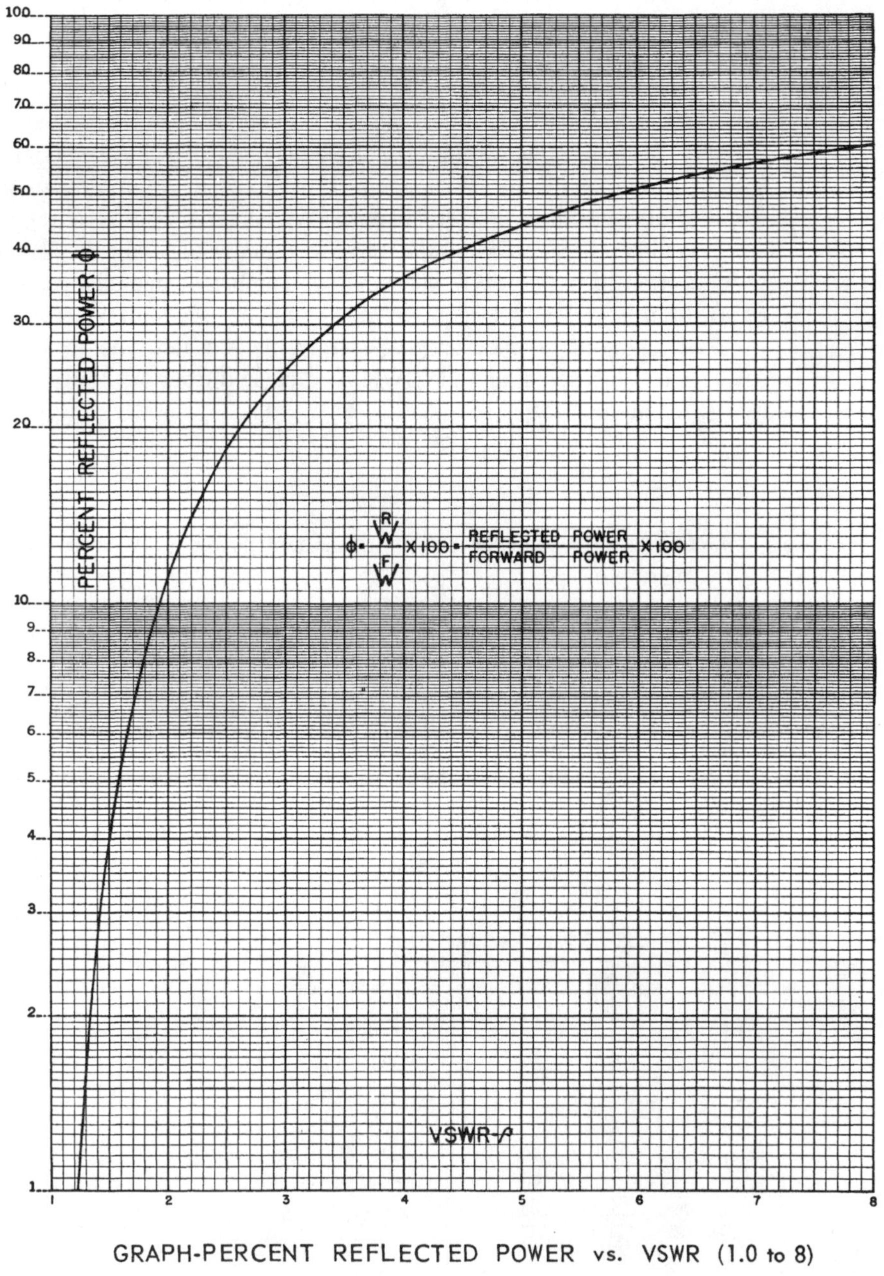

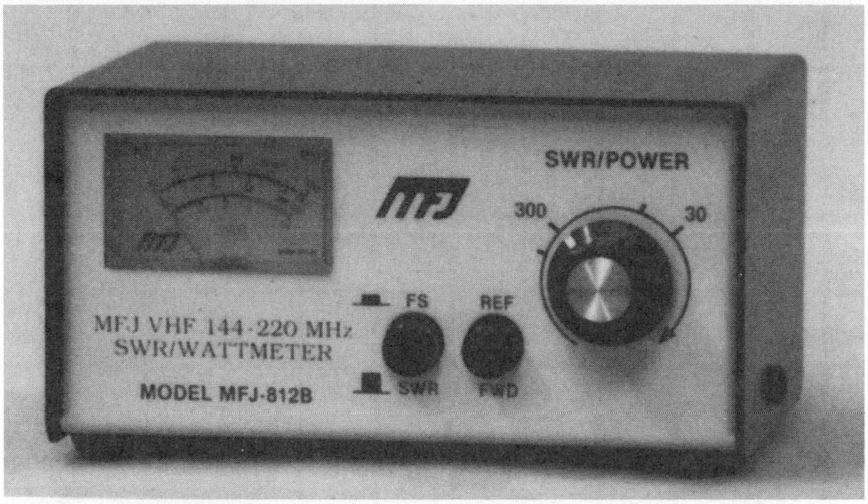

Figure 16.35
SWR meter
MFJ Enterprises, Inc., P.O. Box 494, Mississippi State, MS 39762.

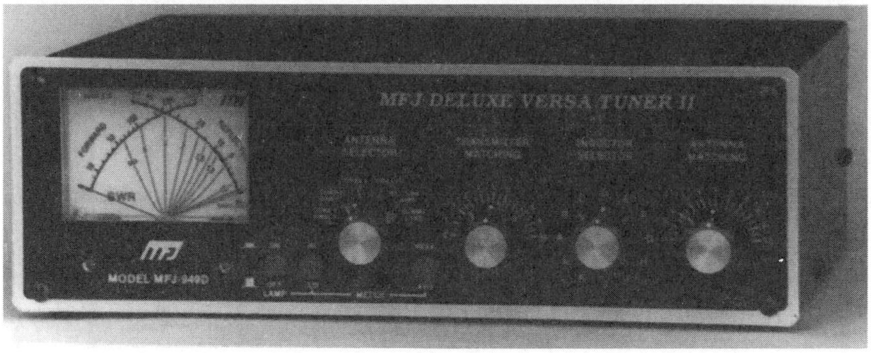

Figure 16.36
Cross-needle SWR meter
MFJ Enterprises, Inc., P.O. Box 494, Mississippi State, MS 39762.

An easier way to calculate the SWR using a directional wattmeter is to use a nomograph, such as the one in Figure 16.34. Redoing the previous example with the nomograph will yield the same result. Try it.

An even more convenient method for measuring SWR uses two directional couplers, one each for forward and reflected power. This allows the use of a meter calibrated directly in SWR. Figure 16.35 is a photograph of a typical instrument of this type. The meter is set for full-scale deflection in the "forward" position of the switch; then the switch is moved to the "reverse" position, and the SWR is read directly from the meter. When any adjustment is made to the line, the forward reading must be readjusted to the set point. Meters of this type are suitable only for SWR measurements; they cannot be used to measure actual power levels.

The ultimate in convenience is the cross-needle SWR meter shown in Figure 16.36. This device responds to forward and reverse power simultaneously, with a separate meter for each. The SWR is found from the point at which the two meter needles intersect. No switching is required, and no readjustment is needed to compensate for changes in forward power as adjustments are made to the line or the load.

Summary

Here are the main points to remember from this chapter.

1. Any pair of conductors can act as a transmission line. The methods presented in this chapter need to be used whenever the line is longer than approximately one-sixteenth the wavelength at the highest frequency in use.
2. Any transmission line has a characteristic impedance determined by its geometry and its dielectric. If a transmission line is terminated by an impedance that is differ-

ent from its characteristic impedance, part of a signal traveling down the line will be reflected. Usually, this is undesirable.

3. Reflections on lines are characterized by the reflection coefficient and the standing-wave ratio. The latter is easier to measure. A matched line has a reflection coefficient of 0 and an SWR of 1.
4. Lines that are terminated by either a short or an open circuit can be used as reactances or as either series- or parallel-resonant circuits, depending on their length.
5. Lines can be matched using lumped constants, transmission line transformers, or shorted stubs.
6. The Smith chart allows graphical transmission line calculations and shows transmission line parameters in an intuitive way.
7. Most transmission line calculations are now done using special computer programs.
8. Specialized test equipment used with transmission lines includes slotted lines, time-domain reflectometers, and SWR meters.

Important Equations

$$Z_0 = \sqrt{\frac{L}{C}} \tag{16.2}$$

$$Z_0 \approx 276 \log \frac{D}{r} \tag{16.3}$$

$$Z_0 \approx \frac{138}{\sqrt{\epsilon_r}} \log \frac{D}{d} \tag{16.4}$$

$$v_f = \frac{v_p}{c} \tag{16.6}$$

$$v_f = \frac{1}{\sqrt{\epsilon_r}} \tag{16.7}$$

$$T = \frac{L}{v_p} \tag{16.8}$$

$$\Gamma = \frac{V_r}{V_i} \tag{16.9}$$

$$\Gamma = \frac{Z_L - Z_0}{Z_L + Z_0} \tag{16.16}$$

$$v_p = f\lambda \tag{16.19}$$

$$\phi = (360)\frac{L}{\lambda} \tag{16.20}$$

$$\text{SWR} = \frac{V_{max}}{V_{min}} \tag{16.21}$$

$$\text{SWR} = \frac{1 + |\Gamma|}{1 - |\Gamma|} \tag{16.24}$$

$$|\Gamma| = \frac{\text{SWR} - 1}{\text{SWR} + 1} \tag{16.25}$$

$$\text{SWR} = \frac{Z_L}{Z_0}, \text{ if } Z_L > Z_0 \tag{16.26}$$

$$\text{SWR} = \frac{Z_0}{Z_L}, \text{ if } Z_0 > Z_L \tag{16.27}$$

$$P_r = \Gamma^2 P_i \quad (16.28)$$

$$P_L = P_i(1 - \Gamma^2) \quad (16.29)$$

$$P_L = \frac{4\text{SWR}}{(1 + \text{SWR})^2} P_i \quad (16.30)$$

$$Z = Z_0 \frac{Z_L \cos\theta + jZ_0 \sin\theta}{Z_0 \cos\theta + jZ_L \sin\theta} \quad (16.31)$$

$$Z = Z_0 \frac{Z_L + jZ_0 \tan\theta}{Z_0 + jZ_L \tan\theta} \quad (16.32)$$

$$Z = jZ_0 \tan\theta \text{ for shorted line} \quad (16.33)$$

$$Z_0' = \sqrt{Z_0 Z_L} \text{ for quarter-wave shorted stub} \quad (16.36)$$

$$z = \frac{Z}{Z_0} \quad (16.37)$$

$$y = \frac{Y}{Y_0} \quad (16.38)$$

$$\text{SWR} = \frac{1 + \sqrt{P_r/P_i}}{1 - \sqrt{P_r/P_i}} \quad (16.40)$$

Glossary

balun a device for coupling balanced and unbalanced lines

characteristic impedance the ratio between voltage and current on an infinitely long transmission line

coaxial line a transmission line containing concentric conductors

open-wire line a transmission line containing parallel conductors separated by spacers

propagation velocity the speed at which signals travel down a transmission line

quarter-wave transformer a section of transmission line, electrically a quarter-wavelength in length, that is used to change impedances on a transmission line

reflection coefficient the ratio of reflected to incident voltage on a transmission line

Smith chart a graphical transmission line calculator

standing-wave ratio (SWR) the ratio of maximum to minimum voltage on a transmission line

stub a short section of line, usually short-circuited at one end, used for impedance matching

surge impedance see *characteristic impedance*

transmission line any pair of conductors used to conduct electrical energy

velocity factor ratio of the speed of propagation on a line to that of light in free space

Questions

1. Why is an ordinary extension cord not usually considered a transmission line, while a television antenna cable of the same length would be?
2. Explain the difference between balanced and unbalanced lines, and give an example of each.
3. Draw the equivalent circuit for a short section of transmission line, and explain the physical meaning of each circuit element.
4. What is meant by the characteristic impedance of a transmission line?
5. Define the velocity factor for a transmission line, and explain why it can never be greater than one.
6. Explain what is meant by the SWR on a line. What is its value when a line is perfectly matched?
7. Why is a high SWR generally undesirable?
8. Why are shorted stubs preferred to open ones for impedance matching?
9. Draw a sketch showing how the impedance varies with distance along a lossless shorted line.
10. What would be the effect of placing a shorted, half-wave stub across a matched line?
11. What are the major contributors to transmission line loss?

12. How does transmission line loss vary with frequency? Why?
13. How does the SWR at the source differ from that at the load with a lossy line? Why?
14. Draw circuits for a balun using a transformer and one using a transmission line segment. Which of these has greater bandwidth? Which is likely to have a larger ratio of power-handling capacity to cost?
15. Draw a sketch showing how a quarter-wave transformer can be used for impedance matching. Is the transformer itself matched?
16. Compare single- and double-stub matching, giving one advantage of each.
17. Why is a Smith chart circular?
18. Why are admittances rather than impedances used for stub-matching calculations?
19. Explain how TDR can be used to find a defect in an underground cable.
20. Describe two ways in which the standing-wave ratio on a transmission line can be measured.

Problems

SECTION 16.3

21. A 12 V dc source is connected to a 93 Ω lossless line through a 93 Ω source resistance at time $t = 0$. The line is 85 m long and is terminated in a resistance.
    (a) What is the voltage across the input of the line immediately after $t = 0$?
    (b) At time $t = 1.0$ μs, the voltage at the input end of the line changes to 7.5 V, with the same polarity as before. What is the resistance that terminates the line?
    (c) What is the velocity factor of the line?
22. A 2 V dc source is connected through a 50 Ω resistor to one end of a 50 Ω line at time $t = 0$. The line is 100 m long and lossless, with a velocity factor of 0.8. Draw a sketch showing how the voltage at the source end of the line varies with time, if the line has a 20 ohm load.
23. A 10 V positive-going pulse is sent down 50 m of lossless 50 Ω cable with a velocity factor of 0.8. The cable is terminated with a 150 Ω resistor. Calculate the length of time it will take the reflected pulse to return to the start and the amplitude of the reflected pulse.
24. Calculate the characteristic impedance of an open-wire transmission line consisting of two wires with diameter 1 mm and separation 1 cm.
25. An open-wire transmission line is made from two wires, each with a diameter of 2 mm, located 15 mm apart.
    (a) Calculate the characteristic impedance of the line.
    (b) What happens to the impedance if the wire spacing is increased?
26. Calculate the characteristic impedance of a coaxial line with a polyethylene dielectric if the diameter of the inner conductor is 3 mm and the inside diameter of the outer conductor is 10 mm.

SECTION 16.4

27. A transmitter delivers 50 W into a 600 Ω lossless line that is terminated with an antenna that has an impedance of 275 Ω, resistive.
    (a) What is the coefficient of reflection?
    (b) How much of the power actually reaches the antenna?
28. A 75 Ω lossless line is terminated in error with a 93 Ω resistor. A generator sends 100 mW down the line.
    (a) What is the SWR on the line?
    (b) What is the reflection coefficient?
    (c) How much power is dissipated at the load?
    (d) What happens to the rest of the power?
29. A generator is connected to a short-circuited line 1.25 wavelengths long.
    (a) Sketch the waveforms for the incident, reflected, and resultant voltages at the instant the generator is at its maximum positive voltage.
    (b) Sketch the pattern of voltage standing waves on the line.
    (c) Sketch the variation of impedance along the line. Be sure to note whether the impedance is capacitive, inductive, or resistive at each point.
30. A transmission line is to be used to shift the phase of a 10 MHz signal by 90°. If RG-8/U with solid polyethylene dielectric is used, how long should the line be?
31. An open-circuited line is 0.75 wavelength long.
    (a) Sketch the incident, reflected, and resultant voltage waveforms at the instant the generator is at its peak negative voltage.
    (b) Draw a sketch showing how RMS voltage varies with position along the line.
32. A 75 Ω source is connected to a 50 Ω load (a spectrum analyzer) with a length of 75 Ω line. The source produces 10 mW. All impedances are resistive.
    (a) Calculate the SWR.

(b) Calculate the voltage reflection coefficient.
(c) How much power will be reflected from the load?

SECTION 16.5

33. A properly matched transmission line has a loss of 1.5 dB/100 m. If 10 W is supplied to one end of the line, how many watts reach the load, 27 m away?
34. A transmitter sends 100 watts down 13 m of matched line with a loss of 2 dB/100 m.
(a) How much power is dissipated at the load?
(b) What happens to the rest of the power?
35. A receiver requires 0.5 μV of signal for satisfactory reception. How strong (in microvolts) must the signal be at the antenna if the receiver is connected to the antenna by 25 m of matched line having an attenuation of 6 dB per 100 m?
36. A transmitter with an output power of 50 W is connected to a matched load by 32 m of matched coaxial cable. It is found that only 35 W of power is dissipated in the load. Calculate the loss in the cable in decibels per 100 meters.

SECTION 16.6

37. A 75 Ω transmission line is terminated with a load having an impedance of $45 - j30\ \Omega$. Use the Smith chart to find:
(a) the distance (in wavelengths) from the load to the closest place at which a quarter-wave transformer could be used to match the line
(b) the characteristic impedance that should be used for the quarter-wave transformer
38. Draw a dimensioned sketch of a transmission line balun that could be used to match RG-59/U type coaxial cable, with foam dielectric, to television twin-lead. Design it to work in the center of the FM broadcast band (98 MHz).
39. Find the length and position of the transformer in Problem 37, given that the operating frequency is 20 MHz and the velocity factor for both lines is 0.66.
40. A transmitter supplies 100 W to a 50 Ω lossless line that is 5.65 wavelengths long. The other end of the line is connected to an antenna with a characteristic impedance of $150 + j25\ \Omega$.
(a) Use the Smith chart to find the SWR and the magnitude of the reflection coefficient.
(b) How much of the transmitter power reaches the antenna?
(c) Use the Smith chart to find the best place to insert a shorted matching stub on the line. (Give the answer in wavelengths from the load.)
(d) Use the Smith chart to find the proper length for the stub (in wavelengths).
41. Use the Smith chart to solve this problem. A 75 Ω transmission line has a load impedance of $40 + j30\ \Omega$. Find a suitable place for a shorted stub to match this line, and calculate the length of the stub. The frequency is 45 MHz, and the line has a velocity factor of 0.95.
42. A 50 Ω transmission line is terminated with a resistive load of 120 Ω impedance. A generator puts a signal with a frequency of 70 MHz and a power level of 20 dBm into the other end of the line. The line has a velocity factor of 0.8.
(a) Calculate the reflection coefficient at the load.
(b) Calculate the voltage standing-wave ratio at the load.
(c) How much power is dissipated by the load?
(d) Design a quarter-wave transformer that could be installed at the load to match it to the line. Sketch the transformer and calculate its length and characteristic impedance.

SECTION 16.7

43. A slotted line having an impedance of 50 Ω and a velocity factor of 0.95 is used to conduct measurements with a generator and a resistive load, which is known to be more than 50 Ω. It is found that the maximum voltage on the line is 10 V and the minimum is 3 V. The distance between two minima is measured as 75 cm.
(a) What is the wavelength on the line?
(b) What is the frequency of the generator?
(c) What is the standing-wave ratio?
(d) What is the load resistance?
44. A time-domain reflectometer produces a 1 V step input. Sketch the signal that would be viewed if the line is lossless, is 50 m in length, has a velocity factor of 0.8, and is terminated in an open circuit.

COMPREHENSIVE

45. A transmission line of unknown impedance is terminated with two different resistances, and the SWR is measured each time. With a 75 Ω termination, the SWR measures 1.5. With a 300 Ω termination, it measures 2.67. What is the impedance of the line?
46. Lumped constants can be used instead of transmission-line stubs to match lines. Solve Problem 40 in this way, by calculating the correct value of inductance or capacitance that could be placed across the line, instead of using a shorted stub. The frequency is 20 MHz.

17 Radio-Wave Propagation

Objectives After studying this chapter, you should be able to:

1. Describe the nature and behavior of radio waves and compare them with other forms of electromagnetic radiation
2. Calculate power density and electric and magnetic field intensity for waves propagating in free space
3. Explain the meaning of wave polarization and differentiate between vertical, horizontal, circular, and elliptical polarization
4. Calculate free-space attenuation
5. Describe reflection, refraction, and diffraction and calculate angles of reflection and refraction
6. Describe the most common methods of terrestrial propagation, decide on the most suitable method for a given frequency and distance, and perform the necessary calculations to determine the communication range

17.1 Introduction

Radio waves are one form of electromagnetic radiation. Other forms include infrared, visible light, ultraviolet, X rays, and gamma rays. Figure 17.1 shows the place of radio waves at the low end of the frequency spectrum.

Scientists believe that electromagnetic radiation has a dual nature. Under some circumstances, it acts like a set of waves, while at other times its behavior is more easily explained by considering it to be a stream of particles called **photons**. Which description is better depends on frequency; at radio frequencies, the wave model is generally more appropriate, while light is sometimes better modeled by photons.

Electromagnetic radiation, as the name implies, involves the creation of electric and magnetic fields in free space or in some physical medium. The waves that propagate are known as *transverse electromagnetic* (TEM) waves—this means that the electric field, the magnetic field, and the direction of travel of the wave are all mutually perpendicular. The sketch in Figure 17.2 is an attempt to represent this three-dimensional process in two dimensions.

Electromagnetic radiation can be generated by many different means, but all of them involve the movement of electrical charges. In the case of radio waves, the charges are electrons moving in a conductor, or set of conductors, called an *antenna*. Antennas are the subject of the next chapter.

Once launched, electromagnetic waves can travel through **free space** (that is, through a vacuum) and through many materials. Any good dielectric will pass radio waves; the material does not have to be transparent to light. Electromagnetic waves do not travel well through lossy conductors, such as seawater, because the electric fields cause currents that dissipate the energy of the wave very quickly. Radio waves reflect from good conductors,

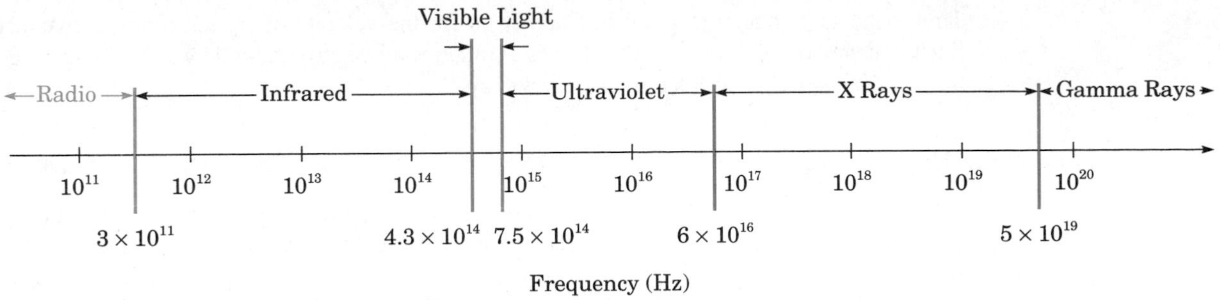

Figure 17.1 Electromagnetic spectrum

such as copper or aluminum, and are refracted as they pass from one medium to another, just as light is.

The speed of **propagation** of an electromagnetic wave in free space is the same as that of light, approximately 3×10^8 m/s. This should be no surprise, since the two forms of energy are very similar. In other media, the velocity is lower. The propagation velocity is given by

$$v = \frac{c}{\sqrt{\epsilon_r}} \tag{17.1}$$

where

v = propagation velocity in the medium
c = 3×10^8 m/s, the propagation velocity in free space
ϵ_r = relative permittivity of the medium

Figure 17.2 Transverse electromagnetic waves

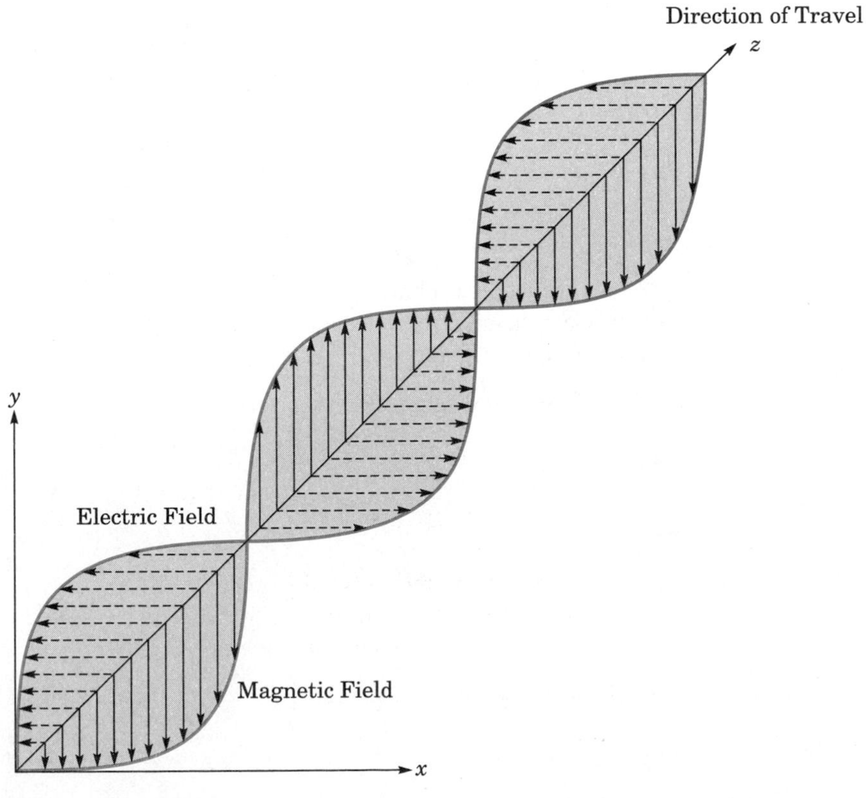

This is the same equation given in Chapter 16 for the velocity of signals along a coaxial line. One way of thinking of such a line is to imagine an electromagnetic wave propagating through the dielectric, guided by the conductors.

17.2 Electromagnetic Waves

In this chapter, we will look at transverse electromagnetic waves propagating through space and other media. Remember that these waves are characterized by frequency and wavelength.

Wavelength was discussed in Chapter 16. It is related to frequency and propagation velocity by a simple equation, repeated here for easy reference:

$$v = f\lambda \tag{17.2}$$

where

v = propagation velocity in meters per second
f = frequency in hertz
λ = wavelength in meters

Radio waves differ from other waves found in nature, such as water waves and sound waves, in that there is no physical motion of the medium, and in fact, there is no need of any physical medium. The waves consist only of time-varying electric and magnetic fields. This makes them somewhat harder to visualize, but it is necessary to make the attempt in order to understand the way they propagate and how they are generated and intercepted by antennas.

17.2.1 Electric and Magnetic Fields

An electromagnetic wave propagating through space consists of electric and magnetic fields, perpendicular both to each other and to the direction of travel of the wave (see Figure 17.2). The fields vary together, both in time and in space, and there is a definite ratio between the electric field intensity and the magnetic field intensity. This ratio is called the *characteristic impedance* of the medium and is expressed in ohms.

The relationship between the electric and magnetic field intensities is analogous to the relation between voltage and current in circuits using lumped constants. For circuits, we have the familiar Ohm's law:

$$Z = \frac{V}{I}$$

where

Z = impedance in ohms
V = electromotive force in volts
I = current in amperes

Ohm's Law for electromagnetic waves is very similar:

$$\mathscr{Z} = \frac{\mathscr{E}}{\mathscr{H}} \tag{17.3}$$

where

$\mathscr{Z}$ = impedance of the medium in ohms
$\mathscr{E}$ = electric field strength in volts per meter
$\mathscr{H}$ = magnetic field strength in amperes per meter

For a lossless medium, this is equivalent to

$$\mathscr{Z} = \sqrt{\frac{\mu}{\epsilon}} \tag{17.4}$$

where

μ = permeability of the medium in henrys per meter
ϵ = permittivity of the medium in farads per meter

For free space,

$$\mu_0 = 4\pi \times 10^{-7} \text{ H/m} \qquad \text{and} \qquad \epsilon_0 = 8.854 \times 10^{-12} \text{ F/m}$$

so the impedance of free space is

$$\mathcal{Z}_0 = \sqrt{\frac{\mu_0}{\epsilon_0}} = 377\ \Omega \tag{17.5}$$

For most media in which electromagnetic waves can propagate, the permeability is the same as that of free space. The permittivity is likely to be given as a dielectric constant, which is simply the permittivity of the medium relative to that of free space, that is,

$$\epsilon_r = \frac{\epsilon}{\epsilon_0} \tag{17.6}$$

where

ϵ_r = dielectric constant (relative permittivity)
ϵ = permittivity of the medium
ϵ_0 = permittivity of free space

It follows from Equations (17.4), (17.5), and (17.6) that the impedance of a nonmagnetic medium is

$$\begin{aligned}\mathcal{Z} &= \sqrt{\frac{\mu}{\epsilon}} \\ &= \sqrt{\frac{\mu_0}{\epsilon_r\epsilon_0}} \\ &= \sqrt{\frac{\mu_0}{\epsilon_0}} \times \frac{1}{\sqrt{\epsilon_r}} \\ &= \frac{377}{\sqrt{\epsilon_r}}\end{aligned} \tag{17.7}$$

Example 17.1 Find the characteristic impedance of polyethylene, which has a dielectric constant of 2.3.

Solution From Equation (17.7),

$$\begin{aligned}\mathcal{Z} &= \frac{377}{\sqrt{\epsilon_r}} \\ &= \frac{377}{\sqrt{2.3}} \\ &= 249\ \Omega\end{aligned}$$

17.2.2 Power Density

In lumped-constant circuits, power is given by the equation

$$P = \frac{V^2}{R}$$

where

P = power in watts
V = voltage in volts
R = resistance in ohms

The analogous equation for electromagnetic waves is

$$P_D = \frac{\mathscr{E}^2}{\mathscr{Z}} \tag{17.8}$$

where

P_D = power density in watts per square meter
$\mathscr{E}$ = electric field strength in volts per meter
$\mathscr{Z}$ = impedance of the medium in ohms

The **power density** can also be expressed in two other forms that are analogous to equations from circuit theory:

$$P_D = \mathscr{H}^2 \mathscr{Z} \tag{17.9}$$

$$P_D = \mathscr{E} \mathscr{H} \tag{17.10}$$

In physical terms, power density in space is the amount of power that flows through each square meter of a surface perpendicular to the direction of travel. To find the power in the whole wave, it would be necessary to integrate (that is, sum) the power density over the surface area.

Example 17.2

The **dielectric strength** of air is about 3 MV/m. Arcing is likely to take place at field strengths greater than that. What is the maximum power density of an electromagnetic wave in air?

Solution

From Equation (17.8),

$$\begin{aligned} P_D &= \frac{\mathscr{E}^2}{\mathscr{Z}} \\ &= \frac{(3 \times 10^6)^2}{377} \\ &= 23.9 \text{ GW/m}^2 \end{aligned}$$

Power densities of this order of magnitude can be found very close to antennas that are physically small and operated at very high power. Some radar antennas fit this description. Such densities can also be found in resonant cavities and waveguides, which will be discussed in Chapter 19.

17.2.3 Plane and Spherical Waves

Conceptually, the simplest source of electromagnetic waves would be a point in space. Waves would radiate equally from this source in all directions. A *wavefront*, that is, a surface on which all the waves have the same phase, would be the surface of a sphere. Such a source, called an **isotropic radiator**, is shown in Figure 17.3.

Of course, an actual point source is not a practical possibility, but at distances from a real source that are large compared with the dimensions of the source, this approximation is good. This is usually the case with radio propagation at reasonable distances from the antenna.

If we look at only a small area on the sphere shown in Figure 17.3 and if the distance from its center is large, the area in question will resemble a plane. (In the same way, we

Figure 17.3
Isotropic radiator

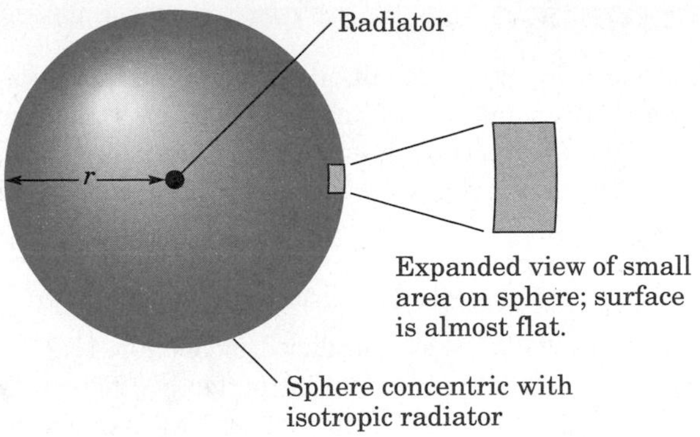

experience the earth as flat, though we know it is roughly spherical.) Consequently, many practical cases of wave propagation can be studied in terms of plane waves, which are often simpler to deal with than spherical waves. Reflection and refraction are examples of phenomena whose study is simplified by assuming plane waves.

17.2.4 Polarization

The **polarization** of a plane wave is simply the direction of its electric field vector (see Figure 17.2). If this is unvarying, the polarization is described as linear. It is also customary to refer the polarization axis to the horizon. Polarization is important because the polarization of the receiving antenna must be the same as that of the wave for best reception.

Sometimes the polarization axis rotates as the wave moves through space, rotating 360° for each wavelength of travel (see Figure 17.4). In that case, the polarization is circular if the field strength is equal at all angles of polarization and elliptical if the field strength varies as the polarization changes. The wave can rotate in either direction—it is called *right-handed* if it rotates in a clockwise direction as it recedes. The wave shown in Figure 17.4 has right-handed circular polarization. Circularly polarized waves can be received reasonably well by antennas using either horizontal or vertical polarization, as well as by circularly polarized antennas. For example, FM radio broadcasting commonly uses circular polarization, producing a signal that can be received by either horizontal or vertical antennas.

Figure 17.4
Circular polarization

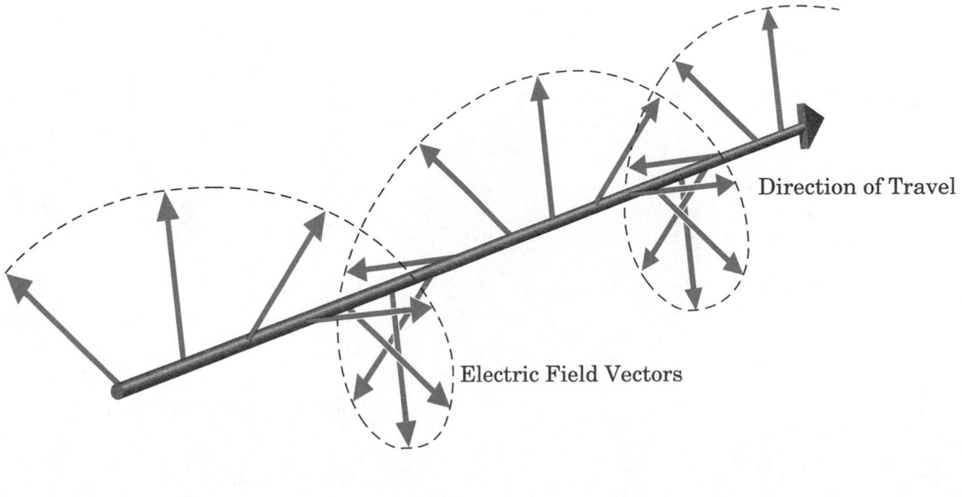

17.3 Free-Space Propagation

Radio waves, like light, propagate through free space in a straight line with a velocity of 3×10^8 m/s. There is no loss of energy in free space, but there is attenuation due to the spreading of the waves. Though the earth's atmosphere is definitely not free space, it can sometimes be treated as if it were. Free-space propagation is also of interest in satellite communications.

17.3.1 Attenuation of Free Space

Consider an isotropic radiator as introduced in Section 17.2.3, that is, an antenna that radiates equally well in all directions and is perfectly efficient. While not exactly a practical construction project, the isotropic radiator has a simple and predictable radiation pattern that will be used in this and the next chapter as an aid to understanding. A real situation can often be approximated either by an isotropic radiator or by some simple modification to it.

An isotropic radiator would produce spherical waves like those shown in Figure 17.5. If a sphere were drawn at any distance from the source and concentric with it, all the energy from the source would pass through the surface of the sphere. Since no energy would be absorbed by free space, this would be true for any distance, no matter how large. However, the energy would be spread over a larger surface as the distance from the source increased.

Since the isotropic radiator radiates equally in all directions, the power density, in watts per square meter, would be simply the total power divided by the surface area of the sphere. Put mathematically,

$$P_D = \frac{P_t}{4\pi r^2} \tag{17.11}$$

where

P_D = power density in watts per square meter
P_t = total power in watts
r = distance from the antenna in meters

Not surprisingly, this is the same *square-law* attenuation that applies to light and sound and, in fact, to any form of radiation. It is important to realize that this attenuation is not due to any loss of energy in the medium but only to the spreading out of the energy as it moves further from the source. Any actual losses will be in addition to this.

Figure 17.5
Spherical waves

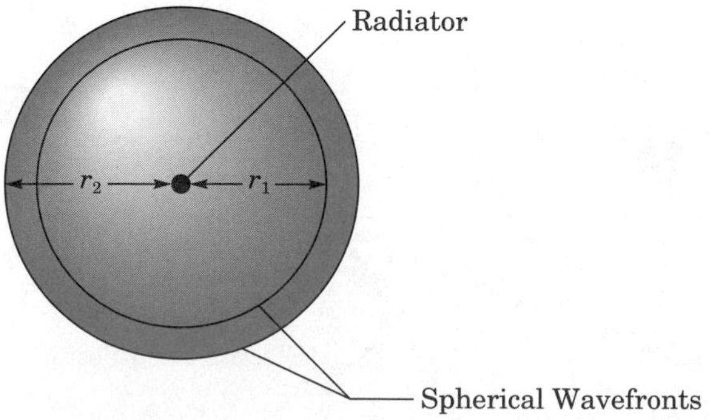

Example 17.3 A power of 100 W is supplied to an isotropic radiator. What is the power density at a point 10 km away?

Solution From Equation (17.11),

$$\begin{aligned} P_D &= \frac{P_t}{4\pi r^2} \\ &= \frac{100\ \text{W}}{4\pi\ (10 \times 10^3\ \text{m})^2} \\ &= 79.6\ \text{nW/m}^2 \end{aligned}$$

By the way, this is actually quite a strong signal, in radio terms. Radio, while very useful for transmitting information, is obviously not an efficient method of power transmission.

The strength of a signal is more often given in terms of its electric field intensity rather than power density, perhaps because the former is easier to measure. It is quite easy to derive an equation relating electric field strength at a distance to radiated power. From Equation (17.8),

$$P_D = \frac{\mathscr{E}^2}{\mathscr{Z}}$$

and from Equation (17.11),

$$P_D = \frac{P_t}{4\pi r^2}$$

Equating these two expressions for power density gives

$$\frac{\mathscr{E}^2}{\mathscr{Z}} = \frac{P_t}{4\pi r^2} \tag{17.12}$$

$$\mathscr{E}^2 = \frac{P_t \mathscr{Z}}{4\pi r^2}$$

$$\mathscr{E} = \sqrt{\frac{P_t \mathscr{Z}}{4\pi r^2}}$$

For free space, $\mathscr{Z} = 377\ \Omega$ and the equation simplifies to

$$\mathscr{E} = \sqrt{\frac{377 P_t}{4\pi r^2}} \tag{17.13}$$

$$\mathscr{E} = \frac{\sqrt{30 P_t}}{r}$$

where

$\mathscr{E}$ = electric field strength in volts per meter
P_t = total power in watts
r = distance from the source in meters

Example 17.4 Find the electric field strength for the signal in the previous example.

Solution From Equation (17.13),

$$\begin{aligned}\mathscr{E} &= \frac{\sqrt{30P_t}}{r} \\ &= \frac{\sqrt{30 \times 100}}{10 \times 10^3} \\ &= 5.48 \text{ mV/m}\end{aligned}$$

17.3.2 Transmitting Antenna Gain

In a practical communication system, it is very important to know the signal strength at the receiver input. It depends on the transmitter power and the distance from transmitter to receiver, of course, but there are two other very important determinants: the transmitting and receiving antennas. The details of antenna design will be dealt with in the next chapter, but it is useful to look at two important antenna characteristics at this point, because they have an effect on propagation calculations. These characteristics are *gain* for a transmitting antenna, and **effective area** for a receiving antenna.

Until now, we have been assuming an isotropic antenna, that is, one that radiates equally in all directions. Many practical antennas are designed to radiate more power in some directions than others. They are said to have *gain* in those directions in which the most power is radiated. This is not gain in the sense in which amplifiers have gain. The antenna is a passive device, so the total output power cannot be more than the power input. It is, in fact, somewhat less because of losses. The gain in some directions is more than compensated for by a loss in others. The antenna can nonetheless be thought of as having gain in its direction(s) of maximum radiation, when compared with an isotropic source.

Antenna gain can be visualized by making an analogy with light. Consider a flashlight bulb suspended by its connecting wires. The amount of light radiated is approximately equal in most directions, ignoring the light that is blocked by the base of the bulb. Now put a reflector behind the bulb, as in a flashlight. The bulb emits the same amount of light as before, but its intensity is greater in the beam of the flashlight. Of course, this is compensated for by the fact that no light at all is radiated to the back of the flashlight.

If the transmitting antenna has gain in a given direction, then the power density in that direction is increased by the amount of the gain, and the equation for power density becomes

$$P_D = \frac{P_T G_T}{4\pi r^2} \tag{17.14}$$

where

P_D = power density in watts per square meter
P_T = total transmitter power in watts
G_T = gain of the transmitter antenna
r = distance from transmitter to receiver, in meters

Another way of looking at the gain of a transmitting antenna is to note that in a given direction, the power density is the same as it would be if the transmitting antenna were replaced with an isotropic radiator and the transmitter power multiplied by the antenna gain. From a distant point, it makes no difference whether a signal comes from a powerful transmitter and an isotropic radiator or from a less powerful transmitter used with an antenna with gain. A common practice is to speak of the *effective isotropic radiated power* (EIRP), which is very easily found:

$$\text{EIRP} = P_T G_T \tag{17.15}$$

17.3.3 Receiving Antenna Gain

A receiving antenna absorbs some of the energy from radio waves that pass it. Since the power in the wave is proportional to the area through which it passes, it seems reasonable that a large antenna intercepts more energy than a smaller one, because it intercepts a larger area. It also seems logical that some antennas are more efficient at absorbing power from some directions than from others. For instance, a satellite dish would not be very efficient if it were pointed at the ground instead of the satellite. In other words, receiving antennas have gain, just as transmitting antennas do. In fact, the gain is the same whether the antenna is used for receiving or transmitting.

Thus the power extracted from a wave by a receiving antenna ought to depend both on its physical size and on its gain. The effective area of an antenna can be defined as

$$A_{eff} = \frac{P_R}{P_D} \tag{17.16}$$

where

A_{eff} = effective area of the antenna in square meters
P_R = power delivered to the receiver in watts
P_D = power density of the wave in watts per square meter

Equation (17.16) simply tells us that the effective area of an antenna is the area from which all the power in the wave is extracted and delivered to the receiver. Combining Equation (17.16) with Equation (17.14) gives

$$\begin{aligned} P_R &= A_{eff} P_D \\ &= \frac{A_{eff} P_T G_T}{4\pi r^2} \end{aligned} \tag{17.17}$$

It can be shown that the effective area of a receiving antenna is

$$A_{eff} = \frac{\lambda^2 G_R}{4\pi} \tag{17.18}$$

where

G_R = antenna gain, as a power ratio
λ = wavelength of the signal

17.3.4 Path Loss

Combining Equation (17.18) with Equation (17.17) gives an expression for receiver power in terms of the gains of the two antennas and the wavelength:

$$\begin{aligned} P_R &= \frac{A_{eff} P_T G_T}{4\pi r^2} \\ &= \frac{\lambda^2 G_R P_T G_T}{(4\pi)(4\pi r^2)} \\ &= \frac{\lambda^2 P_T G_T G_R}{16\pi^2 r^2} \end{aligned} \tag{17.19}$$

It is more common to express this equation in terms of the **attenuation of free space**, that is, the ratio of received power to transmitter power:

$$\frac{P_R}{P_T} = \frac{\lambda^2 G_T G_R}{16\pi^2 r^2} \tag{17.20}$$

While accurate, this equation is not very convenient. Gain and attenuation are usually expressed in decibels rather than directly as power ratios; the distance between transmitter and receiver is more likely to be given in kilometers than meters; and the frequency of the signal, in megahertz, is more commonly used than its wavelength. It is quite easy to perform the necessary conversions to arrive at a more useful equation. The work involved is left as a problem for the reader; the solution follows:

$$\begin{aligned}\frac{P_R}{P_T}\,(\text{dB}) = {} & [G_T\,(\text{dBi})] + [G_R\,(\text{dBi})] \\ & - (32.44 + 20 \log d + 20 \log f)\end{aligned} \tag{17.21}$$

where

d = distance between transmitter and receiver in kilometers
f = frequency in megahertz

The term *dBi* indicates that the antenna gains are given with respect to an isotropic radiator, for instance,

G_T (dBi) = 10 log (power density in the direction of the receiver from the transmitting antenna divided by power density in the same direction from an isotropic radiator with the same power input)

All the other quantities are as defined in the previous equations.

Equation (17.21) is expressed as a decibel gain between transmitting and receiving antennas. Of course, the received signal is weaker than the transmitted signal, so this gain is always negative. Negative gains are more commonly called *losses*, and Equation (17.21) can be written as a loss by simply changing the signs. The loss thus found is called *free-space loss* or **path loss**:

$$\begin{aligned}L_{fs} = {} & 32.44 + [20 \log d\,(\text{km})] + [20 \log f\,(\text{MHz})] \\ & - [G_T\,(\text{dBi})] - [G_R\,(\text{dBi})]\end{aligned} \tag{17.22}$$

where

$$L_{fs} = 10 \log \frac{P_T}{P_R}$$

and everything else is as in Equation (17.21). Note that P_T and P_R are the power levels at the transmitting and receiving antennas, respectively. Attenuation due to transmission line losses or mismatch is not included, but these losses can be found separately and added to the result given above (assuming all the losses are found in decibels).

For those working in miles rather than kilometers, the equivalent equation is:

$$\begin{aligned}L_{fs} = {} & 36.58 + [20 \log d\,(\text{mi})] + [20 \log f\,(\text{MHz})] \\ & - [G_T\,(\text{dBi})] - [G_R\,(\text{dBi})]\end{aligned} \tag{17.23}$$

The following examples show the use of Equation (17.22) in calculating received signal strength. We will first consider a very straightforward application, then one that is a little more complex, involving transmission line loss and mismatch.

Example 17.5

A transmitter has a power output of 150 W at a carrier frequency of 325 MHz. It is connected to an antenna with a gain of 12 dBi. The receiving antenna is 10 km away and has a gain of 5 dBi. Calculate the power delivered to the receiver, assuming free-space propagation. Assume also that there are no losses or mismatches in the system.

Solution

From Equation (17.22),

$$\begin{aligned}L_{fs} &= 32.44 + [20 \log d\ (\text{km})] + [20 \log f\ (\text{MHz})] \\ &\quad - [G_T\ (\text{dBi})] - [G_R\ (\text{dBi})] \\ &= 32.44 + 20 \log 10 + 20 \log 325 - 12 - 5 \\ &= 85.7 \text{ dB}\end{aligned}$$

$$10 \log \frac{P_T}{P_R} = 85.7$$

$$\log \frac{P_T}{P_R} = \frac{85.7}{10}$$

$$\frac{P_T}{P_R} = \text{antilog} \frac{85.7}{10}$$

$$\begin{aligned}P_R &= \frac{P_T}{\text{antilog } (85.7/10)} \\ &= \frac{150 \text{ W}}{372 \times 10^6} \\ &= 404 \times 10^{-9} \text{ W} \\ &= 404 \text{ nW}\end{aligned}$$

Example 17.6

A transmitter has a power output of 10 W at a carrier frequency of 250 MHz. It is connected by 10 m of a transmission line having a loss of 3 dB/100 m to an antenna with a gain of 6 dBi. The receiving antenna is 20 km away and has a gain of 4 dBi. There is negligible loss in the receiver feedline, but the receiver is mismatched: the antenna and line are designed for a 50 Ω impedance, but the receiver input is 75 Ω. Calculate the power delivered to the receiver, assuming free-space propagation.

Solution

When dealing with a relatively complex situation like this, it is a good idea to sketch the system. Then each loss can be added to the sketch as it is found, and a quick check will determine whether anything has been left out. We begin with the sketch, which is shown in Figure 17.6.

Next, we must find the loss at each stage of the system. If we find them all in decibels, we only need to add them to get the total loss.

First, we use Equation (17.22) to find the path loss.

$$\begin{aligned}L_{fs} &= 32.44 + [20 \log d\ (\text{km})] + [20 \log f\ (\text{MHz})] - [G_T\ (\text{dBi})] \\ &\quad - [G_R\ (\text{dBi})] \\ &= 32.44 + 20 \log 20 + 20 \log 250 - 6 - 4 \\ &= 96.42 \text{ dB}\end{aligned}$$

Figure 17.6

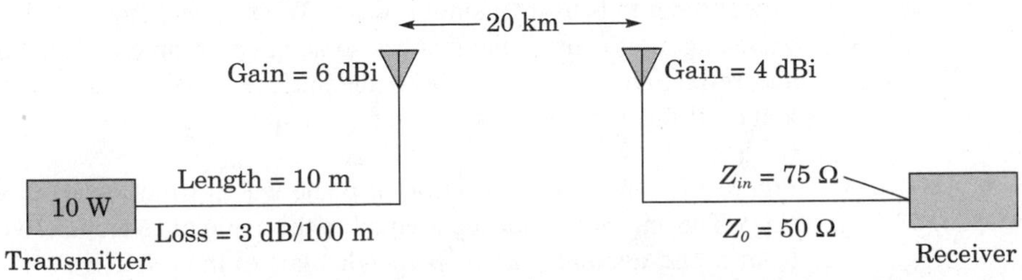

For the transmitter feedline, the loss is

$$L_{TX} = 10 \text{ m} \times 3 \text{ dB}/100 \text{ m}$$
$$= 0.3 \text{ dB}$$

The receiver feedline is lossless, but some of the power is reflected from the receiver back into the antenna due to the mismatch. This power is reradiated by the antenna and never reaches the receiver. Therefore, for our purposes it is a loss. Remember from Chapter 16 that the proportion of power reflected is the square of the reflection coefficient, which is given by

$$\Gamma = \frac{Z_L - Z_0}{Z_L + Z_0}$$
$$= \frac{75 - 50}{75 + 50}$$
$$= 0.2$$
$$\Gamma^2 = 0.2^2$$
$$= 0.04$$

The proportion of the incident power that reaches the load is

$$1 - \Gamma^2 = 0.96$$

In decibels, the loss due to mismatch is

$$L_{RX} = -10 \log 0.96$$
$$= 0.177 \text{ dB}$$

The total loss is

$$L_t = L_{fs} + L_{TX} + L_{RX}$$
$$= 96.42 \text{ dB} + 0.3 \text{ dB} + 0.177 \text{ dB}$$
$$= 96.9 \text{ dB}$$

This represents a power ratio of

$$\frac{P_T}{P_R} = \text{antilog} \frac{96.9}{10}$$
$$= 4.90 \times 10^9$$

So

$$P_R = \frac{10 \text{ W}}{4.90 \times 10^9}$$
$$= 2.04 \text{ nW}$$

17.4 Reflection, Refraction, and Diffraction

The three properties listed in the title of this section should be familiar from the behavior of light. Radio waves are identical to light waves except for frequency, and we should expect them to behave in similar ways. Their lower frequency is associated with a longer wavelength, of course, and this has some effect in practical situations. For both reflection and refraction, it is assumed that the surfaces involved are much larger than the wavelength. If this is not the case, diffraction will occur.

17.4.1 Reflection

Figure 17.7 shows the reflection of plane waves from a smooth surface (*specular reflection*). The angle of incidence is equal to the angle of reflection, with both angles measured from a line **normal** (that is, perpendicular) to the reflective surface. In the case of radio

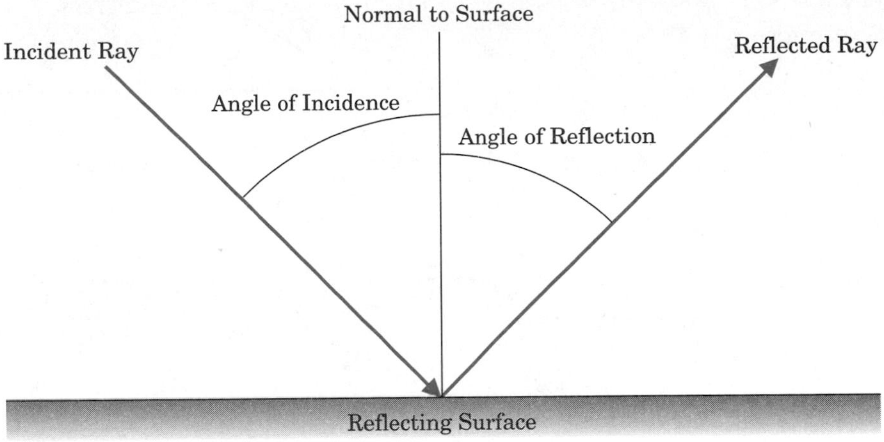

Figure 17.7
Specular reflection

waves, the reflection will be complete if the reflector is an ideal conductor. This is never quite the case, of course, and some of the wave will propagate for a short distance into a lossy material before being completely absorbed.

Reflection can also take place from dielectrics. Usually, some of the energy is reflected and some refracted, but sometimes all the energy is reflected. This is really a special case of refraction, and it will be considered in that context.

The reflecting surface does not have to be a single plane. Figure 17.8 shows a corner reflector, for instance. This scheme is often used for antennas.

Also very useful in antenna design, of course, is the parabolic reflector, shown in Figure 17.9. Any plane wave entering along the axis of the antenna will be reflected in such a way that all of its energy passes through a single point called the *focus* of the reflector. Both the corner reflector and parabolic reflector will be considered in more detail later in this book, when we look at antennas.

Reflections do not require a smooth surface. Radio waves often reflect from the earth, for example. Figure 17.10 shows what happens with these diffuse reflections. The

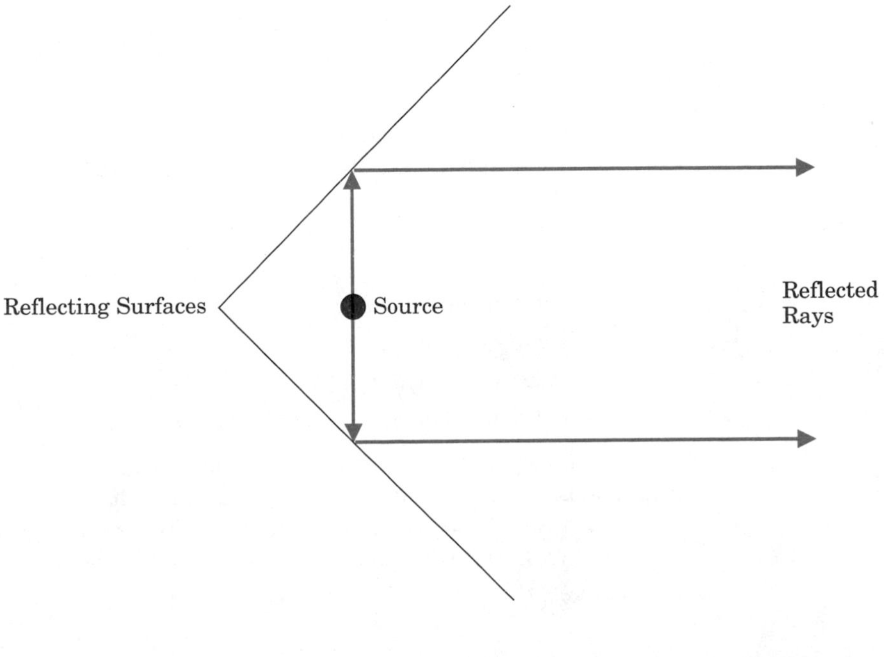

Figure 17.8
Corner reflector

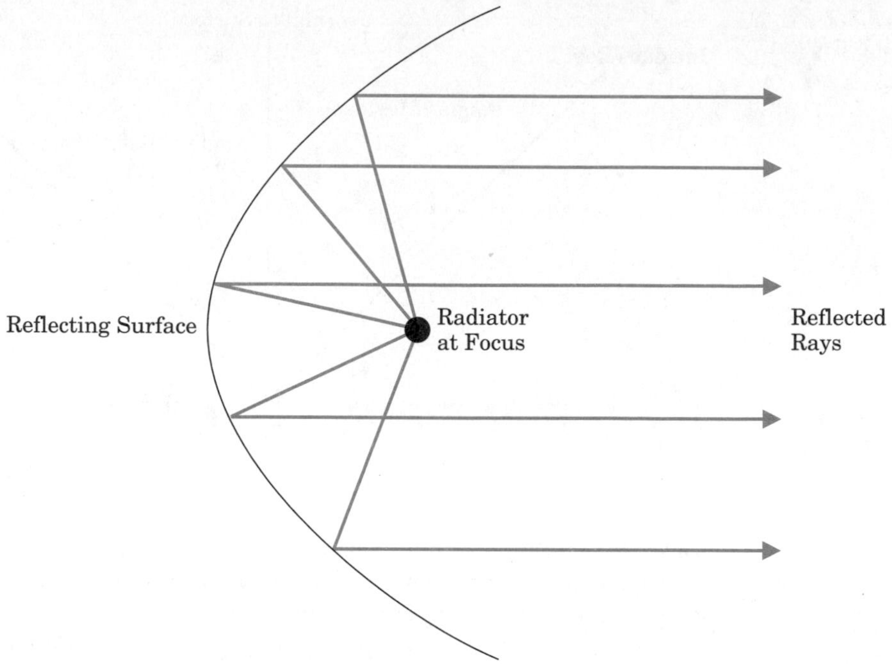

Figure 17.9
Parabolic reflector

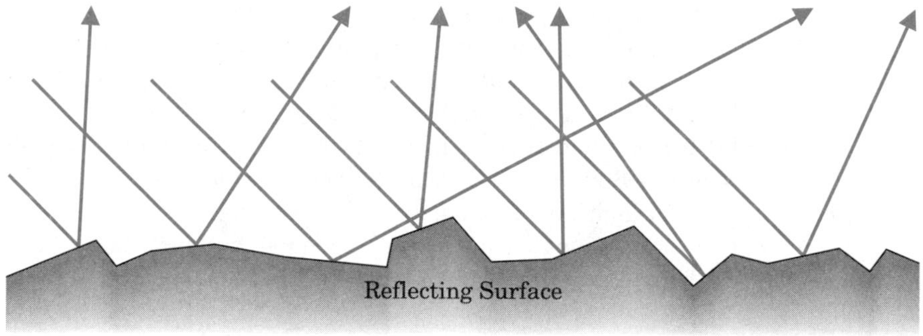

Figure 17.10
Diffuse reflection

angles of incidence and reflection for each part of the surface are equal, as before, but since each part has a different orientation, the reflected wave is scattered.

17.4.2 Refraction

A transition from one medium to another often results in the bending, or refraction, of radio waves, just as it does with light. In optics, the angles involved are given by Snell's law:

$$n_1 \sin \theta_1 = n_2 \sin \theta_2 \tag{17.24}$$

where

θ_1 = angle of incidence
θ_2 = angle of refraction
n_1 = index of refraction in the first medium
n_2 = index of refraction in the second medium

Figure 17.11 shows how the angles are measured. Once again, they are measured from a line normal to the surface and not from the surface itself.

Now if Snell's law works for radio waves (it does), how do we find the **index of refraction**? This is actually quite simple: for a given medium,

$$n = \sqrt{\mu_r \epsilon_r} \tag{17.25}$$

Figure 17.11
Refraction

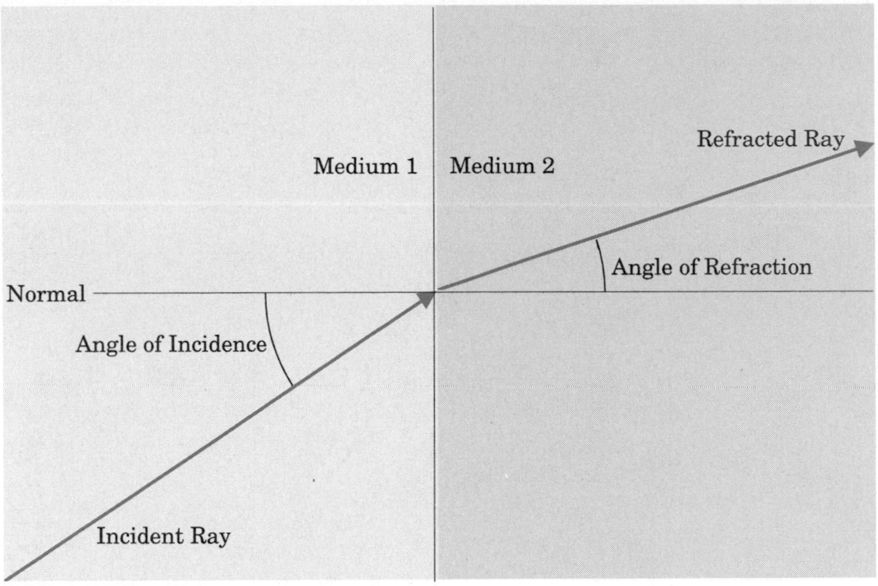

where

μ_r = relative permeability of the medium
ϵ_r = relative permittivity of the medium

Since μ_r is almost always 1 for the medium of interest, in practical terms

$$n = \sqrt{\epsilon_r} \tag{17.26}$$

Substituting Equation (17.26) into Equation (17.24) gives

$$\frac{\sin \theta_1}{\sin \theta_2} = \frac{\sqrt{\epsilon_{r2}}}{\sqrt{\epsilon_{r1}}} = \sqrt{\frac{\epsilon_{r2}}{\epsilon_{r1}}} \tag{17.27}$$

Example 17.7 A radio wave moves from air ($\epsilon_r = 1$) to glass ($\epsilon_r = 7.8$). Its angle of incidence is 30°. What is the angle of refraction?

Solution From Equation (17.27),

$$\frac{\sin \theta_1}{\sin \theta_2} = \sqrt{\frac{\epsilon_{r2}}{\epsilon_{r1}}}$$

$$\sin \theta_2 = \frac{\sin \theta_1}{\sqrt{\frac{\epsilon_{r2}}{\epsilon_{r1}}}} = \frac{\sin 30°}{\sqrt{\frac{7.8}{1}}} = 0.179$$

$$\theta_2 = \arcsin 0.179 = 10.3°$$

The result is sketched in Figure 17.12.

Figure 17.12

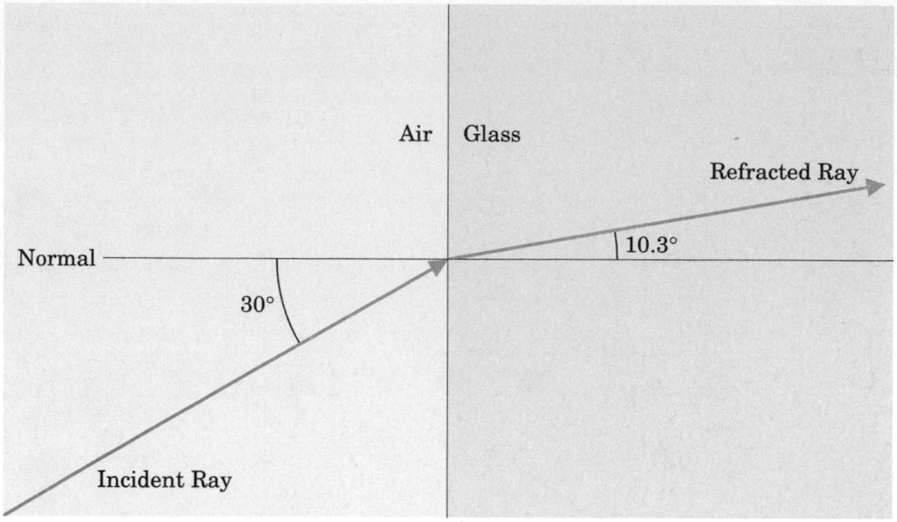

It can be seen from this example that when a wave enters a region with a higher dielectric constant and thus a lower propagation velocity, it bends toward the normal. This can be remembered by thinking of a wavefront perpendicular to the direction of travel, as shown in Figure 17.13. When this wavefront enters a slower medium, the part that enters the medium slows down, causing the whole front to swing toward the normal, as shown in Figure 17.13(a). On the other hand, if the second medium has a lower dielectric constant (and a greater propagation velocity), the wave will be deflected away from the normal, as shown in Figure 17.13(b).

In extreme cases, where the angle of incidence is large and the wave travels into a region of considerably lower dielectric constant, the angle of refraction can be greater than 90°, so that the wave comes out of the second medium and back into the first. Under these circumstances, refraction becomes a form of reflection called *total internal reflection*. Figure 17.14 shows an example. The angle of incidence that results in an angle of

Figure 17.13
Refraction as a function of dielectric constant

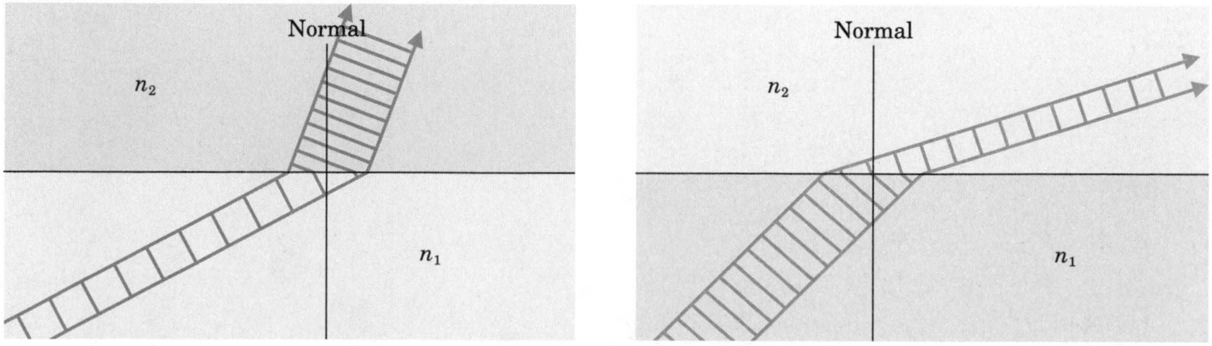

(a) $n_1 < n_2$ (b) $n_1 > n_2$

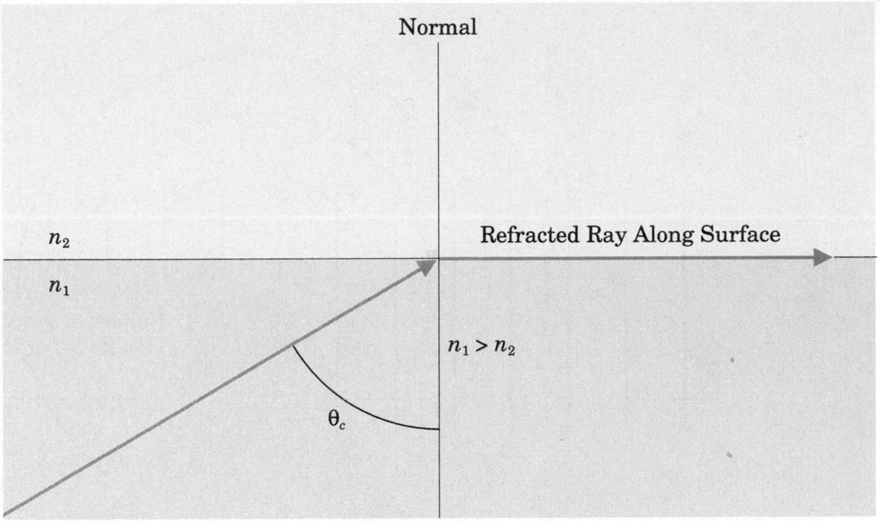

Figure 17.14 Total internal reflection

Critical angle ($n_1 > n_2$)

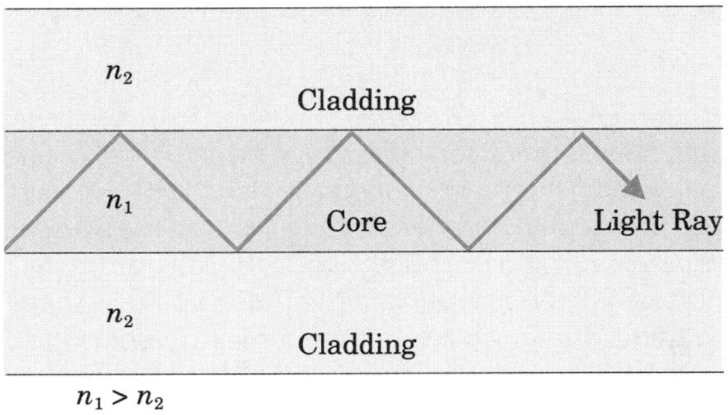

Figure 17.15 Optical fiber ($n_1 > n_2$)

refraction of exactly 90° (so that the wave propagates along the boundary between the two media) is known as the *critical angle* and is given by

$$\theta_c = \arcsin \frac{n_2}{n_1} \tag{17.28}$$

Total internal reflection is found in optical fibers. The light reflects from the boundary between the core of the fiber and a cladding material with lower refractive index. Figure 17.15 shows the basic idea, which will be developed in more detail in Chapter 22.

17.4.3 Diffraction

It is a common saying that light travels in straight lines, but it does appear to go around corners occasionally. This phenomenon is a result of diffraction, and there are numerous examples of its use in optics (for instance, diffraction gratings). Diffraction also occurs for radio waves and can, for example, allow reception from a transmitting antenna on the far side of a mountain.

Figure 17.16 illustrates diffraction. We can describe the effect of diffraction by assuming that each point on a wavefront acts as an isotropic source of radio waves. Thus some of the wavefronts that pass beside or above an obstruction can radiate into the area behind it.

Diffraction is more pronounced when the object that causes it has a sharp edge, that is, when its dimensions are small in comparison with the wavelength.

Figure 17.16
Diffraction

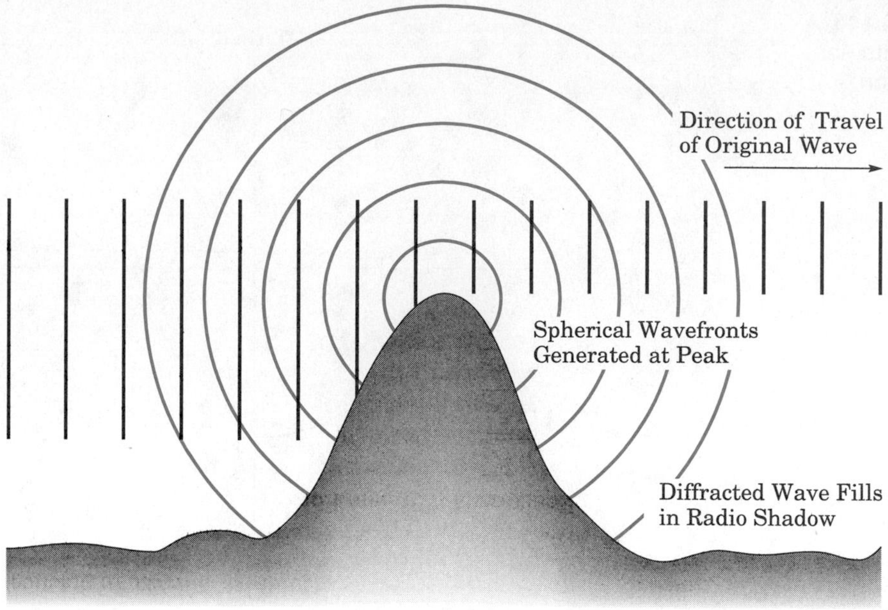

Early Experiences with Propagation

Radio propagation was predicted mathematically by James C. Maxwell in 1865 but was first demonstrated experimentally by Heinrich R. Hertz in 1887. Hertz demonstrated transmission from one spark gap to another and also showed the presence of standing waves. The frequency he used is not known, but the size of his laboratory is—in order to set up standing waves, he must have been operating in the VHF part of the spectrum.

When radio communication became practical around the turn of the century, the frequencies used were much lower. In fact, the terms "medium," "low," and "high" frequency date from these early days. Experimenters began in the medium-frequency range and went lower and lower in frequency in search of more reliable long-distance propagation using ground waves. Frequencies as low as 60 kHz were popular. These low frequencies had the additional advantage that they could be generated by mechanical alternators, which were more efficient than spark gaps. (Vacuum tubes were not yet available.)

In the early days of radio, frequencies in the HF range and up were considered useless for long-distance communication.

17.5 Terrestrial Propagation

Most of the time, the radio waves we use are not quite in free space. Their behavior near the earth is influenced by factors that include the properties of both the earth and the various layers of the atmosphere. Most of these factors are frequency-dependent. We will look briefly at some of the more important **terrestrial propagation** modes, proceeding roughly in order of frequency, from low to high. In addition to direct, straight-line radiation from transmitter to receiver (called *line-of-sight* or sometimes **space-wave** propagation), we will discuss **ground waves**, which follow the surface of the earth, and **sky waves**, which refract from ionized layers in the atmosphere. We will also look at some other less common but still important modes of propagation.

17.5.1 Ground-Wave Propagation

At frequencies up to approximately 2 MHz, the most important method of propagation is by *ground waves*. Ground waves are vertically polarized waves that follow the ground and can therefore follow the curvature of the earth to propagate far beyond the horizon.

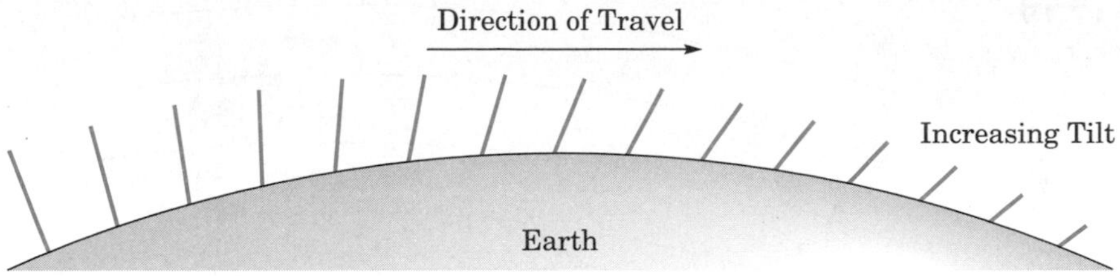

Figure 17.17
Ground-wave propagation

They must have vertical polarization to minimize currents induced in the ground itself, which result in losses. Nonetheless, there is a tendency for the waves to "tilt" toward the horizontal, increasing losses, as the distance from the transmitter increases. Ground-wave propagation is illustrated in Figure 17.17.

At close range, some energy is also transmitted directly from the transmitter to the receiver through the air. However, this signal is often nearly cancelled by another signal reflected from the ground, as shown in Figure 17.18. At low frequencies, the earth is highly reflective and also provides a 180° phase shift. The transmitting and receiving antennas are likely to be close to the ground, so the direct path from transmitter to receiver is almost equal in length to the path involving a reflection from the earth. Since the direct and reflected waves are out of phase, partial cancellation occurs. This leaves the ground wave as the main mode of propagation at low frequencies.

Ground waves provide very reliable communication that is almost independent of weather and solar activity, both of which can greatly affect some other propagation modes. With sufficient power at a low enough frequency, round-the-world communication is possible, and some military transmitters operate at frequencies as low as 15 kHz. There is a United States government time-and-frequency station, WWVL, at 60 kHz, and an international navigation system called LORAN-C operates at 100 kHz. In addition, of course, the standard AM broadcast band relies mainly on ground-wave propagation.

Ground waves are attenuated very quickly above about 2 MHz, so that the usable spectrum for this propagation mode has only one-third the width of a single television channel. This lack of spectrum space is one of the main disadvantages of ground-wave propagation. Ground waves require relatively high power, and their low frequencies are associated with long wavelengths that require physically large antennas for good efficiency. Ground-wave propagation will certainly remain in use, but it has little or no room for expansion.

17.5.2 Ionospheric Propagation

Many people, including the author of this book, first became interested in radio communications when they listened to a high-frequency (shortwave) radio receiver and heard stations from all over the world. Long-range communication in the high-frequency band is possible because of refraction in a region of the upper atmosphere called the **ionosphere**, where some of the air molecules are ionized by solar radiation (see Figure 17.19). Note that Figure 17.19 is not drawn to scale: the ionospheric layers vary in height from about 60 to 400 km above the earth's surface, while the radius of the earth is approximately 6400 km.

The ionosphere can be divided into three regions known as the *D, E*, and *F* layers. The *F* layer itself is divided into two parts, called F_1 and F_2. The level of ionization in-

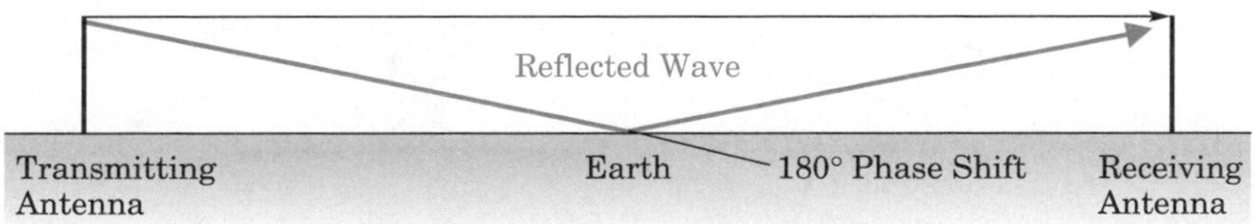

Figure 17.18
Cancellation due to ground reflections

Figure 17.19
Ionospheric layers

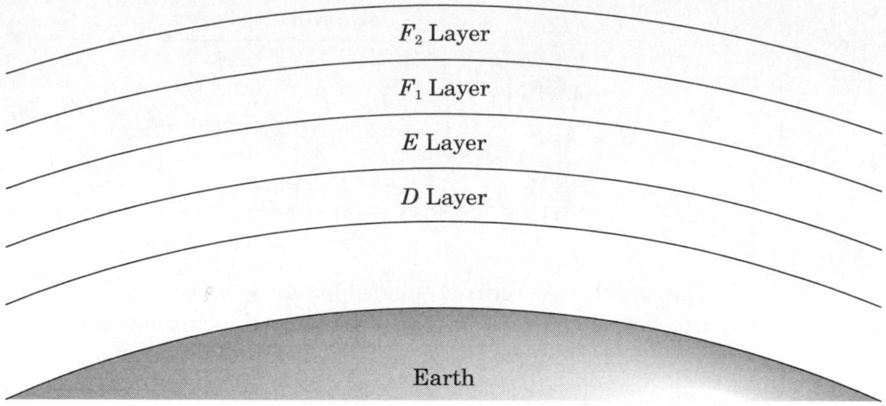

creases with height above the earth and is greater in the daytime. At night, when solar radiation is not received, the D and E regions disappear and the F_1 and F_2 layers combine into a single F layer. The F layer remains during the night because the atmosphere is so rarified at this height that ions take longer to recombine.

Ionization levels change with the amount of solar activity, which varies greatly over an 11-year cycle known as the *sunspot cycle*. Peaks in this cycle are associated with greater activity in the ionosphere and generally more favorable conditions for ionospheric propagation, especially toward the high end of the HF region of the spectrum. The most recent peak occurred in 1990.

The signal is returned from the ionosphere by a form of refraction. The ionized air contains free electrons, which can move in the presence of electromagnetic waves. The interaction is complex, but the net effect is an effective decrease in the dielectric constant, which causes the waves to bend toward the earth. The higher the frequency, the more ionization is necessary for refraction. If the signal is not refracted enough to reach the earth, it may be absorbed or may pass right through the atmosphere into space.

In the daytime, the D and E layers absorb frequencies below around 8 or 10 MHz. Frequencies above this, up to about 30 MHz, are refracted by the F_1 and F_2 layers and may return to earth. At night, the D and E layers virtually disappear, allowing lower frequencies to reach the F layer without being absorbed. These frequencies are refracted by the F layer, and thus propagation at the lower frequencies is better at night than during the day. The higher frequencies pass right through all the layers at night, so propagation at frequencies above about 10 MHz tends to be better during the daylight hours.

It is possible for the signal to reflect from the ground and make two or more "hops" before reaching its final destination, as shown in Figure 17.20, but each reflection from the ground or refraction in the ionosphere greatly reduces its strength.

Figure 17.20
Double-hop propagation

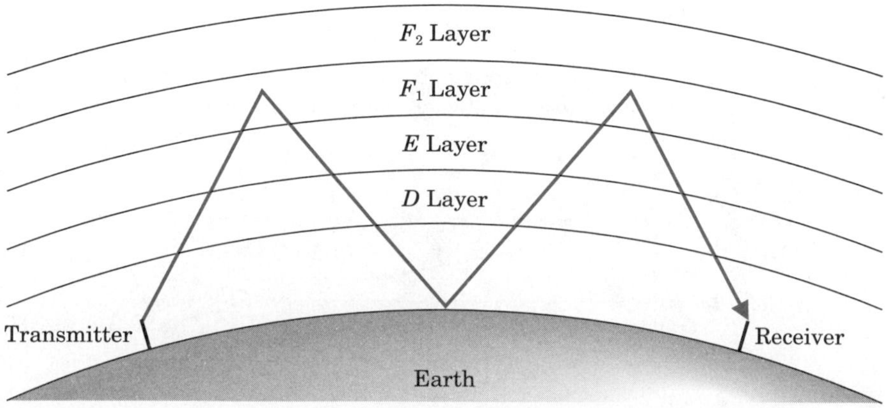

Figure 17.21
Ionospheric
sounding

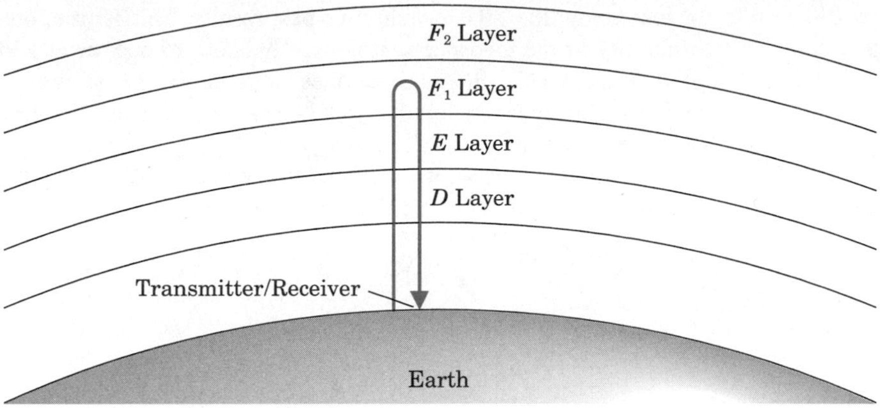

If all this sounds rather vague and approximate, it is because the ionosphere is constantly changing, from hour to hour, month to month, and year to year. It is generally possible to find some frequency at which the ionosphere returns signals in the desired direction and at the correct distance, but any predictions are approximate.

Large users of high-frequency communications, such as shortwave broadcasters and the military, employ *frequency diversity*, that is, they transmit on a number of different frequencies that span the HF spectrum, in the hope that at least one of these will work at a given time. Of course, they also use the kind of general observations that have just been made to reduce the amount of energy expended. Usually, there is no point in trying to use 3 MHz in the daytime or 30 MHz at night.

Ionospheric sounding, shown in Figure 17.21, can be used to make some measurements on which to base propagation predictions. A signal is sent straight up, and the frequency is gradually increased. At low frequencies, the signal will be absorbed. As the frequency is increased, the signal will be returned to earth and can be picked up by the receiver. As the frequency is increased still further, a frequency will be reached at which the signal ceases to return to earth but instead passes right through the ionosphere. The highest frequency that is returned to earth in the vertical direction is called the *critical frequency*.

Practical communication takes place over a path more like that shown in Figure 17.22. Since the angle at which the signal meets the ionosphere is smaller, the signal does not have to refract as much to return to earth, and we would expect that a higher frequency would be usable. We would also expect that longer paths would be operable at higher frequencies, since they require smaller angles. The highest frequency that returns to earth over a given path is called the **maximum usable frequency** (MUF). Since absorption decreases with increasing frequency, it would be expected that a frequency at

Figure 17.22
Ionospheric
propagation

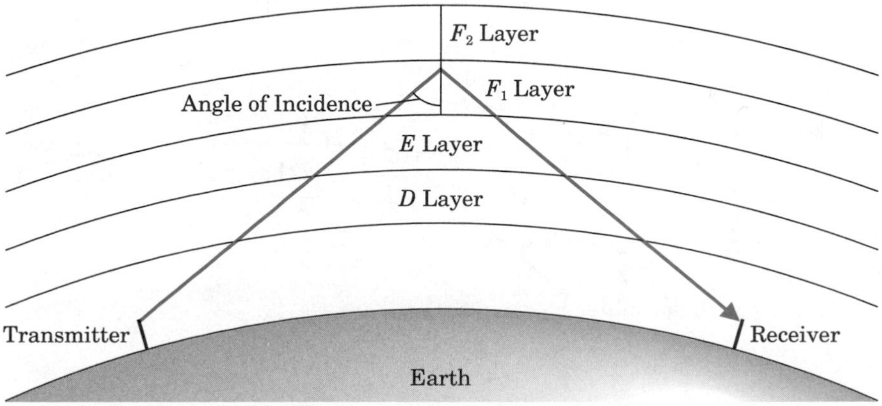

or just below the MUF would give best results. This is true, but because of the general instability of the ionosphere, it is usually better to operate at a lower frequency, perhaps 85% of the MUF—this is sometimes called the *optimum working frequency* (OWF). Of course, a particular station does not have access to every frequency and must choose one of those for which it is authorized.

In theory, at least, it is possible to predict the MUF from the critical frequency. The equation is sometimes called the *secant law*:

$$f_m = f_c \sec \theta_1 \tag{17.29}$$

where

f_m = MUF
f_c = critical frequency
θ_1 = angle of incidence, as shown in Figure 17.22

A more convenient way to use this law, since the secant is not directly available on most calculators, is to rewrite it as

$$f_m = \frac{f_c}{\cos \theta_1} \tag{17.30}$$

Example 17.8

The critical frequency at a particular time is 11.6 MHz. What is the MUF for a transmitting station if the required angle of incidence for propagation to a desired destination is 70°?

Solution

The MUF will be, from Equation (17.30),

$$\begin{aligned} f_m &= \frac{f_c}{\cos \theta_1} \\ &= \frac{11.6 \text{ MHz}}{\cos 70^\circ} \\ &= 33.9 \text{ MHz} \end{aligned}$$

Figure 17.22 hints at an interesting phenomenon. As the angle of elevation becomes larger, the distance covered becomes smaller and the MUF becomes lower. To put it another way, for any given frequency that is greater than the critical frequency, there will be a maximum angle of elevation above the horizon for which the signal will be reflected. Thus for frequencies above the critical frequency, there may be a region relatively close to the transmitter where the signal cannot be received, even though it can easily be picked up at much greater distances. This region, called the *skip zone*, is pointed out in Figure 17.23.

Figure 17.23
Skip zone

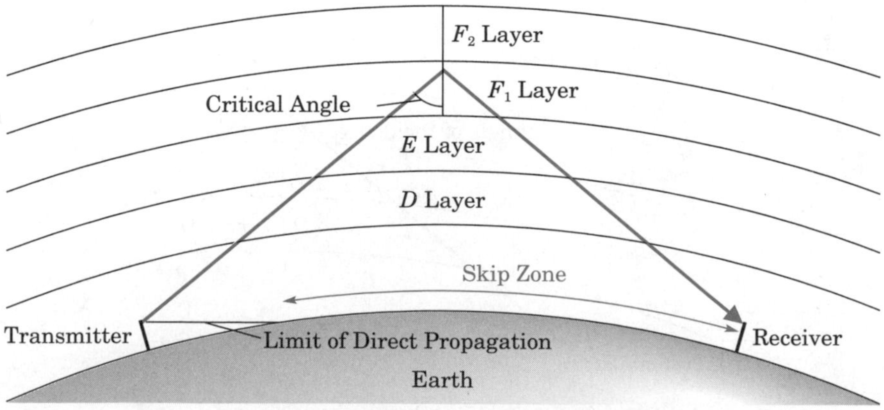

Ionospheric propagation allows communication over great distances with relatively simple equipment and reasonable power levels. Most commercial transmitters range in power from about 100 W to a few kilowatts. Broadcasters are the exception, often using very large power levels (into the megawatt range) in an attempt to overwhelm the competition. Radio amateurs have achieved worldwide communications with very low power levels, sometimes less than 1 W.

On the other hand, HF communication via the ionosphere is noisy and uncertain. It is also prone to phase shifting and frequency-selective fading. For instance, the phase shift and signal attenuation may be different for the upper and lower sidebands of the same signal. This makes high-fidelity music transmission impossible and restricts data transmission to very low rates, typically under 100 bits per second. There is little doubt that ionospheric propagation will remain in use, but it is being superseded for many applications by other technologies, for example, communications satellites and optical-fiber cables. It is still in use to a limited extent for telephony and quite extensively for ship and aircraft communications, international newswire services, military communication links, communication with outlying settlements (in the far north, for instance), and amateur radio. In addition, of course, there is shortwave broadcasting, which is of interest mainly to hobbyists in North America but is an important source of information programming in much of the world.

17.5.3 Line-of-Sight Propagation

Signals in the VHF range and higher are not usually returned to earth by the ionosphere, though during peaks of solar activity, frequencies in the low VHF range (even as high as television channel 2 (54–60 MHz)) have been propagated that way. Most terrestrial communication at these frequencies uses direct radiation from the transmitting antenna to the receiving antenna. There may be reflection from the ground, but that is more likely to cause problems than to increase the signal strength. This type of propagation is variously referred to as *space-wave, line-of-sight*, or *tropospheric propagation* (because the lowest layer of the atmosphere is known as the **troposphere**). Line-of-sight propagation is also the basis for satellite communication, which will be discussed in detail in Chapter 21.

The practical communication distance for terrestrial line-of-sight propagation is limited by the curvature of the earth. In spite of the title of this section, the maximum distance is actually greater than the eye can see because refraction in the atmosphere tends to bend radio waves slightly toward the earth. The dielectric constant of air usually decreases with increasing height, because of the reduction in pressure, temperature, and humidity with increasing distance from the earth. The effect varies with weather conditions, but it usually results in radio communications being possible over a distance approximately one-third greater than the visual line of sight.

Just as a person can see further from a high place, the height above average terrain of both the transmitting and receiving antennas is very important in calculating the maximum distance for space-wave radio communications. Note the differences between space-wave propagation and the other types of propagation we have studied so far. Antennas for ground-wave propagation are usually vertically polarized arrays with one end on the ground. For sky-wave (ionospheric) propagation, the antenna height is important only insofar as reflections from the ground may change the radiation pattern of the antenna. With space-wave propagation, however, height above ground is important, and the higher the better. Figure 17.24 shows the advantage of increased antenna height.

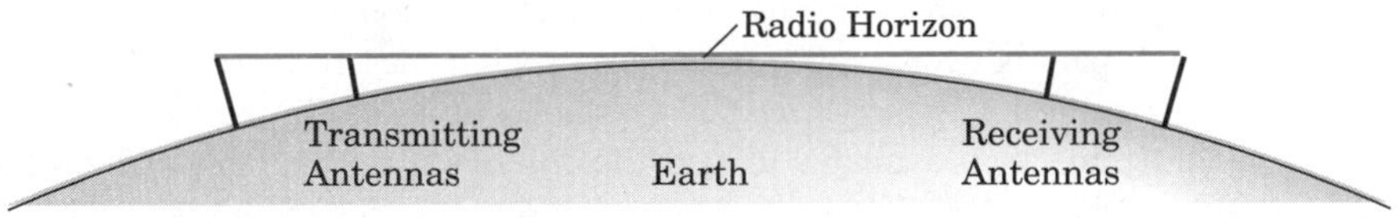

Figure 17.24 Line-of-sight propagation (antenna heights are greatly exaggerated)

An approximate value for the maximum distance between transmitter and receiver, over reasonably level terrain, is given by the following equation:

$$d = \sqrt{17h_T} + \sqrt{17h_R} \tag{17.31}$$

where

d = maximum distance in kilometers
h_T = height of the transmitting antenna in meters
h_R = height of the receiving antenna in meters

Example 17.9 A taxi company uses a central dispatcher, with an antenna at the top of a 15 m tower, to communicate with taxicabs. The taxi antennas are on the roofs of the cars, approximately 1.5 m above the ground. Calculate the maximum communication distance:

(a) between the dispatcher and a taxi
(b) between two taxis

Solution (a)

$$\begin{aligned} d &= \sqrt{17h_T} + \sqrt{17h_R} \\ &= \sqrt{17 \times 15} + \sqrt{17 \times 1.5} \\ &= 21.0 \text{ km} \end{aligned}$$

(b)

$$\begin{aligned} d &= \sqrt{17h_T} + a\sqrt{17h_R} \\ &= \sqrt{17 \times 1.5} + \sqrt{17 \times 1.5} \\ &= 10.1 \text{ km} \end{aligned}$$

The line-of-sight range can sometimes be extended by diffraction, particularly if there is a relatively sharp obstacle, such as a mountain peak, in the way. Diffraction, however, greatly reduces the signal strength, requiring more powerful transmitters and more sensitive receivers than are required for line-of-sight communications.

The attenuation of free space, as expressed by Equation (17.22), is usually the most important factor in determining the signal power at the receiver. Other factors, however, arise because propagation takes place near the ground and because the actual medium is air.

Although line-of-sight propagation uses a direct path from transmitter to receiver, the receiver can also pick up signals that have been reflected or diffracted. For instance, the signal can be reflected from the ground, as discussed in section 17.5.1. If the ground is rough, the reflected signal will be scattered and its intensity will be low in any given direction. If, on the other hand, the reflecting surface is relatively smooth—a body of water, for instance—the reflected signal at the receiver can have a strength comparable to that of the incident wave and the two signals will interfere. Whether the interference is constructive or destructive depends on the phase relationship between the signals: if they are in phase, the resulting signal strength will be increased, but if they are 180° out of phase, there will be partial cancellation. When the surface is highly reflective, the reduction in signal strength can be 20 dB or more. This effect is called *fading*. The exact phase relationship depends on the difference, expressed in wavelengths, between the lengths of the transmission paths for the direct and reflected signals. In addition, there is usually a phase shift of 180° at the point of reflection.

In a practical situation where the transmitter and receiver locations are fixed, the effect of reflections can often be reduced by carefully surveying the proposed route and adjusting the transmitter and receiver antenna heights so that any reflection takes place in wooded areas or rough terrain; the reflection will then be diffuse and weak. If most of the path is over a reflective surface such as desert or water, fading can be reduced by using either *frequency diversity* or *spatial diversity*. In the former method, more than one frequency is available for use; the difference, in wavelengths, between the direct and reflected

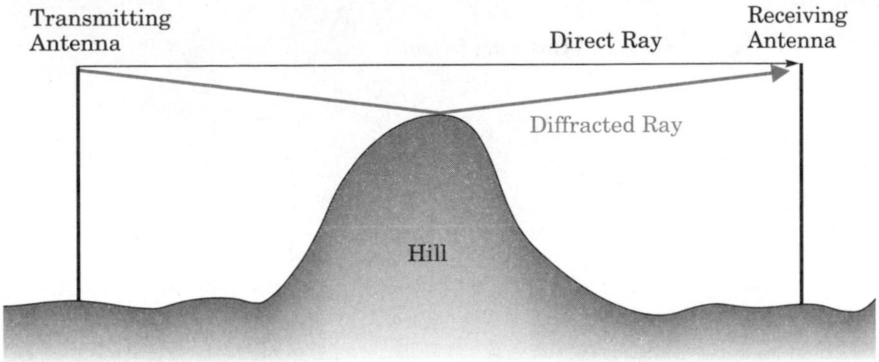

Figure 17.25
Fading due to diffraction

path lengths is different for the two frequencies. For spatial diversity, two antennas are usually mounted one above the other on the same tower, so that the difference between the direct and reflected path length is different for the two antennas.

Diffraction from obstacles in the path can also be a problem for line-of-sight radio links. Earlier, the beneficial effects of diffraction in allowing reception on the far side of an obstruction were described. Diffraction can also cause problems when the direct wave and the diffracted wave have opposite phase and tend to cancel, as shown in Figure 17.25.

The solution to the problem of interference due to diffraction is to arrange for the direct and refracted signals to be in phase. Again, this requires a careful survey of the proposed route and adjustment of the transmitting and receiving antenna heights to achieve this result.

Problems can also occur when the signal reflects from large objects like hills or buildings, as shown in Figure 17.26. There may be not only phase cancellation but also significant time differences between the direct and reflected waves. These can cause a type of distortion called (not surprisingly) **multipath distortion** in FM radio reception. The "ghosts" that appear in television reception have the same cause. Directional receiving antennas aimed in the direction of the direct signal can reduce the problem of reflections for fixed receivers, but they are not very practical on vehicles. In the case of mobile radio, the path length changes constantly, so the transmitted power must be sufficient to achieve usable signal strength at the receiver, even with cancellation due to reflections. Spatial diversity, with two or more antennas mounted on one vehicle, is occasionally used to reduce multipath distortion in mobile radio reception.

There is also some absorption of the signal by the atmosphere. At frequencies under 20 GHz, this absorption is quite small compared with the square-law attenuation of free space.

Figure 17.26
Multipath reception

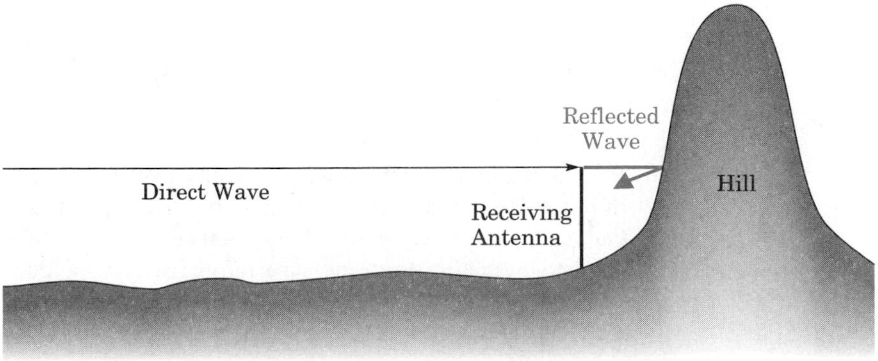

Figure 17.27
Tropospheric scatter

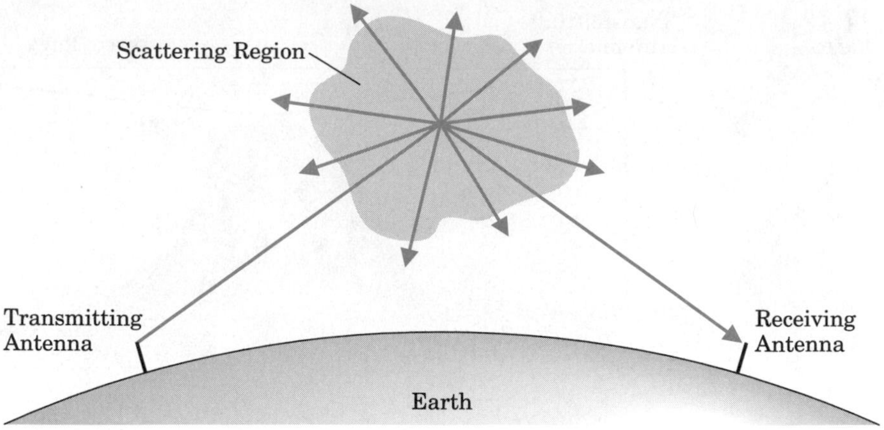

17.5.4 Other Propagation Modes

The types of propagation discussed so far represent the majority of communication systems. There are, however, some other methods that, while not used as commonly as the others, still represent useful ways of getting radio waves—and the information they carry—from one place to another.

Tropospheric Scatter The *troposphere* is the lowest layer of the atmosphere. We have already seen that it refracts radio waves slightly, providing communication at distances somewhat beyond the visual line of sight. The troposphere also absorbs radiation at some frequencies. What is not quite so well known is that irregularities in the troposphere can also scatter radio waves. Exactly what causes these irregularities is not known, but one theory is that they are caused by the effect of variations in temperature on the water vapor content of the atmosphere.

Figure 17.27 shows the basic idea behind tropospheric scatter (often called *troposcatter*). The transmitting antenna is aimed in the direction of the receiver, but the receiver is over the horizon. Most of the transmitted energy simply continues on into space, but a small portion of it is scattered, and a small fraction of the scattered energy reaches the receiver.

Troposcatter can give reliable communication over distances of about 80 to 800 km at frequencies from about 250 MHz to around 5 GHz. It is an inefficient system, requiring larger transmitter power, antennas with higher gain, and more sensitive receivers than line-of-sight systems. It is also subject to fading, which can be reduced by using spatial diversity at both ends, that is, each station has at least two antennas separated by 100 wavelengths or more.

On the other hand, troposcatter can operate at a much greater range than line-of-sight communication, thus reducing the requirement for repeater stations. This is a great benefit when the communications path is over water or over difficult terrain such as mountains or when a foreign, possibly unfriendly government controls the territory between ends of the link. Government and commercial point-to-point radio links are the main users of tropospheric scatter.

Ducting Under certain conditions, especially over water, a *superrefractive* layer (one with a lower refractive index) can form in the troposphere and return signals to earth. The signals can then propagate over long distances by alternately reflecting from the earth (or water) and refracting from the superrefractive layer. Figure 17.28(a) illustrates this phenomenon.

A related condition involves a thin tropospheric layer with a high refractive index and a layer above it with a low refractive index, so that a *duct* forms, as shown in Fig-

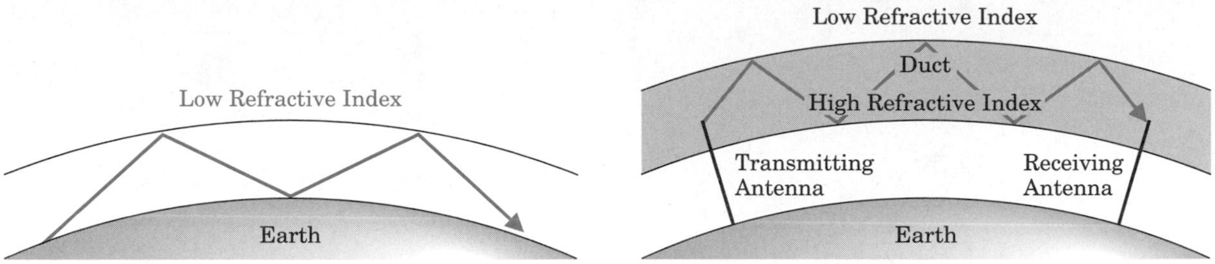

(a) At ground level (b) Between tropospheric layers

Figure 17.28 Tropospheric ducting

ure 17.28(b). This too will propagate waves for long distances, provided that both the transmitting and receiving antennas are in the duct.

Unfortunately **ducting** is not reliable enough for commercial use. In fact, it is more likely to cause problems. By carrying signals far from their intended destinations, this phenomenon can cause fading of desired signals and interference from signals at great distances.

Meteor-Trail Propagation Meteors are constantly entering the earth's atmosphere and being destroyed because of the heat generated by friction with the air. A few of the larger ones are visible, and occasionally enough material survives to strike the earth with enough force to make a crater. Most meteors are the size of dust particles, however, and their paths are not visible. They do leave a trail of ionized air behind, and it is possible to use it for communication, though only for a few minutes. Although it may seem far-fetched, meteor-trail propagation is in use for data communications, especially in the North. It is not suitable for voice because of its intermittent nature.

Summary

Here are the main points to remember from this chapter.

1. Radio waves are transverse electromagnetic waves, very similar, except for frequency, to light waves.
2. The impedance of a medium is the ratio of the electrical and magnetic field strengths in that medium. The impedance of free space is 377 Ω.
3. The polarization of a radio wave is the direction of its electric field vector.
4. As a radio wave propagates in space, its power density reduces with distance, due to spreading of the waves. Directional antennas can partially offset this effect.
5. Reflection, refraction, and diffraction occur with radio waves just as they do with light; the rules are similar to those for light.
6. Ground-wave propagation gives good results over long distances for signals in the MF range and below.
7. Ionospheric propagation is relatively unreliable and noisy but achieves worldwide communication in the HF range.
8. Line-of-sight propagation is the system most commonly used at VHF and above. The radio range is actually somewhat greater than the visible range; it is very dependent on the height of the transmitting and receiving antennas.

Important Equations

$$v = \frac{c}{\sqrt{\epsilon_r}} \tag{17.1}$$

$$\mathcal{Z} = \frac{\mathcal{E}}{\mathcal{H}} \tag{17.3}$$

$$\mathcal{Z}_0 = 377\ \Omega \tag{17.5}$$

$$\epsilon_r = \frac{\epsilon}{\epsilon_0} \tag{17.6}$$

$$\mathcal{Z} = \frac{377}{\sqrt{\epsilon_r}} \tag{17.7}$$

$$P_D = \frac{\mathcal{E}^2}{\mathcal{Z}} \tag{17.8}$$

$$P_D = \mathcal{H}^2\,\mathcal{Z} \tag{17.9}$$

$$P_D = \mathcal{E}\mathcal{H} \tag{17.10}$$

$$P_D = \frac{P_t}{4\pi r^2} \tag{17.11}$$

$$\mathcal{E} = \frac{\sqrt{30P_t}}{r} \tag{17.13}$$

$$P_D = \frac{P_T G_T}{4\pi r^2} \tag{17.14}$$

$$\text{EIRP} = P_T G_T \tag{17.15}$$

$$A_{eff} = \frac{P_R}{P_D} \tag{17.16}$$

$$P_R = \frac{A_{eff} P_T G_T}{4\pi r^2} \tag{17.17}$$

$$A_{eff} = \frac{\lambda^2 G_R}{4\pi} \tag{17.18}$$

$$L_{fs} = 32.44 + [20 \log d\ (\text{km})] + [20 \log f\ (\text{MHz})] - [G_T\ (\text{dBi})] - [G_R\ (\text{dBi})] \tag{17.22}$$

$$n_1 \sin\theta_1 = n_2 \sin\theta_2 \tag{17.24}$$

$$n = \sqrt{\epsilon_r} \tag{17.26}$$

$$\frac{\sin\theta_1}{\sin\theta_2} = \sqrt{\frac{\epsilon_{r2}}{\epsilon_{r1}}} \tag{17.27}$$

$$\theta_c = \arcsin\frac{n_2}{n_1} \tag{17.28}$$

$$f_m = \frac{f_c}{\cos\theta_1} \tag{17.30}$$

$$d = \sqrt{17h_T} + \sqrt{17h_R} \tag{17.31}$$

Glossary

attenuation of free space the reduction in signal strength due to spreading of the waves at a distance from the transmitter

dielectric strength the magnitude of the electric field required to cause breakdown and arcing in a dielectric

ducting a means of propagation in which the waves are confined within a refractive region of the troposphere or between such a region and the ground
effective area the area from which a receiving antenna can be considered to extract all the energy in an electromagnetic wave
electric field intensity (or strength) the ratio of the electric force on a charge to the charge, at a given point (units are volts per meter)
free space a vacuum that allows radio waves to propagate without any obstruction
ground wave a vertically polarized electromagnetic wave that propagates along the surface of the earth
index of refraction the ratio of the phase velocity of a wave in free space to that in the medium under consideration
ionosphere the ionized region of the earth's atmosphere
isotropic radiator a hypothetical antenna having zero physical size and no loss and radiating equally in all directions
magnetic field intensity (or strength) magnitude of the magnetic field vector (units are amperes per meter)
maximum usable frequency (MUF) the highest frequency that will be returned by the ionosphere at a given point
multipath reception a situation in which a signal arrives at a receiving antenna via two or more paths (usually one of these paths is direct from the transmitting antenna and the other(s) involves reflections)
normal a line drawn perpendicular to the interface between two media
path loss the ratio between the signal appearing at the transmitting antenna terminals and that at the receiving antenna terminals
photon a quantum of electromagnetic radiation
polarization the direction of the electric field vector of an electromagnetic wave
power density the power flowing through a unit cross-sectional area normal to the direction of travel of an electromagnetic wave
propagation the process by which waves travel through a medium
sky wave an electromagnetic wave that is returned to earth by the ionosphere
space wave an electromagnetic wave that propagates directly from the transmitting to the receiving antenna
terrestrial propagation propagation along or near the surface of the earth
troposphere the region of the atmosphere closest to the earth

Questions

1. What are the similarities between radio waves and light waves?
2. What is meant by the characteristic impedance of a medium? What is the characteristic impedance of free space?
3. State the difference between power and power density. Explain why power density decreases with the square of the distance from a source.
4. A radio wave propagates in such a way that its magnetic field is parallel with the horizon. What is its polarization?
5. What is an isotropic radiator? Could such a radiator be built? Explain your answer.
6. State three factors that determine the amount of power extracted from a wave by a receiving antenna.
7. Distinguish between specular and diffuse reflection. For wavelengths on the order of 1 m, state which type is more likely from:
   (a) a calm lake
   (b) a field strewn with large boulders
8. For waves passing from one medium to another, what is meant by the critical angle of incidence? What happens when the angle of incidence exceeds the critical value?
9. What phenomenon accounts for the fact that radio waves from a transmitter on one side of a mountain can sometimes be received on the other side?
10. Why do stations in the AM standard broadcast band always use vertically polarized antennas?
11. Why is the ionosphere more highly ionized during the daylight hours than it is at night?
12. When the critical frequency is 12 MHz, what will happen to a 16 MHz signal that is radiated straight up? What will happen to a 10 MHz signal?
13. Sometimes an HF radio transmission can be heard at a distance of 1000 km from the transmitter but can't be heard 100 km away. Explain why.
14. Why is antenna height much more important for an FM broadcast-band antenna than for one designed for the AM broadcast band?
15. State two undesirable effects that can be caused by reflections in line-of-sight communications, and explain how they arise.

16. How can multipath interference be reduced?
17. Explain what is meant by diversity, and describe the various types of diversity.
18. Why are AM broadcast stations often received at greater distances during the night than during the day? Is this always an advantage?
19. Why does tropospheric scatter require high-powered transmitters?
20. Explain how tropospheric ducting works and why it can cause problems.
21. State which mode of propagation is normally used for each of the following services and explain why.
    (a) FM radio broadcasting
    (b) shortwave radio broadcasting
    (c) cellular telephones (frequency of about 800 MHz)
    (d) LORAN-C navigation beacons (frequency 100 kHz)
22. At noon a station can transmit from New York City to Miami at a frequency of 20 MHz but not at 5 MHz. At midnight the situation is reversed. Explain why.

Problems

SECTION 17.1

23. Find the propagation velocity of radio waves in glass, which has a relative permittivity of 7.8.

SECTION 17.2

24. Find the wavelength, in free space, of radio waves at each of the following frequencies:
    (a) 50 kHz
    (b) 1 MHz
    (c) 23 MHz
    (d) 300 MHz
    (e) 450 MHz
    (f) 12 GHz
25. Find the characteristic impedance of glass, which has a relative permittivity of 7.8.
26. An isotropic source radiates 100 W of power in free space. Calculate the power density and the electric field intensity at a distance of 15 km from the source.
27. What power density is required to produce an electric field strength of 100 volts per meter in air?
28. A signal has a power density of 50 mW/m^2 in free space. Calculate its electric and magnetic field strengths.

SECTION 17.3

29. A certain antenna has a gain of 7 dB with respect to an isotropic radiator.
    (a) What is its effective area if it operates at 200 MHz?
    (b) How much power would it absorb from a signal with a field strength of 50 μV/m?
30. A transmitter has an output power of 50 W. It is connected to its antenna by a feedline that is 25 m long and properly matched. The loss in the feedline is 5 dB/100 m. The antenna has a gain of 8.5 dBi.
    (a) How much power reaches the antenna?
    (b) What is the EIRP in the direction of maximum antenna gain?
    (c) What is the power density 1 km from the antenna, in the direction of maximum gain, assuming free-space propagation?
    (d) What is the electric field strength at the location in part (c)?
31. A satellite transmitter operates at 4 GHz with an antenna gain of 40 dBi. The receiver 40,000 km away has an antenna gain of 50 dBi. If the transmitter has a power of 8 W, find (ignoring feedline losses and mismatch):
    (a) the EIRP in dBW
    (b) the power delivered to the receiver
32. What would be the effect on the received signal level of making each of the following changes (separately)? Give your answers in decibels.
    (a) double the transmitter power
    (b) reduce the distance by half
    (c) increase the gain of the receiving antenna by 10 dB
33. A cellular radio transmitter has a power output of 3 W at 800 MHz. It uses an antenna with a gain of 3 dBi. The receiver is 5 km away, with an antenna gain of 12 dBi. Calculate the received signal strength in dBm, ignoring any losses in transmission lines.
34. Suppose that the situation is the same as in Problem 33 except that the transmission line losses are 1 dB at the transmitter and 2 dB at the receiver. What is the received power in dBm under these conditions?
35. Suppose the situation is the same as in Problem 33 except that the transmitter power is reduced to 750 mW. What is the received power in dBm under these conditions?

SECTION 17.4

36. Sketch the path of the reflected waves in each of the diagrams in Figure 17.29.

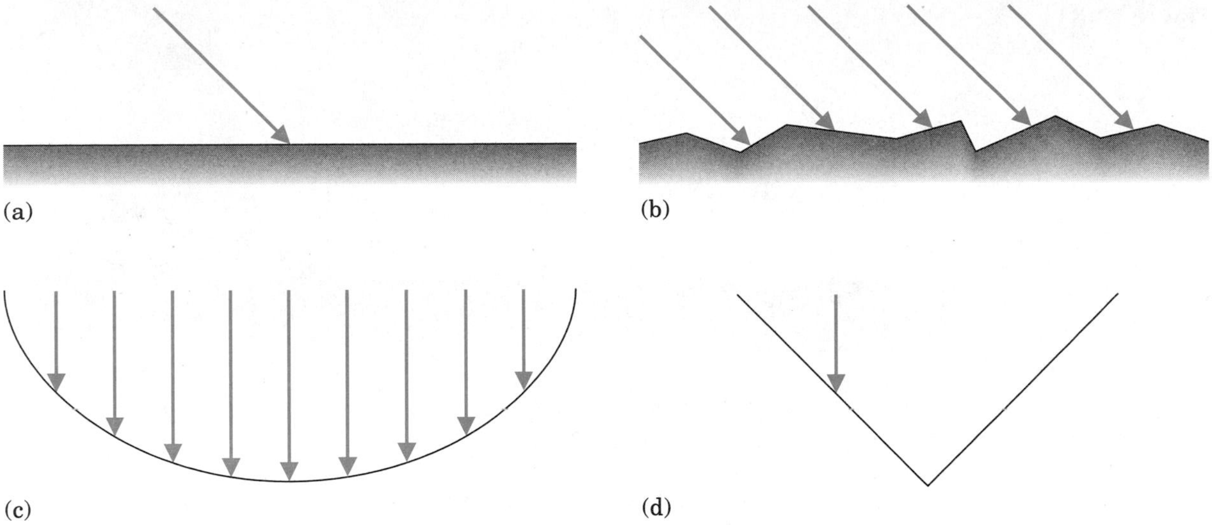

Figure 17.29

37. Find the critical angle when a wave passes from glass, with $\epsilon_r = 7.8$, into air.
38. Sketch the path of the refracted waves in each of the diagrams in Figure 17.30.
39. A radio signal moves from air to glass. The angle of incidence is 20°. Calculate the angle of refraction and sketch the situation. The relative permittivity of the glass is 7.8.

SECTION 17.5

40. If the critical frequency is 10 MHz, what is the OWF at an angle of incidence of 60°?
41. At a certain time, the MUF for transmissions at an angle of incidence of 75° is 17 MHz. What is the critical frequency?
42. If the critical frequency is 12 MHz, what is the critical angle at 15 MHz?

Figure 17.30

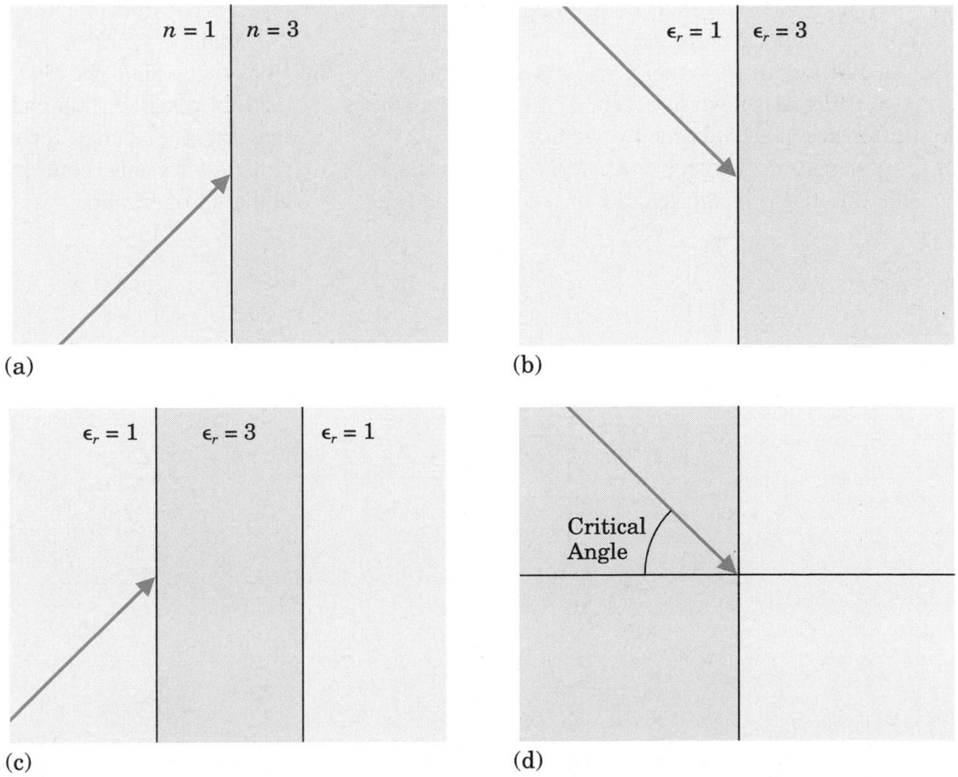

Figure 17.31

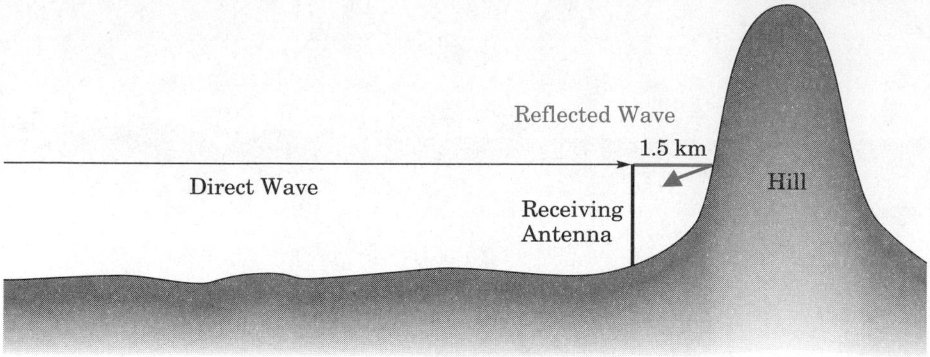

43. An FM broadcast station has a transmitting antenna located 50 m above average terrain. How far away could the signal be received:
    (a) by a car radio with an antenna 1.5 m above the ground?
    (b) by a rooftop antenna 12 m above the ground?
44. A boat is equipped with a VHF marine radio, which it uses to communicate with other nearby boats and shore stations.
    (a) Name the mode of propagation.
    (b) If the antenna on the boat is 2.3 m above the water, calculate the maximum distance for communication with:
        (i) another similar boat
        (ii) a shore station with an antenna on a tower 22 m above the water level
        (iii) another similar boat, using the shore station as a repeater
45. An FM broadcast signal arrives at an antenna via two paths, as shown in Figure 17.31. Calculate the difference in arrival time for the two paths.
46. A transmitter and receiver are separated by water. If the difference in the lengths of the direct and reflected paths is 2 m, calculate the phase difference at 400 MHz.

COMPREHENSIVE

47. Explain each of the following:
    (a) AM radio broadcast stations must often reduce power at night to avoid interference.
    (b) CB radio at 27 MHz is intended for local communication but can often communicate for hundreds of kilometers.
    (c) HF communications are often severely disrupted when the aurora borealis (northern lights) is visible.
    (d) Sometimes VHF signals appear and cause interference hundreds of kilometers away from their intended route.
    (e) It is possible to communicate somewhat further than the visible horizon using UHF signals.
48. Construct a table showing all the propagation methods discussed in this chapter, with the frequency ranges and distances for which they are useful. Include comments concerning economy and reliability of operation.

18 Antennas

Objectives After studying this chapter, you should be able to:

1. Explain the basic principles of operation of antenna systems
2. Define antenna gain and beamwidth and find the gain and beamwidth of an antenna from a plot of its radiation pattern
3. Calculate effective isotropic radiated power and effective radiated power for an antenna-transmitter combination and explain the difference between the two terms
4. Calculate the effective area for a receiving antenna and use it to calculate the power delivered to a receiver
5. Calculate the dimensions of simple practical antennas for a given frequency
6. Identify common types of antennas and antenna arrays, explain their operation, and sketch their approximate radiation patterns
7. Calculate the gain and beamwidth for parabolic antennas

18.1 Introduction

So far, we have considered how electromagnetic waves can propagate along transmission lines and through space. The **antenna** is the interface between these two media and is thus a very important part of the communications path. In this chapter, we will study the basic operating principles of antennas and look at some of the parameters that describe their performance. Some representative examples of practical antennas will be analyzed.

Before we begin, you should have two ideas firmly in mind. First, antennas are passive devices. Therefore, the power radiated by a transmitting antenna cannot be greater than the power entering from the transmitter. In fact, it will be less because of losses. We will speak of antenna gain, but remember that gain in one direction results from a concentration of power and is accompanied by a loss in other directions. Antennas achieve gain the same way a flashlight reflector increases the brightness of the bulb: by concentrating energy.

By the way, you may hear someone talk about an **active antenna** for a high-frequency communications receiver or an FM or television broadcast receiver. This term simply describes the combination of a receiving antenna with a low-noise preamplifier. The antenna part of the combination is still a passive device.

The second concept to keep in mind is that antennas are reciprocal, that is, the same design works equally well as a transmitting or a receiving antenna and in fact has the same gain. That does not mean that transmitting and receiving antennas are necessarily identical. For instance, the conductors in a transmitting antenna must be sized to handle larger currents. However, the designs are quite similar, and many of the calculations are identical.

Essentially, the task of a transmitting antenna is to convert the electrical energy traveling along a transmission line into electromagnetic waves in space. This process, while difficult to analyze in mathematical detail, should not be hard to visualize. The energy in the transmission line is contained in the electric field between the conductors and in the magnetic field surrounding them. All that is needed is to "launch" these fields (and the energy they contain) into space.

At the receiving antenna, the electric and magnetic fields in space cause current to flow in the conductors that make up the antenna. Some of the energy is thereby transferred from these fields to the transmission line connected to the receiving antenna.

18.2 Simple Antennas

In order to understand the operation of antennas, we will first look at two simple antennas. The idea of the **isotropic radiator** was introduced in the previous chapter. Though merely a theoretical construct, it serves as a way to describe the functions of an antenna and as a reference for other antennas. The *half-wave dipole antenna*, on the other hand, is very practical and in common use. An understanding of the half-wave dipole is important both in its own right and as a basis for the study of more complex antennas.

18.2.1 The Isotropic Radiator

An ideal isotropic radiator would radiate all the electrical power supplied to it and would do so equally in all directions. It would also be a point source, that is, it would have zero size.

It would not be a good idea to try to build such an antenna. Of course, the zero size is not physically realizable, and neither are the complete losslessness and complete nondirectionality of the isotropic antenna. Nonetheless, we will use the isotropic antenna as a standard for comparison because even though this antenna cannot be built and tested, its characteristics are simple and easy to derive.

18.2.2 The Half-Wave Dipole

A more practical construction project is the **dipole**, shown in Figure 18.1. The word *dipole* simply means it has two parts, as shown. A dipole antenna does not have to be one-half wavelength in length like the one shown in the figure, but this length is handy for impedance matching, as we shall see. Actually, in practice its length will be slightly less than one-half the free-space wavelength to allow for capacitive effects. A half-wave dipole is sometimes called a Hertz antenna, though strictly speaking, the term *Hertzian dipole* refers to a dipole of infinitesimal length. This, like the isotropic radiator, is a theoretical construct; it is used in the calculation of antenna radiation patterns.

Typically, the length of a half-wave dipole, assuming that the conductor diameter is much less than the length of the antenna, is 95% of one-half the wavelength measured in free space. The free-space wavelength is given by

$$\lambda = \frac{c}{f} \tag{18.1}$$

where

λ = free-space wavelength in meters
c = 3×10^8 m/s
f = operating frequency in hertz

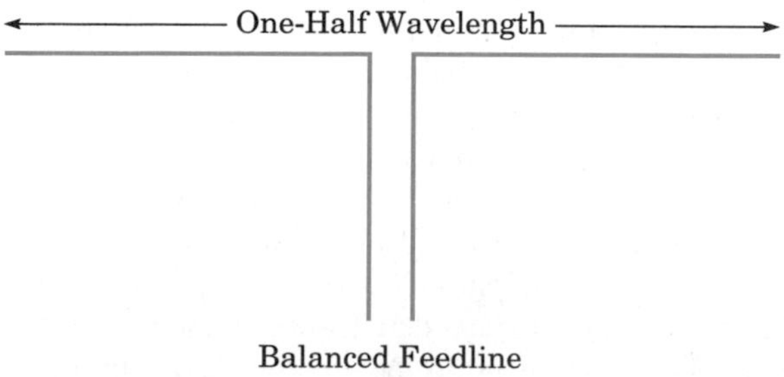

Figure 18.1 Half-wave dipole antenna

Therefore, the length L of a half-wave dipole, is (in meters):

$$L = 0.95 \times 0.5\left(\frac{c}{f}\right) \tag{18.2}$$

$$= .475\left(\frac{c}{f}\right)$$

$$= \frac{.475(3 \times 10^8)}{f}$$

$$= \frac{142.5 \times 10^6}{f}$$

In the above equation, L is the length in meters and f is the frequency in hertz. Very often, megahertz are a more convenient unit for frequency, in which case Equation (18.2) becomes

$$L = \frac{142.5}{f} \tag{18.3}$$

where

L = length of a half-wave dipole in meters
f = operating frequency in megahertz

For length measurements in feet, the equivalent equation is

$$L = \frac{468}{f} \tag{18.4}$$

where

L = length of a half-wave dipole in feet
f = operating frequency in megahertz

Example 18.1 Calculate the length of a half-wave dipole for an operating frequency of 20 MHz.

Solution From Equation (18.3),

$$L = \frac{142.5}{f}$$

$$= \frac{142.5}{20}$$

$$= 7.13 \text{ m}$$

One way to think about the half-wave dipole is to consider an open-circuited length of parallel-wire transmission line, as shown in Figure 18.2(a). The line will have a voltage maximum at the open end, a current maximum one-quarter wavelength from the end, and a very high standing-wave ratio (SWR). In fact, we noted in our discussion of transmission lines in Chapter 16 that the SWR for this open-circuited stub would be infinite, except for the fact that there would be some radiation from the open end. That is, the stub acts as an antenna, though a very inefficient one.

Now, suppose that the two conductors are separated at a point one-quarter wavelength from the end, as in Figure 18.2(b). The drawing shows how the electric field seems to stretch away from the wires. If the process continues, as in Figure 18.2(c), some of the field detaches itself from the antenna and helps to form electromagnetic waves that propagate through space.

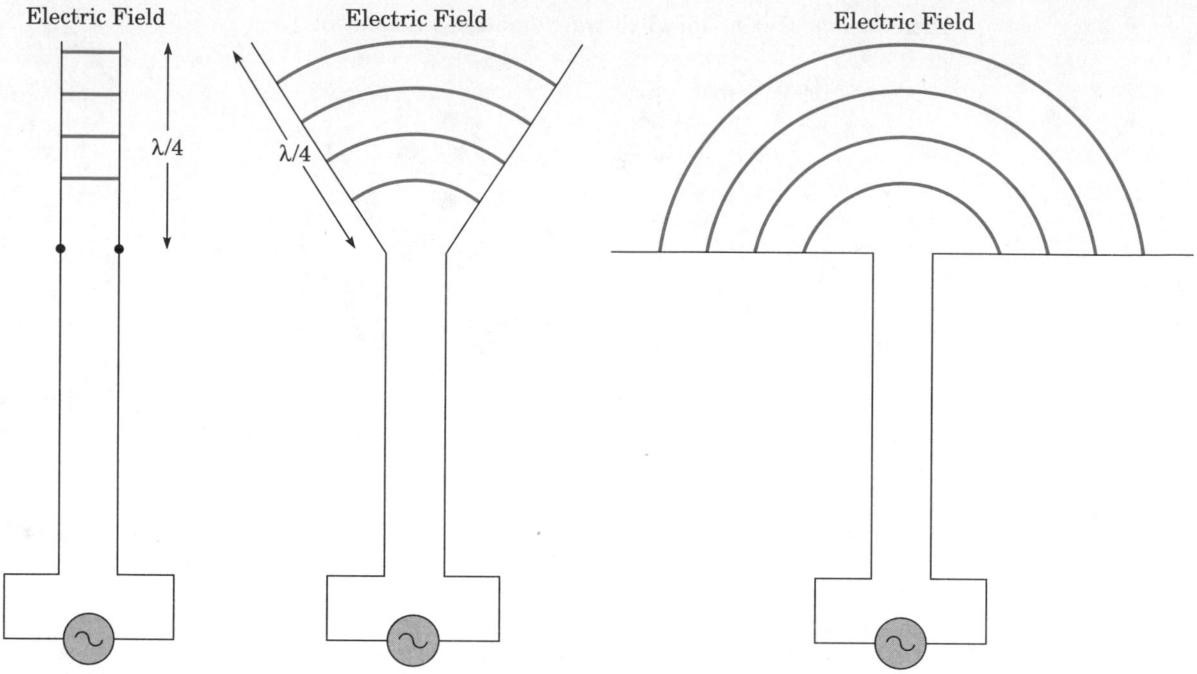

(a) Open-circuited line (b) Spreading of line (c) Dipole

Figure 18.2
Development of the half-wave dipole

18.2.3 Radiation Resistance

The radiation of energy from a dipole is quite apparent if we measure the impedance at the feedpoint in the center of the antenna. An actual open-circuited lossless line, as described in Chapter 16, would look like a short circuit at a distance of one-quarter wavelength from the open end. At distances slightly greater than or less than one-quarter wavelength, the line would appear reactive. There would never be a nonzero resistive component to the feedpoint impedance, since an open-circuited line has no way of dissipating power.

The half-wave dipole does not dissipate power either, assuming the material of which it is made is lossless, but it will radiate power into space. The effect on the feedpoint impedance is the same as if a loss had taken place. Whether power is dissipated or radiated, it disappears from the antenna and therefore causes the input impedance to have a resistive component. The half-wave dipole, for instance, looks like a resistance of about 70 Ω at its feedpoint.

The portion of an antenna's input impedance that is due to power radiated into space is known, appropriately, as the **radiation resistance**. It is important to understand that this does not represent losses in the conductors that make up the antennas.

The idealized antenna just described will radiate all the power supplied to it into space. A real antenna will, of course, have ohmic losses in the conductor and will therefore have an efficiency less than 1. This efficiency can be defined as

$$\eta = \frac{P_r}{P_T}$$

where

P_r = radiated power

P_T = total power supplied to the antenna

Recalling that $P = I^2R$, we have

$$\eta = \frac{I^2R_r}{I^2R_T} = \frac{R_r}{R_T} \tag{18.5}$$

where

R_r = radiation resistance, as seen from the feedpoint
R_T = total resistance, as seen from the feedpoint

Example 18.2 A dipole antenna has a radiation resistance of 67 Ω and a loss resistance of 5 Ω, measured at the feedpoint. Calculate the efficiency.

Solution From Equation (18.5),

$$\eta = \frac{R_r}{R_T} = \frac{67}{67 + 5} = 0.93 \quad \text{or} \quad 93\%$$

The half-wave dipole does not radiate uniformly in all directions. The field strength is at its maximum along a line at a right angle to the antenna and is zero off the ends of the antenna.

18.3 Antenna Directional Characteristics

Now that two simple antennas have been described, it is already apparent that antennas differ in the amount of radiation they emit in various directions. This is a good time to introduce some terms that describe and quantify the directional characteristics of antennas and to demonstrate methods of graphing some of these characteristics. These terms will be applied to the isotropic and half-wave dipole antennas here and will be applied to other antenna types as they are introduced.

18.3.1 Radiation Pattern

The diagrams used in this book follow the three-dimensional coordinate system shown in Figure 18.3. As shown in the figure, the x–y plane is horizontal, and the angle **phi** (ϕ) is measured from the x axis in the direction of the y axis. The z axis is vertical, and the angle **theta** (θ) is measured from the horizontal plane toward the **zenith**. This is not quite the same as the standard polar coordinate system used in geometry texts. In the "standard" system, θ is measured down from the z axis. Antenna manufacturers generally prefer to measure the vertical angle upward from the ground, rather than downward from the zenith, so that is the method that will be used here. The vertical angle, measured upward from the ground, is called the *angle of elevation*.

Figure 18.4 shows two ways to represent the radiation pattern of a dipole. The three-dimensional picture in Figure 18.4(a) is useful in showing the general idea and in getting a feel for the characteristics of the antenna. The two views in Figures 18.4(b) and (c), on the other hand, are less intuitive but can be used to provide quantitative information about the performance of the antenna. They are also much easier to draw. It is the latter type of

Figure 18.3
Three-dimensional coordinate system

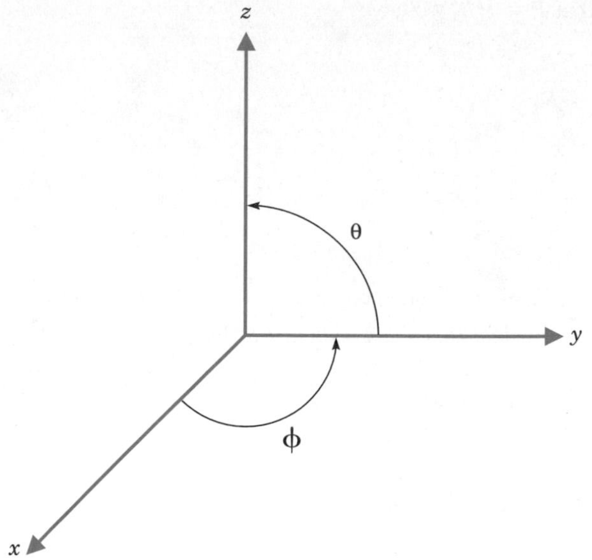

antenna pattern that is usually found in manufacturers' literature, for instance, and it is this type of depiction that will be used in this book.

Notice that polar graph paper is used in Figures 18.4(b) and (c). The angle is measured in one plane from a reference axis. Because the three-dimensional space around the antenna is being represented in two dimensions, at least two views are required to give the complete picture.

It is important to choose the axes carefully to take advantage of whatever symmetry exists. Very often, the horizontal and vertical planes are used; usually the antenna is mounted either vertically or horizontally, so the axis of the antenna itself can be used as one of the reference axes.

The two graphs in Figures 18.4(b) and (c) can be thought of as slices of the three-dimensional pattern of Figure 18.4(a). If the axes are well chosen, two such slices will usually be sufficient to describe the three-dimensional radiation pattern.

In Figure 18.4, the dipole itself is drawn to help you visualize the antenna orientation. Note, however, that these radiation patterns are valid only in the **far-field region,** that is, an observer must be far enough away from the antenna that any local capacitive or inductive coupling is negligible. In practice, this means a distance of at least several wavelengths, and generally an actual receiver is at a much greater distance than that. From this distance, the antenna would be more accurately represented as a dot in the center of the graph. The area close to the antenna is called the **near-field region** and does not have the same directional characteristics.

The directions in Figure 18.4 follow the coordinate system of Figure 18.3. The antenna is considered to be in free space, so its radiation is plotted in all directions. In many practical situations, radiation below the horizon is not plotted. In that case, the bottom half of Figure 18.4(b) would be omitted.

The distance from the center of the graph represents the strength of the radiation in a given direction. The scale is usually in decibels with respect to some reference. Here, there are a lot of choices. For instance, the scale can be arbitrary, with the outside circle representing the maximum radiation from the antenna. Often, the reference is an isotropic radiator, as is the case in Figure 18.4. Note that the point furthest from the center of the graph is at 2.14 dB; that is, we can say that the gain of a lossless dipole, in its direction of maxi-

Figure 18.4 Radiation pattern of horizontal half-wave dipole

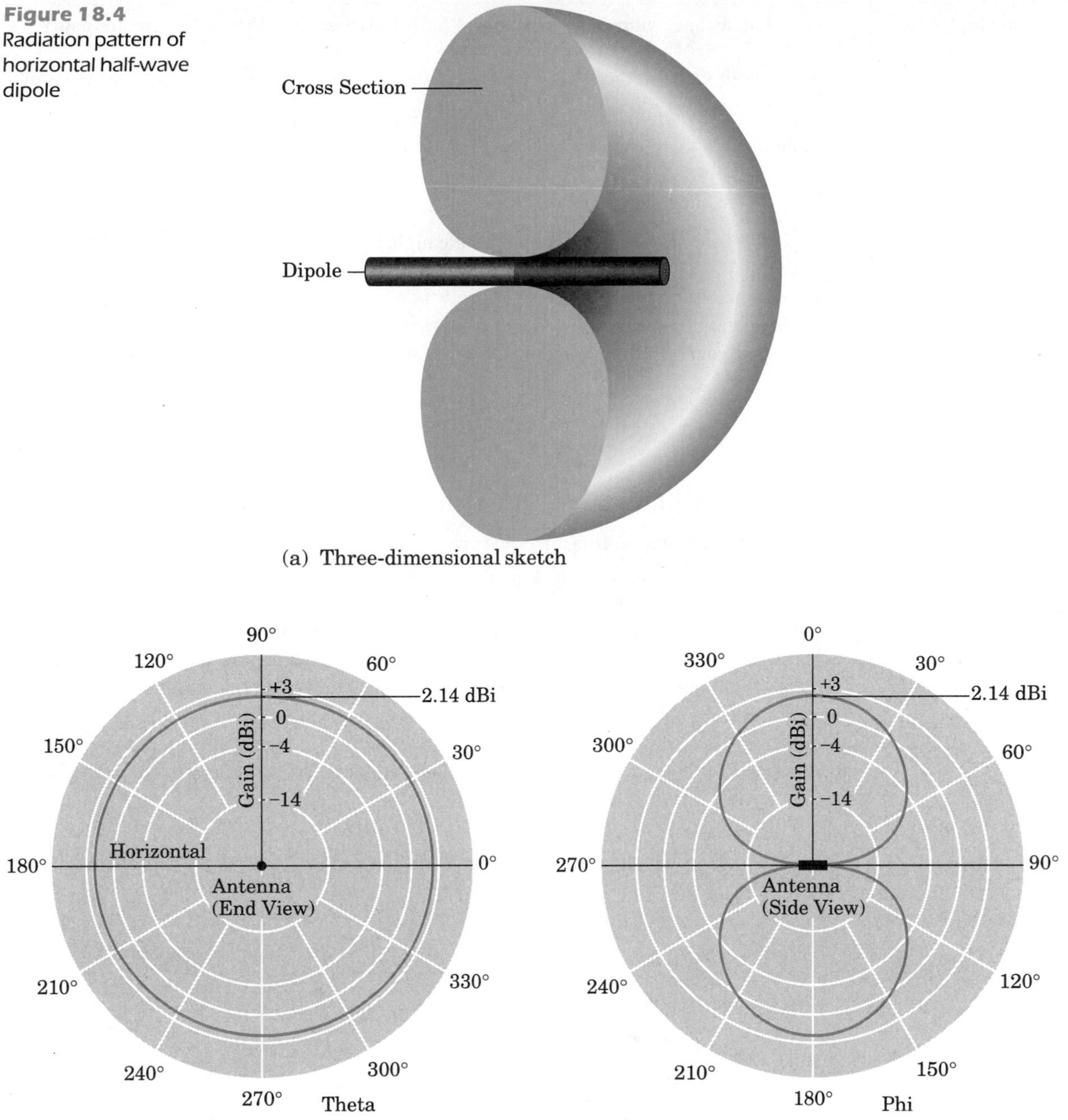

(a) Three-dimensional sketch

(b) Vertical plane

(c) Horizontal plane

mum radiation, is 2.14 dB with respect to an isotropic radiator. This gain is usually expressed as 2.14 **dBi**.

The half-wave dipole itself is sometimes used as a reference, especially for high-frequency antennas. In that case, the gain of an antenna may be expressed in decibels with respect to a half-wave dipole or **dBd** for short. Since the gain of the dipole is known to be 2.14 dBi, the gain of any antenna in dBd is 2.14 dB less than the gain of the same antenna expressed in dBi. Obviously, when comparing antennas, it is important to know which reference antenna was used in gain calculations.

Example 18.3 Two antennas have gains of 5.3 dBi and 4.5 dBd respectively. Which has greater gain?

Solution Convert both gains to the same standard. In this case, let us use dBi. Then, for the second antenna,

$$\begin{aligned} \text{gain} &= 4.5 \text{ dBd} \\ &= 4.5 + 2.14 \text{ dBi} \\ &= 6.64 \text{ dBi} \end{aligned}$$

Therefore, the second antenna has the higher gain.

As mentioned in Chapter 17, the signal power delivered by a receiving antenna depends not only on its gain but also on its effective physical size. The signal power available is simply the power density multiplied by the **effective area**, and the effective area is given by

$$A_{eff} = \frac{\lambda^2 G_R}{4\pi} \tag{18.6}$$

where

A_{eff} = effective area
G_R = antenna gain, as a power ratio
λ = wavelength of the signal

18.3.2 Gain and Directivity

The sense in which a half-wave dipole antenna can be said to have gain can be seen from Figure 18.5. This sketch shows the pattern of a dipole, from Figure 18.4(c), superimposed on that of an isotropic radiator. It can be seen that while the dipole has a gain of 2.14 dBi in certain directions, in others its gain is negative. If the antenna were to be enclosed by a sphere that would absorb all the radiated power, the total radiated power would be found to be the same for both antennas. Remember that for antennas, power gain in one direction is at the expense of losses in others.

Sometimes the term **directivity** is used. Directivity is not quite the same as gain—it is the gain calculated assuming a lossless antenna. Real antennas have losses, and gain is simply the directivity multiplied by the efficiency of the antenna.

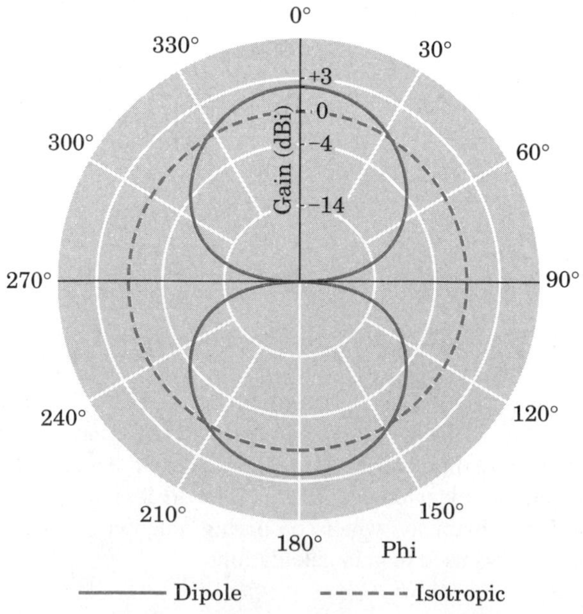

Figure 18.5 Isotropic and dipole antennas

Just as a flashlight emits a beam of light, a directional antenna can be said to emit a beam of radiation in one or more directions. The width of this beam is defined as the angle between its half-power points. These are also the points at which the power density is 3 dB less than it is at its maximum point. An inspection of Figure 18.4(b) and (c) will show that the half-wave dipole has a **beamwidth** of about 78° in one plane and 360° in the other. Many antennas are much more directional than this, with a narrow beamwidth in both planes.

When an antenna is used for transmitting, the total power emitted by the antenna is somewhat less than that delivered to it by the feedline. In fact,

$$P_x = \eta P_T \tag{18.7}$$

where

P_x = total radiated power
P_T = power supplied to the antenna
η = antenna efficiency

The figure of 2.14 dBi we have been using for the gain of a lossless dipole is also the directivity for *any* dipole. To find the gain of a real (lossy) dipole, it is necessary first to convert the decibel directivity to a power ratio and then to multiply by the efficiency.

Example 18.4 A dipole antenna has an efficiency of 85%. Calculate its gain in decibels.

Solution The directivity of 2.14 dBi can be converted to a power ratio:

$$\begin{aligned} D &= \text{antilog}\,\frac{2.14}{10} \\ &= 1.638 \end{aligned}$$

Now, find the gain:

$$\begin{aligned} G &= D\eta \\ &= 1.637 \times 0.85 \\ &= 1.39 \end{aligned}$$

If required, this gain can be converted to decibels:

$$\begin{aligned} G\,(\text{dB}) &= 10 \log 1.39 \\ &= 1.43 \text{ dBi} \end{aligned}$$

(The "i" in "dBi" is a reminder that the gain is with respect to an isotropic radiator.)

18.3.3 Effective Isotropic Radiated Power and Effective Radiated Power

In a practical situation, we are usually more interested in the power emitted in a particular direction than in the total radiated power. Looking from a distance, it is impossible to tell the difference between a high-powered transmitter using an isotropic antenna and a transmitter of lower power working into an antenna with gain. In Chapter 17, we defined an **effective isotropic radiated power** (EIRP), which is simply the actual power going into the antenna multiplied by its gain with respect to an isotropic radiator:

$$\text{EIRP} = P_T G_T \tag{18.8}$$

Another similar term that is in common use is **effective radiated power** (ERP), which represents the power input multiplied by the antenna gain measured with respect to a half-wave dipole. Since an ideal half-wave dipole has a gain of 2.14 dBi, the EIRP is 2.14 dB greater than the ERP for the same antenna-transmitter combination.

Figure 18.6
Variation of dipole reactance with frequency

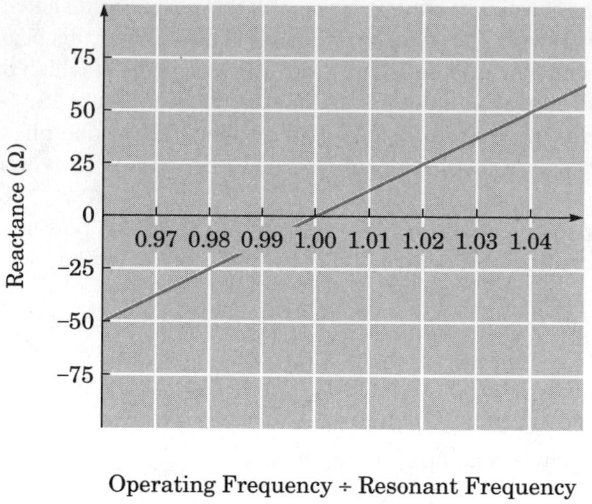

18.3.4 Impedance

The radiation resistance of a half-wave dipole situated in free space and fed at the center is approximately 70 Ω. The impedance will be completely resistive at resonance, which takes place for a physical length of about 95% of the calculated free-space half-wavelength value. The exact length depends on the diameter of the antenna conductor relative to the wavelength. If the frequency is above resonance, the feedpoint impedance will have an inductive component; if the frequency is lower than resonance, the antenna impedance will be capacitive. Another way of saying the same thing is that an antenna that is too short will appear capacitive, while one that is too long will be inductive. Figure 18.6 shows graphically how reactance varies with frequency.

A center-fed dipole is a balanced device and should be used with a balanced feedline. If coaxial cable is used, a *balun* (balanced-to-unbalanced) transformer should be connected between the cable and the antenna. (Balun transformers were described in Chapter 16.)

A half-wave dipole does not have to be fed at its midpoint. It is possible, for instance, to connect the feedline at one end, as shown in Figure 18.7. As you might expect from our earlier analogy with an open-circuited transmission line, this results in a higher impedance at the feedpoint. It is also possible to feed the antenna at some distance from the center in both directions, as shown in Figure 18.8. This system, called a *delta match*, allows the impedance to be adjusted to match a transmission line. Just as for a center-fed antenna, these two systems will result in a resistive impedance only if the antenna length is accurate.

18.3.5 Polarization

The **polarization** of a half-wave dipole is easy to determine: it is the same as the axis of the wire. That is, a horizontal antenna produces horizontally polarized waves, and a vertical antenna gives vertical polarization.

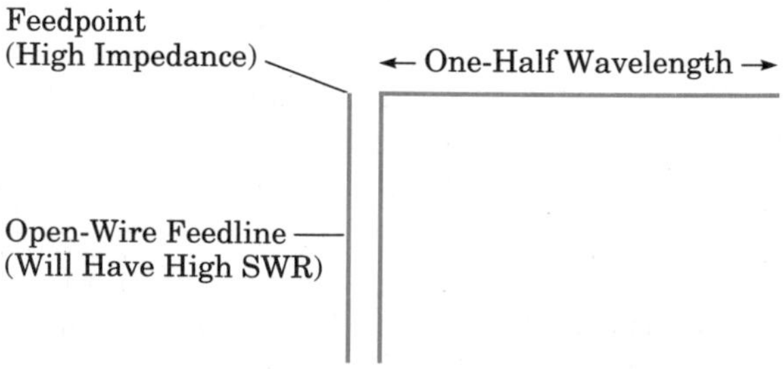

Figure 18.7
End-fed Half-wave dipole

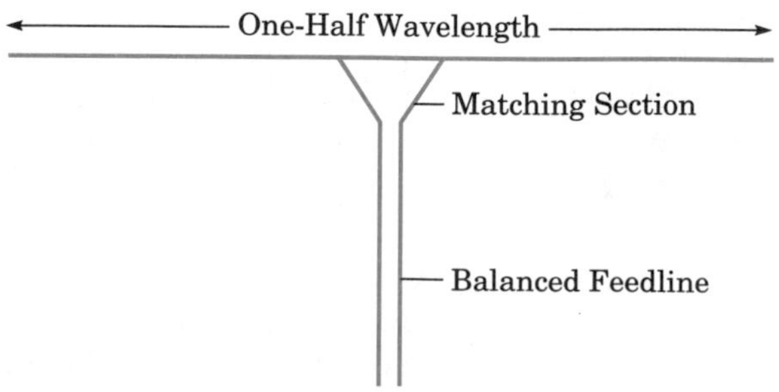

Figure 18.8
Delta match

The choice of polarization is sometimes dictated by the method of propagation, as explained in Chapter 17. For instance, ground-wave propagation requires vertical polarization. For HF communication, polarization is not very important because the ionosphere will randomize it; horizontal polarization is more common because a horizontal wire is usually easier to install. At VHF and above, it is important only that the polarization be the same at the two ends of a communications path.

18.3.6 Ground Effects

When an antenna is installed within a few wavelengths of the ground, the earth acts as a reflector and has a considerable influence on the radiation pattern of the antenna. Ground effects are important up through the HF range. At VHF and above, the antenna is usually far enough above the earth that reflections from the ground near the antenna are not significant. Reflections at a considerable distance from the antenna can still be very important, however, since they can cause *fading* (discussed in Chapter 17).

Ground effects are complex because the characteristics of the ground are so variable. In particular, the conductivity varies over a wide range. Any detailed analysis of ground effects requires a computer, and there are several programs that will do the job. An intuitive understanding of the process, without calculations, can be gained, however, by considering the earth to be a perfectly conductive sheet under the antenna. It thus imparts a 180° phase shift to the reflected wave, and the reflected wave will have the same amplitude as the incident wave. Just as with an ordinary mirror, it is possible to imagine an "image" of the real antenna and use it to understand the changes that ground reflections make in the antenna pattern.

Figure 18.9 shows the effects of average ground on the vertical radiation pattern of a horizontal dipole. Figure 18.9(a) shows the dipole in free space. This figure is the same as Figure 18.4(b) except that the radiation below the horizon has not been shown. Figure 18.9(b) shows the dipole one-quarter wavelength above the ground. Radiation in the upward direction is increased because the reflected wave reinforces the incident wave. On

Figure 18.9 Effect of ground on radiation pattern

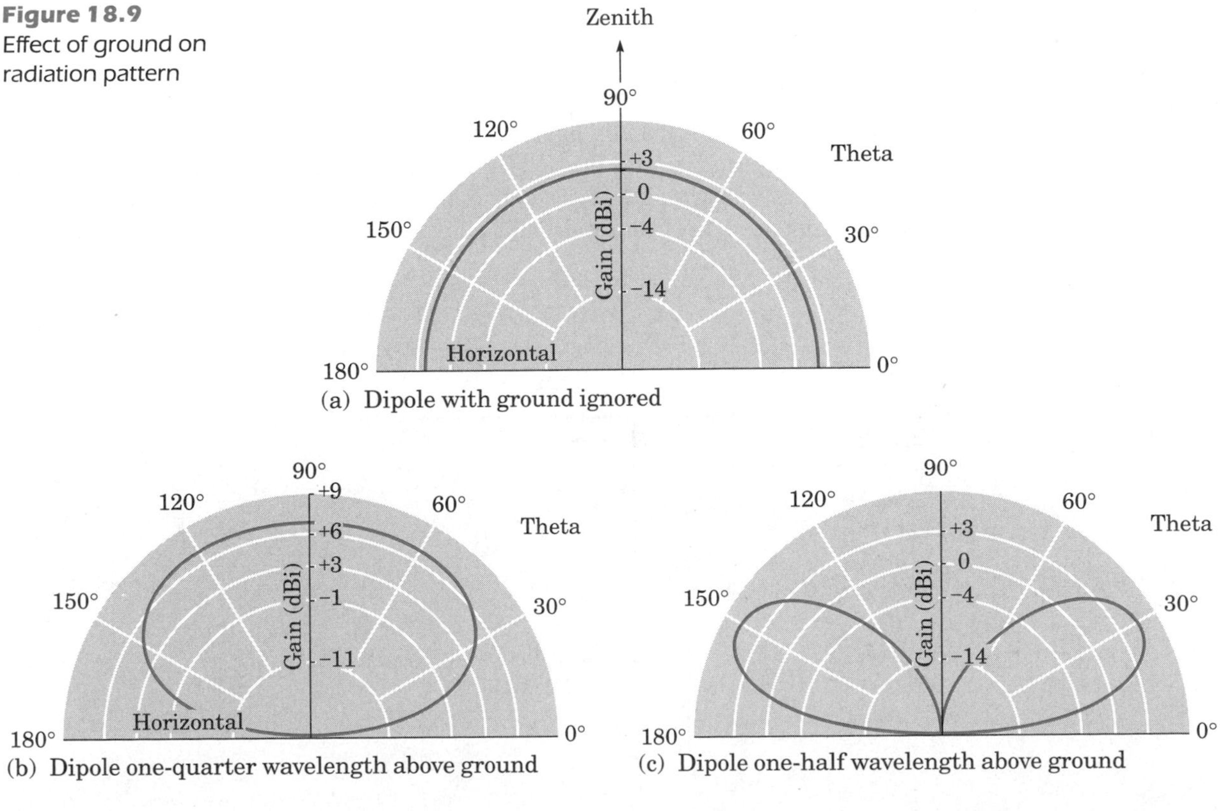

the other hand, the situation in Figure 18.9(c), where the antenna is one-half wavelength above the ground, results in cancellation of radiation toward the zenith, with a corresponding increase in low-angle radiation. The latter is generally more useful than high-angle radiation for HF communications.

18.4 Other Simple Antennas

The half-wave dipole is simple, useful, and very common, but it is by no means the only type of antenna in use. In this section, some other simple antennas will be introduced. Later, ways of combining antenna elements into arrays with specific characteristics will be examined.

18.4.1 The Folded Dipole

Figure 18.10 shows a folded dipole. It is the same length as a standard half-wave dipole, but it is made with two parallel conductors, joined at both ends and separated by a distance that is short compared with the length of the antenna. One of the conductors is broken in the center and connected to a balanced feedline.

The folded dipole differs in two ways from the ordinary half-wave dipole described above. It has a wider bandwidth, that is, the range of frequencies within which its impedance remains approximately resistive is larger than for the single-conductor dipole. For this reason, it is often used—alone or with other elements—for television and FM broadcast receiving antennas. It also has approximately four times the feedpoint impedance of an ordinary dipole. This accounts for the extensive use of 300 Ω balanced line (known as *twin-lead*) in TV and FM receiving installations.

It is easy to see why the impedance of a folded dipole is higher than that of a standard dipole. First, suppose that a voltage V and a current I are applied to an ordinary, nonfolded dipole. Then, at resonance when the feedpoint impedance is resistive, the power supplied is

$$P = VI \tag{18.9}$$

where

P = average power
V = RMS voltage
I = RMS current

Now reconsider the folded dipole of Figure 18.10. Looking at the center of the dipole, the length of the path from the center of the lower conductor to the center of the upper conductor is one-half wavelength. Therefore, the currents in the two conductors will be equal in magnitude. If the points were one-half wavelength apart on a straight transmission

Figure 18.10
Folded dipole

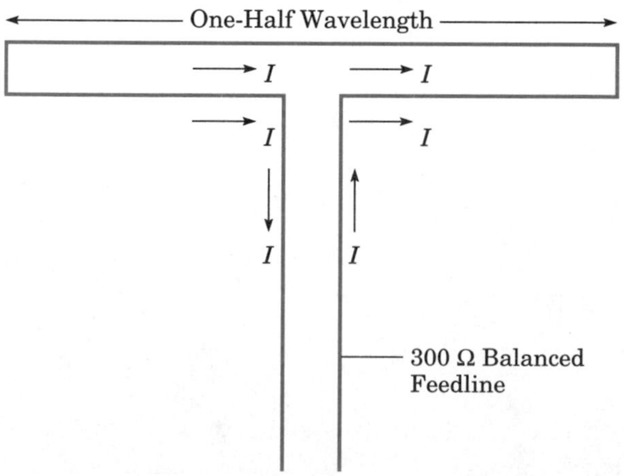

line, we would say that they were equal in magnitude but out of phase. Here, however, because the wire has been folded, the two currents, flowing in opposite directions with respect to the wire, actually flow in the same direction in space and contribute equally to the radiation from the antenna.

If a folded dipole and a regular dipole radiate the same amount of power, the total current in each must be the same. However, the current at the feedpoint of a folded dipole is only one-half the total current. If the feedpoint current is reduced by one-half, yet the power remains the same, the feedpoint voltage must be doubled. that is,

$$\begin{aligned} P &= VI \\ &= 2V\left(\frac{I}{2}\right) \end{aligned}$$

Assuming equal power is provided to both antennas, the feedpoint voltage must be twice as great for the folded dipole.

The resistance at the feedpoint of the ordinary dipole is, by Ohm's law,

$$R = \frac{V}{I}$$

For the folded dipole, it will be

$$\begin{aligned} R' &= \frac{2V}{I/2} \\ &= \frac{4V}{I} \\ &= 4R \end{aligned}$$

Since the current has been divided by two and the voltage multiplied by two, the folded dipole has four times the feedpoint impedance as the regular version.

It is also possible to build folded dipoles with different-size conductors and with more than two conductors. In this way, a wide variety of feedpoint impedances can be produced.

18.4.2 The Monopole Antenna

For low- and medium-frequency transmission, it is necessary to use vertical polarization to take advantage of ground-wave propagation. A vertical half-wave dipole would be possible, of course, but rather long. For instance, at 1 MHz, the wavelength is 300 m. Similar results can be obtained by using a quarter-wave **monopole** antenna, fed at one end with an unbalanced feedline, with the ground conductor of the feedline connected to a good earth ground. In practice, this usually means a fairly extensive array of **radials** (conductors buried in the ground and extending outward from the antenna). Such an antenna is often called a *Marconi antenna* (though there is some doubt as to whether Marconi was actually the first to use it) and usually takes the form of a guyed tower. The tower can be insulated from the ground, as shown in Figure 18.11; alternatively, it can be grounded and fed at a point above ground using a *gamma match*, as in Figure 18.12. Moving the feedline connection higher up on the tower increases the feedpoint impedance.

When connected to an ideal ground (or a good radial system), the radiation pattern of a quarter-wave monopole in the vertical plane has the same shape as that of a vertical half-wave dipole in free space, except that only one-half the pattern is present, since there is no underground radiation. The vertical pattern is shown in Figure 18.13. In the horizontal plane, of course, a vertical monopole will be omnidirectional. Since, assuming no losses, all of the power is radiated into one-half the pattern of a dipole, this antenna has a power gain of two (or 3 dB) over a dipole in free space.

The impedance at the base of a quarter-wave monopole is one-half that of a dipole. This can be explained as follows: with the same current, the antenna produces one-half the

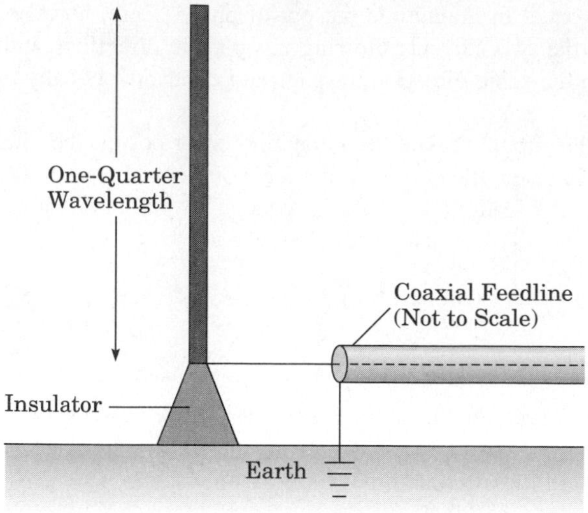

Figure 18.11
Monopole antenna using insulated tower

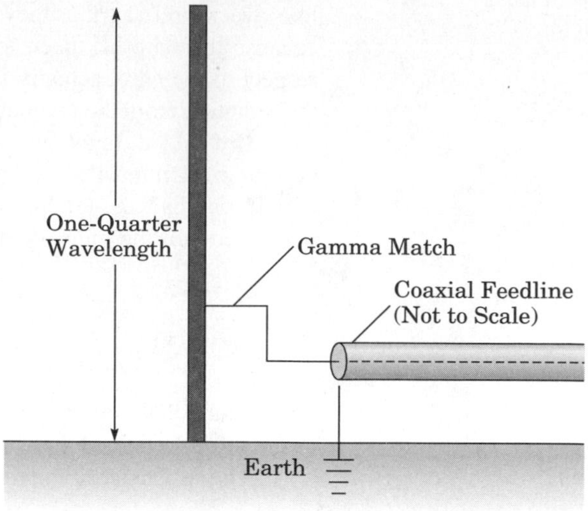

Figure 18.12
Monopole antenna using grounded tower

radiation pattern of a dipole and therefore one-half the radiated power. Assuming there are no losses, the radiated power is given by

$$P_r = I^2 R_r$$

where

P_r = radiated power
I = antenna current at the feedpoint
R_r = radiation resistance measured at the feedpoint

If the radiated power decreases by a factor of two for a given current, then so must the feedpoint radiation resistance.

18.4.3 Ground-Plane Antennas

Vertical antennas need not be mounted at ground level. At VHF and above, where line-of-sight propagation is the most common method, antenna height is very important. An antenna mounted at ground level would be ineffective. It is possible to preserve the simplicity and low radiation angle of the ground-mounted monopole by, in effect, constructing an artificial ground at the base of the antenna. This **ground plane** can be a conductive sheet, but it is more likely to be constructed of four or more metal rods radiating outward from the base of the antenna and made at least as long as the antenna itself. Ground-plane antennas are often seen with CB base stations. Since the wavelength in the 27 MHz CB band is about 11 m, the quarter-wave antenna needs to be almost 3 m long. Such an antenna is

Figure 18.13
Vertical monopole

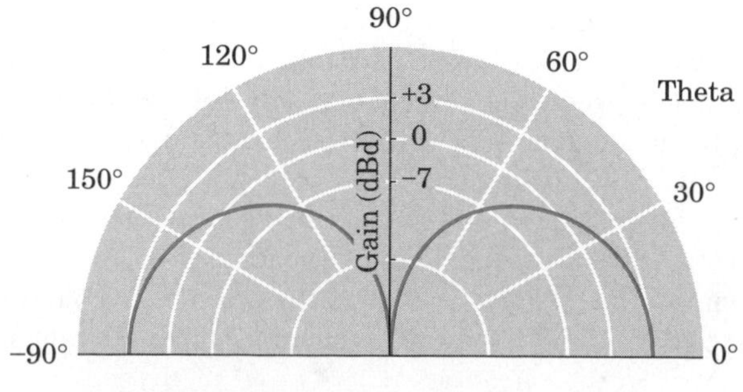

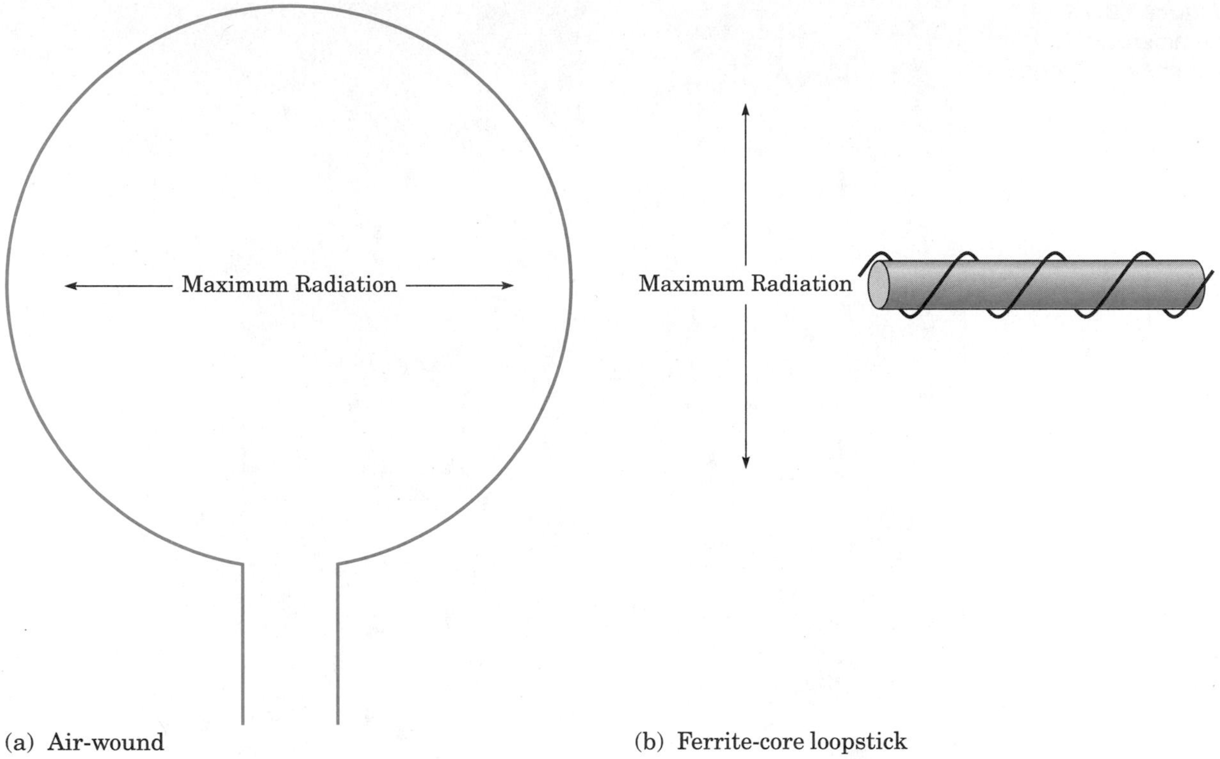

Figure 18.14
Loop antennas

obviously difficult enough to mount on a tower; a full-length dipole at twice the length would be much clumsier.

Mobile antennas are usually ground-plane antennas, with the car itself acting as the ground plane. Thus, a simple whip antenna on an automobile would be expected to be one-quarter wavelength in length. This is quite practical in the FM broadcast band, where the wavelength is about 3 m, but it is rather impractical (though sometimes done) in the 27 MHz CB band and out of the question in the AM broadcast band, where the wavelength is on the order of 300 m.

18.4.4 Loop Antennas

The lengths of the antennas studied so far are all an appreciable fraction of a wavelength. Sometimes, particularly for receiving, a much smaller antenna is required. These antennas are not be very efficient but can perform adequately for such tasks as the reception of local AM broadcasts. They are also used for marine radio direction finders.

Figure 18.14 shows two versions of the popular loop antenna. The one in Figure 18.14(a) is an older design, using an air-wound coil. This antenna is bidirectional, with its greatest sensitivity in the plane of the loop, as shown by the arrow. This would be a multiturn coil for the AM broadcast band; only one turn is shown for clarity. Figure 18.14(b) is a representation of a ferrite "loopstick" antenna, such as is found in practically every AM broadcast receiver (except those in automobiles). Here, the directionality is still in the plane of the individual coil turns, but this is broadside to the axis of the ferrite core. As we saw in Chapter 5, the loopstick antenna usually doubles as the coil in the input tuned circuit of a receiver.

18.4.5 The Five-Eighths Wavelength Antenna

This antenna is often used vertically as either a mobile or base antenna in VHF and UHF systems. Like the quarter-wave vertical antenna, it has omnidirectional response in the horizontal plane. However, the radiation is concentrated at a lower angle, resulting in gain in the horizontal direction, which is often more useful. In addition, it has a higher feedpoint impedance and so does not require as good a ground, because the current at the feedpoint

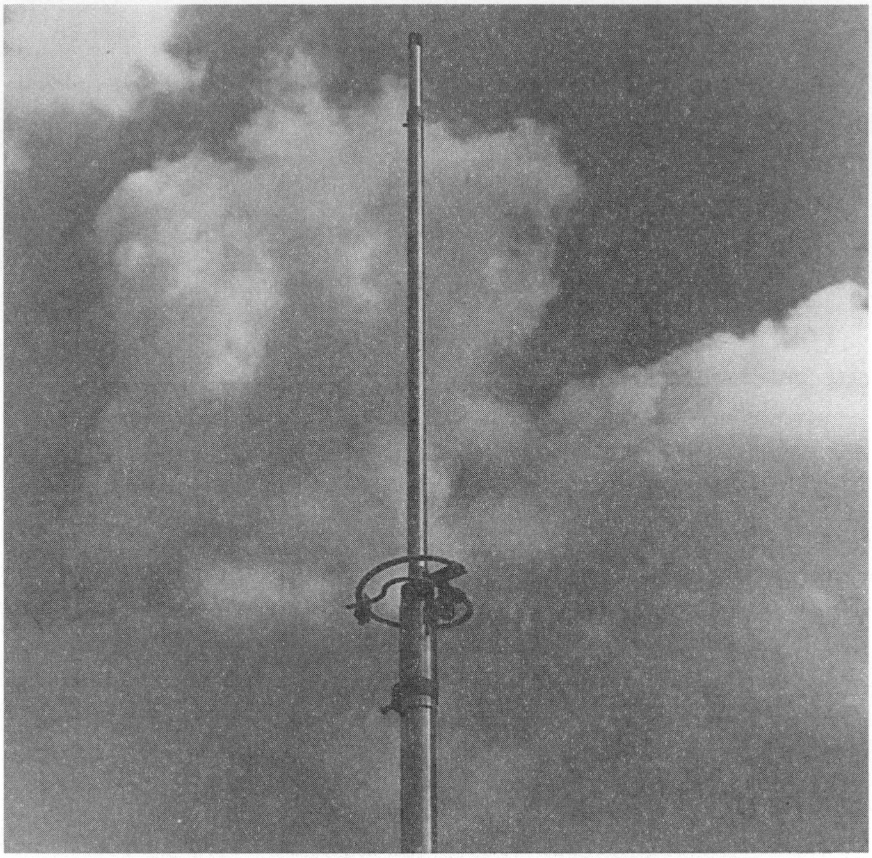

Figure 18.15 Commercial five-eighths wavelength antenna
Courtesy Cushcraft Corporation.

is less. The impedance is typically lowered to match that of a 50 Ω feedline by the use of an impedance-matching section. The circular section at the base of the antenna in Figure 18.15 is an impedance-matching device.

18.4.6 The Discone Antenna

The rather unusual-looking antenna shown in Figure 18.16 is known appropriately as the *discone*. It is characterized by very wide bandwidth, covering approximately a 10:1 frequency range, and an omnidirectional pattern in the horizontal plane. The signal is vertically polarized, and the gain is comparable to that of a dipole. The feedpoint impedance is approximately 50 Ω; the feedpoint is located at the intersection of the disk and the cone. The disk-cone combination acts as a transformer to match the feedline impedance to the impedance of free space, which is 377 Ω. Typically, the length measured along the surface of the cone is about one-quarter wavelength at the lowest operating frequency.

The wide bandwidth of the discone makes it a very popular antenna for general reception in the VHF and UHF ranges. It is a favorite for use with *scanners*. These receivers can tune automatically to a large number of channels in succession and are often used for monitoring emergency services. The discone can be used for transmitting but seldom is. Most transmitting stations operate at one frequency or over a narrow band of frequencies. Simpler antennas with equivalent performance and equally elaborate antennas with better performance are available when wide bandwidth is not required.

18.4.7 The Helical Antenna

There are actually several different types of antennas that are described as *helical*. A **helix**, of course, is simply a spiral. A quarter-wave monopole antenna can be shortened and wound into a helix—this is the common *rubber ducky* antenna used with many handheld transceivers. Sometimes it is called a *helical antenna*, and it certainly is helical in shape.

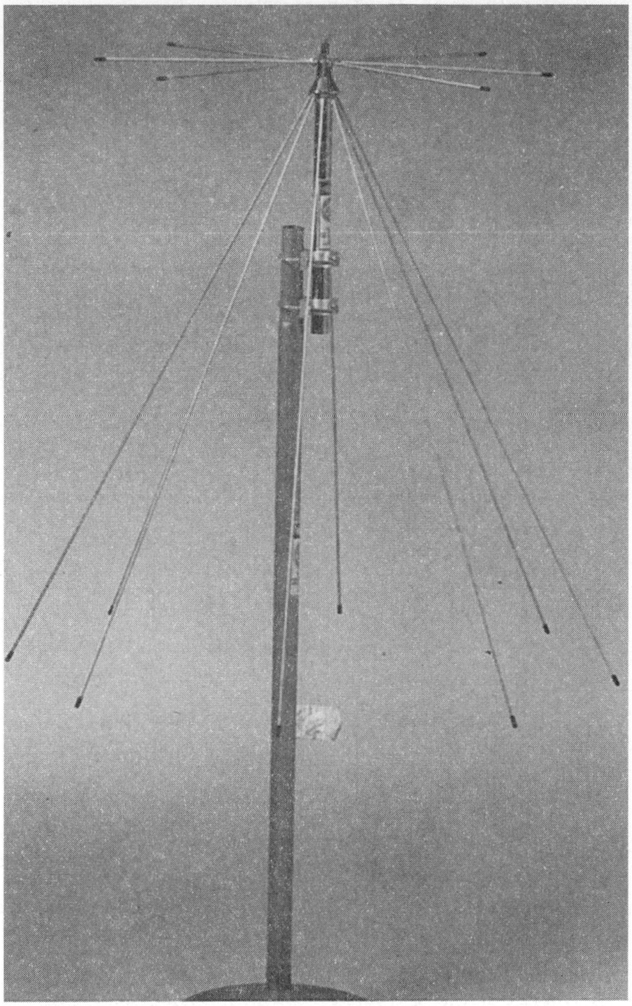

Figure 18.16
Commercial discone antenna
Courtesy of Radio Shack, a division of Tandy Corporation

The antenna that is usually referred to as helical, however, is much longer, usually several wavelengths long. Such an antenna is shown in Figure 18.17. It is used with a plane reflector, as shown, to improve its directional characteristics. Typically, the circumference of each turn is about one wavelength, and the turns are about one-quarter wavelength apart.

Helical antennas of the type shown in Figure 18.17 produce circularly polarized waves whose sense is the same as that of the helix. A helical antenna can be used to receive

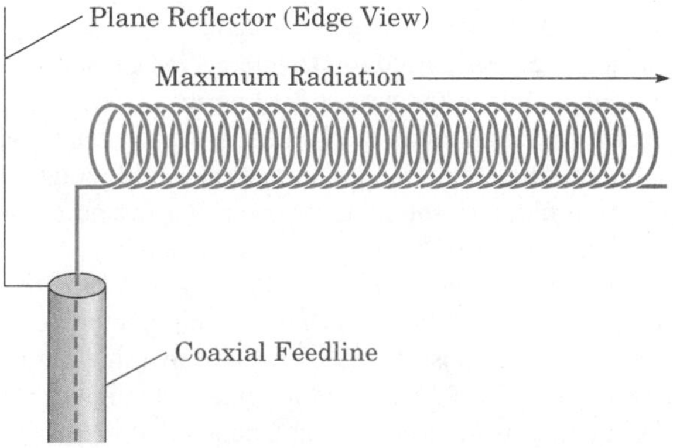

Figure 18.17
Helical antenna

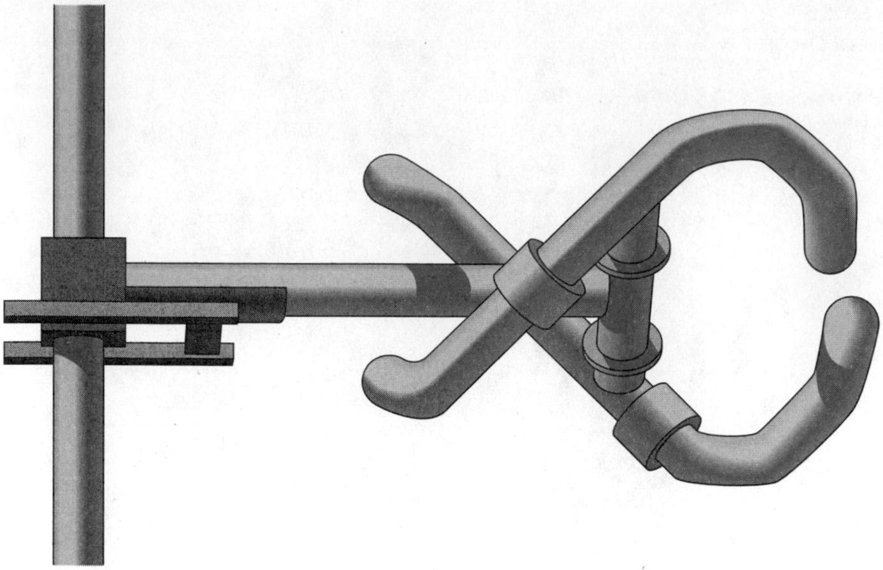

Figure 18.18 Commercial FM broadcasting transmitting antenna
Courtesy Harris Corp.

circularly polarized waves with the same sense and can also receive plane-polarized waves with the polarization in any direction. The gain is proportional to the number of turns and can be several decibels greater than a dipole.

Helical antennas are often used with VHF satellite transmissions. Since they respond to any polarization angle, they avoid the problem of Faraday rotation, which makes the polarization of waves received from a satellite impossible to predict.

There is yet another type of antenna that can be described as helical. The antenna illustrated in Figure 18.18 resembles a half-wave dipole, except that the ends are bent. This antenna is used to produce both horizontally and vertically polarized waves simultaneously, with one-half the input power going into each polarization. Such antennas are very common in FM broadcasting and enable the signal to be received with both vertical antennas (such as those on cars) and horizontal antennas (like the television antennas installed on the roofs of houses). Mounted as shown, the antenna is approximately omnidirectional in the horizontal plane.

18.5 Antenna Matching

Sometimes a resonant antenna is too large to be convenient. The question then arises as to what can be done to facilitate reasonably efficient transmission and/or reception with a shorter antenna. In other situations, one antenna may be required to work at several widely different frequencies and cannot be expected to be of resonant length at all of them: at some frequencies it may be too short, at others, too long.

First note that there is no need for an antenna to be resonant for it to radiate or receive signals—there only needs to be current flowing in the antenna. However, a nonresonant antenna will not be matched to its feedline, and the mismatch will result in a high SWR and a good deal of reflected power. The antenna will still radiate the power that is supplied to it, but it will be difficult to supply that power. Furthermore, the radiation resistance for short antennas is very low, resulting in reduced efficiency.

The problem of mismatch can be rectified by matching the antenna to the feedline. Perhaps the most obvious solution is an *LC* matching network, which should be installed as close to the antenna as possible, because any feedline between the network and the antenna will have a high SWR. Transmission line matching techniques using shorted stubs can also be used (these were described in Chapter 16). Whether lumped- or distributed-constant techniques are used is largely a function of frequency. At VHF and higher, the

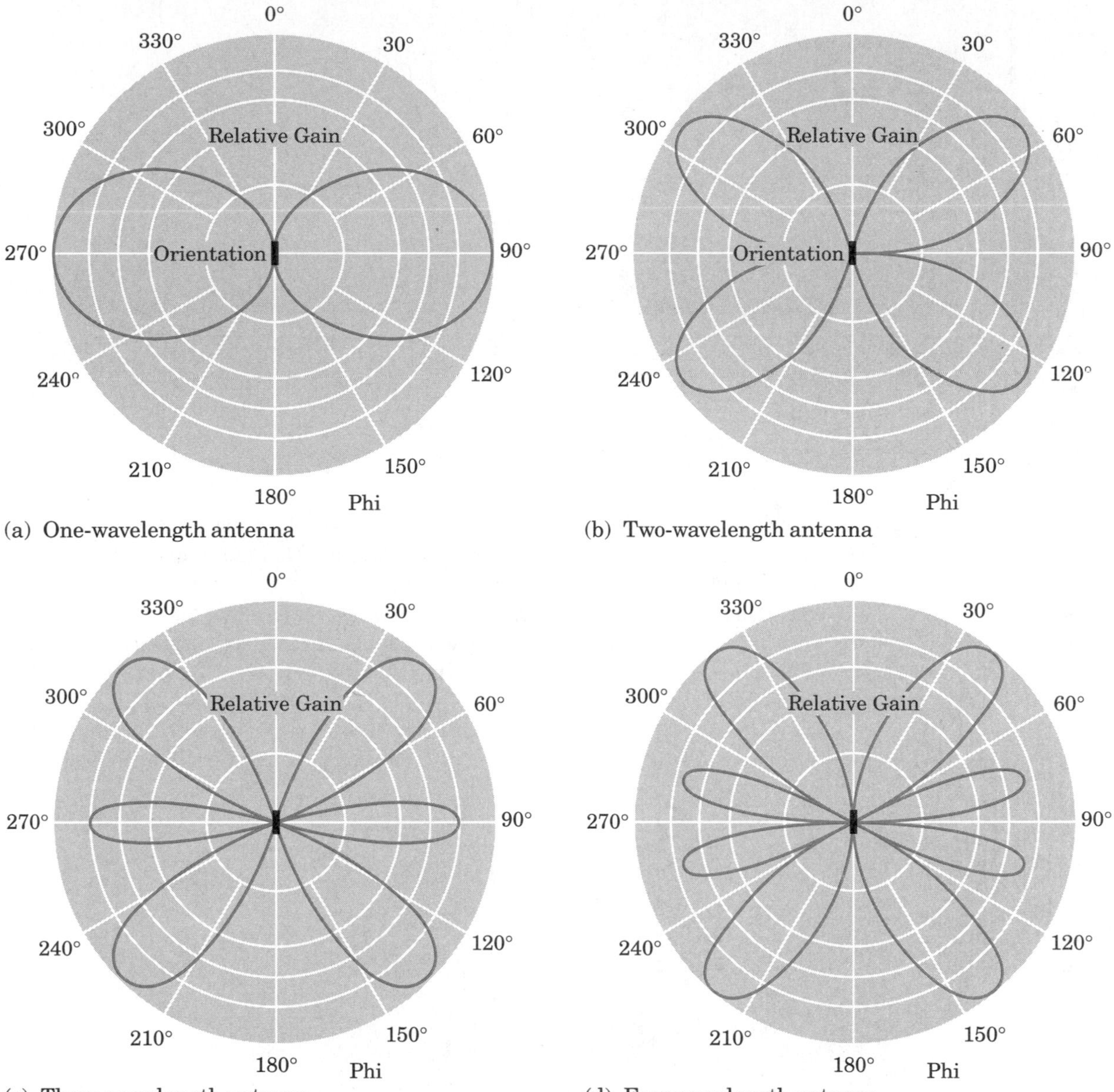

(a) One-wavelength antenna
(b) Two-wavelength antenna
(c) Three-wavelength antenna
(d) Four-wavelength antenna

Figure 18.19
Long-wire antennas

lumped-constant approach may involve impractically small component values, while at low frequencies, transmission line techniques may require inconveniently long stubs.

The antenna-matching network is separate from the transmitter output (or receiver input) circuit. The latter matches the transmitter (or receiver) to the transmission line, and the former matches the line to the antenna.

With the proper matching network, it is possible to use any random length of wire as an antenna. The directional properties of such an antenna are not the same as those of a half-wave dipole, however. As a rough rule of thumb, the longer the wire, the more **lobes**, and the closer the radiation maximum is to the axis of the antenna. Figure 18.19 shows the patterns in the plane of the antenna for a few selected *long-wire* antennas.

18.5.1 Inductive and Capacitive Loading

One simple but effective technique for matching a short antenna to a feedline is to increase its electrical length. For instance, a mobile whip antenna that is less than one-quarter wavelength can have an inductance added at its base, as shown in Figure 18.20. This inductance, called a **loading** coil, cancels the capacitive effect of the too-short antenna and, when care-

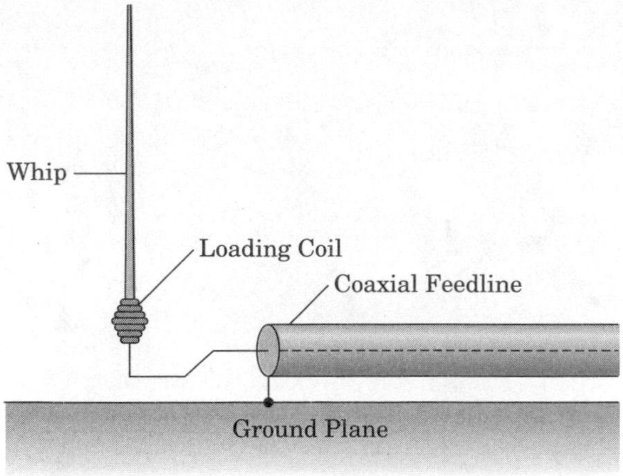

Figure 18.20
Loaded whip antenna

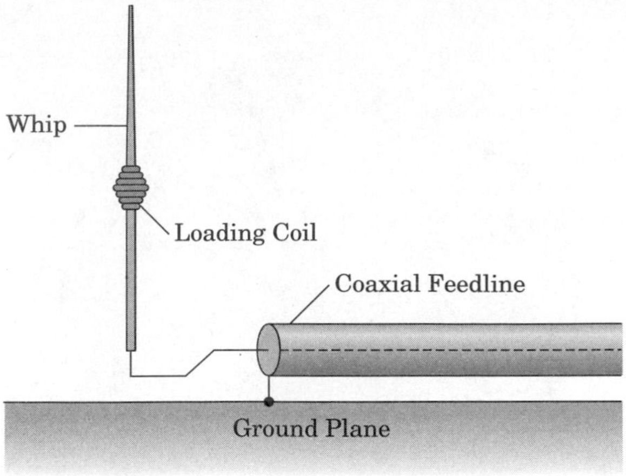

Figure 18.21
Center-loaded whip antenna

fully adjusted, can result in an antenna that looks electrically like a quarter-wave monopole. There are two drawbacks to this solution: the coil raises the Q of the antenna and narrows its bandwidth, and the coil resistance increases the losses.

The problem of coil losses is made more serious by the fact that the coil is installed at the point in the antenna where the current is maximum. In addition, short antennas have very low radiation resistance, so they require high antenna currents, which lead to significant losses.

The I^2R losses in the coil could be reduced by moving it away from the feedpoint, as shown in Figure 18.21. Unfortunately, this also reduces the effectiveness of the coil, re-

Figure 18.22
Commercial rubber-duckie antenna for portable transceiver
Icom America Inc.

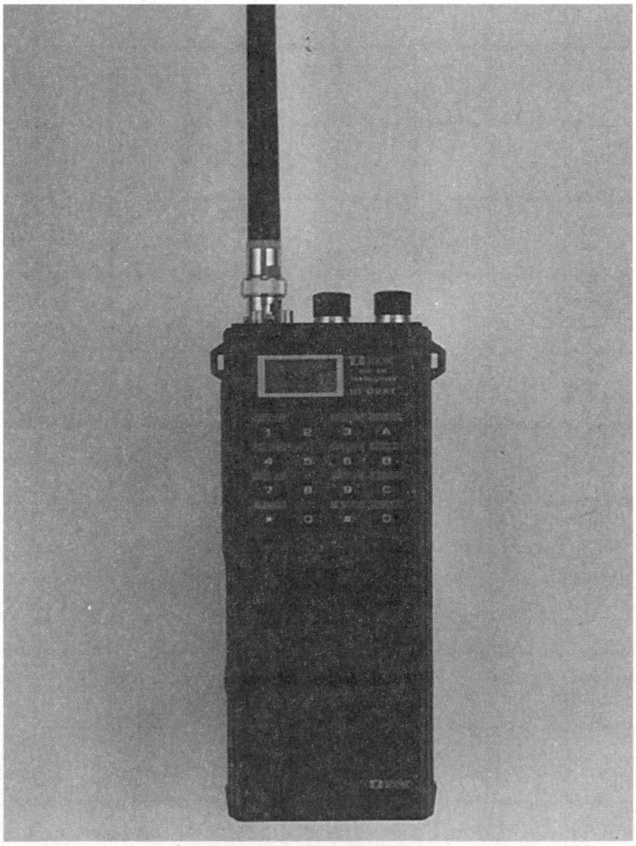

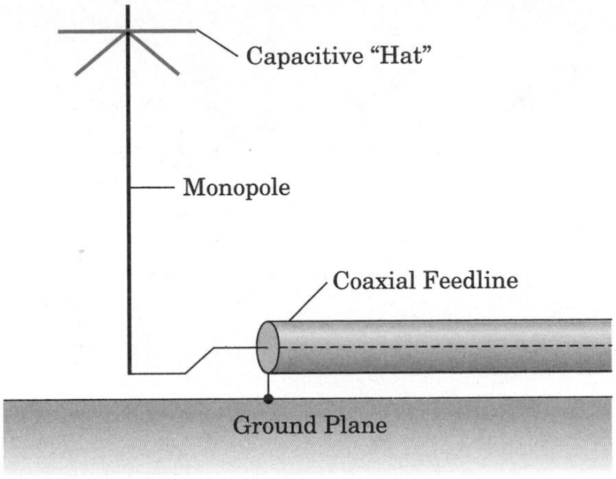

Figure 18.23
Capacitive loading

quiring a greater inductance. Nonetheless, this center-loading technique is often seen with mobile antennas.

Another way to employ inductive loading is to construct the whole antenna in the form of a coil, or helix. The rubber-ducky antennas often found on handheld transceivers are generally of this type. Figure 18.22 shows an example.

It is also possible to increase the antenna's electrical length by adding capacitance to the end away from the feedpoint. This solution takes the form of a "hat" consisting of a metallic disk or a collection of rods or wires, as shown in Figure 18.23. Though not suitable for mobile antennas because it would increase wind resistance, this technique can be seen on fixed antennas (for instance, those used in AM broadcasting).

18.6 Antenna Arrays

The simple elements described above can be combined to build a more elaborate antenna. The radiation from the individual **elements** will combine, resulting in reinforcement in some directions and cancellation in others to give greater gain and better directional characteristics. For instance, it is often desirable to have high gain in only one direction, something that is not possible with the simple antennas previously described.

Arrays can be classified as *broadside* or *end-fire*, according to their direction of maximum radiation. If the maximum radiation is along the main axis of the antenna (which may or may not coincide with the axis of its individual elements), the antenna is an end-fire array. If the maximum radiation is at right angles to this axis, the array has a broadside configuration. Figure 18.24 shows the axes of each type.

Antenna arrays can also be classified according to how the elements are connected. In a *phased array*, all the elements are connected to the feedline. There may be phase-shifting, power-splitting, and impedance-matching arrangements for individual elements, but all receive power from the feedline (assuming a transmitting antenna). Since the transmitter can be said to *drive* each element by supplying power, such arrays are also called *driven arrays*. In some arrays, only one element is connected to the feedline, while the

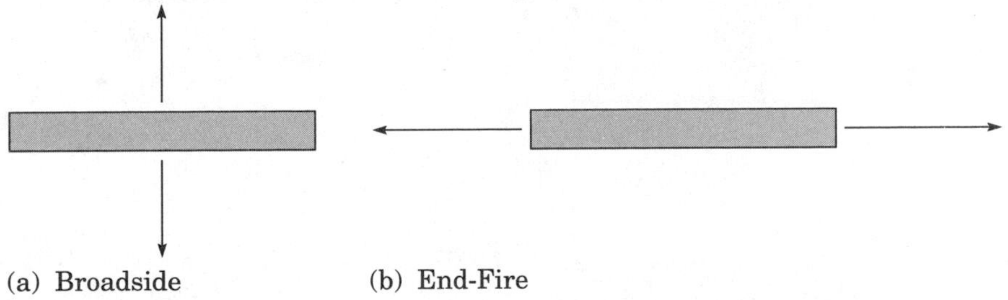

Figure 18.24
Broadside and end-fire arrays. Arrows represent direction(s) of maximum radiation.

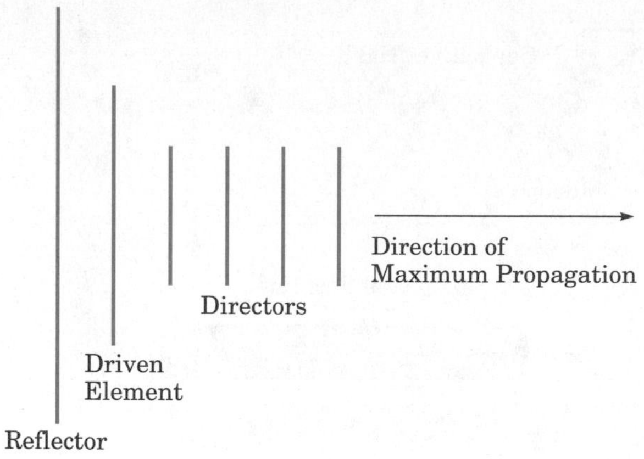

Figure 18.25
Yagi array

others work by absorbing and reradiating power radiated from the driven element. These elements are called *parasitic elements*, and the antennas are known as *parasitic arrays*.

The following sections will describe several common types of arrays and give some general characteristics of each. Detailed analysis of antenna arrays is usually done with the aid of a computer.

18.6.1 The Yagi Array

The Yagi array shown in Figure 18.25 is a parasitic end-fire array. It has one driven element, one *reflector* behind the driven element, and one or more *directors* in front of the driven element. The driven element is a half-wave dipole or folded dipole. The reflector is slightly longer than one-half wavelength, and the directors are slightly shorter. The spacing between elements varies but is typically about 0.2 wavelength. The Yagi antenna is more formally referred to as the *Yagi-Uda array*.

The Yagi antenna is unidirectional, with a single **main lobe** in the direction shown in Figure 18.25, as well as several **minor lobes**. The antenna pattern for a typical Yagi with eight elements—one driven, one reflector, and six directors—is shown in Figure 18.26. Yagis are often constructed with five or six directors for a gain of about 10 dBi, but higher gains, up to about 16 dBi, can be achieved by using more directors.

The antenna pattern shown in Figure 18.26 has a beamwidth of approximately 40° for the main lobe, at the 3 dB down points. It also has six **sidelobes** at the sides of the pattern and a lobe to the back of the pattern. In addition to their gain and beamwidth,

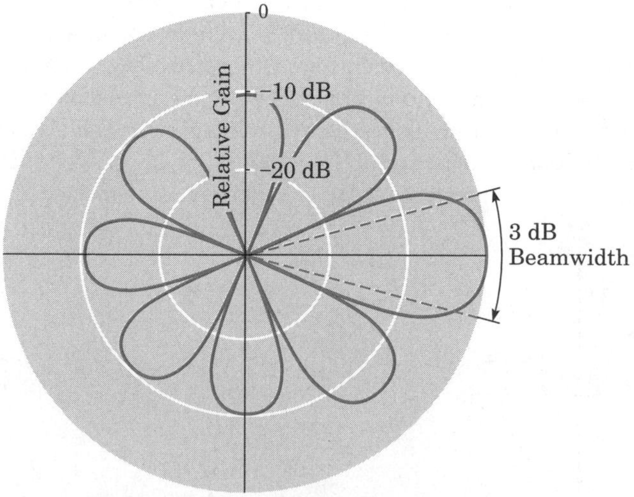

Figure 18.26
Radiation pattern for eight-element Yagi

antennas of this type are characterized by their **front-to-back ratio**, which is the ratio (in decibels) between the power density in the direction of maximum radiation and the power density radiated in a direction 180° away from it. For the pattern in Figure 18.26, this ratio is approximately 11 dB.

The Yagi is a relatively narrow-band antenna. When optimized for gain, its usable bandwidth is only about two percent of the operating frequency. Wider bandwidth can be obtained by varying the length of the directors, making them shorter as the distance from the driven element increases. This is necessary, for instance, when the Yagi is used for television reception. In fact, it is common to build two or three Yagis on one support (called a *boom*), with the elements interspersed, for the low-VHF, high-VHF, and/or UHF television bands. A folded dipole is generally used for the driven element in a TV antenna because its bandwidth is wider than that of an ordinary dipole.

18.6.2 The Log-Periodic Dipole Array

The log-periodic antenna derives its name from the fact that the feedpoint impedance is a periodic function of the operating frequency. Although log-periodic antennas take many forms, perhaps the simplest and most common is the dipole array, illustrated in Figure 18.27. The log-periodic dipole array (LPDA) is probably the most common antenna for television reception.

The elements are dipoles, with the longest one at least one-half wavelength in length at the lowest operating frequency and the shortest one less than one-half wavelength at the highest. The ratio between the highest and lowest frequencies can be 10:1 or more. A balanced feedline is connected to the narrow end, and power is fed to the other dipoles via a network of crossed connections as shown. The operation is quite complex, with the dipoles that are closest to resonance at the operating frequency doing most of the radiation. The gain is reasonable, typically about 8 dBi, but not as good as that of a well-designed Yagi with the same number of elements.

The design of a log-periodic antenna is based on several equations. A parameter τ is chosen, with a value that must be less than 1 and is typically between 0.7 and 0.9. A value toward the larger end of the range gives an antenna with better performance but more elements. The value τ is the ratio between the lengths and spacing of adjacent elements, that is,

$$\tau = \frac{L_1}{L_2} = \frac{L_2}{L_3} = \frac{L_3}{L_4} = \ldots \tag{18.10}$$

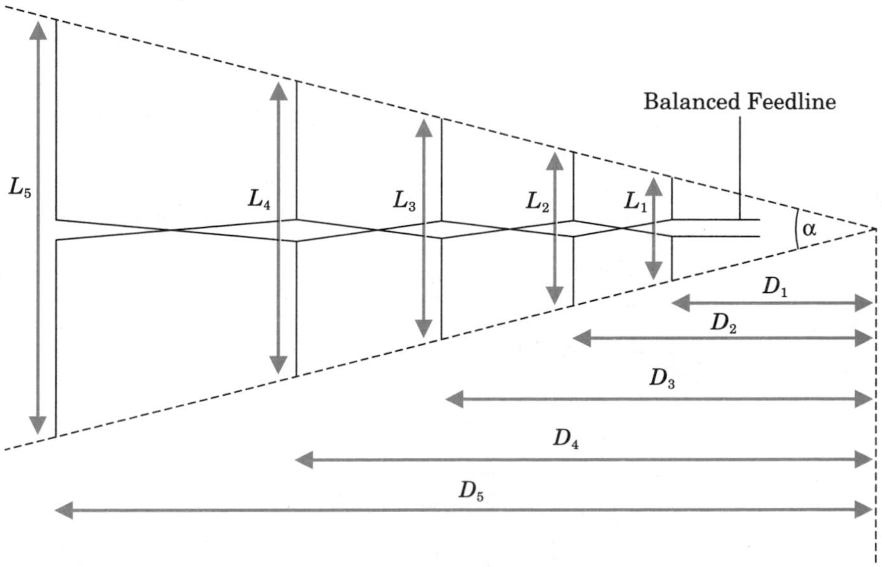

Figure 18.27 Log-periodic dipole array

where

$L_1, L_2, \ldots$ = lengths of the elements, in order from shortest to longest

and

$$\tau = \frac{D_1}{D_2} = \frac{D_2}{D_3} = \frac{D_3}{D_4} = \ldots \qquad (18.11)$$

where

$D_1, D_2, \ldots$ = spacings between the elements and the apex of the angle enclosing them, in order from shortest to longest

This angle is usually designated α and is typically about 30°. From simple trigonometry it can be shown that

$$\frac{L_1}{2D_1} = \tan\frac{\alpha}{2} \qquad (18.12)$$

The following example shows how an LPDA can be designed. Computer programs are available to automate the procedure.

Example 18.5 Design a log-periodic antenna to cover the frequency range from 100 to 300 MHz. Use $\tau = 0.7$ and $\alpha = 30°$.

Solution In order to get good performance across the frequency range of interest, it is advisable to design the antenna for a slightly wider bandwidth. For the longest element, we can use a half-wave dipole cut for 90 MHz, and for the shortest, one designed for 320 MHz.

From Equation (18.3),

$$L = \frac{142.5}{f}$$

For a frequency of 90 MHz,

$$L = \frac{142.5}{90} = 1.58 \text{ m}$$

For 320 MHz,

$$L = \frac{142.5}{320} = 0.445 \text{ m}$$

Because of the way the antenna is designed, it is unlikely that elements of exactly these two lengths will be present, but we can start with the shorter one and simply make sure that the longest element has at least the length calculated above.

Starting with the first element and using Equation (18.12),

$$\frac{L_1}{2D_1} = \tan\frac{\alpha}{2}$$

We have specified $L_1 = 0.445$ m and $\alpha = 30°$, so

$$D_1 = \frac{L_1}{2\tan\frac{\alpha}{2}} = \frac{0.445}{2\tan 15°} = 0.830 \text{ m}$$

From Equation (18.10),

$$\tau = \frac{L_1}{L_2} = \frac{L_2}{L_3} = \frac{L_3}{L_4} = \ldots$$

τ and L_1 are known, so L_2 can be calculated.

$$\begin{aligned} L_2 &= \frac{L_1}{\tau} \\ &= \frac{0.445}{0.7} \\ &= 0.636 \text{ m} \end{aligned}$$

We continue this process until we obtain an element length that is greater than 1.58 m.

$$\begin{aligned} L_3 &= \frac{L_2}{\tau} \\ &= \frac{0.636}{0.7} \\ &= 0.909 \text{ m} \end{aligned}$$

Similarly, $L_4 = 1.30$ m and $L_5 = 1.85$ m. This is longer than necessary, so the antenna will need five elements.

The spacing between the elements can be found from Equation (18.11):

$$\tau = \frac{D_1}{D_2} = \frac{D_2}{D_3} = \frac{D_3}{D_4} = \ldots$$

Since τ and D_1 are known, it is easy to find D_2 and then the rest of the spacings, in the same way as the lengths were found above. We get $D_2 = 1.19$ m, $D_3 = 1.69$ m, $D_4 = 2.42$ m, and $D_5 = 3.46$ m.

18.6.3 The Turnstile Array

Antenna arrays are not always designed to give directionality and gain. The turnstile array illustrated in Figure 18.28(a) is a simple combination of two dipoles designed to give omnidirectional performance in the horizontal plane, with horizontal polarization. The dipoles are fed 90° out of phase.

The gain of a turnstile antenna is actually about 3 dB less than that of a single dipole in its direction of maximum radiation, because each element of the turnstile receives only one-half the transmitter power. Figure 18.28(b) shows the radiation pattern for a typical turnstile antenna.

Turnstile antennas are often used for FM broadcast reception, where they give reasonable performance in all directions without the need for a rotor. Variations of the turnstile are also used for television and FM broadcasting.

18.6.4 The Monopole Phased Array

A single vertical quarter-wave monopole antenna has an omnidirectional radiation pattern in the horizontal plane. Often, however, the AM broadcast stations that use this type of antenna must have a directional pattern to avoid interference with other stations. In fact, it is common for a station to be required to use two different radiation patterns, one for day and one for night. (Interference is more likely at night because of ionospheric propagation.)

Two or more monopole antennas can be arranged in an array. One possibility, using only two antennas, is shown in Figure 18.29. Both elements are driven, and the radiation pattern depends on the phase angle between the elements. Two possibilities are illustrated in the figure. When the antennas are fed in phase and are one-half wavelength apart as shown, the radiation will be at a maximum broadside to a line drawn between the two towers. On the other hand, if the antennas are fed 180° out of phase, the radiation will cancel broadside to the array and will be at a maximum off the ends. Figure 18.30 shows these patterns. A larger number of possible radiation patterns can be created by varying the

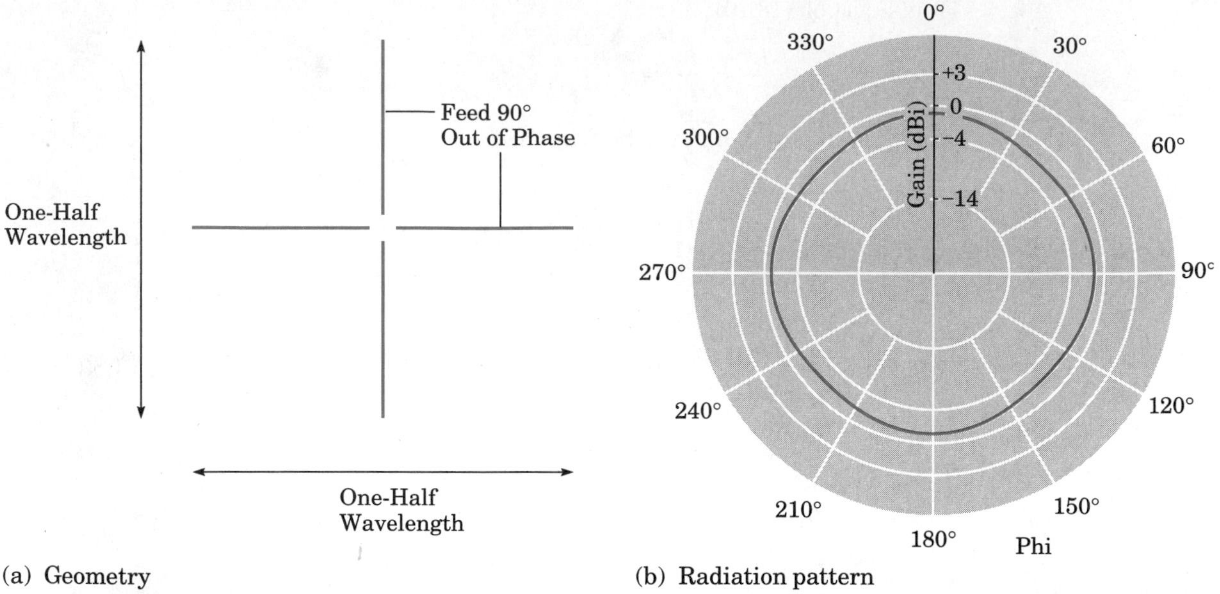

Figure 18.28
Turnstile antenna

number, spacing, and phase angle of the elements. In addition, the current distribution among the elements also affects the pattern. The element with the largest current will radiate the most power.

18.6.5 Other Phased Arrays

Phased arrays can be made by connecting any of the simple antenna types already discussed. Depending on the geometry of the array and the phase and current relationships between the elements, the array can be either broadside or end-fire.

Figure 18.31 shows one type of broadside array using half-wave dipoles. This is called a *collinear array*, because the axes of the elements are all along the same line. Suppose that the collinear antenna in this figure is used for transmitting, and imagine a receiving antenna placed along the main axis of the antenna. None of the individual elements radiates any energy in this direction, so of course there will be no signal from the array in this direction. Now move the hypothetical receiving antenna to a point straight out to one side of the antenna. All of the individual elements radiate a signal in this direction, but we must determine whether these signals add constructively or destructively. If the elements are in phase, the signals will add. A quick check shows that this is indeed the

Figure 18.29
Monopole array

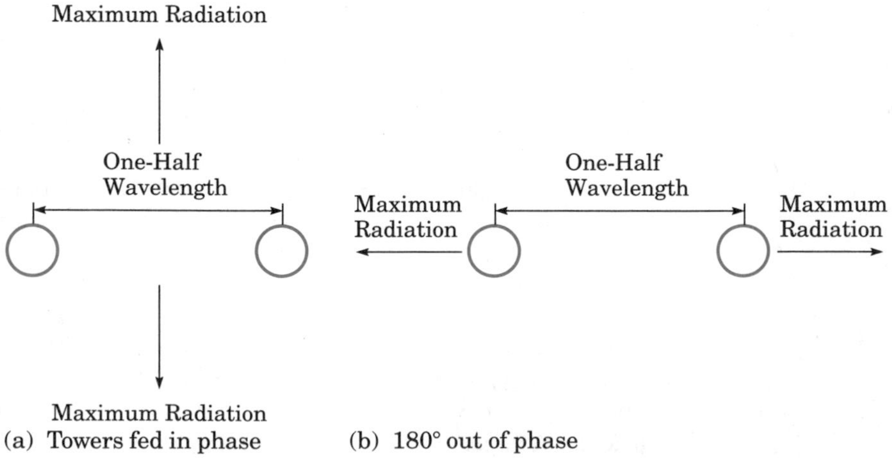

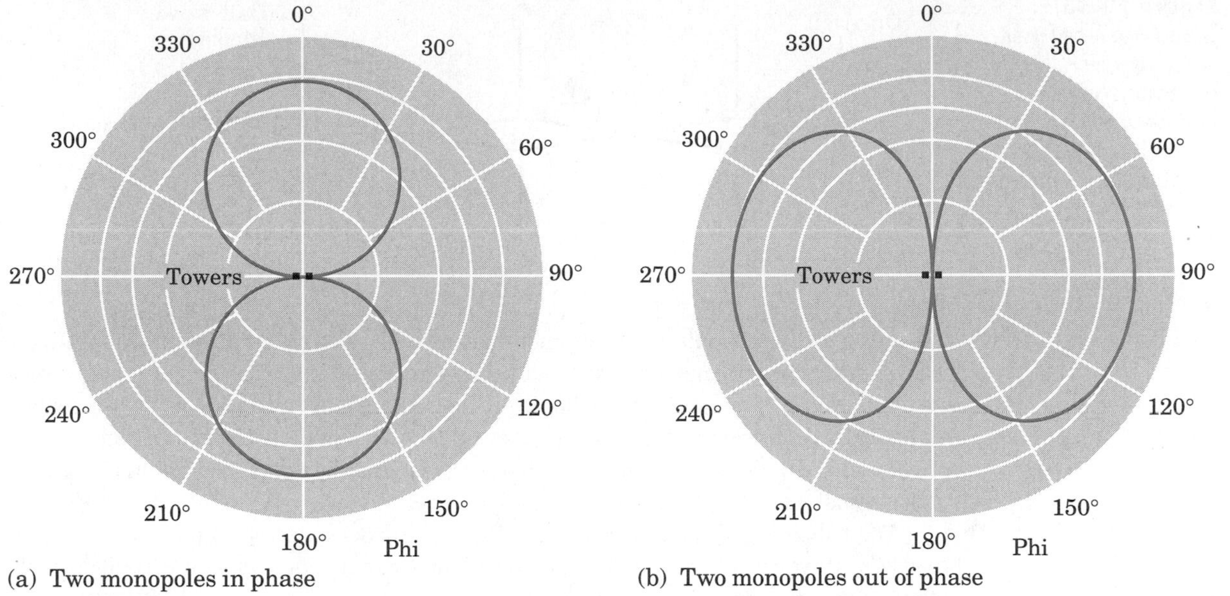

Figure 18.30
Radiation patterns for monopole arrays

case. The half-wave dipoles are linked by quarter-wave transmission line sections that cause a phase reversal between adjacent ends. Therefore, all the dipoles will be in phase.

Collinear antennas are often mounted with the main axis vertical. They will then be omnidirectional in the horizontal plane but will have a narrow angle of radiation in the vertical plane. Thus, they make good base-station antennas for mobile radio systems.

Another broadside array using dipoles is shown in Figure 18.32. This time, the elements are not collinear, but they are still in phase. That may not be obvious because of the crossed lines connecting them, but notice that the separation between the elements is one-half wavelength. This separation causes a 180° phase shift as the signal travels along the feedline from one element to the next. This phase shift is cancelled by the crossing of the transmission line section that joins the dipoles.

Although this antenna, like the previous one, is a broadside array, its pattern is not identical. There is no radiation off the end of any of the elements, so of course there is no radiation in the equivalent direction from the array. If this antenna were erected with its

Figure 18.31
Collinear array (all elements one-half wavelength, fed in phase)

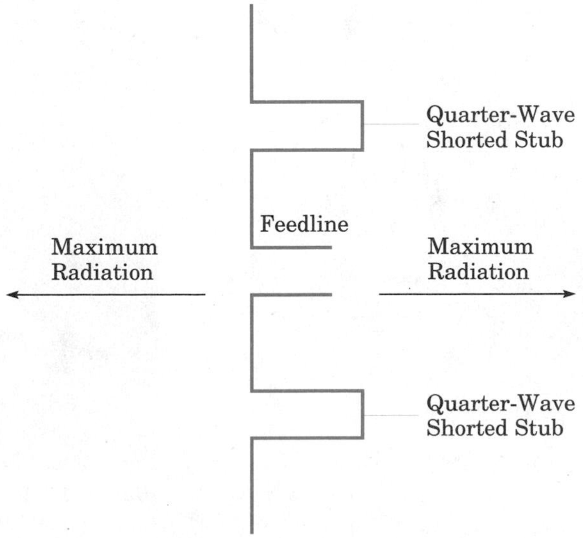

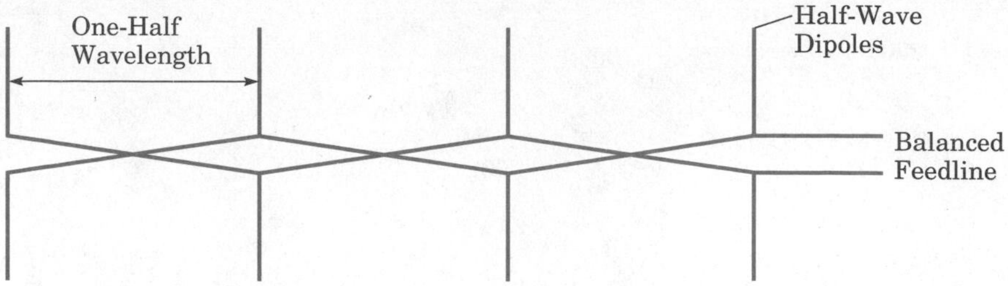

Figure 18.32
Broadside array (half-wave dipoles separated by one-half wavelength; maximum radiation is into and out of the paper)

main axis vertical, it would *not* be omnidirectional in the horizontal plane. It would have a narrow vertical pattern and a bidirectional pattern in the horizontal plane. As for the radiation off the end of the antenna, each dipole provides a signal in this direction, but the signals from adjacent elements cancel due to the one-half wavelength difference in path lengths between adjacent elements. Figure 18.33 shows a comparison of the radiation patterns in the horizontal plane for these two types of antennas.

The helical dipoles described above for FM broadcast use are often combined in a broadside array. Figure 18.34 is an example of such an antenna. Gains of several decibels in the horizontal plane can be achieved in this way.

Figure 18.35 shows an end-fire array using dipoles. It is identical to the broadside array in Figure 18.32 except for the fact that the feedline between elements is not crossed. This causes alternate elements to be 180° out of phase, and the radiation from one element therefore cancels that from the next in the broadside direction. Off the end of the antenna, however, the radiation from all the elements will be additive, since the 180° phase shift between adjacent elements is neatly cancelled by the one-half wavelength physical separation between them.

Phased arrays are not restricted to combinations of simple dipoles. For example, Figure 18.36 shows a pair of "stacked" Yagis, that is, the Yagis are mounted one above the other and fed in phase. This combination gives better directivity than a single Yagi. Ideally, the gain will be 3 dB greater than for a single Yagi of the same type. Of course, it is also

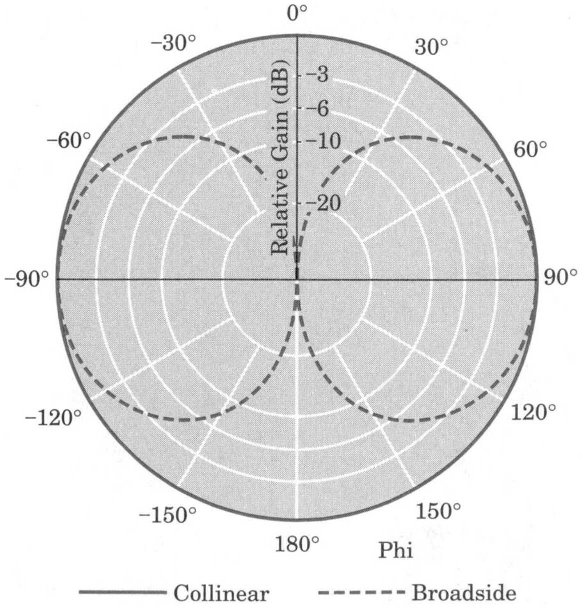

Figure 18.33
Collinear/broadside comparison

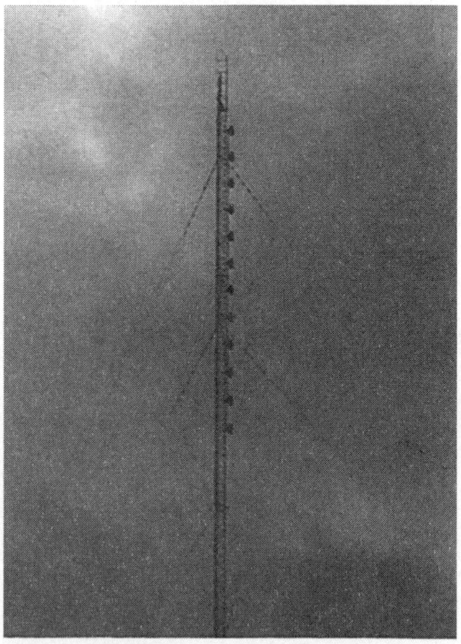

Figure 18.34
Commercial FM broadcast array

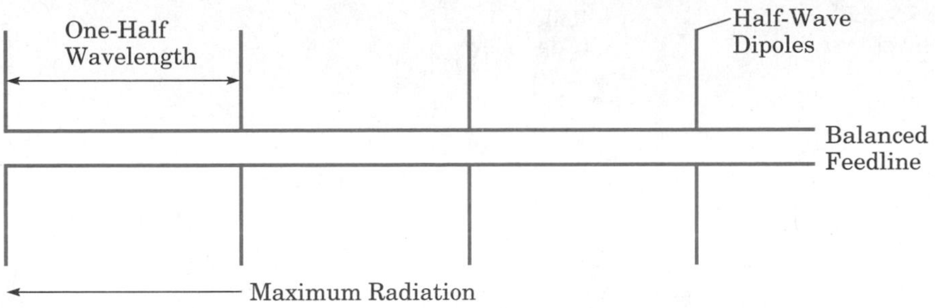

Figure 18.35
End-fire array (half-wave dipoles separated by one-half wavelength)

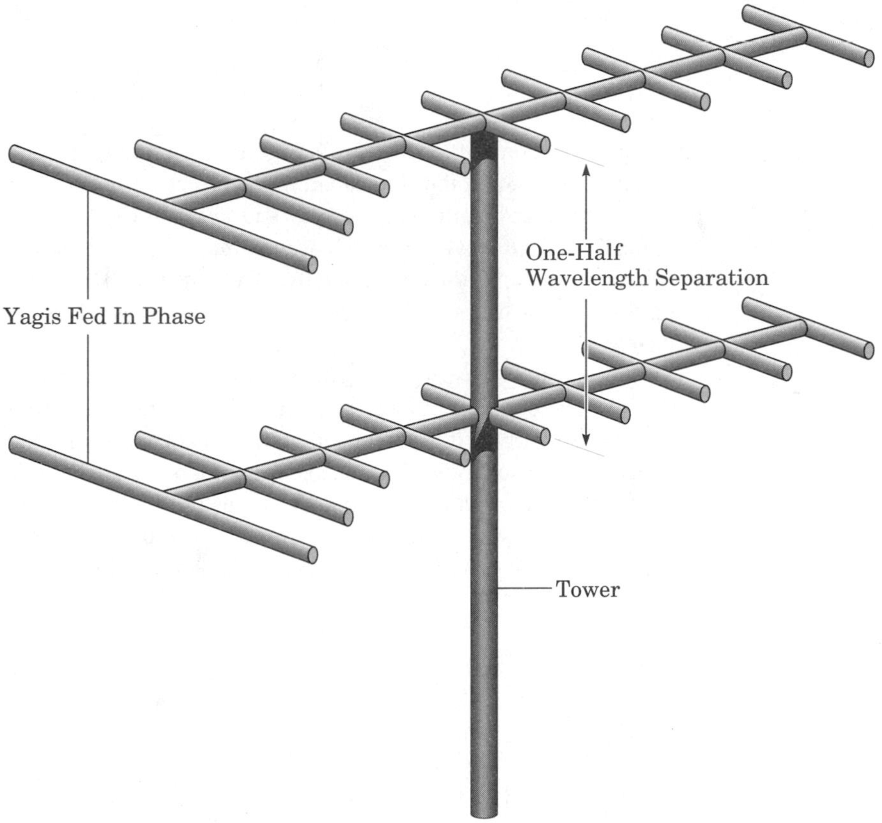

Figure 18.36
Stacked Yagis

possible to combine larger numbers of Yagis, and similar arrays can be constructed from other types of antennas, such as log-periodic arrays or helical antennas.

18.7 Reflecting Surfaces

We have seen that a Yagi antenna contains a reflecting element. It is also possible to construct a conductive surface that reflects antenna power in the desired direction. The surface may consist of one or more planes or may be parabolic in shape.

18.7.1 Plane and Corner Reflectors

A plane reflector acts in a similar way to an ordinary mirror. Like a mirror, its effects can be predicted by supposing that there is an "image" of the antenna on the opposite side of the reflecting surface at the same distance from it as the source. Reflection changes the phase angle of a signal by 180°. Whether the image antenna's signal aids or opposes the signal from the real antenna depends on the spacing between the antenna and the reflector

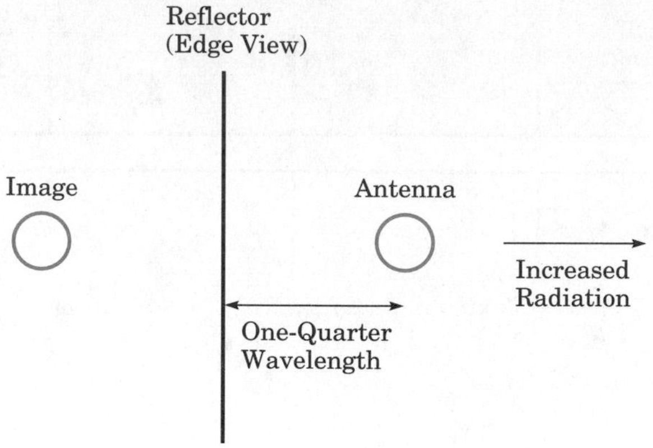

Figure 18.37 Plane reflector and image

and on the location of the receiver. In Figure 18.37, the antenna is one-quarter wavelength from the reflector, and the signals aid in the direction shown. The reflected signal experiences a 180° phase shift on reflection and another 180° shift because it must travel an additional one-half wavelength to reach the receiver. The magnitude of the electric field in the direction shown is thus increased by a factor of two. The power density in this direction is increased by a factor of four, or 6 dB, because power is proportional to the square of voltage.

It is possible to use a plane reflector with almost any antenna. For example, a reflector can be placed behind a collinear antenna, as shown in Figure 18.38. The antenna becomes directional in both the horizontal and vertical planes. Base antennas for cellular radio systems are often of this type. The plane does not have to be solid. In fact, it is often made of wire mesh to reduce wind resistance. It can also be made as a series of metal rods or tubes oriented the same way as the antenna. Reflectors of this type work well provided the separation between the components of the reflector is much less than a wavelength at the operating frequency.

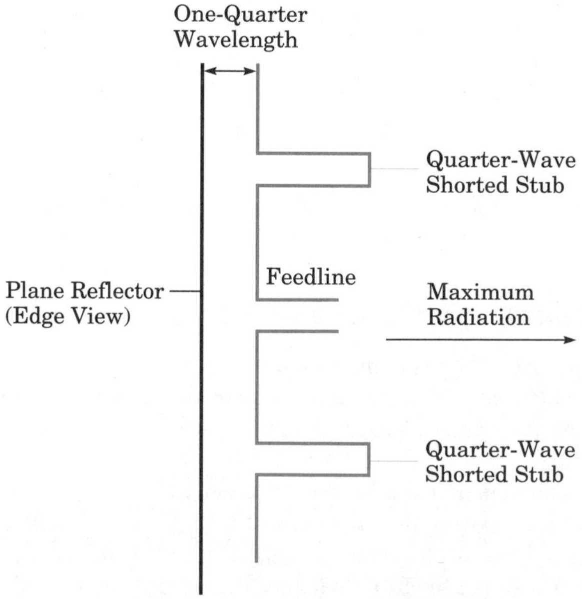

Figure 18.38 Collinear array with plane reflector (all elements one-half wavelength, fed in phase)

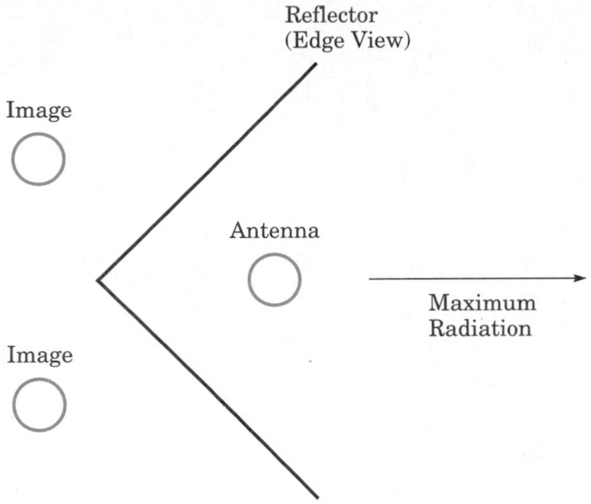

Figure 18.39
Corner reflector and images

The corner reflector creates two images, as shown in Figure 18.39, for a somewhat sharper pattern. Corner reflectors are often combined with Yagi arrays in UHF television antennas.

18.7.2 The Parabolic Reflector

Parabolic reflectors have the useful property that any ray that originates at a point called the *focus* and strikes the reflecting surface will be reflected parallel to the axis of the parabola, that is, a *collimated* beam of radiation will be produced. The parabolic "dish" antenna, familiar from backyard satellite-receiver installations, consists of a small antenna at the focus of a large parabolic reflector, which focuses the signal in the same way as the reflector of a searchlight focuses a light beam. Figure 18.40 shows a typical example. Of

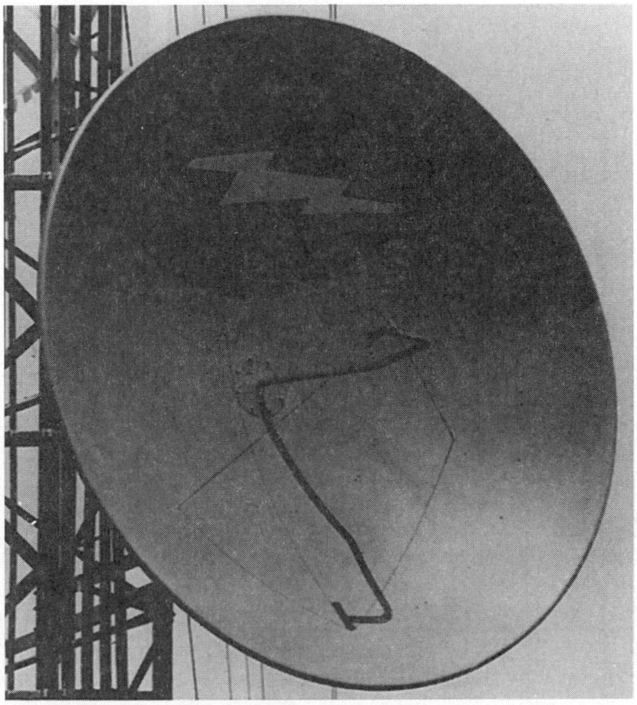

Figure 18.40
Parabolic antenna
Courtesy Andrew Corporation.

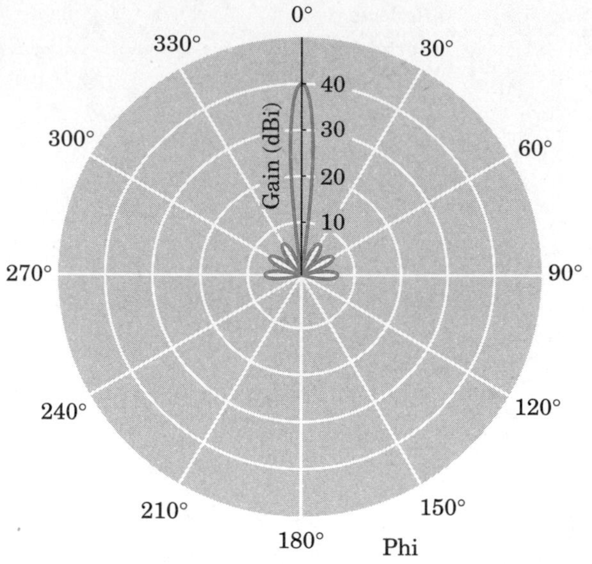

Figure 18.41 Polar pattern for typical parabolic antenna

course, the antenna is reciprocal: radiation entering the dish along its axis will be focused by the reflector.

Ideally, the antenna at the feedpoint should illuminate the entire surface of the dish with the same intensity of radiation and should not spill any radiation off the edges of the dish or in other directions. If that were the case, the gain and beamwidth of the antenna could easily be calculated. The equation for beamwidth is

$$\theta = \frac{70\lambda}{D} \tag{18.13}$$

where

θ = beamwidth in degrees at the 3 dB points
λ = free-space wavelength
D = diameter of the dish

The radiation pattern for a typical parabolic antenna is shown in Figure 18.41. The width of the beam measured between the first nulls is approximately twice the 3 dB beamwidth.

For gain, the equation is

$$G = \frac{\pi^2 D^2}{\lambda^2} \tag{18.14}$$

where

G = gain as a power ratio (not in decibels)
D = diameter of the dish
λ = free-space wavelength

The gain is reduced by uneven illumination of the antenna, losses, and any radiation spilling off at the edges. To include these effects in gain calculations, it is necessary to include a constant η, which is known as the efficiency of the antenna. This constant can theoretically have a value between zero and one, but is between 0.5 and 0.7 for a typical antenna. When this constant is included, Equation (18.14) becomes

$$G = \frac{\eta \pi^2 D^2}{\lambda^2} \tag{18.15}$$

where

G = gain as a power ratio (not in decibels) with respect to an isotropic radiator
D = diameter of the dish
λ = free-space wavelength
η = efficiency

Example 18.6 A parabolic antenna has a diameter of 3 m, an efficiency of 60%, and operates at a frequency of 4 GHz. Calculate its gain and beamwidth.

Solution The free-space wavelength is

$$\begin{aligned} \lambda &= \frac{c}{f} \\ &= \frac{3 \times 10^8}{4 \times 10^9} \\ &= 0.075 \text{ m} \end{aligned}$$

Substituting this into Equation (18.13), the beamwidth is

$$\begin{aligned} \theta &= \frac{70\lambda}{D} \\ &= \frac{70 \times 0.075}{3} \\ &= 1.75° \end{aligned}$$

The gain is given by Equation (18.15):

$$\begin{aligned} G &= \frac{\eta \pi^2 D^2}{\lambda^2} \\ &= \frac{0.6\pi^2 \times 3^2}{0.075^2} \\ &= 9475 \end{aligned}$$

In decibels (relative to an isotropic radiator), the gain is

$$\begin{aligned} G &= 10 \log 9475 \\ &= 39.8 \text{ dBi} \end{aligned}$$

Any type of antenna can be used with a parabolic reflector. In the microwave portion of the spectrum, where parabolic reflectors are most useful because they can have a practical size, a *horn antenna* provides a simple and efficient method to feed power to the antenna. Because horn antennas are essentially an extension of a waveguide, their operation will be described in the next chapter.

Besides the simple horn feed shown in Figure 18.40, there are several other ways to get power to the parabolic reflector. For example, in the Gregorian feed shown in Figure 18.42(a), a feedhorn in the center of the dish itself radiates to a reflector at the focus of the antenna. This reflects the signal to the main parabolic reflector. By removing the feedhorn from the focus, this system allows any waveguide or electronics associated with the

Figure 18.42
Parabolic antenna variations
Courtesy Andrew Corporation.

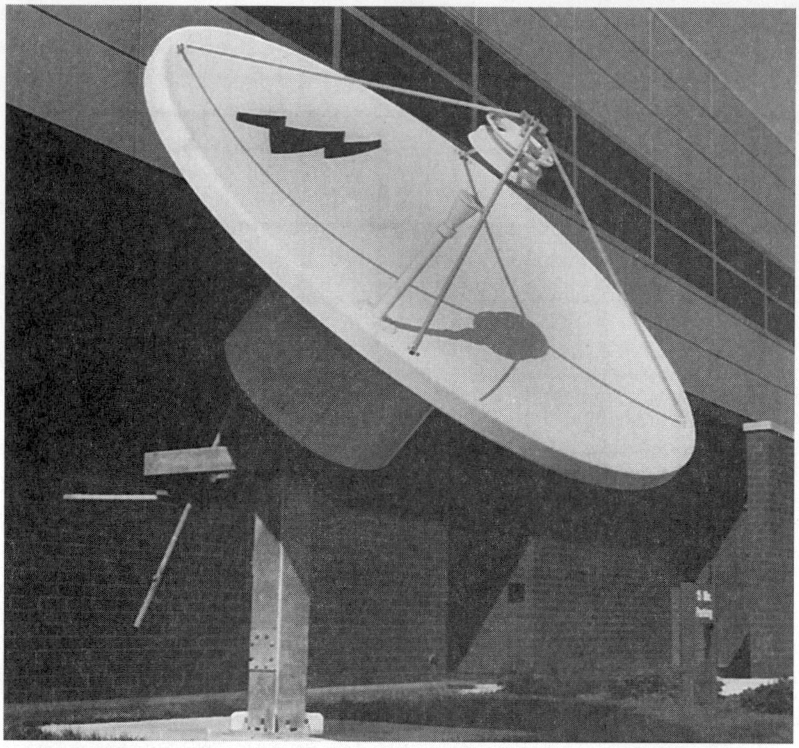

(a) Gregorian feed

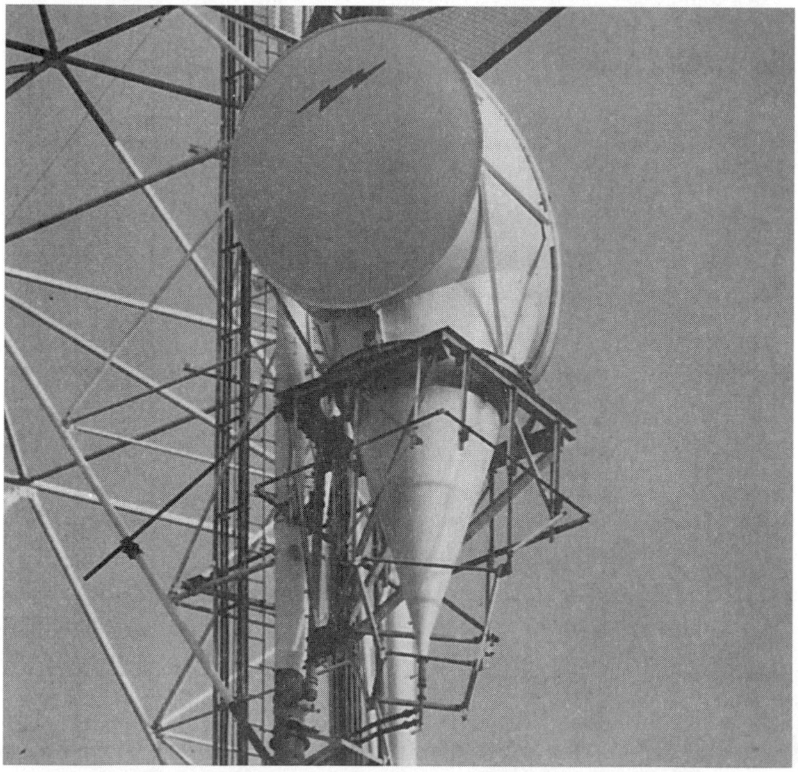

(b) Hog-horn antenna

feedpoint to be placed in a more convenient location. The strange-looking antenna shown in Figure 18.42(b) is a combination of horn and parabolic antennas called a *hog-horn;* it is often used for terrestrial microwave links.

18.8 The Anechoic Chamber

Measurement of antenna gain and directivity requires the ability to set up an antenna in a location that is free from reflecting surfaces and large enough that a receiving antenna can be located in the far field of the antenna under test. This requires a spacing of at least several wavelengths. It is also desirable to electrically shield the whole test setup from the outside world to avoid interference from nearby radio transmitters.

These requirements can be difficult or impossible to meet when tests of large, low-frequency antennas are required. Sometimes an open field called an *antenna range* is used. For higher frequencies, however, it is possible to build an enclosed space called an **anechoic chamber** that will meet all the requirements listed above.

The required size of an anechoic chamber is dictated by the physical size of the antennas to be tested and the frequency of operation. The size of the chamber can be reduced by scaling down the antenna and scaling up the frequency, as long as the effects of ground do not have to be considered. Two representative samples of anechoic chambers are shown in Figure 18.43. Figure 18.43(a) shows a large commercial installation, while the one in Figure 18.43(b) was built as a student project at Niagara College in Welland, Ontario. Though small, the student-built chamber gives quite satisfactory results, at least for demonstration purposes, at frequencies of 1 GHz and above.

An anechoic chamber must be lined with a material that absorbs radio waves. Figure 18.44 shows the material used: a foam plastic that is impregnated with carbon. It absorbs any radio waves originating within the chamber. If these waves were reflected from the walls, they could cause false readings. The absorbent material has the additional advantage of preventing signals originating outside the chamber from reaching the receiving antenna inside.

Inside the chamber, there must be provisions for mounting and rotating the antenna under test and connecting it to a signal source, which can be a laboratory RF generator. There must also be a receiving antenna, which can be a simple arrangement such as a dipole connected to a receiver, and a readout device, such as a chart recorder.

It is easy to obtain a plot of relative gain by simply rotating the antenna under test while observing the output from the receiving antenna. Actual measurements of gain with respect to an isotropic antenna or a dipole are a little more difficult. Of course, an isotropic antenna is impossible to build, but a half-wave dipole is quite simple. The signal strength from the dipole can be measured and compared with that from the test antenna. The only difficulty is to make sure that the dipole and test antennas have the same impedance (using matching devices, if necessary) so that each will absorb the same amount of power from the generator. If a matching network is used, the loss in the network must also be taken into account.

Figure 18.45 shows a Yagi antenna used for tests, and Figure 18.46 shows a plot of its output made in the anechoic chamber in Figure 18.43(b). Notice that this plot is not on polar graph paper because it was made with a conventional x–y plotter. The x axis represents the antenna position in degrees, and the y axis shows output power on a logarithmic scale. This plot can easily be transferred to polar graph paper if desired.

A method similar to the one just described can be used to measure the performance of an antenna system in real-world conditions. The antenna under test is used as a transmitting antenna and driven with a source of known power output. A device called a field-strength meter is then used to measure the field strength in a number of directions at a distance from the antenna that is great enough to ensure that the far-field performance is being measured. A field-strength meter is basically a calibrated antenna-receiver combina-

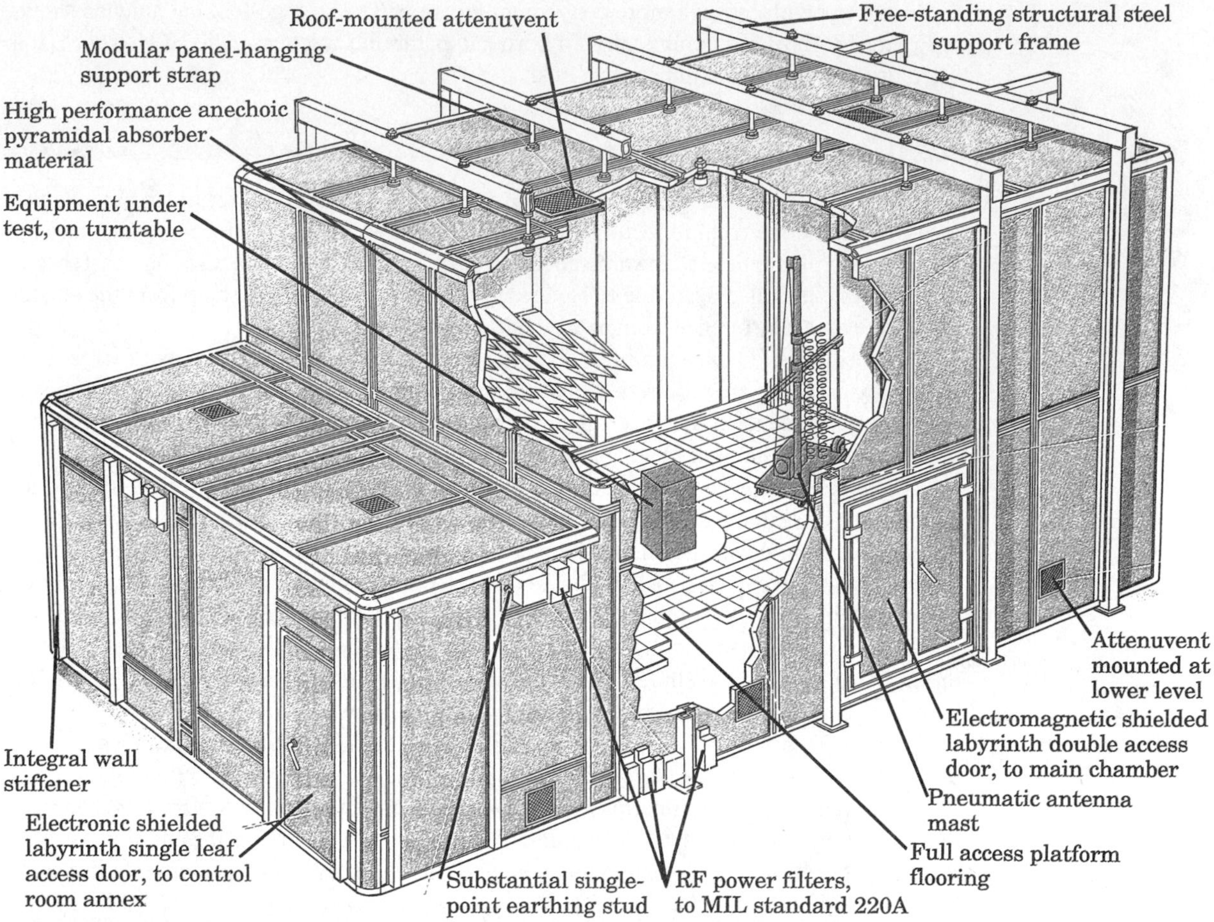

(a) Commercial

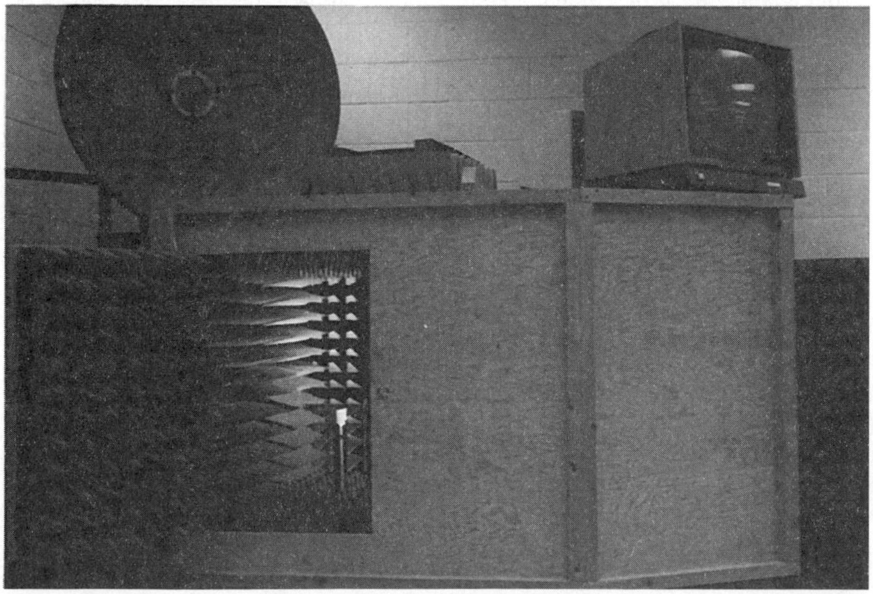

(b) Student-Built

Figure 18.43

Anechoic chambers

(a) Courtesy Rainford division of Devtek Corporation.

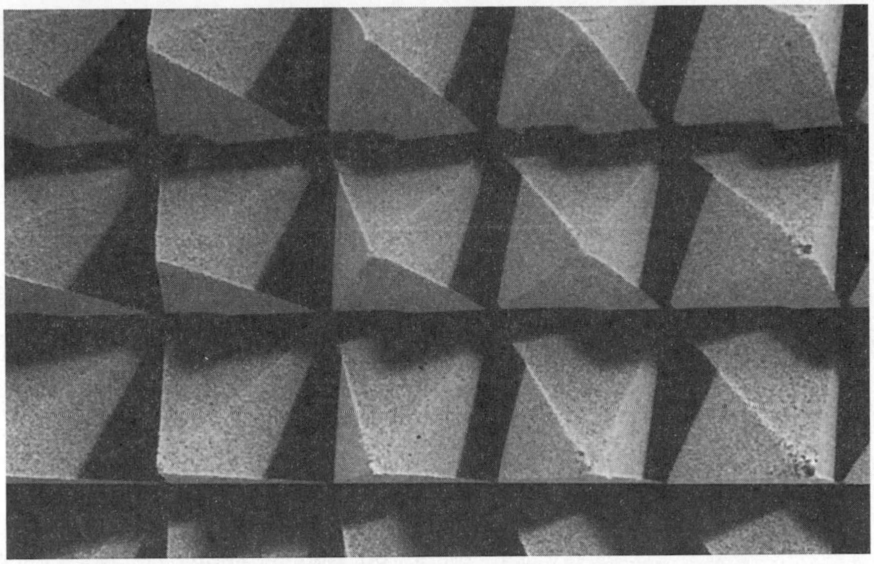

Figure 18.44
Radiation-absorbing material

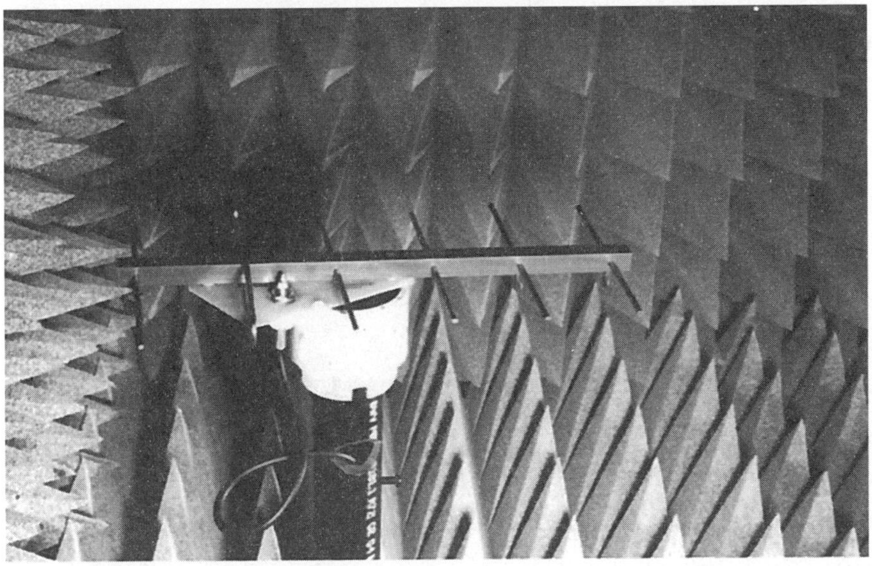

Figure 18.45
Yagi test antenna

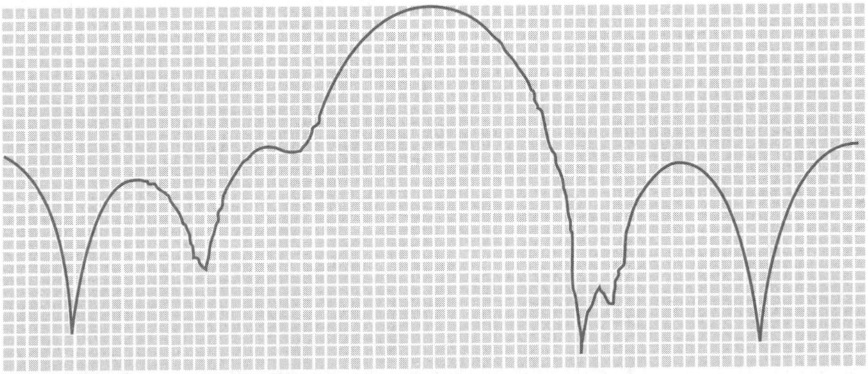

Figure 18.46
Plot from antenna in Figure 18.45
Scales: vertical, 3 dB/division; horizontal, 10°/division

tion. This method is especially useful at lower frequencies, where the wavelength is large and the effect of ground is important. It is also the only practical way to measure the performance of an actual installation, where the effect of obstacles (such as buildings) on field strength may be important.

Summary

Here are the main points to remember from this chapter.

1. Antennas are reciprocal passive devices that couple electrical energy between transmission lines and free space.
2. The isotropic radiator is convenient for calculations because it emits all the energy supplied to it equally in all directions. Antenna gain is usually calculated with reference to an isotropic radiator.
3. Gain is obtained in an antenna when more power is emitted in certain directions than in others.
4. The beamwidth of a directional antenna is the angle between the half-power points of the main lobe.
5. The half-wave dipole is a simple practical antenna with a bidirectional pattern, a small gain over an isotropic radiator, and a feedpoint impedance of about 70 Ω. The dipole is sometimes used instead of an isotropic antenna as a reference for antenna gain.
6. The radiation resistance of an antenna is the equivalent resistance that appears at its feedpoint due to the radiation of energy into space.
7. The polarization of most simple antennas is the same as the axis of the antenna.
8. Antennas must often be matched to the feedline in order to achieve efficient transfer of power. Means for doing this include lumped-constant matching networks, transmission line sections, and loading coils.
9. Simple antennas can be used as elements of an array in order to obtain specified values of gain and directivity.
10. Plane and parabolic reflectors can be used to increase the gain and directivity of antennas.

Important Equations

$$\lambda = \frac{c}{f} \tag{18.1}$$

$$L = \frac{142.5}{f} \tag{18.3}$$

$$\eta = \frac{R_r}{R_T} \tag{18.5}$$

$$A_{eff} = \frac{\lambda^2 G_R}{4\pi} \tag{18.6}$$

$$P_x = \eta P_T \tag{18.7}$$

$$\text{EIRP} = P_T G_T \tag{18.8}$$

$$\tau = \frac{L_1}{L_2} = \frac{L_2}{L_3} = \frac{L_3}{L_4} = \cdots \tag{18.10}$$

$$\tau = \frac{D_1}{D_2} = \frac{D_2}{D_3} = \frac{D_3}{D_4} = \cdots \tag{18.11}$$

$$\frac{L_1}{2D_1} = \tan\frac{\alpha}{2} \tag{18.12}$$

$$\theta = \frac{70\lambda}{D} \tag{18.13}$$

$$G = \frac{\eta\pi^2 D^2}{\lambda^2} \tag{18.15}$$

Glossary

active antenna a receiving antenna with a built-in preamplifier

anechoic chamber an enclosure lined with material that absorbs electromagnetic radiation

antenna a device to radiate or receive electromagnetic radiation at radio frequencies

array an antenna system composed of two or more simpler antenna elements

beamwidth the angle between the points on the major lobe of an antenna at which the radiated power density is one-half its maximum value

dBd a measure of antenna gain: decibels with respect to a lossless half-wave dipole

dBi a measure of antenna gain: decibels with respect to an ideal isotropic radiator

dipole any antenna consisting of a single conductor with zero current only at its two ends

directivity the ratio of the maximum to the average radiation intensity for an antenna

effective area for a receiving antenna, the ratio of the available output power to the power density of the received wave

effective isotropic radiated power (EIRP) the product of the power supplied to a transmitting antenna and the gain of the antenna with respect to an isotropic radiator

effective radiated power (ERP) the product of the power supplied to a transmitting antenna and the gain of the antenna with respect to a lossless half-wave dipole

element in an antenna array, an individual conductor or group of conductors

far-field region a distance far enough from an antenna that local inductive and capacitive effects are insignificant

front-to-back ratio the ratio between the radiation intensity in an antenna's direction of maximum radiation and the intensity at an angle of 180° to this direction

ground plane an artificial ground consisting of a conducting surface or an equivalent (such as wire mesh or a group of wires) at the base of a vertical antenna

helix a spiral

isotropic radiator a hypothetical antenna that would radiate all the energy supplied to it, with equal intensity in all directions

loading the process of increasing the electrical length of an antenna by the addition of inductance or capacitance

lobe the portion of an antenna pattern between two nulls

main lobe the lobe in the direction of maximum radiation

minor lobe a lobe with less intensity than the main lobe

monopole an antenna with a current null at one end and a maximum at the other, with no other nulls in between

near-field region the region close to an antenna, where local inductive and capacitive effects predominate

phi (ϕ) in an antenna pattern, the Greek letter phi denotes the angle in the horizontal plane, from the x axis toward the y axis

polarization the direction of the electric field vector of an electromagnetic wave

radial in a monopole antenna, a wire extending along the surface of the ground or just below it, away from the antenna (a set of radials is used to improve the effective conductivity of the ground)

radiation resistance equivalent resistance at the feedpoint corresponding to the radiation of energy by an antenna

sidelobe a minor lobe at an angle of approximately 90° to the main lobe

theta (θ) in an antenna pattern, the Greek letter theta refers to the angle from the horizontal (x–y) plane toward the zenith, represented by the z axis

zenith the direction straight up from the horizontal plane

Questions

1. Explain how an antenna can have a property called *gain*, even though it is a passive device.
2. Sketch the radiation patterns of an isotropic antenna and of a half-wave dipole in free space, and explain why a dipole has gain over an isotropic radiator.
3. What is the significance of radiation resistance? How would reducing the radiation resistance affect an antenna's efficiency, all other things being equal?
4. Suggest two reasons for using a directional antenna.
5. Distinguish between the near and far fields of an antenna. Why is it necessary to use the far field for all antenna measurements?
6. Draw three types of impedance-matching circuit that can be used to match an antenna to a transmission line.
7. Why is the effect of the ground significant for MF and HF antennas but usually insignificant for UHF and microwave antennas?
8. Monopole antennas have better performance if they are installed on damp ground such as swampland. Sometimes they are even installed in shallow water just off the shore of a lake or other body of water. Explain why.
9. How can the effect of damp ground be simulated (if it is not available,) for a monopole antenna?
10. How does a ground plane differ from an actual ground connection? Why is it used?
11. How can an antenna's electrical length be increased without increasing its physical length?
12. Why are small loop antennas seldom used for transmitting?
13. What advantages does a five-eighths wavelength antenna have over a quarter-wave antenna for mobile use? Does it have any disadvantages?
14. Name two types of antennas that are noted for having particularly wide bandwidth. Which of these is directional?
15. Name one type of antenna that produces circularly polarized waves. Under what circumstances is circular polarization desirable?
16. Distinguish between end-fire and broadside arrays. Name and sketch one example of each.
17. Distinguish between parasitic and phased arrays. Name and sketch one example of each. What is another term for a phased array?
18. What is the purpose of using a plane reflector with an antenna array?
19. What is meant by a collinear antenna? Sketch such an antenna and its radiation pattern, and classify it as broadside or end-fire, phased or parasitic.
20. Describe how an anechoic chamber could be used to measure the front-to-back ratio for an antenna.

Problems

SECTION 18.2

21. Calculate the length of a practical half-wave dipole for a frequency of 15 MHz.
22. An indoor television "rabbit ears" antenna is basically a dipole. To what length should each "ear" be adjusted
    (a) for channel 2
    (b) for channel 13
23. A transmitter with a power output of 100 watts is connected to a dipole antenna with a radiation resistance of 70 Ω and an ohmic resistance of 2 Ω.
    (a) How much power is radiated into space?
    (b) What happens to the rest of the power?
24. Calculate the efficiency of a dipole with a radiation resistance of 68 Ω and a total feedpoint resistance of 75 Ω.

SECTION 18.3

25. Given that a half-wave dipole has a gain of 2.14 dBi, calculate the electric field strength at a distance of 10 km in free space in the direction of maximum radiation from a half-wave dipole that is fed, by means of lossless, matched line, by a 15 W transmitter.
26. A transmitter with a power of 100 W is used with a dipole antenna operating at 90% efficiency. Calculate the power density at a distance of 20 km in the direction in which the antenna has maximum gain. Ignore transmission line losses.
27. A half-wave dipole is mounted horizontally one wavelength above perfectly conducting ground. Will the radiation straight up toward the zenith be increased or decreased by the effect of ground, and by how much?
28. Sketch the radiation pattern, in the vertical plane, for a half-wave dipole at each of the following heights above perfectly conducting ground:
    (a) one-quarter wavelength
    (b) one-half wavelength
    (c) three-quarters wavelength
    (d) one wavelength

29. Refer to the plot in Figure 18.47, and find the gain and beamwidth for the antenna shown.

Figure 18.47

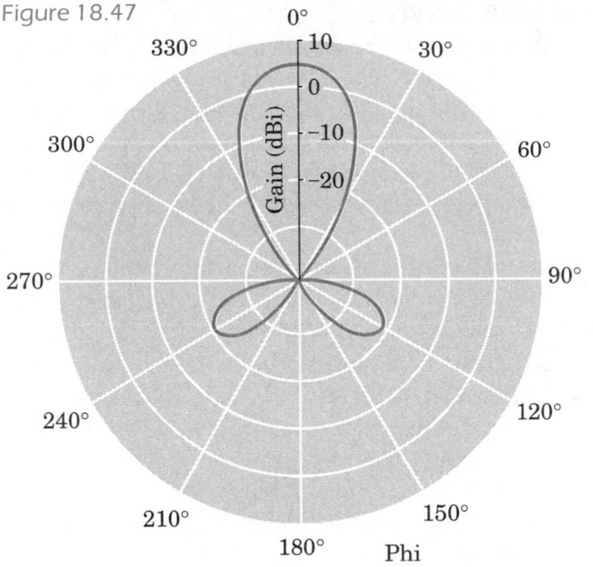

30. Calculate the EIRP in dBW for a 25 W transmitter operating into a dipole with 90% efficiency.

SECTION 18.4

31. Calculate the length of a quarter-wave monopole antenna for a frequency of 1000 kHz.
32. Calculate the optimum length of an automobile FM broadcast antenna, for operation at 100 MHz.
33. The loop antenna shown in Figure 18.48 produces its maximum output when oriented in the direction shown for each of two locations. Locate the transmitting station.

Figure 18.48

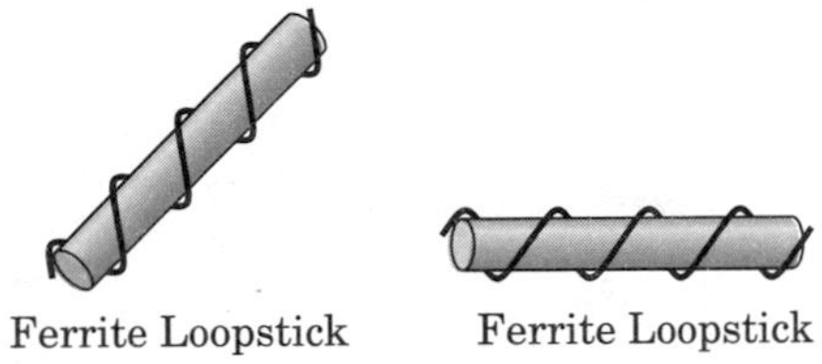
Ferrite Loopstick
Ferrite Loopstick

34. Draw a dimensioned sketch of a discone antenna that will cover the VHF range from 30 to 300 MHz.
35. Calculate the length of a five-eighths wavelength antenna for a cellular phone operating at about 900 MHz
36. An FM broadcasting antenna has the radiation pattern of a vertical dipole but divides its radiation equally between horizontal and vertical polarizations. What is its gain with respect to an isotropic antenna when its signal is received by a vertical antenna at the same height as the transmitting antenna?

SECTION 18.5

37. (a) How could the antenna in Problem 32 be made to work on the 27 MHz CB band?
    (b) Would the modified antenna in part (a) be as efficient as a full-size quarter-wave CB antenna? Explain your answer.

SECTION 18.6

38. Characterize each of the arrays sketched in Figure 18.49 as phased or parasitic and as broadside or end-fire. Indicate the direction(s) of maximum output for each antenna.
39. Design a log-periodic antenna to cover the low-VHF television band, from 54 to 88 MHz. Use $\alpha = 30°$, $\tau = 0.7$, $L_1 = 1.5$ m.
40. A monopole array consists of four towers spaced at one-half-wavelength intervals along a north–south line. Sketch the radiation pattern when the towers are fed:
    (a) in phase
    (b) with adjacent towers 180° out of phase
41. Two horizontal Yagis are stacked, with one of them one-half wavelength above the other and fed in phase. What will be the effect of the stacking on the radiation
    (a) in the horizontal plane
    (b) in the vertical plane

Figure 18.49

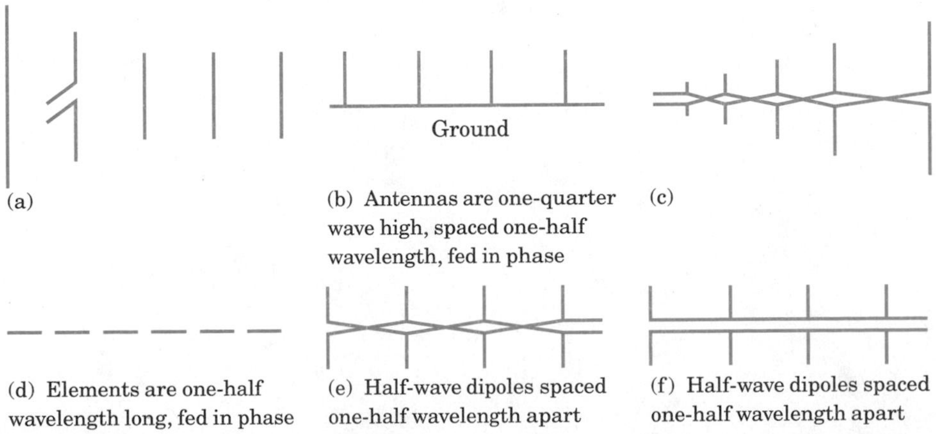
Ground
(a)
(b) Antennas are one-quarter wave high, spaced one-half wavelength, fed in phase
(c)
(d) Elements are one-half wavelength long, fed in phase
(e) Half-wave dipoles spaced one-half wavelength apart
(f) Half-wave dipoles spaced one-half wavelength apart

42. Draw a dimensioned sketch for a collinear antenna designed for 900 MHz. The antenna should have four radiating elements. Indicate on the sketch the directions in which radiation will be maximum.
43. A Yagi antenna for a frequency of 200 MHz has three directors of length 0.4 wavelength, one reflector of length 0.6 wavelength, and a half-wavelength folded-dipole driven element. The spacing is 0.25 wavelength between the driven element and the reflector and 0.2 wavelength from the driven element to the first director and between directors. Sketch the antenna, showing all dimensions.

SECTION 18.7

44. Calculate the minimum diameter for a parabolic antenna with a beamwidth of 2° at a frequency of:
    (a) 4 GHz
    (b) 12 GHz
45. Calculate the gain of each of the antennas in the previous problem, assuming an efficiency of 65%.
46. Calculate the effective area of a 3 m dish with an efficiency of 0.7, at 3 GHz and at 12 GHz. Explain your results.

SECTION 18.8

47. An anechoic chamber is 5 m in length. How far is this in wavelengths at 1 GHz?
48. Redraw the x–y plot in Figure 18.46 on polar graph paper. Then calculate:
    (a) the front-to-back ratio
    (b) the beamwidth
    (c) the strength of the minor lobes compared to the main lobe

COMPREHENSIVE

49. A Yagi antenna has a gain of 10 dBi and a front-to-back ratio of 15 dB. It is located 15 km from a transmitter with an ERP of 100 kW at a frequency of 10 MHz. The antenna is connected to a receiver via a matched feedline with a loss of 2 dB. Calculate the signal power supplied to the receiver if the antenna is:
    (a) pointed directly toward the transmitting antenna
    (b) pointed directly away from the transmitting antenna
50. A lossless half-wave dipole is located in a region with a field strength of 150 μV/m. Calculate the power that this antenna can deliver to a receiver if the frequency is:
    (a) 10 MHz (b) 500 MHz
51. Calculate the gain, with respect to an isotropic antenna, of a half-wave dipole with an efficiency of 90% that is placed one-quarter wavelength from a plane reflector that reflects 100% of the signal striking it.
52. An advertisement describes a "dish" antenna for the VHF television band that sits on top of a television receiver. The picture in the advertisement shows a dish-shaped antenna with a diameter of about 20 cm. Does this really function as a parabolic dish antenna at this frequency? Explain your answer.

19 Microwave Devices

Objectives After studying this chapter, you should be able to:

1. Explain the factors that cause microwave equipment to differ in construction and operation from equipment used at lower frequencies
2. Describe and sketch some of the modes of operation of waveguides
3. Name the dominant mode in a rectangular waveguide and describe its characteristics
4. Calculate the cutoff frequency, phase and group velocity, guide wavelength, and characteristic impedance for the dominant mode in a rectangular waveguide
5. Describe several methods for coupling power into and out of waveguides
6. Explain the operation of passive microwave components, including waveguide bends and tees, resonant cavities, attenuators, and loads
7. Describe and explain the applications of ferrites in microwave systems, particularly their use in circulators and isolators
8. Explain why conventional solid-state diodes and transistors lose efficiency at microwave frequencies, describe the theory of operation of several alternative devices, and give examples of their application
9. Explain why conventional vacuum tubes lose efficiency at microwave frequencies, describe the theory of operation of several alternative devices, and give examples of their application; discuss the relative merits of vacuum-tube and solid-state devices for various applications
10. Describe the theory and operation of specialized microwave antennas
11. Explain the operation of pulse radar systems, and perform signal-strength and range calculations
12. Explain the operation of Doppler radar systems and perform velocity calculations

19.1 Introduction

There is no sharp distinction between **microwaves** and other radio-frequency signals. Conventionally, the lower boundary for microwave frequencies is set at 1 GHz. We will examine microwaves separately because many conventional techniques for generating, amplifying, and transmitting signals become less effective as frequency increases, while other techniques that are impractical at lower frequencies become more useful.

As frequency increases, many of the simplifying assumptions that work at lower frequencies become less accurate. Here are a few examples.

At low frequencies, the inductance and capacitance of component leads can be ignored. At microwave frequencies, even short connecting leads have significant capacitive and inductive reactance, so the physical design of components must change.

At frequencies up to about the UHF range, the time taken for charge carriers to move through such devices as diodes, transistors, and vacuum tubes can usually be ignored. As the period of signals becomes shorter, however, this *transit time* becomes a significant fraction of a complete cycle. Some conventional components have been redesigned to minimize transit time, and other active devices have been specially designed to incorporate transit-time effects in their operation.

Because of the short wavelengths of microwave signals, antennas of reasonable physical size can have very high gain, and parabolic reflectors become practical.

At microwave frequencies, the losses in conventional transmission lines are quite large. Waveguides, which will be introduced in the next section, have much lower losses but are impractical at lower frequencies due to their large size.

19.2 Waveguides

As discussed in Chapter 16, both dielectric and conductor losses in conventional transmission lines increase with frequency. **Waveguides** provide an alternative at microwave frequencies. A waveguide is essentially a pipe through which an electromagnetic wave travels. As it travels along the guide, it reflects from the walls. Figure 19.1 shows the general idea of waveguides and waveguide propagation. Rectangular waveguides of brass or aluminum, sometimes silver-plated on the inside, are most common, but elliptical and circular cross sections are also used.

It is possible to build a waveguide for any frequency, but waveguides operate essentially as high-pass filters, that is, for a given waveguide cross section, there is a cutoff frequency below which waves will not propagate. At frequencies below the gigahertz range, waveguides are too large to be practical for most applications.

As the electric and magnetic fields are completely contained within the guide, waveguides have no radiation loss. Dielectric losses are very small, since the dielectric is usually air. There are some losses in the conductive walls of the waveguide, but because of the large surface area of the walls, these losses are much smaller than the losses in coaxial or open-wire line.

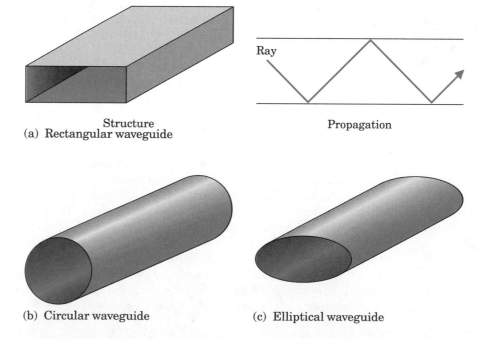

Figure 19.1
Waveguides

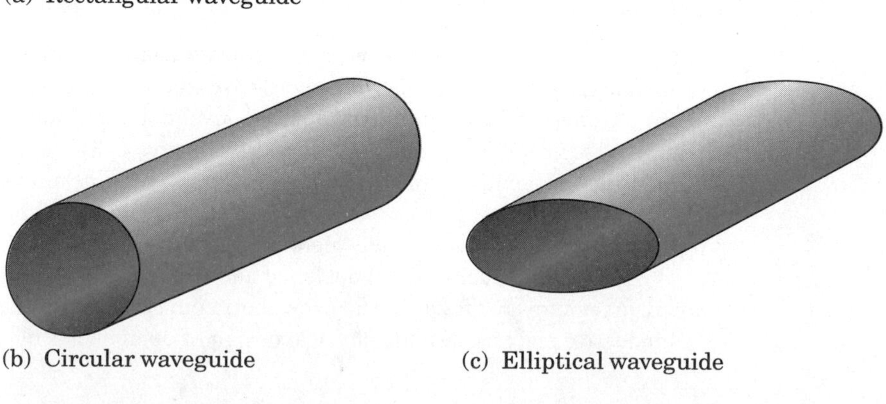

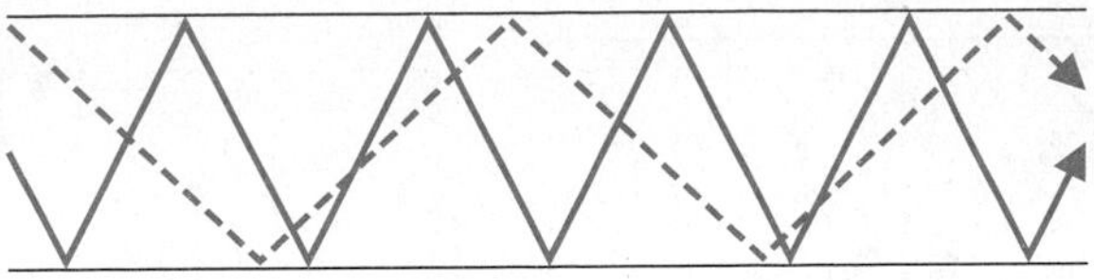

Figure 19.2 Multimode propagation

19.2.1 Modes and Cutoff Frequency

There are a number of ways (called **modes**) in which electrical energy can propagate along a waveguide. All of these modes must satisfy certain boundary conditions. For instance, there cannot be any electric field along the wall of the waveguide (assuming an ideal conductor for the guide). If there were such a field, there would have to be a voltage gradient along the wall, and that is impossible since there cannot be any voltage across a short circuit.

It may help you understand modes if you think of a wave moving through the guide as if it were a ray of light (Figure 19.2 shows the idea). For each different mode, the ray strikes the walls of the waveguide at a different angle. As the angle a ray makes with the wall of the guide becomes larger, the distance the ray must travel to reach the far end of the guide becomes greater. Though propagation in the guide is at the speed of light, the greater distance traveled causes the effective velocity down the guide to be reduced.

It is desirable to have only one mode propagating in a waveguide. To see the effect of *multimode propagation* (more than one mode propagating at a time), consider a brief pulse of microwave energy applied to one end of a waveguide. The pulse will arrive at the far end at several different times, one for each mode. Thus a brief pulse will be spread out over time, becoming longer. If another pulse follows close behind, there may be interference between the two.

The effect just described is called **dispersion**. Dispersion limits the usefulness of waveguides with pulsed signals and other types of modulation. Because of dispersion, it is undesirable to have more than one mode propagating.

Each mode has a cutoff frequency below which it will not propagate. Single-mode propagation can be achieved by using only the mode with the lowest cutoff frequency. This mode is called the *dominant mode*. The waveguide is used at frequencies between its cutoff frequency and that of the mode with the next lowest cutoff frequency.

Modes are designated as transverse electric (TE) or transverse magnetic (TM) according to the pattern of electric and magnetic fields within the waveguide. Recall from Chapter 17 that electromagnetic waves in free space are known as transverse electromagnetic (TEM) waves because both the electric and the magnetic fields are perpendicular to the direction of travel. When these waves travel diagonally along a waveguide, reflecting from wall to wall, only one component—either the electric or the magnetic field—can remain transverse to the direction of travel. The term TE means that there is no component of the electric field along the length of the guide.

Figure 19.3 shows several examples of TE modes in a rectangular waveguide. The electric field strength is represented by arrows, with the lengths of the arrows proportional to the field strength. Note that in all cases the field strength is zero along the walls of the guide. It is possible to have an electric field at the guide surface, but only if the field is perpendicular to the surface.

The field strength varies sinusoidally across the guide cross section. The first number following the TE designation represents the number of half-cycles of the wave along the long dimension (a) of the rectangular guide, and the second represents the number of variations along the short dimension (b). Of course, a and b refer to inside dimensions.

In a rectangular waveguide, TE_{10} is the dominant mode (that is, the mode with the lowest cutoff frequency). In a typical rectangular waveguide with $a = 2b$, the TE_{01} and

Figure 19.3 TE modes in rectangular waveguide (arrows represent electric-field strength)

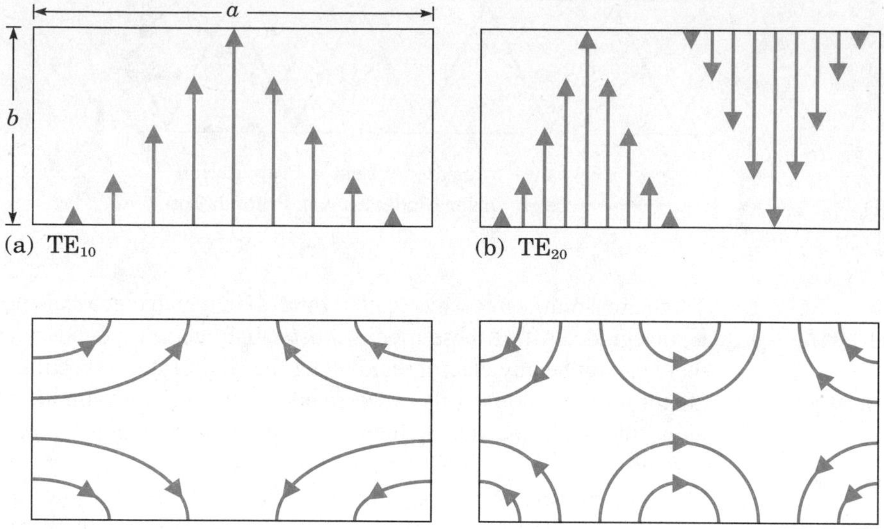

(a) TE_{10} (b) TE_{20} (c) TE_{11} (d) TE_{21}

TE_{20} modes each have a cutoff frequency twice that of the TE_{10} mode, giving an approximate 2:1 frequency range for the waveguide in its dominant mode. In what follows, a rectangular waveguide operating in the TE_{10} mode will be assumed unless otherwise stated.

The cutoff frequency for the TE_{10} mode can easily be found. For this mode to propagate, there must be at least one-half wavelength along the wall. Therefore, at the cutoff frequency,

$$a = \frac{\lambda_c}{2} \tag{19.1}$$

where

a = longer dimension of the waveguide cross section
λ_c = cutoff wavelength in the dielectric material that fills the waveguide (usually air)

Assuming that the waveguide has an air dielectric, the cutoff frequency can be found as follows. From Equation (19.1),

$$a = \frac{\lambda_c}{2}$$

$$\lambda_c = 2a$$

From earlier work we know that for propagation in free space (and as a close approximation, in air dielectric),

$$\lambda = \frac{c}{f}$$

where

λ = free-space wavelength in meters
c = 3×10^8 meters per second
f = frequency in hertz

Therefore, at the cutoff frequency f_c,

$$\lambda_c = \frac{c}{f_c} \tag{19.2}$$

$$2a = \frac{c}{f_c}$$

$$f_c = \frac{c}{2a}$$

Example 19.1

Find the cutoff frequency for the TE_{10} mode in an air-dielectric waveguide with an inside cross section of 2 cm by 4 cm. Over what frequency range is the dominant mode the only one that will propagate?

Solution

The larger dimension, 4 cm, is the one to use in calculating the cutoff frequency. From Equation (19.2), the cutoff frequency is

$$\begin{aligned} f_c &= \frac{c}{2a} \\ &= \frac{3 \times 10^8 \text{ m/s}}{2 \times 4 \times 10^{-2} \text{ m}} \\ &= 3.75 \times 10^9 \text{ Hz} \\ &= 3.75 \text{ GHz} \end{aligned}$$

The dominant mode is the only mode of propagation over a 2:1 frequency range, so the waveguide will be usable to a maximum frequency of $3.75 \times 2 = 7.5$ GHz.

The dominant mode depends on the shape of the waveguide. Figure 19.4 shows some of the modes in a circular guide. Here, the same boundary conditions result in different modes because of the changed geometry. For a circular guide, the dominant mode is the TE_{11} mode, but the TM_{01} mode is also used because it has circular symmetry, which allows its use in rotating joints (necessary with rotating radar antennas). See Figure 19.5 for an example.

19.2.2 Group and Phase Velocity

Assuming an air dielectric, the wave travels inside the waveguide at the speed of light. However, it does not travel straight down the guide but reflects back and forth from the walls. The actual speed at which a signal travels down the guide is called the **group ve-**

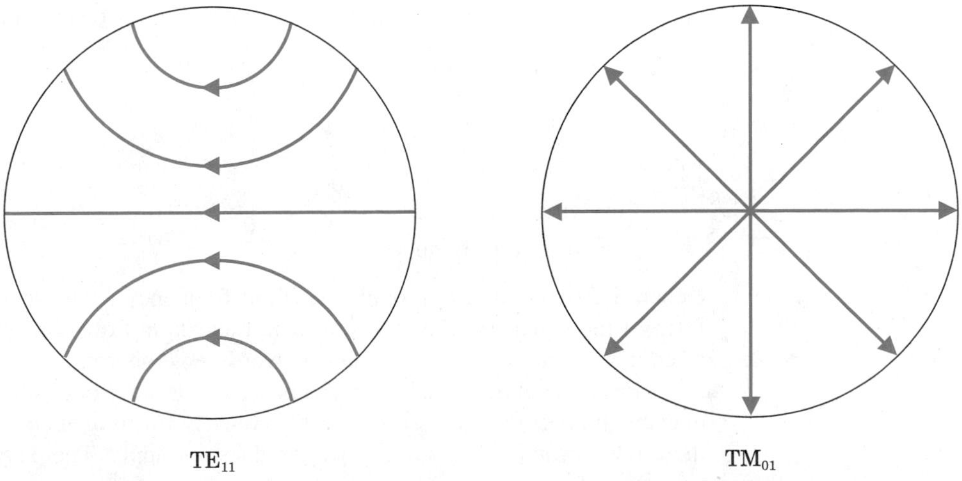

Figure 19.4 Modes in circular waveguide (arrows represent electric field strength)

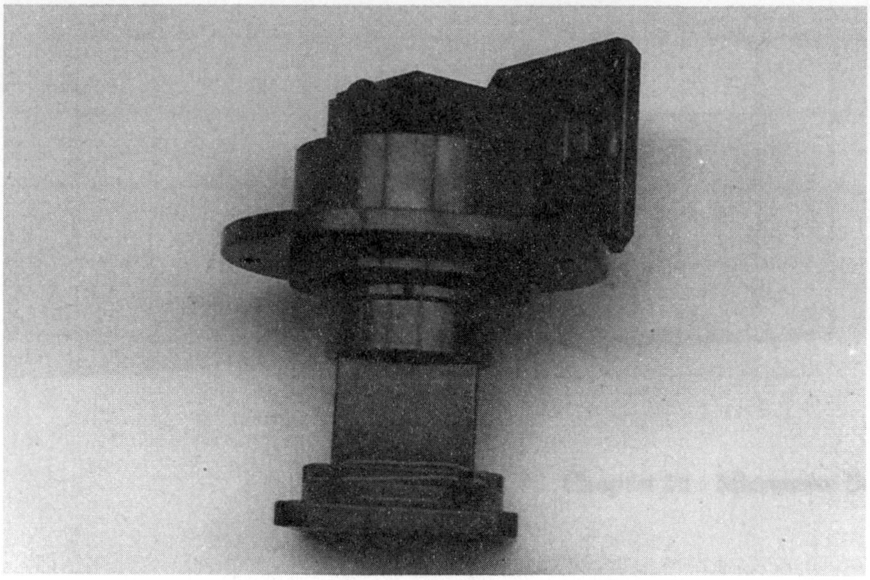

Figure 19.5 Rotating waveguide joint

locity, and it is considerably less than the speed of light. The group velocity in a rectangular waveguide is given by the equation

$$v_g = c\sqrt{1 - \left(\frac{\lambda}{2a}\right)^2} \tag{19.3}$$

where

v_g = group velocity
λ = free-space wavelength
a = larger dimension of the interior cross section

There is an equivalent form of this equation that is more useful in specific situations. Since, from Equation (19.2),

$$f_c = \frac{c}{2a}$$

the wavelength and dimension in Equation (19.3) can be replaced by two frequencies, the operating frequency and the cutoff frequency, giving the equation

$$v_g = c\sqrt{1 - \left(\frac{f_c}{f}\right)^2} \tag{19.4}$$

where

f_c = cutoff frequency
f = operating frequency

Generally, manufacturers provide the cutoff frequency for waveguides, so this is a useful form of the equation. The two equivalent radicals in Equations (19.3) and (19.4) appear often in waveguide calculations and are worth remembering.

From the above equation, it can be seen that the group velocity is a function of frequency and becomes zero at the cutoff frequency. At frequencies below cutoff, of course, there is no propagation, so the equation does not apply. The physical explanation of the variation of group velocity is that the angle the wave makes with the wall of the guide

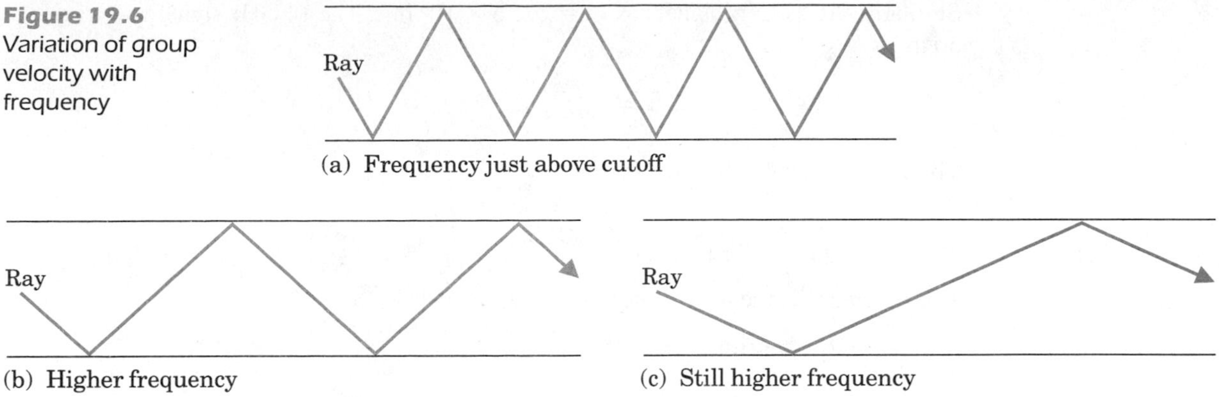

Figure 19.6 Variation of group velocity with frequency

varies with frequency. At frequencies near the cutoff value, the wave moves back and forth across the guide more often while traveling a given distance down the guide than it does at higher frequencies. Figure 19.6 gives a qualitative idea of the effect.

Example 19.2 Find the group velocity for the waveguide in Example 19.1, at a frequency of 5 GHz.

Solution From Example 19.1, the cutoff frequency is 3.75 GHz. Therefore, from Equation (19.4), the group velocity is

$$v_g = c\sqrt{1 - \left(\frac{f_c}{f}\right)^2}$$

$$= (3 \times 10^8 \text{ m/s})\sqrt{1 - \left(\frac{3.75}{5}\right)^2}$$

$$= 1.98 \times 10^8 \text{ m/s}$$

When we first looked at waveguide propagation, we noted that waveguides are generally used with only one mode in order to reduce dispersion (the tendency of signals to spread in time). Since the group velocity varies with frequency, it now appears that some dispersion exists even for single-mode propagation. If two signals of different frequencies start to travel down a guide at the same instant, the higher-frequency signal will arrive at the other end first. This can be a real problem for pulsed and other wideband signals. For instance, the upper sideband of an FM or AM signal travels faster than the lower sideband.

Example 19.3 A waveguide has a cutoff frequency for the dominant mode of 10 GHz. Two signals with frequencies of 12 and 17 GHz propagate down a 50 m length of the guide. Calculate the group velocity for each and the difference in arrival time for the two.

Solution The group velocities can be calculated from Equation (19.4):

$$v_g = c\sqrt{1 - \left(\frac{f_c}{f}\right)^2}$$

For the 12 GHz signal,

$$v_g = (3 \times 10^8 \text{ m/s})\sqrt{1 - \left(\frac{10}{12}\right)^2}$$

$$= 165.8 \times 10^6 \text{ m/s}$$

Similarly, the 17 GHz signal has $v_g = 242.6 \times 10^6$ m/s. The 12 GHz signal will travel the 50 m in:

$$t_1 = \frac{50 \text{ m}}{165.8 \times 10^6 \text{ m/s}} = 301.6 \text{ ns}$$

The 17 Ghz signal will take less time:

$$t_2 = \frac{50 \text{ m}}{242.6 \times 10^6 \text{ m/s}} = 206.1 \text{ ns}$$

The difference in the travel times for the two signals is:

$$t_1 - t_2 = 301.6 \text{ ns} - 206.1 \text{ ns} = 95.5 \text{ ns}$$

It is often necessary to calculate the wavelength of a signal in a waveguide. For instance, it may be required for impedance matching. It might seem that the wavelength along the guide could be found using the group velocity, in much the same way that the velocity factor of a transmission line is used. However, this common-sense approach does not work because what is really important in impedance-matching calculations is the change in phase angle along the line. Figure 19.7 shows how the angle varies along the guide. The guide wavelength shown represents 360° of phase variation. The guide wavelength is always larger than the free-space wavelength. Interestingly, the more slowly the waves in the guide propagate along it, the more quickly the phase angle varies along the guide.

It is possible to define a quantity called **phase velocity** to describe the variation of phase along the wall of the guide. Phase velocity is the rate at which the wave *appears* to move along the wall of the guide, based on the way the phase angle varies along the walls. Surprisingly, the phase velocity in a waveguide is always greater than the speed of light. Of course, the laws of physics prevent anything from actually moving that fast (except in science-fiction stories). A phase velocity greater than the speed of light is possible because phase velocity is not really the velocity of anything. Phase velocity is used when calculating the wavelength in a guide, but the group velocity must be used to determine the length of time it takes for a signal to move from one end to the other.

A similar effect can be seen with water waves at a beach. If the waves approach the shore at an angle, as in Figure 19.8, the crest of the wave will appear to run along the shore

Figure 19.7 Variation of phase angle along a waveguide

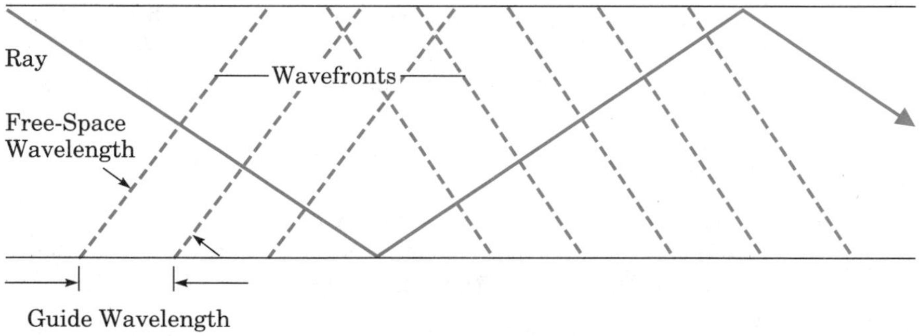

Figure 19.8 Water waves at a shoreline

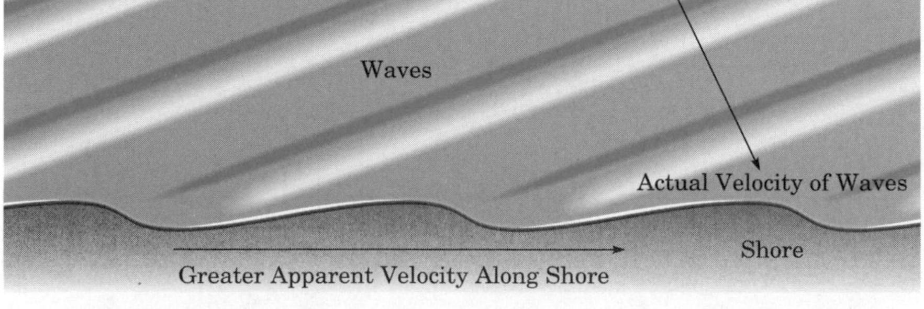

at a faster rate than that at which the waves approach the beach. Once again, there is nothing physically moving at the faster velocity.

The relationship between phase velocity and group velocity is very simple. The speed of light is the geometric mean of the two, that is,

$$v_g v_p = c^2 \tag{19.5}$$

where

v_p = phase velocity

It is easy to modify the equations given earlier for group velocity, so that the phase velocity can be calculated without first finding the group velocity. The derivation is left as an exercise; the results are given below.

$$v_p = \frac{c}{\sqrt{1 - \left(\frac{\lambda}{2a}\right)^2}} \tag{19.6}$$

$$= \frac{c}{\sqrt{1 - \left(\frac{f_c}{f}\right)^2}} \tag{19.7}$$

Example 19.4 Find the phase velocity for the waveguide used in Examples 19.1 and 19.2, at a frequency of 5 GHz.

Solution The answer can be found directly from Equation (19.7):

$$v_p = \frac{c}{\sqrt{1 - \left(\frac{f_c}{f}\right)^2}}$$

$$= \frac{3 \times 10^8 \text{ m/s}}{\sqrt{1 - \left(\frac{3.75}{5}\right)^2}}$$

$$= 4.54 \times 10^8 \text{ m/s}$$

Alternatively, it can be found from Equation (19.5), since the group velocity is already known:

$$v_g v_p = c^2$$

$$v_p = \frac{c^2}{v_g}$$

$$= \frac{(3 \times 10^8 \text{ m/s})^2}{1.98 \times 10^8 \text{ m/s}}$$

$$= 4.54 \times 10^8 \text{ m/s}$$

19.2.3 Impedance and Impedance Matching

Like any transmission line, a waveguide has a characteristic impedance. Unlike wire lines, however, the waveguide impedance is a function of frequency. As has been pointed out, the impedance of free space is 377 Ω. It might be expected that the impedance of a waveguide with an air dielectric would have some relationship to this value, and that is true. The actual impedance Z_0 of a waveguide is given by

$$Z_0 = \frac{377}{\sqrt{1 - \left(\frac{\lambda}{2a}\right)^2}}\ \Omega \tag{19.8}$$

or

$$Z_0 = \frac{377}{\sqrt{1 - \left(\frac{f_c}{f}\right)^2}}\ \Omega \tag{19.9}$$

Example 19.5 Find the characteristic impedance of the waveguide used in the previous examples, at a frequency of 5 GHz.

Solution From Equation (19.9),

$$\begin{aligned} Z_0 &= \frac{377}{\sqrt{1 - \left(\frac{f_c}{f}\right)^2}}\ \Omega \\ &= \frac{377}{\sqrt{1 - \left(\frac{3.75}{5}\right)^2}}\ \Omega \\ &= 570\ \Omega \end{aligned}$$

Smith charts and the computer programs that emulate them can be used with waveguides in much the same way as they can with conventional transmission lines. There are only a few differences. There are two different velocities in a waveguide, and both change with frequency. For calculating wavelength in the guide, the phase velocity must be used. To do this, we can start with the basic wave equation

$$v = f\lambda$$

Here, the velocity in question is the phase velocity v_p, and the wavelength is the guide wavelength λ_g. Rearranging the equation allows λ_g to be found:

$$\lambda_g = \frac{v_p}{f} \tag{19.10}$$

Another way to find the guide wavelength, given the free-space wavelength, is to use the equivalent equations

$$\lambda_g = \frac{\lambda}{\sqrt{1 - \left(\frac{\lambda}{2a}\right)^2}} \tag{19.11}$$

and

$$\lambda_g = \frac{\lambda}{\sqrt{1 - \left(\frac{f_c}{f}\right)^2}} \tag{19.12}$$

where

λ = free-space wavelength
λ_g = guide wavelength

Example 19.6 Find the guide wavelength for the waveguide used in the previous examples.

Solution Since the phase velocity is already known, the easiest way to solve this problem is to use Equation (19.10).

$$\begin{aligned}\lambda_g &= \frac{v_p}{f}\\ &= \frac{4.54 \times 10^8 \text{ m/s}}{5 \times 10^9 \text{ Hz}}\\ &= 0.0908 \text{ m}\\ &= 9.08 \text{ cm}\end{aligned}$$

Techniques for matching impedances using waveguides also differ from those used with conventional transmission lines. Shorted stubs of adjustable length can be used, but a simpler method is to add capacitance or inductance by inserting materials into the guide. In practice, a **tuning screw** (see Figure 19.9) is used. As the screw is inserted further into the guide, the effect is first capacitive, then series-resonant, and finally inductive.

19.2.4 Coupling Power into and out of Waveguides

Now that the propagation of energy down a waveguide has been described, it is time to consider how power can be put into and taken out of the guide. This is not as simple as it is with conventional transmission lines: any attempt to connect wires to the guide merely results in a short circuit.

There are three basic ways to launch a wave down a guide. Figure 19.10 shows all three. The method shown in Figure 19.10(a) uses a probe resembling a quarter-wave monopole antenna. The probe couples to the electric field in the guide, and it therefore should be located at an electric-field maximum. For the TE_{10} mode, it should be in the center of the a (wide) dimension. The probe launches a wave along the guide in both directions. If propagation in only one direction is desired (the usual case), it is only necessary to place the probe a quarter-wavelength from the closed end of the guide. This closed end represents a short circuit, and following the same logic as for transmission lines, there will be an

Figure 19.9 Tuning screw

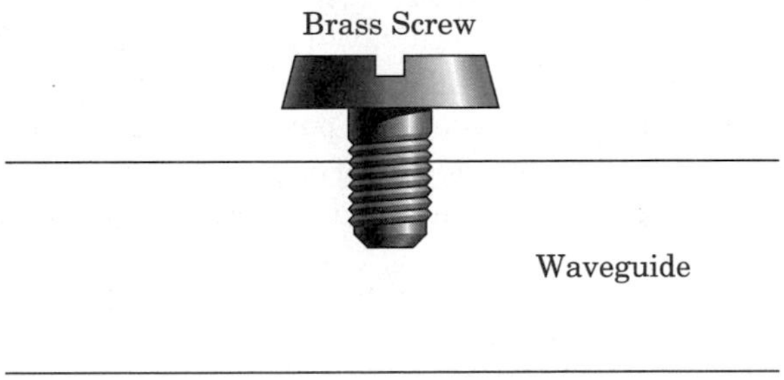

Figure 19.10 Coupling power to a waveguide

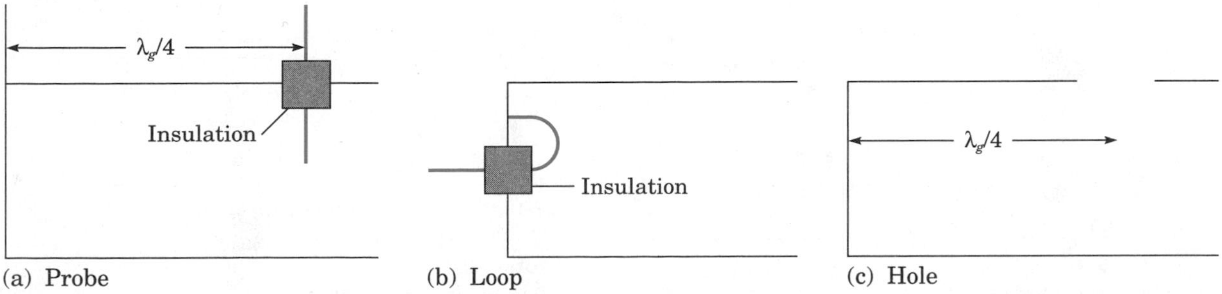

electric-field maximum at a distance of one-quarter wavelength from the end of the guide. This is the waveguide equivalent of the voltage maximum one-quarter wavelength from the shorted end of a transmission line.

The wave emitted by the probe will reflect from the shorted end of the guide. As usual, there will be a 180° phase shift at the reflecting surface, and this, combined with the 180° shift due to the total path length of one-half wavelength, will cause the reflected wave to be in phase with and add to the direct wave in the direction along the length of the guide. The one-quarter wavelength to the end of the guide will have to be calculated using the phase velocity described earlier, of course.

Figure 19.10(b) shows another way to couple power to a guide. A loop is used to couple with the magnetic field in the guide. It is placed in a location of maximum magnetic field, which for the TE_{10} mode occurs close to the end wall of the guide. It may help you understand this if you think of the magnetic field in a waveguide as equivalent to current in a conventional transmission line and think of the electric field as the equivalent of voltage. There is, of course, a current maximum at the short-circuited end of a transmission line.

A third way of coupling energy, shown in Figure 19.10(c), is simply to put a hole in the waveguide, so that electromagnetic energy can propagate into or out of the guide from the region exterior to it.

Like transmission lines and antennas, waveguides are reciprocal devices, that is, the same means can be used to couple power into and out of a waveguide.

Variations on the above methods are also possible. For instance, it is possible to use two probes spaced one-quarter wavelength apart to launch a wave in only one direction. Figure 19.11 shows how this works. The probes are fed 90° out of phase. The waves from the two probes add in phase to the right, because the phase shift between probe 1 and probe 2 along the transmission line is equal to the phase shift along the waveguide and is in the same direction. The signal from probe 1 arrives at probe 2 in phase with the wave generated at probe 2. The signals cancel to the left because the phase shifts add to 180°. Similarly, if this arrangement is used to transfer power from a waveguide to a transmission line, the line will accept only energy traveling along the guide from right to left. Devices of this type are called **directional couplers**.

Holes are often used to make directional couplers. In the example shown in Figure 19.12, a signal moving down the main guide in the direction of the arrow will be coupled to the secondary guide, where it will again move in the direction indicated by the arrow. A wave propagating down the main guide in the opposite direction will also be coupled to the secondary guide, but it will propagate in the opposite direction and will be absorbed by the resistive material in the end of the secondary guide. Though only two holes are shown, in practice there is often a greater number to provide more coupling between the guides.

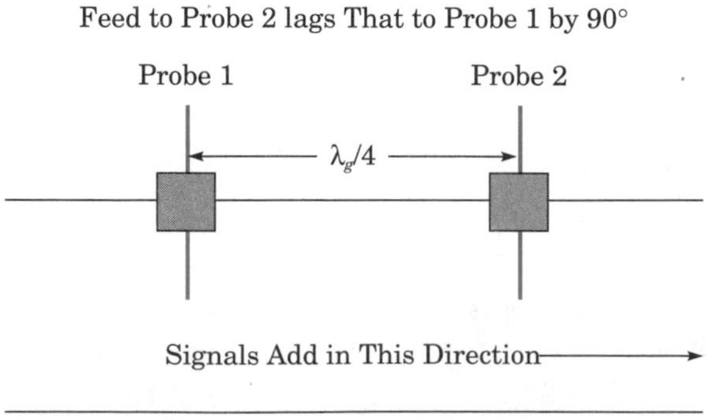

Figure 19.11
Directional coupler using probes

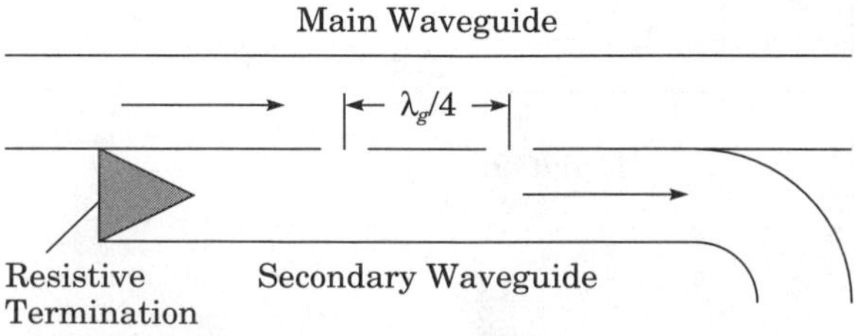

Figure 19.12
Two-hole directional coupler

Directional couplers are characterized by their insertion loss, coupling, and directivity. All three are normally specified in decibels. The *insertion loss* is the amount by which a signal in the main guide is attenuated. The *coupling* specification gives the amount by which the signal in the main guide is greater than that coupled to the secondary waveguide. The *directivity* refers to the ratio between the power coupled to the secondary guide, for signals traveling in the two possible directions along the main guide.

Example 19.7 A signal with a level of 20 dBm enters the main waveguide of a directional coupler in the direction of the arrow. The coupler has an insertion loss of 1 dB, coupling of 20 dB, and directivity of 40 dB. Find the strength of the signal emerging from each guide. Also, find the strength of the signal that would emerge from the secondary guide if the signal in the main guide were propagating in the other direction.

Solution The signal level in the main guide is

$$20 \text{ dBm} - 1 \text{ dB} = 19 \text{ dBm}$$

The signal level in the secondary guide is

$$20 \text{ dBm} - 20 \text{ dB} = 0 \text{ dBm}$$

If the signal direction in the main guide were reversed, the signal level in the secondary guide would be reduced by 40 dB to

$$0 \text{ dBm} - 40 \text{ dB} = -40 \text{ dBm}$$

19.2.5 Striplines and Microstrips

Striplines and **microstrips** are transmission lines that can be constructed on a printed-circuit board. Figure 19.13 shows the construction of these two types of lines in cross section. They are very useful for making interconnections between components. They are included here because only in the UHF and microwave regions of the spectrum must circuit-board traces be considered as transmission lines.

Striplines and microstrips, like waveguides, have a critical frequency. Below the critical frequency, they resemble conventional transmission lines and their impedance depends only on the geometry and the dielectric constant of the line. Above the critical frequency, their characteristic impedance becomes a function of frequency, as it is for a waveguide.

These lines are usually used below the critical frequency. Since the fields are not entirely enclosed within the line, radiation is a problem as the frequency increases.

Figure 19.13 Stripline and microstrip lines: cross sections

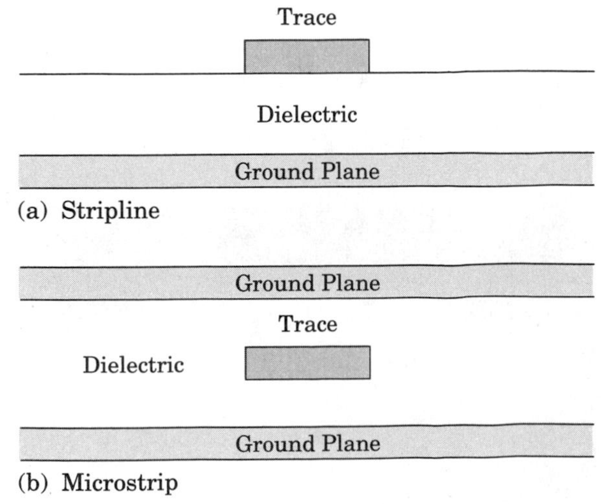

19.3 Passive Components

The use of waveguides requires redesign of some of the ordinary components that are used with feedlines. The lowly tee connector is an example. In addition, other components, such as resonant *cavities*, are too large to be practical at lower frequencies. We will also describe several other varieties of microwave passive components.

19.3.1 Bends and Tees

Anything that changes the shape or size of a waveguide has an effect on the electric and magnetic fields inside. If the disturbance is sufficiently large, there will be a change in the characteristic impedance of the guide. However, as long as any bend or twist is gradual, the effect is minimal. Figure 19.14 shows two examples of bends, which are designated as E-plane bends (Figure 19.14(a)) and H-plane bends (Figure 19.14(b)). Since the rectangular guide shown normally operates in the TE_{10} mode, the electric field lines are perpendicular to the long direction. Therefore, the E-plane bend changes the direction of the electric-field lines, and similar logic holds for the H-plane bend.

Rigid waveguide, with its carefully designed gradual bends, resembles plumbing and is just as tricky to install. Flexible waveguide, as shown in Figure 19.15, is used for awkward installations.

One of the more common components used with ordinary transmission line is the tee, which allows one line to branch into two. A typical tee for coaxial cable is shown in

Figure 19.14
Waveguide bends

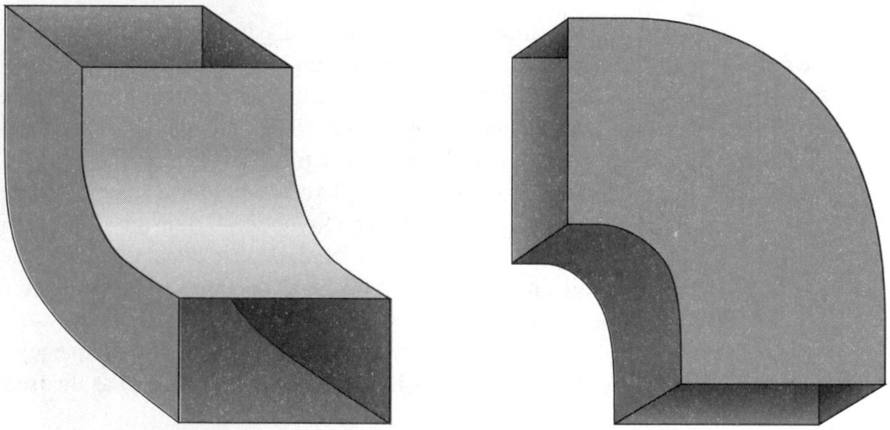

(a) E-plane bend (b) H-plane bend

Figure 19.15
Flexible waveguide

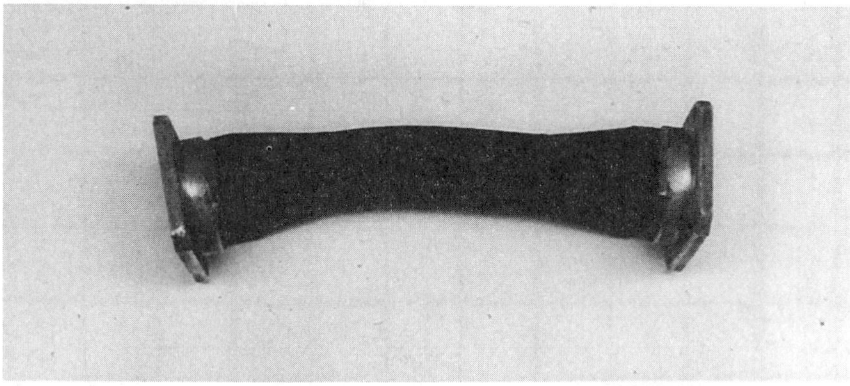

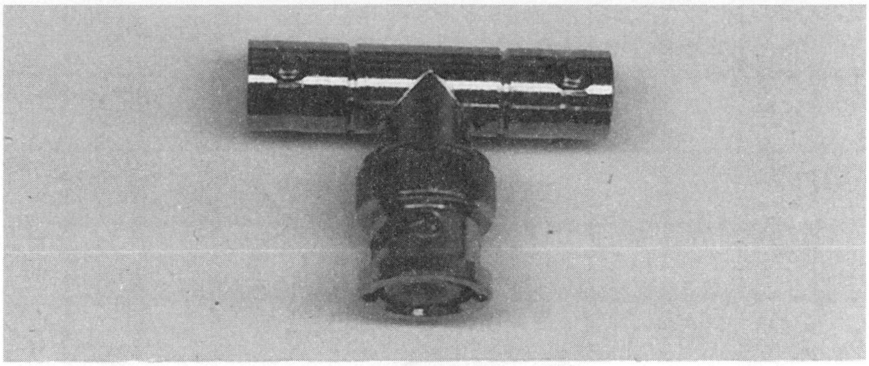

Figure 19.16
Coaxial tee

Figure 19.16. Of course, when using tees, special attention must be paid to impedance matching. Connecting a 50 Ω load to each of the branches of the tee in Figure 19.16, for example, results in a 25 Ω load being applied to the cable.

Tees can also be built for waveguides. Figure 19.17 shows E-plane and H-plane tees, named in the same way as the bends described above. A signal applied to port 1 appears at each of the other ports. For the H-plane tee, the signal will be in phase at the two outputs, while the E-plane tee produces two out-of-phase signals. Sometimes the H-plane tee is referred to as a *shunt tee*, and the E-plane tee is called a *series tee*.

The **hybrid** or **magic tee** shown in Figure 19.18 is a combination of E-plane and H-plane tees. It has some interesting features. In particular, it can provide isolation between signals. An input at port 3 will result in equal and in-phase outputs at ports 1 and 2 but no

Figure 19.17
Waveguide tees

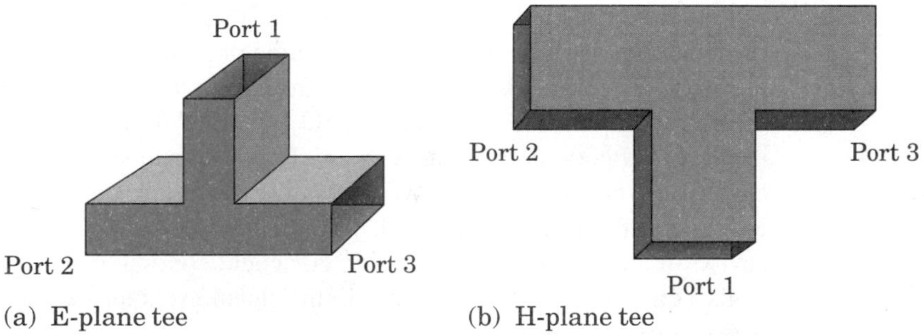

Figure 19.18
Hybrid tee

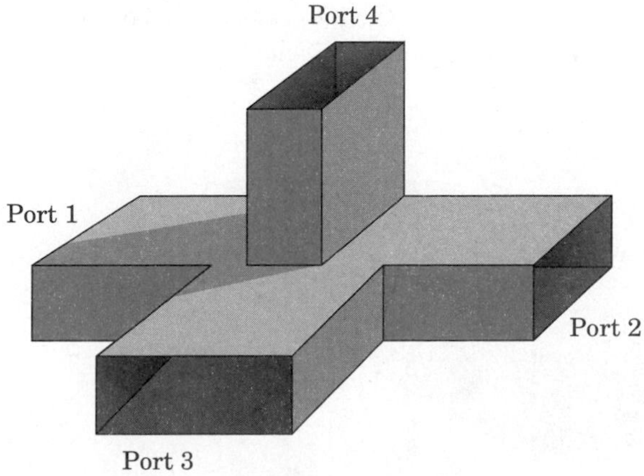

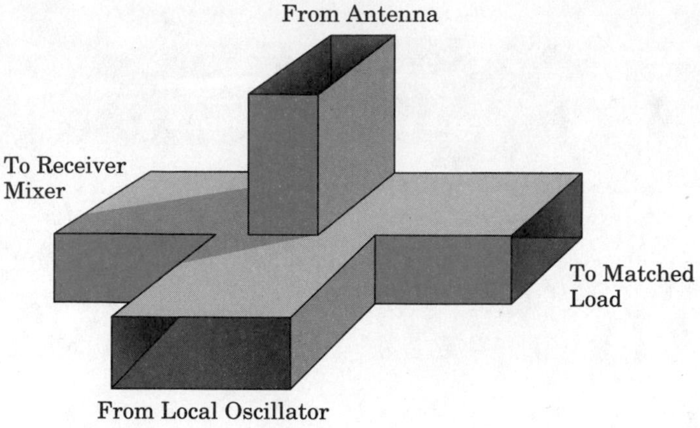

Figure 19.19
Receiver front end using hybrid tee

output at port 4. On the other hand, a signal entering via port 4 will produce equal and out-of-phase outputs at ports 1 and 2 but no output at port 3.

One example of the application of the hybrid tee is the receiver front end shown in Figure 19.19. In any receiver, it is desirable to mix the received signal and the local oscillator signal without allowing the local oscillator signal to reach the antenna. If it does, it may be radiated, causing interference to other communication services. The system in Figure 19.19 allows the incoming signal at port 4 to be combined with the local oscillator signal that is supplied to port 3. The combined output appears at ports 1 and 2. One of these ports will be connected to the mixer input, and the other must be terminated with a matched load to prevent reflections. This means that the input signal suffers a 3 dB loss in the tee.

19.3.2 Cavity Resonators

In our discussion of waveguides, we noted that the waves reflect from the walls as they proceed down the guide. Suppose that instead of using a continuous waveguide, we were to launch a series of waves in a short section of guide called a **cavity**, as illustrated in Figure 19.20. The waves would, of course, reflect back and forth from one end to the other. A cavity of random size and shape would have random reflections with a variety of phase angles and a good deal of cancellation. However, suppose that the cavity had a length of exactly one-half wavelength. Waves would reflect from one end to the other and would be in phase with the incident signal. There would be a buildup of field strength within the cavity. In a perfectly lossless cavity, this could continue forever. Of course, losses in the walls of a real cavity would result in the signal eventually dying out unless sustained by new energy input.

This description should sound familiar: it is a description of resonance. Like any other resonant device, a waveguide cavity has a Q; the Q for resonant cavities is very high, on the order of several thousand. Cavities can be tuned by changing their size (for instance, by moving a short-circuiting plate at one end).

Figure 19.20
Rectangular cavity

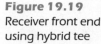

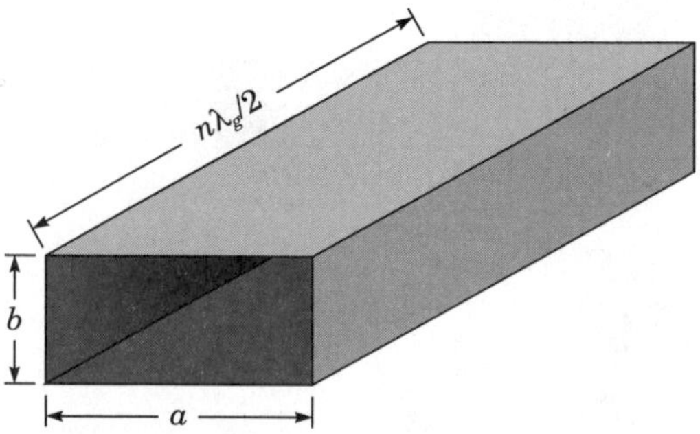

Figure 19.21
Resonant cavities

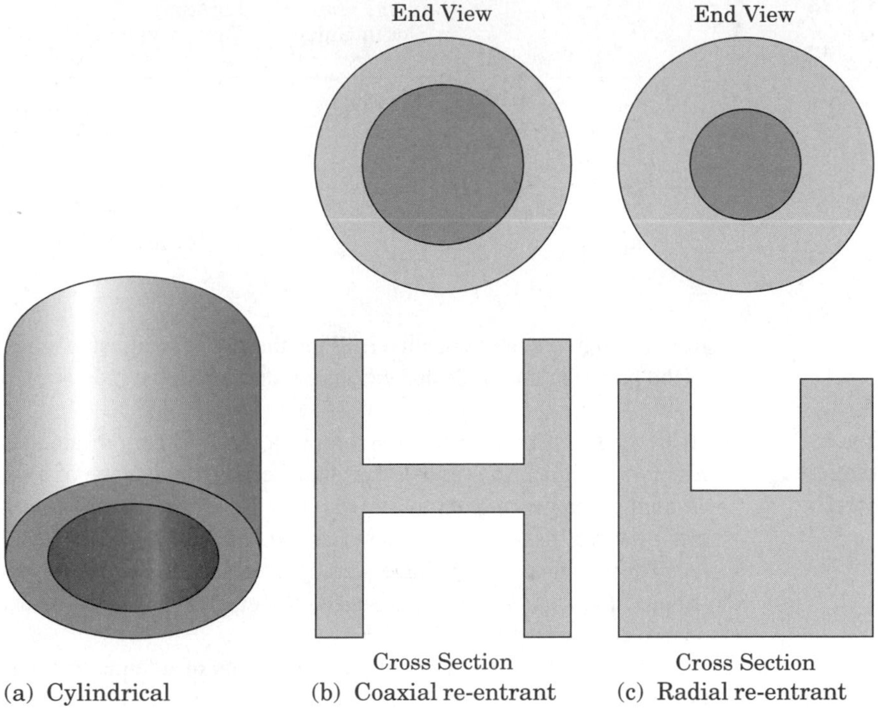

The rectangular cavity described above is not the only possible type. Figure 19.21 shows several others. Resonant cavities of various sorts are found in many types of microwave devices (for example, the **magnetron** and **klystron**, which will be described later).

One useful application of the resonant cavity is the wavemeter shown in Figure 19.22. The wavemeter is a type of frequency meter that consists of a cavity with an adjustable plunger. The adjusting screw is attached to a frequency scale. This cavity is attached to a waveguide through which the signal of unknown frequency passes as it travels

Figure 19.22
Wavemeter

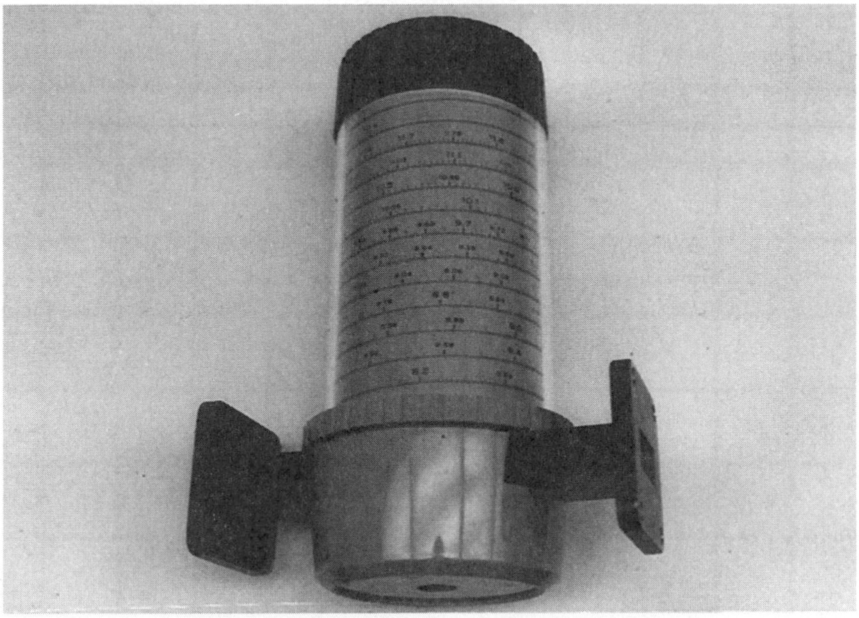

Figure 19.23 Waveguide attenuators

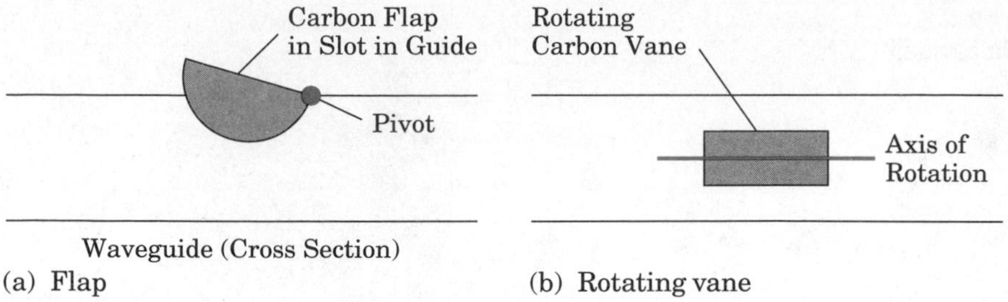

from its source to a power meter. When the cavity is adjusted to resonance, it absorbs some of the power in the guide and causes a reduction in the indication of the power meter.

19.3.3 Attenuators and Loads

At lower frequencies, a load is simply a resistor, and an attenuator is a combination of resistors designed to preserve the characteristic impedance of a system while reducing the amount of power applied to a load. Once again, this lumped-constant approach becomes less practical as frequency increases. At microwave frequencies, an ordinary carbon or metal-film resistor would have a complex equivalent circuit involving a good deal of distributed inductance and capacitance. Nonetheless, the idea of using resistive material to absorb energy is still valid.

Figure 19.23 shows waveguide versions of attenuators. The flap attenuator in Figure 19.23(a) uses a carbon flap that can be inserted to a greater or lesser extent into the waveguide. The fields inside the guide are, of course, present in the carbon as well. A current flows in the flap, causing power loss.

The attenuator shown in Figure 19.23(b) uses a rotating vane. When the vane is rotated so that the electric field is perpendicular to its surface, little loss occurs, but when the field runs along the surface of the vane, a much larger current is induced, causing greater loss. Both fixed and variable vane attenuators are available with a variety of attenuation values.

Figure 19.24 illustrates a terminating load for a waveguide. The carbon insert is designed to dissipate the energy in the guide without reflecting it.

19.3.4 Circulators and Isolators

Isolators and **circulators** are useful microwave components that generally use ferrites in their operation. The theory of their operation will be described shortly; first, a brief description of their operating characteristics is in order.

An *isolator* is a device that allows a signal to pass in only one direction. In the other direction, it is greatly attenuated. An isolator can be used to shield a source from a mismatched load. Energy will still be reflected from the load, but instead of reaching the source, the reflected power is dissipated in the isolator. Figure 19.25 illustrates this application of an isolator.

The *circulator*, shown schematically in Figure 19.26, is a very useful device that allows the separation of signals. The three-port circulator of Figure 19.26 allows a signal introduced at any port to appear at, and only at, the next port in counterclockwise rotation. For instance, a signal entering at port 1 appears at port 2 but not at port 3. Circulators can have any number of ports from three up, but three- and four-port versions are the most common.

Figure 19.24 Waveguide termination

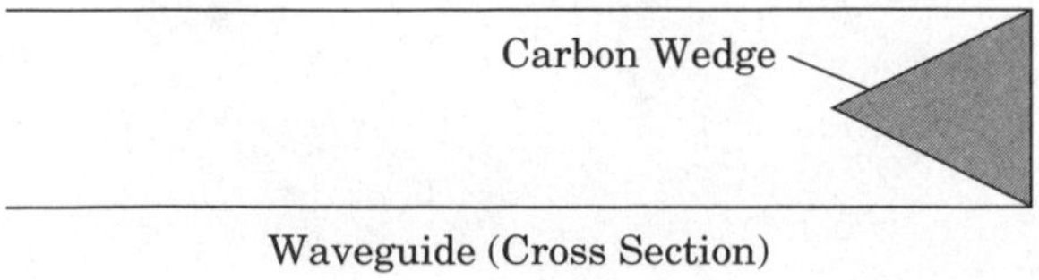

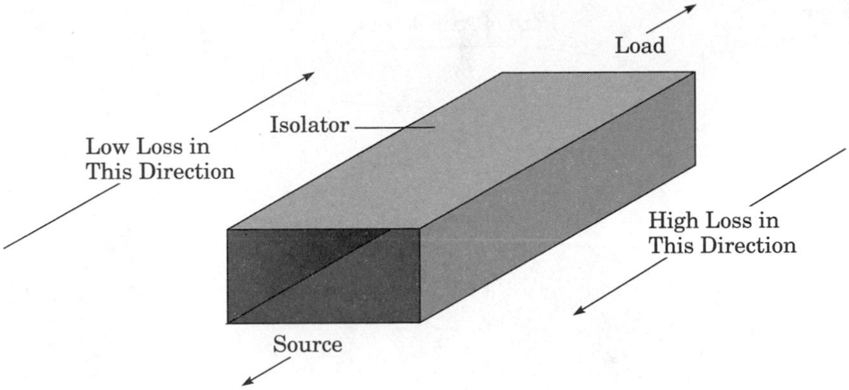

Figure 19.25
Operation of isolator

One simple example of an application for a circulator is as a transmit-receive switch. Figure 19.27 shows the idea. The transmitter output is connected to port 1, the antenna to port 2, and the receiver input to port 3. The transmitter output signal is applied to the antenna, and a received signal from the antenna reaches the receiver. The transmitter signal does not reach the receiver, however; if it did, it would probably burn out the receiver front end. Another feature of this setup is that any local oscillator signal that leaks out the receiver input appears only at the transmitter port, where it will do no harm. It does not reach the antenna, where it could cause undesired radiation and possible interference.

The operation of both the isolator and the circulator is based on the magnetic properties of ferrites. These ceramic compounds of iron oxide with other metals are ferromagnetic but nonconductive. They are very commonly used as cores for radio-frequency inductors and transformers, where they provide a permeability greater than that of air. Microwave applications of ferrites also use the ferromagnetic properties of the material, but in a different way.

In all materials, the electrons spin about their axes while orbiting around the nucleus of an atom. The magnetic properties of ferrites result from the fact that most of their electrons spin in the same direction.

If a ferrite is subjected to a magnetic field from a permanent magnet, the electron spins experience a phenomenon called **precession**: the axis about which an electron spins also begins to spin. Precession can be demonstrated with a child's top, as shown in Fig-

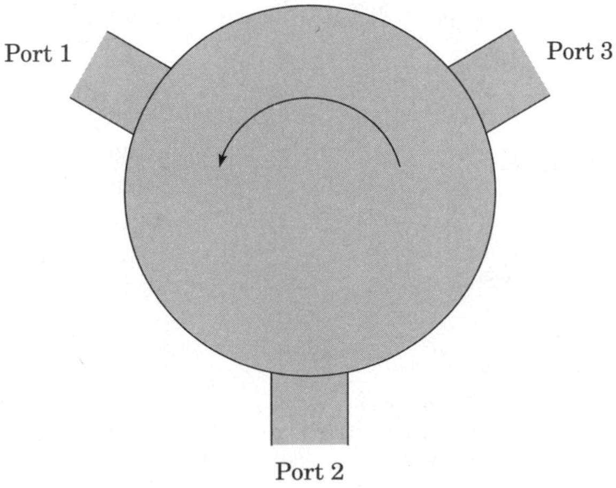

Figure 19.26
Circulator (top view)

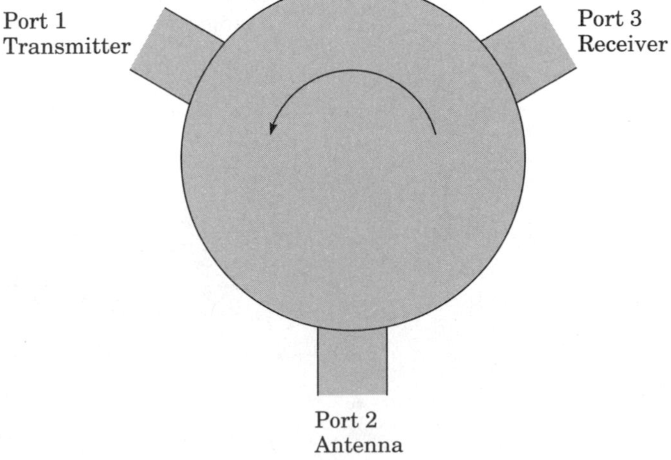

Figure 19.27
Circulator transmit-receive switch

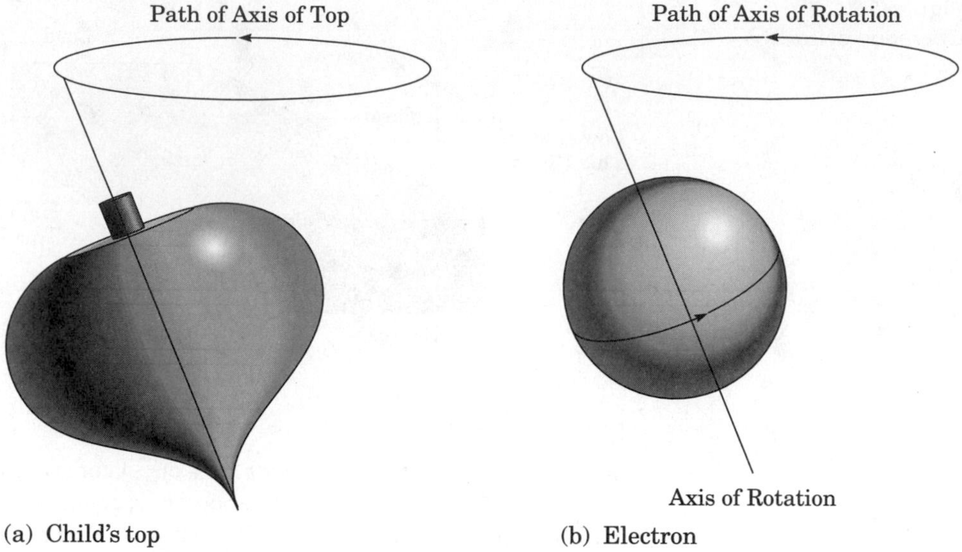

Figure 19.28 Precession

ure 19.28. With the top, however, precession is caused by gravity rather than by a magnetic field. Adjusting the magnetic field can cause the precession in a ferrite to occur at a frequency in the microwave range. If an electromagnetic wave is allowed to propagate through the ferrite, and if the frequency of the wave is equal to the precession frequency, the precession can be either increased or decreased by the magnetic field associated with the signal. Which occurs depends on the relative directions of the signal and the fixed magnetic field. If the interaction increases the precession, energy is extracted from the signal; if the interaction goes in the other direction, little or no loss occurs. These characteristics meet the requirements for an *isolator*, which can consist of a slab of ferrite located in a waveguide and subjected to a constant magnetic field from a permanent magnet, as shown in Figure 19.29.

The theory behind the ferrite *circulator* is more complex, so we will not go into it in detail. The interaction between an electromagnetic wave and the ferrite results in a phase shift as the wave propagates through the material. This shift is called *Faraday rotation*. (It is the same phenomenon that causes the polarization of waves of certain frequencies to change as they move through the ionosphere, as described in Chapter 21.) The amount of phase shift depends on the length of the ferrite and the strength of the dc magnetic field to which it is subjected. The shift is nonreciprocal, that is, it depends on the direction of

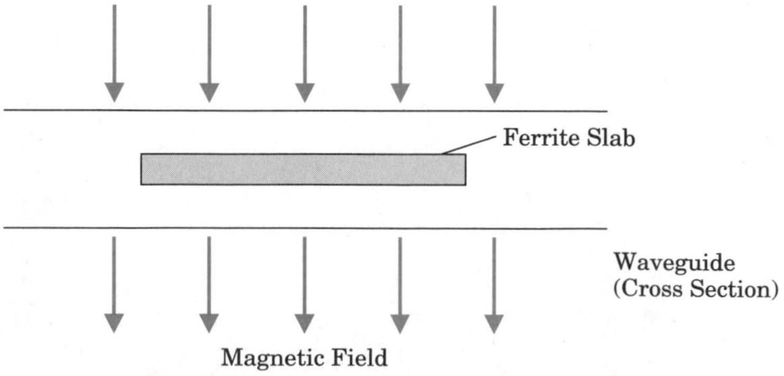

Figure 19.29 Ferrite isolator

propagation of the signal. By careful choice of phase shifts, a circulator can be devised that will achieve addition of the waves along two paths at the desired port and cancellation at the other port or ports.

19.4 Microwave Solid-State Devices

When transistors were first invented, they were restricted to low-frequency applications. The author can remember hybrid AM broadcast-band car radios, which used tubes in the RF section and transistors for the audio portion. The usefulness of transistors has gradually been extended through the radio-frequency spectrum, and there are now transistors that are usable in the microwave range. At the same time, other solid-state devices have been designed to take advantage of some of the factors that reduce transistor efficiency as frequency increases. In this section, we will first take a brief look at the ways in which conventional transistors can be adapted for microwave use, and then we will examine some of these ingenious alternatives.

19.4.1 Microwave Transistors

Conventional silicon transistors suffer from two major problems as frequency increases. First, all the components of the transistor, including the leads and the silicon elements themselves, exhibit stray capacitance and inductance. As the frequency increases, the inductive reactances get larger and the capacitive reactances get smaller, so that the transistor begins to look like the circuit in Figure 19.30. Eventually, there is so much feedback from collector to base that the transistor is useless.

It is impossible to eliminate stray capacitance and inductance, but they can be reduced. Figure 19.31 shows a typical microwave transistor. Note from the photograph in Figure 19.31(a) that the wire leads have been replaced with metal strips that can be soldered directly to striplines. Figure 19.31(b) is a sketch of the internal structure. The "interdigital" design, in which the emitter and base are formed from a series of "fingers," is designed to reduce capacitance to a minimum. The collector is formed from the substrate.

A more fundamental problem with any conventional transistor (and with many other types of devices) is transit time. In order for a transistor, diode, or other similar device to operate, charge carriers must move from one region to another. This takes a small but finite amount of time, depending on the type of carrier, the **mobility** of the carriers in the material, and the distance they have to move. If the transit time is of the same order as the signal period, it will result in phase shifts that can seriously degrade performance. In general, free electrons move more quickly than holes—this accounts for the preponderance of NPN bipolar transistors and N-channel field-effect transistors in high-frequency applications. Gallium arsenide is "faster" than silicon and is preferred in microwave applications.

Figure 19.30 Equivalent circuit of transistor at high frequency

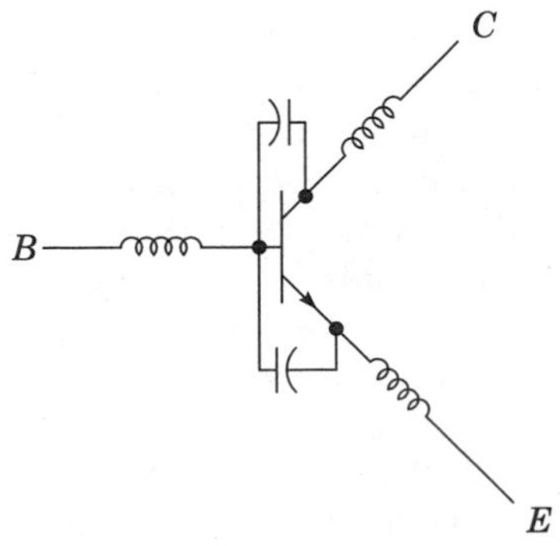

Figure 19.31
Microwave transistor
Copyright by Motorola. Used by permission.

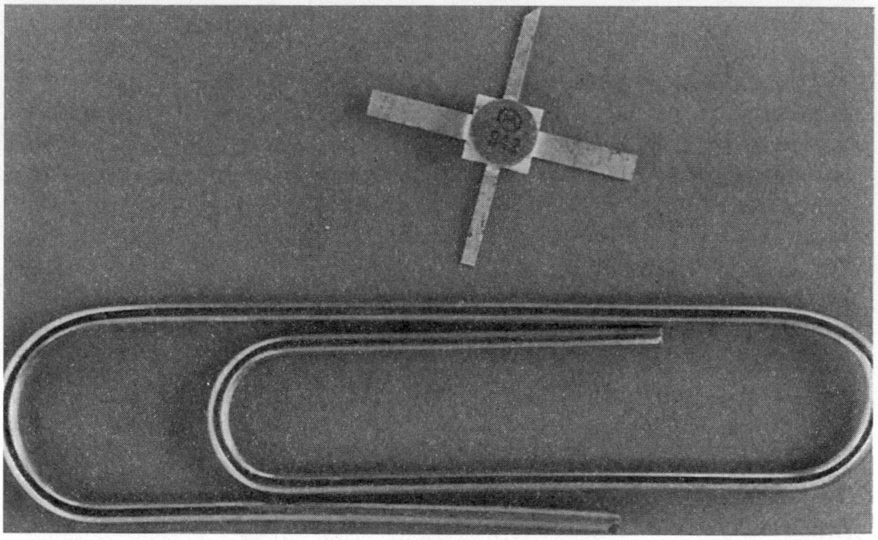

(a) Transistor

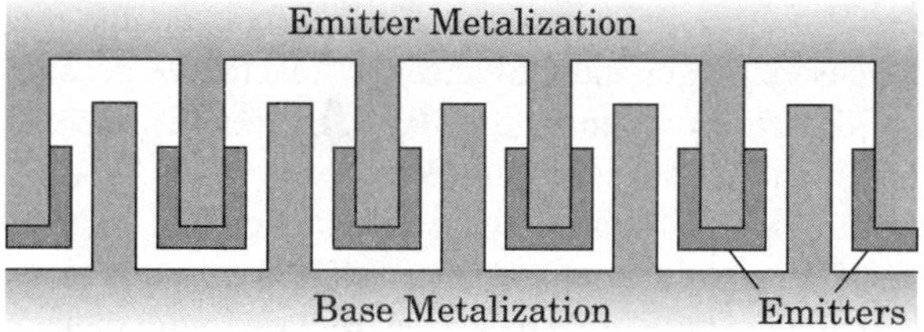

(b) Internal structure

Transistors modified for microwave use are quite common in the lower part of the microwave spectrum. For instance, the low-noise amplifier (LNA) that is the first stage of a satellite-reception system is practically always a gallium arsenide field-effect transistor (GaAsFET).

19.4.2 Gunn Devices

One solution to the problem of transit time appears in a good many microwave devices and can be loosely described as "if you can't beat it, join it." That is, rather than attempting to keep the transit time as small as possible to minimize its undesired effects, the designer includes a phase shift due to transit time as one of the device characteristics. The **Gunn device**, also known as the **transferred-electron device** (TED), is one of the simpler transit-time devices.

The Gunn device is sometimes called a *Gunn diode* because it has two terminals. It has no junction, however, but is just a slab of N-type gallium arsenide, as shown in Figure 19.32. Gallium arsenide has the interesting and unusual property that the mobility of the electrons actually decreases as the electric field strength increases over a certain range. Normally, we would expect that the larger the electric field strength, the faster electrons would move through the conductor, and that is in fact the case with most conductors and semiconductors. With gallium arsenide and a few other semiconductors, such as indium arsenide and gallium phosphide, there is a region where the mobility actually decreases as the field increases. The graph in Figure 19.33 shows this phenomenon.

A curve like the one in Figure 19.33 is said to have a negative resistance region. This does not actually mean that the material has less than zero resistance—it means that in

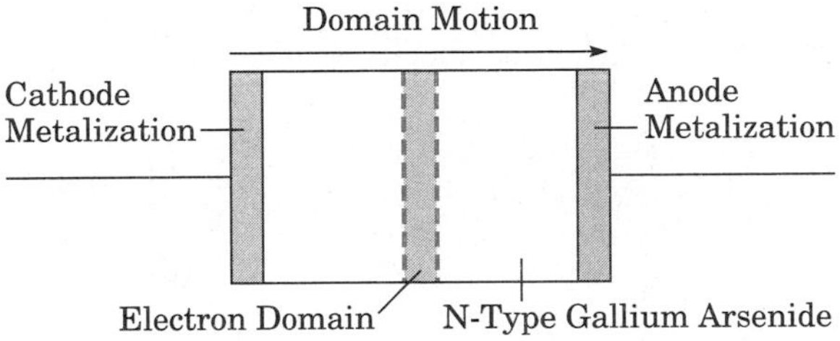

Figure 19.32
Gunn device (cross section)

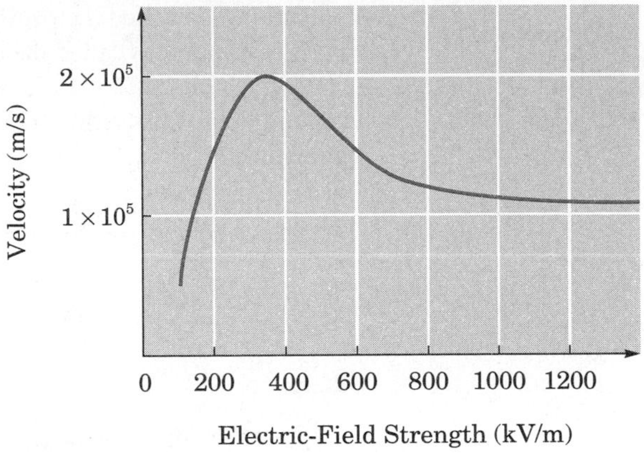

Figure 19.33
Drift velocity in gallium arsenide

some part of the curve, the current decreases as the voltage decreases. A differential resistance could be defined as

$$r = \frac{\Delta v}{\Delta i} \tag{19.13}$$

and it is this differential resistance that is negative over a certain range of field strength in gallium arsenide. A device with negative resistance can be made to work as an oscillator or, with a bit more work, as an amplifier.

This negative-resistance effect was discovered by J. B. Gunn (b. 1928) in 1963. He found that when a voltage was placed across a small slab of gallium arsenide, as shown in Figure 19.34, oscillations occurred. The frequency depended on the thickness of the slab and seemed to have a period equal to the time required for an electron to pass through the slab at a drift velocity of about 1×10^5 m/s. Gunn eventually discovered that the reason for the oscillations was that a **domain** with a large electric field formed in the material and moved toward the positive terminal. The domain forms when there is a small region with an electric field greater than in the rest of the material, possibly due to some local impurity. When this field reaches the threshold value, the electrons in the region move more slowly than elsewhere. A concentration of charge builds up, forming a *domain*, which moves through the device, collecting most of the charge as it does so. When the domain reaches the positive terminal, all of its electrons leave at once, causing a pulse of current. Simultaneously, another domain forms and the process continues. For obvious reasons, this mode

Figure 19.34
Gunn-effect oscillator

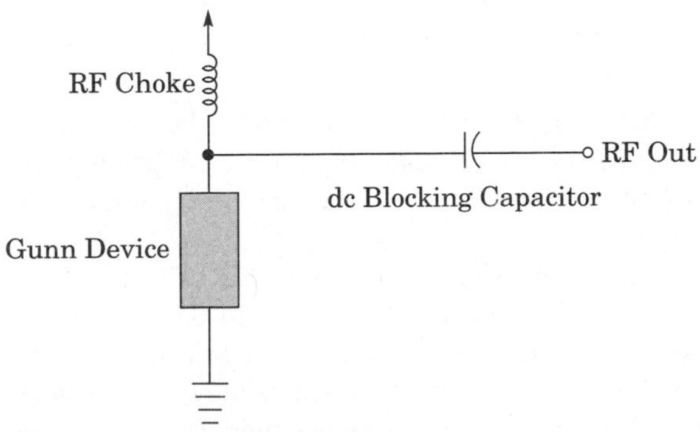

of operation is called the *transit-time mode*. It can also be referred to as the *domain mode*, and it is sometimes called the *Gunn mode*.

Example 19.8 A Gunn device has a thickness of 7 μm. At what frequency will it oscillate in the transit-time mode?

Solution The transit time of an electron can be found from the basic velocity relationship

$$v = \frac{d}{t}$$

where

v = velocity in meters per second
d = distance in meters
t = time in seconds

Rearranging gives

$$\begin{aligned} t &= \frac{d}{v} \\ &= \frac{7 \times 10^{-6}\ \text{m}}{1 \times 10^{5}\ \text{m/s}} \\ &= 7 \times 10^{-11}\ \text{s} \end{aligned}$$

This will be the period T of the oscillation, so the frequency is

$$\begin{aligned} f &= \frac{1}{T} \\ &= \frac{1}{7 \times 10^{-11}\ \text{s}} \\ &= 14.3 \times 10^{9}\ \text{Hz} \\ &= 14.3\ \text{GHz} \end{aligned}$$

The Gunn mode provides oscillations at a frequency that depends on the device geometry and not on the external circuit. There are several other modes of operation that allow the device to be tuned by placing it in a resonant cavity. The effect of the cavity is to remove, or *quench*, the domain before it reaches the anode terminal, causing the device to operate at a higher frequency than it would otherwise. The limited-space-charge-accumulation (LSA) mode is the most common. In this mode, the device is biased in the negative resistance region, but the voltage swing is such that the device moves out of this region once per cycle, so that the domain is quenched.

The Gunn device can also be used as an amplifier, but like all two-terminal negative-resistance devices, it requires a circulator to separate the input from the output. This mode of operation is illustrated in Figure 19.35.

19.4.3 IMPATT Diodes

The acronym **IMPATT** stands for *imp*act *a*valanche and *t*ransit *t*ime. Unlike the Gunn device, the IMPATT has a P–N junction; in fact, it is often a four-layer device, as shown in Figure 19.36. As suggested by its name, the device operates in the reverse-breakdown (avalanche) region.

The width of the intrinsic region in the IMPATT diode is made such that the transit time of an electron across this region is equal to one-half of the period at the operating frequency. The device is reverse biased at just below its breakdown voltage. The ac waveform superimposed upon this bias causes breakdown to occur once per cycle. When the junction breaks down, a surge of electrons enters the intrinsic region, emerging one-half

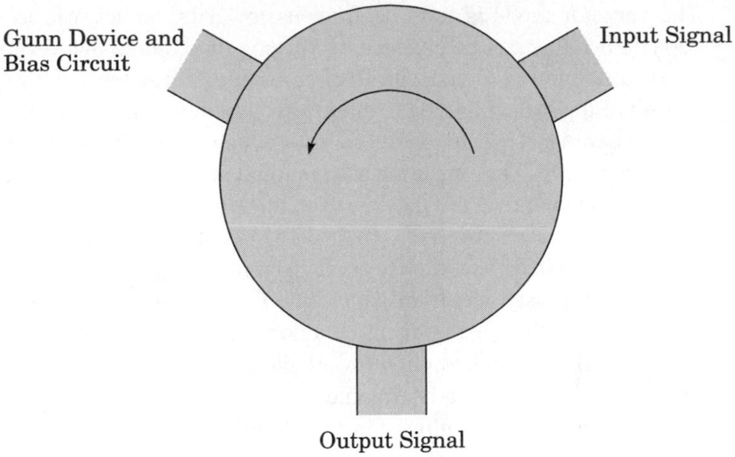

Figure 19.35
Use of Gunn device as an amplifier

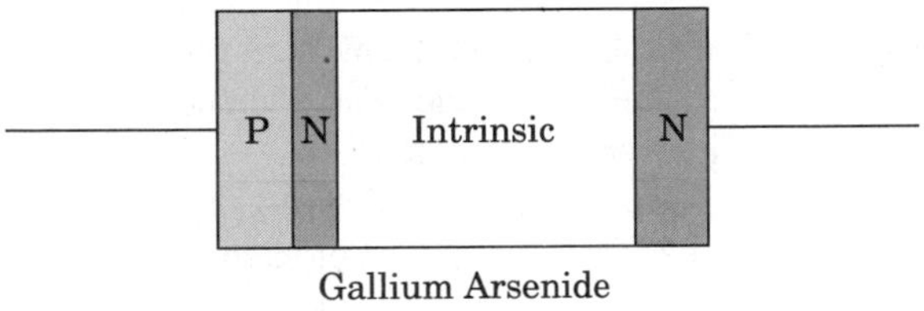

Figure 19.36
IMPATT diode

cycle later. This has an effect similar to that of the domains in the Gunn device, and it results in oscillation at a frequency that depends on the dimensions of the IMPATT diode and the resonant frequency of the cavity in which it is installed.

IMPATT diodes can achieve greater efficiency and higher power levels than Gunn devices (on the order of 10 W compared to 1 W for a Gunn device), but their use of avalanche breakdown causes them to be considerably noisier. A variation of the IMPATT called the **TRAPATT** (for *tra*pped *p*lasma *a*valanche *t*riggered *t*ransit) can operate with still higher power (about 100 W in pulsed operation) and efficiency levels, but it is expensive, requires complex circuitry, and is noisier than the IMPATT.

19.4.4 PIN Diodes

The structure of a P-intrinsic-N (**PIN**) diode, shown in Figure 19.37, is similar to that of an IMPATT diode, but its operation is quite different. This device is used as an electronic switch and an attenuator. When reverse biased, it represents a large resistance in parallel with a small capacitance, but forward biasing the diode results in a variable resistance that can approach zero when the diode is completely turned on. Ordinary silicon diodes are used as switches in just this way at lower frequencies. The PIN diode is preferred in the microwave region of the spectrum because of its small capacitance when reverse biased. At microwave frequencies, there will be enough stored charge in the intrinsic region, when the diode is forward biased, for good conduction.

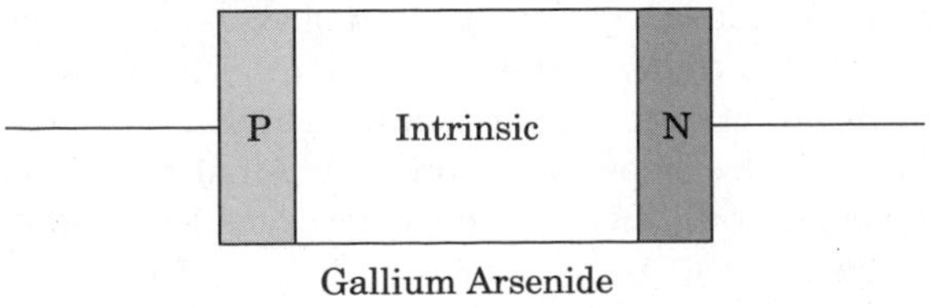

Figure 19.37
PIN diode

19.4.5 Varactor Diodes

The varactor diode is familiar from its lower-frequency use as a means of providing a capacitance that can be changed by varying the voltage that reverse biases the diode. Varactors are found in such devices as frequency synthesizers and frequency modulators. They can serve the same function at microwave frequencies.

There is, however, another use to which varactors can be put in the microwave part of the spectrum. They are often used in frequency multipliers, which should also be familiar from earlier chapters, but they are even more useful at microwave frequencies. They can be used to generate microwave signals with crystal-controlled stability, something that is otherwise difficult to accomplish at such high frequencies. The frequency multipliers described previously used transistor Class C amplifiers, but of course this is not necessary. All that is required is that there be a nonlinear circuit that will generate harmonics of the input signal and a tuned circuit to emphasize the desired harmonic and reject the fundamental and any undesired harmonics.

All diodes are nonlinear in the "knee" region of their characteristic curve, but varactors, due to their variable capacitance, can be used with reverse bias as nonlinear devices, achieving greater efficiency. A variation of the varactor, known as the *step-recovery* or *snap diode*, can also be used as a multiplier. It is operated with forward bias for most of the cycle of the fundamental. During this period, a charge is stored in the diode. When the diode becomes reverse biased at the input cycle peak, the charge flows back across the junction as a brief pulse with a high harmonic content.

Since varactor multipliers are passive devices, they obviously have a gain of less than unity. They can be quite efficient, though: an efficiency of 50% is not unusual. This compares quite favorably with generating a microwave signal directly.

19.4.6 Yttrium-Iron-Garnet Devices

Yttrium-iron-garnet (**YIG**) is a type of ferrite. A YIG sphere can be used in place of a resonant cavity as a microwave resonant circuit. Such a sphere is often used as the frequency-determining circuit in a microwave oscillator. Since the resonant frequency of a ferrite depends on the magnetic field to which it is subjected, the frequency of the oscillator can be changed by varying the dc current to an electromagnet that generates the field. This method of frequency setting is much more convenient than varying the physical size of a resonant cavity.

19.4.7 Dielectric Resonators

Another innovative frequency-determining technique that is often used in solid-state microwave sources is the dielectric resonator. This is essentially a resonant cavity that is made of a solid slab of a dielectric material such as alumina. Microwaves propagate within the medium and are reflected at the interface between the dielectric and air. The result is a stable, low-cost resonant device. Compared to the YIG sphere, it has the disadvantage of being fixed in frequency, but it is inexpensive and needs no magnet.

19.5 Microwave Tubes

As at lower frequencies, vacuum tubes are still the preferred technology when large amounts of microwave power are required. Many of the problems that reduce the efficiency of transistors, such as transit time and stray reactance, also affect tubes. Lead inductance can be reduced by making connections to rings on the tube rather than to pins at one end. Transit time can be reduced by making tubes smaller, but this increases stray capacitance and reduces the tube's ability to dissipate power. Such "fixes" work only up to a point. This section will look at some vacuum-tube designs that avoid the shortcomings of conventional tubes by such techniques as incorporating transit time into the design. There are many other types not considered here, but these examples illustrate the logic behind microwave tube designs.

19.5.1 Magnetrons

The *magnetron*, invented in 1921, is the oldest microwave tube design. The magnetron and its variations are high-power, fixed-frequency oscillators, not noted for stability or ease of modulation but simple, rugged, and relatively efficient (about 40% to 70%). Magnetrons are commonly used in radar transmitters, where they can generate peak power levels in the megawatt range. The smaller ones used in microwave ovens produce several hundred watts continuously.

Figure 19.38 shows a cross-sectional view of a typical cavity magnetron. As might be expected from its name, the magnetron needs a magnet: a powerful permanent magnet. The tube itself is situated between the poles of the magnet, and only the magnetic field lines are shown.

A hot cathode is surrounded by an anode in which there are a number of resonant cavities. Were it not for the magnetic field, electrons emitted from the cathode would simply move radially across to the anode. However, the interaction between the fixed magnetic field and the magnetic field generated by a moving electron produces a force that causes the electron to travel in a curved path. The size of the loop made by the electrons depends on the relation between the magnetic field strength and that of the electric field between anode and cathode. For a large anode-cathode voltage, as in Figure 19.38(b), the electrons reach the anode and anode current flows. If the voltage is reduced, as in Figure 19.38(c), the electrons will circle around and return to the cathode. In this case, there is no anode current and the tube is cut off. The rate at which electrons move around the cathode is called the *cyclotron frequency*. Since the magnetic and electric fields applied to the tube are at right angles, magnetrons are also known as **crossed-field tubes**.

Now let us consider the function of the cavities around the outside of the magnetron. The motion of electrons moving past the opening of a cavity starts oscillations that will be sustained if the conditions are right for the continual transfer of energy from the electron stream to the fields in the cavities. In most magnetrons, this involves the presence of a traveling wave that moves around the tube, with one-half wavelength representing the distance between adjacent cavities. This requires movement of the wave around the tube at a rate much slower than the speed of light, and for this reason the circular arrangement of resonant cavities is called a **slow-wave structure**.

A slow-wave structure is required because the electrons must move around the structure at almost the same rate as the electromagnetic wave in order for the electrons to give up energy to the wave. Of course, the electrons in the tube move with a velocity much less than that of light, so the propagation velocity of the wave must be reduced.

Figure 19.38
Cavity magnetron

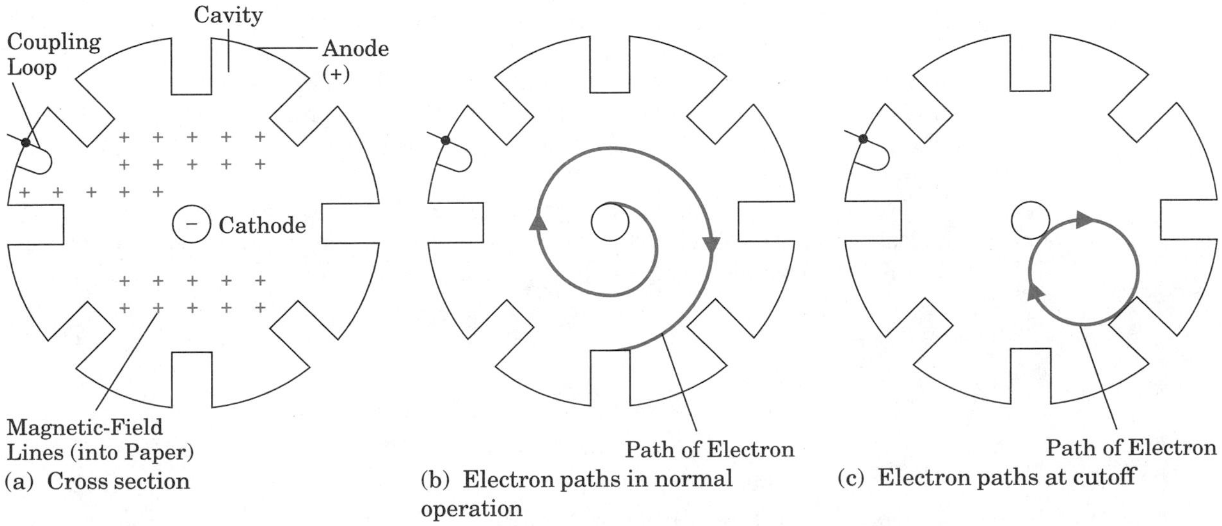

Microwave Ovens

There is very likely a magnetron in your kitchen, since every microwave oven contains one. The magnetron produces several hundred watts at 2.45 GHz, a frequency that is chosen because water molecules are resonant at that frequency. If subjected to a microwave field at that frequency, water molecules absorb power and become hot, causing any food containing water to cook. The output of the magnetron is coupled to a resonant cavity, the oven itself. The several kilovolts required by the tube are supplied by a step-up transformer and a voltage-doubler rectifier circuit. In variable-power ovens, a thyristor circuit pulses the power so that the magnetron switches rapidly on and off.

Now that you know something about microwaves, some of the well-known cautions about microwave ovens should be easy to understand. Operating the oven without food is undesirable, since this is equivalent to operating a transmitter without a load—most of the power will be reflected back to the tube. Similarly, the use of metal containers is frowned on because currents will be induced in the metal, heating it and distorting the field in the oven. Some special microwave cookware contains resistive material designed to absorb power from the microwaves and become hot, as an aid to browning certain foods. Some microwaves rotate the food or use a "stirrer" to disturb the fields in the cavity. This is done because the oven, like any resonant cavity, has standing waves, so the electric-field strength and therefore the cooking ability are not evenly distributed.

Lastly, a knowledge of microwaves can be applied to safety concerns. It is important that the door seal properly, and that the interlocks (there are usually two, for backup) disconnect the power when the door is opened. The microwaves that can cook your food can cook you, too, if allowed to escape. (The eyes are especially vulnerable, because they have poor circulation that does not allow heat to be removed quickly by the blood.) But no, the food itself is not "radioactive" in any way. We are dealing with radio waves, at frequencies far below those of ordinary light. Giving radioactivity to substances requires photons with much more energy. "Nuking" food in the microwave is just a figure of speech!

In order to get power from the magnetron, one of the cavities must be coupled electrically to the outside. The tube shown uses a loop for this. The energy provided by the tube comes from the wave traveling around the tube, and that energy in turn comes from the electron stream. The electrons, of course, get their energy from the applied anode-to-cathode potential. For the tube to work, this voltage must be adjusted so that most of the electrons cycle around the tube. As they spiral from cathode to anode, they gain energy from the dc electric field and give up most of it to the traveling wave. Finally, they land on the anode. Some electrons, on the other hand, will cycle around and land on the cathode—this *back-bombardment* of the cathode reduces the life of the tube and lowers efficiency.

The magnetron described above is essentially a fixed-tuned oscillator. Changing the frequency requires changing the physical size of the cavities, which can only be done mechanically. Voltage-controlled versions are available, but they have much lower efficiencies. There is also an amplifier tube that works on similar principles and is known as the *crossed-field amplifier* (CFA). The operating bandwidth is very limited because of the high-Q resonant cavities.

Many magnetrons are used to generate pulses of radio-frequency energy, for instance, in radar transmitters. Typically these are low-duty-cycle applications. Recall that the duty cycle for any device is simply

$$D = \frac{T_{on}}{T_T} \tag{19.14}$$

where

D = duty cycle as a decimal fraction
T_{on} = on time per operating cycle
T_T = total time per operating cycle

Duty cycle can also be expressed as a percentage, by multiplying D from Equation (19.14) by 100. The power averaged over time for a magnetron (or any other device that generates pulses) is the power in the pulses, which is generally called (somewhat misleadingly) *peak power*, multiplied by the duty cycle, that is,

$$P_{avg} = P_P D \tag{19.15}$$

where

P_{avg} = average power
P_P = pulse power
D = duty cycle

Equation (19.15) is equally applicable to input power, output power, or power dissipation.

Example 19.9 A pulsed magnetron operates with an average power of 1.2 kW and a peak power of 18.5 kW. One pulse is generated every 10 ms. Find the duty cycle and the length of a pulse.

Solution From Equation (19.15),

$$\begin{aligned} P_{avg} &= P_P D \\ D &= \frac{P_{avg}}{P_P} \\ &= \frac{1.2}{18.5} \\ &= 0.065 \quad \text{or} \quad 6.5\% \end{aligned}$$

From Equation (19.14),

$$\begin{aligned} D &= \frac{T_{on}}{T_T} \\ T_{on} &= DT_T \\ &= 0.065 \times 10 \text{ ms} \\ &= 0.65 \text{ ms} \end{aligned}$$

19.5.2 Klystrons

The *klystron* is the preferred tube for high-power, high-stability amplification of signals at frequencies from UHF to about 30 GHz. In fact, it is commonly found in UHF television transmitters. There are actually two quite different types of klystrons: the reflex klystron, a small tube used as an oscillator, and the multicavity klystron, which is employed as a power amplifier. The reflex klystron is nearly obsolete, having been replaced in most applications by the solid-state techniques described above, so this section will deal with the multicavity klystron, some examples of which are pictured in Figure 19.39(a).

Like the magnetron, the klystron works by transferring energy from a stream of electrons to an electromagnetic wave moving in a slow-wave structure. In the klystron, the electrons are formed into a beam that moves in a straight line past a series of resonant cavities. For this reason, the klystron, like the traveling-wave tube to be described next, is known as a **linear-beam tube**.

The electron beam is produced in the conventional manner by a hot cathode. An electron gun, aided by one or more external magnets, focuses the electrons into a beam. As shown in Figure 19.39(b), the beam passes two or more resonant cavities before reaching the anode, which in this tube is called the **collector**.

The input signal is applied to the cavity closest to the cathode. The beam is velocity-modulated, or **bunched**, by the microwave field in the input cavity, which is also called the **buncher**. The output signal is taken from the cavity closest to the collector, which is sometimes called the **catcher**. In between, there may be one or more intermediate cavities.

Figure 19.39
Multicavity klystron tubes
Photograph courtesy of Communications & Power Industries.

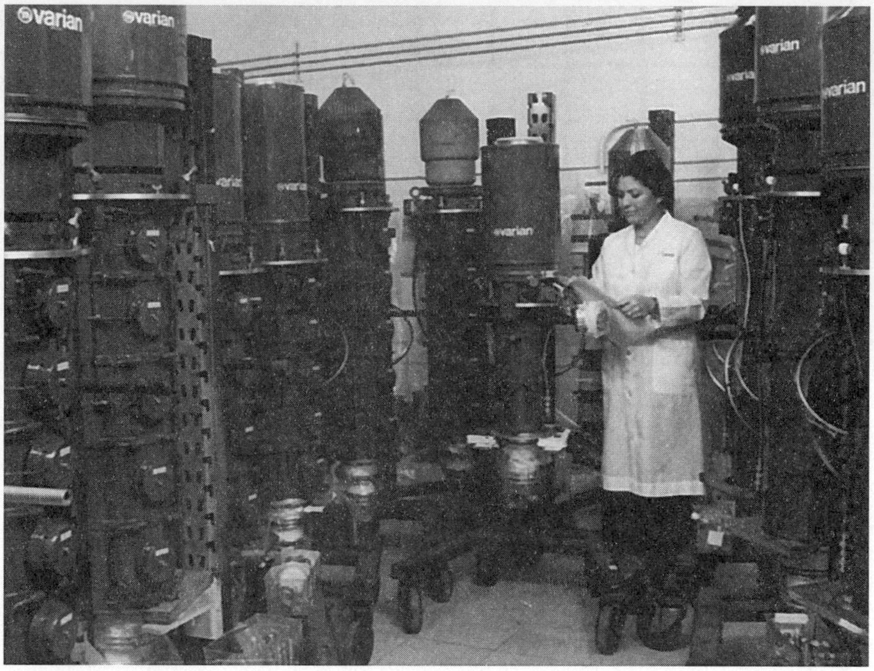

(a) Klystron tubes

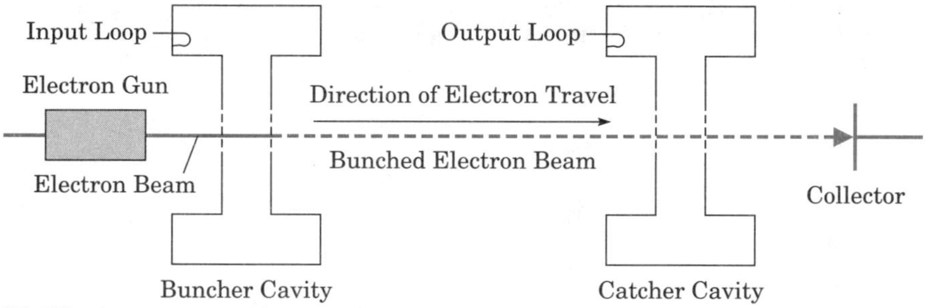

(b) Klystron cross section

The velocity modulation, or bunching, is caused by the interaction between the electric field at the input cavity and the electrons in the beam. Figure 19.40 shows the process. When the electric field due to the input signal is in the same direction as the accelerating field from the main power supply, it causes the velocity of the electrons to increase. On the other hand, when the two fields are in opposition, the beam is slowed. The result is that at a certain distance from the cavity, the electrons bunch together, as the figure shows. A

Figure 19.40
Velocity modulation

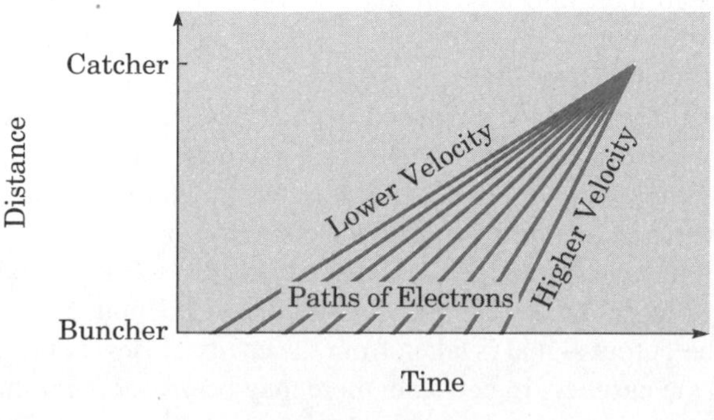

Figure 19.41
Traveling-wave tube

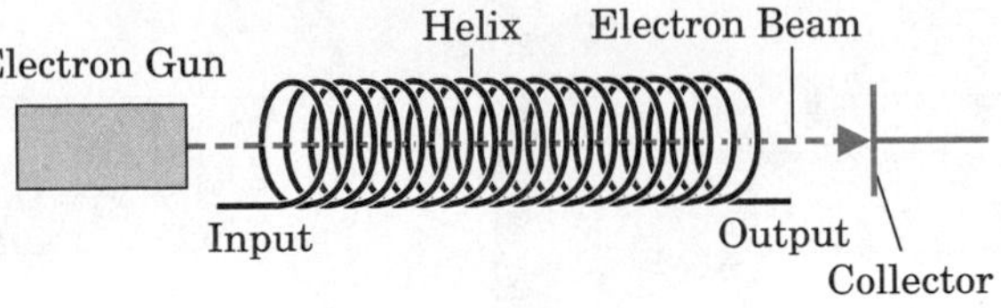

simple analogy would be cars at a traffic light: some cars are stopped at the light, some are let through, and the result is that the cars tend to travel in groups.

The bunched electrons produce an alternating microwave field at the output cavity. This field has the same frequency as the input wave but a greater amplitude, thus fulfilling the requirements for an amplifier. The electrons move on to the collector, but not before much of their energy has been given up. Klystrons can produce very high powers, up to the megawatt range.

19.5.3 Traveling-Wave Tube

The **traveling-wave tube** (TWT) can be used as a moderate-power amplifier or, with modifications, as an oscillator. It is distinguished by its wide bandwidth. The transmitters in communications satellites and their ground stations, for example, often use TWT amplifiers. Like the klystron and unlike the magnetron, the TWT is a linear-beam tube. Like both of these tubes, the TWT has a slow-wave structure. It may consist of a series of coupled cavities, reminiscent of the klystron, but it often has the form of a helix, as shown in Figure 19.41. The helix can be thought of as a sort of waveguide along which a wave travels by going around and around the turns of the spiral. The progress of the wave along the length of the helix is much slower than its speed as it moves around the turns. An electron beam travels from the cathode down the center of the helix to the collector, giving up some of its energy to the wave as it does so. Since the helix, unlike a series of cavities, is nonresonant, the bandwidth of the TWT can be much greater than that of the klystron. On the other hand, it is more difficult to remove heat from the helix than from the cavities of the klystron, so the helix TWT is a low- to medium-power device, for power levels up to about a kilowatt.

Like the klystron, the TWT needs a magnetic structure to focus the electron beam so that it passes within the helical slow-wave structure.

19.6 Microwave Antennas

There is no theoretical difference between microwave antennas and those for lower frequencies. The differences are practical: at microwave frequencies it is possible to build elaborate, high-gain antennas of reasonable physical size. Certainly dipoles, Yagis, and log-periodic antennas are also possible, but this section will look at antennas whose construction would be impractical at lower frequencies.

The parabolic reflector has already been described in Chapter 18. Though occasionally used at lower frequencies (for radio telescopes, for instance), it is most commonly employed at microwave frequencies. The parabolic dish is not really an antenna: it is a reflector, and it needs an antenna to provide it with a signal. The most common feed antenna for use with a parabolic dish is the horn, described below.

19.6.1 Horn Antennas

Horn antennas, like those shown in Figure 19.42, can be viewed as impedance transformers that match waveguide impedances to that of free space. The examples in the figure are the most common types. The E- and H-plane sectoral horns are named for the plane in which the horn flares; the pyramidal horn flares in both planes. With a circular waveguide, the conical horn is most appropriate.

Figure 19.42
Horn antennas

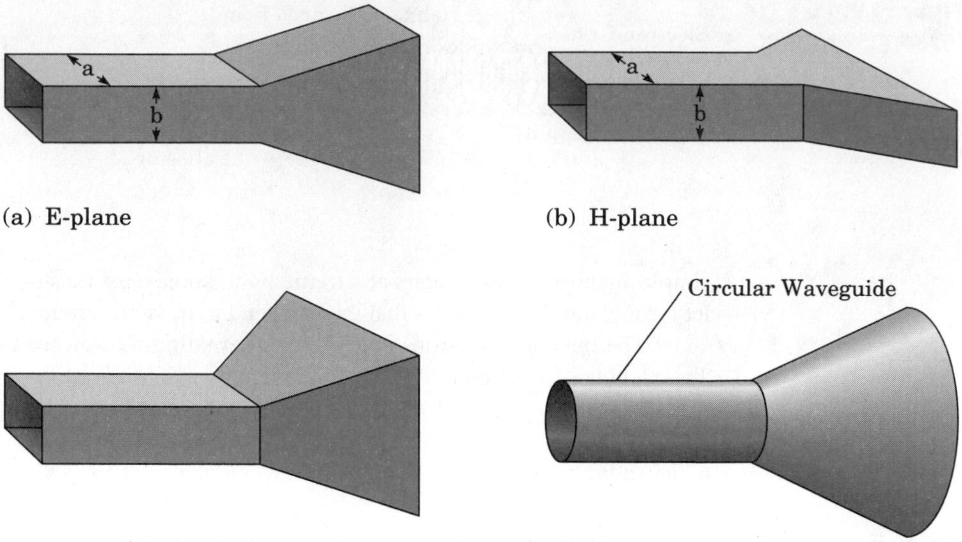

(a) E-plane (b) H-plane (c) Pyramidal (d) Conical

The gain and directivity of horn antennas depend on the type of horn and its dimensions. Gain is often in the vicinity of 20 dBi for practical horns. The bandwidth is about the same as that of the associated waveguide, that is, it works over a frequency range of approximately 2 : 1.

As well as being used with a parabolic reflector, the horn antenna can be—and often is—used alone as a simple, rugged antenna with moderate gain.

19.6.2 Other Microwave Antennas

There are many other types of microwave antennas in common use. We will give a brief introduction to some of them here.

Figure 19.43 shows a slot antenna, which is actually just a waveguide with a hole. The length of the slot is generally one-half wavelength. Its radiation pattern is similar to that of a dipole with a plane reflector behind it. It therefore has much less gain than, for instance, a horn antenna. It is seldom used alone but is usually combined with many other slots to make a phased array. Phased arrays with several elements were discussed in Chapter 18; at microwave frequencies, a great many slot antennas can be used to form a narrow, high-gain beam whose direction can be changed electronically by changing the phase of the signals to the individual elements. Such antennas are especially useful for airborne radar. In this case, the slots would be filled with a dielectric material to present a smooth surface to the air. Large aircraft actually have several slot arrays, for communications as well as for radar.

Of course, phased arrays can also be constructed from more conventional antenna elements, such as dipoles. The dipoles are sometimes assembled in a slightly unconventional way by laying them out on a printed-circuit board in the form of a stripline antenna.

Figure 19.43
Slot antenna

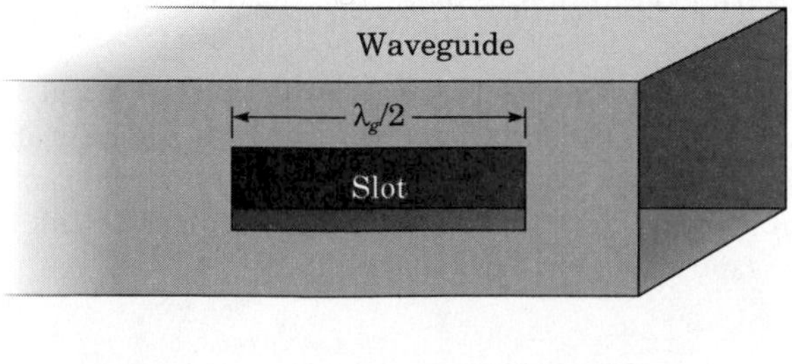

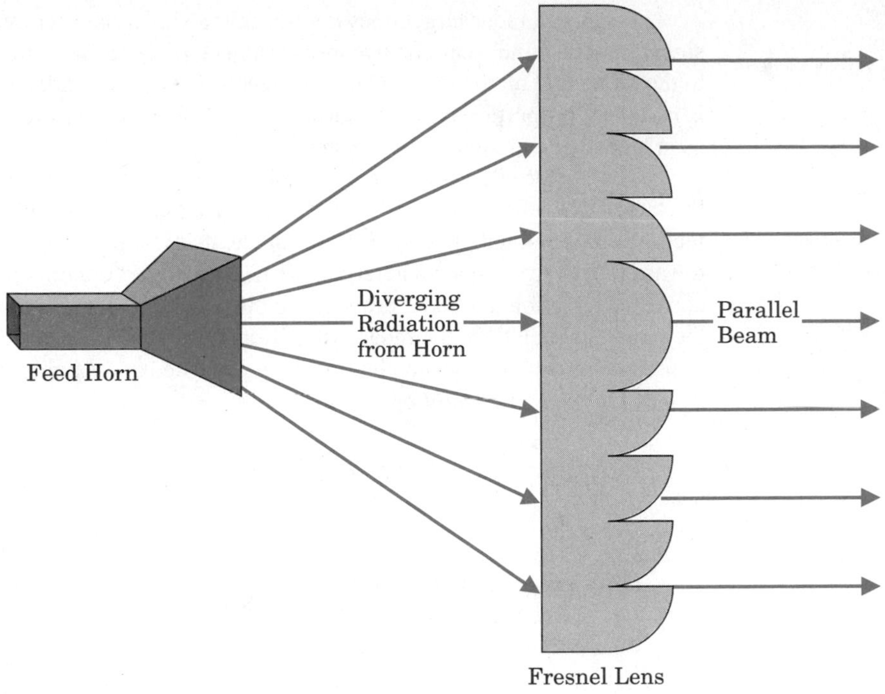

Figure 19.44
Dielectric lens

Just as the optical property of reflection can be put to good use in microwave antennas, so can refraction. At microwave frequencies, it is practical to build a lens for radio waves. The lens does not have to be optically transparent, but it must be made from a good dielectric—teflon is popular. To reduce the physical size of the lens, the **Fresnel lens** is usually chosen (see Figure 19.44). The lens, like the parabolic dish, is not really an antenna; it must be fed by a radiating element such as a horn antenna. One common application of the dielectric lens is in police radar systems.

19.7 Radar

Radar is an acronym for *ra*dio *d*etection *a*nd *r*anging, and it had its origins during World War II. Essentially, radar requires that a transmitter emit a signal using a directional antenna toward some object called the **target**. The signal reflects from the target back to the source, where it is received and interpreted. Information about the target can be obtained by analyzing the reflection, or echo.

Early radars often used signals in the HF or VHF regions of the spectrum. Microwaves are more common now, however. They can be focused into narrower beams, and they can detect smaller targets, since the target must be relatively large compared with the wavelength to provide a good reflection. An exception is long-range over-the-horizon radar, which uses frequencies in the HF range to take advantage of ionospheric propagation.

Radar equipment can be divided into two main categories. Pulse radar works by transmitting a short burst of microwaves called a *pulse*, which reflects from the target and is received at a later time. The time between the transmitted and received pulses gives distance information. Continuous wave (CW) radar transmits continuously and compares the frequency of the received echo with that of the transmitted signal. Relative motion between the radar and the target can cause a change of frequency (the **Doppler effect**), from which velocity information can be obtained. There are also radars that combine pulse and Doppler techniques.

Because a radar target does not transmit a signal but merely reflects it, the returning signal may be quite weak. Its strength is subject to square-law attenuation both as it travels to the target and on the return trip to the receiver. Thus, for radar, the received signal power is inversely proportional to the fourth power of the distance, rather than the second power as in a normal communications system.

The power of a return signal is also affected by the size, shape, and composition of the target. Targets are assigned a **radar cross section** that is defined as the area of a perfectly conducting flat plate, facing the source, that would reflect the same amount of power toward the receiver. Since real targets are neither perfect conductors nor flat planes with the correct orientation, the radar cross section is smaller than the actual cross-sectional area of the target, as seen from the radar installation.

The information given above can be quantified and expressed in a propagation equation called the *radar equation*.

$$P_R = \frac{\lambda^2 P_T G^2 \sigma}{(4\pi)^3 r^4} \tag{19.16}$$

where

P_R = received power in watts
λ = free-space wavelength
P_T = transmitted power in watts
G = antenna gain as a power ratio
σ = radar cross section of the target in square meters
r = range (distance to the target) in meters

Note the fourth-power variation of received signal strength with distance. The transmitting and receiving antennas are normally the same for a radar system, so their gains have been combined in the equation.

Example 19.10 A radar transmitter has a power of 10 kW and operates at a frequency of 9.5 GHz. Its signal reflects from a target 15 km away with a radar cross section of 10.2 m². The gain of the antenna is 20 dBi. Calculate the received signal power.

Solution First, since Equation (19.16) requires wavelength, rather than frequency, and antenna gain must be expressed as a power ratio, some preliminary calculations are necessary:

$$\begin{aligned}\lambda &= \frac{c}{f} \\ &= \frac{3 \times 10^8}{9.5 \times 10^9} \\ &= 0.0316 \text{ m} \\ G &= \text{antilog} \frac{\text{dB}}{10} \\ &= \text{antilog} \frac{20}{10} \\ &= 100\end{aligned}$$

Next, substituting into Equation (19.16):

$$\begin{aligned}P_R &= \frac{\lambda^2 P_T G^2 \sigma}{(4\pi)^3 r^4} \\ &= \frac{0.0316^2 (10 \times 10^3)(100^2)(10.2)}{(4\pi)^3 (15 \times 10^3)^4} \\ &= 10.1 \times 10^{-15} \text{ W} \\ &= 10.1 \text{ fW}\end{aligned}$$

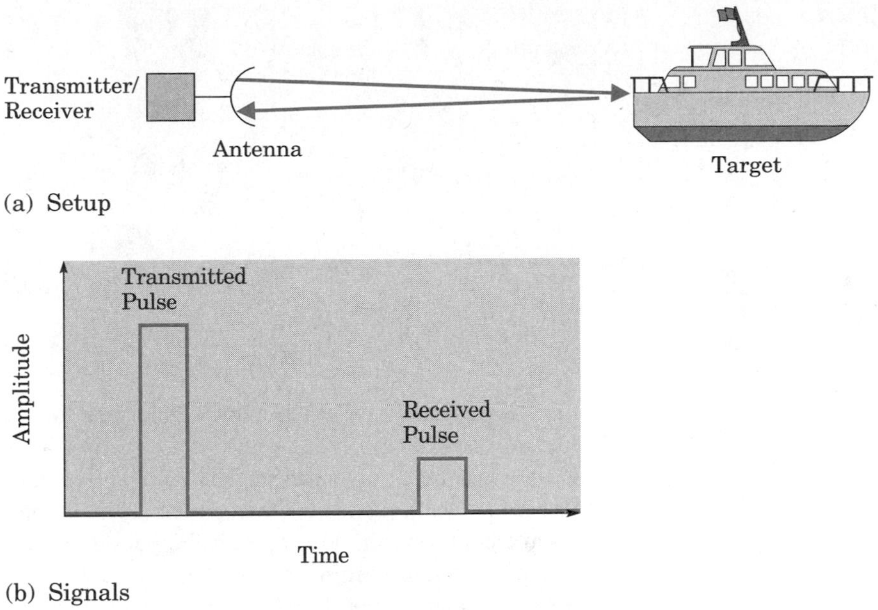

Figure 19.45 Radar range determination

19.7.1 Pulse Radar

Determining the direction from the radar antenna to a target can be very simple. Either the antenna is moved physically, or, in the case of a phased-array antenna, its radiation pattern is moved electronically. When the received signal is strongest, the main lobe of the antenna is pointed at the target. The type of antenna and the way in which it is moved depend on the target characteristics. For instance, radar on a ship intended to locate shoreline features and other ships need only scan in azimuth (that is, horizontally), while radar on an airplane designed to track other airplanes have to scan in both azimuth and elevation. The angular resolution depends on the beamwidth of the antenna.

The distance, or range, to the target can be found by a method analogous to that used with time-domain reflectometry (see Chapter 16). A brief pulse of radio-frequency energy is emitted, and the time that elapses before its return is measured. From this elapsed time, the range can easily be calculated (see Figure 19.45). Of course, the measured time is that for a round trip to the target and back, so the one-way distance is found by dividing by two. Assuming free-space propagation,

$$R = \frac{ct}{2} \tag{19.17}$$

where

R = distance to the target
c = velocity of light
t = time taken for the echo to return

Example 19.11 A pulse sent to a target returns after 15 μs. How far away is the target?

Solution From Equation (19.17),

$$\begin{aligned} R &= \frac{ct}{2} \\ &= \frac{(3 \times 10^8)(15 \times 10^{-6})}{2} \\ &= 2250 \text{ m} \\ &= 2.25 \text{ km} \end{aligned}$$

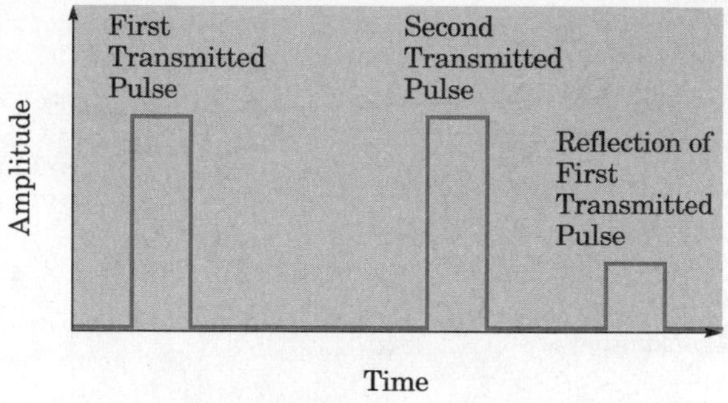

Figure 19.46 Radar range ambiguity

Of course, some of the time taken may be time spent by the signal getting from the electronics to the antenna and back. If significant, this time would have to be subtracted from the total time to get the time for the signal to travel from the antenna to the target and back. It is this net time that is called for in Equation (19.17).

A problem can arise with pulse radar if the period between pulses is less than the time taken for a pulse to return from the target. Figure 19.46 illustrates the problem. There is an ambiguity: the radar cannot distinguish between this target and a much closer one. This means that any pulse radar has a maximum unambiguous range that is limited by the pulse repetition rate. It is easy to calculate this range: it is the distance a signal can travel between pulses, divided by two, of course.

$$R_{max} = \frac{cT}{2} \tag{19.18}$$

where

R_{max} = maximum unambiguous range
c = velocity of light
T = pulse period

Often, the pulse repetition rate (frequency) is used instead of the period. Equation (19.18) then becomes

$$R_{max} = \frac{c}{2f} \tag{19.19}$$

where

f = pulse repetition rate

There is another possible difficulty. The transmitted pulse has a finite pulse duration, also called a *pulse width*. If the echo returns while the pulse is still being transmitted, it will not be detected by the receiver. The minimum usable range for the radar, then, is one-half the distance the signal can travel during the time it takes to transmit the pulse:

$$R_{min} = \frac{cT_P}{2} \tag{19.20}$$

where

R_{min} = minimum usable range
c = velocity of light
T_P = pulse duration

Example 19.12 A pulse radar emits pulses with a duration of 1 μs and a repetition rate of 1 kHz. Find the maximum and minimum range for this radar.

Solution From Equation (19.19), the maximum unambiguous range is

$$\begin{aligned} R_{max} &= \frac{c}{2f} \\ &= \frac{3 \times 10^8}{2 \times 1000} \\ &= 150 \times 10^3 \text{ m} \\ &= 150 \text{ km} \end{aligned}$$

Of course, this range is subject to the signal being able to propagate that far. If this radar were installed on a ship for the purpose of observing objects on the water, the range would actually be limited to the line-of-sight communication distance.

The minimum range is found from Equation (19.20):

$$\begin{aligned} R_{min} &= \frac{cT_P}{2} \\ &= \frac{(3 \times 10^8)(1 \times 10^{-6})}{2} \\ &= 150 \text{ m} \end{aligned}$$

Obviously, the use of short pulses improves the performance of radar at short range. It also improves the ability of the system to separate targets that are in the same direction but at different ranges, as illustrated in Figure 19.47. Short pulses also result in a lower duty cycle for the transmitter, reducing the average power requirement for a given pulse power. The disadvantage of short pulses is that they increase the bandwidth of the signal. A wider signal bandwidth requires a wider receiver bandwidth, reducing the signal-to-noise ratio, and also increases the congestion of the spectrum.

19.7.2 Doppler Radar

Pulse radar can measure velocity only in an indirect way: by finding the position of a target at two different times and calculating how far it has moved in a given time interval. It is possible to use radar to measure velocity directly, with certain limitations. Radars that do this use the Doppler effect, which causes the frequency of an echo to differ from that of the transmitted signal when there is relative motion between the radar and the target, along a line joining the two. Motion that closes the gap between source and target raises the frequency of the reflection, and motion in the other direction reduces it.

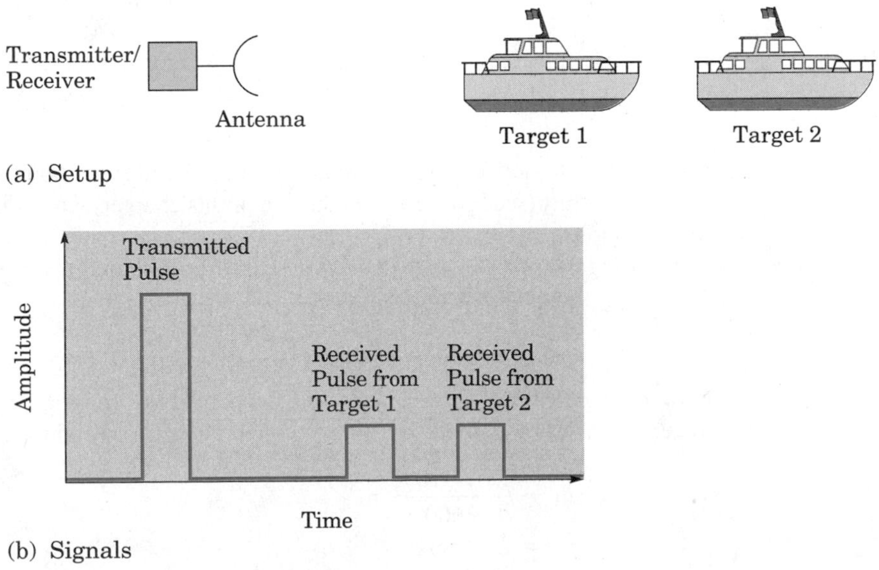

Figure 19.47 Discrimination of targets in same direction

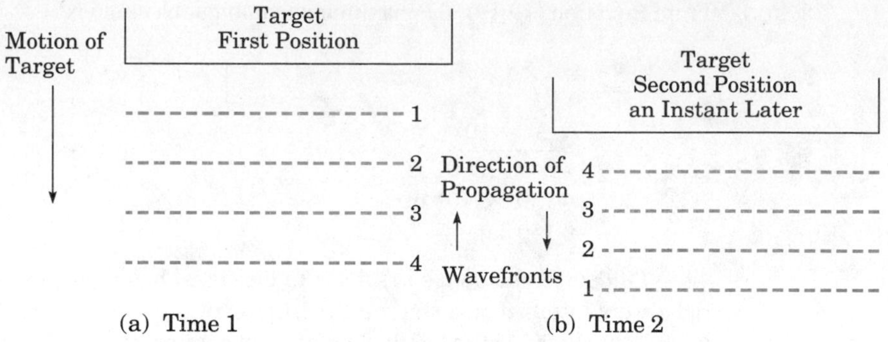

Figure 19.48 Doppler effect

To understand the Doppler effect, consider a plane wave emitted from a stationary source impinging on a reflecting plane that is moving toward the source, as sketched in Figure 19.48. The wavefronts are shown as dashed lines. As soon as front 1 reaches the surface, it reflects. Since the surface is moving in the opposite direction to that of the wave, peak 2 reaches the surface after a shorter interval than it would have had the target been stationary. As soon as peak 2 reaches the target, it reflects. Thus, the period of the reflected wave has been reduced. Since the speed of propagation does not change, this means that the frequency of the reflected wave has correspondingly increased. A similar logic shows that the frequency of the reflection will decrease if the target is moving away from the source.

The equation governing the Doppler effect is quite simple:

$$f_D = \frac{2v_r f_i}{c} \tag{19.21}$$

where

f_D = Doppler shift in hertz
v_r = relative velocity of the source and target in meters per second along a line between them; it is positive if the two are closing (getting closer)
f_i = incident frequency in hertz
c = velocity of light in meters per second

Example 19.13 Find the Doppler shift caused by a vehicle moving toward a radar at 60 mph, if the radar operates at 10 GHz.

Solution It is possible to derive equations relating speed in miles per hour and kilometers per hour to Doppler shift (see Problem 39 at the end of this chapter), but let us do this example from first principles.

A logical first step is to convert miles per hour to kilometers per hour:

$$\begin{aligned}60 \text{ mph} &= 60 \times 1.6 \text{ km/h}\\ &= 96 \text{ km/h}\end{aligned}$$

Now, convert kilometers per hour to meters per second (multiply by 1000 to convert kilometers to meters and divide by 3600 to convert hours to seconds):

$$\begin{aligned}96 \text{ km/h} &= \frac{96 \times 1000}{3600} \text{ m/s}\\ &= 26.7 \text{ m/s}\end{aligned}$$

Finally, the Doppler shift can be found from Equation (19.21),

$$
\begin{aligned}
f_D &= \frac{2v_r f_i}{c} \\
&= \frac{2(26.7)(10 \times 10^9)}{3 \times 10^8} \\
&= 1.778 \text{ kHz}
\end{aligned}
$$

Doppler radar can measure only the velocity component along a line that joins the source and the target. This is, of course, not always the true velocity. For instance, Figure 19.49 shows a police radar setup at the side of a road. It is apparent that the speed measured by the radar is actually less than the true speed of the vehicle. In fact, a little trigonometry will show that

$$x = y \cos \theta \tag{19.22}$$

where

x = component of velocity measured by the radar
y = total velocity
θ = angle between the direction of travel and the direction of a line from the target to the radar

Since the transmitted and received frequencies are different, Doppler radars can use either CW or pulse techniques. Police speed radars, for instance, are CW. The transmitted and received signals are separated by a circulator, as shown in Figure 19.50. A small portion of the transmitted signal is allowed to mix with the received signal to produce a difference signal at the Doppler frequency. A frequency counter measures this frequency, and the result is converted into a speed readout in kilometers or miles per hour. Circuitry can also be provided to take the speed of the vehicle in which the radar is mounted into account, since of course it is the relative speed between police and target vehicle that is measured by the radar.

When applied to pulse radar, Doppler techniques allow the velocity of a target, as well as its position and direction, to be estimated. They also allow the elimination of stationary objects called *clutter* from the display.

19.7.3 Transponders

Radar echoes are generally simple reflections of the original pulse, not by any means images of the target. By using narrow beams, frequency sweeps during the pulse, and a large amount of computing power, it is sometimes possible to make some inferences about the target from its radar *signature*. In the case of friendly targets, however, the whole process can be made much more efficient by installing a transmitter-receiver, called a **transponder**, on the target, usually an airplane. The transponder responds to a radar pulse by transmitting a signal that can identify the aircraft, even down to its flight number and destination. In the

Figure 19.49 Doppler radar used in a speed trap

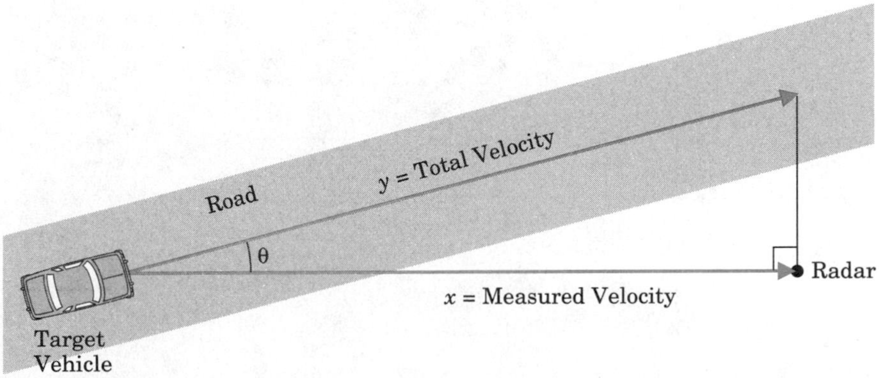

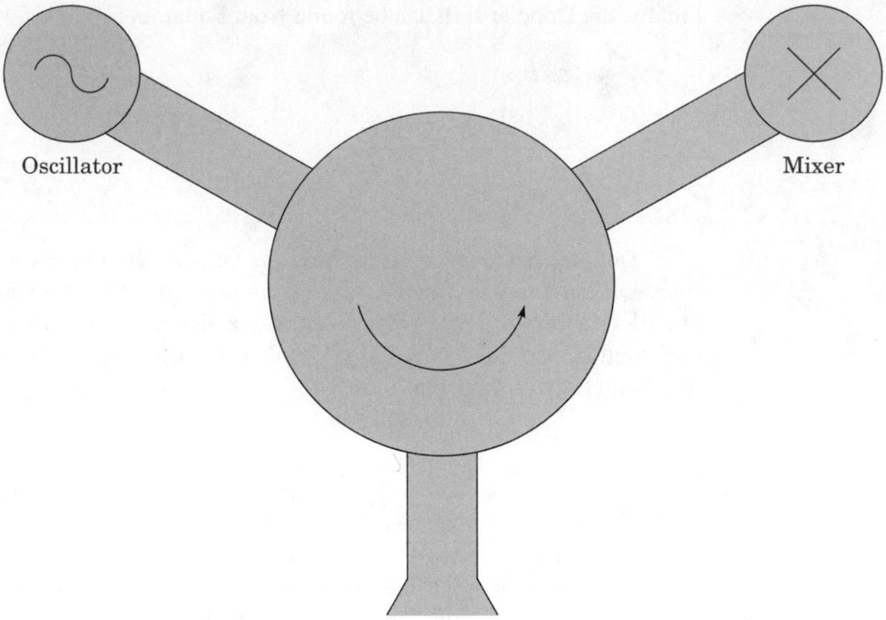

Figure 19.50
CW Doppler radar

case of military aircraft, transponders are part of an "identification friend or foe" (IFF) system designed to prevent friendly aircraft from being attacked.

19.7.4 Stealth

There are times when it would be better for an airplane not to appear on radar, for example, if the airplane is military and the radar belongs to an enemy anti-aircraft battery. The techniques for avoiding detection by radar are collectively known as *stealth*, and the details are, of course, kept secret. Still, the general ideas behind stealth are well known and easily understood.

The two basic ways of escaping radar detection are, first, to absorb radar waves rather than reflect them, and, second, to scatter any reflected signals as widely as possible to avoid returning a strong signal to the hostile radar. Achieving the first result involves the use of resistive materials in the body of the aircraft; sharp angles help to achieve the second. Figure 19.51 shows a stealth fighter developed by the United States military. Note the rather ungainly angular appearance.

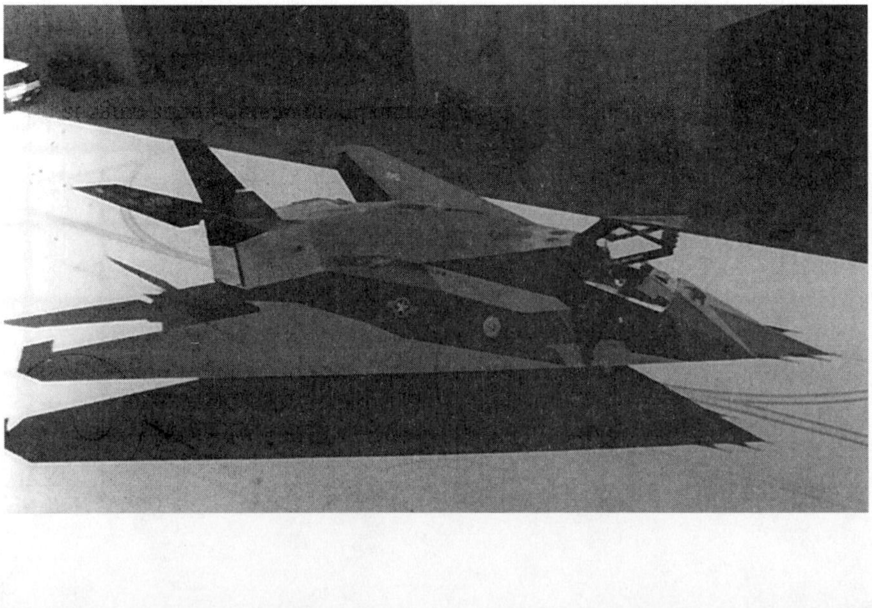

Figure 19.51
Stealth fighter
AP/Wide World Photos

In practice, stealth also involves reducing the *heat signature* of the aircraft, since some missiles use the hot exhaust gas from a jet engine to track their target.

Summary

Here are the main points to remember from this chapter.

1. As frequency increases into the gigahertz range, distributed capacitance and inductance become very important everywhere in a circuit.
2. Waveguides are a very practical means of transmitting electrical energy at microwave frequencies, as they have much lower losses than coaxial cable. They are not very useful at lower frequencies because they must be too large in cross section.
3. Waveguides are generally useful over only a 2:1 frequency range. They have a lower cutoff frequency that depends on their dimensions, and they exhibit dispersion due to multimode propagation at high frequencies.
4. Two velocities must be calculated for waveguides. The group velocity, which is lower than the speed of light, is the speed at which signals travel down the guide. The phase velocity, which is greater than the speed of light, is used for calculating the wavelength in the guide.
5. Power can be coupled into and out of waveguides using probes, loops, or holes in the guide.
6. Ferrites have special properties that make them just as useful at microwave frequencies as at lower frequencies. Applications of ferrites include circulators, isolators, attenuators, and resonators.
7. At microwave frequencies, conventional active components such as tubes and transistors suffer from excessive carrier transit time as well as stray capacitance and inductance. Solutions to these problems include redesigning standard component types as well as innovative solutions designed to include transit time as part of the device operation.
8. The short wavelengths in the microwave region allow antennas to be built with high gain and narrow beamwidth. The combination of a horn antenna with a parabolic reflector is very popular.
9. Radar systems are of two basic types. Pulse radar works by reflecting pulses from a target, whose distance is gauged from the time taken for pulses to return. Doppler radar estimates the velocity of a target by measuring the difference in frequency between transmitted and received signals.

Important Equations

$$a = \frac{\lambda_c}{2} \tag{19.1}$$

$$f_c = \frac{c}{2a} \tag{19.2}$$

$$v_g = c\sqrt{1 - \left(\frac{\lambda}{2a}\right)^2} \tag{19.3}$$

$$v_g = c\sqrt{1 - \left(\frac{f_c}{f}\right)^2} \tag{19.4}$$

$$v_g v_p = c^2 \tag{19.5}$$

$$v_p = \frac{c}{\sqrt{1 - \left(\frac{f_c}{f}\right)^2}} \tag{19.7}$$

$$Z_0 = \frac{377}{\sqrt{1 - \left(\frac{f_c}{f}\right)^2}}\ \Omega \tag{19.9}$$

$$\lambda_g = \frac{v_P}{f} \tag{19.10}$$

$$\lambda_g = \frac{\lambda}{\sqrt{1 - \left(\frac{f_c}{f}\right)^2}} \tag{19.12}$$

$$D = \frac{T_{on}}{T_T} \tag{19.14}$$

$$P_{avg} = P_P D \tag{19.15}$$

$$P_R = \frac{\lambda^2 P_T G^2 \sigma}{(4\pi)^3 r^4} \tag{19.16}$$

$$R = \frac{ct}{2} \tag{19.17}$$

$$R_{max} = \frac{cT}{2} \tag{19.18}$$

$$R_{max} = \frac{c}{2f} \tag{19.19}$$

$$R_{min} = \frac{cT_P}{2} \tag{19.20}$$

$$f_D = \frac{2v_r f_i}{c} \tag{19.21}$$

$$x = y \cos \theta \tag{19.22}$$

Glossary

buncher in a klystron, a cavity that velocity-modulates the electron beam

bunching velocity modulation of an electron beam

catcher in a klystron, a cavity that removes some of the energy from the electron beam and transfers it in the form of microwave energy to the output

cavity a space in which microwaves can resonate by means of in-phase reflections from the walls

circulator a device with three or more ports that allows an input to one port to emerge only at the next port in order

collector in a klystron or traveling-wave tube, the element that receives the electron beam; in a conventional tube, this element is called the anode

crossed-field tube a microwave tube in which the electric and magnetic fields are at right angles

directional coupler a device that launches or receives a wave in a transmission line or waveguide, in one direction only

dispersion variation of velocity as a function of frequency in a waveguide or medium

domain in a semiconductor, a concentration of charge

Doppler effect the change in frequency that occurs when a wave reflects from a moving object

Fresnel lens a lens that is stepped in order to reduce its size

group velocity the speed of transmission of a signal along a waveguide

Gunn device a slab of N-type gallium arsenide that can operate as an oscillator or amplifier by means of domain formation

hybrid tee a combination of E-plane and H-plane tees

IMPATT diode a junction device that can operate as an oscillator or amplifier, by means of avalanche breakdown

isolator a waveguide device that has low loss in one direction and high loss in the other

klystron a type of linear-beam microwave tube that uses velocity modulation of the electron beam
linear-beam tube a microwave tube in which electrons travel in a straight line down the length of the tube
magic tee see *hybrid tee*
magnetron a crossed-field microwave-tube oscillator in which electrons circle around the cathode under the influence of a magnetic field
microstrip a microwave transmission line constructed on a printed-circuit board, consisting of a single conductor on one side of the board and a ground plane on the other side
microwave conventionally, electromagnetic radiation in the range above approximately 1 GHz
mobility the speed of electron drift in a conductor or semiconductor
mode in a waveguide, a specific configuration of electric and magnetic fields that allows a wave to propagate
phase velocity the apparent speed of propagation along a waveguide based on the distance between wavefronts along the walls of the guide
PIN diode a three-layer diode (P-intrinsic-N) that can be used as a switch and an attenuator at microwave frequencies
precession in a ferrite, rotation of the axis of rotation of the electrons
radar cross section the equivalent size of a radar target, in terms of a perfectly conducting flat plate oriented toward the receiver
slow-wave structure in a microwave tube, any device that causes a wave to propagate at less than the speed of light, so that the electron beam and the wave move at approximately the same speed
stripline a microwave transmission line that consists of a conductor inside a circuit board, working against two ground planes, one on the top and one on the bottom of the board
target in radar, the object whose range, direction, and/or velocity is to be measured
transferred-electron device (TED) see *Gunn device*
transponder a transmitter-receiver combination
TRAPATT diode a variation of the IMPATT designed for high-power operation
traveling-wave tube (TWT) a linear-beam microwave tube in which an electron beam gives up energy to a slow-wave structure
tuning screw a metal object threaded into a waveguide to add capacitance or inductance
waveguide a hollow structure that has no center conductor but allows waves to propagate down its length
YIG yttrium-iron-garnet, a type of ferrite

Questions

1. Give the approximate limiting frequencies for the microwave portion of the spectrum.
2. Explain what is meant by the dominant mode for a waveguide and why the dominant mode is usually the one used.
3. Name the dominant mode for a common rectangular waveguide, and explain what is meant by the numbers in the designation.
4. Explain the difference between phase velocity and group velocity in a waveguide. Which of these is greater than the speed of light? Explain how this is possible.
5. State what is meant by dispersion, and show how it can arise in two different ways in a waveguide.
6. Give one difference between the use of characteristic impedance in a coaxial line and a waveguide.
7. Draw diagrams showing the correct positions in which to install a probe and a loop to launch the dominant mode in a rectangular waveguide.
8. Explain the function of a directional coupler, and draw a sketch of a directional coupler for a waveguide.
9. Compare stripline and microstrip construction techniques, and draw a sketch of each type.
10. Explain the operation of a hybrid tee.
11. The wavemeter shown in Figure 19.52 is based on a resonant cavity. Which way should the plunger be moved to tune it to a higher frequency?

Figure 19.52

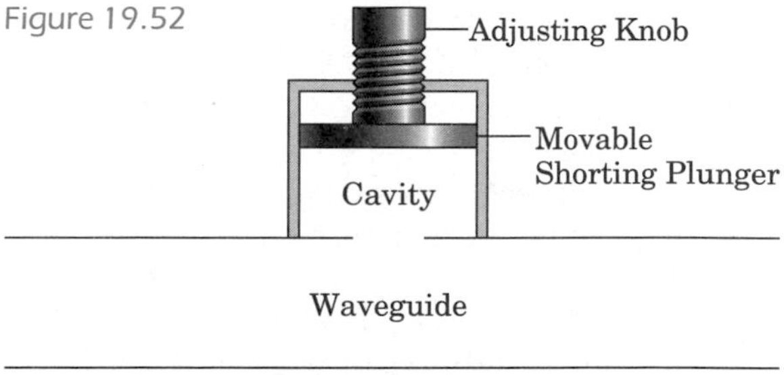

12. Sketch a four-port circulator and show what happens to a signal entering at each port.
13. Microwave transistors are always NPN if bipolar and N-channel if FETs. Why?
14. Explain what is meant by negative resistance, and give two examples of solid-state microwave devices that have this characteristic.
15. Give two applications each for PIN diodes and varactor diodes.

16. Characterize the microwave tubes studied with respect to the following factors: linear-beam or crossed-field, two or three terminals, amplifier or oscillator, and narrow- or wideband.
17. Explain how ambiguities can arise when measuring distance with a pulse radar installation.
18. Assuming that a Doppler radar has been correctly calibrated for targets moving directly toward the antenna, will its readings be high or low for targets moving toward it at an angle? Explain your answer.
19. How can velocity be measured with pulse radar?
20. How can a vehicle increase its visibility to radar? When would this be desirable?
21. How can a vehicle decrease its visibility to radar? When would this be desirable?

Problems

SECTION 19.2

22. RG-52/U waveguide is rectangular with an inner cross section of 22.86 by 10.16 mm. Calculate the cutoff frequency for the dominant mode and the range of frequencies for which single-mode propagation is possible.
23. For a 10 GHz signal in an RG-52/U waveguide (described in Problem 22), calculate:
    (a) the phase velocity
    (b) the group velocity
    (c) the guide wavelength
    (d) the characteristic impedance
24. A waveguide has a cutoff frequency for the TE_{10} mode of 3 GHz.
    (a) What is the cutoff frequency for the next higher mode?
    (b) Why is it desirable to use this waveguide above 3 GHz and below the frequency you found in part (a)?
    (c) A pulse of RF energy at a frequency of 5 GHz is sent down a 10 m length of this waveguide. How long does it take this pulse to get to the other end?
    (d) A designer wants to match a load to this waveguide. At a frequency of 5 GHz, what should the load impedance be?
25. A rectangular waveguide has an inner cross section of 8 by 16 mm. Find:
    (a) the cutoff frequency for the TE_{10} mode
    (b) the waveguide impedance at a frequency of 15 GHz
    (c) the guide wavelength at a frequency of 15 GHz
    (d) the group velocity at a frequency of 15 GHz
26. Explain how you could measure the insertion loss, coupling, and directivity of a directional coupler. Draw a block diagram and give step-by-step procedures. Include sample results for a source with an output of 0 dBm (measured at the input to the directional coupler) and specifications as follows:
    insertion loss: 0.5 dB
    coupling: 25 dB
    directivity: 35 dB

SECTION 19.3

27. An isolator has a loss of 1 dB in the forward direction and 20 dB in the reverse direction. If a power of 100 mW is applied to its input, calculate the amount of power that will reach a matched load.
28. Suppose the load in Problem 27 is short-circuited. How much power will be returned to the source?
29. What is the shortest length of a rectangular cavity made from a section of RG-52/U waveguide (see Problem 22) that will resonate at 10 GHz?
30. (a) Draw a sketch showing how a hybrid tee could be used to connect a transmitter and a receiver to the same antenna in such a way that the transmitter power would not reach the receiver.
    (b) Draw a sketch showing how a circulator could be used to do the same thing.
    (c) What advantage does the circulator version of this device have over the hybrid-tee version?

SECTION 19.4

31. What would be the approximate thickness of a Gunn device intended to operate in the Gunn mode at a frequency of 20 GHz?
32. A crystal oscillator operates at a frequency of 10 MHz with a stability of ±0.002%. A series of varactor multipliers is used to multiply this frequency to a nominal frequency of 6 GHz. By how much could the output frequency vary?

SECTION 19.5

33. A radar transmitter uses a magnetron to generate pulses with a power level of 1 MW. The pulses have a duration of 2 μs and the pulse repetition rate is 500 Hz. The magnetron has an efficiency of 60%. Calculate:
    (a) the duty cycle of the tube
    (b) the peak power input to the magnetron
    (c) the average power output from the magnetron
    (d) the average power input to the magnetron
    (e) the average power dissipated in the magnetron

SECTION 19.7

34. Calculate the time between the emission of a pulse and the detection of its reflection for a target at a distance of:

    (a) 100 m (b) 3 km (c) 75 km

35. Find the maximum and minimum useful ranges for a radar with a pulse duration of 50 μs and a pulse repetition rate of 200 Hz.

36. A radar transmits a pulse with a power of 50 kW, which is reflected from a target that is 20 km away and has a radar cross section of 3 m^2. The signal frequency is 20 GHz, and the antenna has a gain of 25 dB. Calculate the received signal strength.

37. A Doppler radar at 15 GHz has a return signal with a frequency 50 kHz higher than the transmitted signal. Assume that the radar installation is stationary.
    (a) What is the component of the target velocity along a line joining the radar and the target?
    (b) Is the target moving toward the radar or away from it?

38. Find the required transmission frequency for a Doppler radar if the frequency difference in hertz is to be equal to the target velocity in meters per second.

39. Rewrite Equation (19.21) so that the frequency difference in hertz is expressed in terms of the target velocity in kilometers per hour and the radar frequency in gigahertz.

40. A Doppler radar is set up beside a highway as shown in Figure 19.53. It measures the speed of a vehicle as 110 km/hr. What is the vehicle's actual speed

COMPREHENSIVE

41. An RG-50/U waveguide is rectangular with an interior cross section of 34.9 by 15.8 mm.
    (a) What is the name of the dominant mode in this waveguide?
    (b) Over what frequency range is the dominant mode the only mode of propagation?
    (c) What length of this guide is required to achieve a phase shift of 180° at 6 GHz?
    (d) Draw a dimensioned cross-sectional sketch showing how a probe can be used to couple power to this waveguide.
    (e) A signal with a bandwidth of 80 MHZ and a center frequency at 6 GHz is transmitted down 100 m of this waveguide. Calculate the dispersion, that is, the difference in the length of time it takes for the highest and lowest frequencies to propagate down the guide.

42. Convert the radar equation (Equation 19.16) to a more practical form: antenna gain should be expressed in dBi, distance in kilometers, and frequency in gigahertz.

43. An isolator has a forward loss of 0.7 dB and a return loss of 26 dB. A source provides 1 W to the isolator, and the load is resistive with an SWR of 3.
    (a) How much power is dissipated in the load?
    (b) How much power returns to the source?

44. Which of the active devices studied in this chapter would be most appropriate for each of the following applications? Why?
    (a) a radar transmitter with pulse power of 100 kW
    (b) a low-noise preamplifier operating at 4 GHz
    (c) a low-cost oscillator for low-power applications
    (d) an amplifier with a frequency range of 10 to 20 GHz and a power output of 100 W.

45. A pulse radar system has the block diagram shown in Figure 19.54. The specifications of the components are as follows:

    transmitter pulse power: 2 kW

Figure 19.53

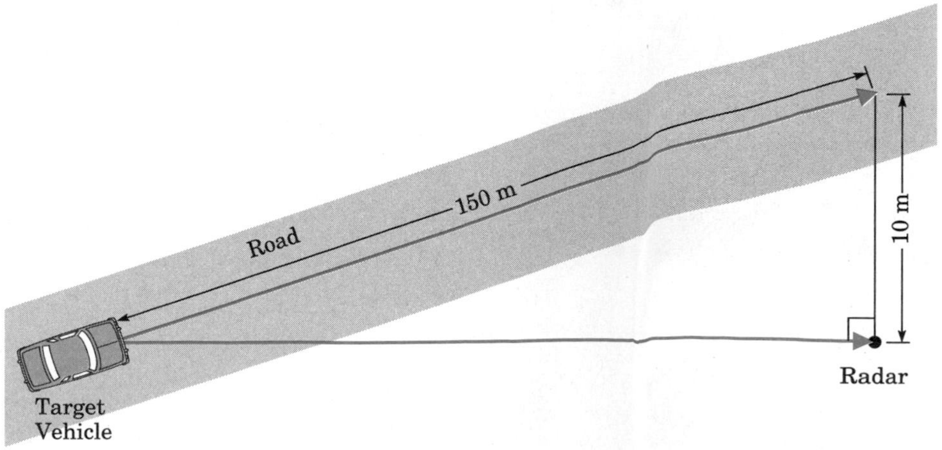

Figure 19.54

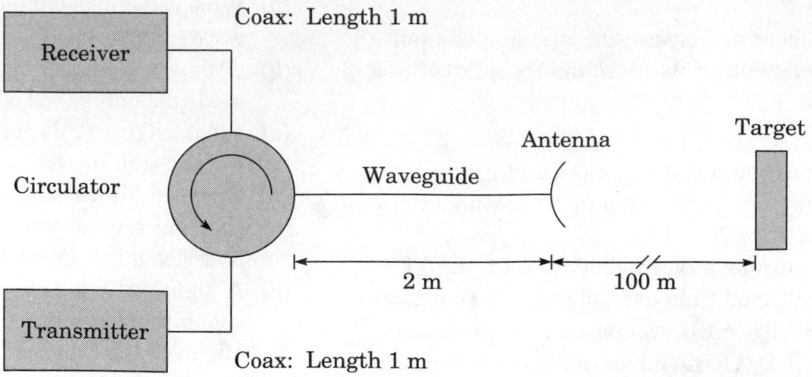

operating frequency: 15 GHz
coaxial cable: velocity factor 0.8, loss 0.5 dB/m
waveguide: internal dimensions 8 mm × 16 mm, loss 0.1 dB/m
antenna: gain 15 dBi
target radar cross section: 15 m^2
circulator loss 1 dB (each way)

(a) Calculate the time delay and loss (round trip) in the coax.
(b) Calculate the time delay and loss (round trip) in the waveguide.
(c) Calculate the time delay and loss (round trip) in free space.
(d) Calculate the total time delay from transmitter to receiver.
(e) Calculate the signal strength at the receiver, in dBm.

Calculate all losses in dB.

20 Terrestrial Microwave Systems

Objectives After completing this chapter, you should be able to:

1. Describe the basic structure and uses of microwave radio links
2. Explain the methods used in choosing sites for repeaters
3. Calculate the signal strength at the receiver for a variety of transmitter, antenna, and terrain configurations
4. Calculate the required clearance of a microwave path from an obstacle
5. Calculate the noise temperature and carrier-to-noise level of a microwave system
6. Calculate the energy per bit per noise density ratio for a microwave system
7. Explain fading and describe the diversity schemes that are used to overcome it
8. Describe the transmitting and receiving equipment used for FM, SSB, and digital systems
9. Describe the types of repeaters used for analog and digital systems and perform frequency calculations for these systems

20.1 Introduction

Point-to-point microwave radio links have many uses. They can be used as *studio-to-transmitter (STL) links* for radio and television broadcasting stations, and they can also link the *head-ends* (antenna sites) of many cable television installations to their distribution systems. Another very common application of microwave links is as part of a communications network involving telephone, data, or television signals.

Fiber-optic systems are being installed in preference to microwave radio for some fixed point-to-point services. The bandwidth available with fiber-optic systems is greater than with radio, and they require less maintenance. On the other hand, microwave relays are needed only at intervals of approximately 40 km, so microwave systems are easier to install in difficult terrain, such as in mountainous or wilderness areas. There is no doubt that terrestrial microwave systems will continue to be part of the evolving communications grid.

Some microwave systems use only one link or **hop**, while others are multihop systems that use **repeaters** to extend the system beyond the line-of-sight range of a single link. Figure 20.1 shows the two types of systems.

Microwave systems can also be classified by the modulation technique used. Older systems are analog; most employ frequency modulation, though some use single-sideband AM. Many of the newer systems use digital modulation schemes, generally quadrature amplitude modulation (QAM). All of these techniques have been discussed in previous chapters. The analog and digital systems are similar in many respects, since the radio propa-

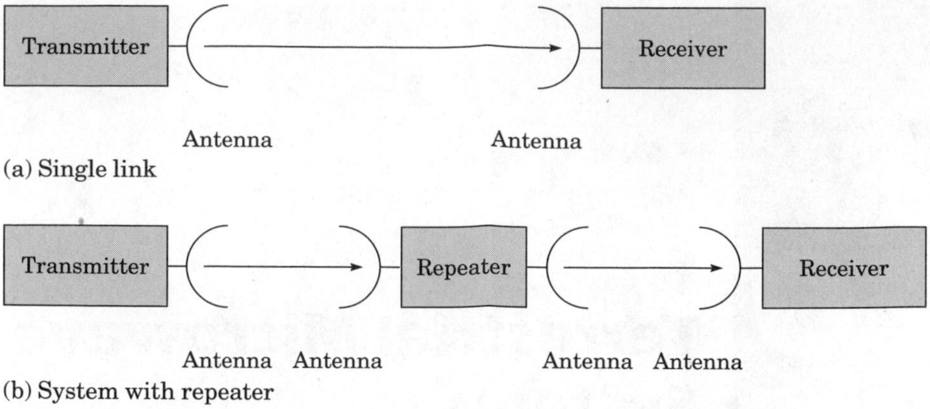

Figure 20.1 Microwave communication systems

gation part of the system is the same for both. The modulation and demodulation techniques and the design of repeaters are the main differences between the two types.

This chapter will first review microwave propagation and then discuss the components used in microwave links, both analog and digital. Because these systems are used in commercial applications where reliability is of utmost importance, we will also consider the means of ensuring the greatest possible reliability. Practical systems often have reliability in the range of 99.99%, that is, the system may be "down" for about one hour per year, or even less.

20.2 Terminal and Repeater Siting

In general, a microwave system should use as few repeaters as possible. Repeaters cost money, of course, and each one increases the chances of an equipment breakdown that can disable the link. More importantly, additional links contribute to noise levels in analog systems and increase the **jitter** in digital ones.

On the other hand, repeater stations must not be located beyond line-of-sight propagation range from each other, and all sorts of practical considerations can prevent certain sites from being used for repeaters. Land must be acquired, access and electrical power must be arranged, and the topography must be inspected for the best repeater sites, preferably on high points of the terrain.

Terrestrial microwave systems use relatively low power transmitters with high-gain parabolic or hog-horn antennas. By concentrating the transmitted power into a narrow beam, these antennas increase the effective power and reduce interference to and from other systems. There are limits to antenna gain, however. Gain figures greater than 45 dB should be avoided because antennas with this much gain have such narrow beamwidth (less than one degree) that mounting requirements are severe—slight motion of an antenna tower due to wind can be sufficient to cause signal loss with such narrow antenna beams.

Feedlines between transmitters and/or receivers and antennas are almost always constructed from waveguide at frequencies above 2 GHz, to reduce losses. At lower frequencies, coaxial cable may be used.

20.2.1 Path Calculations

Terrestrial microwave links generally use line-of-sight (LOS) propagation (described in Chapter 17). The name is not quite accurate, as radio waves actually travel about one-third farther than visible light under normal conditions. The maximum distance between two stations depends on the height of the transmitting and receiving antennas as well as on the nature of the terrain between them. As previously discussed, for average atmospheric conditions and level terrain, the distance over which line-of-sight propagation is possible is given by:

$$d = \sqrt{17\,h_T} + \sqrt{17\,h_R} \tag{20.1}$$

where

d = maximum distance in kilometers
h_T = height of the transmitting antenna in meters
h_R = height of the receiving antenna in meters

Example 20.1 Suppose that the transmitter and receiver towers have equal height. How high would they have to be to communicate over a distance of 40 km?

Solution From Equation (20.1),

$$d = \sqrt{17h_T} + \sqrt{17h_R}$$

Let $h_T = h_R = h$ and $d = 40$ km. Then

$$d = 2\sqrt{17h}$$

$$h = \frac{d^2}{68} = \frac{40^2}{68} = 23.5 \text{ m}$$

Each tower must be at least 23.5 m in height.

Equation (20.1) gives approximate results only. For accurate calculations, it is necessary to obtain **topographic maps** of the area and note any obstacles in the path between transmitter and receiver. The path is then plotted on graph paper. There are two common methods of doing this. The first is to use ordinary graph paper and adjust the heights of obstacles and the terrain to compensate for the curvature of the earth. The second way is to use special graph paper known as *4/3 earth paper*, on which the horizontal lines are curved to represent the earth's curvature. As mentioned, for radio waves the earth looks about one-third larger than it really is, hence the term *4/3 earth paper*. An example is shown in Figure 20.2.

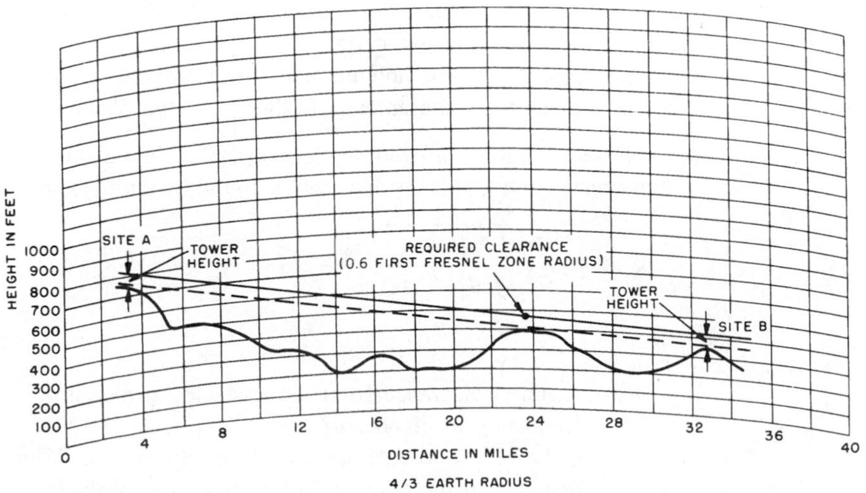

Figure 20.2
4/3 earth paper
From *Radio System Design for Telecommunications,* © 1987 by Roger L. Freeman. Reprinted by permission of John Wiley & Sons, Inc.

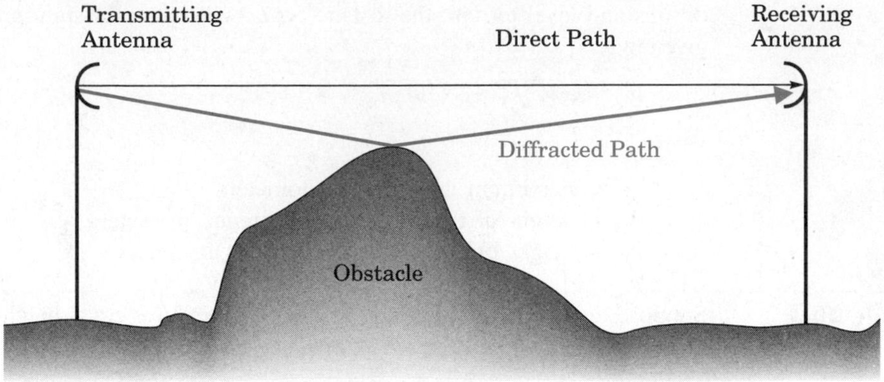

Figure 20.3 Interference due to diffraction

20.2.2 Diffraction

It is obviously necessary to avoid obstruction of the signal by large obstacles such as mountains, but it is also important to avoid **diffraction**, which occurs if the radio beam is partly obstructed by an object on the ground. Diffraction can cause a second signal to appear at the receiver, and the two signals, depending on their relative phase angles, may cancel each other out to some extent, resulting in **fading** of the signal.

The effects of diffraction can be reduced by making sure that the path from transmitting antenna to receiving antenna clears an obstacle by at least 60% of a distance known as the first **Fresnel zone**. Fresnel zones derive from the Huygens-Fresnel wave theory, which states that an object that diffracts waves acts as if it were a second source of those waves. The direct and refracted waves add, but due to the difference in path length for the two rays, the interference is sometimes constructive (when the direct and diffracted signals are in phase), and sometimes destructive (when they are out of phase). The areas where the interference is constructive are called Fresnel zones. As the distance between the direct path and the refracting object increases, the strength of the diffracted signal decreases and the interference becomes less pronounced. Figure 20.3 shows how interference takes place.

The distance from an obstacle to any Fresnel zone, in the direction at right angles to the propagation path, is given by

$$R_n = 17.3\sqrt{\frac{nd_1d_2}{f(d_1 + d_2)}} \tag{20.2}$$

where

R = distance to the Fresnel zone, in meters
n = number of the Fresnel zone
f = frequency in gigahertz
d_1 = distance to the antenna nearer the obstacle, in kilometers
d_2 = distance to the antenna farther from the obstacle, in kilometers

As a general rule, it is sufficient to clear obstacles by a distance corresponding to 60% of the first Fresnel zone. This distance can be found by modifying Equation (20.2) slightly:

$$R = 10.4\sqrt{\frac{d_1d_2}{f(d_1 + d_2)}} \tag{20.3}$$

where

R = required clearance from the obstacle, in meters
f = frequency in gigahertz
d_1 = distance to the antenna nearer the obstacle, in kilometers
d_2 = distance to the antenna farther from the obstacle, in kilometers

Example 20.2 A line-of-sight radio link operating at a frequency of 6 GHz has a separation of 40 km between antennas. An obstacle in the path is located 10 km from the transmitting antenna. By how much must the beam clear the obstacle?

Solution From Equation (20.3),

$$R = 10.4\sqrt{\frac{d_1 d_2}{f(d_1 + d_2)}}$$

$$= 10.4\sqrt{\frac{10 \times 30}{6\ (10 + 30)}}$$

$$= 11.6\text{m}$$

20.3 System Gain Calculations

Once the path for a microwave link has been determined, it is necessary to ensure that the received signal power is sufficient for the required signal-to-noise ratio. The variables involved here include the power of the transmitter, the receiver noise figure, and the gain of their respective antennas.

The reader should recall from the basic theory of electromagnetic propagation (Chapter 17) that the power density of a signal under free-space propagation conditions is given by the equation

$$P_D = \frac{\text{EIRP}}{4\pi r^2} \tag{20.4}$$

where

P_D = power density in watts per square meter
$EIRP$ = effective isotropic radiated power in watts
r = distance from antenna in meters

A more convenient version of the same equation is

$$\frac{P_R}{P_T}\ (\text{dB}) = G_T\ (\text{dBi}) + G_R\ (\text{dBi}) - (32.44 + 20 \log d + 20 \log f) \tag{20.5}$$

where

P_R/P_T (dB) = ratio of received to transmitted power, expressed in decibels
G_T (dBi) = gain of transmitting antenna in decibels with respect to an isotropic radiator
G_R (dBi) = gain of receiving antenna in decibels with respect to an isotropic radiator
d = distance between transmitter and receiver in kilometers
f = frequency in megahertz

Example 20.3 A transmitter and a receiver operating at 6 GHz are separated by 40 km. How much power (in dBm) is delivered to the receiver if the transmitter has an output power of 2 W, the transmitting antenna has a gain of 20 dBi, and the receiving antenna has a gain of 25 dBi?

Solution From Equation (20.5),

$$\frac{P_R}{P_T}\text{ (dB)} = G_T\text{ (dBi)} + G_R\text{ (dBi)} - (32.44 + 20\log d + 20\log f)$$
$$= 20 + 25 - (32.44 + 20\log 40 + 20\log 6000)$$
$$= -95\text{ dB}$$

$$P_T\text{ (dBm)} = 10\log\frac{2\text{ W}}{1\text{ mW}} = 33\text{ dBm}$$

$$P_R\text{ (dBm)} = 33\text{ dBm} - 95\text{ dB} = -62\text{ dBm}$$

20.3.1 Carrier-to-Noise Ratio

Received signal power is calculated to determine whether the system noise performance is satisfactory. For analog microwave systems, satisfactory performance is normally defined as a **carrier-to-noise ratio** that exceeds a given number of decibels. The carrier-to-noise ratio is simply the signal-to-noise ratio measured before the signal is demodulated, that is, while it still has a carrier. With frequency modulation, you should recall, the signal-to-noise ratio can actually be greater after detection than before.

In order to determine the carrier-to-noise ratio, we need the signal power and the noise power. The signal power can be found using Equation (20.5). The noise consists mainly of thermal noise, either received by the antenna (space noise) or generated in the antenna, transmission line, or receiver. The easiest way to combine these sources is to find the corresponding noise temperature for each, referenced to the receiver input, add the noise temperatures, and then find the noise power from the equation given in Chapter 1:

$$P_N = kTB \tag{20.6}$$

where

P_N = noise power in watts
k = Boltzmann's constant, 1.38×10^{-23} joules/kelvin (J/K)
T = absolute temperature in kelvins (K)
B = noise power bandwidth in hertz

Let us examine the antenna first. The antenna receives noise from the sky and perhaps from the earth as well, depending on whether the beam of the antenna includes the ground. The noise temperature of the sky depends on the angle of elevation of the antenna, the frequency, and atmospheric conditions. For a terrestrial microwave link with an antenna elevation of 0° and a frequency of 6 GHz, a typical sky noise temperature would be 120 K. Resistive losses in the antenna and its feedline must also be taken into account to get an equivalent noise temperature at the receiver input. Assuming the feedline is at the reference temperature of 290 K (17° C), the equivalent noise temperature at the receiver input is given by:

$$T_a = \frac{(L - 1)\,290 + T_{sky}}{L} \tag{20.7}$$

where

T_a = effective noise temperature of antenna and feedline, in kelvins, referenced to receiver antenna input
L = loss in feedline and antenna as a ratio of input to output power (not in decibels)
T_{sky} = effective sky temperature, in kelvins

Example 20.4 In a microwave system, the antenna sees a sky temperature of 120 K, and the antenna feedline has a loss of 2 dB. Calculate the noise temperature of the antenna/feedline system, referenced to the receiver input.

Solution First it is necessary to change the feedline loss into a power ratio:

$$L = \text{antilog}\ (2/10) = 1.58$$

Now we can find the noise temperature from Equation (20.7):

$$\begin{aligned} T_a &= \frac{(L - 1)\,290 + T_{sky}}{L} \\ &= \frac{(1.58 - 1)\,290 + 120}{1.58} \\ &= 182\ \text{K} \end{aligned}$$

Once the equivalent noise temperature at the receiver input (due to the antenna and feedline) has been found, it is only necessary to add the noise temperature of the receiver. Sometimes receiver noise temperature is specified, but often the noise performance of a receiver is given as a noise figure. In the latter case, it is easy to convert noise figure to noise temperature, as explained in Chapter 1, by using the following equation:

$$T_{\text{eq}} = 290\ (NF - 1) \tag{20.8}$$

where

T_{eq} = equivalent noise temperature in kelvins (K)
NF = noise figure as a ratio (not in dB)

Example 20.5 A receiver has a noise figure of 2 dB. Calculate its equivalent noise temperature.

Solution First it is necessary to convert the noise figure to a ratio:

$$NF = \text{antilog}\ (2/10) = 1.58$$

Now we can use Equation 20.8:

$$\begin{aligned} T_{\text{eq}} &= 290\ (NF - 1) \\ &= 290\ (1.58 - 1) \\ &= 168\ \text{K} \end{aligned}$$

Once the noise temperatures of the antenna-feedline combination and the receiver have been determined, they can be summed to find the noise temperature of the system. From that information and the system bandwidth, it is easy to determine the noise power.

Example 20.6 The antenna and feedline combination from Example 20.4 is used with the receiver from Example 20.5. Calculate the thermal noise power in dBm, referred to the receiver input, if the receiver has a bandwidth of 20 MHz.

Solution First add the noise temperatures:

$$\begin{aligned} T_N(\text{system}) &= T_a + T_{eq} \\ &= 182\ \text{K} + 168\ \text{K} \\ &= 350\ \text{K} \end{aligned}$$

Now use Equation (20.6) to calculate the noise power:

$$\begin{aligned} P_N &= kTB \\ &= 1.38 \times 10^{-23}\ \text{J/K} \times 350\ \text{K} \times 20\ \text{MHz} \\ &= 96.6\ \text{fW} \end{aligned}$$

Convert to dBm:

$$P_N \text{ (dBm)} = 10 \log (96.6 \text{ fW} / 1 \text{ mW})$$
$$= -100 \text{ dBm}$$

Given the signal power at the antenna input, the carrier-to-noise ratio can be found.

Example 20.7 Calculate the carrier-to-noise ratio, in decibels, for the signal in Example 20.3, received by the installation in Example 20.6.

Solution From Example 20.3,

$$P_R = -62 \text{ dBm}$$

From Example 20.6,

$$P_N = -100 \text{ dBm}$$

Therefore,

$$C/N = -62 - (-100) = 38 \text{ dB}$$

20.3.2 Energy per Bit per Noise Density Ratio

In digital communication schemes, it is common to specify the noise performance of a system in terms of the ratio, in decibels, of the energy per received information bit to the noise density. Recall that the noise density is the noise power in one hertz of the spectrum. The energy per bit per noise density is directly related to the bit error rate for the system.

The **energy per bit** is the energy received in the time taken to transmit one bit. That is, it is the received signal power multiplied by the period for one bit. Since period is the reciprocal of frequency, it is easy to calculate the energy per bit if the received signal power and the bit rate are given:

$$E_b = \frac{P_R}{f_b} \tag{20.9}$$

where

E_b = energy per bit in joules
P_R = received signal power in watts
f_b = bit rate in bits per second

The noise power density is given simply by

$$N_0 = kT \tag{20.10}$$

where

N_0 = noise power density in watts per hertz
k = Boltzmann's constant, 1.38×10^{-23} joules/kelvin (J/K)
T = temperature in kelvins

Example 20.8 The system in Example 20.7 operates at a bit rate of 40 Mb/s. Calculate the energy per bit to noise density ratio, in decibels.

Solution First, the received power must be expressed in watts rather than in dBm:

$$P_R = \text{antilog } (-62/10) \text{ mW}$$
$$= 631 \text{ pW}$$

From Equation (20.9),

$$\begin{aligned} E_b &= \frac{P_R}{f_b} \\ &= \frac{631 \text{ pW}}{40 \text{ Mb/s}} \\ &= 15.8 \times 10^{-18} \text{ J} \end{aligned}$$

The noise power density is found from Equation (20.10):

$$\begin{aligned} N_0 &= kT \\ &= 1.38 \times 10^{-23} \times 350 \\ &= 4.83 \times 10^{-21} \text{ W/Hz} \end{aligned}$$

The energy per bit to noise density ratio, in decibels, is then:

$$\frac{E_b}{N_0} = 10 \log \frac{15.8 \times 10^{-18}}{4.83 \times 10^{-21}} = 35.1 \text{ dB}$$

This value would likely give satisfactory performance. Required E_b/N_0 ratios for a bit error rate of 1×10^{-4} vary from about 10 to 20 dB for practical digital modulation schemes.

20.3.3 Fading

Any microwave system must include an allowance for **fading** in its system gain calculations. Fading is a reduction in signal strength below its nominal level. It has a number of causes, including

- *multipath reception*, in which a direct signal is partially cancelled by reflections from ground or water
- *attenuation due to rain*, mainly at frequencies above 10 GHz
- *ducting*, in which signals are deflected by layers of different temperature and humidity in the atmosphere
- *aging or partial failure* of transmitting or receiving equipment

There are two basic methods for dealing with fading. One is to overbuild the system, that is, to make sure that the system has sufficient gain to achieve the desired signal-to-noise ratio or bit error rate, even when fading takes place. The additional gain, called *fade margin*, can be obtained by increasing transmitter power, antenna gain, or receiver sensitivity.

Unfortunately, fading due to multipath reception can reduce the received signal strength by 20 decibels or more. A brute force approach to obtaining a satisfactory fade margin could therefore require increasing the transmitter power by a factor of 100. However, another approach, called **diversity**, can reduce fading without resorting to extreme power levels.

Often fading is *frequency-selective*, so changing the frequency slightly can eliminate the problem. For instance, suppose the signal reaches the receiving antenna by two paths, one direct and the other reflected from water, as shown in Figure 20.4. Fading occurs when the path lengths are such that there is a 180° phase difference between the two signals. One might think that the path could simply be adjusted by moving either the transmitting or receiving antenna to produce constructive interference. However, variations in atmospheric conditions can easily change the path length enough, at microwave wavelengths, to change the interference from constructive back to destructive.

Fading due to multipath propagation can be avoided by slightly changing the frequency (and therefore the wavelength) so that the phase difference between the direct and reflected signals is no longer 180°. This technique is called *frequency diversity*.

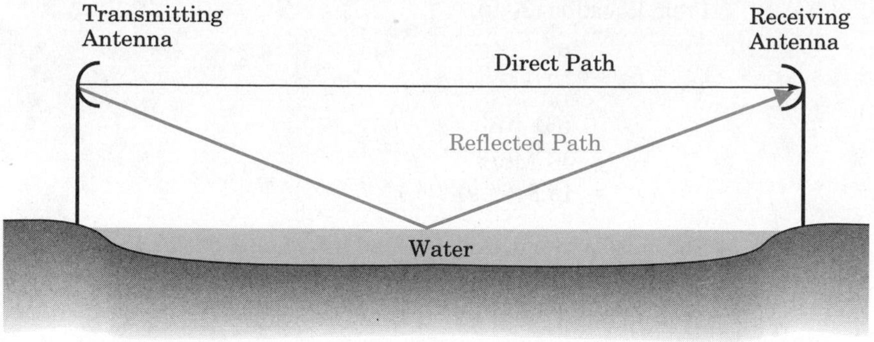

Figure 20.4
Multipath fading

To protect against fading on a moment-to-moment basis, frequency diversity requires two transmitters and two receivers, separated in frequency. Ideally the separation should be at least 5%, though in practical situations a shortage of spectrum space often limits the separation to 2%. Providing dual transmitters and receivers is expensive, but it also allows *hot standby* protection—if one transmitter or receiver should fail, communication can continue uninterrupted. Figure 20.5 illustrates frequency diversity.

Sometimes spectrum requirements are such that an extra channel cannot be obtained to provide frequency diversity. Another way to prevent multipath fading is to change the path length by moving either the transmitting or the receiving antenna. This technique, called *space diversity*, generally involves placing two antennas one above the other on the same tower. The two antennas should be separated by 200 wavelengths or more. At 6 GHz, a separation of 10 meters is required. Since the lower of the two antennas must be high enough for reliable line-of-sight communication, space diversity requires taller towers as well as more antennas. Space diversity is illustrated in Figure 20.6.

Figure 20.5
Frequency diversity with hot standby protection

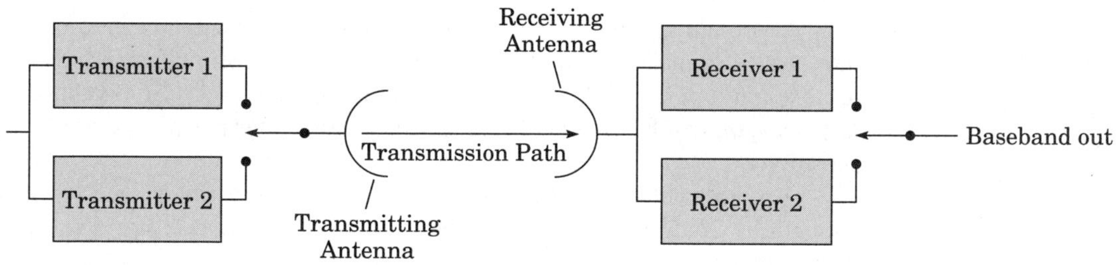

Figure 20.6
Space diversity

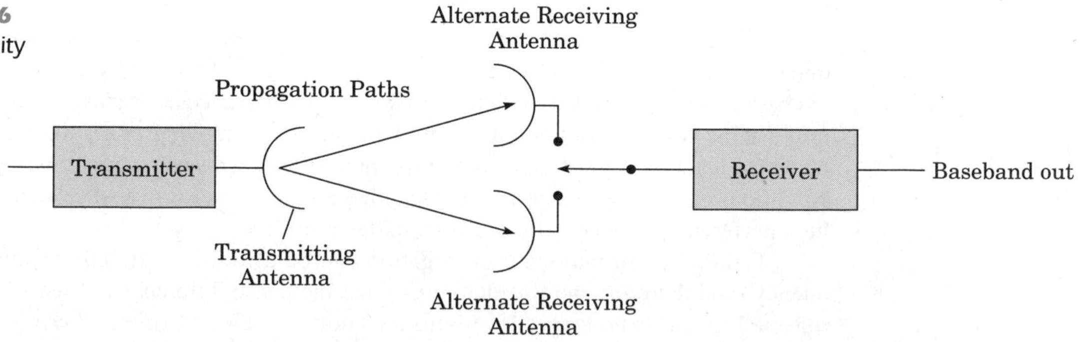

20.4 Transmitting and Receiving Equipment

Terrestrial microwave links can be either analog or digital systems. Most new systems are digital, but there are still many analog installations in use. Analog systems use either frequency modulation (FM) or single-sideband suppressed-carrier amplitude modulation (SSB). Digital systems use phase-shift keying or quadrature amplitude modulation (QAM). We shall examine each of these methods in turn.

20.4.1 FM Systems

The transmitter and receiver topologies shown in Figure 20.7 closely resemble those in use at lower frequencies. Typically the IF is 70 MHz and narrowband FM is used, so that the transmitted bandwidth is only slightly more than twice that of the baseband signal. The baseband signal shown at the transmitter input and receiver output can consist of a single broadcast-quality television signal or a number of telephone signals combined using frequency-division modulation. Typical FDM systems in the 6 GHz band carry from 300 to 600 channels. In the 12 GHz band, from 600 to 1800 voice channels (one to three mastergroups) can be carried per FM carrier.

The transmitter block diagram in Figure 20.7(a) is quite conventional. The FM signal is generated at a relatively low frequency, then mixed up to the required transmit carrier frequency. The local oscillator shown can be a microwave oscillator such as a reflex klystron or a Gunn device, or it can be a VHF transistor oscillator followed by a frequency multiplier using varactors. Output powers are generally under 10W, with 1–2W being typical. A traveling-wave tube (TWT) is commonly used in the output stage, though for power levels near the low end of the above range, a solid-state power amplifier using field-effect transistors can be used.

The receiver block diagram in Figure 20.7(b) shows an RF amplifier before the mixer, but this is not always present, particularly in older designs. Until recently it was easier to build a low-noise mixer (often using a Schottky diode) than a low-noise microwave amplifier. Low-noise amplifiers involved exotic technology such as parametric amplifiers and masers and were not economical enough for everyday use. However, most recent microwave receiver designs use GaAsFET transistors in a low-noise amplifier (LNA) before the mixer.

The intermediate frequency is generally 70 MHz, but the IF bandwidth depends on the bandwidth of the signal, which varies with the baseband signal bandwidth.

Figure 20.7 Microwave transmitter and receiver

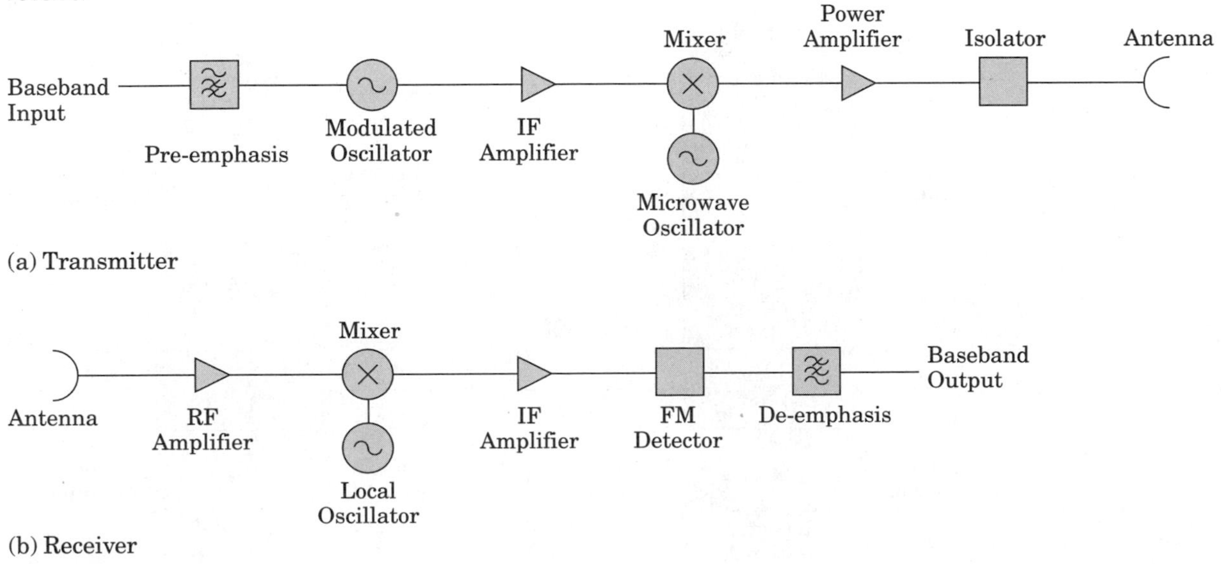

A single installation can use more than one carrier, depending on the required traffic capacity and the available spectrum space. For instance, a 12 GHz system could use 12 carriers, with a 20 MHz separation between carriers to allow room for sidebands.

Example 20.9 In the receiver in Figure 20.7 (b), the received carrier frequency is 6870 MHz and the IF is 70 MHz. Calculate the local oscillator frequency if the receiver uses low-side injection.

Solution This is a standard superheterodyne receiver. Subtracting the local oscillator frequency from the incoming frequency should give the IF. Therefore, the local oscillator frequency is

$$f_{LO} = 6870 \text{ MHz} - 70 \text{ MHz} = 6800 \text{ MHz}$$

Digital data can also be transmitted with FM systems, using external modems. This is a logical approach when digital signals must be translated over existing analog systems. When new systems are designed especially for digital signals, it is more efficient to use digital modulation schemes, such as those described in section 20.4.3.

Repeaters are necessary in any microwave system that extends for more than the line-of-sight distance. A repeater receives from one direction on one frequency and simultaneously transmits in a different direction on another frequency. Different frequencies must be used for transmission and reception to avoid the possibility of feedback due to signals leaking from the transmit to the receive antenna.

Most microwave systems are bidirectional, so they transmit and receive in two directions. The transmit and receive frequencies at a relay site must be different to prevent feedback, but both transmitters can use the same frequency, as can the two receivers. Adjacent links in the chain can alternate between two pairs of frequencies, as illustrated in Figure 20.8.

Two basic types of repeaters are used in analog systems. Both are shown in Figure 20.9. In Figure 20.9(a), the repeater moves the received signal, by mixing, to an intermediate frequency (IF) to amplify and filter it and then moves it to the new transmitting frequency, all without demodulating it. The frequency of the shift oscillator is equal to the difference between the input and output frequencies of the repeater, that is,

$$f_{SO} = | f_i - f_o | \tag{20.11}$$

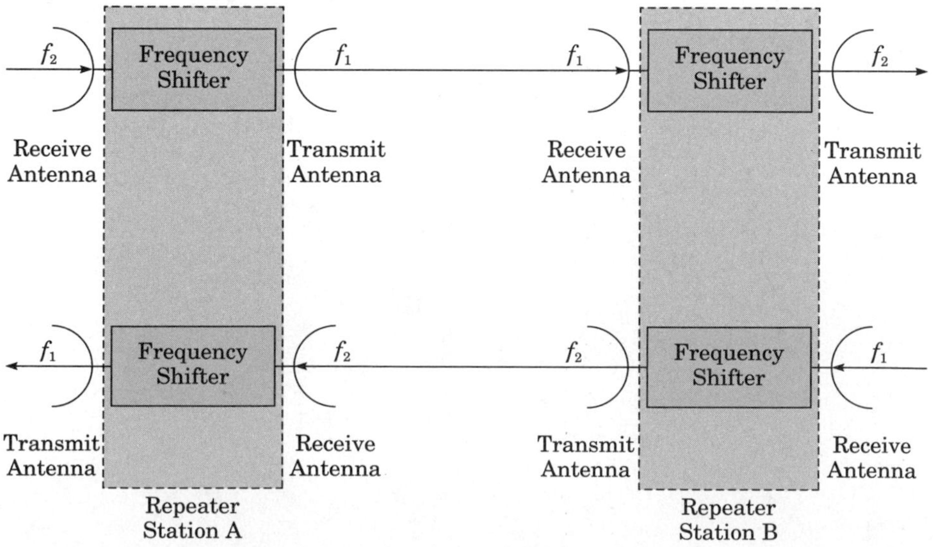

Figure 20.8
Bidirectional system

Figure 20.9 Repeaters

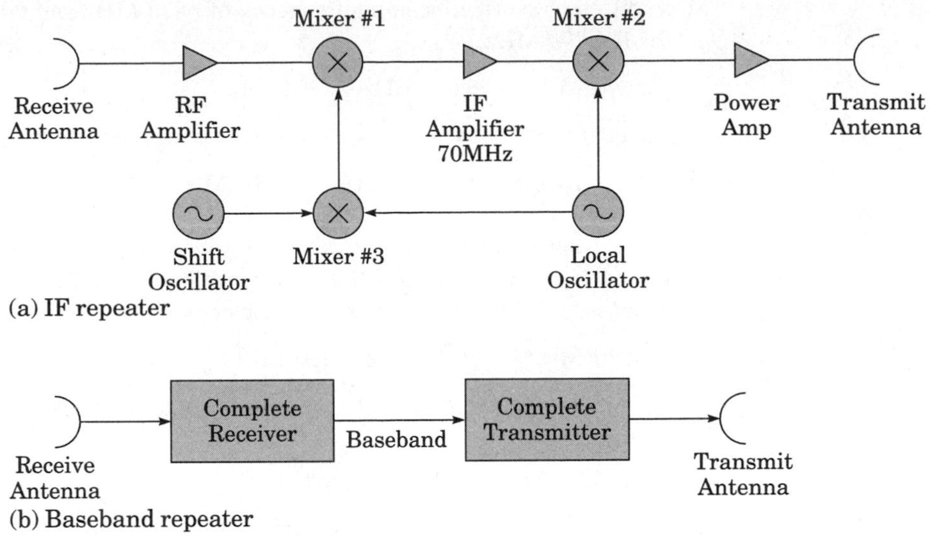

where

$$f_{SO} = \text{shift oscillator frequency}$$
$$f_i = \text{input (receive) frequency}$$
$$f_o = \text{output (transmit) frequency}$$

The local oscillator frequency is set so that the sum of the local oscillator frequency and the intermediate frequency is equal to the output carrier frequency, that is,

$$f_{LO} + 70 \text{ MHz} = f_o$$

Therefore,

$$f_{LO} = f_o - 70 \text{ MHz} \tag{20.12}$$

The shift oscillator frequency is added to or subtracted from the local oscillator frequency in mixer #3 in order to lower or raise, respectively, the signal frequency as the signal moves through the repeater. An example should make this clear.

Example 20.10 A microwave repeater has the block diagram shown in Figure 20.9 (a). The received signal has a carrier frequency of 6870 MHz, and the transmitted signal is to have a carrier frequency of 6710 MHz. The IF is 70 MHz. What should be the frequencies of the local oscillator and the shift oscillator? To what frequency should the output of mixer #3 be tuned? Verify these results by following the signal through the repeater.

Solution The shift oscillator should be set to the difference between the transmit and receive frequencies:

$$f_{SO} = 6870 \text{ MHz} - 6710 \text{ MHz} = 160 \text{ MHz}$$

The local oscillator frequency should be:

$$\begin{aligned} f_{LO} &= f_o - 70 \text{ MHz} \\ &= 6710 \text{ MHz} - 70 \text{ MHz} \\ &= 6640 \text{ MHz} \end{aligned}$$

Mixer #3 will produce an output equal to the sum of the two oscillator frequencies, that is,

$$6640 \text{ MHz} + 160 \text{ MHz} = 6800 \text{ MHz}$$

Mixer #1 combines the incoming frequency of 6870 MHz and the output of Mixer #2 to get the IF of 70 MHz:

$$6870 \text{ MHz} - 6800 \text{ MHz} = 70 \text{ MHz}$$

This frequency is added to the local oscillator frequencies to get

$$f_o = 70 \text{ MHz} + 6640 \text{ MHz} = 6710 \text{ MHz}$$

The same oscillator frequencies could be used to go the other way, that is, to change a received frequency of 6710 MHz to a transmitted frequency of 6870 MHz. The local oscillator would have to be retuned to a frequency of

$$6870 \text{ MHz} - 70 \text{ MHz} = 6800 \text{ MHz}$$

and the output of mixer #3 would have to be tuned to the difference between the two oscillator frequencies, which is

$$6800 - 160 \text{ MHz} = 6640 \text{ MHz}$$

When this frequency is subtracted from the incoming frequency in mixer #1, we get an IF of

$$6710 \text{ MHz} - 6640 \text{ MHz} = 70 \text{ MHz}$$

When it is necessary to connect to the baseband signal, for instance, to add or drop groups of telephone signals, a *baseband repeater* like the one shown in Figure 20.9(b) is required. In a baseband repeater, the signal is demodulated, then retransmitted. This type of repeater is just a receiver with its output connected to the modulation input of a transmitter.

If baseband access is not required, IF repeaters add less noise to the signal than do baseband repeaters. The noise performance of repeaters is very important in analog systems, because noise is cumulative throughout the chain. Once added, noise cannot be removed from an analog system.

20.4.2 Single-Sideband Systems

Some recent analog microwave systems use SSB in order to reduce the bandwidth required. The reader should recall from basic communications theory that the bandwidth required for SSB transmission is equal to the baseband bandwidth, while the bandwidth needed for FM is always more than twice as great as the baseband bandwidth. This bandwidth-conserving property of SSB allows a very considerable increase in traffic-handling capacity along busy routes. For instance, FM microwave links in the 6 GHz band typically carry either 1800 or 2400 voice channels in a bandwidth of 29.65 MHz. The same bandwidth can support 6000 voice channels (ten mastergroups) using SSB.

It is interesting to note that analog voice channels are already multiplexed using SSB before they reach the microwave transmitter, whether the transmitter itself uses FM or SSB. SSB microwave transmission simply continues the bandwidth efficiencies established when the FDM signals were created. (See Chapter 10 for a description of FDM telephony.)

SSB microwave equipment resembles that used for FM, except that the IF frequency is usually 74.13 MHz instead of 70 MHz and great care must be taken to preserve linearity in all amplifiers. Like any amplitude-modulation system, SSB requires linear amplification if distortion of the baseband signal is to be avoided. As in FM systems, most SSB microwave transmitters use traveling-wave tubes for the transmitter power amplifier stage, though there are also solid-state designs using GaAsFET power amplifiers.

SSB radio links can be used to transmit digital signals, just as in FM systems, and this will become more important as the telephone system gradually converts to digital operation. A typical SSB system that can carry 6000 analog voice channels (ten mastergroups) can carry a 135 Mb/s digital bit stream (three DS-3 signals) instead, using 64-

QAM (quadrature amplitude modulation with eight phase angles and eight amplitudes, for a total of 64 possibilities, or six bits per symbol).

20.4.3 Microwave Digital Radio

Digital techniques have two main advantages in microwave radio links. First, the accumulation of noise as the signal travels through many links can be avoided by regenerating the signal at each repeater. This requires demodulating the signal, decoding the data, then recoding and remodulating on a new carrier. As long as the signal-to-noise ratio for each single link is high enough to avoid errors, there will be no increase in error rates as the signal progresses through the system. In practice, a zero error rate is not obtainable, but the error rates will add from link to link, rather than multiplying as in analog systems.

The second major reason for employing digital modulation techniques in microwave radio is to maintain compatibility with digitally-coded baseband signals. Digital signals using time-division multiplexing can pass through digital microwave systems without alteration.

The only disadvantage of digital transmission of voice and video signals is that it requires more bandwidth than analog FM or SSB. However, as better compression algorithms are devised, this is becoming less of a problem, at least in theory. In practice, most systems still use 64 kb/s for each voice channel.

To be transmitted by microwaves, a digital signal must modulate a microwave-frequency carrier. Two modulation schemes are used for digital microwave radio: phase-shift keying (PSK) and quadrature amplitude modulation (QAM), which involves both amplitude and phase shifts. These systems were discussed in Chapter 13—the emphasis there was on telephone modems, where the channel bandwidth is only about 3 kHz, the carrier frequency is on the order of 1 kilohertz, and the data rate is a few kilobits per second. Microwave systems, of course, have very much higher carrier frequencies, bandwidths, and data rates, but the underlying principles are the same.

In general, the more recent and the more sophisticated the system, the more bits per symbol are used. Recall that a symbol is transmitted whenever the transmitter changes state: the state changes can involve amplitude, phase, or both (frequency shifts are also possible but are not used in modern microwave systems). The number of bits of information transmitted with each symbol is a function of the number of possible states:

$$N_B = \log_2 N_S \tag{20.13}$$

where

N_B = number of bits per symbol

N_S = number of possible states per symbol

Recall too that the baud rate is the number of symbols per second.

Transmitting more bits per symbol yields a higher data rate for a given RF bandwidth. On the other hand, transmitting more information per symbol requires greater precision at both transmitter and receiver, so that small amplitude and phase changes can be generated and detected reliably. A better signal-to-noise ratio is also required for a given error rate with more bits per symbol.

Most digital radio systems use 16-QAM or 64-QAM. The first uses four amplitudes and four phase angles for a total of 16 possible states or four bits per symbol; the second has eight amplitudes and eight phases, giving 64 states or six bits per symbol. The current state of the art is 256-QAM, which transmits eight bits per symbol. Experimental use has been made of 1024-QAM. All QAM schemes have the disadvantage, common to all AM systems, of requiring linear amplifiers in the transmitter. FSK and PSK schemes do not require this linearity, so the transmitters can be more efficient. (See Chapter 2 for a discussion of linear and nonlinear amplifiers.) On the other hand, QAM has greater noise immunity than FSK or PSK for a given transmitter power and data rate, and it requires less bandwidth.

Example 20.11 A typical microwave digital radio system uses 16-QAM. It has a bit rate of 90.524 Mb/s (two DS-3 signals or 1344 voice channels, plus overhead—see Chapter 11 for a discussion of digital signals in telephony). Calculate:

(a) the number of bits per symbol
(b) the baud rate

Solution This problem should remind you of some we did in Chapter 13; only the numbers have changed.

(a) There are 16 possibilities, so each symbol can transmit a total of $\log_2 16 = 4$ bits.
(b) Therefore the baud rate is one-quarter of the bit rate or 22.631 Mbaud.

Aside from the modulator, transmitters and receivers for digital microwave radio resemble those for analog systems. Repeaters, however, must demodulate the signal to baseband in order to achieve the advantages of digital transmission. That is, they must be regenerative repeaters, as shown in Figure 20.10.

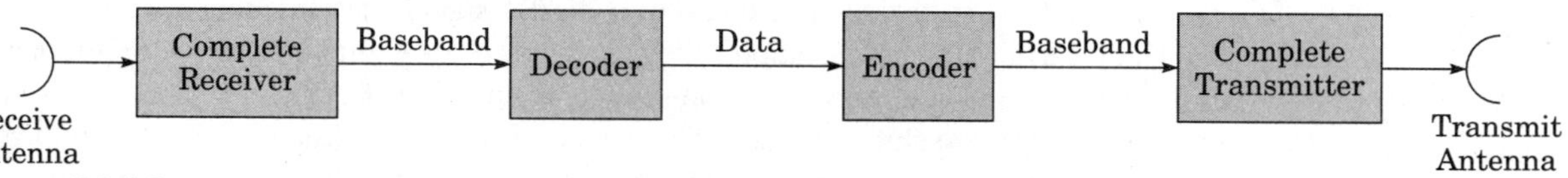

Figure 20.10 Regenerative repeater

Such repeaters can restore a signal that has been distorted or corrupted by noise to its original state, as long as the signal is still readable at the receiver.

Summary

Here are the main points to remember from this chapter:

1. Terrestrial microwave links are useful for short-range communication systems and, when repeaters are used, for long-range systems spanning thousands of kilometers.
2. Microwave systems use line-of-sight propagation and require repeaters approximately every 40 km, depending on the terrain.
3. Care must be taken to keep the microwave beam well above obstacles on the ground to avoid signal loss due to diffraction.
4. Transmitter power and antenna gain must be chosen to give a satisfactory carrier power-to-noise ratio for analog systems. In digital systems, the equivalent specification is the energy per bit per noise density ratio.
5. Fading is common with microwave signals. It can be overcome by using either frequency or space diversity.
6. Analog microwave systems use either FM or SSB modulation. Digital systems use PSK or QAM.

Important Equations

$$d = \sqrt{17h_T} + \sqrt{17h_R} \tag{20.1}$$

$$R_n = 17.3\sqrt{\frac{nd_1d_2}{f(d_1 + d_2)}} \tag{20.2}$$

$$R = 10.4\sqrt{\frac{d_1d_2}{f(d_1 + d_2}} \tag{20.3}$$

$$\frac{P_R}{P_T}\text{ (dB)} = G_T\text{ (dBi)} + G_R\text{ (dBi)} - (32.44 + 20 \log d + 20 \log f) \quad (20.5)$$

$$P_N = kTB \quad (20.6)$$

$$T_a = \frac{(L - 1)\,290 + T_{sky}}{L} \quad (20.7)$$

$$T_{eq} = 290\,(NF - 1) \quad (20.8)$$

$$E_b = \frac{P_R}{f_b} \quad (20.9)$$

$$N_0 = kT \quad (20.10)$$

$$N_B = \log_2 N_S \quad (20.13)$$

Glossary

carrier-to-noise ratio the signal-to-noise ratio in a receiver at a point before the detector

diffraction the deviation of a wave as it passes an obstacle or passes through a small aperture

diversity use of more than one frequency or transmission path, to improve system reliability in the presence of fading

energy per bit energy received in the time taken to transmit one bit

fading variation in received field strength over time due to changes in propagation conditions

Fresnel zone a region near an object in which diffraction effects are significant

hop a single transmission path from transmitter to receiver

jitter abrupt variations in the timing of a digital signal

repeater a receiver-transmitter combination that amplifies and retransmits a signal

topographic map a map showing surface features, including the elevation of the terrain

Questions

1. Briefly compare microwave and fiber-optic links, in terms of bandwidth, cost, and need for right-of-way.
2. What are repeaters? Why are they needed in long-haul terrestrial microwave systems?
3. Approximately how does the line-of-sight distance for radiowave propagation differ from that for visible light?
4. What type of feedline is typically used for microwave systems? Why?
5. Describe 4/3 earth graph paper and explain its use.
6. What is a Fresnel zone? What is its importance in microwave communication systems?
7. Why is it advisable in a terrestrial microwave system to keep the antenna gain under about 40 dB?
8. Why is the term *carrier-to-noise level*, rather than *signal-to-noise level*, used for path calculations with FM microwave systems?
9. What is the equivalent to carrier-to-noise level for a digital microwave radio system? How is it calculated?
10. What causes fading in microwave radio systems?
11. Name two types of diversity, and describe how and why they are used.
12. Which type of diversity is preferred when spectrum space is at a premium? Why?
13. What is a hot standby system? Why is it a desirable feature of a communication system?
14. What range of power output levels is typical for the transmitter in a point-to-point microwave link?
15. Why did microwave receivers lack an RF amplifier stage until recently?
16. How can digital data be communicated using analog microwave systems?
17. What advantage does SSB have over FM for microwave systems?
18. What advantages does digital transmission have over analog for microwave systems?
19. How do repeaters for digital systems differ from those used in analog systems?
20. What modulation schemes are typically used with digital microwave radio?

Problems

SECTION 20.2

21. How far from the transmitter could a signal be received if the transmitting and receiving antennas were 40 m and 20 m, respectively, above level terrain?
22. A transmitter site is on a hill 40 m above average terrain and uses a tower 20 m in height. How far above average terrain would the receiving antenna have to be for reliable communication over a distance of 45 km?
23. Suppose there is an obstacle midway between the transmitter and receiver in the previous question. By how much must the path between the towers clear the obstacle in order to avoid diffraction at a frequency of 11 GHz?
24. How many repeaters would be required in a system spanning 2000 km if the towers are on average 40 km apart?
25. How many repeaters could be eliminated from the system in the previous question if the repeater spacing could be increased to 45 km?

SECTION 20.3

26. Suppose the transmitter in Problem 22 has an output power of 2 W. Its feedline has a loss of 1 dB, and the antenna gain is 28 dBi. Calculate the power density at the receiving antenna.
27. The transmitter in the previous problem works with a receiving installation having an antenna gain of 32 dBi, a feedline loss of 1.5 dB, and a receiver noise figure of 2.5 dB. The bandwidth is 20 MHz at a carrier frequency of 6 GHz.
    (a) What is the power at the receiver input, in dBm?
    (b) Calculate the antenna noise temperature, referred to the receiver input, assuming a sky temperature of 130 K.
    (c) Calculate the receiver noise temperature.
    (d) Calculate the system noise temperature, referred to the receiver input.
    (e) Calculate the noise power, referred to the receiver input.
    (f) What is the carrier-to-noise ratio at the receiver, in decibels?
28. What effect would each of the following changes have on the carrier-to-noise ratio at the receiver in Problem 27?
    (a) doubling the transmitter power
    (b) reducing the required bandwidth by 50%
    (c) improving the receiver noise figure by 0.5 dB
    (d) increasing the gain of the antennas by 3 dB each
29. What is the noise density at the receiver in Problem 27?
30. If the system in Problem 27 is digital, operating at 100 Mb/s, calculate:
    (a) the energy per bit at the receiver
    (b) the energy per bit per noise density ratio at the receiver, in decibels
31. Calculate the effect on the energy per bit per noise density ratio for the system in the previous problem if
    (a) the bit rate is doubled with no change in transmitted bandwidth
    (b) the bit rate is doubled but the modulation scheme remains unchanged
    (c) the transmitter power is doubled
32. What is the fade margin for the system in Problem 27, if the minimum required carrier-to-noise ratio for satisfactory operation is 30 dB?
33. By how much should two antennas be separated for frequency diversity in the 11 GHz band?

SECTION 20.4

34. An IF repeater has the block diagram shown in Figure 20.9(a). Redraw this diagram, showing all the frequencies present if the repeater is used to receive a signal at 1980 MHz and retransmit it at 1900 MHz with an IF of 70 MHz.
35. If a voice channel occupies 4 kHz, how many channels could be transmitted using SSB in a bandwidth of 29.65 MHz? Compare your result with the data in the text and suggest reasons for the difference.
36. Suppose a digital modulation scheme uses 32 different amplitudes and 32 phase angles.
    (a) What would this scheme be called?
    (b) How many total possibilities would there be for each symbol?
    (c) How many bits per symbol would be transmitted?
37. Suppose a signal with a bit rate of 100 Mb/s is transmitted. What would be the baud rate with
    (a) 16-QAM
    (b) 64-QAM
    (c) the system described in the previous problem

COMPREHENSIVE

38. A microwave system has a transmitter power of 5 W at a carrier frequency of 6870 MHz, a bandwidth of 30 MHz, and antennas with gain of 30 dBi at each end of the link. The loss in the waveguides at each end is 1 dB. The receiving antenna sees a sky noise temperature of 100 K. The receiver itself has a noise temperature of 80 K.

(a) Draw a block diagram of the system.
(b) Calculate the carrier-to-noise ratio, in decibels, for a 25 km path.

39. A microwave digital radio system requires an E_b/N_0 ratio of 12 dB for a satisfactorily low error rate. If it operates at 90.524 Mb/s using a bandwidth of 30 MHz, what transmitter power will be required for a communications range of 35 km? The antennas at each end of the path have 20 dB gain, and the sky noise temperature is 120 K. The receiver has a noise temperature of 65 K. Ignore losses in the feedlines. The carrier frequency is 6.5 GHz.

21 Satellite Communication

Objectives After completing this chapter, you should be able to:

1. Explain the advantages of satellites over terrestrial radio communication
2. Compare satellite communication with terrestrial microwave, coaxial cable, and fiber-optic communication systems
3. Distinguish between geostationary satellites and those in lower orbits and explain the advantages and disadvantages of each
4. Calculate the velocity and orbital period for a satellite at a given distance from the earth
5. Adjust a polar mount so that a satellite antenna can track all geostationary satellites that are visible from a given location
6. Calculate the path length and path loss for a signal from a geostationary satellite to a given point on earth
7. Calculate noise figure, noise temperature, and G/T for satellite installations
8. Calculate the propagation time required for a signal to reach a satellite and return
9. Explain the difference between "bent-pipe" and processing satellites
10. Describe the operation of typical geostationary satellites used for television broadcasting
11. Explain how satellites are used to relay telephone communications, using both analog and digital techniques
12. Describe several ways in which satellites can be used for data communication
13. Describe several applications for satellites in low earth orbit
14. Discuss the use of satellites in mobile telephony

21.1 Introduction

One of the main obstacles to long-distance radio communication has always been the curvature of the earth. Early radio systems used low frequencies to take advantage of ground waves, which follow the earth's curve. Later, high-frequency communication systems used the ionosphere as a passive reflector of radio waves. However, ionospheric propagation is sporadic and unpredictable and is suitable only for narrow-bandwidth applications that can tolerate a good deal of noise and distortion. Furthermore, the portion of the radio spectrum to which ground-wave and sky-wave propagation can be applied amounts to only about 30 MHz.

In the previous chapter we discussed using repeaters to extend communication beyond line-of-site range for frequencies greater than those that are returned by the ionosphere. The higher above ground level the repeater can be located, the farther the signal can travel between repeaters (see Figure 21.1).

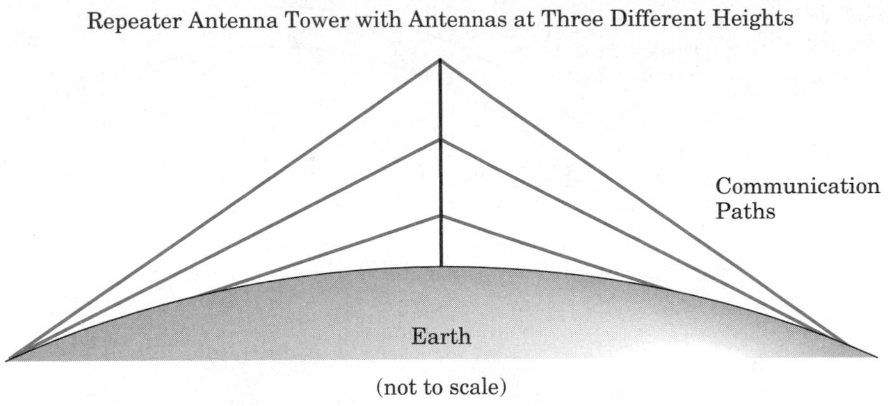

Figure 21.1 Variation in communication distance with repeater height

It follows then, that a very great repeater height, higher than is practicable with towers, could permit communication over very large distances. Aircraft have been used for this purpose (the use of unmanned, remotely piloted aircraft as radio repeaters is quite common for military communications), but the use of **artificial satellites** is more common.

Satellites can be passive reflectors. The first such satellite was Echo, a metallized balloon 30 m in diameter that was launched by the United States in 1960 and orbited at a height of about 1600 km. However, much stronger return signals can be achieved by putting a repeater, complete with transmitter and receiver, on a satellite, and all current satellite communication systems use this technique. The signal path from the earth-station transmitter to the satellite receiver is called the **uplink**, and the path from the satellite to earth is known as the **downlink**.

The choice of orbit is an important consideration in satellite-system design. Most communication satellites use the **geosynchronous orbit**, that is, they occupy a circular orbit above the equator at a distance of 35,784 km above the earth's surface. At this height, the satellite's orbital period is equal to the time taken by the earth to rotate once, that is, approximately 24 hours. If the direction of the satellite's motion is the same as that of the earth's rotation, the satellite appears to remain almost stationary above one spot on the earth's surface. The satellite is then said to be in a **geostationary orbit**, as illustrated in Figure 21.2. Although theoretically an orbit can be geosynchronous without being geostationary (if the satellite orbits in the direction opposite to the earth's rotation), in practice all geosynchronous satellites are also geostationary and the two terms are used as synonyms.

In most cases it is necessary to aim an earth-station antenna at a geostationary satellite only once. Satellites in other orbits require the antenna to follow or **track** the satellite (which will not always be above the horizon). Therefore, nongeostationary satellites (sometimes called **orbital satellites**, though, of course, all satellites are in orbit) cannot be used singly for continuous communication.

Geostationary satellites are far enough from the earth that one satellite can cover about 40% of the earth's surface and three can provide worldwide communication, except for the polar regions.

The first artificial earth satellite, Sputnik I, was launched in 1957, but it was in a much lower orbit, transmitted only telemetry data, and operated for only three weeks. The first geostationary communications satellites were launched in 1962. A brief chronology is presented in the accompanying box.

Though geostationary satellites are very convenient and are, in fact, the basis for most satellite communication systems, they do have some disadvantages. They are very far from earth, so signals suffer a great deal of free-space attenuation on the way to and from the satellite. This is not too serious for fixed applications, where large parabolic antennas

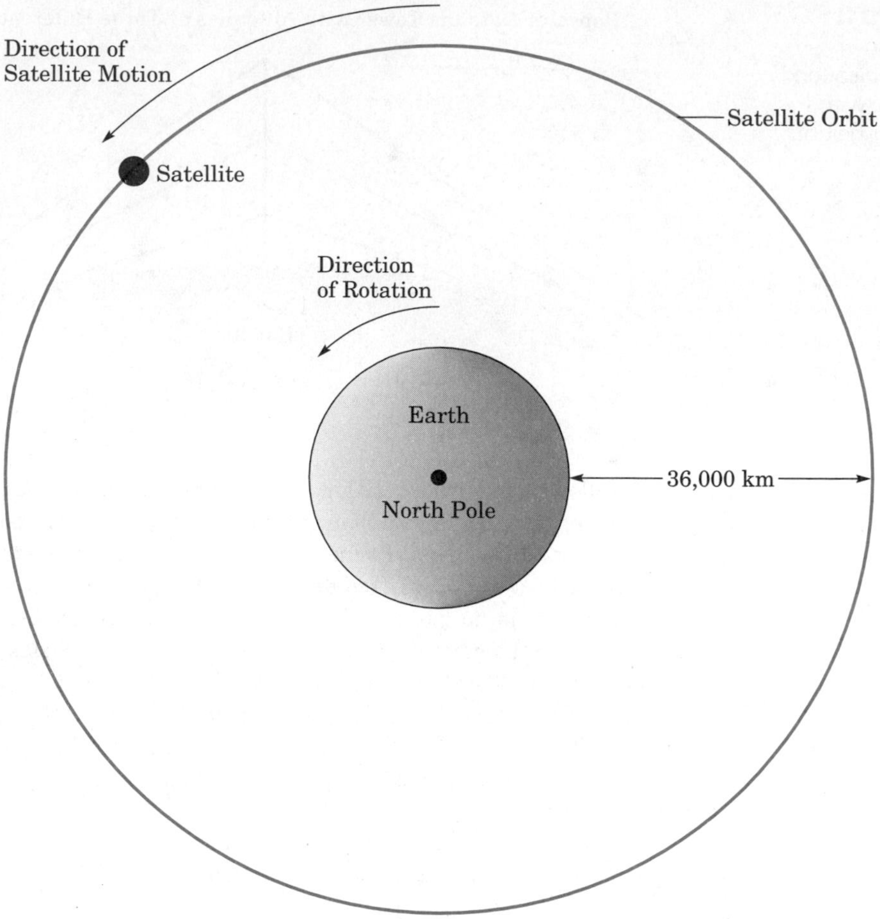

Figure 21.2 Geostationary satellite orbit

are practical, but it makes geostationary satellites awkward for mobile communications. It also causes a considerable time delay (about half a second for a round trip) that is a nuisance in telephone conversations. Furthermore, all the satellites are in the same orbit, so there is a problem with congestion above the most desirable spots on the earth. Since all satellites must be directly above the equator, communication in extreme northern and southern latitudes (above about 80°) is difficult, and it is impossible at the poles (see Figure 21.3). As the location of a transmitter or receiver moves further toward one of the poles, the required angle of elevation of the antenna becomes lower, making the system more susceptible to blocking of the antenna beam by trees, buildings, hills, and so forth. The

Early Geostationary Satellites

1962 Telstar I (AT&T) operated for only a few weeks due to radiation damage
1963 Telstar II (AT&T) first transatlantic video transmission
1963 Syncom II (Syncom I was lost)
1964 Syncom III used to transmit 1964 Olympics from Tokyo
1965 Early Bird I first Intelsat satellite
1969 Intelsat satellites placed over the Atlantic, Pacific, and Indian oceans, allowing worldwide satellite communications
1972 Anik A-1 (Canada) first commercial domestic communications satellite
1972 Westar 1 and 2 (Western Union) first commercial U.S. domestic communications satellites

Figure 21.3
Variation in satellite-antenna elevation angle with latitude

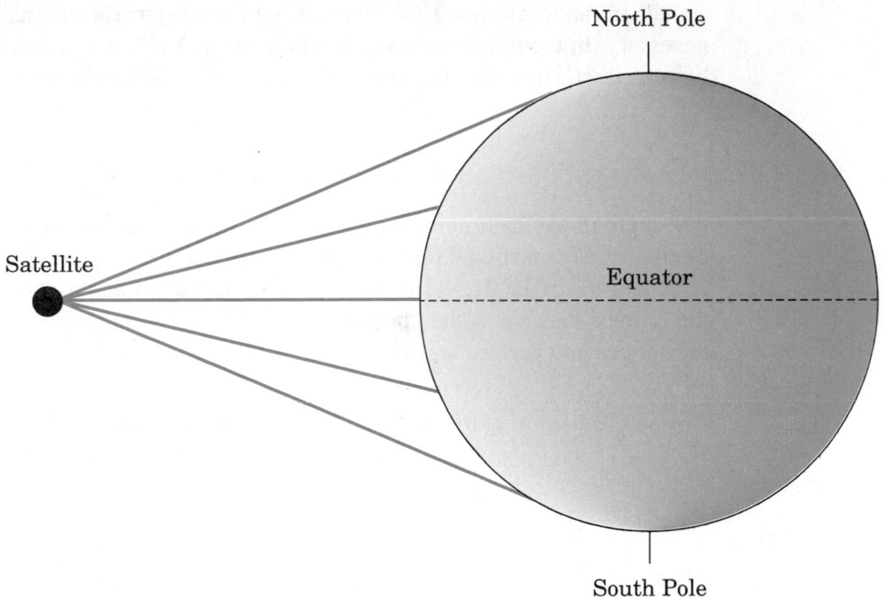

signal must also travel a greater distance through the atmosphere, where it is subject to absorption and scattering, and the beam pattern of a satellite antenna tends to spread out at higher latitudes, as shown in Figure 21.4. This makes the design of the satellite antenna more complex, and the resulting *footprint* (the outline of the antenna pattern on the earth) may be larger than desired.

Therefore, other types of satellites are used in specialized applications such as far-northern and mobile communications. This chapter will emphasize geostationary systems, but we will also look at representative examples of nongeostationary satellites.

Signal propagation between earth and satellite is very similar to terrestrial line-of-sight propagation. Since signals travel through the ionosphere, frequencies below the VHF region are not generally suitable (though a few amateur-radio satellites operate at 28 MHz, just below the VHF region). At VHF and UHF, the ionosphere rotates the polarization of

Figure 21.4
Spreading of satellite beam at high latitudes

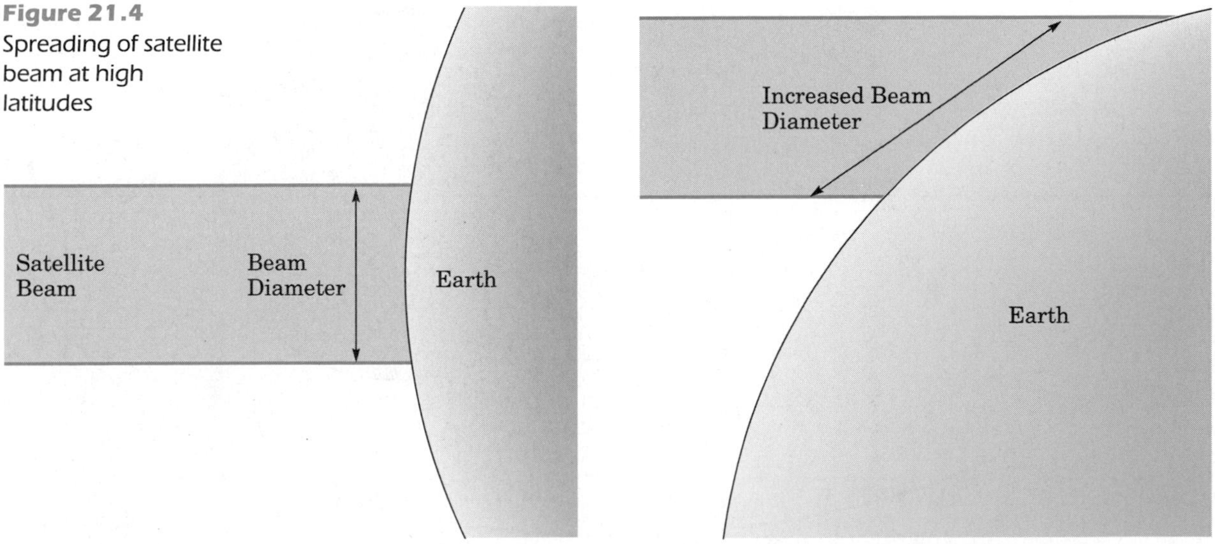

(a) Beam at equator

(b) Beam at northern latitude

signals in random ways. This effect, called **Faraday rotation**, makes circular polarization necessary. In the microwave region where most satellites operate, the ionosphere has negligible effect, and either linear or circular polarization can be used.

21.2 Satellite Orbits

All satellites are held in orbit by a balance between the inertia due to their motion (commonly called centrifugal force) and the centripetal force of gravity. All orbits are elliptical, as shown in Figure 21.5. The distance of closest approach to the earth is the **perigee**, and the farthest distance is the **apogee**. A circular orbit is a special case of an ellipse in which the apogee and perigee are equal.

21.2.1 Orbital Calculations

Any satellite orbiting the earth must satisfy this equation:

$$v = \sqrt{\frac{4 \times 10^{11}}{(d + 6400)}} \tag{21.1}$$

where

v = velocity in meters per second
d = distance above earth's surface in km

Several important points can be noted from Equation 21.1. First, the farther a satellite is from the surface of the earth, the slower it travels. Second, since a satellite that is farther from the earth obviously has farther to go to complete an orbit, and since it also travels slower than one closer to earth, then the orbital period of a distant satellite must be longer than the period of one closer to the earth. An example will illustrate this.

Figure 21.5
Elliptical orbit

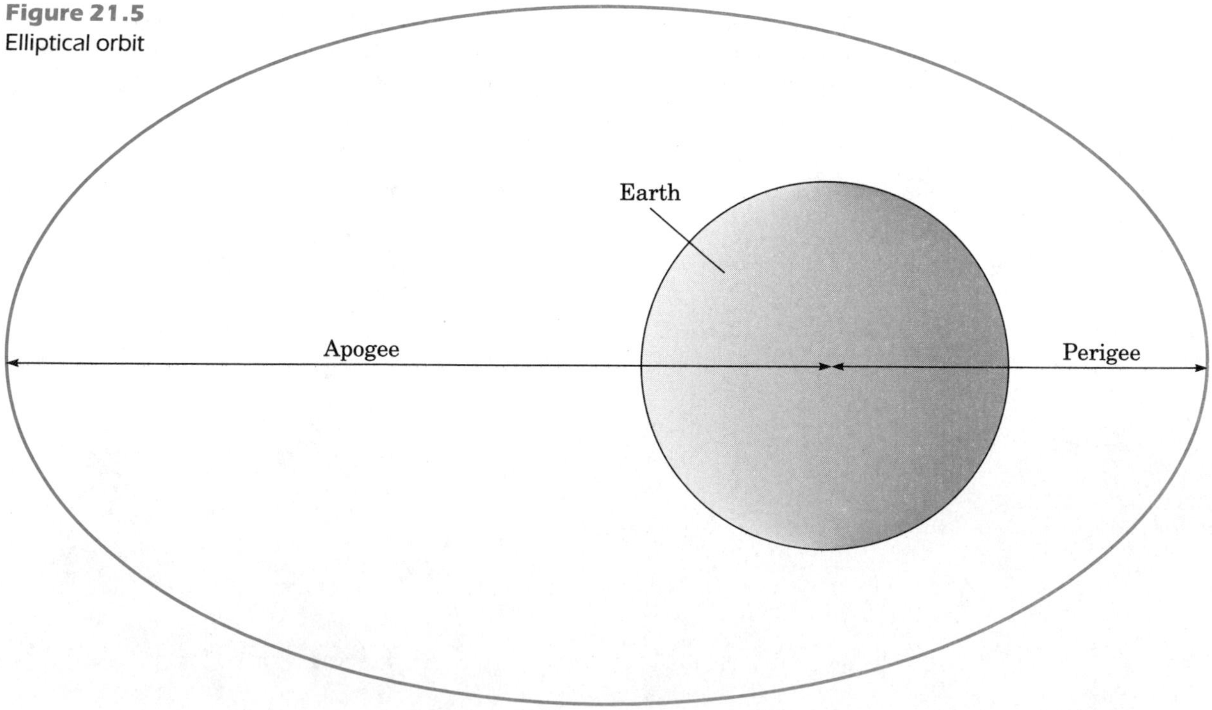

Example 21.1 Find the velocity and the orbital period of a satellite in a circular orbit

(a) 500 km above the earth's surface
(b) 36,000 km above the earth's surface (approximately the height of a geosynchronous satellite)

Solution (a) From Equation 21.1

$$\begin{aligned} v &= \sqrt{\frac{4 \times 10^{11}}{(d + 6400)}} \\ &= \sqrt{\frac{4 \times 10^{11}}{(500 + 6400)}} \\ &= 7.6 \text{ km/s} \end{aligned}$$

The circumference of the orbit can be found from its radius, which is that of the earth (6400 km) plus the distance of the satellite from the earth. In this case the total distance is

$$r = 6400 \text{ km} + 500 \text{ km} = 6900 \text{ km}$$

and the circumference of the orbit is

$$\begin{aligned} C &= 2\pi r \\ &= 2\pi \times 6900 \text{ km} \\ &= 43.4 \text{ Mm} \end{aligned}$$

The period of the orbit can be found by dividing the circumference by the orbital velocity,

$$\begin{aligned} T &= \frac{C}{v} \\ &= \frac{43.4 \times 10^6 \text{ m}}{7.6 \times 10^3 \text{ m/s}} \\ &= 5.71 \times 10^3 \text{ s} \\ &= 1.6 \text{ hours} \end{aligned}$$

(b) Again, start with Equation 21.1:

$$\begin{aligned} v &= \sqrt{\frac{4 \times 10^{11}}{(d + 6400)}} \\ &= \sqrt{\frac{4 \times 10^{11}}{(36{,}000 + 6400)}} \\ &= 3.07 \text{ km/s} \end{aligned}$$

Note that the speed is less than before. The new radius is

$$\begin{aligned} r &= 6400 \text{ km} + 36{,}000 \text{ km} \\ &= 42.4 \text{ Mm} \end{aligned}$$

Note that the orbit for a geostationary satellite has a radius more than five times as large as that of the earth.

The circumference of the orbit is, as before,

$$\begin{aligned} C &= 2\pi r \\ &= 2\pi \times 42.4 \text{ Mm} \\ &= 266.4 \text{ Mm} \end{aligned}$$

and the period of the orbit is

$$T = \frac{C}{v}$$
$$= \frac{266.4 \times 10^6 \text{ m}}{3.07 \times 10^3 \text{ m/s}}$$
$$= 86.8 \times 10^3 \text{ s}$$
$$= 24 \text{ hours}$$

Note that the satellite is, at least approximately, geosynchronous. If the orbit is above the equator and the satellite travels in the same direction as the earth's rotation, it will also be geostationary.

When a satellite has an elliptical orbit in which the earth is at one focus, it spends more time in that part of the orbit that takes it farthest from earth. This is in agreement with Kepler's second law, which states that such a satellite sweeps out equal areas in space in equal times. See Figure 21.6 for an illustration. Such orbits have their advantages. The Russian Molniya series of communications satellites, for instance, uses highly elliptical polar orbits arranged so that the satellites spend about 11 hours of a 12 hour orbit over the Northern Hemisphere, thus facilitating communication in regions near the North Pole. Geostationary satellites, which must be above the equator, are at or below the horizon in these regions.

Figure 21.6
Kepler's second law

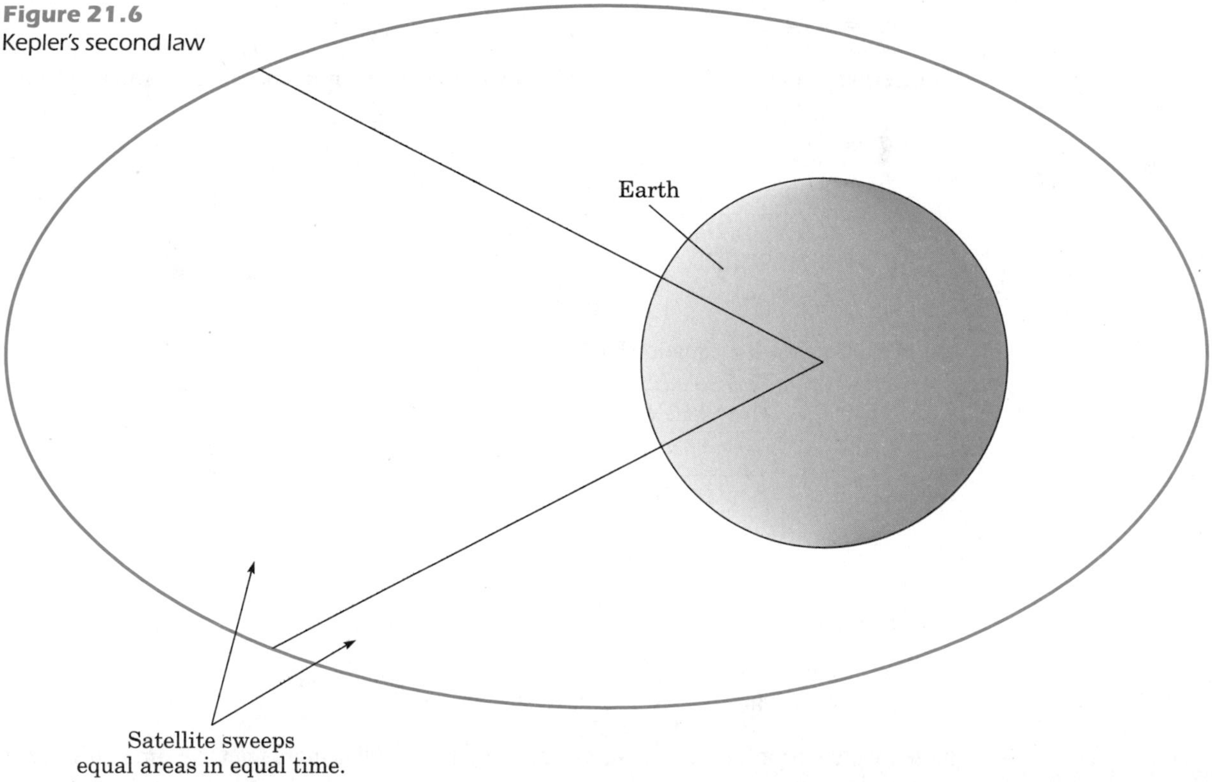
Earth
Satellite sweeps
equal areas in equal time.

21.3 Geostationary Satellites

As we have already mentioned, most communications satellites in current use are geostationary, so we will study geostationary satellites in some detail. We begin by considering the siting and adjustment of antennas; next, we look at signal losses and **propagation time** over the very large distances between satellite and earth; and finally, we discuss the satellites themselves and the earth stations that communicate with them.

21.3.1 Antenna Siting and Adjustment

All geostationary satellites orbit above the equator; therefore antennas in the Northern Hemisphere point south and those in the Southern Hemisphere point north. We will discuss only the Northern Hemisphere in the following paragraphs.

In general, an antenna is aimed at a satellite by adjusting its *azimuth* (horizontal position) and *elevation* (vertical angle with the ground). (Of course, it is necessary to have a clear path from the antenna to the satellite.) These adjustments would seem to require motion in two directions and, for automatic positioning, two motors. This is indeed the requirement for tracking orbital satellites. However, a good deal of simplification is possible when an antenna is designed for use only with geostationary satellites. As seen from any point in the Northern Hemisphere, all geostationary satellites lie along an arc from southeast to southwest. Each satellite appears almost stationary as seen from the earth, so an antenna intended to communicate with only one satellite need have its azimuth and elevation adjusted only once, then it can be fixed in place. This method is used by commercial installations, such as cable-television head-ends, which generally have a separate antenna for each satellite they use, and it is also used with the new direct-to-home digital television systems. The azimuth and elevation can be calculated, based on the longitude of the satellite and the latitude and longitude of the earth station. Usually, however, approximate coordinates can be found from manufacturer-supplied tables, and the fine adjustment can be made by adjusting for maximum signal from the satellite.

If, as is often the case, the same antenna is to be used with more than one geostationary satellite, it is possible to use a very simple means to move the antenna. If the antenna is correctly mounted, it need be rotated along only one axis to aim at any geostationary satellite. The mount can be a simple hingelike affair called a polar mount, and only one motor is needed. Figure 21.7 shows a typical polar mount with actuator.

In order for a polar mount to work properly, it must be carefully installed. At the equator, the axis of rotation of the antenna would be horizontal, running in a north-south direction, and the antenna would rotate from east to west, passing through a vertical position. A similar arc would be followed anywhere in the Northern Hemisphere, except that the antenna's axis would be tilted southward. First it must be raised from the horizontal by an angle equal to the latitude of the earth station, so that the axis of rotation of the antenna is parallel to the earth's axis, pointing towards the North Star rather than the earth's North Pole. This adjustment is usually accomplished by using a compass, properly compensated to find true north, and an *inclinometer*, which is a device incorporating a level that can measure the angle of the antenna axis from the horizontal. It can also be performed by sighting on the North Star, if one doesn't mind the inconvenience of working at night.

If the antenna were left as just described, it would point straight out into space, as shown in Figure 21.8 (a). It must now be moved a little farther, so that it points at the satellites above the equator. The amount by which the antenna axis is offset from the earth's axis is called the *declination*. Figure 21.8(b) shows how the addition of a declination adjustment allows a polar mount to track satellites above the equator. The declination can be calculated from the earth-station latitude as follows:

$$\theta = \arctan\left(\frac{R \sin L}{H + R(1 - \cos L)}\right) \tag{21.2}$$

where

R = radius of the earth (6370 km)
H = height of the satellite above the earth (35,784 km)
L = earth-station latitude

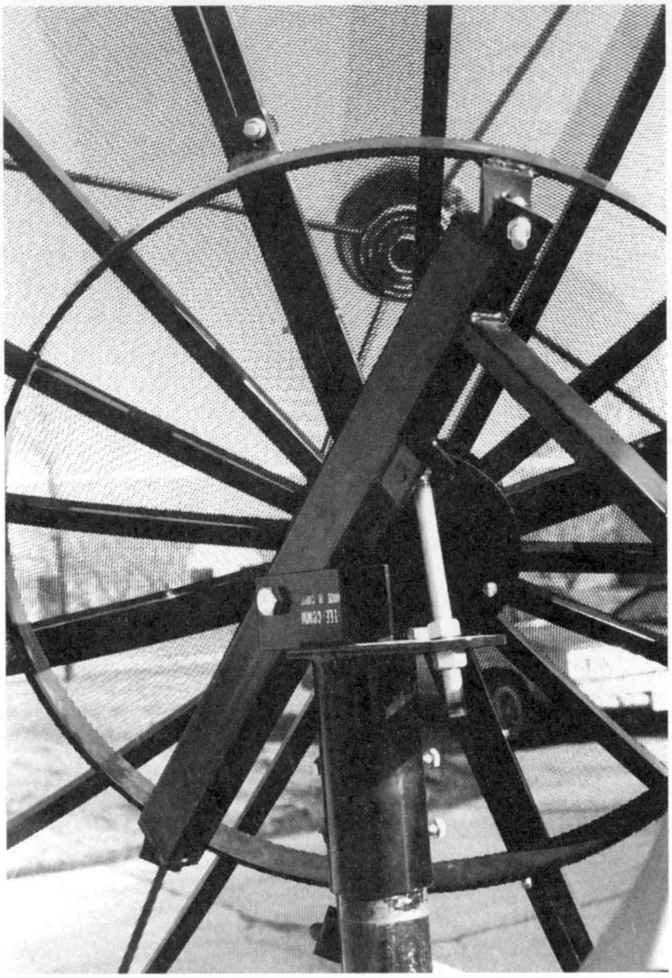

Figure 21.7
Polar mount
Tee-Comm Electronics, Inc., 775 Main St. E., Milton, ON.

Example 21.2 Calculate the angle of declination for an antenna using a polar mount at a latitude of 45°.

Solution From Equation (2.2),

$$\theta = \arctan\left(\frac{6370 \sin 45°}{35784 + 6370(1 - \cos 45°)}\right)$$
$$= 6.82°$$

Once the axis of rotation has been accurately aligned, the antenna will follow the arc of satellites from one horizon to the other (provided that the actuator can move it far enough). Of course, only those satellites that are above the horizon at a particular location can be used.

21.3.2 Path Loss Calculations

The standard equation for path loss, which was discussed in Chapter 17, applies here:

$$\frac{P_R}{P_T}\text{(dB)} = G_T\text{ (dBi)} + G_R\text{ (dBi)} - (32.44 + 20 \log d + 20 \log f) \tag{21.3}$$

where

P_R/P_T(dB) = ratio of received to transmitted power, expressed in decibels
G_T (dBi) = gain of transmitting antenna in decibels with respect to an isotropic radiator

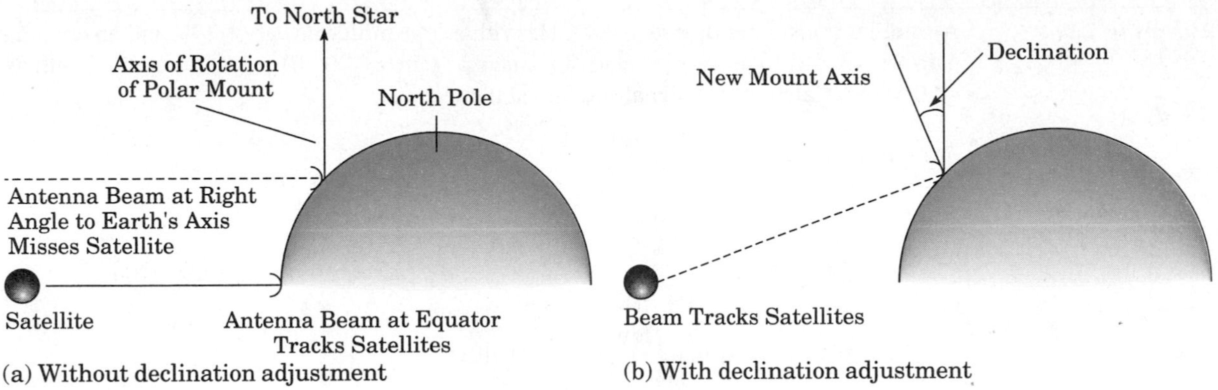

Figure 21.8
Use of polar mount

G_R (dBi) = gain of receiving antenna in decibels with respect to an isotropic radiator
d = distance between transmitter and receiver in kilometers
f = frequency in megahertz

In terms of loss, the only difference between satellite communication and terrestrial line-of-sight propagation is the much greater distance involved in satellite systems. The path length from the earth to a geostationary satellite is about 36,000 km for an earth station at the equator, at the same longitude as the satellite, and it increases to as much as 41,700 km for an earth station close to either the North or South Pole. The actual path length can be found from the equation:

$$d = \sqrt{(r + h)^2 - (r \cos \theta)^2} - r \sin \theta \tag{21.4}$$

where

d = distance to the satellite in km
r = radius of the earth in km (6400 km)
h = height of satellite above equator (35780 km)
θ = angle of elevation to satellite at antenna site

The angle of elevation is, of course, 90° for an antenna at the equator and at the same longitude as the satellite, decreasing to zero at about 80° latitude.

Example 21.3 Calculate the length of the path to a geostationary satellite from an earth station where the angle of elevation is 30°.

Solution From Equation 21.4,

$$\begin{aligned} d &= \sqrt{(r + h)^2 - (r \cos \theta)^2} - r \sin \theta \\ &= \sqrt{(6400 + 35780)^2 - (6400 \cos 30°)^2} - 6400 \sin 30° \\ &= 38{,}614\text{km} \end{aligned}$$

The increased distance causes greater attenuation and requires antennas with high gain, coupled with sensitive receivers. On the ground, transmitters with several kilowatts of output power are common. On the satellite, where power must come from solar cells, transmitter power is strictly limited, generally to about 10 W for satellites intended for use with commercial ground stations. For C-band satellite communications, another limiting factor is the need to avoid interference with terrestrial microwave links, which share the same frequency band. Ku-band satellites that transmit directly to home receivers have power outputs on the order of 50–240 W per transponder. An example will give you some idea of the received signal levels to be expected.

Example 21.4 A satellite transmitter operates at 4 GHz with a transmitter power of 7 W and an antenna gain of 40 dBi. The receiver has an antenna gain of 30 dBi, and the path length is 40,000 km. Calculate the signal strength at the receiver.

Solution From Equation (21.3),

$$\frac{P_R}{P_T}\text{ (dB)} = G_T\text{ (dBi)} + G_R\text{ (dBi)} - (32.44 + 20\log d + 20\log f)$$
$$= 40 + 30 - (32.44 + 20\log(40\times 10^3) + 20\log 4000)$$
$$= -126.5\text{ dB}$$
$$P_T\text{ (dBm)} = 10\log\frac{7\text{ W}}{1\text{ mW}} = 38.5\text{ dBm}$$
$$P_R\text{(dBm)} = 38.5\text{ dBm} - 126.5\text{ dB} = -88\text{ dBm}$$

Because the received signals are very weak, geostationary satellite systems require low-noise receivers and high-gain antennas for both the satellite and the earth station. A **figure of merit** called ***G/T*** has evolved to measure the combination of gain and equivalent noise temperature. *G/T* is defined as:

$$G/T\text{(dB)} = G_R\text{(dBi)} - 10\log(T_a + T_{eq}) \tag{21.5}$$

where

G/T(dB) = a figure of merit for the receiving system
G_R(dBi) = the gain of the antenna system with respect to an isotropic radiator
T_a = the noise temperature of the antenna
T_{eq} = the equivalent noise temperature of the receiver

The gain and noise temperatures should all be taken at the same reference point. The gain required is the antenna gain less any losses up to the reference point. Usually receiving antennas for use with satellites have the first receiver stage, called the low-noise amplifier (LNA), right inside the dish, connected directly to the feedhorn by a short piece of waveguide. In that case, the reference point for antenna gain and noise temperature calculations is usually the LNA input.

Gain and noise temperature calculations have been described previously. Recall that the antenna noise temperature depends on the angle of elevation of the antenna: when the antenna beam includes the ground, the noise level increases because of radiation from the ground itself. Luckily, this is seldom the case with satellite ground stations, except in very high latitudes where the satellite is just above the horizon. Thus noise entering the antenna originates mainly from extraterrestrial sources (stars, for instance) and from the atmosphere. Occasionally the sun passes through the main lobe of the antenna pattern; the sun is a very powerful noise source and makes communication impossible for the few minutes it takes to pass through the antenna beam. Otherwise the sky noise temperature for an earth-station receiving antenna is quite low, typically 20 K or less.

Losses in the antenna system also contribute to its noise temperature. In the previous chapter we noted that the noise temperature of an antenna system at the far end of a feedline is given by

$$T_a = \frac{(L-1)\,290 + T_{sky}}{L} \tag{21.6}$$

where

T_a = effective noise temperature of antenna and feedline, referenced to receiver antenna input, in kelvins
L = loss in feedline and antenna as a ratio of input to output power (not in decibels)
T_{sky} = effective sky temperature, in kelvins

Example 21.5 A receiving antenna with a gain of 40 dBi looks at a sky with a noise temperature of 15 K. The loss between the antenna and the LNA input, due to the feedhorn, is 0.4 dB, and the LNA has a noise temperature of 40 K. Calculate G/T.

Solution First, we find G in dB—it is simply the antenna gain less any losses up to the reference point. In this example,

$$\begin{aligned} G &= 40 \text{ dBi} - 0.4 \text{ dB} \\ &= 39.6 \text{ dBi} \end{aligned}$$

The calculation of T is a little more difficult. We have to convert the feedhorn loss into a ratio, as follows:

$$L = \text{antilog } (0.4/10) = 1.096$$

Substituting into Equation (21.6), we get

$$\begin{aligned} T_a &= \frac{(L - 1)290 + T_{sky}}{L} \\ &= \frac{(1.096 - 1)290 + 15}{1.096} \\ &= 39 \text{ K} \end{aligned}$$

The receiver noise temperature is given with respect to the chosen reference point, so it can be used directly. Therefore,

$$\begin{aligned} G/T(\text{dB}) &= G_R(\text{dBi}) - 10 \log (T_a + T_{eq}) \\ &= 39.6 - 10 \log (39 + 40) \\ &= 20.6 \text{ dB} \end{aligned}$$

Sometimes receiver performance is specified in terms of noise figure rather than noise temperature. Typically, for instance, C-band consumer equipment uses temperature and Ku-band equipment uses decibels. It is easy to convert from one to the other. Recall from basic theory (see Chapter 1) that

$$T_{eq} = 290(NF - 1) \tag{21.7}$$

where

T_{eq} = receiver equivalent noise temperature in kelvins, referred to its input
NF = receiver noise figure as a ratio, not in dB

Example 21.6 A receiver has a noise figure of 1.5 dB. Find its equivalent noise temperature.

Solution First find the noise figure as a ratio:

$$\begin{aligned} NF &= \text{antilog } (1.5/10) \\ &= 1.41 \end{aligned}$$

Now use Equation (21.7) to determine the equivalent noise temperature:

$$\begin{aligned} T_{eq} &= 290(NF - 1) \\ &= 290(1.41 - 1) \\ &= 119 \text{ K} \end{aligned}$$

Once G/T has been found, it can be used in the calculation of the carrier-to-noise ratio for a system. The method is similar to that used in the previous chapter, except that this time antenna gain is considered separately from path loss. We can use the following equation:

$$\frac{C}{N}(\text{dB}) = EIRP(\text{dBW}) - FSL(\text{dB}) - L_{misc} + G/T - k(\text{dBW}) - 10 \log B \qquad (21.8)$$

where

$C/N(\text{dB})$ = carrier-to-noise ratio in decibels
$EIRP(\text{dBW})$ = effective isotropic radiated power in dBW
$FSL(\text{dB})$ = free space loss in decibels, without compensation for antenna gains
L_{misc} = miscellaneous losses, such as feedline losses, in dB
G/T = figure of merit as given in Equation 21.5
$k(\text{dBW})$ = Boltzmann's constant expressed in dBW (−228.6 dBW)
B = bandwidth in hertz

The free-space loss for this application is simply the loss given in Equation (21.3) with the antenna gains left out, that is

$$FSL(\text{dB}) = 32.44 + 20 \log d + 20 \log f \qquad (21.9)$$

where d = path length in km
f = frequency in MHz

Example 21.7 The receiving installation whose G/T was found in Example 21.5 is used as a ground terminal to receive a signal from a satellite at a distance of 38,000 km. The satellite has a transmitter power of 50 watts and an antenna gain of 30 dBi. Assume losses between the satellite transmitter and its antenna are negligible. The frequency is 12 GHz. Calculate the carrier-to-noise ratio at the receiver, for a bandwidth of 1 MHz.

Solution The earth station was found to have $G/T = 20.6$ dB. The satellite transmitter power, in dBW, is

$$P_T(\text{dBW}) = 10 \log 50 = 17 \text{ dBW}$$

The EIRP in dBW is just the transmitter power in dBW plus the antenna gain in dBi, less any feedline losses, which are negligible here (see Chapter 17 if you need to review this). Here, we have

$$\text{EIRP(dBW)} = 17 \text{ dBW} + 30 \text{ dBi} = 47 \text{ dBW}$$

Next we find FSL(dB). Using Equation (21.9),

$$\begin{aligned} FSL(\text{dB}) &= 32.44 + 20 \log d + 20 \log f \\ &= 32.44 + 20 \log 38{,}000 + 20 \log 12{,}000 \\ &= 205.6 \text{ dB} \end{aligned}$$

There are no miscellaneous losses, since the receiver feedline loss was already considered in the calculation of G/T. Now we can find C/N. From Equation 21.8,

$$\begin{aligned} C/N(\text{dB}) &= EIRP(\text{dBW}) - FSL(\text{dB}) - L_{misc} + G/T - k(\text{dBW}) - 10 \log B \\ &= 47 \text{ dBW} - 205.6 \text{ dB} + 20.6 \text{ dB} + 228.6 \text{ dBW} - 10 \log (1 \times 10^6) \\ &= 30.6 \text{ dB} \end{aligned}$$

Since the EIRP of a satellite depends on the gain of its antenna, it will be different at different points on earth. Satellite operators often publish maps showing the footprint of a satellite on the earth. These maps show the effective EIRP, in dBW, for the satellite at each point on the earth where reception is possible. Satellite antennas can range from high-gain antennas whose *spot beams* cover only a small portion of the earth to antennas with much lower gain and wider patterns that cover almost a hemisphere of the earth, but with much lower EIRP. Figure 21.9 shows an example of a satellite footprint.

Figure 21.9
Satellite footprint for Anik E-1 satellite
Courtesy Telesat Canada.

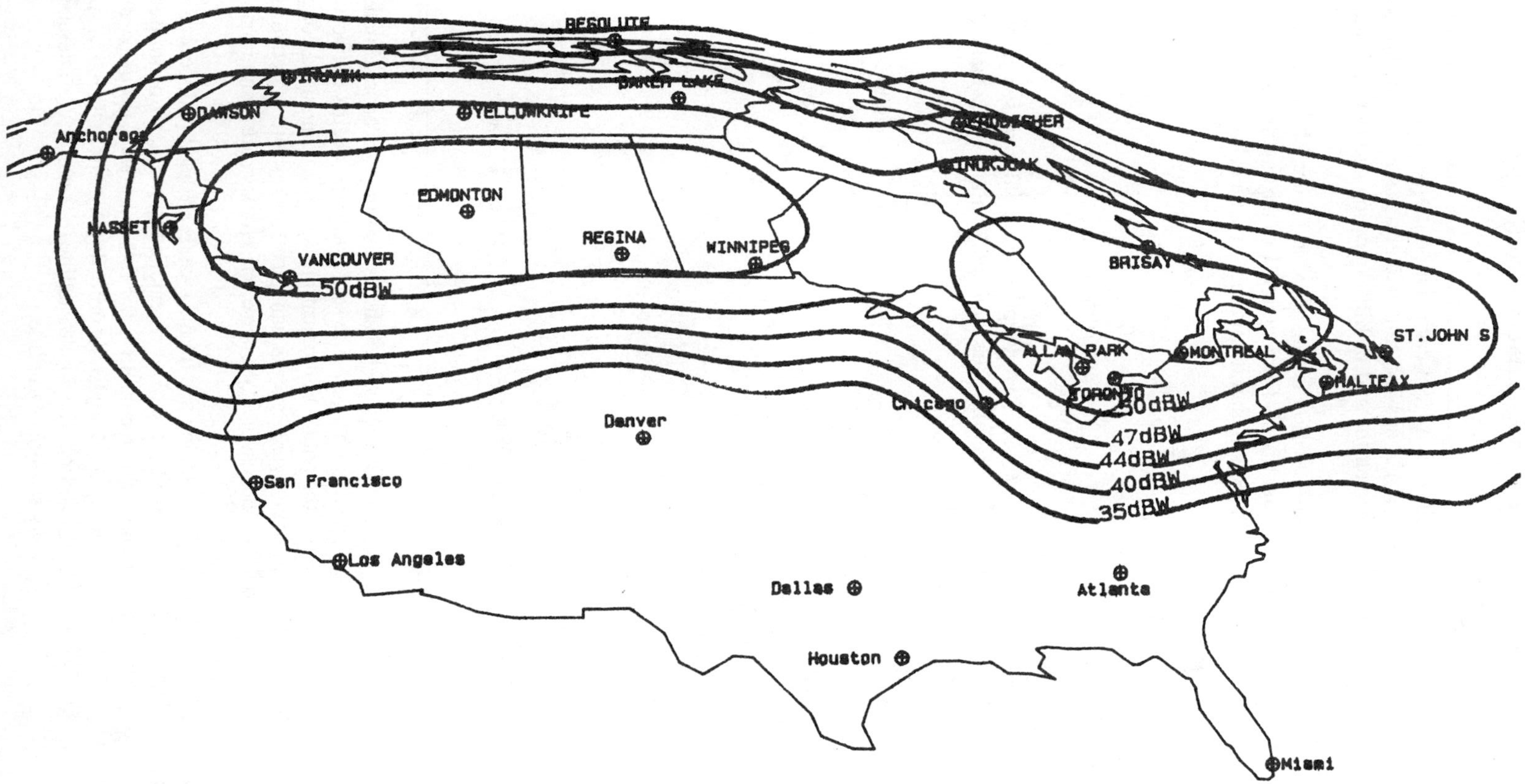

21.3.3 Propagation Time

One disadvantage of geostationary satellites (compared with such techniques as terrestrial microwave links or fiber-optic cable) is the considerable amount of time it takes for a signal to make the round trip from earth to satellite and back. Even though signals travel at the speed of light, the distance is so great that the propagation time is significant for such applications as telephony and full-duplex data communications. A quick calculation will give us an estimate of this delay.

Example 21.8 Telephone communication takes place between two earth stations via a satellite that is 40,000 km from each station. Suppose Bill, at station 1, asks a question and Sharon, at station 2, answers immediately, as soon as she hears the question. How much time elapses between the end of Bill's question and the beginning of Sharon's reply, as heard by Bill?

Solution This situation is illustrated in Figure 21.10. Sharon will not hear Bill's question until the signal travels from station 1 to the satellite and back to earth at station 2. This distance is 80,000 km, and it represents a time delay given by

$$\begin{aligned} t &= \frac{d}{v} \\ &= \frac{80 \times 10^6 \text{ m}}{300 \times 10^6 \text{ m/s}} \\ &= 0.26 \text{ s} \end{aligned}$$

Once Sharon replies, there will be a second time delay before Bill hears the answer. Therefore the total delay will be twice that calculated above, or 0.52 second.

Figure 21.10

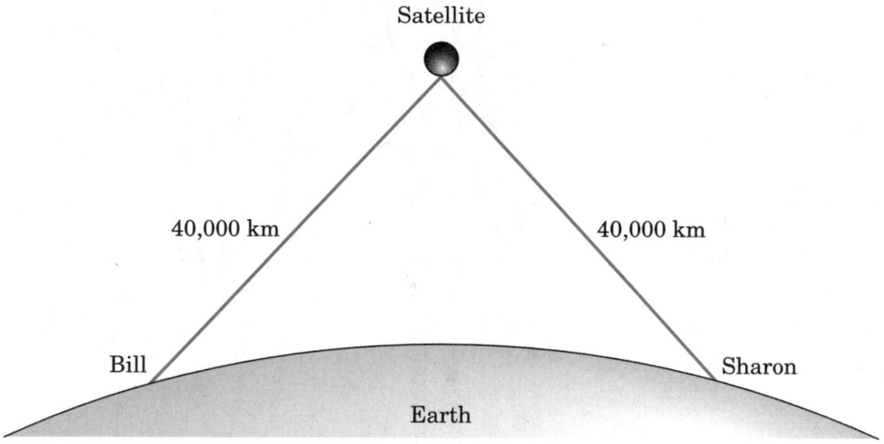

Example 21.8 shows that there is approximately a half-second delay each time the speaker changes. The same situation occurs in two-way data communications. Depending on the data-communications scheme being used, long delays in the receipt of transmission acknowledgements can greatly slow the effective data rate of the communication. For broadcast use, of course, these delays are insignificant.

21.3.4 Satellite and Transponder Types

Both analog and digital signals are used with geostationary satellites. The satellite *payload* consists of several repeaters, known as **transponders**, mounted on a platform called the *bus*, which contains solar cells for power, telemetry and control systems, and small jets that are used to adjust the orbit slightly. Often there is a spinning wheel to stabilize the satellite; sometimes the whole satellite (except for the antennas) spins. Figure 21.11 shows a typical geostationary satellite.

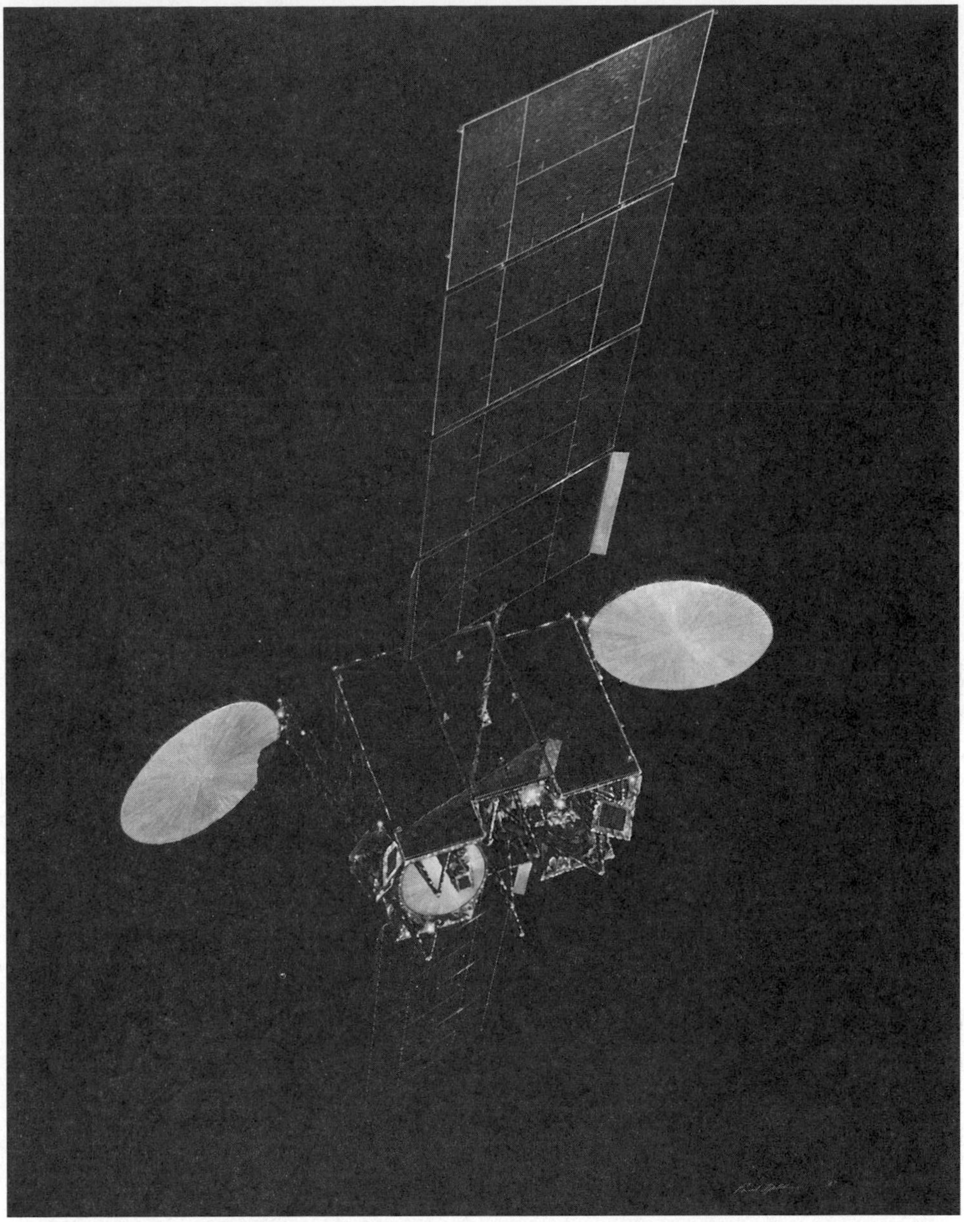

Figure 21.11
Anik E-1 satellite
Courtesy Telesat Canada.

Even though a satellite is in a geosynchronous orbit, it does not remain perfectly stationary above one position on the earth without help—small forces such as the gravitational attraction of the sun and moon and the solar wind cause its orbit to change slowly over time. To counteract this effect, small gas jets controlled from the ground are used for **station-keeping**. Once the fuel for these jets is exhausted, the satellite drifts away from its correct position and becomes useless, so this fuel supply is one of the primary limitations on the operational life of a satellite, which is typically eight to ten years.

The most common type of satellite transponder is colloquially known as the **bent-pipe configuration**. Figure 21.12 shows why.

Signals received by a satellite are received at one frequency, amplified, and moved to another frequency for retransmission. The transmit and receive frequencies are quite widely separated to avoid interference. The most popular frequency ranges are shown in Table 21.1. There are two major frequency assignments. One is in the C-band, with the uplink (transmission from ground to satellite) at approximately 6 GHz and the downlink

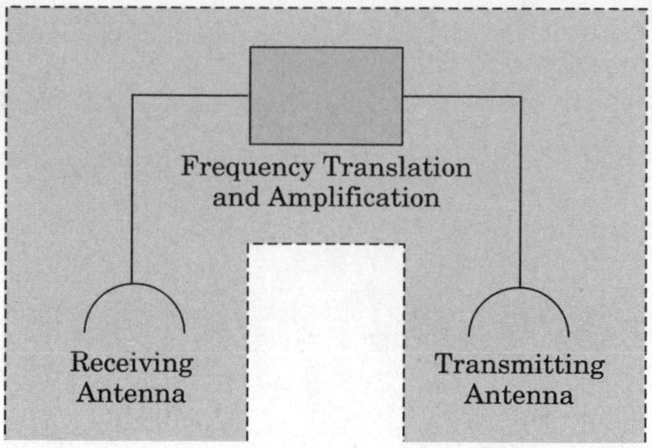

Figure 21.12
Bent-pipe satellite transponder

(from satellite to ground) at about 4 GHz. The other is in the Ku-band, with the uplink and downlink in the range of 14 GHz and 12 GHz, respectively. The fact that good gain and narrow beamwidth can be obtained at 12 GHz with smaller antennas than those used at 4 GHz makes the Ku-band popular for the newer direct-to-home (*direct-broadcast satellite* or *DBS*) systems. The C-band was established first and is widely used for such purposes as television network feeds and transmissions to cable-television head-ends, as well as for some DBS services. The letter band designations in Table 21.1 are those commonly used in North America; their origins are in U.S. military communications.

The structure of satellite transponders varies somewhat depending on the age, frequency band, and application of the satellite. We will first look at a typical North American domestic satellite operating in the C-band in order to get an understanding of satellite transponders. Then we will look at some of the differences found in Ku-band satellites and in satellites designed for international communications.

Figure 21.13 is a block diagram of one transponder. There are 24 transponders on a typical North American C-band satellite. Each transponder is assigned a bandwidth of 36 MHz, with a 4 MHz guard band between transponders. A transponder is essentially a linear amplifier and frequency translator—it reproduces at its output whatever type of modulation appears at the input. Many different types of signals can be used with such a

Table 21.1
Commonly-used geostationary satellite frequencies

Band	Uplink (GHz)	Downlink (GHz)	Use
S	1.6265–1.6605	1.53–1.559	Marine (Inmarsat)
C	5.925–6.425	3.7–4.2	Commercial
X	7.9–8.4	7.25–7.75	Military
Ku	14–14.5	11.7–12.2	Commercial
K	27.5–30.5	17.7–20.2	Commercial
K	30–31	20.2–21.2	Military

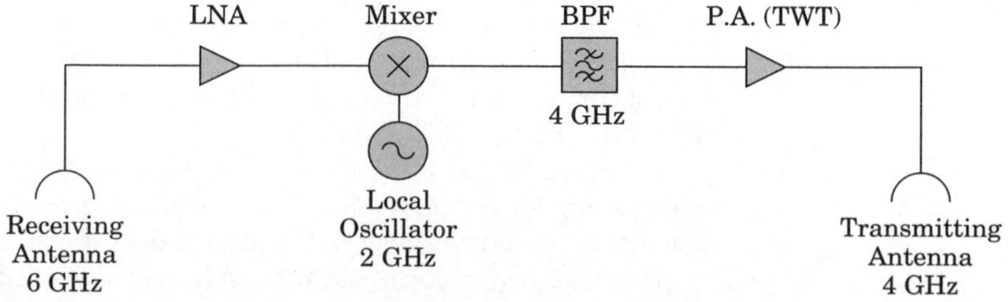

Figure 21.13
Transponder block diagram (C-band)

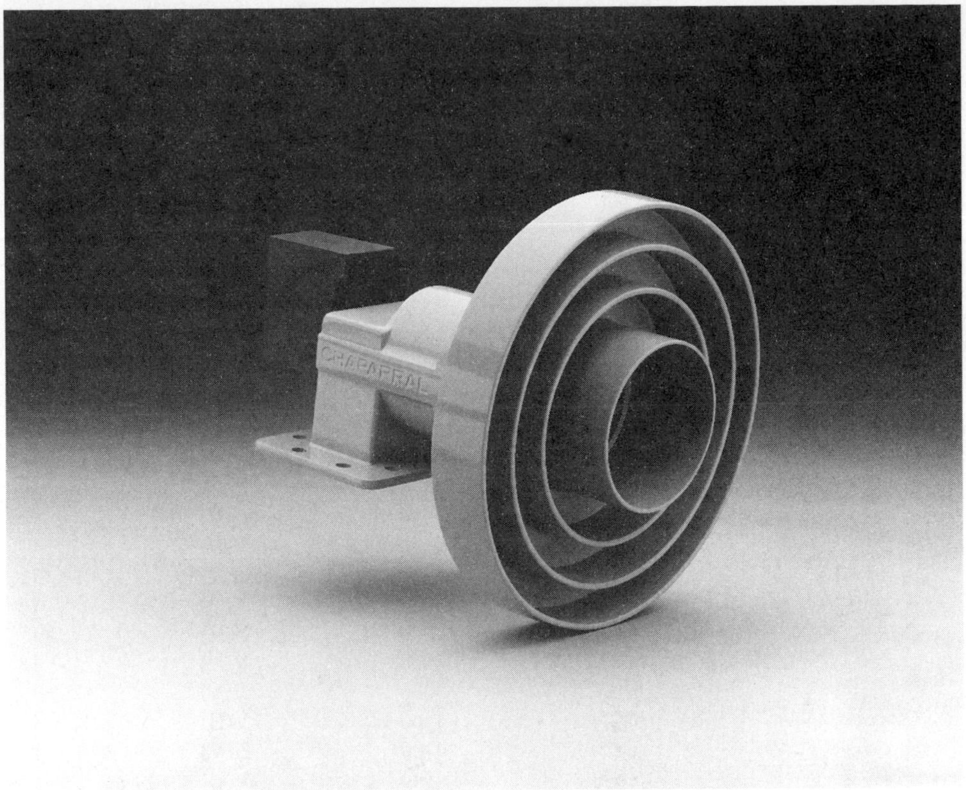

Figure 21.14
C-band feedhorn
Tee-Comm Electronics, Inc., 775 Main St. E., Milton, ON.

transponder, and some will be discussed later in this chapter. Perhaps the most common type of signal consists of a single analog television signal, transmitted using frequency modulation, with stereo sound signals on subcarriers.

There is not enough spectrum space for a satellite to use 24 transponders unless some form of frequency reuse is employed. The method used here is for half the transponders to use vertical polarization and half, horizontal polarization, with overlapping frequencies, as shown in Table 21.2. Depending on the satellite, either the first or the second (bracketed) polarization scheme is used. Receiving antennas on the ground use switchable polarization to separate out the transponders. This is usually accomplished by physically rotating a pickup element in the feedhorn, as illustrated in Figure 21.14, though it is also possible to use two pickups and switch between them.

Aside from the fact that weight and power consumption have to be minimized, the design of a satellite transponder is quite straightforward. The received signal is amplified by a low-noise amplifier (LNA), then translated in frequency using a mixer and a local oscillator. No demodulation takes place. After frequency translation, the signal is amplified by a power amplifier, usually employing a traveling-wave tube (TWT). When linear operation is required, the tube must be operated at several decibels less than its maximum power output in order to reduce intermodulation distortion to a reasonable level. This power reduction is called *backoff*.

The type of transponder discussed so far is equally usable with analog and digital signals, since it simply amplifies signals and shifts their frequencies without demodulating the signals or performing any other processing. In fact, such a transponder is equally capable of handling one broadband signal using the whole transponder bandwidth or a number of narrowband signals, each with its own carrier. For purely digital signals, it is possible to design a transponder as a regenerative repeater. That is, the transponder can demodulate the signal to baseband, decode the digital information, then remodulate the signal on a

Table 21.2
C-band transponder frequencies

Uplink frequency (MHz)	Downlink frequency (MHz)	Polarization	Channel number	Alternate numbering system
5945	3720	H(V)	1	1A
5965	3740	V(H)	2	1B
5985	3760	H(V)	3	2A
6005	3780	V(H)	4	2B
6025	3800	H(V)	5	3A
6045	3820	V(H)	6	3B
6065	3840	H(V)	7	4A
6085	3860	V(H)	8	4B
6105	3880	H(V)	9	5A
6125	3900	V(H)	10	5B
6145	3920	H(V)	11	6A
6165	3940	V(H)	12	6B
6185	3960	H(V)	13	7A
6205	3980	V(H)	14	7B
6225	4000	H(V)	15	8A
6245	4020	V(H)	16	8B
6265	4040	H(V)	17	9A
6285	4060	V(H)	18	9B
6305	4080	H(V)	19	10A
6325	4100	V(H)	20	10B
6345	4120	H(V)	21	11A
6365	4140	V(H)	22	11B
6385	4160	H(V)	23	12A
6405	4180	V(H)	24	12B

carrier with a different frequency. This allows the satellite repeater to regenerate the signal while removing accumulated noise and distortion.

Regenerative repeaters are particularly useful in situations where there are many amplifiers in cascade, for example, in long coaxial or optical cable runs and in the terrestrial microwave networks discussed in the previous chapter. Regeneration is not as important in satellite systems, since there is usually only one repeater, and conventional analog transponders are often used for data signals. Digital regenerative transponders have been designed, however. The block diagram of a typical digital transponder is shown in Figure 21.15.

More elaborate switching techniques can be used with satellites. For instance, the output from a transponder can be switched to any one of several antennas to allow signals from one region to be sent to any of several other regions within the satellite coverage area. Figure 21.16 illustrates beam switching, which is mainly applicable to digital signals, which can use TDM (time-division multiplexing) and can function adequately with access to the satellite only part of the time.

For wider coverage, it is also possible to transmit signals directly between satellites. These cross-links allow signals to be transmitted to areas outside the coverage area of one

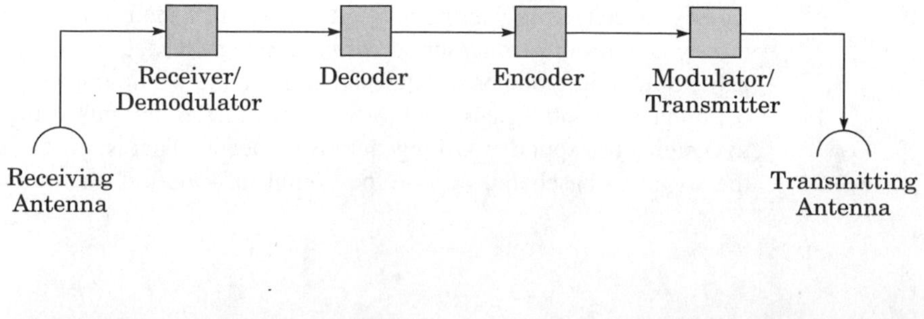

Figure 21.15
Digital transponder block diagram

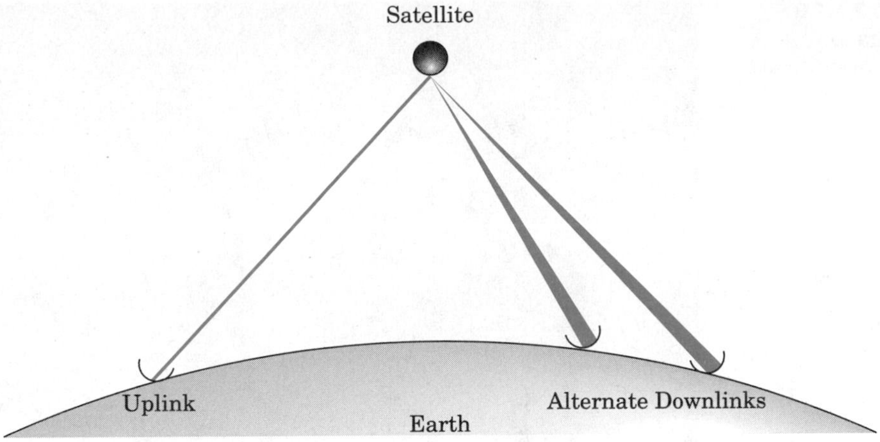

Figure 21.16
Beam switching

satellite, without using an extra earth station in between. A cross-link is illustrated in Figure 21.17. The frequency range of 58–62 GHz has been assigned to these intersatellite links, which are quite common in military satellite systems.

21.3.5 Earth Stations

Earth stations vary greatly in design, depending on whether they are used to transmit to a satellite as well as to receive its signals, the type of signals in use (television, data, etc.), and the strength of the received signal. For instance, Intelsat satellites have antenna beams that cover nearly half the earth, though at a very low received power density. Domestic communications satellites, on the other hand, often use spot beams, which cover only one country or part of a large country like the United States or Canada. Since a spot beam concentrates the available transmitter power into a smaller area, the antenna gain is higher and the power density at the receiver antenna is much greater. Receiving stations for these satellites do not have to be as sophisticated as those used with Intelsat hemispheric beams.

As we pointed out earlier, geostationary satellites are really only approximately stationary as viewed from earth. Small station-keeping jets are used to prevent a satellite from wandering very far away from its assigned position, but even so, a satellite seen from earth appears to move in a small pattern shaped like a figure eight. Most antennas on the ground have beamwidths large enough to include the range of apparent movement and need be moved only to change from one satellite to another. Some very large antennas used as part of Intelsat type A earth stations have beamwidths of only about 0.1 degree and must be moved to track geostationary satellites; they are used with the hemispheric beams described above. Receivers for these extremely weak signals must employ exotic low-noise parametric amplifiers that use cryogenic cooling to reduce the noise level further.

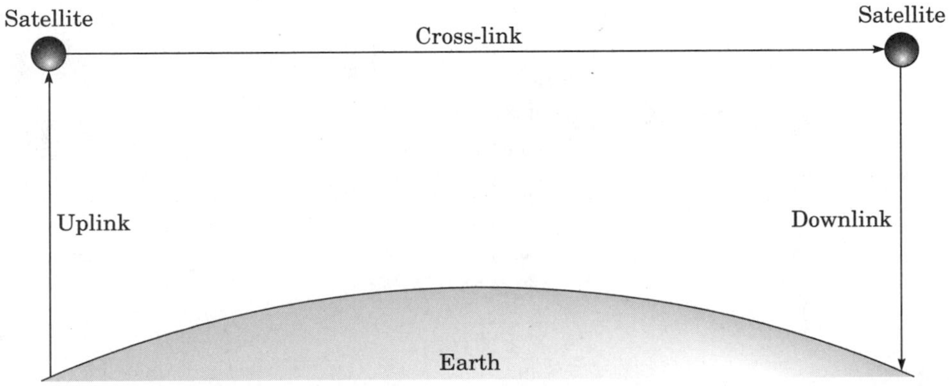

Figure 21.17
Cross-links

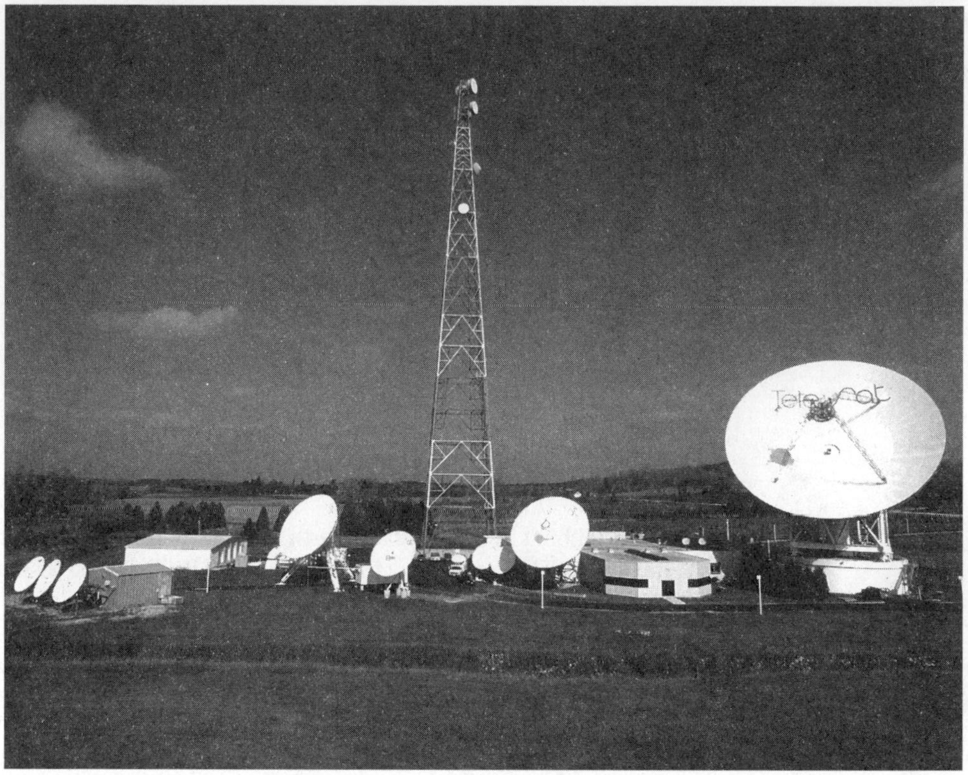

Figure 21.18 Earth Stations
Courtesy Telesat Canada.

You should recall that both gain and beamwidth for a parabolic reflector are functions of the diameter of the dish. For less critical applications, such as television receive-only (TVRO) setups for use with domestic communications satellites, the dish diameter can be reduced—this reduces the gain while also increasing the beamwidth.

Figure 21.18 shows examples of large and small earth stations.

Example 21.9 A typical TVRO installation for use with C-band satellites (downlink at approximately 4 GHz) has a diameter of about 3 m and an efficiency of about 55%. Calculate its gain and beamwidth.

Solution Recall from earlier work that the gain of a parabolic antenna is given by

$$G = \frac{\eta \pi^2 D^2}{\lambda^2}$$

where

G = power gain (as a ratio, not in decibels)
η = efficiency
D = diameter of the dish
λ = free-space wavelength

At 4 GHz the wavelength is given by

$$\begin{aligned}\lambda &= \frac{c}{f} \\ &= \frac{3 \times 10^8 \text{ m/s}}{4 \times 10^9 \text{ Hz}} \\ &= 7.5 \text{ cm}\end{aligned}$$

Substituting into the previous equation, we get

$$G = \frac{0.55\ \pi^2 \times 3^2}{0.075^2}$$
$$= 8.69 \times 10^3$$
$$= 39\text{dB}$$

For the beamwidth, we can use the equation

$$\theta = \frac{70\ \lambda}{D}$$
$$= \frac{70 \times 0.075}{3}$$
$$= 1.75°$$

The receiver circuitry is also much less exotic in a TVRO system. A block diagram of a typical system is shown in Figure 21.19. The low-noise amplifier uses gallium arsenide field-effect transistors (GaAsFETs) and is located right at the antenna. Also included in the same package is a block downconverter that reduces the frequency from the microwave to the UHF region, generally 950–1450 MHz, to reduce losses on the cable to the receiver proper. This receiver contains another converter, the IF amplification and demodulation. Typical noise performance is about 35 kelvins (0.5 dB) at 4 GHz and 60 kelvins (0.8 db) at 12 GHz.

Figure 21.19
TVRO block diagram

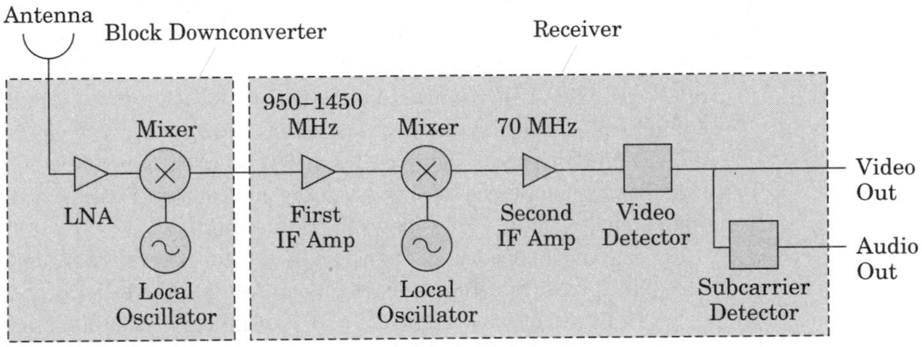

21.4 Applications of Geostationary Satellites

Though geostationary satellites are not the only type of communications satellite, they are certainly the most important at the present time, and they are used in many different ways. This section will examine some of the most important uses for geostationary satellites, namely, television and radio broadcasting, telephony, and data transmission.

21.4.1 Television and Radio Broadcasting via Satellite

One of the most important applications of geostationary satellites is the transmission of television programming. Some of this is international and intercontinental, but much satellite television is domestic. In North America, for instance, satellites are used to carry network program feeds to affiliates, to transmit pay-television and other cable-only channels to cable-television company head-ends, and to transmit remote news feeds from anywhere in the world to network headquarters. Radio broadcasts are also carried, usually as subcarriers on the same transponders that carry television signals. There are also a considerable number of individuals with television receive-only (TVRO) installations.

At the present time, the most common way of transmitting television signals by satellite is to use wideband frequency modulation (WBFM). With a typical C-band North

Figure 21.20
Generation of a television signal with multiplexed stereo audio

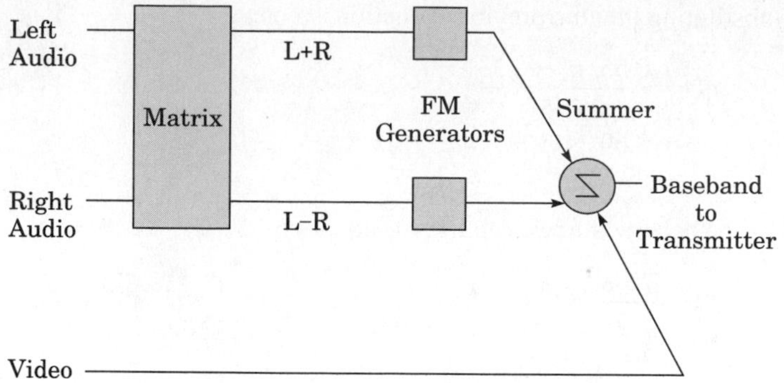

American domestic satellite, this FM signal occupies the entire transponder bandwidth of 36 MHz, which allows for a 4 MHz guard band between transponders. The baseband signal used to modulate the main carrier includes the composite video, complete with both luma and chroma signals, and one or more audio channels transmitted on subcarriers. The most common frequency for the program audio subcarrier is 6.8 MHz for North American domestic satellites. The audio signal is used to frequency-modulate a 6.8 MHz carrier, and the resulting signal is added to the baseband video signal, which contains frequencies from about 30 Hz to 4.2 MHz. The combined video and multiplexed audio signal is then used to frequency-modulate the main carrier. Figure 21.20 shows the signal and a simplified block diagram of the way in which it is created.

Stereo television audio is quite common and can be transmitted via satellite in any of several ways. The simplest method is to use two separate subcarriers, one each for the left and right channel signals. Usually these are at 5.58 and 5.76 MHz, with a combined left-plus-right (mono) signal also transmitted at 6.8 MHz for the benefit of users without stereo equipment. Another method is to transmit the left-plus-right signal on 6.8 MHz and a left-minus-right difference signal on another subcarrier, typically at 5.8 MHz. The stereo audio signal can be recovered by simply adding and subtracting the left-plus-right and left-minus-right signals. Figure 21.21 shows these possibilities.

A transponder has enough bandwidth for several additional subcarriers, which are sometimes used to carry second-language audio, FM radio broadcasts, or digital signals.

Until recently, virtually all video transmitted by satellite used analog FM modulation, as described above. Lately, however, there has been a great deal of interest in digital video. Digital video has been late in coming because without compression, it requires a very high data transmission rate, on the order of 100 Mb/s. Recent advances in data-compression algorithms have allowed this rate to be reduced to a range of about 3 to 7.5 Mb/s using MPEG-2 encoding (described in Chapter 11). The advent of practical, low-cost, direct-to-home satellite transmission is causing some consternation among cable-television companies, who refer to these direct-broadcast satellites as "death stars." Of course, there have

Figure 21.21
Stereo television audio via satellite

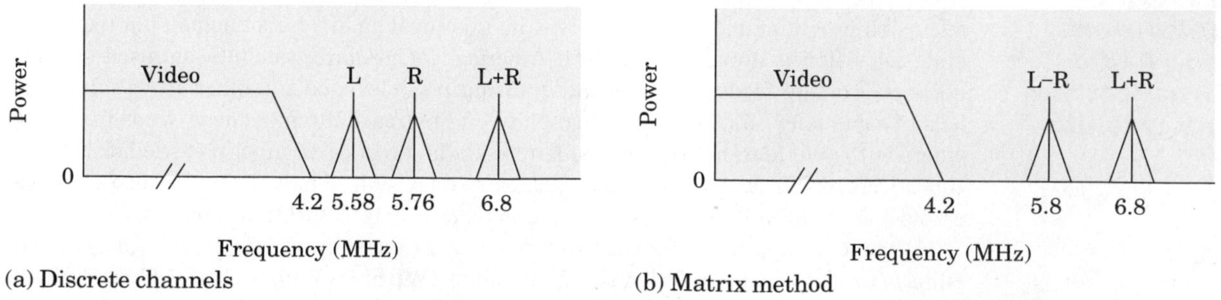

been home satellite receivers for years, but the size, complexity, and cost of a typical installation have kept the satellite receiver from becoming a truly mass-market item.

At this writing, the most popular digital direct-to-home television system is the Digital Satellite System (DSS) (a trademark of Hughes Communications). It uses an 18″ diameter dish antenna to receive signals from three satellites, designated DBS-1, -2, and -3, all operating in the Ku-band and located close enough to each other to make it unnecessary to move the dish to receive signals from all three satellites. A typical DSS receiver and antenna are pictured in Figure 21.22. The antenna has a gain of about 34 dBi and a beamwidth of approximately 3.5°. It uses an offset feed. Though it does not appear so, the feedhorn is at the focus of the antenna, which is a section of a larger paraboloid. Installation is very simple, since the antenna does not need to rotate. The approximate azimuth and elevation angles for the dish can be read from a program built into the receiver once the

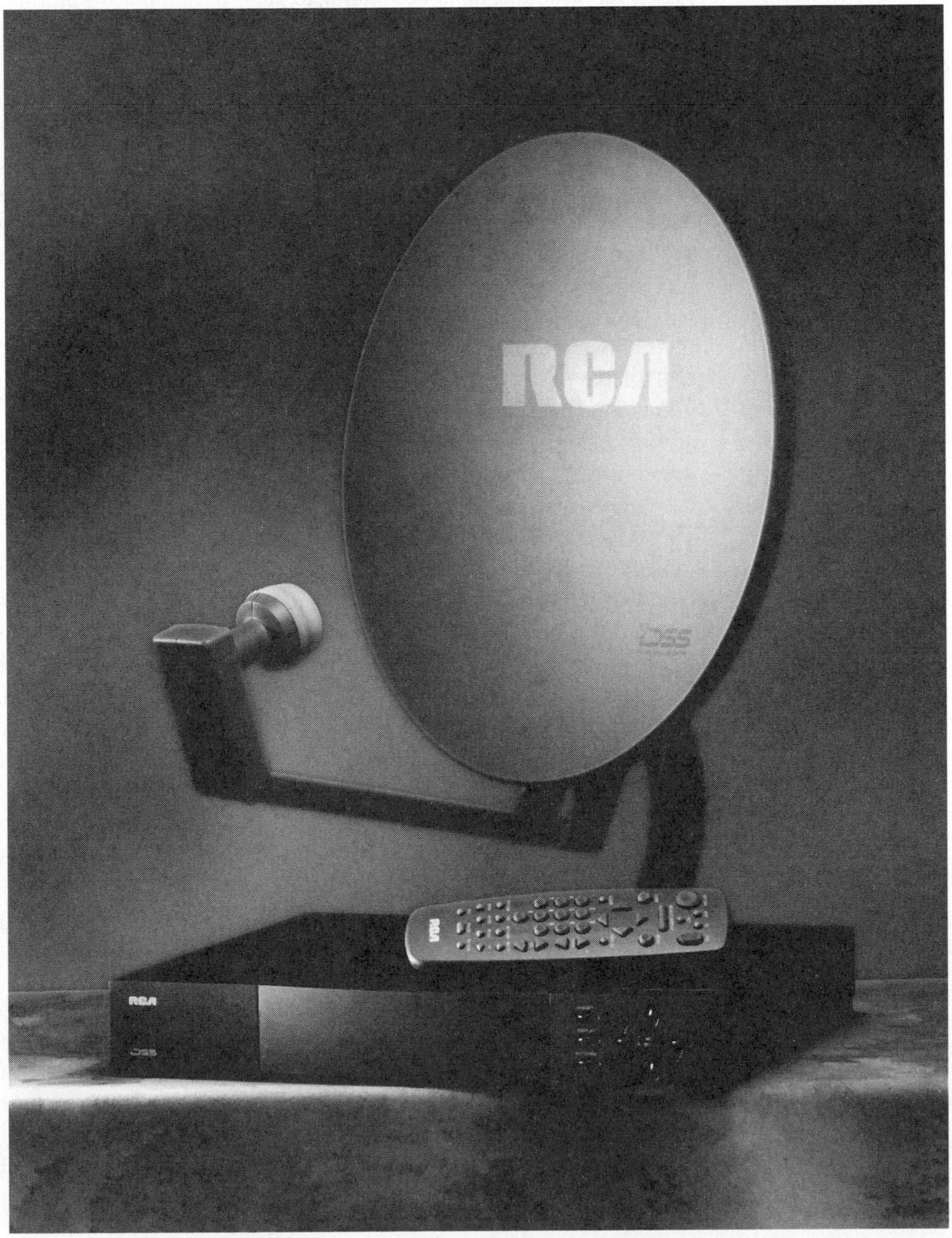

Figure 21.22 RCA DSS satellite system
From Thomson Consumer Electronics.

receiver location is entered. Fine adjustments can be made by using the receiver to monitor signal strength as the antenna is moved.

Two of the three satellites are fully utilized. The third has spare transponders so that it can act as a partial backup for the other two. Altogether, the three satellites carry approximately 175 television channels. Each satellite can have 16 Ku-band transponders transmitting 120 watts each. DBS-1 does use 16 transponders, but DBS-2 and DBS-3 are configured to use 8 transponders, each with 240 watts of power.

Each transponder on the DBS-1 satellite has a *payload data rate*, that is, a useful bit rate after error correction, of approximately 23 Mb/s; the other two satellites have payload data rates of about 30 Mb/s. (The raw data rate before error correction is much higher, about 40 Mb/s.) Typically six programs share a transponder, with the bit rate for each being allocated as needed. This process, called *statistical multiplexing*, allows a greater bit rate to be allocated to a program that has a great deal of motion, while a lower bit rate is allocated to programs with more static scenes. Of course, this technique does not provide any benefit if several programs on the same transponder have a lot of motion at the same time.

Programming for the DSS system is provided by two companies, DirecTV and United States Satellite Broadcasting (USSB). Most of the material on these satellites is delivered to the uplink transmitter by conventional analog C-band satellites, as described earlier. The quality of the eventual DSS picture is thus no better than that from an analog satellite, but it is usually much greater than that delivered by cable television. Some programming is provided via digital videotape, which can achieve better quality. MPEG-2 decoders include a frame store, so that programs derived from film can be shown at 24 frames per second and converted to 30 frames per second in the receiver. This process reduces the bit rate for these programs.

Most DSS programs use 544 pixels per line and 480 lines per frame—this compares favorably with broadcast television, which typically has about 449 details per line and 480 visible lines. The MPEG-2 system is quite capable of higher resolution if required, and it could easily be adapted to high-definition television. Of course, doing so would require an increase in the bit rate per channel and reduce the number of programs that could be transmitted per transponder.

There are currently two other direct-to-home satellite television systems in the U.S. The Echostar DISH (*di*gital *s*ky *h*ighway) began service in March 1996 with one high-power satellite and approximately one hundred channels. It is scheduled to expand to about two hundred channels, using a second satellite scheduled for launch in 1996 and a third planned for 1997. Its technology is very similar to that of the DSS system.

Primestar, another direct-to-home satellite service, uses a medium-power satellite, Satcom K1, and so does not officially qualify as a DBS service. In fact, its purpose is to transmit directly to homes using a 1 meter diameter dish. Its transmissions are digital, but they use a proprietary compression technology. The company plans to switch to MPEG-2 at a later date.

Canada will eventually get direct broadcast satellites. Two companies were licensed, but one, Power DirecTv (affiliated with DirecTv) has decided not to enter the market at present. Expressvu, the other licensee, has had technical problems and was not yet operating as of August 1996. Meanwhile, there has been a flourishing "grey market" in Canada for receivers for the U.S. systems, especially DSS.

21.4.2 Telephony via Satellite

Geostationary satellites are not ideal for long-distance telephony because of the time delays involved. Cable (either copper or fiber-optic) and terrestrial microwave radio have much less delay. The use of satellites for telephony went through a period of rapid increase when geostationary satellites were first deployed, but recently there has been little increase in this use of satellites, largely because of the proliferation of fiber-optic cables. However, for reaching remote areas where it would be too costly to install cable or microwave relays, satellite telephony is of great benefit. It is much more reliable than high-frequency radio, which was previously the most common way of providing telephone communications to these areas. Telephony via satellite is also important for ships at sea, which can be connected to the telephone network via Inmarsat satellites, and for transoceanic communications, since undersea cables are still very costly, even with fiber optics, and there are not yet enough of them to handle transoceanic telephone traffic at peak hours.

Telephony via satellite can be accomplished in several ways, and both analog and digital means are employed. We will begin with analog, which was first historically and is probably still more common than digital transmission using satellites. Since one satellite transponder has room for many telephone channels, some form of frequency-division multiplexing is always used for analog telephony. The methods used are grouped under the generic title of frequency division multiple access (FDMA) schemes. The most important FDMA schemes are frequency division multiplexing–frequency modulation (FDM–FM) and single channel per carrier (SCPC).

Frequency Division Multiplexing–Frequency Modulation The simplest type of satellite telephony is known as frequency division multiplexing–frequency modulation (FDM–FM). An uplink station takes a number of groups or supergroups from the FDM–SSB telephony hierarchy, which was discussed earlier in this book, and modulates the whole package onto a carrier using frequency modulation. Recall from the previous section that a transponder can carry a video signal with baseband frequencies ranging up to 4.2 MHz, plus its associated audio subcarriers at frequencies up to about 8 MHz. Therefore, any other signal with a maximum frequency of about 8 MHz or less can be transmitted by satellite using FM. Figure 21.23 shows how this is accomplished. The FDM–SSB signal is formed in the usual way, then used to modulate the main carrier for the transponder using wideband FM. Demodulation, like modulation, takes place in two stages. First the signal is demodulated by an FM detector, then the SSB signals are demodulated.

With FDM–FM telephony, all the signals on a transponder have the same carrier, so they must be uplinked by the same earth station. They do not all have to go to the same destination, however. Each receiving station must demodulate the entire FM signal, but

Figure 21.23
FDM–FM telephony

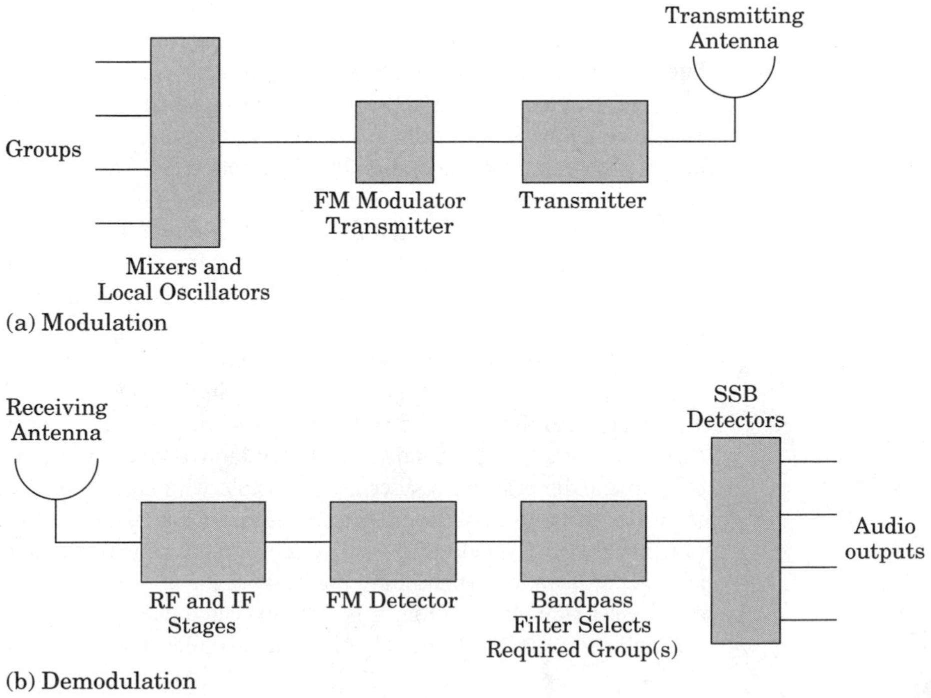

then it only needs to demodulate the SSB signals for whichever groups are intended for it. Conversations are split, with the two sides on different transponders.

Single Channel Per Carrier FDM–FM telephony works well when a large number of voice channels originate at a single point, for example, from a large city. Tying up a whole transponder for a light-traffic route requiring only a few voice channels, however, is very wasteful of satellite resources and spectrum space. Where there are a number of sources and destinations, each of which has a need for only one or a few (typically less than twelve) voice channels, the transponder bandwidth can be divided into a number of channels. Each channel is associated with one carrier and one voice channel. Channel spacings vary, with 45 kHz per channel being typical. Voice signals are modulated using FM.

The major difference between FDM–FM and SCPC is that with the latter, there is no need for all the voice channels on a transponder to originate from one location. There can be many uplinks, each one providing one or a few carriers. Channels can be assigned either permanently or on the basis of demand. The latter technique is called demand assignment multiple access (DAMA), and it allows higher traffic densities.

One drawback of SCPC is that the multiplicity of carriers on one transponder makes the system susceptible to intermodulation distortion, which is caused by mixing of the multiple signals in the power amplifier. When FM is used with a single main carrier, as is the case for television and for FDM–FM telephony, linearity in the power amplifier is not important, and the amplifying device on the satellite (usually a traveling wave tube) can be run at full power (saturation). The multiple signals of SCPC require a linear amplifier, which means that the power input to the traveling-wave-tube amplifier on the satellite must be reduced ("backed off") to less than its maximum value to reduce distortion. Typical backoffs are five to ten decibels.

Though originally an analog system, SCPC can also be used with digital signals. Any of the usual digital modulation schemes can be used to modulate each of the individual carriers.

Time-Division Multiple Access (TDMA) You will recall from basic communications theory that whereas each baseband signal is assigned a part of the communications channel bandwidth for continuous use in FDM, time-division multiplexing (TDM) assigns the whole channel bandwidth to each signal, but only for a brief time slot that recurs at regular intervals. TDM is useful for data communications and also for voice, provided the voice signal is sampled. Analog TDM is possible using analog pulse modulation, but all practical systems employ digital coding using either pulse-code modulation (PCM) or delta modulation. TDMA on satellites resembles landline TDM, except that a time slot is assigned to each of several earth stations. The number of earth stations using a transponder depends on the amount of traffic generated by each. More than one hundred earth stations can use a single transponder when each has only light traffic. Satellite TDMA interfaces very nicely with TDM digital landlines, just as FDM–FM on satellites works with the analog FDM hierarchy.

TDMA has several advantages over analog FDM–FM and SCPC systems. Since only one signal appears on the transponder at any given time, intermodulation distortion is not a problem, and thus satellite and ground-station power amplifiers can be operated at full power. The transponder bandwidth is used more efficiently, allowing about twice as many telephone calls per transponder as is possible with analog systems. In addition, there are the usual advantages of digital transmission, such as signal regeneration and error correction. On the other hand, the digital system is more complex, and precise timing is required at each earth station. This timing must take the propagation time between the earth station and the satellite into account. It is to be expected that as the earthbound telephone system gradually converts to all-digital operation, a similar process will take place in space.

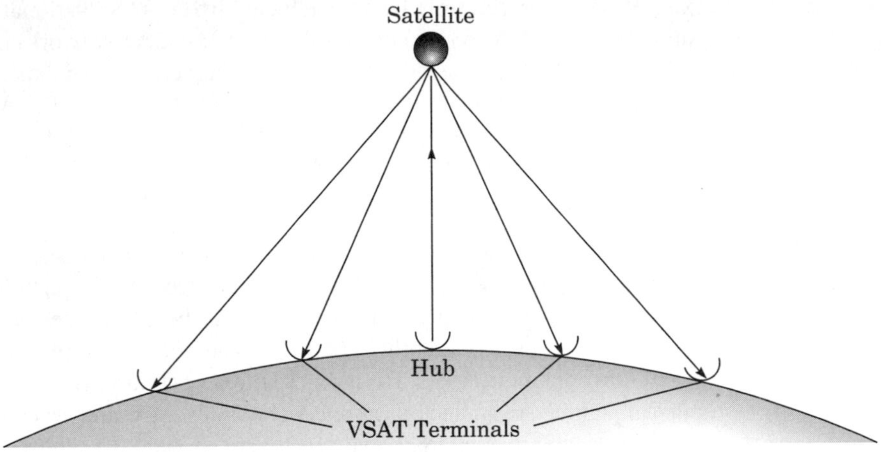

Figure 21.24
VSAT data network

21.4.3 Data via Satellite

Much data traffic uses satellites as an extension of the telephone system. The satellites used for telephony are transparent to data, that is, the fact that a satellite forms part of a telephone link has no effect on the data (except, of course, for introducing an additional time delay).

Large data users can lease a transponder or part of a transponder for a point-to-point or point-to-multipoint link. One type of data network that is commonly used with satellites has a central hub that communicates with many remote sites. The remote sites use very small aperture terminals (VSATs), which are low-cost installations with small antennas and low-power transmitters.

Figure 21.24 shows a typical VSAT network. As you can see, it is basically a star network. The hub uses a relatively expensive earth station with a large antenna and powerful transmitter and transmits a high-speed data stream to the outlying terminals. The terminals use small dishes and low-power transmitters to transmit data at a much lower rate to the hub. Typical applications for VSAT networks are inventory control for large companies with many branches, point-of-sale data transmission in retail chains, and banking. All of these businesses need relatively low data rate transmission from many branches to a central site such as a head office and usually transmission at a greater data rate from the head office to the branches. VSAT networks can satisfy these needs while remaining competitive with landline systems.

A typical VSAT system uses antennas of about 1.2 to 2.4 m diameter at the remote terminals and about 5 to 7 meters at the hub. Data rates from terminals to hub range from 12 to 19.2 kb/s, and the hub usually transmits at 256 to 512 kb/s. The star configuration reduces costs by making it unnecessary for the terminals to have the level of sophistication required to transmit to another similar terminal.

To some extent, an analogy can be made between VSAT systems and ordinary terrestrial broadcasting, in which one rather expensive, high-power transmitter communicates with many receivers, which can be made very cheaply because the signal levels from the transmitter are so high. Indeed, VSAT systems can be used for one-way transmission, for instance, wire service news feeds and stock market reports can be sent this way. VSAT networks typically share an ordinary bent-pipe transponder with other VSAT networks or with SCPC telephony. Current systems generally use satellites in the Ku-band (14 GHz uplink, 12 GHz downlink).

It is possible to build mesh-type VSAT systems, sometimes called M–VSAT, in which each terminal can communicate directly with any other terminal, but the terminals are much more expensive than for the star configuration.

At present mesh systems are not competitive with landline networks, where these exist. Research is currently being conducted by the U.S. National Aeronautics and Space Administration (NASA) into the feasibility of M-VSAT networks using advanced satellites with narrow spot beams and onboard signal processing. An experimental satellite known as the Advanced Communication Technology Satellite (ACTS) was launched in 1993 to test advanced satellite systems. It uses the 30 GHz and 20 GHz frequency bands, narrow spot beams, and onboard signal processing.

21.4.4 Inmarsat Marine Service

Communication between ships at sea and between ship and shore stations has historically depended on high-frequency radio, with lower frequencies also used for distress signals and radio navigation. High-frequency radio has the advantages of simplicity and low cost, but it cannot offer the reliability and voice quality (or data rate, for data communications) of satellite communications. HF radio is still very common at sea, but as time passes, more ships are taking advantage of satellite technology. A system especially designed for marine use is operated by an international organization called the International Maritime Satellite Organization (Inmarsat). This organization began by leasing satellite capacity from various other organizations, such as Intelsat and the European Space Agency, but now it has its own dedicated satellites as well. The first of these went into service in 1990. Most developed nations now have Inmarsat shore stations, which can interconnect between ships at sea and can link ships to the country's domestic telephone system. The Inmarsat system offers worldwide coverage except near the North and South Poles (since geostationary satellites are used).

The primary Inmarsat service is known as Inmarsat-A. It is used by ships at sea and also by aircraft. The ships use small antennas, on the order of 1.3 m in diameter, while diameters of about 13 m are more typical of shore stations. Portable Inmarsat-A stations are available and have been used from land, for example, by news and aid organizations at remote locations or during crises that disrupt normal communications.

Another service called Inmarsat-C offers a low-data-rate message forwarding service. Terminals for Inmarsat-C are relatively low in cost and are used on smaller craft such as fishing boats.

21.5 Satellites in Low Earth Orbits

So far our discussion has centered on geostationary satellites, which are most convenient to use because of the reduced need for tracking; up till now, geostationary satellites have been by far the most widely used satellites for communications. Low-earth-orbit (sometimes abbreviated LEO) satellites have certain advantages, however, and are beginning to be used in communications and related systems.

Because a low-orbit satellite is closer to the earth, it will (all other things, such as transmitter power and antenna gain, being equal) produce a greater power density at the earth's surface. Thus it can be used with less sophisticated earth stations. Similarly, earth-station transmitters will produce greater power density at the satellite for a given EIRP. This makes low-orbit satellites especially attractive for communication with mobile and portable equipment, where high-gain parabolic antennas are impractical.

Low-orbit satellites tend to be less expensive than geostationary satellites because of relaxed power and antenna requirements, and they are also cheaper to launch. This is fortunate, because several low-orbit satellites are required to replace one geostationary satellite. Low satellites are constantly disappearing over the horizon, and a system designer must ensure that there is always at least one satellite in view if reliable, real-time communication is required. (Sometimes this requirement can be relaxed for data communication, if data can be stored until a link is available.)

21.5.1 Orbital Characteristics

Unlike geostationary satellites, low-earth-orbit (LEO) satellites can have a wide variety of orbits. They can be circular or elliptical (remember that a circle is actually a special case of an ellipse) and can have any orientation with respect to the equator and the poles. The Russian Molniya series, already mentioned, has an elliptical polar orbit designed to keep it over Russia for as much of its orbit as possible. On the other hand, weather satellites often have circular polar orbits so that they will pass over most of the earth's surface within a relatively short time period.

LEO satellites can also be any distance from the earth, at least in theory, as long as they are far enough away (about 300 km) to avoid most of the drag of the atmosphere. The further out they are, the longer they remain within sight of a given point on the earth. Thus with larger orbits, fewer satellites are required for continuous communications. On the other hand, smaller orbits, which put the satellite closer to the earth stations with which it communicates, allow the use of less complex equipment, in particular, much smaller antennas, in the earth stations. This is a great advantage for portable and mobile use. However, very low orbits cause satellites to have a reduced operational lifetime, because the greater atmospheric drag causes their orbits to decay quickly.

Besides making it more difficult to track the satellite, lower orbits make it necessary to follow the signal as it changes in frequency due to the Doppler effect (discussed in connection with radar in Chapter 19). Because the distance of a nongeostationary satellite from the earth station is constantly changing, there is a Doppler shift in the frequency of its transmissions: the frequency is shifted upward when the satellite is approaching the earth station and downward when it is receding. Transmissions from the earth to the satellite are affected the same way. Doppler shift is more of an inconvenience than a serious problem, but it must be allowed for.

21.5.2 Mobile Radio via Satellite

At first glance, satellites seem an inappropriate medium for mobile radio. Antennas for use with satellites are traditionally quite large and require aiming at the satellite. Obviously, keeping an antenna on a moving vehicle pointed in a given direction is a nontrivial task. On the other hand, while terrestrial systems such as cellular radio are very practical where there is a large amount of traffic, they are not cost-effective in remote regions.

Several schemes have been proposed for the use of satellites for mobile communications. Most rely on a relatively large number of low-earth-orbit satellites in order to make it possible to use simple, nondirectional antennas on vehicles. However, such schemes are very complex and expensive, and the first operational mobile satellite telephone systems will use geostationary satellites. One example of a geostationary system is the joint United States and Canadian MSAT system (see the accompanying box).

At the 1992 World Administrative Radio Conference (WARC '92), held in Torremolinos, Spain, the members of the International Telecommunication Union (ITU) agreed to an increase in the amount of spectrum allocated to the mobile satellite service (MSS). Frequency allocations in the areas of 150 MHz, 300 MHz, 400 MHz, 800 MHz, and 1.5–2.5 GHz are currently set aside for the MSS. See Table 21.3 for the details. Some of the bands shown in Table 21.3 are shared with other services.

Many systems using LEOs and MEOs (medium earth orbit satellites) for the mobile satellite service have been proposed. Table 21.4 shows the important features of some of

Table 21.3 Frequencies (in MHz) for mobile satellite service

137–138	148–150.05	312–315	387–390
400.15–401	806–840	856–890	1492–1544
1555–1559	1610–1631.5	1656.6–1660.5	1675–1710
1930–2010	2120–2200	2483.5–2520	2670–2690

The MSAT Mobile Satellite System

The MSAT system, which is currently under development, is unusual among proposed mobile satellite communication systems in that it is designed to operate with geostationary satellites. The main problem with such operation has always been the difficulty of mounting and tracking a high-gain directional antenna on a vehicle. Most proposed mobile satellite systems use simple omnidirectional antennas and must therefore use low-orbit satellites in order to achieve strong enough signals. In contrast, the MSAT system uses an advanced phased-array antenna developed by the Communications Research Centre in Ottawa, Canada. This antenna mounts on the roof of a vehicle and tracks the geostationary antenna without moving parts. Figure 21.25 shows a vehicle equipped with an MSAT antenna.

The first of two MSAT satellites was launched in April 1995, and commercial service began in January 1996. A second satellite was launched in the spring of 1996. MSAT satellites use spot beams, each covering a small area of North America. Spot beams achieve a greater power density on the ground than does a single antenna with a wide beam, and they also allow the same frequencies to be reused in more than one beam, increasing the number of conversations that can be carried in a given spectrum allocation. MSAT uses frequencies in the L-band part of the spectrum, around 1.5 GHz.

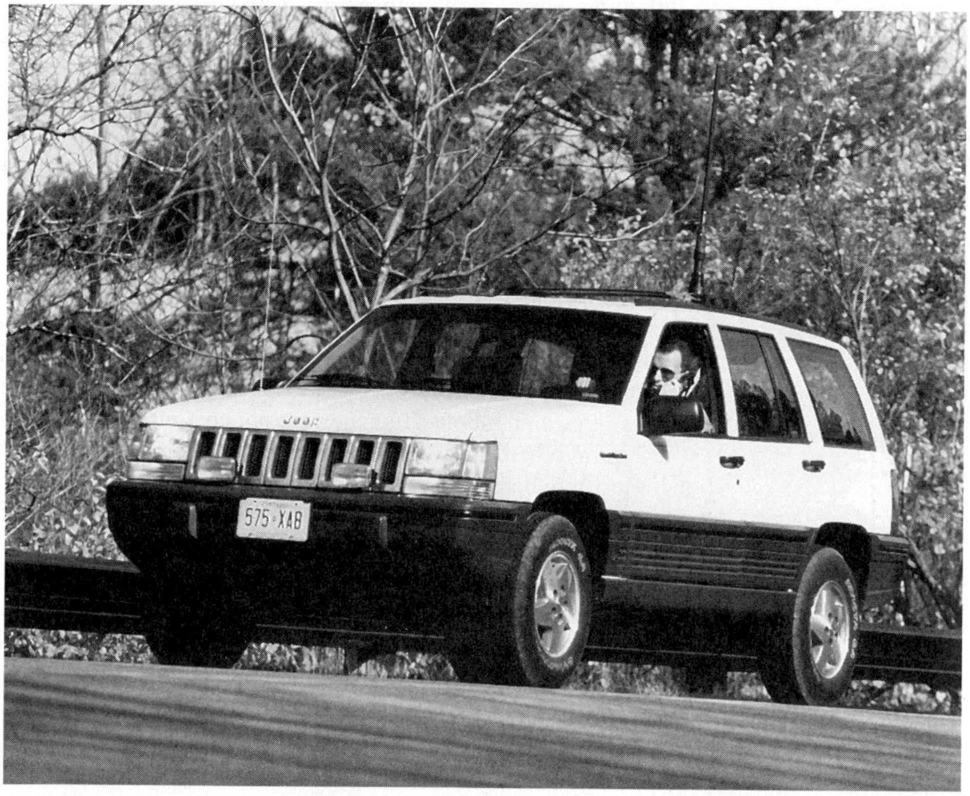

Figure 21.25
Photo of mobile mast antenna

Provided courtesy of TMI Communications, owner and operator of the MSAT® network and a world leader in mobile satellite communications. MSAT is a registered trademark of TMI Communications.

these systems. As of this writing, only the Iridium system has been funded and is under development. In Table 21.4, CDMA stands for Code Division Multiple Access, a form of spread-spectrum communications that will be discussed in more detail in Chapter 24.

Table 21.4 shows mobile satellite systems with full voice capability as well as data and tracking. There is another set of proposals for a less sophisticated class of mobile satellite systems that will operate at lower frequencies and be useful for such operations as keeping track of truck fleets, but not as a cellular-phone substitute. Sometimes these sys-

Table 21.4
Proposed mobile satellite systems

	Name					
	Iridium	Odyssey	Ellipsat	Global Star	Teledesic	Aries
Sponsor	Motorola	TRW	Ellipsat Corp.	Loral/ Qualcomm	MicroSoft/ McCaw Cellular	CCI
Number of satellites	66	12	6	24	840	48
Class	LEO	MEO	LEO	LEO	LEO	LEO
Orientation	Circular	Circular	Elliptical	Elliptical	Circular	Circular
Altitude (km)	755	10,600	2903/426	1390	700	1000
Service markets	Voice Positioning Paging Messaging Data	Voice Positioning Paging Messaging Data	Voice Positioning Paging Messaging	Voice Positioning Paging Messaging	Voice Data Paging	Voice Positioning Paging Messaging
Uplink bands	L-band (1616.5–1626.5 MHz)	L-band	L-band	L-band	Ka-band (26–40 GHz)	L-band
Downlink bands	L-band	S-band (2483.5–2500 MHz)	S-band	L-band S-band	Ka-band	S-band
Access method	FDMA	CDMA	CDMA	CDMA	FDMA	CDMA

tems are called "little LEOs," as compared with the "big LEOs" that operate at frequencies greater than 1 GHz. Table 21.5 shows some of these systems.

21.5.3 Global Positioning System

"Tracking" is one of the functions of several of the mobile satellite schemes described in the previous section. This does not mean that the satellite itself can determine the location of a mobile unit, such as a truck. The satellite simply relays a digital message from the mobile to the base station, reporting the position of the mobile. Determining that position is another problem.

That problem, among many others, can be solved by the Global Positioning System (GPS). This satellite navigation system was developed by the U.S. military for its own use, but has been made available to the public (although the military version has greater accuracy). The GPS system uses satellites in low earth orbit. GPS receivers must receive the signals from several satellites simultaneously to generate accurate position information. Three satellites are required for accurate determination of latitude and longitude, and four

Table 21.5
Proposed mobile satellite systems operating at frequencies under 1 GHz

	Name		
	Orbcomm	Starnet	Vitasat
Sponsor	Orbital Comm. Corp.	Starsys	Volunteers in Technical Assistance (VITA)
Number of satellites	20	24	2
Class	LEO	LEO	LEO
Orientation	Circular	Circular	Circular
Altitude (km)	970	1300	815
Service markets	Tracking Paging Messaging	Tracking Paging Messaging	Data transfer
Uplink bands	VHF (148–149.9 MHz)	VHF (148–149.9 MHz)	VHF (148–149.9 MHz)
Downlink bands	VHF (137–138 MHz) UHF (400.075–400.125 MHz)	VHF (137–138 MHz)	VHF (137–138 MHz)
Access method	FDMA	CDMA	FDMA

are needed if altitude information is also required, as it is for aircraft. A portable GPS receiver is pictured in Figure 21.26.

Summary

1. Satellite communication has advantages over terrestrial modes when mobile or point-to-multipoint communication is required.
2. Most communications satellites are geostationary in order to simplify tracking requirements, but low-earth-orbit satellites are coming into increasing use, especially for mobile communications, because the shorter distance to the satellite results in stronger signals.
3. Most communications satellites are analog repeaters and can be used for a wide variety of both analog and digital signals. Common applications include television, telephony, and data communications.
4. When a satellite is to be used with digital signals only, improved performance can be obtained by using signal processing on the satellite.
5. Digital direct-to-home television broadcasting via satellite is rapidly increasing in importance.

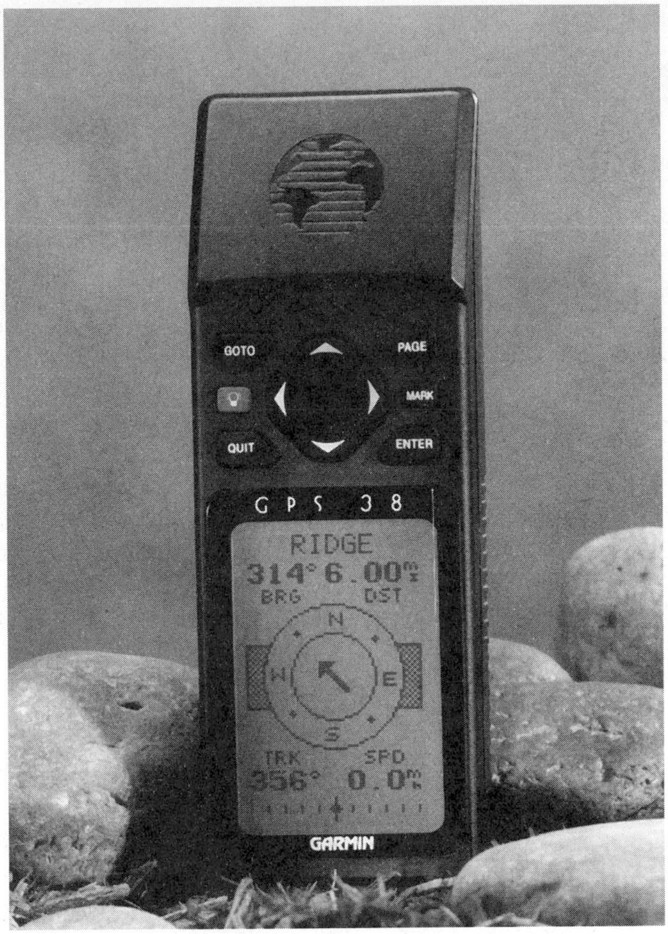

Figure 21.26 Portable GPS Receiver
Courtesy of Garmin International.

Important Equations

$$v = \sqrt{\frac{4 \times 10^{11}}{(d + 6400)}} \tag{21.1}$$

$$\theta = \arctan\left(\frac{R \sin L}{H + R(1 - \cos L)}\right) \tag{21.2}$$

$$\frac{P_R}{P_T}\,(\text{dB}) = G_T\,(\text{dBi}) + G_R\,(\text{dBi}) - (32.44 + 20 \log d + 20 \log f) \tag{21.3}$$

$$d = \sqrt{(r + h)^2 - (r \cos \theta)^2} - r \sin \theta \tag{21.4}$$

$$G/T(\text{dB}) = G_R(\text{dBi}) - 10 \log (T_a + T_{eq}) \tag{21.5}$$

$$T_a = \frac{(L - 1)290 + T_{sky}}{L} \tag{21.6}$$

$$T_{eq} = 290(NF - 1) \tag{21.7}$$

$$\frac{C}{N}(\text{dB}) = EIRP(\text{dBW}) - FSL(\text{dB}) - L_{misc} + G/T - k(\text{dBW}) - 10 \log B \tag{21.8}$$

$$FSL(\text{dB}) = 32.44 + 20 \log d + 20 \log f \tag{21.9}$$

Glossary

apogee the point farthest from earth in a satellite orbit

artificial satellite a structure that orbits the earth and was built by humans (the moon is a natural satellite); usually referred to as just a *satellite* when the context is clear

bent-pipe configuration a satellite transponder design that receives signals and retransmits them at higher power and at a different frequency, without any other processing

downlink transmission of signals from a satellite to an earth station

figure of merit for a microwave receiver, the ratio, expressed in decibels, of gain to noise temperature

Faraday rotation the change in the direction of polarization of signals passing through the ionosphere

G/T see *figure of merit*

geostationary orbit an orbit in which a satellite appears to remain in one position as seen from the earth; a geostationary orbit is also *geosynchronous*

geosynchronous orbit an orbit in which a satellite remains above the equator and has a period equal to one day. If the satellite moves in the same direction as the earth's rotation, the orbit is also *geostationary*. Since in practice all geosynchronous satellites are geostationary, the two terms are often used as synonyms.

orbital satellite any artificial satellite that is not geostationary

perigee the point closest to earth in a satellite orbit

propagation time the time taken for a signal to travel through space from transmitter to receiver

station-keeping the process of adjusting the orbit of a geostationary satellite so that it appears to remain stationary above a point on earth

track to keep an antenna pointed at a satellite as it moves with respect to the antenna

transponder a receiver-transmitter combination on a satellite

uplink transmission of signals from an earth station to a satellite

Questions

1. Compare the following in terms of cost and practical communications distance:
   (a) microwave repeaters on towers
   (b) satellites in low earth orbit
   (c) geostationary satellites
2. Why is it necessary to use circular polarization with satellites that operate at frequencies below the microwave region of the spectrum?
3. How does the orbital period of a satellite change as it moves farther from the earth?
4. Sketch the earth and the orbit of a geosynchronous satellite, approximately to scale.
5. Why do all geostationary satellites orbit the earth at the same distance and above the equator?
6. Why are geostationary satellites unusable from earth stations in the polar regions?
7. What factors limit the power density levels (on the ground) of the radiation from a C-band geostationary satellite?
8. What factors allow Ku-band satellite reception with smaller antennas than are needed for C-band satellites?
9. List two types of mounts for use with earth-station antennas. Explain the uses of each.
10. Describe the operation of a polar mount. Explain why it is usable only with geostationary satellites.
11. What is meant by a "bent-pipe" satellite transponder?
12. What method is used on geostationary satellites to allow more transponders to operate than would fit into the available bandwidth?
13. Name several uses for subcarriers on satellite transponders that carry broadcast television.
14. Describe two ways in which stereo television audio can be transmitted via satellite.
15. Describe and compare three currently available direct-to-home satellite systems.
16. Have any new direct-to-home systems become available since this book was written? Find out and compare them with the systems listed in the text.
17. Explain the difference between FDM–FM and SCPC telephony via satellite. Suggest suitable applications for each.
18. What is meant by the term *backoff* with respect to a traveling-wave tube? Why is backoff necessary when the transponder is shared by several different signals?
19. Explain the operation of TDMA telephony. Compare it with FDM–FM and SCPC.
20. What is a VSAT data network? Compare VSAT networks with those using conventional telephone lines.
21. What satellite-communications system is especially designed for use by ships at sea?
22. How do the mobile antennas used with the MSAT system track a satellite?

23. What are the advantages and disadvantages of satellites in low earth orbit, compared with geostationary satellites, for mobile communication systems?
24. Explain the relevance of the Doppler effect to satellite communications. Why is the Doppler effect unimportant in systems using geostationary satellites?
25. What position accuracy is available with the Global Positioning System?

Problems

SECTION 21.2

26. Find the orbital velocity and period of a satellite that is 1000 km above the earth's surface.
27. The moon orbits the earth with a period of approximately 28 days. How far away is it?
28. What velocity would a satellite need to have to orbit just above the surface of the earth? Why would such an orbit be impossible in practice?

SECTION 21.3

29. An uplink transmitter operates at 6 GHz with a transmitter power of 10 kW and an antenna gain of 50 dBi. The receiver on the geostationary satellite has an antenna gain of 40 dBi. The elevation angle to the satellite from the ground station is 45°. Calculate the received signal strength at the satellite.
30. A receiving installation has a *G/T* of 30 dB. If its antenna gain is 50 dB and the antenna noise temperature is 25 K, what is the equivalent noise temperature of the receiver?
31. How could the *G/T* of the system described in the preceding problem be improved to 35 dB? Give two ways, and perform calculations for each.
32. Find the noise temperature of a receiver with a noise figure of 2 dB.
33. Find the noise temperature of an antenna on a satellite if it looks at the earth (giving it a "sky" noise temperature of 290 K) and is coupled to the reference plane by a waveguide with a loss of 0.3 dB.
34. Compare the time delay for a signal transmitted by geostationary satellite to that for a signal transmitted via terrestrial microwave radio link over a distance of 2000 km. The angle of elevation for both earth station antennas is 30°. For both systems, ignore any time delay in the electronics.
35. A satellite-dish owner has a 3 meter dish designed for C-band (4 GHz) operation. The owner wants to use the same dish, with a new feedhorn, for Ku-band (12 GHz) satellites. What effect will the change in frequency have on the gain and beamwidth of the antenna?

SECTION 21.4

36. How many one-way telephone channels could theoretically be carried on a satellite transponder using FDM–FM if the allowable baseband spectrum extends from dc to 8 MHz and each channel occupies 4 kHz?
37. How many one-way FDM–FM telephone channels can actually be carried on a satellite transponder like the one in the previous question if the transponder carries two U600 mastergroups?
38. How many one-way telephone channels can be carried on a transponder with a bandwidth of 36 MHz, if 40 kHz of bandwidth is allocated to each channel in an SCPC system?

SECTION 21.6

39. Compare the signal strengths from two satellites as received on the ground. One is geostationary, with a path length of 40,000 km; the other is in low earth orbit, with a path length of 500 km. Assume all other factors are equal. Which signal is stronger, and by how many decibels?

COMPREHENSIVE

40. It is often claimed that the Ku-band (12 GHz) is better than the C-band (4 GHz) for TVRO reception because a parabolic antenna of a given diameter has higher gain at the higher frequency. Though the gain is undoubtedly higher, check the difference in path loss at the two frequencies to find out if there is really a net improvement in the signal strength obtained for a given receiving antenna diameter and transmitter EIRP. Are there any other advantages of the Ku-band over the C-band for satellite television broadcasting?
41. It is possible to communicate between points on earth by using the moon as a passive reflector. In fact, it is done routinely by radio amateurs. Calculate the time delay and round-trip path loss at 1 GHz using the moon as a reflector (the distance to the moon was calculated in Problem 27). Note that your path loss calculation ignores losses in the reflection at the moon's surface. What can you conclude about the likely commercial possibilities of this idea?

22 Fiber Optics

Objectives After studying this chapter, you should be able to:

1. Give reasons for the use of optical fiber in preference to wire cable and suggest suitable applications for fiber optics
2. Explain the operation of the three major types of optical fiber and compare their performance
3. Perform calculations involving critical angle, numerical aperture, and bandwidth-distance product for fiber
4. Calculate the required core diameter for single-mode operation at a given wavelength
5. State the requirements for a good splice or connector
6. Use the quantum theory of light to explain the operation of light-emitting diode and laser-diode sources
7. Explain the operation of optical detectors and perform calculations involving responsivity and spectral response

22.1 Introduction

An optical fiber is essentially a waveguide for light (usually infrared). As shown in Figure 22.1(a), it consists of a **core** and a **cladding** that surrounds the core. Both are made of transparent material, either glass or plastic, but the **index of refraction** of the cladding is less than that of the core, causing rays of light leaving the core to be refracted back into the core, so that the light propagates down the fiber.

Figure 22.1(b) shows how an optical fiber can be used for communications by applying a modulated light source to one end and connecting a photodetector to the other. A light-emitting diode (LED) or a **laser diode** (LD) can be used for the source. Any of the modulation schemes studied so far can be used, but digital transmission is more common than analog. The light source can be turned on and off or switched between two different power levels for digital transmission. Analog signals are usually transmitted digitally using PCM, though analog AM and FM schemes are sometimes employed.

Though its operating principle is that of a waveguide, optical fiber is used in practice as a substitute for a copper cable (either coaxial or twisted-pair) or a point-to-point microwave radio link. It is found in many applications—a few random examples include telephone cables, point-to-point transmission of television signals, and computer networks.

Optical fiber has several advantages over electrical cable for communications. In general, fiber has greater bandwidth than coaxial cable, giving it the ability to handle greater data rates. This increased bandwidth allows more signals to be multiplexed. Optical fibers can be built with lower loss than copper cables, increasing the allowable distance between repeaters, and the cable itself can be less expensive.

Figure 22.1
Optical fiber and communications system

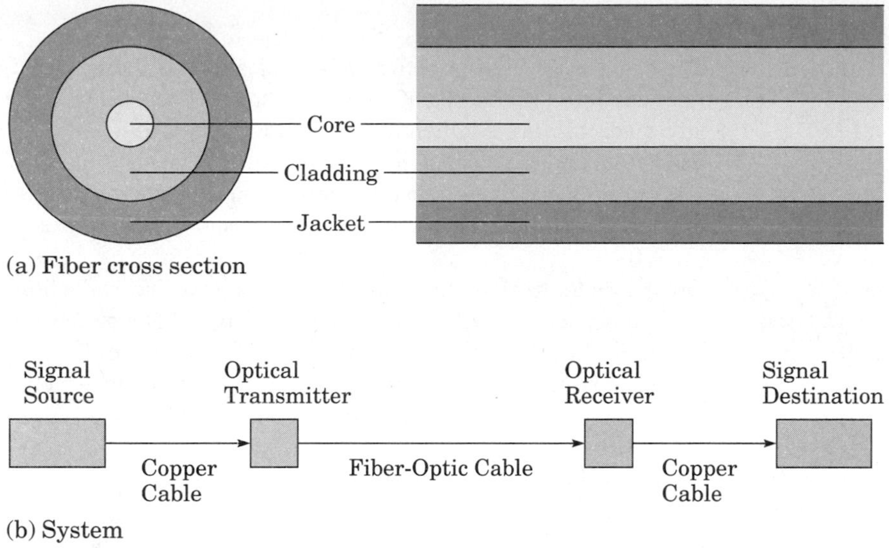

The nonelectrical nature of the signals on optical fiber also confers certain advantages. Since optical cables do not carry electric currents, they are immune to **crosstalk** between cables and to electromagnetic interference (EMI) from other sources. For the same reason, optical fiber is useful where secure communications are concerned, because it is impossible to "tap" the cable without first breaking it and causing an obvious disruption. Electrical cables can often be monitored by inductive coupling from outside, and, of course, radio signals can easily be picked up by unauthorized persons.

The lack of an electrical connection is also useful in explosive environments where a spark hazard may exist and in applications where one circuit must be isolated electrically from another. Fiber links are sometimes used to control high-voltage circuits in power stations, for example. In medical electronics—a very different area—fiber optics is often used to isolate the patient from circuits connected to the electrical power line and thus avoid shock hazards.

Radio links as well as copper cables can be replaced with fiber optics. The two types that are obvious competitors are point-to-point microwave links and geostationary satellite channels. Radio links have the advantage of not needing to have cable laid. This advantage is especially attractive in the case of satellites, which require no access to the terrain between source and destination, which can be separated by thousands of kilometers. On the other hand, the delay of about one-half second between a question and the response to it is a considerable nuisance in telephony via satellite. Optical fibers have greater bandwidth than either type of channel, and, of course, they are much more private.

Physically, optical fibers have size and weight advantages over copper cables. The cost factor depends on the application: optical fiber itself can be less expensive than coaxial cable, but it is more expensive than twisted-pair. For wide-bandwidth long-distance applications, the cost accounting is in favor of fiber, but for short-distance low-bandwidth applications, copper is still cheaper, especially when the cost of converting an electrical signal to and from optical form is taken into account.

One application in which optical fibers cannot substitute for copper cable or microwave waveguide is the transmission of power. They cannot be used to connect a transmitter to an antenna, for instance. Fiber optics are used with power levels in the milliwatt range and are strictly for the transmission of information, not energy.

22.2 Optical Fiber

The actual fiber used in optical communications is a thin strand of glass or plastic. This fiber has very little mechanical strength, so it is enclosed in a protective jacket that is usually made of plastic. Often, two or more fibers are included in one cable for increased bandwidth and redundancy in case one fiber breaks. It is also easier to build a full-duplex system using two fibers, one for transmission in each direction, than to send signals in both directions along the same fiber. Figure 22.2 shows several examples of fiber-optic cables.

22.2.1 Total Internal Reflection

Optical fibers work on the principle of **total internal reflection**. We discussed this principle in connection with radio-wave propagation, but a review seems in order at this point. Figure 22.3(a) shows the situation in which a wave moves from one medium to another one in which its velocity is greater. Rather than specifying the velocity in the medium directly, with light it is usual to list the refractive index, which is given very simply by

$$n = \frac{c}{v} \tag{22.1}$$

where

n = index of refraction of a medium
c = velocity of light in free space
v = velocity of light in the medium

You may recall that the velocity of propagation of any electromagnetic wave in a medium is closely related to the dielectric constant.

$$\epsilon_r = \left(\frac{c}{v}\right)^2 \tag{22.2}$$

where

ϵ_r = relative permittivity, also called the *dielectric constant*, of the medium

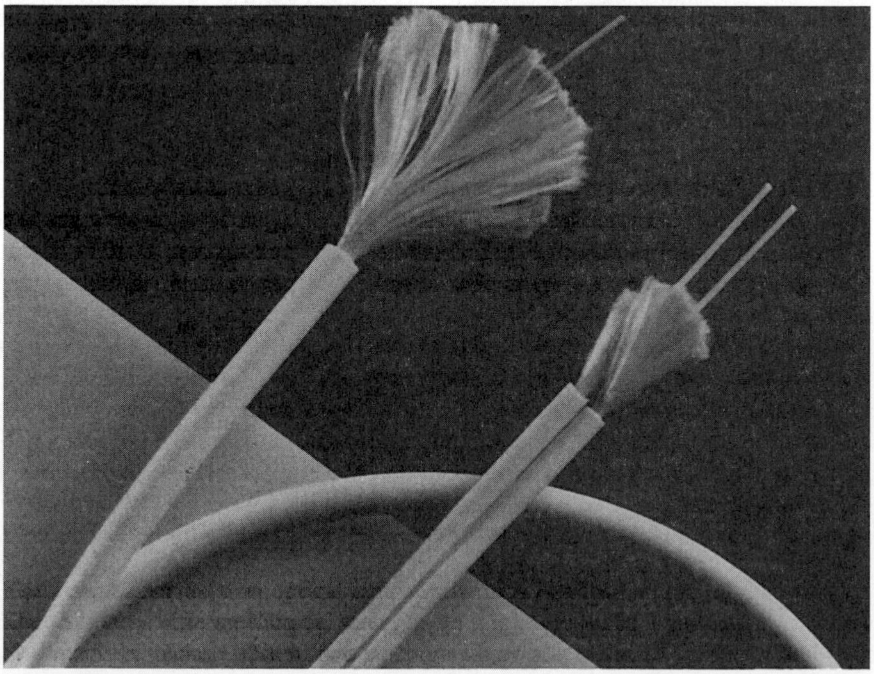

Figure 22.2
Fiber-optic cables
Belden, Inc.

Figure 22.3
Refraction and total internal reflection($n_1 > n_2$)

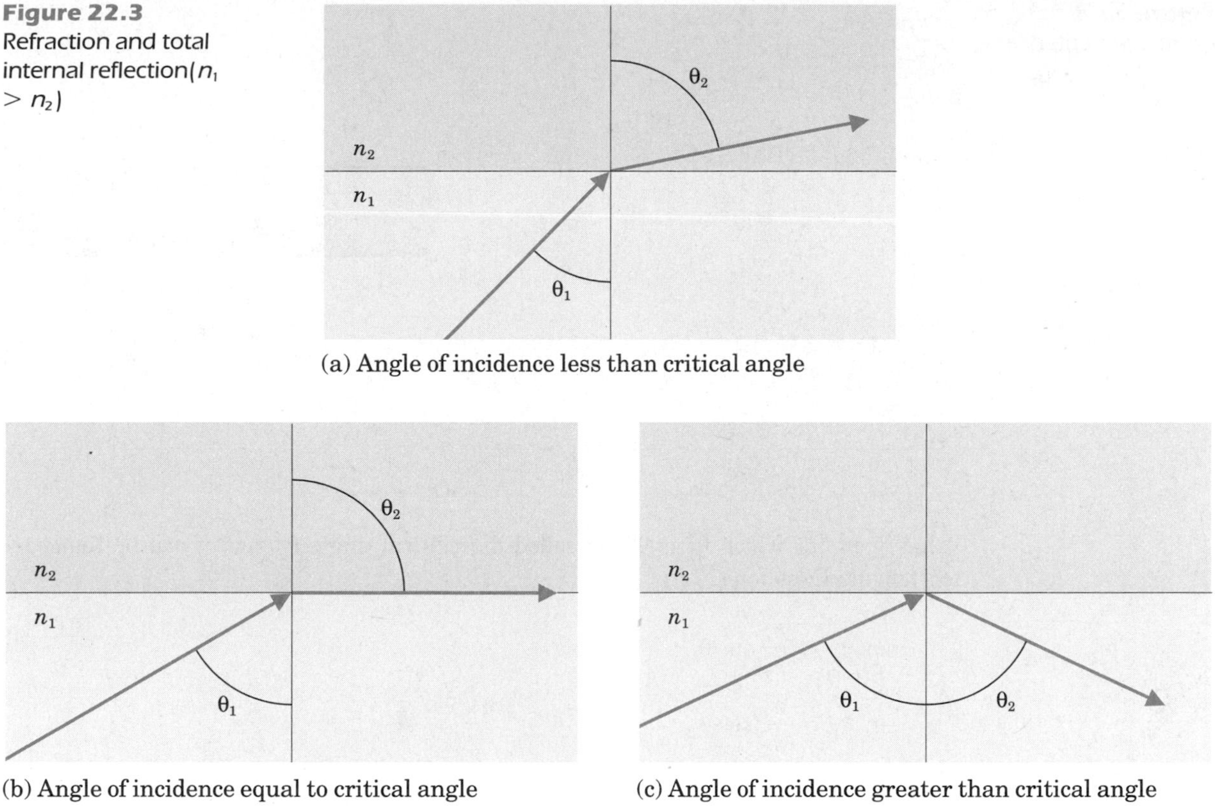

Combining these two equations gives a simple relationship between the index of refraction and the dielectric constant:

$$n = \sqrt{\epsilon_r} \tag{22.3}$$

The **angle of refraction** at the interface between two media is governed by Snell's law:

$$n_1 \sin \theta_1 = n_2 \sin \theta_2 \tag{22.4}$$

where

n_1 = index of refraction in the first medium
n_2 = index of refraction in the second medium
θ_1 = angle of incidence
θ_2 = angle of refraction

In each case, the angle is measured between the ray of light and the **normal**, which is a line perpendicular to the interface between the media.

By rearranging Equation (22.4), we can get an expression for the angle of refraction:

$$\sin \theta_2 = \frac{n_1}{n_2} \sin \theta_1 \tag{22.5}$$

For an optical fiber, the first medium is the core, the second is the cladding, and n_1 is always greater than n_2. There will be some value of θ_1, less than 90°, for which $\sin \theta_2$ will be 1, that is, θ_2 will be 90° and the refracted ray will lie along the interface between the two media, as shown in Figure 22.3(b). For any greater value of θ_1, the ray will be reflected rather than refracted. In this case, Snell's law ceases to operate, and the **angle of reflection** is equal to the **angle of incidence**, as in any reflection situation (see Figure 22.3(c)). The

Figure 22.4
Cone of acceptance

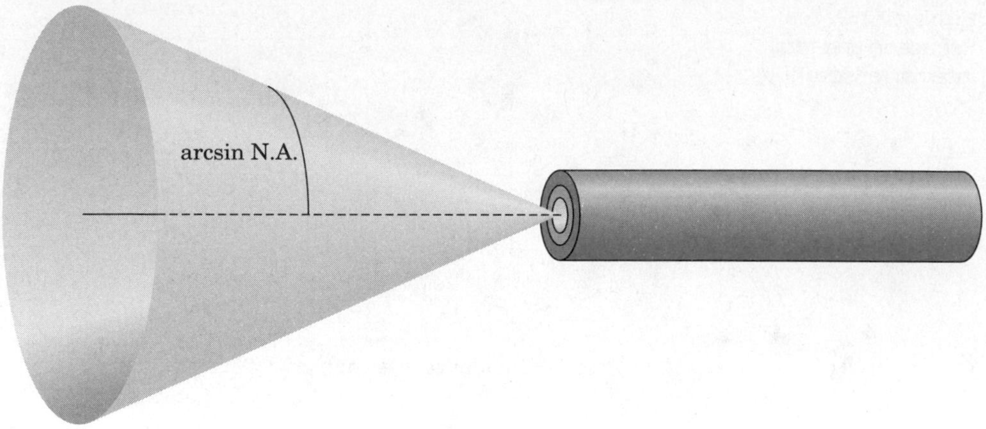

value of θ_1 for which θ_2 is 90° is called the **critical angle** θ_c, and it can be found by rearranging Equation (22.5):

$$\sin\,\theta_2 = \frac{n_1}{n_2}\sin\,\theta_1$$

$$\sin\,\theta_1 = \frac{n_2}{n_1}\sin\,\theta_2$$

For $\theta_2 = 90°$, $\sin\,\theta_2 = 1$ and $\theta_1 = \theta_c$:

$$\sin\,\theta_c = \frac{n_2}{n_1}\sin\,90° \qquad \textbf{(22.6)}$$

$$= \frac{n_2}{n_1}$$

$$\theta_c = \arcsin\frac{n_2}{n_1}$$

where

θ_c = critical angle of incidence
n_1 = index of refraction of the first medium (core)
n = index of refraction of the second medium (cladding)

Example 22.1 A fiber has an index of refraction of 1.6 for the core and 1.4 for the cladding. Calculate:

(a) the critical angle
(b) θ_2 for $\theta_1 = 30°$
(c) θ_2 for $\theta_1 = 70°$

Solution (a) From Equation (22.6),

$$\theta_c = \arcsin\frac{n_2}{n_1}$$

$$= \arcsin\frac{1.4}{1.6}$$

$$= 61°$$

(b) Since 30° is less than the critical angle, Equation (22.5) applies:

$$\begin{aligned} \sin \theta_2 &= \frac{n_1}{n_2} \sin \theta_1 \\ &= \frac{1.6}{1.4} \sin 30° \\ &= 0.571 \\ \theta_2 &= \arcsin 0.571 \\ &= 34.8° \end{aligned}$$

(c) Since 70° is greater than the critical angle, total internal reflection will take place and the angle of reflection will be equal to the angle of incidence, that is, 70°.

22.2.2 Numerical Aperture

The **numerical aperture** of a fiber is closely related to the critical angle and is often used in specifications for optical fiber and the components that work with it. The numerical aperture is defined as the sine of the maximum angle a ray entering the fiber can have with the axis of the fiber and still propagate by internal reflection. Using trigonometry, it can be shown that

$$\text{N.A.} = \sqrt{n_1^2 - n_2^2} \tag{22.7}$$

where

N.A. = numerical aperture

Equation (22.7) assumes that light enters the fiber from free space. (As a practical matter, air can be considered equivalent to free space.) Note that the total **angle of acceptance** is twice that given by the numerical aperture (see Figure 22.4). Any light entering the cone of acceptance illustrated will be reflected internally and may propagate down the fiber. Light entering from outside the cone of acceptance will merely be refracted into the cladding and will not propagate. Numerical apertures for optical fiber vary from about 0.1 to 0.5.

Example 22.2 Calculate the numerical aperture and the maximum angle of acceptance for the fiber described in Example 22.1.

Solution From Equation (22.7),

$$\begin{aligned} \text{N.A.} &= \sqrt{n_1^2 - n_2^2} \\ &= \sqrt{1.6^2 - 1.4^2} \\ &= 0.775 \end{aligned}$$

This corresponds to a maximum angle from the fiber axis of

$$\arcsin 0.775 = 50.8°$$

The total width of the cone of acceptance will be twice this, or 101.6°.

22.2.3 Modes and Materials

Since an optical fiber is in reality a waveguide, light can propagate in a number of specific modes. If the diameter of a fiber is relatively large, light entering at different angles will excite different modes. On the other hand, a fiber that is sufficiently narrow may support only one mode.

Multimode propagation will cause **dispersion**, just as it does in waveguides. Dispersion results in the spreading of pulses and limits the usable bandwidth of the fiber. **Single-mode fiber** has much less dispersion, but it is more expensive to manufacture and its small diameter, coupled with the fact that its numerical aperture is less than that of **multimode fiber**, makes it more difficult to couple light into and out of the fiber.

The maximum allowable diameter for a single-mode fiber varies with the wavelength of the light and is given by the following equation:

$$r_{max} = \frac{0.383\lambda}{\text{N.A.}} \tag{22.8}$$

where

r_{max} = maximum radius of the core
λ = wavelength
N.A. = numerical aperture

Example 22.3 A single-mode fiber has a numerical aperture of 0.15. What is the maximum core diameter it could have for use with infrared light with a wavelength of 820 nm?

Solution From Equation (22.8), the maximum radius is

$$\begin{aligned} r_{max} &= \frac{0.383\lambda}{\text{N.A.}} \\ &= \frac{0.383 \times 820 \times 10^{-9}\ \text{m}}{0.15} \\ &= 2.1 \times 10^{-6}\ \text{m} \\ &= 2.1\ \mu\text{m} \end{aligned}$$

Of course, the maximum diameter is twice this radius, or 4.2 μm.

Both types of fiber described so far are known as **step-index fibers** because the index of refraction changes sharply between core and cladding. A compromise is possible using **graded-index fiber**. Graded-index fiber is a multimode fiber, but the index of refraction gradually decreases away from the center of the core, so light travels faster near the outside. Graded-index fiber has less dispersion than a multimode step-index fiber, though single-mode fibers are still preferred for the most demanding applications. Figure 22.5 shows typical cross sections for all three types of fiber.

Figure 22.5 Types of optical fiber (dimensions are typical)

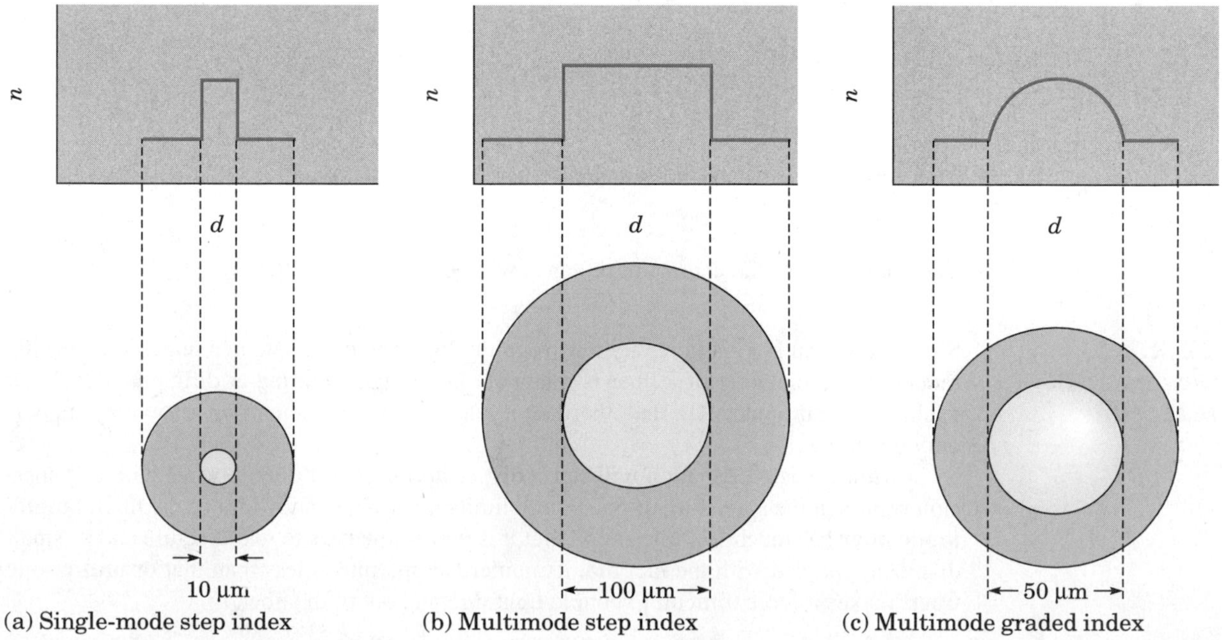

As might be expected, most optical fibers are made of high-quality glass chosen for its very great transparency in order to reduce losses. Some low-cost multimode fibers designed for short-distance applications (such as optical links in consumer electronics and control-signal lines in automobiles) are made of acrylic plastic. The losses in these fibers are very much greater than in glass fiber, but this is of no importance when the distances involved are a few meters or less.

22.2.4 Dispersion

Dispersion due to the propagation of more than one mode has already been mentioned. Just as in microwave waveguides, dispersion in optical fiber results from the fact that in multimode propagation, the signal travels farther in some modes than it would in others. Figure 22.6 shows the effect of multimode propagation on the time it takes for a signal to get from one end of the fiber to the other.

Single-mode fibers are free of modal dispersion effects. Graded-index multimode fibers reduce dispersion by taking advantage of the fact that signals propagating in higher-order modes spend more of their time near the outside of the core than do the low-order modes (this should be clear from Figure 22.6). The reduction of the refractive index toward the outside of the core results in increased velocity in this region, so the high-order-mode components of the signal, which have farther to travel, propagate more quickly. The result is that dispersion is greatly reduced, though not eliminated.

Even single-mode fibers exhibit some dispersion, which is known as *intramodal dispersion* because it takes place within one mode of propagation. It is also called *chromatic dispersion*, because it results from the presence of different wavelengths of light. The wavelength of visible light corresponds to its color. In optical communication, infrared light is much more common than visible light, but the term "chromatic dispersion" is used anyway.

One form of intramodal dispersion is called *material dispersion* because it depends on the material of the core. A beam of light, even **laser** light, consists of more than one wavelength, and different wavelengths propagate at different velocities in glass. This is the same effect that causes white light to break up into a spectrum as it passes through a prism. The wavelengths corresponding to the various colors of the spectrum have different velocities in the glass, resulting in different indices of refraction. They are therefore refracted at different angles, as illustrated in Figure 22.7. In an optical fiber, the difference in propagation velocity for different wavelengths causes them to take varying amounts of time to propagate down the fiber.

Figure 22.6 Multimode propagation

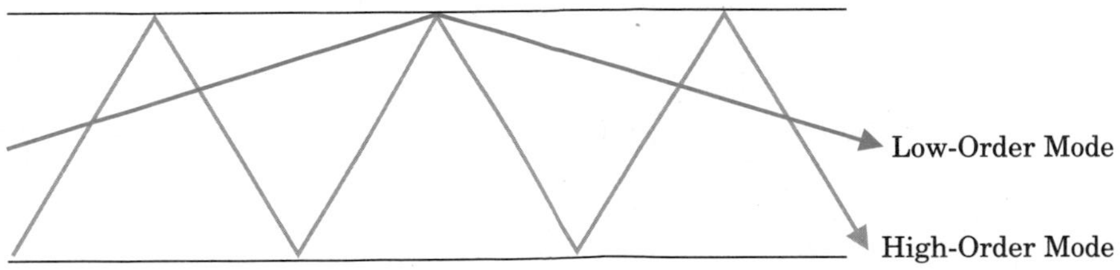

Figure 22.7 Chromatic dispersion in a prism

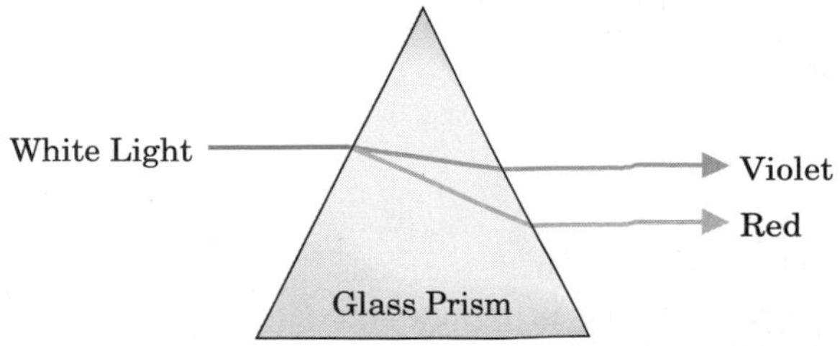

Another source of dispersion in single-mode fiber is called *waveguide dispersion*. It results from the fact that some of the energy propagates in the cladding rather than in the core. This is not obvious from our rather brief study of total internal reflection, but it would be revealed by a more detailed analysis. Since the core and cladding have different indices of refraction, light propagates at different velocities in the two.

These two sources of dispersion are not necessarily additive. In fact they can and do cancel each other out at some wavelength. For a normal glass fiber, this wavelength is about 1.3 μm, in the infrared region.

Dispersion increases with the bandwidth of the light source. (For LEDs and lasers, the term *linewidth* is generally used rather than *bandwidth*.) Where minimum dispersion is important, the laser diode is greatly preferred over the LED because of its much narrower linewidth.

The effects of dispersion increase with the length of the fiber, since a difference in the velocity of two signals produces a difference in the time required to reach the end of the fiber, and this difference is proportional to the length of the fiber. The greater the bandwidth of the signal, the greater the effect of a given amount of dispersion. Some optical fibers are rated according to the product of distance and bandwidth. Multimode fibers have the lowest bandwidth-distance products, and single-mode fibers have the highest.

Example 22.4 An optical fiber has a bandwidth-distance product of 500 MHz-km. If a bandwidth of 85 MHz is required for a particular mode of transmission, what is the maximum distance that can be used between repeaters?

Solution

$$\text{bandwidth} \times \text{distance} = 500 \text{ MHz-km}$$

$$\text{distance} = \frac{500 \text{ MHz-km}}{\text{bandwidth}} = \frac{500 \text{ MHz-km}}{85 \text{ MHz}} = 5.88 \text{ km}$$

Dispersion for single-mode fiber is often expressed directly in units of time rather than bandwidth. The dispersion is then the duration of the output pulse when an infinitesimally short pulse of light is applied to the input. Since single-mode fiber has only intramodal (chromatic) dispersion, the linewidth of the source is very important in determining the dispersion for this fiber. Dispersion is proportional to source linewidth and also, of course, to the length of the fiber. Dispersion in single-mode fiber is given in picoseconds per nanometer of source linewidth per kilometer of fiber length (abbreviated either ps/nm/km or ps/(nm-km), where the - indicates multiplication, not subtraction!). We can define a chromatic dispersion constant D_c, and the dispersion per kilometer is then

$$D = D_c \Delta\lambda \tag{22.9}$$

where

D = dispersion per kilometer in ps/km
D_c = chromatic dispersion in ps/nm/km
$\Delta\lambda$ = linewidth of the source

The total dispersion is then

$$\Delta t = Dl \tag{22.10}$$

where

Δt = total dispersion in picoseconds
D = fiber dispersion in ps/(nm-km)
l = length of fiber in km

Example 22.5 A fiber has a dispersion rating of 6 ps/(nm-km) and a length of 50 km. Calculate the dispersion when this fiber is used with a laser diode source having a linewidth of 5 nm.

Solution From Equation (22.10),

$$\begin{aligned}\Delta t &= Dl \\ &= 6 \text{ ps/(nm-kn)} \times 5 \text{ nm} \times 50 \text{ km} \\ &= 1500 \text{ ps} = 1.5 \text{ ns}\end{aligned}$$

Converting between dispersion and bandwidth specifications is not an exact process, but an approximate idea can be obtained by assuming that a square wave signal is applied to the fiber, as shown in Figure 22.8. This corresponds to NRZ data with alternating ones and zeros. When the dispersion is great enough that a pulse merges into the next bit period, we have intersymbol interference and have reached the limit of the fiber's capacity for most applications. This interference takes place when the dispersion is one-half the period of the square wave. Since it is only necessary to transmit the fundamental component of the square wave in a digital application, the dispersion can be said to equal half the period of a signal at the maximum bandwidth of the fiber, that is,

$$\Delta t = T_{min}/2$$

where T_{min} is the period of a signal at the maximum frequency transmitted by the fiber. Rearranging this equation gives

$$T_{min} = 2\Delta t$$

Since for any signal,

$$f = 1/T$$

where f is the frequency and T is the period, then for the fiber,

$$f_{max} = 1/T_{min}$$

where

f_{max} = the maximum signal frequency

Obviously this is the same as the bandwidth B, so we can say

$$B = \frac{1}{2\Delta t} \tag{22.11}$$

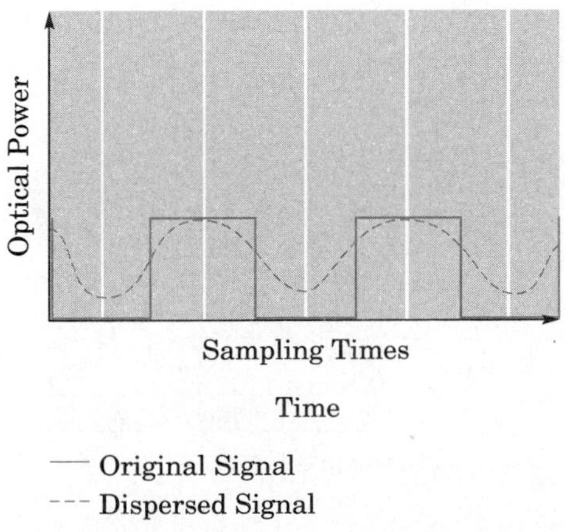

Figure 22.8 Dispersion applied to a square wave

Example 22.6 Find the bandwidth and bandwidth-distance product for the fiber in Example 22.5.

Solution For the fiber given in the example, $\Delta t = 1.5$ ns, so

$$\begin{aligned} B &= \frac{1}{2\Delta t} \\ &= \frac{1}{2 \times 1.5 \text{ ns}} \\ &= 333 \text{ MHz} \end{aligned}$$

The bandwidth-distance product is

$$333 \text{ MHz} \times 50 \text{ km} = 16.7 \text{ GHz-km}$$

It must be recognized that the foregoing analysis is only approximate. It is generally best to work with either bandwidth-distance product or dispersion as given in the manufacturer's data. The determination of bandwidth or bit rate for the whole system will be covered in the next chapter.

22.2.5 Losses

Losses in optical fibers result from attenuation in the material itself and from scattering, which causes some of the light to strike the core-cladding boundary at angles less than the critical angle. This light is refracted into the cladding and is lost to the fiber. Figure 22.9 shows approximately how the loss varies with wavelength for typical glass fiber. The sharp peak at about 1400 nm is due to the presence of water ions, and operation near this wavelength should be avoided. Loss in glass fibers is lowest at wavelengths around 1550 nm, in the infrared region of the spectrum, while plastic fibers perform best at about 660 nm, which corresponds to visible red light. Losses are an important factor limiting the distance between repeaters, since even with laser diodes the light that can be coupled into the fiber is on the order of a milliwatt.

Optical fibers are not always used at the wavelength of lowest loss. Many short-range systems use a wavelength of about 820 nm, as LEDs for this wavelength are less costly than those designed for longer wavelengths. On the other hand, long-distance high-data-rate links may be limited more by dispersion than by power loss. In this case, they can operate at the lowest-dispersion wavelength of 1300 nm. Another possibility is to modify the fiber by changing its composition or the way in which the refractive index varies to bring its zero-dispersion wavelength closer to the minimum-loss wavelength.

Bending an optical fiber too sharply can also increase losses by causing some of the light to meet the cladding at less than the critical angle. Figure 22.10 shows how this can happen.

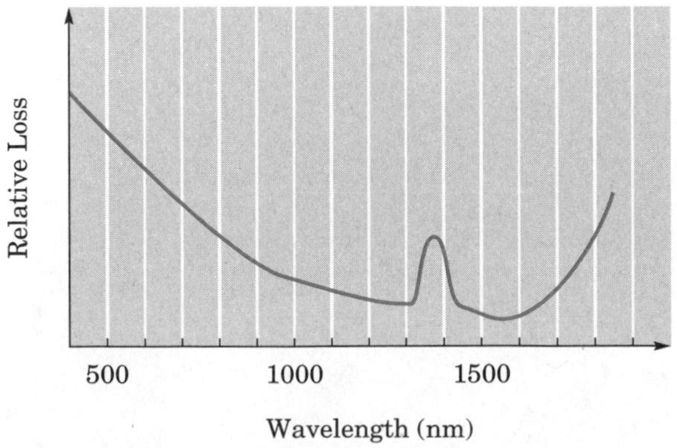

Figure 22.9
Variation of loss with wavelength in glass fiber

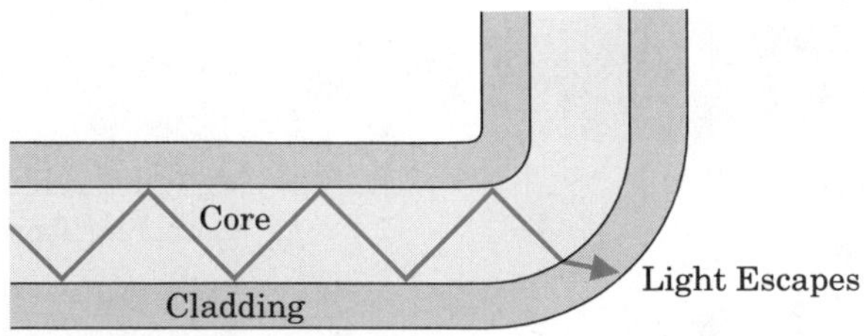

Figure 22.10
Losses due to bending

Table 22.1 Typical specifications for optical fiber

	Multimode	Graded-index	Single-mode
Core diameter	50–100 μm	50–85 μm	8–10 μm
Loss	plastic: 100+ dB/km glass: 8–10 dB/km	1–3dB/km	0.16–1 dB/km
Bandwidth × distance	12–20 MHz-km	1 GHz-km	20+ GHz-km
Repeater spacing	100 m	a few km	40–130 km
Light source	LED	LED or laser diode	laser diode

Table 22.1 compares the specifications of typical examples of the different types of optical fiber.

22.3 Fiber-Optic Cables

Although strong for their size, optical fibers are very thin. They need protection from external forces, both while being installed and during use. Damage from corrosive environments must also be avoided, and it is often desirable to run several fibers together. For these reasons, individual optical fibers, like their copper counterparts, are usually found as part of cables that include protective jackets and strength members.

Most optical fibers, other than those with plastic cladding, have a plastic coating applied over the cladding when the fiber is manufactured. This coating provides some protection against damage in handling. The coated fiber is then installed in a cable, which very often is made by a different company from the one that manufactures the fiber.

There are two basic types of fiber cable. The difference is in whether the fiber is free to move inside a tube with a diameter much larger than the fiber or is inside a relatively tight-fitting jacket. The two types are usually called loose-tube and tight-buffer cables, respectively, and both are illustrated in Figure 22.11.

Both methods of construction have advantages. In loose-tube cables, all the stress of cable pulling is taken up by the cable's strength members. The fiber is also free to expand and contract with temperature at a different rate from the rest of the cable. On the other hand, tight-buffer cables are smaller, cheaper, and generally easier to use. Loose-tube cables tend to be relegated to applications where their greater strength is important, such as telephone cables that have to be pulled for long distances through ducts.

Figure 22.11 Fiber-optic cable construction

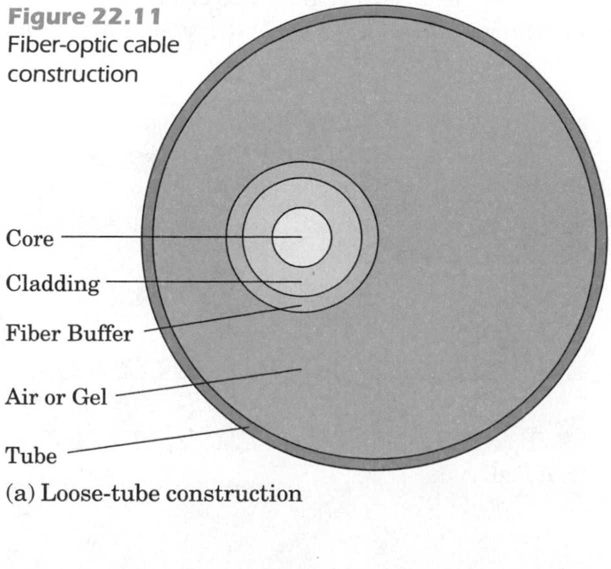

(a) Loose-tube construction

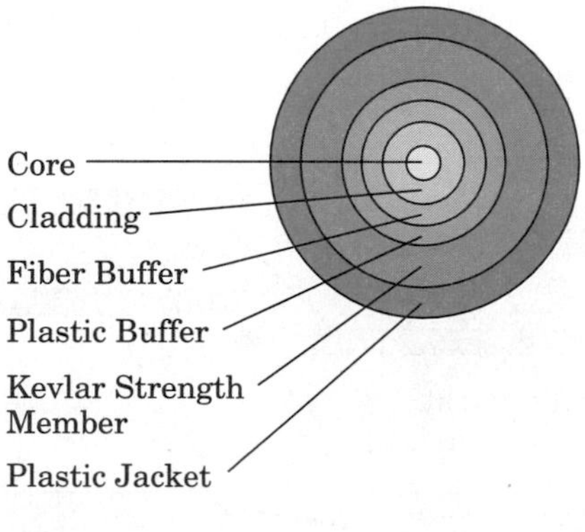

(b) Tight-buffer construction

Since the fiber itself is brittle, cables generally have strength members. Kevlar and steel are popular. Kevlar is generally used in small cables, where it has the advantage of flexibility. It also allows the cable to be completely nonmetallic, which is an advantage in some environments. The strands of plastic material that are visible in Figure 22.2 are Kevlar. Note that the figure shows a single-fiber and a duplex (two-fiber) cable.

Fiber cables, like the copper cables they replace, come in a wide variety of types. There are flat cables to install under carpets, armored and waterproof cables for underwater use, and fiber cables with copper pairs included for communication during installation or for the provision of power to repeaters.

22.4 Splices and Connectors

In an optical communication system, the losses in splices and connectors can easily be more than in the cable itself. Losses result from axial or angular misalignment of the fibers, air gaps between the fibers (which lead to spreading of light), or rough surfaces at the ends of the fibers (which allow light to escape at various angles). Figure 22.12 illustrates these problems. For simplicity, only the cores of the fibers are shown in the figure.

Coupling the fiber to sources and detectors creates losses as well, especially when it involves mismatches in numerical aperture or in the size of optical fibers. For instance, a light source may come with a *pigtail*, a short length of fiber that carries the light away from the source. If this pigtail is joined to a fiber with a smaller diameter or lower numerical aperture, some of the light will be lost. In the case of coupling to a smaller fiber, some of the light from the first fiber never enters the second. However, if the second fiber has equal or greater diameter but a smaller numerical aperture, some of the light that enters the second fiber will fail to propagate. Figure 22.13 shows how this happens. Figure 22.13(a) shows a ray entering the second fiber, then escaping from the core. Figure 22.13(b) illustrates the problem by showing the cone of acceptance for each fiber. The first fiber will emit light over a larger angle than the second fiber can accept.

Good connections are more critical with single-mode fiber, due to its small diameter and small numerical aperture. The low loss of single-mode fiber also makes connector loss more significant. For example, a connector loss of 1 dB is more significant in a system in which the cable loss is 1 dB/km than in one where it is 1 dB/m, as it is in some cheap plastic cables.

The terms *splice* and *connector* are related but not equivalent. Generally, a splice is a permanent connection, while connectors are removable. Splices are necessary where sections of cable are joined. For practical reasons, the length of a spool of cable is limited to about 10 km, so longer spans between repeaters require splices to be made in the field. Connectors are needed between sources and detectors and the fiber cable. Generally, the loss in a properly made splice is less than that in a connector. Splices can have losses of 0.2 dB or less, while connector losses are often about 1 dB. One of the reasons for this is

Figure 22.12
Coupling losses

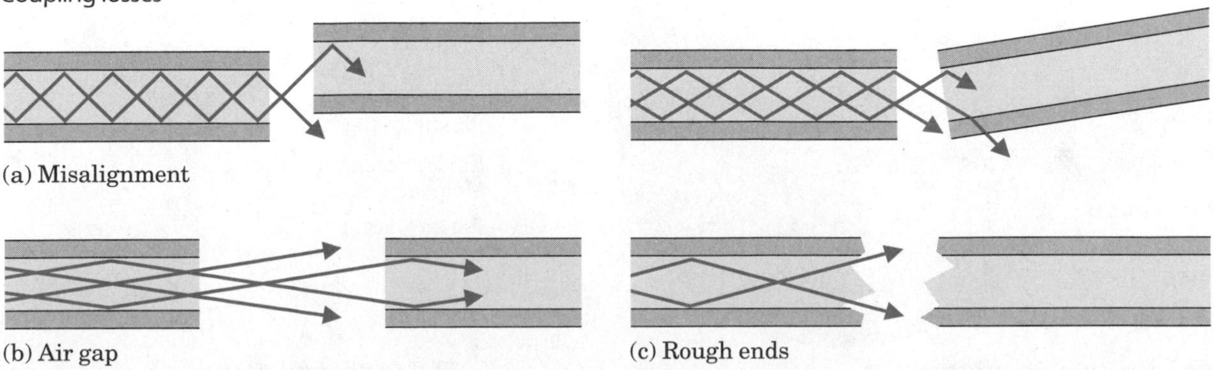

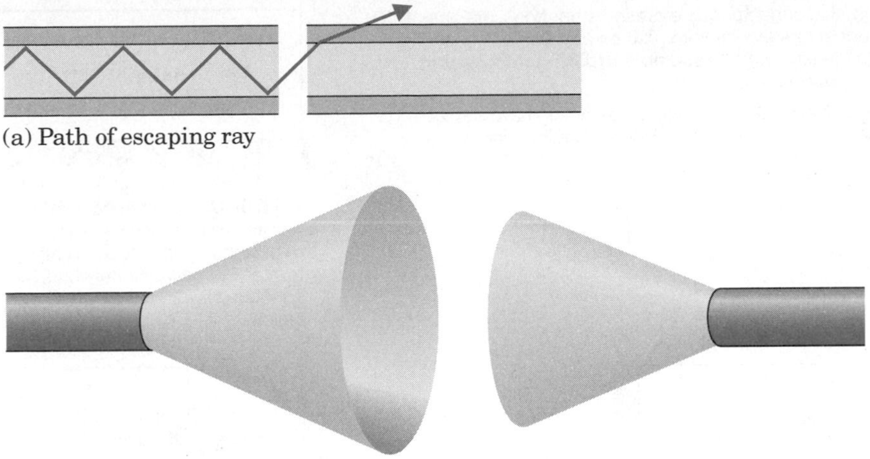

Figure 22.13
Loss due to coupling to fiber with smaller numerical aperture

(a) Path of escaping ray

(b) Different cones of acceptance

that the ends of the fibers touch in a well-made splice, while in a connector there is a small air gap so that the polished fiber surfaces will not be damaged during the process of connecting or disconnecting.

Figure 22.14 shows some typical commercial connectors. Note the care that is taken to align the fibers. Splices are usually created by cleaving the glass ends and then fusing them together with an electric arc, but some field splices are made using specially formulated adhesives.

A fusion splicer is illustrated in Figure 22.15.

22.5 Optical Couplers and Switches

Just as with coaxial cable and microwave waveguide, it is possible to build power splitters and directional couplers for fiber-optic systems. However, it is considerably more complex (and expensive) to tap or split a fiber system than one using copper wire. Sometimes this is an advantage: for instance, a fiber-optic communication system is more secure against unauthorized wiretapping than one using copper, which can often be tapped by placing a detector in the magnetic field surrounding the wire, with no need to cut it. However, when it is necessary to tap into a fiber, as in some types of local-area networks, for instance, taps and splitters can be obtained.

Optical couplers are often categorized as either star couplers, with many inputs and outputs, or tee couplers, which have only one input and two outputs. The construction of these two types is similar, however, and it is more useful to describe couplers as directional or bidirectional, depending on whether light can travel through the coupler in both directions. A typical directional coupler is shown schematically in Figure 22.16(a).

22.5.1 Coupler Construction

Optical couplers can be made in many different ways. Two of the most common types will be described. A number of fibers can be fused together as shown in Figure 22.16(b) to make a transmissive coupler. Light coming in on any fiber on one side of the coupler will exit by all the fibers on the other side. There appears to be a requirement that the number of ports on each side be equal, but if this is not desired, one or more of the fibers on one side can be ignored. For instance, one input can be split among two outputs to create a tee connector. By its nature, the transmissive coupler is directional. In the example shown in Figure 21.16(a), for instance, light entering port 1 will leave by ports 2 and 3, but any light entering port 2 will exit by port 1 and not by port 3.

Another common style of coupler is the reflective type illustrated in Figure 22.16(c). Here, a signal entering on any fiber will exit on all the other fibers, so the coupler is bidirectional.

This fiber optic connector guide presents brief descriptions of popular connectors available for use on any Belden® fiber optic cable assembly. **For assembly ordering information see catalog page 29.**

For further assistance and connector selection for a Belden cable assembly, contact Belden Product Engineering at **1-317-983-5200.**

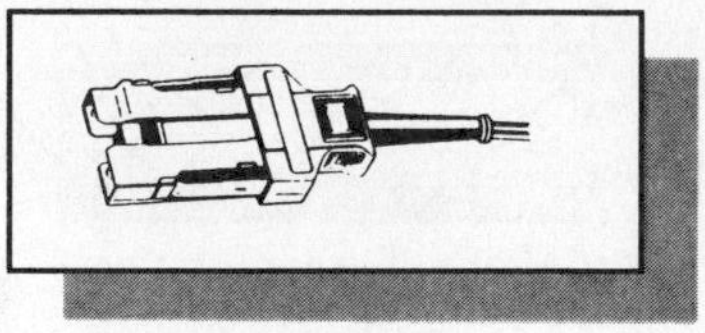

■ **FDDI.** Duplex fiber optic connector system with ceramic ferrule, fully compatible with ANSI FDDI PMD document. For data communcations applications, including FDDI backbone, frontend or backend networks and IEEE 802.4 token bus. Dry connection, with positive latching mechanism. Low insertion loss.

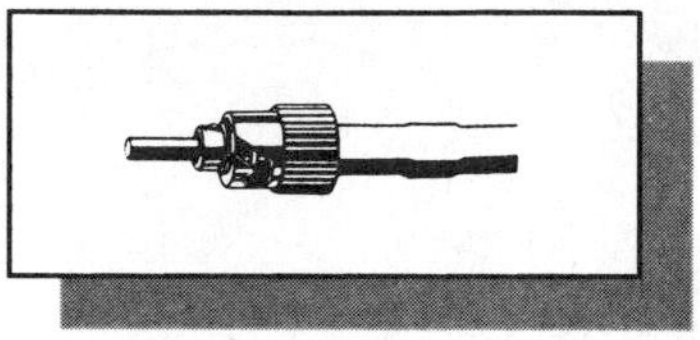

■ **ST Compatible.** Small size connector with keyed bayonet coupling for simple ramp latching or disconnect; dry connection. Available in multimode and single mode versions. Fully compatible with existing ST hardware. For data processing, telecommunications and local area networks, premise installations, instrumentation and other distribution applications. Low insertion and return loss.

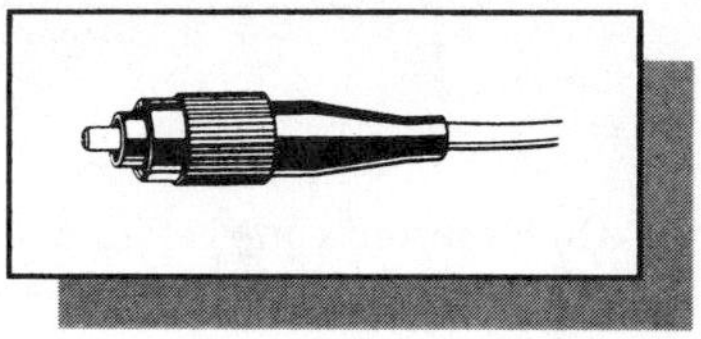

■ **FC.** One piece connector design for easy termination. Compatible with NTT-FC and NTT-D3 hardware. Dry connection with screw type strength member retention. Available in multimode and single mode versions. Applicable for telecommunications and data communications networks, premise installations and instrumentation. Low insertion and return loss.

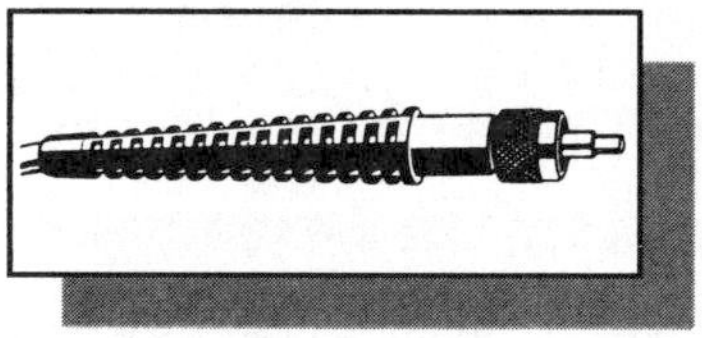

■ **SMA.** Small size connector with SMA coupling nut; dry connection. For use with multimode cables in data communications applications such as local area and data processing networks, premise installations and instrumentation. Low insertion loss. Fully compatible with all existing SMA hardware.

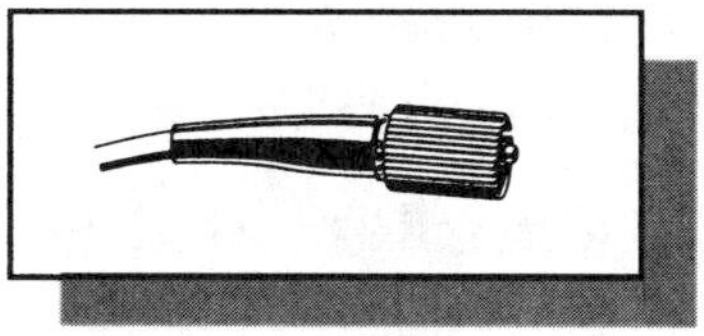

■ **D4.** Compatible with NTT-D4 hardware. Ferrule alignment key for consistent remating. Rugged construction for long life and durability. Low insertion and return loss.

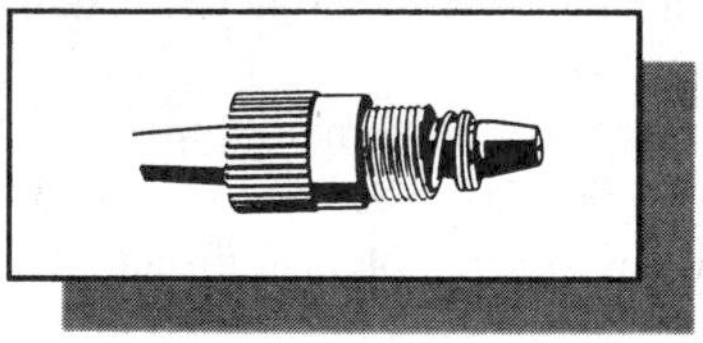

■ **Biconic.** Small size connector with screw thread, cap and spring loaded latching mechanism. Low insertion and return loss. Compatible with all biconic hardware.

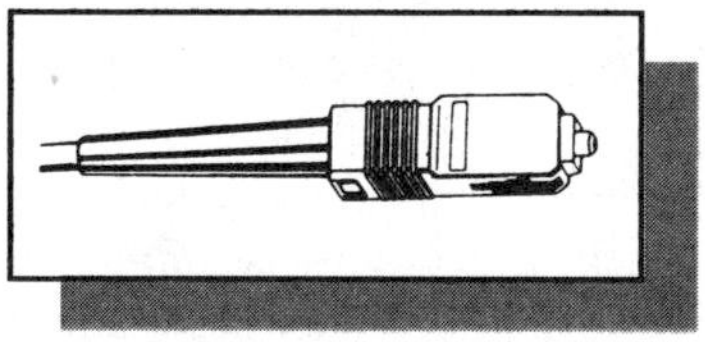

■ **SC.** Square design for high packing density. Push-pull operation simplifies connections. Available in single mode and multimode versions. Low insertion and return loss.

Figure 22.14
Fiber-optic connectors
Belden, Inc.

NOTE: Information on connectors is presented for selection assistance only. Belden Wire and Cable does not assume any liability or responsibility for the accuracy of descriptive or performance data herein. Product and performance specifications should be verified with the connector manufacturer.

22.5.2 Coupler Specifications

Passive optical couplers like those just described divide the input power among two or more outputs. In addition, there is some **excess loss**, usually less than 1 dB, in the coupler itself, so the sum of the output power levels will not quite equal the input power. Some couplers, like the transmissive coupler shown in Figure 22.16(b), are directional, so there will be a

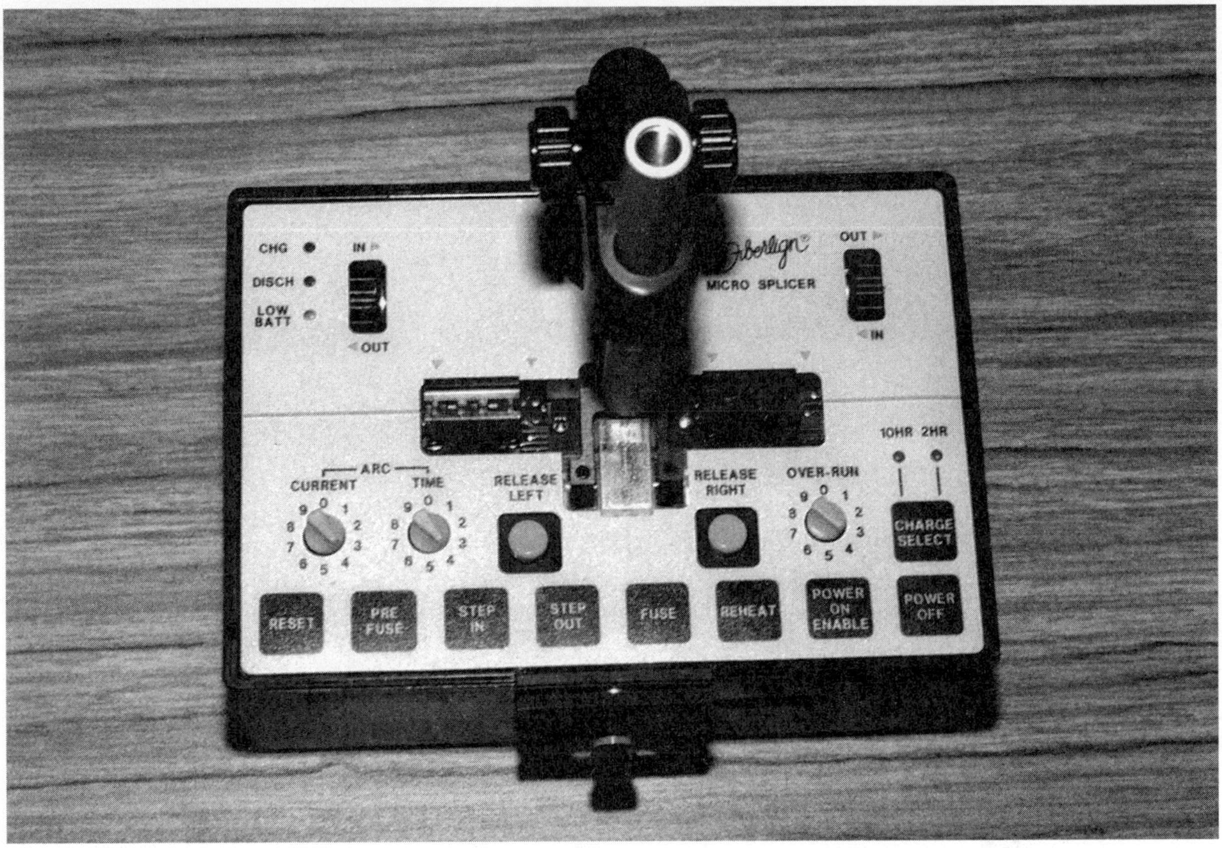

Figure 22.15
Fusion splicer

Figure 22.16
Optical couplers

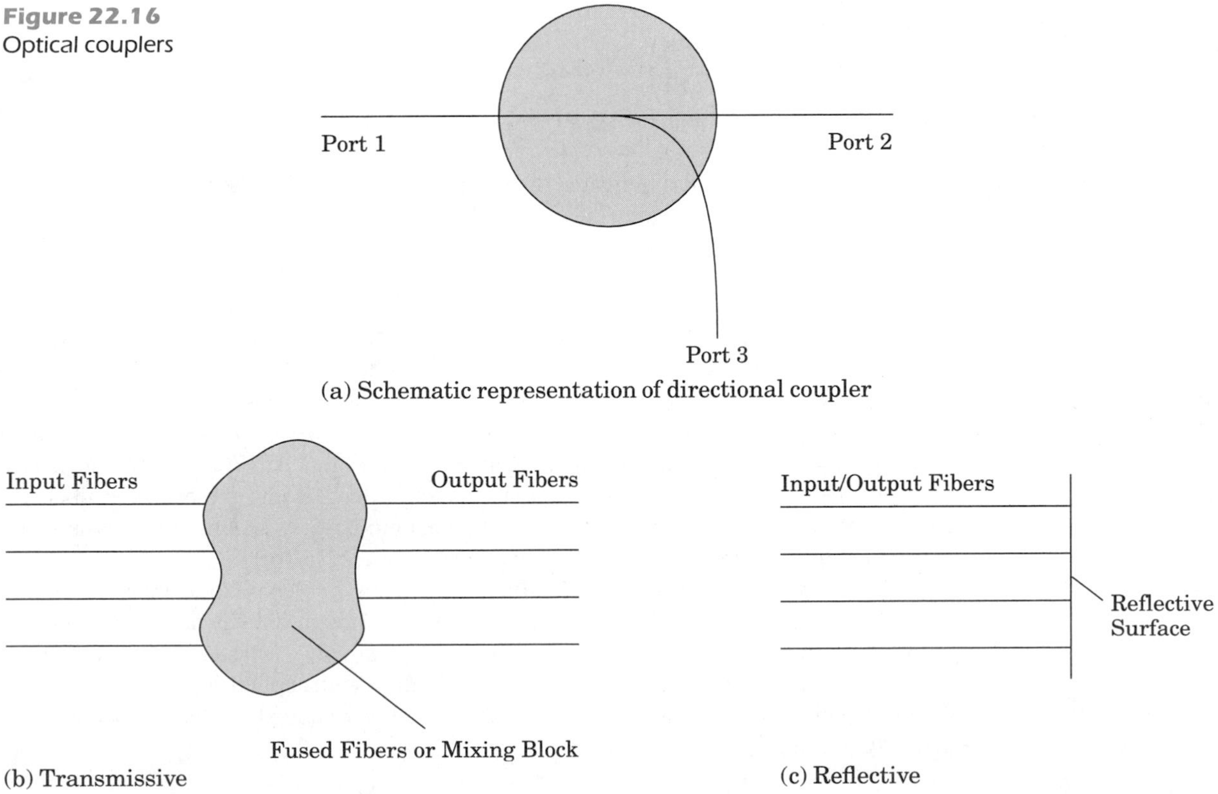

directivity specification as well. Output power at each port can be specified as a number of decibels down from the input power or as a percentage of the input power. Specifications in decibels are usually more convenient.

Example 22.7

An optical coupler has one input (port 1) and two outputs (ports 2 and 3). Its specifications are given below. Even though negative signs are not used for the coupling specifications, they are implied. Since this is a passive coupler, the output power must always be less than the input power.

Inport port	Outport port	Coupling dB
1	2	3
1	3	6
2	1	40
2	3	40
3	1	40
3	2	40

Find:

(a) the percentage of the input power that emerges from each of ports 2 and 3 when the input is at port 1
(b) the directivity
(c) the excess loss in decibels

Solution

(a) The proportion of input power emerging at port 2 is:

$$\frac{P_2}{P_{in}} = \log^{-1}\left(\frac{-3}{10}\right) = .501 = 50.1\%$$

For port 3 the result is:

$$\frac{P_3}{P_{in}} = \log^{-1}\left(\frac{-6}{10}\right) = 0.251 = 25.1\%$$

(b) In the reverse direction, the signal is 40 dB down for all combinations, so the directivity is 40 dB.
(c) The total power output, compared to the input, is 50.1% + 25.1% = 75.2%. The excess loss in decibels is $-10 \log .752 = 1.24$ dB

22.5.3 Optical Switches and Relays

Occasionally it is necessary to switch optical signals from one fiber to another. This can be done electrically, by converting the signal from light to electrical form and then back to light, but it is simpler to use a switch that can work directly with the light.

The simplest type of optical switch moves fibers so that an input fiber can be positioned next to the appropriate output fiber, as shown in Figure 22.17(a). Another approach is to direct the incoming light into a prism, which reflects it into the outgoing fiber. By moving the prism, the light can be switched among output fibers. Lenses are necessary with this approach to avoid excessive loss of light (see Figure 22.17(b)).

Either of these methods can be automated by using a solenoid to operate the moving parts, as in a mechanical relay for electrical circuits. Though reliable and fairly efficient, with losses on the order of 1 dB, these relays are rather slow. A faster though lossier approach is to use crystals whose refractive index changes with applied voltage. Relays based on this principle have no moving parts and can switch very quickly (in nanoseconds, compared with milliseconds for mechanical relays).

Figure 22.17(a)
Optical switch

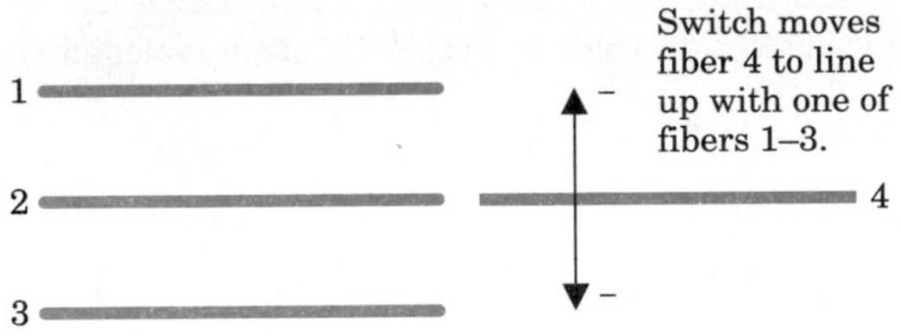

Figure 22.17(b)
Using movable prism

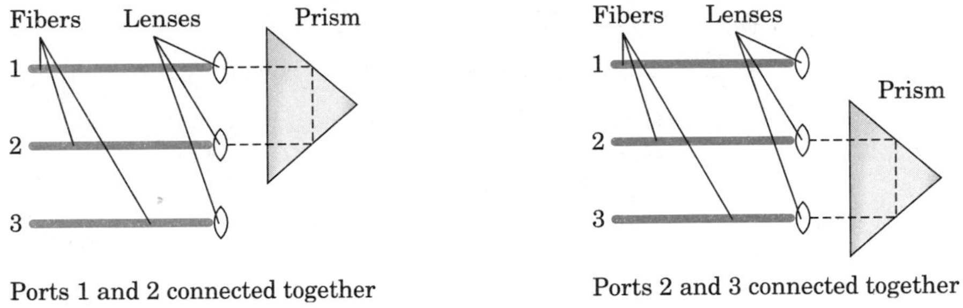

22.6 Optical Emitters

Electromagnetic energy can appear only in multiples of a discrete amount known as a **quantum**. These quanta are called **photons** when the energy is radiated. A *photon* is an entity that is somewhere between a wave and a particle. It has a wavelength corresponding to the radiation and an amount of energy equal to one quantum, but no mass. The details are perhaps best left to physicists, but the existence of photons is important to the operation of the emitters and detectors used with optical fiber.

The amount of energy that corresponds to a quantum and is found in a photon is given by a simple equation:

$$E = hf \tag{22.12}$$

where

E = energy in one quantum in joules
h = Planck's constant, 6.626×10^{-34} joule-seconds
f = frequency of the radiation in hertz

From Equation (22.12), you can see that the energy in one photon varies directly with frequency. For radio frequencies, the energy per photon is so small and the number of photons present in even the weakest signals is so large that it is seldom necessary to consider photons at all. With light, however, the frequency is much higher and the amount of energy in a photon is sometimes sufficient to create a free electron when a photon collides with an atom. Photons can also trigger the movement of an electron from one valence state to another. Similarly, the movement of an electron from a higher energy level to a lower one can cause the emission of a photon.

When talking about energy levels in atoms, it is usual to use the **electron-volt** (eV), rather than the joule, as the unit of energy. Converting between the two is simple. By definition, the electron-volt is the energy gained or lost by one electron moving through a potential difference of one volt. A volt, of course, is one joule per coulomb, so the energy given to an electron is

$$\begin{aligned} 1 \text{ eV} &= (1 \text{ joule/coulomb})(1.6 \times 10^{-19} \text{ coulomb}) \\ &= 1.6 \times 10^{-19} \text{ joule} \end{aligned} \tag{22.13}$$

Example 22.8 Find the energy, in electron-volts, in one photon at a wavelength of 1 μm.

Solution First find the frequency.

$$c = f\lambda$$
$$f = \frac{c}{\lambda}$$
$$= \frac{3 \times 10^{8} \text{ m/s}}{1 \times 10^{-6} \text{ m}}$$
$$= 3 \times 10^{14} \text{ Hz}$$

Next, find the energy in joules (J). From Equation (22.12),

$$E = hf$$
$$= (6.626 \times 10^{-34})(3 \times 10^{14})$$
$$= 1.99 \times 10^{-19} \text{ J}$$

This can easily be converted to electron-volts. From Equation (22.13),

$$1 \text{ eV} = 1.6 \times 10^{-19} \text{ J}$$

so

$$1 \text{ J} = \frac{1}{1.6 \times 10^{-19}} \text{ eV}$$
$$= 6.25 \times 10^{18} \text{ eV}$$

and the energy of our photon, in electron-volts, is

$$E = (1.99 \times 10^{-19})(6.25 \times 10^{18})$$
$$= 1.24 \text{ eV}$$

22.6.1 Light-Emitting Diodes

A light-emitting diode (LED) is a form of junction diode that is operated with forward bias. The recombination of electron-hole pairs in this or any junction diode causes energy to be released. With an ordinary silicon diode, this energy is in the form of heat, but in an LED, a significant proportion of the energy is radiated as photons of either visible or infrared light. Sometimes an infrared LED is called an *infrared-emitting diode* (IRED).

Several types of structures can be used for LEDs in optical communications. Figure 22.18 is a cross-sectional sketch of one of these, the Burrus etched-well LED. This is a surface-emitting **heterojunction** diode. *Surface-emitting* means that the light is emitted from the flat surface of the junction, rather than its edge. A *heterojunction* is a junction between two different materials, in this case, between P-type gallium arsenide (GaAs) and

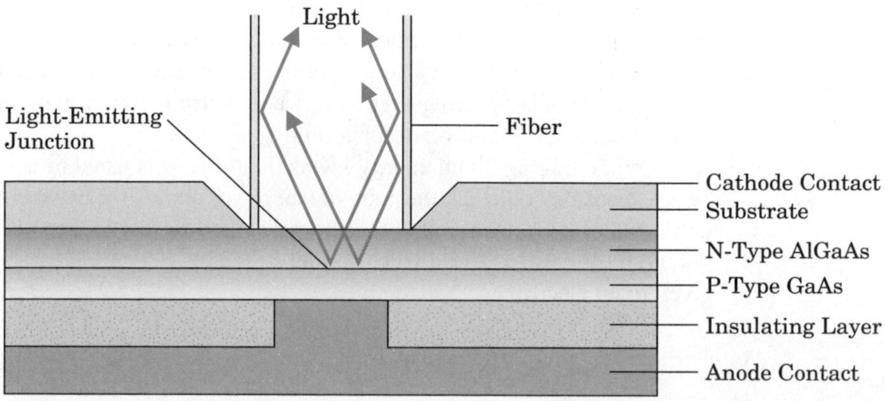

Figure 22.18 Simplified cross section of Burrus etched-well LED

N-type aluminum gallium arsenide (AlGaAs). The diode's geometry is such that a high current density is created in a small region of the junction, producing a small light source that can more easily be coupled to the fiber. The etched well helps to confine the light. The output can be coupled to a fiber *pigtail* as shown or emitted through a lens inserted into the well.

Since the variation of light output with current is reasonably linear for an LED, it can be used with analog amplitude modulation. It is much more common, however, to use pulse transmission, in which the diode is simply switched on and off. Analog pulse techniques like PDM are sometimes used, though digital techniques are more common.

An idea of the specifications and physical characteristics of a typical LED designed for fiber optics can be gained by studying the data sheet in Figure 22.19. This LED emitter

Figure 22.19 Specifications for a typical LED
Copyright of Motorola. Used by permission.

MFOE71

FIBER OPTICS

Low Cost System FLCS Infrared-Emitting Diode

FLCS LINE
FIBER OPTICS
AlGaAs LED

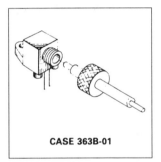

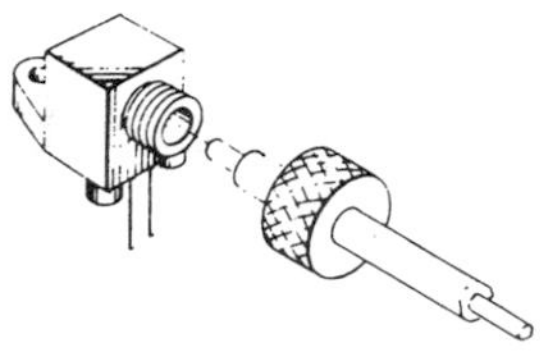

CASE 363B-01

. . . designed for low cost, medium frequency, short distance Fiber Optics Systems using 1000 micron core plastic fiber.

Typical applications include: high isolation interconnects, disposable medical electronics, consumer products, and microprocessor controlled systems such as coin operated machines, copy machines, electronic games, industrial clothes dryers, etc.

- Fast Response — > 10 MHz
- Spectral Response Matched to FLCS Detectors: MFOD71, 72, 73
- FLCS Package
  - — Low Cost
  - — Includes Connector
  - — Simple Fiber Termination and Connection
  - — Easy Board Mounting
  - — Molded Lens for Efficient Coupling
  - — Mates with 1000 Micron Core Plastic Fiber (DuPont OE1040, Eska SH4001)

MAXIMUM RATINGS

Rating	Symbol	Value	Unit
Reverse Voltage	V_R	6	Volts
Forward Current	I_F	150	mA
Total Power Dissipation @ T_A = 25°C Derate above 25°C	P_D(1)	150 2.5	mW mW/°C
Operating and Storage Junction Temperature Range	T_J, T_{stg}	−40 to +85	°C

(1) Measured with the device soldered into a typical printed circuit board.

ELECTRICAL CHARACTERISTICS (T_A = 25°C unless otherwise noted)

Characteristic	Fig. No.	Symbol	Min	Typ	Max	Unit
Reverse Breakdown Voltage (I_R = 100 μA)	—	$V_{(BR)R}$	2	4	—	Volts
Forward Voltage (I_F = 100 mA)	—	V_F	—	1.5	2	Volts

OPTICAL CHARACTERISTICS (T_A = 25°C unless otherwise noted)

Characteristic	Fig. No.	Symbol	Min	Typ	Max	Unit
Power Launched (I_F = 100 mA)	4, 5	P_L	110	165	—	μW
Optical Rise and Fall Time (I_F = 100 mA)	2	t_r, t_f	—	25	35	ns
Peak Wavelength (I_F = 100 mA)	1	λ_P	—	820	—	nm

For simple fiber termination instructions, see the MFOD71, 72 and 73 data sheet.

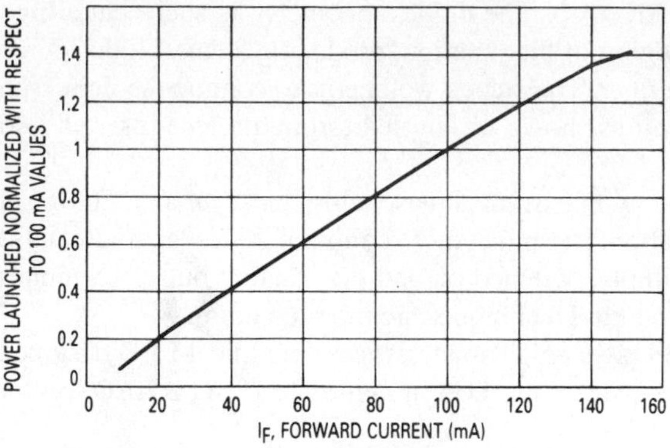

Figure 1. Normalized Power Launched versus Forward Current

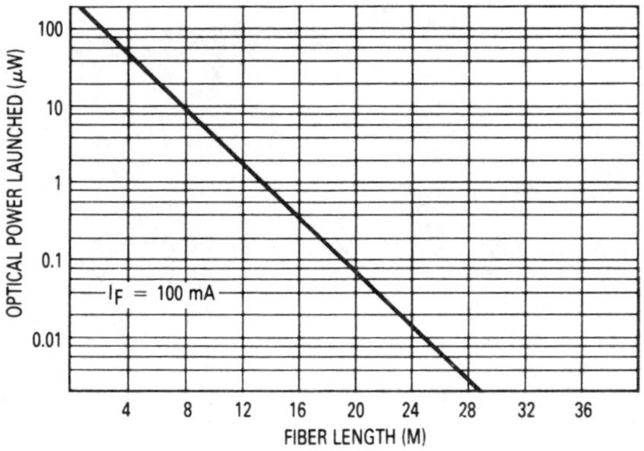

Figure 2. Power Launched versus Fiber Length

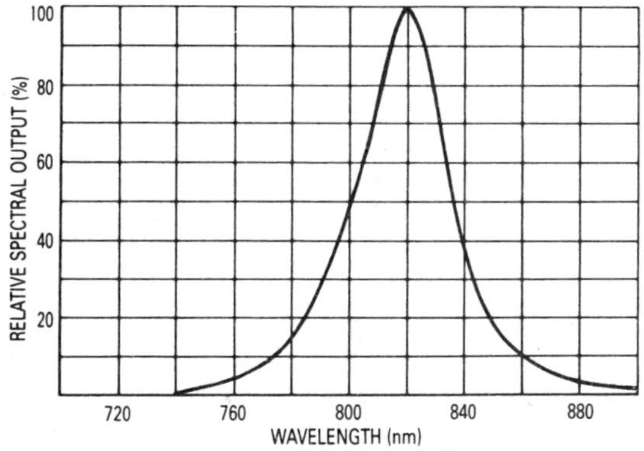

Figure 3. Typical Spectral Output versus Wavelength

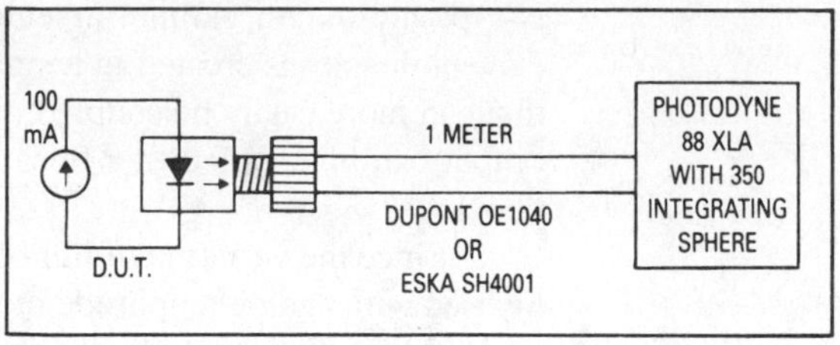

Figure 4. Power Launched Test Set

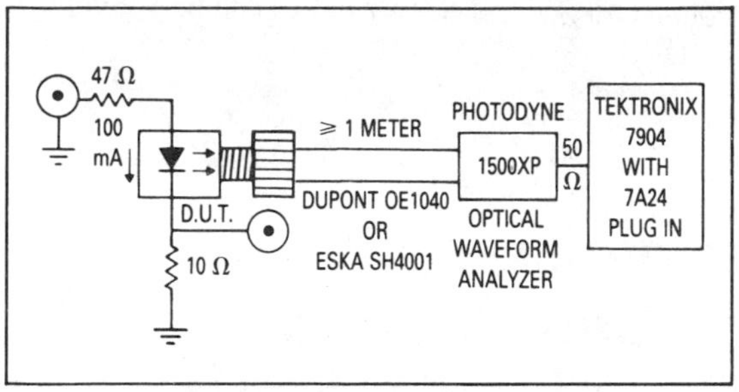

Figure 5. Optical Rise and Fall Time Test Set (10%–90%)

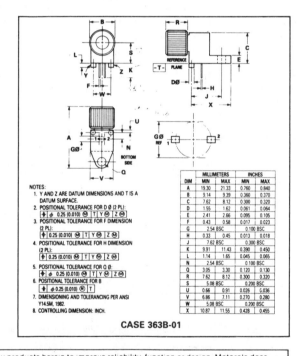

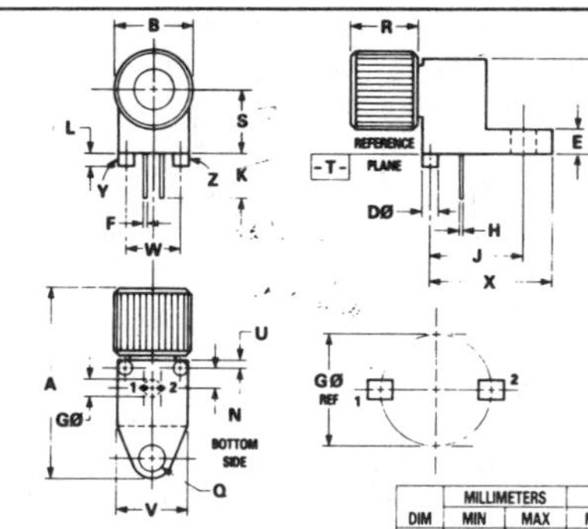

NOTES:
1. Y AND Z ARE DATUM DIMENSIONS AND T IS A DATUM SURFACE.
2. POSITIONAL TOLERANCE FOR D Ø (2 PL):
⌖ | ϕ 0.25 (0.010) Ⓜ | T | Y Ⓜ | Z Ⓜ
3. POSITIONAL TOLERANCE FOR F DIMENSION (2 PL):
⌖ | 0.25 (0.010) Ⓜ | T | Y Ⓜ | Z Ⓜ
4. POSITIONAL TOLERANCE FOR H DIMENSION (2 PL):
⌖ | 0.25 (0.010) Ⓜ | T | Y Ⓜ | Z Ⓜ
5. POSITIONAL TOLERANCE FOR Q Ø:
⌖ | ϕ 0.25 (0.010) Ⓜ | T | Y Ⓜ | Z Ⓜ
6. POSITIONAL TOLERANCE FOR B
⌖ | ϕ 0.25 (0.010) Ⓜ | T
7. DIMENSIONING AND TOLERANCING PER ANSI Y14.5M, 1982.
8. CONTROLLING DIMENSION: INCH.

	MILLIMETERS		INCHES	
DIM	MIN	MAX	MIN	MAX
A	19.30	21.33	0.760	0.840
B	9.14	9.39	0.360	0.370
C	7.62	8.12	0.300	0.320
D	1.55	1.62	0.061	0.064
E	2.41	2.66	0.095	0.105
F	0.43	0.58	0.017	0.023
G	2.54 BSC		0.100 BSC	
H	0.33	0.45	0.013	0.018
J	7.62 BSC		0.300 BSC	
K	9.91	11.43	0.390	0.450
L	1.14	1.65	0.045	0.065
N	2.54 BSC		0.100 BSC	
Q	3.05	3.30	0.120	0.130
R	7.62	8.12	0.300	0.320
S	5.08 BSC		0.200 BSC	
U	0.66	0.91	0.026	C.036
V	6.86	7.11	0.270	0.280
W	5.08 BSC		0.200 BSC	
X	10.87	11.55	0.428	0.455

CASE 363B-01

Figure 22.19
Continued

is designed for low-cost fiber-optics systems. Its maximum output is at a wavelength of 820 nm, which is a good match for the silicon photodetectors with which it is usually used. Notice the inefficiency of the device. It typically operates with a forward voltage of 1.5 V and a current of 100 mA, for a total input power of 150 mW. The optical power launched into the fiber is typically 165 μW for an efficiency of about 0.1%. It is easy to see why fiber optics is used to transmit information, not energy.

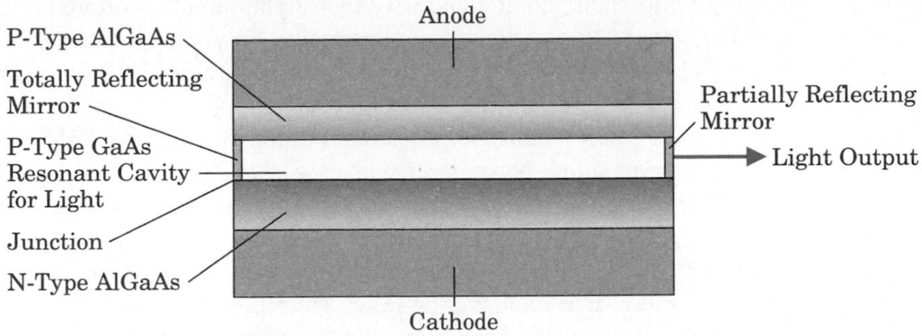

Figure 22.20 Simplified cross section of laser diode

22.6.2 Laser Diodes

The word *laser* is really an acronym for *light amplification by stimulated emission of radiation.* In operation, a laser acts as an optical oscillator, generating phase-coherent light with a much narrower bandwidth than is possible with other sources. A typical laser diode has an emission linewidth of about 1 nm, compared with 40 nm for a typical LED, and some narrow-bandwidth lasers have much better performance still. Laser diodes also produce more light in the narrow cone required for coupling to optical fibers. This advantage is particularly great with single-mode fibers, which are very narrow and also have a small numerical aperture.

The construction of a laser diode is rather similar to that of an LED, and in fact, at low currents the laser diode operates as an LED. Figure 22.20 is a simplified cross-sectional diagram of a common type of laser diode known as a *striped heterojunction injection laser*. This diode has a junction surrounded by material that has a lower refractive index than the material used for the junction. Many of the photons produced at the junction remain in the plane of the junction; if they move away from it, they will be reflected in the same way that light is contained within an optical fiber. The ends of the junction have mirrored surfaces, with one of the mirrors being partly transmissive so that light can escape in one direction only. The whole junction area forms a resonant cavity at the operating frequency.

Laser diodes operate with high current densities of 11–30 kA/cm^2. The actual current is typically only about 100 mA; the high current density is produced by masking off most of the junction so that the area through which the current flows is very small. The high current density causes many electron-hole pairs to be generated. The recombination of these pairs generates photons that stimulate other pairs to combine. Laser operation, sometimes called *lasing*, occurs only above a certain threshold current. Below that current, the laser diode acts as a conventional LED. The graph in Figure 22.21 shows the threshold and the variation of light output with current.

Laser diodes are more difficult to operate than LEDs: their current must be carefully controlled, and the required current varies with temperature. With too little current, lasing stops, but too much will destroy the device. Laser diodes often use thermoelectric coolers

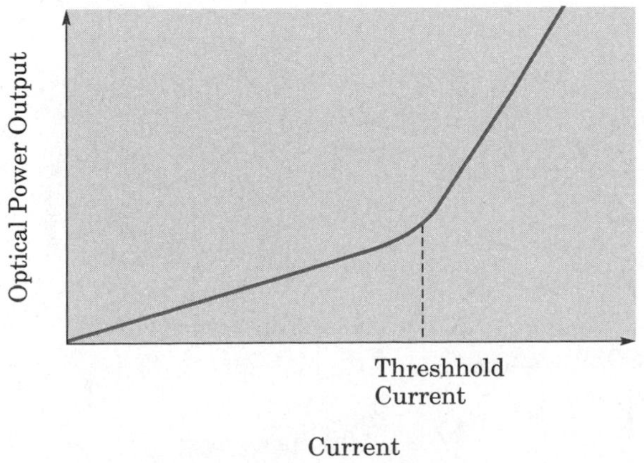

Figure 22.21 Light output as a function of current for a laser diode

for temperature control and have their light output monitored by a photodetector to allow the current to be adjusted automatically. Lasers are usually modulated by changing the current from a value just below the lasing threshold to some considerably higher value. This can be done at rates exceeding 1 GHz.

A safety note about lasers is in order. Because the light emitted is coherent and intense, even a low-power laser such as a laser diode can damage the eye, which may focus the light onto a small spot on the retina. Infrared lasers are especially dangerous because the light is invisible and it is not obvious that the laser is operating. Anyone working with lasers, even the very low-powered ones used in communications, should be very careful not to look into an operating laser. More powerful lasers can cause other damage due to heating effects: they can burn skin, start fires, and so on. These lasers are not found in communications applications, however, and are not within the scope of this book.

22.7 Optical Detectors

The most common types of detectors for optical communications are PIN diodes and avalanche photodiodes. The PIN diode, shown in cross section in Figure 22.22, was previously discussed in connection with its uses as a microwave switch and attenuator. As a photodetector, the PIN diode takes advantage of its wide depletion region, in which photons can create electron-hole pairs. If a photon has enough energy to move an electron from the valence to the conduction band, it can, by colliding with an electron, create a free electron. The crystal lattice then contains a hole where the electron used to be. Since the PIN diode is reverse-biased, the free electron and hole migrate in opposite directions, creating a current that is proportional to the light power. The low junction capacitance of the PIN diode, which results from the wide depletion region, allows for very high-speed operation, into the gigahertz range in some cases.

The avalanche photodiode (APD), shown in Figure 22.23, is also operated with reverse bias. In this case, however, the device is biased just below its reverse-breakdown voltage. The creation of an electron-hole pair due to the absorption of a photon of incoming

Figure 22.22
PIN-diode photodetector

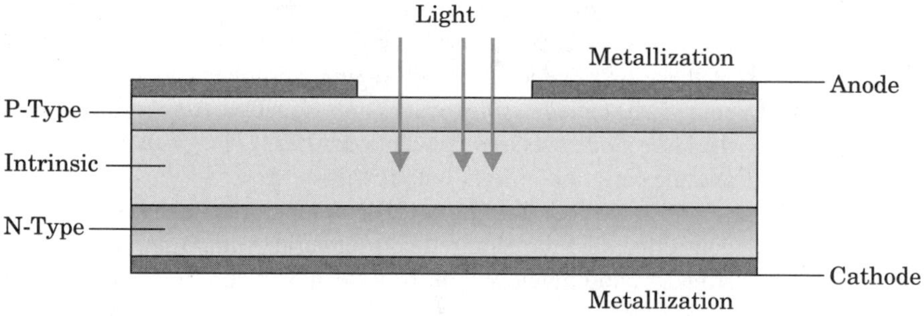

Figure 22.23
Avalanche photodiode

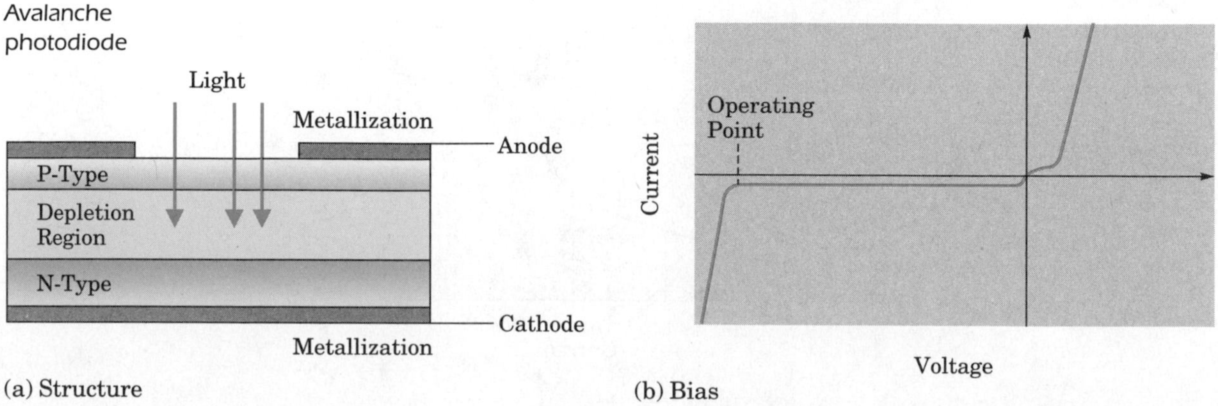

light may set off avalanche breakdown, creating up to 100 more pairs. This multiplying effect gives the APD high sensitivity, about 5 to 7 dB greater than the PIN diode. On the other hand, the operation of an avalanche photodiode requires a carefully controlled bias voltage to keep the diode biased on the verge of avalanche breakdown.

Figure 22.24 is a data sheet for the Motorola MFOD71, a low-cost PIN diode photodetector, which could be used with the LED shown earlier. A brief study of this data sheet

Figure 22.24 Specifications for a PIN-diode photodetector

FIBER OPTIC LOW COST SYSTEM
FLCS DETECTORS

. . . designed for low cost, short distance Fiber Optic Systems using 1000 micron core plastic fiber.

Typical applications include: high isolation interconnects, disposable medical electronics, consumer products, and microprocessor controlled systems such as coin operated machines, copy machines, electronic games, industrial clothes dryers, etc.

- Fast PIN Photodiode: Response Time <5.0 ns
- Standard Phototransistor
- High Sensitivity Photodarlington
- Spectral Response Matched to MFOE71 LED
- Annular Passivated Structure for Stability and Reliability
- FLCS Package
  - — Includes Connector
  - — Simple Fiber Termination and Connection (Figure 4)
  - — Easy Board Mounting
  - — Molded Lens for Efficient Coupling
  - — Mates with 1000 Micron Core Plastic Fiber (DuPont OE1040, Eska SH4001)

FLCS LINE

FIBER OPTICS

DETECTORS

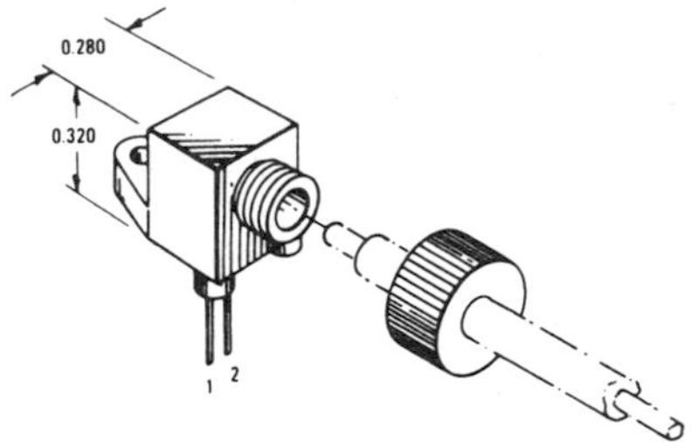

MAXIMUM RATINGS (T_A = 25°C unless otherwise noted)

Rating		Symbol	Value	Unit
Reverse Voltage	MFOD71	V_R	100	Volts
Collector-Emitter Voltage	MFOD72 MFOD73	V_{CEO}	30 60	Volts
Total Power Dissipation @ T_A = 25°C MFOD71 Derate above 25°C MFOD72/73 Derate above 25°C		P_D	 100 1.67 150 2.5	 mW mW/°C mW mW/°C
Operating and Storage Junction Temperature Range		T_J, T_{stg}	−40 to +85	°C

FIGURE 1 — RELATIVE SPECTRAL RESPONSE

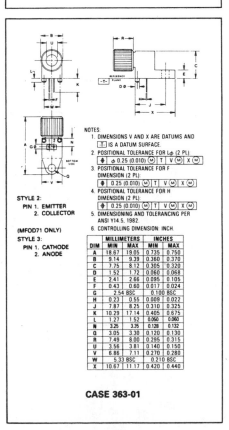

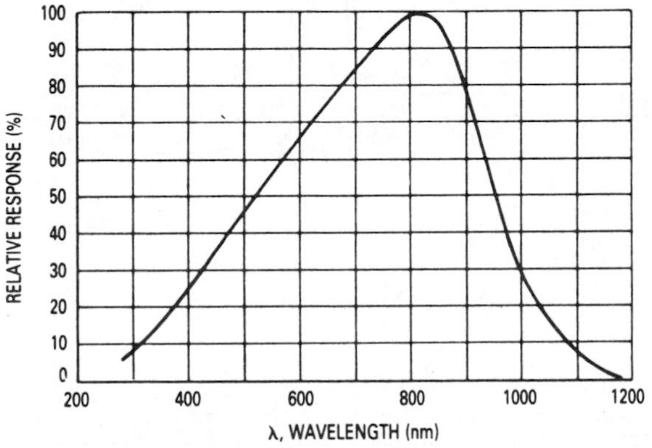

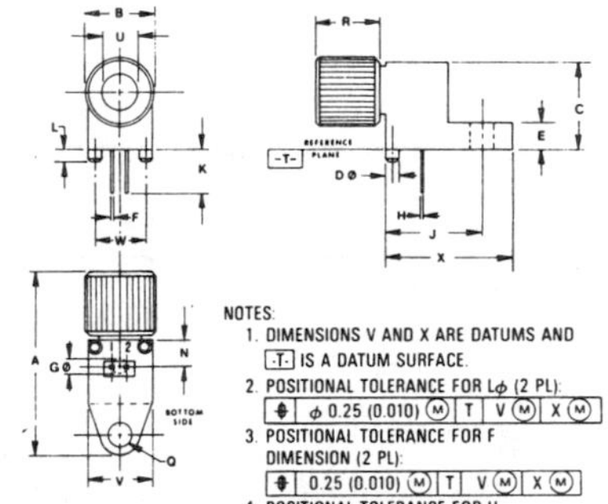

STYLE 2:
PIN 1. EMITTER
2. COLLECTOR

(MFOD71 ONLY)
STYLE 3:
PIN 1. CATHODE
2. ANODE

NOTES:
1. DIMENSIONS V AND X ARE DATUMS AND -T- IS A DATUM SURFACE.
2. POSITIONAL TOLERANCE FOR Lϕ (2 PL): ⌖ | ϕ 0.25 (0.010) Ⓜ | T | V Ⓜ | X Ⓜ
3. POSITIONAL TOLERANCE FOR F DIMENSION (2 PL): ⌖ | 0.25 (0.010) Ⓜ | T | V Ⓜ | X Ⓜ
4. POSITIONAL TOLERANCE FOR H DIMENSION (2 PL): ⌖ | 0.25 (0.010) Ⓜ | T | V Ⓜ | X Ⓜ
5. DIMENSIONING AND TOLERANCING PER ANSI Y14.5, 1982.
6. CONTROLLING DIMENSION: INCH.

DIM	MILLIMETERS		INCHES	
	MIN	MAX	MIN	MAX
A	18.67	19.05	0.735	0.750
B	9.14	9.39	0.360	0.370
C	7.75	8.12	0.305	0.320
D	1.52	1.72	0.060	0.068
E	2.41	2.66	0.095	0.105
F	0.43	0.60	0.017	0.024
G	2.54 BSC		0.100 BSC	
H	0.23	0.55	0.009	0.022
J	7.87	8.25	0.310	0.325
K	10.29	17.14	0.405	0.675
L	1.27	1.52	0.050	0.060
N	3.25	3.35	0.128	0.132
Q	3.05	3.30	0.120	0.130
R	7.49	8.00	0.295	0.315
U	3.56	3.81	0.140	0.150
V	6.86	7.11	0.270	0.280
W	5.33 BSC		0.210 BSC	
X	10.67	11.17	0.420	0.440

CASE 363-01

MFOD71

STATIC ELECTRICAL CHARACTERISTICS (T_A = 25°C unless otherwise noted)

Characteristic	Symbol	Min	Typ	Max	Unit
Dark Current (V_R = 20 V, R_L = 1.0 MΩ) T_A = 25°C T_A = 85°C	I_D	— —	0.06 10	10 —	nA
Reverse Breakdown Voltage (I_R = 10 μA)	$V_{(BR)R}$	50	100	—	Volts
Forward Voltage (I_F = 50 mA)	V_F	—	—	1.1	Volts
Series Resistance (I_F = 50 mA)	R_s	—	8.0	—	ohms
Total Capacitance (V_R = 20 V; f = 1.0 MHz)	C_T	—	3.0	—	pF

OPTICAL CHARACTERISTICS (T_A = 25°C)

Characteristic	Symbol	Min	Typ	Max	Unit
Responsivity (V_R = 5.0 V, Figure 2)	R	0.15	0.2	—	μA/μW
Response Time (V_R = 5.0 V, R_L = 50 Ω)	$t_{(resp)}$	—	5.0	—	ns

MFOD72/MFOD73

STATIC ELECTRICAL CHARACTERISTICS

Characteristic		Device	Symbol	Min	Typ	Max	Unit
Collector Dark Current (V_{CE} = 10 V)			I_D	—	—	100	nA
Collector-Emitter Breakdown Voltage (I_C = 10 mA)		MFOD72 MFOD73	$V_{(BR)CEO}$	30 60	— —	— —	Volts

OPTICAL CHARACTERISTICS (T_A = 25°C unless otherwise noted)

Characteristic		Device	Symbol	Min	Typ	Max	Unit
Responsivity (V_{CC} = 5.0 V, Figure 2)		MFOD72 MFOD73	R	80 1,000	125 1,500	— —	μA/μW
Saturation Voltage (λ = 820 nm, V_{CC} = 5.0 V) (P_{in} = 10 μW, I_C = 1.0 mA) (P_{in} = 1.0 μW, I_C = 2.0 mA)		 MFOD72 MFOD73	$V_{CE(sat)}$	 — —	 0.25 0.75	 0.4 1.0	Volts
Turn-On Time	R_L = 2.4 kΩ, P_{in} = 10 μW, λ = 820 nm, V_{CC} = 5.0 V	MFOD72	t_{on}	—	10	—	μs
Turn-Off Time			t_{off}	—	60	—	μs
Turn-On Time	R_L = 100 Ω, P_{in} = 1.0 μW, λ = 820 nm, V_{CC} = 5.0 V	MFOD73	t_{on}	—	125	—	μs
Turn-Off Time			t_{off}	—	150	—	μs

TYPICAL COUPLED CHARACTERISTICS

FIGURE 2 — RESPONSIVITY TEST CONFIGURATION

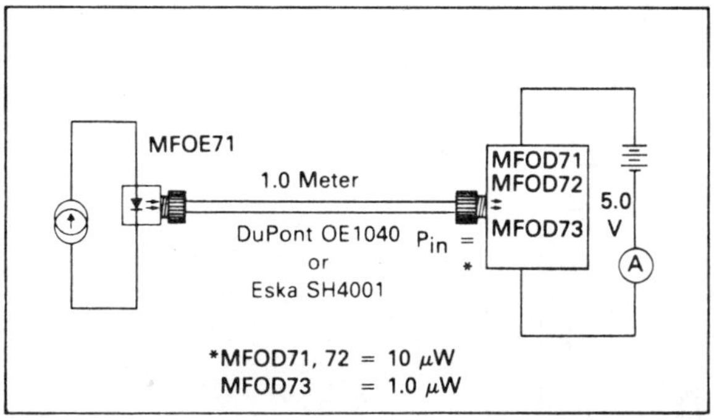

FIGURE 3 — DETECTOR CURRENT versus FIBER LENGTH

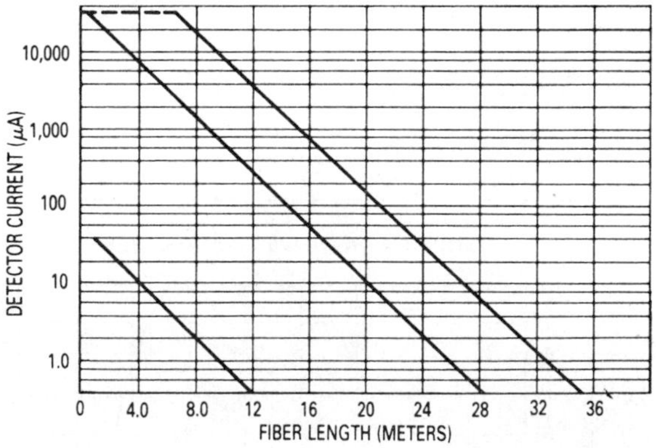

Figure 22.24
Continued

Continued

FLCS WORKING DISTANCES

The system length achieved with a FLCS emitter and detector using the 1000 micron core fiber optic cable depends upon the forward current through the LED and the Responsivity of the detector chosen. Each emitter/detector combination will work at any cable length up to the maximum length shown.

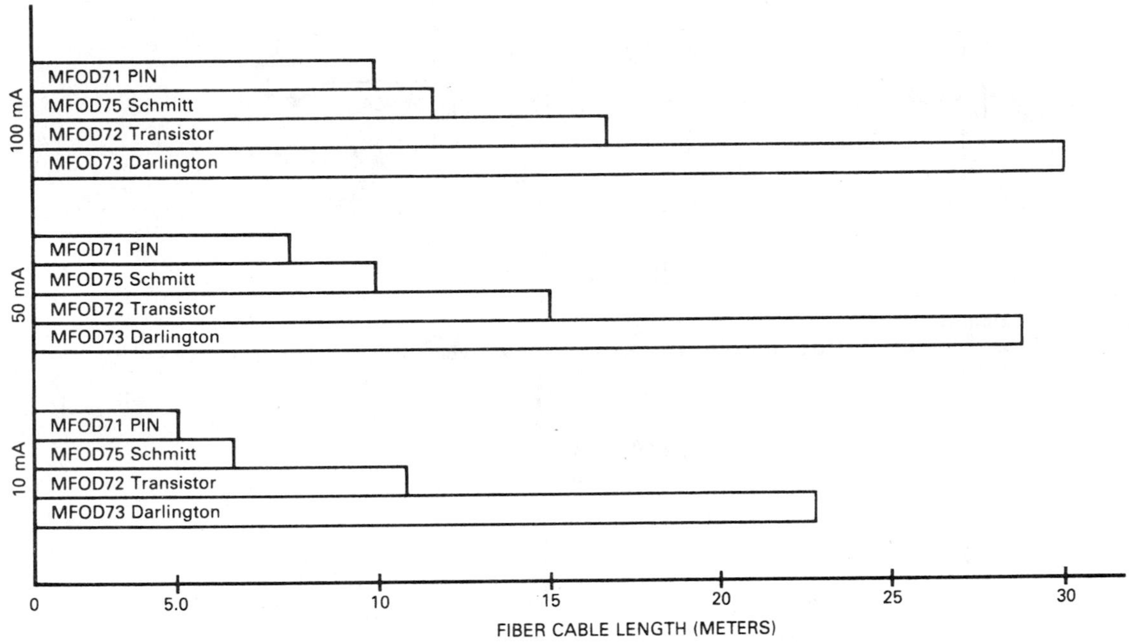

FIGURE 4 — FO CABLE TERMINATION AND ASSEMBLY

Cross Section of FLCS Package

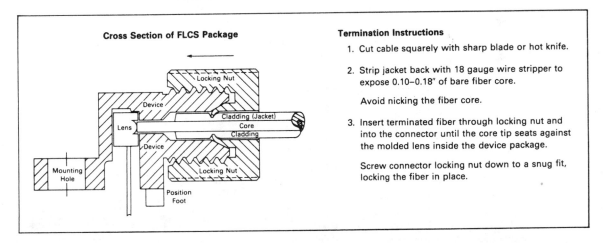

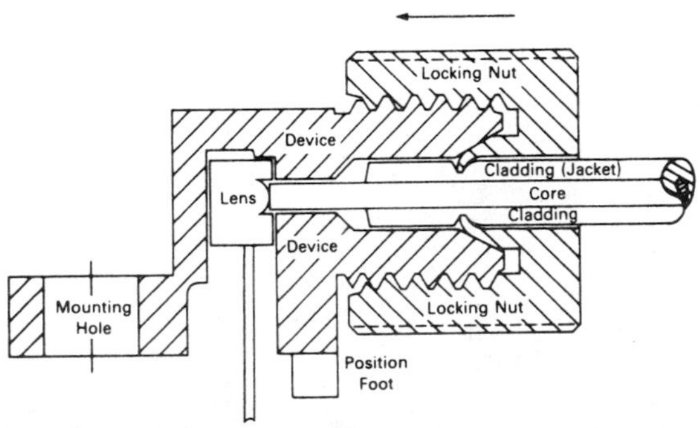

Termination Instructions

1. Cut cable squarely with sharp blade or hot knife.
2. Strip jacket back with 18 gauge wire stripper to expose 0.10–0.18″ of bare fiber core.

   Avoid nicking the fiber core.
3. Insert terminated fiber through locking nut and into the connector until the core tip seats against the molded lens inside the device package.

   Screw connector locking nut down to a snug fit, locking the fiber in place.

Figure 22.24
Continued

Input Signal Conditioning

The following circuits are suggested to provide the desired forward current through the emitter.

TTL TRANSMITTERS

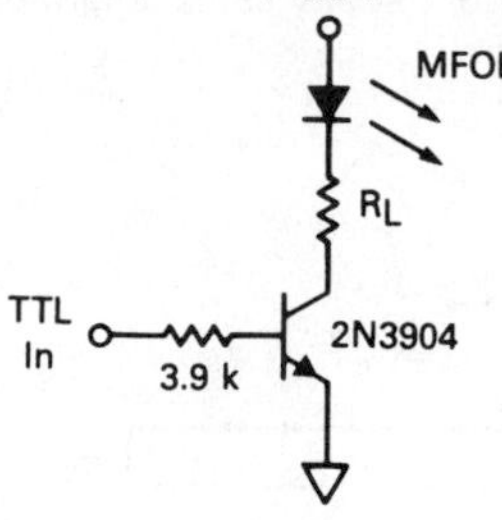

Noninverting

I_F	R_L
10 mA	330
50 mA	68
100 mA	33

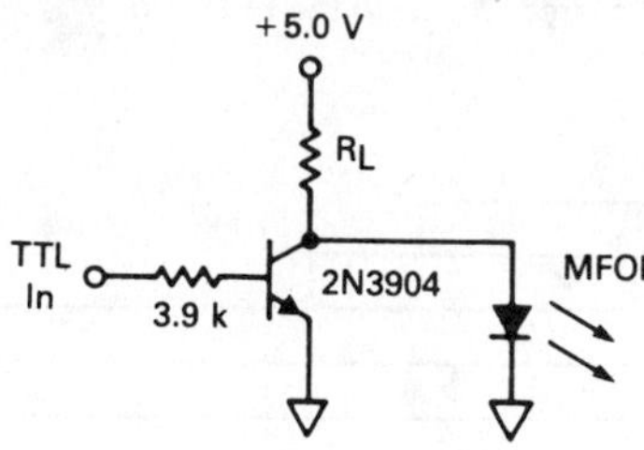

Inverting

Output Signal Conditioning

The following circuits are suggested to take the FLCS detector output and condition it to drive TTL with an acceptable bit error rate.

TTL RECEIVERS

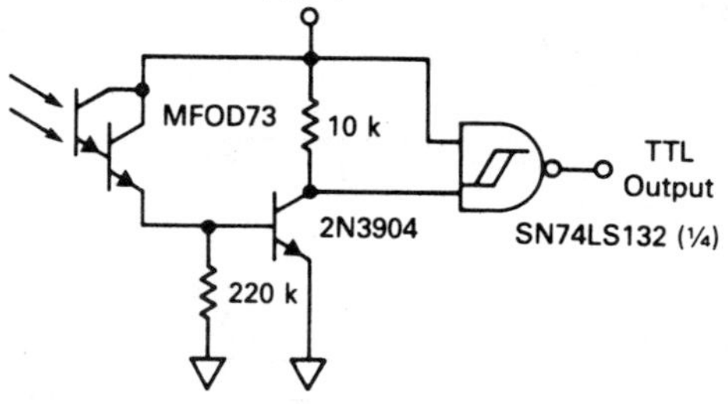

1.0 kHz Darlington Receiver

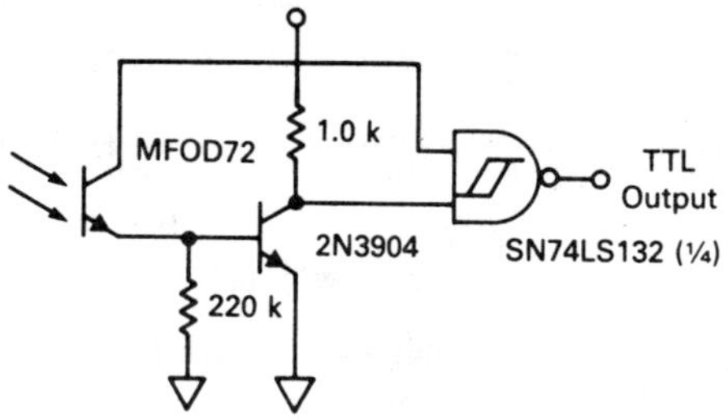

5.0 kHz Transistor Receiver

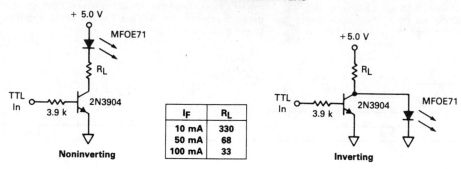

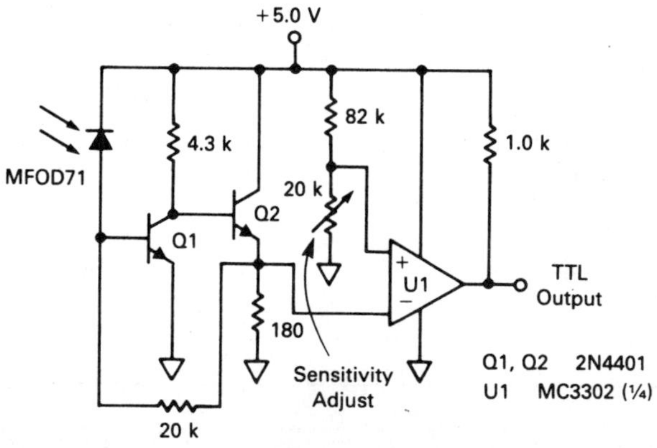

1.0 MHz PIN Receiver

Figure 22.24
Continued

will familiarize the reader with photodetector specifications. The most important of these are described below.

Dark current is the leakage current in the absence of light, that is, it is the normal reverse current of the diode. The dark current sets a sensitivity floor. Since the dark current is due to electron-hole pairs created by thermal activity, it increases with temperature, just as for any diode. For this diode, typical dark-current values range from 0.06 nA at 25°C to 10 nA at 85°C.

Responsivity is a measure of the sensitivity of the detector to light. Its basic units are amperes per watt of light; in the case of the MFOD71, a typical value is given as 0.2 μA/μW. (The practical unit is actually the same as the basic one, since the numerator and denominator have both been divided by 10^6.)

The responsivity is given for the wavelength of light to which the detector is most sensitive. If the light source in use has its maximum output at some other wavelength, it is necessary to consider the spectral response of the detector, that is, the variation of detector sensitivity with wavelength. The spectral response curve provides a percentage or decibel factor that modifies the responsivity according to the source wavelength.

Response time is a very important specification in high-speed systems, as it limits the maximum data rate that can be used.

Example 22.9 A typical Motorola MFOD71 photodetector is used with an LED having its maximum output at 600 nm. Calculate the output current for a power input of 50 μW.

Solution The responsivity of 0.2 μA/μW will have to be multiplied by the relative spectral response at 600 nm. This can be found from the graph in Figure 22.24 as approximately 65% or 0.65. Therefore, the output current will be

$$I_o = 50\ \mu\text{W} \times 0.2\ \mu\text{A}/\mu\text{W} \times 0.65$$
$$= 6.5\ \mu\text{A}$$

The data sheet in Figure 22.24 also has suggested circuits for simple optical transmitters and receivers designed to work with TTL-level digital signals.

Summary

Here are the main points to remember from this chapter.

1. Optical fiber has many advantages over copper cable for communications, including larger bandwidth, greater distance between repeaters, lower weight and smaller size, immunity from electrical interference, and (often) lower cost.
2. Optical fibers are waveguides for light. Several qualities are available, depending on the application. Single-mode fibers have greater bandwidth and lower losses, while multimode fibers are cheaper and easier to use.
3. Light sources for fiber optics include LEDs for low-cost systems and laser diodes for applications where long distances and wide bandwidth are important.
4. Detectors in fiber-optics systems are usually PIN diodes or avalanche photodiodes.
5. Connectors and splices are important in fiber systems. Even a well-made splice can have as much loss as 1 km of fiber.

Important Equations

$$n = \frac{c}{v} \tag{22.1}$$

$$n = \sqrt{\epsilon_r} \tag{22.3}$$

$$n_1 \sin \theta_1 = n_2 \sin \theta_2 \tag{22.4}$$

$$\theta_c = \arcsin \frac{n_2}{n_1} \tag{22.6}$$

$$\text{N.A.} = \sqrt{n_1^2 - n_2^2} \tag{22.7}$$

$$r_{max} = \frac{0.383\lambda}{\text{N.A.}} \tag{22.8}$$

$$D = D_c \Delta\lambda \tag{22.9}$$

$$\Delta t = Dl \tag{22.10}$$

$$B = \frac{1}{2\Delta t} \tag{22.11}$$

$$E = hf \tag{22.12}$$

$$1 \text{ eV} = 1.6 \times 10^{-19} \text{ joule} \tag{22.13}$$

Glossary

angle of acceptance the maximum angle between the axis of an optical fiber and a ray of light entering the fiber

angle of incidence the angle an incident ray makes with the normal to a reflecting or refracting surface

angle of reflection the angle a reflected ray makes with the normal to a reflecting surface

angle of refraction the angle a refracted ray makes with the normal to a refracting surface

cladding in an optical fiber, the material of lower refractive index that surrounds the core

core in an optical fiber, the central part of the fiber in which the light propagates

critical angle the maximum angle of incidence for which refraction takes place

crosstalk interference between signals on separate cables in close proximity

dark current in a photodetector, the current that flows in the absence of light

directivity a measure of how well a coupler or similar device rejects power passing through it in the reverse direction

dispersion variation of propagation velocity with wavelength

electron-volt (eV) the energy given to or absorbed by an electron that moves through a potential difference of one volt (1 eV = 1.6×10^{-19} joule)

excess loss the proportion of the power entering a coupler that is lost inside the coupler

graded-index fiber an optical fiber in which the index of refraction of the core decreases gradually with increasing distance from the center

heterojunction a P–N junction in which the two sides of the junction are made of different materials

index of refraction the ratio between the velocity of light in free space and that in a given medium

laser acronym for *l*ight *a*mplification by *s*timulated *e*mission of *r*adiation; a laser is an optical oscillator that produces phase-coherent light with narrow bandwidth

laser diode (LD) a low-power laser resembling an LED in its construction

multimode fiber a fiber that allows light to travel along it in more than one waveguide mode, at different velocities

normal a line perpendicular to a reflecting or refracting surface

numerical aperture for an optical fiber, the sine of the angle of acceptance

photon a quantum of electromagnetic radiation

quantum the smallest amount in which energy can exist; the size of a quantum depends on the wavelength of the energy

responsivity the relationship between output current and input light power for a photodetector

single-mode fiber an optical fiber whose core is sufficiently narrow that only one waveguide mode can propagate

step-index fiber an optical fiber that has one index of refraction for the core and a second, lower index for the cladding, with a sharp transition between them

total internal reflection reflection at the boundary between two media when the angle of incidence is greater than the critical angle

Questions

1. List four possible advantages of optical fiber over wire cables, and explain each.
2. Give one advantage and one disadvantage of fiber compared with a geostationary satellite radio link.
3. Explain briefly the difference between single-mode and multimode fiber. Which gives better performance? Why?
4. What are the advantages of using a laser diode with single-mode fiber instead of an LED?
5. State Snell's law and illustrate it.

6. What is total internal reflection? Under what circumstances does it occur?
7. What is meant by the numerical aperture of an optical fiber? What happens if light moves from one fiber to another with a lower numerical aperture?
8. List the three types of optical fiber discussed in this chapter, and order them in terms of dispersion and loss.
9. Describe the mechanisms by which dispersion takes place in optical fibers. Which of these mechanisms apply to single-mode fiber?
10. Which type of fiber has the highest bandwidth-distance product? Why?
11. How does dispersion limit the maximum data rate that an optical fiber can carry?
12. Explain why the maximum data rate that can be used with a fiber decreases as the length of the fiber increases.
13. Explain the difference between loose-tube and tight-buffer cables. Give an application in which each would be preferred.
14. Why do fiber-optic cables require strength members? What materials are most commonly used to add strength to cables?
15. Explain the differences between transmissive and reflective optical couplers.
16. How can a tee connector be made from a four-port transmissive coupler?
17. What is meant by excess loss in a coupler?
18. What advantage does optical switching have over electrical switching in a fiber-optic system?
19. What is a photon?
20. Draw a cross section of an LED, and describe its operation.
21. Draw a cross section of a laser diode, and describe its operation.
22. For what is the term *laser* an acronym?
23. What safety precautions are necessary with infrared laser diodes? Why?
24. Name and compare two types of photodetectors that are used in optical communications.

Problems

SECTION 22.1

25. Calculate the frequency that corresponds to each of the following wavelengths of light. Assume free-space propagation.
    (a) 400 nm (violet)
    (b) 700 nm (red)
    (c) 900 nm (infrared)
26. Calculate the wavelengths of each of the light sources in Problem 25 in glass with a refractive index of 1.5.
27. What is the critical angle for light moving from glass with a refractive index of 1.5 to free space?

SECTION 22.2

28. The indices of refraction of the core and cladding of an optical fiber are 1.5 and 1.45, respectively.
    (a) Calculate the dielectric constants for the core and the cladding.
    (b) Calculate the speed of light in the core and the cladding.
29. For the fiber in Problem 28,
    (a) what is the critical angle for a ray moving from the core to the cladding?
    (b) what is the numerical aperture?
    (c) what is the maximum angle (from the axis of the fiber) at which light will be accepted?
30. For a cable with indices of refraction of 1.46 and 1.41 for the core and cladding, respectively, calculate the maximum diameter the core could have for single-mode propagation at a wavelength of 1.5 μm.
31. A commonly used wavelength for optical communication is 1.55 μm.
    (a) What type of light is this?
    (b) Calculate the frequency corresponding to this wavelength, assuming free-space propagation.
32. A fiber-optic cable has a bandwidth-distance product of 500 MHz-km. What bandwidth can be used with a cable that runs 30 km between repeaters?
33. A fiber is installed over a distance of 15 km, and it is found experimentally that the maximum operating bandwidth is 700 MHz. Calculate the bandwidth-distance product for the fiber.
34. From the chart labeled "Power Launched versus Fiber Length" on the MFOE71 data sheet in Figure 22.19, find the loss of the fiber used, in dB/km.

SECTION 22.5

35. An optical tee coupler has the following specifications: With input at port 1, the output at ports 2 and 3 is 3.6 dB (each) down from the input power level. Calculate the excess loss of the coupler.
36. A ten-port reflective star coupler has an excess loss of 1 dB. Each of the ports has the same coupling. If 500 microwatts of power is put into one of the ports, how much comes out of each of the others?

37. A three-port directional coupler has a directivity of 35 dB between ports 2 and 3. If 250 microwatts is put into port 2, what power appears at port 3?
38. A four-port coupler has 500 microwatts put into port 1. The outputs are:
port 2: 200 microwatts
port 3: 100 microwatts
port 4: 150 microwatts
Calculate the coupling, in decibels, to each port and the excess loss, in decibels, for the coupler.

SECTION 22.6

39. Calculate the energy in one photon of a light wave at a wavelength of 400 nm. Express the result in both joules and electron-volts.
40. An LED has a forward voltage of 1.5 volts at a current of 40 mA. It produces a usable light output of 200 microwatts. How efficient is the diode?
41. A single-mode fiber has a dispersion of 10 ps/(nm-km). Calculate the dispersion over a 10 km distance using
(a) an LED with a linewidth of 40 nm
(b) an LD with a linewidth of 5 nm

SECTION 22.7

42. Figure 22.25 is a graph relating light output to input current for a laser diode. Find the threshold current level and the power output for a current of 120 mA.

Figure 22.25

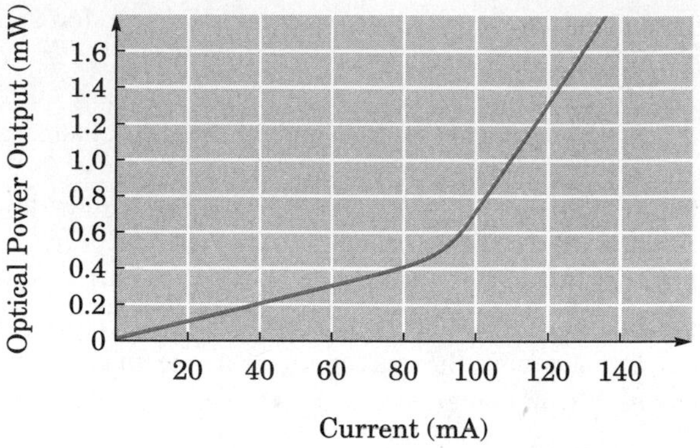

43. A Motorola MFOD71 photodiode is connected in series with a resistor of 10 kΩ and reverse biased. Light with a wavelength of 700 nm and a power of 75 μW is applied. Calculate the current through the diode and the output voltage across the resistor.

23 Fiber-Optic Systems

Objectives After studying this chapter, you should be able to:

1. Prepare a loss budget for a fiber-optic system
2. Calculate the maximum possible data rate for a fiber-optic system
3. Explain and calculate the rise time and the pulse-spreading constant in a fiber-optic system
4. Describe repeaters and optical amplifiers and explain their function and uses
5. Explain how wavelength-division multiplexing can be used to increase the maximum data rate for a fiber-optic system
6. Describe the uses of optical fiber in submarine cables
7. Explain the SONET optical communications standard and outline its relation to earlier digital communications standards
8. Discuss the use of fiber optics in local-area networks and describe the FDDI standard
9. Discuss the present and future uses of fiber optics in telephone and cable television systems
10. Explain how solitons and heterodyne reception can be used to extend the capabilities of optical fiber
11. Describe the operation of optical time-domain reflectometry and explain how it differs from TDR with copper cables

23.1 Introduction

Fiber-optic systems are becoming very important in modern communications. Most new telephone cables of any great length, whether on land or under water, employ fiber optics rather than the coaxial cable previously used. Cable television systems use fiber optics for new trunk lines. Many data networks are routed over optical fiber, and it appears to be only a matter of time before fiber is used for direct residential connections for telephone and/or television service—in fact, a single fiber would have plenty of bandwidth for both. Consequently, it is important to understand how fiber-optic networks are designed and constructed and how they are connected to more conventional systems.

In the previous chapter we discussed the basics of fiber optics, including the theory of light transmission through single-mode and multimode fiber and the properties of simple optical transmitters and receivers. This chapter will emphasize the use of optical components in systems and some of the factors that limit overall system performance. Some new developments that look very promising for the future of fiber optics will also be discussed.

23.2 Basic Fiber-Optic Systems

Let us first examine a simple fiber link involving a transmitter, a receiver, and a length of optical fiber, which can be either single-mode or multimode. The transmitter can be either a light-emitting diode (LED) or a laser diode (LD), and the receiver can be either a PIN diode or an avalanche photodiode (APD). We will consider the distance over which data can be transmitted with this system. We will need to look at the effect of fiber and other losses on the received power, and we must also consider the relationship between the length of the fiber and the permissible bandwidth or data rate.

In general, short-range systems, such as those within offices, factories, or automobiles, use LED emitters and multimode fiber, while long-range systems (for example, telephone trunk lines) use laser diodes and single-mode fiber. Early single-mode systems used a wavelength of 1.3 μm for lowest loss, but currently the zero-dispersion wavelength of 1.55 μm is more commonly used. Bit rates of 2.488 Gb/s are common in high-speed systems, and even higher rates are used in the newest undersea telephone cables.

23.2.1 Loss Budget

The most basic limitation on the length of a fiber-optic link is loss in the fiber and at connectors and splices. These losses cause the received power to be less than the transmitted power. If the length of the fiber is too great, the optical power level at the receiver will be insufficient to produce an acceptable signal-to-noise ratio. Given the optical power output of the transmitter and the signal level required at the receiver, it is easy to draw up a **loss budget** that shows where the losses occur and the amount of power delivered to the receiver. If this power level is greater than that required by the receiver, the system will work. If not, it will be necessary to do one or more of the following: increase transmitter power, increase receiver sensitivity, reduce the length of the cable, or reduce losses by using better cable and/or connectors and splices with less loss. Another possibility, though not generally a very practical one, is to reduce the data rate, since the power needed at the receiver is roughly proportional to the data rate.

The simplest way to prepare a loss budget is to start with the transmitter power in dBm, subtract from it all the losses, in decibels, and compare the result with the required power at the receiver. Any excess power is called the *system margin*. A margin of five to ten decibels is generally required to allow for the deterioration of components over time and the possibility that additional splices will be needed (if the cable is accidentally cut, for instance).

Example 23.1

A fiber-optic link extends for 40 km. The laser-diode emitter has an output power of 1.5 mW, and the receiver requires a signal strength of -25 dBm for a satisfactory signal-to-noise ratio. The fiber is available in lengths of 2.5 km and can be spliced with a loss of 0.25 dB per splice. The fiber has a loss of 0.3 dB/km. The total of all the connector losses at the two ends is 4 dB. Calculate the available system margin.

Solution

It is much easier to use decibels throughout to solve this type of problem, so let us begin by converting the input power to the system to dBm:

$$\begin{aligned} P_{in}\text{ (dBm)} &= 10 \log P_{in}\text{(mW)} \\ &= 10 \log 1.5 \\ &= 1.76 \text{ dBm} \end{aligned}$$

The connector and fiber losses are quite straightforward. The only problem in finding the loss in the splices is determining the number of splices. There will be one less splice than the number of fiber spans, as the sketch in Figure 23.1 should make clear.

Consider each length of fiber to be attached to one splice at its output end. Then there is one splice per length, except for the last one. The number of lengths of fiber can be found

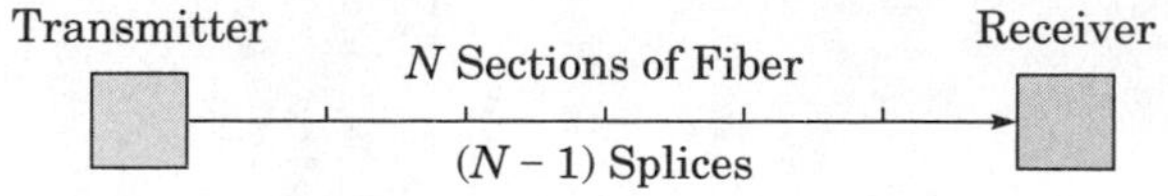

Figure 23.1

by dividing the total length by the maximum length for a fiber section, then rounding upward. In this case the number of lengths is found from

$$\text{span length/fiber length} = 40\ \text{km}/2.5\ \text{km} = 16$$

Therefore there will have to be 15 splices.

Now we need to find the total losses. These can be calculated as follows:

Connector losses	4 dB
Fiber loss: 40 km × 0.3 dB/km	12 dB
Splice loss: 15 splices × 0.25 dB/splice	3.75 dB
Total loss	19.75 dB

Next, find the output level

$$P_{out} = 1.76\ \text{dBm} - 19.75\ \text{dB} = -17.99\ \text{dBm}$$

The receiver requires a power of −25 dBm, so the available system margin is

$$\text{system margin} = -17.99\ \text{dBm} - (-25\ \text{dBm}) = 7.01\ \text{dB}$$

This system should be satisfactory in terms of signal strength.

23.2.2 Rise Time Budget

In the previous chapter we noted that the data rate that can be used with a length of fiber is limited by dispersion. Dispersion is greatest with step-index multimode fibers and least with single-mode fibers, particularly when they are used at a wavelength of about 1.55 μm. However, even single-mode fibers have some dispersion.

The effects of dispersion increase with the length of the fiber, since a difference in the time two signals of different velocity require to reach the end of the fiber is proportional to the length of the fiber. The effect of dispersion is also proportional to the bandwidth of the information signal, and for single-mode fibers, the dispersion is proportional to the source linewidth. Most of the dispersion in multimode fibers is due to the numerous modes, so source linewidth has little effect on dispersion in multimode fibers.

The fiber itself is not the only part of the system that limits bandwidth and data rate. Both transmitters and receivers have finite **rise times** that limit their bandwidth, and their effects must be included when calculating the maximum data rate.

In Chapter 22 we looked at the effect of dispersion on a square wave information signal in order to find a rough equivalence between dispersion and bandwidth. Another way to look at dispersion is to consider its effect on a short pulse. As a pulse of light propagates down the fiber, its duration increases. This effect, known as *pulse spreading*, is illustrated in Figure 23.2. The amount of pulse spreading is proportional to the length of the fiber and to its dispersion per kilometer, which is also called its *pulse-spreading constant* or *pulse-stretching constant.* The pulse-spreading constant is given in nanoseconds or picoseconds per kilometer, and the total spread of an infinitesimally short pulse for a given fiber length is then:

$$\tau = Dl \tag{23.1}$$

Figure 23.2
Pulse spreading

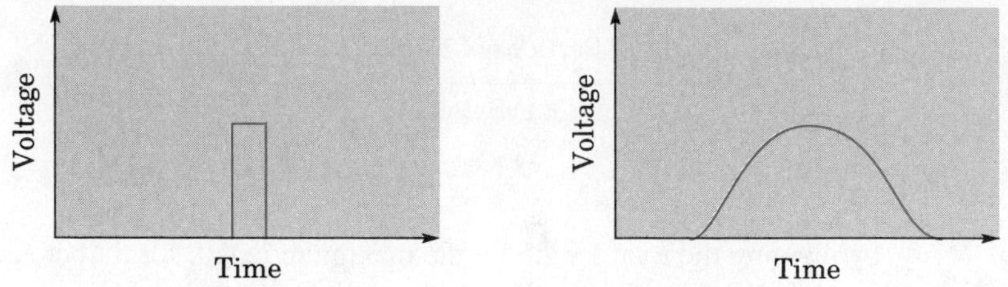

(a) Original pulse (b) Spread pulse

where

τ = total pulse spreading in nanoseconds
D = pulse-spreading constant in ns/km
l = fiber length in km

Example 23.2 A 45 km length of fiber must not lengthen pulses by more than 100 ns. Find the maximum permissible value for the pulse-spreading constant.

Solution From Equation (23.1),

$$D = \frac{\tau}{L} = \frac{100 \text{ ns}}{45 \text{ km}} = 2.22 \text{ ns/km}$$

When a square pulse of finite length is applied to the fiber, the effect of dispersion is to make the rise and fall times of the pulse approximately equal to the dispersion.

The pulse-spreading effect of the fiber itself combines with the rise times of the transmitter and receiver. These rise times combine as follows:

$$T_{RT} = \sqrt{T^2_{Rtx} + T^2_{Rrx} + T^2_{Rf}} \tag{23.2}$$

where

T_{RT} = total rise time
T_{Rtx} = transmitter rise time
T_{Rrx} = receiver rise time
T_{Rf} = fiber rise time

The cumulative effect of these rise times is to limit the rate at which pulses can be transmitted, and this in turn limits the data rate.

There are two major digital transmission modes for fiber. The laser diode or light-emitting diode is switched on and off, or sometimes, when a laser diode is used, between low and high power levels. The line code may be a non-return-to-zero (NRZ) type, as shown in Figure 23.3(a). For this code, the amount of pulse stretching must be less than one bit period.

Alternatively, if a return-to-zero code, as illustrated in Figure 23.3(b), is used, the amount of pulse stretching must be less than half a bit period. Thus the maximum bit rate can be found from the dispersion, as follows:

For RZ code:

$$f_b = \frac{1}{2T_{RT}} \tag{23.3}$$

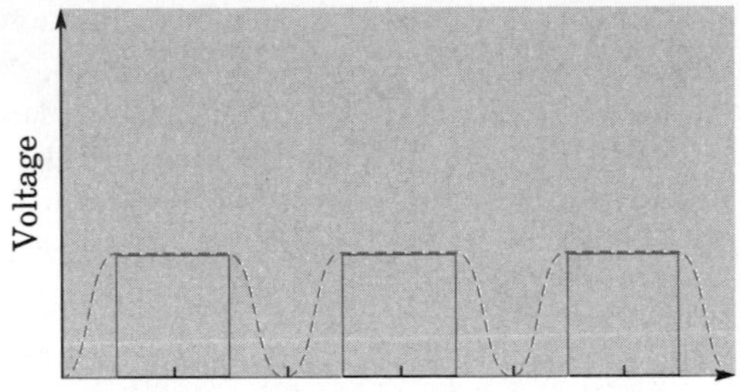

(a) NRZ

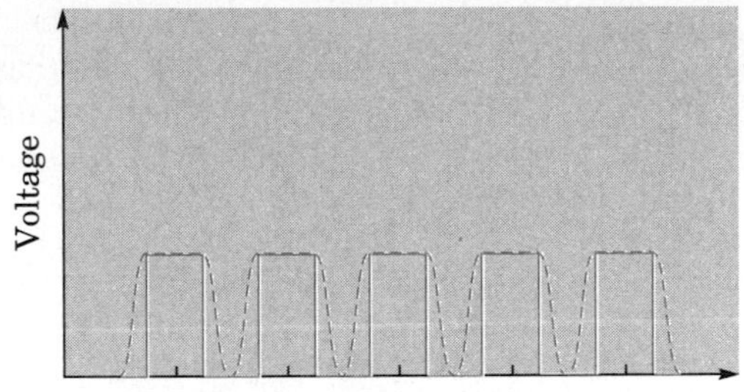

(b) RZ

Figure 23.3
Line codes and pulse spreading

For NRZ code:

$$f_b = \frac{1}{T_{RT}} \tag{23.4}$$

where

f_b = maximum data rate in bits per second
T_{RT} = total rise time in seconds

Now we can easily prepare a rise time budget for a system.

Example 23.3 Calculate the maximum data rate for the 45 km fiber system in the previous example when it is used with a transmitter having a rise time of 50 ns and a receiver having a rise time of 75 ns, if the line code is

(a) NRZ
(b) RZ

Solution The first step is to calculate the total rise time for the system. From Equation 23.2,

$$\begin{aligned} T_{RT} &= \sqrt{T^2_{Rtx} + T^2_{Rrx} + T^2_{Rf}} \\ &= \sqrt{50^2 + 75^2 + 100^2} \\ &= 135 \text{ ns} \end{aligned}$$

Now we can find the maximum bit rate for each scheme.

(a)
$$\begin{aligned} f_b &= \frac{1}{\tau} \\ &= \frac{1}{100 \text{ ns}} \\ &= 10 \text{ MHz} \end{aligned}$$

(b)
$$\begin{aligned} f_b &= \frac{1}{2\tau} \\ &= \frac{1}{2 \times 100 \text{ ns}} \\ &= 5 \text{ MHz} \end{aligned}$$

Since the bandwidth-distance product and the pulse-stretching constant are two different ways of describing the same property of a fiber, namely, its dispersion, it seems logical that the two ratings should be related for a given fiber. In Chapter 22 we derived an approximate relationship by using a square wave. If we use the rise time of a short pulse instead, we can look at the situation another way. In general, in any system that can be modeled as a first-order low-pass filter, the relation between rise time and electrical bandwidth is given by:

$$B_{el} = \frac{0.35}{T_R} \tag{23.5}$$

where

T_R = rise time in seconds
B_{el} = electrical bandwidth in hertz

A fiber is not exactly a first-order low-pass system, but an analysis based on that assumption gives conservative results. For a fiber, T_R is approximately equal to the dispersion. Therefore, for a 1 km length of fiber we should have:

$$B_{el} = \frac{0.35}{D}$$

The bandwidth given in fiber bandwidth-distance product specifications is optical bandwidth. Since current in optical transmitters and receivers is proportional to power in the fiber, the traditional half-power (-3 dB) bandwidth limit in the fiber corresponds to half-current or one-quarter power (-6 dB) in the electrical system. Thus the electrical bandwidth at the -3 dB point is smaller than the optical bandwidth. This can be compensated for by multiplying the electrical bandwidth by $\sqrt{2}$ to get the optical bandwidth.

The optical bandwidth for the fiber is then

$$B = \sqrt{2}\, B_{el} = \frac{\sqrt{2} \times 0.35}{\Delta t} \approx \frac{0.5}{\Delta t} = \frac{1}{2\Delta t} \tag{23.6}$$

where

B = optical bandwidth of the fiber
B_{el} = electrical bandwidth of the fiber
Δt = dispersion of the fiber

This is the same result we got in Chapter 22, where we used a different approach. Note that for analog signals, the electrical bandwidth should be used. Also note that these numbers are approximate and that the bandwidth of a fiber, being due to dispersion and not attenuation, is not really directly comparable to that of an amplifier or similar low-pass system. As an analog signal approaches the fiber bandwidth, distortion increases, so it would not be wise to approach the limit too closely.

Now it is easy to relate the bandwidth-distance product and the pulse-spreading constant. Since the bandwidth-distance product is equal to the bandwidth of a 1 km length and the pulse-spreading constant is equal to the dispersion in a 1 km length, we can simply substitute into Equation (23.6):

$$Bl = \frac{1}{2D}$$

where

Bl = bandwidth-distance product in Hz-km
D = pulse-spreading constant in seconds/km

These are not practical units, of course. If the bandwidth-distance product is given in MHz-km and the pulse-spreading constant is in ns/km, the relation becomes

$$Bl = \frac{1}{2D \times 10^{-3}} = \frac{500}{D} \tag{23.7}$$

where

Bl = bandwidth-distance product in MHz-km
D = pulse-spreading constant in ns/km

Example 23.4 A fiber is rated as having a gain-bandwidth product of 500 MHz-km. Find its dispersion in ns/km, and find the rise time of a pulse in a 5 km length of this cable.

Solution From Equation (23.7),

$$Bl = \frac{500}{D}$$

$$D = \frac{500}{Bl} = \frac{500}{500} = 1 \text{ ns/km}$$

Since the fiber is 5 km long, the total rise time is 5 ns.

Now let us look at a slightly more complex example.

Example 23.5 A fiber-optic system uses a detector with a rise time of 3 ns and a source with a rise time of 2 ns. If an RZ code is used with a data rate of 100 Mb/s over a distance of 25 km, calculate the maximum acceptable dispersion for the fiber and the equivalent bandwidth-distance product.

Solution First we have to find the maximum permissible dispersion. For an RZ code, we can use Equation (23.3):

$$f_b = \frac{1}{2T_{RT}}$$

so

$$\begin{aligned} T_{RT} &= \frac{1}{2f_b} \\ &= \frac{1}{2 \times 100 \times 10^6 \text{ b/s}} \\ &= 5 \text{ ns} \end{aligned}$$

Next, we need to rearrange Equation (23.2) to get fiber rise time as the unknown:

$$\begin{aligned} T_{RT} &= \sqrt{T^2_{Rtx} + T^2_{Rrx} + T^2_{Rf}} \\ \sqrt{T^2_{Rtx} + T^2_{Rrx} + T^2_{Rf}} &= T_{RT} \\ T^2_{Rtx} + T^2_{Rrx} + T^2_{Rf} &= T^2_{RT} \\ T^2_{Rf} &= T^2_{RT} - T^2_{Rtx} - T^2_{Rrx} \\ T_{Rf} &= \sqrt{T^2_{RT} - T^2_{Rtx} - T^2_{Rrx}} \\ &= \sqrt{5^2 - 2^2 - 3^2} \text{ ns} \\ &= 3.464 \text{ ns} \end{aligned}$$

This value is for the whole fiber length, of course. We now have to divide by 25 to find the dispersion per kilometer:

$$\tau = \frac{3.46 \text{ ns}}{25 \text{ km}} = 0.1386 \text{ ns/km}$$

Expressed in terms of a bandwidth-distance product, the cable rating would be

$$Bl = \frac{500}{0.1386} = 3608 \text{ MHz-km} = 3.61 \text{ GHz-km}$$

In addition to the optical components, the electronic parts of the system, such as amplifiers and analog-to-digital converters, also have rise-time specifications. Therefore, specifying the rise time, rather than the bandwidth, for a given length of fiber is very useful in calculating the response of the whole system.

23.3 Repeaters and Optical Amplifiers

As a result of either loss or dispersion, there is always a limit to the length of a single span of fiber-optic cable. Historically, the trend has been to longer spans at higher data rates, made possible by the continuing development of fiber with lower loss coupled with lower dispersion. Still, when distances are great, some form of gain must be provided. There are two basic ways to do this. One is to change the signal to electrical form, amplify it, regenerate it if it is a digital signal, and then convert it back to optical form. The other, newer technique, is to amplify the optical signal directly, without converting it to an electrical signal. Let us look briefly at each of these techniques.

23.3.1 Regenerative Repeaters

In its most common form, a repeater converts the signal from optical to electrical energy, amplifies it, then converts it back to optical form. Figure 23.4 shows a block diagram of a typical repeater.

One of the advantages of using digital techniques in communications is the fact that regenerative repeaters can be used, that is, a distorted or noisy received signal can be decoded into ones and zeros, from which a new, perfect pulse train can be reconstructed. As long as repeaters are spaced closely enough to keep the error rate for one span very low, digital systems can avoid the accumulation of noise and distortion that plagues analog systems. Though most fiber-optic systems are digital, analog systems are still used for some applications. One example is the use of analog fiber trunks in cable television systems. The use of analog technology here allows easier interconnection between the fiber and coaxial-cable portions of the system.

Figure 23.4 Optical communication systems

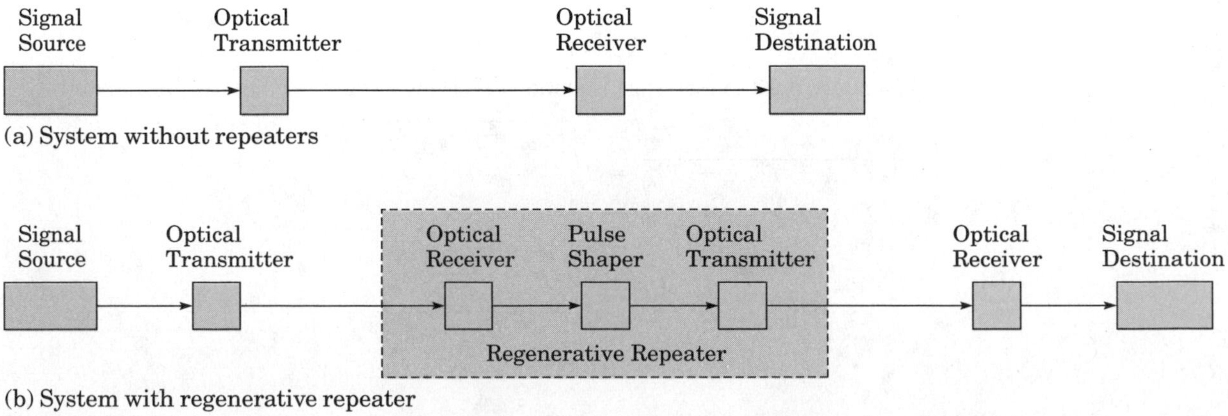

Figure 23.5
Uses of optical amplifiers

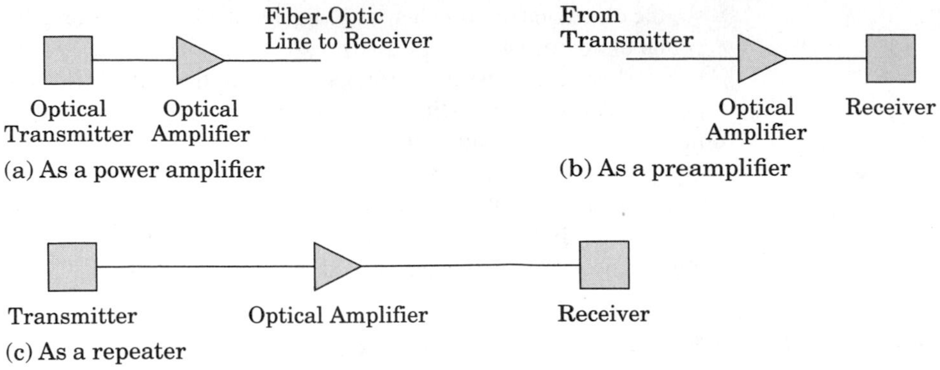

Repeaters need electrical power, of course, and it is preferable to have them accessible for maintenance. Consequently, the trend in optical system design is to minimize the number of repeaters by increasing the bandwidth-distance product and reducing the losses of the fiber.

23.3.2 Erbium-Doped Fiber Amplifiers

In situations where fiber loss, not dispersion, is the limiting factor on the length of a fiber span, it is possible to amplify the optical signal directly, without the trouble and expense of converting it to electrical form and back. Furthermore, an optical amplifier can work with any type of signal, analog or digital, whether multiplexed or not.

The construction of optical amplifiers is based on principles similar to those that govern the operation of lasers. Figure 23.5 shows three possible uses for optical amplifiers. In Figure 23.5(a), the amplifier is installed at the output of the optical emitter and is used as a power amplifier. In Figure 23.5(b), the amplifier is shown just ahead of the receiver—it is used as a preamplifier. Finally, in Figure 23.5(c), the amplifier is used in the middle of the communications link, instead of using a repeater. There are several ways to build optical amplifiers, but the one that currently shows the most promise is the erbium-doped fiber amplifier sketched in Figure 23.6. This amplifier is a very new development, with commercial versions first becoming available in 1990. It is most efficient at wavelengths in the vicinity of 1550 nm, which in any case is the most popular wavelength for current system designs.

As Figure 23.6 shows, an erbium-doped fiber amplifier consists of several meters of doped fiber, with an arrangement to couple the signal into and out of the fiber and to couple the light from a semiconductor **pump laser** into the fiber. The pump laser provides the power that is used to amplify the input signal. The pump is a laser diode operating at a shorter wavelength than that of the signal to be amplified. In erbium-doped fiber amplifiers, the pump generally operates at either 980 or 1480 nm. A shorter wavelength (820 nm) has also been used. The amplifier is not as efficient when pumped at that wavelength, but gallium arsenide laser diodes are available at 820 nm, and they have much higher power output than the longer-wavelength lasers. The extra pumping power can more than make up for the reduction in efficiency and can result in greater output power from the amplifier.

The pump excites electrons in the fiber into a higher energy level. Light from the signal to be amplified then causes the electrons to drop back to a lower level. As they do

Figure 23.6
Erbium-doped fiber amplifier

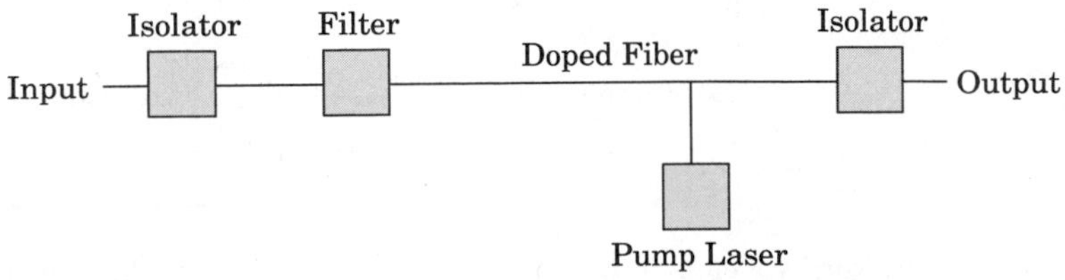

so, the electrons release photons of light at the same wavelength as, and in phase with, the signal to be amplified.

There are some disadvantages to using fiber amplifiers rather than regenerative repeaters. As is the case with any analog amplifier, the signal-to-noise ratio of a fiber amplifier is worse at its output than at its input. Some electrons that have been excited by the pump laser spontaneously drop to a lower energy level without being subjected to an input signal, and the light thus produced is noise, which can accumulate from one amplifier to the next, as it does in any analog system. Since fiber amplifiers have no way to reshape pulses, the effects of dispersion also accumulate.

23.4 Wavelength-Division Multiplexing

Most optical-fiber systems use time-division multiplexing to take advantage of the available bandwidth using one LED or laser diode. This bandwidth, which is usually limited by dispersion, is really only a small fraction of the actual bandwidth available on a fiber, however. There is no theoretical reason why there cannot be several light sources, each operating at a different wavelength and coupled into the same fiber. At the receiving end, the different wavelengths would be sent to separate receivers.

A diagram of this scheme, called **wavelength-division multiplexing (WDM)**, is given in Figure 23.7. Practical implementation requires lasers with narrow bandwidth. WDM is really a form of frequency-division multiplexing (FDM)—think of the laser diodes as carrier oscillators, each with a different carrier frequency and separately modulated.

One difference between FDM on radio-frequency carriers and WDM on optical fibers is that for FDM the separation between carriers is limited by the sidebands created by modulation, while with even the best lasers, the width of the carrier signal itself is such that it, rather than the modulation rate and type, determines the signal bandwidth. That is, the laser can be considered an optical oscillator, but by radio-frequency standards it is a rather unstable oscillator with plenty of spurious signals. This is changing, however, with improvements in laser technology.

One obvious use for WDM is to implement full-duplex communications on a single fiber. Figure 23.8 shows how this can be done. At present, however, it is still usually more cost-effective to use two fibers and dedicate each to service in one direction.

Figure 23.7
Wavelength-division multiplexing

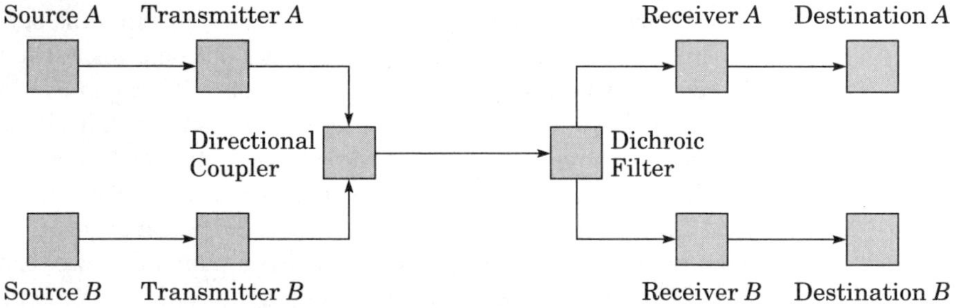

Figure 23.8
Full-duplex communication using wavelength-division multiplexing

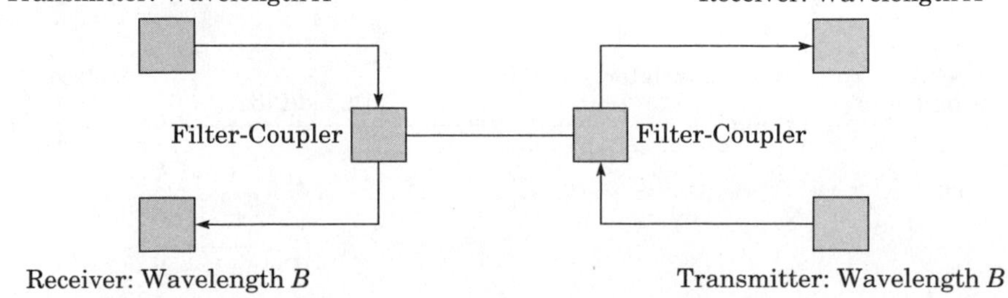

23.5 Submarine Cables

The use of fiber optics for underwater telephone cables is a very logical application of the technology. Coaxial cable has traditionally been used for such cables, but its bandwidth is much less than that of fiber and the number of repeaters required is much greater. Reducing the number of repeaters is of special importance in undersea cables, since gaining access to the repeaters for maintenance is very difficult and expensive once the cable has been installed.

Short fiber-optic cables with lengths under about 100 km are generally built without repeaters. Putting all the electronics on shore makes maintenance and upgrading much easier. Cables of this sort are used to communicate across narrow straits, such as from a mainland to offshore islands, and sometimes to link coastal cities without the need to purchase land rights-of-way.

Cables for longer distances, such as across oceans, currently need repeaters, though their eventual elimination is one of the main reasons for research into solitons, which will be discussed later in this chapter. As technology has improved, the data rate and repeater separation have both increased.

For example, the first fiber-optic transatlantic undersea fiber cable, designated TAT-8, was completed in December 1988. The repeater spacing is 70 km, and 109 repeaters are required altogether. Its primary data rate is 295.6 Mb/s, which is equivalent to 40,000 simultaneous telephone calls when digital compression is taken into account. Of course the system can handle other types of communications (for example, video and data) as well. Two single-mode fibers are used, one for communication in each direction. Laser diode emitters and PIN diode detectors operating at a wavelength of 1.3 μm are used.

The second generation of fiber transatlantic cables, exemplified by TAT-10 (1992), uses a wavelength of 1.55 μm. It operates at roughly double the data rate of TAT-8, and the repeater spacing is more than 100 km.

A third generation of transatlantic cables that operates at 5 Gb/s in each direction (two STM-16 signals, described in the next section) and uses optical amplifiers instead of repeaters is now entering service. The first segment of the TAT-12/13 system began carrying traffic in October 1995. When the system is complete, it will have the ring configuration shown in Figure 23.9. Each cable has one *service* (active) pair of fibers (one fiber for communication in each direction) and one *restoration* (spare) pair. If a cable is cut, service can be restored using the restoration pair in the intact segments. Meanwhile the restoration pairs can be used for lower-priority traffic or as backup for other cables. The optical amplifiers under the Atlantic Ocean have gains of about 10 dB each and are spaced at 45 km intervals.

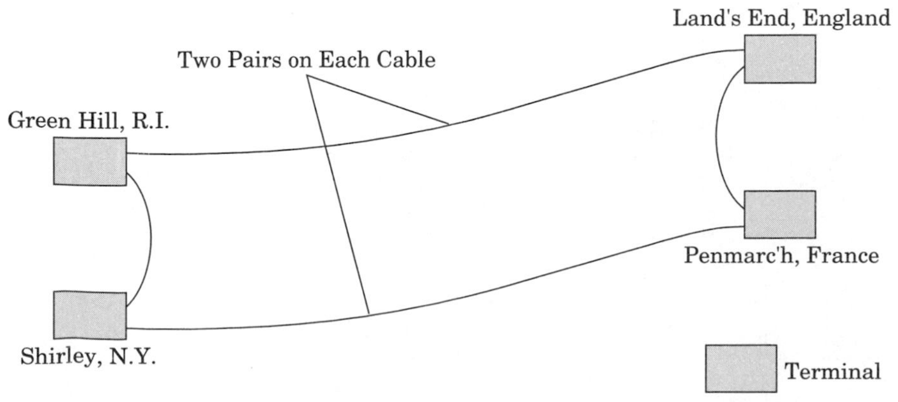

Figure 23.9 TAT-12/13 transatlantic fiber optic cables

23.6 The Synchronous Optical Network (SONET)

The very high data rates possible with fiber optics require new standards for digital transmission. The familiar DS signals and T-series carriers, starting with T-1 at 1.544 Mb/s, were developed for coaxial cable and microwave transmission and do not take full advantage of the bandwidth available with fiber.

The **synchronous optical network (SONET)** standard was especially designed for the high data-rate capability of fiber-optic transmission. SONET is a North American standard; the European equivalent is called the **synchronous digital hierarchy (SDH)** and is so similar that the abbreviation SONET/SDH is often used to describe both. The basic SONET data rate is 51.840 Mb/s, and any multiple of this rate is possible. A signal at the basic rate is called a *synchronous transport signal level one* or STS-1 signal. When transmitted on a fiber, the same signal is referred to as *optical carrier one* or OC-1. This terminology is analogous to the DS-1 signal and T-1 carrier previously described.

Currently popular SONET rates are shown in Table 23.1. STM stands for *synchronous transport module*, and the STM rates are those used for synchronous transmission.

The lowest speed in the hierarchy is designed to accommodate both the North American DS-3 data rate of 44.736 Mb/s and the European E-3 rate of 34 Mb/s. SONET/SDH is designed to be an international standard.

The last line in Table 23.1 shows a data rate of almost 2.5 Gb/s, but this is well within the present capability of optical links using single-mode fiber and laser diodes.As technology progresses, additional multiples of the SONET basic rate can easily be employed. The TAT-12/13 system, which was discussed in the previous section, already multiplexes two STM-16 signals.

In order to use the available data rate more efficiently, the SONET standard differs in some other ways from the older digital transmission scheme. Like the DS-1 signal, the STS-1 signal has 8,000 frames per second, corresponding to the sampling rate for digital telephony. Recall that one frame of the DS-1 signal contains 193 bits: 8 bits for one sample from each of 24 channels, plus 1 framing bit. Every sixth frame uses the least-significant bit from each channel for signaling. The SONET STS-1 signal, in contrast, has 6480 bits (810 eight-bit bytes) per frame. The frame can be thought of as being arranged in the rows and columns illustrated in Figure 23.10. Transmission of a frame begins in the upper left-hand corner and ends in the lower right-hand corner, 125 μs later.

There are nine rows in the table in Figure 23.10, and each row has 90 bytes. Higher levels in the hierarchy would still have nine rows, but the number of columns would increase. An STS-3 signal, for instance, has 270 bytes in each row, making a frame three times as big (2430 bytes). Since the larger frame is still transmitted in 125 μs, the bit rate for STS-3 is three times that for STS-1. The STS-3 signal is formed by transmitting one byte from each of three STS-1 signals in turn.

Table 23.1
SONET line rates

STS-level	OC-level	STM-level	Line rate (Mb/s)
STS-1	OC-1		51.84
STS-3	OC-3	STM-1	155.52
STS-9	OC-9		466.56
STS-12	OC-12	STM-4	622.08
STS-18	OC-18		933.12
STS-24	OC-24		1244.16
STS-36	OC-36		1866.24
STS-48	OC-48	STM-16	2488.32

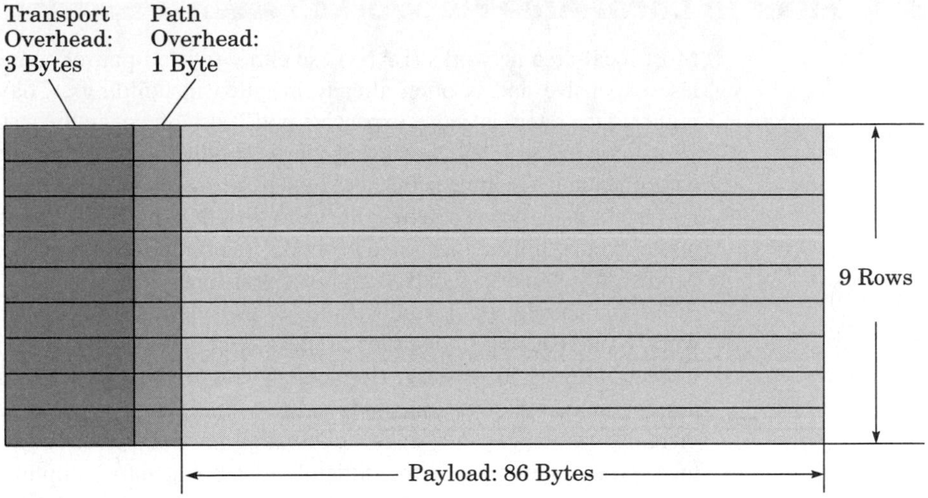

Figure 23.10
SONET DS-1 frame

The first three columns of the STS-1 signal, 27 bytes in all, are designated as transport overhead. One column (9 bytes) is dedicated to line overhead, and the other two columns are for section overhead. The remaining 87 columns are referred to as the STS-1 synchronous payload envelope (SPE). The first column (one byte in each of the nine columns) is designated as path overhead (POH), but the remainder can be used for the payload information. Thus, of the total of 810 bytes in a frame, 36 bytes (4.4%) are used for overhead.

The SONET frame thus has considerably more overhead than the DS series it replaces. On the other hand, DS-1 signals are not capable of transmitting nearly as much information about signal routing and setup, and what little information they do contain is carried at the expense of creating errors in the least-significant bit of some of the voice samples.

Until the development of SONET, there were timing problems in multiplexing digital signals. Suppose two signals are to be multiplexed—even if the clock rates for both signals are exactly the same (that is, they are synchronous), the phase is almost certain to be different. Samples from the two signals will not begin and end at the same time. Even if the signals were aligned at some point, differences in the lengths of their transmission paths would soon cause phase errors. Recall, for instance, that even at the speed of light, signals travel only 300 meters per microsecond. Thus signals traveling different distances are practically guaranteed to be out of phase. In the early days of digital switching, when most transmission was still analog, the signals to be switched had often just been digitized in the switching equipment itself, so synchronization was not a common problem. Later, the introduction of digital transmission caused these phase differences to become a more serious problem.

The earlier DS series of digital signals solved the phase problem by adding bits to one of the signals to synchronize the two. This technique was called **bit stuffing**. Of course, it added to the overhead of the signal, and it also required some arrangement for removing ("destuffing") the excess bits at the signal's destination. SONET, on the other hand, uses a pointer to denote the starting position of a frame of information. The information frame does not necessarily coincide with the SONET frame—one frame of information can easily straddle two SONET frames. In order to indicate where each information frame starts, SONET frames include as part of the transport overhead a payload pointer that indicates the beginning of the synchronous payload envelope (SPE). The pointer value can be changed to account for phase and frequency differences between signals.

23.7 Fiber in Local-Area Networks

Most local-area networks (LANs) use either twisted-pair or coaxial cable. Twisted pair is less expensive and is often already installed in buildings. Coaxial cable, which allows higher data rates, is more expensive and must be specially installed. Fiber optics have seen some use in LANs because the large bandwidth of fiber allows it to operate at higher data rates and for greater distances than the other media. (Fiber also has lower loss, but loss is not usually a problem at the short distances normally found in LANs.) The trend to higher bandwidths in local-area networks is certain to continue as more use is made of such bandwidth-intensive signals as digitized real-time video.

You should recall from Chapter 14 that there are three main ways to connect a network—the star, ring, and bus topologies—which are shown again in Figure 23.11. Of these three topologies, the easiest to implement with fiber optics is the ring, since it involves only direct connections from one node to the next. One ring can be used, as shown, but often there are two rings, with signals circulating in opposite directions. The bus system is the most problematic, due to the difficulty of tapping into an optical fiber at many different points along its length. Optical tee connectors, which were described in Chapter 22, are available, but they are relatively expensive, and each connector contributes about a decibel of loss to the signal moving along the bus. Furthermore, most fiber systems are unidirectional and thus not well-suited to a bus topology. The star system is of intermediate difficulty: optical power splitters are available (see Chapter 22), but they are lossy. Separate fiber connections can be made for signals moving to and from the central hub, so that each fiber carries signals in only one direction.

Currently most fiber-optic LANs use a ring topology and a standardized interface known as the **fiber distributed data interface (FDDI)**. The system uses multimode fiber operating at a wavelength of 1.3 μm. LED transmitters and PIN diode receivers are used for low cost. The short distances and modest data rates used with FDDI make laser transmitters unnecessary. The data rate is 100 Mb/s, which is high for a LAN but relatively low for fiber. In fact, this data rate can also be transmitted on coaxial cable, and the FDDI data standard has been used with copper as well as fiber connections.

The FDDI system uses two token rings that carry signals in opposite directions. Usually only one is used, and the second is held in reserve for backup. Each node decodes the signal coming along the fiber. If the message is for the connected terminal, it is sent to that

Figure 23.11
Network Topologies

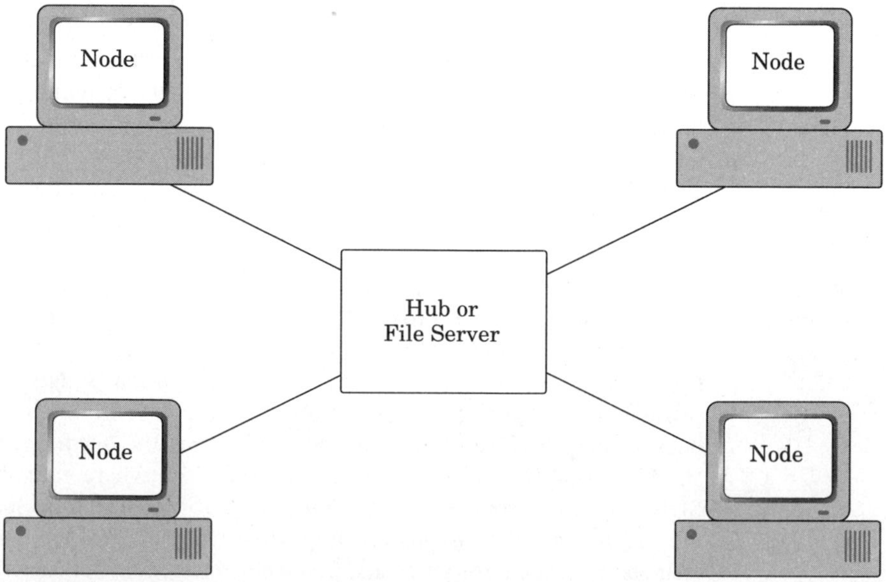

(a) Star network

Figure 23.11
Continued

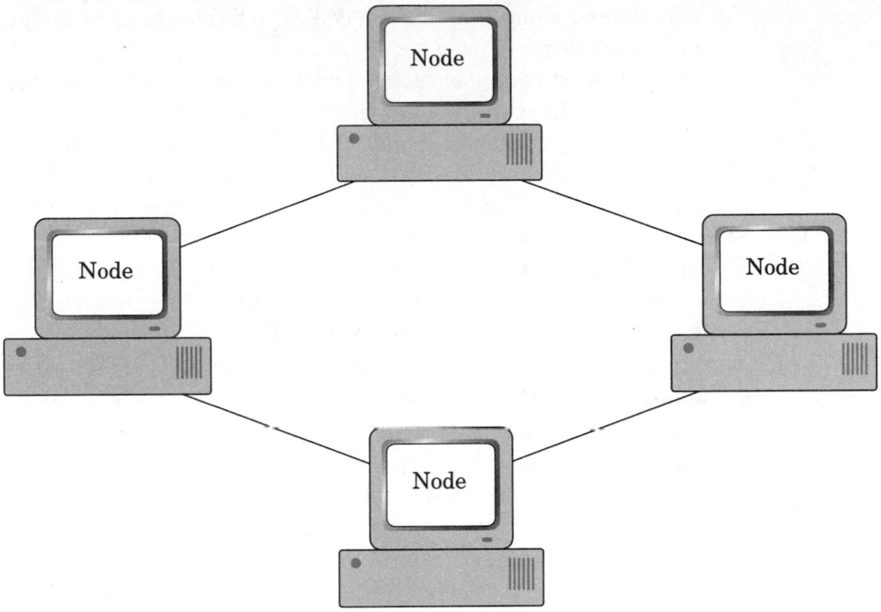

(b) Ring network

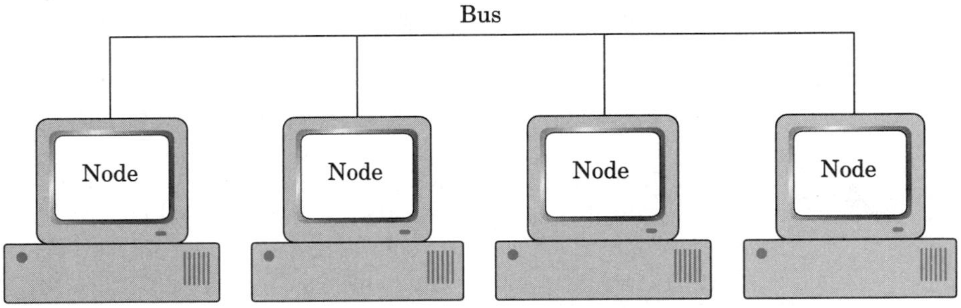

(c) Bus network

terminal; otherwise it is re-encoded and retransmitted. Thus each node in the network acts as a regenerative repeater. Because of this regeneration and the low loss of fiber, distances in FDDI networks can be much greater than in other LANs: 2 km between nodes with a total length of 200 km. In fact, because of its high data rate and long-distance capability, FDDI is often used to connect copper-based LANs together.

23.8 Local Telephone Applications

Telephone companies were among the first communications users to recognize the advantages of fiber optics. Practically all new trunk cable for long-distance telephony is now fiber. Most fiber trunks use single-mode fiber operating at 1.3 μm, though some of the latest installations use 1.55 μm. However, the local loop from the central office to the individual subscriber remains almost entirely twisted-pair copper.

There are several good reasons why local loops still rely on copper. Fiber's most important advantage over copper is its much greater bandwidth, but so far most local loops need a bandwidth of only about 3 kHz for one voice channel. Even ISDN, with its requirement for two 64 kb/s channels and one 16 kb/s channel, is easily accommodated using twisted-pair lines. There has been much talk lately about combining telephone, cable television, and some as-yet ill-defined data services on one cable, and there have been some experiments along those lines. If and when this happens on a large scale, fiber will certainly be a contender for that cable.

At least two other factors impede the development of fiber in local loops (often referred to as **fiber in the loop**, or **FITL**). One is the fact that most local loops are already installed and are quite functional using copper. The telephone companies (telcos) at present have no compelling reason to replace billions of dollars worth of installed copper plant. Another issue is that each fiber local loop would require an optical-to-electrical (and probably digital-to-analog) conversion device to be installed at the customer's premises. In addition to its extra expense, this device would have to obtain power from the customer's electrical service. While this in itself would not seem to be a problem, since having the customer supply the power would actually reduce the telco's costs, it could seriously compromise the reliability of the telephone system. At present, power comes from the central office, which has battery and diesel standby power in case of power failures. Consequently, most electrical power interruptions now have no effect on telephone service. That could change with fiber in the loop, unless a backup battery supply were installed at each customer's premises—at a further increase in cost.

If there has not been very much progress to date in FITL, **fiber to the curb (FTTC)** is a different matter. For some time, it has been common in new construction to run a fiber optic cable from the central office to a concentrator that consists of an optical-to-electrical interface and a multiplexer/demultiplexer at a central location from which a number of customers—a few hundred houses in a subdivision, perhaps—can be served. Many signals can be digitally multiplexed on a single pair of fibers, saving money compared with multipair copper cable. The optical-to-electronic interface can be powered by a separate copper cable from the central office, which can also carry the power for the conventional local loops that connect the subscribers' homes. Alternatively, for larger installations, the concentrator can be powered from the local electrical utility, with battery backup power also available.

In the future it seems likely that fiber will approach the home even more closely than this, probably to a connection that serves a small number of residences within a radius of a few hundred meters. The remaining distance could operate at quite a large bandwidth, even with twisted-pair cable, without the need for optical-to-electrical and digital-to-analog conversion at each customer's premises.

23.9 Cable-Television Applications

Community-antenna television systems (CATV or, informally, cable television) were described in Chapter 9. They have traditionally used coaxial cable to carry a number of analog television and FM radio signals from a single antenna location, known as the head-end, to a large number of subscribers. A typical CATV system has a tree structure, as shown in Figure 23.12.

The use of coaxial cable for CATV trunk lines requires closely spaced amplifiers, generally separated by less than a kilometer. Many systems now use fiber-optic cable for trunk lines to avoid the necessity for repeaters. Except in a few experimental systems, the drop lines to individual subscribers are still coaxial copper cable. As in telephone systems, using copper for the subscriber connection avoids the expense of providing optical-to-electrical conversion at each subscriber's premises.

Cable television systems are still almost completely analog, and their fiber-optic components are no exception. Though most fiber-optic systems use digital schemes, cable television trunks are generally analog, using frequency-division multiplexing of the various television and radio broadcast signals. The resulting wideband electrical signal can either be applied directly to a coaxial cable or used to modulate a fiber-optic transmitter. The optical receiver reproduces this analog FDM signal, which can pass through the copper part of the system with no change. This system makes the requirements for the interfaces between the optical and electrical parts of the system much simpler.

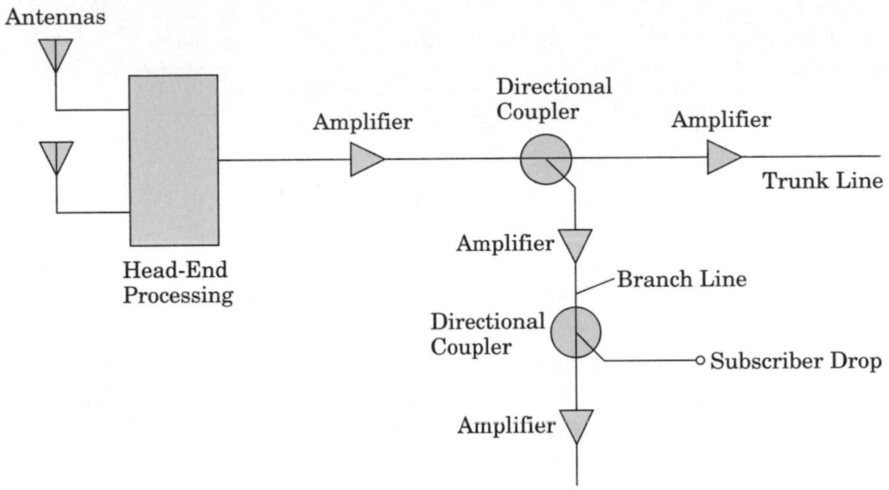

Figure 23.12 Cable-television system

In the near future it is expected that CATV systems will begin to carry television signals in a compressed digital format. Two such formats have already been described: the Grand Alliance high-definition television system was examined in Chapter 9 and the MPEG-2 compression system was discussed in Chapter 11. Either of these schemes would be well-suited to fiber: the very wide bandwidth and suitability of fiber for digital transmission could make the much-touted "500-channel universe" a practical possibility. Since such a scheme would likely require a digital-to-analog converter at each customer's premises, it might not add very much to the cost to run fiber to each residence. Doing so would provide the subscriber with almost unlimited potential bandwidth.

23.10 Experimental Techniques

Fiber optics is still in its infancy. In fact, it could be considered to be about where radio communication was around 1920. The LED, with its broad spectrum, can be compared to a spark transmitter and the laser diode to a rather primitive oscillator-only transmitter. Erbium-doped amplifiers are as significant an advance as vacuum-tube amplifiers were in radio communication. The PIN diodes and avalanche photodiodes commonly used today are quite reminiscent of the early "crystal" receivers, which also used a diode as their main component.

There is still much work to be done in fiber optics. In this section we look at two of the many developments that will shortly leave the laboratory and enter everyday use.

23.10.1 Solitons

As we have mentioned, the zero-dispersion wavelength for ordinary silica fiber is about 1300 nm. Even at that wavelength, there is some broadening of pulses as they travel long distances because the index of refraction of the fiber increases slightly with the optical power level. This nonlinearity causes the center part of a pulse, which normally has the greatest amplitude, to propagate more slowly than the lower-intensity portions at its beginning and end. The flat top of a pulse contains more low-frequency and less high-frequency energy than its leading and trailing edges. The resulting spreading of the pulse is called *chromatic dispersion*.

Solitons are created by using the fiber at a wavelength slightly greater than the zero-dispersion value. At such a wavelength, the higher-frequency components at the leading and trailing edges of each pulse propagate more quickly than the lower-frequency components in the center, canceling the effect of the chromatic dispersion. Pulses so created can propagate for very long distances (thousands of kilometers) without dispersion, provided the pulse amplitude remains large enough. Soliton propagation works only for relatively strong pulses, so amplifiers (such as erbium-doped fiber amplifiers) must be used at intervals of about 25 to 50 km.

Figure 23.13
Optical version of superheterodyne receiver

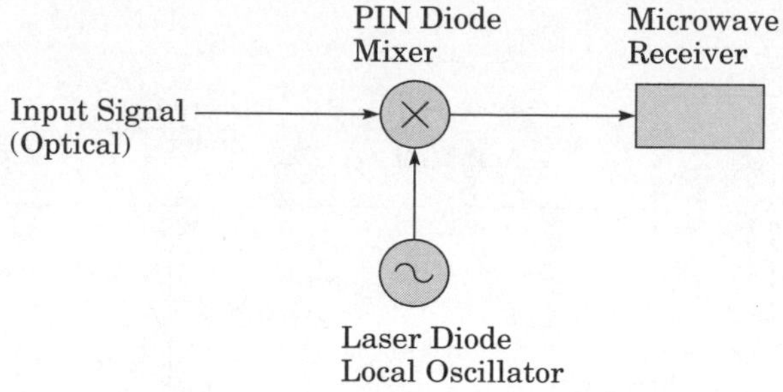

23.10.2 Heterodyne Reception

In view of the analogy we made with crystal receivers, one might wonder whether an optical equivalent to the superheterodyne receiver could be built. The answer is yes, and it has been done. Figure 23.13 shows how: a laser diode is used as a local oscillator, a photodiode as a mixer, and the intermediate frequency is a microwave signal. This kind of device is still largely experimental, but it shows the direction in which technology is moving. If the twentieth century has been dominated by radio and its variations, it appears that the twenty-first may be the century of lightwave communication.

23.11 Optical Time-Domain Reflectometry

You should recall that time-domain reflectometry is a useful technique for finding faults in copper cables. A pulse or a voltage step is sent down a cable, and the reflection, if any, is analyzed. A perfectly matched cable has no reflection. If a reflection exists, information can be gained about the type and position of the termination.

Figure 23.14
Optical time-domain reflectometry
Photo courtesy of Hewlett-Packard Company.

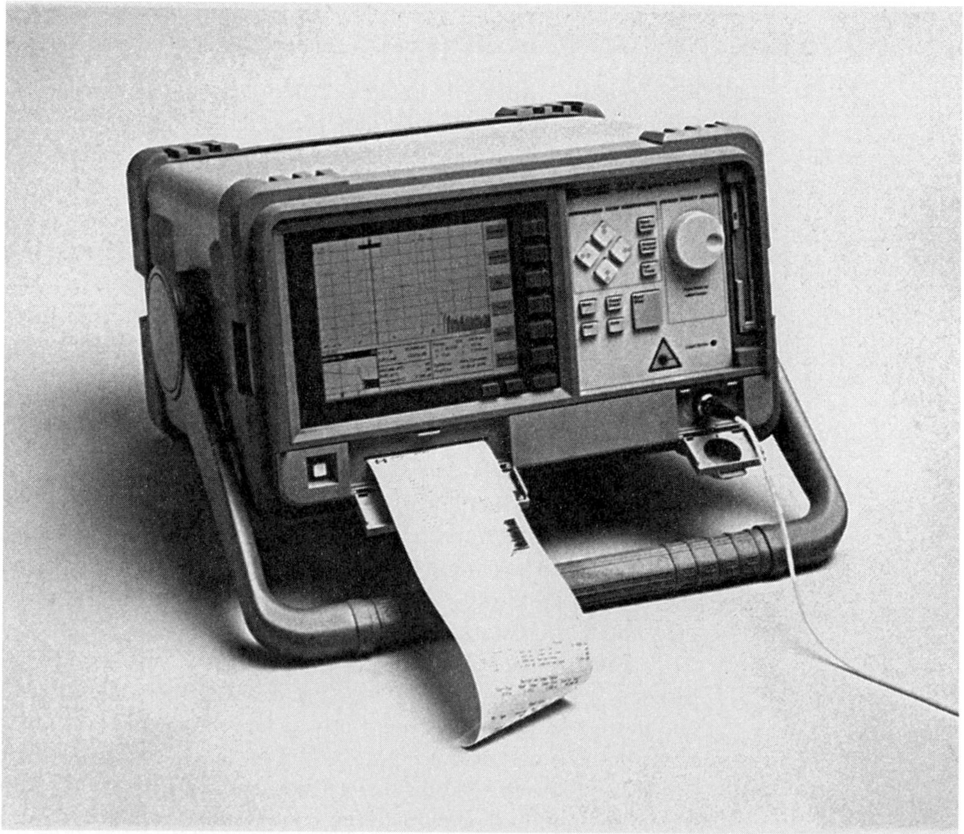

(a) Equipment

Figure 23.14
Continued

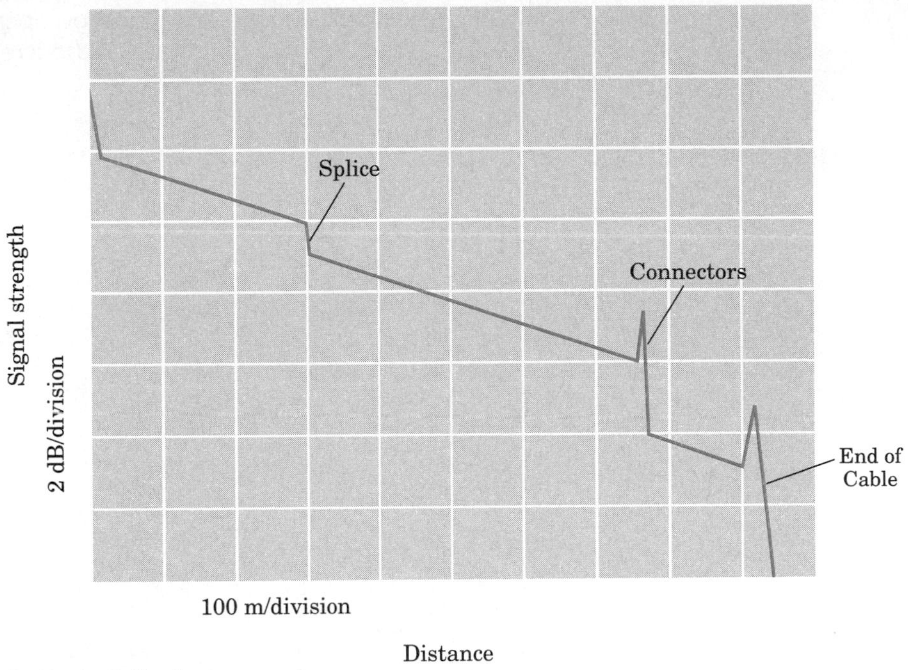

(b) Typical display

Optical time-domain reflectometry is similar to its electrical equivalent, but there are important differences. Because of back-scattering in the fiber, there is always a reflected signal. As a pulse propagates down the fiber, the reflected signal gradually becomes weaker due to attenuation: the signal strength is reduced as the distance increases on both the forward and return legs.

Figure 23.14 shows a typical optical time-domain reflectometry (OTDR) setup. Figure 23.14(a) shows the equipment, and Figure 23.14(b) is a typical display. Note the sloping line representing scattering. Also note that losses due to splices and connectors are clearly visible. The position of the loss can be found from the time the signal takes to travel to the loss and back, just as in a conventional TDR setup. The amount of the loss can be determined from the amount by which the scattering level drops at the lossy element.

Summary

Here are the main points to remember from this chapter.

1. The maximum distance that can be used with a fiber link can be limited by either losses or dispersion. Limitations due to losses are expressed in a loss budget, while dispersion is expressed in terms of either gain-bandwidth product or rise time.
2. Regenerative repeaters can be used in digital systems to extend the communication distance indefinitely.
3. Optical amplifiers can amplify an optical signal without converting it into electrical form.
4. Wavelength-division multiplexing can be used in conjunction with time-division multiplexing to increase the amount of data that can be carried on a fiber.
5. Fiber optics has greatly increased both the capacity and the reliability of submarine cables.
6. The SONET data standard unites North American and European digital hierarchies in one international high-speed fiber-optic data standard.
7. Fiber optics are increasingly being used in telephony and cable television, though it may be some time before fiber comes right to the home.

8. The use of solitons and heterodyne reception may allow fiber optics to operate with much higher data rates and over longer distances without repeaters than any other system now in use.

Important Equations

$$\tau = Dl \tag{23.1}$$

$$T_{RT} = \sqrt{T^2_{Rtx} + T^2_{Rrx} + T^2_{Rf}} \tag{23.2}$$

$$f_b = \frac{1}{2T_{RT}} \tag{23.3}$$

$$f_b = \frac{1}{T_{RT}} \tag{23.4}$$

$$B = \frac{1}{2\Delta t} \tag{23.6}$$

$$Bl = \frac{500}{D} \tag{23.7}$$

Glossary

bit stuffing synchronization of digital signals by adding extra bits to one signal

fiber distributed data interface (FDDI) a 100 Mb/s LAN signaling standard intended for use with fiber optics but also used with coaxial cable

fiber in the loop (FITL) use of fiber-optic cable for telephone subscriber connections

fiber to the curb (FTTC) use of fiber for all of a telephone system except for the subscriber loop

loss budget calculation of received power in a system in order to compare it with the power required for satisfactory performance

pump laser a laser used as an energy source to excite electrons into a higher energy state

rise time the time required for the voltage level at the beginning of a pulse to increase from 10% to 90% of its maximum value

solitons single pulses that can travel through a medium with no dispersion

synchronous digital hierarchy (SDH) the European standard for synchronous transmission over a fiber-optic network

synchronous optical network (SONET) the North American standard for synchronous transmission over a fiber-optic network

wavelength-division multiplexing (WDM) the use of two or more light sources at different wavelengths, each separately modulated, with the same fiber

Questions

1. What types of emitters, detectors, and fiber are most common for short-range fiber systems?
2. What types of emitters, detectors, and fibers are most common in long-distance, high-data-rate systems?
3. What are the most commonly used wavelengths for short- and long-range fiber-optic links?
4. What is meant by a loss budget for a fiber-optic system?
5. What can be done to improve a system that does not have sufficient power at the receiver?
6. What is system margin? Why is it necessary?
7. Why do even single-mode fibers have some dispersion?
8. Explain the relationship between bandwidth-distance product and rise time.
9. Explain why the permissible amount of pulse stretching is different for RZ and NRZ codes.
10. What advantage does a regenerative repeater have over an amplifier as a means of extending the length of a fiber-optics system?
11. Describe the construction and operation of an erbium-doped fiber amplifier.
12. Does a fiber amplifier have any advantages over a regenerative repeater? If so, state them.
13. How can full-duplex communication be implemented on a single optical fiber?
14. Compare copper and fiber optics as media for submarine communications cables.

15. Why is the SONET system better adapted to optical fiber than the T-carrier system?
16. How does SONET deal with problems in synchronization among the systems with which it must interface?
17. How is a SONET STS-3 signal constructed from STS-1 signals?
18. Why are fiber-optic cables more popular in telephone trunk applications than for local loops?
19. How can the equivalent of a superheterodyne receiver be implemented in an optical communication system?
20. Discuss the advantages and disadvantages of the use of fiber optics in local-area networks.
21. Which network topology is used by the FDDI standard? Why?
22. What is the standard data rate for FDDI?
23. What other medium, in addition to optical fiber, can be used with FDDI?
24. Distinguish between FITL and FTTC, and comment on the relative practicality of each.
25. Why do cable television systems still use analog signals with their fiber-optic trunk lines?
26. What are solitons? How does their use improve the performance of a fiber-optic system?
27. Describe the fiber-optic equivalent of a superheterodyne receiver.

Problems

SECTION 23.2

28. A communications link has a length of 50 km. The transmitter power output is 3 mW, and the losses are as follows:

    Connector loss (total): 5 dB
    Splice loss: 0.3 dB per splice (splices are 2 km apart)
    Fiber loss: 1.5 dB/km

    Calculate the power level at the receiver, in dBm.
29. A laser diode emits a power of 1 mW. It is to be used in a fiber-optic system with a receiver that requires a power of at least 1 μW for the required bit error rate. Determine whether the system will work over a 10 km distance. Assume that it will be necessary to have a splice every 2 km of cable. The losses in the system are as follows:

    Coupling and connector losses, transmitter to cable: 10 dB
    Cable loss: 0.5 dB per km
    Splice loss: 0.2 dB per splice
    Connector loss between cable and receiver: 2 dB
30. A fiber-optic cable has a bandwidth-distance product of 500 MHz-km. What bandwidth can be used with a cable that runs 50 km between repeaters?
31. A fiber has a loss of 0.5 dB/km and a bandwidth-distance product of 1 GHz-km. Which of these specifications limits the distance over which a signal with a bandwidth of 50 MHz can be transmitted, with a loss of 20 dB or less?
32. An optical communications link is to be built using fiber rated for 5000 MHz-km. The light source is a laser diode producing 0.25 mW. The fiber has losses of 0.4 dB/km and is available in 2 km lengths. It can be spliced with a loss of 0.2 dB per splice, and there will be a 5 dB loss in the connectors throughout the system. The receiver used has a sensitivity of −30 dBm. The source rise time is 2 ns, and the receiver rise time is 4 ns.
    (a) Prepare a loss budget for a link 20 km long, and calculate the system margin in dB.
    (b) What would be the maximum bandwidth that could be used with the system in (a)?
33. A fiber-optic cable has a rise time of 2 ns/km and a length of 40 km. It is used with a receiver with a rise time of 20 ns and a transmitter with a rise time of 50 ns. Calculate the total rise time and bandwidth of the system.
34. Over what distance could a signal be transmitted using the cable in Problem 33 and a data rate of 100 Mb/s with RZ code? Neglect the transmitter and receiver rise times, and consider only the cable.
35. What is the maximum total system rise time that would be acceptable for a 10 Mb/s signal using NRZ code?
36. Calculate the gain-bandwidth product for a cable with a rise time of 0.3 ns/km.

SECTION 23.6

37. Sketch one frame of a SONET STS-3 signal.
38. Approximately how many telephone calls could be carried by an STM-16 signal?

SECTION 23.9

39. Draw a block diagram of a cable television system, indicating areas that are likely candidates for conversion to fiber optics.

SECTION 23.10

40. Draw a block diagram showing how an optical heterodyne system can be implemented.

24 Cellular Radio and Personal Communication Systems

Objectives After studying this chapter, you should be able to:

1. Explain the concept of a personal communication system
2. Describe the ways in which personal communication systems have been attempted in the past
3. Describe the operation and limitations of citizens band radio and cordless telephones and explain how experience with these systems has been used in the development of more modern techniques
4. Describe the IMTS system for trunked mobile telephony and explain how its deficiencies led to the creation of cellular radio systems
5. Show how the use of cellular technology can make more efficient use of the radio spectrum than is achieved with conventional trunked radio systems and explain how the use of microcells can improve spectrum utilization still further
6. Describe the AMPS cellular radio system used in North America and compare it with newer digital standards
7. Describe and compare the two most common systems for implementing spread-spectrum radio and explain the advantages and disadvantages of spread-spectrum as compared with narrowband techniques
8. Explain code-division multiple access and compare it with frequency-division and time-division systems
9. Describe the use of spread-spectrum communications in wireless local-area networks, wireless modems, and personal communications systems
10. Explain the differences between cellular telephones and the newly emerging personal communication systems

24.1 Introduction

Most of this book deals with communication between stations rather than directly between people. This concept has been so much a part of electronic communications for so long that it is taken for granted—we are used to having a home phone and an office phone, each with its own number.

Portable and mobile transceivers have been used from almost the earliest days of radio (one of the first practical uses of radio was for ship-to-ship and ship-to-shore communication). Mobile and portable radio communication systems have become more complex and sophisticated as time goes on. Until recently, however, most of these techniques have been separate from the general communications infrastructure, communicating with the landline telephone system, if at all, through primitive gateways. Like fixed stations, mobile and portable telephones have been identified as stations, not people. The newer forms of communication, such as cellular telephones, are more integrated with terrestrial communication, but they are still organized more by location than by person—in addition to our home and office phone numbers, we now need a third number for the car phone.

Some people feel that a shift in this way of communicating is just beginning to take place. In the future, they speculate, people, not places, will have telephone numbers. A call to your number will reach you wherever you are. If you are at home, you may answer a more-or-less conventional telephone, or you may use a pocket-sized cordless phone. If you are at work, the system will find you there, perhaps using the same portable phone, but accessed from a base station in your office building. If you are out, your portable phone will be contacted by the system. The portable phone may receive signals by radio from a transmitter in the same building (in a shopping mall, for instance), from a cellular system (in a metropolitan area or along a major highway), or even by satellite (in rural areas or at sea, for example).

We are still a long way from this situation, and no one really knows when, or if, this model of telecommunications will become practical for the majority of people.

In the meantime, we do have a wide variety of systems that can be loosely grouped under the heading of personal communication systems, that is, systems designed to give mobility to their users, freeing them from dependence on permanently installed facilities. Personal communication systems can be as simple as a cordless phone or as complex as a satellite communication system involving dozens of low-earth-orbit satellites. In this chapter we will sample some of the developments in personal communication systems (PCS) and try to get a sense of the directions in which this relatively new and somewhat confusing field may be headed.

Part of the confusion surrounding personal communication systems derives from the fact that they have developed from different sources and in different ways. Cordless phones, for instance, were intended as very short-range extensions of a single, already installed telephone connection. The technology was quite primitive in the beginning and is still relatively unsophisticated. Cellular radio, on the other hand, developed from earlier mobile telephone systems as an extension of the telephone network itself, rather than of one node of that network. It is not surprising, therefore, that attempts to marry the two technologies are awkward and hard to standardize.

24.2 Early PCS

Before leaping into the somewhat confusing world of PCS, it will be instructive to look at systems that have existed for some time in order to see what works and what should be improved. To some extent, all personal communication systems are outgrowths of previous systems.

24.2.1 Citizens Band Radio

Citizens band radio is probably the earliest true personal communication system. Introduced in the United States in the 1960s, it enjoyed great popularity in the 1970s, followed by an almost equally steep decline as its limitations became better known.

CB radio was intended to do some of the same things that are envisaged by more recent personal communication systems. In fact, it can be thought of as an early form of PCS. Its relatively low frequency, 27 MHz, made transceivers affordable when CB radio was introduced, and the absence of any test for a license made it easy for anyone to get involved. The power limit of four watts for full-carrier AM or twelve watts PEP for SSB is designed to reduce interference by restricting the communications range, which should not be a problem since CB radio is intended for local communications. The fact that most CB operation is between mobile units with low antenna heights and no repeaters also limits the effective range. Selection among the 40 available channels is made by the operator, who is supposed to listen and manually switch to a clear channel before transmitting. Since anyone can listen in on any channel, there is no privacy.

Example 24.1

Two drivers communicate with each other via CB radio at 27 MHz. Each has a vertical antenna with its center 1.5 m above level ground, and each uses 4 W of carrier power output. The antennas are loaded verticals with a gain of 1 dBi. Ignoring feedline losses, calculate

(a) the maximum range for communications
(b) the signal strength in microvolts delivered to the receiver at that distance (assume the receiver has a 50Ω input impedance)

Solution

(a) Recall from basic propagation theory (Chapter 17) that for line-of-sight propagation,

$$d = \sqrt{17h_T} + \sqrt{17h_R} \tag{24.1}$$

where

$$\begin{aligned} d &= \text{maximum distance in kilometers} \\ h_T &= \text{height of the transmitting antenna in meters} \\ h_R &= \text{height of the receiving antenna in meters} \end{aligned}$$

It follows that the maximum distance for communication is

$$\begin{aligned} d &= \sqrt{17h_T} + \sqrt{17h_R} \\ &= \sqrt{17 \times 1.5} + \sqrt{17 \times 1.5} \\ &= 10.1 \text{ km} \end{aligned}$$

(b) Recall that the signal loss for free-space propagation is given by

$$L_{fs} = 32.44 + 20 \log d + 20 \log f - G_T - G_R \tag{24.2}$$

where

$$\begin{aligned} L_{fs} &= \text{free-space loss in decibels} \\ d &= \text{distance in km} \\ G_T &= \text{transmitting antenna gain in dBi} \\ G_R &= \text{receiving antenna gain in dBi} \end{aligned}$$

In this case, G_T and G_R are each 1 dBi, and d is 10.1 km. This gives a path loss of

$$\begin{aligned} L_{fs} &= 32.44 + 20 \log d + 20 \log f - G_T - G_R \\ &= 32.44 + 20 \log 10.1 + 20 \log 27 - 1 - 1 \\ &= 79.2 \text{ dB} \end{aligned}$$

The transmitter power is 4 watts. In dBW this is

$$\begin{aligned} P_T \text{ (dBW)} &= 10 \log 4 \\ &= 6.02 \text{ dBW} \end{aligned}$$

At the receiver, the power is

$$\begin{aligned} P_R \text{ (dBW)} &= 6.02 \text{ dBW} - 79.2 \text{ dB} \\ &= -73.18 \text{ dBW} \end{aligned}$$

In watts, the power will be

$$\begin{aligned} P_R &= \text{antilog } (-73.18/10) \\ &= 48.1 \text{ nW} \end{aligned}$$

The corresponding input voltage at the receiver can now be found from

$$P = \frac{V^2}{R}$$

$$V = \sqrt{PR}$$

$$= \sqrt{48.1 \text{ nW} \times 50 \ \Omega}$$

$$= 1.55 \text{ mV}$$

This is quite a strong signal, so in this case, the limiting factor is the height of the antennas.

What happened to CB radio is instructive. There was a tremendous boom in CB for a few years, resulting in major interference problems. The fact that 27 MHz transmissions are susceptible to ionospheric propagation, sometimes over distances of hundreds or thousands of kilometers, certainly did nothing to alleviate the interference problem. Indeed, some people make a hobby of long-distance CB communication, often using higher-than-legal transmitter power levels.

If CB radio were begun today, no doubt a higher frequency would be used. Current technology would allow the construction of low-cost transceivers operating in the VHF range. In fact, there are frequency ranges available in the 46 and 49 MHz regions, but they are currently used for low-power services like cordless phones, which will be described shortly.

Partly as a result of interference levels that often made communication impossible, and partly because CB radio is inherently a short-range system, lacking radio repeaters or any convenient access to the telephone system, CB use went into a sharp decline. Ironically, there is at present very little problem with interference at most times and at most locations, yet the service remains underused. In many areas it is used mainly for a few specialized applications, such as for conversations between truck drivers on the highway.

Some of the problems with CB radio are caused by the same features that gave it its initial cost advantage: the relatively low frequency and very informal structure. Efficient antennas for 27 MHz are relatively large—too large for convenient use with handheld portable transceivers. The power level needed for reliable communication over a few kilometers, without repeaters, is a watt or more, and some handheld transceivers have up to four watts output. These power levels require handheld CB transceivers to have relatively large and heavy batteries. See Figure 24.1 for a typical example. Even so, portable or mobile communication is limited in range and unreliable at best in the absence of repeaters.

It appears, then, that to avoid the fate of CB radio, a PCS should have some form of automatic frequency management to avoid interference and provide a measure of privacy, should be usable with small handheld transceivers and small antennas, and should connect to the telephone system. In addition, the use of radio repeaters would improve reliability and reduce interference.

24.2.2 Cordless Telephones

Early cordless phones were intended as simple wireless extensions to ordinary telephone service. For best results, a telephone, cordless or otherwise, should operate in full-duplex mode, that is, it should be capable of transmitting and receiving at the same time. Thus a cordless phone needs two radio channels, separated widely enough in frequency to avoid interference between them. Early designs (and some of the cheaper models available today) had only a single channel for each direction, so the transmitter power levels and range had to be kept very small to minimize interference. Nonetheless, consumers found that a telephone that could be carried freely throughout a house and its grounds was very useful, and cordless phones have been very popular. A typical modern example is shown in Figure 24.2.

Current models typically use FM with either 10 or 25 selectable pairs of channels to minimize interference, though single-channel models are still available. Most current designs use 10 channels with frequencies between 46.61 and 46.97 MHz for the base-unit

Figure 24.1 Handheld CB transceiver

Courtesy of Radio Shack, a division of Tandy Corporation.

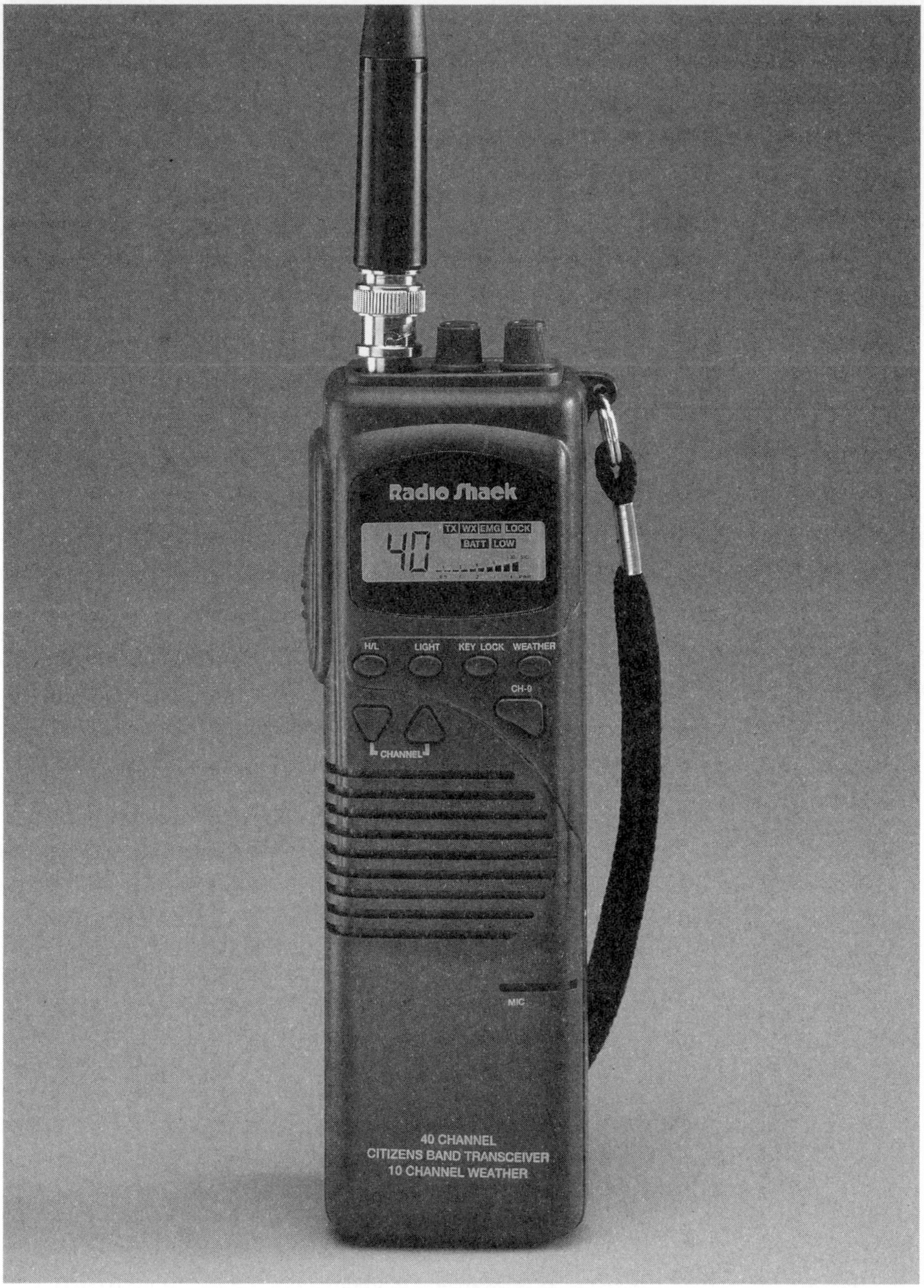

transmitter and frequencies between 49.67 and 49.97 MHz for the handset. Recently, 15 more channel pairs have been added between 43.72 MHz and 44.48 MHz for the base and between 48.76 and 49.5 MHz for the handset. Many newer cordless phones can use any of the 25 channel pairs. These new designs are also capable of scanning the channels to find a clear frequency before transmitting. A few of the most recent (and most expensive) designs operate in the frequency range between 902 and 928 MHz, and older units use AM at just under 2 MHz (carried by the power line) for the base unit's transmitter.

Cordless telephones share much of the simplicity of CB radio. There are no license

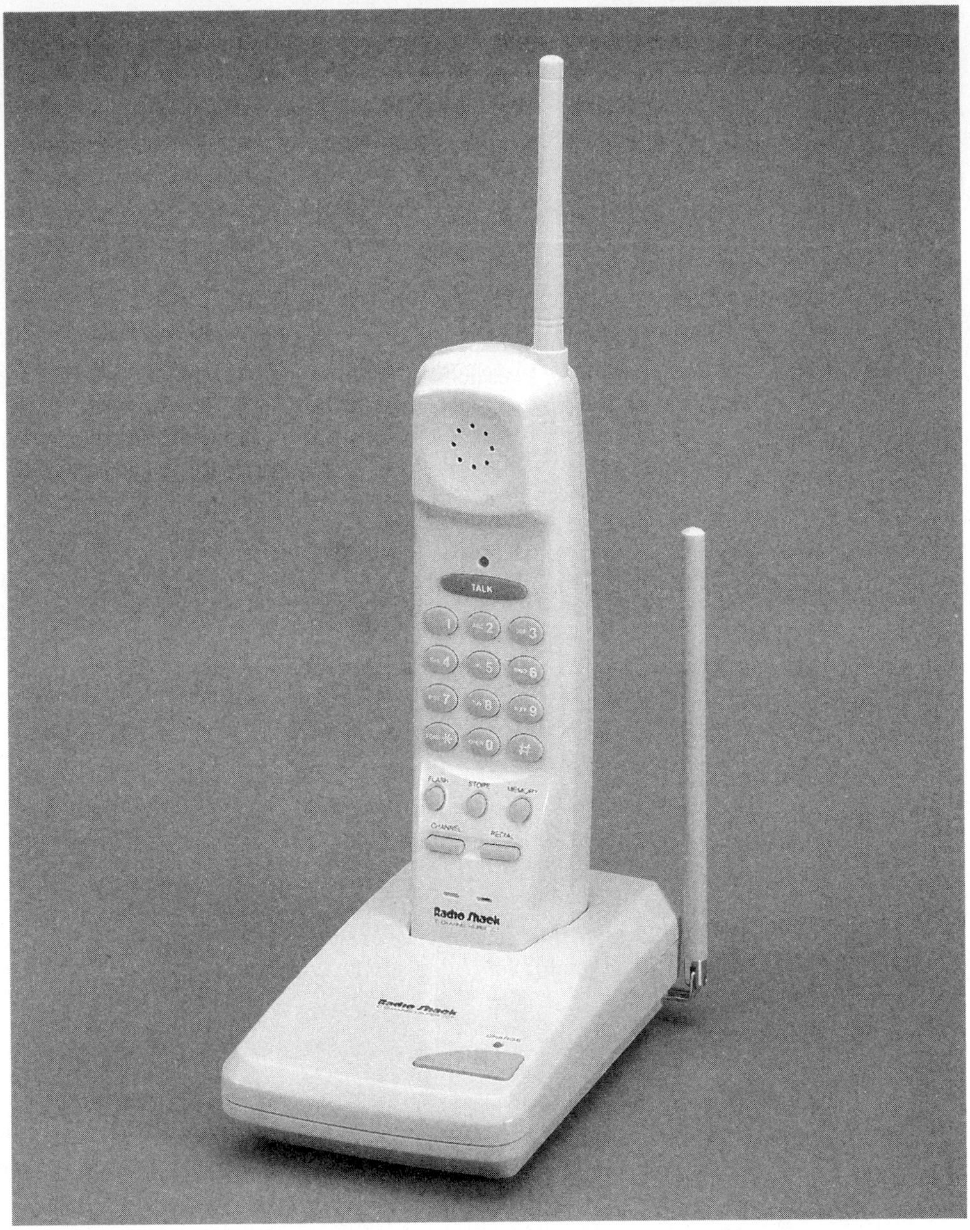

Figure 24.2
Cordless phone
Courtesy of Radio Shack, a division of Tandy Corporation.

requirements, and there is no official coordination of frequencies. Users simply try to choose a channel that is not used by their neighbors. Many of the newer cordless phones use digital access codes to prevent unauthorized persons from dialing the phone and possibly making toll calls, but it is still not possible to use two nearby phones on the same channel at the same time. The use of FM does provide some protection from interference: due to the capture effect, the desired signal has only to be a few decibels stronger than the interfering signal in order to reduce interference to a reasonable level. As with CB radio, privacy is almost nonexistent.

Example 24.2

A cordless phone is in use 50 m from its base station. How much farther away does a second, interfering phone (with the same transmitter power) have to be to have a signal 3 dB weaker at the receiver input?

Solution An interfering signal that is 3 dB weaker has half the power density as the desired signal at the receiving antenna. We know that in free space, power density obeys the square-law equation:

$$P_D = \frac{P_T}{4\pi r^2} \tag{24.3}$$

In the cluttered home environment, with many surfaces and structures to reflect and absorb radio waves, this attenuation estimate is very conservative—the signal strength may actually fall off as the fourth power of distance in such environments. However, let us use the conservative estimate to do this example.

We do not have to know the transmitter power as long as the transmitters have equal power. Let the power density of the desired transmission be P_{D1} and that due to the interference be P_{D2}. Then the ratio of their signal strengths is

$$\frac{P_{D1}}{P_{D2}} = \frac{\dfrac{P_T}{4\pi r^2{}_1}}{\dfrac{P_T}{4\pi r^2{}_2}}$$

$$= \frac{P_T}{4\pi r^2{}_1} \times \frac{4\pi r^2{}_2}{P_T} = 2$$

$$\frac{r^2{}_2}{r^2{}_1} = 2$$

$$r^2{}_2 = 2r^2{}_1$$

$$r_2 = \sqrt{2}r_1$$

$$= \sqrt{2} \times 50 \text{ m}$$

$$= 70.7 \text{ m}$$

Despite their interference problems and severely limited range, cordless phones have been (and remain) very popular with consumers.

24.3 Cellular Radio

Strictly speaking, cellular radio is not a "personal" communication system—it is basically designed for mobile use, and the telephone number is still associated with a station rather than a person. Still, it is a very popular system, important in its own right, and it provides the model for many newer ideas in PCS.

24.3.1 Trunked Radio-telephone Systems

The familiar cellular radiotelephone system has its origins in much earlier systems that used a few widely spaced repeaters with powerful transmitters and antennas mounted as high as possible to obtain wide coverage. The mobile transceivers likewise used relatively high power, on the order of 30 watts. Very similar systems are still widely used in dispatching systems, such as those for taxicabs and ambulances. Figure 24.3 illustrates a system using a central repeater.

The most common type of mobile telephone system, from its introduction in the mid-1960s until the coming of cellular radio in the early 1980s (the first commercial cellular system became operational in Chicago in 1983), was known as improved mobile telephone service (IMTS). IMTS is a **trunked system**, that is, radio channels are assigned to mobile users by the system as needed, rather than having one channel, or pair of channels, permanently associated with each user. Narrowband FM technology is used.

IMTS is capable of assigning channels automatically, by the rather simple means of transmitting a tone from the base station on unoccupied channels. The receiver in the mobile unit scans channels until it detects the tone.

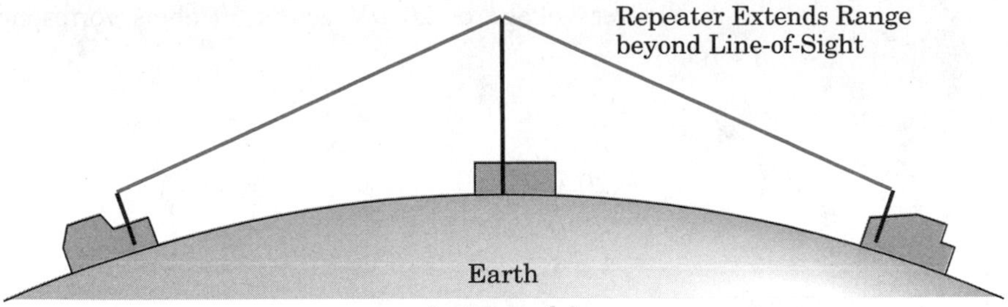

Figure 24.3 Mobile telephone system with central repeater

IMTS is capable of full-duplex operation using two channels per telephone call. Direct dialing is also possible, so that using a mobile phone is almost as simple as using an ordinary telephone at home.

The main problem with IMTS and similar systems is that whatever bandwidth is made available to a single repeater is tied up for a radius of perhaps 50 km or even more, depending on the height of the antenna and the power of the transmitter at the base station. Any attempt to reuse frequencies within this radius is likely to result in harmful interference. Simple systems like this also suffer from fading and interference near the edges of their coverage areas. For instance, suppose two similar trunked systems with identical repeaters are located 50 km apart. Then, at a location midway between the two, a receiver would receive equally strong signals from each. Communication would be impossible if the two repeaters used the same frequencies.

As the demand for mobile telephony grew, it became obvious that another way had to be found to accommodate more users without greatly increasing the spectrum allocation.

Example 24.3 Suppose an IMTS system has to communicate over a radius of 50 km.

(a) How high would its antenna have to be, assuming level terrain and communication with automobiles whose antennas are at a height of 1.5 m? Assume a frequency of 900 MHz.

(b) What EIRP would be required at the repeater transmitter if the signal has to have a strength of 20 μV at the mobile receiver? As before, assume a receiver antenna gain of 1 dBi.

Solution (a) We can use Equation (24.1) again. This time we know h_R and d, so we rearrange the equation in terms of h_T.

$$\sqrt{17h_T} = d - \sqrt{17h_R}$$

$$17h_T = (d - \sqrt{17h_R})^2$$

$$h_T = \frac{(d - \sqrt{17h_R})^2}{17}$$

$$= \frac{(50 - \sqrt{17 \times 1.5})^2}{17}$$

$$= 119 \text{ m}$$

(b) Since we are looking for EIRP, we can neglect the gain of the transmitting antenna. We will modify Equation 24.2 by removing G_T and find the path loss. Then we can find the power that corresponds to a signal of 20 μV at the receiver input. If we find that power in dBW and add the path loss in dB, we will get the required power in dBW. We can convert this to watts to get our answer.

A receiver voltage of 20 μV across 50 ohms corresponds to a receiver input power of

$$P = \frac{V^2}{R} = \frac{(20\ \mu V)^2}{50\ \Omega} = 8\ \text{pW}$$

In dBW, this is

$$P_R(\text{dBW}) = 10 \log (8\ \text{pW}) = -111\ \text{dBW}$$

The path loss can be found from Equation 24.2:

$$\begin{aligned} L_{fs} &= 32.44 + 20 \log d + 20 \log f - G_T - G_R \\ &= 32.44 + 20 \log 50 + 20 \log 900 - 1 \\ &= 124.5\ \text{dB} \end{aligned}$$

Now, the transmitter EIRP in dBW will be

$$P_T\ (\text{dBW}) = -111 + 124.5 = 13.5\ \text{dBW}$$

Convert to watts and we're finished:

$$P_T = \text{antilog}\ (13.5/10) = 22.4\ \text{W}$$

24.3.2 Basic Cellular Concepts

Cellular radio goes a long way toward relieving the congestion described above by essentially reversing the conventional wisdom about radio systems using repeaters. Instead of trying to achieve long range by using high power, cellular repeaters are deliberately restricted in range by using low power. A reasonable elevation is still required to minimize radio shadows (behind buildings, for instance). Similarly, the mobile radios use low power (no more than 3 W for mobiles and 700 mW or less for portable phones), and in fact, the transmitter power is automatically limited by the system to the minimum required for reliable communication. Furthermore, instead of one repeater, there are many, located in a grid pattern like that shown in Figure 24.4. Each repeater is responsible for coverage in a small cell. The cells are shown as hexagons, but of course in a real situation the antenna patterns do not achieve this precision—the actual cells are more likely to be approximately circular, with some overlap.

Since each transmitter operates at low power, it is possible to reuse frequencies over relatively short distances. Typically a repeating pattern of either 12 or 7 cells is used, and

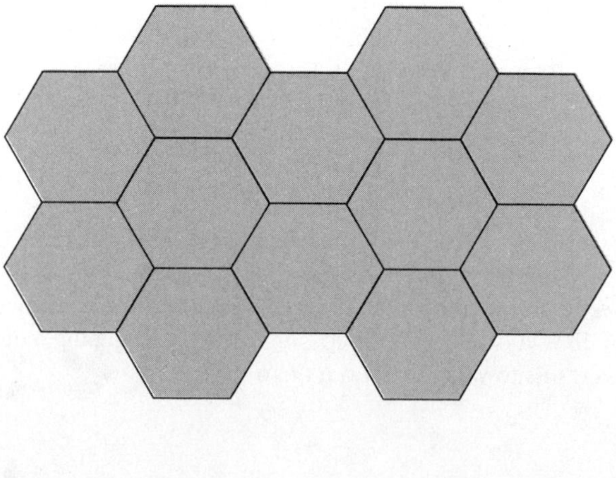

Figure 24.4
Basic cellular pattern

Figure 24.5

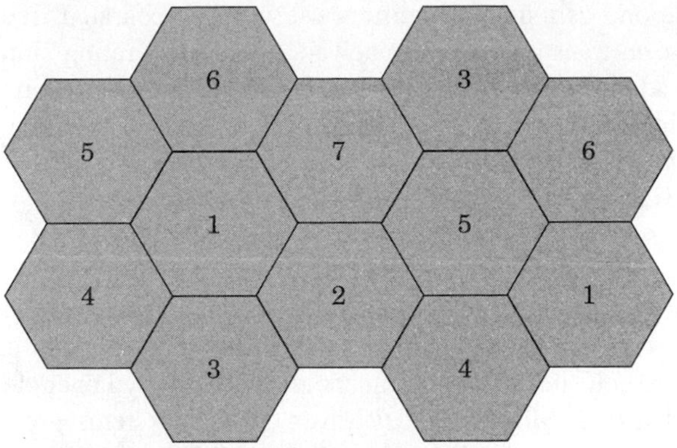

(a) Seven-cell repeating pattern

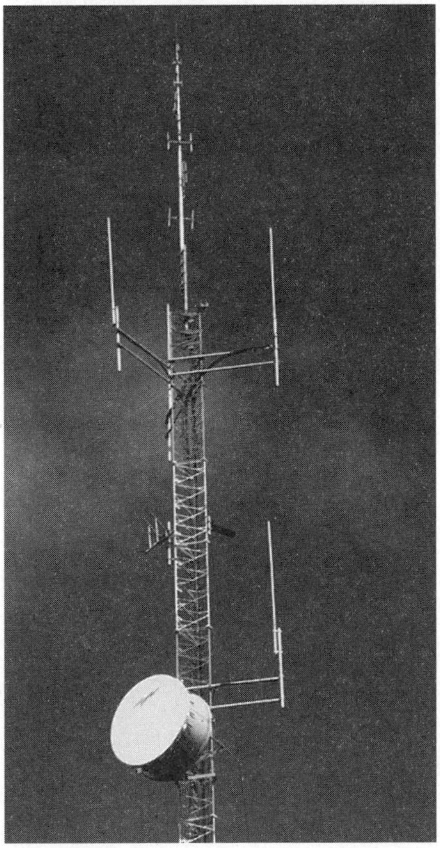

(b) Cell site
Courtesy Bell Mobility

the available bandwidth is divided among these cells. The frequencies can then be reused in the next pattern. A pattern of 12 cells is needed when the cell site uses a single omnidirectional antenna in the center of the cell. Alternatively, if three antennas are used, each spanning an angle of 120°, the number of cells in a pattern can be reduced to 7, with different channels being assigned to each of the three antennas in a cell. In the current North American system, there are 395 duplex voice channels consisting of a channel in each direction. Therefore each cell can have an average of 57 channels in each direction. See Figure 24.5 for an example of how the cells can be arranged and a photograph of a typical antenna installation.

Mobile and cell-site transmitters use widely separated frequencies to avoid interference between transmitters and receivers that are operating simultaneously at the same location. In addition, there are two distinct frequency allocations to allow for two independent cellular systems in any given locality. Each carrier also needs frequencies for setting up calls. The whole system uses 832 channels in two frequency ranges, from 869–894 MHz (receive) and 824–849 MHz (transmit).

With a seven-cell repeating pattern (as in Figure 24.5), frequencies can be reused at a distance of 4.6 times the cell radius. It can be shown that this distance gives a minimum signal-to-interference ratio of about 18 dB, which gives acceptable results when narrowband FM is used. As the number of users increases, cell sizes can be made smaller by installing more cell sites and the frequencies can be reused over closer intervals. This technique is called **cell-splitting**, and it gives cellular systems great flexibility to adapt to changes in demand over both space and time. Cells typically range from one to several kilometers in radius, depending on the traffic density. As traffic increases in a given area, cell-splitting allows the system capacity to increase along with it.

Let's see how cell-splitting works. The number of cells required for a given area is given by:

$$N = \frac{A}{a} \tag{24.4}$$

where

N = number of cells
A = total area to be covered
a = area of one cell

This equation assumes that all cells are of equal size and that there is no overlap between cells. Nevertheless, it serves as a reasonably good estimate.

If we assume that the cells are circular, then the area of an individual cell is given by

$$a = \pi r^2 \tag{24.5}$$

where

a = area of one cell
r = radius of a cell

The number of cells is then approximately

$$N = \frac{A}{\pi r^2} \tag{24.6}$$

Equation (24.6) is an approximation because there will have to be some overlap between cells, since the cells are assumed to be circular. If hexagonal cells are used, there need be no overlap, at least in theory. (In practice, the cells are not perfectly hexagonal.) The area of a regular hexagon (one in which all six sides are equal) is given by the equation

$$A = 3.464r^2 \tag{24.7}$$

where

A = the area of the hexagon
r = the radius of a circle inscribed in the hexagon, as in Figure 24.6

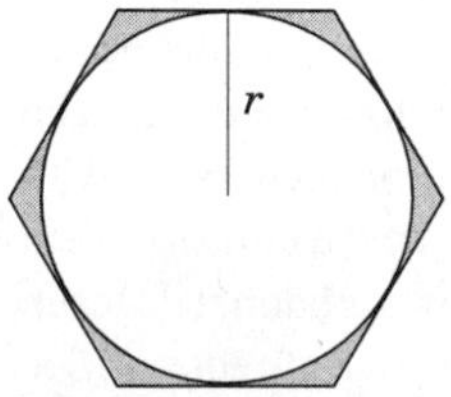

Figure 24.6
Area of hexagon

Using hexagonal cells slightly changes the calculation in Equation 24.6. With hexagonal cells, the required number of cells is:

$$N = \frac{A}{a} = \frac{A}{3.464r^2} \tag{24.8}$$

In either case, however, the number of cells is inversely proportional to the square of the cell radius. Reducing the cell radius by a factor of two increases the number of cells and therefore the system capacity by a factor of four. Unfortunately, reducing the cell radius by a factor of two also increases the number of cell sites required by a factor of four. As cell sizes are reduced, the available spectrum is used more efficiently, but cell-site equipment is not, since the maximum number of users per cell remains constant.

Example 24.4 A metropolitan area of 1000 square km is to be covered by cells with a radius of 2 km. Find the number of cell sites that would be required assuming

(a) circular cells
(b) hexagonal cells

Solution (a) From Equation (24.8),

$$N = \frac{A}{\pi r^2} = \frac{1000}{\pi \times 2^2} = 80 \text{ cells}$$

(b) From Equation (8.8),

$$N = \frac{A}{3.464 r^2} = \frac{1000}{3.464 \times 2^2} = 73 \text{ cells}$$

Users of earlier mobile-radio systems could use the same frequencies throughout a conversation, but since the cells in a cellular radio system are relatively small, many calls from or to moving vehicles must be transferred or *handed off* from one cell site to another as the vehicles cross cell boundaries. Handing off requires a change in frequency, since frequencies are not reused in adjacent cells. The mobile-unit frequencies are controlled from the cell site with the aid of a computer-controlled switching center variously called a mobile telecommunications switching office (MTSO) or a mobile switching center (MSC). All cell sites within a region are linked to each other and to the MTSO by telephone cables or microwave radio links that are separate from those belonging to the ordinary ("wireline") telephone system. The computer uses the cell receiving the strongest signal from the mobile unit. A separate control channel tells the mobile unit which channels and what power level to use.

The most common cellular technology in North America is known as advanced mobile phone service (AMPS). It uses narrowband analog FM with a maximum frequency deviation of 12 kHz and a channel spacing of 30 kHz for voice transmission. It can also be used for data, including fax, if modems are used at both ends. It is not as satisfactory for data as the wireline telephone system, however, because there is often a momentary interruption of service when a handoff between cell sites occurs. For this reason, data transmission is usually more successful from a stationary location than from a moving vehicle.

Because of the delays due to handoffs and the relatively high noise level, specially designed modems achieve better performance than those designed for normal landline use. Some cellular systems use modem pools equipped with these modems, which allow a cellular user with a specialized cellular modem to communicate with another user with a standard modem. The modems in the pool translate between cellular and landline standards.

24.3.3 Microcells

As we have seen, cell-splitting can be used to increase the capacity of a cellular system. At a certain point, however, diminishing returns set in. Cell sites are very expensive (on the order of a million dollars), and the increase in capacity does not justify the increase in cost for very small cells. Another problem is that real-estate costs are highest in the areas where demand is greatest, and the use of small cells means less choice in cell-site location and thus higher costs for access to cell sites.

In high-demand areas, **microcells** have been used to help relieve the congestion at relatively low cost. A microcell site is a very small unit that can mount on a streetlight pole

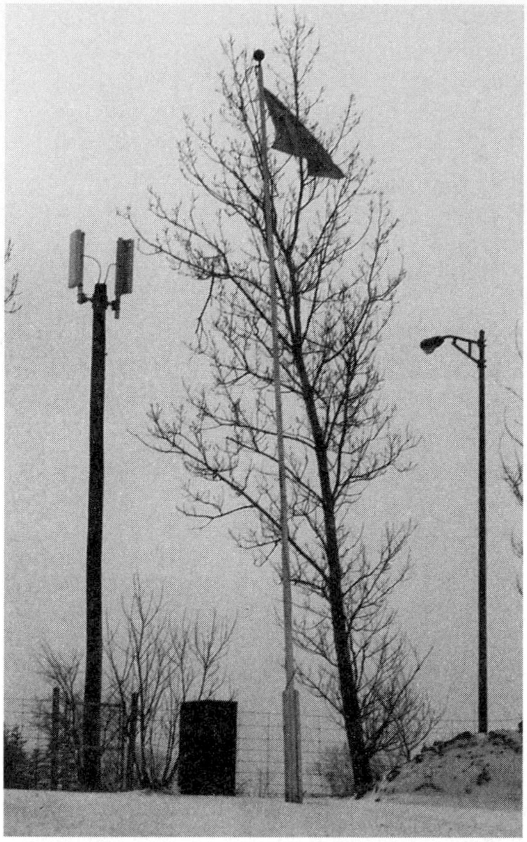

Figure 24.7
Microcell site
Courtesy Bell Mobility

(see Figure 24.7). It is under the control of a conventional cell site, with which it usually communicates by microwave radio. The microcell may itself be divided into several zones. Figure 24.8 illustrates how microcells and conventional cells can be arranged to work together.

In order to save costs, many microcell sites are not true transceivers but are only amplifiers and frequency translators (see Figure 24.9). The main cell site up-converts the whole transmitted spectrum to microwave frequencies. The transmitter at the microcell site simply down-converts the block of frequencies to the cellular-radio band and amplifies it. No modulation or channel switching is required at the microcell site. Similarly, the micro-

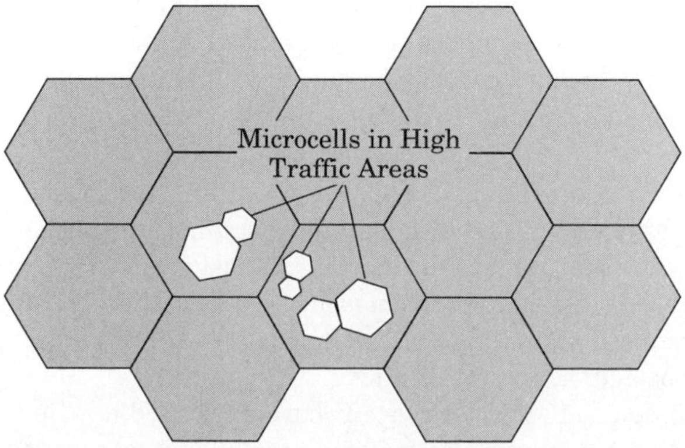

Figure 24.8
Integration of microcells and conventional cells

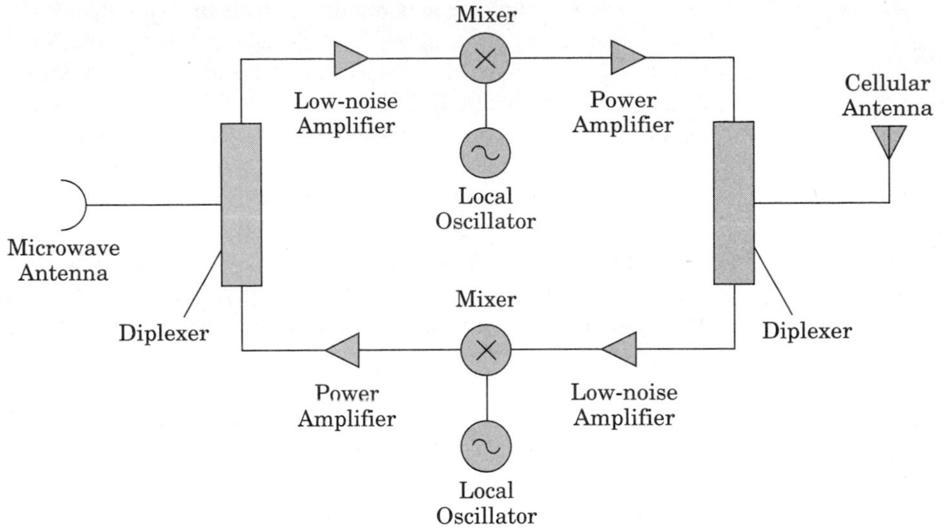

Figure 24.9 Microcell block diagram

cell's receiver consists of a low-noise amplifier that amplifies the entire frequency range in use at that site plus an up-converter to convert the signal to microwave frequencies. All demodulation is handled at the main cell site.

24.3.4 Cellular Phone Operation

The cellular radio system is designed to be as transparent to the user as possible. Thus using a cellphone is more like using an ordinary phone than it is like using most other radio communication systems. Simplicity of operation is made possible by the complex but hidden structure of the cellular phone system.

Each area in North America can be served by two companies. One, the *B* carrier, is affiliated with the local *wireline* company, that is, the company that provides ordinary local phone service. The other, the *A* carrier, is not. This system ensures a certain amount of competition, and in any given area, each company generally offers slightly different rate packages. The rates are typically composed of a fixed monthly cost plus "airtime" for the minutes the phone is actually in use plus any applicable long distance charges. Different companies offer different combinations of rates and usually offer some free airtime as part of the monthly rate. Value-added services like voice mail and call forwarding are available at extra cost. Most cellphone users subscribe to only one service, but it is possible to subscribe to both using one phone. In that case, the phone has two phone numbers, one for each service, and the desired service is chosen by the user by activating the appropriate *number assignment module (NAM)* in the phone.

In most cases, all that is necessary to make a cellular phone call is to turn on the phone, enter the number, and press the *send* key. To hang up, the *end* key is pressed. Partly because mobile use makes entering numbers difficult, cellphones tend to have many memories for often-used numbers. Answering the phone is just as simple: the user presses *send* to answer and *end* to hang up.

This apparent simplicity hides the sophistication of the system. When *send* is pressed for an outgoing call, the phone has to communicate its unique identification number, called the *electronic serial number (ESN)*, and its phone number to the system, so that the call can be billed correctly. (By the way, the electronic serial number makes the theft of cellphones rather pointless, as the phone can be deactivated as soon as the owner discovers the theft; thereafter, that phone is useless.) The system has to assign a pair of frequencies and a power level and make the connection. For an incoming call, the system locates the telephone, first checking in its local area and then farther afield if necessary, identifies it, and makes the connection. If the phone moves, the system follows it from cell to cell.

When a cellphone user ventures outside of the company's local area, he or she is said to be *roaming*. In some areas (all of Canada and part of the United States) the system will automatically find the phone if a call is made to it. This is known as National Call Delivery (Canada) or Automatic Call Delivery (US). With this system, no action is required on the part of the cellphone user to receive calls, other than turning on the phone. Some parts of the United States use follow-me roaming, which requires the user to announce his or her presence to the system daily so that incoming calls can be routed directly.

Both types of roaming tend to be costly. A cheaper method is available in many cities. Callers who know which area the roamer is in can dial a number in that city, from which the system will contact the cellphone locally, avoiding extra roaming charges. Of course this system is less convenient, as the traveler has to give his or her itinerary to anyone who is likely to call.

24.3.5 Digital Cellular Radio

In some metropolitan areas, FM cellular radio has already reached its maximum capacity. Cell-splitting has been used to its maximum practical extent, and microcells have been added. The system reaches a point of diminishing returns, however, where a further increase in the number of cells provides very little additional capacity at a great deal of additional cost. At this point, capacity can be increased only by enlarging the spectrum allotted to cellular radio or by reducing the bandwidth required by each telephone conversation.

The first of these solutions is actually more political than technical. There is considerable competition for frequencies in the 900 MHz region currently used for cellular radio. Spectrum is available in the microwave region for new applications such as mobile satellite services and short-range personal communication systems, but for the time being, there does not seem to be much chance of an increase in the frequency allocation for mobile cellular telephony.

There are several possibilities for reducing the bandwidth used by a conversation. AMPS technology uses conventional narrowband FM with a channel spacing of 30 kHz. FM is famous for trading bandwidth for signal-to-noise ratio. Moving to a more spectrum-efficient modulation scheme, such as single-sideband suppressed-carrier AM, would save a considerable amount of bandwidth. However, FM has the advantage of the capture effect, which causes a desired signal only a few decibels stronger than an interfering signal to take over the receiver. Under these conditions, the signal-to-interference ratio at the receiver output is much greater than the ratio at the input. Cellular radio, with its emphasis on frequency reuse over a small area, relies on the capture effect for its effectiveness. Moving to a single-sideband AM system would reduce the bandwidth needed per conversation but would also reduce the amount of frequency reuse and would result in no net improvement.

Another possibility would be to use **spread-spectrum** techniques, which will be described in the next section. These schemes are already in use for wireless local-area networks, and it appears at this writing that they will be very popular for short-range personal communications (for instance, within buildings and small groups of buildings). There is also a spread-spectrum standard for cellular telephony called IS-95. However, at the present time these techniques are not widely used for mobile cellular radio in North America, partly because the equipment required is radically different from that already in place.

The current North American standard for digital cellular telephony, known variously as TDMA-30 and IS-54, fits three time-division multiplexed (TDM) digital signals into each existing 30 kHz channel. Phase-shift keying with four levels is used, and special speech encoding limits the bit rate to 9.6 kb/s for each speech channel. (Compare this with the 64 kb/s used for conventional PCM telephony.) Each channel also has 3.4 kb/s allocated to error control and 3.2 kb/s to signaling, resulting in a total data rate of 16.2 kb/s per voice channel or 48.6 kb/s for the 30 kHz wide RF channel. Note that this data rate is less than 2 b/s per hertz of bandwidth. This is quite conservative, necessarily so because of the radio environment, which is subject to noise, interference, and very deep fading.

Eventually it is planned to increase the capacity of the digital system to six voice channels per 30 kHz radio channel by using vocoders to reduce the bit rate for one voice channel to 4.8 kb/s.

Because it works with a 30 kHz RF channel, the new digital standard can use much of the existing cell-site equipment, though it will be necessary to redesign mobile transceivers. At this writing, most mobile equipment is still analog, although some is capable of using both schemes. It is expected that digital and analog cellular telephony will coexist for some time, with some channels in each cell remaining analog in urban areas, as will all channels in less-busy areas.

24.4 Spread-Spectrum Radio

As radio communication systems proliferate and traffic increases, interference problems become more severe. Interference is nothing new, of course, but it has been managed reasonably successfully in the past by careful regulation of transmitter locations, frequencies, and power levels. There are some exceptions to this, however, notably CB radio and cordless telephones. In fact, wherever government regulation of frequency use is informal or nonexistent, interference is likely to become a serious problem. Widespread use of personal communication systems by millions of people obviously precludes the tight regulation associated with services such as broadcasting and makes bothersome interference almost inevitable. One approach to the interference problem, used by cellular radio systems, is to employ a complex system of frequency reuse, in which computers choose the best channel at any given time. However, this system too implies strong central control, if not by government then by one or more service providers, each with exclusive rights to certain radio channels. That is, in fact, the current situation with respect to cellular telephony, but problems can arise when several widely different services use the same frequency range. The 49 MHz band, for instance, is currently used by cordless phones, baby monitors, remote controlled models, and various other users in an almost completely unregulated way.

Another problem with channelized communication, even when tightly controlled, is that the number of channels is strictly limited. If all available channels are in use in a given cell of a cellular phone system, the next attempt to complete a call will be **blocked**, that is, the call will not go through. Service does not degrade gracefully as traffic increases; rather, it continues normally until the traffic density reaches the limits of the system, and then it ceases altogether for new calls.

Spread-spectrum communication is another way to reduce interference without requiring strong central control. This technique has been used for some time in military applications, where interference often consists of deliberate jamming of signals by an enemy. Such interference, of course, is not under the control of the communicator, nor is it subject to government regulation.

Military communication systems also need to avoid unauthorized eavesdropping on confidential transmissions, and this is another problem that can be alleviated by using spread-spectrum techniques. Privacy is also a concern for personal communication systems, but many current systems, such as cordless and cellular telephone systems, have nonexistent or very poor protection of privacy at present.

For these reasons, and also because the availability of large-scale integrated circuits has reduced the costs involved, there has recently been a great deal of interest in the use of spread-spectrum technology in personal communication systems.

The basic idea in spread-spectrum systems is, as the name implies, to spread the signal over a much wider portion of the spectrum than usual. A simple audio signal that would normally occupy only a few kilohertz of spectrum can be expanded to cover many megahertz. Thus only a small portion of the signal is likely to be masked by any interfering signal. Of course, the average power density in watts per hertz of bandwidth is also reduced, and this often results in a signal-to-noise ratio of less than one (that is, the signal power in any given frequency range is weaker than the noise power in the same bandwidth). It may seem at first glance that this would make the signal almost impossible to detect, and this is true unless special techniques are used to "despread" the signal while at the same time spreading the energy from interfering signals. In fact, the low average power density of a spread-spectrum signal is responsible for its relative immunity to both interference and eavesdropping.

Example 24.5 A cellular telephone signal normally occupies a channel 30 kHz wide. Suppose a spread-spectrum system is used to increase its bandwidth to 10 MHz. If the signal has a total signal power of −110 dBm at the receiver input and the system noise temperature referred to the same point is 500 K, calculate the signal-to-noise ratio for both systems, ignoring any gain during demodulation.

Solution Recall that thermal noise power is given by

$$P_N = kTB$$

where

P_N = noise power in watts
k = Boltzmann's constant (1.38×10^{-23} J/K)
T = absolute temperature in kelvins
B = noise power bandwidth in hertz

In general, the noise power bandwidth for a system is approximately equal to the receiver bandwidth. For the original signal with a bandwidth of 30 kHz and a noise temperature of 500 K,

$$\begin{aligned} P_N(30\text{ kHz}) &= 1.38 \times 10^{-23}\text{ J/K} \times 500\text{ K} \times 30 \times 10^3\text{ Hz} \\ &= 207 \times 10^{-18}\text{ W} \end{aligned}$$

In dBm, this is

$$P_N(30\text{ kHz}) = 10 \log \frac{207 \times 10^{-18}\text{ W}}{1 \times 10^{-3}\text{ W}} = -127\text{ dBm}$$

When the signal bandwidth increases to 10 MHz, the signal is spread over a much wider region of the spectrum, and a receiver designed to receive the whole signal bandwidth would need a bandwidth of 10 MHz as well. It would receive a noise power equal to

$$\begin{aligned} P_N(10\text{ MHz}) &= 1.38 \times 10^{-23}\text{ J/K} \times 500\text{ K} \times 10 \times 10^6\text{ Hz} \\ &= 69 \times 10^{-15}\text{ W} \\ &= -102\text{dBm} \end{aligned}$$

With both signal and noise in dBm, we can subtract to get the S/N:
For the 30 kHz bandwidth,

$$S/N = -110\text{ dBm} - (-127\text{ dBm}) = 17\text{ dB}$$

For the 10 MHz bandwidth,

$$S/N = -110\text{ dBm} - (-102\text{ dBm}) = -8\text{ dB}$$

Spread-spectrum communication is especially effective in a portable or mobile environment. Cancellation due to the reflection of signals often causes very deep fading, called *Rayleigh fading* over a narrow frequency range. When one of these fades happens to coincide with the frequency in use, the signal can be lost completely. With spread spectrum, only a small part of the communication would be lost, and this can usually be made up, at least in digital schemes, by using error-correcting codes.

Figure 24.10 Frequency-hopping transmitter

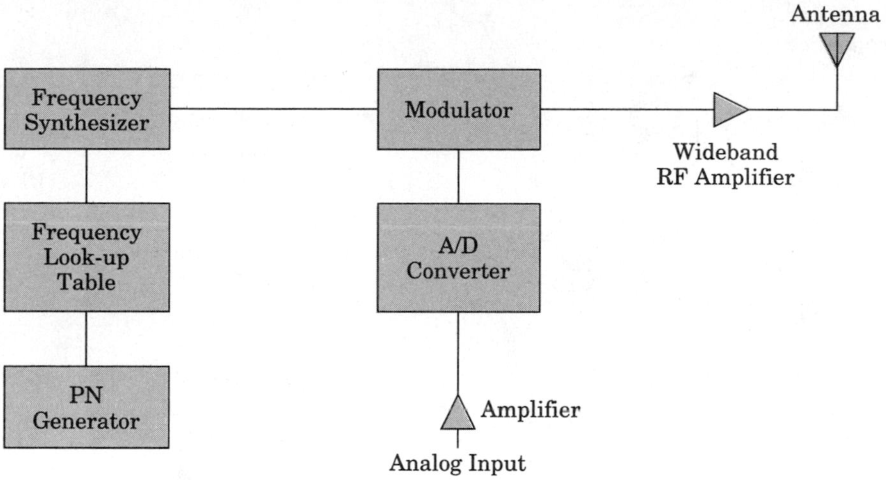

There are several ways of generating spread-spectrum signals. The most common ones are known as **frequency hopping** and **direct sequence**.

24.4.1 Frequency Hopping

Frequency hopping is perhaps the simplest spread-spectrum technique to understand. A frequency synthesizer is used to generate a carrier in the ordinary way. However, instead of operating at a fixed frequency, the synthesizer changes frequency many times per second according to a preprogrammed sequence of channels. This sequence is known as a **pseudo-random noise (PN)** sequence because to an outside observer who has not been given the sequence, the transmitted frequency appears to hop about in a completely random and unpredictable fashion. In reality, however, the sequence is not random at all, and a receiver that has been programmed with the same sequence can easily follow the transmitter as it hops and can decode the message normally.

The block diagram in Figure 24.10 shows how a frequency-hopping transmitter operates. Note the digitally-controlled synthesizer; otherwise, this is a fairly conventional transmitter design.

Since the frequency-hopping signal typically spends only a few milliseconds or less on each channel, any interference from a signal on that frequency will be of short duration. If an analog modulation scheme is used for voice, the interference will appear as a click and may well pass unnoticed. If the spread-spectrum signal is modulated using digital techniques, the system can use an error-correcting code that will allow these brief interruptions in the received signal to be ignored, and the user will probably not experience any signal degradation at all. Thus reliable communication can be achieved in spite of interference.

Example 24.6 A frequency-hopping spread-spectrum system hops to each of 100 frequencies every ten seconds. How long does it spend on each frequency?

Solution The amount of time spent on each frequency is

$$\begin{aligned} t &= 10 \text{ seconds}/100 \text{ hops} \\ &= 0.1 \text{ second per hop} \end{aligned}$$

If the frequency band used by the spread-spectrum system contains known sources of interference, such as carriers from other types of services, the frequency-hopping scheme can be designed to avoid these frequencies entirely. Otherwise, the communication system

Figure 24.11
Direct sequence transmitter

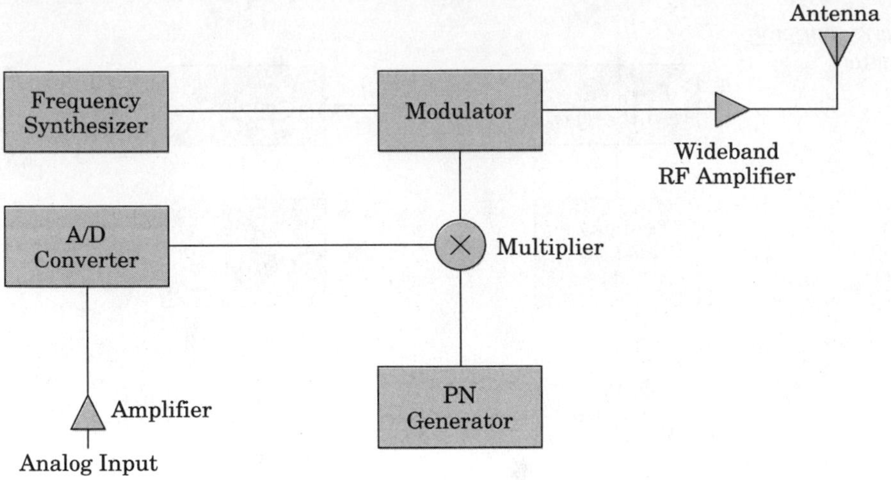

will degrade gracefully as the number of interfering signals increases, since each new signal will simply increase the noise level slightly. There will be no point at which calls become blocked, but eventually, if the traffic is too great, the quality of communication will degrade to the point where people keep their calls short or decide to try again later.

24.4.2 Direct Sequence

The direct-sequence form of spread-spectrum communication is commonly used with digital modulation schemes. The basic idea is to modulate the transmitter with a bit stream consisting of pseudo-random noise (PN) that has a much higher rate than the actual data to be communicated. Here the term *pseudo-random* means that the bit stream appears at first glance to be a random sequence of zeros and ones but is actually generated in such a way as to repeat exactly from time to time. The data to be transmitted is combined with the PN. One common technique is to invert all the bits of the PN stream during the time the "real" data is represented by a one and to leave the PN bit stream unchanged when a data zero is to be transmitted. The extra bits transmitted in this way are called **chips**, and the resulting bit rate is known as the **chipping rate**. Most direct-sequence spread-spectrum systems use a chipping rate at least ten times as great as the bit rate of the actual information to be transmitted. Figure 24.11 shows a block diagram of a direct-sequence transmitter. The chipping bits and the data signal are summed before being transmitted. The rest of the transmitter is conventional, except that it is designed for a much higher bit rate (and therefore a wider bandwidth) than would be normal for the data rate in use.

Using a high-speed PN sequence results in an increase in the bandwidth of the signal, regardless of what modulation scheme is used to encode the bits into the signal. Recall from Hartley's Law (Chapter 1) that for any given modulation scheme, the bandwidth is proportional to the bit rate, that is,

$$I = ktB \tag{24.9}$$

where

I = the amount of information to be transmitted, in bits
k = a constant that depends on the modulation scheme used
B = bandwidth in hertz

It follows from Hartley's Law that a direct-sequence system that transmits a total of ten bits for each information bit will use ten times as much bandwidth as a narrowband signal with the same type of modulation and the same information rate. That is, the side-

Figure 24.12
Spectrum of direct-sequence spread-spectrum signal

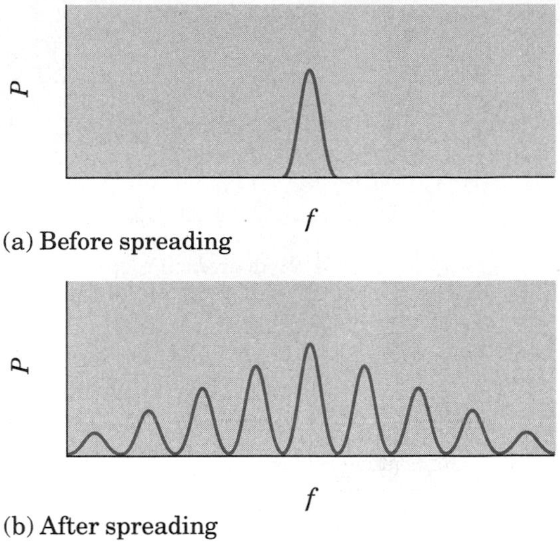

bands will extend ten times as far from the carrier, as shown in Figure 24.12. Direct-sequence spread-spectrum schemes typically use some form of phase-shift keying (PSK).

Example 24.7

A digital communication scheme uses DPSK with four possible phase shifts. It is to transmit a compressed PCM audio signal that has a bit rate of 16 kb/s. The chipping rate is 10 to 1. Calculate the number of signal changes (symbols) that must be transmitted each second.

Solution

The total bit rate, including the chips, is 10 times the data rate, or 160 kb/s. Since there are four signal states, each state represents two bits. Therefore the symbol rate is

$$160/2 = 80 \text{ kilobaud.}$$

Expanding the bandwidth by a factor of ten while keeping the transmitted power constant decreases the received signal-to-noise ratio by the same factor. As before, the pseudo-random sequence is known to the receiver, which has to separate the information signal from the chips.

24.4.3 Reception of Spread-Spectrum Signals

The type of receiver required for spread-spectrum reception depends on how the signal is generated. For frequency-hopped transmissions, what is needed is a relatively conventional narrowband receiver that hops in the same way as and is synchronized with the transmitter. The receiver must be given the frequency-hopping sequence, and there must be some form of synchronizing signal (similar to the signal usually sent at the start of a data frame in digital communications) to keep the transmitter and receiver synchronized. Some means must also be provided to allow the receiver to detect the start of a transmission, since, if this is left to chance, the transmitter and receiver will most likely be on different frequencies as a transmission begins.

One way to synchronize the transmitter and receiver is to have the transmitter transmit a tone on a prearranged channel at the start of each transmission, before it begins hopping. The receiver can synchronize by detecting the end of the tone and then beginning to hop according to the prearranged PN sequence. Of course, this method fails if there happens to be an interfering signal on the designated synchronizing channel at the time synchronization is attempted.

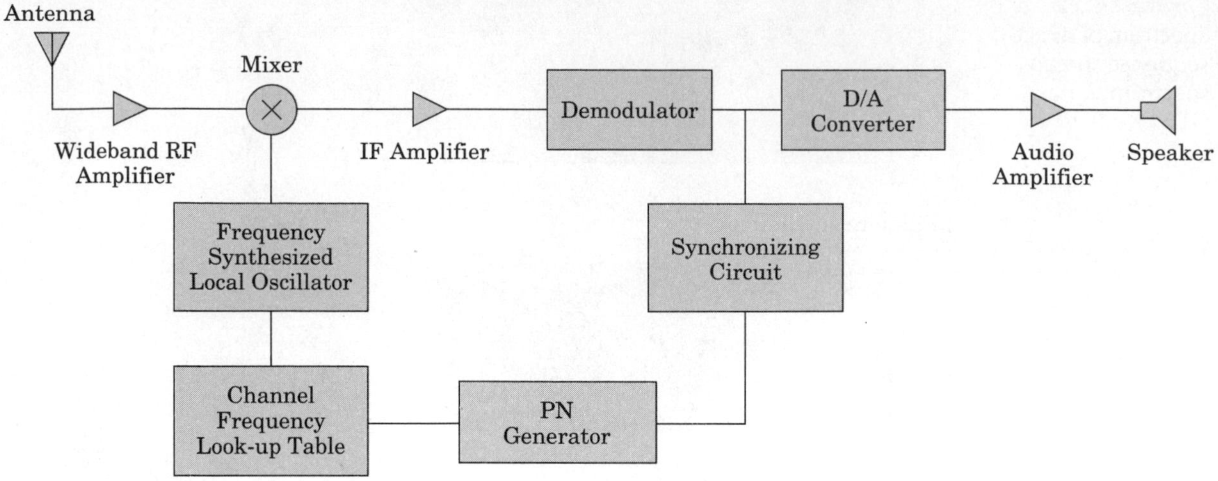

Figure 24.13 Frequency-hopping receiver

A more reliable method of synchronizing frequency-hopping systems is for the transmitter to visit several channels in a prearranged order before beginning a normal transmission. The receiver can monitor all of these channels sequentially, and once it detects the transmission, it can sample the next channel in the sequence for verification and synchronization. Aside from the need for synchronization and a synthesizer that is capable of frequency-hopping for the local oscillator, the receiver can be a conventional narrowband design. A block diagram of a frequency-hopping receiver is shown in Figure 24.13.

Direct-sequence spread-spectrum transmissions require different reception techniques. Narrowband receivers will not work with these signals, which occupy a wide bandwidth on a continuous basis. A wideband receiver is required, but a conventional wideband receiver would output only noise. In order to distinguish the desired signal from noise and interfering signals, which over the bandwidth of the receiver are much stronger than the desired signal, a technique called *autocorrelation* is used. Essentially this involves multiplying the received signal by a signal generated at the receiver from the PN code. When the input signal corresponds to the PN code, the output from the autocorrelator is large; at other times, this output is very small. Once again, the transmitter and receiver will probably not be synchronized at the start of a transmission, so the transmitter must send a preamble signal that is a prearranged sequence of ones and zeros to let the receiver synchronize with the transmitter. See Figure 24.14 for a block diagram of a direct-sequence receiver.

24.4.4 Code-Division Multiple Access (CDMA)

Traditionally, most analog communication systems have used frequency-division multiplexing and digital systems have used time-division multiplexing in order to combine many information signals into a single transmission channel. When the signals originate from different sources, these two methods become frequency-division multiple access (FDMA) and time-division multiple access (TDMA), respectively.

Spread-spectrum communication allows a third method for multiplexing signals from different sources—**code-division multiple access (CDMA)**. All that is required is for each transmitter to be assigned a different PN sequence. If possible, **orthogonal sequences** should be chosen, that is, the transmitters should never be in the same place at the same time. The PN sequence for the transmitter is given only to the receiver that is to operate with that transmitter. This receiver will then receive only the correct transmissions, and all other receivers will ignore these signals. This technique, which is applicable to both frequency-hopping and direct-sequence transmissions, allows many transmissions to share the same spread-spectrum channel.

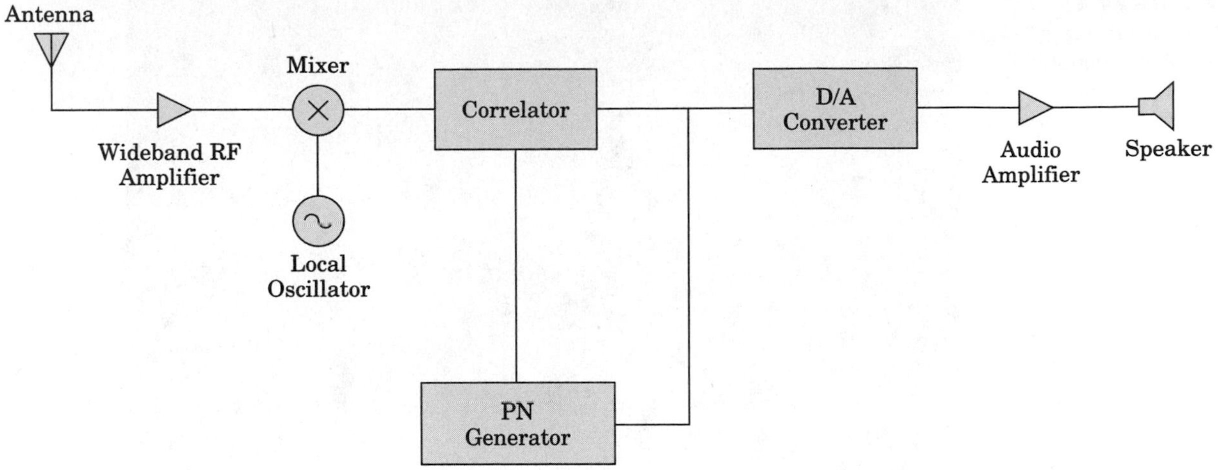

Figure 24.14
Direct-sequence receiver

24.5 Wireless Local-Area Networks

Recently there has been a good deal of interest in networking computers within a building without a physical connection. Such a system can be useful for offices that are subject to frequent rearrangement, for people who work from more than one office, and for setting up temporary networks for meetings or conferences. In general, the data rate for a wireless LAN connection is much less than for a fixed connection—typical wireless LAN connections communicate at 1–2 Mb/s, compared with 10 Mb/s for a typical wired Ethernet connection.

Most wireless LANs use either direct-sequence or frequency-hopping spread-spectrum systems. (The rest use infrared light, which is less expensive, but since it is unable to go through walls or around corners, it is usually less satisfactory.) In general, direct-sequence systems are more effective but also more expensive than frequency-hopping designs.

Wireless LANs currently use frequencies between 902 and 928 MHz and between 2.4 and 2.484 GHz. These bands are designated for industrial, scientific, and medical (ISM) uses, and no license is required for low-power spread-spectrum operation. In North America, frequency-hopping spread-spectrum LANs are restricted to a total bandwidth of 500 kHz in the 900 MHz band and 1 MHz in the 2.4 GHz band. They must not spend more than 400 milliseconds on any one channel in any 20-second period in the 900 MHz band or more than 400 ms in any 30-second period in the 2.4 GHz band. It appears likely that in the future, frequency-hopping transmissions will be permitted only in the 2.4 GHz band, with the 900 MHz band reserved for direct-sequence systems.

Wireless LANs that use direct-sequence spread-spectrum transmission must use at least 10 redundant bits or "chips" for each data bit, and many use 11. Thus the spread-spectrum signal must have a bandwidth at least ten times as great as would be needed to transmit the same data rate using conventional modulation schemes.

24.5.1 Wireless Modems

Related to wireless LANs (but not identical) are radio modems intended to link computers that are more widely separated—by perhaps a few kilometers. Current models, like wireless LANs, use spread-spectrum techniques in the ISM bands at about 900 MHz and 2.4 GHz. Figure 24.15 shows a typical example. This model uses frequency-hopping at 915 MHz and, with a total power output of 1 W, has a communications range of up to

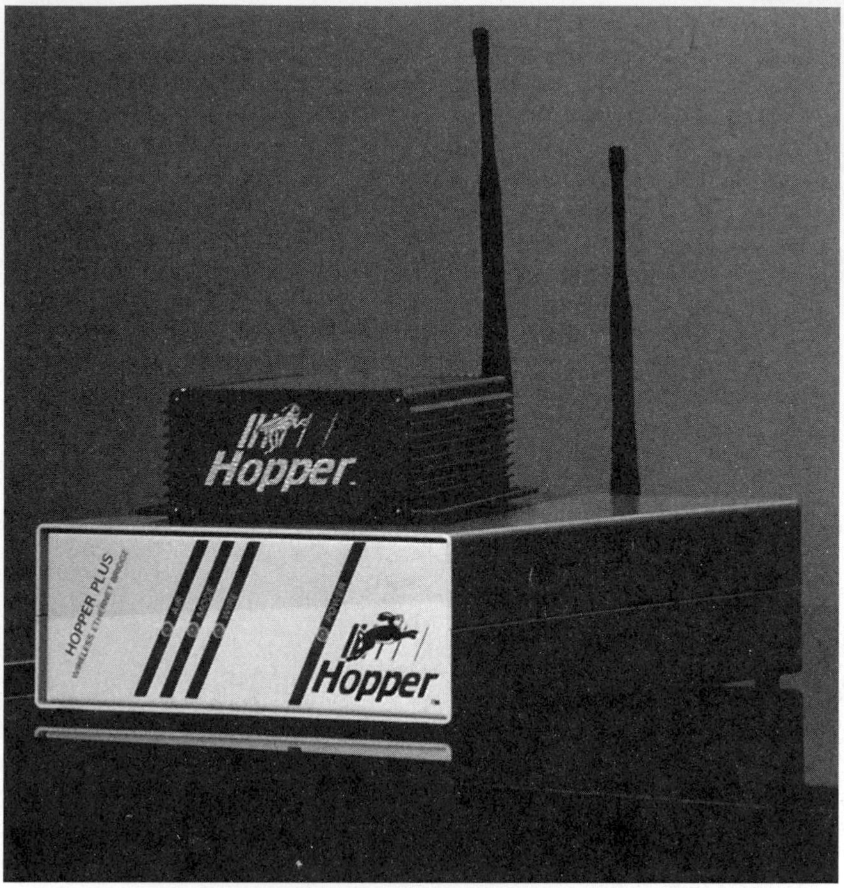

Figure 24.15 Wireless modem (top) and wireless LAN bridge
Photo by Ashraf Khoury, used by permission of Wi-Lan, Inc.

approximately 30 km, depending on radio propagation conditions, at a maximum data rate of 19,200 bits per second.

24.6 Personal Communication Systems

It is still not clear exactly what will happen to personal communications in the future. The most likely situation, for the near term at least, seems to be something like this: Current cellular systems will continue to expand, mainly for mobile use, with a gradual switch to digital technology. Likewise, there will continue to be cordless telephones and wireless LANs operating in frequency bands that do not require licensing. There will be an increasing trend toward digital, spread-spectrum technology for cordless phones, both to reduce interference and to increase privacy (wireless LANs already use spread-spectrum).

Satellite systems will be used in locations that are not served by cellular radio, and they are likely to be more expensive than cellular.

The gap between cordless phones and existing cellular radio systems will be filled by new systems with more range than the cordless phone and a lower cost structure than cellular radio. These PCS will initially be installed in high traffic areas, for instance, outdoors in city business districts and indoors in office complexes and shopping malls. They will operate with small handheld transceivers, which may also be capable of operating as cordless phones and/or using the existing cellular system when there is no available PCS.

The current cellular system is optimized for traffic that is moving fairly quickly, is outdoors, and has relatively high-powered transmitters. Recently, more portable than mobile cellular phones have actually been sold, showing that there is a market for PCS. The cellular system as it stands is not well suited to pedestrian use, especially indoors. The new PCS are expected to be aimed mainly at pedestrian traffic. They will use portable transceivers with very low transmitter power (a few milliwatts) to conserve battery power and will have many closely spaced, low-power repeaters to reduce fading and allow use with low-powered transmitters.

At this writing it is not clear exactly what forms the new PCS will take. In Europe, cordless phones and PCS are moving to common digital standards. In North America, where low-cost FM cordless phones have been popular for years, this does not seem as likely. In the United States, the Federal Communications Commission (FCC) is currently in the process of assigning licenses for PCS, largely by auction. A total bandwidth of 120 MHz has been reserved for new spectrum allocations in the 2 GHz region (seven different blocks located between 1850 and 2200 MHz) for "wideband" personal communications services. Wideband services include voice, data, and possibly, at some later date, video. Allocations are also being auctioned in 3 MHz of the 900 MHz region for "narrowand" PCS, mainly paging and messaging services. In addition, 40 MHz in the 2 GHz region (1890–1930 MHz) has been allocated for unlicensed use; this region contains the ISM bands, where no license is required. Canada has set aside the same frequency bands as in the United States but is allocating them by license hearings rather than by auction.

It is expected that the new PCS services will be digital. Both CDMA and TDMA systems have been proposed, and it seems at this writing that TDMA will be more popular. Designers have already produced prototype transceivers that can work with both PCS and the existing FDMA-FM cellular system. This duality is important because, at least at first, PCS will be available only in urban areas, and most users will probably want conventional cellular service as well, for backup. In fact, portable cellphones have become so small and so economical of battery power, at least when on standby, that some people wonder if PCS operators will have a hard time convincing customers that they need anything beyond conventional cellular service. PCS can be expected to get off to a slow start; in fact, as of May 1996, several auction winners had already defaulted on their license payments and lost their frequency allocations.

Eventually the seamless web described at the beginning of the chapter may appear—but not immediately. A logical first step would be to use common multiple-access and coding systems for all PCS, but even that seems unlikely for the present. The next step would be to make cellular systems and PCS compatible, except for operating frequency. Eventually the goal of a single telephone number and perhaps even a single telephone that would work with all systems may be realized.

Summary

Here are the main points to remember from this chapter.

1. Personal communication systems will eventually represent a radical new way of communicating. They will emphasize communication between people rather than places.
2. Early attempts at creating personal communication systems include citizens' band radio and cordless telephones.
3. Cellular radio telephony is currently the closest thing to a personal communication system in common, commercial operation. The current AMPS system developed from an earlier trunked system known as IMTS.
4. Most current cellular radio systems in North America use analog FM technology, but there is a gradual shift to digital systems. Most of the digital systems in current use employ time-division multiplexing, but there is also great interest in code-division multiple-access systems using spread-spectrum technology.
5. Practical spread-spectrum systems use either frequency-hopping or direct-sequence techniques. Frequency-hopping systems visit each of a number of frequencies in

sequence, while direct-sequence systems transmit many extra bits in order to increase the signal bandwidth.

6. Many spread-spectrum signals can occupy the same spectrum by means of code-division multiplexing, which requires that each signal use an orthogonal bit sequence to direct the hopping or chipping.
7. Even with no coordination to ensure orthogonal bit streams, spread-spectrum systems can tolerate a good deal of interference. In the presence of interference, the signal-to-noise level and the performance of the system degrade gradually.
8. Spread-spectrum technology is currently used in some cellular telephone systems. It is also used in most wireless modems and in the emerging personal communication systems.
9. The personal communication systems being developed bridge the gap between cordless phones and cellular radio. Typically they use spread-spectrum technology and are concentrated in areas of high pedestrian traffic.

Important Equations

$$d = \sqrt{17h_T} + \sqrt{17h_R} \tag{24.1}$$

$$L_{fs} = 32.44 + 20 \log d + 20 \log f - G_T - G_R \tag{24.2}$$

$$P_D = \frac{P_T}{4\pi r^2} \tag{24.3}$$

$$N = \frac{A}{\pi r^2} \tag{24.6}$$

$$N = \frac{A}{3.464 r^2} \tag{24.8}$$

$$I = ktB \tag{24.9}$$

Glossary

blocked a telephone call that cannot be connected because of a shortage of equipment or channels is said to be blocked

cell-splitting reducing the size of cells to increase their number and increase the capacity of a cellular phone system

chipping rate the total bit rate for a direct-sequence spread-spectrum system

chips extra bits added to the signal in a direct-sequence spread-spectrum system

code-division multiple access (CDMA) multiplexing of spread-spectrum signals using the same frequency range but different frequency-hopping or chipping codes

direct sequence spread-spectrum a technique in which a pseudo-random noise sequence is modulated by the information signal

frequency hopping a spread-spectrum technique in which the transmitter moves quickly, in a predetermined way, among a number of frequencies

microcells small cells, often remote-controlled from standard cells, that are added to a cellular phone system to increase capacity

orthogonal sequences pseudo-random chipping or hopping sequences that do not result in two transmitted signals occupying the same frequencies at the same time

pseudo-random noise (PN) a signal that moves between binary zero and one in a way that appears to be random but actually follows a preset pattern

spread-spectrum any technique that involves increasing the bandwidth of a transmitted signal greatly beyond its normal bandwidth

trunked system a radiotelephone system where channels are assigned to users on an as-needed basis

Questions

1. Explain how the concept of a personal communication system differs from that of a conventional telephone system.
2. What factors limit the range that can be obtained with CB radio?
3. Why are handheld CB transceivers typically larger and heavier than those for other personal communication systems?
4. How is interference between cordless telephones prevented?
5. Explain how a trunked mobile radio system differs from the systems used to assign frequencies to CB transceivers and cordless phones.
6. Describe how the IMTS system assigns frequencies to mobile telephones.
7. What factors limit the range of a mobile telephone using the IMTS system?
8. What is meant by frequency reuse? Why is frequency reuse desirable?
9. What factors limit frequency reuse with the IMTS system?
10. Explain how cellular systems allow frequency reuse, and discuss the factors that limit the distance at which a given frequency is reusable.
11. Briefly describe the difference between microcells and conventional cells.
12. What modulation scheme is used for the advanced mobile phone system (AMPS)?
13. Describe the procedure for placing and receiving phone calls using cellular radio.
14. What is *roaming*? Describe two different ways in which it can be implemented.
15. How can data be transmitted using a cellular telephone system operating under the AMPS standard?
16. What techniques have been proposed for reducing the bandwidth required by a single voice channel on a cellular phone system? Which of these are most popular?
17. Explain how direct-sequence and frequency-hopping spread-spectrum systems differ from each other and from narrowband techniques.
18. How do spread-spectrum signals avoid the problem of interference? Compare the different methods used by frequency-hopping and direct-sequence systems to avoid interference.
19. Compare code-division multiple access with frequency-division and time-division multiple access systems.
20. What methods are commonly used to implement wireless local-area networks?
21. What is the difference between a wireless LAN connection and a wireless modem?
22. What are the main uses proposed for new personal communication systems (PCS)? Compare them with existing services such as cordless phones and cellular phones.

Problems

SECTION 24.2

23. A CB user wants to communicate between a base station and a mobile unit that has an antenna height of 1.5 m and is 30 km away.
    (a) How high must the base station antenna be?
    (b) What will be the minimum signal strength at the receiver under the following conditions:
    Transmitter power: 4 W
    Antenna gains: fixed: 3 dBi
    mobile: 1 dBi
    Feedline losses: fixed: 2 dB
    mobile: negligible
24. Amateur radio clubs often operate repeaters, usually in frequency bands near 144 and 430 MHz. They do not have the same power restrictions as CB radio. Calculate the minimum repeater height and transmitted EIRP for a repeater that must communicate with mobile stations within a 50 km radius. A minimum electric field strength of 50 μV/m must be maintained at the mobile receiver, which has an antenna height of 1.5 m.
25. What is the maximum possible distance between two mobile stations with the repeater setup of problem 24?
26. Suppose that cordless telephone manufacturers double the power output of their transmitters in an attempt to increase range.
    (a) If the usable range was 50 m before the power increase, what is it after the increase?
    (b) What effect will the power increase have on interference from other portable telephones
    (i) operating at the old (low) power level?
    (ii) operating at the new (high) power level?
    (c) From your answer, what can you conclude

about the effect of increasing power in an environment where transmission range is limited by
(i) signal-to-noise ratio?
(ii) signal-to-interference ratio?

27. An old cordless phone works well in the house but not in the backyard. What is the most likely problem?

SECTION 24.3

28. A cellular radio system covers an area of 200 square kilometers using hexagonal cells with a 2 km radius. How many cells are required?
29. An area of 300 square kilometers is to be served by no more than 50 cells. Assuming the cells are hexagonal in shape, what is the maximum distance between cell sites?
30. Calculate the percent increase in the number of possible simultaneous phone calls with a 7 cell rather than a 12 cell repeating cell pattern.
31. By what percentage can the number of possible simultaneous phone calls be increased by converting a cellular radio system from analog to an IS-54 digital system?

SECTION 24.4

32. An FM system with a signal bandwidth of 20 kHz and a signal-to-noise ratio of 30 dB is converted to spread-spectrum, occupying a bandwidth of 5 MHz. What is the new, predetection signal-to-noise ratio?
33. What is the hopping rate for a frequency-hopping transmitter that visits 200 frequencies in 5 seconds?
34. A direct-sequence spread-spectrum scheme is to be used to spread a signal that formerly occupied 50 kHz to occupy one MHz. How many chips are needed per signal bit?

SECTION 24.5

35. A frequency-hopping LAN terminal must spend no more than 400 ms in any 30 second period on one channel in the 2.4 GHz band. Calculate the minimum number of channels needed by such a system.

COMPREHENSIVE

36. Compare analog FDM, digital FDM/TDM, and digital CDM using spread-spectrum techniques as possible ways to transmit cellular telephone signals.
37. (a) Calculate the length of a quarter-wave vertical antenna for each of the following services: CB radio at 27 MHz, cordless phone at 49 MHz, cellular radio at 900 MHz, and PCS at 2 GHz.
(b) What can you conclude about the ease of designing portable transceivers for each of these services?

APPENDIX A

Decibels

The decibel is a logarithmic way of expressing the ratio of two power levels or, sometimes, voltage levels. Decibels are used in almost every part of electronic communications. Since the decibel is based on logarithms, a brief review of logarithms is in order before beginning to discuss it.

A.1 Review of Logarithms

The *logarithm* (log) of a number to a given base is the power to which the base must be raised to give the number. Although any number can be used as the base, the base 10 is used in decibel calculations. Logarithms to the base 10 are called *common logarithms* and are abbreviated $\log_{10}$ or just log. For example, the log of 100 to the base 10 is 2, because 10 raised to the power of 2 equals 100. Expressed in mathematical notation,

$$\log_{10} 100 = 2$$

Similarly,

$$\log_{10} 0.01 = -2$$

Negative numbers do not have logs, because no matter what power 10 is raised to, the result is always positive.

The *antilog* of a number to a given base is simply the base raised to that number. For example, using base 10,

$$\begin{aligned} \text{antilog } 2 &= 10^2 \\ &= 100 \end{aligned}$$

The antilog is the inverse operation to the log of a number; that is, the antilog of the log of a number is the number itself. The antilog of x can also be written as inverse log x, $\log^{-1} x$, or 10^x (assuming the base 10 is being used).

A.1.1 Simple Operations with Logarithms

The following results can easily be proved by going back to the definition of a logarithm. It is assumed that all logs are to the same base.

$$\log ab = \log a + \log b \tag{A.1}$$

$$\log \frac{a}{b} = \log a - \log b \tag{A.2}$$

$$\log a^b = b \log a \tag{A.3}$$

A.2 Decibels

The decibel, in its simplest form, expresses the ratio of two power levels, logarithmically. If P_1 and P_2 are two power levels, P_2 can be said to be greater than P_1 by a number of decibels given by

$$\frac{P_2}{P_1}\ (\text{dB}) = 10 \log \frac{P_2}{P_1} \tag{A.4}$$

From now on, to reduce clutter in the equations, the base 10 will be assumed. Note that, if P_1 happens to be greater than P_2, the result will be negative.

Example A.1 Find the ratio between P_2 and P_1, in decibels, if

(a) $P_1 = 2$ W, $P_2 = 3$ W (b) $P_1 = 3$ W, $P_2 = 2$ W

Solution

(a)
$$\begin{aligned}\frac{P_2}{P_1}\ (\text{dB}) &= 10 \log \frac{P_2}{P_1}\\ &= 10 \log \frac{3}{2}\\ &= 1.76\ \text{dB}\end{aligned}$$

(b)
$$\begin{aligned}\frac{P_2}{P_1}\ (\text{dB}) &= 10 \log \frac{P_2}{P_1}\\ &= 10 \log \frac{2}{3}\\ &= -1.76\ \text{dB}\end{aligned}$$

A.2.1 Decibel Gain and Loss

If P_o is the output power of a device, and P_i is the input power, then the power gain in decibels is

$$A_P\ (\text{dB}) = 10 \log \frac{P_o}{P_i} \tag{A.5}$$

A negative result simply means that the output power is less than the input and there is a power loss in the device. In decibels, a gain of $-x$ dB is equivalent to a loss of x dB.

Example A.2 An amplifier has an input of 100 mW and an output of 4 W. Find its gain in decibels.

Solution

$$\begin{aligned}A_P\ (\text{dB}) &= 10 \log \frac{P_o}{P_i}\\ &= 10 \log \frac{4\ \text{W}}{0.1\ \text{W}}\\ &= 16\ \text{dB}\end{aligned}$$

Example A.3 An attenuator has a loss of 26 dB. If a power of 2 W is applied to the attenuator, find the output power.

Solution Equation (A.5) can be rearranged quite easily to solve this.

$$A_P\ (\text{dB}) = 10 \log \frac{P_o}{P_i}$$

$$\frac{P_o}{P_i} = \text{antilog} \frac{A_P\ (\text{dB})}{10}$$

$$P_o = P_i \text{ antilog} \frac{A_P\ (\text{dB})}{10}$$

Since the 26 dB is given as a loss, its sign must be changed before it can be used in this equation.

$$P_2 = 2\text{ W antilog} \frac{-26}{10}$$

$$= 5.02 \times 10^{-3}\text{ W}$$

$$= 5.02\text{ mW}$$

A.2.2 Use of Reference Power Levels: dBm, dBW, dBf

The decibel expresses the ratio between two power levels, but there is no requirement for both of these two signals to exist physically. For instance, we could ask, by how many decibels is the power in the circuit greater than one milliwatt? This does not imply that we actually have a power of 1 mW somewhere in the circuit. Power levels expressed in this way are said to be in **dBm**.

$$P\ (\text{dBm}) = 10 \log \frac{P}{1\text{ mW}} \tag{A.6}$$

Power is often measured in dBm in both radio- and audio-frequency applications.

Example A.4 Convert a power level of 0.5 W to dBm.

Solution

$$P\ (\text{dBm}) = 10 \log \frac{P}{1\text{ mW}}$$

$$= 10 \log \frac{500\text{ mW}}{1\text{ mW}}$$

$$= 27\text{ dBm}$$

Note that the ratio in Equation (A.6) must be dimensionless; that is, a power level given in some unit other than milliwatts must be converted to milliwatts before the log is found.

Other reference power levels can be used whenever it is more convenient. Two common references are the femtowatt and the watt.

$$P\ (\text{dBf}) = 10 \log \frac{P}{1\text{ fW}} \tag{A.7}$$

$$P\ (\text{dBW}) = 10 \log \frac{P}{1\text{ W}} \tag{A.8}$$

The femtowatt reference is useful where power levels are very low, as in the calculation of receiver sensitivity. For relatively large power levels, such as the output power of transmitters, the watt is a more useful reference.

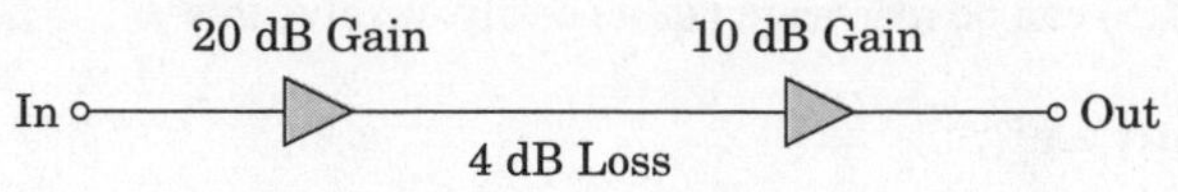

Figure A.1
Decibel Gains and Losses

A.2.3 Operations with Decibels

These follow directly from the properties of logarithms and the fact that decibels are logarithmic.

1. Decibel gains add.

   Example: An amplifier with 20 dB gain is connected to another with 10 dB gain by means of a transmission line with a loss of 4 dB. (See Figure A.1.) The total system gain is

   $$\begin{aligned} A_P\ (\text{dB}) &= 20 - 4 + 10 \\ &= 26\ \text{dB} \end{aligned}$$

   Note that a loss of 4 dB is equivalent to a gain of −4 dB, as explained above.
2. Gains in decibels can be added directly to powers expressed in dBm, dBf, dBW, etc., giving an output in the same unit as the input.

   Example: If a signal with a power level of −12 dBm were applied to the system of Figure A.1, the output would be

   $$\begin{aligned} P_o &= -12\ \text{dBm} + 26\ \text{dB} \\ &= 14\ \text{dBm} \end{aligned}$$

   It may seem that we are adding different quantities in the above example, but in fact we are not. Both the quantities are actually the logarithms of power ratios, and thus dimensionless. Decibels and dBm are not units like amperes or volts: the "dB" indicates the operation that has been performed on a ratio, and the "m" keeps track of a reference level.
3. To change reference levels, express the old reference level in terms of the new one, then add this amount to every value.

   Example: To convert dBW to dBm, we note that

   $$0\ \text{dBW} = 1\ \text{W} = 30\ \text{dBm}$$

   so any power given in dBW can be converted to dBm by adding 30 dB. For example, 2 dBW = 32 dBm, and −10 dBW = 20 dBm.

A.2.4 Decibels and Voltages

Until now, we have dealt only with power ratios. If signal voltages are compared instead, the ratio of the voltages will be the square of the power ratio, provided that the voltages are developed across identical impedances. This is very often the case in radio-frequency amplifiers like that shown in Figure A.2. In this case, both the input impedance and the load impedance are 50 Ω, so

$$\frac{P_o}{P_i} = \left(\frac{V_o}{V_i}\right)^2$$

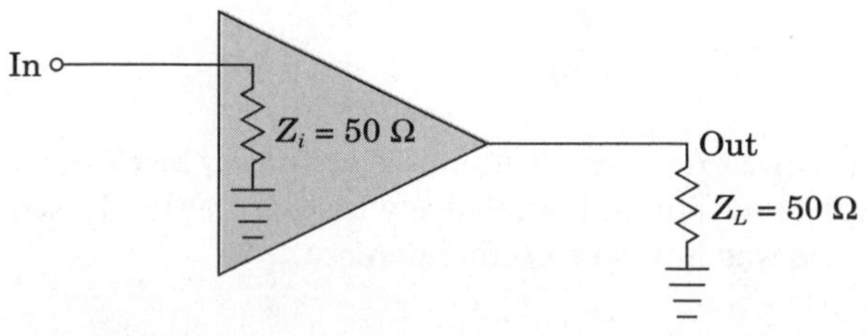

Figure A.2
RF Amplifier

Using the fact that

$$\log a^b = b \log a$$

it is easy to show that

$$A_P \text{ (dB)} = 20 \log \frac{V_o}{V_i}$$

Please note carefully that this is only true when both voltages are measured across the same impedance. It would not be true for a typical audio power amplifier, for instance. Here the input impedance Z_i might be 10 kΩ, while the load resistance Z_L is commonly about 8 Ω. It is possible to define a decibel voltage gain as follows:

$$A_V \text{ (dB)} = 20 \log \frac{V_o}{V_i} \tag{A.9}$$

A_V (dB) and A_P (dB) will be equal only when $Z_i = Z_L$. In practice, this seldom causes problems, because it is usually clear from the context which type of decibel gain is being referred to.

Just as for power, it is possible to use decibels to compare a voltage with a reference level. For example, the term dBmV is used in cable-television systems. It is a measure of a voltage level compared to 1 mV; that is,

$$V \text{ (dBmV)} = 20 \log \frac{V}{1 \text{ mV}} \tag{A.10}$$

Example A.5

A signal in a cable-television system has an amplitude of 3 mV in 75 Ω (resistive). Calculate its level in

(a) dBmV (b) dBm

Solution

(a) From Equation (A.10),

$$\begin{aligned} V \text{ (dBmV)} &= 20 \log \frac{V}{1 \text{ mV}} \\ &= 20 \log \frac{3 \text{ mV}}{1 \text{ mV}} \\ &= 9.54 \text{ dBmV} \end{aligned}$$

(b) Since dBm is a measure of power, we first find the power in the signal.

$$\begin{aligned} P &= \frac{V^2}{R} \\ &= \frac{(3 \times 10^{-3} \text{ V})^2}{75 \ \Omega} \\ &= 120 \times 10^{-9} \text{ W} \end{aligned}$$

Now, Equation (A.6) can be used to calculate the power in dBm:

$$\begin{aligned} P \text{ (dBm)} &= 10 \log \frac{P}{1 \text{ mW}} \\ &= 10 \log \left(\frac{120 \times 10^{-9} \text{ W}}{1 \times 10^{-3} \text{ W}} \right) \\ &= -39.2 \text{ dBm} \end{aligned}$$

A.2.5 Decibel Measurements

Since decibels simply express power or voltage ratios in a logarithmic way, it is not necessary to have specialized equipment to measure decibels. Power and voltage levels can be measured in any of the usual ways, and then decibel quantities can be calculated as explained above. Many meters, however, have scales calibrated directly in decibels. These include, among others, output-level meters on RF generators, audio voltmeters, and, sometimes, the ac ranges on multimeters.

As always, of course, the decibel measurements provided by these meters must be with respect to some reference, either stated or implied. A typical audio voltmeter, for instance, will have two scales, one labeled dBV and the other dBm. The dBV scale is self-explanatory, but one might ask how a voltmeter measures power in order to find a level in dBm. It does not actually measure power, of course: it measures voltage and calculates power, based on an assumption about the impedance across which the voltage appears. The most common impedance value used for audio voltmeters is 600 Ω, but some multimeters allow this to be changed. RF meters, on the other hand, use 50 Ω, or sometimes 75 Ω, as the reference impedance.

APPENDIX B

CB Channel Frequencies

Channel	Frequency (MHz)	Channel	Frequency (MHz)
1	26.965	21	27.215
2	26.975	22	27.225
3	26.985	23	27.255
4	27.005	24	27.235
5	27.015	25	27.245
6	27.025	26	27.265
7	27.035	27	27.275
8	27.055	28	27.285
9	27.065	29	27.295
10	27.075	30	27.305
11	27.085	31	27.315
12	27.105	32	27.325
13	27.115	33	27.335
14	27.125	34	27.345
15	27.135	35	27.355
16	27.155	36	27.365
17	27.165	37	27.375
18	27.175	38	27.385
19	27.185	39	27.395
20	27.205	40	27.405

APPENDIX C

Television Channel Frequencies

VHF: Low Band

Channel	Frequency (MHz)
2	54–60
3	60–66
4	66–72
5	76–82
6	82–88

VHF: High Band

Channel	Frequency (MHz)
7	174–180
8	180–186
9	186–192
10	192–198
11	198–204
12	204–210
13	210–216

UHF

Channel	Frequency (MHz)	Channel	Frequency (MHz)	Channel	Frequency (MHz)
14	470–476	34	590–596	54	710–716
15	476–482	35	596–602	55	716–722
16	482–488	36	602–608	56	722–728
17	488–494	37	608–614	57	728–734
18	494–500	38	614–620	58	734–740
19	500–506	39	620–626	59	740–746
20	506–512	40	626–632	60	746–752
21	512–518	41	632–638	61	752–758
22	518–524	42	638–644	62	758–764
23	524–530	43	644–650	63	764–770
24	530–536	44	650–656	64	770–776
25	536–542	45	656–662	65	776–782
26	542–548	46	662–668	66	782–788
27	548–554	47	668–674	67	788–794
28	554–560	48	674–680	68	794–800
29	560–566	49	680–686	69	800–806
30	566–572	50	686–692	70	806–812
31	572–578	51	692–698	71	812–818
32	578–584	52	698–704	72	818–824
33	584–590	53	704–710		

APPENDIX D

Cable-Television Frequencies

Note: There are several schemes in use with different designations. The list below includes the most popular variations. Channels are sometimes designated by numbers and sometimes by letters, and there is more than one system for each! Not all systems carry all these channels. In addition to the channels listed, cable-television systems also use the normal VHF channels 2 through 13, at the same frequencies as for broadcast television. Some systems also have a *hyperband* extending past channel 64, with additional channels at 6 MHz intervals.

Midband Channels (Between Broadcast Channels 6 and 7)

Letter	Number	Frequency Range (MHz)	Letter	Number	Frequency Range (MHz)
A-5	95	90–96 *	C	16	132–138
A-4	96	96–102*	D	17	138–144
A-3	97	102–108*	E	18	144–150
A-2	98	108–114	F	19	150–156
A-1	99	114–120	G	20	156–162
A	14	120–126	H	21	162–168
B	15	126–132	I	22	168–174

*These channels cannot be used when the cable carries FM broadcasting.

Superband Channels (Above Broadcast Channel 13)

Letter	Number	Frequency Range (MHz)	Letter	Number	Frequency Range (MHz)
J	23	216–222	Q	30	258–264
K	24	222–228	R	31	264–270
L	25	228–234	S	32	270–276
M	26	234–240	T	33	276–282
N	27	240–246	U	34	282–288
O	28	246–252	V	35	288–294
P	29	252–258	W	36	294–300

Hyperband Channels

Letter	Letter + Number	Number	Frequency Range (MHz)
AA	W+1	37	300–306
BB	W+2	38	306–312
CC	W+3	39	312–318
DD	W+4	40	318–324
EE	W+5	41	324–330
FF	W+6	42	330–336
GG	W+7	43	336–342
HH	W+8	44	342–348
II	W+9	45	348–354
JJ	W+10	46	354–360
KK	W+11	47	360–366
LL	W+12	48	366–372
MM	W+13	49	372–378
NN	W+14	50	378–384
OO	W+15	51	384–390
PP	W+16	52	390–396
QQ	W+17	53	396–402
RR	W+18	54	402–408
SS	W+19	55	408–414
TT	W+20	56	414–420
UU	W+21	57	420–426
VV	W+22	58	426–432
WW	W+23	59	432–438
XX	W+24	60	438–444
YY	W+25	61	444–450
	W+26	62	450–456
	W+27	63	456–462
	W+28	64	462–468

APPENDIX E

Waveguide Table

The following table consists of selected rectangular waveguides. Dimensions are internal. The useful ranges and cutoff frequencies (f_c) given are for single-mode propagation using the TE_{10} mode.

U.S. RG-Number						
Brass	Aluminum	European IEC Number	Width (mm)	Height (mm)	f_c (GHz)	Useful Range (GHz)
69/U	103/U	R 14	165	82.6	0.91	1.14–1.73
104/U	105/U	R 22	109	54.6	1.38	1.72–2.61
112/U	113/U	R 26	86.4	43.2	1.74	2.17–3.30
48/U	75/U	R 32	72.1	34.0	2.08	2.60–3.95
49/U	95/U	R 48	47.6	22.1	3.16	3.94–5.99
50/U	106/U	R 70	34.9	15.8	4.29	5.38–8.17
51/U	68/U	R 84	28.5	12.6	5.26	6.57–9.99
52/U	67/U	R 100	22.9	10.2	6.56	8.20–12.5
91/U	107/U	R 140	15.8	7.9	9.49	11.9–18.0
53/U	121/U	R 220	10.7	4.3	14.08	17.6–26.7

APPENDIX F Answers to Odd-Numbered Problems

CHAPTER 1

21. 750 THz, 429 THz

23.

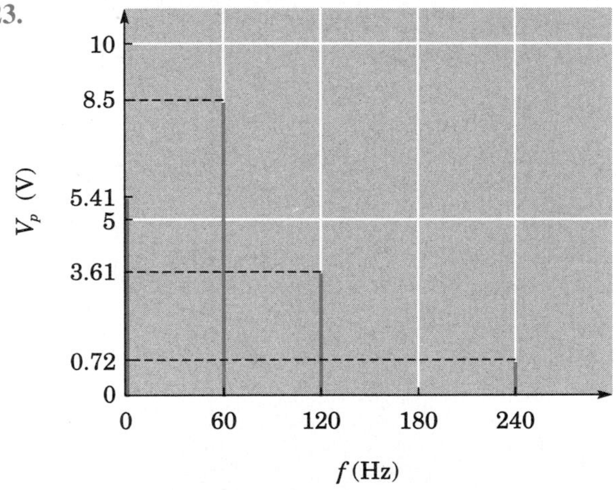

25. **(a)**

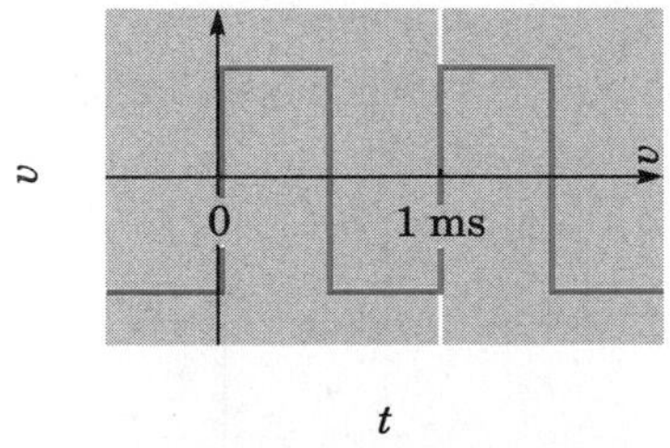

(b)

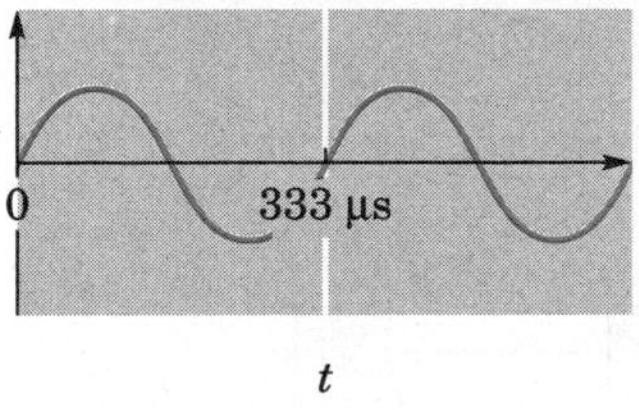

(c)

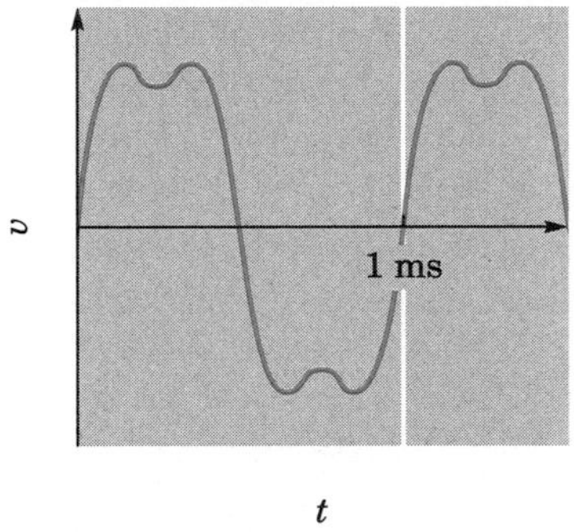

27. Average dc value is zero.

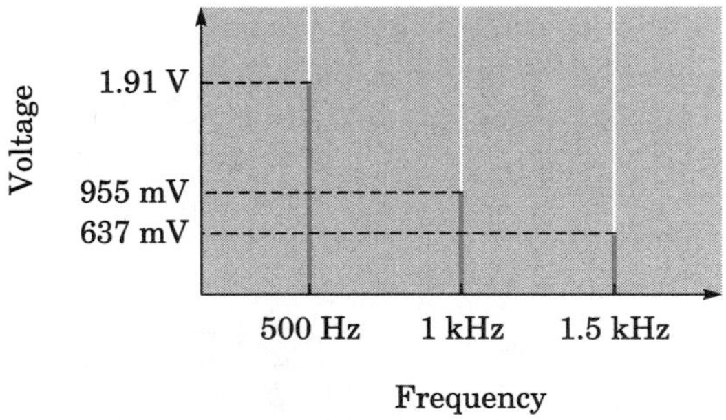

29. **(a)** 90.1 nV **(b)** 2.2 µV
31. **(a)** 2.7 db **(b)** 249 K
33. **(a)** 1.29 dB
    (b) Better: lower *NF* means less noise contribution and better *S/N*
35. $A_T = 6000 = 37.8$ dB, $NF = 3.32 = 5.2$ dB
37. **(a)** $f = 867$ MHz, $P = -50$ dBm $= 10$ nW, $V = 707$ µV

(b) $f = 80$ MHz, $P = -19$ dBm $= 12.5$ μW, $V =$ 25 mV
(c) $f = 260$ kHz, $P = 12$ dBm $= 15.8$ mW, $V =$ 890 mV

39. (a) 10 dB (b) 0.56 dB (c) reduced by 15 K
41. $B = 50$ MHz, $f_u = 235$ MHz, $f_l = 185$ MHz
43. 27 dB
45. S/N decreases by 3 dB.

CHAPTER 2

27. **(a)** 17.6 pF **(b)** 600 kHz
29. **(a)** Class C: zero bias with no input **(b)** 0
(c)

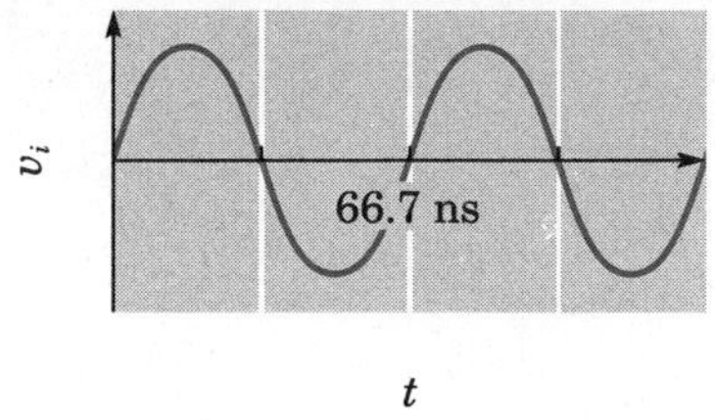

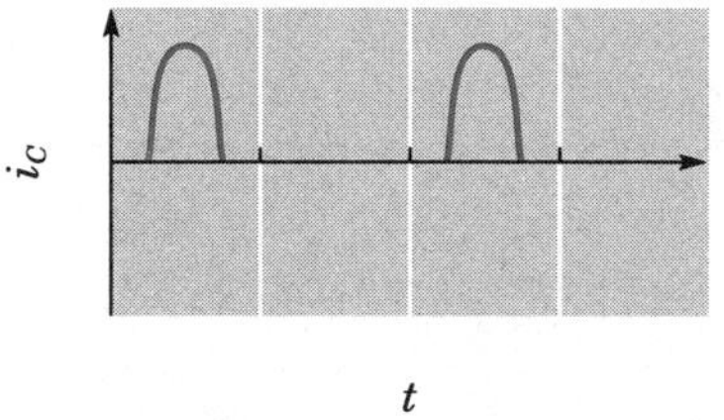

(d) 3.75 V peak
31. **(a)** 10.7 W **(b)** 100 W
33. 1.4 μH
35. **(a)** Pierce
(b)

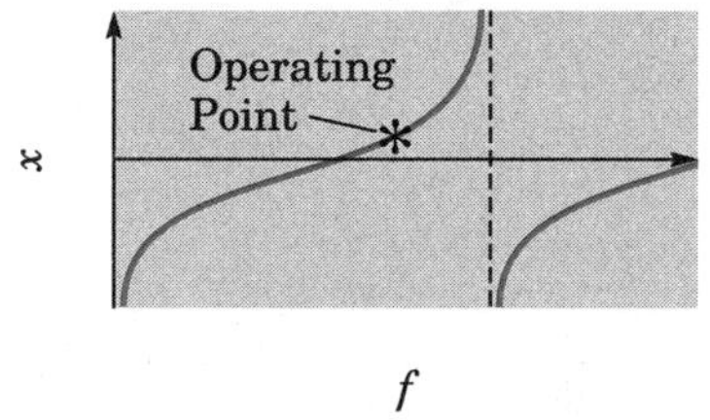

37. 6.24 MHz to 10.2 MHz
39. 5.79 ppm
41.

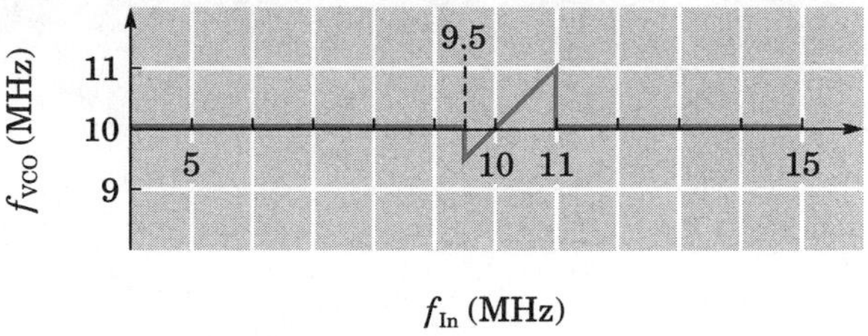

43. **(a)** 2–4 MHz **(b)** 20 kHz
45. 500 MHz, 100 kHz
47. 15 MHz
49. **(a)** 10.00125 MHz, 9.99875 MHz
(b) 45.005625 MHz, 44.994375 MHz
51. One solution is shown below. Others are possible.

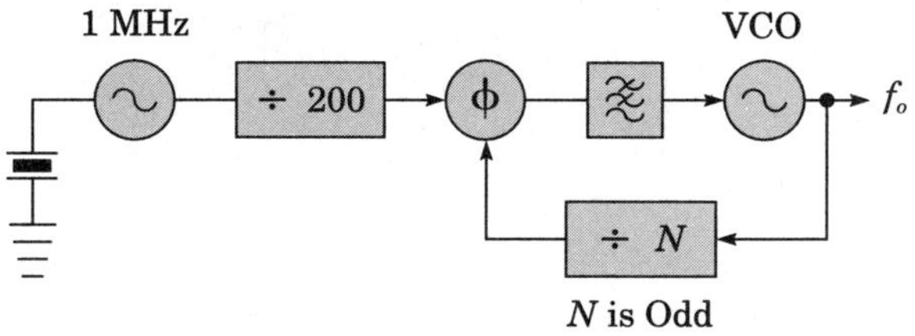

53. One solution is shown below. Others are possible.

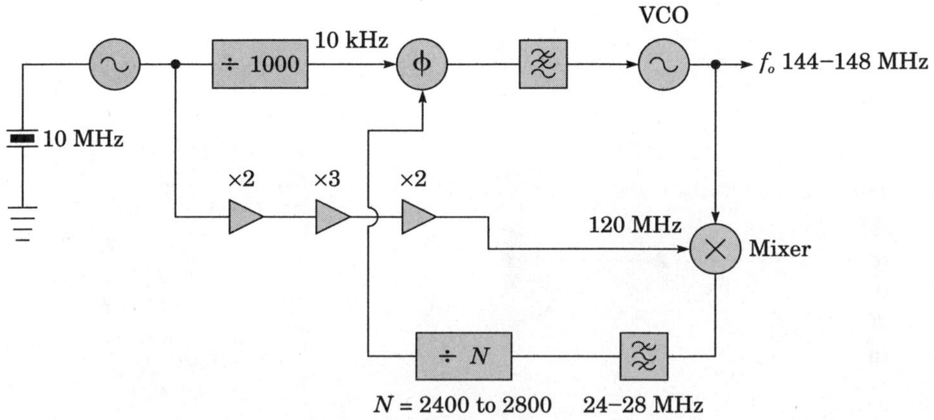

CHAPTER 3

19. $v(t) = [5 + 2 \sin (3.14 \times 10^3 t)] \sin (18.8 \times 10^6 t)$, $m = 0.4$
21. 0.56
23. $m = 0.75$, $V_c = 283$ V RMS
25. 7.2 MHz, 7.2015 MHz, 7.203 MHz, 7.1985 MHz, 7.197 MHz
27. **(a)** 0.8 **(b)** 6 kHz **(c)** 20 V
29. **(a)**

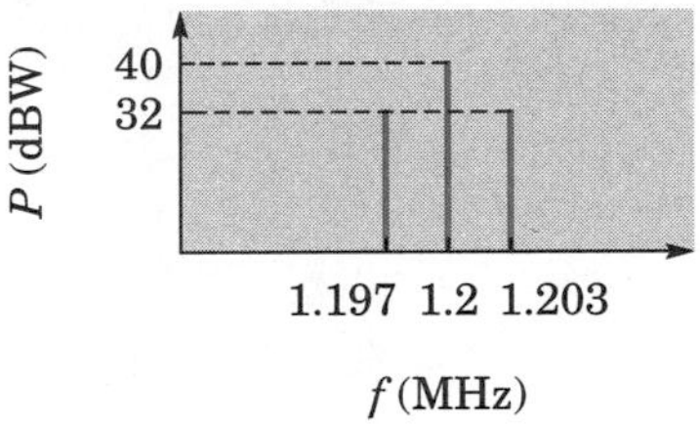

(b) 13.2 kW = 41.2 dBW **(c)** 812 V
(d) 1.8 kV

31. **(a)** 50% **(b)** 283 V, 533 Ω
(c)

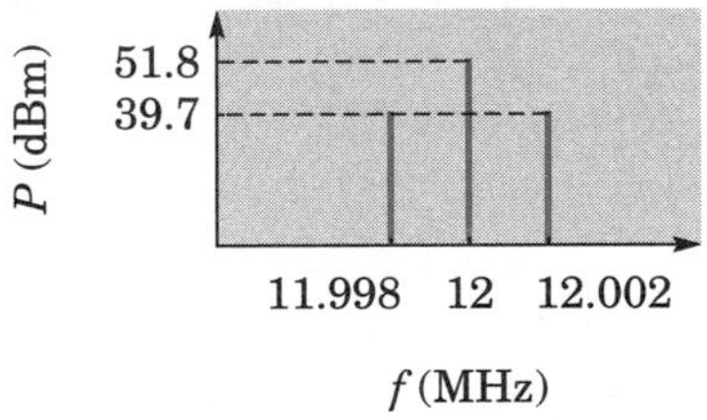

33.

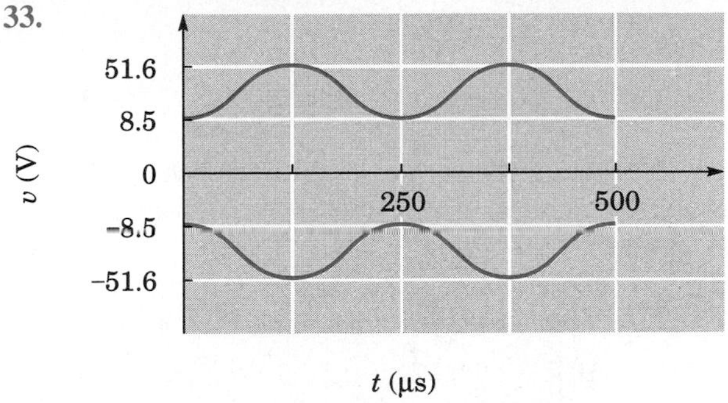

35. $v(t) = 3.57(1 + 0.4 \sin 18.85 \times 10^3 t)\sin 31.4 \times 10^6 t$
37. **(a)**

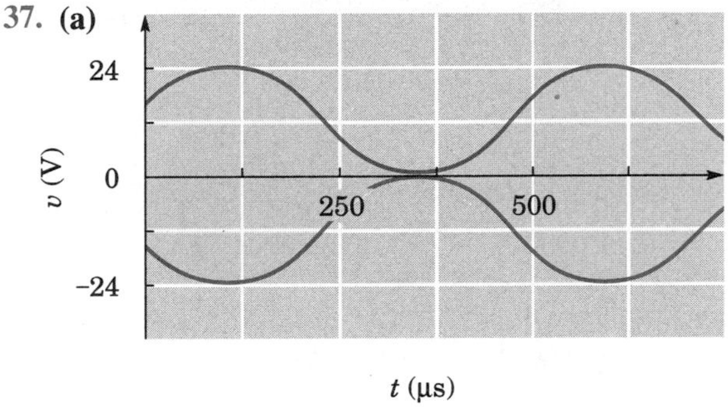

(b) (i) 3 MHz, (ii) 2 kHz (iii) 2.16 W (iv) 4 kHz

CHAPTER 4

21. 27.2064 MHz, 27.2036 MHz
23. 63.7%
25.

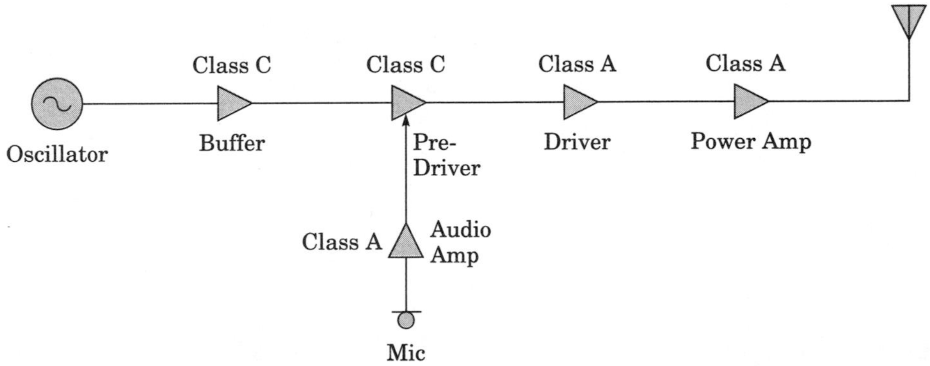

27. 17.9 W
29. **(a)** 857 mA **(b)** 58.3 Ω **(c)** 21.4 W
(d) 200 V
31. 0.38 dB
33. 20 W
35. **(a)** LED glows when collector voltage goes negative.
(b) LED will not glow until collector voltage exceeds about −2 V.
37. **(a)**

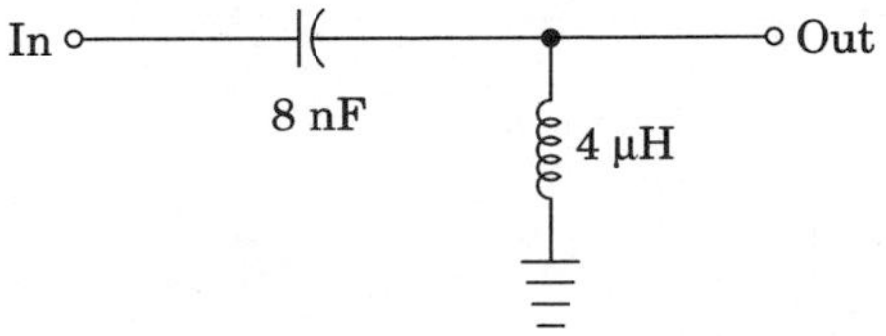

(b)

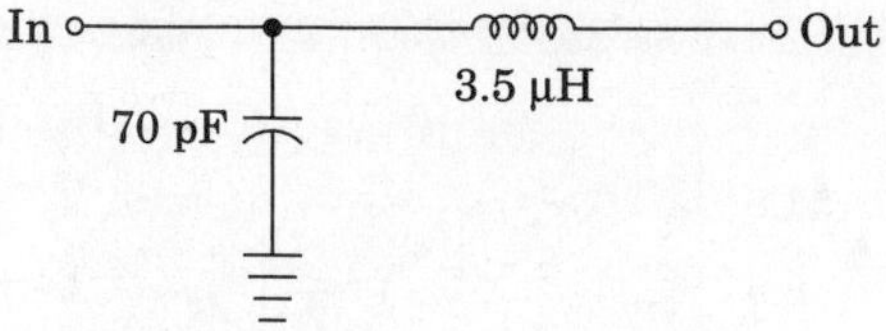

(b)

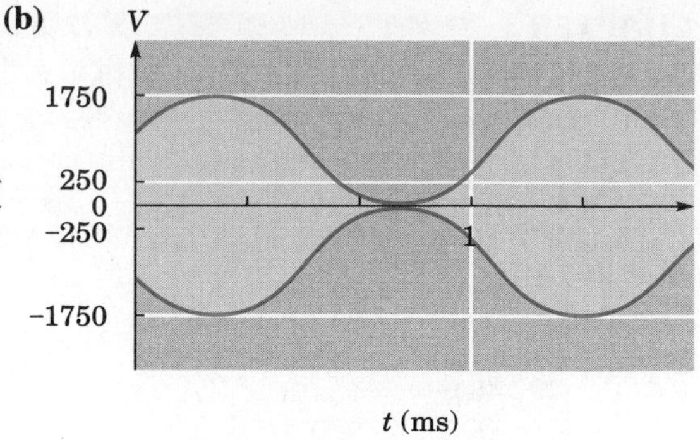

39. See Figure 4.4(c). Classes: buffer, driver, RF power amp: C; audio amp: A for earlier stages, AB or B for output stage.
41. **(a)**

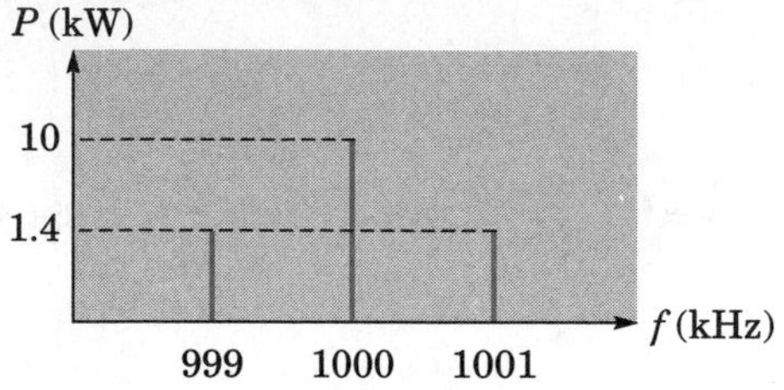

CHAPTER 5

25. **(a)** 1.65 MHz **(b)** High-side injection
27. 98.7 MHz to 118.7 MHz
29. 108 dB
31. **(a)** 17.5 MHz **(b)** 14 MHz
33. 15 kHz
35. **(a)** 58.6 dB **(b)** 9.08 dB
37. 359 kHz
39. **(a)** 180 nH **(b)** 14.4 pF
41. **(a)** 1st: 130–230 MHz, 2nd: 29 MHz
    (b) See Figure 5.15.
43. **(a)** S-7 **(b)** S-9 + 6 dB
    (c) S-9 + 32 dB
45. **(a)** Audio sine wave **(b)** AM at IF
    (c) AM at IF **(d)** AM at signal frequency

CHAPTER 6

23. **(a)**

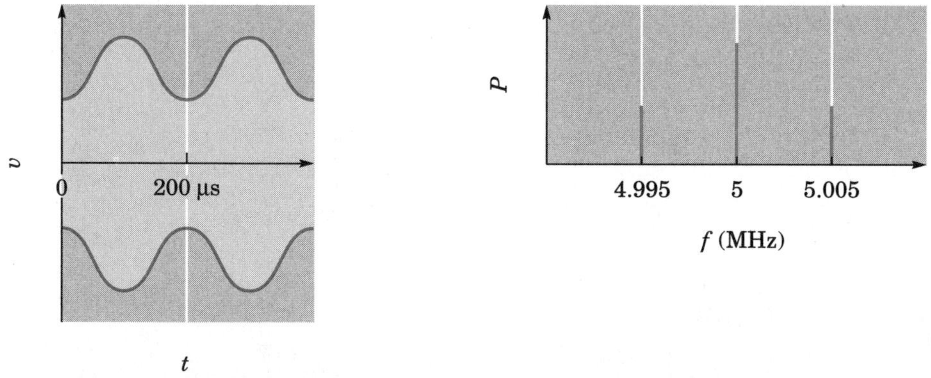

(b)

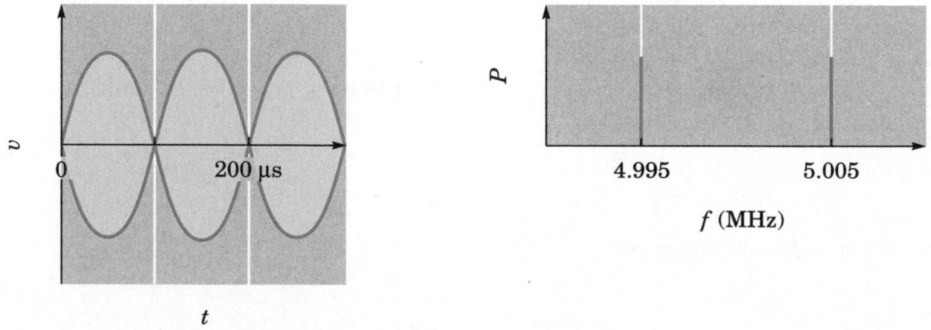

(c)

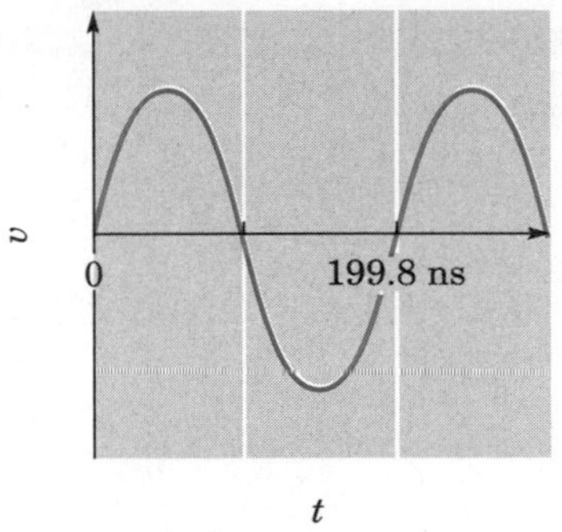

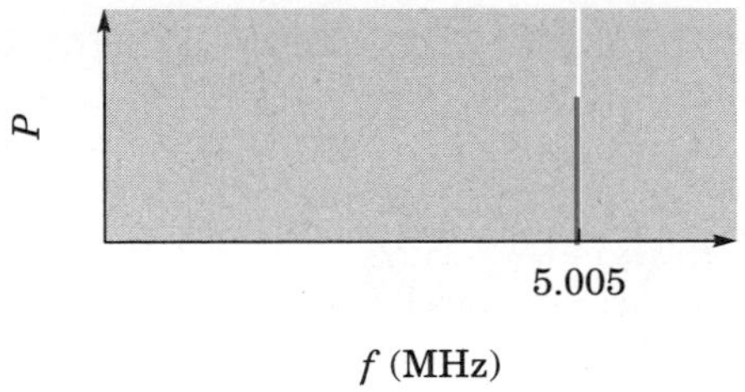

25. **(a)** 16 W **(b)** 8 dB
27. 80
29.

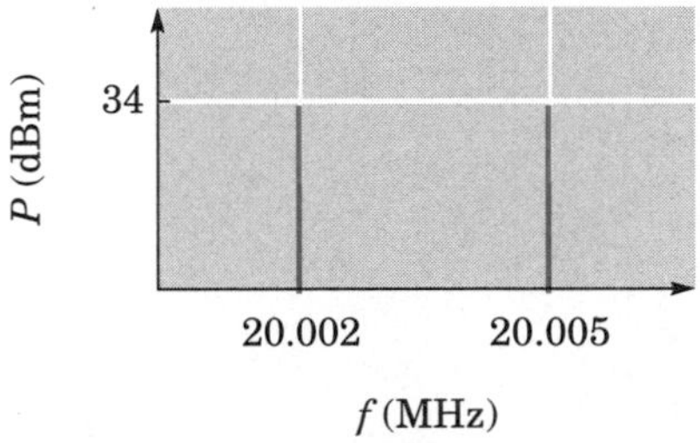

31.

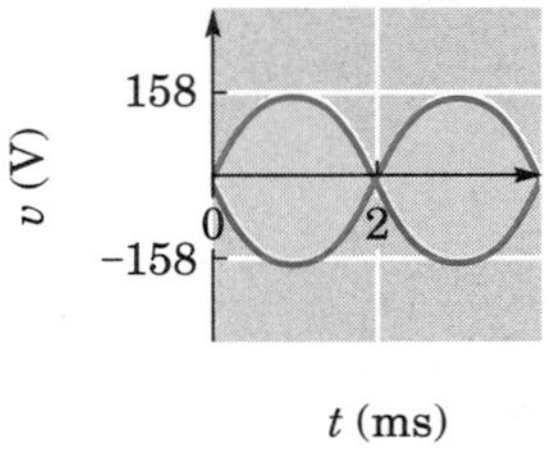

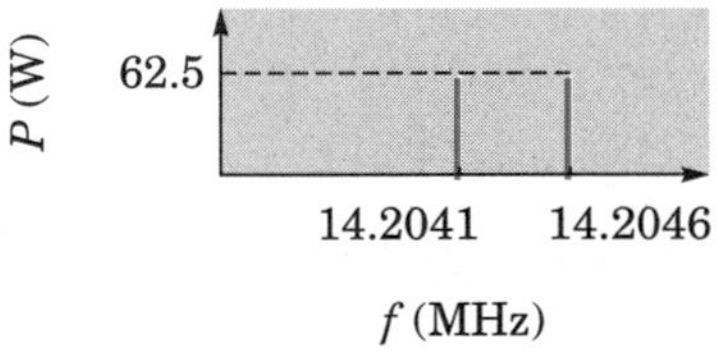

33. 4.99837 MHz
35. **(a)** 8.0015 MHz **(b)** 38.0015 MHz
    (c)

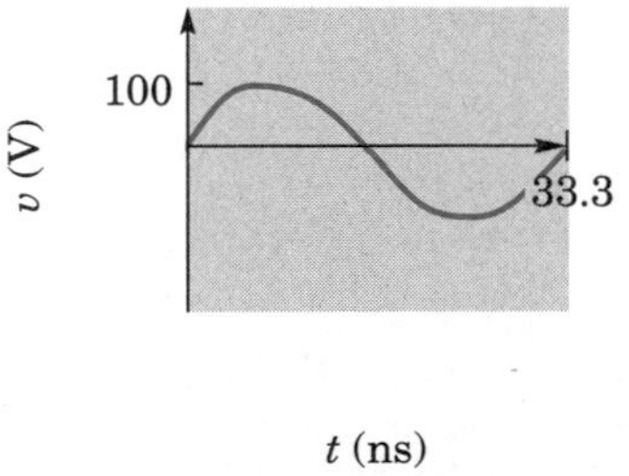

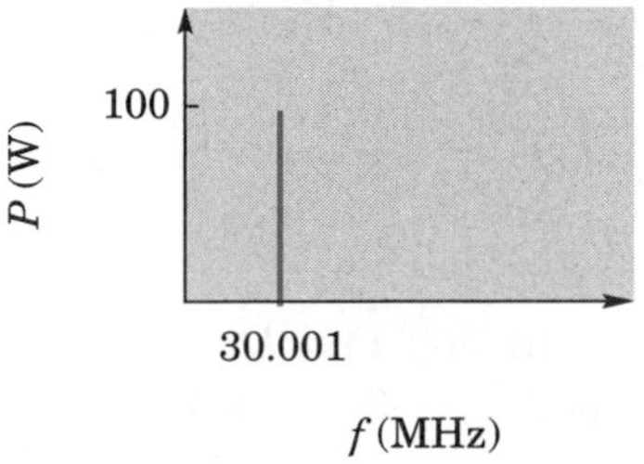

37. **(a)** LSB **(b)** 26.5015 MHz
    (c) Change BFO frequency to 8.9985 MHz, change LO frequency to 17.497 MHz
39. The signal can be received with an SSB receiver adjusted so that the BFO is separated by a convenient audio frequency from the carrier, as it appears in the IF. It doesn't matter which sideband is used, and the receiver should be tuned a few hundred hertz above or below 7.100 MHz.
41. Yes. Audio frequency response will be limited. A wider IF filter should be used for best results. BFO should be turned off if possible. If not, receiver must be tuned so that BFO is at exactly the same frequency as carrier to avoid an audible tone at the difference between the two receivers.
43. **(a)** 500 Hz, 1.75 kHz **(b)** USB **(c)** 4 W
    (d) 50 dB **(e)** 40 dB, 60 dB
45. **(a)** 86.6 V **(b)** 37.5 W
    (c)

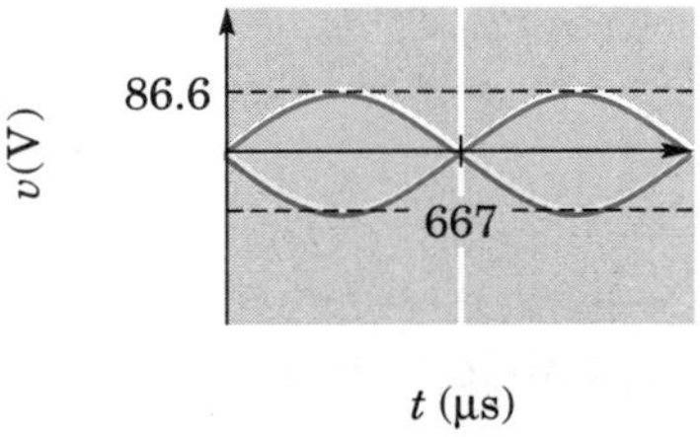

(d)

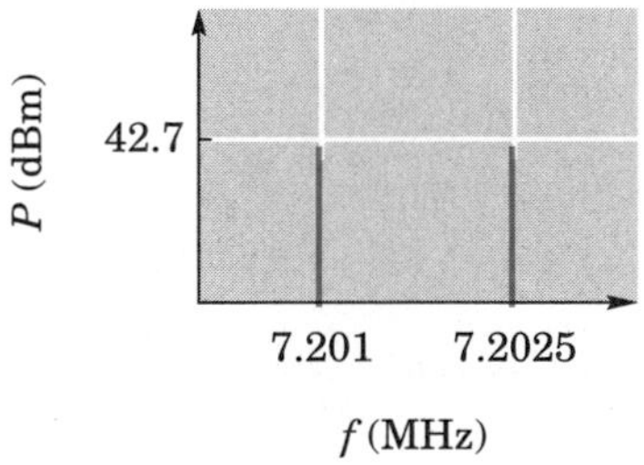

CHAPTER 7

21. 5

23.

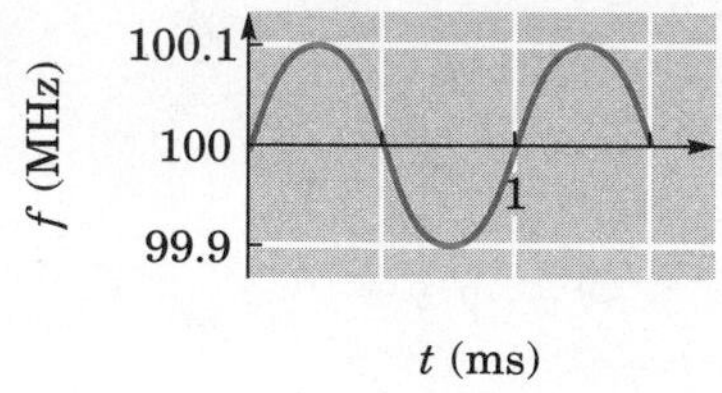

25. −20 kHz/V
27. 17
29. 500 Hz
31. 85.9 degrees
33. **(a)** 2 **(b)** 605 mW **(c)** 8.75 V(RMS) **(d)** 50 kHz
35. **(a)** 15 kHz
(b)

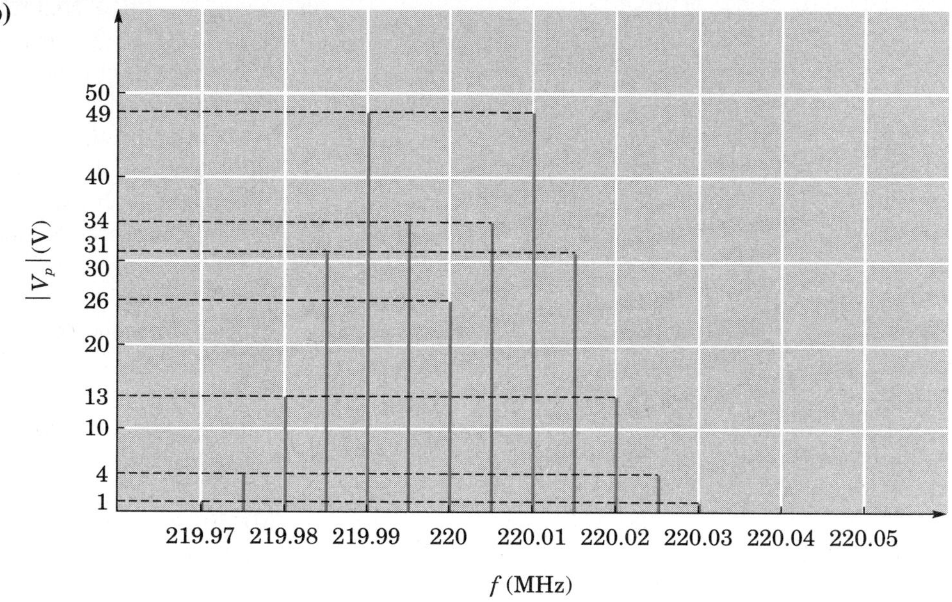

(c) 70 kHz **(d)** 40 kHz

37. **(a)** 16 kHz **(b)** 156 kHz
39. **(a)** De-emphasis, $R = 7.5\ \text{k}\Omega$
(b) Pre-emphasis, $C = 75$ nF
41. 12 dB
43. V (not to scale)

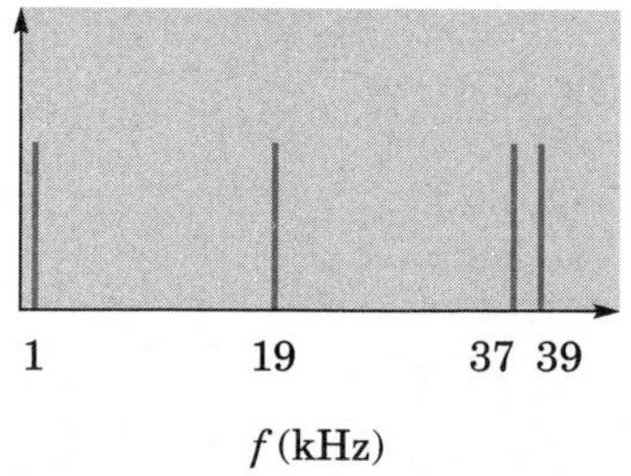

45. Use $f_m = 31.25$ kHz for first null or 13.6 kHz for second null.
47. For full-carrier AM, sideband power is in addition to carrier power, which remains constant. For SSB, the carrier power is zero.

CHAPTER 8

21. $f_c = 216$ MHz, $\delta = 54$ kHz
23. **(a)** 258 degrees
(b) Use a frequency multiplier, multiplying frequency by at least 6 times. See Figure 8.8.
25. **(a)** 16.7 MHz **(b)** 0.21 rad **(c)** Converts PM to FM
27. **(a)** 911 **(b)** 26.5 kHz/V
29. **(a)** 1.5 V (peak) **(b)** 60 kHz

31. **(a)** 563×10^3 or 115 dB **(b)** Yes, but the noise advantage of FM would be lost.
33. **(a)** 156.7 MHz, 139.8 MHz
(b) Transmitter is direct-FM using PLL.
(c) 16 kHz (from Carson's rule)
35. 4 μV
37. Increase signal level slightly and measure SINAD again.
39. Approach **(b)**, because modulator output is proportional to deviation.

CHAPTER 9

23. **(a)** 417 μs **(b)** 6.6
25. **(a)** $Y = 0.498$, $I = 0.304$, $Q = 0.022$
(b) 53.6 IRE
27. **(a)** 669.25 MHz **(b)** 673.75 MHz
(c) 672.83 MHz
29.

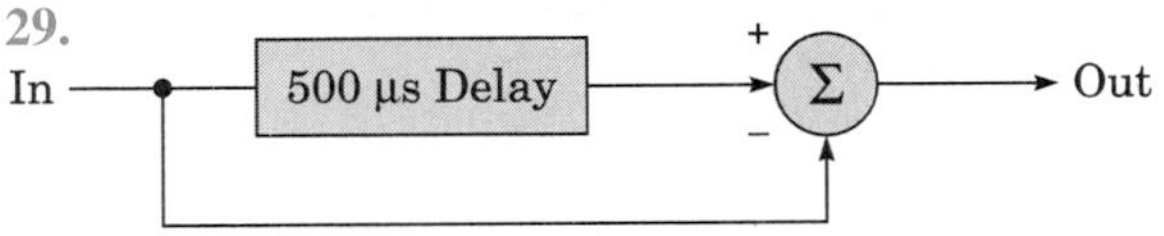

31. **(a)** 1.68 mV **(b)** 37.6 nW
33. **(a)** No sound or picture on VHF or UHF
(b) No picture or noisy picture, noisy sound
(c) No horizontal or vertical sync
(d) No raster, probably no sound (due to lack of low-voltage power supply)
35. **(a)** AFPC
(b) Color killer, color demodulator, color oscillator
(c) CRT driver for red, CRT
37. **(a)** 674 **(b)** 6.3 MHz
39. 20%
41. 12.9 MHz
43. $R = Y + 0.9482I + 0.624Q$

CHAPTER 10

31. 2600 Ω
33. 4
35. See section 10.3.2.
37. 50.4 dB. No: the received signal would be inaudible.
39. 74 dBrnc
41. 79.6 kHz

CHAPTER 11

21. **(a)** 48 Mb/s **(b)** 24 dB
23. See Figure 11.11. Make the time intervals 250 μs and the peak voltage 1 V in part (a).
25. 75.2%
27. **(a)** 256 **(b)** 80 Mb/s **(c)** 49.9 dB
(d) Quantizing noise
29. 156.256 mV
31. **(a)** Unipolar NRZ

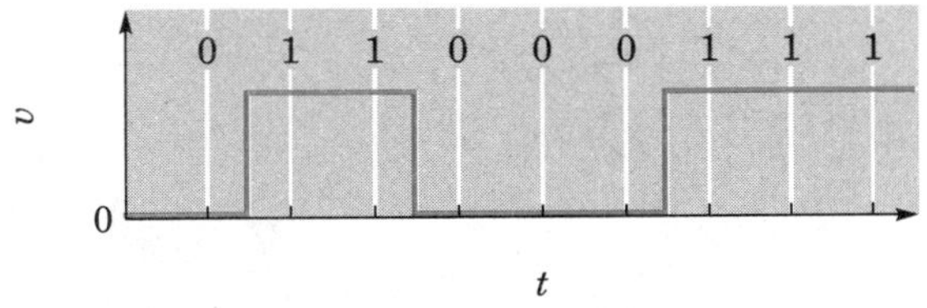

(b) AMI

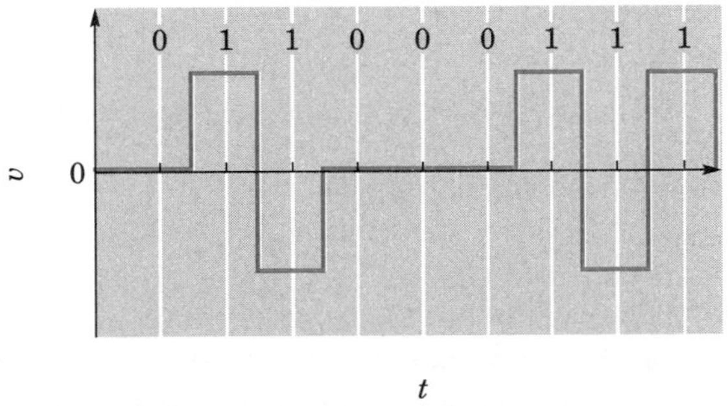

(c) Manchester

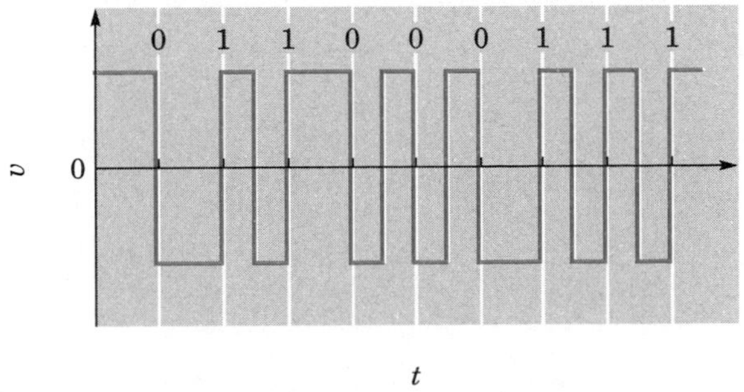

33. 100100

35. 2.048 Mb/s

37.

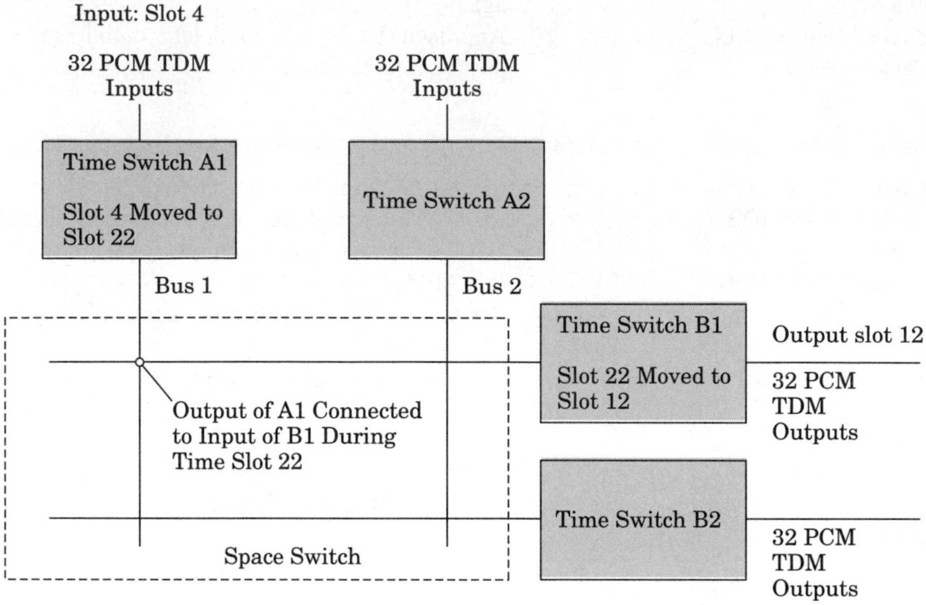

39. 35:1

CHAPTER 12

23. **(a)** BE HERE59,853 ?6 8
(b) BE HERE59,853 BY 8

25. As written (MSB first): 10100 00001 10010 10010 11000 00100 10000 10100 00001 01010 00001

27. As transmitted: (LSB first, with start and stop bits): Gaps between characters are for clarity, would not be transmitted.
0000100101 0101001101 0001101101
0001101101 0111101101 0000001001
0001010101 0000101101 0101001101
0010011101 0101001101 0100001001

29. **(a)** η = data bits/total bits = 8/10 = 0.8 = 80%
(b) 960 characters/second, 9600 words/minute
(c) Increase; Less overhead

31. **(a)** Unipolar NRZ

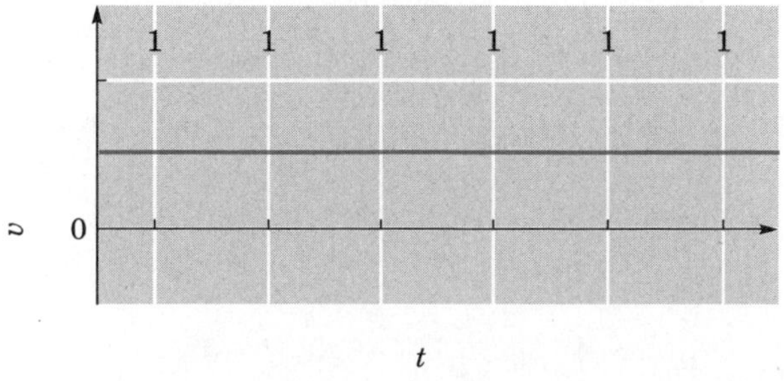

(b) Bipolar NRZ

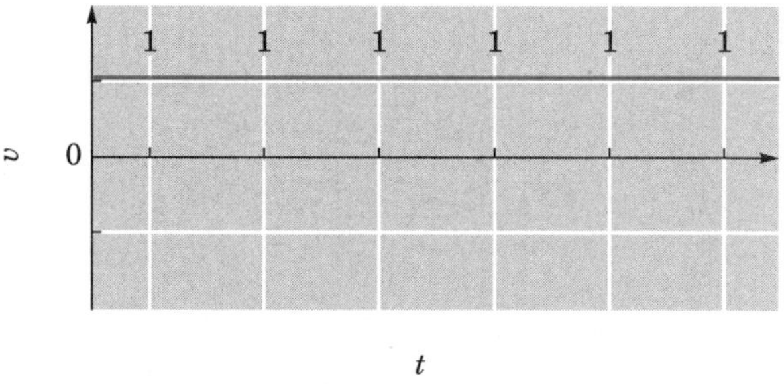

(c) AMI

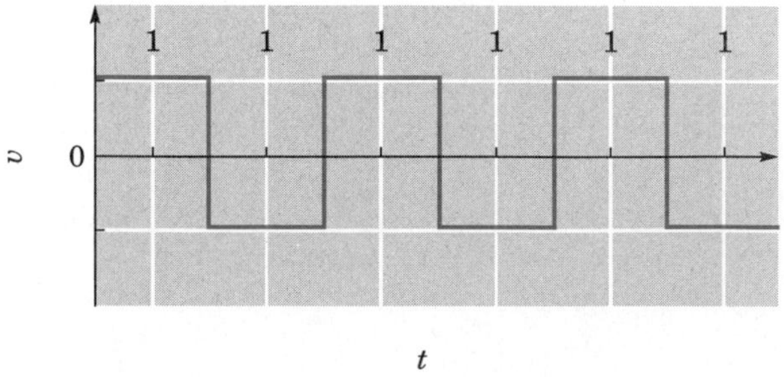

(d) Manchester

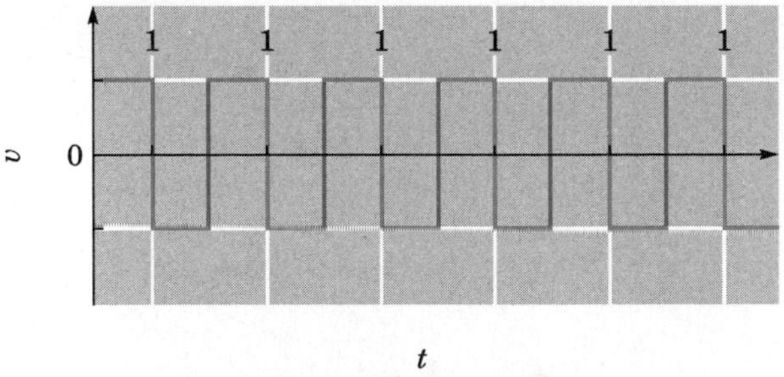

(c) and **(d)** are suitable for synchronous transmission

33. See Figure 12.6.
35. **(a)** BEL
    (b) Receiver would ignore it if not preceded by DLE.
    (c) Receiver would ignore it because it would look only for flags within a block.
37. LRC: 0111010 VRC: 11100011010
39. The third and fourth characters are illegal. The message is KEY

CHAPTER 13

25. **(a)** 16-level QAM **(b)** 2400 baud
    (c)

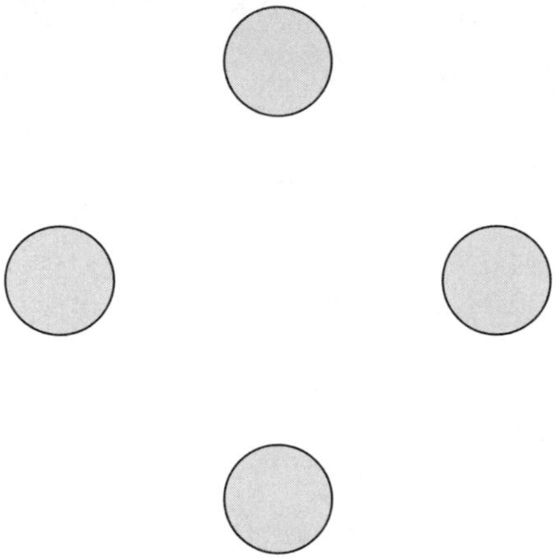

    (d) Fallback would be used when the signal-to-noise ratio was lower than normal.
27.

25 pin connector	9 pin connector
2	3
3	2
4	7
5	8
6	6
7	5
8	1
20	4
22	9

29. ATE1L3M1N1&C1&D2
31. **(a)** 8 **(b)** 640 Mb/s

CHAPTER 14

19. **(a)** 15 **(b)** 30

CHAPTER 15

29. See Figure 15.3
31. **(a)** repeater **(b)** bridge
    (c) router **(d)** gateway

CHAPTER 16

21. **(a)** 6 V **(b)** 155 Ω **(c)** 0.567
23. 417 ns, 5 V
25. **(a)** 325 Ω **(b)** increases
27. **(a)** −0.371 **(b)** 43.1 W

29. **(a)**

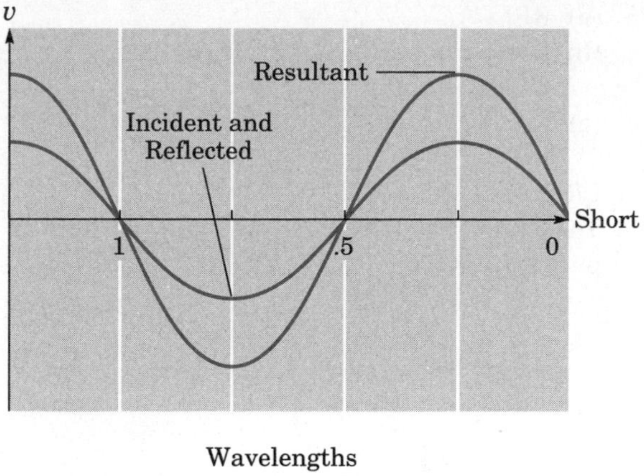

(b)

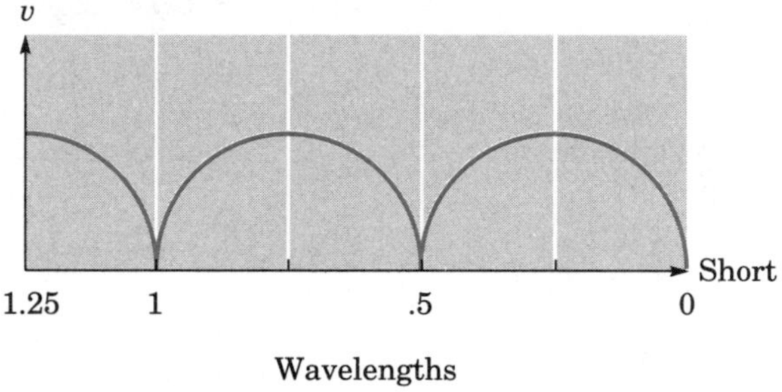

(c)

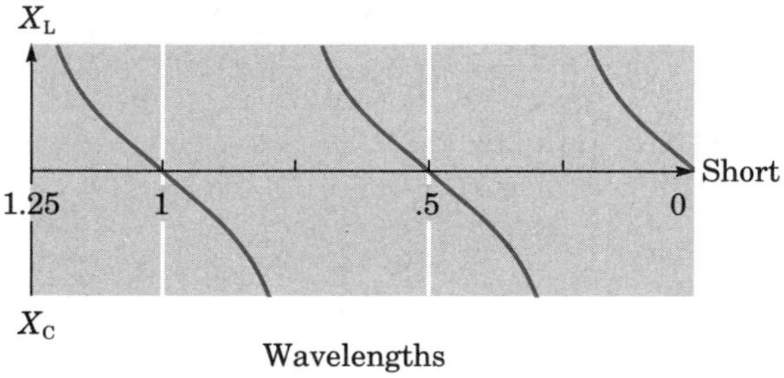

31. **(a)**

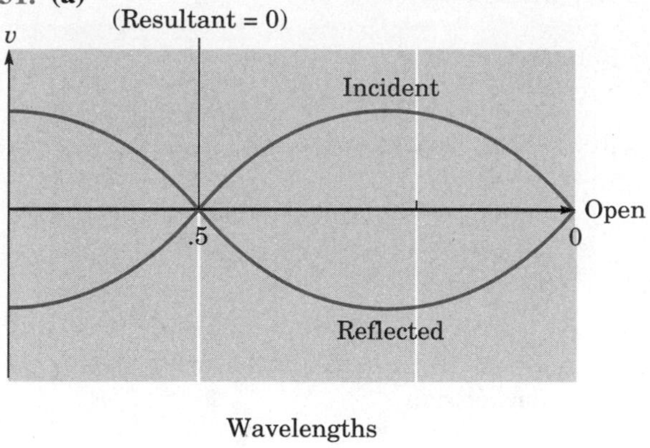

(b)

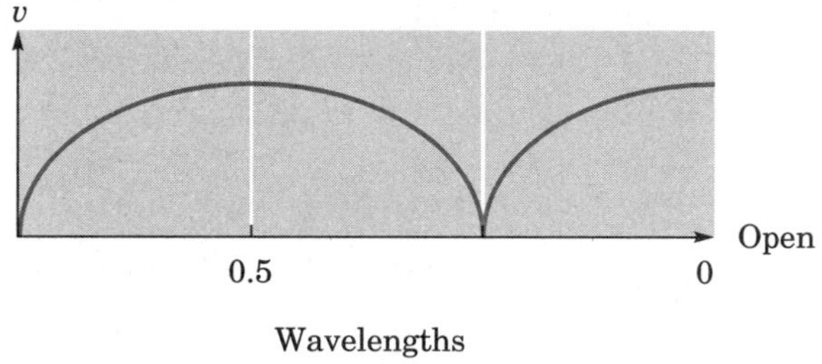

33. 9.1 W
35. 0.594 μV
37. **(a)** 0.082 λ **(b)** 52.5 Ω
39. Length 2.48 m, position 0.81 m from load
41. Length 0.88 m, position 2.08 m from load
43. **(a)** 1.5 m **(b)** 190 MHz
    (c) 3.33 **(d)** 167 Ω
45. 112 Ω

CHAPTER 17

23. 107×10^6 m/s
25. 135 Ω
27. 26.5 W/m²
29. **(a)** 0.897 m² **(b)** 5.95 pW
31. **(a)** 49 dBW **(b)** 178 pW
33. −54.7 dBm
35. −60.7 dBm
37. 21°
39. 7.04°. See Figure 17.12.
41. 4.4 MHz approx.
43. **(a)** 34.2 km **(b)** 43.4 km
45. 10 μs
47. **(a)** Ionospheric propagation is more likely at these frequencies at night.
    (b) Ionospheric propagation is common at 27 MHz, especially near peaks of the solar cycle.
    (c) This phenomenon causes disturbances in the ionosphere.
    (d) The likely cause is tropospheric ducting.
    (e) There is refraction in the troposphere.

CHAPTER 18

21. 9.5 m
23. **(a)** 97.2 W
    (b) dissipated as heat in the antenna
25. 2.71 mV/m
27. Cancels completely
29. Gain = 5 dBi, beamwidth = 20°

31. 71.3 m
33. Draw lines at right angles to each antenna's axis; find point of intersection.
35. Approximately 20 cm
37. **(a)** Add base or center inductive loading.
    (b) No, because of losses due to low radiation resistance and presence of coil.
39. $L_2 = 2.14$ m, $L_3 = 3.06$ m, $D_1 = 2.8$ m, $D_2 = 4$ m, $D_3 = 5.7$ m
41. **(a)** 3 dB increase in gain **(b)** Cancellation
43.

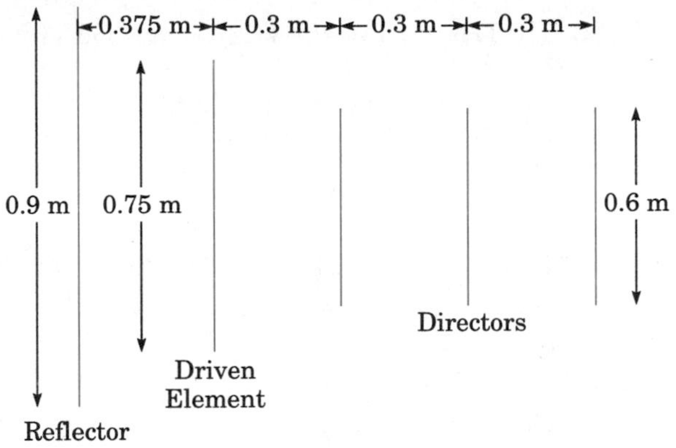

45. **(a)** 39 dBi **(b)** 39 dBi
47. 16.7λ
49. **(a)** 26.2 mW **(b)** 828 μW
51. 7.68 dBi

CHAPTER 19

23. **(a)** 397×10^6 m/s **(b)** 226×10^6 m/s
    (c) 39.7 mm **(d)** 500 Ω
25. **(a)** 9.375 GHz **(b)** 483 Ω **(c)** 25.6 mm
    (d) 2.34×10^8 m/s
27. 79.4 mW
29. 19.9 mm
31. 5 μm
33. **(a)** 0.001 **(b)** 1.67 MW **(c)** 1 kW
    (d) 1.67 kW **(e)** 670 W
35. Max. 750 km, min. 7.5 km
37. **(a)** 500 m/s **(b)** Toward
39. $f_D = 1.85 v_r f_i$
41. **(a)** TE_{10} **(b)** 4.3 to 8.6 GHz
    (c) 35.9 mm **(d)** See Figure 19.10(a). Distance from probe to end of guide is 17.9 mm.
    (e) 6.72 ns
43. **(a)** 638 mW **(b)** 535 μW
45. **(a)** 8.33 ns, 1 dB **(b)** 17.08 ns, 0.4 dB
    (c) 667 ns, 104.76 dB **(d)** 692 ns
    (e) −43.15 dBm

CHAPTER 20

21. 44.5 km
23. 10.5 m
25. 5
27. **(a)** −50.6 dBm **(b)** 177 K **(c)** 226 K
    (d) 402 K **(e)** 111 fW **(f)** 49 dB
29. 5.55×10^{-21} W/Hz
31. **(a)** 3 dB decrease **(b)** 6 dB decrease
    (c) 3 dB increase
33. 5.5 m
35. 7412 channels. Text has 6000; difference is due to guard bands.
37. **(a)** 25 MBd **(b)** 16.7 MBd **(c)** 10 MBd
39. 15.2 dBm
41. 2.62s, 210 dB

CHAPTER 21

27. 384 Mm
29. −39.5 dBm
31. Increase antenna gain to 55 dB or reduce receiver noise temperature to 6.6 K.
33. 290 K
35. Gain increases by 9.54 dB, beamwidth decreases to one-third its former value.
37. 1200
39. LEO signal is stronger by 38 dB

CHAPTER 22

25. **(a)** 750 THz **(b)** 429 THz **(c)** 333 THz
27. 41.8°
29. **(a)** 75° **(b)** 0.384 **(c)** 22.6°
31. **(a)** infrared **(b)** 194 THz
33. 10.5 GHz-km
35. 0.6 dB
37. 79.1 nW
39. 497×10^{-21} J, 3.11 eV
41. **(a)** 4 ns **(b)** 500 ps
43. 12.75 μA, 127.5 mV

CHAPTER 23

29. Yes
31. Dispersion
33. 96.4 ns, 3.6 MHz
35. 100 ns
37. See Figure 23.10
39.

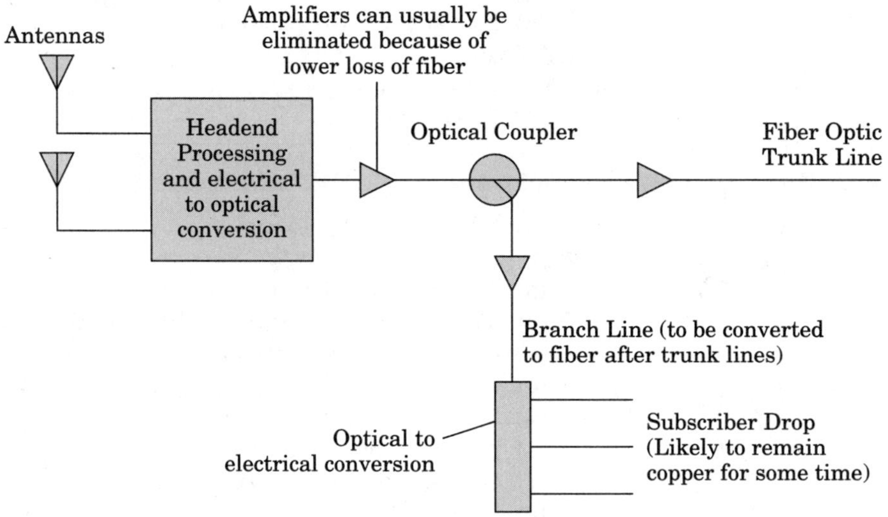

CHAPTER 24

23. **(a)** 36.6 m **(b)** −52.6 dBm
25. 100 km
27. Older designs use the power line to carry signals in one direction. These signals are usable only close to the line. Probably there are no power lines in the backyard.
29. 2.6 km
31. 200%
33. 40 hops per second
35. 75
37. **(a)** 5.3 m; 2.9 m; 16 cm; 7 cm
    (b) Higher frequency makes antenna design easier.

Index

B

C

P

Q

R

S

T